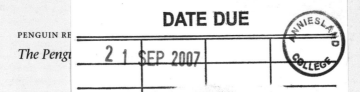

Michael Thain was born in Hampstead in 1946 and went to University College School and to Keble College, Oxford, where he read Zoology and then Human Biology. He studied History and Philosophy of Science at University College London, Philosophy at Birkbeck College and History of Technology at Imperial College, London. A Fellow of the Institute of Biology, he is currently completing a *Dictionary of Human Biology*, also for Penguin. He joined the staff of Harrow School as a biology teacher in 1969 and was Head of Biology for eleven years before retiring in 2003. A keen naturalist, he managed a small nature reserve in the school grounds. He now teaches at Davies, Laing and Dick Independent College, London, and tries to devote more time to writing. He has one stepdaughter and one son and lives in Northwood, Middlesex.

Michael Hickman was born in Worcestershire in 1943. He gained a First Class Honours BSc in Biology (Botany) from the University of Western Ontario, Canada, in 1966, and a PhD in Phycology/Freshwater Ecology from the University of Bristol, England, in 1970. He has been Professor of Botany (now Biological Sciences) at the University of Alberta, Canada, since 1981, having been initially an Assistant and then Associate Professor in the same department since 1970. He has published numerous papers on his research activities into the ecology of benthic and planktonic algae in lakes and rivers of Alberta, and has been retained as a consultant by various organizations, both private and governmental. His current research interests lie in the field of palaeoecology, specifically palaeolimnology, and how climatic changes affect lakes and their biota. He lives in Edmonton with his wife and two sons.

The Penguin Dictionary of

Biology

M. Thain
M. Hickman

ELEVENTH EDITION
being the Eleventh Edition of the work originally compiled by
M. ABERCROMBIE, C. J. HICKMAN, M. L. JOHNSON

Diagrams by
RAYMOND TURVEY

PENGUIN BOOKS

For Avril, Harry and Margaret

PENGUIN BOOKS

Published by the Penguin Group
Penguin Books Ltd, 80 Strand, London WC2R 0RL, England
Penguin Group (USA) Inc., 375 Hudson Street, New York, New York 10014, USA
Penguin Group (Canada), 90 Eglinton Avenue East, Suite 700, Toronto, Ontario, Canada M4P 2Y3
(a division of Pearson Penguin Canada Inc.)
Penguin Ireland, 25 St Stephen's Green, Dublin 2, Ireland (a division of Penguin Books Ltd)
Penguin Group (Australia), 250 Camberwell Road, Camberwell, Victoria 3124, Australia
(a division of Pearson Australia Group Pty Ltd)
Penguin Books India Pvt Ltd, 11 Community Centre, Panchsheel Park, New Delhi – 110 017, India
Penguin Group (NZ), cnr Airborne and Rosedale Roads, Albany, Auckland 1310, New Zealand
(a division of Pearson New Zealand Ltd)
Penguin Books (South Africa) (Pty) Ltd, 24 Sturdee Avenue, Rosebank, Johannesburg 2196, South Africa

Penguin Books Ltd, Registered Offices: 80 Strand, London WC2R 0RL, England

www.penguin.com

First published as *The Penguin Dictionary of Biology*, 1951
Second edition 1954; Third edition 1957; Fourth edition 1961; Fifth edition 1966; Sixth edition 1973;
Seventh edition 1980; Eighth edition, entitled *The New Penguin Dictionary of Biology*, 1990
Reprinted with amendments 1991, 1992
Reprinted, with amendments, as *The Penguin Dictionary of Biology*, 1992
Ninth edition 1994
Reprinted with minor revisions 1995, 1996
Tenth edition 2000
Reprinted with minor revisions 2001
Eleventh edition 2004
6

The acknowledgements on pages ix–vii constitute an extension of this copyright page

Set in 7/9 pt PostScript ITC Stone
Typeset by Rowland Phototypesetting Ltd, Bury St Edmunds, Suffolk
Printed in England by Clays Ltd, St Ives plc

ISBN-13: 978–0–141–01396–1
ISBN-10: 0–141–01396–6

Introduction to the Eleventh Edition

In the Introduction to the Tenth Edition we claimed that barely a page of the previous edition remained unchanged. Although the Eleventh Edition has not involved such an all-embracing revision, there have been some major achievements in biology since the eve of the millennium which have added considerably to much of this dictionary. These include publication in February 2001 of the bulk of the human genome, and of complete genomes of a growing number of other species too. The debate surrounding genetically modified organisms and food continues to polarize just as the science warns against catch-all 'for' or 'against' attitudes, each such case of applied genetics probably having different consequences for the environment. This is surely a technology that needs to involve the public far better than it does.

There have been significant advances in the study of DNA repair mechanisms, chromatin remodelling and nanobiotechnology, and in our appreciation of the biological roles of RNA. Several non-coding RNA species have emerged whose involvement in developmental biology promise to be of great interest. The fields of RNA and gene silencing have certainly yielded surprises. It is also becoming clearer that many intracellular signalling pathways involve protein complexes that are linked to form networks of considerable complexity. Indeed, 'complexity' itself is a term attracting more attention as bioinformatics develops into a discipline in its own right.

We have made a stab at introducing the topic 'plant disease and defences', in which exciting progress is being made, not least in our appreciation of the significance of Toll-like receptors. And we have made changes to the 'recombination' and 'eukaryote' entries in the hope that these reflect mainstream thinking better. There are also some new but long overdue entries, including 'artificial intelligence' and 'Bayesian statistics', which we have deliberately kept brief but which we hope to develop further in a subsequent edition.

MT
MH
February 2004

Introduction to the Tenth Edition

In the Preface to the Ninth Edition we noted the need for biologists to be alert to advances in fields which have not traditionally been considered *de rigueur*. Six years on, this need is even more strongly evident. It is now probably more desirable than ever before for a trainee biologist to establish an interest in the earth and physical sciences, and in mathematics and information technology. Here, we can only flag up this belief and hope that, in these times of educational parsimony, schools and colleges will encourage their students to read well beyond what their official syllabus demands. In addition to providing up-to-date information on biological issues, we hope that this dictionary stimulates further reading and most entries include suggestions (in SMALL CAPITALS) for further consultation within its pages.

In our efforts to keep the dictionary up to date, barely a page of the Ninth Edition remains unaltered and thereby – we earnestly hope – unimproved. A significant minority of new and modified entries deal with biological classification – a perennially unstable minefield where controversies within molecular systematics will continue to ebb and flow, and where each new fossil discovery (see, e.g., *Psarolepis* in the entry on FISH) has its own, sometimes seismic, implications for taxonomy. Especially problematic is plant taxonomy – so we accept that those taxa we have chosen to include will not meet with universal approval. The important point here is that the lists of organisms making up taxa, and especially plant taxa, are unusually fluid at present. In part, this is due to the different emphases which different research groups place upon the various forms of evidence used in taxonomy. Some rely purely on anatomical features, others argue there is greater objectivity in the interpretation of DNA sequences. Still others try to combine the two approaches and this reflects to some extent the continuing debate between the 'evolutionary taxonomy' and 'cladistic' schools of biological classification.

To illustrate the controversies within plant taxonomy, here are just two of many current issues, both involving flowering plants. The recent tradition of flowering-plant taxonomy has been to split this division into monocotyledons (now termed Liliopsida) and dicotyledons (now termed Magnoliopsida) – a partition justified by the clustering of several observable characters, including cotyledon number. But some research groups evaluating DNA sequences suggest that the real split among flowering plants lies elsewhere. One such group (at the Royal Botanic Gardens, Kew) identifies the major morphological distinction in the flowering plants not with cotyledons but with the anatomy of the basic pollen type. They conclude that magnolia – traditionally a 'primitive dicot', but with a single pollen grain pore rather than the three pollen grain furrows of dicots – has closer evolutionary links with traditional monocots such as bamboos and lilies than it does with other plants having two seed leaves. This would lead to a change in the criteria for membership of the two major flowering plant divisions. As a second illustration, four independent groups of DNA sequencers in the USA have, to majority surprise, produced flowering plant phylogenies implicating *Amborella*, a genus of New Caledonian shrub, as the closest living relative of the first flowering plants. Both examples, if valid, will necessitate reconstruction of current plant classifications.

It is a common complaint among researchers who use DNA sequence data in phylogenetic reconstruction that many classical taxa, such as 'monocots' and 'dicots', fail to group plants accurately. We have tried to reflect such criticisms in our minor revision to the LINNAEUS entry in this edition. Some researchers would even prefer to name plants according to clade – a system termed the PhyloCode. Even some of the more traditionally minded admit that botanists may soon have to cope with two systems of nomenclature, one Linnaean, the other cladistic.

Given such welcome diversity of opinion on taxonomic matters, it is pleasing to be able to recommend the reader to a companion volume, the *Penguin Dictionary of Plant Sciences*, whose taxonomic entries can usefully be read alongside ours.

MT
MH
September 1999

Acknowledgements

For Figures

We gratefully acknowledge the following publishing houses for permission to reproduce material as indicated below. Figures may have been modified slightly to fit the contexts of entries, and the legends will generally have been rewritten. In several cases, authors' permissions were also received. While it is very much hoped that all necessary permission for work reproduced here will have been obtained at the time of going to press, the authors sincerely apologize for use of any uncredited material and will be happy to hear from any such interested parties.

American Association for the Advancement of Science: Fig. **2** is derived from Constance Holden, *Science 273*:47 (1996); Fig. **6** is derived from Jean-Philippe Vielle Calzada, Charles F. Crane & David M. Stelly, *Science 274*:1322 (1996); Fig. **40** is derived from Fig. 3, Gerard Evan & Trevor Littlewood, *Science 281*:1320 (1998); Fig. **52** is from S. L. Baldauf, *Science 300*: 1705 (2003); Fig. **96** is derived from Fig. 1, B. Bruce Nicklas, *Science 275*:632 (1997); Fig. **116** is from R. Keller, *Science 298*: 1951 (2002); Fig. **129** is derived from Fig. 2, Gerard Evan & Trevor Littlewood, *Science 281*:1320 (1998); Fig. **138b** is derived from Fig. 3, Richard F. Kay, Callum Ross & Blythe A. Williams, *Science 275*:799 (1997); Table **8** is from 'Some emerging plant pathogens', *Science 292*: 2272 (2001).

Benjamin/Cummings Publishing Company: Watson, J. D., *Molecular Biology of the Gene* (3rd edn, 1976), for Fig. **94**.

Blackwell Scientific Publications Ltd: May, R. M. (ed.), *Theoretical Ecology* (2nd edn, 1981), for Table **6**; Lewin, Roger, *Human Evolution* (2nd edn, 1989), for Fig. **82**; M. Begon, J. L. Harper & C. R. Townsend, *Ecology* (2nd edn, 1990), for Fig. **156**.

The Blakiston Company (McGraw-Hill Book Co.): Kingley, J. S., *Outline of Comparative Anatomy of Vertebrates* (1928), with permission of McGraw-Hill, Inc., for Fig. **68a**.

British Museum (Natural History): Charig, A., *A New Look at the Dinosaurs* (1979), for Fig. **161**.

The British Ornithologists Union: Landsborough Thompson, A. (ed.), *A New Dictionary of Birds* (1964), for Fig. **57**.

Butterworth & Company (Publishers): Cohen, J., *Reproduction* (1977), for Fig. **160**.

Cambridge University Press: Willmer, P., *Invertebrate Relationships* (1990), for Table **1** and Fig. **15**; Lowrie, Pauline and Wells, Sue, *Microorganisms, Biotechnology & Disease* (student's book) (1991), for Figs. **10b** and **14**; D. O. Hall and K. K. Rao, *Photosynthesis* (4th edn, 1987) for Fig. **19**; Austin, C. R. and Short, R. V., *Reproduction in Mammals, Book 2: Embryonic and Fetal Development* (2nd edn, 1982), for Figs. **54** and **135**; Slack, J. M. W., *From Egg to Embryo* (1983), for Fig. **67**; S. Jones, R. Martin and D. Pilbeam, *The Cambridge Encyclopedia of Human Evolution* (1995), for Fig. **87**; van Emden, H. F., *Pest Control* (1992), for Fig. **98**.

Chapman and Hall Ltd: Mather, K., *Genetical Structure of Populations* (1973), for Fig. **66**; Martin Ingrouille, *Diversity and Evolution of Plants* (1992), for Fig. **137**.

The Company of Biologists Ltd, Cambridge: Grimstone, A. V., Harris, H. and Johnson, R. T., *Prospects in Cell Biology* (1986), for Fig. **92**.

Current Science: Fig. **1** is modified from Fig. 1, P. J. Goldschmidt-Clermont, et al., *Current Biology*, 2:669 (1992); Fig. **6** is derived from Miranda Robertson, *Current Biology*, 8:597 (1998); Fig. **22c** is from Fig. 1, Andrew W. Murray, *Current Biology*, 3:291 (1993); Fig. **40** is derived from M. Javed Aman and Waren J. Leonard, *Current Biology*, 7:784 (1997); Fig. **41a** is from Fig. 1, G. I. Evan and T. D. Littlewood, *Current Opinion in Genetics & Development*, 3:46 (1993); Fig. **82** is derived from Fig. 2, Caro-Beth Stewart and Todd R. Disotell, *Current Biology*, 8:584 (1998); Fig. **86a** is derived from Miranda Robertson, *Current Biology*, 8: 596 (1998); Fig. **91** is from Fig. 1, Jacopi Meldolesi, *Current Biology*, 3:911 (1993); Fig. **108a** is derived from Michael J. Novacek, *Current Biology*, 7:490 (1997); Fig. **143a** is from Fig. 1, T. Pawson and W. Schlessinger, *Current Biology*, 3:435 (1993); Fig. **150** is from Fig. 1, F-U. Hartl and M. Weidmann, *Current Biology*, 3:86 (1993); Fig. **157** is from Fig. 2, Grace Gill, *Current Biology*, 2:566 (1992); Fig. **158** is from Fig. 1, Anthony L. DeFranco, *Current Biology*, 2:478 (1992); Fig. **159** is from Fig. 1, Brigitte T. Huber, *Current Biology*, 2:495 (1992).

Elsevier Trends Journals, Cambridge, UK: Fig. **7** is from Fig. 1, Robert Foley, *Trends in Ecology and Evolution*, 8:196 (1993); Fig. **77** is modified from David Schubert, *Trends in Cell Biology*, 2:65 (1992); Fig. **106a** is modified from Fig. 1, Robert DeMars and Thomas Spies, *Trends in Cell Biology*, 2:82 (1992); **133a** is from Fig. 1, Deborah Brown, *Trends in Cell Biology*, 2:338 (1992); Fig. **143b** contains information originally published in Fig. 1, Stephen P. Jackson, *Trends in Cell Biology*, 2:105 (1992); Fig. **120** is from M-A. Selosse and F. Le Tacon, *Trends in Ecology and Evolution*, 13:16 (1998); Fig. **123** is from E. D. Brodie III, A. J. Moore and F. J. Janzen, *Trends in Ecology and Evolution*, 10:313 (1995).

The English Universities Press Ltd: John T. Price, *The Origin and Evolution of Life* (1971) for Fig. **35**.

W. H. Freeman & Co. Ltd: John Maynard Smith and Eörs Szathmáry, *The Major Transitions in Evolution* (1995), for Fig. **8**; Eckert, *Animal Physiology* (4th edn, D. Randall, W. Burggren, K. French, 1997), for Fig. **17**; Anthony J. F. Griffiths, et al., *An Introduction to Genetic Analysis* (6th edn, 1996), for Figs. **46**, **61** and **117**; James D. Watson, Michael Gilman, Jan Witkowski and Mark Zoller, *Recombinant DNA* (2nd edn, 1992) for Fig. **63**; *Life, Death & the Immune System* (1994), for Figs. **81** and **106b**.

Garland Publishing Inc.: Alberts, B., et al., *Molecular Biology of the Cell* (1st edn, 1983), for Figs. **25**, **28a**, **31**, **45**, **65**, **105** and **153**; Alberts, B., et al., *Molecular Biology of the Cell* (2nd edn, 1989) for Figs. **22**, **75**, **85**, **112**, **125**, **139** and Fig. **16**, modified from a drawing by Jane Richardson; Fig. **112**, which includes data supplied by Linda Amos; Alberts, B., et al., *The Molecular Biology of the Cell* (3rd edn, 1994), for Figs. **92** and **121**; Alberts, B., et al., *Molecular Biology of the Cell* (4th edn, 2002), for Figs. **38**, **46**, **62**, **85**, **144a** and **144b**; Fig. **143c**, from N. H. Boke, *American Journal of Botany*, 36: 535–547 (1949).

Gower Medical: Staines, N., Brostoff, J. and James, K., *Introducing Immunology* (1985), for Fig. **104**.

Harcourt Brace Jovanovich: Frobisher, M., et al., *Fundamentals of Microbiology* (9th edn, 1974), for Figs. **20c** and **76**; Hopkins, C. R., *Structure and Function of Cells* (1978), for Figs. **24a** and **24b**.

HarperCollins: Tortora, G. J. and Anagnostokos, N. P., *Principles of Anatomy and Physiology* (4th edn, 1984), for Figs. **72**, **79**, **97**, **111**, **119**, **130**, **145** and **155a**; Tortora, G. J. and Anagnostokos, N. P., *Principles of Anatomy and Physiology* (6th edn, 1990), for Figs. **9**, **20a** and **128**.

Harvard University Press: Hughes, G. M., *Comparative Physiology of Vertebrate Respiration* (1963), for Figs. **68b** and **68c**.

Heinemann: Freeman, W. H., and Bracegirdle, B., *An Advanced Atlas of Histology* (1976), for Fig. **69**.

Hodder Headline plc: Chapman, R. F., *The Insects* (2nd edn, 1971), for Fig. **118**.

CBS College Publishing (Holt-Saunders): Goodenough, U., *Genetics* (3rd edn, 1984), for Fig. **93**; Romer, A. S., *The Vertebrate Body* (5th edn, 1977), for Figs. **130b** and **131**.

IRL Press: Fig. **4** is from Fig. 1.2, *Immune Recognition*, M. J. Owen & J. R. Lamb (1988); Fig. **28b** is from Fig. 3.4, *Gene Structure & Transcription*, T. Beebee & J. Burke (1988); Fig. **34** is from Fig. 5.9, *Proton Targeting & Secretion*, B. M. Austen & O. M. R. Westwood (1991); Fig. **60** is from Fig. 3.2, *Gene Structure & Transcription*, T. Beebee & J. Burke (1988); Fig. **74a** is from Fig. 1.2, *Protein Targeting & Secretion*, B. M. Austen & O. M. R. Westwood (1991).

S. Karger AG, Basel: Fig. **170** is from *Intervirology*, *12*:3–5 (1979).

Longman Group Ltd: Lewis, K. R., and John, B., *The Matter of Mendelian Heredity* (2nd edn, 1972), for Fig. **100(i)**; J. S. Garrow, W. P. T. James, *Human Dietetics & Nutrition* (9th edn, 1993), for Fig. **127**.

McGraw-Hill Inc.: Katz, B., *Nerve, Muscle and Synapse* (1966), for Fig. **88**; K. V. Kardong, *Vertebrates* (2nd edn, 1998), for Figs. **101** and **163**.

Macmillan Magazines Ltd: Fig. **12** is from Fig. 3, Paulo Sassone-Corsi, *Nature*, *392*:874 (1998); Fig. **21** is modified from *Nature*, *401*: 535 (1999); Fig. **47** is from Fig. 2, Regis B. Kelly, *Nature*, *364*:488 (1993); Fig. **48** is from Fig. 2, W. Ford Dolittle, *Nature*, *392*:16 (1998); Fig. **51** is from Fig. 1, Richard C. Strohman, *Nature Biotechnology*, *15*:197 (1997); Fig. **53** is from Fig. 2, William Martin and Miklós Müller, *Nature*, *392*:39 (1998); Fig. **64**, is from *Nature 348*: 109 (1990); Fig. **74b** is from Fig. 1a, Regis B. Kelly, *Nature*, *364*:487 (1993); Fig. **83a** is from Box 4, Bernard Wood, *Nature*, *355*:789 (1992); Fig. **84** is from *Nature 418*: 134 (2002); Fig. **107** is from Fig. 4, Peter R. Crane, Else Marie Friis and Kaj Raunsgaard Pedersen, *Nature*, *374*:31 (1995); Fig. **108b** is from Fig. 1, Timothy Rowe, *Nature*, *398*:283 (1999); Fig. **126** is from Figs. 1 and 2, Erich A. Nigg, *Nature*, *386*:780–1 (1997); Fig. **135** is from Fig. 1, Robert D. Martin, *Nature*, *363*:224 (1993); Fig. **140** is from *Nature 416*: 499 (2002); Fig. **148** is from Fig. 1, Gordon G. Carmichael, *Nature*, *418*: 380 (2002); Fig. **149** is from Fig. 1, Michael P. Robertson and Andrew D. Ellington, *Nature*, *395*:224 (1998); Fig. **151** is derived from Sean E. Egan and Robert A. Weimberg, *Nature*, *365*:782 (1993); Fig. **158** is from *Nature 418*: 380 (2002); Fig. **160** is from Fig. 1, Alan P. Wolffe, *Nature*, *387*:17 (1997); Fig. **169** is from *Nature 414*: 409 (2001).

Medical Research Council: B. Everitt 'From pleasure to compulsion', *MRC News*, Summer 1996, for Fig. **141**.

John Murray (Publishers) Ltd: Clegg, C. J. and Cox, G., *Anatomy and Activities of Plants* (1988), for Fig. **165**.

Thomas Nelson & Sons, Ltd: G. H. Harper, T. J. King, M. B. V. Roberts, *Biology Advanced Topics* (1987) for Fig. **39**; Barrington, E. J. W., *Invertebrate Structure and Function* (2nd edn, 1979), for Fig. **55**; Roberts, M. B. V., *Biology: A Functional Approach* (3rd edn, 1982), for Fig. **132**.

Philip Allen Publishers: Fig. **80** is from *Biological Sciences Review*, 6:13 (September 1993).

Prentice-Hall, Inc.: Brock, *Biology of Microorganisms* (8th edn, eds M. T. Madigan, J. M. Martinko and J. Parker, 1997), for Figs. **10a**, **90**, **102** and **154**.

Scientific American Inc.: Fig. **113** is from Douglas C. Wallace, *Scientific American*, August 1997.

Sinauer Associates Inc.: Scott F. Gilbert, *Developmental Biology* (3rd edn, 1991) – Fig. **11** is from Fig. 3, p. 653, and Fig. 16, p. 663, Fig. **24c** is from Fig. 17 (A), p. 537, Fig. **24d** is from Fig. 17 (B), p. 537, Fig. **164** is from Fig. 7, p. 416; Scott F. Gilbert, *Developmental Biology* (5th edn, 1997) for Fig. **18b**; Jeffrey M. Camhi, *Neuroethology* (1984), for Fig. **124**.

University of London: ULSEB (1990), Human Biology A Level Specimen Paper, for Fig. **13**.

Viking Penguin Inc.: Hartman, P. E., and Suskind, S. R., *Gene Action* (1965), for Fig. **32**.

Wadsworth Inc.: Salisbury, F. B. and Ross, C. W., *Plant Physiology* (4th edn, 1992), for Figs. **20b** and **134**.

Worth Publishers Inc.: Lehninger, A. L., *Biochemistry* (2nd edn, 1975), for Figs. **36**, **70**, **71b**, **73**, **122** and **167**.

Fig. **108** is reproduced here from Prof. R. D. Martin's 2nd edition of *Primate Origins and Evolution*, Chapman Hall: London.

For Assistance

It is with pleasure that we record with gratitude those who have made writing this new edition easier. Our editor, Martin Toseland, has been his characteristically supportive and accommodating self, and Ellie Smith and Patricia Croucher have done splendid editing jobs on the manuscript. Steve West and Adrian Harwood helped us with certain entries and Helen Piwnica-Worms generously allowed us to use unpublished material in Figure 21. But, inevitably, the largest debt of gratitude is owed to the many other unnamed research scientists upon whose work the authors of a dictionary such as this depend. Any errors remaining in this text lie solely at the doors of the authors, who would welcome having them pointed out.

MT
MH
February 2004

Units and Symbols

> greater than
< less than
~ approximately

Prefixes

M = mega = 10^6
K = kilo   = 10^3
d = deci   = 10^{-1}
c = centi = 10^{-2}
m = milli = 10^{-3}
μ = micro = 10^{-6}
n = nano = 10^{-9}

Suffixes

°C = degrees Celsius
      (0°C = 273° Kelvin)
J = Joule
m = metre
s = second
h = hour
g = gram
mole = 6.023×10^{23} particles
      (ions, atoms, molecules)
dm^3 = 1,000 cm^3 or litre (l)
ha = $10^4 m^2$
ppm = parts per million
yr = year
BP = before present
Ig = immunoglobulin
spp. = species (plural)
atm = atmosphere
N = Newton
bar = 10^5 N m^{-2}

ABA See ABSCISIC ACID.

A-band See STRIATED MUSCLE.

abaptation A process in which the present match between organisms and their environment, and the constraints on this match, have been determined by evolutionary forces acting upon ancestors. The prefix 'ab-' emphasizes that the heritable traits of an organism are consequences of the past and not anticipation of the present or future. See ADAPTATION.

abaxial Facing away from the main axis of an organ or organism; e.g. of a leaf surface, typically the lower surface, facing away from the stem. Compare ADAXIAL.

abdomen (1) Vertebrate body region containing viscera (e.g. intestine, liver, kidneys) other than heart and lungs; bounded anteriorly in mammals but not other classes by a diaphragm. (2) Posterior arthropod trunk segments, exhibiting TAGMOSIS in insects but not in crustaceans.

abducens nerve Sixth vertebrate CRANIAL NERVE. Mixed, but mainly motor, supplying external rectus eye-muscle.

abductor Muscle moving skeletal element, e.g. a bone, away from the midline.

aberrant chromosome behaviour Departures from normal mitotic and meiotic chromosome behaviour, often with a recognized genetic basis. Includes (1) *achiasmate meiosis*, where chiasmata fail to form (e.g. in *Drosophila* spermatogenesis; see SUPPRESSOR MUTATION); (2) *amitosis*, where a dumbbell-like constriction separates into two the apparently 'interphase-like', but often

highly polyploid, ciliate macronucleus prior to fission of the cell; (3) *chromosome extrusion* or loss, as with X-chromosomes in egg maturation of some parthenogenetic aphids (see SEX DETERMINATION); and in *Drosophila* where gynandromorphs may result; but notably in some midges (e.g. *Miastor*, *Heteropeza*) where paedogenetic larvae produce embryos whose somatic cells contain far fewer chromosomes than GERM LINE cells, owing to selective elimination during cleavage (see WEISMANN). In some scale insects, males and females develop from fertilized eggs, but males are haploid because the entire paternal chromosome set is discarded at cleavage (see HETEROCHROMATIN, PARASEXUALITY, GYNOGENESIS). Infection of the haplodiploid wasp *Nasonia vitripennis* by the bacterium *WOLBACHIA* is now known to induce extrusion of the entire paternal chromosome set through delayed nuclear envelope breakdown of male pronuclei. Such a mechanism may promote reproductive isolation and rapid speciation. The bacterium *Wolbachia*, infecting perhaps 16% of all insect species, is known to cause diploidy in the wasp *Trichogramma* by disrupting the egg's first division. This causes eggs to develop as females, condemning them to perpetual asexual reproduction by infectious thelytoky, reversible by exposure to antibiotics; (4) *meiotic drive*, where segregation of chromosomes during meiosis is non-Mendelian (transmission ratio distortion, TRD), one member of a pair being inherited more frequently than the other owing to differential effects on gamete development or survival. If sex chromosomes are involved, offspring of a cross may be all, or nearly all, of the same sex. Two loci are usually involved, one gene (the driving gene) encoding information to kill gametes (e.g. the segregation distorter, *Sd*, in male *Drosophila*), the other (*Responder*, *Rsp*, in *Drosophila*) determining sensitivity of the cell to the killer gene product. In *Drosophila*, sensitive *Rsp* sperm fail to develop. *Rsp* comprises a chain of AT-rich 120-base-pair repeats, and the longer the chain, the more sensitive *Rsp* is to *Sd*, which has several inversions, preventing *Rsp* from crossing over into the *Sd* chromosome during meiosis (which would cause suicide of *Sd* spermatids). The *Sd* protein product is a defective GAP, expressed in the fly's testes, and appears to interfere with nuclear transport in spermatids carrying a sensitive *Rsp* gene. Most cases of meiotic drive involve either non-homologous regions of sex chromosomes or inverted repeat sequences on autosomes (high copy number of the sequence brings insensitivity to the driving gene) when linkage disequilibrium between driving and insensitivity genes allows the trait to spread (crossing-over would provide an antidote to such spread, and may have contributed to its selective advantage). The *t* haplotype in mice and the Spore killer system in the mould *Neurospora* are functionally analogous to *Sd*, but each apparently operates through a different molecular mechanism. Several polymorphic drive systems in natural populations appear to have lower frequencies of driving genes than might be expected, suggesting a general strong selection acting against drive systems. See SELFISH DNA/GENES, SEX; (5) *premeiotic chromosome doubling* (see AUTOMIXIS); (6) ENDOMITOSIS, where chromosomes replicate and separate but the nucleus and cell do not divide; (7) POLYTENY, where DNA replication occurs but the strands remain together to form thick, giant chromosomes.

abiotic Environmental features, such as climatic and EDAPHIC factors, that do not derive directly from the presence of other organisms. See BIOTIC.

abomasum The 'true' stomach of RUMINANTS.

aboral Away from, or opposite, the mouth.

abscisic acid (ABA, abscisin, dormin) Inhibitory plant GROWTH SUBSTANCE, produced in plastids from carotenoids (see ISOPRENOIDS). A sesquiterpene present in a variety of plant organs: leaves, buds, seeds, fruits, roots and tubers. Promotes senescence and abscission in leaves; that produced in the core cells of the root cap has been implicated in the response of roots to gravity; induces dormancy in buds and seeds; induces stomata to close, helping reduce water loss from the plant. Antagonizes the influences of growth-promoting substances. Believed to inhibit protein and nucleic acid synthesis while apparently acting as a calcium agonist, causing influxes of the SECOND MESSENGER calcium (Ca^{2+}) into a wide variety of cells, from cyanobacteria to plants and animals, suggesting genetic conservation of ABA receptors and calcium transport mechanisms during evolution. Inhibits fecundity and egg-viability in several insects and has been located in mammalian brain tissue.

abscission layer Layer at base of leaf stalk in woody dicotyledons and gymnosperms, in which the parenchyma cells become separated from one another through dissolution of the middle lamella before leaf-fall.

absorption Uptake of water (by osmosis) and/or solutes (e.g. by simple DIFFUSION, FACILITATED DIFFUSION, ACTIVE TRANSPORT or ENDOCYTOSIS) by cells, often those forming part of a specialized epithelium. Not to be confused with ADSORPTION. Compare ASSIMILATION.

absorption spectrum Graph of light absorption versus wavelength of incident light. Shows how much light (measured as quanta) is absorbed by a pigment (e.g. plant pigments) at each wavelength. Compare ACTION SPECTRUM.

abyssal Inhabiting deep water, roughly below 1,000 metres.

abzyme Monoclonal antibody with catalytic activity.

Acanthodii Class of primitive, usually minnow-sized, fossil fish abundant in early

Devonian freshwater deposits. Earliest known gnathostomes. Bony skeletal tissue. Fins supported by very stout spine; several accessory pairs of fins common. Rows of spines between pectoral and pelvic fins. Heterocercal tail. Relationships with osteichthyan fishes uncertain, but probably not directly ancestral. See FISH, PLACO-DERMI.

Acanthopterygii Spiny-rayed fish. Largest superorder of (teleost) fishes. Spiny rays in their fins consist of solid pieces of bone (and not numerous small bony pieces); are unbranched and pointed at their tips. Radial bones of each ray are sutured or fused, preventing relative lateral movement. Often have short, deep bodies, and relatively large fins, making these fish very manoeuvrable. See TELEOSTEI.

Acari (Acarina) Order of ARACHNIDA including mites and ticks. External segmentation much reduced or absent. Larvae usually with three pairs of legs, nymphs and adults with four pairs. Of considerable economic and social importance as many are ectoparasites and vectors of pathogens.

accessory bud A bud generally situated above or on either side of main axillary bud.

accessory chromosome See SUPERNUMER-ARY CHROMOSOME.

accessory molecules Membrane glycoproteins of the Ig superfamily on certain T CELL surfaces, additional to the T CELL RECEP-TOR itself (see Fig. 159), regulating ADHESION between T cell and ANTIGEN-PRESENTING CELLS (APCs). These include many 'cluster of differentiation' (CD) antigens, first identified using monoclonal antibodies. The CD4 (or T4) gene product assists helper T cells (T$_H$) bind MHC Class II-bound antigens on the APC surface and binds a p56 tyrosine kinase on the cytoplasmic side, while the heterodimer CD8 (or T8) formed from CD8α and CD8β helps cytotoxic T cells (T$_C$) bind MHC Class I-bound antigens externally and binds a p56 tyrosine kinase on the cytoplasmic side. CD8 binds the α3 domain of MHC Class I molecules; CD2 glycoprotein binds

leucocyte functional antigen-3, promoting adhesion of target cell prior to TCR engagement. CD4 is a receptor for HIV-1 (AIDS virus).

accessory nerve Eleventh cranial nerve of tetrapod vertebrates, unusual in originating from both brainstem and spinal cord. A mixed nerve, whose major motor output is to muscles of throat, neck and viscera.

accessory pigment Pigment that captures light energy and transfers it to chlorophyll *a*, e.g. chlorophyll *b*, carotenoids, phycobiliproteins.

acclimation Usually reversible regulatory biochemical and/or physiological compensation enabling increased tolerance to changes in an animal's external or internal environment as studied under laboratory conditions (contrast ACCLIMATIZATION). Thermal acclimation, e.g., increases tolerance to temperature change; such animals are *thermally acclimated*.

acclimatization Naturally induced, and usually reversible, regulatory biochemical and/or physiological compensation enabling increased tolerance to changes in animal's external or internal environment (contrast ACCLIMATION). Thermal acclimatization, e.g., increases tolerance to temperature change; such animals are *thermally acclimatized*.

accommodation Changing the focus of the eye. In man and a few other mammals occurs by changing the curvature of the lens; at rest lens is focused for distant objects and is focused for near objects by becoming more convex with contraction of the ciliary muscles in the CILIARY BODY. See EYE, OCULO-MOTOR NERVE.

acellular Term sometimes applied to organisms or their parts in which no nucleus has sole charge of a specialized part of the cytoplasm, as in unicellular organisms. Applicable to coenocytic organisms (e.g. many fungi), and to tissues forming a SYNCY-TIUM. Sometimes preferred to 'unicellular'. See MULTICELLULARITY.

acentric (Of chromosomes) chromatids or their fragments lacking any CENTRO-MERES.

acetabulum Cup-like hollow on each side of hip girdle into which head of femur (thigh bone) fits, forming hip joint in tetrapod vertebrates. See PELVIC GIRDLE.

acetification See FERMENTATION.

acetylation (ethanoylation) Replacement of a hydrogen atom by an acetyl ($-COCH_3$) group, often by an acetyltransferase. Of considerable biological interest in the modification of tubulin and of histones.

acetylcholine (ACh) NEUROTRANSMITTER of many interneural, neuromuscular and other *cholinergic* effector synapses. Formed in mitochondria, it relays electrical signals in chemical form, with transduction back to electrical signal at the postsynaptic membrane. Initiates depolarization of postsynaptic membranes to which it binds, but hyperpolarizes vertebrate cardiac muscle membranes (slowing heart rate). Different ACh receptors mediate these effects: the cardiac receptor is a non-channel-linked RECEPTOR (*muscarinic*, activated by the fungal toxin muscarine); channel-linked ACh receptors are *nicotinic* (activated by nicotine). Nicotine receptors (ganglia, neuromuscular junctions and maybe some brain and spinal cord regions) are blocked by curare, muscarinic receptors (peripheral autonomic interneural synapses) by atropine. An anaesthetist can relax skeletal muscles by administering curare, leaving heart muscle unaffected. ACh relaxes smooth muscle and blood vessels by causing endothelial release of nitric oxide (NO), a relaxing factor, which in turn activates a secondary messenger CASCADE involving cGMP. Stored in SYNAPTIC VESICLES inside axon terminals and released there in quantal fashion in response to calcium ion uptake on arrival of an ACTION POTENTIAL (the greater the Ca^{2+} entry, the greater the ACh release). It diffuses across the synaptic cleft and binds to receptor sites on the postsynaptic membrane, whereupon these ion channels open and allow appropriately sized positive ions to enter cell, initiating membrane depolarization. Hydrolysis to choline and acetate by cholinesterase in the synaptic cleft ensures that the chemical signal is appropriately brief (see SUMMATION). Its hydrolysis is blocked by organophosphates (see PEST MANAGEMENT), which bind to active site of cholinesterase and cause accumulation of ACh in synaptic cleft and prolonged receptor activation. ACh is found in some protozoans. Compare ADRENERGIC.

acetylcholinesterase See CHOLINESTERASE.

acetyl coenzyme A See COENZYME A.

achene Simple, dry, one-seeded fruit formed from a single carpel, without any special method of opening to liberate seed; seed coat is not adherent to the pericarp; may be smooth-walled (e.g. buttercup), feathery (e.g. traveller's joy), spiny (e.g. corn buttercup), or winged (when termed a *samara*) as in sycamore and maple.

achiasmate Of meioses lacking chiasmata. One form of ABERRANT CHROMOSOME BEHAVIOUR. See SUPPRESSOR MUTATION.

achlamydeous (Of flowers) lacking petals and sepals; e.g. willow.

acicular (Of leaves) needle-shaped.

acid dyes Dyes consisting of an acidic organic compound (anion) which is the actively staining part, combined with an inorganic cation, e.g. eosin. Stain particularly cytoplasm and collagen. See BASIC DYES.

acid growth Term denoting the fact that growing plant cell walls typically extend faster at low pH, which requires active wall proteins called extensins (see CELL GROWTH) whose action is termed 'wall loosening'.

acid hydrolase Any hydrolytic enzyme whose optimum pH of activity is in the acidic range. Many different examples occur in LYSOSOMES. Pepsin is an acid protease.

acid phosphatase One of several acid hydrolases located in LYSOSOMES and concentrated in the trans-most cisternae of the GOLGI APPARATUS.

acid rain Phrase introduced in 1872 by the first British air pollution inspector, Robert Angus Smith, when he discovered that rain falling around Manchester contained sulphuric acid. Today, 'acid rain' is used more generally to describe the total deposition of atmospheric pollutants on vegetation, soils and aquatic ecosystems. This deposition can take three possible forms: (1) *Wet deposition* – rain, hail or snow; (2) *dry deposition* – where pollutants as gases or particles are absorbed on to plant and soil surfaces or settle on water surfaces; and (3) *occult deposition* – where pollution is incorporated into cloud droplets which are deposited on vegetation when enveloped in the cloud. Generally, rainwater is considered acid if the pH value falls below 5.6. Rain dissolves carbon dioxide, forming carbonic acid, giving it a normal pH of 5.6. This arbitrary pH value does not allow for natural sources of sulphur and nitrogen. Gaseous sulphur compounds of terrestrial and marine origin (e.g. dimethyl sulphide, carbonyl sulphide, hydrogen sulphide) can be oxidized to sulphur dioxide. Similarly, the release of nitric acid by natural processes of DENITRIFICATION can contribute nitric acid to the precipitation. So, the pH of unpolluted rainwater varies between 4.6 and 5.6, with a mean value of 5.0. Pollutants, attributable to human activities, can further reduce the pH of rainwater, while occult precipitation can act as a further concentrating mechanism, especially in regions where orographic cloud sits for considerable periods over upland regions. Polluted air rising into clouds can leave the pollutants dissolved in water droplets which then deposit a concentrated solution on surfaces with which they come into contact. Similarly, rainfall as drizzle possesses a greater total drop surface area than the equivalent amount of heavy rain; pH values between 2.5 and 3.0 can result from such light precipitation.

Consequences of acid precipitation include leaching of cations, particularly calcium, from soils, causing knock-on Ca^{2+} depletion in lakes and rivers – a natural process which has been accelerated this century by POLLUTION from strong acids by industry, and by car exhausts. The result (especially in Scandinavia, northern Scotland and eastern North America) has been the progressive transfer of sulphuric acid and acidity into surface waters, so that many lakes cannot support fish populations. High acidity alone interferes with fish reproduction by inhibiting softening of the hard egg case, preventing hatching. Fish are also killed by a secondary consequence of acid run-off from the soil by soluble aluminium, which upsets gill operations by injuring epithelial cells; gills become clogged with mucus and the oxygen content of the blood falls. Aluminium toxicity is greatest when Ca^{2+} concentrations are low. Damage to forest trees leading to forest decline has also occurred. Owing to leaching of Mg^{2+} ions from the soil, leaves become chlorotic and growth is reduced. Sensitivity of trees to this pigment loss may depend upon the overall nutritional status of the soil.

acinar cells See ACINI.

acinar gland (Zool.) Multicellular gland (e.g. seminal vesicle) with flask-like secretory portions.

acini (Zool.) Cells lining tubules of pancreas and secreting digestive juices. Their secretory vesicles (*zymogen granules*) concentrate the enzymes and fuse with the apical portion of the plasmalemma under the stimulus of ACETYLCHOLINE or CHOLECYSTOKININ, releasing their contents into the lumen of the duct. Much used in the study of secretion.

acoelomate Having no COELOM. Refers to some lower animal phyla, e.g. coelenterates, platyhelminths, nemerteans and nematodes.

acoustic nerve See VESTIBULOCOCHLEAR NERVE.

acoustico-lateralis system See LATERAL LINE SYSTEM.

acquired characteristics, inheritance of See LAMARCKISM.

acquired immune deficiency syndrome (AIDS) A disorder caused by the human immunodeficiency virus (HIV), which may be acquired congenitally. In 2003, 70% of the 40 million people with HIV/AIDS infection were concentrated in countries with dysfunctional health systems. Pilot studies in Africa suggest that DNA-based diagnostic methods are valuable in identifying genotypes that are resistant to the drugs used in delaying onset of AIDS in HIV-positive individuals.

acquired immune response Secondary antibody response to presence of antigen and differing from the initial response (which may precede it by a matter of years) in that it appears more quickly, achieves a higher antibody titre (concentration) in the blood and in that the principal IMMUNO-GLOBULIN species present is IgG rather than IgM. See ANTIBODY, IMMUNITY.

Acrania See CEPHALOCHORDATA.

Acrasiomycota Cellular slime moulds (Kingdom PROTISTA) comprising four genera and twelve species found living on a wide range of decaying plant material, soil and dung. These slime moulds may exist as separate amoebae (myxamoebae), which retain their original identities within the pseudoplasmodium (the 'slug') formed by swarming. A myxamoeba's pseudopodia are globose. The sporocarp is sessile (some have simple supportive stalks), independent, multispored (spores in chains or delimited sori) and divides when vegetative. Flagellated cells are usually absent, and sexual reproduction is unknown. Compare MYCETOZOA.

acrocentric Of chromosomes and chromatids in which the CENTROMERE is close to one end.

acromion Point of attachment of clavicle to scapula in mammals and mammal-like reptiles. A bone process.

acropetal (Bot.) Development of organs in succession towards apex, the oldest at base, youngest at tip (e.g. leaves on a shoot). Also used in reference to direction of transport of substances within a plant, i.e. towards the apex. Compare BASIPETAL.

acrosome Specialized penetrating vesicular organelle, formed from GOLGI APPARATUS and part of the nuclear envelope at the tip of a spermatozoon. It contains HYALURONIDASE, several lytic enzymes and acid hydrolases released when the sperm cell membrane fuses at several points with the acrosome during the *acrosome reaction*, dissolving the jelly-like zona pellucida so that the sperm head can fuse with the ovum membrane (see CELL FUSION). Some sperm discharge an *acrosomal process* composed of rapidly polymerizing ACTIN which punctures the egg membranes prior to fusion with ovum (e.g. in some echinoderms).

ACTH (adrenocorticotropic hormone, corticotropin) A polypeptide of thirty-nine amino acids secreted by anterior lobe of the pituitary, involved in the growth and secretory activity of adrenal cortex. Has a minor positive effect on aldosterone secretion, but an important role in glucocorticoid secretion. Both stress and low blood glucocorticoid levels cause release from the hypothalamus of *corticotropin releasing factor* (*CRF*) which initiates ACTH release. See ADRENAL GLAND, CORTISOL.

actin Eukaryotic protein, varying in sequence slightly from species to species, which binds to many other proteins in performance of its function. Actin has a 20% sequence similarity to the bacterial protein FtsA – both probably having diverged from a common ancestral protein – and appears to have diverged very slowly among eukaryotes: perhaps 10% in 10^9 years. Filamentous actin (*F-actin*) is composed of 43Kd globular protein monomers (*G-actin* molecules) polymerized to form long fibrous molecules, two of which coil round one other in the thin actin filaments of muscle and other eukaryotic cells, where they are termed *microfilaments*. Each *G-actin* molecule binds one calcium ion and one ATP or ADP molecule, when it polymerizes to form *F-actin* with ATP hydrolysis. F-actin formation begins with nucleation, when three or four G-actin monomers cluster;

then polymerization (elongation) produces filament growth. Nucleation is the rate-limiting step and is speeded up by the PROTEIN COMPLEX Arp2/3. Like MICROTUBULES, the opposite ends of actin filaments grow and depolymerize at different rates and play a vital role in CYTOSKELETON structure, under the influence of the G PROTEINS Rho and Rac (see CELL LOCOMOTION). Stress fibres are bundles of actin filaments and other proteins at the lower surfaces of cells in culture dishes and will contract if exposed to ATP *in vitro* (see CONTRACTILE RING). Microfilaments are involved in the building of FILOPODIA, microspikes and MICROVILLI where, as in stress fibres, they form paracrystalline bundles. Filaments of actin and MYOSIN are capable of contracting together as ACTO-MYOSIN in both muscle and non-muscle cells, e.g. in the contractile ring of dividing cells, in belt DESMOSOMES and in CYTOPLASMIC STREAMING. Numerous actin-related proteins (ARPs), with at least 40–50% sequence identity to actin, have been discovered. See CELL POLARITY.

actinomorphic (Of flowers) regular; capable of bisection vertically in two (or more) planes into similar halves, e.g. buttercup. Such flowers are also said to exhibit RADIAL SYMMETRY.

Actinomycetes 'Ray fungi'. A group of usually filamentous Gram-positive bacteria that have sometimes been classified as mitosporic fungi. Morphologically diverse, actinomycetes are usually saprotrophs (e.g. in soil). A few are pathogens, infecting humans, other animals and plants. Others form lichen-like associations with green algae. Some are important sources of ANTI-BIOTICS (e.g. *Streptomyces*).

actinomycin D Antibiotic derived from species of the bacterial genus *Streptomyces*. Binds to DNA between two G–C base pairs and prevents movement of RNA polymerase, so preventing transcription in both prokaryotes and eukaryotes. Penetrates into intact cells. See ANTIBIOTIC.

Actinopterygii Ray-finned fishes. Generally regarded as a subclass of OSTEICHTHYES, and includes all common fish except sharks, skates and rays. Earliest forms (chondrosteans) represented in the Devonian by the palaeoniscoids and today by e.g. *Polypterus*; later forms (holosteans) were predominantly Mesozoic fishes but represented today by e.g. *Lepisosteus* (gar pikes); teleosts are the dominant fish of the modern world and represent the subclass in almost every part of the globe accessible to fish. Internal nostrils absent; SCALES ganoid in primitive forms, but reduced or even absent in teleosts. The paired fins are webs of skin braced by horny rays (like ribs of a fan), each a row of slender scales, there being no fleshy fin lobes except in the most primitive forms. A swim bladder is present and the skeleton is bony. Internal groupings given here probably represent GRADES rather than CLADES. See TELEOSTEI, ACANTHOPTERYGII.

Actinozoa (Anthozoa) Sea anemones, corals, sea pens, etc. A class of Coelenterata (Subphylum Cnidaria). The body is a polyp, there being no medusoid stage in the life cycle. Polyp more complexly organized than that of other coelenterates; coelenteron divided by vertical mesenteries. May have external calcareous skeleton as in well-known corals, but some forms have internal skeleton of spicules in mesogloea.

action potential Localized reversal and then restoration of electrical potential between the inside and outside of a nerve or muscle cell (or fibre) which marks the position of an impulse as it travels along the cell. See IMPULSE, RESTING POTENTIAL, ACTIVATION.

action spectrum Plot of the quanta of different wavelengths required for a photochemical response against the wavelength of light used. Its reciprocal indicates photochemical efficiency.

activation (Of eggs) when the membrane of the sperm ACROSOME fuses with the egg plasma membrane, an *activation reaction* passes over the egg surface involving an ACTION POTENTIAL of longer duration than in nerve or muscle. It signifies the onset of embryological development and may be achieved merely by pricking of some eggs (e.g. frog). See CALCIUM SIGNALLING.

activators Proteins binding to ENHANCER sites (see Fig. 49) of genes and to coactivators (which bind transcription factors), determining which genes are transcribed as well as their speed of transcription. See GENE EXPRESSION.

active site Part of an enzyme molecule in its natural hydrated state which, by its three-dimensional conformation and charge distribution, confers upon the enzyme its substrate specificity. It binds to a substrate molecule, forming a transient enzyme–substrate complex. Enzymes may have more than one active site and so catalyse more than one reaction. Competitive inhibitors of an enzyme reaction bind reversibly to the active site and reduce its availability for normal substrate. Active sites may only take on their appropriate conformation after the enzyme has combined at some other site with an appropriate modulator molecule. Some active sites require metal ions as prosthetic groups (e.g. human carboxypeptidase requires a zinc atom). See ENZYME.

active transport The energy-dependent carriage of a substance across a cell membrane, accumulating it on the other side in opposition to chemical or electrochemical gradients (i.e. 'uphill'). The process involves 'pumps' composed of protein molecules in the membrane (often traversing it) which carry out the transport. Requires an appropriate energy supply, commonly ATP, or a gradient of protons across the membrane itself usually generated by redox, photochemical or ATP-hydrolysing reactions. Collapse of this gradient drives proton-linked symports or antiports (see TRANSPORT PROTEINS). Alternatively, a membrane potential arising from ion asymmetry across the membrane may drive specific ions through special transport systems. Solute transport (commonly by facilitated diffusion) across a cell membrane is often only possible because active transport is occurring elsewhere in the cell. In this case, passage across the first membrane is termed *secondary active transport* (as with sodium/glucose co-transport into cells of the proximal convoluted tubule in the vertebrate kidney).

Probably all cells engage in active transport. See SODIUM PUMP, ELECTRON TRANSPORT SYSTEM, FACILITATED DIFFUSION.

active zone A specialized region of a pre-synaptic membrane to which synaptic vesicles fuse prior to release of transmitter.

activin A member of the transforming growth factor β (TGF-β) superfamily of secreted growth factors. Has strong mesoderm-inducing properties in the early *Xenopus* embryo and can induce gene activity in distant cells in a concentration-dependent manner (i.e. activin may be a MORPHOGEN). Activins include some gonadal peptides produced by Sertoli cells, which stimulate FSH release. See TGFS, INHIBINS.

actomyosin Complex formed when the pure proteins ACTIN and MYOSIN are mixed, resulting in increased viscosity of the solution. Actomyosin undergoes dissociation in the presence of ATP and magnesium ions (Mg^{2+}), when ATP hydrolysis occurs. Completion of this hydrolysis results in reaggregation of the two proteins. Live muscle cells have an absolute requirement for calcium ions (Ca^{2+}) before myosin and actin filaments will interact, and when Ca^{2+} is removed the actin and myosin dissociate. Such interactions form the basis of many biological force-generating events, notably during MUSCLE CONTRACTION, CYTOPLASMIC STREAMING, CELL LOCOMOTION and blood clot contraction.

acute-phase proteins Serum proteins which build up rapidly in concentration (*c*. 100-fold) following infection (the acute phase response). Include C-reactive protein (CRP), which binds to C-protein of pneumococci, in turn promoting COMPLEMENT binding (and hence phagocytosis) when attached to bacteria. See MACROPHAGE.

adaptation (1) Evolutionary. Some property of an organism is normally regarded as an adaptation (i.e. fits the organism in its environment) if (a) it occurs commonly in the population, and (b) the cause of its commonness was NATURAL SELECTION in its favour. Adaptations are not, therefore, fortuitous benefits, the implication being that they have a genetic basis, since selection operates only upon genetic differences between individuals. Alternatively, we often in practice identify an adaptation by its effects rather than its causes. Learned abilities which improve an individual's FIT-NESS or inclusive fitness, but without clear genetic causation, are cases in point. See TELEOLOGY. (2) Physiological. A change in an organism, resulting from exposure to certain environmental conditions, allowing it to respond more effectively to them. (3) Sensory. A change in excitability of a sense organ through continuous stimulation, increasingly intense stimuli being required to produce the same response.

adaptive enzyme Inducible enzyme. See ENZYME.

adaptive immune response See IMMUNITY, APOPTOSIS.

adaptive radiation Evolutionary diversification from a single ancestral (prototype) population of descendant populations into more and more numerous ADAPTIVE ZONES and ecological NICHES. When a new higher taxon emerges, there is often a geologically rapid diversification into different adaptive zones. Such evolution occurred in the Cretaceous period, when early mammals evolved into carnivores, primates and ungulates. In the Eocene, ~50 Myr BP, there was a burst of mammalian evolution, including many new ungulates which experienced a range of diversity of upper second molars (linked to diet). In the Oligocene, ungulate genera crashed to pre-Eocene levels; but the group expanded again in the Miocene, 25 Myr later, although this time there was far less tooth diversity. The implication is that not every aspect of morphology automatically radiates in parallel with taxonomic diversification.

Experimental introduction of the anolid lizard *Anolis sagrei* on to fourteen very small Bahamian islands where no lizards occurred before, indicated that in just fourteen years the within-species adaptative differentiation mirrored in miniature some aspects of the larger-scale adaptive radiations of anolids in the Greater Antilles over

geological time. May involve both ANA-GENESIS and CLADOGENESIS. See CHARACTER DISPLACEMENT.

adaptive zone A more or less distinctive set of ecological niches established and occupied by an evolutionary lineage with time.

adaptors Molecules which link a cell surface receptor to an intracellular signalling pathway. See SIGNAL TRANSDUCTION.

adaxial Facing towards the main axis of an organ or organism; e.g. of a leaf surface, typically the upper surface, facing the stem. Compare ABAXIAL.

addiction For addiction to drugs, see PSYCHOACTIVE DRUGS.

addition mutation See MUTATION.

adductor Muscle moving a bone closer to the midline. Contrast ABDUCTOR.

adenine A purine base of DNA, RNA, some nucleotides and their derivatives. Precursor of plant CYTOKININS. See ADENOSINE.

adenohypophysis See PITUITARY GLAND.

adenosine Purine nucleoside breakdown product of ATP at some synapses, acting as an inhibitory sympathetic neurotransmitter and neuromodulator where purine receptors are present. Reduces heart muscle and vas deferens contractility, and is released from nerve plexuses connecting with smooth muscle of the gut and from dorsal root ganglion cells synapsing in the dorsal horn of the spinal cord. Adenosine receptors are linked to G PROTEINS and some are involved in anti-inflammatory responses in immune cells. CAFFEINE blocks adenosine receptors. Presynaptic receptors for adenosine have been described.

adenosine deaminase Enzyme converting adenosine to inosine. Once 'hyperedited' in this way, double-stranded RNAs resulting from antisense transcription are bound tightly to a protein complex and retained within the nucleus; but messages resulting from modification of short duplex components of pre-RNAs, with only a few inosines, may be delivered to the cytoplasm. See RNA EDITING.

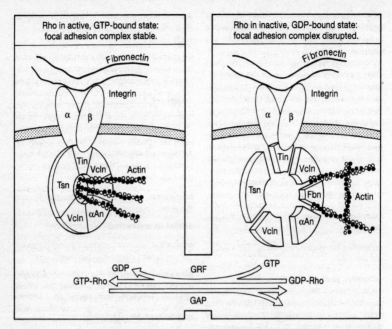

FIG. 1 *A model for focal* ADHESION *regulation by the GTPase, Rho. The complex lipid and protein structure is stabilized by GTP-Rho. Some of the proteins involved are shown: the heterodimer integrin; Tin = talin; Tsn = tensin; Vcln = vinculin; Fbn = fimbrin; aAn = α-actinin; GRF = guanine nucleotide release factor; GAP = GTP-ase activity protein; Pi = inorganic phosphate. The shapes and stoichiometry of molecules are not to be taken literally.*

adenosine diphosphate See ADP.

adenosine monophosphate See AMP.

adenosine triphosphate See ATP.

adenovirus One kind of DNA tumour virus of animals. See VIRUS.

adenylyl cyclase (adenylate cyclase) A membrane receptor-bound, or freely soluble, enzyme converting ATP to CYCLIC AMP, nearly all forms of which are regulated by one or other arm of the PHOSPHOLIPASE C pathway, intimately associated with sites of calcium entry into the cell. Rise in intracellular Ca^{2+} ion level commonly feeds back negatively on adenylyl cyclase activity, causing oscillations in intracellular cAMP levels (see CALCIUM SIGNALLING, GABA). It is the target for the A_1 subunit of cholera enterotoxin,

which activates the enzyme and causes active secretion of chloride and hydrogen carbonate ions from the mucosa, the resulting loss of water from the ileum exceeding uptake by the colon, leading to dehydration. See ORAL REHYDRATION THERAPY. Many peptide hormones and local chemical signals operate through activation of this enzyme. See RECEPTOR.

ADH See ANTIDIURETIC HORMONE.

adhesion Mechanisms by which cells are able, by forming focal contacts (focal adhesions, see Fig. 1) with one another and/ or with an appropriate substratum, to link the EXTRACELLULAR MATRIX to the intracellular ACTIN network. This GTP-dependent process is often made possible through the several affinities of their cell-surface receptors, often acting as components of INTERCELLULAR JUNCTIONS. Involved in MULTICELLULARITY, MORPHOGENESIS and the functions of the immune system.

Several families of adhesion receptors have been recognized (see ACCESSORY MOLECULES). Cell–cell adhesion molecules (CAMs) are either calcium-dependent or calcium-independent proteins. The former (which include CADHERINS, INTEGRINS and SELECTINS) seem more important in tissue-specific cell–cell adhesion in vertebrate embryogenesis, as when blastomeres become tightly joined by desmosomes and adhesion belts; each cadherin is encoded by a separate gene and most are transmembrane linker proteins which link the cytoskeletons of the adhering cells. Expression of E-cadherin is sufficient to cause cell–cell adhesion even in non-adherent cell lines; under-expression promotes invasiveness (a feature of transformed CANCER CELLS). Most integrins, however, seem to bind cells to their EXTRACELLULAR MATRIX. Selectins are more ephemeral in their action, e.g. binding some white blood cells to the endothelial lining of a small blood vessel and enabling extravasation (creeping out) into the tissue fluid during INFLAMMATION. Calcium-independent CAMs include the neural cell adhesion molecules (N-CAMs), immunoglobulin-like proteins whose various forms are arrived at by alternative RNA SPLICING from a single gene. Many N-CAMs contain negatively charged sialic acid residues in their extracellular domains which inhibit cell–cell adhesion and so modify their adhesive function. Adhesive interactions between animal epithelial cells and the basement membrane or extra-cellular matrix determine cell shape through changes in cytoskeletal reorganization through mediation of CYCLIC NUCLEOTIDES and divalent ions. Such changes may modulate a cell's response to GROWTH FACTORS. Some adhesion molecules are exploited as VIRUS RECEPTORS. Only cells circulating in the blood are adapted to survive without anchorage.

adiabatic expansion Expansion that takes place without heat entering or leaving a system (e.g., air expands adiabatically as atmospheric pressure falls with increasing altitude).

adipose tissue A connective tissue. (1) Brown adipose tissue (brown fat) comprises cells whose granular cytoplasm is due to high concentration of cytochromes and whose function appears to be release of heat in the neonatal mammal. Brown adipocytes have mitochondria with proton-leaky mitochondrial membranes, causing heat and not ATP to be evolved from electron transport. Distributed around neck and between scapulae in these and hibernating mammals but not otherwise extensively in adults. Richly innervated and vascularized. (2) White adipose tissue is distributed widely in animal bodies, comprising large cells (fat cells) each with a single large fat droplet inside a thin rim of cytoplasm. This depot fat is composed largely of triglyceride. ADRENALINE, GLUCAGON, GROWTH HORMONE and ACTH all stimulate release of fatty acids and glycerol via activation of intrinsic lipases via CYCLIC AMP. It is less richly innervated than brown adipose tissue. Fat cells also produce hormones which directly affect other tissues' responses to insulin (see LEPTIN).

In humans, adipocyte precursors (fibroblast-like cells) divide through the first year of life, after which increase in tissue volume is by HYPERTROPHY accentuated by giving high-fat diets, and as they enlarge adipocytes start losing their ability to regulate the daily influx of dietary fat. Fat is deposited in other tissues (e.g. skeletal muscle and liver), the adipocytes become less sensitive (more resistant) to insulin, and increasing adipose tissue volume leads to increasing risk of developing DIABETES mellitus. Cases of genetic obesity may be due to prolongation of the period for which adipocyte precursors can divide. See HIBERNATION.

adjuvant A substance non-specifically enhancing the immune response to an antigen; e.g. lipopolysaccharide of Gram-negative bacterial cell walls and spirochaete peptidoglycan. In laboratory use, Complete Freund's adjuvant comprises killed *Mycobacterium tuberculosis* suspended in oil and emulsified with the aqueous antigen preparation. See MUTATION, VACCINES.

ADP (adenosine diphosphate) A nucleoside diphosphate found universally inside cells. Phosphorylated to ATP during energy-yielding catabolic reactions and produced in turn when ATP itself is hydrolysed.

adrenal gland (suprarenal g., epinephric g.) Endocrine gland of most tetrapod vertebrates lying paired on either side of the mid-line, one atop each kidney. Each is a composite of an outer cortex derived from coelomic mesoderm, making up the bulk of the gland, and an inner medulla derived from neural crest cells of the ectoderm. Rarely found as a composite gland in fish. Cortex comprises three zones, the outermost secreting aldosterone which promotes water retention by kidneys by increasing renal potassium excretion and sodium retention; the middle zone secretes cortisol and other glucocorticoids under ACTH control and enhancing GLUCONEOGENESIS; the innermost zone secretes sex (mainly male) hormones. The medulla comprises *chromaffin cells*, effectively postganglionic sympathetic nerve cells that have lost their axons and secrete adrenaline (epinephrine) and a little noradrenaline (norepinephrine) into the large surrounding blood-filled sinuses. These mimic effects of the sympathetic nervous system (SEE AUTONOMIC NERVOUS SYSTEM), release being under hypothalamic control via the splanchnic nerve. They promote liver and muscle glycogenolysis via CYCLIC AMP, lipolysis in ADIPOSE TISSUE, vasodilation in skeletal and heart muscle and brain, and vasoconstriction in skin and gut. They relax bronchi and bronchioles and increase rate and power of heart beat, raising blood pressure. All adrenal hormones are known as 'stress' hormones, those of the cortex responding to internal physiological stress such as low blood temperature or volume, while medullary hormones are released in response to stress situations (often auditory or visual) outside the body. See L-DOPA.

adrenaline (In USA, EPINEPHRINE.) Hormone derivative of amino acid tyrosine (hence a catecholamine) secreted by chromaffin cells of ADRENAL GLAND and to a lesser extent by sympathetic nerve endings. May enhance memory during periods of excitement. Effect on cells is mediated by ADENYLYL CYCLASE membrane receptors.

adrenergic Of a motor nerve fibre secreting at its end noradrenaline (norepinephrine) or, less commonly, adrenaline. Characteristic of postganglionic sympathetic nerve endings. Compare CHOLINERGIC.

adrenocorticotropic hormone (ACTH) See ACTH, ADRENAL GLAND.

adsorption Attachment by chemical bond or by van der Waals forces of one substance to the surface of another (e.g. of gas on to solid). Important in heterogeneous catalysis, when the catalyst is in a different phase from the reactants.

adventitious Arising in an abnormal position; of roots, developing from part of the plant other than roots (e.g. from stem or leaf cutting); of buds, developing from part of the plant other than a leaf axil (e.g. from a root).

aecium A cup-shaped structure in which binucleate (dikaryotic) spores (aeciospores) of rust fungi (BASIDIOMYCOTA, Order UREDINALES) are produced. The aeciospores develop in chains and are dispersed by wind; they only infect wheat or related grasses. See PYCNIUM, UREDIUM, UREDINALES, TELIUM.

aerenchyma Secondary spongy tissue of some aquatic plants, with intercellular air spaces formed by the activity of a CORK cambium or phellogen. May develop in a lesser way from the lenticels of land plants such as willow (*Salix*), and poplar (*Populus*) if partially submerged. Seems to function mainly in a flotation capacity rather than as a respiratory aid.

aerobic Requiring free (gaseous or dissolved) oxygen. In most cases the oxygen is utilized in aerobic respiration, but a few enzymes (oxygenases) insert oxygen atoms directly into organic substrates. See RESPIRATION.

aerobic anoxygenic photosynthesis Method of PHOTOSYNTHESIS, employed by some marine α-proteobacteria and probably by many other marine microbes, in which oxygen is not evolved but is used instead to synthesize chlorophyll and enable growth. Involves possibly some 10% of all open ocean bacteria, in addition to those in beach sands and the surfaces of seaweeds.

aerobic exercise For aerobic training, see MUSCLE CONTRACTION.

aerobic respiration See AEROBIC, RESPIRATION.

aerosols (1) See GLOBAL WARMING; (2) in inoculating media with microorganisms; see AIRBORNE INFECTIONS.

aesthetic injury level The level of pest abundance above which aesthetic or sociological considerations suggest PEST MANAGEMENT measures should be taken.

aestivation (Bot.) Arrangement of parts in a flower-bud. (Zool.) DORMANCY during summer or dry season as, e.g., in lungfish (dipnoans). See HIBERNATION.

aethalium See MYCETOZOA.

afferent Leading towards, as of arteries leading to vertebrate gills or of nerve fibres (sensory) conducting an input towards the central nervous system. Opposite of EFFERENT.

affinity chromatography See CHROMATOGRAPHY.

affinity labelling Method by which a label is used to identify the active site of an enzyme. The label resembles the enzyme's normal substrate, but then forms a covalent bond with a group on the enzyme in close proximity to the active site and remains stably attached.

affinity maturation Introduction of point mutations in immunoglobulin (Ig) V genes (somatic hypermutation, see ANTIBODY DIVERSITY) within a population of memory B cell precursors followed by the selective proliferation of those B cells having mutations causing them to bind antigen with increased affinity. COGNATE help by T cells and cross-linking of the Ig surface are required *in vitro*.

aflatoxins EXOTOXINS produced by the fungus *Aspergillus flavus*, which grows on improperly stored grain. Aflatoxins are highly toxic and induce tumours, especially in birds feeding on contaminated grain. See POISONS.

A-form helix (A DNA, DNA-A) Right-handed, double helical form of DNA. Less hydrated and consequently more compact and stable than the more common B-FORM HELIX. Unlike that form, the axis of the molecule does not pass through the hydrogen bonds holding the base pairs together, so the minor groove is narrower. See Fig. 42a.

after-ripening Dormancy exhibited by certain seeds (e.g. hawthorn, apple) which, although embryo is apparently fully developed, will not germinate immediately seed is formed. Even when removed from seed coat and provided with favourable conditions, the embryo has to undergo certain chemical and physical changes before it can grow. Possibly associated with delay in production of required growth substances, or with gradual breakdown of growth inhibitors. See DORMANT.

agamospecies See SPECIES.

agamospermy Any plant APOMIXIS in which embryos and seeds are formed but without prior sexual fusion. Excludes vegetative reproduction (vegetative apomixis). Occurs widely in higher plants, both ferns and flowering plants. Unknown in gymnosperms. See PSEUDOGAMY.

agamospory Asexual formation of an embryo and the subsequent development of a seed.

agar The gelatinous fraction of the cell wall of certain red algae (Rhodophyta), comprising the polysaccharide *agarose* (β-1,3-linked D-galactose and 1,4-linked anhydro-L-galactose) and *agaropectin* (a sulphated galactan mixture) in a ratio of about 7:3. Once solidified, agar gel does not melt below 100°C. A 1.5% solution is clear and forms a solid but elastic gel on cooling to 32–39°C, not dissolving again at a temperature below 85°C. Comprises two polysaccharides, agarose and agaropectin, and is commercially extracted from such red algal species as *Gelidium*, *Pterocladia*, *Acanthopeltis*, *Ahnfeltia* and *Gracilaria*. Greatest use of agar occurs in the food and pharmaceutical industries for gelling and thickening purposes (e.g. canning of fish and meat, processed cheese, mayonnaise, puddings, jellies), emulsions, ointments and lotions. It is also widely used as a solidifying agent for culture media in bacteriology, mycology and phycology. Compare ALGINATES.

agarose Polysaccharide used as gel (see AGAR) in column chromatography and in electrophoresis. See NORTHERN BLOT TECHNIQUE, SOUTHERN BLOT TECHNIQUE.

ageing (senescence) Progressive deterioration in activity of cells, tissues, organs, etc., as a function of the time or number of cell generations since that activity commenced in the individual. Ageing cells in culture may remain viable after a definable number of replications, but they will then stop growing. Cells that have undergone irreversible cessation of cell division are said to exhibit 'cellular senescence'.

T cells from older humans are fewer in number (see THYMUS) and respond less well to mitogens. Virgin T cells are replaced by memory cells, and cells with signal transduction defects increase. B cells also tend to respond less well to antigen in mounting a secondary immune response. Senescence may in part be due to inability of body cells to respond to GROWTH FACTORS and other extrinsic factors. TELOMERE shortening has been linked to some aspects of cellular ageing. In humans, the normal cross-linking of proteins through diketone bridges caused over time by exposure to blood sugar (*glycation*) is accelerated in those with sugar DIABETES, and such modified proteins are thought to represent tissue damage at the molecular level. Advanced glycation has also been implicated in the pathology of ALZHEIMER'S DISEASE. Many cell surfaces bear the protein RAGE ('receptor for advanced glycation end-products'), a member of the IMMUNOGLOBULIN SUPERFAMILY of proteins, which recognizes potentially damaged, glycated proteins that accumulate during diabetes. By dividing indefinitely, bacteria and many protozoans avoid ageing; higher plants often seem capable of unlimited vegetative propagation. However, even young cultures may be provoked to senesce within days of experiencing the effects of an introduced *RAS* oncogene. As with APOPTOSIS, senescence represents a short-term response that cells may activate to prevent their own further proliferation. Regeneration and renewal in many simple invertebrates seem to permit escape from senescence. In the nematode *Caenorhabditis elegans*, four genes, when mutated, can make the worms use energy more efficiently, feed and swim at a slower pace, and live many times their normal life-span. In 1956, Harman proposed that the metabolic by-products known as reactive oxygen species ('ROS', see SUPEROXIDES) continually damage cellular macromolecules such as DNA, proteins and lipids, and that imperfect repair of this damage would lead to age-related deterioration. This '*free radical attack theory*' is supported by several observations, not least, that from yeast to rodents, dietary restriction delays ageing and age-related accumulation of genetic lesions, and extends *lifespan*. In the worm *C. elegans*, the transcription factor DAF-16 switches on several minor genes (of small individual effect) which promote longevity – apparently in an additive way – and switches off others associated with reduced lifespan. The tyrosine kinase receptors InR in *Drosophila* and DAF-2 in *Caenorhabditis* have also been shown to regulate lifespan and have homologues in the mammalian insulin receptor and insulin-like growth factor type-1 receptor (IGF-1R). Heterozygous knockout *Igf1r*[+/-] mice display greater resistance to oxidative stress than do their wild-type littermates and the IGF-1 receptor may be a central regulator of mammalian lifespan (see DNA REPAIR MECHANISMS, INSULIN/INSULIN-LIKE GROWTH FACTOR PATHWAY, p53). Many of the genes influencing longevity are involved with hormonal signals, and mechanisms promoting such survival often oppose the progress of ageing. Some of these genes encode enzymes that protect against, or repair, oxidative damage. There may be an internal clock which puts into synchrony everything with a temporal component in the animal. Mutations in the *clk-1* gene, which is also a MATERNAL EFFECT gene, cause individual cells to divide more slowly so that the worm spends more time in each phase of its life cycle, such mutations being overridden by proteins in the egg produced by any mother containing at least one normal *clk-1* gene.

GERM LINES of sexual metazoa are potentially immortal (see WEISMANN); and cells with certain genetic defects can become immortal and never stop dividing – a common characteristic of cultured CANCER CELLS. Expressed as disintegration of somatic tissue, ageing may be due to gradual accumulation of somatic mutations or to late expression of genes not subject to strong selection. In yeast, accumulation of extra-chromosomal ribosomal DNA (rDNA) circles has been implicated in ageing, accompanied by progressive enlargement and fragmentation of the nucleolus. Elsewhere, loss of DNA methylation has been held responsible. In the population context, it may be due to inbreeding or to some other factor reducing genetic variation. In many plants, particularly annuals and biennials, senescence and death of the whole organism follow flowering and fruiting (CYTO-KININS may delay senescence until fruiting is over). New leaves often induce senescence in older leaves, which may be reversed if the former are removed. Several GROWTH SUB-STANCES, including ETHYLENE, ABSCISIC ACID and AUXINS are known to affect senescence in one or more plant organs. See APOPTOSIS.

age pyramids Figures showing consecutive age-classes (cohorts) of a population as percentages of the total, with the two sexes represented on either side of the midline. The resulting shape can indicate whether the population is likely to remain stable, decline or increase. See Fig. 2, DEMOGRAPHY, SURVIVORSHIP CURVES.

agglutination Sticking together or clumping; as of bacteria (an effect of antibodies), or through mismatch of AGGLUTINOGENS of red blood cells and plasma AGGLUTININS in blood transfusions. See LECTIN.

agglutinins (isoantibodies) Plasma and cell-surface proteins that by interacting with AGGLUTINOGENS (antigens) on foreign cells can cause cell clumping (AGGLUTINA-TION). Commonly LECTINS.

agglutinogen Proteins acting as cell-surface antigens of red blood cells and interacting with AGGLUTININS to cause red cell clumping and possible blockage of blood vessels. Genetically determined, and the basis of BLOOD GROUPS.

aggregated distribution The distribution of organisms in which individuals are closer together than they would be if they were randomly or evenly distributed.

aggregate fruit Fruit which develops from several separate carpels of a single flower (e.g. magnolia, raspberry, strawberry).

aggregative response The response of a predator which results in it spending more time in habitat patches with higher densities of prey, leading to higher densities of predators in patches with higher densities of prey.

aging See AGEING.

Agnatha Class of Subphylum Vertebrata (if treated as a class, other vertebrates form Superclass Gnathostomata). Modern forms (cyclostomes) include lampreys (Subclass Mororhina) and hagfishes (Subclass Diplorhina), but fossil forms included anaspids, osteostracans and heterostracans. Jawless vertebrates. Buccal chamber acts as muscular pump sucking water in, serving

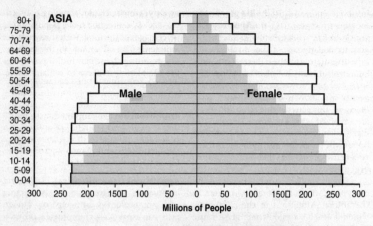

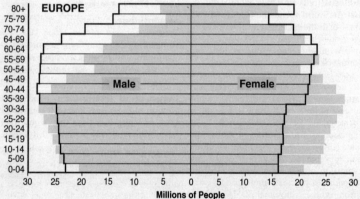

FIG. 2 *Human age distributions in Asia and Europe compared, for 1995 (dark) and projected 2025 (outlined). Asia is starting to resemble Europe in its demography. See* AGE PYRAMIDS.

for filter-feeding in lamprey larvae as well as ventilating gills – an advance over ciliary mechanisms. Paired appendages almost unknown. Earliest forms (heterostracans) appear in the late Cambrian.

agonist A substance binding to and activating a receptor, thereby mimicking the effect of a natural neurotransmitter or hormone. Thus phenylephrine activates α_1 adrenergic receptors, constricting blood vessels in the nasal mucosa and relieving symptoms of the common cold. Compare ANTAGONIST.

agonistic behaviour Intraspecific behaviour normally interpreted as attacking, threatening, submissive or fleeing. Actual physical injury tends to be rare in most apparently aggressive encounters.

agrestrials Non-crop plants that are adapted to agricultural practices and can often be associated with a particular crop; these are of economic importance, for they compete for space, nutrients, light and water. See WEED.

Agrobacterium Bacterial genus noted for crown gall tumour-inducing ability.

Oncogenic strains are host to a tumour-inducing (Ti) PLASMID which can be transmitted between species. A segment (T) of the Ti plasmid is transmitted to the plant host cell and is the immediate agent of tumour induction. It is not known to infect monocotyledonous plants, which include the economically important cereals. See ONCOGENE, PROTOPLAST, *RHIZOBIUM*.

ahnfeltan A complex phycocolloid substance occurring in the cell walls of some red algae (Rhodophyta).

AIDS See ACQUIRED IMMUNE DEFICIENCY SYNDROME.

airborne infections Only those pathogens which can resist drying are commonly transmitted from host to host aerially, often by infectious droplets (AEROSOLS) originating in the respiratory tract – although these often do not remain airborne for long. INFLUENZA virus and the common cold virus are often spread in this way. See ANTHRAX.

air bladder See GAS BLADDER.

air pollution See POLLUTION.

air sacs (1) Expanded bronchi in abdomen and thorax of birds, initially in five pairs but one or more pairs fusing to form thin-walled passive sacs with limited vascularization. Ramify throughout the body and within bones. Connected to lung by small tubes whose relative diameters are probably crucially important in establishing a unidirectional passage of air from lung to sacs and back to lung. The avian ventilation system lacks a tidal rhythm characteristic of mammals. (2) Expansions of insect tracheae into thin-walled diverticulae whose compression and expansion assist VENTILATION.

akinete Vegetative cell which becomes transformed into a thick-walled, resistant spore. Formed by certain Cyanobacteria and some algae (e.g. some Chlorophyta).

Akt See PROTEIN KINASE B.

albinism Failure to develop pigment, particularly melanin, in skin, hair and iris. Resulting *albinos* light-skinned with white hair and 'pink' eyes due to reflection from choroid capillaries behind retina. In mammals, including humans, usually due to homozygous autosomal recessive gene resulting in failure to produce enzyme tyrosine 3-mono-oxygenase.

albumen Egg-white of birds and some reptiles comprising mostly solution of ALBUMIN with other proteins and fibres of the glycoprotein *ovomucoid*. Contains the dense rope-like CHALAZA and with yolk supplies protein and vitamins to embryo, but is also major source of water and minerals.

albumin Group of several small proteins produced by the liver, forming up to half of human plasma protein content, with major responsibility for transport of free fatty acids, for blood viscosity and OSMOTIC POTENTIAL. If present in low concentration oedema may result, as in kwashiorkor. See IMMUNOASSAYS.

albuminous cells Ray and parenchyma cells in gymnosperm phloem, closely associated morphologically and physiologically with sieve cells.

alcohols Organic compounds with at least one (monohydric) or more (dihydric, trihydric, etc.) hydroxyl groups (-OH) attached directly to carbon atoms. Aliphatic dihydric alcohols are known as glycols; cyclic alcohols include phenols. Alcohols ending -ol are named after the corresponding paraffin hydrocarbon, e.g. ethanol for ethyl alcohol. Monohydric alcohols react with metals such as sodium, calcium and aluminium to give alkoxides and give esters on reacting with acids. On oxidation, primary alcohols (RCH_2OH) give aldehydes; secondary alcohols ($RR'CHOH$) give ketones, while tertiary alcohols ($RR'R''COH$) give a mixture of carboxylic acids with fewer carbon atoms than the original alcohol. Many of the toxic and addictive effects of alcohol (e.g. ethanol C_2H_5OH) are due to the production of ethanal (CH_3CHO) by alcohol dehydrogenase in the cytosol, by the microsomal ethanol oxidizing system in the ENDOPLASMIC RETICULUM, and by catalase in cell peroxisomes. Ethanal is very reactive, forming covalent bonds with functional groups such as –SH and –NH_2 groups of biologically active compounds (e.g. in enzyme active sites, membrane proteins and cytoskeletal microfilaments). It also inhibits the metabolism of biogenic amines such as NORADRENALINE. Ethanol's anaesthetic effect appears to result from its inhibition of some synaptic ion channels and potentiation of others. The earliest chemical evidence of wine comes from a pottery jar at a Neolithic site in Iran between 7.4–7 Ka BP (see FERMENTATION) and ethanol was first used as a motor fuel in 1890. Fuel shortages in World War II encouraged this technology, and at present most industrial alcohol is produced as a by-product of the petrochemical industry, some of it mixed with petrol to form 'gasohol'. Brazil leads the field in the use of industrial fermenters to produce sufficient alcohol to replace its petrol needs, *Zygomonas mobilis* rather than *Saccharomyces* species being used as a more efficient converter of sugar to ethanol. When immobilized to an inert support column, *Zygomonas* cells are particularly good for continuous culture production (see BIOREACTOR). Up to one third of hospital patient admissions in the UK are for alcohol-related problems, placing a severe burden on National Health Service funds. See CORONARY HEART DISEASE.

Alcyonaria Order of coelenterates within the Class Actinozoa. Sea pens, soft corals, etc. Have eight pinnate tentacles and eight mesenteries. Polyps colonial, with continuity of body wall and enteron. Skeleton, often of calcareous spicules, within mesogloea and occasionally externally.

aldosterone Hormone of ADRENAL cortex. See OSMOREGULATION.

aleurone grains Membrane-bound granules of storage protein occurring in the outermost cell layer of the endosperm of wheat and other grains.

aleurone layer Metabolically active cells of outer cereal endosperm (in contrast to metabolically inactive cells of most of the endosperm) containing *aleurone grains*, several hydrolytic enzymes and reserves of *phytin* (releasing inorganic phosphate and inositol on digestion by phytase). During germination, aleurone cells secrete α-amylase into the endosperm, initiating its digestion. Recent work suggests that the synthesis of enzymes by aleurone cells may not be as specifically induced by gibberellins from the embryo axis as was once thought, although these growth substances are certainly implicated in the control of endosperm digestion.

aleuroplast Colourless plastid (leucoplast) storing protein; found in many seeds, e.g. brazil nuts.

algae An informal grouping of a diverse array of mainly photosynthetic organisms within the Kingdom PROTISTA (= Protoctista), all possessing chlorophyll *a* as their primary photosynthetic pigment, and lacking a sterile covering of cells around reproductive cells. The body comprises a thallus (i.e. lacks roots, stems and leaves). Algae live in the oceans and seas to a depth of about 150 m, depending upon the transparency of the water. They occur on shores and coasts, attached to the bottom (benthic species) or live suspended in the open water (planktonic species). Algae populate freshwater, as well as terrestrial habitats in soil, among bryophytes, on damp rock faces, and on tree trunks. A few live endolithically in deserts, relying upon night-time dew for their source of water; others grow on melting snow, or attached to the under-surface of floating ice. Algae are extremely important ecologically, as they probably account for more than half the total primary production worldwide (and virtually all aquatic animals are dependent on this production). Some algae, particularly species of red algae (RHODOPHYTA) and brown algae (PHAEOPHYCEAE), are harvested and eaten as a vegetable, or the mucilages are extracted for use as sizing, gelling and thickening agents (see AGAR, CARRAGEENAN).

Within the algae are forms not necessarily closely related but included with the other algae by most phycologists (e.g. CYANOBACTERIA and PROCHLOROPHYTA). These belong to the kingdom BACTERIA which, together with the Archaebacteria, comprise the PROKARYOTES. The formal eukaryotic divisional taxon 'Algae' has been aban-

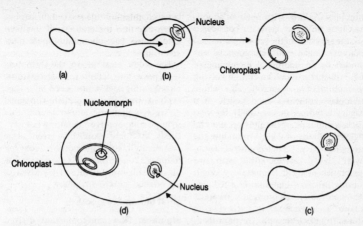

FIG. 3 *The possible steps involved in the origin of the cryptophyte* ALGAE *such as* Cryptomonas. *Two endosymbiotic events are required: first, a photosynthetic prokaryote* (a) *is engulfed by eukaryote I* (b) *which is subsequently engulfed by a second eukaryote* (c) *to form the present* Cryptomonas *cell* (d). *The membranes surrounding the chloroplast in* (d) *form part of the* CHLOROPLAST ENDOPLASMIC RETICULUM, *in which the nucleomorph is trapped.* (From *Recombinant DNA* (2nd edn) by Watson, Gilman, Witkowski, and Zoller. Copyright © James D. Watson, Michael Gilman, Jan Witkowski and Mark Zoller, 1992. Reprinted with permission of W. H. Freeman and Company.)

doned, component groups being considered sufficiently distinct to merit divisional or class status on the basis of comparisons between pigments, assimilatory (storage) products, flagella, cell wall chemistry and structure, and aspects of ultrastructure and biological molecular information. Electron microscopy has revealed the presence and structure of flagella, their flimmers, swellings and scales; chloroplasts; chloroplast endoplasmic reticulum (CER) (see Fig. 3); thylakoids; phycobilisomes; external organic and inorganic scales; silica deposition vesicles; theca; nuclear structure and division, and cell division – all are employed in algal systematics. Simultaneously, biochemists analysed the molecular details of pigments, assimilatory products and cell walls. More recently it has become possible to use molecular biological techniques to examine genomes of organisms. These techniques are now being applied to algae with examination of nucleotides in their RNA and DNA or similar sequences of amino acids in their proteins.

Today, four main evolutionary groups are recognized as 'algae': (i) Prokaryotic algae, the first to evolve (e.g. Cyanobacteria, Prochlorophyta); (ii) Eukaryotic algae with a chloroplast envelope of two membranes. These probably evolved via an evolutionary event involving capture of a prokaryotic cell in a food vesicle by a phagocytic, non-photosynthetic protozoan. Eventually, the plasmalemmas of the endosymbiont and food vesicle became the inner and outer chloroplast envelope membranes, respectively. Chloroplasts of certain algae have retained more of the characteristics of blue-green algae than typical ones. These can be interpreted as being intermediate between chloroplasts and blue-green algae. The chloroplasts of *Glaucocystis* and *Cyanophora* (GLAUCOPHYTA) resemble unicellular blue-green algae and are referred to as cyanelles. These cannot survive outside their host. Recent genome structure studies have confirmed their intermediate status. The Rhodophyta (red algae), and Chlorophyta (green algae), both represent completion of this evolutionary process, resulting in chloroplasts, along with their offshoots. The Chlorophyta are generally considered to be the ancestors of the (green) plants; (iii) Eukaryotic algae with chloroplasts surrounded by an additional membrane of CER. Probably arose via an event involving the capture of a eukaryotic alga whose chloroplast entered a food vesicle of a phagocytic protozoan; eventually, the food vesicle membrane of the host became the single membrane of CER surrounding the chloroplast (e.g., Euglenophyta, Dinophyta); (iv) Eukaryotic algae with two membranes of CER. Evolutionary origin probably involved ingestion of a red alga into a protozoan's food vesicle where it remained as a symbiont, its nucleus being reduced to a nucleomorph. The protozoan, with its algal symbiont, was probably taken into the food vesicle of another phagocytic protozoan where it remained as an endosymbiont. The nucleus of the first protozoan took over the functioning of the cell, while that of the second protozoan was lost, along with the food vesicle and the outer nuclear membrane of the second and first protozoans, respectively. Thus, two membranes of CER resulted (e.g. CRYPTOPHYTA, HETEROKONTOPHYTA, CHRYSOPHYCEAE, BACILLARIOPHYCEAE, XANTHOPHYCEAE, EUSTIGMATOPHYCEAE, RAPHIDOPHYCEAE, PHAEOPHYCEAE, PRYMNESIOPHYTA). Other evidence used to support the endosymbiotic theory comes from organisms that contain endosymbiotic bluegreen algae serving as chloroplasts (e.g. *Geosiphon pyriforme* is a filamentous fungus found in the autumn growing in clay soils together with the liverwort *Anthoceros*). This fungus forms vesicular cells ~1 mm in diameter, which contain filaments of the photosynthetic blue-green alga *Nostoc*. Chloroplasts and mitochondria are autonomous to some extent. CHLOROPLASTS normally divide independently of the nucleus and possess their own DNA. Same is true for MITOCHONDRIA. These are interpreted as relic features from the time when these organelles were free-living prokaryotes. Also, comparisons of molecular sequences in ribosomal RNAs of mitochondria, chloroplasts and prokaryotes have confirmed that there is a close phylogenetic link between mitochondria and certain aerobic, heterotrophic bacteria, and between chloroplasts and photoautotrophic bluegreen algae.

The oldest 'algal' group, the Cyanobacteria, possesses a definite fossil STROMATOLITE record dating back almost 3,000 Myr. They evolved when the atmosphere contained no oxygen; methane, ammonia and other reduced compounds dominated. Photosynthesis by these early blue-green algae eventually increased atmospheric oxygen to today's concentrations. Eukaryotic algae did not evolve until about 700–800 Myr ago. See INTRODUCTION to dictionary.

algal bloom See BLOOM.

alginates Non-toxic compounds derived from members of the PHAEOPHYCEAE, used extensively in industry because of their colloidal properties. They make up the greater part of the cell wall of brown algae and are salts of alginic acid. This is a polymer of two sugar acids, D-mannuronic acid and L-guluronic acid, which are connected by (β1,4 linkages). In the food and pharmaceutical industries, alginates are used to stabilize emulsions and suspensions. Products containing them include ice cream, jam, sausage, ointments, lotions, medicine capsules and slimming foods. They are also used in the manufacture of dyes, building materials, glue and paper.

alimentary canal The gut; a hollow sac with one opening (an enteron) or a tube (said to be 'entire' since it opens at both mouth and anus) in whose lumen food is digested, and across whose walls the digestion products are absorbed. The epithelium lining the lumen is endodermal in origin, but the bulk of the organ system in higher forms is mesodermal, and is muscularized and vascularized. There are usually many associated glands.

Alismatidae A subclass of the LILIOPSIDA that comprises about 500 species of plants that are considered to be the most archaic group of the Liliopsida. Characteristically they are either with apocarpous (sometimes monocarpous) flowers, or more or less aquatic, or very often both. Plants are always herbaceous and their vascular systems are not strongly lignified, often reduced, with vessels confined to the roots or lacking. The endosperm is also mostly lacking, nor starchy if present. Pollen is trinucleate. See INTRODUCTION to dictionary.

alkaline phosphatase Broad specificity enzyme, hydrolysing many phosphoric acid esters with an optimum activity in the basic pH range. Breaks down pyrophosphate in vertebrate blood plasma, enabling bone mineralization. In the yeast *Saccharomyces cereviseae*, it releases phosphates in the periplasmic space between plasma membrane and cell wall, in which (like invertase) it is somehow anchored.

alkaloids Wide group of basic nitrogenous organic compounds, many of them lipids, produced by a few families of flowering plants (e.g. Solonaceae, Papaveraceae); often derivatives of pyridine, pyrimidine, quinoline or isoquinoline. Include atropine (anti-cholinergic; application may double human heart rate; generally blocks effect of MUSCARINE on vagus nerve effectors); CAFFEINE (stimulant of central nervous system); cocaine (local anaesthetic), morphine (analgesic); NICOTINE (general stimulant) and QUININE. See ERGOT, CYANOBACTERIA, GLYCOSIDE.

alkylating agent A substance introducing alkyl groups (e.g. $-CH_3$, $-C_2H_5$, etc.) into either hydrocarbon chains or aromatic rings. Alkylation of DNA residues important in regulating transcription. See DNA REPAIR MECHANISMS, DNA METHYLATION.

allantois Stalk of endoderm and mesoderm which grows out ventrally from the posterior end of embryonic gut in AMNIOTES, expanding in reptiles and birds into a large sac underlying and for much of its surface attached to the CHORION. May represent precocious development of ancestral amphibian bladder. One of the three EXTRA-EMBRYONIC MEMBRANES. A richly vascularized organ of gaseous exchange within cleidoic eggs, also functioning as a bladder to store embryo's nitrogenous waste. In higher primates and rodents, persists into later life as the urinary bladder.

alleles (allelomorphs) Different sequences of genetic material occupying the same gene LOCUS, and therefore on different chromosomes; said to be alleles of (allelomorphic to) one another – a relational property. In classical genetics, alleles were ascribed to the same gene on the basis of two criteria: (i) failure to recombine with one another at meiosis, as if occupying the same locus, and (ii) failure, when mutant, to exhibit COMPLEMENTATION when present together in a diploid. Alleles of the same gene differ by MUTATION at one or more nucleotide sites within the same length of DNA, and back-mutation from one to another may occur. There may be many alleles of a gene in a population, but normally only two in the same diploid cell. See GENE, MULTIPLE ALLELISM, ALLELIC COMPLEMENTATION.

allelic complementation Interaction between individually defective mutant alleles of the same gene to give a phenotype more functional than either could produce by itself. Due to interaction (hybridization) of protein products. A source of confusion in the delineation of CISTRONS. See COMPLEMENTATION.

allelic exclusion Process whereby successful expression of one allele prevents expression of its homologue.

allelopathy Inhibition of one species of plant by chemicals produced by another plant (e.g. by *Salvia leucophylla* – purple sage).

Allen's rule States that the extremities (tail, ears, feet, bill) of ENDOTHERMIC animals tend to be relatively smaller in cooler regions of a species range. See BERGMANN'S RULE.

allergic reactions Pointless immune response to some non-threatening foreign protein, initiated when IgE-bearing B CELLS are activated by antigen to secrete IgE antibody. These bind MAST CELLS and BASOPHILS resulting in degranulation and release of HISTAMINE and other potent mediators causing allergic symptoms such as long-term inflammatory effects. See MILK (I).

alloantibody Antibody introduced into an individual but produced in a different member of the same species.

alloantigen (isoantigen) Antigen stimulating antibody response in genetically different members of the same species.

allochronic Of species or species populations that are either sympatric at different times of the year or otherwise have non-overlapping breeding seasons (e.g. different flowering seasons in anthophytes). See ALLO-PATRY, SYMPATRIC.

allochthonous Originating somewhere other than where found.

allogamy (Bot.) Cross-fertilization.

allogeneic (allogenic) Genetically different, but from the same species. Compare AUTOGENEIC, XENOGENEIC.

allogenic succession A temporal SUC-CESSION of species at a location that is driven from an adjacent terrestrial system. Estuaries, for example, are subject to quite fast silt deposition. This results in formation of a salt marsh, which gradually extends seawards, while terrestrial vegetation invades the landward limits of the marsh. See AUTO-GENIC SUCCESSION.

allograft (homograft) See GRAFT.

allometry Derived from the Greek 'other measure', allometry is often thought to describe cases of differential growth, or relationships between size and shape; but it is also applied wherever size affects organic form or process: i.e. to the effects of scale. Organisms do not grow isometrically; rather proportions change as size changes. Thus juvenile mammals have relatively large heads, while limb proportions of arthropods alter in successive moults. In mathematical terms, such relationships are usually dealt with by means of a power function (see POWER LAWS) of the form $y = bx^a$, where y = size of structure at some stage, b = a constant for the structure, x = body size at the stage considered and a = allometric constant (unity for isometric growth). The analysis is open to multivariate generalization. See HETEROCHRONY.

If we consider some three-dimensional objects of different sizes but of exactly the same shape, choosing cubes for simplicity, if the length is L, then the volume, V, is L^3 and the total surface area, A, of the six sides is $6L^2$. So

$$L = V^{1/3} = (A/6)^{1/2}$$

so that

$$A \propto V^{2/3}$$

Therefore, if a cube's mass is M, and all cubes have the same density, we find that

$$A \propto M^{2/3}$$

Now, if two mammals differ in body mass by a factor of 1,000 but are otherwise similar in most respects, e.g. in shape and body temperature, this last equation tells us that their surface areas should differ by a factor of $1,000^{2/3} = 100$. If heat production is a function of their total tissue mass, and if heat is lost largely through their body surfaces, then if both animals have the same METABOLIC RATE per unit body mass, the absolute metabolic rate of the larger mammal is 1,000 times that of the smaller – which would cause its blood to boil! In fact, larger mammals have lower metabolic rates per unit body mass than do smaller mammals: metabolism, Mb, scales with body mass, M, according to $Mb \propto M^{3/4}$ (Kleiber's Law), and heat production is actually proportional not to body mass but to surface area. It follows from Kleiber's Law that a limited quantity of available energy per unit area of ecosystem will sustain a larger number of individuals of a small-sized species than of a larger species – and both are expected to conform to a power law of the form $N \propto M^{-3/4}$, where N = population density. Available estimates confirm this expectation.

allopatric Geographical distribution of different species, or subspecies or populations within a species, in which they do not occur together but have mutually exclusive distributions. Populations occupying different vertical zones in the same geographical area may still be fully allopatric. See ALLOCHRONIC, SYMPATRIC.

allopolyploid Typically, a TETRAPLOID organism derived by chromosome doubling from a hybrid between diploid species whose chromosomes have diverged so much that little or no synapsis occurs between them at meiosis, so that only bivalents are formed (e.g. New World cottons, *Gossypium* spp.). This clearly distinguishes the term from AUTOPOLYPLOID, but some polyploids do not fall readily into either category. Allopolyploids may backcross with one or other diploid parent stock; hence allotetraploids, which are generally themselves fully fertile (since they form bivalents at meiosis), behave in effect as new reproductively isolated species. However, if the original diploid progenitors were closely related species, or even ecotypes of the same species, then MULTIVALENTS may arise in meioses, which then resemble meioses in typical autopolyploids. Nevertheless, as a result of their greater fertility classical allopolyploids have been more significant in evolution than have classical autopolyploids. Many new plant species have arisen this way. Cultivated WHEAT (*Triticum aestivum*) is an allohexaploid, combining doubling in a triploid hybrid between an allotetraploid and a diploid.

all-or-none response Ability of certain excitable tissues, under standardized conditions, to respond to stimuli of whatever intensity in just two ways: (a) no response (stimulus sub-threshold), or (b) a full-size response (stimulus at or above threshold). ACTION POTENTIALS of nerve and muscle membranes are characterized by all-or-none behaviour. Where thresholds of different units in a response differ, as in the many motor fibres of the sciatic nerve, or the various MOTOR UNITS of an entire muscle, an increase in stimulus intensity may bring progressively more units to respond. In muscle, this constitutes spatial SUMMATION. Nerve signals cannot use such amplitude variations.

allosteric Of those molecules (typically proteins) whose three-dimensional conformations alter in response to their environmental situation, normally registered by a change in molecule function. Often the key to regulation of critical biochemical pathways, serving as a feedback monitoring

device in cybernetic circuits both inside and outside cells (see REGULATORY ENZYMES). At least as significant is allosteric control of GENE EXPRESSION by regulatory proteins. Among non-enzyme proteins, the haemoglobin molecule is allosteric under different blood pH values, with marked effects upon its oxygen saturation curve (see BOHR EFFECT). For allosteric inhibition and induced fit of enzymes, see ENZYME. See RECEPTOR (2).

allotetraploid An ALLOPOLYPLOID derived by doubling the set of chromosomes resulting from fusion between haploid gametes from more or less distantly related parental species. In classical cases, there is no meiotic SYNAPSIS between the chromosomes of different origin, and more or less complete fertility is achieved. Far more common in plants than animals, probably through comparative rarity of vegetative habit and/or parthenogenesis in the latter, in which it is difficult to rule out autopolyploidy as the source. See POLYPLOIDY.

allotopic Of closely related sympatric populations, whose distributions are such that both occupy the same geographical range, but each occurs in a different habitat within that range.

allotype Genetic variant within a LOCUS of a given species population, such as allelic forms within a BLOOD GROUP SYSTEM or variants of heavy chain constant regions of ANTIBODY molecules. See IDIOTYPE, ISOTYPE.

allozymes Forms of an enzyme encoded by different allelic genes.

alpha-actinin (α-actinin) An accessory protein of muscle, anchoring actin filaments at the Z-disc and cross-linking adjacent sarcomeres; also cross-links actin in many other cells to contribute to the CYTOSKELETON.

alpha blocker Drug blocking ADRENERGIC alpha receptors, preventing activity of the sympathetic neurotransmitter NORADRENALINE.

alpha helix (Of proteins) a common secondary structure, in which the chain of amino acids is coiled around its long axis. Not all proteins adopt this conformation, it depending upon the molecule's primary structure. When adopted there are about 3.6 amino acids per turn (corresponding to 0.54 nm along the axis), amino acid R-groups pointing outwards. Hydrogen bonds between successive turns stabilize the helix. The α-helix may alternate with other secondary structures of the molecule such as β-sheets or 'random' sections. See PROTEIN.

alpha receptors ADRENERGIC membrane receptor site binding NORADRENALINE in preference to ADRENALINE. May be excitatory or inhibitory, depending on the tissue. As with BETA RECEPTORS, effects are mediated through an adenylate cyclase molecule adjacent in the membrane. The commonest receptors on postsynaptic membranes of postganglionic cells of sympathetic system. See CHOLINERGIC, AUTONOMIC NERVOUS SYSTEM, ALPHA BLOCKER.

alpha-richness Number of species present in a small, local, homogeneous area. See DIVERSITY.

alternation of generations Either (1) *metagenesis*, a life cycle alternating between a generation reproducing sexually and another reproducing asexually, the two often differing morphologically; or (2) the alternation within a life cycle of two distinct cytological generations, one being haploid and the other diploid. See LIFE CYCLE.

Metagenesis occurs in a few animals, e.g. CNIDARIA and parasitic flatworms, where both generations are normally diploid. The alternation of distinct cytological generations is clearest in some algae and ferns, where the two generations (gametophyte and sporophyte) are independent and either identical in appearance (alternation of *isomorphic* generations) or quite dissimilar (alternation of *heteromorphic* generations). In mosses and liverworts the dominant (vegetative) plant is the gametophyte while the sporophyte (the capsule) is more or less nutritionally dependent on the gametophyte. In flowering plants, the male (micro-) and female (macro-) gametophytes are reduced to microscopic pro-portions, the male gametophyte being shed as the pollen grain and the female gametophyte (embryo sac) being retained on the sporophyte in the ovule. A clearcut alternation of physically distinct plants is avoided here, although alternating cytological phases are still discernible. In vascular plants generally, the sporophyte generation is the vegetative plant itself, be it a fern, herb, shrub or tree.

alternative splicing Production of different polypeptides from the same pre-mRNA transcript by splicing it in different ways. See RNA PROCESSING.

altricial Animals born naked, blind and immobile (e.g. rat and mouse pups, many young birds). See NIDICOLOUS.

altruism Behaviour benefiting another individual at the expense of the agent. Widespread and apparently at odds with Darwinian theory, which predicts that any genetic component of such behaviour should be selected against. Theories of altruism in biology tend to be concerned with cost-benefit analysis, as dictated by the logic of natural selection. One component of Darwinian FITNESS may be the care a parent bestows upon its offspring, although this is not usually considered altruism. HAMILTON'S RULE indicates the scope for evolutionary spread of genetic determinants of altruistic character traits, compatibly with Darwinian theory, and explains the evolution of parental care, while showing that *reciprocal altruism* can evolve even in the absence of relatedness between participants (e.g. members of different species). Some models are exploring the possibilities of a co-operative behaviour evolving if discrimination between individuals (for reciprocity) is based not upon classical, genetically inherited, traits but upon social ones. MULTI-CELLULARITY may afford opportunities for sacrifice of somatic cells for a genetically related germ line harbouring the potentially immortal UNITS OF SELECTION. See ARMS RACE.

Alu A TRANSPOSON, and the only active member of the SINE family in the human genome, in which it is abundant and amplified. It lacks coding for either a reverse transcriptase or an endonuclease. Alu sequences appear to have arisen from a 7SL RNA gene (see SIGNAL RECOGNITION) and have changed far more slowly in mammalian evolution than have SATELLITE DNAS. There appears to have been strong selection favouring retention of Alu elements in GC-rich regions of the human genome, and that these 'selfish' elements may confer advantages to their hosts. See REPETITIVE DNA; compare LINES.

alveolus (1) Minute air-filled sac, grouped together as *alveolar sacs* to form the termini of bronchioles in vertebrate lungs. Their thin walls are composed of squamous epithelial and surfactant-producing cells. A rich capillary network attached to the alveoli supplies blood for gaseous exchange across the huge total alveolar surface. The total human alveolar surface is 140 m^2, with 125 m^2 of capillary surface. A surfactant (lecithin) layer reduces surface tension, keeping alveoli open from birth onwards, and provides an aqueous medium to dissolve gases. Without surfactant, premature babies tend to suffer from respiratory stress syndrome: their lungs are hard to inflate and become congested with tissue fluid. Macrophages in the alveolar walls remove dust and debris. (2) Expanded sac of secretory epithelium forming internal termini of ducts of many glands, e.g. mammary glands. (3) Bony sockets into which teeth fit in mandibles and maxillae of jawed vertebrates, lying in the *alveolar processes* of the jaws. (4) An elongated chamber on the cell wall of some diatoms (Bacillariophyta) from the central axis to the margin, and opening to the inside of the cell wall.

Alzheimer's disease Neurodegenerative disease currently affecting nearly half of those over 85 years, with huge social and economic cost. A full cellular and molecular explanation is not currently available, but genetic predisposition seems probable. Involves loss of cholinergic neurones in the cerebral cortex as well as those using other neurotransmitters, e.g. glutamate, dopamine (compare PARKINSON'S DISEASE) and serotonin, all of which are targets for potential therapies. Symptoms include memory loss, agitation and emotional outbursts. Possible causes include brain inflammatory changes; deposition in the brain of PLAQUES of a small neurotoxic protein, β-amyloid, or Aβ peptide (see CHAPERONES); excess phosphorylation of another brain neuroprotein, tau; possession of a genetic variant of the cholesterol-carrying protein apo-E4 which may inhibit growth of neural spines involved in nerve communication; and missense mutations in genes encoding the two presenilins (1 and 2), which alter the cleavage site of the intramembrane protease γ-secretase (later identified as presenilin) and which when mutated produces a longer, more toxic and more easily precipitated Aβ peptide from its larger precursor (APP) than normal. This fragment is released extracellularly and aggregates to form the amyloid plaques (compare PRION). Initial palliative drugs inhibiting acetylcholine breakdown by acetylcholinesterase had intolerable side effects at effective doses though reversed symptoms by a month to a year. One type of glutamate receptor (the AMPA receptor thought to be needed for the long-term potentiation linked to memory formation) is the target of drugs called ampakines, which reverse memory loss in rats. Antioxidants such as vitamin E, anti-inflammatory drugs such as prednisone, and oestrogen supplements and the NMDA-glutamate ion receptor blocker memantine are each being given clinical trials. Efforts to preserve brain neurons with nerve growth factors like NGF (which in animals improves cholinergic neuron survival, memory and learning) are worth investigating although this protein is itself too large to cross the BLOOD–BRAIN BARRIER. See AGEING, APOPTOSIS, PRION.

amacrine cell One of the three classes of neuron in mid-layer of vertebrate retina. Conducts signals laterally without firing action potentials.

Amastigomycota In older fungal classifications, that division of fungi lacking a motile stage. See ZYGOMYCOTA, BASIDIOMYCOTA, ASCOMYCOTA, DEUTEROMYCOTA.

amber mutation One of three mRNA CODONS not recognized by transfer RNAs commonly present in cells, and bringing about normal polypeptide chain termination. Its triplet base sequence is UAG. *Missense* or *stop mutation*. See OCHRE and OPAL MUTATIONS, GENETIC CODE.

amenorrhoea Disruption of the MENSTRUAL CYCLE. See PROLACTIN.

amensalism Interaction in which one animal is harmed and the other unaffected. See SYMBIOSIS.

Ames test Test assessing mutagenic potential of chemicals. Strains of the bacterium *Salmonella typhimurium* having qualities such as permeability to chemicals, inability to repair DNA damage, or ability to convert DNA damage into heritable mutations, are made AUXOTROPHIC for histidine. After mixing with potential mutagen prior to plating, increase in normal (PROTOTROPHIC) colonies indicates mutagenicity.

Ametabola Primitively wingless insects (APTERYGOTA).

amino acids Amphoteric organic compounds of general structural formula

$$H_2N-\underset{\underset{R}{|}}{\overset{\overset{H}{|}}{C}}-C\overset{O}{\underset{OH}{\diagup}}$$

(where R may be one of twenty atomic groupings) occurring freely within organisms, and polymerized to form dipeptides, oligopeptides and polypeptides. Amino acids differ in their R-groups (seven of which are readily ionizable) and the amino acid sequence in a polypeptide determines not only its charge sequence but also its conformation in solution (the distribution of acidic, basic and non-polar R-groups depends upon R-group composition, the pH and the microenvironment). Relative molecular masses of the common forms vary from 75 (glycine) to 204 (tryptophan). Only three commonly contain sulphurous R-groups: methionine, cysteine and cystine (formed from two oxidized cysteines, providing 'sulphur bridges'). During PROTEIN SYNTHESIS the carboxy- and amino-terminal ends of adjacent amino acids condense to form peptide bonds, leaving only the N-terminal and C-terminal ends of the protein and appropriate R-groups ionizable. About 20 amino acid radicals occur commonly in proteins, encoded by the GENETIC CODE. Their modification after attachment to a transfer RNA molecule may result in rare nonencoded amino acids occurring in proteins. Some amino acids (e.g. ornithine) never occur in proteins. Most naturally occurring amino acids have a free carbonyl group and (except proline) are alpha amino acids, bearing a free amino group on the α carbon atom (that adjacent to the free carboxyl group). Aspartic acid is required for pyrimidine synthesis, as glutamine is for purine synthesis. *Essential amino acids* are required by an organism from its environment, due to inability to synthesize them adequately from precursors (see VITAMINS, which they are not); there are ten in humans, eight of which we cannot synthesize at all (isoleucine, leucine, lysine, methionine, phenylalanine, threonine, tryptophan and valine), and two of which (arginine and histidine) are inadequately synthesized during childhood. Non-essential amino acids can be synthesized by transamination (e.g. on to pyruvic acid or another acid of the KREBS CYCLE; see GLUCONEOGENESIS). CEREALS are relatively poor in lysine but adequate in methionine, so cereals and legumes eaten together provide a good balance (see COMPLEMENTARY RESOURCES). Amino acids are not stored, and protein synthesis is prevented if appropriate amino acids are lacking from a cell. Different amino acids within a protein often serve as targets for different PROTEIN KINASES in SIGNAL TRANSDUCTION pathways – a fact of inestimable significance in cell biology.

amitochondriate Term used for those eukaryotic cells lacking mitochondria. Many microbial eukaryotes previously thought to lack mitochondria have actually retained the organelle, although normal markers may be absent (see MICROSPORIDIA). Protists such as *Giardia lamblia* and *Entamoeba histolytica* are type I amitochondriate eukaryotes and are thought to lack mitochondria and other 'core metabolism' organelles primitively but *Entamoeba* contains an organelle (a mitosome, or crypton) which now seems to be mitochondrially derived; type II amitochondriate eukaryotes, e.g. *Trichomonas*, harbour HYDROGENOSOMES; while cells such as mammalian erythrocytes lack organelles of core metabolism secondarily (having inherited, but then lost, their mitochondria). See Figs. 52 and 53.

amitosis See ABERRANT CHROMOSOME BEHAVIOUR (2).

ammocoete Filter-feeding larva of lamprey, capable of attaining lengths of over 10 cm if conditions for metamorphosis do not prevail.

ammonification Decomposition of amino acids and other nitrogenous organic compounds; results in production of ammonia (NH_3) and ammonium ions (NH_4^+). Bacteria involved are *ammonifying bacteria*. See NITROGEN CYCLE.

ammonites Group of extinct cephalopod molluscs (Subclass Ammonoidea, Order Ammonitida) dominating the Mesozoic cephalopod fauna. Had coiled shells, with protoconch (calcareous chamber) at origin of the shell spiral. Of great stratigraphic value.

ammonotelic (Of animals) whose principal nitrogenous excretory material is ammonia. Characterizes aquatic, especially freshwater, forms. See UREOTELIC, URICOTELIC.

amniocentesis See AMNION.

amnion Fluid-filled sac in which AMNIOTE embryo develops. An EXTRAEMBRYONIC MEMBRANE (Fig. 54) formed in reptiles, birds and some mammals by extraembryonic ectoderm and mesoderm growing up and over embryo, the (amniotic) folds overarching and fusing to form the amnion surrounding the embryo, and the CHORION surrounding the amnion, ALLANTOIS and YOLK SAC. The amnion usually expands to meet the chorion. In humans and many other mammals the amnion originates by rolling up of some of the cells of the INNER CELL MASS during GASTRULATION. Amniotic fluid (amounting to

about one dm³ at birth in humans) is circulated in placental mammals by foetal swallowing, enabling wastes to pass to the placenta for removal. Provides a buffering cushion against mechanical damage, helps stabilize temperature and dilate the cervix during birth. In *amniocentesis*, amniotic fluid containing cells from the human foetus is withdrawn surgically at around the sixteenth week of pregnancy for signs of congenital disorder, such as DOWN'S SYNDROME.

amniote Reptile, bird or mammal. Distinguished from anamniotes by presence of EXTRAEMBRYONIC MEMBRANES in development.

amniotic egg Egg type characteristic of reptiles, birds and PROTOTHERIA (much modified in placental mammals). Shell leathery or calcified; ALBUMEN and yolk typically present. EXTRAEMBRYONIC MEMBRANES occur within it during development. See CLEIDOIC EGG.

Amoeba Genus of sarcodine protozoans. Protists of irregular and protean shape, moving and feeding by use of PSEUDOPODIA. Some slime mould cells are also loosely termed 'amoebae', while any CELL LOCOMOTION resembling an amoeba's is termed 'amoeboid'.

amoebocyte Cell (haemocyte) capable of active amoeboid locomotion found in blood and other body fluids of invertebrates; in sponges, an amoeboid cell type implicated in mobilization of food from the feeding CHOANOCYTES and its conveyancing to non-feeding cells in absence of true vascular system.

amoeboid Describing cells resembling those of the genus *AMOEBA*.

amoebozoans One of the major groups of eukaryotes; with separate (not fused) DHFR and TS genes (see EUKARYOTE) and comprising mostly naked amoebae (lacking tests), often with lobose pseudopodia at some stage in their life cycle. Include dictyostelid, protostelid and plasmodial slime moulds.

AMP Adenosine monophosphate. Nucleotide component of DNA and RNA (in deoxyribosyl and ribosyl forms respectively), and hydrolytic product of ADP and ATP. Converted to CYCLIC AMP by ADENYLATE CYCLASE, intracellular concentrations of cAMP rising rapidly in response to extracellular (esp. hormonal) signals and falling rapidly due to activity of intracellular phosphodiesterase. Its level dictates rates of many biochemical pathways, depending upon cell type. See CASCADE, SECOND MESSENGER, G-PROTEIN, GTP, RECEPTOR.

amphetamines   Powerful behavioural stimulants which augment the neurotransmitters NORADRENALINE, SEROTONIN and DOPAMINE – as does cocaine. See PSYCHOACTIVE DRUGS.

Amphibia   Class of tetrapod vertebrate, its first fossil representatives being Devonian ichthyostegids and its probable ancestors rhipidistian crossopterygian fishes. A POLYPHYLETIC origin has not been ruled out. Many early forms had scaly skins, almost entirely lost in the one modern Subclass (Lissamphibia) of three orders: Apoda, legless caecilians; Urodela, salamanders and newts; Anura, toads and frogs. Compared with their mainly aquatic ancestors, the more terrestrialized amphibians have: vertebrae with larger, more articulating neural arches and larger intercentra (see VERTEBRAL COLUMN); greater freedom of the PECTORAL GIRDLE from the skull, allowing some lateral head movement; PELVIC GIRDLE composed of three paired bones (pubis, ischium and ilium) with some fusion to form the rigid PUBIC SYMPHYSIS; eardrums (homology with part of the spiracular gill pouch of fish) and a single middle ear ossicle, the columella, homologous with the hyomandibular bone of fish. Fertilization is internal or external (but intromittant organs are lacking). Most return to water to lay anamniote eggs, although some are viviparous. The skin is glandular for gaseous exchange. Modern forms specialized and not representative of the Carboniferous amphibian radiation.

amphicribral   (Bot.) Type of vascular arrangement where phloem surrounds the xylem. Compare AMPHIPHLOIC.

amphidiploid   See ALLOTETRAPLOID.

amphimixis   Normal sexual reproduction, involving meiosis and fusion of haploid nuclei, usually borne by gametes. See AUTOMIXIS, APOMIXIS, PARTHENOGENESIS.

Amphineura   Class of MOLLUSCA, including the chitons. Marine, mostly on rock surfaces; head reduced and lacking eyes and tentacles; mantle all round head and foot; commonly eight calcareous shell plates over visceral hump; nervous system primitive, lacking definite ganglia.

amphioxus   Lancelets (Subphylum CEPHALOCHORDATA). Widely distributed marine filter-feeding burrowers up to 5 cm long. Two genera (*Branchiostoma*, *Asymmetron*). Giant larva resulting from prolonged pelagic life once given separate genus (*Amphioxides*) and develops premature gonads, providing support for the evolutionary origin of vertebrates by PROGENESIS.

amphipathic   Of protein molecules with one surface containing hydrophilic and the other hydrophobic amino acid residues. See LDL.

amphiphloic   (Bot.) Type of vascular arrangement where phloem is on both sides of the xylem. Compare AMPHICRIBAL.

Amphipoda   Order of Crustacea (Subclass Malacostraca). Lack carapace; body laterally flattened. Marine and freshwater forms; about 3,600 species. Very important detritus feeders and scavengers. Includes gammarids.

amphistylic   Method of upper jaw suspension in a few sharks, in which there is support for the jaw both from the hyomandibular and the brain case. See AUTOSTYLIC, HYOSTYLIC.

amphoteric   A substance capable of donating and accepting protons, so capable of acting as an acid or a base. See AMINO ACID, PROTEIN.

amphoterin   A heparin-binding protein essential for normal neurite growth. Abundant in the extracellular regions of the developing brain and other organs. Its interactions with the membranes of neuronal surfaces seem to be required for extension of neuronal processes. See NEURON.

AMPK AMP-activated protein kinase. See LEPTIN.

ampulla (Of inner ear) see VESTIBULAR APPARATUS.

amygdala (amygdaloid bodies or **nuclei)** See LIMBIC SYSTEM.

amylases (diastases) Group of enzymes hydrolysing starches or glycogen variously to dextrins, maltose and/or glucose; α-amylase (in saliva and pancreatic juice) yields maltose and glucose; β-amylase (in malt) yields maltose alone. Present in germinating cereal seeds (see ALEURONE LAYER), where only α-amylase can digest intact starch grains, and produced by some microorganisms.

amylopectin Highly branched polysaccharide component of the plant storage carbohydrate STARCH. Consists of homopolymer of α[1,4]-linked glucose units, with α[1,6]-linked branches every 30 or so glucose radicals. Like GLYCOGEN it gives a red-violet colour with iodine/KI solutions. See AMYLASES.

amyloplast Colourless plastid (leucoplast) storing STARCH; e.g. found in cotyledons, endosperm and storage organs such as potato tubers.

amylose Straight-chain polysaccharide component of STARCH. Comprises α[1,4]-linked glucose units. Forms hydrated micelles in water, giving the impression of solubility. Gives a blue colour with iodine/KI solutions. Hydrolysed by AMYLASES to maltose and/or glucose.

anabolism Enzymatic synthesis (build-up) of more complex molecules from more simple ones. Anabolic processes include multi-stage photosynthesis, nucleic acid, protein and polysaccharide syntheses. ATP or an equivalent needs to be available and utilized for the reaction(s) to proceed. See CATABOLISM, GROWTH HORMONE, METABOLISM.

anadromous Animals (e.g. lampreys, salmon) which must ascend rivers and streams from the sea in order to breed. See OSMO-REGULATION.

anaerobic (Of organisms) ability to live *anoxically*, i.e. in the absence of free (gaseous or dissolved) oxygen. (Of processes) occurring in the absence of such oxygen. *Anaerobic respiration* is the enzyme-mediated electron-transferring process by which cells (or organisms) liberate energy by oxidation of substances but without involving molecular oxygen. This involves less complete oxidation of substrates, with less energy released per g of substrate used, enabling *anaerobes* to exploit environments unavailable to obligate aerobes. *Facultative anaerobes* can switch metabolism from aerobic to anaerobic under anoxic conditions, as required of many internal parasites of animals, some yeasts and other microorganisms. GLYCOLYSIS is anaerobic but may require aerobic removal of its products to proceed. Relatively anoxic environments include animal intestines, rumens, gaps between teeth, sewage treatment plants, polluted water, pond mud, some estuarine sediments and infected wounds. See OXYGEN DEBT, RESPIRATION. For anaerobic training, see MUSCLE CONTRACTION.

anaesthetics Molecules, often volatiles, which depress impulse conduction and synaptic transmission by raising the threshold for excitation and cause partial or complete loss of feeling. See ALCOHOLS, ENKEPHALINS.

anagenesis (1) Process by which characters change during evolution within species, by NATURAL SELECTION or GENETIC DRIFT. (2) Any non-branching speciation in which species originate along a single line of descent yet only one species represents the lineage after any speciation event (contrast CLADOGENESIS). Gradual anagenetic speciation is not possible within the *biological species concept*, for reproductive isolation is never completed between ancestral and descendant species. CLADISTICS excludes anagenetic speciation by definition, but the term is retained in the context of characters. See SPECIES.

analgesics Pain-killing drugs. See PSYCHO-ACTIVE DRUGS.

analogous A structure present in one evolutionary lineage is said to be analogous to a structure, often performing a similar function, within the same or another evolutionary lineage if their phyletic and/or developmental origins were independent of one another; i.e. if there is HOMOPLASY. Tendrils of peas and vines and eyes of squids and vertebrates are pairs of analogous structures. See CONVERGENCE, HOMOLOGY, PARALLEL EVOLUTION.

anamniote (Of vertebrates) more primitive than the AMNIOTE grade. Includes agnathans, all fish, and amphibians.

anandrous (Of flowers) lacking stamens.

anaphase Stage of mitosis and meiosis during which either bivalents (meiosis I) or sister chromatids (mitosis, meiosis II) separate and move to oppositive poles of the cell. See ANAPHASE-PROMOTING COMPLEX, SPINDLE.

anaphase-promoting complex (cyclosome) Large protein complex which initiates chromatid segregation and exit from mitosis by degrading anaphase inhibitors and mitotic cyclins.

anaphylaxis A type of hypersensitivity to antigen (allergen) in which IgE antibodies attach to mast cells and basophils. May result in circulatory shock and asphyxia. See ALLERGIC REACTION.

anatropous (Of ovule) inverted through 180°, micropyle pointing towards placenta. Compare CAMPYLOTROPOUS.

ancient DNA See DNA.

androdioecious Having male and hermaphrodite flowers on separate plants. Compare ANDROMONOECIOUS.

androecium A collective term referring to the stamens of a flower. Compare GYNOECIUM.

androgen Term denoting any substance with male sex hormone activity in vertebrates, but typically steroids produced by vertebrate testis and to a much lesser extent by adrenal cortex. See TESTOSTERONE.

androgenesis Process by which an embryo (an *androgenone*) is produced solely from sperm-derived chromosomes. Occurs when the female pronucleus is absent (or removed) from an egg and the sperm chromosomes duplicate on entry, restoring the diploid number (similar to production of hydatidiform mole in mammals). *Gynogenones* are equivalent, but totally female-derived, embryos. Experimental work in this field, including pronuclear transplantation, suggests that sperm-derived genes are required for normal chorion development in mammals, while egg-derived genes are required for normal development of the embryo itself. Expression of *Xist* RNA (see DOSAGE COMPENSATION) from paternal X chromosomes, X^p, during preimplantation development leads to repression of genes near the *Xist* gene; but whereas XY androgenones only shut off these genes and can form blastocysts, XX androgenones inactivate both X chromosomes completely and die before the blastocyst stage. See CHROMOSOMAL IMPRINTING.

androgenone Diploid embryo with two paternal genomes produced by pronuclear transplantation (see ANDROGENESIS).

andromonoecious Having male and hermaphrodite flowers on the same plant. Compare ANDRODIOECIOUS.

anemophily Pollination of flowers by the wind. Compare ENTOMOPHILY.

anergy (adj. **anergic**) Situation in which, if a T cell receives only one of the two signals it needs for making an appropriate immune response, it becomes unable to divide or produce growth factors (esp. IL-2).

aneuploid (heteroploid) Of nuclei, cells or organisms having more or less than an integral multiple of the typical haploid chromosome number. See EUPLOID, MONOSOMY, TRISOMY, NULLISOMY.

angiogenesis Process by which pre-existing blood vessels (both in embryo and adult) send out capillary sprouts to produce new blood vessels. Involves exquisite regulation by paracrine signals, e.g. chemoattractants, many of which are protein ligands which bind RTKs. The proteins vascular endothelial growth factor (VEGF, secreted in HYPOXIA) and transforming growth factor (TGF) are involved, respectively, in endothelial precursor cell division and differentiation, and smooth muscle cell differentiation from endothelium. Platelet-derived growth factor (PDGF) induces smooth muscle precursors to migrate to branching endothelial channels. A common progenitor for both endothelial and smooth muscle cells has been discovered. See HYPOXIA.

angiosperm Literally, a seed borne in a vessel; thus one of a group of plants, namely the flowering plants, whose seeds are borne within a mature ovary (fruit). This term is used to refer to flowering plants but has no taxonomic status. Angiosperm genomes appear to have undergone an episode of gene duplication ~80 Myr ago, at a time in the Cretaceous when fermentable fruits became prominent; indeed overgrazing by dinosaurs may have been a factor promoting flowering plant diversity. Yeast (which ferment fruits) and fruit fly (whose larvae eat yeast) genomes appear to have undergone similar bouts of gene duplication at this time. See MAGNOLIOPHYTA (Magnoliopsida), MAGNOLIOPSIDA (Magnoliidae), LILIOPSIDA (Liliidae), INTRODUCTION to dictionary.

angiotensins *Angiotensin I* is a decapeptide produced by action of kidney enzyme, renin, on the plasma protein angiotensinogen when blood pressure drops. It is in turn converted by a plasma enzyme in the lung to the octapeptide *angiotensin II*, an extremely powerful vasoconstrictor which raises blood pressure and also results in sodium retention and potassium excretion by kidney. See OSMOREGULATION.

Ångström unit (Å) Unit of length, 10^{-10} metres (0.1 nm); 10^{-4} microns. Not an SI unit.

Animalia Animals. Kingdom containing those heterotrophic eukaryotes lacking cell wall material and having a BLASTULA stage in their development. One proposal for the defining character of the taxon (its autapomorphy, or ZOOTYPE) is the presence of *HOX* GENE clusters and their characteristic mode of expression during development. For earliest fossils, see CAMBRIAN.

animal pole Point on surface of an animal egg nearest to nucleus, or extended to include adjacent region of cell. Often marks one end of a graded distribution of cytoplasmic substances. The cells derived by CLEAVAGE from this pole region are termed 'animal cells' and are not usually nutrient-rich; by contrast, the 'vegetal cells' derived from the vegetal pole region usually are so. See CELL POLARITY, CENTROSOME.

anisogamy Condition in which gametes which fuse differ in size and/or motility. In OOGAMY, gametes differ in both properties. Significantly, the sperm often contributes the sole centriole for the resulting zygote. See FERTILIZATION, ISOGAMY, PARTHENOGENESIS.

Annelida (Annulata) Soft-bodied, metamerically segmented coelomate worms with, typically, a closed blood system; excretion by NEPHRIDIA; a central nervous system of paired (joined) nerve cords ventral to the gut, and a brain comprising paired ganglia above the oesophagus, linked by commissures to a pair below it. Cuticle collagenous, not chitinous. Chitin present in CHAETAE, which may be quite long, bristle-like and associated laterally with fleshy parapodia (e.g. ragworms, Class Polychaeta) or shorter and not housed in parapodia (e.g. earthworms, Class Oligochaeta). Leeches (Class HIRUDINEA) have 34 segments, confused by surface annulations. CLITELLUM present in both oligochaetes and leeches. Septa between segments often locally or entirely lost. The coelom acts as a hydrostatic skeleton against which longitudinal and circular muscle syncytia (and diagonal muscles in leeches) contract. Cephalization most pronounced in polychaetes (largely marine); eyes and mandibles often well developed but oligochaetes lack specialized head structures. Gametes leave via COELOMODUCTS. Oligochaetes and leeches are typically hermaphrodite, polychaetes frequently dioecious.

annual Plant completing its life cycle, from seed germination to seed production followed by death, within a single season, regardless of the number of times reproduction occurs (see SEMELPAROUS). Compare BIENNIAL, EPHEMERAL, PERENNIAL. See DESERT, R-SELECTION.

annual ring Annual increment of secondary wood (xylem) in stems and roots of woody plants of temperate climates. Because of sharp contrast in size between small wood elements formed in late summer and larger elements formed in spring the limits of successive annual rings appear in a cross-section of stem as a series of concentric lines.

annular thickening In protoxylem, internal thickening of a xylem vessel or tracheid wall, in rings at intervals along its length. Provides mechanical support, permitting longitudinal stretching as neighbouring cells grow.

annulus (1) Ring of tissue surrounding the stalk (stipe) of fruit bodies of certain Basidiomycota (e.g. mushrooms); (2) line of specialized cells involved in opening moss capsules and fern sporangia to liberate spores.

anoestrus Period between breeding seasons in mammals, when OESTROUS CYCLES are absent. See OESTRUS.

Anoplura See SIPHUNCULATA.

anoxia Deficiency or absence of free (gaseous or dissolved) oxygen. See HYPOXIA.

antagonist A substance which binds to, and blocks, a receptor, preventing a natural neurotransmitter or hormone from exerting its effect. Thus, atropine blocks muscarinic ACETYLCHOLINE receptors. Compare AGONIST.

antagonistic resources A pair of substitutable resources are antagonistic when a species requires proportionately more resource to maintain a given rate of increase when two resources are consumed together than when consumed separately (e.g.

D,L-pipecolic and djenkolic acids, believed to have defensive roles in some seeds, had no significant effect on the growth of the seed-eating larvae of a bruchid beetle if consumed together; however, if the seed of one species contained one of these compounds, and another seed the second compound, then a mixed diet would produce an adverse effect on growth).

antenna Paired, preoral, tactile and olfactory sense organ developing from second or third embryonic somites of all arthropod classes other than Onychophora and Arachnida. Usually much jointed and mobile. In some crustaceans locomotory or for attachment, a pair of ANTENNULES (often regarded as antennae) typically occurring on the segment anterior to that with antennae. ONYCH-OPHORA have a pair of cylindrical preantennae on first somite. See TENTACLES.

antenna complex Clusters of several hundred chlorophyll and accessory molecules fixed to the thylakoid membranes of chloroplasts by proteins in such a way as to harvest light energy falling on them, relaying it to a special chlorophyll molecule in an associated PHOTOSYSTEM. See PHOTOSYNTHESIS and Fig. 45b.

antennapedia complex (Ant-C) Complex of HOMEOTIC and segmentation loci in *Drosophila* which, when homozygously mutant, may result in conversion of antennal parts into leg structures. Intensely studied in contexts of MORPHOGENESIS, and POSITIONAL INFORMATION. Some loci in the complex appear to be expressed only in specific embryonic COMPARTMENTS. The *Drosophila* Antennapedia and Bithorax genes form one (split) cluster of *HOX* GENES (four in each), while vertebrate antennapedia-like HOMEOGENES are homologues.

antenna pigments See ANTENNA COMPLEX.

antennule Paired and most anterior head appendages of crustaceans; uniramous, whereas antennae like most appendages in the class are biramous.

anther Terminal portion of a STAMEN, containing pollen in *pollen sacs*.

antheridiophore In some liverworts, a stalk that bears the antheridia.

antheridium 'Male' sex organ (gametangium) of fungi, and of plants other than seed plants (e.g. algae, bryophytes, lycophytes, sphenophytes and pterophytes).

antherozoid Synonym of SPERMATOZOID.

Anthocerotopsida Hornworts. Class of BRYOPHYTA. Small, widely distributed group, especially in tropical and warm temperate regions, growing in moist, shaded habitats. Plant a thin, lobed, dorsiventral THALLUS, anchored by rhizoids. Each cell usually has a single large chloroplast rather than the many small discoid ones found in cells of other bryophytes and vascular plants; and each chloroplast possesses a PYRENOID, all features suggesting algal affinities. Some (e.g. *Anthoceros*) contain Cyanobacteria (e.g. *Nostoc* spp.), supplying fixed nitrogen to their host plants.

anthocyanins Group of water-soluble, flavonoid pigments (glycosides) occurring in solution in vacuoles in flowers, fruits, stems and leaves. Change colour, depending on acidity of solution. Responsible for most red, purple and blue colours of plants, especially in flowers; contribute to autumn (fall) colouring of leaves and tint of young shoots and buds in spring. Colours may be modified by other pigments, e.g. yellow flavonoids.

Anthophyta See MAGNOLIOPHYTA (Magnoliopsida), INTRODUCTION to dictionary.

Anthozoa See ACTINOZOA.

anthrax Disease of cattle, and occasionally of humans, caused by the bacterium *Bacillus anthracis*, whose spores can remain potent in soil for decades. The cutaneous form of the disease is not as dangerous as the pulmonary form (caused by inhaling spores). The toxin produced by the bacterium has three components: edema factor (EF) prevents macrophages from engulfing the bacteria; lethal factor (LF) kills macrophages, and eventually the host; protective antigen (PA) assists the other two components to enter the cell. Once seven PA molecules have formed a ring-like complex on the cell surface, they bind an EF/LF complex and become engulfed by the membrane. Once inside, PA molecules form a pore in the membrane of an ENDOSOME, pierce it, and allow EF and LF into the cell proper to cause damage to and death of the macrophage.

anthropoid apes Members of what used to be the primate Family Pongidae (see PONGINAE): the so-called 'great apes'. Include orangutan, chimpanzees and gorillas. Common ancestor of pongids and hominids ('men') probably Miocene in age. Gibbons (Family Hylobatidae) are in same suborder (Anthropoidea) as 'great apes' (pongids) and occasionally included in the term 'anthropoid ape'. Much ape anatomy may be regarded as adaptation to brachiation. Fundamentally quadrupedal, with limited (but famous) tendency to bipedalism. Markedly prognathous, with diastemas. All are Old World forms.

Anthropoidea (Simii) Suborder of PRIMATES (see Figs. 129a and b). Three living superfamilies: Ceboidea (New World monkeys); Cercopithecoidea (Old World monkeys); Hominoidea (gibbons, great apes and man). Eyes large and towards front of face; brain expansion associated with relative expansions of frontal, parietal and occipital bones of skull; great manual dexterity. The earliest fossils bearing clear anthropoid (simian) characters now seem to be from the late Eocene of Algeria; but the most numerous are from the Fayum (now dated in the Lower Oligocene) of Egypt – once dated forest, now desert. The skull and postcranial skeleton of the monkey-sized *Aegyptopithecus* from Fayum suggest a proto-hominoid of DRYOPITHECINE affinities, whereas the squirrel-sized *Apidium* and Parapithecus appear closer to prosimians and New World monkeys. There may have been an early simian radiation before the end of the Eocene, for *Algeripithecus minutus* – a putative simian from the Glib Zegdou deposits of Algeria – may be of middle or even early Eocene age. A date of 55 Myr BP rather than 35 Myr BP for simian origins now seems likely.

antiauxins Chemicals which can prevent the action of AUXINS in plants, e.g. 2,6-dichlorophenoxyacetic acid; 2,3,5-triiodobenzoic acid.

antibiotic Diverse group of generally low molecular mass organic compounds (see SECONDARY METABOLITE). Produced by spore-forming soil microorganisms (esp. moulds such as *Penicillium*, filamentous bacteria, e.g. *Streptomyces*, and other bacteria of the genus *Bacillus*) during or just after sporulation (noticeable in culture). Although they are often regarded as growth inhibitors of potentially competing microorganisms (bacteristatic when reversibly so; bactericidal when irreversibly), this is not totally convincing and may obscure their original role. They may be relics from the prebiotic era (prior to the evolution of cells) possibly interacting with RIBOZYME-like RNA molecules ('RNA-life'), as some do within cells today. Certainly many antibiotics serve mankind well by interfering with pathogen protein synthesis by adhering to ribosomal RNA, a property for which they have been artificially selected; but these effects may be due to overproduction of substances normally present in much smaller quantities within the cell. *Streptomycins* affect DNA, RNA and protein synthesis; *penicillins* (produced by *Penicillium chrysogenum*; see MITOSPORIC FUNGI) prevent cross-linking of the glycan chains of peptidoglycan molecules in bacterial cell walls (cell growth producing wall-deficient or wall-less cells, which burst); ACTINOMYCIN prevents RNA transcription; *puromycins* prevent translation; *anthracyclins* block DNA replication and RNA transcription. These effects generally involve the antibiotic complexing with, or inserting into, a nucleic acid. Antibiotics have been widely used as clinical drugs against bacterial pathogens, but are ineffective against viruses. It was Sir Alexander Fleming who first drew attention to antibiotics. In 1928, he observed that a strain of *Penicillium*, which had contaminated a culture of the bacterium *Staphylococcus* growing upon nutrient agar, completely halted the growth of the bacterium. Ten years later, Howard Florey and Ernst Chain at the University of Oxford, purified penicillin. It was Florey's tireless efforts which eventually led to its mass-production in America.

Many antibiotics are now used, or are potentially useful, as antitumour agents (e.g. mitomycin), immunosupressive agents (e.g. cyclosporin), as blood cholesterol-reducing agents (e.g. lovastatin), enzyme inhibitors (e.g. clavulanic acid), antimigraine agents (ergot alkaloids), herbicides (glufosinate), antiparasitic agents, ruminant growth promoters and bioinsecticides (e.g. *Bacillus thuringensis* toxin). Their use has put new selection pressures on the target microorganisms, resulting in the troublesome spread of ANTIBIOTIC RESISTANCE ELEMENTS in what resembles an ARMS RACE. See MITOSPORIC FUNGI (for a little history), PLASMIDS. Compare ANTIMICROBIAL PEPTIDES.

antibiotic resistance element Genetic element, composed of DNA and often borne on a TRANSPOSON, conferring bacterial resistance to an antibiotic. Often with INSERTION SEQUENCES at either end, when capable of moving between PLASMID, viral and bacterial DNA and selecting insertion sites, sometimes turning off expression of genes it inserts into or next to. Able to spread rapidly across species and other taxonomic boundaries, making design of new antibiotic drugs even more urgent. All pathogens usually found in hospitals are involved, plus mycobacteria, pneumococci and Enterobacteriaceae, so that treatment with even our most powerful drugs is sometimes unsuccessful, viz. vancomycin-resistant *Enterococcus* (VRE) and methicillin-resistant *Staphylococcus aureus* (see MRSA). Over-use of antibiotics on human patients selects for resistant strains, and a similar situation arises in animal husbandry where antibacterials are used in prophylaxis, chemotherapy and growth promotion. Worldwide differences in use and licensing of antibiotics are large. Non-homologous recombination occurs between plasmids, bacterial plasmid R1 conferring resistance to chloramphenicol, kanamycin, streptomycin, sulphonamide and ampicillin. If use of an antibiotic is stopped, over time resistance to it appears to reverse so that some antibiotics may become useful again through prudent long-term monitoring and use. Resistance could be minimized if antibiotics were used only for serious diseases, not at all when they would be useless in any case, and given in sufficiently high doses that bacterial populations are reduced before resistant mutants begin to appear. Unless patients complete their course of antibiotics, premature fall in antibiotic levels may allow the bacterial population to rise, increasing the numerical potential for newly mutating resistant strains to emerge and crowd out the wild-type strains (the equivalent process may occur in anti-retroviral treatments, which do not, however, involve antibiotics: see HIV). See NICIN PLASMID, TRANSFORMATION (1), NATURAL GENETIC COMPETENCE.

antibody Member of the IMMUNOGLOBULIN SUPERFAMILY of glycoproteins secreted by mature vertebrate B CELLS, binding selectively to epitopes of antigens and clumping them (agglutination) prior to phagocytic engulfment. Antibodies travel via the lymph to the blood, and viruses coated in antibody cannot enter host cells. In humans, antibodies in breast milk are not absorbed into the baby's blood but only act in the gut; but in cattle they are absorbed into the calf's blood by pinocytosis.

Five major classes differ principally in their type of heavy protein chain, and the degree to which the molecule is a polymer of immunoglobulin 'monomers'. Each immunoglobulin unit comprises two identical H- (heavy) and two identical L- (light) polypeptide chains forming mirror images of each other and joined by a flexible hinge region involving disulphide bridges, the hinge allowing the molecule to adapt slightly to different shapes when binding cell-surface antigens. They bind to antigen at specific antigen-binding regions provided uniquely by the combination of H- and L-chain amino-terminal portions (see Fig. 4), which are extremely variable in their amino acid sequences between different antibodies, in contrast to constant regions at their carboxy-terminal portions. Only about 20–30 amino acids of the variable regions of H- and L-chains contribute to the antigen-binding site, these being located in three short hypervariable regions of each variable region. These lie themselves within relatively invariant 'framework regions' of the variable regions. The other biological properties of the molecule are determined by the constant domains of the heavy chains.

Digestion of antibody with papain produces two identical Fab (antigen-binding) fragments and one Fc (crystallizing) fragment. The latter region in the intact Ig (immunoglobulin) molecule is responsible for determining which component of the immune system the antibody will bind to (for Fc receptor, see RECEPTOR (2)). The Fc region of IgG may bind phagocytes and the first component of COMPLEMENT. Only the IgG antibody can cross the mammalian placenta. IgM is the major Ig type secreted in a primary immune response, but IgG dominates in secondary immune responses (see B CELL).

Transformation of B cells into differentiated antibody-producing plasma cells generally requires both antigen-presenting cells and a signal from a helper T cell (see T CELL). Because B cells have only a few days' life in culture they are not suitable for commercial antibody production; however, if an antibody-producing B cell from an appropriately immunized mouse is fused to an appropriate mutant tumour B cell, the hybrid cell formed may continue dividing and producing the particular antibody required. The resulting HYBRIDOMA can be sub-cloned indefinitely, giving large amounts of antibody. Initial isolation of the appropriate B cell follows discovery of the required antibody in the growing medium. The purity of the resulting *monoclonal antibody* (mAb) and its production in response to what is possibly a minor component of an impure antigen mixture are both desirable features of the technique (see ABZYME). All the ACCESSORY MOLECULES known to participate in T cell recognition of their targets were first identified by monoclonal antibodies raised against these cell-surface markers. See ANTIBODY DIVERSITY, ANTIGEN–ANTIBODY REACTION, ANTITOXINS, IgA–IgM, OPSONINS.

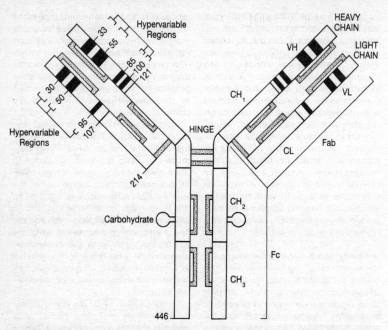

FIG. 4 *Diagram of the basic IgG structure. The hypervariable regions (amino acid residues numbered) are the antigen-binding sites. Disulphide bridges are indicated within and between light and heavy by stippling. Fab regions are those released by popain digestion, along with the complement-fixing Fc portions, which also bind Fc receptors. See* ANTIBODY.

antibody diversity (a. variation) Production of different ANTIBODY molecules by different B CELLS (see Figs. 4 and 5 for symbols). Light and heavy chains are encoded by different gene clusters. In humans, light chain genes lie on chromosomes 2 and 22, heavy chain genes on chromosome 14, and the light chains of a particular immunoglobulin molecule are encoded either by chromosome 2 or chromosome 22, not both. There are different constant (C) regions of the molecule for different immunoglobulin classes and two forms of light chain ($\varkappa$ and λ). Each light chain V region is encoded by a DNA sequence originating in two separate gene segments, a long V gene segment and a short J (joining) segment. But three gene segments are involved in the assembly of each heavy chain V (hypervariable) region:

the variable (V), joining (J) and diversity (D) elements, all occurring at distinct locations in germ line cells but becoming re-assorted into a continuous EXON during a process known as $V(D)J$ rearrangement during B cell and T cell maturation, in which the DNA-dependent protein kinase (DNA-PK) plays a major role (see also AFFINITY MATURATION). Any of 40 V segments in the human $\varkappa$ light-chain gene-segment pool can be recombined to any of the five J segments, so that at least 200 $\varkappa$-chain V regions alone can be produced from this pool; and any of the 51 V segments of the heavy-chain pool can be joined by any of the six J segments and any of the 27 D segments to produce at least 8,262 different heavy-chain V (V_H) regions. But a human can also produce 316 different V_L regions (200 $\varkappa$ and 116 λ). So these could,

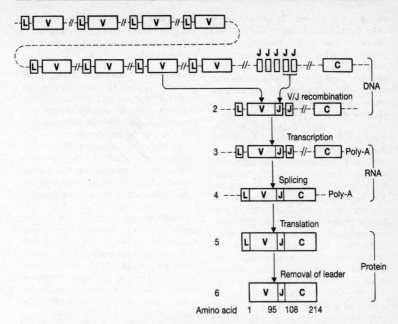

FIG. 5 *The formation of an active light chain of an antibody. The DNA is organized into families of related elements: one copy of the gene for the constant (C) region, five repeats of a joining (J) sequence, and up to 200 related but different variable (V) sequences, each separated by a leader (L) sequence. Through recombination, one of the V elements with its L sequence is inserted into a J segment. The entire V–J–C complex is transcribed into RNA, and excess J and linking sequences to C are excised as introns. The mature mRNA is translated, and the amino acids coded by the L portion, which are necessary to secret the polypeptide, are removed from the protein. Active heavy chains arise by similar rearrangements. Poly-A indicates polyadenosine tail. See* ANTIBODY DIVERSITY. *(From Introduction to Genetic Analysis (5th edn) by Griffiths, Miller, Suzuki, Lewontin and Gelbert. Copyright © W. H. Freeman and Company, 1993. Reprinted with permssion.)*

in principle, be combined to produce ~2.6 × 10⁶ (316 × 8,262) different antigen-binding sites (see ANTIBODY).

Each *V, D* and *J* element is flanked by a recombination signal sequence (rss). One of a number of *V* elements can rejoin with any *J* or *D* element, creating a wide repertoire of different immunoglobin and T CELL RECEPTOR genes. The recombination process is initiated by the introduction of a site-specific double-stranded (ds) DNA break at the border between the rss and its adjacent coding sequence. The two signal ends rejoin precisely head-to-head,

but the coding elements rejoin less precisely, involving deletions of 10 base pairs or less, as well as nucleotide insertions. This defined imprecision provides an additional mechanism for diversifying immunoglobin and T cell receptor genes, making the number of possible antibodies far greater than the total number of B cells (10¹²) in a human.

Somatic HYPERMUTATION in the IgV genes resulting from the generation of uracil from cytosine introduces point mutations within a population of dividing B cell precursors (see AFFINITY MATURATION) prior to clonal selection, a process in which DNA-dependent protein kinase DNA-PK plays a major role. The enzymes bringing together genes from different parts of a chromosome are performing a form of RECOMBINATION. Diversity arises from the randomness with which particular genes from heavy and light chain clusters are brought together. In addition, extra short nucleotide sections (N segments) get inserted, probably in some rule-following way, into the DNA encoding the antigen-binding regions of the molecule, and this together with variation in RNA PROCESSING of the hnRNA transcript increases still further the total antibody diversity, often classified as follows: (1) Allotypic: variation in the C_HI, C_H3 and C_L antibody regions caused by allelic differences between individuals at one or more loci for a subclass of immunoglobulin chains; (2) Idiotypic: variation in the V_L and V_H regions (especially in the hypervariable regions) that are generally characteristic of a particular antibody clone, and therefore not present in all members of a population; (3) Isotypic: variation in the C_L, and in the C_HI–3 antibody regions, determined by loci whose representative alleles are shared by all healthy members of a population. See AFFINITY MATURATION.

anticlinal (Bot.) Alignment of the plane of cell division approximately at right angles to the outer surface of the plant part. Compare PERICLINAL. See Fig. 23.

anticoagulant Any substance preventing blood clotting. Blood naturally contains such substances: fibrin and antithrombin III absorb much of the thrombin formed in the clotting process and HEPARIN inhibits conversion of prothrombin to thrombin. Blood-sucking animals (leeches, insects, bats, etc.) frequently produce anticoagulants in their saliva. Artificial anticoagulants (e.g. dicumarol) are either helpful to patients, or prevent blood samples from clotting in blood banks (e.g. EDTA). Sodium citrate precipitates calcium ions from plasma (as calcium citrate) and prevents clotting. The rat poison warfarin is an anticoagulant. See BLOOD CLOTTING.

anticodon The triplet sequence of tRNA nucleotides capable of base-pairing with a codon triplet of an mRNA molecule. See PROTEIN SYNTHESIS.

antidiuretic hormone (ADH, vasopressin) Ring-structure octapeptide hormone produced by hypothalamic neurosecretory cells and released into posterior pituitary circulation if blood water potential drops below the homeostatic norm. Has marked vasoconstrictor effects on arterioles, raising blood pressure, and increases water permeability of collecting ducts and distal convoluted tubules to the 10–20% of the initial glomerular filtrate still remaining (see KIDNEY), resulting in water retention. It acts via cAMP; its release is inhibited by alcohol, which therefore stimulates urine production; its concentration in blood depends on the rate of urine production and is zero when urine production is at its maximum rate (1 $dm^3.hr^{-1}$ in humans). Its release follows a circadian rhythm (see BIOLOGICAL CLOCK). See NICOTINE, OSMOREGULATION, OXYTOCIN.

antigen Any substance (often protein or glycoprotein) that can be recognized by an already induced immune response and initiate production of further specific ANTIBODY, to which the latter binds at a specific conformational domain of the antigen molecule called the antigenic determinant, or epitope. Not every antigen is immunogenic, i.e. capable of initial induction of an immune response. Mistaken recognition of AUTOANTIGENS (self-antigens) is the cause of AUTOIMMUNE REACTIONS. See antigen references and cross-references below.

antigen–antibody reaction Noncovalent bonding between antigenic determinant of ANTIGEN and antigen-combining site on an immunoglobulin molecule (see ANTIBODY). Several such bonds form simultaneously. The reactions show high specificity but cross-reactivity may result if some determinants of one antigen are shared by another. Antibodies seem to recognize the three-dimensional configuration and charge distribution of an antigen rather than its chemical make-up as such.

Such reactions form the basis of humoral and of many cell-mediated immune responses. See AGGLUTININ, COMPLEMENT, PRECIPITIN, IMMUNITY.

antigenic variation Ability of some pathogens, notably viruses (see HAEMAGGLUTININ), bacteria and protozoa, to change their coat antigens during infection. Trypanosomes and some stages in the malarial life-cycle achieve it, making the search for vaccines to some devastating human diseases very difficult. Can result from GENE CONVERSION. *Antigenic drift* involves minor changes in INFLUENZA virus antigens due to gene mutation; *antigenic shift* involves major changes in influenza virus antigens due to gene reassortment, often by recombination with a different strain.

antigen presentation See ANTIGEN-PRESENTING CELL.

antigen-presenting cell (APC) Few antigens bind directly to antigen-sensitive T CELLS or B CELLS but are generally 'presented' to these lymphocytes on the surfaces of other cells, the antigen-presenting cells. First identified in 1868 by Paul Langerhans, who initially mistook them for nerve endings, dendritic cells (DCs) are major antigen-presenting cells with large surface areas for this end. They make up only 0.2% of circulating white blood cells and are rarer still in most tissues (e.g. the spleen and the epidermis of the skin) except mucous membranes (e.g. the vaginal mucosa). Like macrophages, they develop from monocytes. There are several DC subtypes, but once an antigen is ingested they mature and present its peptide fragments on their surface, then travel to the spleen (via the blood) or to lymph nodes (via the lymph) where they mobilize their antigen-laden MHC class II molecules (see MAJOR HISTOCOMPATIBILITY COMPLEX) along microtubules and present them at the cell surface to naïve T helper cells at an IMMUNOLOGICAL SYNAPSE. These T cells are now activated to clonal expansion, differentiating into either T_H1 or T_H2 subclasses depending on the cytokines the DC produces. T_H2 cells assist B cells with the same MHC-bound antigen as themselves to produce antibodies specific to that antigen. DCs also activate T killer cells (see T CELLS). Some of the cells which DCs activate develop into memory cells, making them crucial to IMMUNOLOGICAL MEMORY and IMMUNE TOLERANCE. But DCs also activate phagocytic cells (neutrophils and macrophages) in innate immune responses. Antigen presentation to previously activated T cells is also a function of macrophages and B cells; but dendritic cells are unique in presenting antigenic peptide fragments to naïve T cells. Some cancer therapies may soon depend on VACCINES based on dendritic cells. See Figs. 98, 146, 147, IMMUNITY, PROTEASOMES.

antigen receptors Membrane glycoproteins on T CELL and B CELL surfaces. Those on B cells are membrane-bound antibodies; those on T cells are antibody-like heterodimers (see T CELL RECEPTOR).

antigestogen See CONTRACEPTION.

antigibberellins Organic compounds of varied structure causing plants to grow with short, thick stems or with appearance opposite to that obtained with GIBBERELLIN, which can reverse the action of most of these compounds. Of agricultural importance, they include phosphon and maleic hydrazide (retarding growth of grass, reducing frequency of cutting).

antimicrobial peptides Insects lack lymphocytes and antibodies and plant seeds germinate in microbe-infested soils, made possible in large part by their production of broad-spectrum antimicrobial peptides which target the susceptibility of the microbial (*contra* the eukaryotic) cell membrane. Many have AMPHIPATHIC structures, in which clusters of cationic and hydrophobic amino acid residues occupy distinct sectors of the molecule. Mammals have several gene families encoding such peptides (acting as 'natural antibiotics'), such as β-defensins and cathelicidins, expressed on epithelial surfaces and in neutrophils. See PHYTOALEXINS, PLANT DISEASE AND DEFENCES.

antioxidants See SUPEROXIDES.

antipodals Three (sometimes more) cells of the mature EMBRYO SAC, located at the end opposite the micropyle.

antiport See TRANSPORT PROTEINS.

antisense Originally applied to those artificially synthesized single-stranded DNA oligomers which bound to complementary mRNA, thereby preventing their translation. But the term is now of wider application, referring to any single-stranded DNA or RNA which complements another, binds to it and blocks its function. They have been explored for several years as experimental and therapeutic tools. Antisense RNA, for instance, can bind a complementary mRNA molecule and produce double-stranded RNA (dsRNA). It now seems that perhaps as many as 8% of human genes are transcribed from both strands and encode antisense RNA, and most if not all naturally occurring antisense transcripts are made in the nucleus, not in the cytoplasm. See RNA SILENCING.

antisense interference A GENE SILENCING effect by which artificially provided single-stranded RNA complementary to the target gene blocks its activity. See dsRNA-TRIGGERED INTERFERENCE.

antiseptic Substance used on a living surface (e.g. skin) to destroy microorganisms and sterilize it. Ethyl and isopropyl alcohol, diluted 70% with sterile water, destroy vegetative bacteria and some viruses, but not spores of bacteria or fungi. Iodine (dissolved with potassium iodide in 90% ethanol) is rapidly bactericidal, killing both vegetative cells and spores. It does, however, stain. See DISINFECTANT, AUTOCLAVE.

antiserum SERUM containing antibodies with affinity for a specific antigenic determinant (see ANTIGEN) to which they bind. May result in cross-reactivity (see ANTIGEN–ANTIBODY REACTION) within recipient.

antitoxin Antiserum containing an ANTIBODY specifically binding and neutralizing a toxin; e.g. diphtheria and tetanus antitoxins. See TOXOIDS.

antler Bony projection from skull of deer. Unlike HORN (which is matted hair) they are often branched, are shed annually, and are confined to males (except in reindeer).

Anura (Salientia) Frogs and toads. An order of the Class AMPHIBIA. Hind legs modified for jumping and swimming; no tail; often vocal.

anus The opening of the alimentary canal to the exterior through which egested material, some excretory material and water may exit. When present, the gut is said to be entire. Absent from coelenterates and platyhelminths. See PROCTODAEUM.

aorta Term applied to some major vertebrate arteries. See AORTA, DORSAL; AORTA, VENTRAL; AORTIC ARCH.

aorta, dorsal Major vertebrate (and cephalochordate) artery through which blood passes to much of body, supplying arteries to most major organs. In sharks a single dorsal aorta collects oxygenated blood from the gills, but in bony fish paired dorsal aortae on either side in the head region perform this task before uniting as a single median vessel. Oxygenated blood then passes backwards to the body; but in fish too blood flowing up through the third AORTIC ARCH tends to pass anteriorly through the aorta(e) rather than posteriorly (see CAROTID ARTERY). In adult tetrapods, those parts of the single or paired dorsal aortae between the third and fourth aortic arches tend to disappear, blood from the fourth (systemic) arch(es) passing back within two uniting dorsal aortae (terrestrial salamanders, lizards) or within a single dorsal aorta (most reptiles, birds and mammals) derived from the right arch (mammals). Protected throughout in vertebrates by proximity of bone above (typically vertebrae).

aorta, ventral Large median artery of fish and embryonic amniotes leading anteriorly from ventricle of heart, either giving off branches to gills or running uninterrupted as AORTIC ARCHES to dorsal aorta(e). In lungfish, branches differ in this respect. In living amphibians it has disappeared, while in other tetrapods it serves merely as a channel supplying blood to aortic arches III, IV and VI.

aortic arches Paired arteries (usually six, but up to fifteen in hagfishes) of vertebrate embryos connecting ventral aorta with dorsal aorta(e) by running up between gill slits or gill pouches on each side, one in each VISCERAL ARCH. The study of their comparative anatomy in embryos and, where they persist, in adults provides striking support for macroevolutionary change. Each is given a Roman numeral, beginning anteriorly. Arches I and II do not persist in postembryonic tetrapods, but arch II at least is present in sharks, some bony fish and lungfish. Arch III usually serves (with parts of the dorsal aortae) as the tetrapod carotid arteries, but in fish is usually interrupted by gills; arch IV is separated from the anterior arches in most tetrapods and becomes the systemic arch (see AORTA, DORSAL); arch V is absent from adult tetrapods other than urodeles, but serves as the DUCTUS ARTERIOSUS in development prior to lung function; arch VI then shifts to supply the lungs.

APC protein (adenomatous polyposis coli (APC) protein) A protein which is mutated in about 70% of colorectal CANCERS. The normal protein has highly conserved nuclear export signals, although these are lost in the mutant protein. The ability of APC protein to leave the nucleus seems to be important to its tumour-suppressing function. See TUMOUR SUPPRESSOR GENE.

ape General term for HOMINOID primates of families Hylobatidae (gibbons, siamangs) and Pongidae ('great apes'). See ANTHROPOID APES.

apetalous Lacking petals, e.g. flower of wood anemone.

Aphaniptera See SIPHONAPTERA (fleas).

aphid Green fly or black fly. Homopteran insect (Superfamily Aphidoidea) notorious for sucking plant juices, for transmitting plant viral diseases, and for phenomenal powers of increase by viviparous PARTHENO-GENESIS.

aphyllous Leafless.

apical dominance (Bot.) Influence exerted by a terminal bud in suppressing growth of lateral buds. See AUXINS.

apical meristem Growing point (zone of cell division) at tip of root and stem in vascular plants. Responsible for postembryonic development, positional control signals from more mature cells guiding the pattern of meristem cell differentiation. An apical meristem has its origin in a single cell (*initial*), e.g. PTEROPHYTA, or in a group of cells (initials), e.g. Anthophyta. In the latter, the growing point apex (*promeristem*) consists of actively dividing cells. Behind this, division continues and differentiation begins, becoming progressively greater towards mature tissues. One (older) concept of growing point organization in flowering plants recognizes differentiation into three regions (*histogens*): *dermatogen*, a superficial cell layer giving rise to the epidermis; *plerome*, a central core of tissue giving rise to the vascular cylinder and pith; and *periblem*, tissue lying between dermatogen and plerome, that gives rise to cortex. It is now evident that respective roles assigned to these histogens are by no means universal; nor can periblem and plerome always be distinguished, especially in the shoot apex. Becoming widely accepted is the *tunica-corpus* concept, an interpretation of the shoot apex recognizing two tissue zones in the promeristem: *tunica*, consisting of one or more peripheral layers, in which the planes of cell division are predominantly anticlinal, enclosing the *corpus* or central tissue of irregularly arranged cells in which the planes of cell division vary. No relation is implied between cells of these two regions and differentiated tissue behind the apex as in the histogen concept. Although epidermis arises from the outermost tunica layer, underlying tissue may originate in tunica or in corpus, or in both, in different plant species.

Stem apical meristem allows growth in length of main axis and is site of origin of leaf and bud primordia. In roots, two types of apical meristem occur, one in which vascular, cylinder, cortex and root cap can be traced to distinct layers of cells in the promeristem, and a second type in which all tissues have a common origin in one group of promeristem cells. In contrast to those of stems, apical meristems of roots provide only for growth in length, lateral roots originating some distance from apex and, endogenously, from pericycle. See INTRODUCTION to dictionary.

Apicomplexa Class of parasitic protozoans, many intracellular, some with alternate hosts. Mature stages lack locomotor organelles, but young may be amoeboid or flagellated. Most characteristic structure is the *apical complex*, a device required for entry to the host cell. Food is generally not ingested but is rather absorbed in soluble form, as in fungi and many prokaryotes. Large numbers of young are produced, either naked or in spores, by multiple fission after syngamy, when transmission to another host may occur. Includes plasmodia, e.g. *Plasmodium* (see MALARIA), *Cryptosporidium* (agent of water-borne dysentery) and COCCIDIA, e.g. *Toxoplasma* (causing toxoplasmosis). CYTOCHALASINS are usually effective in preventing host cell invasion.

aplanospore Non-motile algal spore, produced through a subdivision of the protoplast of a cell (sporangium). They have no flagella, but possess certain of the characteristics of flagellate cells, such as contractile vacuoles.

apocarpous (Of the gynoecium of flowering plants) having separate carpels, e.g. buttercup. See FLOWER.

apocrine gland Type of gland in which only the apical part of the cell from which the secretion is released breaks down during secretion, e.g. mammary gland. Compare HOLOCRINE GLAND, MEROCRINE GLAND.

Apoda (Gymnophiona) Caecilians. Order of limbless burrowing amphibians with small eyes and, sometimes, a few scales buried in the dermis of the skin, and a pair of tentacle-like structures in grooves above the maxillae.

apoenzyme The protein component of a holoenzyme (enzyme-cofactor complex) when the COFACTOR is removed. It is catalytically inactive by itself.

apogamy See APOMIXIS.

apogeotropic Growth of roots away from the earth and from the force of gravity (i.e. into the air).

apomict Plant produced by APOMIXIS.

apomixis Most common in botanical contexts. (1) AGAMOSPERMY, reproduction which has the superficial appearance of ordinary sexual reproduction (amphimixis) but occurs without fertilization and/or meiosis. Affords the advantages of the seed habit (dispersal, and survival through unfavourable conditions) without risks in achieving pollination. Recurrent apomixis occurs in more than 40 angiosperm families, including those containing grasses, sunflowers and roses and is often genetically equivalent to asexual reproduction. Diploid pollen of the Mediterranean cypress tree *Cupressus dupreziana* produces quite naturally an embryo without fertilization which is nourished in the seed tissues of a surrogate mother, *C. sempervirens*. This seems to be the first known case of paternal apomixis in plants, and may be a reproductive strategy by *C. dupreziana* in response to the threat of extinction. Various genes are being investigated which may determine whether a plant reproduces by normal sexual processes, or apomictically. It may be possible to use these genes to develop new strains of apomictic crop plants while offering a way of avoiding the degeneration of breeding stocks of vegetatively propagated plants, such as potato and cassava, which accumulate pathogens through repeated usage. See Fig. 6, PARTHENOGENESIS. (2) Vegetative apomixis; ASEXUAL methods of propagation such as by rhizomes, stolons, runners and bulbils.

apomorphous In evolution, of a character derived as a novelty from pre-existing (plesiomorphous) character. The two form a homologous pair of characters, termed an *evolutionary transformation series* in CLADISTICS. See SYNAPOMORPHY.

apoplast Cell wall continuum of plant or organ; movement of substances in the cell

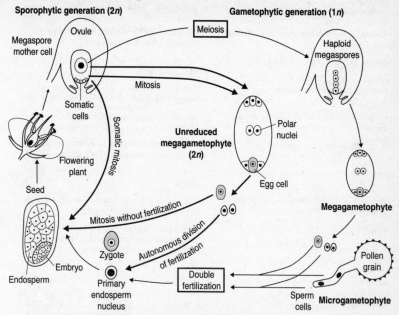

Sporophytic generation (2n)   **Gametophytic generation (1n)**

Meiosis

Megaspore mother cell

Ovule

Haploid megaspores

Mitosis

Somatic cells

Somatic mitosis

Unreduced megagametophyte (2n)

Polar nuclei

Flowering plant

Seed

Egg cell

Mitosis without fertilization

Megagametophyte

Zygote

Autonomous division of fertilization

Pollen grain

Embryo

Endosperm

Primary endosperm nucleus

Double fertilization

Sperm cells   **Microgametophyte**

FIG. 6 *The normal angiosperm sexual cycle (outer cycle) compared with the apomictic short-circuit (inner cycle) in which embryos arise from a cell lineage lacking both meiotic reduction and fertilization, and in which the entire maternal genotype is transmitted to the next generation in the form of a seed-propagated clone.* See APOMIXIS.

walls is termed apoplastic movement or transport. Compare SYMPLAST.

apoprotein A protein without its characteristic PROSTHETIC GROUP.

apoptosis A manner of cell death which, unlike cell necrosis, releases no cell debris onto adjacent cells. In animals, mitochondria are involved in integration of diverse cellular stress signals and initiate a death execution pathway, and a similar role for mitochondria in plants now appears probable, although the molecular regulators identified in animals are absent from the *ARABIDOPSIS* genome. It is sometimes brought on by a lack of environmental SURVIVAL FACTORS (e.g. in the EXTRACELLULAR MATRIX). Plants probably use other regulators to control their programmed cell death (see PLANT DISEASE AND DEFENCES). Apoptosis is a key feature of normal development, tissue

homeostasis and disease progression in animals, abnormal resistance to which causes malformations, autoimmune disease or cancer through persistence of superfluous, self-specific, or mutated cells, respectively. The molecular mechanisms involved are highly conserved between nematodes and humans, but more complex in the latter. The process is characterized by DNA fragmentation, chromatin condensation, membrane blebbing, nuclear and cytoplasmic shrinkage and disassembly into membrane-enclosed vesicles, followed by phagocytosis by macrophages – without any leakage of cell contents, so avoiding inflammation. See Table 1.

Too much apoptosis may lead to stroke damage or the neurodegeneration of ALZHEIMER'S DISEASE. Work on the nematode *Caenorhabditis elegans* has identified three key genes in the process: *ced-3*, *ced-4* and

Assumption	Invalidation
Apoptosis involves the expression of 'killer genes'.	Apoptosis can be induced in all cells in the presence of cycloheximide.
Apoptosis is an abortive cell cycle.	Apoptosis can be induced during any phase of the cell cycle.
Apoptosis is a nuclear process.	Cytoplasts (enucleated cells) can undergo receptor-mediated, Bcl-2-regulated cell death.
Apoptosis is mediated by reactive oxygen species.	Apoptosis can be induced in/by the absence of oxygen, in a Bcl-2-regulated fashion.
Apoptosis involves an elevation of Ca^{2+} in some subcellular compartment.	Ca^{2+} depletion can induce apoptosis. Nuclear apoptosis can be induced in Ca^{2+}-free media.
Apoptosis is due to cytosolic acidification.	In some models, acidification inhibits cell death.
Apoptosis always involves specific proteases (caspases).	Bax and Bak overexpression induces death in the presence of a caspase inhibitor.
Apoptosis does not involve mitochondria.	Cell-free systems of apoptosis require mitochondria or mitochondrial products.

TABLE 1 *A few incorrect assumptions that have marred* APOPTOSIS *research.*

ced-9. Expression of the first two is required for induction of cell death, while *ced-9* product (CED-9) binds to the CED-3/CED-4 protease complex, an expression necessary to inhibit cell death and protect cells from inappropriate activation of the 'death programme'.

The terms 'pro-apoptotic' and 'anti-apoptotic' are used of molecules (e.g. signal molecules, receptors) which respectively make apoptosis more, or less, likely. Thus, a pro-apoptotic dependence RECEPTOR, such as Patched-1 (Ptc-1), requires binding by its ligand Sonic hedgehog (Shh) in the chick neural tube for survival of some ventral cells, and cleavage of this signal molecule by caspase-3 induces cell death, which regulates numbers of cells in several central nervous regions, notably the cerebellum. Many, if not all, human CANCER cells exhibit resistance to apoptosis and cells from many tumour types have an increased ratio of anti-apoptotic (e.g. BCL-2) to pro-apoptotic (e.g. Bax, Bak) proteins.

One route for apoptosis is via the cell membrane receptors TNFR1, TNFR2 and Fas which, on binding their respective ligands (TNF, FasL), transmit 'death orders' via other proteins to a protein-cutting enzyme (ICE, interleukin-1β converting enzyme), a CASPASE related to CED-3. A second, and probably not independent, route for apoptosis onset involves precipitous collapse of the cell's mitochondrial membrane potentials and release from mitochondria of apoptosis-initiating factors (AIFs, or Apafs) which trigger latent caspases into action, bringing about most of the terminal effector events of apoptosis (see APOPTOSOME). Mitochondrial membranes contain Bcl-2 anti-apoptotic proteins which probably regulate flow of ions and molecules between this compartment and the cytosol. It seems that the mitochondrion is an amplifier of cellular stress signals arriving from different subcellular compartments, including: the proteinaceous steroid hormone receptor TR3 from the nucleus; Bcl-2 and Bax from the endoplasmic reticulum; caspase-8 from the plasma membrane, on activation by the death receptor CD95; Bim from the cytoskeleton and Bax, Bak and Bid from the cytosol. These proteins modify permeability of the outer mitochondrial membrane (OMM), causing release of the cell-death activators mentioned above, as well as inhibitors and inhibitor depressors. Release of the protein CYTOCHROME c regulates the activity of the initiator caspase procaspase-9 and involves the adaptor protein Apaf-1 (see PLANT DISEASE AND DEFENCES). A third key to induction of apoptosis involves the onco-suppressor protein p53, which commits to death cells that have sustained DNA damage from mutagens (purging tissues of founder CANCER CELLS). It is possible that of the several genes induced by p53, some may activate mitochondrial disruption, releasing death agonists (Bax, Ced-4). See C-*MYC*, FAS.

apoptosome One of a series of PROTEIN COMPLEXES whose three-dimensional structures have been determined by CRYO-ELECTRON TOMOGRAPHY. It comprises seven molecules each of the proteins cytochrome *c* and Apaf-1 and is a major player in APOPTOSIS. When free from inner mitochondrial membranes, cytochrome *c* binds Apaf-1 to create the apoptosome, a complex which activates the cell's procaspase-9 which breaks up the cell components for recycling.

aposematic Colour, sound, behaviour or other quality advertising noxious or otherwise potentially harmful qualities of an animal. See MIMICRY.

apospory See APOMIXIS, PARTHENOGENESIS.

apostatic selection Selection by visual predators, held to increase the divergence between morphs in a polymorphic population by favouring any variation departing from the predator's SEARCH IMAGE.

apothecium Cup or saucer-shaped fruit body (ASCOCARP) of certain ASCOMYCOTA and LICHENS); lined with a hymenium of asci and paraphyses. They are sessile, often brightly coloured, varying from a few mm to more than 40 cm across.

appendage A functional projection from an animal surface; termed *paired appendages* if bilaterally symmetrical. Two such pairs (e.g. limbs, fins) generally occur in gnathostome vertebrates. Primitively one pair per segment in arthropods (walking legs, mouthparts, antennae) and polychaetes (parapodia).

appendicular skeleton See ENDOSKELETON.

appendix, vermiform Small diverticulum of human caecum, of many other primates, and of rodents, containing lymphoid tissue. Not a vestigial structure, contrary to common belief.

appetite For control in mammals, see HYPOTHALAMUS.

appetitive behaviour Behaviour (e.g. locomotory activity) variable with circumstances, increasing the chances of an animal satisfying some need (e.g. for food, nesting material) usually through a more stereotyped CONSUMMATORY ACT, such as eating. To this extent it is *goal-oriented*.

apposition (Bot.) Growth in thickness of cell walls by successive deposition of material, layer upon layer. Compare INTUS-SUSCEPTION.

aptamer See OLIGONUCLEOTIDES.

apterous Wingless; either of insects which are polymorphic for winged and wingless forms, e.g. aphids, many social insects; or of insects which have discarded wings, as do some ants and termites; or of primitively wingless (apterygotan) insects.

Apterygota (Ametabola) Subclass of primitively wingless insects. Comprises orders Thysanura (bristletails, silverfish), Collembola (springtails), Protura and Diplura. Probably a polyphyletic assemblage. Some abdominal segments in members of all four orders have small paired lateral appendages, another primitive characteristic. Metamorphosis slight or absent. See PTERYGOTA.

aquaporins Members of the MIP (major intrinsic protein) family functioning as water channels in CELL MEMBRANES. The first to be sequenced was CHIP28 (channel-forming integral protein of molecular mass 28,000), having six membrane-spanning regions allowing water molecules (0.3 nm diameter) through but able to exclude charge-bearing ions and uncharged molecules (e.g. urea) of similar size. Bacterial glycerol facilitator (GlpF) is a channel homologous with the human water channel aquaporin-1 (AQP1) but selecting for glycerol in preference to water. Both belong to a large family of integral membrane proteins, the aquaglyceroporin family. See OSMOSIS.

aqueous humour Fluid filling the space between cornea and VITREOUS HUMOUR of vertebrate EYE. The iris and lens lie in it. Much like cerebrospinal fluid in composition. Continuously secreted by ciliary body, and drained by canal of Schlemm into blood. Much less viscous than vitreous humour. Links circulatory system to lens and cornea, neither having blood vessels for optical reasons; also maintains intraocular pressure.

Arabidopsis thaliana Common thale cress, a cruciferous weed (Tribe Sisymbriae) of no economic value but of increasing importance as an experimental plant for molecular and developmental biology on account of its (1) very small genome size (7×10^7 base pairs; about five times that of the yeast *Saccharomyces cerevisiae*) in five chromosome pairs, over 60% of it encoding protein; (2) rapid life cycle (6–8 weeks); (3) small size (15–30 cm height). Because of the small amount of repetitive DNA, cloning procedures involving chromosome walking and gene tagging are applicable.

The genome of *A. thaliana*, sequenced in 2000, is the principal reference genome for dicotyledons. Comparison with the nearly completed genome of the distantly related RICE reveals that they have much in common: 85% of *Arabidopsis* genes have high sequence similarity to rice genes. However, there is only a limited degree of conserved gene order, since extensive chromosomal and wholesale genome duplications and selective gene loss have occurred in ANGIOSPERM evolution.

arachidonic acid A twenty-carbon fatty acid; precursor of EICOSANOIDS and product of PLA$_2$ (phospholipase A$_2$) activity. See LEUKOTRIENES.

Arachnida Class of chelicerate arthropods. Most living forms terrestrial, using lung books (scorpions), lung books and tracheae (spiders), tracheae alone (e.g. pseudoscor-

pions, larger mites), or just the body surface (smaller mites) for gaseous exchange. Usually there is TAGMOSIS into a *prosoma* of eight adult segments anteriorly and an *apisthosoma* of thirteen segments posteriorly. No head/thorax distinction. Prosoma lacks antennae and mandibles; first pair of appendages clawed and prehensile chelicerae; second pair (pedipalps) may be prehensile and sensory, copulatory or stridulatory devices. Remaining four pairs of prosomal appendages are legs. Bases (gnathobases) of second and subsequent pairs of appendages are often modified for crushing and 'chewing' (in absence of true jaws). Includes orders: Acari (mites and ticks), Araneae (spiders), Scorpiones (scorpions), Pseudoscorpiones (false scorpions), Palpigrada (palpigrades), Solifugae (solfugids) and Opiliones (harvestmen). Xiphosura (king crabs), and the predatory and extinct Eurypterida are usually placed as subclasses of the MEROSTOMATA.

arachnoid membrane One of the MENINGES around vertebrate spinal cord and brain.

Araneae (Araneida) Order of ARACHNIDA. Spiders. Abdomen (opisthosoma) almost always without any trace of segmentation and joined to prosoma (cephalothorax) by 'waist'; silk produced from two to four spinning glands (spinnerets); pedipalps in male modified as intromittant organs for copulation; ends of chelicerae modified as poison fangs.

arbuscule See MYCORRHIZA.

Archaea (adj. **archaeal**) In taxonomic systems that regard the prokaryote/eukaryote distinction as unnatural and failing to represent an inner phylogenetic division within prokaryotes, an alternative to the more traditional ARCHAEBACTERIA. The first archaeon to be genomically sequenced (1996) was *Methanococcus jannaschii*, from a Pacific deep sea hydrothermal vent. Its genome contained 1,738 genes, 56% of which were entirely new to science. The DNA-replicating and RNA translation machinery of *M. jannaschii* have far more in common with those of yeast than of bacteria; and their histones (unlike bacterial 'histone-like' proteins) form part of a nucleosome-like chromatin. *M. jannaschii* appears to lack four tRNA-binding enzymes common to all other forms of life; but those that are present are eukaryote-like, while archaeon tRNA introns are very similar to those of eukaryotes. The LIPIDS present in archaeon membranes have ether linkages between glycerol and their hydrophobic side chains (sometimes covalently bonded to form a lipid monolayer rather than a bilayer); hyperthermophilic archaeon membranes lack fatty acids, but instead have repeating units of the hydrocarbon isoprene (see ISOPRENOIDS) and phytane chains linked to glycerolphosphate. All this supports C. R. Woese's (1990) three-DOMAIN view of life: Bacteria (formerly Eubacteria), Archaea and Eukarya, the bacteria branching first from the common trunk of future archaeons/eukaryons. Work on archaeon ribosomal RNA (e.g. 16S sequences) suggests three subdivisions (kingdoms): very early hyperthermophilic forms (Korarchaeota); later hyperthermophiles (Crenarchaeota); and methanogens and extreme halophiles (Euryarchaeota). Crenarchaeota account for 30% of microbes in the Antarctic oceans and are not restricted to hot springs; indeed, they occur in lake and marsh sediments, and in soil.

In 2002, small, spherical structures (~400 nm) – an organism named *Nanoarchaeum equitans* – were discovered attached to cells of a new species of the archaeal genus *Ignicoccus*. It has a very small genome and its small subunit rRNA sequence places it in the archaeal domain, but not within any of the three subdivisions just mentioned. A new group, Nanoarchaeota, has been postulated to include it – not to be confused with the very much smaller nanobacteria. See ENDOSYMBIOSIS and Fig. 48.

Archaean (Archaeozoic) Geological division preceding PROTEROZOIC; earlier than about 2,500 Myr BP. See Appendix.

Archaebacteria Ancient lineage of bacteria distinct from other bacteria (eubacteria) and from eukaryotes (but see ARCHAEA). Many live in hot acidic conditions (i.e. they are thermophilic and acidophilic), growing best at temperatures approaching 100°C. Formerly in two groups, either aerobic (Sulfolobales) or anaerobic (Thermoproteales). Facultative anaerobic forms are now known. Many unusual biochemical characteristics including possession of a novel 16S-like ribosomal RNA component in the small ribosome subunit, which with their peculiar membrane composition indicates that there may be a deep divide among prokaryotes between archaebacteria and eubacteria. Halophiles, methanogens and sulphur-dependent thermophiles occur.

Archaeopteryx Most ancient recognized fossil bird (late Jurassic, 150–145 Myr BP). Exhibits mixture of reptilian and bird-like characters, having feathered wings and tail (impressions clear in limestone) and furcula (fused clavicles and interclavicles); but with teeth, bony tail, and claws on three digits of forelimbs. Only known representative of Subclass Archaeornithes of Class AVES.

archegoniophore In some liverworts, a stalk bearing archegonia.

archegonium 'Female' sex organ of liverworts, mosses, ferns and related plants, and of most gymnosperms. Multicellular, with neck composed of one or more tiers of cells, and swollen base (venter) containing egg-cell.

archenteron Cavity within early embryo (at gastrula stage) of many animals, communicating with exterior by BLASTOPORE. Formed by invagination of mesoderm and endoderm cells at gastrulation; becomes the gut cavity.

archesporium Cells or cell from which spores are ultimately derived, e.g. in developing pollen sac, fern sporangium.

Archezoa Taxon posited to house ancestral AMITOCHONDRIATE, but nonetheless eukaryotic, cells (i.e. nucleated, cytoskeletal, and endomembranous). The taxon was postulated to have derived from hypothetical common ancestors of archaebacteria and eukaryotes, a descendant of which was conjectured to have engulfed a bacterium (giving rise to mitochondria), others remaining amitochondriate. It now appears that loss of mitochondria may have been a secondary event, and that a revised theory of EUKARYOTE origins may make the taxon redundant. Included diplomonads (e.g. *Giardia*), trichomonads and microsporidians.

Archonta Mammalian superorder employed by some authors, who generally regard it as monophyletic, containing primates, dermopterans, tree shrews (Order Scandentia) and bats (see CHIROPTERA). Interpretation of morphological similarities between these groups is questionable; no consistent picture emerges from molecular comparisons.

archosaurs 'Ruling reptiles'; the Subclass Archosauria. Originating with thecodonts in the Triassic, it includes the bipedal carnivorous dinosaurs (saurischians) and the bird-like dinosaurs (ornithiscians). Crocodiles and alligators are living representatives. Birds are descendants. See DINOSAUR.

Ardipithecus ramidus A Pliocene hominine. See AUSTRALOPITHECINE.

Arecidae A subclass of morphologically and ecologically diverse plants belonging to the LILIOPSIDA comprising about 5,600 species. More than half of these species belong to a single order, the Arecales, which includes only the family Arecaceae (palms). Characteristically, flowers of this group are mostly numerous, usually small and subtended by a prominent spathe (a large bract subtending and often enclosing an inflorescence) or several spathes. Flowers are often aggregated into a spadix (a spike with small, crowded flowers on a thickened fleshy axis); inflorescence much reduced in the Lemnaceae. Plants are either arborescent or with relatively broad leaves that do not have the typical parallel venation or both (but sometimes lacking both of these characteristics, and in the Lemnaceae they are even thalloid and free-floating). Vessels generally present in all vegetative organs (except among the Arales) and the endosperm is not starchy (except in some Arales). See INTRODUCTION to dictionary.

aril Accessory seed covering, often formed from an outgrowth at the base of the OVULE (e.g. yew); often brightly coloured, aiding dispersal by attracting animals that eat it and carry seed away from the parent plant.

arista See AWN.

arms race Term sometimes used to express the dialectical changes in selection pressure that occur when regular, often unavoidable, conflicts of interest between two or more 'ways of life' favour an adaptation for one party which creates a fresh 'challenge' for the other to respond by adapting to. Such conflicts are common: predator/prey; parasite/host; parent/offspring and male/female. It has been argued that selection will be the stronger where one party has more to lose by 'not evolving' and minimizing the probability of losing the conflict. Consequence to a prey organism of losing a predator/prey conflict is probably more serious than to a predator on any occasion. Much depends on how likely such conflict encounters are as to whether selection will favour whatever 'costs' may be involved in evolving a ploy to avoid or win the conflict. Conflicts are best generalized as conflicts of 'ways of life', or strategies, rather than between individuals *per se*. The caterpillar *Helicoverpa zea*, which feeds on the tobacco plant *Nicotiana tabacum*, produces GLUCOSE OXIDASE as its principal salivary enzyme. *N. tabacum* produces NICOTINE in response to herbivory by caterpillars, but produces much less if caterpillar regurgitant is applied to its leaves. Glucose oxidase is responsible for suppressing this induced resistance and represents one component in the evolutionary arms race between plants and herbivores (SEE PLANT DISEASE AND DEFENCES). Some conflicts of interest may resolve in favour of one party through inability of the genetic system to 'represent' the other in the arms race. There have been few convincing demonstrations of the 'arms race principle'; but some experimental work suggests that even in apparently unchanging sexual interactions, males and females may, in evolutionary terms, be 'running frantically to stand still' (SEE RED QUEEN HYPOTHESIS). See ALTRUISM, COEVOLUTION, LD$_{50}$, RNA SILENCING.

arousal General causal term (and factor) invoked to account for the fact that animals are variably alert and responsive to potential stimuli. There may be a general 'sleeping/waking' difference; but it is less clear that there is a continuum of levels of awareness or responsiveness during either of these states. The phenomenology of arousal may be correlated with neural activity in the RETICULAR FORMATION of the medulla, hypothalamus and cortex of the vertebrate brain. Physiological processes which facilitate certain behaviours include hormone release and endogenous rhythms. Both may then be said to be arousal mechanisms, or to affect motivation.

arrhenotoky See MALE HAPLOIDY.

arteriole A very small, almost microscopic, artery delivering blood to capillaries. The tunica interna comprises an endothelial lining covered by a basement membrane and elastic lamina; the tunica media is a patchy affair of smooth muscle with very few elastic fibres; and the tunica externa comprises elastic and collagen fibres. Crucially important in regulating blood supply to capillaries through VASOCONSTRICTION and VASODILATION. See METARTERIOLES.

artery Any relatively large blood vessel carrying blood (not necessarily oxygenated) from the heart towards the tissues. Vertebrate arteries have thick elastic walls of smooth muscle and connective tissue (larger ones have capillaries in them), damping blood pressure changes. Their innermost layer is endothelium, as with all vertebrate blood vessels. They divide repeatedly to form arterioles. *Elastic arteries* (aka *conducting arteries*) contain a high proportion of elastic fibres with relatively thin walls in relation to overall diameter. They absorb energy from the blood after ventricular contraction by stretching, and release it again during elastic recoil as blood pressure falls as a result of ventricular relaxation, helping to force blood along the artery. These effects damp the oscillations in blood pressure which would otherwise have far greater amplitude. The aorta, brachiocephalics, common carotids, subclavian, pul-monary and iliac arteries are of this kind. *Muscular arteries* (aka *distributing arteries*), such as the brachial and radial arteries, are medium-sized arteries whose tunica media contain more smooth muscle and fewer elastic fibres than do elastic arteries. They are therefore capable of greater vasoconstriction and vasodilation and have more influence on the rates of blood flow. See SCLEROSIS.

Arthropoda The largest phylum in the animal kingdom in terms of both number of taxa and biomass. Bilaterally symmetrical and metamerically segmented coelomates, with appendages on some or all segments (somites). A chitinous cuticle provides the exoskeleton, flexible to provide joints. Haemocoele is the main body cavity (coelom reduced). They lack true nephridia and cilia (onychophorans have the latter); with an annelid-like central nervous system and one or more pairs of coelomoducts acting as gonoducts or excretory ducts. Taxonomy varies. Thirteen classes are widely recognized, including: Onychophora (peripatids), Myriapoda (centipedes and millipedes), Insecta (insects), Trilobita (trilobites, extinct), Merostomata (king, or horseshoe crabs and extinct eurypterids), Arachnida (scorpions, spiders, harvestmen, solfugids, mites and ticks), Crustacea (crabs, prawns and shrimps, water-fleas). However, onychophorans are sometimes excluded from extant arthropods, while merostomatans and arachnids are often grouped into the CHELICERATA (see Table 2). Some take the view that Crustacea, Insecta, Myriapoda and Chelicerata are phyla in their own rights within an arthropod superphylum. See BIRAMOUS APPENDAGE for diagram of limbs. The extent and patterning of TAGMOSIS reflect the locomotory method, while appendages have proved marvellously adaptable and account in large measure for the success of the group. There appear to be three major evolutionary lineages: the *Onychophora-Myriapoda-Insecta* group, the *Merostomata-Arachnida-Trilobita* group, and the *Crustacea*. The phylum may be regarded as a GRADE, a polyphyletic origin not yet discounted. It seems that three new *HOX*

	Appendages			Tagmosis	Other features
	Antennae	Legs	Mouthparts		
Crustacea	2 pairs	*n* pairs, biramous	mandibles maxillae 1 maxillae 2	variable	nauplius larva
Insecta	1 pair	3 pairs, appear uniramous	mandibles maxillae 1 maxillae 2	3: head thorax abdomen	2 pairs wings (usually)
Myriapoda	1 pair	*n* pairs, appear uniramous	mandibles maxillae 1 (+ variable)	2: head trunk	
Chelicerata	0	4 pairs, appear uniramous	chelicerae (+ pedipalps)	2: prosoma opisthosoma (= cephalothorax, abdomen)	

TABLE 2 *A simple diagnosis of major extant* ARTHROPOD *groups.*

GENES arose early in the ancestral onychophoran/arthropod clade, before their divergence: *Ubx*, *abd-A* and *ftz* genes may represent synapomorphies for this joint clade and help as phylogenetic tools to resolve the relationship of arthropods to other taxa.

articular cartilage Cartilage providing the articulating surfaces of vertebrate joints.

artificial chromosome See YEAST ARTIFICIAL CHROMOSOME, HUMAN ARTIFICIAL CHROMOSOME.

artificial insemination Artificial injection of semen into female reproductive tract. Much used in animal breeding.

artificial intelligence In 1950, Claude Shannon wrote the first chess-playing program for a computer, using a 'minimax' algorithm permitting the computer to choose a good move without evaluating all the states in the intervening moves – a task which, if done by humans, would be said to involve intelligence. Neural networks, first suggested in the 1940s, consist conventionally of layers of cells (artificial neurons) between which connections in adjoining layers are made through 'weights' (representing actions of synapses in real biological systems), the body of the cell summating inputs causing it to 'fire' at an output (the axon) when the weighted incoming signal exceeds some mathematically specified gradient function. Learning occurs through alteration of the weights. By analogy with natural selection, a neural network which performs its goal best can be made to act as the source of networks with slightly modified weight functions, a process which can, if pursued iteratively, generate improved problem-solving systems. See COMPLEXITY.

artificial key Any IDENTIFICATION KEY not based upon evolutionary relationships but rather upon any convenient distinguishing characters. See CLASSIFICATION.

artificial selection Directional selection imposed by humans, deliberately or otherwise, upon wild or domesticated organisms. Crop plants originated in many cases from such deliberate crosses, sometimes involving one or more polyploid stocks. Procedures employed in harvesting these crops commonly involve unintentional (but still artificial) selection upon plants growing with the crops, favouring weed properties (see WEEDS). The phenomenon was well known to DARWIN and examples of conscious human selection provided an analogy through which his readers could grasp the theory of NATURAL SELECTION. See ANTIBIOTIC.

Artiodactyla Order of eutherian mammals. Ungulates (hoofed), with even number of toes upon which they walk. Includes pigs and hippopotamuses (four toes) and camels, deer, antelopes, etc. (two toes – the cloven-hoof). See PERISSODACTYLA, RUMINANT.

Aschelminthes A phylum of pseudocoelomate animals, probably representing a grade rather than a clade. The body cavity is a persistent blastocoele. Many are worm-like, and most either aquatic, soil-dwelling or parasitic. They are bilaterally symmetrical, unsegmented, and frequently minute, composed of remarkably few cells in view of their anatomical complexity. The classes (often regarded as phyla) include: NEMATODA, ROTIFERA, KINORHYNCHA, GASTROTRICHA and NEMATOMORPHA. See PSEUDOCOELOM.

ascidian Sea squirt, or tunicate; member of Class Ascidiacea of Subphylum URO-CHORDATA.

ascocarp Fruiting body of ASCOMYCOTA; consists of tightly interwoven hyphae. Many are macroscopic and may be open and cup- or saucer-shaped (APOTHECIUM); closed and spherical (CLEISTOTHECIUM); or flask-shaped with a small pore through which the ASCOSPORES escape (PERITHECIUM). Asci are usually borne on the inner surface of the ascocarp.

ascogenous hyphae Hyphae containing haploid 'male' and 'female' nuclei, which develop from ascogonia (the oogonia or female gametangia of the ASCOMYCOTA); eventually give rise to asci.

ascogonium Oogonium or 'female' gametangium of the ASCOMYCOTA.

Ascomycota Division of the Kingdom Fungi. This is the largest group of fungi for which the ascus is the diagnostic character with about 3,266 genera and 32,260 species. Ascomycota include terrestrial and aquatic fungi. This group includes saprotrophs, parasites (especially of plants) and lichen-forming species. Most of the blue-green, red and brown moulds that cause food spoilage are members of the Ascomycota, including Neurospora, which has played a central role in modern genetics. Many yeasts are also members of the Ascomycota, as are the edible morels and truffels. The presence of lamellate hyphal walls with a thin electron-dense outer layer and a relatively thick electron-transparent inner layer also appears diagnostic. Sexual reproduction involves formation of a characteristic cell, the ascus; meiosis and spore formation occur in the ascus. Ascus formation takes place within a complex structure composed of tightly interwoven hyphae, the ascocarp. Asexual reproduction takes place through production of non-motile spores (conidia).

ascorbic acid (vitamin C) A water-soluble sugar acid; a VITAMIN for a few arthropods and vertebrates (including humans), readily undergoing oxidation without loss of vitamin activity, Especially abundant in citrus fruits, potatoes and tomatoes. Hydrolysis of its lactone ring destroys its vitamin activity, as often occurs in cooking. Mild dietary deficiency affects connective tissue as it appears to be required for collagen formation; its complete dietary absence in humans results in symptoms of scurvy, with poor wound-healing.

ascospore Fungal spore of ASCOMYCOTA.

ascus In some FUNGI (i.e. ASCOMYCOTA) a spherical, cylindrical or club-shaped cell in which fusion of haploid nuclei (KARYOGAMY) occurs during sexual reproduction, followed by meiosis and formation of, usually, eight ascospores.

aseptate Those algal filaments and fungal hyphae devoid of cross-walls (septa).

asexual In non-zoological contexts, indicating reproduction involving spore production, but in which meiosis, gamete production, fertilization (leading to genome or nuclear union), transfer of genetic material between individuals and PARTHENO-GENESIS do not occur. In this sense, therefore, not synonymous with vegetative reproduction (SEE LIFE CYCLE). In zoological contexts, the stipulation that spores be produced does not apply; but the other criteria included above do. Often employed, with vegetative reproduction and partheno-genesis, as a means of rapidly increasing progeny output during a favourable period (these having practically uniform geno-type); hence common in internal parasites (see POLYEMBRYONY). The basis of natural cloning (artificially imposed in the propagation of plants by cuttings). May alternate with sexual phase in LIFE CYCLE (see ALTERNATION OF GENERATIONS). Some organisms (e.g. *Amoeba*, trypanosomes) are obligately asexual, and this raises questions about the evolutionary and ecological significance of SEX, not least because some asexual forms are so because of infection with *WOLBACHIA*.

A-site (Of ribosome) binding site on ribosome for charged (amino-acyl, hence A for acyl) tRNA molecule in PROTEIN SYNTHESIS. See P-SITE.

aspirin See SALICYLIC ACID.

assay See BIOASSAY.

assemblage (Of plants) collection of populations; merely a collective term and is best used when no attempt is made to define dominance. See ASSOCIATION.

assimilation Absorption of simple substances by an organism (i.e. across cell membranes) and their conversion into more complicated molecules which then become its constituents.

association (Of plants) climax plant community dominated by a particular species and named according to them (e.g. oak–beech association of a deciduous forest). It can be defined as a segment of the BIOCOENOSIS. The association is an assemblage of species that recurs under comparable ecological conditions in different places; keystone of ecological studies, since it forms a recognizable entity which can be described and within which species are in intimate interaction.

assortative mating (a. breeding) Non-random mating, involving selection of breeding partner, usually based on some aspect of its phenotype. This 'choice' (consciousness not implied) may be performed by either sex, and may be positively assortative (choice like self in some respect) or negatively assortative (disassortative, choice unlike self in some respect). Likely to have consequences for degree of in-breeding and maintenance of POLYMORPHISM. Assortative mating by size may be due to mate choice (large individuals choosing large mates is common), to mate availability or to physical mating constraints. Natural selection and sexual selection may be involved. Some INCOMPATIBILITY mechanisms in plants are analogies. See SEXUAL SELECTION, SPECIATION.

aster (1) (Zool.) A star-like perinuclear structure containing an MTOC and appearing as a system of cytoplasmic striations radiating from the centriole, comprising MICROTUBULES. Often conspicuous during cleavage of egg, or during fusion of nuclei at fertilization. Also probably present in many other animal cells during division. Absent from higher plants. (2) (Bot.) The common name applied to a family of flowering plants, the Asteraceae (Compositae), possessing evolutionarily specialized flowers. Individual flowers are subordinated to the overall display of the head, which functions as a single large flower in attracting pollinators. Individual flowers are epigynous, rather small, and possess an inferior ovary comprising two fused CARPELS within a single OVULE in a locule. Stamens are reduced to five; usually fused to one another and to the COROLLA. Petals, also five, are fused to one another and to the ovary. Sepals absent or reduced to a series of bristles or scales (the pappus). Pappus often aids in seed dispersal by wind (e.g. dandelion, *Taraxacum*); may be barbed so becoming attached to passing animals. In many members each flower head comprises two flower types: (a) *disk flowers* – make up the central portion of the head; (b) *ray flowers* – arranged on the outer periphery; often carpellate but sometimes completely sterile. Flower head matures over several days with individual flowers opening serially in a centripetal spiral. Very successful family with about 22,000 species; probably the largest family of flowering plants.

Asteridae (Asteriflorae) The most advanced, and possibly the most recently evolved, subclass of the Magnoliopsida comprising about 60,000 species. About one third of the Asteridae belong to the family Asteraceae, which is the largest family of dicotyledons and one of the largest families of plants. Characteristically, flowers are sympetalous in which the stamens are isomerous and alternate with or fewer than the corolla lobes. The nucellus comprises a single layer of cells but has a single massive integument (there are exceptions). More than in any other subclass, they exploit specialized pollinators and specialized means of presenting their pollen. The rise of the Asteridae seems closely correlated with the evolution of insects capable of recognizing complex floral patterns. Chemically, the Asteridae are noteworthy for the frequent occurrence of iridoid compounds, the usual absence of ellagic acid and proanthocyanins, and absence of betalains, mustard oils and benzyl-isoquinoline alkaloids. See ASTER (2).

Asteroidea Class of ECHINODERMATA. Starfishes. Star-shaped; arms, containing projections of gut, not sharply marked off from central part of body; mouth downwards; suckered TUBE FEET; spines and pedicillariae. Carnivorous (some notoriously on oysters or corals).

astrocyte One type of GLIAL CELL of central nervous system. Star-shaped, with numerous processes, they provide mechanical support by ensheathing synaptic junctions, associate with nodes of Ranvier and clear cellular debris and secrete trophic factors in response to disease and injury. Some span the entire width of the brain radially from the hollow ventricles to the pial surface, providing scaffolding along which neurons migrate during foetal life. Others link blood capillaries to neurons, transporting ions and other substances. Astrocyte–astrocyte network signalling via gap junctions can produce a chain reaction, and impinges on synaptic signalling. Culture studies indicate that astrocytes play a powerful role in regulating the formation and maintenance of synaptic connections between neurons in development. Ca^{2+} levels in astrocyte cytoplasm respond to many external influences, including extracellular ATP and the excitatory neurotransmitter GLUTAMATE. Astrocytes are larger than previously thought, and their stellate processes do not overlap extensively. Thus, astrocytes divide parts of the brain such as the hippocampus, where memories are formed, into different compartments, each the sole prerogative of an individual astrocyte. The significance of this is currently unknown. See OLIGODENDRO-CYTES.

asymmetric cell division See CELL DIVISION, MATING TYPE.

atavistic structures Structures occurring inappropriately during development and thought (sometimes erroneously) to reflect ancestral morphology. See HOMOLOGY, *HOX* GENES.

atherosclerosis See SCLEROSIS.

atlas First VERTEBRA, modified for articulation with skull. Modified further in amniotes, which have freer head movement, than in amphibians. Consists of simple bony ring, while a peg (odontoid process) of the next vertebra (the AXIS) projects forward into the ring (through which the spinal cord also runs). This peg represents part of the atlas (its centrum) which has become detached and fused to the axis. Nodding the head takes place at the skull–atlas joint; rotation of head at atlas–axis joint.

ATM See DNA REPAIR MECHANISMS.

A toxins See POISONS.

ATP Adenosine triphosphate. Adenyl nucleotide diphosphate. The common 'energy currency' of all cells, whose hydrolysis accompanies and powers most cellular activity, be it mechanical, osmotic or chemical. Its two terminal phosphate groups have a more negative STANDARD FREE ENERGY of hydrolysis than phosphate compounds below it on the thermodynamic scale (e.g. sugar phosphates), and a less negative one than those higher (e.g. phosphocreatine, phosphoenolpyruvate), but this varies with intracellular concentrations of ATP, ADP and free phosphate as well as pH. A HIGH-ENERGY PHOSPHATE (see Fig. 80), it tends to lose its terminal phosphate to substances lower on the scale, provided an appropriate enzyme is present, and its mid-position on the scale enables it to serve as a common intermediate in the bulk of enzyme-mediated phosphate-group transfers in cells. Its relationship with ADP and AMP may be summarized:

$$ATP + H_2O \rightleftharpoons AMP + PP_i - 10 \text{ kcal mol}^{-1}$$

$$ATP + AMP \rightleftharpoons ADP + ADP$$

$$ATP \rightleftharpoons ADP + P_i - 7.3 \text{ kcal mol}^{-1}$$

The energy values are STANDARD FREE ENERGY *changes at pH 7, standard temperature and pressure, at 25°C.*
1 kcal = 4.184 kJ.

Cells normally contain about ten times as much ATP as ADP and AMP, but when metabolically active the drop in the ATP/(ADP + AMP) ratio results in acceleration of GLYCOLYSIS and aerobic respiration (see RESPIRATION), the signal being detected by ALLOSTERIC enzymes in these pathways whose modulators (see ENZYME) are ATP, ADP or AMP. ATP is not a reservoir of chemical energy in the cell but rather a transmitter or carrier of it. The bulk of ATP in eukaryotic cells is provided by mitochondria, where these are present. Some extra-mitochondrial ATP is produced anaerobically in the cytosol, and chloroplasts produce it but do not export it. ATP hydrolysis is used to transfer energy when work is done in cells.

ATPase activity is found in MYOSIN (e.g. MUSCLE CONTRACTION) and DYNEIN (e.g. ciliary/flagellar beating). Membrane ion pumps (e.g. sodium and calcium pumps) and macromolecular syntheses of all kinds involve ATPase activity. Ultimately the energy source for ATP formation in the biosphere is solar energy trapped by autotrophs in photosynthesis – plus some lithotrophy. All heterotrophs depend upon respiratory oxidation of these organic compounds to power their own ATP synthesis. ATP is, like the other common nucleoside triphosphates in cells (CTP, GTP, TTP, UTP), a substrate in nucleic acid synthesis, and its hydrolysis provides the energy needed to

build the resulting AMP monomer into the growing polynucleotide chain. These other triphosphates may participate in some other energy transfers; but ATP has by far the major role. ATP (and its breakdown product adenosine) can act as neurotransmitters, but only where purine receptors (often presynaptic) are present. See AMP, ADP, PHOSPHAGEN, BACTERIORHODOPSIN.

ATPase Any member of several families of enzymes bringing about either (i) orthophosphate (P_i) cleavage of ATP yielding ADP and inorganic phosphate, or (ii) pyrophosphate (PP_i) cleavage of ATP to yield AMP and pyrophosphate. The latter provides a greater decrease in free energy and is involved where a 'boost' is needed for an enzyme reaction. ATPase activity is found in cell motor molecules such as kinesins and dyneins (both microtubule-associated), myosins (actin-associated), and in chloroplast thylakoids and inner mitochondrial membranes (as ATP synthetases). See MITOCHONDRION, CHLOROPLAST, BACTERIORHODOPSIN.

atrium (1) Chamber, closed except for a small pore, surrounding gill slits of Amphioxus and urochordates. (2) A type of heart chamber of vertebrate chordates synonymous with 'auricle'; receives blood from major vein and passes it to ventricle. Walls not as muscular as those of ventricle. Fishes have single atrium, but tetrapods, breathing mainly or entirely by lungs, have two: one (the left) receives oxygenated blood from lungs, the other (the right) receives deoxygenated blood from the body. Much of the blood flow through the atria is passive (see HEART CYCLE). Non-chordates may have an atrial component of the heart, e.g. some polychaete worms and most molluscs, in which the term 'auricle' is sometimes preferred. (3) A space or cavity in some invertebrates (e.g. platyhelminths, some molluscs) known as the genital atrium, which houses the penis and/or opening of the vagina, and into which these may open.

atrophy Diminution in size of a structure, or in the amount of tissue of part of the body. Generally involves destruction of cells, and may be under genetic and hormonal control, as is frequently the case in metamorphosis. May also result from starvation. Compare HYPERTROPHY.

atropine See ACETYLCHOLINE, ALKALOIDS.

attenuation (Of pathogenic microorganisms) a procedural term referring to the selection and isolation, or enrichment, of a mutant or recombinant having a lower VIRULENCE than its parent. See PASTEUR, VACCINE.

auditory (otic) capsule Part of vertebrate skull, enclosing auditory organ.

auditory nerve See VESTIBULOCOCHLEAR NERVE.

auditory organ Sense organ detecting pressure waves in air ('sound'), in vertebrates represented by the cochlea of the inner ear, but the term often intended to include VESTIBULAR APPARATUS detecting positional and vectorial information as well. See LATERAL LINE SYSTEM.

Auerbach's plexus That part of the autonomic nervous system in vertebrates (mostly from the vagus nerve) lying between the two main muscle layers of the gut and controlling its peristaltic movements.

auricle (Zool.) (1) Often used synonymously with atrial heart chamber (see ATRIUM). (2) External ear of vertebrates. (Bot.) Small ear- or claw-like appendage occurring one on each side at the bases of leaf-blades in certain plants.

Australian region ZOOGEOGRAPHICAL REGION consisting mainly of Australia, New Guinea and the Celebes; demarcated from south-east Asia by WALLACE'S LINE.

australopithecine Member of genus *Australopithecus*, now extinct; of the primate family Hominidae (see HOMINID). It appears to have been a long extant genus (at least 4.4–1.0 Myr BP), with perhaps as many as eight African species. *Australopithecus ramidus*, renamed *Ardipithecus ramidus*, is the oldest to date at 4.4 Myr BP. *A. afarensis* (4.0–3.0 Myr BP) appears to have been a conservative species near the common ancestry of later forms, with chimp-like thin dental enamel and strongly-built arms but with a foramen magnum resembling later hominids and inhabiting closed-canopy woodland; *A. anamensis* (4.2–3.9 Myr BP) from Turkana had thicker dental enamel and larger molars than *Ardipithecus*; *A. bahrelghazali* from Chad (*c*. 3.5 Myr BP) is a long way west of other australopithecine fossils in the Rift Valley; *A. aethiopicus* and *A. africanus* were later contemporaries (3–2.1 Myr BP) and possibly sister species with *A. afarensis* as common ancestor; *A. robustus* and *A. boisei* were later still (approx. 2–1.2 Myr BP) and shared several derived features (synapomorphines). *A. aethiopicus*, *A. robustus* and *A. boiseri* were all 'robust' (megadontic) forms, with heavy skulls and facial features; but megadonty may have evolved in *A. aethiopicus* earlier than, and independently of, its appearance in the other two species. *A. africanus* had more 'gracile' features and may have been ancestral to *Homo*, although some cladistic analysis suggests it has strong links with neither *Homo* nor the robust australopithecines and was possibly too specialized to have been either ancestral to one of these clades or, as was once thought, the common ancestor of all later hominines (see *HOMO*, HOMININI). One possible phylogeny, showing the hypothetical independent evolution of megadontic characters (darker shading), is illustrated in Fig. 7. Cranial capacities of typical australopithecines were 400–600 cm^3 (modern humans average 1,400 cm^3), although there is now some concern that these australopithecine values may be too high; BROCA'S AREA was seemingly absent. They had bipedal posture (and, from the footprints at Laetoli 3.6 Myr BP, were capable of bipedal gait). They are not generally thought to have been responsible for the stone choppers of the Oldowan tool culture of *c*. 1.75 Myr BP (*Homo habilis* preferred). Physiologically, bipedalism would have reduced thermal stress and this in turn would have reduced water requirement; it may not have been an adaptation to locomotion in a savannah environment since modern chimpanzees are usually only

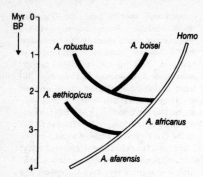

Myr 0
BP

A. robustus   A. boisei   Homo

A. aethiopicus

A. africanus

A. afarensis

FIG. 7 *A possible phylogeny of early African homin-ines as implied by some cladistic analysts. The darker shading refers to megadontic lineages, the paler shading to more gracile lineages as indicated in the* AUSTRALOPITHECINE *entry.*

bipedal when static and feeding under trees. *A. afarensis* may have been such a postural biped; hominid locomotory bipedalism may not have appeared until *Homo erectus* or *H. ergaster* (see *HOMO*). Further revisions of phylogenetic status are to be expected as new fossils and improved dating techniques occur. All fossil material currently comes from Africa.

autapomorphy In CLADISTICS, one of the character states uniquely defining a taxon.

autecology Ecology of individual species, as opposed to communities (synecology).

autoantibodies Antibodies raised against self-antigens, as in certain AUTOIMMUNE REACTIONS.

autoantigen Molecular component of organism, normally regarded as 'self' by its immune system, but here recognized as 'non-self' and eliciting an AUTOIMMUNE REACTION.

autocatalyst Any molecule catalysing its own production. The more produced, the more catalyst there is for further, non-linear, production. Most likely, some such process was involved in the origin of pre-biological systems which, once enclosed in a membrane, might be called

'living'. The chirality of sodium chlorate crystals precipitating from a stirred (but not unstirred) supersaturated solution depends on an autocatalytic effect arising from the rapid production of secondary 'nuclei' from a single primary nucleating crystal: in effect, a chiral template. The Calvin cycle of photosynthesis is an example of symmetrical autocatalysis. RIBOZYMES are almost, but not quite, autocatalytic; but an autocatalytic RNA will probably soon be engineered. Population growth is inherently autocatalytic. See Fig. 8, ANTI-BIOTIC, ORIGIN OF LIFE.

autochthonous (1) Of soil microorganisms whose metabolism is relatively unaffected by increase in organic content of soil. See ZYMOGENOUS. (2) The earliest inhabitant or product of a region (aboriginal). In this sense contrasted with *alloch-thonous* (not native to a region).

autoclave Widely-used equipment for heat-sterilization. Commonly air is either pumped out prior to introduction of steam, or simply replaced by steam as the apparatus is heated under pressure. Material being sterilized is usually heated at 121°C and 138–172 kNm pressure for 12 minutes, which destroys vegetative bacteria, all bacterial spores and viruses; but these figures will vary with the size and nature of material.

autocolony (Of algae) a daughter colony (daughter COENOBIUM) formed within the cell of the parent colony (mother coenobium) and resembling a miniature version of the parent colony.

autocrine signals Molecules which act upon the cell secreting them.

autoecious Of rust FUNGI (BASIDIOMYCOTA) having different spore forms during the life cycle, all produced upon one host species, as in mint rust. Compare HETEROECIOUS.

autogamy Fusion of nuclei derived from the same zygote but from different meioses. Includes all forms of self-fertilization. See AUTOMIXIS.

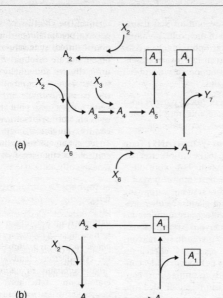

FIG. 8 *Autocatalysis: (a) An asymmetrical autocatalytic cycle; (b) A symmetrical autocatalytic cycle.*
A_i = cycle intermediates; X_i = raw materials; Y_i = waste.

autogeneic Genetically identical to donor or recipient. See GRAFT. Compare ALLOGENEIC, XENOGENEIC.

autogenic succession A SUCCESSION that occurs on newly exposed landforms, and in the absence of gradually changing abiotic influences. If the landform has not previously been influenced by a community, the sequence of species is termed primary succession (e.g. freshly formed sand dunes, substrata left behind by melting glaciers). Then, in instances where the vegetation of an area has been partially or completely removed, but where well-developed soil, and seeds and spores remain, the subsequent sequence of species is termed a secondary succession (e.g. after fire, disease). See ALLOGENIC SUCCESSION.

autograft Tissue grafted back on to the original donor. See ALLOGRAFT, ISOGRAFT, XENOGRAFT.

autoimmune reaction Response by an organism's immune system to molecules normally regarded as 'self' but which act as antigens. Quite often the thyroid gland, adrenal cortex or joints become damaged by the action there of antibodies or sensitized lymphocytes. Autoimmunity seems to arise from autoantibody production by defective (mutant) B CELLS rather than from any defect in helper T CELLS. Autoimmune diseases include insulin-dependent (Type I) DIABETES, rheumatoid arthritis, myasthenia gravis (antibodies produced against nicotinic acetylcholine receptors, e.g. at neuromuscular junctions, causing skeletal muscle weakening) and possibly multiple sclerosis (a demyelinating disease).

autologous Used of cells which are transplanted from one part of the body to another in the same individual. Contrast ALLOGENEIC.

autolysis (1) Self-dissolution that tissues undergo after death of their cells, or during metamorphosis or atrophy. Involves LYSO-SOME activity within cell. (2) Prokaryotic self-digestion, dependent upon enzymes of cell envelope. See APOPTOSIS.

automixis Fusion of nuclei derived both from the same zygote and from the same meiosis. See AUTOGAMY, PARTHENOGENESIS.

autonomic nervous system (ANS) Term sometimes referring to the entire vertebrate visceral nervous system, but more often restricted to the efferent (motor) part of it (the visceral motor system), supplying smooth muscles and glands. Neither anatomically nor physiologically autonomous from the central nervous system. Sometimes termed the 'involuntary' nervous system; but here again its effects (in humans) can largely be brought under conscious control with training. For convenience the ANS can be subdivided into two components: the *parasympathetic* and *sympathetic* systems. See Fig. 9.

Parasympathetic nerve fibres are CHOL-INERGIC and in mammals are found as motor components of CRANIAL NERVES III, VII, IX and X, as well as of three spinal nerves in sacral segments 2–4. Most of its effects are brought about by its distribution in the vagus (CNX), serving the gut (see AUERBACH'S PLEXUS), liver and heart among other organs.

Fibres of the sympathetic system originate within the spinal cord of the thoracic and lumbar segments, but beyond the vertebrae each departs from the cord and turns ventrally in a short white ramus (rami communicantes) to enter a sympathetic ganglion, a chain of which lies on either side of the mid-line. In the sympathetic ganglia many of the preganglionic fibres relay with postganglionic fibres that innervate target organs (e.g. mesenteries and gut); others pass straight through as splanchnic nerves and meet in plexi (collectively termed the *solar plexus*) beneath the lumbar vertebrae. From here postganglionic fibres innervate much of the gut, liver, kidneys and adrenals. Postganglionic fibres usually liberate catecholamines, particularly noradrenaline. Preganglionic parasympathetic fibres are

relatively long and usually synapse in a ganglion on or near the effector, postganglionic fibres being short. In general ANS preganglionic fibres are myelinated, postganglionic unmyelinated and usually (there are exceptions) where the sympathetic stimulates, the parasympathetic inhibits, and vice versa; but organs are not always innervated by both. Both are, however, under central control, notably via the hypothalamus. Together they afford homeostatic nervous control of the internal organs, often reflexly.

autophagy Occurring in plant, animal and fungal cells, the process by which portions of cytoplasm are sequestered for degradation within a double-membrane vesicle, termed an *autophagosome* (in yeast) or *autophagic vacuole* (in mammals). This forms in the cytosol from membranes of unknown origins (probably, in mammals at least, the endoplasmic reticulum) and is then delivered by microtubules to, and fuses with, a LYSOSOME or vacuole where breakdown of the contents occurs. It is not understood how specific substrates are recognized as cargo for sequestration, but its occurrence is linked to the sensing of a starvation signal by the cell. See PROTEASOMES.

autopolyploid In classical cases, a POLY-PLOID (commonly a tetraploid) in which all the chromosomes are derived from the same species, frequently the same individual. Compare ALLOPOLYPLOID.

autoradiography Method using the energy of radioactive particles taken up by cells, tissues, etc., from an artificially enriched medium and localized inside them, to expose a plate sensitive to the emissions, thus indicating where radioactive atoms lie. Much used in tracing pathways within cells, DNA SEQUENCING and SOUTHERN BLOT TECHNIQUE. See LABELLING.

autosome Chromosome that is not a SEX CHROMOSOME.

autospore (Of algae), non-motile spore produced within a parent cell, which develops the same shape as the parent cell at an early stage, before release.

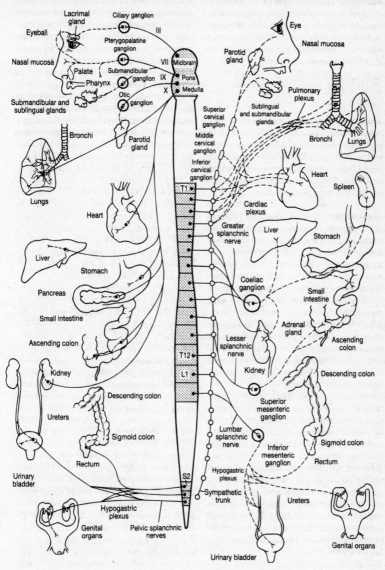

FIG. 9 *The human* AUTONOMIC NERVOUS SYSTEM. *Parasympathetic components are shown only on the left side, sympathetic components only on the right (in the body, both occur on both sides). Cranial nerve components have roman numerals. Thoracic and lumbar segments are indicated by T1–T12, L1, etc., and sacral segments by S2, etc. Preganglionic fibres = solid lines, postganglionic fibres = broken lines.*

autostylic jaw suspension The method of upper jaw suspension of modern chimaeras and lungfishes; presumed to have been that employed by earliest jawed fishes, in which the hyomandibular bone has no role in the suspension. The upper jaw (palatoquadrate) attaches directly to the cranium. See HYOSTYLIC, and AMPHISTYLIC JAW SUSPENSION.

autotomy Self-amputation of part of the body. Some lizards can break off part of the tail when seized by a predator, muscular action snapping a vertebra. Both here and in many polychaete worms which can shed damaged parts of the body, REGENERATION restores the autotomized part.

autotrophic Of those organisms independent of external sources of organic carbon (compounds) for provision of their own organic constituents, which they can manufacture entirely from inorganic material. Autotrophs may be phototrophic or chemotrophic as regards energy supply, using solar or chemical energies respectively. Most chlorophyll-containing plants are autotrophic, manufacturing organic material from water, carbon dioxide and mineral ions (nitrates, phosphates, sulphates, metal ions, etc.) and using solar energy phototrophically (see PHOTOSYNTHESIS). Some bacteria and the blue-green algae are also phototrophic; other bacteria are chemotrophic, using energy released from inorganic oxidations (e.g. of hydrogen sulphide by sulphur bacteria, or of hydrogen by hydrogen bacteria). The term *lithotroph* is applied to those autotrophs which employ inorganic oxidants to oxidize inorganic reductants (e.g. the nitrifying bacteria *Nitrosomonas* and *Nitrobacter*, of the NITROGEN CYCLE). All other organisms, HETEROTROPHIC in one form or another, depend upon the synthetic activities of autotrophic organisms for their energy and nutrients. See ECOSYSTEM.

autotrophic succession A temporal SUCCESSION of species at a location principally involving plants (e.g. where new habitat opens up for invasion by plants).

auxins Group of plant GROWTH SUBSTANCES produced by many regions of active cell division and enlargement (e.g., growing tips of stems and young leaves) that regulate many aspects of plant growth and development. Auxin signalling seems to take place by targeting for degradation negative regulators such as the auxin/indole-3-acetic acid (Aux/IAA) proteins in an auxin-dependent manner via the ubiquitin-related protein degradation machinery, so de-repressing a network of genes which guide growth and development. Promote growth by increasing the plasticity of the cell wall. When the cell wall softens, the cell enlarges because of turgor pressure. As the turgor pressure is reduced by cell expansion, the cell takes up more water, and the cell continues to enlarge until it encounters sufficient resistance from the wall. Cell wall softening results from changes in the cell's metabolism that causes a rapid pumping of protons across the plasmalemma, pumping H^+ ions out of the cell. The resulting acidification of the cell wall leads to hydrolysis of restraining bonds within the wall, and consequently, to cell elongation driven by the cell's turgor pressure. This is termed the *acid growth hypothesis*. But it cannot account for the continued effect of auxin treatment upon the plant. It appears that auxin has two different effects upon cell elongation, a rapid short-term effect caused by acid growth, and a second continued effect due to the regulation of gene expression. Auxin causes expression of at least ten specific genes, all presumably involved in the growth effect, the step affected by auxin being transcription: auxin's effect on gene expression in plants parallels that of several animal hormones. Auxins are transported basipetally in shoots at a rate of about 1 cm/hr^{-1}, act synergistically with GIBBERELLIN in stem elongation and with CYTOKININ in control of buds behind the apical bud (apical dominance). Effects of auxin on cell growth include curvature responses such as GEOTROPISM and PHOTOTROPISM. They may also have mitogenic effects, as in the initiation of cambial activity in association with cytokinins (and the differentiation of vascular tissue in association with sucrose), and adventitious root formation. Interactions between auxins and gibberellins determine rates of production of secondary phloem and secondary xylem. Auxins are implicated in flower initiation, sex determination, fruit growth (via production of auxin, or a factor promoting its synthesis, by pollen tube), delays in leaf fall and fruit drop.

Naturally occurring auxins include indole-3-acetic acid (IAA) and indole-3-acetonitrile (IAN). IAA has been isolated from such diverse sources as corn, endosperm, fungi, bacteria, human saliva and, the richest source, human urine. In addition to naturally occurring auxins, many substances with plant growth regulatory activity (*synthetic auxins*) have been developed. Some are used on a very large scale for regulating growth of agriculturally and horticulturally important plants, as in inhibition of sprouting in potato tubers, prevention of fruit drop in orchards, achievement of synchronous flowering (hence fruiting) in pineapple, in parthenocarpic fruit production (as in tomatoes, so avoiding risk of poor pollination). At increased, yet still relatively low, concentrations, auxins inhibit growth, sometimes resulting in death. Some synthetic auxins have differential toxicity in different plants: toxicity of 2,4-dichlorophenoxyacetic acid (2,4-D) to dicotyledonous and non-toxicity to monocotyledonous plants is perhaps the best known example, being exploited successfully in control of weeds in cereal crops and lawns. The synthetic 'rooting powders' containing IBA (indole-3-butyric acid) or NAA (naphthalene-acetic acid) are commonly used to induce formation of adventitious roots in cuttings, and to reduce fruit drop in commercial crops. See CYCLIC AMP.

auxospore See BACILLARIOPHYTA.

auxotroph Mutant strain of bacterium, fungus or alga requiring nutritional supplement to the MINIMAL MEDIUM upon which the wild-type strain can grow. See PROTOTROPH, REPLICA PLATING.

Aves Birds. A vertebrate class. Usually regarded as derived from small theropod

ARCHOSAURS (T. S. Huxley was first, in 1870, to suggest the connection, his evidence including that they shared three functional toes and hollow bones). Several groups working in the Liaoning Province of northern China have discovered a remarkable trove of fossils from the early Cretaceous (~124–128 Myr BP), including several THEROPOD dinosaurs showing fully modern feathers as well as a variety of primitive feather structures. This has precipitated a reconsideration of what it means to be a bird. The conclusion is that birds are a group of feathered theropod dinosaurs that evolved the capacity for flight. Members of Class Aves are distinguished by having a reversed first toe and fewer than 26 tail vertebrae. At least one theropod reptile (*Oviraptor*) had a furcula (see WISHBONE) formed from fused clavicles, as in modern birds. Cladistic analysis includes in the Class Aves the ancestor of ARCHAEOPTERYX and all its other descendants. This clade is part of the larger THEROPOD clade Maniraptora, which is itself a part of the theropod clade Tetanura, descended in turn from the basal theropods. The two superorders with living representatives in the Subclass Neornithes are the Palaeognathae (ratites) and Neognathae (20 major orders; about half the 2,900 living species, including songbirds, belonging to the Order Passeriformes, or 'perching' birds). Distinctive features include: FEATHERS; forelimbs developed as wings. Bipedal and homeothermic, laying cleidoic eggs and (excluding *Archaeopteryx* and some other fossil genera) lacking teeth, but with the skin of the jaw margins cornified to form a beak (bill), whose diversity of form is in large part responsible for the Cretaceous, and subsequent, adaptive radiation of the group. New fossils will bring review of the above, almost certainly displacing *Archaeopteryx* from its position as the earliest-known fossil bird. See FLIGHT.

avirulence factors (*Avr* factors) See RESISTANCE GENES.

***avr* genes** See RESISTANCE GENES.

awn (arista) Stiff, bristle-like appendage occurring frequently on the flowering glumes of grasses and cereals.

axenic (Of cultures of organisms) a pure culture.

axial skeleton See ENDOSKELETON.

axial system In secondary xylem and secondary phloem, collective term for those cells originating from fusiform cambial initials. Long axes of these cells are orientated parallel with the main axis of the root or stem.

axil (Of a leaf) the angle between its upper side and the stem on which it is borne; the normal position for lateral (axillary) buds.

axillary Term used to describe buds, etc., occurring in the AXIL of a leaf.

axis, axis specification (axiation) (1) Very generalized plane of symmetry in animals. There are generally three such: antero-posterior, dorso-ventral and medio-lateral (left–right), established early in development, and sometimes by the CELL POLARITY of the egg, and may involve materials deposited inside the egg cell by nurse cells (or other accessory cells of the germ line), specifying POSITIONAL INFORMATION. Specification involves transient organizing processes, in the oocyte or a later multicellular stage, producing the initial and minimal level of spatial organization (usually in two dimensions) bearing reliable correspondence to the eventual antero-posterior and dorso-ventral axes (or animal-vegetal axis) of the body, which are normally at right angles to each other. Vertebrate left–right asymmetry, however, begins during gastrulation when the anticlockwise beating of cilia of the mid-line node (see HENSEN'S NODE), brought about by the left–right DYNEIN in their axonemes, wafts morphogen-like molecules specifically to the left side of the embryo. Members of the same phylum often display very different axis-forming processes despite all going through the same PHYLOTYPIC STAGE

(e.g. the chordate pharyngula or the arthropod segmented germ band stage). This diversity may have been selected for its role in accommodating diverse reproductive specializations: shells, nutrient sources, extra-embryonic tissues, speed of development and larval stages. Such specializations increase the demand on axis specification and on other organizing processes up to the phylotypic stage. In the diploblastic animal *Hydra*, it appears that WNT SIGNALLING is involved in axis formation, suggesting that this involvement may have been central in the evolution of axial differentiation in early animals (see BODY PLANS). Axis specification may involve self-organization, in that organization is generated where none was present before. In ascidians, the end of the meiotic spindle which embeds in the cortex of the oocyte biases the position of the animal pole. In amphibians, the site of sperm entry biases the equatorial position furthest from the entry point to become the dorsal side of the embryo. In birds, the antero-posterior axis is biased (but not determined) by the orientation of the egg in the oviduct in relation to gravity at the 20,000-cell blastodisc stage. (2) Second amniote VERTEBRA, modified for supporting the head.

axon The long process which grows out from the cell bodies of some neurons towards a specific target with which it connects and carries impulses away from the cell body. See GROWTH CONE, NEURON, NERVE FIBRE.

axoneme Complex microtubular core of CILIUM and flagellum. Contains actin in *Chlamydomonas*.

axopod PSEUDOPODIUM of some sarcomastigophoran protozoans in which there is a thin skeletal rod of siliceous material upon which streaming of the cytoplasm occurs. They may bend to enclose prey items.

B7 A group of cell-surface proteins on macrophages and dendritic cells whose production is required, along with presented antigen fragments, as a co-stimulatory signal for T cell activation and proliferation. One signal for B7 and cytokine production is activation of the TOLL-LIKE RECEPTOR TLR4. See Fig. 158, ANTIGEN-PRESENTING CELLS.

BACs, and BAC clones See BACTERIAL ARTIFICIAL CHROMOSOMES.

Bacillariophyceae (Diatomophyceae)

Diatoms. Microscopic algae in the Division Heterokontophyta (*c.* 10,000 spp.) accounting for about 20% of all primary production. All species are eukaryotic, unicellular or colonial coccoid algae; pigmented and photosynthetic containing in their chloroplasts chlorophylls *a*, c_1 and c_2, together with the major carotenoid fucoxanthin, which gives the cells their characteristic brown colour. Plastid structure is similar to other members of the Heterokontophyta, but often larger and more elaborate. Chloroplast DNA is organized into a ring-shaped nucleoid. Oil droplets, reserve polysaccharides (chrysolaminarin) and other inclusions occur within the cytoplasm, often visibly. Some species can live heterotrophically in the dark, if there is a source of organic carbon. Less than ten species are obligate heterotrophs and all colourless (apochlorotic) forms are found in two genera (*Nitzschia*, *Hantzschia*).

Each diatom cell is encased by a characteristic and highly differentiated cell wall, almost always heavily impregnated with silica ($SiO_2.nH_2O$), whose uptake and deposition require less energy than the formation of an equivalent organic wall. The cell wall comprises two largely intricately sculptured overlapping halves (valves), together with several thinner linking structures (girdle elements), and mostly comprises quartzite or hydrated silica. The valves lie at the end of the cell, and the girdle elements surround the region in between. Wall components (the frustule) fit together closely, so movement of materials occurs mainly via pores or slits in them. Probably all diatoms secrete polysaccharides, some of which may diffuse into the surrounding medium, while some remain as a gelled capsule around the cell or as threads, pads, stalks or tubes. Each frustule possesses one valve formed just after the last cell division, and an older valve, which may have existed for several generations since its own formation. Each element of the siliceous wall is formed within the cytoplasm in a silica deposition vesicle.

Mitosis is open, the nuclear envelope breaking down before metaphase, and the telophase spindle is persistent. The spindle is formed outside the nucleus, between two darkly staining polar plates, and then sinks down into the nucleus as the nuclear envelope disperses. It comprises two overlapping and interdigitating sets of microtubules (half spindles); each associated with one polar plate. In the region of overlap, the microtubules of the two half spindles slide over each other, thus pushing the poles apart.

Some diatoms form thick ornamented resting spores, usually in response to stress; others form resting cells, identical morphologically to vegetative cells but possessing no protective layer. Auxospores represent another way for re-establishing the original size of a cell; they result from sexual reproduction, which can involve uniflagellate male and non-motile female gametes or the fusion of amoeboid gametes. The only flagelleted cells produced by diatoms are the male gametes of the centric diatoms (Order Centrales). Each gamete has a single pleuronematic flagellum, which is apically inserted and lacks the central two microtubules of the axoneme. The transition zone of the flagellum also lacks a transitional helix. Each species that reproduces sexually has a diplontic life cycle, with gametic meiosis.

Diatoms exhibit several features typical of the Heterokontophyta (e.g. girdle lamellae in the chloroplasts; the presence of chloroplast endoplasmic reticulum; Golgi bodies lying with their forming faces appressed to the nuclear envelope; presence of typical stiff tubular hairs on the pleuronematic flagellum of the male gamete, and the presence of a periplastidial reticulum in the narrow space between the nuclear envelope – where this is continuous with and replaces the CER – and the chloroplast). The class Bacillariophyceae comprises two major groups, which are sometimes recognized as two orders, the Pennales and Centrales. These are differentiated from one another by differences in cell wall structure.

Many diatoms, particularly those possessing a raphe (longitudinal slit in the valve of some pennate diatoms) or labiate processes (appendage at a valve's periphery through which mucilage is secreted) can move over a substratum. Movement is through secretion of material from the raphe or labiate process; the motile force being generated by interaction between actin filaments and transmembrane structures, which are free to move within the cell and raphe but fixed to the substratum at their distal ends. The transmembrane structure probably includes ATPase and a protein able to make translational movements with the plasmalemma. The transmembrane structure is itself connected to filaments of acid glycosaminoglycan, which can become attached to the substratum at their distal ends. Thus, transmembrane structures are moved along the raphe by their interactions with actin filament bundles beneath, and the cell moves relative to the substratum.

Epipelic and epipsammic diatoms exhibit endogenous vertical migration rhythms. Diatoms are abundant in both the benthos and plankton of both fresh and marine water. Their fossil record extends back to the early Cretaceous (120 Myr BP). The oldest diatoms were marine and they belong to the Centrales. The first marine pennate diatoms appear by the end of the Cretaceous (70 Myr BP); however, diatoms do not seem to have achieved their present pre-eminence until the beginning of the Miocene (24 Myr BP). The oldest known freshwater diatoms come from the early Tertiary (60 Myr BP) and are pennate forms. Past deposits of countless numbers of diatom frustules have formed siliceous or diatomaceous earths; industrial uses of such earths are diverse (e.g. as a mild abrasive in toothpaste and metal polishes, as an absorbent for nitroglycerin in making dynamite, and in the filtration of liquids, especially in sugar refineries).

Diatoms are important and ecologically sensitive microfossils, being used extensively in palaeolimnology, enabling both qualitative and quantitative reconstruction of past lake histories, palaeoclimates, pH values, nutrient loading, water depth and palaeosalinity. Very important microfossils in determining responses of lakes to acid deposition and anthropogenic effects upon lakes.

bacillus General term for any rod-shaped bacterium. Also a genus of bacteria: *Bacillus*.

Bacillus thuringiensis See GMO.

backbone See VERTEBRAL COLUMN.

backcross Cross (mating) between a parent and one of its offspring. Employed in CHROMOSOME MAPPING, when the parent is homozygous and recessive for at least two character traits, and to ascertain genotype of offspring (i.e. whether homozygous or heterozygous for a character), the parent used being the homozygous recessive. Where the organism of known genotype in the cross is not a parent of the other, the term *testcross* is often used.

Bacteria Now generally regarded as a DOMAIN of life, these PROKARYOTES exhibit a huge range of genetic and metabolic variation (see Fig. 10a). They may be unicellular, filamentous or mycelial, some forming BIO-FILMS, and some bacterial species communicate by extracellular signals (see QUORUM SENSING). They are among the simplest living organisms (see CELL for diagram). Cyanobacteria are nowadays included in the bacterial domain. It is not known how many bacterial species exist, not even the correct order of magnitude, since bacteria cannot be differentiated under the microscope; but a phylogenetic system based on 5S and 16S rRNA sequences is being developed and is a useful tool in estimating bacterial species richness. Multiplication is by simple fission; other forms of asexual reproduction, e.g. production of aerially dispersed spores or flagellated swarmers, occur in some bacteria. As prokaryotes, they have no meiosis or syngamy, but genetic recombination occurs in many of them (see F FACTOR, PILI, PLASMID, RECOMBINATION). Endosymbiotic PROTEOBACTERIA are now firmly implicated in the origin of EUKARYOTES, and bacterial BIOFILMS play an important role in the settlement and development of algal communities.

Bacteria are ubiquitous, occurring in a large variety of habitats. 1 g of soil may contain from a few thousand to several million (see BIOFILMS); 1 cm^3 of sour milk, many millions. Most are saprotrophs or parasites; but a few are autotrophic, obtaining energy either by oxidation processes or from light (in the presence of bacteriochlorophyll). In soil, their activities are of the utmost importance in the decay of dead organic matter and return of minerals for higher plant growth (see DECOMPOSERS, CARBON CYCLE, NITROGEN CYCLE). In the surface waters of oceans, α-proteobacteria are thought to be responsible for 5% of all primary production, by AEROBIC ANOXYGENIC PHOTOSYN-THESIS. As they also utilize organic solutes, their net role in recycling surface CO_2 levels in the open ocean awaits resolution. Some are sources of ANTIBIOTICS (e.g. *Streptomyces griseus* produces streptomycin). Some resilient bacteria (EXTREMOPHILES) survive in the Earth's crust to the depths of the temperature-limited biosphere, notably in conti-

nental aquifer sediments, fractured and porous rocks, deep ocean sediments and arctic permafrost. Oil drillers at Taylorsville in Virginia have successfully cultured 'Hell's bacterium' (*Bacillus infernus*) discovered 2.8 km down in a deep rift at over 60°C and without oxygen. The water that carried these to such depths may have left the surface 30 Kyr ago.

Bacterial metabolism in anoxic environments is particularly interesting. Members of the Geobacteriaceae can couple oxidation of organic compounds to reduction of insoluble Fe(III) oxides. Thus the marine *Desulfuromonas acetoxidans* can grow anaerobically by oxidizing acetate with associated reduction of elemental sulphur or Fe(III). Electrical current can be generated from anoxic marine sediments by embedding a graphite electrode (the anode) in it and connecting it appropriately to a similar electrode (the cathode) in the surface waters. This is due to the growth of *D. acetoxidans* and others, and may hold promise for the bioremediation of organic contaminants or energy harvesting for electronic instruments (see BIOFUEL CELLS). Acid mine drainage sites afford excellent opportunities for the study of bacterial mineral solubilization, where bacteria can account for at least 70% of Fe(II) released by pyrite (FeS_2) dissolution. Only about 1% of all known bacteria can be cultured at present, as few of them grow on cultures in Petri dishes; however, better success has been recently achieved by providing them with the chemical components of their natural environment in diffusion chambers and incubating in an aquarium in association with other microorganisms to simulate their natural settings (see also POLYMERASE CHAIN REACTION for culture-independent PCR). In natural habitats, different physiological types of microorganisms interact closely, and studies from pure cultures are often of limited relevance. L-form bacteria are those lacking CELL WALLS. As agents of plant disease bacteria are less important than fungi; but they cause many diseases in animals and humans (e.g. diphtheria, tuberculosis, typhoid, some forms of pneumonia) (See Fig. 10b). See ANTISEPTIC,

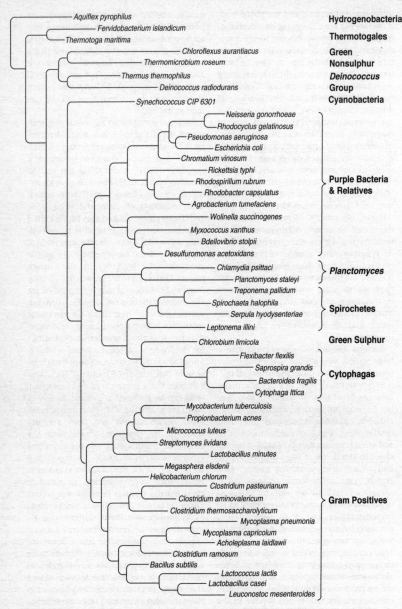

FIG. 10a *One classification of* BACTERIA, *based on 16S ribosomal RNA sequences (data from Carl R. Woese).*

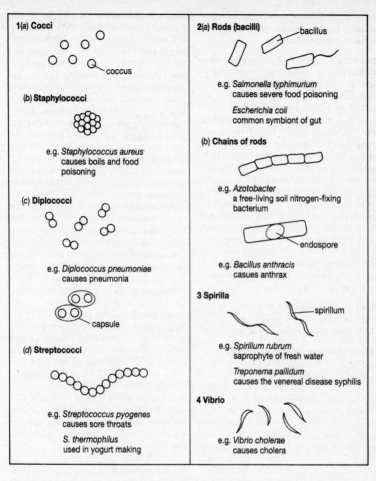

1(a) Cocci

coccus

(b) Staphylococci

e.g. *Staphylococcus aureus*
causes boils and food
poisoning

(c) Diplococci

e.g. *Diplococcus pneumoniae*
causes pneumonia

capsule

(d) Streptococci

e.g. *Streptococcus pyogenes*
causes sore throats

S. thermophilus
used in yogurt making

2(a) Rods (bacilli)

bacillus

e.g. *Salmonella typhimurium*
causes severe food poisoning

Escherichia coli
common symbiont of gut

(b) Chains of rods

e.g. *Azotobacter*
a free-living soil nitrogen-fixing
bacterium

endospore

e.g. *Bacillus anthracis*
casues anthrax

3 Spirilla

spirillum

e.g. *Spirillum rubrum*
saprophyte of fresh water

Treponema pallidum
causes the venereal disease syphilis

4 Vibrio

e.g. *Vibrio cholerae*
causes cholera

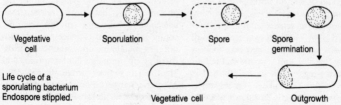

Vegetative
cell

Sporulation

Spore

Spore
germination

Life cycle of a
sporulating bacterium
Endospore stippled.

Vegetative cell

Outgrowth

FIG. 10b *Some bacterial forms and diseases caused by particular species in humans.*

BACTERICIDAL, BACTERIORHODOPSIN, BIOLUMIN-
ESCENCE, *ESCHERICHIA COLI*, GRAM'S STAIN, MYCO-
PLASMAS, SPIROCHAETES.

**bacterial artificial chromosomes
(BACs)** These are derived by inserting DNA
into *E. coli* F plasmids, which are found in
just one or two copies per cell. They can
accept large DNA inserts (from 300,000 to
1 million nucleotide pairs), making them
very suitable as vectors for carrying a genomic DNA insert; but in practice the
sequences are usually only 100–200 kb in
length when used for cloning purposes and
for building GENE LIBRARIES. Each clone (a
genomic DNA clone) carries just one of the
DNA fragments produced by endonuclease
action on the whole organism's genome and
the complete set of colonies forms the genomic DNA library. BAC inserts form fewer
hybrid inserts than do those of YEAST ARTI-
FICIAL CHROMOSOMES and bacterial technology is somewhat simpler to use. See CONTIG,
SHOTGUN SEQUENCING.

bactericidal Substances which kill bacteria. Include many ANTIBIOTICS. For *colicins*,
see PLASMID. Compare BACTERIOSTATIC.

bacteriophage (phage) A VIRUS parasitiz-
ing bacteria. The genetic material is always
housed in the centre of the phage particle
and may be DNA or RNA – the former either
double- or single-stranded, the latter
double-stranded, RNA phages are very
simple; the more complex T-even DNA
phages have a head region, collar, tail and
tail fibres; some (e.g. φX174) are entirely
spherical; others (e.g. M13) are filamentous.
Some (e.g. PM2) contain a lipid bilayer
between an outer and inner protein shell.
The complex phage T4 has a polyhedral
head about 70 nm across containing double-
stranded DNA, enclosed by a coat (capsid)
of protein subunits (capsomeres); a short
collar or 'neck' region; a cylindrical tail (or
tail sheath) region and six tail fibres. The
particle (virion) of T4 phage is quite large –
about 300 nm in length. It initiates infection
by attachment of tail fibres to the bacterial
cell wall at specific receptor sites. This is
followed by localized lysis of the wall by
previously inert viral enzymes in the tail

and by contraction of the viral tail sheath,
forcing the hollow tail core through the host
cell wall to inject the phage DNA. *Virulent
phages* (such as T4, T7, φX174) engage in a
subsequent *lytic cycle* inside the host cell.
Their circular genomes are transcribed and
translated, causing arrest of host macromolecule synthesis and production of virion
DNA and coat proteins. Lysis of the cell
releases the virions. *Filamentous phages* (e.g.
M13, fl, fd) do not lyse the host cell, but
even permit it to multiply. Eventually they
get extruded through the cell wall. Some
temperate phages (e.g. LAMBDA PHAGE, λ) can
insert their genome into their host's and be
replicated with it; others (e.g. P1) replicate
within a bacterial plasmid. This non-lethal
infective relationship does not involve transcription of phage genome, and is termed
lysogeny, the phage being termed *prophage*.
Lysogenic bacteria (those so infected) can
produce infectious phage, but are immune
to lytic infection by the same or closely
related phage (superinfection immunity).
Conversion from lysogenic state to lytic
cycle (induction) may be enhanced by UV
light and other mutagens. A replication
cycle from adsorption to release of new
phage takes ≈15–20 mins. *Transduction*
occurs when temperate phage from one
lysogenic culture infects a second bacterial
culture, taking with it a small amount of
closely-linked DNA which remains as a
stable feature of the recipient cell. Antibiotic
resistance can be transferred this way (see
ANTIBIOTIC RESISTANT ELEMENT). See PHAGE CON-
VERSION, PHAGE RESTRICTION.

bacteriorhodopsin Conjugated RECEPTOR
protein of the 'purple membrane' of the
photosynthetic bacterium *Halobacterium
halobium*, forming a proton channel
composed of seven α-helices, each spanning the membrane once, and whose prosthetic group (retinal) is light-absorbing and
identical to that of vertebrate rod cells.
Allosteric change on illumination of the pigment results in proton ejection from the
cell, the resulting proton gradient being
used to power ATP synthesis. Its integration
into LIPOSOMES along with mitochondrial ATP synthetase showed that the

latter is also a proton channel, driven by a proton gradient. Bacteriorhodopsin absorbs strongly at about 570 nm (in the green region of the spectrum), and if chlorophyll-based photosynthesis evolved after a rhodopsin-based variety then absence of green-absorption in chlorophyll's action spectrum may be due to selection in favour of a pigment avoiding competition with abundant rhodopsin-based forms. See CELL MEMBRANES, CHLOROPLAST, ELECTRON TRANSPORT SYSTEM, MITOCHONDRION, ROD CELL.

bacteriostatic Inhibiting growth of bacteria, but not killing them. See BACTERICIDAL.

BAD A pro-apoptotic protein, and BCL-2 family member, encoded by *Bad* and regulated by phosphorylation in response to GROWTH FACTORS and SURVIVAL FACTORS. Glucose deprivation results in dephosphorylation of BAD, and to BAD-dependent cell death (see APOPTOSIS). Its phosphorylation helps regulate glucokinase activity.

balanced lethal system Genetic system operating when the two homozygotes at a locus represented by two alternative alleles each produce a lethal phenotype, yet the two alleles persist in the population through survival of heterozygotes, which thus effectively breed true. Compare HETEROZYGOUS ADVANTAGE.

balance of nature Phrase glossing observations that in natural ecosystems, communities and the biosphere at large, herbivores do not generally overgraze, predators do not generally over-predate nor parasites decimate host populations; that populations of all appear to be held roughly in equilibrium, and that drastic (sometimes trivial) disturbance of this harmony between organisms and the physical environment will have inevitable and generally unfavourable consequences for mankind. Causal processes involved in the complex systems with which ecologists deal are increasingly amenable to computer simulation. Theories of food web dynamics involve the terms 'connectedness' (the degree to which all species are directly connected in trophic relations) and 'interaction strength' (the change in one unit of a species' growth rate effected by a one unit increase in the density of another'. Theoretical studies suggest that (i) increasing species number, interaction strength and connectedness diminishes the proportion of randomly created webs that are stable, and (ii) species-rich, highly connected food webs have to be carefully 'constructed' if their component species are to persist at a stable equilibrium. There is broad agreement between simple food web models predicted to be stable and those which are actually statistically common; and these include relatively short food chains and those with little omnivory. Initial field work suggests that those parts of the food web where the trophic connections are most complex are those most important to its stability. Natural food webs have parameter values and interaction patterns closer to those predicted to be stable than would be expected by chance, and to this extent food webs are 'engineered', like a complex piece of machinery. Ecologists have long realized that the loss of a species doing a specific ecological 'job' can have wave-like effects across an ecosystem. Recent work shows that, in South America, the set of probably doomed primates spans ecological niches; but in Asia, Africa and especially Madagascar, entire GUILDS of primates are threatened with extinction (e.g. great apes, which eat massive amounts of foliage and disperse seeds; or fruit-eating lemurs, which also disperse seeds). Such losses may hasten extinction of trees dependent on primates for dispersal. See CHAOS, DENSITY-DEPENDENCE, GAIA HYPOTHESIS, HOMEOSTASIS.

balancing selection See NATURAL SELECTION.

Balbiani ring See PUFF.

Baldwin effect Reinforcement or replacement of individually acquired adaptive responses to altered environmental conditions, through selection (artificial or natural) for genetically determined characters with similar functions. Not regarded as evidence of Lamarckism. See GENETIC ASSIMILATION, LAMARCK.

ballistospore Of fungi, a spore that is forcibly discharged. See STATISMOSPORE.

barbiturates Group of compounds of varied action and low selectivity, but all capable of producing behavioural depression – usually affecting polysynaptic pathways in the central nervous system. Commonly fat-soluble (i.e. lipids), crossing the BLOOD–BRAIN BARRIER with ease. Usually induce TOLERANCE, more of the drug being needed for the same effect. Induce enzymes non-selectively. See PSYCHOACTIVE DRUGS.

barbs See FEATHER.

barbules See FEATHER.

bark Protective corky tissue of dead cells, present on the outside of older stems and roots of woody plants, e.g. tree trunks. Produced by activity of cork cambium. Bark may consist of cork only or, when other layers of cork are formed at successively deeper levels, of alternating layers of cork and dead cortex or phloem tissue (when it is known as *rhytidome*). Popularly regarded as everything outside the wood.

baroreceptor (pressoreceptor) Receptor for hydrostatic pressure of blood. In man and most tetrapods, located in carotid sinuses, aortic arch and wall of the right atrium. Basically a kind of stretch receptor. When stimulated, those in the first two locations activate the cardio-inhibitory centre and inhibit the CARDIO-ACCELERATORY CENTRE, while those in the atrium stimulate the cardio-acceleratory centre, helping to regulate blood pressure.

Barr body Heterochromatic X chromosome occurring in female placental mammals. Paternally- or maternally-derived X chromosomes may behave in this way, often a different X chromosome in different cell lineages. The nuclear RNA molecule *Xist* appears to be at least one RNA that forms a 'cage' around the X chromosome from which it is transcribed and is necessary and sufficient for mammalian X chromosome inactivation. See DOSAGE COMPENSATION, HETEROCHROMATIN.

basal body Structure indistinguishable from centriole, acting as an organizing centre (nucleation site) for eukaryotic cilia and flagella, unlike which its 'axoneme' is a ring of nine triplet microtubules, each comprising one complete microtubule fused to two incomplete ones. It is a permanent feature at the base of each such flagellum or CILIUM. See CENTRIOLE for details.

basal ganglia (striatum) Lateroventral component of the subpallial region of the vertebrate telencephalon (see Fig. 18), controlling sequences of actions in complex movements. Receives sensory input from pallium and from the substantia nigra of the midbrain. Comprises the *caudate nucleus* (the most ventral part being the *nucleus accumbens*, which communicates with the LIMBIC SYSTEM); the *putamen*; the *globus pallidus*, and the *amygdala*. Disruption of the basal nuclei leads to involuntary motions (dyskinesias) characteristic of PARKINSON'S DISEASE.

basal metabolic rate (BMR) The energy budget over a specified time period (e.g. 24 hr) of an endothermic animal which is fasting (not digesting or absorbing a meal), lying motionless and is within its THERMONEUTRAL ZONE. *Standard metabolic rate* (SMR) is an animal's resting and fasting metabolic rate at a given body temperature and is applicable to ectotherms. Can be measured directly by calorimetry, or indirectly by oxygen uptake (once RESPIRATORY QUOTIENTS have been taken into account). See SPECIFIC DYNAMIC EFFECT.

basal placentation (Bot.) Condition in which ovules are attached to the bottom of the locule in the ovary.

base Either a substance releasing hydroxyl ions (OH^-) upon dissociation, with a pH in solution greater than 7, or a substance capable of acting as a proton acceptor. In this latter sense, the nitrogenous cyclic or heterocyclic groups combined with ribose to form nucleosides are termed bases. See PURINE, PYRIMIDINE.

basement membrane Combination of BASAL LAMINA with underlying reticular fibres and additional glycoproteins, situated between many animal epithelia and connective tissue.

base pairing Hydrogen bonding between appropriate purine and pyrimidine bases of (antiparallel) nucleic acid sequences, as during DNA synthesis, mRNA transcription and during translation (see PROTEIN SYNTHESIS). If two DNA strands align then an adenine in one strand pairs with a thymine and a guanine with a cytosine (i.e. A:T, G:C); but if a DNA strand aligns with an RNA strand, then adenines in the DNA pair align with uracils in the RNA (i.e. A:U, G:C). Without these 'rules' there could be no GENETIC CODE or heredity as we know it. See DNA HYBRIDIZATION.

base ratio The (A + T):(G + C) ratio in double-stranded (duplex) DNA. It varies widely between different sources (i.e. from different species). The amount of adenine equals the amount of thymine; the amount of guanine equals the amount of cytosine. See BASE PAIRING.

basic dyes Dyes consisting of a basic organic grouping of atoms (cation) which is the actively staining part, usually combined with an inorganic acid. Nucleic acids, hence nuclei, are stained by them and are consequently referred to as BASOPHILIC.

Basidiomycota A fungal division comprising about 1,400 genera and 22,200 species of diverse forms (e.g. jelly fungi, bracket fungi, mushrooms, toadstools, stinkhorns, puffballs, smuts and rusts). Most of these common names refer to the visible part of the fungus, the conspicuous reproductive or fruiting structure (basidiocarp) supported by an extensive assimilative mycelium. The presence of a basidium bearing basidiospores is the diagnostic feature of this division. A typical basidium is aseptate and has four one-celled haploid basidiospores (ballistospores or statismospores) dispersed by air currents. The basidium may be transversely or longitudinally septate and the number of spores (which may be statismospores) are occasionally fewer than or more than four. Nuclear fusion occurs in the young basidium and meiosis takes place before basidiospore development. The mycelium is always septate; when a spore germinates, a primary haploid mycelium is formed, which is initially multinucleate but septa soon form such that the mycelium becomes divided into uninucleate (monokaryotic) cells. A secondary mycelium is produced by fusion of primary hyphae from the same (homokaryotic) or different mating types (heterokaryotic) resulting in the formation of a binucleate (dikaryotic) mycelium. Apical cells of secondary mycelium usually divide by formation of highly characteristic clamp connections, which are another diagnostic feature of the division along with DOLIPORE SEPTA and a double-layered wall, lamellae and electron-opacity in electron microscopy. The primary mycelium may also bear 'oidia'. The mycelium may also be perennial in soil or wood (e.g. fairy rings, sclerotia, rhizomorphs or mycorrhizas). Although typically mycelial, some of these fungi are yeasts or have a yeast-like state. These yeasts are distinguished from those of the ASCOMYCOTA by the morphology of the bud scars, giving a red colour with diazonium blue.

basidiospore Characteristic propagative spore produced by members of the BASIDIOMYCOTA. It is typically a ballistospore but in the gasteromycetes it is a statismospore. Spores contain one or two haploid nuclei produced externally, after meiosis, on a basidium. The colour, form and ornamentation of the basidiospore are important taxonomic characteristics.

basidium Specialized reproductive cell of BASIDIOMYCOTA; often club-shaped, cylindrical or divided into four cells. Nuclear fusion and meiosis occur within the basidium, resulting in the formation of four BASIDIOSPORES, borne externally on minute stalks or sterigmata.

basipetal (Bot.) (Of organs) developing in succession towards the base, oldest at the apex, youngest at the base. Also used of the direction of transport of substances within a plant: away from apex. Compare ACROPETAL.

basophil Class of granulocytic white blood cell involved primarily in inflammatory allergic responses and parasitic (as opposed to bacterial) infections. Originating in the bone marrow, they circulate in the blood (comprising less than 0.2% of leucocytes) and are recruited wherever MAST CELLS are active. They respond chemotactically to COMPLEMENT factor C5a and express defined adhesion molecules and receptors for C3. Granules contain histamine and peroxidase, released on degranulation. Basophils and mast cells bear high-affinity Fc receptors for IgE and effective allergens cross-link these IgE molecules.

basophilic Staining strongly with basic dye. Especially characteristic of nucleic acids, and hence of nucleus (during division of which the condensed chromosomes are strongly basophilic), and of cytoplasm when actively synthesizing proteins (due to rRNA and mRNA).

Bateman principle The theory, proposed on the basis of experiments with fruit flies by A. J. Bateman in 1948, that male promiscuity is more valuable to the reproductive success of males than female promiscuity is to the reproductive success of females. His explanation was that sperm are small and cost little to produce, so that a male is better able to spread them widely; whereas a female invests relatively much more in each egg and needs one good male to maximize her reproductive output. R. Trivers expanded Bateman's principle in 1972 to include the entire investment a parent makes in the calculation, the sex investing more being expected to be the more discriminating, while that investing less being expected to court more mates and show readiness to defend that situation against competitors. Again, S.Hrdy proposed in 1981 that in some primates promiscuity might have an advantage for females through confusing paternity and enlisting support in rearing offspring from males with whom she has mated. Then again, female prairie dogs (*Cynomys gunnisoni*) mating with multiple males produce larger litters and healthier pups.

Batesian mimicry See MIMICRY.

Bayesian statistics Methods of inference first described in the 1700s (Thomas Bayes, d.1761), and later developed by Laplace (1790s), for assessing how new information influences the chances that a current belief continues to be correct. Its use is undergoing a renaissance with researchers in fields as otherwise unrelated as astrophysics, new drug testing and the setting of limits for fish catches. It was first used in assessing the relative likelihoods of different evolutionary trees in 1996 (an optimality criterion being a rule used to decide between competing trees). Evolutionary biologists are now using LIKELIHOOD models within a Bayesian statistical approach to estimate ancestral character states and test for character correlations.

Bayes' Theorem is:

$$P(H \mid D) = P(D \mid H) \times P(H)/P(D)$$

which says that the probability of a hypothesis, given the data, is equal to the probability of the data, given that the hypothesis is correct, multiplied by the probability of the hypothesis before obtaining the data divided by the averaged probability of the data. See OPTIMALITY THEORY, PARSIMONY.

B cell (B lymphocyte) A LEUCOCYTE, derived from a LYMPHOID TISSUE stem cell which has migrated from foetal liver to bone marrow and has not entered the thymus but has settled either in a lymph node or in the spleen. Immature B cells are versatile. If deprived of Pax5 transcription factor, they do not proceed down the path to mature antibody production but can instead develop into a T CELL, NATURAL KILLER CELL or take on the functions of other bone marrow-derived cell types such as antigen-presenting dendritic cells, macrophages, osteoclasts or even neutrophils. B lymphocytes must integrate signals from two classes of membrane receptor before they can proliferate and mature into antibody-secreting plasma cells. This is crucial to the immune system's ability to differentiate between foreign and self 'antigens'. The first signal occurs when an antigen binds directly to a specific immunoglobulin molecule (B cell receptor) integrated into the cell surface. This activates PI3-KINASES, which indirectly signal the B cell to survive and proliferate. The second signal often takes the form of CYTOKINE proteins or cell-surface growth factors made by helper T cells that recognize the same antigen. This dual molecular signal recognition and the involvement of two classes of immune cells helps to reduce the risk of AUTOIMMUNE REACTIONS. The second signal may, however, be specific molecules on the pathogen surface, e.g. LIPOPOLYSAC-CHARIDE in the case of Gram-negative bacteria, which bind to the TOLL-LIKE RECEPTOR 4, present on all B cell surfaces. Mature B cells express a specific immunoglobulin (Ig, or ANTIBODY) on their plasma membranes. They usually first encounter antigens in systemic lymphoid organs such as the spleen (responsive to blood-borne antigens) and lymph nodes (responsive to antigens coming from the tissues via the lymphatic system), at a time when they express germline IgM antibodies (Igμ) on their surfaces. Pre-B cells can bind circulating antigen at their receptor, internalize it and present the processed peptides bound to MHC class II molecules (see MAJOR HISTOCOMPATIBILITY COMPLEX), particularly important in secondary immune responses. On encountering antigen, B cells express new receptors allowing them to respond to cytokines from T cells, whereupon they differentiate along any of several pathways, into memory cells or antibody-forming cells (plasma cells). One pathway somehow induces somatic HYPER-MUTATION of the two variable regions (V_H and V_L) of the genes encoding antibody. This may lead to programmed cell death, but B cells can be rescued from death by antigen presented by dendritic cells within the germinal centres of lymphoid tissue. Secondary B cells leave these centres as either memory cells or plasma cell precursors. Further encounters with antigen (cross-linking the B cell receptors) select cells (clones with higher affinity Ig receptors on their membranes) to develop to the plasma cell stage at which immunoglobulin is

secreted. This is usually brought about by a chromosomal rearrangement that brings the V_H exon adjacent to a new C_H, followed by secretion of a new immunoglobulin type, such as IgG or IgA (see ANTIBODY DIVERSITY). Antibody produced by such a plasma cell will almost always be of one class only, or isotype, (see IgM, IgD, IgG, IgA, IgE). Memory cells may circulate and persist for several years, capable of clonal expansion and rapid Ig secretion (a secondary immune response) if activated by the initial antigen. See CLONAL SELECTION THEORY.

B-chromosome See SUPERNUMERARY CHROMOSOME.

Bcl-2 A family of proteins whose members, when activated, suppress APOPTOSIS by their involvement in survival factor signalling pathways. One member, located on the cytoplasmic side of mitochondrial outer membrane, endoplasmic reticulum and nuclear envelope, possibly registers cell damage to these organelles by modifying movements across them of small molecules or proteins. Can function both as an ion channel and as a docking protein. Its association with nuclear pore complexes might enable Bcl-2 to catch proteins as they cross the envelope. Prevents escape of cytochrome c from mitochondria. Conformations of some resemble those of COLICINS. See P53, ULTRAVIOLET IRRADIATION.

BCR A B CELL antigen receptor. See CLONAL SELECTION THEORY.

belt desmosome See DESMOSOME.

Benedict's test A modification of Fehling's test for sugar using just one solution. Contains sodium citrate, sodium carbonate and copper sulphate dissolved in water in the ratio $1.7 : 1.0 : 0.17$ g : 30 cm^3 water. Five drops of test solution are added to 2 cm^3 Benedict's solution. If a REDUCING SUGAR is present then a rust-brown cuprous oxide precipitate forms on boiling. This reaction occurs on account of the presence of aldehyde or ketone groups of sugars, which occur in the 'open' but not the 'ring' arrangement of the molecule; but since these forms are interconvertible in solution, all sugar is eventually oxidized.

Bennettitales See CYCADOPHYTA.

benthos General term originally introduced to describe those animals found living on the sea bottom; now used to describe organisms of aquatic ecosystems that live associated with a substratum (crawling or burrowing there, or attached to rocks as with seaweeds and sessile animals). Many schemes have been proposed for subdividing the benthos. With respect to higher aquatic plants and algae, the benthos can be subdivided into four categories which have characteristic communities. (1) RHIZOBENTHOS, vegetation rooted via roots, rhizomes or rhizoids into the submerged sediments, e.g. emergent and submersed higher plants such as *Typha*, *Scirpus*, *Potamogeton*, *Nuphar*, and the algae *Chara* and *Nitella*; (2) Haptobenthos, communities of algae attached to a solid substratum, e.g. epilithon, attached to rocks and stones, EPIPHYTON, attached to plant surfaces, EPIZOON, attached to animals, and EPIPSAMMON, attached to sand grains; (3) Herpobenthos, algae living on and or moving through the very top few mm of submerged sediment, mostly motile but also some non-motile forms, e.g. EPIPELON; (4) endobenthos, living within, and often boring into, a substratum, e.g. endolithon, within pores of rocks, ENDOPELON, within sediments, ENDOZOON, within animals, and ENDOPHYTON, within plants. Organisms feeding primarily upon the benthos are termed benthophagous. See PELAGIC, NEKTON, PLANKTON.

Bergmann's rule States that in geographically variable species of ENDOTHERMIC animals, body size tends to be larger in cooler regions of a species range, selection at low temperatures allegedly favouring individuals with small surface area/body mass ratios on account of their reduced heat loss. This explanation remains controversial. See ALLEN'S RULE.

berry Many-seeded succulent fruit, in which the wall (*pericarp*) consists of an outer skin (*epicarp*) enclosing a thick fleshy *mesocarp* and inner membranous *endocarp*, as in gooseberry, currant, tomato. Compare DRUPE.

beta blocker Substance, such as the drug propranolol, selectively blocking BETA RECEPTORS. Clinical use is to slow heart rate and lower blood pressure so that there is less hydraulic stress and consequent wear and tear on the walls of the arteries. They reduce the incidence of strokes and heart attacks. See OPIATES.

beta cells See PANCREAS.

beta-globulins A class of vertebrate plasma proteins including certain lipoproteins, TRANSFERRIN and plasminogen (precursor of fibrinolysin, see FIBRINOLYSIS).

beta-oxidation (β-oxidation) An iterative series of reactions shortening a fatty acid chain by two carbon atoms at a time. Has been shown to occur in PEROXISOMES. See FATTY ACID OXIDATION.

beta receptors ADRENERGIC receptors (e.g. β_1 and β_2 receptors) binding different CATECHOLAMINES, with different affinities. A G PROTEIN links the β receptor to ADENYLYL CYCLASE. Cardiac muscle cells have β_1 receptors and bind noradrenaline, raising heart rate. β_2 receptors are found in smooth muscle fibres in airways and in blood vessels serving the heart wall, skeletal muscle, adipose tissue, liver and walls of visceral organs; on binding catecholamine, the muscle relaxes. See ALPHA RECEPTOR, BETA BLOCKER, AUTONOMIC NERVOUS SYSTEM.

beta-sheets (beta-conformation) One type of protein secondary structure. See PROTEIN, AMINO ACID.

B-form helix (B DNA, DNA-B) The (paracrystalline) form of DNA believed to be adopted most frequently in living cells, as in aqueous media generally. Compared with the less common A-FORM HELIX, the molecule is more hydrated (hence less stable and compact) and the axis of the molecule passes through the hydrogen bonds of the base pairs, producing a wider minor groove. See Z-FORM HELIX.

***bicoid* gene (*bcd*)** (See Fig. 11.) A pattern-specifying MATERNAL GENE in *Drosophila*,

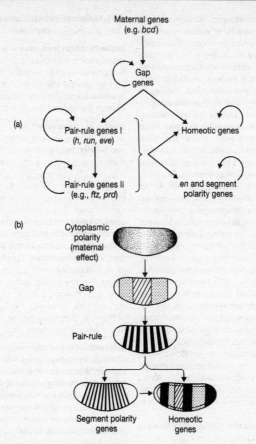

FIG. 11 *BICOID* GENES: (a) *Proposed cascade of segmentation gene activity in* Drosophila *and* (b) *the pattern formation in the fly resulting from this. The pattern is established by maternal effect genes which set up morphogenic gradients of proteins as described in the entry for* GAP GENES. *Italicized acronyms refer to the relevant genes. See also* PAIR-RULE GENES.

whose mRNA transcript passes from maternal ovary nurse cells to anterior poles of developing oocytes where it becomes localized, apparently trapped by the oocyte cytoskeleton. Its eventual protein product bcd (or bicoid) is a transcription activator (with a DNA-binding homeodomain), is located in nuclei of the syncytial blastoderm, and is translocated half-way along the embryo from the anterior pole. The bcd product resembles a MORPHOGEN except that

it is a transcription factor that spreads through the syncytial *Drosophila* embryo – unlike normal morphogens, which are extracellular and spread through multicellular tissue. Different threshold concentrations of bcd activate different target genes, and bcd also appears to bind a specific messenger RNA and act as a translational repressor generating an opposing gradient on the homeodomain protein caudal (cad). With a similar translocation of nanos gene product (see *OSKAR* GENE) in the opposite direction, two opposing gradients provide quantitative POSITIONAL INFORMATION which the embryo genome converts into qualitative phenotypic differences. Genes involved in this conversion process, i.e. *Krüppel* (*Kr*), *hunchback* (*hb*) and *knirps* (*kni*), belong to the GAP GENE class, and all three encode proteins with DNA-binding FINGER DOMAINS. Bicoid product is responsible for production of a normal head and thorax, structures absent in embryos from females lacking functional *bcd*, and both it and the oskar product are examples of MORPHOGENS.

By repressing transcription of *Kr* in anterior and posterior embryo regions, bicoid and oskar morphogens only permit *Kr* expression in mid-embryo. High levels of *bcd* product activate *hb* transcription, while low levels permit *Kr* transcription. However, their abilities to elicit such zygotic (embryo) gene expression depend upon the affinities of their respective DNA-binding sites for them, and these are known to vary. The *hb* and *Kr* products are mutual repressors of each others' transcription, so that their 'domains' of product expression become stable and restricted. These *gap gene domains* delimit the domains of HOMEOTIC GENE expression but also enable position-specific regulation of the PAIR-RULE GENES, whose expression leads to metameric segmentation of the embryo.

bicollateral bundle See VASCULAR BUNDLE.

bicuspid (1) A tooth with two cones, or cusps, on the biting surface. Whereas in *Australopithecus afarensis* the first premolar is sectorial (front surface long and blade-like for shearing), in all later fossil hominids, and in modern humans, the first premolar is a bicuspid tooth. All remaining premolars in monkeys, apes and hominids are also bicuspid. (2) Alternative term for the mitral valve of the mammalian heart.

biennial Plant requiring two years to complete life cycle, from seed germination to seed production and death. In the first season, biennials store up food which is used in the second season to produce flowers and seed. Examples include carrot and cabbage. Compare ANNUAL, EPHEMERAL.

bikonts A major eukaryote grouping, considered to be monophyletic, distinguished by being ancestrally biciliate, or biflagellate, and by having the DHFR-TS gene fusion (see EUKARYOTE, Fig. 52). Compare HETEROKONTS.

bilateral symmetry Property of most metazoans, having just one plane in which they can be separated into two halves, approximately mirror images of one another. This is usually the plane separating left and right halves of the embryo, located at the interface of the antero-posterior and dorsoventral axes. The vertebrate body plan is asymmetric in certain respects however – notably the heart and parts of the gut – a phenomenon arising at the gastrula stage (see AXIS SPECIFICATION). Major metazoan phyla excluded are coelenterates and echinoderms (having radial symmetry as adults). In flowers, the condition is termed zygomorphy. Diatoms (BACILLARIOPHYCEAE) of the order Pennales display bilateral symmetry when observed in valve view.

Bilateria One of the two super-phyletic divisions of animals employed in early phylogenies (the other being the Radiata). Such schemes are losing much support, radially symmetrical animals (ctenophores and cnidarians) no longer being set apart from other animals. However, bilaterally symmetric animals (unlike radial ones) do tend to be triploblastic, and recognition of the three major tissue layers remains useful in interpreting later development, even though these layers (mesoderm especially) may not be homologous across groups.

bile Fluid produced by vertebrate liver cells (hepatocytes), containing both secretory and excretory products, and passed through bile duct to duodenum. Contains (i) BILE SALTS (e.g. sodium taurocholate and glycocholate) which emulsify fat, increasing its surface area for lipolytic activity. Also form micelles for transport of sterols and unsaturated fatty acids towards intestinal villi (see CHYLOMICRON); (ii) bile pigments, breakdown products of haemoglobin such as bilirubin, which are true excretory wastes and colour faeces; (iii) LECITHIN and CHOLESTEROL as excretory products. Bile is aqueous and alkaline due to NaHCO₃, providing a suitable pH for pancreatic and subsequent enzymes. Stored in gall bladder. See MICELLE.

bile duct Duct from liver to duodenum of vertebrates, conveying BILE. See GALL BLADDER.

bile salts Conjugated compounds of bile acids (derivatives of CHOLESTEROL and taurine and glycine), forming up to two thirds of dry mass of hepatic BILE.

bilharzia Schistosomiasis. See *SCHIS-TOSOMA*.

binary fission Vegetative reproduction occurring when a single cell divides into two equal parts. Compare MULTIPLE FISSION.

binocular vision Type of vision occurring in primates and many other active, predatory vertebrates; eyeballs can be so directed that an image of an object falls on both retinas. Extent to which eyes converge to bring images on to the foveas of each retina gives proprioceptive information needed in judging distance of object. Stereoscopic vision (perception of shape in depth) depends on two slightly different images of an object being received in binocular vision, the eyes viewing from different angles.

binomial nomenclature The existing method of naming organisms scientifically and the lasting contribution to taxonomy of LINNAEUS. Each newly described organism (usually in a paper published in a recognized scientific journal) is placed in a genus, which gives it the first of its two (italicized) Latin names – its generic name – and is always given a capital first letter. Thus the genus *Canis* would probably be given to any new placental wolf discovered. Within this genus there may or may not be other species already described; in any case, the new species will receive a second (specific) name to follow its generic name, this time with a lower case first letter. The wolf found in parts of Europe, Asia and North America belongs to the species *Canis lupus*; jackals,

coyotes and true dogs (including the domesticated dog) also belong to the genus *Canis*, but each species is further identified by a different specific name: the domestic dog, in all its varieties, is *C. familiaris* (often only the first letter of the genus is given, if it is contextually clear what the genus is). If a previously described species is subsequently moved into another genus, it takes the new generic name, but carries the old specific name. The author who published the original description of a species often receives credit in the form of an abbreviation of their name after the initial mention of the species in a paper or article. Thus one might see *Canis lupus* Linn., since Linnaeus first described this species. The specimen upon which the initial published description was based is termed the type specimen, or 'type', and is probably housed in a museum for comparison with other specimens. If sufficient variation exists within a species range for SUBSPECIES or varieties to be recognized, then a trinomial is employed to identify the subspecies or variety. The British wren rejoices under the trinomial *Troglodytes troglodytes troglodytes*.

bioassay Quantitative estimation of biologically active substances by measurement of their activity in standardized conditions on living organisms or their parts. A 'standard curve' is first produced, relating the response of the tissue or organism to known quantities of the substance. From this the amount giving a particular experimental response can be read. See BIOSENSOR.

biochemical oxygen demand (biological oxygen demand) Amount of dissolved oxygen (in mg/dm³ water) which disappears from a water sample in a given time at a certain temperature, through decomposition of organic matter by microorganisms. Used as an index of organic POLLUTION, especially sewage.

biochemistry The study of the chemical changes within, and produced by, living organisms. Includes molecular biology, or molecular genetics.

biocoenosis Totality of organisms living in a biotope, itself defined as any segment of the environment which has convenient but arbitrary upper and lower limits, characterized by dominant or characteristic species. Virtually synonymous with COMMUNITY.

biodegradable (Of substances) which can be broken down and recycled as part of a FOOD CHAIN (see DECOMPOSER), e.g. sewage and oil pollutants. Non-biodegradable pollutants include the elements lead, mercury and cadmium and, among compounds, polychlorinated biphenyls (see PCBs), those with radioactive atoms (radioactivity itself is not lost, although the compounds containing it may otherwise be biodegradable), dioxins (found in many plastics and herbicides), DDT and other organochlorine pesticides (see PEST MANAGEMENT).

biodiversity The level of global or local biological diversity, often quantified crudely as numbers of species or of higher taxa; increasingly taken as indicative of genetic diversity level (see GENETIC VARIATION, GENOME). Total species estimates for Earth vary from 3×10^6 to 30×10^6, but these are almost entirely terrestrial estimates. Terrestrial insects account for more than half of all named and recorded species, and it has been argued that marine invertebrates alone may provide a further 10×10^6 species. In terms of diversity of basic body plan (phyla), more than 80% of all phyla are found only in the sea, and most of these in sediments. Genotyping techniques that sample and compare genotypes from different organisms are demonstrating that the bulk of genetic diversity lies not with plants and animals but with prokaryotes (see BACTERIA, ARCHAEA). The most widely used technique depends on sequencing the DNA of genes encoding the 16S subunit of ribosomal RNA (see RIBOSOMES), which evolves very slowly.

There is a growing consensus that species numbers of predators and parasites within a community are not simply summative: weighted (ordinal) measures of diversity contrast with summative (cardinal) ones. Weightings often reflect either the functional importance of taxa within communities or their degree of taxonomic variability, species number within a community (its richness) being a poor indicator of its structure. Much interest surrounds TROPICAL RAIN FORESTS where, e.g., an individual tree may harbour as many species and genera of ants as are found in the entire British Isles. Few of the species present in a community are usually abundant (i.e. have a large population size, large biomass, productivity, or some other measure of importance). The large number of rare species (see RARITY) mainly determines the species diversity, and the ratios between the number of species and their 'importance values' are termed *species diversity indices*. The particular index employed depends on the type of community studied. The Shannon–Weiner diversity index, derived from information theory, tries to describe community diversity in terms of the amount of information it contains. It is given by the equation:

$$H = -\sum_{i=1}^{s} (p_i)(\log_e p_i)$$

where H is the value of the index, S is the number of species and p_i is the proportion of individuals of the ith species.

Emerging infectious diseases (EIDs) of humans and domesticated animals pose a problem for terrestrial wildlife. Thus, wild dogs became extinct in the Serengeti in 1991 at the same time as an epidemic of canine distemper occurred in local domestic dogs. Such 'spill-over' is a particular threat to endangered species. The movement around the world of wildlife (for breeding and conservation purposes), and of domesticated animals, risks transport of their pathogens too. One cost of human domination of the Earth's biota is the increasing biogeographical homogeneity caused by introduction of non-native flora and fauna to new areas. Such 'biological pollution' causes serious loss of biodiversity, particularly on oceanic islands. Recent increasingly intensive agricultural systems have been linked to declines in plant, invertebrate and bird diversity and there is much debate as to whether the introduction of GM crops will accelerate this trend (see GMO). See BIOGEOGRAPHY, SEED BANK, SPECIES RICHNESS.

bioengineering See GENE MANIPULATION.

biofilms Large aggregations of BACTERIA surrounded by slime, resisting chemical disinfectants, antibiotics and the immune system. This is the commonest growth form adopted by bacteria outside laboratories (where they are normally grown as broth suspensions or as colonies on agar). Researchers are designing new antibiotics to combat biofilms. See QUORUM SENSING.

biofuel cells Submillimetre fuel cells, currently in the research-and-development stage. One variety uses GLUCOSE OXIDASE tethered to the anode to remove electrons from hydrogen atoms in the gluconolactone substrate. The protons produced travel to the cathode where another tethered enzyme, laccase, forms water molecules by reacting them with oxygen. Osmium-containing polymers transport the electrons between the electrodes and their enzymes – when they can be channelled off to do work. Once the problems of immune rejection are overcome, such biofuel cells may breed a new generation of implanted machines powered by the body's own glucose.

biogas Mixture of methane (CH_4) and carbon dioxide (CO_2) produced anaerobically during a two-stage bacterial breakdown of (i) carbohydrates, fats and proteins to fatty acids, alcohols, CO_2 and hydrogen gas (H_2), followed by (ii) production by methanogenic bacteria (e.g. *Methanococcus, Methanospirillum*) of methane by reduction of CO_2 by H_2 and by the splitting of acetic acid to methane and CO_2. Domestic biogas generators capable of using organic wastes, excreta and leafy vegetable crop remains can make a significant contribution to the energy budget of a country while reducing pollution. See GASOHOL.

biogenesis The theory that all living organisms arise from pre-existing life forms. The works of Redi (1688) for macroorganisms, and of Spallanzani (1765) and PASTEUR (1860) in particular for microorganisms, stand as landmarks in the overthrow of the theory of SPONTANEOUS GENERATION. Since their time it has become generally accepted that every individual organism has a genetic

ancestry involving prior organisms. The first appearance of living systems on the Earth (see ORIGIN OF LIFE) is still problematic, but there is no reason in principle to dispense with faith in natural and terrestrial causation, and every reason to pursue that line of inquiry.

Biogenetic Law Notorious view propounded by Ernst Haeckel in about 1860 (a more explicit formulation of his mentor Muller's view) that during an animal's development it passes through ancestral adult stages ('ontogenesis is a brief and rapid recapitulation of phylogenesis'). Much of the evidence for this derived from the work of embryologist Karl von Baer. It is now accepted that embryos often pass through stages resembling related *embryonic*, rather than adult, forms. Such comparative embryology provides important evidence for EVOLUTION. See BODY PLANS, PAEDOMORPHOSIS.

biogeography A major branch of geography, concerned with the study and interpretation of geographical distributions of animals (ZOOGEOGRAPHY), plants (PLANT GEOGRAPHY), and man (cultural biogeography), both extant and extinct. One approach (*dispersal biogeography*) stresses the role of dispersal of animals and plants from a point of origin or across pre-existing barriers. An alternative approach, *vicariance biogeography*, studies the relation between continental drift and biogeography. In cases where it leads to different predictions from dispersal biogeography, it may be possible to decide between the two theories.

Whereas large animals have biogeographies, microbial species – being ubiquitous – do not. Current work suggests that the transition from ubiquity to biogeography takes place in the species size range of 1–10 mm. Free-living microbial EUKARYOTES all have body sizes less than ~2 mm. See BIODIVERSITY.

bioinformatics Algorithmic and other computational techniques employed in arguing from the data of GENOMICS and PROTEOMICS towards conclusions about issues as wide-ranging as structural biology (e.g. 3D molecular interactions), drug design, GENE

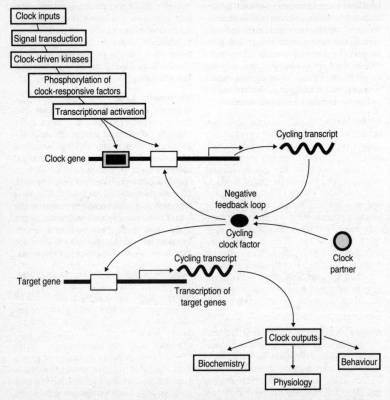

FIG. 12 *One molecular model of autoregulatory feedback loops involved in* BIOLOGICAL CLOCK *function. 'Clock factors' may bind their own promoters (rectangles in figure), feeding back either positively or negatively on gene expression. Short-lived mRNA transcripts can cause cyclic oscillations of their own, or of a target gene's, activity. Additional proteins (clock partners) may be required in this feedback.*

MANIPULATION and phylogeny. See COMPLEXITY.

biological clock In its widest sense, any form of biological timekeeping, such as heart beat or ventilatory movements; more often used in context of PHOTOPERIODISM, DORMANCY and DIAPAUSE, but mostly associated with physiological, behavioural, etc., rhythms relating to environmental cycles, notably CIRCADIAN, tidal, lunar-monthly and annual. Mechanisms counting the number of cell divisions would also be included,

i.e. developmental clocks (see TELOMERES). In higher plants, assembly of the photosynthetic apparatus begins (adaptively) in the hours before dawn. Membrane regulation of ions between cellular compartments, and of energy metabolism by energy-transducing membranes of chloroplasts and mitochondria, are implicated in the clock mechanism. Any clock, or oscillator, needs to be entrained (synchronized) to the solar cycle, running at precisely 24 hours with a defined phase relationship to the light–dark cycle, and be able to both measure and show time. The oscillator can be reset in free-running individuals (those held in darkness, or in dim light), such that bright light presented at a time normally perceived as day (subjective day) may have little effect; but if presented at the start of subjective night, it delays the clock and its rhythms; and if presented at the end of subjective night it advances the clock and its rhythms. Clocks have three conceptual components: an input pathway linking the internal cycle to external light–dark cycles; an autonomous pacemaker generating the circadian oscillation (measuring time); and an output pathway producing the circadian rhythm physiology itself (showing time). In fish, reptiles and birds, the pineal gland contains photosensitive cells that have pacemaker properties and influence hormonal secretory cycles; in mammals the clock centre is the SUPRACHIASMATIC NUCLEI (SCN) in the HYPOTHALAMUS of the brain, whose neurons secrete neuropeptides (e.g. ANTIDI-URETIC HORMONE) with a circadian rhythm while others send rhythmic signals to the pineal gland, effecting circadian output of MELATONIN. Transplantation of the SCN can restore circadian activities in mammals that have had their SCN removed.

Two effects are key tests of a clock-driven circadian rhythm: dramatic swings in mRNA levels over a 24-hr period, even in free-running organisms; and alteration of gene expression by changing the light–dark cycle experienced. In the mould fungus *Neurospora* (involving cycles of asexual sporulation) and the fruit fly *Drosophila* (involving cycles of rhythmic locomotor activity) spectacular success has been achieved in understanding the negative feedback involved between the initiation of transcription of genes (*frequency* (*frq*) in *Neurospora*, and *period* (*per*) and *timeless* (*tim*) in *Drosophila*) in the absence of their product, and their repression by the product. Hamsters heterozygous for the mutation *tau* show a 22-hr period in locomotory activity whereas those homozygous for *tau* have a 20-hr periodicity. Even individual cells seem capable of supporting their own clocks, independently of a 'master clock', e.g. cells from the retina of hamsters, and in the antennae, proboscis and Malpighian tubules of *Drosophila*. Cells outside the retina lack the photosensitive pigments found in the eye, and it may be that proteins sensitive to blue light (like those involved in DNA repair, but different from them) are involved as photoreceptors for activating the circadian clock. Animal biological clocks are temperature insensitive, with Q_{10}s of 1.0. See Fig. 12.

biological control Artificial control of pests and parasites by use of organisms or their products; e.g. of mosquitoes by fish and aquatic insectivorous plants which feed on their larvae; or of the prickly pear (*Opuntia*) in Australia by the moth *Cactoblastis cactorum*. There is increasing use of PHERO-MONES in attracting pest insects, which may then be killed or occasionally sterilized and released. Sometimes attempts are made to encourage spread of genetically harmful factors in the pest population's gene pool, although this needs great care. For *Bacillus thuringiensis*, see PEST MANAGEMENT. See also PESTS. Success depends on a thorough grasp of relevant ecology, not least ensuring that insects, especially parasitoids, used as control agents do not prey on other species than that requiring control, and that a control organism does not turn to other prey once the pest has been controlled.

biology Term coined by LAMARCK in 1802. The branch of science dealing with properties and interactions of physico-chemical systems of sufficient complexity for the term 'living' (or 'dead') to be applied. These are usually cellular or acellular in organization; but since viruses share some of the same polymers (nucleic acid and protein) as cells, and moreover are parasitic, they are regarded as biological systems but not usually as living organisms. See LIFE.

bioluminescence Emission of light by 'photogenic' organisms, including bacteria, fungi, dinoflagellates, jellyfish, brittle stars, worms, fireflies, molluscs and fish. Bacteria often form symbiotic associations with animals such as fish, which thereby become luminous (see PHOTOPHORE). Some animals are self-luminous, having photogenic organs containing photocysts. In the coelenterate *Obelia* these are scattered through the endoderm. Two types of light emission occur in living organisms: (1) bioluminescence (chemiluminescence), in which energy from an exergonic chemical reaction is transformed into light energy; and (2) photoluminescence, which is dependent on the prior absorption of light.

Dinoflagellates are the main contributors to marine bioluminescence, emitting a bluish-green (maximum wavelength 474 nm) flash of light of 0.1 s duration when the cells are stimulated. The compound responsible for the bioluminescence is *luciferin*, which is oxidized with the aid of the enzyme luciferase. The chemistry of luciferin varies with the organisms, e.g., in bacteria, insects, and dinoflagellates it is a flavin, a (benzo)thiazole nucleus and a tetrapyrrole respectively. Luciferases share the feature of being oxygenases (enzymes that add oxygen to compounds). In the dinoflagellate *Gonyaulax polyedra* bioluminescence is rhythmic and circadian. Most luminescent bacteria are marine forms, and usually associated with fish. In culture, they produce a specific *autoinducer* whose concentration builds up until at a critical value induction of the luciferase occurs. Cultures of luminous bacteria at low densities are therefore not luminous but become so at high densities – termed *quorum sensing* on account of the density-dependent nature of the phenomenon. Symbiotic bacteria, as in the light organs of fish, do reach high densities, but it is not clear why bioluminescence is density dependent in free-living forms.

Flashing that occurs in luminous organs of many animals (often serving in mate or prey attraction) is often under nervous control, the organs having a rich supply of nerve terminals. Bioluminescent flashing in dinoflagellates has been suggested to serve as an antipredation device. Feeding upon such dinoflagellates by copepods produces a bright spot of light making the copepod conspicuous to its predators. Hence such copepods feeding upon bioluminescent dinoflagellates would be at a selective disadvantage.

biomagnification The increasing concentration of a compound in the tissues of organisms as the compound passes up a food chain, usually as a result of food intake. This results from the accumulation of the compound at each trophic level prior to its consumption by organisms at the next trophic level (e.g. chlorinated hydrocarbons in PESTICIDES are very susceptible to biomagnification). An animal's fat level will probably be a major factor in accumulation of fat-soluble organic residues, but not all persistent pollutants in aquatic food chains do bioaccumulate.

biomass The total quantity of matter (the non-aqueous component frequently being expressed as dry mass) in organisms, commonly of those forming a trophic level or population, or inhabiting a given region. See STANDING CROP, PYRAMID OF BIOMASS.

biome Major terrestrial, climatically controlled, regional ecological complex or set of ecosystems possessing characteristic vegetation, and among which there is exchange of water, nutrients and biotic components (e.g. TROPICAL RAIN FOREST, coral reefs, grasslands). Biomes occupy large areas of the land surface, and typically occur on more than one continent. The actual distribution of biomes results from (1) geological factors, e.g., proximity of mountain ranges, their height and orientation; (2) direction in which the prevailing moisture-laden winds are blowing, hence global climate patterns; and (3) seasonal changes in insolation arising from the sun.

biometry Branch of biology dealing specifically with application of mathematical techniques to the qualitative study of varying characteristics of organisms, populations, etc.

biomonitoring The use of living organisms to track environmental POLLUTION.

biophysics Fields of biological inquiry in which physical properties of biological systems are of overriding interest. Biophysics departments tend to work in such areas as crystal structures of macromolecules and neuromuscular physiology.

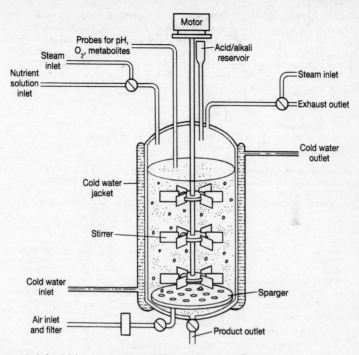

FIG. 13 *An industrial fermenter suitable for continuous cultures. See* BIOREACTOR.

biopoiesis The generation of living from non-living material. See ORIGIN OF LIFE.

bioreactor A chamber within which the activity of a biocatalyst can be optimally controlled, often for industrial production. They include fermenters such as the CHEMO-STAT (Fig. 13). In *batch fermentation*, the nutrients and microorganism are put into a closed reactor, nothing is added while fermentation occurs and nothing except waste gases leaves the chamber, and most environmental conditions except temperature are normally allowed to vary unchecked. The product is separated from the microorganism when nutrients have been utilized; the exponential growth phase usually lasts only a short time, and secondary metabolites such as antibiotics (e.g. penicillin) are produced only towards the end of the EXPONENTIAL GROWTH phase, risks of con-

tamination being reduced compared with continuous culturing methods. In the *fed-batch method*, the first stage involves the above batch method but then controlled feeding is applied in order to reduce growth rate compatibly with the oxygen supply, resulting in good quality protein production. *Continuous fermentation* is less often employed than batch fermentation – as in vinegar production, and production of glucose isomerase by *Bacillus coagulans* (see INDUSTRIAL ENZYMES). Here open reactors (chemostats) are used and nutrients added continuously at a rate which balances their removal rate. Microorganisms are thus kept in their exponential growth phase, for which monitoring and constancy of environmental variables (pH, temperature, O_2, substrate, product and waste levels) are essential. The difficulty of achieving such steady-state conditions

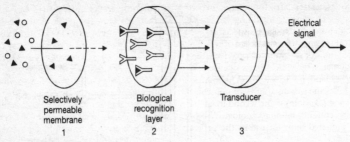

(a) *The parts of a biosensor*

1 The substance to be detected passes through the membrane

2 The substance binds to the recognition layer which may contain antibodies or enzymes; a reaction takes place.

3 The product of the reaction passes to a transducer and this produces an electrical signal

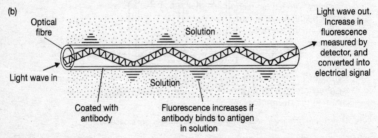

FIG. 14 (a) *The principal components of a* BIOSENSOR, *illustrated in (b) by an optical fibre on which antibodies to a particular antigen have been tagged with a fluorescent molecule. Light travelling along the fibre is reflected back off the parts where antigen has bound, so increasing light transmitted along the fibre.*

could be offset by the greater cost-effectiveness achievable. When fermentation is completed, the product is extracted, concentrated (or purified) and made commercially presentable, a combination of procedures termed *downstream processing.* See SINGLE CELL PROTEIN.

biorhythm See BIOLOGICAL CLOCK, CIRCADIAN RHYTHM.

biosensor (See Fig. 14.) Device in which a bioactive layer lies in intimate contact with a transducer whose responses to changes in the bioactive layer generate electronic signals for subsequent interpretation. The bioactive layer may consist of membrane-bound enzymes, antibodies or receptors. The membrane itself may be synthetic (e.g. acetyl cellulose) and coat an inorganic chip. The potential for this fusion of electronics and biotechnology includes the direct assay of clinically important substrates (e.g. blood glucose) and of substances too unstable for storage or whose concentrations fluctuate rapidly. New designs incorporate stable gated ion channels which amplify a signal, perhaps a single molecular interaction, so mimicking natural bioamplifiers.

biosphere That part of Earth and its atmosphere inhabited by living organisms. See GAIA HYPOTHESIS.

biosystematics (1) In botany, experimental taxonomy including genetical, cytological and ecological aspects. (2) Biological systematics – the study of relationships and CLASSIFICATION of organisms and the process by which the organism has evolved and by which it is maintained (includes the sub-disciplines of nomenclature and taxonomy). The second usage has a wide currency amongst zoologists and is the preferred term for use in the general context.

biota The flora and fauna together; all the living organisms at a location, at a time.

biotechnology Application of discoveries in biology to large-scale production of useful organisms and their products. Centres on development of enzyme technology in industry and medicine, and of GENE MANIPULATION, often in the service of plant and animal breeding (e.g. see PROTOPLAST). Together, these constitute biomolecular engineering. Branches include fermentation technology, waste technology (e.g. recovery of metals from mining waste) and renewable resources technology, such as the use of LIGNOCELLULOSE to generate more usable energy sources. Organisms involved tend to be microorganisms, and their traditional involvement in the brewing, baking and cheese/yoghurt industries is also affected by the new technology. See BIOREACTOR, BIOSENSOR, DNA CHIP, INDUSTRIAL ENZYMES, MICROPROPAGATION, NANOBIOTECHNOLOGY.

biotic factors Those features of the environments of organisms arising from the activities of other living organisms; as distinct from such abiotic factors as climatic and edaphic influences.

biotic potential Overall reproductive output. Its value will depend upon the numbers of eggs laid, the relative importance of asexual and sexual reproduction, and upon generation time. Parasites tend to have huge biotic potentials.

biotin VITAMIN of the B-complex made by intestinal bacteria, so difficult to prevent its uptake and assimilation. However avidin, a component of raw egg white, binds tightly to biotin and prevents its uptake from the gut lumen, resulting in biotin deficiency in those eating raw eggs too avidly. A modified biotin when bound to the enzyme propionyl-CoA carboxylase acts as a coenzyme in the conversion of pyruvic acid to oxaloacetic acid (see KREBS CYCLE) and in the synthesis of fatty acids and purines, again facilitating carboxylation and decarboxylation reactions.

biotrophic (Of parasites) which cannot live on a dead host; e.g. most parasitic worms, lice and fleas. Contrast NECROTROPHIC; see PLANT DISEASE AND DEFENCES.

biotype Naturally occurring group of individuals of the same genotype. See INFRASPECIFIC VARIATION.

bipedalism For hominid bipedalism – postural and locomotory – see AUSTRALOPITHECINES.

bipolar neuron See NEURON.

biramous appendage Paired crustacean appendage branching distally from a basal region (protopodite) to form two rami, the exopodite and endopodite, the exopodite often being more slender and flexible. Subject to considerable adaptive radiation, serving varied functions, such as locomotion, feeding and gaseous exchange. Trilobite limbs also had a biramous structure, but the origin of the second ramus is different. *Stenopodia* has been the term used for crustacean appendages in which one or more processes, epipodites, lie on the outer sides of protopodites; *phyllopodia* are broader and flatter than most stenopodia and may be unjointed, bearing lobes known as endites and exites. See UNIRAMOUS APPENDAGE, and Fig. 15.

birds See AVES.

bisaccate Pollen grains possessing two air sacs or bladders; mainly coniferous pollen, e.g. pine pollen.

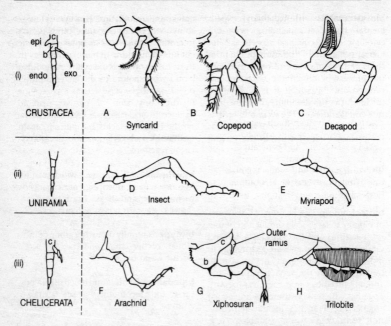

FIG. 15 *A diagram indicating the main features of adult limb structures in arthropods. (i) biramous, (ii) uniramous, (iii) primitively biramous, with the exite/outer ramus more proximally placed than in crustaceans. epi = epipodite; endo = endopodite; exo = exopodite; b = basipodite; c = coxopodite.*

bisexual See HERMAPHRODITE.

bisporangiate cone (Bot.) Cone containing both megaspores and microspores (e.g. in the lycopsid *Selaginella*).

bisporic embryo sac Embryo sac developing from the inner spore of a diad of spores produced by incomplete meiosis (as in the onion, *Allium*).

bithorax complex (BX-C) Three HOMEO-GENES in *Drosophila* involved in the PATTERN FORMATION of embryonic PARASEGMENTS 5–14. The proteins encoded by all three genes (*Ultrabithorax*, *Ubx*; *abdominal-A*, *abd-A*; *Abdominal-B*, *Abd-B*) vary through alternative splicing during RNA PROCESSING. Each works in a particular parasegmental domain within the embryo, sometimes in a combinatorial way.

biuret reaction Reaction often used as a test for presence of proteins and peptides (as a result of peptide bonds; but also works for any compound containing two carbonyl groups linked via a nitrogen or a carbon atom). A few drops of 1% copper sulphate solution and sodium hydroxide solution are added to the test sample, when a purple Cu^{2+}-complex is produced if positive; solution stays blue otherwise.

bivalent Pair of homologous chromosomes during pairing (synapsis) at the first meiotic prophase and metaphase. See MEIOSIS.

bivalve Broadly speaking, any animal with a shell in two parts hinged together, e.g. bivalve molluscs (with which the term is often equated) and BRACHIOPODA. Occasionally ostracod crustaceans are said to have a bivalve carapace, although this is not a shell. See BIVALVIA.

Bivalvia (Lamellibranchia, Pelecypoda) Class of MOLLUSCA, with freshwater, brackish and marine forms with greatly reduced head (no eyes, tentacles or radula), and body laterally compressed and typically bilaterally symmetrical. Shell composed of two hinged. valves (both lateral) under which lie two large gills (*ctenidia*) used for gaseous exchange and generally for filter-feeding. Sexes nearly always separate; trochosphere and veliger larvae, but occasionally a parasitic glochidium. Considerable radiation, including fixed forms (e.g. mussels, clams) and burrowers (e.g. shipworm, razorshell). See BRACHIOPODA.

bladder (1) Urinary bladder. Part of the anterior ALLANTOIS of embryonic amniotes, which persists into adult life and receives the openings of the ureters either directly (mammals) or via a short part of the cloaca (lower tetrapods). In fish the urinary ducts themselves may enlarge as 'bladders', or fuse for part of their length forming a single sac, receiving also the oviducts in females, as a urogenital sinus. In all cases the bladder serves for urine retention, and in tetrapods is distensible with a thick smooth-muscle wall. In lower tetrapods it opens into the cloaca, but in mammals its contents leave via the urethra. (2) Other sac-like fluid-filled structures in animals termed bladders include GALL BLADDER and GAS BLADDER.

bladderworm The CYSTICERCUS stage of some tapeworms.

blastema Mass of undifferentiated tissue which forms, often at a site of injury or amputation, and from which regenerating parts regrow and differentiate. See REGENERATION.

blastocoele Animal primary body cavity, arising as fluid-filled cavity in the BLASTULA and persisting as blood and tissue fluids where these occur; otherwise obliterated during GASTRULATION. Much expanded in most arthropods as the HAEMOCOELE. Compare COELOM, PSEUDOCOELOM.

blastocyst Stage of mammalian development at which implantation into the uterine wall occurs, the INNER CELL MASS spreading inside the blastocoele as a flat disc.

blastoderm Sheet of cells, usually one or just a few cells thick, surrounding the blastocoele in non-yolky eggs, or covering it in yolky ones. Consists usually of small, tightly packed cells.

blastomere Cell produced by cleavage of an animal egg, up to the late blastula stage. In yolky eggs especially this results in smaller *micromeres* at the animal pole, with larger yolky *macromeres* at the vegetal pole where the rate of division is slower. See CLEAVAGE.

blastopore Transitory opening on surface of gastrula by which the internal cavity (archenteron) communicates with the exterior; produced by invagination of superficial cells during GASTRULATION. It becomes the mouth in PROTOSTOMES and the anus in DEUTEROSTOMES. The *dorsal lip* (= Spemann organizer) of amphibian blastopores is famous for INDUCTION of overlying tissue in gastrulation. Transplant experiments indicate that a region up to about 60° around the original blastopore material is the source of a dorsoventral gradient responsible for regional subdivision of embryonic mesoderm since it is a source of the dorsalizing proteins noggin and chordin which bind the ventralizing protein BMP-4 (see BMPS, TGFs). See POSITIONAL INFORMATION, ORGANIZER, PRIMITIVE STREAK.

blastula Stage of animal development at or near the end of cleavage and immediately preceding gastrulation. May consist of a hollow ball of cells (*blastomeres*), especially in non-yolky embryos.

blight For 'potato blight', caused by *Phytophthora infectans*, see OOMYCOTA.

blind spot Region of vertebrate retina at which optic nerve leaves; devoid of rods and cones and hence 'blind'.

blood Major fluid transport medium of many animal groups, including nemerteans, annelids, arthropods, molluscs, brachiopods, phoronids and chordates. Derived from the BLASTOCOELE. Comprises an aqueous mixture of substances in solution (e.g. nutrients, wastes, hormones, gases and osmotically active compounds) in which are suspended cells (haemocytes) functioning either in defence (e.g. phagocytes) or oxygen transport (e.g. RED BLOOD CELLS). Blood is moved by muscle contraction in some of the vessels it passes through. Hearts are such specialized vessels, the hydrostatic pressure generated being employed in filtration processes (in capillaries generally, and kidney glomeruli in particular), in locomotion (e.g. many bivalve molluscs) and other activities besides solute translocation. Arthropod blood (haemolymph) is hardly confined to vessels (open circulation), insects and onychophorans having least vascularization, the haemocoele being much expanded. When blood is confined to vessels, the circulation is said to be closed, as it is throughout annelid and vertebrate bodies, although expanded blood sinuses are a feature of lower vertebrates. Respiratory pigment (absent in most insects) is in simple solution in invertebrate blood, but confined to red blood cells (erythrocytes) in vertebrates. See TISSUE FLUID.

blood–brain barrier Phrase indicating the relative resistance to diffusion of molecules across (i) the unfenestrated capillaries of the brain, whose cells have tight junctions, and (ii) the fatty glial cell (astrocyte) sheath surrounding the capillaries. The latter obstructs polar solutes more than it does such lipid-soluble solutes as PSYCHOACTIVE DRUGS.

blood clotting (b. coagulation) Adaptive response to haemorrhage involving local conversion of liquid blood to a gel, which plugs a wound. In vertebrates, blood does not clot in normal passage through vessels since the smooth endothelial lining prevents damage to platelets. The altered surface of damaged vessel walls plus turbulence of blood flow results in ADP release from platelets, and the exposed collagen in vessel walls causes adherence of platelets, which release Ca^{2+} and cause more aggregation. Vasoconstriction, stemming blood flow, results from SEROTONIN released when platelets fragment. A temporary plug of platelets is made permanent by a CASCADE of enzymic reactions caused by release of phospholipid and the protein thromboplastin (thrombokinase) by damaged cells and platelets. Ten clotting factors in addition to the above have been isolated; but the main sequence may be summarized as: tissue thromboplastin (Factor III) and plasma thromboplastin component (Factor IX) convert prothrombin (Factor II) to thrombin, which converts fibrinogen (Factor I) to an insoluble fibrin meshwork, trapping erythrocytes, platelets and plasma to form clot, stabilized by Factor XIII in presence of Ca^{2+} (Factor IV). Fibrin absorbs 90% of the thrombin formed and prevents the clot from spreading away from the damaged area. Several genetic disorders cause poor clotting. Factor VIII, one of the substances required for thromboplastin formation, is absent in classical X-linked haemophilia. See ANTICOAGULANT, FIBRINOLYSIS, THROMBOSIS, VITAMIN K, and Fig. 16.

blood group Either a group of people bearing the same antigen(s) on their red blood cells within a particular BLOOD GROUP SYSTEM, or the blood characteristic used to distinguish groups of individuals within a blood group system. Main clinical significance lies in blood transfusion and Rhesus incompatibility between mother and foetus. Without due matching of donor and recipient blood, death or severe illness of a recipient may result from a single transfusion (in ABO system) or after repeated transfusion (in Rhesus system). Causes of death include haemolysis (red cell rupture), agglutination (red cell clumping) with blockage of capillaries etc., and tissue damage. Danger in transfusion comes when donor antigens meet antibodies to them in recipient plasma and elsewhere resulting in INNATE IMMUNE RESPONSE. Antibodies to antigens of the ABO system occur naturally in plasma; those to Rhesus antigens occur in plasma only as a result of immunization, during pregnancy or transfusion. People with an antigen of the ABO system on their red cells automatically lack the antibody to it in their plasma,

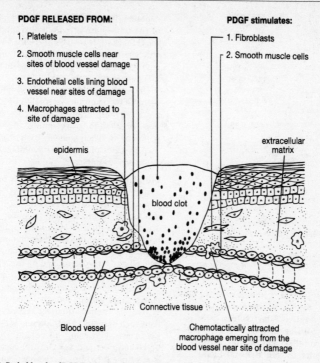

PDGF RELEASED FROM:

1. Platelets
2. Smooth muscle cells near sites of blood vessel damage
3. Endothelial cells lining blood vessel near sites of damage
4. Macrophages attracted to site of damage

PDGF stimulates:

1. Fibroblasts
2. Smooth muscle cells

epidermis

extracellular matrix

blood clot

Connective tissue

Blood vessel

Chemotactically attracted macrophage emerging from the blood vessel near site of damage

FIG. 16 *Probable role of PGDF (platelet-derived* GROWTH FACTOR) *in wound healing. Its release stimulates division of fibroblasts and smooth muscle cells, the former producing extracellular matrix material, while PGDF attracts macrophages.*

	Blood Group			
	A	B	AB	O
Antigen(s) on red cells	A	B	A & B	—
Antibody (ies) in plasma	anti-B	anti-A	—	anti-A anti-B

TABLE 3a *Relationship of antigen and antibody distributions in the ABO* BLOOD GROUP *system.*

		Donor Blood Group			
		A	B	AB	O
Recipient Blood Group	A	√	×	×	√
	B	×	√	×	√
	AB	√	√	√	√
	O	×	×	×	√

TABLE 3b *Success [√] or failure [×] of blood transfusions from specific donor blood groups to specific recipient blood groups.*

whereas lack of an ABO antigen on the red cells is coupled to presence of antibody to it in plasma (Table 3a).

Success or failure of transfusion is determined by the data outlined in Table 3b.

People with O-type are universal donors; those with AB-type blood are universal recipients. The Rhesus antibody may be

Blood Group [phenotype]	Possible Genotypes
A	$I^A I^A$ or $I^A I^O$
B	$I^B I^B$ or $I^B I^O$
AB	$I^A I^B$
O	$I^O I^O$

TABLE 4a *Genetics of ABO system.*

P_1 Phenotypes	A		×	O
	Case 1	Case 2		
P_1 genotypes	$I^A I^A$	$I^A I^O$	×	$I^O I^O$
Gametes	I^A	I^A & I^O	×	I^O
F_1 genotypes	$I^A I^O$	$I^A I : I^O I^O$		
F_1 phenotypes	all A	A : O		

TABLE 4b *Example showing expected offspring from marriage between group A and group O people.*

found in plasma of Rhesus negative (Rh−) women who have been pregnant with Rhesus positive (Rh+) babies, who have Rhesus antigen on their red cells. This is because mothers may become sensitized to the Rhesus antigen if foetal blood leaks across the placenta during birth of a Rh+ baby. Later Rh+ foetuses are at risk if anti-Rh antibody crosses the placenta from the mother, who had produced it as an immune response to the earlier leaked foetal blood. Foetal haemolysis can be averted by injecting an RH -mother with anti-Rh antibody (anti-D) prior to delivery of any Rh+ offspring; immediate blood transfusion of babies ('blue babies') born with haemolytic anaemia is an alternative. Blood groups are a classic instance of POLYMORPHISM in man. The genetics involved is often complex (there are at least three loci responsible for the Rh system), but the ABO system is a simple multiple allelism, as shown in Tables 4a and 4b, where blood groups A and B are codominant and O recessive. The antigens of the ABO system are found on, and appear to have evolved earlier than, cell types (mainly epithelial cells) other than red blood cells and should therefore be termed histo-blood group antigens. The I^O allele contains a frame-shift mutation causing loss of the transferase activity which modifies the cell surface H antigen into either A or B antigens. Adenocarcinomas of most organs of O-type individuals are less frequent than in A-type individuals.

The ratios of different blood groups within a blood group system differ geographically and indicate either NATURAL SELECTION or GENETIC DRIFT among different populations. Their genetic basis may be used in cases of contested parentage. See MENDELIAN HEREDITY.

blood group system A person's blood groups are genetically determined, each person belongs to several, and most are determined by loci situated on autosomes. The alleles at these loci determine antigens on a person's red blood cell membranes. A blood group system refers to the range of red cell antigens determined by the alleles at one such locus, or sometimes a group of closely linked loci. Some non-human primates share certain human blood group systems. In humans there are at least fourteen such systems (ABO, Rhesus, MNS, P, Kell, Lewis, Lutheran, Duffy, Kidd, Diego, Yt, I, Dombrock and Xg). Only the Xg locus is sex-linked. No linkage between any of these systems is apparent, other than between Lewis and Lutheran.

blood plasma Clear yellowish fluid of vertebrates, clotting as easily as whole blood and obtained from it by separating out suspended cells by centrifugation. An aqueous mixture of substances, including plasma proteins. See BLOOD.

blood platelets See PLATELETS, BLOOD CLOTTING.

blood pressure Usually refers to pressure in main arteries; in humans, normally between 120 mm Hg at SYSTOLE and 80 mm Hg at DIASTOLE (i.e. 120/80). Pressure drops most rapidly in arterioles, falling further in capillaries and more slowly in venules and veins. In the venae cavae it is 2 mm Hg, and 0 mm Hg in the right atrium. Homeostatically controlled (see BARORECEPTOR) at a high level in the mammalian systemic circulation because of the filtration needs of the kidney; low in pulmonary circuit because the extra tissue fluid produced would interfere with gaseous exchange (oedema). An important diagnostic indicator (see CORONARY HEART DISEASE). See NITRIC OXIDE for role in blood pressure stabilization.

blood serum Fluid expressed from clotted blood or from clotted blood plasma. When blood clots, platelets in the clot release the contents of their secretory vesicles, platelet-derived growth factor among them. This accounts for the ability of serum (*contra* plasma) to promote fibroblast, smooth muscle and neuroglial cell division in culture. Serum contains far lower levels of fibrinogen or other clotting proteins than does plasma. See MITOGEN.

blood sugar Glucose dissolved in blood. Homeostatically regulated in humans at about 90 mg glucose/100 cm^3 blood. Hormones affecting level include: INSULIN, GLUCAGON, ADRENALINE, GROWTH HORMONE and THYROXINE. For glycation effects of blood sugar on proteins, see AGEING.

blood system (blood vascular system) The system of blood vessels (in sequence: arteries, arterioles, metarterioles, capillaries, venules and veins) and/or spaces (often sinuses) through which blood flows in most animal bodies. See BLOOD for other details.

bloom A term used to describe the dense growth of planktonic algae which imparts a distinct colour to the water. Commonly, blue-green algae (Cyanobacteria) form such blooms in EUTROPHIC lakes. See EUTROPHICATION.

blue-green algae See CYANOBACTERIA.

BMPs (bone morphogenetic proteins) Serine proteases; dimeric proteins of the transforming growth factor-β superfamily (see TGFs). Binding to the extracellular matrix (e.g. heparin sulphate, heparin and type IV collagen), they are also ligands of cell membrane BMP receptors where they are pleiotropic regulators of a bone morphogenesis cascade. In normal limb development, BMPs are crucial to the formation of the cartilage that dictates where bones will form. Thus inhibition of BMP-7 by the Noggin protein antagonist causes joint fusions (dysplasias). Acting in a concentration-dependent manner, BMPs also specify cell death between digits, so producing finger shape (see GROWTH FACTORS), are involved in dorso-ventral AXIS formation and are chemotactic for human lymphocytes.

BOD See BIOCHEMICAL OXYGEN DEMAND.

body cavity (1) Primary body cavity; see BLASTOCOELE. (2) Secondary body cavity; see COELOM.

body-mass index (BMI) See OBESITY.

body plans At its simplest, a 'body plan' is what results after the main body axes are established in the early embryo (see AXIS). The 'EVO-DEVO' school of developmental biologists, fascinated by the implications of developmental genetics for evolutionary studies, would distinguish three historical ideas in the concept: (i) adult body plan, Richard Owen's 'archetype' or (Woodger, 1945) 'bauplan'– a homologous structural plan underlying evolutionary transformations within a taxonomic group; (ii) early developmental stages, or larval body plans, often dramatically distinct from the adult in form (see the discredited BIOGENETIC LAW); and (iii), which builds on (ii), the actual developmental processes which generate the adult structure. Thus, despite their varying conditions of growth, fishes and mammals are said to share the vertebrate PHYLOTYPIC STAGE (the pharyngula) – introducing the controversial idea that constraints of development might permit no other way for a vertebrate to be built.

The invention of DNA PROBES revealed the large numbers of developmental genes shared by widely different animal groups. In 1994, De Robertis and Sasai discovered in frogs a gene, *chordin*, that helps establish the dorso-ventral axis; and in the same year, F. and E. Bier discovered a gene, *sog*, with a similar DNA sequence, achieving a similar effect in the fruit fly *Drosophila*, but reversing the axis polarity: where *chordin* is active, frog cells become dorsal, whereas where *sog* is active, fly cells become ventral. This is evidence that the *chordin/sog* precursor helped establish a dorso-ventral axis in a common ancestor of insects and vertebrates and that changes in early development somehow 'flipped' the axis and inverted the expression pattern of the gene (see HOM-OLOGY, *Hox* GENES). The vertebrate/invertebrate 'inversion' theory was postulated by E. Geoffroy Saint-Hilaire in 1822 but was regarded as 'totally preposterous' at the time. See MORPHOGENESIS, WNT SIGNALLING.

body size Body and organ size are regulated by three processes: CELL GROWTH, CELL DIVISION and cell death. Extracellular signal molecules affecting the first two of these are MITOGENS, GROWTH FACTORS and SURVIVAL FAC-TORS, the last of which, in turn, suppress APOPTOSIS (aka *programmed cell death*), an essential process in organ development and size control. Body size has important implications for metabolic rate and ecological parameters (see ALLOMETRY), as well as for BIOGEOGRAPHY.

Bohr effect The effect of dissolved carbon dioxide on the oxygen equilibria curves or respiratory pigments (see HAEMOGLOBIN), whereby increased carbonic acid level (through raised pCO_2) shifts the curve to the right (lactic acid has same effect), decreasing the percentage O_2-saturation of the pigment for a given pO_2, increasing the rate of oxygen unloading in regions of high respiratory activity (the tissues) and loading of oxygen in regions of low pCO_2 (e.g. lungs, gills). Brought about by ALLOSTERIC effect of pH on haemoglobin molecule.

bone Vertebrate connective tissue laid down by specialized mesodermal cells (osteocytes) lying in LACUNAE within a calcified matrix which they secrete, containing about 65% by weight inorganic salts (mainly hydroxyapatite, providing hardness); remainder largely organic and comprising mostly COLLAGEN fibres providing tensile strength. *Compact bone* forms outer cylinder of shafts of long bones of limbs, and is typified by HAVERSIAN SYSTEMS; *spongy bone* forms vertebrae, most 'flat' bones (e.g. skull), and ends of long bones (epiphyses), typified by presence of TRABECULAE. Bone is a living tissue, supplied with nerves and blood vessels. Its constitution changes under hormone

influence (see CALCITONIN, PARATHYROID HOR-MONE). Beside its skeletal role in providing levers for movement and support for soft parts of the body, it protects many delicate tissues and organs. See OSSIFICATION, HAEMO-POIESIS, CARTILAGE, and Fig. 128.

bony fish See OSTEICHTHYES.

bootstrapping See PHYLOGENETIC TREES.

botany Branch of biological science concerned with plants. Since animal life depends for its existence on the process of PHOTOSYNTHESIS, its importance is self-evident. Its roots lie deep in antiquity, for our interest in plants as sources of food and medicine goes back to the very origins of human society; but the scientific study of plants (a requirement of which is the ability to distinguish and record plants in question) did not begin until the sixth century BC in Asia Minor and the Greek-speaking cities of Ionia. Modern botany commenced in the mid to late nineteenth century when some splendid textbooks were published and research schools became established at universities.

boundary elements (insulating elements) DNA sequences regulating the interaction between PROMOTER and ENHANCER elements.

bouton Knob-like terminal of nerve axon, containing synaptic vesicles. See SYNAPSE.

Bowman's capsule Cup-like receptacle composed of two epithelial layers, formed by invagination of a single layer, surrounding a glomerulus and forming part of a vertebrate KIDNEY nephron. The visceral layer, lying on top of the basement membrane of the glomerular capillaries, consists of podocytes specialized for filtering the blood. See Fig. 17.

brachial Relating to the arm; e.g. brachial plexus, the complex of interconnections of spinal nerves V–IX supplying the tetrapod forelimb.

brachiation The arm-over-arm locomotion adopted by many arboreal primates.

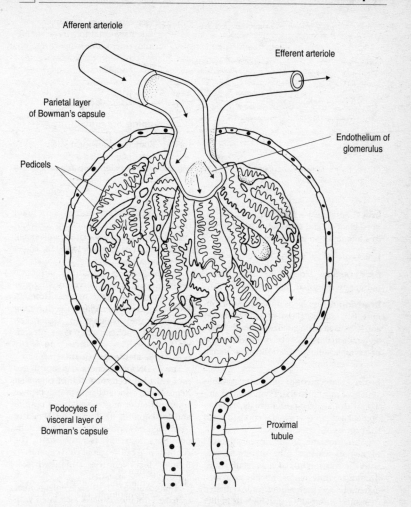

Afferent arteriole

Efferent arteriole

Parietal layer
of Bowman's capsule

Endothelium of
glomerulus

Pedicels

Podocytes of
visceral layer of
Bowman's capsule

Proximal
tubule

FIG. 17 *Renal corpuscle comprising* BOWMAN'S CAPSULE *and glomerular capillary knot. The inner (visceral) layer of the capsule comprises podocytes involved in filtration; the outer (parietal) layer forms the capsule itself.*

Brachiopoda Lamp shells. A small phylum of marine coelomate bivalve invertebrates, valves being dorsal and ventral (see BIVALVIA); with teeth and sockets along the hinge in articulate, but absent from inarticulate forms. Ciliated LOPHOPHORE serves for feeding and gaseous exchange. Of enormous value in dating rock strata, appearing in Lower Cambrian, with major extinctions in Devonian and Permian and expansions in the Ordovician, Carboniferous and Jurassic. One of the few surviving genera, *Lingula* (inarticulate) is almost unchanged since the Ordovician.

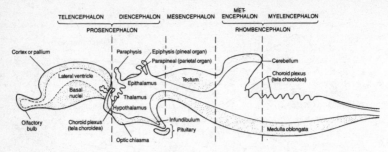

FIG. 18a *Diagram showing median section of typical embryonic vertebrate brain, illustrating the principal divisions and structures.*

bract Small leaf with relatively undeveloped blade in axil of which arises a flower or a branch or an inflorescence.

bracteole Small BRACT.

bradycardia Slowing of heart (hence pulse) rate. Compare TACHYCARDIA.

bradykinin Nonapeptide hormone of submaxillary salivary gland, inducing dilation and increased permeability of blood vessels and smooth muscle contraction; released by parasympathetic stimulation; implicated in allergic reactions.

brain Enlargement of the central nervous system of most bilaterally symmetrical animals with an antero-posterior axis. Its development anteriorly is a major component of the process of CEPHALIZATION.

(1) In vertebrates, an anterior enlargement of the hollow neural tube, sharing features with the spinal cord, such as relative positioning of white matter (myelinated axons) externally and grey matter (unmyelinated neurones) internally – although in higher forms there is migration of grey matter cell bodies above the white matter to form a third layer, reaching its zenith in the human cerebrum. Three dilations of the neural tube give rise to forebrain, midbrain and hindbrain.

The *forebrain* comprises the diencephalon anteriorly and the telencephalon posteriorly. Primitively, this is olfactory but in higher vertebrates the telencephalon roof is expanded to form the paired cerebral hemispheres and dominates the rest of the brain in its sensory and motor function. The telencephalon retains its association with smell, forming a pair of olfactory lobes. See Fig. 18 and LIMBIC SYSTEM.

The *midbrain* is primitively an optic centre, the pair of optic nerves entering after DECUSSATION at the optic chiasma. Although terminations of these fibres enter the visual cortex of the cerebrum in higher vertebrates, the midbrain still retains integrative functions (SEE BRAINSTEM) and is the origin of some CRANIAL NERVES.

The *hindbrain* generally has its anterior roof enlarged to form a pair of cerebellar hemispheres associated with proprioceptive coordination of muscle activity in posture and locomotion. Its floor is thickened to form the pons anteriorly and the medulla oblongata posteriorly (SEE BRAINSTEM). The central canal of the spinal cord expands to form the brain ventricles, and the whole is surrounded by MENINGES.

(2) In invertebrates, there is enlargement of paired anterior GANGLIA associated with cephalization, those above the gut often fusing to form complex integrative centres; but in segmented forms, segmental ganglia and local reflexes are just as important in nervous integration. In molluscs 'brain functions' may be divided between several pairs of distinct ganglia or, as in cephalopods (octopus, squid, etc.), as many as 30 integrated brain lobes, enabling a complexity of behaviour on a par with the lower vertebrates.

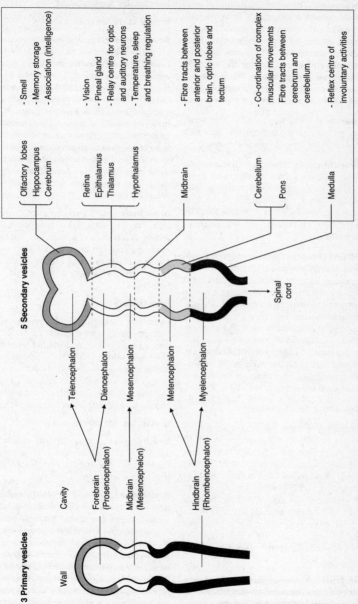

3 Primary vesicles

Wall

Cavity

Forebrain (Prosencephalon) → Telencephalon

Diencephalon

Midbrain (Mesencephalon) → Mesencephalon

Hindbrain (Rhombencephalon) → Metencephalon

Myelencephalon

5 Secondary vesicles

Olfactory lobes
Hippocampus
Cerebrum
- Smell
- Memory storage
- Association (intelligence)

Retina
Epithalamus
Thalamus
Hypothalamus
- Vision
- Pineal gland
- Relay centre for optic and auditory neurons
- Temperature, sleep and breathing regulation

Midbrain
- Fibre tracts between anterior and posterior brain, optic lobes and tectum

Cerebellum
Pons
- Co-ordination of complex muscular movements
Fibre tracts between cerebrum and cerebellum

Medulla
- Reflex centre of involuntary activities

Spinal cord

FIG. 18b *Early human* BRAIN *development. Three primary vesicles become subdivided to form the adult derivatives, shown on the right with some of their functions.*

brainstem The vertebrate midbrain, pons and medulla oblongata. It links the cerebrum to the pons and spinal cord, has reflex centres for control of eyeballs, head and trunk in response to auditory and other stimuli, and houses part of the RETICULAR FORMATION. The respiratory, VASOMOTOR and CARDIAC CENTRES are here, associated with appropriate cranial nerves. See VENTILATION.

branchial Relating to gills. The aortic arches serving the gills in fish (the third arch onwards) are termed *branchial arches*.

Branchiopoda CRUSTACEA with at least four fairly uniform pairs of phyllopodial trunk limbs (see BIRAMOUS APPENDAGE) used in gaseous exchange, and a variable but usually large number of somites. Primitive. Includes water fleas (e.g. *Daphnia*) and brine shrimps (e.g. *Artemia*).

BRCA1 (Brca1) A tumour suppressor protein strongly associated when mutated with breast and ovarian cancer and implicated in a range of nuclear roles including genomic integrity, transcription regulation, chromatin remodelling and cell cycle checkpoint control. It co-localizes with, and seems to direct localization of, Xist RNA (see DOSAGE COMPENSATION) and loss of BRCA1 results in loss of Xist from the inactive X chromosome, leading to re-activation of at least one gene on the inactive X. See *recA*.

breathing See VENTILATION.

breeding system All factors, apart from mutation, affecting the degree to which gametes which fuse at fertilization are genetically alike. Includes population size, and levels of INBREEDING and OUTBREEDING controlled by variable selfing, incompatibility mechanisms, assortative mating, heterostyly, dichogamy, arrangement and distribution of sex organs (e.g. size of plant and inflorescence number are factors), mechanism of sex-determination, etc. See GENETIC VARIATION, RARITY.

brefeldin A An antibiotic from fungi, blocking transport of proteins from the endoplasmic reticulum to the GOLGI APPARATUS; can cause the latter's disintegration.

bristletails See APTERYGOTA.

brittle stars See ECHINODERMATA.

Broca's area A language-associated centre, visible as a bump on the left side of the frontal lobe of the brain in modern humans; apparently also present in all members of the genus *Homo* (from endocranial casts), but absent from *Australopithecus*. Receiving auditory input from WERNICKE'S AREA, it is involved in the motor output of communication and stringing of sounds into words. Injury causes slow, laboured and simplified speech. Visual information about a written word goes from the association cortex straight to Broca's area without entering Wernicke's area.

bronchiole Small air-conducting tube (less than 1 mm in diameter) of tetrapod lung, arising as a branch of a bronchus and terminating in alveoli. Smooth muscle abundant in walls, controlling lumen diameter. Lacks cartilage and mucous glands of bronchi.

bronchitis See LUNG.

bronchus Large air tube of tetrapod lung. One per lung, connecting it to trachea; divided into smaller and smaller bronchi, and finally into BRONCHIOLES. Each bronchus has cartilaginous plates and smooth muscle in its walls, and mucus from glands traps bacteria and dust, the whole being beaten by cilia up to the pharynx for swallowing.

brown fat See ADIPOSE TISSUE.

brush border Animal cell surface (often apical), often a whole epithelial surface, covered in MICROVILLI and serving for absorption and/or DIGESTION.

Bryophyta Division of the plant kingdom with over 15,000 species, making it the second largest green land plant division. It comprises the HEPATICOPSIDA (liverworts), BRYOPSIDA (mosses) and ANTHOCEROTOPSIDA (hornworts). However, some scientists would argue that this division is polyphyletic and these three groups should be recognized as separate divisions. Habitats various, e.g. wet banks, on soil, rock surfaces; some epiphytic, others aquatic. Small plants, flat, prostrate or with a central stem up to 30 cm in length, bearing leaves. Vascular tissue may be poorly developed; attached to substratum by RHIZOIDS. Reproducing sexually by fusion of macro- and microgametes produced in multicellular sex organs; antheridia liberate microgametes, motile by flagella; archegonia contain a single macrogamete (egg cell). Sexual reproduction is followed by development of capsule containing spores giving rise to PROTONEMATA to new plants. A well-marked ALTERNATION OF GENERATIONS: the leafy vegetative plant is the gametophyte generation; the capsule (and seta) comprise the sporophyte generation, partially dependent nutritionally upon the gametophyte. Origins linked to CHAROPHYCEAE.

Bryopsida (Musci) Mosses. Class of BRYOPHYTA, having fairly conspicuous PROTONEMATA and multicellular RHIZOIDS. The sporophyte develops from an apical cell, and the capsule has a complex opening mechanism. ELATERS are absent. Cosmopolitan, occurring in damp habitats (e.g. moist humus, peat, wet cliff faces, dry boulders, etc.) and as epiphytes on branches and tree trunks. Sex organs (antheridia and archegonia) borne either terminally or in lateral bud-like structures (perigonia and perichaetia), either on separate or the same plants. Fertilization is followed by development of the sporophyte, the capsule elevated on a seta and covered when young by a gametophytic CALYPTRA. Spores are dispersed by a unique structure termed the peristome, and germinate to produce a filamentous or thalloid protonema, giving rise to new moss plants from lateral buds.

Bryopsidophyceae Class of CHLOROPHYTA. Almost entirely marine, most genera tropical, or extend from the tropics into warm temperate waters. There are about 25 genera and 150 species. Members show only one level of thallus organization, and this is siphonaceous. Each individual comprises essentially a single cell, which contains a vast central vacuole with a thin layer of cytoplasm around it. The cell is coenocytic. Parts are sometimes sealed off by plugs of cell wall material (e.g., during formation of

gametangia, as in *Codium*, or sporangia as in *Derbesia*). In many bryopsidophycean algae the multinucleate siphons combine, forming quite complex tissues. The principal polysaccharides found in the fibrillar, structural fraction of the cell wall are mannan, xylan and glucan ('cellulose'). These occur in varying proportions.

Chloroplasts are numerous, fusiform or ellipsoidal. Members of this class can be divided into two groups according to the kinds of plastids they contain: *heteroplastic* members contain amyloplasts as well as chloroplasts, while *homoplastic* members have only chloroplasts. Siphonoxanthin and siphonein are the accessory pigments in the chloroplasts.

Zoospores have two or four flagella, and the flagella apparatus has a CRUCIATE 11 O'CLOCK–5 O'CLOCK configuration, the basal bodies exhibiting a marked overlap. Some members produce stephanokont zoospores (many flagella arranged in a whorl around the anterior end of the cell). Mitosis is closed, centric or acentric (i.e. with or without centrioles at the poles) and there is a prominent persistent telophase spindle, which gives the telophase nucleus a dumbbell shape. Sexual reproduction seems usually to have a haplontic life cycle, in which a multinucleate, haploid gametophyte alternates with a microscopic zygote stage, containing a giant diploid nucleus. The zygote may germinate releasing stephanokont zoospores. These then develop into gametophytes or it may itself bud off gametophytes directly. Sexual fusion is anisogamous.

Bryozoa See ECTOPROCTA.

***Bt* maize** A strain of maize developed to express an insect toxin gene from the bacterium *Bacillus thuringiensis*. See GMO, MAIZE.

buccal cavity The mouth cavity, lined by ectoderm of stomodaeum, leading into the PHARYNX.

bud (Bot.) (1) Compact embryonic plant shoot comprising a short stem bearing crowded overlapping immature leaves. (2) Vegetative outgrowth of yeasts and some bacteria serving for vegetative reproduction. See BUDDING. (Zool.) Outgrowth of organism that may detach, as in BUDDING; or a morphogenic feature of a growing region, as in vertebrate limb buds (see REGENERATION).

budding (Bot.) (1) Grafting in which grafted part is a bud. (2) Asexual reproduction in which a new cell is formed as an outgrowth of a parent cell, e.g. in yeast. Compare BINARY FISSION. (Zool.) ASEXUAL method of reproduction common in many invertebrates such as sponges, coelenterates (e.g. *Hydra*), ectoproctans and urochordates, but also a feature of the HYDATID CYSTS of tapeworms. Rarely known as gemmation. See POLYEMBRYONY.

buffer A solution resisting pH change on addition of acid (i.e. H^+) or alkali (OH^-), absorbing protons from acids and releasing them on addition of alkali. Usually consists of a mixture of a weak acid and its conjugate base, or vice versa. Intracellular and extracellular buffers may differ; thus the commonest intracellular buffer is the acid-base pair $H_2PO_4^- - HPO_4^{2-}$ and such organic phosphates as ATP, but the bicarbonate buffer system ($H_2CO_3 - HCO_3^-$) is common extracellularly, as in vertebrate blood plasma, where plasma proteins are also a major buffer system. HAEMOGLOBIN acts as a buffer during the CHLORIDE SHIFT. If pH varies, PROTEIN shape and function may be affected. See KIDNEY.

Buffon, G. L. L. de French naturalist (1707–1788), with influence comparable with LINNAEUS on contemporaries. His great work is *Histoire Naturelle* (1749–1788), a natural history in 44 volumes. Espoused NOMINALISM with regard to species and other taxa, but had noted that species seemed to 'breed true'. He held that the environment had important effects on animal characteristics, amounting in time to a sort of degeneration from original types rather than being in any way creative. He favoured the theory of the GREAT CHAIN OF BEING, indicating his distance from the later DARWINISM.

bug See HEMIPTERA.

bulb Organ of vegetative reproduction; modified shoot consisting of very much shortened stem enclosed by fleshy scale-like leaves (e.g. tulip) or leaf-bases (e.g. onion). See BULBIL. Compare CORM.

bulbil Dwarf shoot occurring in place of a flower (e.g. *Saxifraga, Festuca, Allium*), borne either in lower part of the inflorescence or in axils of leaves (lesser celandine) and serving as an organ of vegetative reproduction. See APOMIXIS.

bulk flow Overall movement of water or some other liquid induced by gravity or pressure or the interplay of both.

bulla, auditory Bony projection of middle ear cavity in most placental, but not most marsupial, mammals. Absent also in earliest mammals.

bundle scar The scar from a vascular bundle remaining on a leaf scar after ABSCISSION.

bundle sheath Layer of cells which surrounds a vascular bundle, comprising cells of PARENCHYMA, SCLERENCHYMA, or both. See KRANZ ANATOMY.

Burgess Shale Early Palaeozoic (Middle CAMBRIAN, *c*. 530 Myr BP) sedimentary strata discovered in 1909 by Charles Walcott, just north of Fields, British Columbia, in the Canadian Rocky Mountains. Characterized by presence of exquisitely preserved invertebrate fossils whose decay was somehow prevented, revealing soft parts. Previously unobserved specimens of problematic phylogeny include the lobopod *Hallucigenia sparsa* and *Anomalocaris* (now generally regarded as arthropods). The two other most important soft-bodied faunas of Cambrian age are the Lower Cambrian Sirius Passet fauna of North Greenland (with the Burgess Shale, once a part of the ancient continent Laurentia) and Lower Cambrian Chengjiang fauna of South China which, although on a different tectonic plate from the other two and probably then separated by tens of thousands of km, shows faunal similarities to the Burgess Shale, including a lobopod *Microdictyon*. See CHENGJIANG FAUNA.

bursa of Fabricius Thymus-like LYMPHOID organ found in birds, but not in mammals, developing dorsally from CLOACA. Like the mammalian THYMUS, it is a specialized site for B CELL (i.e. bursa-derived cell) maturation. Contains ANTIGEN-PRESENTING CELLS. See T CELL.

buttress root An ADVENTITIOUS root on a stem, functioning in support; found mainly in monocotyledonous plants.

C

c- Genes which are cellular counterparts of ONCOGENES are prefixed by c-, as in c-*fos* – the cellular counterpart of the oncogene v-*fos*. Such genes are also termed c-oncs (cellular oncogenes).

cadherins Cell-type-specific adhesion glycoproteins, attached to the cytoskeleton by β-CATENIN and occurring transiently or permanently on the cell surfaces of probably all vertebrate cells at some stage in their differentiation. Transmembrane molecules, most have five folded extracellular domains, four of them homologous and calcium-binding. Because they link up in homophilic bonds when Ca^{2+}-activated, cadherins hold cells of the same type together to form tissues. Switches in the type of cadherin expressed on a cell surface can affect cell adhesion and separation during morphogenesis. See ADHESION, INTERCELLULAR JUNCTIONS.

caducous (Bot.) Not persistent. Of sepals, falling off as flower opens (e.g. poppy); of stipules, falling off as leaves unfold (e.g. lime).

caecum Blind-ending diverticulum, commonly of gut. One or two may be present at junction of vertebrate ileum and colon, housing cellulose-digesting bacteria. Thin-walled, sometimes with spiral valve for increased surface area, terminating in vermiform APPENDIX. Mesenteric (midgut) caeca in some annelids (e.g. leeches) and many arthropods are secretory and absorptive, and may be generally 'liver-like' en masse.

Caenorhabditis elegans A small (1 mm) transparent, hermaphroditic, bacteria-feeding soil nematode of relatively small genome (about 3,000 genes) and few cell types, whose 16-hour embryogenesis can be achieved in a petri dish, making it highly suitable for the study of developmental and behavioural genetics. Normally dying after 14 days, they have become classic organisms for studying the genetics of AGEING and longevity (see APOPTOSIS). One such gene, *daf-16*, which encodes the transcription factor DAF-16, is a powerful regulator of its lifespan and is activated by a hormonal signalling pathway similar to that activated by the mammalian insulin and insulin-like growth factor proteins (IGF-1; see INSULIN/INSULIN-LIKE GROWTH FACTOR PATHWAY). Microarray techniques permit discovery of the activities of most of the genes in the animal's sequenced genome and comparison of their expression patterns in normal worms and those in which a gene is mutated. In 1998 it became the first multicellular organism to have its genome sequenced (97 megabases; > 19,000 genes). See SEX DETERMINATION.

caffeine A powerful ALKALOID stimulant of the central nervous system inducing increased mental alertness, clearer flow of thought, wakefulness and restlessness; in consequence, a psychoactive drug. Present in coffee, tea, cocoa, cola drinks and chocolate. An average cup of coffee contains 100 mg caffeine; a cup of tea contains about 50 mg. About 80% of adult Americans consume 3–5 cups of coffee per day on average. It owes its psychostimulant effects to blockage of adenosine (A_{2A}) receptors, increasing the state of phosphorylation of a dopamine- and cyclic AMP-regulated phosphoprotein (DARPP-32) and not by stimulation of cyclin-dependent kinase 5 (Cdk5)-catalysed phosphorylation. These receptors are located in the CNS and peripheral nervous system, and as a result caffeine prevents the sedating effect of adenosine but also reduces asthma and migraine. It is only a slight heart stimulant. High doses cause *caffeinism*: nervousness, irritability, muscle hyperactivity and twitching, raised body temperature and ventilation, heart palpitations and arrhythmias. Persons prone to panic attacks may be particularly sensitive to even moderate doses (4–5 cups of coffee). Total caffeine intake does not seem to increase risk of coronary artery disease or increase cancer; however, it appears to induce physiological dependence, with withdrawal symptoms including headache, irritability, fatigue, muscle pain and stiffness. It does not appear to be a MUTAGEN, but should be regarded as possibly hazardous to the human foetus and neonate since it crosses the placenta and is concentrated in breast milk. See NICOTINE.

CAK (cdk-activating kinase) See CDKs.

calciferol See VITAMIN D.

calcitonin (CT) Polypeptide hormone of parafollicular cells (*C-cells*) of thyroid gland. Lowers plasma calcium and phosphate levels by inhibiting bone degradation and stimulating their uptake by bone. May have evolved alongside conquest of land by vertebrates, given its role in regulating plasma ion levels. Antagonized by PARATHYROID HORMONE. See OSSIFICATION.

calcium pump An ATP-driven TRANSPORT PROTEIN. Calcium ions (Ca^{2+}) act as SECOND MESSENGERS in the cell cytosol and their changing concentration there, particularly in eukaryotes, is significant. Although total cell Ca^{2+} concentration approximates to that of the environment, it is unevenly distributed and Ca^{2+} pumps in the plasma membrane expel Ca^{2+} when it enters. Much is accumulated by pumps in MITOCHONDRIA and other organelles, causing a thousandfold drop in Ca^{2+} concentration across the plasma membrane, a gradient down which the ion moves. The calcium pump in the sarcoplasmic reticulum of striated muscle accumulates Ca^{2+} from the cytosol,

enabling MUSCLE CONTRACTION. In plants, calcium pumps exist at the plasmalemma (see GUARD CELLS), tonoplast and endoplasmic reticulum, keeping Ca^{2+} levels low until an appropriate signal. Cadmium ions block voltage-gated calcium pumps. See CALCIUM SIGNALLING, CALMODULINS and Fig. 91.

calcium signalling Calcium ions (Ca^{2+}) can enter the cell through voltage-gated or receptor-gated calcium channels in the cell membrane, be sequestered by CALCIUM PUMPS and stored mostly in the endoplasmic or sarcoplasmic reticulum (lesser amounts in mitochondria) and released into the cytosol when a calcium channel opens. These calcium releases, involving either INOSITOL 1,4,5-TRIPHOSPHATE receptors (see Fig. 91) or ryanodine receptors, serve as signals (see ADENYLYL CYCLASE, SECOND MESSENGER) which either activate very localized cellular responses adjacent to the channel (e.g. opening K^+ channels in the smooth muscle cell membrane causing relaxation), or set in train more global responses when channels communicate with each other (e.g. smooth muscle contraction deeper in the cell in response to a large-scale Ca^{2+} wave). Many signal transduction pathways require changes in the signal molecule (or ion) to be very precisely localized. Although Ca^{2+} is very mobile in water, because of the large number of high-affinity binding sites for it on proteins it diffuses far more slowly than the very mobile messenger cAMP. Very localized Ca^{2+} signals are thus made possible, still more so if the distribution of calcium channels in a membrane (e.g. endoplasmic reticulum) is non-random and some distance from the site of initial receptor interaction. In these ways, Ca^{2+} can be used as a messenger to control many different pathways simultaneously in the same cell.

The versatile and multi-subunit calcium/calmodulin-dependent PROTEIN KINASE CaM-kinase II switches to an active state when it encounters Ca^{2+}/calmodulin and remains active in phosphorylating target proteins even when the Ca^{2+} signal has passed. This 'memory mechanism' allows it to 'decode' intracellular Ca^{2+} oscillations since its catalytic activity increases as a function of Ca^{2+} pulse frequency. The resulting frequency-dependent (sometimes non-oscillatory) cellular responses are a form of CELL MEMORY. Transcription of different sets of genes in the same cell can be activated by different frequencies of Ca^{2+} wave (see CALMODULINS). ACTIVATION of the mammalian egg at fertilization involves a Ca^{2+} oscillation produced by repetitive Ca^{2+} pulses lasting several hours and triggers development by activating the enzymes controlling the cell division cycle (see *CDC* GENES, CELL CYCLE). Calcium is pivotal in the function of neurons, not only in release of neurotransmitters (e.g. ACETYLCHOLINE) but probably in the synaptic plasticity leading to short-term memory. Involved in ATP breakdown, and cAMP production, through calmodulins. Abnormally high Ca^{2+} levels in mitochondria triggers APOPTOSIS. See IMMUNOPHILINS.

callose A complex branched polysaccharide associated with the SIEVE AREAS of sieve elements. May form in reaction to injury of these and parenchyma cells and be deposited so that their activity is impaired or finished, permanently or seasonally.

callus (Bot.) Superficial tissue developing in woody plants, usually through cambial activity, in response to wounding, protecting the injured surface. Often used in tissue culture, when the effects of hormones upon cell differentiation can be studied. (Zool.) Fibrocartilage produced at bone fracture, developing into bone as blood vessels grow into it and pressure and tension are applied. Also applied to the thickened stratum corneum of mammalian epidermis, forming where there is regular abrasion.

calmodulins Small multiply-allosteric proteins required for Ca^{2+}-dependent activities of many cellular (esp. membrane-bound) enzymes. Said to be *activated* when bound to Ca^{2+}. Ubiquitous cellular component related to tropinin C (see MUSCLE CONTRACTION) which, once activated can in turn bind to several cell proteins (e.g. adenylate cyclase, some ATPases and membrane pumps) and regulate their activities and is a component of muscle PHOSPHORYLASE KINASE, accounting for its Ca^{2+}-dependence.

Calmodulin activation is facilitated by Ca^{2+} concentration, a 10-fold increase in Ca^{2+} causing a 50-fold increase in activation. Many effects of Ca^{2+} flux in animal cells are mediated by Ca^{2+}/calmodulin-dependent kinases (see PROTEIN KINASES), one multifunctional form being the autophosphorylating CaM-kinase II which comprises 2% of the cell's protein mass in some brain regions and when activated by Ca^{2+}/calmodulin becomes very concentrated at SYNAPSES where its ability to act as a frequency decoder of Ca^{2+} oscillations is thought to play a role in SYNAPTIC PLASTICITY (see CALCIUM SIGNALLING). Present in all plant species so far tested and implicated in geotropic response of roots. Its presence in plant cell walls probably affects the level there of free Ca^{2+}, which in turn seems to influence both growth and development (see INOSITOL 1,4,5-TRIPHOSPHATE). Calmodulin gene expression in tobacco seedlings is greatly increased by touch. See CALCIUM SIGNALLING, SIGNAL TRANSDUCTION.

Calvin cycle Series of enzymic photosynthetic reactions in which carbon dioxide is reduced to 3-phosphoglyceraldehyde, while the carbon dioxide acceptor ribulose-1,5, bisphosphate is regenerated. For every six molecules of carbon dioxide that enter the cycle, a net gain of two molecules of glyceraldehyde-3, phosphate results. Employed by phototrophic purple bacteria and most other phototrophs; but not by green sulphur bacteria, which fix carbon by a reversal of steps in the Krebs cycle (the 'reverse', or 'reductive', citric acid cycle). See PHOTOSYNTHESIS, Fig. 19.

calyptra Hood-like covering of moss and liverwort capsules, developing from the archegonial wall.

calyptrogen Layer of actively dividing cells formed over apex of growing part of roots in many plants, giving rise to ROOT CAP.

calyx Outermost part of a flower, consisting usually of green, leaf-like members (sepals)

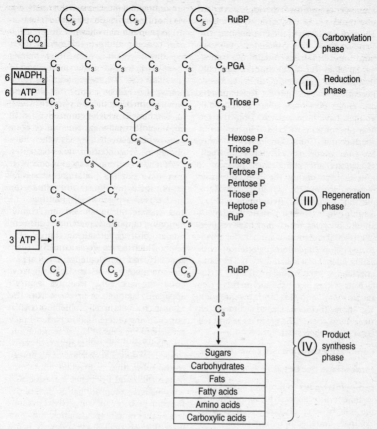

FIG. 19 *Three pathways in the* CALVIN CYCLE *by which fixed* CO_2 *(bound to* C_5 *ribulose bisphosphate) enters. PGA = phosphoglycerate; Ru = ribulose. One* C_3 *compound is generated for three* CO_2 *molecules which enter; three ribulose bisphosphate molecules being likewise regenerated for carboxylation.*

that in the bud stage enclose and protect other flower parts. See FLOWER.

CAM (1) See CRASSULACEAN ACID METABOLISM. (2) Cell–cell adhesion molecule. See AD-HESION.

cambium (Bot.) A meristem giving rise to parallel rows of cells. See VASCULAR CAMBIUM, CORK CAMBIUM.

Cambrian The earliest period of the PALAEO-ZOIC era (see Appendix), extending from 543–

505 Myr BP, the start of the Phanerozoic age ('evident animals') marked by the appearance of abundant skeletalized fossils. The dearth of animal fossils below the Precambrian–Cambrian boundary, and the 'sudden' appearance over the next few tens of millions of years of traces of almost all the major animal phyla, has led to the postulation of a 'Cambrian explosion', the main divergence of animal phyla apparently occurring no more than 565 Myr BP. Initial DNA sequencing of living forms indicated

a divergence time twice this long; but a date of around 670 Myr is currently in vogue. The topmost levels of the Precambrian (the Vendian, 680–550 Myr) include the Ediacara sandstones and shales containing a controversial biota (the Vendobionta), generally thought to lack clear animal fossils. Recently, however, minute fossil sponges and embryos of more advanced, bilateral, animals have been reported from the 570 Myr phosphorite rocks of Doushantuo in south-central China. Geological evidence for a succession of near global 'snowball' glaciations alternating with periods of intense heat during the Neoproterozoic between 750 and 590 Myr BP indicates the likelihood of intense 'environmental filters' resulting in a series of genetic 'bottleneck and flush' cycles, which may have resulted in an initial radiation of animals (i.e. metazoans) before the terminal glaciation, with a subsequent Ediacaran radiation. The Cambrian fauna itself includes trilobites, crustaceans, king crabs, eurypterids (see ARTHROPODA), annelids and brachiopods. The flora included bacteria, blue-green algae, large coenocytic green algae and red algae. See BURGESS SHALE.

Cambrian explosion See CAMBRIAN.

camouflage See CRYPSIS.

campylotropous (Of ovule) curved over so that funicle appears to be attached to the side, between chalaza and micropyle. Compare ANATROPOUS, ORTHOPTEROUS.

cancer, cancer cells Cells which have escaped normal controls regulating growth and division, producing clones of dividing daughter cells which invade adjacent tissues and may interfere with their activities. Some cancer cells express self-antigens normally only expressed early in development, and so can be detected by T CELLS. Despite a normal oxygen uptake, cancer cells tend to use several times the normal glucose requirement. In vertebrates they produce lactic acid under aerobic conditions (termed *aerobic glycolysis*). This places a burden on the liver, which must use ATP to get rid of lactate. Cancer cells that proliferate but stay together form benign tumours; those that

not only proliferate but also shed cells, e.g. via the blood or lymphatic system (metastasis) to form colonies elsewhere form malignant tumours, and cancer generally refers to a disease resulting from either. Among these, *carcinomas* are malignant tumours of epithelial cells. *Teratocarcinomas* are carcinomas that can be cultured *in vitro* and serially grafted to other hosts (see TERATOMA); *sarcomas* are cancers of connective tissue; *myelomas* are malignant tumours of bone marrow. Cancer cells are cells which have undergone neoplastic TRANSFORMATION. Cells tend to become cancerous only after they have suffered mutations in several genes, including those controlling the mechanisms whereby cells multiply, die and migrate. Although different combinations of mutations are found in different human patients, producing cancers that respond differently to treatment, there appears to be a step-by-step mutational progression involved, with mutations inactivating the *APC* gene (see APC PROTEIN) apparently among the first to occur. The resulting increase in cell proliferation (without changing differentiation pattern) may predispose the cells to *Ras* gene mutation, associated in culture with aspects of transformed phenotype such as loss of anchorage to the substratum. Subsequent mutations in the *DCC* and *p53* genes are associated with malignant transformation, loss of p53 function probably allowing cells to accumulate further mutations since they now progress through the cell cycle despite being unfit to do so (see CHECKPOINTS). Many such cells therefore undergo complex recombinations, often involving three or more chromosomes. Cancer cells appear more susceptible to effects of PROTEASOME inhibitors than do normal cells; some cancers result from insufficient apoptotic turnover rather than from unrestrained cell proliferation. Cancer cells often display gross changes in the organization of the nuclear matrix likely to affect DNA metabolism and subnuclear organization. Tumorigenesis is a complex and step-by-step process producing a pathologic 'organ' in which host and cancer cells cooperate in the production of an entity which grows and then may invade locally and then spread (metastasize) and ultimately kill its host. This interaction leads to growth of new blood vessels, evasion of host immune recognition and alteration of the microenvironment so that metastases form despite unfavourable growth conditions. Invasion from an epithelium into the surrounding connective tissue involves cross-talk between cancer cells and local fibroblasts and endothelial cells (see EXTRACELLULAR MATRIX). The gene *p53* must be inactivated or bypassed in order for tumour progression to occur. Cancer invasion may be a deregulated form of the normal physiological invasion process required for neuronal growth and wiring in the embryo, tissue remodelling, ANGIOGENESIS and wound healing. The ability of tumour cells to metastasize may be the main reason that certain cancer cell types are often fatal. HYPOXIA caused by highly aggressive tumours is a factor promoting metastasis. Activation of the gene *CXCR4* by oxygen deprivation causes transformed cell migration and targeting to a specific organ set, while increased synthesis of HIF (hypoxia-inducible factor) activates several genes – including that encoding vascular endothelial growth factor (VEGF) – causing angiogenesis and improving oxygen supply. The angiogenesis associated with a tumour often provokes pro-inflammatory signals which can convert immune tolerance to activation by induction of dendritic cells, which can ingest tumour antigens and carry them to lymph nodes. Alternatively, tumour cells may migrate directly to lymph nodes and activate T cells by this route. Immune tolerance of a tumour may arise through 'ignorance' of its foreign nature because it does not generate a new antigen during its pro-inflammatory phase, or by down regulating the MHC or antigen-processing machinery, T cell anergy or deletion. Several types of DNA viruses can initiate tumours in animals, each encoding one or more ONCOGENES whose products promote growth and division of the host cells, some by binding to (sometimes promoting degradation of) negative cell growth regulators such as the proteins Rb (retinoblastoma gene product) and tumour necrosis

factor p53. Genetic alterations affecting the p16 gene product and cyclin D1, both of which govern phosphorylation of the retinoblastoma protein, and control exit from G1 phase of the cell cycle, are very common in human cancers and may be necessary for tumour development. See CYCLINS, *RAS* GENES.

canine tooth Dog- or eye-tooth of mammals; usually conical and pointed, one on each side of upper and lower jaws between incisors and premolars. Missing or reduced in many rodents and ungulates, they are used for puncturing flesh, threat, etc. Sometimes enlarged at tusks (e.g. in wild boar).

CAP (cyclic AMP receptor protein) A prokaryotic transcription activator which binds as a homodimer to a specific DNA sequence in many bacterial promoters upstream of the region recognized by σ factor. Binds to the major groove and imparts a kink, inducing an overall bend of about 90° within two helical turns of DNA. The double helix wraps round the sides of the dimer, creating a stable DNA-protein complex for RNA polymerase holoenzyme to bind.

capacitation Final stage in maturation of, at least, bird and mammal sperm, without which fertilization is impossible. Generally occurs in female tract (sometimes *in vitro*) where substances, perhaps secreted by the ovary or by the uterine lining, must be encountered for the sperm to undergo the acrosome reaction (see ACROSOME). Sperm do 'wait' at specific points on the uterus wall, and may be capacitated then.

capillary (1) (Of blood system) an endothelial tube, one cell thick and 5–20 μm in internal diameter, on a BASAL LAMINA, and linking a narrow metarteriole to a venule. Permits exchange of water and solutes between blood plasma and tissue fluid (hence called exchange vessels). Their walls lack smooth muscle and connective tissue, and their permeability depends on the junctions between the endothelial cells. Three main types: (i) *continuous capillaries* (e.g. in muscle), where just one endothelial cell with overlapping ends forms the whole tightly sealed structure; (ii) *fenestrated capillaries* (as in intestine, endocrine glands), where pores through the cell are closed by just a cell membrane diaphragm, offering little resistance to small solute molecules; or (e.g. in glomeruli) pores occur between the adjacent cells, the basal lamina alone restricting solute passage; (iii) *sinusoids*, discontinuous capillaries (as in liver and spleen), where complete gaps occur between fenestrated endothelial cells, the gaps between the cells being about 4 nm wide; protein hormones cross the endothelium by endocytosis at the inner surface and exocytosis at the outer surface. These incomplete capillaries have the highest permeability and largest diameters of the three, proteins passing through, although few blood cells do. *Precapillary sphincters* at the junctions of capillaries and metarterioles can slow or shut off blood flow in response to pH, carbon dioxide, oxygen, temperature, dilator and constrictor agents (e.g. ADRENALINE, NORADRENALINE). Blood pressure squeezes water and solutes from plasma across capillaries, forming the tissue fluid bathing body cells. Blood cells and plasma proteins are retained on grounds of size, the latter causing the relatively low WATER POTENTIAL of plasma at the venule end of a capillary, returning water to the blood. Solutes diffusing across the endothelium include oxygen, glucose, amino acids and salts (all outwards), carbon dioxide and metabolic wastes (e.g. urea in liver) inwards. Capillaries are absent from animals with open blood systems, but most vertebrate body cells are no further than 50 μm from a capillary. Capillaries are absent from the cornea, lens and fovea of the vertebrate eye. (2) (Of the LYMPHATIC SYSTEM) structurally similar to blood capillaries, but blind-ending and with non-return valves, draining off surplus water from the tissue fluid. See BLOOD, BLOOD–BRAIN BARRIER, INFLAMMATION.

capillitium (Bot.) (1) Tubular protoplasmic threads in fruiting bodies of Myxomycota (slime moulds), assisting discharge of spores in some species by their move-ments in response to changes in humidity. See ELATERS. (2) Sterile hyphae in the fruiting bodies of certain fungi, e.g. puff-balls.

capitulum (Bot.) (1) In flowering plants, inflorescence composed of dense aggregation of sessile flowers. (2) In the Sphagnidae (BRYOPSIDA), a dense tuft of branches at the apex of the gametophyte.

cAPK Cyclic-AMP-dependent protein kinase.

capping (1) See RNA CAPPING. (2) *Cell capping*. Process by which antibodies or other membrane components are attached by cross-linking ligands (see LECTIN) to cell-surface antigens and then swept along the surface to one end (cap) of a motile cell (commonly the rear) where they may be ingested by endocytosis. Unlinked membrane components diffuse fast enough in the membrane to avoid being swept back.

capsid Coat of virus particle, composed of one or a few protein species whose molecules (capsomeres) are arranged in a highly ordered fashion. See VIRUS, BACTERIOPHAGE.

capsomere See CAPSID.

capsule (Bot.) (1) In flowering plants, dry indehiscent fruit developed from a compound ovary; opening to liberate seeds in various ways, e.g. by longitudinal splitting from apex to base, separated parts being known as *valves* (e.g. iris); by formation of pores near top of fruit (e.g. snapdragon) or in the *pyxidium*, by detachment of a lid following equatorial dehiscence (e.g. scarlet pimpernel). (2) In liverworts and mosses, organ within which spores are formed. (3) In some kinds of bacteria, a gelatinous envelope surrounding the cell wall, being a thickened glycocalyx which makes the cell highly resistant to phagocytosis by macrophages. (Zool.) Connective tissue coat of an organ, providing mechanical support.

carapace (1) Bony plates, often fused, beneath the horny scutes of the chelonian dorsal skin (turtles, tortoises). See PLASTRON. (2) Dorsal skin fold of many crustaceans arising from posterior border of head and reaching to varying extent over trunk somites. May enclose whole body (ostracods), the thorax (malacostracans), or be absent altogether (e.g. copepods). May enclose chamber in which gills are housed, embryos protected, etc.

carbohydrate The class of organic compounds with the approximate empirical formula $(C_x(H_2O)_y$, (i.e. literally 'hydrated carbon'), where $y = x$ (monosaccharides) or $y = x - [n - 1]$ (dioligo- and polysaccharides) where n is the number of monomer units in the molecule. Of enormous biological importance both structurally and as energy stores. Sometimes atoms of nitrogen and other elements are also present (e.g. acetylglucosamine). They include CELLULOSE, CHITIN, GLUCOSE, GLYCOGEN, RIBOSE, STARCH and SUCROSE, but also occur as components of GLYCOLIPIDS and GLYCOPROTEINS.

carbon cycle The constant recycling of carbon atoms between inorganic (carbon dioxide, carbonates, bicarbonates) and organic sources. Both abiotic factors (e.g. volcanic activity, rock-weathering) and biotic factors are involved. The major carbon-fixing process is PHOTOSYNTHESIS by plants, phytoplankton, marine and freshwater algae and CYANOBACTERIA, during which they incorporate CO_2 from the atmosphere into organic carbon-containing compounds and release oxygen into the atmosphere. During RESPIRATION by all organisms, these compounds are broken down and CO_2 is again released. These processes result in the cycling of carbon. About 75 billion metric tons of carbon a year are bound into carbon compounds by photosynthesis. Some carbohydrates are used by the autotrophs themselves; plants release CO_2 from their roots, stems and leaves, and algae release carbon dioxide into the water where it maintains an equilibrium with that in the atmosphere. The seas and atmosphere have carbon sinks of some 500 and 700 billion metric tons respectively. Heterotrophs (herbivores, secondary consumers, decay organisms) use the organic products of photosynthesis for their own metabolic processes. CO_2 is released into the reservoir of the air and the oceans by almost all organisms through respiration for re-fixing.

Another even larger carbon reservoir exists below the earth's surface as coal and oil, deposited there some millions of years ago. Another occurs in limestone rocks and as peat in the world's vast tracts of wetlands.

Natural processes of photosynthesis and respiration balance one another out. But since 1850 atmospheric CO_2 concentrations have been increasing dramatically, due in part to the burning of fossil fuels (coal, oil), the ploughing of the soil and the destruction of forests, particularly in the tropics, where, however, in contrast to the northern hemisphere, they are being replanted. Associated with carbon recycling is massive recycling of associated oxygen and hydrogen. See ECOSYSTEM, FOOD WEB, GREENHOUSE EFFECT.

carbonic anhydrase Zinc-containing metalloenzyme of vertebrate red blood cells, brush borders of kidney proximal convoluted tubule, and other body cells. Essential in catalysing the reaction: $CO_2 + H_2O \leftrightarrow H^+ + HCO_3^-$ (reversibly under different blood pH conditions in lungs and tissues), speeding CO_2 transport. Involved in H^+-secretion by oxyntic cells of gastric pits, Cl^- following passively to produce hydrochloric acid. Also involved in blood pH regulation by kidney. Has very high MOLECULAR ACTIVITY. See RED BLOOD CELL.

Carboniferous A PALAEOZOIC period, lasting from 350–286 Myr BP, notable for its coal measures, with lycopods (*Lepidodendron*) dominating along with sphenophytes (*Calamites*). During it, the lycopods influenced the evolution of many other groups of organisms. The multilayered forest ecosystems, dominated by lycopods, cordaites and calamites, produced a shaded environment which probably influenced the selection of plants tolerant of such conditions. This environment would have provided shelter for many organisms intolerant of more exposed conditions present at that time. Also, thick limestone deposits formed, rich in brachiopods. The present-day continents were in greater contact than today, but PANGAEA had yet to form. Amphibians radiated during it, and reptiles appeared in its lowest deposits.

carboxylase Enzyme fixing carbon dioxide or transferring COO⁻ groups. Important carboxylases occur in both respiration and photosynthesis.

carboxypeptidases Pancreatic exopeptidases, hydrolysing peptide bonds adjacent to different C-terminal amino acids.

carboxysome Structure in some bacteria (e.g. chemoautotrophic *Thiobacillus*), housing the CO_2-fixing enzyme ribulose-1,5, bisphosphate carboxylase.

carcinogen Any factor resulting in transformation of a normal cell into a CANCER CELL. The AMES TEST assesses carcinogenicity of a substance. Gut bacteria and other fermenting organisms often produce carcinogens as by-products, many of them glycosides (sugar-containing). When bacterial glycosidase cleaves the sugar group, these become mutagenic and potentially carcinogenic. Red wine and tea appear more carcinogenic than white wine and coffee. Benzene is carcinogenic to humans, but we face greater risks from this through CIGARETTE SMOKING than from car exhausts, while household toilet cleaners and deodorants often contain paradichlorobenzene, a carcinogen of laboratory animals. Ultraviolet and X-radiation and mustard gas are classic carcinogens, their effects usually being attributable to mutation. See MUTAGEN.

carcinoma See CANCER, CANCER CELLS.

cardiac Of the HEART; hence cardiac cycle (see HEART CYCLE), CARDIAC MUSCLE, cardiac sphincter (at the junction of oesophagus and stomach, near the heart).

cardiac centres See CARDIO-ACCELERATORY CENTRE.

cardiac muscle One of three vertebrate MUSCLE types; restricted to the heart walls. Striated, and normally involuntary. Myogenic (see PACEMAKER). Most obvious structural distinctions from skeletal muscle are its anastomosing (branching and rejoining) fibres, and the periodic irregularly thickened sarcolemma, forming *intercalated discs* which appear dark in most stained light microscope preparations. Unlike skeletal muscle, cardiac muscle tissue is not a multinucleate syncytium; each fibre is uninucleate and limited by its sarcolemma. Cardiac muscle tissue has a longer REFRACTORY PERIOD than skeletal muscle and consequently does not fatigue (see MUSCLE CONTRACTION for effects of training on cardiac output). Both pacemaker and accompanying Purkinje fibres are modified cardiac tissue, but with neuron-like properties. See BETA RECEPTORS.

cardiac output Volume of blood leaving heart via the aorta per minute (i.e. stroke volume × heart rate per minute). Increased venous return of blood increases cardiac output 2 to 3-fold without additional sympathetic nervous stimulation (see CARDIO-ACCELERATORY CENTRE), although such stimulation raises this further, esp. in trained athletes.

cardinal veins Paired veins dorsal to the gut of fish and tetrapod embryos, taking blood towards the heart from the head/front limb region (anterior cardinals) to join posterior cardinals (from trunk), forming a common cardinal (Cuvierian) duct which enters the sinus venosus. Replaced by venae cavae in adult tetrapods.

cardio-acceleratory and cardio-inhibitory centres (cardiac centres) Association centres in medulla oblongata (see BRAINSTEM) of homeothermic vertebrates, with reciprocal effects on heart rate. The former employs sympathetic nerves, the latter parasympathetic (vagus). Regulated by hypothalamus and cerebrum. Adjustments of heart rate involve BARO-RECEPTORS. See CAROTID SINUS, PACEMAKER, VASO-MOTOR CENTRE.

caretakers Genes whose products repair DNA and whose inactivation leads to genetic instability. Compare GATEKEEPERS. See CANCER, CANCER CELLS.

carinate Of those birds (the majority apart from RATITES) with a keel (*carina*) on the sternum. The group so formed is not now regarded as more than a GRADE.

carnassial teeth Modified last premolars in each upper half-jaw and corresponding to first lower molars of carnivorous mammals.

Between them they shear and slice, e.g. tendons and bones, when jaw closes.

Carnivora Order including all living carnivorous mammals. Fossil CREODONTS, also carnivorous, form a separate order. Two suborders: FISSIPEDIA (dogs, cats, weasels, civets), and PINNIPEDIA (seals, walrus). Canine and carnassial teeth and retractile claws usually present. Compare CARNIVORE.

carnivore Any meat-eater, usually deriving most of its glucose from protein metabolism. Sometimes indicates a member of the CARNIVORA.

carotenoids Group of yellow, orange and red lipid-soluble pigments found in all chloroplasts, CYANOBACTERIA and some bacteria and fungi, and chromoplasts of higher plants. Chromoplasts lack chlorophyll but synthesize and retain carotenoids, which are responsible for the colour of many fruits and roots (e.g. carrots). Like chlorophylls, carotenoids are embedded in thylakoid membranes. Carotenoids are not essential to photosynthesis but are involved as accessory pigments in the capture of light energy, absorbing photons of different energy which is then transferred to chlorophyll *a*; they cannot substitute for chlorophyll *a* in photosynthesis. Chemically, carotenoid pigments are long-chain compounds (tetraterpenes) and include carotenes (oxygen-free hydrocarbons) and XANTHOPHYLLS (oxygenated derivatives of carotenes). The most widespread carotene is β-carotene, a principal source of vitamin A required by humans and other animals. A large number of xanthophylls exist, some unique to particular groups of ALGAE and important in their systematics (e.g. *myxoxanthophyll*, *aphanizophyll* and *oscillaxanthin* are unique to the CYANOBACTERIA). In several photosynthetic and non-photosynthetic bacteria, carotenoids protect cells from photochemical damage, and some (e.g. lycopene) appear to protect animal cell membrane phospholipids from nitrogen dioxide radical ($NO_2^•$) damage.

carotid artery Major paired vertebrate artery, one on each side of neck, supplying oxygenated blood to head from heart. Derived from third AORTIC ARCH. Each common carotid branches into internal and external carotids; their origins from the aorta vary.

carotid body Small neurovascular structure near branch of internal and external carotids (near carotid sinus); supplied by vagus and glossopharyngeal nerves (see CRANIAL NERVES). Sensitive to oxygen content of blood, also monitoring pH fall and CO_2 rise, since blood pH falls as CO_2 concentration rises (see VENTILATION).

carotid sinus Small swelling in internal carotid artery (therefore paired) in whose walls lie BARORECEPTORS innervated by the glossopharyngeal nerve (cranial nerve IX). Increase in arterial pressure and sinus stimulation causes reflex homeostatic drop in heart rate and vasodilation, involving the cardio-inhibitory and vasomotor centres of the BRAINSTEM.

carpal bones Bones of proximal part of hand (roughly the wrist) of vertebrates. Compact group of primitively 10–12 bones, reduced to 8 in man. Articulate with radius and ulna on proximal side, and with metacarpals on distal side. See PENTADACTYL LIMB.

carpel Female reproductive organ (megasporophyll) of flowering plants. Consists of ovary containing one or more OVULES (which become seeds after fertilization), and a STIGMA, a receptive surface for pollen grains. Often borne at apex of a stalk, the STYLE. See FLOWER.

carpellate See PISTILLATE.

carpogonium Female sex organ of red algae (RHODOPHYTA). Consists of swollen basal portion containing the egg, and an elongated terminal projection (trichogyne) receiving the microgamete.

carpospore In red algae (RHODOPHYTA), the single diploid protoplast found within a containing cell (the carposporangium). Formed after fertilization and borne at the end of an outgrowth of the mature carpogonium.

carpus Region of vertebrate fore-limb containing carpal bones. Approximates to wrist in man.

carrageenan The gelatinous fraction of the cell wall in certain red algae (Rhodophyta, e.g. *Chondrus crispus*, *Gigartina stellata*), which consists of a sulphated complex polysaccharide composed mainly of 1,3- and 1,4-linked D-galactopyranose units. It is similar to AGAR but has a higher ash content and requires a higher concentration to become a gel. Carrageenan is used extensively for many of the same purposes as agar; however, it provides a lower gel strength than agar. Used for stabilization of emulsions in cosmetics, paints, pharmaceutical preparations, stiffening of ice cream, instant puddings, sauces and creams.

carrier (1) An individual HETEROZYGOUS for a recessive character and who does not therefore express it, but half of whose gametes would normally contain the allele for the character (sex linkage excepted). (2) An individual infected with a transmissible pathogen and who may or may not suffer from the disease.

carrier molecule See IONOPHORE, PERMEASE.

carrying capacity An idealized concept: *K*, the maximum population size that can be supported indefinitely by a given environment, at which intraspecific COMPETITION has reduced the per capita net rate of increase to zero. See *K*-SELECTION.

cartilage With BONE, the most important vertebrate skeletal connective tissue. Cells (chondroblasts) derive from mesenchyme and become *chondrocytes* when surrounded within lacunae by the ground substance they secrete. This amorphous matrix (chondrin) contains glycoproteins, basophilic chondroitin and fine collagen fibres, varying proportions of which determine whether it is hyaline (gristle), elastic or fibrocartilage. Matrix gels formed from highly hydrophilic glycosaminoglycans can resist compressive forces of several hundred atmospheres (*contra* collagen, the main bone protein, which resists stretching forces). The surface of cartilage is surrounded by irregular connective tissue forming the perichondrium. Growth may be interstitial (endogenous) resulting from chondrocyte division and matrix deposition within existing cartilage; or appositional (exogenous) resulting from activity of deeper cells of the perichondrium. Lacks blood vessels or nerves. Cartilage is more compressible than bone and in the form of intercostal cartilage absorbs stresses generated throughout the vertebral column during locomotion, lifting, etc.; costal cartilage caps the articulating bone surfaces of JOINTS. The trachea is kept open by rings of hyaline cartilage; the pinnae of ears and auditory tubes contain elastic cartilage. In some kinds of OSSIFICATION cartilage is destroyed and replaced by bone. The CHONDRICHTHYES have entirely cartilaginous skeletons.

cartilage bone See OSSIFICATION.

caruncle Warty outgrowth on seeds of a few flowering plants, e.g. castor oil; obscures MICROPYLE.

Caryophyllidae (Caryophylliflorae) A subclass of the MAGNOLIOPSIDA comprising 11,000 species. About 90% of the species belong to the order Caryophyllales. The remaining 10% belong to Polygonales and Plumbaginales. Each of these orders is distinctive; no one feature is characteristic of the subclass. However, it comprises those dicotyledons that have ovules with two integuments, the nucellus is several cells thick (at least at the micropylar end) and either have betalains instead of anthocyanins or have free-central or basal placentation in a compound ovary. Most species are herbaceous, and those species that have woody growth usually have anomalous secondary growth or otherwise anomalous stem-structure. Stamens, when numerous, originate in centrifugal sequence, and pollen grains are usually trinucleate. See INTRODUCTION to dictionary.

caryopsis A simple, dry, single-seeded indehiscent fruit. An ACHENE with ovary wall (pericarp) firmly united with seed coat (testa). Characteristic of grasses (Fam. Poaceae).

cascade Biological process by which progressive amplification of a signal via a sequence of biochemical/physiological events results in a very localized response. Such a sequence might involve a hormone or other ligand binding to a membrane receptor site, activation of membrane adenylate cyclase producing many cAMP molecules, each activating many kinase molecules, which in turn activate many enzyme molecules, each producing quantities of product. Activation of COMPLEMENT, BLOOD CLOTTING, FIBRINOLYSIS, RHODOPSIN activity and embryonic acquisition of POSITIONAL INFORMATION all result from cascades (see *BICOID* GENE and RECEPTOR, Figs. II, 143C.)

casein Conjugated milk protein. A phosphate ester of serine residues. RENNIN and calcium precipitate it to produce curd; also a major component of cheese.

Casparian strip See ENDODERMIS.

caspases (cysteine-dependent aspartate-specific proteases) A family of proteases, each having a cysteine residue in their active site, and an integral part of the machinery of APOPTOSIS in animals at least. Synthesized as latent zymogens (procaspases), they are activated by proteolytic cleavage at aspartate residues by other caspases, thus amplifying a signal in a CASCADE, similar to a kinase cascade. Some then cleave nuclear lamins and result in breakdown of the nuclear lamina (see NUCLEUS), while another cleaves a protein normally preventing a DNAse from degrading the nuclear DNA, so causing chromosome fragmentation. Caspases include interleukin-Iβ-converting enzyme (ICE) and the *ced-3* product CED-3.

cassettes Alternative functional DNA (or amino acid) sequences which tend to be substituted and recombined in a modular fashion within a larger molecule. See EXPRESSION SIGNALS, MATING TYPE, MUTAGEN, REGULATORY PROTEINS.

caste In EUSOCIAL insects, a structurally and functionally specialized individual: a MORPH. Caste determination may depend upon the state of ploidy (e.g. haploid bees,

ants and wasps are male), or a combination of ploidy (the number of haploid chromosome sets) and environmental factors (e.g. worker and queen bees are diploid and female, but only queens are fed royal jelly as larvae); in lower termites at least it appears to be non-genetic, pheromones produced by king and queen controlling differentiation of caste. Hymenopteran castes are: queen, worker (some ant species having soldier and non-soldier sub-castes of worker) and drone; termite castes include: primary reproductives (king, queen, supplementary reproductives, workers and soldiers. See POLYMORPHISM.

catabolism The sum of enzymatic breakdown processes, such as digestion and respiration in an organism. Opposite of ANABOLISM.

catabolite Metabolite broken down enzymatically.

catabolite repression Suppression by a fuel molecule, or one of its breakdown products, of synthesis of inducible enzymes which would make use of alternative fuel molecules in the cell. Glucose represses production of galactosidase and some respiratory enzymes (the glucose effect) in bacteria. This involves gene repression, a glucose breakdown product combining with the cell's cyclic AMP (cAMP), reducing the amount available for transcription of the operon. See GENE REGULATION, PASTEUR EFFECT.

catalase Haem enzyme of PEROXISOMES of many eukaryotic cells. Converts hydrogen peroxide, produced by certain dehydrogenases and oxidases, to water and oxygen. Used commercially in converting latex to foam rubber and in removing hydrogen peroxide from food.

catalyst Substance speeding up a reversible chemical reaction without altering its equilibrium point. Biological catalysts are ENZYMES and RIBOZYMES.

cataphyll Small scale-like leaf in flowering plants, often serving for protection.

catarrhines Old World monkeys and apes, and all humans; i.e. all cercopithecoids and hominoids (the anthropoid Infra-order Catarrhini). Characterized by narrow nasal septum, thirty-two teeth (two premolars in each jaw quadrant) and by menstrual cycle. No prehensile tail. Typically dry-nosed, contra the typically wet-nosed platyrrhines. See Fig. 82 (HOMINOID) for possible phylogeny; also ANTHROPOIDEA.

catastrophe A major change in the environment that causes extensive damage and usually widespread death, and occurs so infrequently that the effects of natural selection by similar events in the past do not remain in the 'genetic memory' of the species (e.g. volcanic eruptions). See DISASTER.

catecholamines See ADRENALINE, PSYCHOACTIVE DRUGS.

β-catenin A multifunctional protein (called Armadillo in flies) functioning in both cell–cell ADHESION and as a latent gene regulatory protein (see COACTIVATORS). It undergoes proteolysis when Wnt SIGNALLING pathways dependent on the Frizzled receptor protein and Disshevelled adaptor are activated. β-catenin helps link cadherin molecules to the actin cytoskeleton, where most of it is normally located. Any not located there is degraded by ubiquitinization and proteasome action unless a Wnt protein binds to Frizzled, when β-catenin enters the nucleus and induces transcription of Wnt target genes, among which is c-*myc* (see entry). See HEDGEHOG.

caterpillar Larval stage of Lepidoptera, Mecoptera and some Hymenoptera, bearing abdominal prolegs in addition to thoracic legs. Generally poorly sclerotized and inactive, living close to food.

Catharanthus Genus of flowering plant; the rosy periwinkle (*C. roseus*), native to Madagascar, is the natural source of two highly effective anti-cancer drugs: *vinblastine* (used to treat Hodgkin's disease) and *vincristine* (used in cases of acute leukaemia).

cathelicidins Important native components of the innate immune system which, in mice at least, offer protection against necrotic skin infection caused by some *Streptococcus*. See ANTIMICROBIAL PEPTIDES.

cathepsins A group of proteolytic enzymes occurring in LYSOSOMES.

catheter Tube, often plastic, inserted into gut, blood vessels, etc. for withdrawal/introduction of material. Balloon catheters have an inflatable tip and may be used to dilate blocked vessels (e.g. the coronary artery).

cation A positively charged ion.

caudal (Of the tail) caudal vertebrae are tail vertebrae, the caudal fin of a fish is its tail fin.

cauline Belonging to the stem, or arising from it.

cauline bundle A VASCULAR BUNDLE forming part of the stem tissue.

caveolae Flask-shaped invaginations of the cell surface membrane, their shape conferred by CAVEOLIN, and present on many cells, demarcating cholesterol- and sphingolipid-rich domains in which many diverse signalling molecules and receptors are located. With CLATHRIN, important in ENDOCYTOSIS.

caveolin A dimeric protein which binds cholesterol, inserts a loop in the inner leaflet of the plasma membrane and self-assembles to form a striated caveolin coat on the surface of the membrane invaginations (see POTOCYTOSIS).

cavitation Occurrence of air pockets and/or bubbles in xylem vessels when tension exerted on the water column exceeds that enabling cohesion. It may occur during water stress. An alternative route for the TRANSPIRATION stream would be needed, bypassing the blockage; however, air pockets so formed may be squeezed out again by ROOT PRESSURE.

CD2, CD4, CD8, etc. (cluster of differentiation) proteins See ACCESSORY MOLECULES.

***cdc* genes** Cell-division-cycle genes of yeasts, conserved in all eukaryotes so far examined, forming a regulatory network involved in the timing of mitosis. Their products are sometimes involved in CAS-CADES. Mutations in *cdc* genes are often temperature-sensitive. See CELL CYCLE.

cdk Cyclin-dependent KINASE gene, encoding CDK protein product, which is regulated by transphosphorylation by CDK-activating kinases. See CELL CYCLE, CYCLINS, C-*MYC*.

cDNA (complementary DNA) DNA complementary to RNA and produced by REVERSE TRANSCRIPTASE activity. Initially single-stranded, can be converted to double-stranded cDNA by DNA POLYMERASE activity. cDNA complementary to mRNA lacks INTRON sequences, useful when cloning functional DNA. If a single cDNA clone is required, a far more rapid method than searching through cDNA libraries is to identify the end-points of the relevant mRNA and then design primers for retrieving the cDNA using reverse-transcriptase PCR analysis.

CED proteins See APOPTOSIS.

cell Mass of protoplasm made discrete by an enveloping plasma membrane (plasmalemma). Any cell wall material is, strictly speaking, extracellular (e.g. in most plants and fungi); but distinctions between intracellular and extracellular may be arbitrary (see GLYCOCALYX).

The two basic types of cell architecture are those of PROKARYOTES and EUKARYOTES. See Fig. 20. In the former, cells consist entirely of cytoplasm (lacking nuclei); in the latter, cells have (or had) in addition one or more nuclei. Eukaryotic cells have greater variety of organelles, many enclosed in one or more membranes (see CELL MEMBRANES). They are further distinguished by the presence of distinctive proteins, particularly ACTIN, MYOSIN, TUBULIN and HISTONE, that have very significant uses and are entirely absent from prokaryotic cells. Actin is paramount in the structure of the eukaryotic CYTOSKELETON; tubulin is fundamental in cilium and flagellum structure, and in mitotic and meiotic spindles – none of these being found in prokaryotes, whose flagella are rigid and of a completely different structure. These and other features indicate how similar even

such apparently dissimilar cells as those of plants and animals are when compared with those of prokaryotes (bacteria and blue-green algae). Basic eukaryotic cell architecture is elaborated upon in many ways, notably by fungi, where true cells are commonly absent in much of the vegetative body, organization being COENOCYTIC. A similar multinucleate situation, without intervening cell membranes, arises where eukaryotic cells fuse to form a SYNCYTIUM. Both may be termed ACELLULAR. The plasmodesmata uniting many plant cells may be regarded as producing an intermediate condition. See MULTICELLULARITY, ORIGIN OF LIFE.

cell body (perikaryon) Region of a neuron containing the nucleus and its surrounding cytoplasm. Generally swollen compared with rest of cell. Some ganglia consist of aggregations of cell bodies.

cell–cell adhesion molecule (CAM) See ADHESION.

cell centre Alternative term for CENTROSOME.

cell cycle The period during which events involved in successful eukaryotic nuclear and cell reproduction are completed (see Fig. 22 for some of the details). In proliferative cells this includes all the events taking place between the completion of one round of mitosis and cytokinesis and the next. Cells which commence upon their differentiation pathway have generally left the cell cycle for good. The molecular details of the cell cycle have been the subject of intense research in recent decades, not least because of their implications for our understanding of the origins of many cancers (see CANCER CELLS).

The onset of DNA REPLICATION at S-phase (within interphase) and of mitosis (M-phase), are the cycle's two key events (some embryonic cleavage divisions dispense with G1 and/or G2). The controls of entry to both may be common to all eukaryotic cells, involving successive waves of CYCLIN-dependent kinase (CDK) activity leading to phosphorylation of certain key proteins (e.g. cyclins, H1 histone, LAMINS, ELONGATION FACTORS and RNA polymerase II; see DNA

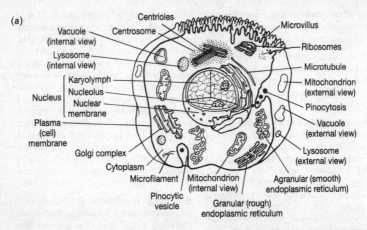

(a)

Centrioles
Centrosome
Vacuole (internal view)
Lysosome (internal view)
Nucleus
  Karyolymph
  Nucleolus
  Nuclear membrane
Plasma (cell) membrane
Golgi complex
Cytoplasm
Microfilament
Pinocytic vesicle
Mitochondrion (internal view)
Granular (rough) endoplasmic reticulum
Microvillus
Ribosomes
Microtubule
Mitochondrion (external view)
Pinocytosis
Vacuole (external view)
Lysosome (external view)
Agranular (smooth) endoplasmic reticulum

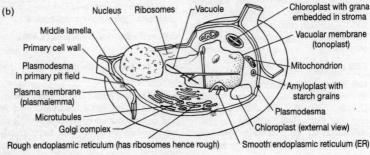

(b)

Nucleus
Ribosomes
Vacuole
Middle lamella
Primary cell wall
Plasmodesma in primary pit field
Plasma membrane (plasmalemma)
Microtubules
Golgi complex
Rough endoplasmic reticulum (has ribosomes hence rough)
Chloroplast with grana embedded in stroma
Vacuolar membrane (tonoplast)
Mitochondrion
Amyloplast with starch grains
Plasmodesma
Chloroplast (external view)
Smooth endoplasmic reticulum (ER)

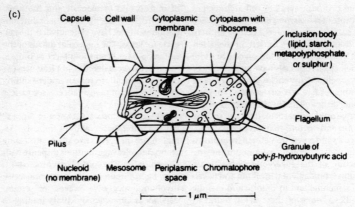

(c)

Capsule
Cell wall
Cytoplasmic membrane
Cytoplasm with ribosomes
Inclusion body (lipid, starch, metapolyphosphate, or sulphur)
Flagellum
Granule of poly-β-hydroxybutyric acid
Pilus
Nucleoid (no membrane)
Mesosome
Periplasmic space
Chromatophore

— 1 μm —

FIG. 20 (a) *Generalized animal* CELL *structure on electron microscope observations (approx. diameter 50 μm). (b) Generalized photosynthetic plant cell (approx. diameter 100 μm). (c) Generalized prokaryotic (bacterial) cell structure (approx. length 1–2 μm).*

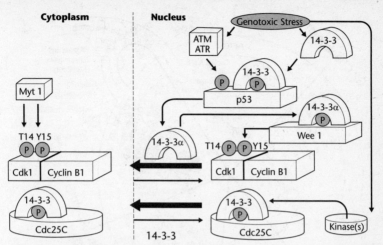

FIG. 21 *Some of the cellular responses to genotoxic stress (see DNA REPAIR MECHANISMS). The cell-division cycle is blocked after DNA damage, allowing time for repair. After damage, Cdk1 is maintained in an active state by phosphorylation on threonine 14 (T14) and tyrosine 15 (Y15). This is done by Wee1, a nuclear kinase that phosphorylates Cdk1 on Y15, and by Myt1, which is cytoplasmic and phosphorylates Cdk1 on both T14 and Y15. Cdk1/Cyclin B1 complexes continually shuttle between the nucleus and the cytoplasm. Cdc25C, which activates Cdk1/Cyclin B1 complexes by dephosphorylating Cdk1, also shuttles between the nucleus and cytoplasm, and the 14-3-3-binding proteins (although not 14-3-3σ) contribute to the nuclear exclusion of Cdc25C. DNA damage activates the nuclear kinases which maintain Cdc25C in a 14-3-3-bound form and also results in the nuclear protein kinase complex ATM/ATR combining with 14-3-3 to stabilize p53. This leads to transcriptional activation of the 14-3-3σ gene, which contributes to nuclear exclusion of the CdK1/Cyclin B1 complexes. Although not illustrated here, the large ATM/ATR protein complex also leads indirectly to proteolysis and removal from the nucleus of 14-3-3-bound phosphorylated prot ein phosphatase Cdc25A. See p. xii for assistance received.*

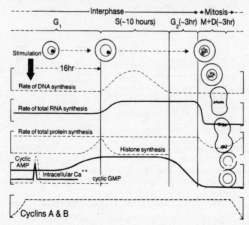

FIG. 22a *Breakdown of the phases of the eukaryotic cell cycle. Times are only approximate and vary for different systems.*

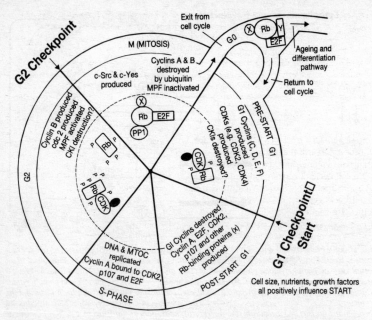

FIG. 22b *The major mammalian cell-cycle stages, with some of their prominent molecular influences and accompaniments. G, in G0, G1, G2 = 'gap'; CDK = cyclin-dependent kinase; CKI = cyclin-dependent inhibitors; Rb = retinoblastoma protein, its phosphorylation states (p) indicated by figures; E2F = transcription factor E2F; c-Src and c-Yes are non-receptor protein kinases (cellular oncogenes). The table below indicates some of the complexes required for passage into, or through, critical cell-cycle stages in mammalian cells. Cyclin equivalents (Cln, Clb) in yeasts are shown in parenthesis.*

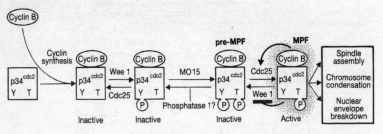

FIG. 22c *Figure to show mechanism of* MATURATION PROMOTING FACTOR (MPF) *activation. Y = tyrosine 15; T = threonine 161 (amino acid residues of p34^{cdc2}). MO15 is a kinase. Proteins Cdc25 and Wee1 modulate p34^{cdc2} in opposite directions. Molecules phosphorylated on both Y and T residues accumulate in G2 and are converted to active MPF by Cdc25, in turn accompanied by Cdc25 activation and Wee1 inactivation.*

Stage in cell cycle	Progress requires these complexes
	cdk4/cyclin D (Cln3)
G1→S	cdk2/cyclin E (Cln1/2)
S	cdk2/cyclin A (Clb5/6)
G2→M	cdc2/cyclin A (Clb5/6)
M	cdc2/cyclin B (Clb1/2)

FIG. 22d

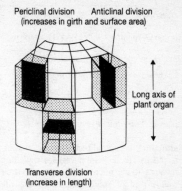

FIG. 23 *The three main planes in which plant cells can divide.*

REPLICATION for roles of CDKs and proteasomes). Towards the end of G1 (just prior to entry to S phase), and at the end of G2 there are cell cycle CHECKPOINTS at which brakes can be applied on further commitment to cell cycle completion if feedback signals indicate failures in preparation, or if external cues (e.g. growth factors) mitigate. Members of the PROTEIN KINASE p34 family (also variously known as $p34^{cdc2}$, $p34^{cdc2/CD28}$, phosphoprotein pp34, or cdk1) encoded by the *cdc2* gene (see *CDC* GENES) are key players in these events. Activity of $p34^{cdc2}$ kinase is itself regulated by phosphorylation at three or more sites, by phosphatases and by cyclins, with which it must complex prior to activation. The timing of ubiquitination and PROTEASOME degradation of several key proteins is crucial for successful transit through the cell cycle, that of 'mitotic cyclin' potentiating the $p34^{cdc2}$ kinase inactivation required for exit from M-phase. Other phases of the cell cycle are mediated by other key regulatory protein kinases (e.g. the Abl cytoplasmic tyrosine kinase family) and their inhibitors. The c-Abl protein binds both actin and specific nuclear DNA sequences and is regulated by a cdc2-mediated phosphorylation (c-Abl has three phosphorylated serine/threonine sites in interphase cells, and seven more in mitotic cells). See ANAPHASE-PROMOTING COMPLEX, CKIS, CENTROSOME, MATURATION PROMOTING FACTOR, p53 (in cell cycle arrest).

cell division Process by which a cell divides into two; commonly apparently symmetric, but asymmetric divisions occur in both bacteria (e.g. sporulating *Bacillus subtilis*) and eukaryotes (see CELL POLARITY, DYNEIN), the plane of the mitotic spindle in the latter being the crucial factor. (a) *Prokaryotic*: one

event achieves separation of both DNA and cytoplasm into daughter cells. Since the two sister chromosomes are attached separately to the cell membrane they can become separated by the cleavage furrow formed between them as the cell membrane invaginates. Fission occurs as membrane intuckings fuse. No microtubules occur in prokaryotes, so there is no mitosis or meiosis; but there might be dedicated machinery to align chromosomes with their origins of replication towards the poles of the predivisional cell and direct chromosome movement to progeny cells, mere chromosome–membrane attachment being insufficient (there is speculation that there may be a bacterial cytoskeleton, and accompanying motor proteins). (b) *Eukaryotic*: nuclear and cytoplasmic divisions are achieved by separate mechanisms. In higher eukaryotes the nuclear membrane breaks down and chromosomes attach to microtubules of the spindle by their KINETOCHORES (see p.53 for its role in spindle microtubule assembly); cytoplasmic division (*cytokinesis*) usually starts in mitotic anaphase and proceeds by a furrowing of the plasmalemma in the plane of the metaphase plate, achieved by a contractile ring of ACTIN filaments. Fusion of the invaginating plasmalemma then occurs. In plant cells with walls the new wall is built upon a CELL PLATE. The plane of plant cell division may be

Low Speed	Medium Speed	High Speed	Very High Speed
(1,000 G, 10 min)	(20,000 G, 20 min)	(80,000 G, 1 h)	(150,000 G, 3 h)
nuclei	mitochondria	microsomes	ribosomes
whole cells	lysosomes	rough & smooth ER	viruses
cytoskeletons	peroxisomes	small vesicles	large macro-molecules

TABLE 5 *Centrifugation of cell components.*

anticlinal, periclinal or transverse, as illustrated in Fig. 23. Golgi vesicles travelling towards it on microtubules deposit their wall precursor molecules, extending the plate to the cell membranes and pinching off the cell into two. Cells often need to coordinate cell division and cell mass. Cells in early embryos divide repeatedly with little increase in mass, periodicity of cell division being controlled by an internal timer, or oscillator. In somatic cells, M-phase is coupled to completion of S-phase (see CELL CYCLE). In others, rate of mass increase dictates periodicity of cell division. See MITOSIS and MEIOSIS for nuclear division; see CYTOSTATIC FACTORS for control of cell proliferation.

cell-division-cycle genes See CDC GENES, CELL CYCLE.

cell fractionation Process whereby cells are first appropriately buffered (often in sucrose solution) and then disrupted (by osmotic shock, sonic vibration, maceration or grinding with fine glass, sand, etc.); the cell fragments are then spun in a refrigerated centrifuge. Different cell components descend to the bottom of the centrifuge tube at different speeds, and these can be increased progressively. Forces generated may be 500,000 times that of gravity (G). The G-forces and times required to spin down different cell constituents are shown in Table 5.

cell fusion Process involving fusion of plasma membranes of two cells to form one resultant cell. All such membrane fusion events require the release of calcium ions (Ca^{2+}) from intracellular stores and are often initiated by receptor/ligand- and phospho-lipase C-mediated release of INOSITOL 1,4,5-TRIPHOSPHATE (see Fig. 91). One of the last steps in membrane fusion is overseen by a family of proteins called SNARES. Naturally occurring cell fusion may or may not result in hybridization (unity of genomes). Fusion of MYOBLASTS in skeletal muscle development, and other syncytial organizations, does not normally involve hybridization. The processes of PLASMOGAMY and KARYOGAMY are temporally separated in those fungal life cycles where a DIKARYON occurs at some stage. In fertilization, separation of plasmogamy and karyogamy is usually brief. Artificial cell fusion is often achieved by treatment with inactivated viruses, or a glycol. The heterokaryon, with its separate nuclei intact, may then divide, in which case all chromosomes may end up within a single nuclear membrane. Irregular chromosome loss may permit CHROMOSOME MAPPING in tissue culture, as with mouse-human hybrid cells. Techniques resulting in fusion and hybridization of normal and tumour B CELLS have yielded HYBRIDOMAS capable of generating monoclonal antibodies on a commercial or clinical scale. Protoplasts resulting from enzymic digestion of plant cell walls can be encouraged to fuse, and may generate heterokaryons or even fusion hybrids. Appropriate horticulture can generate somatic hybrid plants between species that would not normally hybridize. As with mouse-human somatic hybrids, chromosome loss often prevents a genetically stable product.

cell growth In general, cells need to grow in size before they can divide (but see CLEAVAGE), and this leads to tissue hypertrophy. An organism's BODY SIZE is dependent upon cell growth as well as cell division, but it remains unclear how different cells types in the same organism come to have different sizes. Extracellular GROWTH FACTORS that stimulate cell growth bind initially to receptors linked to signal transduction pathways, one of the most important in animals being activated by the PROTEIN KINASE B and PI $_3$-KINASE pathway which in turn can be activated by the Ras/MAP kinase pathway (see RAS PROTEINS and Figs.143b and 151). Growth factor binding increases the rate of protein synthesis by enhancing the rate at which ribosomes translate certain mRNAs – notably those for the ribosomal proteins themselves. The protein kinase S6 kinase is part of one such pathway, phosphorylating and activating the S6 ribosomal subunit.

During growth, plant cells secrete the protein *expansin*, which unlocks the polysaccharide network of the CELL WALL and allows turgor-driven cell enlargement (see AUXINS).

cell hybridization See CELL FUSION.

cell locomotion There are various methods by which cells move, those of prokaryotes having apparently little in common with those of eukaryotes. For the latter, most mechanisms seem to involve protein tubules or filaments sliding past one another and generating force. The details of how force is transmitted to the substratum are largely unknown, although ADHESION proteins, including integrins (see Fig. 1) are involved. Local concentration gradients of second messengers (e.g. calcium ions, phosphoinositides), of enzymes and motor proteins, along the cell length probably also play a part.

(1) *Bacterial*: H$^+$ gradients across the inner cell membrane provide the motive force for rotation of the FLAGELLUM, whose fixed helix of protein subunits permits clockwise and counter-clockwise rotations, like a corkscrew. This involves an extraordinary 'wheel-like' rotor in the inner membrane, and cylindrical fixed bearing in the outer membrane. Reversal of flagellar rotation alters the behaviour of the cell.

(2) *Eukaryotic*. (*a*) *Ciliary/flagellar*: see CILIUM for structure. Paired outer microtubules

over adjacent pairs in response to forces generated by dynein arms coupled to their ATPase activity. Radial spokes and the inner sheath apparently convert this sliding to bending of the organelle. The axoneme can beat without the cell membrane sheath around it. The dynein arms probably act in an equivalent fashion to myosin heads during MUSCLE CONTRACTION and make contact with adjacent microtubule pairs during their power stroke. Control of ciliary/flagellar beat appears to be independent of Ca^{2+} flux, but may be dependent upon signal relay via proteins of the actual structure. However, reversal of ciliary beat in some ciliates is associated with membrane voltage change brought about by Ca^{2+} influx. It is still uncertain how waves of ciliary beating in cell surfaces are coordinated. (*b*) *'Fibroblastic' crawling*: the leading edge of a cell engaged in this method of locomotion, characteristic of fibroblasts, extends forwards and, after attachment to the substratum, pulls the rest of the cell forward by contraction of actin microfilaments under influence of MYOSIN II. Typical features associated with this method are lamellipodia and microspikes (see CELL MEMBRANES, FILOPODIUM), which both pass backwards in waves along the upper cell surface ('ruffling'), typically when the anterior of the cell has failed to attach to the substratum. Molecular mechanisms include polymerization and depolymerization of actin filaments in lamellipodia under the influence of the GTPases Rho and Rac, and regulation of actin branching by the PROTEIN COMPLEX Arp2/3, thought to provide a strong scaffold capable of pushing the cell membrane forward during extension. Random endocytosis of plasmalemma and its later restricted exocytosis at the anterior of the cell generate a circulation of membrane akin to movement of tank caterpillar tracks. The protein FIBRONECTIN is involved in fibroblast crawling. (*c*) *'Amoeboid' (pseudopodial)*: the cell's outermost layer is gel-like (plasmagel) while the core is a fluid sol (plasmasol). It is possible that contraction of the thick cortical plasmagel squeezes the plasmasol and generates pseudopodial extensions of the cell, at the tips of which sol-to-gel transformation occurs. Gel-to-sol changes accompany this elsewhere in the cell, e.g. as a pseudopod retracts. Just how these CYTOPLASMIC STREAMING events are coupled to locomotion is not clear, but motive force must act against regions where the cell adheres to its substratum (see FIB-RONECTIN). Apparently, surfaces of large amoebae are relatively permanent, undergoing folding and unfolding to accommodate pseudopod extension and retraction. ACTIN is implicated in the process. Characteristic of amoebae and macrophages. Cell migration plays a crucial part in development (see GASTRULATION, MORPHOGENESIS, POSITIONAL INFORMATION). Nematode sperm appear to crawl like amoebae and macrophages; but they appear to have evolved a unique cytoskeleton, containing very little actin, the basis instead being *major sperm protein* (MSP) for which no homologues or analogues have been found in other animals. See CAPPING, DESMID.

cell markers See GENETIC MARKER.

cell-mediated immunity See IMMUNITY.

cell membrane (plasma membrane, plasmalemma) The membrane surrounding any cell. See CELL MEMBRANES.

cell membranes Cells may have a wide variety of membranes (often called 'unit' membranes) varying from 5–10 nm in thickness; but all have a plasma membrane (plasmalemma), the outer limit of the cell proper, which is generally quite distinct from any cell wall material present (which is extracellular). See SELF-ASSEMBLY.

Major membrane functions include: restriction and control of movements of molecules (e.g. holding the cell together) enabling scarce metabolites to reach local concentrations sufficient to enhance enzyme-substrate interactions (see POTO-CYTOSIS); to act as platforms for the spatial organization of enzymes and their cofactors, while carotenoids and chlorophylls act as better energy-couplers in chloroplast thylakoids owing to their loss of mobility in the membranes. Membranes also act as platforms for the spatial organization of enzymes and their cofactors, holding other-

wise scattered molecules in functional contact: there appear to be lipid microdomains formed by clusters of sphingolipids and cholesterol, that move within the fluid bilayer and function as 'rafts' (membrane rafts) for the attachment of proteins; and membranes also separate and localize incompatible reactions. Many eukaryotic organelles have one or two membranes around them, chloroplasts having yet a third system within. The currently accepted structure of most cell membranes is that proposed in the *fluid mosaic model*, the evidence coming from X-ray crystallography, freeze-fracture and freeze-etching electron microscopy (see MICROSCOPE), radiolabelling, electron spin resonance spectroscopy and fluorescence depolarization. The last two involve insertion of *molecular probes* with particular spectroscopic features adding peaks or troughs to the lipid spectrum.

In this model an outer and an inner phospholipid monolayer (major components phosphatidyl ethanolamine and lecithin) lie with their polar phosphate heads in the direction of the water which the bilayer thus separates (see Fig. 24b). The minor phospholipid phosphatidyl inositol is crucial in INOSITOL 1,4,5-TRIPHOSPHATE signalling. Specific, and different, proteins lie in one or other layer or traverse the bilayer, making the membrane *asymmetric*. Proteins which span the membrane (transmembrane proteins) have their topologies established in the endoplasmic reticulum. Their hydrophobic portions (esp. the outer surfaces of α-helices) often span the phospholipid bilayer several times (e.g. BACTERIORHODOPSIN, RHODOPSIN, RECEPTOR TYROSINE KINASES and the acetylcholine receptor), such multi-spanning proteins often acting as ION CHANNELS or ion gates – the hydrophilic inner portions of their α-helices allowing water through (see AQUAPORINS). Such integrated membrane proteins commonly require detergent for their release, but proteins linked by ester bonds to the membrane fatty acids are removed by 1 M hydroxylamine or high pH. Proteins for incorporation into the plasmalemma will have appropriate signal regions targeting them there (see PROTEIN TARGETING).

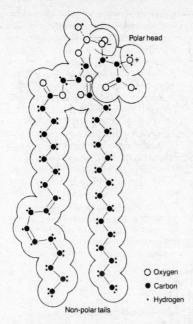

○ Oxygen
● Carbon
· Hydrogen

FIG. 24a *A phospholipid molecule (phosphatidyl-serine).*

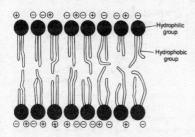

FIG. 24b *A phospholipid bilayer.*

The whole structure has fluid properties resulting from rapid lateral movement of most of its molecular components through thermal agitation ($1\ \mu m.s^{-1}$ for lipids, $10\ \mu m.min^{-1}$ for proteins). Thus fused mouse and human cells, each with differently labelled membrane proteins, exhibit rapid mixing of labels over the entire cell surface. The ionophore gramicidin functions only when the two halves of the molecule,

one in each half of the bilayer, come together – which they do in a quantized way, indicating membrane fluidity. Endocytosis, exocytosis and other processes involving membrane fusion (e.g. fertilization) are made possible by this fluidity. Molecules that occasionally tumble, or 'flip-flop' from one layer of the plasmalemma to the other, are reinserted by 'flippase' enzymes.

The LIPIDS present in the membranes of archaeons (SEE ARCHAEA) are somewhat different from those of bacteria and eukaryons. Sterols are absent from prokaryote membranes, but hopanoids (see Fig. 154) may be found instead. Thermophiles tend to have saturated fatty acids in their membranes, which are more heat-stable than unsaturated fatty acids on account of their stronger hydrophobic bonds. The membranes of HYPERTHERMOPHILES lack fatty acids altogether. The outer leaflet of plasma membranes of plants and animals, the face exposed to the outer world, is composed largely of lipids with no net charge. By contrast, bacterial membranes have an outer leaflet heavily populated by lipids with negatively charged phospholipid headgroups (targeted by ANTIMICROBIAL PEPTIDES). The plasmalemmas of animal cells typically have the oligosaccharide chains of their GLYCOLIPIDS and GLYCOPROTEINS exposed freely on their outer surfaces (see Fig. 24c), playing important roles in immunological responses, in cell–cell ADHESION and identification, and in cell surface changes. Most plasmalemmas comprise about 40–50% lipid and 50–60% protein by weight. The phospholipid bilayer has a non-polar hydrophobic interior, preventing passage of most polar and all charged molecules. Small non-polar molecules readily dissolve in it, and uncharged polar molecules (e.g. H_2O) can also diffuse rapidly across it, possibly assisted by the polar phospholipid heads, or by such ionophores as gramicidin. Lipid bilayers are impermeable to carbohydrates and ions at the diffusion rates needed by cells; but membranes contain various TRANSPORT PROTEINS which speed transfer of metabolites across them so that small and otherwise inaccessible ions and molecules

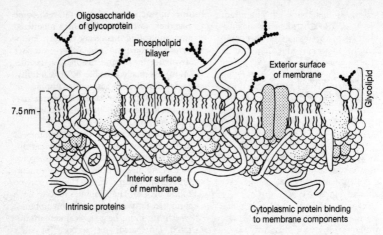

FIG. 24C *Depiction of the fluid mosaic model of cell membranes.*

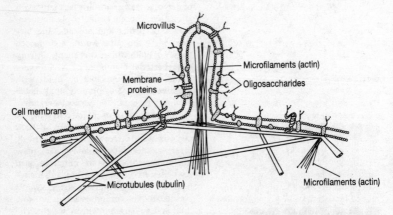

FIG. 24d *Components of the cytoskeleton in the region of a microvillus. Actin microfilaments and tubulin microtubules are linked to each other and to the cell membrane by an array of linker proteins not shown.*

may be carried across cell membranes by GATED CHANNELS, FACILITATED DIFFUSION, IONOPHORES or ACTIVE TRANSPORT. A typical vertebrate cell is estimated to have several hundred different channels and receptors on its surface. One by-product of this activity may be to generate ionic imbalances across the membrane which may be used to power ATP synthesis, or drive symports and antiports.

Large molecules or even solid particles gain access to the cell's geographic interior by pinocytosis and phagocytosis, and may be jettisoned by EXOCYTOSIS. All these involve enclosure of transported molecules within membranous vesicles which fuse only with appropriate cell membranes. This recognition ability probably resides in the specificity of proteins exposed at a membrane surface.

Not all membrane phospholipids are identical, and this prevents their crystallization at low temperatures (CHOLESTEROL has an important role here in animal plasma membranes) as well as permitting local loss of fluidity as at synapses and DESMOSOMES. Poikilothermic animals respond to chronic cold by enzymically increasing the level of unsaturation of their cold-rigidified membranes, improving their fluidity.

The carbohydrate content of plasma membrane glycolipids and glycoproteins may be such as to create a cell coat or GLYCO-CALYX. Other membrane proteins act as *receptor sites* binding specific ligands (e.g. see CASCADE, GATED CHANNELS). The eukaryotic plasma membrane is involved in the structures of CILIA and FLAGELLA, MICROVILLI (see Fig. 24d), LAMELLIPODIA, MICROSPIKES and several sorts of INTERCELLULAR JUNCTION. See appropriate organelles for further membranes. For membrane movement through the cell, see GOLGI APPARATUS, LYSOSOME, SNARES.

cell memory (1) Heritable patterns of gene transcription. Cells often need to pass on their pattern of gene expression to daughter cells, and this can be achieved when, during cell division, each conserved strand of the DNA duplex carries with it the proteins bound to it and engaged in GENE SILENCING through HETEROCHROMATIN formation. These can then recruit newly synthesized proteins to reconstruct the same chromatin remodelling on the new complementary DNA strand. Alternatively, a cell may pass to its daughter cells the transcription factors required for exactly that combination of gene expression which it undertook itself – sometimes involving POSITIVE FEEDBACK of a protein on the promoter region of its own encoding gene. Again, the pattern of DNA METHYLATION, affecting gene expression, can be inherited following DNA REPLICATION. See GENOMIC IMPRINTING. (2) Cells may respond over the medium or long term to far more transient signals. This may be regarded as a form of cell memory, and would include changes of SYNAPTIC STRENGTH. See also CAL-CIUM SIGNALLING, NEUROGENESIS.

cell migration Essentially, any persistent directional CELL LOCOMOTION. Migrating cells shape organs and tissues during animal development (see MORPHOGENESIS, NEURAL CREST), move to the sites of wounds (e.g. MACROPHAGES), generate new blood vessels (see ANGIOGENESIS) and are the cause of tumour malignancy (see CANCER). Stimuli that induce migration by lamellipodia activate the Rho protein Rac at the cell's leading edge, promoting ACTIN polymerization.

cell movement See CELL LOCOMOTION, CELL MIGRATION, MORPHOGENESIS.

cell plate (Bot.) 'Plate' of differentially staining material which appears at telophase in the PHRAGMOPLAST across the equatorial plane of the spindle. Believed to be forerunner of MIDDLE LAMELLA. See CELL WALL.

cell polarity Many structures within cells come to be arranged in such a manner as to give the cell a distinctive and non-random polarity. Epithelial cells are perhaps the most obviously polarized animal cells, the apical and basolateral surfaces being importantly distinctive (see Figs. 74, 92).

Cell polarity is especially important in an animal egg prior to fertilization. Often crucial for the future patterning of the embryo, it is determined largely by the position of the egg in the ovary, its association with nutritive cells, and its attachment to the ovary wall. Cell polarity may change at fertilization or cleavage and is sometimes marked by yolk and/or CENTROSOME distribution. The animal/vegetal axis may dictate the future anterior-posterior AXIS of the embryo. Genes controlling the polarity of bristles on the epidermal cells of *Drosophila* (fruit fly) wings also control the polarized motilities underlying the morphogenetic movements during gastrulation which shape the vertebrate body plan. See ANIMAL POLE, MATERNAL EFFECT.

The term also extends to the establishment of the budding site in budding yeast – microtubule alignment being critical. Work on these and other cells indicates some factors in common with the establishment of animal cell polarity: e.g. the asymmetric distributions of patterning molecules by transport along ACTIN microfilaments of the cytoskeleton into one daughter cell during cleavage; capture of determinants by a region of the cortex inherited by one cell; or association of different mRNA molecules with the CENTROSOMES passing into different cells. Spatial distributions of the cytoskeleton, secretory apparatus, membrane proteins and nucleus also have a role. Thus, in the nematode worm *Caenorhabditis elegans*, establishment of polarity in the fertilized egg involves intimate interaction between the cortical actin cytoskeleton and the sperm-derived asters and developing microtubules, sperm aster position initiating the processes which determine the future posterior of the embryo. Both extrinsic cues (e.g. cell contact, extracellular matrix) and intrinsic cues (e.g. as set up by previous cell cycle) initiate polarity, a spatial cue inducing localized assembly of a signalling network including small GTP-binding proteins (see RAS PROTEINS), kinases, phosphatases, etc., and cytoskeletal proteins resulting in localized assembly of the actin cytoskeletal matrix. 'Targeting patches' in the plasma membrane and secretory membranes for transport vesicles are then established, bringing changes in actin and microtubule distributions – the secretory apparatus aligning itself along an axis which is dependent upon these cues. Work on eggs of *C. elegans* and *Drosophila* indicate common roles for the *par* genes (and their non-secreted products, PAR proteins) in converting transient asymmetries into stably polarized embryonic axes. One PAR protein plays a role in the asymmetric positioning of the first spindle in *C. elegans*, although other gene products are also implicated in spindle positioning. See CELL LOCOMOTION.

cell shape For some of the factors involved, see ADHESION, CELL GROWTH, CYTOSKELETON.

cell size See CELL GROWTH, CYTOSKELETON.

cell theory The theory, first proposed by Schwann in 1839, that organic structure originates through formation and differentiation of units, the cells, by whose divisions

and associations the complex bodies of organisms are formed. Much of the original theory is now untenable. Schleiden's name is also associated with the theory. See VIRCHOW.

cellular immunity and cellular response See IMMUNITY.

cellular memory Inheritance, often through many cell divisions, of an altered cell phenotype even when the original stimulus for the change is no longer present. These changes are usually structural and functional and often involve EPIGENESIS and include chromatin-marking systems (see DNA METHYLATION) and steady-state inheritance systems (where gene products act as positive regulators of their gene's expression).

cellular senescence Irreversible cessation of cell division. See AGEING.

cellulose The most abundant organic polymer. A polysaccharide, occurring as the major structural cell wall material in the plant kingdom. Some fungi have it as a component of their hyphal walls, and it may occur in animal cell coats (see GLYCOCALYX). A long-chain polysaccharide of repeating *cellobiose* units, it may also be considered as a long chain of $\beta[1,4]$-linked glucose units. Hydrogen bonding both within each molecule as well as between parallel molecules (producing crystalline *microfibrils*) gives cellulose its great tensile strength; but microfibrils can be loosened by lowered pH (an effect of AUXINS on the cell) allowing for wall extension in cell growth, when more cellulose may be laid down between existing microfibrils. With LIGNIN, it forms *lignocellulose*. The fibrous texture of cellulose is responsible for its use in textile industries (cotton, linen, artificial silk). See CELL WALL for cellulose distribution.

cell wall Extracellular coat of cells of bacteria, CYANOBACTERIA (blue-green algae), PROCHLOROPHYTA, PLANTAE, FUNGI and many PROTISTA; secreted by the protoplasm, and closely investing it. The bacterial wall is a component of its envelope and contains either a thick layer of PEPTIDOGLYCAN, or

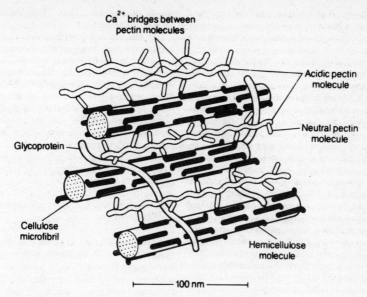

Ca²⁺ bridges between
pectin molecules

Acidic pectin
molecule

Neutral pectin
molecule

Glycoprotein

Cellulose
microfibril

Hemicellulose
molecule

|← 100 nm →|

FIG. 25a *The relative arrangements of molecule types in a primary cell wall.*

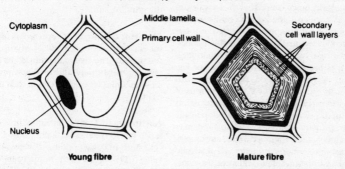

Cytoplasm

Middle lamella

Primary cell wall

Secondary
cell wall layers

Nucleus

Young fibre

Mature fibre

FIG. 25b *Secondary cell wall deposition by a phloem fibre cell, to show different wall layers.*

rather little (SEE GRAM'S STAIN). Mucins may also be present. The cell wall structure of the blue-green algae and prochlorophytes is very similar to the Gram-negative bacteria, but more complex, comprising several layers, while outside is a mucilage layer (sheath or capsule) which is fibrillar. Protruding from the wall of some blue-green algae are fimbriae or pili, which are possibly involved in prokaryotic–eukaryotic interactions (e.g. symbiosis). Comparatively rigid, these and the chitinous walls of fungal cells and hyphae provide mechanical support. Protistan algae may also have cell walls comprising fibrillar and amorphous components – again, the most common fibrillar component is cellulose; but in some siphonaceous green algae and red algae, mannan (polymer of 1,4-linked β-D-mannose) and xylans (of different polymers) replace cellulose. The amorphous mucilaginous components occur in greatest amounts in the brown and red algae, and are commercially exploited (see ALGINIC ACID, FUCOIDIN, AGAR, CARRAGEENAN). Other protista may have incomplete cellulose walls (e.g. LORICA or theca of *Dinobyron, Trachelomonas*); others have exquisitely ornate siliceous walls (e.g. BACILLARIOPHYTA – diatoms). In *Chlamydomonas*, the glycoprotein *extensin* forms the entire cell wall. See CHRYSOPHYTA, PELLICLE, SCALE.

Walls of newly formed plant cells (PLANTAE) are at first very thin and permeable, thickening as cells mature. At plant CELL DIVISION (see Fig. 23), PECTIC COMPOUNDS are laid down in the CELL PLATE across the equatorial plane of the division's spindle forming the MIDDLE LAMELLA, intercellular material cementing adjacent cells together. Each new cell lays down a *primary wall* consisting of CELLULOSE (fibrillar component; polymer of $1,4$-linked β-D-glucose), HEMICELLULOSES, and negatively charged pectins (see CUTIN). Hydrogen bonds bind hemicellulose molecules to cellulose microfibrils, cross-linking them. Pectin molecules, being negatively charged, bind cations such as calcium (Ca^{2+}), and in doing so form a gel-like matrix (amorphous component) filling the interstices between the cellulose microfibrils, holding them together. Glycoprotein molecules probably attach to pectins (see Fig. 25a). At maturity, a cell may remain with just its primary wall (e.g. in some forms of PARENCHYMA); in others, after cell growth has ceased, a *secondary wall* may develop inside the primary wall (see Fig. 25b) and may include LIGNIN (lignified walls not as permeable to water as primary wall), SUBERIN (when impermeable to water) or SPOROPOLLENIN. During deposition of these layers, certain small areas remain largely unthickened, forming PITS. Pits of adjacent cells usually coincide, so that in these areas protoplasts are separated by the PIT MEMBRANE on each side. Through the pit membrane pass the majority of the plasmodesmata, fine protoplasmic connections which are elements of the SYMPLAST. Some walls undergo further modifications, waxy cuticles developing on epidermal cells; others undergo suberization (e.g. cork cells which become impermeable to water). Lignification of fibre

vessels and tracheids gives them more strength and rigidity.

The plant cell wall limits CELL GROWTH (see AUXINS, EXPANSINS), is a barrier to digestion (especially when toughened by aromatic polymers; see TANNINS), glues adjacent cells together and plays an important role in plant morphogenesis. Its stretch-resistance is a major contributory factor to a plant cell's WATER POTENTIAL. Cell walls can contain enzymes which incompletely digest its polysaccharides, releasing oligosaccharides that can act like growth substances and serve in cell-to-cell signalling (see CELLULOSE, CHITIN). They also contain CALMODULIN, regulating the concentration of free Ca^{2+} in the wall, which in turn may affect cell growth and development.

cement (cementum) Modified bone surrounding roots of vertebrate teeth (i.e. below gum), binding them to periodontal ligament by which tooth is attached to jaw. In some herbivorous mammals, occurs between folds of the tooth, forming part of the grinding surface.

Cenozoic (Cainozic) The present geological era; extends from about 65 Myr BP to the present. The 'age of mammals'. Its two periods, the Tertiary and the Quaternary, are sometimes regarded as eras in their own rights.

centimorgan (cM) Unit of relative distance between genes on a chromosome, 1 centimorgan corresponding to a cross-over value (COV) of 1%; in the human genome, this corresponds to about 10^6 base pairs. See HUMAN GENOME PROJECT.

central dogma (Of biology) the proposal, originally in a version by F. H. C. Crick in 1958, that information can flow from nucleic acid to protein and from one nucleic acid to another, but cannot flow from protein to nucleic acid or from one protein to another. Prior to the discovery of REVERSE TRANSCRIPTASE, information was thought to flow from DNA, to RNA, and to protein and never from RNA to DNA.

central mother cells Relatively large vacuolated cells in a subsurface position in the apical meristem of a plant shoot.

central nervous system (CNS) A body of nervous tissue integrating animal sensory and motor functions and providing through-conduction pathways to transmit impulses rapidly, usually medially, along the body. In vertebrates it comprises the BRAIN and SPINAL CORD; in annelids and arthropods a pair of solid ventral nerve chains, each with segmental ganglia, and a pair of dorsal ganglia anteriorly serving as a 'brain', united to the nerve chains by commisures. Impulses travel to and from the CNS via peripheral nerves (vertebrate spinal nerves), while local reflex arcs (vertebrate spinal reflexes) produce adaptive responses to stimuli independently of higher centres (the brain), although these centres initiate and coordinate actions and store memory. See NERVOUS SYSTEM, SPINAL CORD.

centric diatom Common term for a diatom (BACILLARIOPHYTA) which is radially symmetrical when viewed in valve view.

centriole Organelle (probably of endosymbiotic origin) found in cells of those eukaryotic organisms which have cilia or flagella at some stage in their life cycle; hence absent from higher plants. Each comprises a hollow cylinder composed of nine sets of triplet microtubules held together by accessory proteins. Each is 300–500 nm long and 150 nm in diameter. Often functionally interconvertible with BASAL BODY. They occur at right angles to each other near the nucleus, separating at cell division and organizing the spindle microtubules (which arise from material surrounding the centriole, but possibly in turn organized by it). Centrioles generally arise at right angles to existing centrioles. Normally an animal obtains its centrioles from the sperm cell at fertilization; rarely, an egg may form its own (see PARTHENOGENESIS). Centrioles possess their own DNA and appear to be self-replicating, and there may be a link between centriole replication and nuclear DNA replication. Similar or identical structures (BASAL BODIES), possibly functionally interconvertible,

occur at the bases of cilia or flagella in cells which have these. See CENTROSOME.

centrolecithal Of eggs (typically insect) where yolk occupies centre of egg as a yolky core. See TELOLECITHAL.

centromere (spindle attachment) A chromosome region holding sister chromatids together until mitotic or second meiotic anaphase. The position of a centromere defines the ratio between the lengths of the two chromosome arms. Centromeres may be associated with REPETITIVE DNA sequences (not in the yeast *Saccharomyces cerevisae*) and centromeric DNA may be late-replicating. They either include or correspond to KINETOCHORES, which attach to the spindle fibres and by replicating at late metaphase allow the forces pulling sister chromatids apart to operate only if chromosomes are properly aligned. Normally one per chromosome; but chromosomes with 'diffuse' centromeres (e.g. those of many lepidopterans) permit spindle fibre attachment along the whole chromosome length. See ACENTRIC, ACROCENTRIC, METACENTRIC, TELOCENTRIC.

centrosome (cell centre) Amorphous, electron-dense material (in most animal cells surrounding and including the CENTRIOLE pair) situated close to the nuclear envelope, serving as a microtubule-organizing centre (see MTOC). Despite a lack of clear physical boundaries, centrosomes have their own duplication cycle lasting through all phases of the CELL CYCLE. Removal of the centriole itself seems not to prevent expression of a new MTOC but prevents centrosome partitioning and formation of a normal bipolar mitotic spindle. They must be accurately duplicated once per cell cycle, and activation of the $p34^{cdc2}$ kinase controlling M-phase of the cell cycle seems to depend on this. Functional details are expected to emerge as the biochemistry of this enigmatic organelle is better understood; e.g., the actin-related protein centractin is associated both with centrosomes and cytoplasmic dynein. Differences between centrosomes control mRNA targeting. In one mollusc embryo (*Ilyanassa* *obsoleta*) different mRNAs associate with different centrosomes in different cells during embryonic cleavage cycles, moving in a microtubule-dependent manner to the pericentriolar matrix after being diffusely distributed in the cytoplasm. During division, they assemble from the core mitotic centrosome and move by actin transport to the presumptive 'animal' daughter cell cortex (SEE ANIMAL POLE).

centrum Bulky part of a vertebra, lying ventral to spinal cord. In function, as in development, replaces the notochord. Each is firm but flexible, attached to adjacent centra by collagen fibres.

Cephalaspida (Osteostraci) Extinct group of monorhine vertebrates. See AGNATHA.

cephalic index In anthropometry, the value of

$$(B/L) \times 100$$

where B is the maximum breadth of the skull, generally in the vicinity of the parietal eminences, and L is the maximum length of the skull, as measured with spreading calipers from the glabella (bony prominence between eyebrows) to the opisthocranion (most posterior point on occiput). A crude measure of skull shape in the horizontal plane; a continuum, with no clear heritability.

cephalization The tendency, during evolution of animals with an antero-posterior axis, for sense organs, feeding apparatus and nerve tissue to proliferate and enlarge at the anterior end, forming a head.

Cephalochordata (acrania) Subphylum of marine chordates characterized by persistence of notochord in adult, extending (unlike in vertebrates) to the tip of the snout. Metameric segmentation, dorsal hollow nerve cord, gill slits and post-anal tail also present. Amphioxus (*Branchiostoma*) is typical. Compare UROCHORDATA.

Cephalopoda Most advanced class of the phylum MOLLUSCA. All are aquatic, and most marine, possessing a well-developed head surrounded by a ring of prehensile tentacles;

and a muscular siphon derived from the foot through which water is forced from the mantle cavity during locomotion. Primitively (e.g. *Nautilus* and extinct AMMONITES) the animal inhabits the last chamber of an external spiral shell which also serves for buoyancy; in the cuttlefish *Sepia* the shell is internal, while in squids it is much reduced, and absent altogether in *Octopus*. The complexity of cephalopod eyes rivals that of vertebrates (and provides an example of convergent evolution), while the large brain enables powers of learning and shape recognition on a par with simple vertebrates. Much has to be learnt about cephalopod communication; some believe that cuttlefish employ their phenomenal powers of colour and pattern change to this effect.

cephalothorax Term indicating either fusion of, or indistinctness between, head and some or all anterior trunk (thoracic) segments in crustacean and arachnid arthropods.

cercaria The last larval stage of flukes (Order Digenea); produced asexually by POLYEMBRYONY within preceding redia larva inside secondary host, often a snail, from which it emerges and swims with its tail to penetrate skin of primary host (e.g. man in *Schistosoma* causing bilharzia) or to encyst as a metacercaria awaiting ingestion by primary host.

cerci A pair of appendages, often sensory, at the end of the abdomen of some insects. Long in mayflies, short in cockroaches and earwigs (where they are curved).

Cercopithecoidea Old World monkeys. See ANTHROPOIDEA.

cereal ('grain') Any grass (see POACEAE) producing starchy seeds and cultivated as a crop. Includes MAIZE (or corn, *Zea mays*), RICE (*Oryza sativa*), WHEAT (*Triticum aestivum*), oats (*Avena sativa*), barley (*Hordeum* spp), SORGHUM (*Sorghum bicolor*) and millets (e.g. *Eleucine coracana*). Major STAPLE foods in most countries (see AMINO ACID). Cultivated bread wheats have evolved from wild ancestors by interspecific breeding and POLYPLOIDY, and cultivated maize is also polyploid. See COMPLEMENTARY RESOURCES, CROP GENOME PROJECT.

cerebellum Enlargement of the hindbrain of vertebrates, anterior to the medulla oblongata. Coordinates posture (balance) during rest and activity through reflexes initiated by inputs mainly from the VESTIBULAR APPARATUS fed via acoustic regions of the medulla (the lower vertebrate ACOUSTICO-LATERALIS SYSTEM), and from the PROPRIOCEPTORS in muscles and tendons. Compares intended with actual movements and sends impulses to cerebral cortical motor areas to modify behaviour accordingly. Does not directly innervate skeletal muscles. In mammals, covered in a cortex of grey matter. See Fig. 18 (BRAIN).

cerebral cortex (pallium) Layer of GREY MATTER rich in synapses lying atop white matter, covering cerebral hemispheres of amniote and some anamniote vertebrates. In advanced reptiles and all mammals a new association centre, the neopallium, appears in the cortex receiving sensory inputs from the brainstem and initiating actions via motor bundles of the pyramidal tract. Its evolving dominance in mammalian brain involves its reception of increasingly wide ranges of sensory information via the thalamus and the emergence of higher neural (i.e. mental) activities based upon these data. Folding of the cortical surface in mammals provides a large surface area for synaptic association.

cerebral hemispheres (cerebrum) Paired outpushings of vertebrate forebrain, originally olfactory in function, whose evolution has involved progressive movement of grey matter to its surface and an increasing role as an association and motor control centre. The CEREBRAL CORTEX dominates the mammalian brain both physically and functionally. The two hemispheres are anatomically asymmetrical. Although it is often said that in humans the left hemisphere excels in intellectual, rational and verbal reasoning while the right excels in sensory discrimination, emotional, nonverbal and intuitive thought, there is plenty of evidence that each needs inputs from the other, via the corpus callosum, for proper functioning. We cannot clearly demarcate their roles.

cerebroside SPHINGOLIPID of the myelin sheaths of nerves, the commonest being *galactocerebrosides* with a polar head group containing D-galactose. Other tissues contain small amounts of glucose-containing cerebrosides.

cerebrospinal fluid (CSF) Fluid filling the hollow neural tube and subarachnoid space of vertebrates. Secreted continuously into ventricles of the brain by the choroid plexuses and reabsorbed by veins. Clear and colourless fluid, with some white blood cells, supplying nutrients. Serves as shock absorber for the central nervous system. About 125 cm^3 present in humans. See MENINGES.

cerebrum See CEREBRAL HEMISPHERES.

cervical (Adj.) Of the neck; or CERVIX. Cervical vertebrae have reduced or absent ribs; almost all mammals (including giraffe) have seven.

cervix Cylindrical neck of mammalian uterus, leading into vagina. Glands secrete mucus into vagina.

Cestoda Tapeworms. Endoparasites (Class PLATYHELMINTHES) lacking gut and absorbing digested food from host gut lumen across microtriches, minute folds of the surface epithelial cell membranes similar to microvilli. Tapeworms are unsegmented, but body sections (proglottides) budded off from head region (scolex) give segmented appearance. Sequentially hermaphrodite, young proglottides male but become female with age. Self-fertile. Life cycle involves primary and secondary hosts. Larva a six-hooked onchosphere egested in proglottis with faeces of primary host. Sense organs reduced.

Cetacea Whales. Order of placental (eutherian) mammals. Entirely aquatic. Digit symmetry in the vestigial hindlimbs of the Eocene fossil whale *Basilosaurus* supports a common ancestry of cetaceans and artiodactyls with mesonychid CONDYLARTHRA. Morphology convergent with ichthyosaurs, with a dorsal fin, forelimbs developed as flippers, and tail a powerful fluked swimming organ. Eocene fossil *Ambulocetus* (52 Myr BP) represented a 'seal-like' amphibious phase, with well-developed front and hind limbs (hind feet paddle-shaped and huge) and pelvic girdle; toes hoofed; tail long. The later fossil, *Rhodocetus*, was more specialized for swimming. Subcutaneous fat (blubber) for thermal insulation. Dorsal blowhole connects with lungs. Includes Odontoceti (toothed whales, including porpoises and dolphins), Mysticeti (baleen whales) and extinct Archaeoceti; but toothed-whale monophyly has been questioned by molecular studies. See http://www.cetacea.org.

CFCs (chlorofluorocarbons) Halocarbons containing carbon, chlorine and fluorine, all human-derived. Global mean tropospheric concentrations of the two most important CFCs in global warming through the GREENHOUSE EFFECT, i.e. CFC-11 (CFCL$_3$) and CFC-12 (CF$_2$CL$_2$), increased by about 6% between 1977 and 1986. Their main uses are in aerosols, air conditioning and refrigerators, while their major natural sink is photolytic breakdown in the stratosphere at 20–40 km, releasing free chlorine atoms which combine with OZONE, decreasing the latter's concentration and UV-absorbing effect. Their radiative effect per molecule relative to CO$_2$ varies from 11,000–14,000 times.

c-*fos* Ubiquitous 'immediate-early' ONCOGENE, encoding a LEUCINE ZIPPER-type transcription factor (Fos) required for normal cell division and containing large 3'-untranslated regions. Often dimerizes with c-Jun protein, after which it binds DNA more strongly.

chaeta Chitinous bristle characteristic of oligochaete (where few) and polychaete (where many) annelid worms. In polychaetes they are borne on parapodia. Assist in contact with substratum during locomotion. See SETA.

Chaetognatha Arrow-worms. Small phylum of hermaphrodite marine coelomate invertebrates, abundant in plankton. Hermaphrodite. Probably protostomes, on molecular evidence; in which case their radial cleavage is a counterexample to the view that this feature is a deuterostome autapomorphy.

chain elongation Growth of polypeptide on a ribosome during PROTEIN SYNTHESIS. See ELONGATION FACTORS.

chalaza (Bot.) Basal region of ovule, where the stalk (funiculus) unites with the INTEGUMENTS and the NUCELLUS. (Zool.) (Of a bird's egg), the twisted strand of fibrous ALBUMEN; two are attached to the vitelline membrane, one each at opposite poles of the yolk, lying in the long axis of the egg. They stabilize the position of the yolk and early embryo in the albumen.

chalk Variety of limestone, fine-grained and generally friable, largely comprising inorganic remains (shells of foraminiferans and molluscs, tests of sea urchins) of organic organisms, often with bands of silicious flint nodules formed from sponge spicules, dissolved and redeposited.

chalone Term once used for substances now regarded as growth-inhibiting GROWTH FACTORS.

chamaerophytes Class of RAUNKIAER'S LIFE FORMS.

channel proteins See ION CHANNELS.

chaos Biological systems can often be studied at more than one level of organization or complexity and each may exhibit dynamic properties. Symptoms of what has come to be termed 'chaotic behaviour' arise when attempts to model these deterministic systems mathematically fail to predict their future states in all but the very short term, and when the system's characteristics show unexpected sensitivity to slight differences in initial conditions. Importantly, equilibrium points (points of stability) within the system do not recur despite stability of the system as a whole, which is either dominated by POSITIVE FEEDBACK or displays such over-compensating negative feedback that instability arises in just those regions where positive feedback occurs. In ecology, DEN-SITY-DEPENDENCE is central to the notion of chaos, for populations can be affected by direct density-dependence (rapid negative feedback), delayed density-dependence (delayed negative feedback) and inverse density-dependence (positive feedback). Although the jury is still out on whether natural ecosystems exhibit chaotic dynamics, there is concern that they may be made chaotic by human actions. See BALANCE OF NATURE.

chaperones Proteins helping non-covalently in the assembly and folding of a protein during translation without forming part of its final assembly (see PROTEIN SYNTHESIS). Preventing non-productive protein–protein interactions and premature attainment of mature protein conformation, they can enable a protein to pass through a membrane (pore) in open conformation. One class of chaperones is the *heat-shock proteins* (prokaryotic and eukaryotic types occur) whose synthesis and/or activity rate is increased when the cell is subjected to higher-than-optimal temperatures (the appropriate ENHANCER is sometimes itself heat-affected, but heat-induced phosphorylation may be a factor). They are also implicated in ATP-driven protein transport across (e.g. mitochondrial) membranes, which involves temporary loss of tertiary structure. Chaperones provide hydrophobic surfaces which stabilize other proteins and either prevent them from denaturing or promote refolding after heat-induced modification. Examples of the other major chaperone class, *chaperonins*, have been isolated from bacteria, chloroplasts and mitochondria. SIGNAL RECOGNITION particles also have chaperoning functions. Poorly structured native conformations of proteins have been implicated in several neurodegenerative disorders, including ALZHEIMER'S and PARKINSON'S DISEASE, and PRION diseases. Adoption of β-sheet structures leads to formation of protein aggregates (plaques), and one suggestion is that a molecular chaperone may be involved, possibly stabilizing insoluble intermediates rich in β-sheets.

chaperonins A class of molecular CHAPERONES, restricted to bacteria, mitochondria and chloroplasts.

character displacement Evolutionary phenomenon whereby, it is believed, interspecific competition causes two closely related species to become more different in regions where their ranges overlap than in regions where they do not. Such differences are often anatomical, but may involve any aspect of phenotype. Few rigorously documented examples exist where such differences have been shown to be due to competition; but studies on fish in postglacial lakes and lizards on Caribbean islands indicate multiple speciation events accompanied by similar patterns of ecological and morphological divergence involving, apparently, character divergence – implicit as this is in current thinking on ADAPTIVE RADIATION. See SPECIATION.

Charophyceae Stoneworts, e.g. *Chara*, *Nitella*. A class of green algae (CHLOROPHYTA). Most grow in still, clear freshwater (e.g. ditches, ponds and lakes) where they form extensive underwater meadows, anchored into the sediment by their rhizoids. They are particularly abundant in hard, basic waters (ph>7), where the cell walls often become encrusted with calcium carbonate. Sometimes they can be found living in clear, hard-water streams (e.g. in the Canadian Rocky Mountains).

Stoneworts are macroscopic, growing several decimetres tall. They are differentiated into a series of nodes and internodes, which alternate along the axis of the plant. Whorls of branches, each of limited growth, are borne at each node. Arising from the axils of some side branches are others that have unlimited growth like the main axis. Initially cells are uninucleate becoming multinucleate with age. Many round, discoid chloroplasts are present in the peripheral layer of the cytoplasm. These lack pyrenoids, and in cells that are elongating, they divide continuously. There is a large vacuole in the centre of each cell when fully mature. The cytoplasm close to the vacuole streams around the cell incessantly, in a longitudinal direction. The structure part of the cell wall is fibrillar and comprises microfibrils of crystalline cellulose. The cellulose microfibrils are arranged in a helicoidal, 'crossed fibrillar pattern'. As in the ZYGONEMATOPHYCEAE, mosses, liverworts and vascular plants, the microfibrils are synthesized by rosette-like complexes of cellulose synthase molecules, which lie in the plasmalemma. Flagellate cells are formed during sexual reproduction in the form of asymmetrical male gametes, which are covered with tiny organic scales. These are diamond-shaped and resemble the inner body scales of the PRASINOPHYCEAE. The motile cells usually lack an eyespot while the flagellar root comprises a broad band of microtubules and a second smaller microtubular root. A multi-layered structure may be present, but no rhizoplasts. Mitosis is open and the telophase spindle is persistent. Cytokinesis is brought about by formation of a cell plate, which is derived from Golgi vesicles. It lies with a phragmoplast. The cross walls are pierced by plasmodesmata. Sexual reproduction is oogamous. The ovoid oogonium is borne on determinate lateral branches and comprises a large egg cell surrounded by narrow sterile cells. Antheridia are also borne on determinate laterals. They are spherical and contain many multicellular, spermatogenous threads, each cell of which produces a single spermatozoid. The antheridium is surrounded by a covering of sterile cells. Stoneworts are an ancient group of algae. A recent (2001) four-gene analysis using plastid, mitochondrial and nuclear DNA sequences, supports the hypothesis that land plants lie within the Charophyceae (Charophyta) and identifies Charales (stoneworts) as their closest living relatives. Calcified envelopes of the zygotes, called gyrogonites, have been found from the Silurian period dating to 420 Myr. Present day genera date back to the Mesozoic era (130–200 Myr ago).

checkpoints The CELL CYCLE checkpoints, activated by damaged or unreplicated DNA, turn on signalling pathways which block cyclin-dependent kinase (CDK) activity. Cdks, with their cyclin-binding partners, regulate cell cycle progression. Members of the Cdc25 family of phosphatases are crucial in activating several of these Cdk–cyclin complexes, and are variously degraded in response to such stimuli as blocked DNA replication and ionizing radiation such as ultraviolet light, leading to cell cycle arrest.

chela The last joint of an arthropod limb, if it can be opposed to the joint preceding it so that the appendage is adapted for grasping, as in pincers of lobster and some CHELICERAE. Such a limb is termed *chelate*.

chelicerae Paired, prehensile first appendages of CHELICERATA, contrasting with antennae of other groups. Often form CHELAE (when said to be *chelate*).

Chelicerata Probably natural assemblage containing those arthropods with chelicerae. Includes MEROSTOMATA and ARACHNIDA. No true head, but an anterior tagma termed the PROSOMA. Mandibles absent. Probably closed related to trilobites. See ARTHROPODA, BIRAMOUS APPENDAGE (for limb diagram), MOUTHPARTS.

Chelonia (Testudines) Tortoises and turtles. Anapsid reptile order, with bony plates enclosing body or covered by epidermal horny plates. Shoulder and pelvic girdles uniquely within rib cage. Teeth absent.

chemiosmotic theory (chemiosmotic-coupling hypothesis) Hypothesis of P. Mitchell, now generally accepted, that chloroplasts and mitochondria require their appropriate membrane to be intact so that a proton gradient created across it by integral membrane pumps can be coupled to ATP synthesis as protons return across the membrane down their electrochemical gradient. See CHLOROPLAST, CYTOCHROMES, ELECTRON TRANSPORT SYSTEM, MITOCHONDRION.

chemoautotrophic (chemosynthetic) See CHEMOTROPHIC.

chemoheterotrophic See CHEMOTROPHIC.

chemokines Small chemoattractant proteins (currently 30 forms). Almost any nucleated cell can release some form of chemokine in response to infection. RANTES, MIP-1α (macrophage-inflammatory protein 1α) and MIP-1β help attract immune cells to sites of INFLAMMATION during infection. Chemokine receptor 5 (CCR5) is a receptor for these three ligands, and a co-receptor with CD4 in HIV-1 infection (see HIV).

chemoorganotroph An organism obtaining its energy by oxidation of organic compounds (which are therefore electron donors). Compare AUTOTROPHIC, CHEMOSYNTHESIS. See HETEROTROPHIC.

chemoreceptor (1) In bacteria, molecules (also termed chemosensors) located either in the periplasmic space or spanning the plasma membrane (when signal transduces) which, by binding attractor or repellant molecules from the environment, activate a motor system controlling FLAGELLUM behaviour and hence cell direction (chemotaxis). (2) Receptor cell responding to chemical aspects of internal or external environment (aqueous solubility of signal molecule being a prerequisite). Taste and olfaction are chemosenses. See SIGNAL TRANSDUCTION and RECEPTOR for some components of cellular pathways. See CAROTID BODY.

chemostat A BIOREACTOR (see Fig. 13) in which maintenance of microbial culture density is achieved by exhaustion of some limiting substrate (e.g. nutrient). As such, chemostatic control differs from control by optical density of culture (turbidostatic maintenance).

chemosynthesis The synthesis of organic molecules by certain bacteria that obtain their energy by oxidation of specific inorganic molecules or ions, which serve as electron donors (i.e. chemolithotrophs). See AUTOTROPHIC, LITHOTROPH. Compare CHEMOORGANOTROPH.

chemosynthetic See CHEMOAUTOTROPHIC.

chemotaxis Directional response by motile cells to chemical gradient. Important in fertilization, inflammation, diapedesis, wound healing, angiogenesis, metastasis and axonal targeting. Often, but not always, involves G-protein-linked signal transduction pathways. See CHEMOKINES, CHEMORECEPTOR.

chemotrophic Of organisms obtaining energy by chemical reactions independent of light. Reductants obtained from the environment may be inorganic (CHEMOAUTOTROPHIC), or organic (chemoheterotrophic). See AUTOTROPHIC, HETEROTROPHIC, PHOTOTROPHIC.

chemotropism (Bot.) TROPISM in which stimulus is a gradient of chemical concentration, e.g. downward growth of pollen tubes into stigma due to presence of sugars. (Zool.) Rarely used as a synonym of CHEMOTAXIS.

Chengjiang fauna A remarkable assemblage of fossil forms from southern China comparable with, and slightly older than, the BURGESS SHALE (see also CAMBRIAN).

chiasma (pl. chiasmata) (1) The visible effects of the process of genetic CROSSING-OVER between chromosomes which have paired up (i.e. between bivalents) in appropriately stained meiotic cells, and hence indicators of homologous (non-random) RECOMBINATION. Each chiasma may involve either of the two chromatids of each chromosome. Appreciation that chiasmata result from breakage and reciprocal refusion between chromatids during the first meiotic prophase was a major achievement of classical cytogenetics and is due largely to Jannsens and Darlington (their *chiasmatype theory*). The molecular mechanism involved may incorporate enzymes that were formerly part of a DNA REPAIR MECHANISM. Several chiasmata may occur per bivalent, longer bivalents having more on average. Their frequency and distribution are not entirely random and are sometimes under genetic control. See SUPPRESSOR MUTATION. (2) See OPTIC CHIASMA.

chief cells (zymogenic cells) Pepsinogen-secreting cells of the gastric pits of the vertebrate stomach. Also secrete gastric lipase.

Chilopoda Centipedes and their allies. Class (or Subclass) of ARTHROPODA. See MYRIAPODA.

chimaera (1) Usually applied to organisms which (*contra* MOSAICS) comprise cells of two or more distinct genomes, resulting from experimental manipulation (e.g. grafting or aggregation), often early in development (see EMBRYONIC STEM CELLS). Sometimes used in the context of cells and organisms (e.g. CRYPTOPHYTA) whose evolutionary history has involved ENDOSYMBIOSIS, where parental genomes remain more or less intact. (2) A genus of holocephalan fish (*Chimaera*).

Chiroptera Bats. Order of eutherian (placental) mammals; characterized by membranous wing spread between arms, legs, and sometimes tail, generally supported by greatly elongated fingers. Use of echolocation for avoidance of objects and food capture during commonly nocturnal insectivorous feeding. Some are plant pollinators. Diurnal fruit-eating bats (suborder Megachiroptera) share more derived (apomorphous) anatomical characters with euprimates and dermopterans than do the smaller-eyed echolocating bats (suborder Microchiroptera); but immunological evidence suggests that all bats form a cohesive phylogenetic unit. See ARCHONTA.

chi-squared (χ²) test Statistical test for assessing the significance of departures of sets of whole numbers (those observed) from those expected by hypothesis, as when scoring phenotypic classes obtained from a genetic cross. The formula used is

$$\chi^2 = \sum \frac{(n_{obs} - n_{exp})^2}{n_{exp}}$$

The value obtained has to be assessed in relation to the number of *degrees of freedom*, which is the number of classes minus 1, and a χ^2 table will then give the probability (P) of finding as poor a fit with the expected results owing to random sampling error. If, for instance, $P < 0.05$, the data are said to be significantly different from expectation at the 5 per cent level. The χ^2 test becomes seriously inaccurate if any of the expected numbers is less than 5. See NULL HYPOTHESIS.

chitin Nitrogenous polysaccharide found in many arthropod exoskeletons, hyphal walls of many fungi and cell walls of the protistan CHYTRIDIOMYCOTA. Comprises repeated N-acetylglucosamine units (β[1,4]-

linked). Strictly a PROTEOGLYCAN, owing to peptide chains attached to its acetamido groups. Of considerable mechanical strength, hydrogen bonding between adjacent molecules stacked together forming fibres giving structural rigidity; also resistant to chemicals. With lignocellulose, among the most abundant of biological products.

Chlamydia Genus of obligately parasitic bacteria, with specialized loss of metabolic function, giving them possibly the simplest biochemical repertoire of all cells. *C. psittaci* causes psittacosis; *C. trachomatis* causes trachoma. Unrelated to rickettsias and other Gram-negative bacteria, they are Gram-negative despite lacking peptidoglycan in their walls.

chlamydospore Thick-walled fungal spore capable of surviving conditions unfavourable to growth of the fungus as a whole; asexually produced from a cell or portion of a hypha.

chloramphenicol Antibiotic, formed originally by *Streptomyces* bacteria, inhibiting translation of mRNA on prokaryotic ribosomes, eukaryotic translation being unaffected. Its use can thus distinguish proteins synthesized by mitochondrial/chloroplast ribosomes from those manufactured in the rest of eukaryotic cell. See CYCLOHEXIMIDE.

Chlorarachniophyta Division of the ALGAE. Cells are naked, amoeboid, uninucleate and united via filopodia into a net-like plasmodia. Each cell can form a coccoid resting stage or produce uniflagellate zoospores. The zoospores are ovoid and have a single flagellum that bears delicate hairs. The flagellum is inserted a little below the apex of the cell. When the cell is in motion the flagellum wraps around the cell in a downward spiral, lying in a groove along the body of the cell. The zoospore has no eyespot. The chloroplasts are bilobed, discoid, bright green and surrounded by four membranes. The inner two are the two chloroplast membranes, while the outer two are the chloroplast endoplasmic reticulum. Thylakoids are usually in stacks of one to three. The pyrenoid is pear-shaped and around it is a vesicle containing reserve polysaccharide (probably paramylon) situated outside the four membrane chloroplast envelope. The chloroplast has chlorophylls *a* and *b*. A peculiar organelle, the nucleomorph, is located in a pocket of the cytoplasm in a depression in the surface of the pyrenoid, between the chloroplast envelope and the chloroplast endoplasmic reticulum. This nucleomorph contains DNA and a nucleolus-like body, and is surrounded by a porous, double membrane. It is hypothesized to be a vestigial nucleus of a photosynthetic, eukaryotic endosymbiont (probably a chlorophyte) that became incorporated into the heterotrophic, amoeboid ancestor of this division. Further support for this hypothesis comes from the observation that *Chlorarachnion* is able to feed phagotrophically. Lastly, trichocysts are found below the plasmalemma.

Only two representatives of this group have been found so far and both are associated with cultures of tropical or subtropical siphonaceous green algae. As a group they provide further evidence for the hypothesis that many groups of eukaryotic algae have arisen through incorporation of other eukaryotic algae by ENDOSYMBIOSIS with colourless, heterotrophic protozoa.

chlorenchyma Parenchymatous tissue containing chloroplasts.

chloride shift Entry/exit of chloride ions across red blood cell membranes to balance respective exit/entry of hydrogen carbonate ions resulting from CARBONIC ANHYDRASE activity. See BOHR EFFECT.

chlorocruorin Respiratory pigment (green, fluorescing red) dissolved in plasma of certain polychaete worms. Conjugated iron-porphyrin protein resembling haemoglobin.

Chlorophyceae Class of the CHLOROPHYTA – green algae. This class comprises about 355 genera and 2,650 species with the vast majority being in freshwater. Terrestrial and a few brackish-water and marine forms also exist. A few members of the Volvocales are colourless and heterotrophic but have a

similar morphology and ultrastructure to the photosynthetic species. Members of the Chlorophyceae including free-living flagellates, either unicellular or colonial; others are non-motile coccoid or palmelloid, filamentous, thallose or siphonaceous. Flagellate members have cells surrounded by a glycoprotein envelope, while the non-flagellated forms have firm polysaccharide walls. Asexual reproduction occurs via production of zoospores, which have two or four apically inserted flagella. Some members only produce non-flagellate spores (aplanospores), while others produce both. Sexual reproduction is isogamous, anisogamous or oogamous. The life cycle is haplontic, including a hypnozygote (thick-walled resting zygote). Zoospores have a cruciate type of flagellar root system with a 1 o'clock–7 o'clock or 12 o'clock–6 o'clock configuration of the basal bodies. The stephanokont zoospores of the Oedogoniales, which have a crown of flagella around the anterior of the cell, are considered to have been derived from the simpler cruciate types of zoospores. Mitosis is closed, the telophase spindle is not persistent and cytokinesis occurs via the formation of a transverse septum, which develops within a phycoplast. The septum is formed via a cleavage furrow (an invagination of the plasmalemma) or via a coalescence of vesicles within a cell plate.

chlorophyll Green pigment found in all algae and higher plants except a few saprotrophs and parasites. Responsible for light capture in PHOTOSYNTHESIS. Located in CHLOROPLASTS, except in CYANOBACTERIA (blue-green algae) where borne on numerous photosynthetic membranes (thylakoids) dispersed in the cytoplasm at the periphery of the cell. Each molecule comprises a magnesium-containing *porphyrin* group, related to the prosthetic groups of haemoglobin and the cytochromes, ester-linked to a long *phytol* side-chain (see ISOPRENOIDS). Several chlorophylls exist (*a*, *b*, *c*, *d* and *e*), with minor differences in chemical structure. Chlorophyll *a* is the only one common to all plants (and the only one found in blue-green algae). In photosynthetic bacteria, other kinds of chlorophyll (*bacteriochlorophylls*) occur (see also BACTERIORHODOPSIN). Can be extracted from plants with alcohol or acetone and separated and purified by chromatography. See ACCESSORY PIGMENTS, ANTENNA COMPLEX.

Chlorophyta Green algae. This algal division contains about 500 genera and 8,000 species. These algae are primarily found in freshwater habitats; only 10% are marine. If sufficient moisture is present, some species grow aerially (e.g. on the bark of trees, attached to mosses or rocks). These aerial species include unicellular coccoid species (e.g. *Trebouxia*) and filamentous ones (e.g. *Trentepohlia*). Species of these genera are also found as the phycobiont within lichens. One genus, *Cephaleuros*, is a plant parasite responsible for the disease red rust, an economically important disease of tea plants in some regions of the world; it also causes economic losses in other crops (e.g. citrus and peppers). Morphologically diverse, including motile flagellated unicells and coenobia, palmelloid colonies, unbranched and branched filaments, parenchymatous and siphonaceous thalli. Many unicellular and colonial green algae live in the plankton while many others are found in the benthos, growing epilithically or epiphytically. Many filamentous green algae are attached to a substratum during the early stages of their development and later become free floating, form mats or balls composed of many intertwined filaments. Particularly striking are the green growths and mats of green algae that sometimes are found in shallow ponds, ditches, around the edges of small lakes or in quiet backwaters of streams. Along rocky coasts in the upper intertidal zone, rocks are often carpeted with green algae (e.g. *Ulva* – sea lettuce, *Enteromorpha*, *Cladophora*). In Japan both *Enteromorpha* and *Ulva* species are grown for food. They are cultivated in nutrient-rich bays or estuaries, where they are grown attached to racks or nets. Green algae are also found living on sandy or muddy bottoms of tropical lagoons (e.g. *Caulerpa*, *Penicillus*). Green algae can also be found living in snow and ice. *Chlamydomonas nivalis* grows in the Rocky Mountains on permanent snow, imparting a pink to red colour. In this species chlorophyll *a* is masked by 'haematochrome', a mixture of carotenoid pigments.

Chlorophyte cells are characteristically bright green, as the chloroplasts contain chlorophylls *a* and *b*. This characteristic is shared by the PROCHLOROPHYTA, EUGLENOPHYTA, CHLORARACHNIOPHYTA, BRYOPHYTA and vascular plants. The chloroplast of the green algae is enclosed by a double-membraned chloroplast envelope. In this respect these algae resemble the GLAUCOPHYTA, RHODOPHYTA, Bryophyta and vascular plants. Thylakoids of the chloroplast are grouped to form lamellae containing two to six or more thylakoids, pseudograna (stacks formed by partial overlap of thylakoids) or grana (discrete, almost columnar stacks of thylakoids with few connections to each other). The Chlorophyta also have a characteristic set of accessory pigments, including the xanthophylls lutein, zeaxanthin, violaxanthin, antheraxanthin and neoxanthin. Siphonein and siphonoxanthin occur in the Bryopsidophyceae, although siphonoxanthin is also found in some representatives in other classes. Where present the pyrenoids are situated within the chloroplast and are often penetrated by thylakoids. A shell of starch grains surrounds each pyrenoid. Starch grains can also be found scattered through the chloroplast stroma. Chloroplast DNA molecules are circular and are concentrated in numerous small (1–2 μm diameter) nucleoids, which are distributed throughout the chloroplast. The DNA is never organized in a single ring-like nucleoid. Flagellated cells are isokont, which means that the flagella are similar in structure but may differ in length. There are usually two flagella per cell but there can be four or many. Various kinds of delicate hairs and scales may be found on the flagella. Between the flagellar axoneme and the basal body is a stellate transition zone. Asexual reproduction occurs through formation of aplanospores, autospores or zoospores. Sexual reproduction may be isogamous or anisogamous, involving either flagellated or amoeboid gametes (e.g. Zygonematophyceae) or oogamous. Cells may also divide vegetatively, while fragments may also be formed through breakage of filaments.

Advances in electron microscopy, biochemistry and now macromolecular evidence from nuclear, chloroplast or mitochondrial DNA or ribosomal RNA or through comparisons of amino acid sequences in selected proteins have greatly influenced present-day concepts of green algal evolution and systematics. One of the latest classifications recognizes eleven classes (Prasinophyceae, Chlorophyceae, Ulvophyceae, Cladophorophyceae, Bryopsidophyceae, Dasycladales, Trentepohliophyceae, Pleurastrophyceae, Klebsormidiophyceae, Zygnematophyceae, Charophyceae). See entries on these for further details.

chloroplast Chlorophyll-containing plastid of eukaryotic organisms; the organelle within which both light and dark reactions of photosynthesis occur. Present in all PLANTAE (but not usually in all their cells), and most ALGAE (see further details). The similarity of chloroplasts in diverse autotrophs suggests a common origin. Based upon close cytological and biochemical similarities between certain bacteria and chloroplasts, the most accepted theory is that chloroplasts evolved through a series of independent endosymbioses, involving different groups of photosynthetic bacteria. Chloroplasts contain certain DNA (see cpDNA) and protein-synthesizing machinery, including RIBOSOMES, of a prokaryotic type. Hence they are semi-autonomous organelles. Isolated chloroplasts can synthesize RNA but only under the direction of nuclear DNA. The ability to form chloroplasts and associated pigment complexes is largely controlled by nuclear DNA, interacting with chloroplast DNA. Chloroplasts originate by division of pre-existing chloroplasts, in which the prokaryotic protein FtsZ is involved.

Where present, chloroplasts may be numerous per cell or single. In higher plants, they are usually discoid (discshaped), about 4–6 μm in diameter and arranged in a single layer in the cytoplasm but capable of changing both shape and position in relation to light intensity (see CYCLOSIS). In algae, they are variously shaped (cup-shaped, spiral, stellate, reticulate, lobed and discoid). Often accompanied by PYRENOIDS, with which storage products are frequently associated. Those of green algae (CHLOROPHYTA) and plants often contain starch grains and small lipid droplets.

Mature chloroplasts (see Fig. 26) are typically bounded by two outer membranes enclosing a homogeneous *stroma* (where the dark reactions occur). The outer of these, once considered along with the outer mitochondrial membrane to represent the residual phagosomal membrane of the original engulfing eukaryote, is now thought to be a remnant of the Gram-negative outer membrane of the ancestral cyanobacterium (see GRAM'S STAIN). Traversing the stroma are membranes in the form of flattened sacs (*thylakoids*), comprising two membranes. Stacks of thylakoids form the grana. Thylakoid membranes house the photosynthetic pigments and electron transport system involved in the light-dependent reactions of photosynthesis. Thylakoids of the grana are connected to each other by stomal thylakoids (*intergranal thylakoids*). In algae, the chloroplast may or may not be surrounded by one or two membranes of CHLOROPLAST ENDOPLASMIC RETICULUM, and the thylakoid groupings vary: bands of three with a girdle or peripheral band running parallel to the chloroplast envelope (e.g. EUGLENOPHYTA, CHRYSOPHYTA, XANTHOPHYTA, RAPHIDOPHYTA, BACILLARIOPHYTA and PHAEOPHYTA); or bands of two to six, with thylakoids running from one band to the next or free from one another (e.g. RHODOPHYTA). Photosynthetic prokaryotes lack chloroplasts, the numerous thylakoids lying free in the cytoplasm and varying in arrangement and shape between species.

chloroplast endoplasmic reticulum The outer one or two membranes, respectively, where a chloroplast is bounded by either three membranes (DINOPHYTA) or, in the CRYPTOPHYTA, BACILLARIOPHYCEAE, CHRYSOPHYTA, EUMASTIGOPHYCEAE, PHAEOPHYCEAE, HAPTOPHYTA and XANTHOPHYTA, by four. When composed of two membranes, the outer one

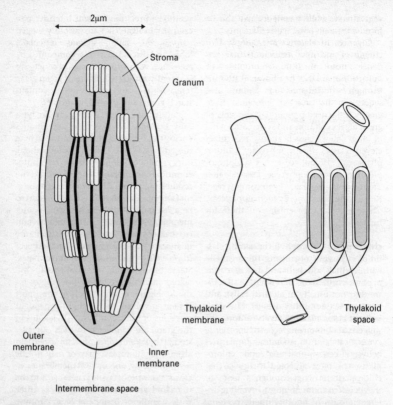

FIG. 26 CHLOROPLAST. *This photosynthetic organelle contains three distinct membranes (the outer membrane, the inner membrane, and the thylakoid membrane) that define three separate internal compartments (the intermembrane space, the stroma, and the thylakoid space).*

is often continuous with the outer membrane of the nuclear envelope. See NUCLEOMORPH, Fig. 3.

chlorosis Disease of green plants characterized by yellow (chlorotic) condition of parts that are normally green; caused by conditions preventing chlorophyll formation, e.g. lack of light or of appropriate soil nutrients.

choanae (internal nares) Paired connections between nasal and oral cavities of typical crossopterygian (lobe-finned) fish, some teleosts, lungfish and higher vertebrates; probably evolved independently in differ-

ent fish groups. Not used for respiratory purposes in any living jawed fish, but providing a passage for ventilation of lungs in tetrapods. Situated near front of roof of mouth, unless false palate (see PALATE) present, when they are at the back. See CHOANICHTHYES, NARES.

Choanichthyes (Sarcopterygii) A probably natural vertebrate clade, containing CROSSOPTERYGII (coelacanths), DIPNOI (lungfishes) and RHIPIDISTIA (including porolepids, osteolepids and tetrapods). New fossils have been shown to link lungfishes and tetrapods to separate extinct rhipidistian groups.

choanocyte (collar cell) Cell with single flagellum generating currents by which sponges (PORIFERA) draw water through their ostia and catch food particles which stick to the outside of cylindrical protoplasmic collar around base of flagellum. Affinities of sponges with the protozoan choanoflagellates problematical.

cholecystokinin-pancreozymin (CCK-PZ) Hormone of mucosa of small intestine, released in response to presence of CHYME. Causes pancreas to release enzymatic juice and gall bladder to eject bile. Promotes intestinal secretion but inhibits gastric secretion. Acts as a NEUROTRANSMITTER in the brain, involved in satiety after eating and experience of fear. See SECRETIN.

cholesterol Sterol lipid derived from squalene, forming a major component of many animal CELL MEMBRANES where it affects membrane fluidity. Absent from higher plants and most bacteria. Precursor of several potent steroid hormones (e.g. corticosteroids, sex hormones) which are in turn converted back to it in liver. Synthesis from acetyl coenzyme A in liver suppressed by dietary cholesterol. A transcription factor (SREBP) inserted into the endoplasmic reticulum and released from the Golgi apparatus causes transcription of genes involved in cholesterol and fatty acid synthesis in mammals, but increased cholesterol and fatty acid production block its production. Can also play important role in covalent modification of proteins during development. Most plasma cholesterol is transported esterified to long-chain fatty acids within a micellar lipoprotein complex. These structures, LOW-DENSITY LIPOPROTEINS (LDL), are about 22 nm in diameter and adhere to plasma membrane receptor sites produced on COATED PITS when a cell needs to make more membranes using the cholesterol in the LDL (see CORONARY HEART DISEASE). Cholesterol ester transfer protein (see PHOSPHOLIPID TRANSFER PROTEINS) transfers a portion of HDL2 to triglyceride-rich lipoproteins while triglyceride is transferred in the opposite direction to modify HDL2. Cholesterol is excreted in BILE, both in native form (as micelles) and conjugated with taurine or glycine as bile salts. See LIPOPROTEIN, CHYLOMICRON.

choline An organic base (formula $OHC_2H_4.N[CH_3]_3OH$); a vitamin for some animals and since 1998, classified as a human vitamin, (humans can synthesize it only in small amounts). Present in high levels in eggs, liver, peanuts and various meats and vegetables. Precursor of lecithin and sphingomyelin, and of such intracellular SECOND MESSENGERS as diacylglycerol (ceramide). Also a precursor of ACETYLCHOLINE and of methionine (important in protein synthesis and transmethylations).

cholinergic Of nerve fibres which secrete, or are stimulated by, ACETYLCHOLINE. In vertebrates, motor fibres to striated muscle, parasympathetic fibres to smooth muscle, and fibres connecting CNS to sympathetic ganglia are cholinergic, as are some invertebrate neurons.

cholinesterase Hydrolytic enzyme anchored to BASAL LAMINA between synapsing membranes of most (especially vertebrate) neuromuscular junctions and of cholinergic synapses. Degrades ACETYLCHOLINE to choline and acetate.

Chondrichthyes Vertebrate class containing cartilaginous fish, first appearing in the Devonian. Includes HOLOCEPHALI (e.g. ratfish, *Chimaera*) and ELASMOBRANCHII (sharks, skates and rays). Cartilaginous skeleton; PLACOID SCALES (denticles), modified to form replaceable teeth; intromittant organs (claspers) formed from male pelvic fins. No GAS BLADDER. See OSTEICHTHYES.

chondrin Matrix material of CARTILAGE.

chondroblast, chondrocyte See CARTILAGE.

chondrocranium Part of the skull first formed in vertebrate embryos as a cartilaginous protection of brain and inner ear. Usually ossified during development to form membrane bones. See OSSIFICATION.

chondroitin Sulphated GLYCOSAMINOGLYCAN composed largely of D-glucuronic acid and N-acetylgalactosamine. Found in cartilage, cornea, bone, skin and arteries.

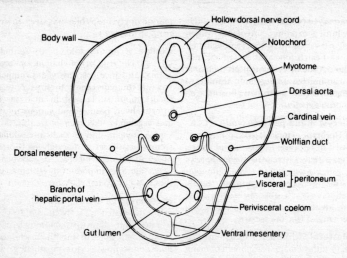

FIG. 27 *Transverse section through embryonic vertebrate, indicating the layout of the trunk region prior to the origin of the skeleton.*

Chondrostei Group (often considered a superorder) of the ACTINOPTERYGII. Includes the primitive Palaeozoic palaeoniscoids represented today by the bichirs (*Polypterus*), paddlefishes and sturgeons. Ganoid scales of bichirs are lost altogether in paddlefish, sturgeons having rows of bony plates lacking ganoine. Ancestral bony internal skeleton largely substituted by cartilage. Primitive heterocercal tail present in sturgeons and paddlefish. Bichirs have lungs, sturgeons a gas bladder.

Chordata Animal phylum, characterized by presence at some stage in development of a NOTOCHORD, by the dorsal hollow nerve cord, pharyngeal gill slits and a post-anal tail. Monophyly of chordates is almost universally accepted. Includes the invertebrate subphyla UROCHORDATA and CEPHALOCHORDATA, and vertebrates (Subphylum VERTEBRATA). Various fossils in the early CAMBRIAN have been interpreted as cephalochordates and one remarkable site (Chengjiang, in southern China) has produced hundreds of 530-Myr-old fine-grained rocks where even soft animal tissue is preserved. One such FOSSIL, *Haikouella lanceolata*, seems to have several key chordate structures, and appears to be even more primitive than the previously most primitive chordate, *Yunnanozoon*, from the same site. See Fig. 27 and VETULICOLIA.

chorion (1) One of three EXTRA-EMBRYONIC MEMBRANES of amniotes. Comprises the TROPHOBLAST with an inner lining of mesoderm, coming to enclose almost the entire complement of embryonic structures. In reptiles and birds it forms with the ALLANTOIS a surface for gaseous exchange within the egg. In most mammals it combines with the allantois to form the PLACENTA. Chorionic villus sampling (CVS) involves removal by catheter, or ultrasound needle probe, of *c.* 30 mg villus tissue to test for chromosomal disorders between 6–10 weeks of pregnancy (earlier than in amniocentesis). See ANDROGENESIS. (2) Egg shell of insects, secreted by follicle cells of ovary, and often sculptured externally.

choroid Mesodermal layer of vertebrate eyeball between outer sclera and retina within. Soft and richly vascularized (supplying nutrition for retina); generally pigmented to prevent internal reflection of light, but reflecting crystals of TAPETUM, part of the choroid, increase retinal stimulation in many nocturnal/deep-water vertebrates. Becomes the CILIARY BODY anteriorly.

choroid plexuses Numerous projections of non-nervous epithelium into ventricles of brain, secreting CEREBROSPINAL FLUID from capillary networks. One plexus occurs in the roof of each of the four ventricles in man.

chromaffin cell, c. tissue Cells derived from NEURAL CREST tissue, which having migrated along visceral nerves during development come to lie in clumps in various parts of the vertebrate body (e.g. the adrenal medulla). They are really postganglionic neurons of the sympathetic nervous system, which have 'lost' their axons and secrete the catecholamines ADRENALINE and NOR-ADRENALINE into the blood, the former more abundantly. Stain readily with some biometric salts (hence name).

chromatid One of the two strands of CHROMATIN, together forming one CHROMOSOME, which are held together after DNA replication during the cell cycle by one or more CENTROMERES prior to separation at either mitotic anaphase or second meiotic anaphase. In mitosis the strands are genetically identical (barring mutation), but in meiosis crossing-over increases the likelihood of dissimilarity.

chromatin (nucleohistone) The material of which eukaryotic CHROMOSOMES are composed. Consists of DNA and proteins, the bulk of them HISTONES, organized into nucleosomes. See CHROMOSOMAL IMPRINTING, EUCHROMATIN, HETEROCHROMATIN.

chromatin remodelling Modification of chromatin structure, as by changes to its structural proteins (especially HISTONES); by enzymes (e.g. multi-subunit ATP-dependent enzymes) which are able to expose or redistribute NUCLEOSOMES, often in a tissue-specific manner; and by interfering RNAs (see RNA INTERFERENCE).

During gametogenesis in humans, the two parental genomes are formatted to respond to the oocyte's cytoplasm so as to proceed through development. Zygotes remodel the paternal genome shortly after fertilization, but before embryonic genome activity commences; and when nuclear transfer techniques are employed in mammalian cloning, any somatic nuclei transferred into oocytes must be quickly reprogrammed so as to express genes required for early development. Although relatively little is known yet about the initial molecular events involved, such remodelling involves changes in chromatin structure, DNA METHYLATION, CHROMOSOMAL IMPRINTING, TELOMERE length adjustment, and X CHROMOSOME INACTIVATION. In humans, there are considerable differences in the extent to which male and female pronuclei are organized epigenetically, the sperm chromatin having been sequentially remodelled, silenced and ultimately compacted with protamines, but the female pronucleus is more transcriptionally repressive and is deficient in generalized transcription factors (e.g. see TATA FACTOR). DNA methylation is also more pronounced during spermatogenesis than during oogenesis, but within hours of fertilization the paternal genome is actively demethylated, in contrast to the maternal genome which appears to be passively demethylated during cleavage. By the blastocyst stage, the embryo's genome is hypomethylated and then undergoes global de novo methylation, giving an apparently uniform pattern across both parental genomes by the time of gastrulation. The expression of imprinted genes is set by parent-of-origin-specific methylation marks during late gametogenesis and, when lost, cannot be reset until passage through the germ line. Genome-wide disruption of imprinted gene expression, as occurs in uniparental embryos and nuclear transfer embryos derived from male primordial germ cells or non-growing oocytes, results in postimplantation lethality. See CLONE, HELICASES.

chromatography Techniques involving separation of components of a mixture in solution through their differential solubilities in a moving solvent (*mobile phase*) and absorptions on, or solubilities in, a *stationary phase* (often gels, e.g. polyacrylamide or agarose; or special paper). In *gel* *filtration*, mixture to be separated (often proteins) is poured into column containing beads of inert gel and then washed through with solvent. Speed of passage depends on relative solubilities in solvent and on ability to pass through the pores in the gel, a function of relative molecular size. Components may then be identified. Development of microparticles (600–800 nm) for the packing material reduces intraparticle mass flow and improves resolution of separation during even high-velocity perfusion. Proteins can be separated by their net charge during *ion-exchange chromatography*. A column of appropriately charged beads is used while the buffered eluting fluid (variable salt concentration) contains metal ions which compete with positively charged groups on the protein for binding to the beads so that proteins with low positive charge tend to emerge first. In *affinity chromatography*, a solution containing a desired protein flows through a column packed with beads on whose surfaces are covalently attached a specific ligand (e.g. antibody, enzyme) which will bind the desired protein. The protein can be separated from the antibodies by washing with concentrated salt solutions, or solutions with extreme pH. See ELECTROPHORESIS, R_f VALUE.

chromatophore (Zool.) Animal cell lying superficially (e.g. in skin), with permanent radiating processes containing pigment that can be concentrated or dispersed within the cell under nervous and/or hormonal stimulation, effecting colour changes. When dispersed, the pigment of groups of such cells is noticeable; when condensed in centre of cells the region may appear pale. Three common types occur in vertebrates: *melanophores*, containing the dark brown pigment melanin; *lipophores*, with red to yellow carotenoid guanine crystals whose light reflection may lighten the region when other chromatophores have their pigments condensed. MELANO-CYTE-STIMULATING HORMONE disperses melanin, while melatonin (see PINEAL GLAND) and adrenaline concentrate it. (Bot.) (1) See CHROMOPLAST. (2) In prokaryotes (bacteria, blue-green algae), membrane-bounded vesicles (thylakoids) bearing photosynthetic pigments. See PROCHLOROPHYTA.

chromatosome A NUCLEOSOME core particle plus a number of adjacent DNA base pairs on either side. Obtained by moderate nuclease digestion of a polynucleosome fibre.

chromocentre Region of constitutive HETEROCHROMATIN which aggregates in interphase nucleus. In *Drosophila* all four chromosome pairs become fused at their centromere regions in POLYTENE nuclei to form a large chromocentre.

chromomeres Darkly staining (heterochromatic) bands visible at intervals along chromosomes in a pattern characteristic for each chromosome. Especially visible in mitotic and meiotic prophases, and at bases of loops of LAMPBRUSH CHROMOSOMES. Probably reflects tight clustering of groups of chromosome loops (see CHROMOSOME). Dark bands of polytene chromosomes are probably due to multiple parallel chromomeres.

chromonema Term usually used for chromosome thread while extended and dispersed throughout nucleus during interphase.

chromophore A light-absorbing pigment; often a component of a larger, complexed, molecule (e.g. II-*cis*-retinal in rhodopsin).

chromoplast (chromatophore) Pigmented plant cell PLASTID of variable shape, lacking chlorophyll but synthesizing and retaining carotenoid pigments; often responsible for the yellow to orange or red colours of many flowers, old leaves, some fruits and roots. May arise from a chloroplast whose internal membrane structure disappears, when masses of carotenoids accumulate. Precise function is not well understood. Often used synonymously with CHLOROPLAST; in older literature a chloroplast that has a colour other than green is often called a chromoplast. See LEUCOPLAST.

chromosomal imprinting A form of GENE SILENCING whereby the same genes can result in different phenotypes depending upon the sex of the parent they are inherited from. It is often held that maternal and paternal parents have conflicting interests in terms of how much the mother should contribute to the growth of an offspring. The *Igf-2* gene encodes a growth factor operating during pregnancy. When expressed, demands are placed on the mother to provide more nutrients; and, in mice and humans, copies of the gene inherited from the mother are switched off, whereas those from the father are not. Other embryonic growth genes behave similarly, providing evidence for a sexual ARMS RACE. Heritable change in gene expressibility possibly brought about by genes becoming heritably silenced in a cell line by incorporation into a HETEROCHROMA-TIN cluster on a chromosome. HOMEOGENE activity/inactivity may depend upon such events. Because there now seem to be many cases of imprinted genes which are difficult to reconcile with embryonic or growth genes, attempts are being made to explain all chromosomal imprinting by theories which do not invoke conflict between the sexes. In studies of TURNER'S SYNDROME, it turns out there is an imprinted gene on the human X chromosome influencing social functioning and related cognitive abilities. Female mice lacking one X chromosome are larger if their one X chromosome was inherited from the father, in contrast with conflict theory prediction. It may, after all, be that chromosomal imprinting is the result of selection for DOSAGE COMPENSATION. See ANDROGENESIS, GENOMIC IMPRINTING, POS-ITION EFFECT.

chromosome It is difficult to think of structures with greater biological impor-tance (see GENETIC VARIATION, RECOMBINATION). Literally, a coloured (i.e. stainable) body; originally observed as threads within eukaryote nuclei during mitosis and mei-osis. Composed of nucleic acid, most com-monly DNA, usually in conjunction with various attendant proteins, in which form the genetic material of all cells is organized. Chromosomes are linear sequences of GENES, plus additional non-genetic (i.e. apparently non-functional) nucleic acid sequences.

Gene sequence is probably never random, being the result of selection for particular LINKAGE groups (but see TRANSPOSABLE ELEMENT). Prokaryotes and eukaryotes differ in the amount of genetic material which needs to be packaged, and in resulting com-plexities of their chromosomes. Thus the absence to date from bacterial chromo-somes of the DNA-binding proteins, his-tones, has some taxonomic value (see ARCHAEA, CHROMATIN). Non-histone proteins (e.g. *protamines*) form part of the structure of all chromosomes, however, and their roles, for example as activators of transcription, are being increasingly elucidated. The DNA of a normal individual chromosome or chromatid is probably just one highly folded molecule.

The bacterial chromosome (usually one main chromosome per cell) is just over 1 mm in length, contains about 4×10^6 base pairs of DNA, is circular and is attached to the cell membrane, at least during DNA repli-cation. It lacks the nucleosome infrastruc-ture of eukaryotic chromatin. Additionally, there may be one or more PLASMIDS, some of which (*megaplasmids*) may constitute more than 2% of the cell's DNA. There is no nuc-leus to contain the chromosome, but the term 'nucleoid' may be used to indicate the DNA-containing region of the cell. The DNA appears to be packaged in a series of loops (see later). Archaeal and eukaryotic chromosomes are made of chromatin and are nucleosomal in organization. Eukary-otic chromatin comprises DNA and five dif-ferent histone species roughly equal in total weight to the DNA; plus various attendant proteins. The fundamental organizational unit is the NUCLEOSOME, a polynucleosome giving rise in turn during nuclease digestion to mononucleosomes (200 DNA base pairs), chromatosomes (165 DNA base pairs) and nucleosome core particles (145 DNA base pairs). See Fig. 28. All eukaryotic chromo-somes have numerous origins of replication (see DNA REPLICATION). Among chromosomal MUTATIONS, INVERSIONS occur in prokaryotes and eukaryotes, usually being larger in the former but more frequent in the latter.

The polynucleosome filament has a diam-eter of about 10 nm, but adopts a tight 30 nm

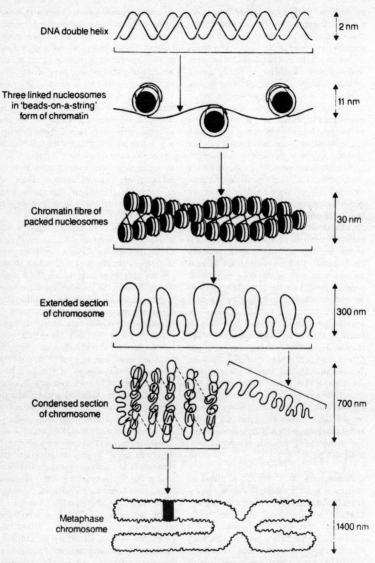

DNA double helix — 2 nm

Three linked nucleosomes in 'beads-on-a-string' form of chromatin — 11 nm

Chromatin fibre of packed nucleosomes — 30 nm

Extended section of chromosome — 300 nm

Condensed section of chromosome — 700 nm

Metaphase chromosome — 1400 nm

FIG. 28a *Possible progressive packing arrangement of a DNA duplex with histones to form nucleosomes and then subsequent packing of these, ultimately to form the chromosomes visible in light microscopy.*

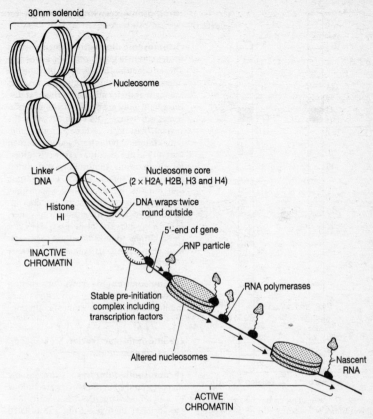

30 nm solenoid

Nucleosome

Linker DNA

Nucleosome core
(2 × H2A, H2B, H3 and H4)

DNA wraps twice
round outside

Histone HI

5'-end of gene

RNP particle

INACTIVE
CHROMATIN

Stable pre-initiation
complex including
transcription factors

RNA polymerases

Altered nucleosomes

Nascent
RNA

ACTIVE
CHROMATIN

FIG. 28b *Model indicating the unwinding of inactive heterochromatin around an active gene, showing the persistence of nucleosomes in transcribed regions.*

helix under physiological ion concentrations. This reduces the DNA length 50-fold and may be the normal interphase state of chromatin. Further looping along a single axis forms a fibre 0.3 μm in width which may in turn form a helix of radially arranged loops about 0.7 μm in diameter, possibly the metaphase chromatin condition. Bands seen in stained mitotic chromosomes probably reflect tight clustering of groups of loops, which stain more densely. Polytene chromosome bands (see POLYTENY) would result from lateral amplification of these tightly clustered loops. The higher orders of chromatin packing are features of HETERO-CHROMATIN such as chromocentres, centromeres and pericentric regions (see CHROMOSOMAL IMPRINTING). Chromosomes can be stained in various ways to reveal different banding patterns. Routinely, G (Giemsa) banding is employed and generates 300–400 alternating light and dark bands in the human karyotype, reflecting differing levels of chromosome condensation. The convention for eukaryotic chromosomes is to call the two arms of a chromosome on either side of the centromere p (short arm) and q (long arm). Prominent bands then subdivide the arms further, each region being further subdivided by the

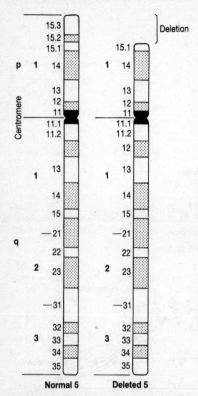

Normal 5   **Deleted 5**

FIG. 28C *The cause of the* cri du chat *syndrome of abnormalities in humans is loss of the tip of the short arm of one of the homologues of chromosome 5. The figure indicates chromosome banding in human chromosome 5, both normal and with the deletion. See* CHROMOSOME *text for banding notation. (From Introduction to Genetic Analysis (5th edn) by Griffiths, Miller, Suzuki, Lewontin and Gelbert. Copyright © W. H. Freeman and Company, 1993. Reprinted with permission.)*

next most prominent bands, and so on. Thus band 5p15.2 is found in the short arm (p) of chromosome 5, in region 1, band 5, sub-band 2 (see Fig. 28C and CHROMOSOME MAPPING). The short arm of human chromosome 21 can vary greatly between individuals on account of the several classes of highly repetitive DNA sequences. See CHROMOSOME PAINTING.

chromosome crawling Alternative for CHROMOSOME WALKING.

chromosome diminution Phenomenon in nematode (e.g. *Paracaris*) eggs whereby after an equatorial first cleavage division the upper (animal) blastomere's chromosomes fragment at their ends during the next division with only a portion of the chromosomes surviving. This contrasts with the vegetal blastomere, whose chromosomes remain normal but whose cell line is such that all but the eventual germ cell undergo chromosome diminution and differentiate into somatic cells. Cytoplasmic determinants in the egg are responsible. Similar events occur in some dipteran insects (e.g. *Wachtiella*), although in others (e.g. *Drosophila*) the GERM PLASM does not cause chromosome diminution. See ABERRANT CHROMOSOME BEHAVIOUR (3), MATERNAL EFFECT.

chromosome engineering Construction of defined deficiencies, inversions and duplications in chromosomes to produce segmental haploids.

chromosome inactivation See BARR BODY, DOSAGE COMPENSATION.

chromosome jumping Chromosome mapping (and DNA sequencing) technique in which quite large DNA fragments (80–150 kb long) are first cut out and isolated by gel separation from very high molecular mass DNA, then circularized and cloned inside a phage or cosmid (see VECTOR) before plating out on bacteria designed to allow only phages or cosmids with the inserted fragments to form plaques. After isolating the inserts, an already mapped DNA sequence is then used as a probe, when those fragments containing the probe sequence will have DNA from a distantly linked site at their other end; intervening sequences can be filled in by CHROMOSOME WALKING. The combined methods have enabled mapping and sequencing of the human cystic fibrosis gene. See Fig. 29.

chromosome landing Alternative term for POSITIONAL CLONING.

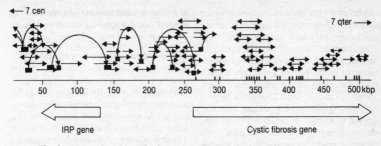

←— 7 cen   7 qter —→

50   100   150   200   250   300   350   400   450   500 kbp

IRP gene   Cystic fibrosis gene

FIG. 29 *The chromosome jumping and subsequent walking involved in the isolation and cloning of the cystic fibrosis (CF) gene. The calibrated base line represents a DNA strand over 500 kilobase pairs long. The centromere on chromosome 7 (7 cen) is to the left of the chromosome's tip (7 qter), off to the right of the diagram. Curved arrows show the length and direction of each jump. Horizontal arrows are overlapping phage and cosmid clones containing DNA isolated from regions at the ends of each jump; arrow directions show directions of cloning (double-headed arrows show clones that went both ways). The CF exons are shown by small vertical bars on the base line.* (From *Recombinant DNA (2nd edn)* by Watson, Gilman, Witkowski, and Zoller. Copyright © James D. Watson, Michael Gilman, Jan Witkowski and Mark Zoller, 1992. Reprinted with permission of W. H. Freeman and Company.)

chromosome map Linear map (circular in bacteria, plasmids, etc.) of the sequence of genes (cistrons or loci) on a chromosome as defined by CHROMOSOME MAPPING techniques. The MAP DISTANCE between two genes does not accurately reflect their physical separation but only their probability of RECOMBINATION.

chromosome mapping Techniques involved in producing either (a) *genetic maps* of chromosomes, mainly through LINKAGE studies involving appropriate breeding routines and scoring of phenotypic ratios; or (b) *physical maps* of chromosomes through analysis of its entire DNA sequence. For most eukaryotes, linkage between two or more loci is normally detected by first obtaining a generation (normally an F1) heterozygous for the two loci concerned (i.e. doubly heterozygous). This is normally achieved by first crossing two stocks, each pure-breeding for *one* of the two mutant phenotypes involved. The F1 stock is then crossed to a doubly mutant stock and the resulting offspring scored for phenotypes. If all four possible phenotypes (assuming complete dominance of wild-type over the mutant phenotype) are present in equal ratio, linkage is not probable; but if there is a departure on the null hypothesis of no

linkage, then this departure can be tested for its significance (using CHI-SQUARED TEST). Where the ratio is obviously non-Mendelian (i.e. departs obviously from 1:1:1:1), with the parental classes outnumbering the recombinants, then a CROSS-OVER VALUE can be determined giving a map distance between the two loci.

When we wish to know whether the loci bearing the alleles for black body and vestigial wing (both recessive characters) in *Drosophila* are linked, then using the symbols

b =   black body
+ =   wild-type body
v =   vestigial wing
+ =   wild-type wing

first pure-breeding black body/wild-type wing flies (bb + +) are mated with pure-breeding wild-type body/vestigial wing flies (+ + vv). F1 offspring are then mated with a double recessive stock (i.e. pure-breeding black body/vestigial wing, $bbvv$) as a TEST CROSS. If all four resulting offspring phenotypes (+ +, + v, b +, bv) occur in equal ratio then, given adequate sample size, linkage is unlikely. If two phenotypic classes (the parental classes, b +, + v) outnumber the other two (the two recombinant classes, bv, + +) then linkage is likely and a provisional map distance can be calculated, equal to the frequency of the recombinant offspring as a percentage of the total number of offspring. (The example is actually more complex, for only when male flies are used as the double recessive in the backcross do four phenotypic classes appear in the F_2 generation. This is because in male *Drosophila* there is no crossing-over during meiosis (see SUP-PRESSOR MUTATION) so the males cited only produce two gamete types, giving only two F_2 phenotypes.) Sex-linked loci would give a different result, suitably modified to take account of the chromosome arrangement of the heterogametic sex. When testing for linkage between mutations for dominant characters, the recessive characters in the method employed above would be wild-type characters.

Chromosome mapping in bacteria can employ transformation, TRANSDUCTION or interrupted mating. In the latter, progress of the donor bacterial chromosome into the recipient cell during conjugation is interrupted, as by shaking (see F FACTOR, for *Hfr* strain). The map of cistrons on the incoming chromosome will be a function of the time allowed for conjugation before interruption, and is deduced from recipient cell phenotypes. The CIS-TRANS TEST may be used to determine whether two mutations lie within the same cistron. In *deletion mapping*, gene sequences can be ascertained by noting whether or not wild-type recombinants occur in appropriate crosses between mutant strains: they will not do so if the part of the chromosome needed for recombination is missing, so that fine mapping of such recombinants can indicate the limits of a deletion and the genes involved in it. Plasmid and viral chromosome maps may be constructed using *restriction fragment mapping* techniques in which different RESTRICTION ENDONUCLEASES digest the chromosome, and electrophoretic patterns of resulting fragments are used to reconstruct the complete nucleotide sequences of the chromosomes. New electrophoretic techniques with infrequently cutting restriction endonucleases now permit restriction fragment mapping of even entire mammalian chromosomes and render chromosome

mapping an extension of DNA SEQUENCING in general. See CELL FUSION, CHROMOSOME WALKING, POSITIONAL CLONING, RFLP, VECTOR.

chromosome painting A technique for resolving complex abnormalities in the structure of chromosomes which are impossible or very difficult to detect by standard histological preparations using brightfield microscopy. It involves fluorescence *in situ* hybridization (FISH), in which a DNA PROBE labelled by addition of a reporter molecule (e.g. a small protein or a fluorescent dye – a fluorophore) hybridizes to target DNA *in situ* (i.e. to chromosomes or nuclei which are already fixed on a slide). Excess dye is washed off, and if the reporter is a protein, a fluorescent antibody to it is then added and the preparation incubated. Hybridized slides are then studied with a fluorescence microscope, when a fluorescent signal indicates the presence and position of the target material. Different fluorophores emit in different wave bands, so that different target sections of chromosomes can be 'painted' different colours.

chromosome puff See PUFF.

chromosome walking Chromosome mapping (and DNA sequencing) technique in which a small DNA sequence from one end of a DNA clone is used as a probe to isolate other DNA sequences containing this and the next adjacent sequence, which is used in turn as a probe, and so on iteratively until an already known sequence is discovered. All intervening sequences will by then have been cloned. Formerly used in conjunction with CHROMOSOME JUMPING (see Fig. 29), the technique is now considered very time consuming and has been largely replaced by POSITIONAL CLONING. See Fig. 30, COSMID, YEAST ARTIFICIAL CHROMOSOME.

chrysalis The PUPA of lepidopterans (butterflies and moths).

chrysolaminarin (leucosin) Polysaccharide storage product in certain algal divisions (CHRYSOPHYCEAE, PRYMNESIOPHYTA, BACILLARIOPHYCEAE). Comprises β-1,3-linked

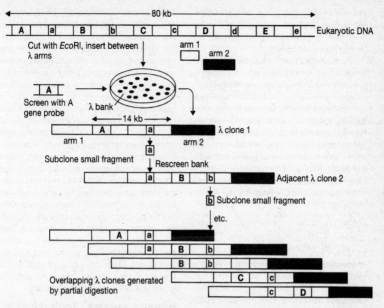

FIG. 30 *Chromosome walking. One recombinant phage obtained from a phage bank made by the partial EcoRI digest of a eukaryotic genome can be used to isolate another recombinant phage containing a neighbouring segment of eukaryotic DNA. In this case the recombinant phage contains the A gene, as detected by the A gene probe. It is cleaved into smaller fragments using different restriction enzymes, and these fragments are subcloned into plasmic VECTORS. One such subclone (a in the example) from the end of the λ phage clone 1 serves as a probe to detect a clone in the original GENE LIBRARY that also contains the sequence defined by probe a, which will have some sequences in common with clone 1 but will also contain new sequences further along the chromosome. This produces a new probe b, and this can be used in the same way as probe a, and so on iteratively.* (From *Introduction to Genetic Analysis (5th edn)* by Griffiths, Miller, Suzuki, Lewonton and Gelbert. Copyright © W. H. Freeman and Company, 1993. Reprinted with permission.)

D-glucose residues with 1–6 glycosidic bonds per molecule. Resides in vesicles outside the chloroplast.

Chrysophyta Golden-brown algae. Protists, whose colour is due to the abundance of carotenoid pigments, including β-carotene, fucoxanthin and other xanthophylls present within the chloroplast, together with chlorophylls a, c_1, and c_2 (Synurophyceae do not have chlorophyll c_2). Reserve assimilatory product is stored as oils and CHRYSOLAMINARIN, a polysaccharide deposited in a vesicle outside of the chloroplast. Many lack a cell wall; when present,

the cell wall is composed of cellulose. LORICAS, silicified, and organic scales occur in some species. Siliceous scales are radially or bilaterally symmetrical, possess species-specific morphology, and are formed by species which are very sensitive to environmental change. The scales preserve in lake sediments and provide yet another microfossil for the palaeolimnologist. Most chrysophytes are freshwater algae occurring in soft water (low in calcium); many freshwater species are planktonic and flagellate (both unicells and colonies); coccoid and filamentous species are mostly found in cold water springs and streams. Marine species also occur. A diverse group, which includes the SILICOFLAGELLATES, having links to other protists (e.g. protozoa, dinoflagellates and brown algae) and fungi.

These algae form characteristic cysts (STATOSPORES, or stomatocysts) asexually or sexually and possess a siliceous wall and one or more pores which, when the protoplast is inside, are closed by an organic plug. Cysts may be spherical, ellipsoidal or ovate and bear species-specific ornamentation used by palaeolimnologists (see HETEROKONT); but only some 5% of the cysts can be related to the algae that actually produce them, so they are described as morphotypes in a standard manner until related to the species forming them.

chyle The milky suspension of fat droplets within LACTEALS and THORACIC DUCTS of vertebrates after absorption of a meal.

chylomicron Plasma LIPOPROTEIN (see Fig. 103) with mean diameter of ~100 nm, containing reconstituted triglycerides, phospholipids and CHOLESTEROL produced by the epithelial cells of intestinal villi after long-chain fatty acids and monoglycerides have diffused across the microvilli. Also act as transport vehicles for dietary lipids within the LACTEALS, LYMPHATIC SYSTEM and BLOOD PLASMA, being absorbed ultimately by the liver. See FAT, LIPOPROTEIN.

chymases Enzymes cleaving proteins after (on the C-terminal side of) aromatic residues (phenylalanine, tyrosine and tryptophan); e.g. chymotrypsin.

chyme Partially digested food as it leaves the vertebrate stomach. See CHOLECYSTOKININ, SECRETIN.

chymotrypsin Proteolytic enzyme secreted as inactive chymotrypsinogen by vertebrate pancreas. An endopeptidase, it converts proteins to peptides and is activated by the enzyme enterokinase. It cleaves peptides on the C-terminal side of tyrosine and phenylalanine residues. See CHYMASES.

Chytridiophyta (chytrids) Predominantly aquatic fungi possessing a coenocytic, holocarpic or eucarpic, monocentric (having one centre of growth and develop-

ment) or polycentric (having several centres of growth and development and more than one reproductive organ) and mycelial thallus. The cell wall is chitinous (at least in the hyphal stages). The mitochondrial cristae are flat. Zoospores have a single (rarely polyflagellate), posteriorly pointing flagellum lacking mastigonemes or scales, with a unique flagellar root system and sometimes rumposomes (an organelle located close to the cell wall and tooth-like in section and honeycomb-like in surface view). The presence of flagellate zoospores has led to these fungi being sometimes included in the algal division HETEROKONTOPHYTA. However, flagella lack mastigonemes, while the cell wall is chitinous, so they have been retained with the fungi. Chytrids are aquatic saprotrophs or parasites, growing upon decaying and living organisms (e.g. nematodes, insects, plants, other chytrids, pollen grains, algae and fungi). They also occur in soils and a few are marine, while others are obligate anaerobes in the guts of herbivores.

cigarette smoking For health-related aspects, see HYPOXIA, LDL, LUNG, NICOTINE, SCLEROSIS, SUPEROXIDES.

ciliary body Anterior part of the fused RETINA and CHOROID of the eyes of vertebrates and cephalopod molluscs; containing ciliary processes secreting the aqueous humour, and ciliary muscles (circular smooth muscle) which may permit ACCOMMODATION of the eye either by altering the focal length of the lens (amniotes), or by moving the lens to and fro (cephalopods, sharks and amphibians). In mammals the lens is suspended from it by ligaments, and the IRIS arises from the same region.

ciliary feeding Variety of feeding mechanism (MICROPHAGY) by which many soft-bodied aquatic invertebrates draw minute water-borne food particles through e.g. gills or the pharyngeal region of the gut, when the particles are frequently trapped in mucus and moved either towards the gut (often by further cilia) or further along it (by peristalsis).

Ciliata Class of Protozoa (Subphylum Ciliophora) containing the most complex cells

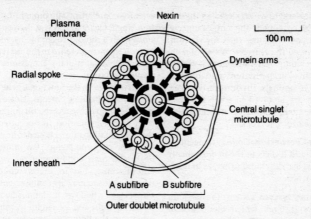

FIG. 31 *Diagram of a* CILIUM *or flagellum in cross-section, as viewed by light microscopy. The microtubule apparatus is termed the* AXONEME.

in the phylum. Covered typically in cilia, with meganucleus, micronucleus, and a cytostome (at the end of a depression, or 'mouth') at which food vacuoles form. Includes familiar *Paramecium* and *Vorticella*, and voracious predatory suctorians. All species probably have a close to cosmopolitan distribution. CONJUGATION and BINARY FISSION both occur, as may AUTOGAMY (see PARTHENOGENESIS).

ciliated epithelium Layer of columnar cells with apices covered in cilia whose coordinated beating enables CILIARY FEEDING, movement of mucus in the respiratory tract, etc.

Ciliophora Protozoan Subphylum containing the solitary Class CILIATA.

cilium Organelle of some eukaryotic cells. See Fig. 31. Tubular extension of the cell membrane, within which a characteristic 9 + 2 apparatus of MICROTUBULES and associated proteins occurs (nine paired outer tubules and a lone central pair). Used either for CELL LOCOMOTION (see for details of ciliary action) or for movement of material past a CILIATED EPITHELIUM; but frequently sensory, especially elongated cilia known as FLAGELLA. Cilia may beat in an organized METACHRONAL RHYTHM, for which KINETODESMATA are probably responsible. Such rows of beating cilia

may fuse to form *undulating membranes*; or several cilia may mat together to beat as one, as in the conical *cirri* of some ciliates used for 'walking'. Factors known to increase ciliary beat include SEROTONIN (mussel gills), nervous stimulation and mucus (mammalian trachea); a cyclic-AMP-dependent signal transduction pathway has been implicated, with phosphorylation of outer arm dynein a key component. Tiny cilia are involved in creating currents on node cells (see HENSEN'S NODE) in the gastrulas of all vertebrates, distributing morphogen-like signals from one side of the embryo to the other and generating left–right asymmetry. No ciliated animal cell can divide (if by 'animal' one adopts the definition taken of ANIMALIA here). For *stereocilium*, see HAIR CELL.

Ciona intestinalis A sea squirt, whose genome is being actively investigated. See UROCHORDATA.

CiPCR Culture-independent POLYMERASE CHAIN REACTION.

circadian rhythm (diurnal rhythm)
Endogenous (intrinsic) rhythmic changes occurring in an organism with a periodicity of approximately 24 h; even persisting for some days in the experimental absence of the daily rhythm of environmental cycles

(e.g. light/dark) to which circadian rhythm is usually entrained, the entraining cue (which may sometimes reset the rhythm) being termed the *zeitgeber*. Widely distributed, including leaf movements, growth movements, sleep rhythms and running activity. In animals, rhythms of hormone secretion are involved in some circadian rhythms, their existence indicating an internal BIOLOGICAL CLOCK.

circinate vernation Coiled arrangement of leaves and leaflets in the bud; gradually uncoils as leaf develops further, as in ferns. See PHYLLOTAXY.

circulatory system System of vessels and/ or spaces through which blood and/or lymph flows in an animal. See BLOOD SYSTEM, LYMPHATIC SYSTEM.

circumnutation See NUTATION.

Cirripedia Barnacles and their relatives. Subclass of CRUSTACEA. Typically marine, sessile and hermaphrodite. Unlike most of the Class in appearance, with a carapace comprising calcareous plates enclosing the trunk region. Usually a cypris larva, which becomes attached to the substratum by its 'head', remaining fixed throughout its adult life and filter-feeding using BIRAMOUS APPENDAGES on its thorax. Several parasitic forms occur.

***cis*-acting control elements** Regulatory genetic elements (e.g. promoters, enhancers), mutations which affect synthesis of an mRNA molecule downstream on the same chromosome. Contrast *TRANS*-ACTING CONTROL ELEMENTS.

cisternae Flattened sac-like vesicles of ENDOPLASMIC RETICULUM and GOLGI APPARATUS intimately involved in transport of materials via vesicles which either bud from or fuse with their membranous surfaces.

***cis-trans* test (complementation test)** Genetic test to discover whether or not two mutations which have arisen on separate but usually homologous chromosomes are located within same CISTRON. See Fig. 32. The two chromosomes, e.g. of phage or prokary-

ote origin, are artificially brought together in a single bacterial cell (e.g. by TRANSDUCTION) or in a diploid eukaryote by a sexual cross. If their co-presence in the cell rectifies their individual mutant expression, then COMPLEMENTATION is reckoned to have occurred between the functional gene products of two distinct cistrons. However, if their resultant expression is still mutant, no such complementation has occurred and the two mutations are reckoned to lie within the same cistronic region. Two mutations lie in the *trans* condition if on separate chromosomes, but in the *cis* condition when on the same. *Trans*-complementation only occurs when two mutations lie in different cistrons, and by careful mapping of mutations the boundary between two cistrons can be located from the results of the *cis-trans* test.

cistron A region of DNA within which mutations affect the same functions by the criterion of the *CIS-TRANS* TEST. In molecular terms, the length of DNA (or RNA in some viruses) encoding a specific and functional product, usually a protein, in which case the cistron is 'read' via messenger RNA; but both ribosomal RNA and transfer RNA molecules have their own encoding cistrons. In modern terminology, 'cistron' is equivalent to 'gene', except that not all putative genes have been fully validated by complementation analysis.

citric acid cycle See KREBS CYCLE.

c-Jun N-terminal protein kinase See JNK SIGNALLING PATHWAY.

CKIs (cyclin-dependent kinase inhibitors) Control elements of the eukaryotic CELL CYCLE which bind to and inactivate cyclin-dependent kinases. Some of them are probably constitutive of the cell cycle, others are thought to be induced by such conditions as DNA damage, starvation or GROWTH FACTORS (see DNA REPLICATION). One 'hydraulic' model of the cell cycle proposes that the peaks of cdks observed at G_1-S and G_2-M are caused by their build-up behind a 'CKI dam' until the excess cyclins somehow trigger CKI destruction or inactivation so

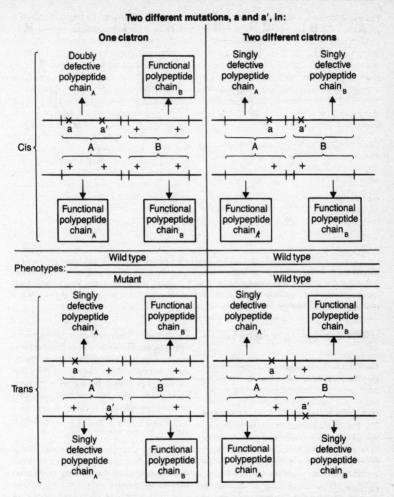

FIG. 32 *Theoretical basis of the* CIS-TRANS TEST. *Where two mutations (→←) occur within the same chromosome (cis configuration), complementation occurs through functional polypeptide production by the complementary cistron or cistrons of the other chromosome. Mutant phenotype only occurs where the two mutations occur within the same cistron but on different chromosomes. This effect enables precise mapping of the physical limits of cistrons within chromosomes.*

that the cell becomes irreversibly committed to S-phase and M-phase respectively, after the latter of which cyclins are destroyed.

clade A truly MONOPHYLETIC group of organisms; i.e. one containing one ancestor not shared by any species outside that group. See CLADISTICS, CROWN CLADE, GRADE.

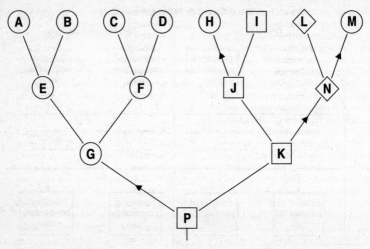

FIG. 33 *A cladogram illustrating terminology employed. Character states are represented by shapes of figures, and a change in these by an arrow. Character state in circles (○) is symplesiomorphous for taxa A & B and C & D but is synapomorphous for taxa E, F and G. It is homoplasous between taxa A – G (being an example of parallel development) and homoplasous between taxa A – H and M, but here it is also an example of convergent development.*

cladist Proponent of, or worker in, CLADISTICS.

cladistics Method of CLASSIFICATION employing genealogies alone in inferring phylogenetic relationships among organisms – disregarding their phenetic adaptations, which may not be due to HOMOLOGY. The resulting diagram of inferred relationships is termed a CLADOGRAM (not the same as a PHYLOGENETIC TREE, which adds non-phylogenetic information, such as times of transitions, from other sources; see Fig. 33, BAYESIAN STATISTICS). A line of descent is characterized by the occurrence of one or more evolutionary novelties (*apomorphies*). Any character found in two or more taxa is homologous in them if their most recent common ancestor also had it. Such a shared homologue may be *symplesiomorphous* (or *-morphic*) in these taxa if it is believed to have originated as a novelty in a common ancestor earlier than the most recent common ancestor, but *synapomorphous* (*-morphic*) if not. HOMOPLASY occurs between characters that share structural aspects but

are thought to have arisen independently, either by PARALLEL EVOLUTION or CONVERGENCE: the cladistic method does not distinguish between these, because it does not need to; nor does it permit what it terms PARAPHYLETIC taxa. For just two examples of how the effect of a cladistic approach can alter classification, see REPTILIA, SARCOPTERYGII. See PHENETICS, PHYLOGENETICS.

Cladocera Order of BRANCHIOPODA. 'Water fleas', including *Daphnia*. Carapace encloses trunk limbs, used for feeding. Antennae used for swimming.

clade (phylloclade) Modified stem, having appearance and function of a leaf, e.g. butcher's broom.

cladogenesis Branching SPECIATION, in which an evolutionary lineage splits to yield two or more lineages. See CLADISTICS.

cladogram See CLADISTICS.

Cladophorophyceae A Class in the algal Division CHLOROPHYTA. Algae having uniseriate filaments that can be branched or unbranched and multinucleate cells. Some members have evolved complex thalli in which the filamentous organization is difficult to recognize. Cells are surrounded by a cellulose wall in which the cellulose is highly crystalline. Zoospores have two or four flagella. The basal bodies have an II o'clock–5 o'clock configuration and exhibit a distinct overlap. Mitosis is closed, while the telophase nucleus has a characteristic dumb-bell shape because of the presence of a prominent, persistent telophase spindle, which forms between polar pairs of centrioles (mitosis is centric). Vegetative cell division is not synchronous with mitosis and the cross walls are not perforated by plasmodesmata.

Each cell of a haploid or diploid thallus is multinucleate and contains irregular angular chloroplasts that form a parietal net. In many chloroplasts there is a single bilenticular pyrenoid divided into two hemispheres by a thylakoid. A bowl-shaped starch grain caps each hemisphere. Life cycles of sexually reproducing species are diplohaplontic and isomorphic, and display isogamy or anisogamy. The haploid plants produce biflagellate gametes. The zygote develops into a diploid sporophyte, which is morphologically identical to the gametophyte. The sporophyte produces quadriflagellate meiospores, which grow into gametophytes. Some reproduce, either additionally or exclusively by means of asexual biflagellate or quadriflagellate zoospores.

cladophyll Branch that resembles a foliage leaf.

clamp connection Lateral connection between adjacent cells of a dikaryotic hypha, found in some of the BASIDIOMYCOTA. Ensures that each cell of the hypha contains two genetically dissimilar nuclei. Compare CROZIER FORMATION.

class A taxonomic category in CLASSIFICATION. Of higher rank (more inclusive) than ORDER but of lower rank (less inclusive) than PHYLUM (or DIVISION). Thus there may be one or more classes in a phylum or division and one or more orders in a class.

In employing any taxonomic category one aim is to ensure that all its members share a common ancestor which is also a member of the taxon, although opinions differ whether the resulting group should include all *descendants* of the common ancestor. Thus evolutionary taxonomists recognize the Class Reptilia, whereas cladists do not.

classification Any method organizing and systematizing the diversity of organisms, living and extinct, according to a set of rules. Belief in the existence of a pre-arranged (divine?) natural order, which it was the role of the scientist to discover, was common until the early 19th century, but has dwindled since the publication of Darwin's *The Origin of Species* in 1859. However, there is still considerable dialogue between ESSENTIALIST and NOMINALIST accounts of biological classification as to whether the groups which different classifications recognize are 'natural' (real) or merely human constructs (artificial). Characters are not randomly distributed among organisms but tend to cluster together with high predictability, suggesting that all the taxonomist has to do is discover the various nested sets of characters to attain a 'natural' classification. However, even those taxa which seem to have most to recommend their objective realities in nature (i.e. species) lack some of the features of NATURAL KINDS.

The charge of arbitrariness (artificiality) over the rule of classification has led to the search for an objective methodology. The solution of *numerical taxonomists* has been to select not just one or a few characters which are given added weight (often apparently arbitrarily) in comparisons between organisms, but to give all phenetic characters equal weight, in the expectation that natural (as opposed to artificial) groups will automatically emerge as clusters through overall phenetic similarity. The taxonomist then arranges these clusters into a rule-governed hierarchy of groups.

But this approach also has arbitrary elements, although favoured by some mathematically minded taxonomists. Critics argue that it fails to achieve a genealogically based classification: they deny that overall similarity alone is a sure guide to recency of common ancestry. Thus, crocodiles and lizards share more common features than either does with birds; yet crocodiles and birds share a more recent common ancestor than either does with lizards. If there is anything like an objectifying principle available to taxonomists it must surely be genealogy. Two principal schools of taxonomy which endeavour to objectify their methods by acknowledging the process of evolution in this way are *cladism* and *evolutionary taxonomy*. The principal difference between them (see CLADISTICS) is that cladists include *all* descendant species along with the ancestral species within taxonomic groups; evolutionary taxonomists hold that different rates of adaptation within the descendant groups of a single ancestor should be reflected in the classification. This may require that those that have diverged more from the ancestral stock are given special (often higher) taxonomic ranking compared with those that have diverged less, and that consequently *not all* descendant species will be included in the taxon of the ancestral species. Much depends on the accepted definition of MONOPHYLETIC. All biological classifications are hierarchical: the higher the taxonomic category, the more inclusive. In descending order of inclusiveness, and omitting intermediate (sub-, infra-, super-, etc.) taxa, the Linnean sequence is: kingdom, phylum (division in botany), class, order, family, genus, species. There are however enormous difficulties in establishing accurate genealogical classifications, not least with taxa which are extinct. Some claim that the Linnean system no longer accommodates modern taxonomy and prefer molecular approaches to classification. One of these, claiming some currency, divides organisms into three great DOMAINS. Molecular phylogeny is full of interpretational pitfalls, but has led some to question long-accepted morphological transformations (see BAYESIAN STATISTICS, OPTIMALITY THEORY). See DNA HYBRIDIZATION, DNA SEQUENCING, IDENTIFICATION KEYS, IMMUNOLOGICAL DISTANCE, LINNAEUS, ORTHOLOGOUS, PARALOGOUS.

clathrin One of the main proteins covering some COATED VESICLES, interconnecting molecules (triskeletons) forming varying numbers of lattice-like pentagonal/hexagonal facets on the cytosolic surfaces of coated pits, causing invagination to form the vesicle. The clathrin coat is shed after the vesicle is formed and internalized. See Fig. 34.

clavicle MEMBRANE BONE of ventral side of PECTORAL (shoulder) GIRDLE of many vertebrates. Collar-bone of man.

clearing Process used to prepare many histological slides in light microscopy. The object is to remove any alcohol used in the dehydration of the material; the preparation is soaked in two or three changes of *clearing agent* (e.g. benzene, xylene, or oil of cloves). Clearing makes the material transparent and permits embedding in paraffin wax (insoluble in alcohol) prior to sectioning.

cleavage (segmentation) Repeated subdivision of egg or zygote cytoplasm associated with, but not always accompanied by, mitoses. In animals it often produces a mass (the blastula) of small cells (blastomeres) which subsequently enlarge. Bilateral (radial) cleavage, in which ANIMAL POLE blastomeres tend to lie directly on top of vegetal blastomeres, occurs in echinoderms and chordates. Spiral cleavage, in which the first four animal blastomeres lie over the junctions of the first four vegetal blastomeres, is characteristic of other animal phyla. But if chaetognaths are protostomes, their radial cleavage is a counterexample. Cleavage may be *deterministic* or *indeterministic*, depending respectively upon whether the fates of blastomeres are already fixed or are plastic. Cleavage is complete (*holoblastic*) in eggs with little yolk; partial (*meroblastic*) in yolky eggs where only the non-yolky portion engages in cell division; *superficial* in centrolecithal eggs, where nuclear division produces many nuclei towards the centre of the cell and which then migrate to the cytoplasmic periphery to become partitioned by cell membranes (see MOSAIC DEVELOPMENT). Mammalian development is unique in involving compaction, whereby cells of the 8-cell stage maximize their contact with each other to form a compact ball stabilized by tight junctions sealing off the sphere's interior prior to division to form the 16-cell morula. See CYTOKINESIS.

cleavage furrow See CYTOKINESIS.

cleidoic egg Egg of terrestrial animal (e.g. bird or insect) enclosed within protective shell, largely isolating it from its surroundings and permitting gaseous exchange and minor water loss or gain. Contrasts with most aquatic eggs, in which exchange of water, salts, ammonia, etc., occurs fairly freely. See URICOTELIC.

cleistocarp (cleistothecium) Completely closed spherical fruit body (ascocarp) of some of the ASCOMYCOTA, e.g. powdery mildews, from which spores are eventually liberated through decay or rupture of its wall.

cleistogamy Fertilization within an unopened flower; e.g. in violet.

climacteric (Bot.) Occurrence of a large increase in cellular respiration in fruit-ripening (e.g. tomatoes, avocadoes, apples, etc.). Such fruits are called *climacteric fruits* (those that display a steady decline or gradual ripening are called *non-climacteric fruits*). Exposure of the fruits to low temperatures stops the climacteric permanently. Fruits can be stored for long periods in a vacuum, and since the available oxygen is minimal, cellular respiration and ethylene are suppressed. After the climacteric, fruits senesce and become susceptible to fungal and bacterial attack. See ETHYLENE.

climax community The concept that communities of organisms will eventually develop with compositions more or less stable and in equilibrium with existing natural environmental conditions (e.g. oak forest in lowland Britain; spruce and aspen forests of the Boreal Forest region). The concept of a climax has a long history. Frederic Clements in 1916 hypothesized that there was only one true climax in any given climatic region (climatic climax community). This rather extreme *monoclimax theory*

became challenged by many ecologists, including Arthur Tansley in 1939 and the *polyclimax* school of thought was born. This recognized that a local climax may be governed by one or more factors so a single climatic area could contain several specific climax types. Later Whittaker in 1953 proposed his climax-pattern hypothesis. It conceives of a continuity of climax types, varying gradually along environmental gradients and not necessarily separable into discrete climaxes. It is difficult to identify a stable climax community. Usually ecologists can do no more than indicate that a rate of change of SUCCESSION slows to a point where any change is imperceptible.

cline Continuous gradation of phenotype or genotype in a species population, usually correlated with a gradually changing ecological variable. See INFRASPECIFIC VARIATION, ECOTYPE.

clisere Succession of CLIMAX COMMUNITIES in an area as a result of climatic changes.

clitellum Saddle-like region of some annelid worms (Oligochaeta, Hirudinea), prominent in sexually mature animals. Contains mucus glands secreting a sheath around copulating worms binding them together; the resultant cocoon houses the fertilized eggs during their development.

clitoris Small erectile organ of female amniotes, homologous to the male's penis; anterior to vagina and urethra.

cloaca Terminal region of the gut of most vertebrates into which kidney and reproductive ducts open. There is only one posterior opening to the body, the cloacal aperture, instead of separate anus and urogenital openings (e.g. placental mammals). Also terminal part of intestine of some invertebrates, e.g. sea cucumbers.

clonal selection theory In immunology, the theory that each antibody-forming cell is genetically committed to producing just one antibody type, distinct from that of other such cells, expressed as a cell-surface receptor, and that proliferation of this cell is initiated only by selection by antigen. Early in development, encounters with self-antigen lead to cell death; but encounters with external antigen cause the cells to respond by clonal expansion and differentiation into mature B CELLS. This theory, associated with Macfarlane Burnet, explains several immunological phenomena (e.g. affinity maturation, self-tolerance, immunological 'memory'). Selection processes also occur for T CELLS.

clone (1) A group of organisms of identical genotype, produced by some kind of ASEXUAL reproduction and some sexual processes, such as haploid selfing, or inbreeding a completely homozygous line. Nuclear transplantation techniques introducing genetically identical nuclei into enucleated eggs can also produce clones in some animals, even *in utero*. The first mammal to be artificially cloned was a sheep named 'Dolly', in 1997. A donor nucleus (G0-arrested) from the udder of a six-year-old ewe was inserted into an enucleated metaphase II-arrested egg. This successful outcome has renewed doubts about the ethics of cloning and, within days of the report's publication, US President Clinton instituted a ban on federal funding of attempts to clone human beings by this method, and requested a report on legal and ethical issues from the National Bioethics Advisory Commission (website http://www.bioethics.gov/). This concluded, *inter alia*, that it is currently morally unacceptable for anyone to attempt to create a child using somatic cell nuclear transfer, and proposed continuation of the moratorium on federal funding. Interestingly, it seems that Dolly's 'genetic age', as indicated by TELOMERE lengths, was greater than her temporal age (see AGEING). The fact that cloned animals may have developmental and morphological irregularities did not surprise botanists, and is often due to imprinting and other epigenetic effects (see CHROMATIN REMODELLING). See EMBRYONIC STEM CELLS, GMO, MICROPROPAGATION. (2) A group of cells descended from the same single parent cell. Often used of sub-populations of multicellular organisms (e.g. see CLONAL SELECTION THEORY) rather than the entire organism, which may in any case be a MOSAIC. (3) Nucleic acid sequences are said to be cloned when they are inserted into VECTORS (e.g. PLASMIDS, YEAST ARTIFICIAL CHROMOSOME) and then copied along with them within host cells.

cloning The production of one individual organism from a nucleus, cell or asexual offshoot of another. Aims and methods differ greatly, especially between the cloning of plants and of higher animals. See CLONE, PROTOPLAST, CUTTING.

club moss See LYCOPHYTA.

clumped distribution The distribution of organisms in which individuals are closer together than if they were distributed at random or equidistant from each other.

c-*myc* Proto-oncogene activated by β-CATENIN believed to be a key regulator of cell growth and differentiation, whose protein product (Myc) is a transcription factor associated with mitogenic activation (cell proliferation) in a variety of cell types. It forms a heterodimer with a partner protein (Max, product of c-*max* gene) and binds another proto-oncogene, *cdc25A*, which triggers APOPTOSIS in the absence of adequate levels of GROWTH FACTORS. Myc stimulates cyclin E- and cyclin D1-dependent kinases (see CDK). Deregulated *myc* expression is frequent in CANCER CELLS, while its ectopic expression is sufficient to send many cells into the cell cycle in the absence of external mitogens. Although, individually, Myc, Ras and Bcl-2 are potent growth promoters, these potentially cancer-promoting activities are not normally actuated because of their corresponding growth-inhibiting properties (apoptosis for Myc and Ras, growth arrest for Ras and Bcl-2). See CYTOSTATIC FACTORS, E1A, Fig. 41.

Cnidaria Subphylum of the COELENTERATA containing hydroids, jellyfish, sea anemones and corals. Gut incomplete (one opening); ectoderm containing CNIDOBLASTS. Two structural forms: (i) attached, sessile *polyp*, (ii) free-swimming *medusa*. Former is a cylindrical sac with mouth and tentacles at opposite end to the attachment; latter is umbrella-shaped, with flattened enteron, and mouth in middle of concave

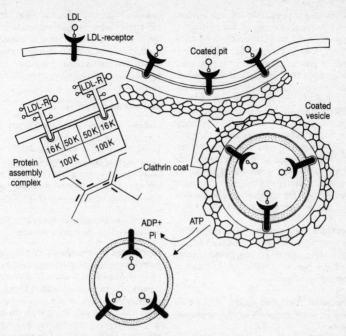

FIG. 34 *Involvement of a* COATED PIT *and* COATED VESICLE *in endocytosis of LDL via a cell-surface LDL-receptor. Details of the protein components of the assembly complex are given (K = kilodalton).*

under-surface. Sometimes the phases alternate in a single life-cycle; sometimes only one phase occurs. Compare CTENOPHORA.

cnidoblast (thread cell) Specialized stinging cell found only in CNIDARIA and a few of their predators which incorporate them. Several kinds exist for, e.g., adhesion, penetration, injection. This cell produces an inert ORGANELLE, a *cnida* (nematocyst), commonly regarded as an *independent effector* (see EFFECTOR). Compare LASSO CELL.

CNS See CENTRAL NERVOUS SYSTEM.

coA See COENZYME A.

coacervate Inorganic colloidal particle (e.g. clay) on to which have been absorbed organic molecules; maybe acted as an important concentrating mechanism in prebiotic evolution. See ORIGIN OF LIFE for discussion of illites and kaolinites.

coactivators Proteins mediating between activators and the TATA-binding protein in eukaryotic transcription (see GENE EXPRESSION). See ENHANCER (Fig. 49); compare CO-REPRESSORS.

coated pit Specialized regions of most eukaryotic CELL MEMBRANES, appearing as depression in electron micrographs before pinching off as a COATED VESICLE to initiate an endocytic cycle. Clathrin-coated pits are the vehicles for RECEPTOR-mediated endocytosis.

coated vesicles Membranous vesicles (c. 50 nm diameter) budded off endocytically from the plasma membrane and some other cell membranes (see PINOCYTOSIS). Either CLATHRIN-coated or non-clathrin-coated; the former arise from coated pits during receptor-mediated ENDOCYTOSIS (e.g. in uptake of CHOLESTEROL as LDL), each vesicle containing a portion of the extracellular fluid and contributing to the bulk-phase aspect of endocytosis (*pinocytosis*); the latter (coat proteins include β-COP, recruited from a protein pool in the cytosol) carry cargo from the endoplasmic reticulum to and through the GOLGI APPARATUS. Clathrin- and non-clathrin-coated vesicles are budded off from the *trans* Golgi network, the former carrying proteins to endocytic organelles and secretory granules. Coat proteins are normally jettisoned soon after vesicle formation and prior to fusion with the target membrane, when the cargo (glycoprotein, neurotransmitter, etc.) is released (see Fig. 34). See CAVEOLAE, CELL MEMBRANES, CELL LOCOMOTION, PHAGOCYTOSIS.

coccidia Apicomplexan protozoa parasitic in guts of vertebrates and invertebrates. Probably ancestral to haemosporidians (e.g. malarial parasites). Give rise to diseases termed *coccidioses*.

coccolith A calcified organic SCALE occurring in some members of the PRYMNESIOPHYTA.

coccyx Fused tail vertebrae. In man comprises two to three bones.

cochlea Diverticulum of the sacculus of inner ears of crocodilians, birds and mammals, usually forming a coiled spiral. Contains the organ of Corti, involved in sound detection and pitch analysis, a longitudinal mound of HAIR CELLS running the length of the cochlea supported on the *basilar membrane* with a membranous flap (*tectorial membrane*) overlaying the hair cells. Outer hair cells are responsible for the sensitivity and frequency-resolving capacity (the 'cochlear amplifier'); inner hair cells convey auditory information to the brain. Vibrations in the round window caused by vibrations in the EAR OSSICLES are transmitted to the perilymph on one side of the basilar membrane, vibrating it and stimulating hair cells of the organ of Corti. The apex of the cochlea (spiral top) is most sensitive to lower frequency vibrations, the base to higher frequencies. See VESTIBULAR APPARATUS.

cocoon Protective covering of eggs, larvae, etc. Eggs of some annelids are fertilized and develop in a cocoon. Larvae of many endopterygotan insects spin cocoons in which pupae develop (cocoon of silkworm moth is source of silk). Spiders may also spin cocoons for their eggs.

codominant See DOMINANCE.

codon Coding unit of MESSENGER RNA, comprising a triplet of nucleotides which base-pairs with a corresponding triplet (*anticodon*) of an appropriate TRANSFER RNA molecule. Most codons encode amino acids (see GENETIC CODE). Some codons signify termination of the amino acid chain (the *nonsense codons* UAG, UAA and UGA, respectively *amber*, *ochre* and *opal*). For role of AUG codon, see PROTEIN SYNTHESIS.

coefficient of selection, s Proportionate reduction in contribution to the gene pool (at a specified time *t*) made by gametes of a particular genotype, compared with the contribution made by the standard genotype, which is usually taken to be the most favoured. If s = 0.1, then for every 100 zygotes produced by the favoured genotype only 90 are produced by the genotype selected against, and the genetic contribution of the unfavoured genotype is 1 − s. See FITNESS.

Coelacanthini Large suborder of CROSSOPTERYGII, mostly fossil (Devonian onwards). Freshwater, but with living marine representatives (*Latimeria*) in the Indian Ocean. Thought to have been extinct since the Cretaceous; but since 1938 several have been found. CHOANAE absent. Living coelacanth posterior dorsal and anal fins (both medial) resemble their (paired) pelvic fins in musculature, innervation and skeletal details. In this they differ from other sarcopterygians and other fishes, and the reasons are as yet unclear. Considerable debate surrounds whether coelacanth α- and β-haemoglobins and mtDNA resemble more closely those of other teleosts or those of larval amphibians.

Coelenterata Phylum of diploblastic and radially symmetrical aquatic animals, comprising the subphyla CNIDARIA and CTENOPHORA. Ectoderm and endoderm separated by *mesogloea* (jelly-like matrix of variable thickness) and enclosing the gut cavity (*coelenteron*), with a single opening to the exterior. Peculiar cell types include MUS-CULO-EPITHELIAL CELLS, and either CNIDOBLASTS (cnidarians) or LASSO CELLS (ctenophores).

coelom Main (secondary) body cavity of many triploblastic animals, in which the gut is suspended. Lined entirely by mesoderm. Principal modes of origin are either by separation of mesoderm from endoderm as a series of pouches which round off enclosing part of the archenteron (ENTEROCOELY), or *de novo* by cavitation of the embryonic mesoderm (SCHIZOCOELY). (See Fig. 35.) Contains fluid (coelomic fluid), often receiving excretory wastes and/or gametes, which reach the exterior via ciliated funnels and ducts (*coelomoducts*). May be subdivided by septa into *pericardial, pleural* and *peritoneal* coeloms enclosing respectively the heart, lungs and gut. Reduced in arthropods, restricted to cavities of gonads and excretory organs, the main body cavity being the HAEMOCOELE, as in Mollusca.

coelomate An animal with a COELOM.

coelomoduct Mesodermal ciliated duct (its lumen never intracellular) originating in the COELOM and growing outwards from the gonad or wall of the coelomic cavity to fuse with the body wall. Sometimes conveys gametes to exterior (see MÜLLERIAN DUCT, WOLFFIAN DUCT); sometimes excretory, e.g. kidneys of molluscs. Compare NEPHRIDIUM. See KIDNEY.

coenobium Type of algal colony where the number of cells is determined at its formation. Individual cells are incapable of cell division, are arranged in a specific manner, are coordinated and behave as a unit; e.g. *Volvox, Pandorina, Pediastrum, Hydrodictyon*.

coenocyte (adj. coenocytic) Multi-nucleate mass of protoplasm formed by division of nucleus, but not cytoplasm, of an original cell with single nucleus; e.g. many fungi and some green algae. Compare SYNCY-TIUM. See ACELLULAR.

coenospecies Group of related species with the potential, directly or indirectly, of forming fertile hybrids with one another.

coenzyme Organic molecule (often a derivative of a mononucleotide or dinucleotide) serving as COFACTOR in an enzyme reaction, but, unlike a PROSTHETIC GROUP, binding only temporarily to the enzyme molecule. Often a recycled vehicle for a chemical group needed in or produced by the enzymic process, reverting to its original form when the group is removed – often by another enzyme in a pathway. Removal by coenzymes of reaction products from the enzyme environment may be essential to prevent end-product inhibition of the enzyme. In heterotrophs they are frequently derivatives of water-soluble VITA-MINS. Many coenzymes, e.g. ATP NAD, FAD, are equipped with nucleotide components, usually catalytically inactive. However, these act and probably originally acted, as handles by which enzymes attach. Amino acids might originally have served preadaptively as coenzymes of RIBOZYMES, where they were equipped with oligonucleotide handles. See COENZYME A, COEN-ZYME Q, FAD, FMN, NAD, NADP.

coenzyme A (CoA, CoA-SH) Mononucleotide phosphate ester of pantothenic acid (a vitamin for vertebrates). Carrier of acyl groups in fatty acid oxidation and synthesis, pyruvate oxidation (see KREBS CYCLE) and various acetylations. When carrying an acyl group, referred to as acetyl coenzyme A (acetyl-CoA) which is converted to malonyl-CoA by acetyl CoA carboxylase (see LEPTIN). See Figs. 36a and b.

coenzyme Q (CoQ, ubiquinone) Lipid-soluble quinone coenzyme transporting electrons from organic substrates to oxygen in mitochondrial respiratory chains. Several forms; but all function by reversible quinone/quinol redox reactions. In plants, the related *plastoquinones* perform similar roles in photosynthetic electron transport. See ELECTRON TRANSPORT SYSTEM.

coevolution Evolution in two or more species of adaptations caused by the selection pressures each imposes on the other.

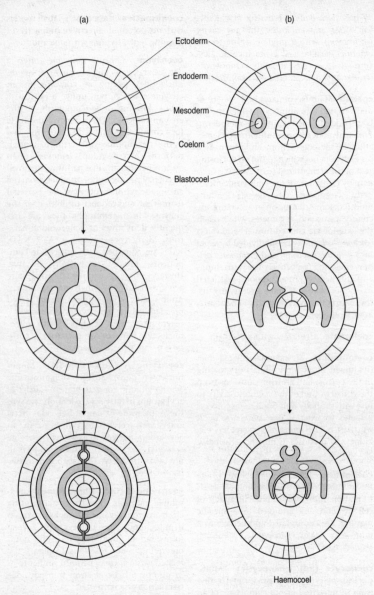

(a)   (b)

Ectoderm

Endoderm

Mesoderm

Coelom

Blastocoel

Haemocoel

FIG. 35 *Cross-sections showing* COELOM *development (schizocoely) in (a) annelids, compared with (b) enlargement of the blastocoele in molluscs and arthropods to form the* HAEMOCOEL *and greatly reduced coelom.*

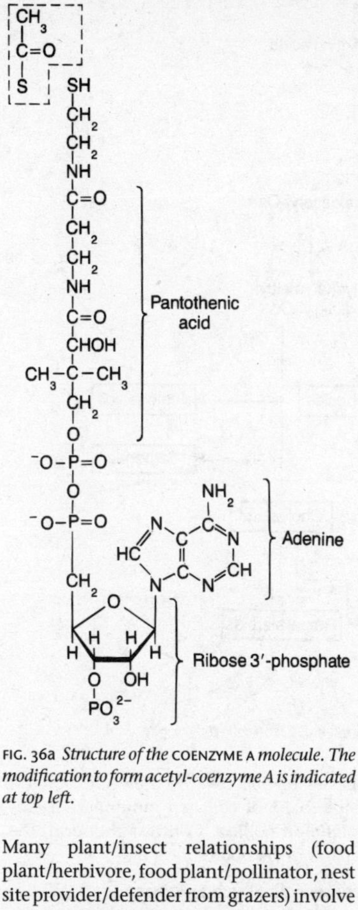

FIG. 36a *Structure of the* COENZYME A *molecule. The modification to form acetyl-coenzyme A is indicated at top left.*

Many plant/insect relationships (food plant/herbivore, food plant/pollinator, nest site provider/defender from grazers) involve reciprocal adaptations. Most host/parasite, predator/prey, cleaner/cleaned relationships, etc., are likely to involve coevolution. See ARMS RACE.

cofactor Non-protein substance essential for one or more related enzyme reactions. They include PROSTHETIC GROUPS and COENZYMES. An enzyme-cofactor complex is termed a *holoenzyme*, while the enzyme alone (inactive without its cofactor) is the *apoenzyme*.

cognate Related, or connected. As of a gene and its *cognate protein*, i.e. the protein it

encodes; or growth factors and their *cognate receptors*.

cohesion-tension theory See TRANSPIRATION STREAM.

cohort (1) In some classifications, a formal taxonomic category between Infraclass and Superorder. (2) An age-class in a population; e.g. those individuals aged between more than one year of age (i.e. >1.0 yr) up to and including those exactly two years of age. See AGE PYRAMIDS.

coiled body Domain within the nucleus containing high concentrations of small nuclear ribonucleoprotein particles (snRNPs), including splicing snRNPs. Sometimes seen close to the edge of the nucleolus. Function uncertain. See RNPs.

colchicine Antimitotic drug, binding to one tubulin molecule and preventing its polymerization. Depolymerization of tubulin then results, disappearance of the mitotic spindle blocking the cell mitosis. The drug taxol stabilizes microtubules, causing free tubulin to polymerize and for this different reason also arrests the cell in mitosis.

Coleoptera Beetles. Huge order of endopterygote insects. Fore-wings (elytra) horny, covering membranous and delicate hind-wings (which may be small or absent) and trunk segments. Biting mouthparts. Larvae may be active predators (*campodeiform*), caterpillar-like (*eruciform*), or grub-like (*apodous*). Many larvae and adults are serious pests of crops, stored produce and timber. Some borers of live wood may transmit fungal disease (e.g. *Scolytus* and Dutch elm disease). Size range is probably greater than that of any other insect order.

coleoptile Protective sheath surrounding the apical meristem and leaf primordia (plumule) of the grass embryo: often interpreted as the first leaf.

coleorhiza Protective sheath surrounding RADICLE in grass seedlings.

colicins Antibiotic toxins determined by plasmid-borne genes in *ESCHERICHIA COLI* (see

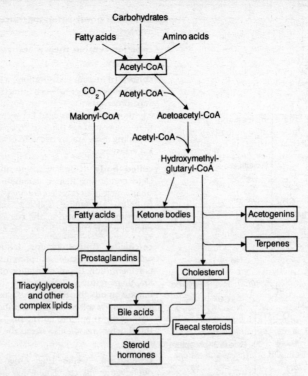

FIG. 36b *Some metabolic pathways in which acetyl-coenzyme A is involved.*

PLASMID). Colicin A is a protein forming voltage-gated channels in bacterial membranes. See BCL-2.

coliform Gram-negative, rod-shaped, non-sporulating and facultatively anaerobic bacteria which can ferment lactose with acid and gas formation within 24 hours at 37°C (or 48 hours at 35°C). *E. coli* is a typical coliform. Sometimes loosely implies any Gram-negative, rod-shaped enteric bacterium.

collagen Major fibrous (structural) protein of CONNECTIVE TISSUE, occurring as *white fibres* produced by fibroblasts. Forms up to one third of total body protein of higher vertebrates. Provides high tensile strength, i.e. resists stretching forces, as in tendon, without much elasticity (unlike ELASTIN). Collagen fibres are composed of masses of *tropocollagen* molecules, each a triple helix of collagen monomers. Yields *gelatin* on boiling. Contrast chondrin, the matrix of CARTILAGE.

collar cell See CHOANOCYTE.

collateral bundle See VASCULAR BUNDLE.

Collembola Springtails. Order of small primitively wingless insects (see APTERY-GOTA). Two caudal furcula fold under the abdomen and engage the hamula (another pair of abdominal appendages) prior to explosive release resulting in the spring. First abdominal segment carries ventral tube for adhesion. Compound eyes absent. No Malpighian tubules, and usually no tracheal system. Immensely abundant in soil, under bark, on pond surfaces, etc., forming vital link in detritus food chain.

collenchyma Tissue composed of collenchyma cells, which provide mechanical support in many young growing plant structures (stems, petioles, leaves), but uncommon in roots. Collenchyma tissue occurs in discrete strands or as continuous cylinders beneath the epidermis, as well as bordering veins in dicotyledon leaves. Cells are typically elongated and contain unevenly thickened, non-lignified primary walls, and are living at maturity. Being primary walls they are readily stretched, offering little resistance to elongation. Compare SCLERENCHYMA.

colloid A substance having particles of about 100–10,000 nm diameter, which remain dispersed in solution. Such *colloidal solutions* are intermediate in many of their properties between true solutions and suspensions. Brownian motion prevents colloid particles from sedimenting under gravity, and in lyophilic colloids (solvent-loving) such as aqueous protein solutions, each particle attracts around it a 'shell' of solvent forming a hydration layer, preventing them from flocculating. Competition for this solvent by addition of strong salt solution will precipitate the colloid.

colon Large intestine of vertebrates, excluding narrower terminal rectum. In amniotes and some amphibians, but not fish, is clearly marked off from small intestine by a valve. In mammals at least, has essential water-absorbing role in preparation of faeces. Bacteria housed within it produce vitamins (esp. VITAMIN K).

colony (1) Several plant and animal organizations where various more or less (and often completely) distinct individuals live together and interact in mutually advantageous ways. Sometimes, as in the alga *Volvox* and certain ciliates, the organization approaches multicellularity and may even have been a transitional stage in its attainment; but there is usually insufficient communication or division of labour between cells for full multicellular status (see COENO-BIUM). In colonial CNIDARIA and ECTOPROCTA there may be considerable POLYMORPHISM between the individuals (termed ZOOIDS)

with some associated division of labour. In all these there are good asexual budding abilities, and it is likely that the colonial habit originated by failure of buds to separate. Not so in colonial insects (e.g. HYMENOPTERA, ISOPTERA), although here too division of labour is the rule (see CASTE). Among vertebrates, several bird and mammal species live and/or breed in colonies, and many behavioural adaptations reflect this. (2) A group of microorganisms (bacteria, yeasts, etc.) arising from a single cell and lying on surface of food source – as in culture on agar.

colony stimulating factors (CSFs) Glycoproteins stimulating mitosis in (i.e. they are mitogens of) bone marrow granulocyte and macrophage progenitor cells. They are also SURVIVAL FACTORS, and promote differentiation in these cells and enhance phagocytosis by mature cells. Lung tissue is a major source. Nomenclature of CSFs is not standardized; e.g., multi-CSF is also called INTERLEUKIN 3 and stimulates multipotential stem cells to produce all the major blood cell types, while the CSF haemopoietin (see HAEMOPOIESIS) is also regarded as a GROWTH FACTOR. CSF-1 binds receptor tyrosine kinase of mononuclear phagocytes, initiating signal transduction with pleiotropic results due to expression of several different genes (e.g. some cyclin genes). Some can force leukaemic cells to differentiate and so stop dividing. Some CSFs form part of the INNATE IMMUNE RESPONSE.

colostrum Cloudy fluid secretion of mammary glands during the first few days after birth of young and before full-scale milk production. Important source of antibodies (passive immunity) too large to cross the placenta. Rich in proteins, but low in fat and sugar.

columella (Bot.) (1) Dome-shaped structure present in sporangia of many Zygomycotina (fungi) of the order Mucorales (pin moulds); produced by formation of convex septum cutting off sporangium from hypha bearing it. (2) Sterile central tissue of moss capsule. (Zool.) The *columella auris*, or stapes. The EAR OSSICLE of land vertebrates (often complex in reptiles and birds) homologous with the HYOMANDIBULAR bone of crossopterygians, and transmitting air vibrations from the ear drum (to which it is primitively attached) directly to the oval window of the inner ear (amphibia, primitive reptiles). In mammals it no longer attaches to the ear drum, articulating with the incus at its outer edge.

columnar epithelium See EPITHELIUM.

Commelinidae (Commeliniflorae) A subclass of the LILIOPSIDA comprising about 15,000 species with more than half belonging to the family Poaceae (grasses). The Poaceae and Cyperaceae (sedges) together account for 80% of the species. Characteristically vessels are found in all vegetative organs, the endosperm is starchy (often with compound starch grains). Pollen is smooth and monoporate. The perianth in the more archaic families is trimerous (parts in sets of three) or wanting. Families with a reduced perianth are typically wind-pollinated. The ovary is always superior (or nude). See INTRODUCTION to dictionary.

commensalism See SYMBIOSIS.

commissure Nerve tissue tract joining two bilaterally symmetrical parts of the central nervous system. In arthropods and annelids, they connect ganglia of the paired ventral nerve cords, and the supraoesophageal ganglia (brain) with the suboesophageal. Commissures unite the vertebrate cerebral hemispheres (see CORPUS CALLOSUM).

common (Of vascular bundles) passing through stem and leaf. Compare CAULINE.

community Term describing an assemblage of populations living in a prescribed area or physical habitat, inhabiting some common environment. An organized unit in possessing characteristics additional to its individual and population components, functioning as a unit in terms of flow of energy and matter. The biotic community is the living part of the ecosystem. It remains a broad term, describing natural assemblages of variable size, from those living upon submerged lake sediments to those of a vast rain forest. See ASSOCIATION, CONSOCIATION, SOCIETY.

compaction See CLEAVAGE.

companion cells Small cells characterized by dense cytoplasm and prominent nuclei, lying side-by-side with sieve tube cells in the phloem of flowering plants and arising with them by unequal longitudinal division of a common parent cell. One of their functions is to transport soluble food molecules into and out of sieve tube elements.

comparative methods Species comparisons are the most commonly used technique for estimating whether and how organisms are adapted to their environments, providing DARWIN with evidence for many of the arguments developed in his works *The Origin of Species* and *The Descent of Man*. Their popularity stems from our inability to do the experiments ideally suited to test evolutionary explanations, although proponents would argue that evolution provides its own natural experiments. Comparative methods essentially aim to determine whether or not different species living in the same environment predictably evolve the same characteristics, and nowadays greater weight is put on the number of times the same characteristic has evolved rather than simply on the number of species having it. In the last decade or so, statistical (and computerized) techniques have helped solve two problems inherent in comparative studies: evolutionary history and alternative explanations. Thus, monogamy and polygyny have evolved several times in birds and mammals; but monogamy (and egg-laying and feathers) predominates in birds and polygyny (and fur and viviparity) in mammals. It is unlikely that monogamy is causally associated with feathers or polygyny with fur, yet the correlations between these characters is highly significant. In order to explain adaptations we need to sift out the correlated but causally unrelated pairs of characters from the causally correlated, and in the case above we can do this only when we use information from polygamous birds and monogamous mammals. This way the merits of alternative phylogenies and adaptive explanations can be compared.

compartmentalization in communities A tendency in communities to be organized into sub-units within which interactions are strong but between which interactions are weak.

compartments Anatomical regions in some (maybe many) animals, their boundaries well defined in development by cell lineage and expression of regulatory genes. In *Drosophila*, a compartment comprises cells forming more than one CLONE (or POLY-CLONE), whose growth respects a compartment boundary even when one of the component clones is a rapid-growing mutant. In vertebrates, the search is on for equivalents of such HOMEOTIC genes as *engrailed* and *wingless* in insects (see SEGMENT POLARITY GENES), which delimit PARASEGMENTS, and the *HOX* GENE *Hox-2.9* seems to be the vertebrate homologue of *engrailed* and is involved as a regulatory gene in compartmentation of the hindbrain (see RHOMBOMERE). Fate of compartment polyclones is associated with the pattern of expression of homeotic genes.

compensation point Light intensity at which rate of respiration by a photosynthetic cell or organ equals its rate of photosynthesis. At this intensity there is no net gain or loss of oxygen or carbon dioxide from the structure. Compensation points for most plants occur around dawn and dusk, but vary with the species.

competence, competent A cell is said to be competent if it is able to respond to an inducing signal (see INDUCTION). This competence may require the presence of appropriate membrane receptors and transcription factors, and sometimes gap junctions with adjacent cells. A cell's competence for a particular response may change with time.

competition The effect (result) of a common demand by two or more organisms upon a limited supply of resource, e.g. food, water, minerals, light, mates, nesting sites,

etc. When *intraspecific*, it is a major factor in limiting population size (or density); when *interspecific*, it may result in local extinction of one or more competing species. An integral factor in Darwinian theory. See COMPETITIVE EXCLUSION PRINCIPLE, DENSITY-DEPENDENCE, EXPLOITATION COMPETITION, INTERFERENCE COMPETITION, OVERGROWTH COMPETITION, NATURAL SELECTION, SCRAMBLE COMPETITION.

competition coefficient In interspecific COMPETITION, a measure of the competitive effect of one species on another relative to the competitive effect of the second on itself.

competitive exclusion principle The empirical generalization, often regarded as axiomatic, probably first enunciated by J. Grinnell (1913) but generally attributed to G. F. Gause (1934), that as a result of interspecific competition, no two ecological niches can be precisely the same (see NICHE); i.e. niches are mutually exclusive, and mutual coexistence of two species will require that their niches be sufficiently different. In plants, it may be important in studying any interacting species pair to compare above-ground and below-ground competition. The competitive exclusion principle implies that any niche differentiation which occurs will tend to permit coexistence between competitors: a limiting similarity may be permitted, species perhaps evolving towards an optimum similarity. In competitive exclusion, one species is eliminated by another from an area or volume of habitat through interspecific COMPETITION. See CHARACTER DISPLACEMENT.

complement Nine interacting serum proteins (beta globulins, C_1–C_9), mostly enzymes, activated in a coordinated way and participating immunologically in bacterial lysis and macrophage chemotaxis. Thus C_9, a lytic component, has structural homology to PERFORINS, which generate pores in target cell membranes. Genetic loci responsible map in the S region of the H-2 COMPLEX in mice, and the HLA-B region of the MHC region in man. The principal event in the system is cleavage of the plasma

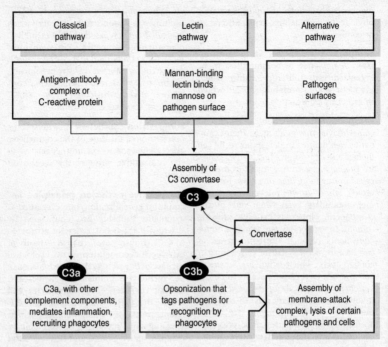

FIG. 37 *The three pathways of* COMPLEMENT *activation and their effects.* C3 *convertase is the enzyme which cleaves* C3.

protein C3, with subsequent attachment of the larger cleavage product (C3b) to receptors present on neutrophils, eosinophils, monocytes, macrophages and B CELLS. C3b combines with immune complexes and causes them to adhere to these white cells, promoting ingestion. C3 cleavage (*fixation*) terminates a CASCADE, itself initiated by immune complexes. See Fig. 37.

complementary DNA See cDNA.

complementary males Males, often degenerate and reduced, living attached to females; e.g. in barnacles (Cirripedia), ceratioid angler fish.

complementary resources A pair of resources for which consumption by the consumer of one resource reduces the consumer's requirements for the other: the combination is often greater than the sum of the independent parts. Thus, humans eating certain kinds of beans together with rice can increase the usable protein content of their food by 40%, because protein synthesis is limited by the first amino acid to become scarce, and unless all dietary amino acids are present in ratios representing those in the proteins found in the organism, protein intake will not lead to protein assimilation unless limiting amino acids are present in adequate quantities. The beans are poor in tryptophan and other sulphur-containing amino acids, but rich in lysine, an essential amino acid in low concentration in rice. Rice, however, is rich in sulphur-containing amino acids that are in low concentration in beans. Maize and wheat are low in isoleucine and lysine but high in other amino acids absent from legumes but required in low levels. See SUB-STITUTABLE RESOURCE.

complementation Production of a phenotype resembling wild-type (normal) when two mutations are brought together in the same cell in the *trans* configuration (on separate chromosomes); it occurs infrequently if the same functional unit (CIS-TRON or GENE) is defective on both chromosomes. The *CIS-TRANS* TEST is the rigorous method for defining the limits of cistrons, but often only the *trans* complementation test is made, the *cis* combination of individually recessive mutations almost always giving a wild phenotype.

complexity Biological systems can be analysed at several structural levels, including the molecular and sub-cellular, cellular, and multicellular levels. Although controversial, the derived condition here is usually treated as the more complex since it adds a structure previously absent. In turn, new (derived) developmental genes, or their new expression patterns, would be sought experimentally to confirm or disconfirm this approach (e.g. see GENE DUPLICATION). Some believe that a complete mathematical description of complex biological systems will come from treating biology as an information science, where genes and proteins are studied in the context of their informational pathways or networks. The functionings of such networks may themselves modulate the physical properties of intracellular and multicellular structures. COMPLEXITY emerges at higher levels in a manner unpredictable from a complete knowledge of lower levels. Biological systems are dynamic and their operations often nonlinear. They are also modular, so it is often fruitful to understand each component before constructing mathematical models to describe their integration into functional systems. If the *modules* are akin to ingredients, parts, subsystems and players, then just as important are *protocols*, which describe the recipes, rules, interfaces and etiquettes permitting functions not attainable by isolated modules. Protocols also facilitate addition of new protocols. A good protocol supplies both robustness and evolvability.

Systems biology may often require a shift in the traditional notion of 'what to look for' in biology. Its success will depend upon our acquisition of the kind of high throughput and accurate measurements that challenge our present level of experimental practice. Computer software will also need to be standardized if models are to be exchangeable.

Very interesting attempts are being made to simulate the evolution of biological complexity using 'digital organisms' – computer programs that self-replicate, mutate, compete and evolve (e.g. see http://myxo.css.msu.edu/papers/nature2003 and ARTIFICIAL INTELLIGENCE). Emerging evidence suggests that organismal complexity (at least in animals) arises from progressively more elaborate regulation of gene expression rather than the invention of new genes. See ENHANCER.

compost and composting See DECOMPOSER.

compound leaf Leaf whose blade is divided into several distinct leaflets.

concentric bundle See VASCULAR BUNDLE.

conceptacle Cavity containing sex organs, occurring in groups on terminal parts of branches of thallus in some brown algae, e.g. bladderwrack, *Fucus vesiculosus*.

condensation reaction Reaction in which two molecules are joined together (often by a covalent bond) with elimination of elements of water (i.e. one H_2O molecule). Occurs in all cells, during polymerization and many other processes. Contrast HYDROLYSIS.

conditioned reflex (conditional reflex) A REFLEX whose original unconditioned stimulus (US) is replaced by a novel conditioned stimulus (CS), the response remaining unaltered. Classically, the CS is presented to an animal either just prior to or in conjunction with the US; they are *associated* (presumably by formation of new neural pathways), and will each elicit the response when presented separately.

Although Ivan Pavlov (1849–1936) is best remembered for conditioning dogs to salivate in response to the sound of a bell (CS) instead of the sight or smell of food (US), his dogs also became excited at the CS and moved to where it or the eventual food was delivered. A simple form of LEARNING. See CONDITIONING.

conditioning Associative learning. There are two broad categories. (1) *Classical conditioning*, in which an animal detects correlations between external events, one of which is either *reinforcing* or *aversive*, modifying its behaviour in such a way that appropriate *consummatory responses* (usually) are elicited. (2) *Instrumental* or *operant conditioning*, in which aspects of consummatory responses are modified as an animal correlates and learns how variations in those responses affect its success in attaining a reinforcing, or avoiding an aversive, stimulus. Both are adaptive, since opportunities for successful instrumental conditioning may depend upon appropriate prior classical conditioning. See CONSUMMATORY ACT, LEARNING, REINFORCEMENT.

Condylarthra Extinct order of Paleocene and Eocene ungulate (hooved) mammals, possibly ancestral to most other ungulates and even to carnivores.

condyle Ellipsoid knob of bone, fitting into corresponding socket of another bone. Condyle and socket form a joint, allowing movement in one or two planes, but no rotation; e.g. condyle at each side of lower jaw, articulating with skull; *occipital condyles* on tetrapod skull, fitting into atlas vertebra.

cone (Bot.) *Strobilus*. Reproductive structure comprising a number of SPOROPHYLLS more or less compactly grouped on a central axis; e.g. cone of pine tree. (Zool.) (1) Light-sensitive RECEPTOR of most vertebrate retinas, though not usually of animals living in dim light. The cone-shaped outer segment appears to be a modified CILIUM, with the 9 + 2 microtubule arrangement where the outer segment joins the rest of the cell. Pigment in the cone (*retinene*) requires bright light before it bleaches; three different classes of cone each contain a combination of retinene with a different genetically determined protein (opsin) bleaching at different wavelengths (red, green, blue). Cones are thus responsible for colour vision. There are about 6 million cones per human RETINA, mostly concentrated in the FOVEA. (2) Any of the cusps of mammalian molar teeth.

conformation The three-dimensional shape (configuration) of a molecule. The final conformation adopted, preparatory to function, is the molecule's *native conformation*.

congenital Of a property present at birth. In one sense a synonym of HERITABLE (e.g. disorders arising from mutation such as DOWN'S SYNDROME, PHENYLKETONURIA). Congenital heart defects (CHD) are the most common group of human live birth defects (~1%). Infectious diseases (e.g. German measles) may be transmitted across the placenta from mother to foetus, as can some venereal diseases (e.g. congenital syphilis) which are otherwise transmitted by sexual intercourse. In some cases, however, the newborn may acquire a venereal disease by infection from the mother at birth; this would be congenital infection.

conidiophore See CONIDIUM.

conidium Asexual spore of certain fungi, cut off externally at apex of specialized hypha (conidiophore).

conifer Informal term for CONE-bearing tree. See CONIFEROPHYTA.

Coniferophyta (Pinopsida) Conifers. The most widespread and numerous of the gymnosperms with about 50 genera and 650 species. Characterized by active cambial growth and simple leaves. Ovules and seeds are exposed; sperm non-flagellated. Conifers range from the giant redwood (*Sequoia sempivirens*), to the pines (*Pinus*), firs (*Abies*) and spruces (*Picea*). Conifer history extends back to the late Carboniferous period; the Cordaites of that period were primitive conifers. Leaves of modern conifers possess drought-resistant features, which may be related to the diversification of the group during the relatively dry, cold Permian period when increasing global aridity would have favoured structural adaptations like those of the conifer leaves. One genus, *Metasequoia*, was abundant during the Tertiary in Eurasia and North America; it was in 1941, then in 1944, that live specimens were found in Sichuan Province in southwestern China. Subsequently, thousands more were found growing in China. See PINACEAE, CUPRESSACEAE, TAXACEAE, INTRODUCTION.

conjugated protein Protein to which a non-protein portion (PROSTHETIC GROUP) is attached.

conjugation (1) Union in which two individuals or filaments fuse together to exchange or donate genetic material. The process can involve a *conjugation tube* as in members of the Zygnematales (e.g. *Spirogyra*), or a *copulation tube* as seen in some BACILLARIOPHYCEAE, or no tube at all as in the Desmidiales. Process also occurs in fungi. In bacterial conjugation only a portion of the genetic material is transferred from the donor cell. Sex pili form (see PILUS), followed by transmission of one or more plasmids from one cell to the other; less commonly the chromosome itself is transferred (see F FACTOR for *Hfr* strain), when conjugation may be employed in CHROMOSOME MAPPING. Conjugation occurs in some animals; in ciliates (e.g. *Paramecium*) interesting varieties of the process occur. In the simplest case, after partial cell fusion, macronuclei disintegrate and each of the two micronuclei undergoes meiosis, after which three of the four nuclei from each cell abort. The remaining nucleus in each cell divides mitotically, one of the two nuclei from each cell passing into the other in reciprocal fertilization. The cells separate, their nuclei each divide mitotically twice, two nuclei reforming the macronucleus while the other two regenerate the micronucleus. The significance of this is obscure, but cultures of just one mating type seem to 'age' and die out sooner than those that can conjugate; CYTOPLASMIC INHERITANCE is probably involved. Compare FERTILIZATION.

conjunctiva Layer of transparent epidermis and underlying connective tissue covering the anterior surface of the vertebrate eyeball to the periphery of the cornea, and lining the inner aspect of the eyelids.

connectin (titanin) The third most abundant protein in a vertebrate skeletal muscle fibre, after actin and myosin. Connects myosin thick filaments to the Z-discs of a sarcomere (SEE MUSCLE CONTRACTION).

connective tissue A variety of vertebrate tissues derived from MESENCHYME and the ground substance these cells secrete. A characteristic cell type is the FIBROBLAST, producing fibres of the proteins *collagen* and *elastin*, providing tensile strength and elasticity respectively. Another protein, *reticulin*, is associated with polysaccharides in the BASEMENT MEMBRANES underlying epithelia and surrounding fat cells in ADIPOSE TISSUE. Loose connective tissue (*areolar tissue*) binds many other tissues together (e.g. in capsules of glands, meninges of the CNS, bone periosteum, muscle perimysium and nerve perineurium). Collagen fibres align along the direction of tension, as in TENDONS and LIGAMENTS. The viscosity of many connective tissues is due to the 'space-filling' *hyaluronic acid*. The PERITONEUM, PLEURA and PERICARDIUM (serous membranes) are modified connective tissues, as are the skeletal tissues BONE and CARTILAGE. Besides supportive roles, connective tissue is defensive, due largely to presence of HISTIOCYTES (macrophages), which may be as numerous as fibroblasts. These and MAST CELLS constitute part of the RETICULOENDOTHELIAL SYSTEM. Connective tissues are frequently well vascularized, and permeated by tissue fluid.

connexins For their role in gap junctions, see INTERCELLULAR JUNCTIONS.

conodonts Extinct group, believed to be primitive vertebrates (AGNATHA), first appearing in Upper Cambrian strata (40 Myr before other vertebrate remains). Most fossils are of phosphatic mouth parts (especially of teeth, judging by the wear); and the presence in some of what appear to be extrinsic eye muscles indicates a level of encephalization comparable with that of lampreys. Possibly the sister groups of anaspids.

consensus sequences The most typical, or 'average', sequence in a group of related DNA, RNA or protein sequences. Such similarity is indicative of functional importance in the sense that, being conserved in evolution, departures from the sequence tend to be selected against. Consensus sequences are consequently often found in regulatory regions of a gene, such as PROMOTER sequences. See CO-OPTION, GENE EXPRESSION.

conserved Term applying to DNA and protein sequences, and to any structures open to comparative study, which have remained relatively unchanged over large expanses of geological time. See MOLECULAR CLOCK.

consociation (Of plants) CLIMAX COMMUNITY of natural vegetation, with an ASSOCIATION dominated by *one* particular species; e.g. beech wood, dominated by the common beech tree.

consomics Chromosome substitution strains.

conspecific Of individuals that are members of the same species.

constitutive enzyme An enzyme synthesized continuously, regardless of substrate availability. Compare INDUCIBLE ENZYME.

constitutive mutant, constitutive phenotype Of mutants, initially studied in the context of the LAC OPERON in *E. coli*, whose continuous and abnormal production of a cell product is caused by a mutation in the cistron encoding the REPRESSOR molecule of the appropriate PROMOTER region, or from a mutation in the promoter region itself. For gain-of-function mutations, see MUTATION.

consumer Indicates HETEROTROPHIC organisms in FOOD CHAINS (and food webs).

consummatory act An act, often stereotyped, terminating a behavioural sequence and leading to a period of quiescence. Compare APPETITIVE BEHAVIOUR. Factors, sometimes very specific, which lead to termination of the sequence are sometimes referred to as *consummatory stimuli*.

contact inhibition Phenomenon in which cells (e.g. fibroblasts) grown in culture in a monolayer normally cease movement at point of contact with another cell (see ADHESION). When all available space is filled by cells, they also cease dividing – a phenomenon known as *contact inhibition of cell division*. Probably a normal self-regulatory device in tissue and organ size. It tends to be lost in transformed CANCER CELLS.

contact insecticide Insecticide whose mode of entry to the body is via the cuticle rather than the gut. DDT and dieldrin are notorious examples, their fat-solubility (often needed to penetrate waxy epicuticle) resulting in accumulation in fat reserves of animals in higher TROPHIC LEVELS.

contig The DNA (or genome) sequence arrived at by joining together overlapping sequences from cloning, e.g. by using BACTERIAL ARTIFICIAL CHROMOSOME clones.

continental drift Theory, widely accepted since about 1953 but better described nowadays as *plate tectonics*, that crustal plates bounded by zones of tectonic activity, with the continents upon them, move slowly but cumulatively relative to one another. Helps explain distribution of many fossil and present-day forms formerly interpreted by invoking supposed land-bridges, and elevations and depressions of the sea bed. See GONDWANALAND, LAURUSSIA, PANGAEA.

continuous culture Procedure employed in BIOREACTORS in which both the nutrients/substrates are added and the cells are harvested at a steady rate.

continuous variation See VARIATION.

contour feather See FEATHER.

contraception Deliberate prevention of fertilization and/or pregnancy, usually without hindering otherwise normal sexual activity. Includes: preventing sperm entry by use of a *condom*, a protective sheath over the penis; preventing access of sperm to the cervix by means of a *diaphragm* placed over it manually before copulation (later removed); preventing IMPLANTATION by means of an *intra-uterine device* (*IUD*), a plastic or copper coil inserted under medical supervision into the uterus; and the CONTRACEPTIVE PILL. Withdrawal of the penis prior to ejaculation (*coitus interruptus*) is *not* an effective method. *Abstention* during the phase in the MENSTRUAL CYCLE when fertilization is likely is a further method. Use of condoms is fairly effective, with the added advantage of reducing the risk of infection by microorganisms during intercourse. IUDs can cause extra bleeding during menstruation and may not be tolerated by some women. Male fertility control currently involves weekly injections of testosterone ethanoate, suppressing pituitary LH and FSH; sperm disappear from the ejaculate after 120 days. Recovery of fertility occurs about 100 days after injections cease. *Sterilization* must be regarded as irreversible. In women, this involves tubal ligation (preferably by endoscopy, e.g. laparoscopy), in which each oviduct is tied or blocked by diathermy (heating), clips or bands. In men, this involves vasectomy (diathermy of the vasa deferentia).

contraceptive pill A pill taken to control fertility. Used by millions of women worldwide, the most popular UK method of CONTRACEPTION is the combined oral contraceptive pill. Each pill contains a mixture of OESTROGENS and progestogens which together prevent ovulation. The additional oestrogen may induce headache and weight gain, and there is a small increased risk of thrombosis and heart attack. Another type of pill (the 'mini pill') consists of pure synthetic PROGESTERONE; while avoiding the side effects of oestrogens it must be taken at the same time each day and may cause frequent and unpredictable uterine bleeding. Antigestogens, such as mifepristone, block progesterone's effects by binding its receptors, disrupting the MENSTRUAL CYCLE, preventing normal thickening of the uterine lining in preparation for pregnancy and inhibiting gene action necessary for normal implantation. Daily doses of 2 mg, less than 1% of that inducing abortion, inhibit ovulation

with very few side effects and without preventing normal oestrogen release. Such doses have the advantage over the mini pill of causing very light menstrual blood loss or none at all. A single large dose of mifepristone is effective as an emergency 'morning after' contraceptive and will prevent ovulation and hence pregnancy in a single dose as low as 10 mg, and may be used by women as an oral 'once-a-month' pill.

In men, synthetic hormones such as those used in the female combined pill can suppress sperm production completely. Large testosterone doses also do this, but can induce side effects such as acne, weight gain and unfavourable blood lipid levels. However, in combination with about 3×10^{-4} g of the oestrogen desogestrel, testosterone loses these side effects without loss of libido or well-being, paving the way for the development of an acceptable 'male pill'.

contractile ring Belt-like bundle of ACTIN and MYOSIN-II filaments assembling just beneath the plasma membrane at anaphase of animal cell mitosis and meiosis, generating the force pulling opposed membrane surfaces together prior to 'pinching-off' and completion of CYTOKINESIS. A highly dynamic structure, its components changing by the minute, regulated by the polo-like family of protein kinases. When fully assembled it contains many proteins besides actin and myosin II, although contraction begins when Ca^{2+}-CALMODULIN activates myosin light chain kinase to phosphorylate myosin II.

contractile root Root undergoing contraction at some stage, causing a change in position of the shoot relative to the ground. Some corm-bearing plants produce large, thick, fleshy roots possessing few root hairs. These roots store large amounts of carbohydrate in the cortex, which may be rapidly absorbed by the plant. The cortex then collapses and the root contracts downwards, pulling the corm deeper into the soil.

contractile vacuole Membrane-bound organelle of many protozoans (especially ciliates, sponge cells, and flagellated algal cells – vegetative cells, zoospores and gametes). In flagellated algae there are usually two anterior contractile vacuoles; however, they can be located posteriorly (e.g. in some CHRYSOPHYCEAE). Contractile vacuoles occur more frequently in freshwater than marine algae, suggesting that they maintain a water balance in cells, since cells in freshwater possess a higher concentration of dissolved substances in their protoplasm than in the surrounding medium, so that there is a net increase of water into the cells. Contractile vacuoles act to expel this excess water. A vacuole will fill with an aqueous solution (*diastole*), and then expel the solution outside of the cell (*systole*); this is repeated rhythmically, and if two contractile vacuoles are present, they usually fill and empty alternately. The process of filling is ATP-dependent, and therefore active. An alternative theory is that the vacuoles remove wastes from the cell. Dinoflagellates (DINOPHYTA) have a similar, but more complex structure called a PUSULE, which functions similarly to a contractile vacuole.

control, controlled experiment See EXPERIMENT.

control action threshold (CAT) The combination of pest density and the densities of the pest's natural enemies beyond which it is deemed necessary to take PEST MANAGEMENT measures preventing the pest population rising to a level at which it will cause economic harm.

control element (regulatory element) Any gene PROMOTER or ENHANCER.

convergence (convergent evolution) The increasing resemblance over time of distinct evolutionary lineages, in one or perhaps several phenotypic respects, increasing their *phenetic* similarity but generally without associated genetic convergence. Usually interpreted as indicating similar selection pressures in operation; i.e. constraints on evolution. Structures coming to resemble one another this way are ANALOGOUS. Convergence of protein or DNA sequences can occur, but is rare; e.g. digestive enzymes of lemurs and cows. Convergence poses problems for any purely phenetic CLASSIFICATION. Compare PARALLEL EVOLUTION. See CLADISTICS.

convergent extension See MORPHOGENESIS.

cooperative behaviour A distinction is usually made between cooperative behaviour between relatives and that between non-relatives. Although RELATEDNESS will generally facilitate the evolution of cooperative behaviour, it is less certain now than it was in the 1980s that KIN SELECTION provides a satisfactory general explanation of cooperative societies. See ALTRUISM, HELPER, RED QUEEN HYPOTHESIS.

co-option The acquisition by a gene, gene expression pattern, embryological pathway, or morphological structure, of a FUNCTION not previously associated with it. Thus, anatomical wings evolved three times within vertebrates and on each occasion the components of the forearm were selected for novel functions. Such events were once previously said to occur through preadaptation; but 'co-option' is preferred because it avoids the semantic overtone of 'foresight' sometimes associated with the former. The theory of co-option, for which there is growing evidential support, underpins the modern concept of HOMOLOGY.

Copepoda Large subclass of CRUSTACEA. No compound eyes or carapace, and most only a few mm long. Usually six pairs of swimming legs on thorax; abdominal appendages absent. Filter-feeders, using appendages on head. Some marine forms, e.g. *Calanus*, occur in immense numbers in plankton and are vital in grazing food-webs. About 4,500 species; some (e.g. fish louse) parasitic.

copia TRANSPOSABLE ELEMENTS in *Drosophila* which resemble integrated retroviral proviruses. Extractable from eggs and cultured cells as double-stranded extrachromosomal circular DNA.

coprolite Fossil dung.

copy number (1) The number of copies of a gene in a genome. (2) The number of copies of PLASMID per host chromosome, typically in a yeast or bacterial cell.

coracoid A cartilage bone of vertebrate shoulder girdle. Meets scapula at glenoid cavity, but reduced to a small process in non-monotreme mammals. See PECTORAL GIRDLE.

coral See ACTINOZOA.

coralline A term referring to some members of the red algae (RHODOPHYTA) which become encrusted with lime (e.g. Corallinaceae).

Cordaitales Order of extinct Palaeozoic gymnosperms that flourished particularly during the Carboniferous. Tall, slender trees, with dense crown of branches bearing many large, simple, elongated leaves. Sporophylls distinct from vegetative leaves, much reduced in size and arranged compactly in small, distinct, male and female cones. Microsporophylls stamen-like, interspersed among sterile scales; megasporophylls similarly borne, each consisting of a stalk bearing a terminal ovule.

co-receptor Any cell-surface receptor acting in cooperation with another in a biological response; e.g. the CHEMOKINE receptors CCR5 and CXCR4 in cooperation with CD4 for HIV entry.

corepressors Pleiotropic REGULATORY PROTEINS that do not bind DNA directly but only indirectly by binding specific repressor molecules that have done so. They can interact with chromatin remodelling complexes, histone-modifying enzymes, the RNA polymerase holoenzyme complex and other transcription factors. See COACTIVATORS.

corium See DERMIS.

cork (phellum) Protective tissue of dead, impermeable cells, formed by the CORK CAMBIUM which, as the diameter of young stems and roots increases, replaces the epidermis. During differentiation of the cork cells, their inner walls are lined with a relatively thick layer of a fatty substance SUBERIN, which makes them impermeable to water and gases. The walls of cork cells may also become lignified. See LIGNIN.

cork cambium (Bot.) One of two lateral meristems in vascular plants which produces secondary tissues, which comprise the secondary plant body. Cork cambium is responsible for the formation of cork, which usually follows formation of both secondary xylem and secondary phloem. Repeated divisions of the cork cambium give rise to radial rows of compactly arranged cells, most of which are cork cells. These cells are formed toward the outer surface of the cork cambium, and the phelloderm is formed toward the inner surface. Together, cork, cork cambium, and phelloderm comprise the periderm. See CORK, VASCULAR CAMBIUM.

corm Organ of vegetative reproduction; swollen stem-base containing food material and bearing buds in the axils of scale-like remains of leaves of previous season's growth; food reserves not stored in leaves (compare BULB). Examples include crocus and gladiolus.

cormophytes Refers to plants that possess a stem, leaf and roots (e.g. ferns, seed plants). Contrast BRYOPHYTA.

cornea Transparent exposed part of the sclerotic layer of vertebrate and cephalopod eyes. Flanked in former by conjunctiva and responsible for most refraction of incident light (a 'coarse focus'), the lens producing the final image on the retina. Composed of orderly layers of COLLAGEN, lacking a blood supply, its nutrients derived via aqueous humour from CILIARY BODY.

cornification (keratinization) Process whereby cells accumulate the fibrous protein *keratin* which eventually fills the cell, killing it. Occurs in the vertebrate epidermis, in nails, feathers and hair. See CYTOSKELETONS.

corolla Usually conspicuous, often coloured, part of a flower within calyx, consisting of a group of petals. See FLOWER.

corolla tube Tube-like structure resulting from fusion of petals along their edges.

coronary heart disease (CHD) Currently assuming epidemic proportions in affluent societies, the basis lies in the formation of atheroma in coronary arteries (see SCLER-OSIS), coupled with further blockage due to thrombus (clot) formation. Lipoproteins, cholesterol, triglycerides, platelets, mono-cytes, endothelial cells, fibroblasts and smooth muscle cells are all involved. The two major clinical (diagnostic) conditions are angina (chest pains due to diseased heart muscle, brought on by exercise or stress) and pain in the left arm and neck.

In 60% of all fatal heart attacks (i.e. myo-cardial infarction and/or coronary throm-bosis), death occurs in the first hour – too soon for effective treatment. The strongest correlation between diet and CHD is with the energy derived from saturated fat, but weaker correlations are found with un-saturated (monounsaturated and polyunsa-turated) fat. Blood pressure and plasma cholesterol levels also show significant associations with CHD, but the high levels of CIGARETTE SMOKING in northern Europe and northern USA made the correlation between saturated fat intake and CHD stronger still (see LDL).

The association of total cholesterol with CHD appears to derive largely from the cor-related high level of plasma LDL. Dietary factors which determine total cholesterol do so largely via an effect on LDL. Overweight and OBESITY are strikingly related to total and LDL-associated cholesterol. Very high intakes of dietary cholesterol have been shown to downregulate cell LDL receptors, which are required for cells to be able to internalize and break down plasma LDL-borne cholesterol. Carbohydrate intake *per se* has little direct effect on total and LDL-associated cholesterol levels, although increasing carbohydrate enables a re-duction in saturated fatty acids, which in turn leads to reduction in this lipoprotein fraction. Very high intakes of soluble fibre (non-starch polysaccharide) may bring down cholesterol (LDL), but the effect is almost certainly less than the effect of diet-ary fat in raising LDL.

There is a striking relation between sys-tolic and diastolic blood pressure and CHD.

Obesity and excessive alcohol intake raise blood pressure, risking damage to endo-thelia and the onset of atheroma and throm-bus formation. Persons with DIABETES mellitus are particularly likely to develop vascular disease, not least atheroma. There may be a link between cardiovascular dis-ease and the presence of *Chlamydia pneu-moniae* in the circulation. Both risk of diabetes mellitus and insulin-dependency are reduced in adults by aerobic exercise.

Aerobic exercise is a sustained level of exertion, usually of the large muscles (walk-ing, jogging, swimming, cycling) which can be maintained for several minutes without excessive breathlessness. It reduces total blood cholesterol, reduces hypertension, increases stroke volume, cardiac output, rate of heart relaxation and filling. It reduces the risk of thrombosis, reduces abdominal fat in males (particularly associated with CHD), and increases flexibility of joints. The exercise is performed with aerobic metab-olism of glycogen and fat without accumula-tion of lactic acid. Aerobic training increases the percentage of fat used. Anaerobic exer-cise, e.g. press-ups, sprints, involves anaer-obic metabolism of glycogen, is usually short-lived, and is limited by accumulation of lactic acid. Muscle strengthening exer-cises which benefit normal health differ from those required for particular sports (e.g. rugby, weight-lifting or body-building). Excessively large muscle mass is only achieved when training involves very heavy resistances. See also MUSCLE CONTRACTION, NIC-OTINE, HYPOXIA.

coronary vessels Arteries and veins of vertebrates carrying blood to and from the heart.

corpora allata Small ectodermal endo-crine glands in the insect head, connected by nerves to the CORPORA CARDIACA (to which they may fuse) and producing juvenile hor-mone (neotenin) which is responsible for maintenance of the larval condition during moulting. Decreasing concentration of their product is associated with progressive sequence of larval stages. Their relative inactivity in final larval stage of ENDOPTERYG-OTA brings about pupation, and their complete inactivity in the pupa is responsible for differentiation into the final adult stage. Removal is termed *allatectomy*.

corpora cardiaca (oesophageal ganglia) Transformed nerve ganglia derived from the insect foregut, usually closely associated with the heart. Connected by nerves to, or fusing with, the CORPORA ALLATA and producing their own hormones; but mainly storing and releasing brain neurosecretory hormones, particularly *thoracotrophic hormone* which stimulates thoracic (prothoracic) glands to secrete ECDYSONE, initiating moulting.

corpus callosum Broad tract of nerve fibres (commissures) connecting the two CEREBRAL HEMISPHERES in mammals in the neopallial region.

corpus luteum Temporary endocrine gland of ovaries of elasmobranches, birds and mammals. In mammals develops from a ruptured GRAAFIAN FOLLICLE, and produces PROGESTERONE (as in elasmobranchs). Responsible in mammals for maintenance of uterine endometrium until menstruation, also during pregnancy. Its normal life (fourteen days in humans) is prolonged in pregnancy by *chorionic gonadotrophin* (hCG in humans). Its initial growth is due to LUTEINIZING HORMONE from the anterior pituitary. See OESTROUS CYCLE.

corpus striatum (striatum) See BASAL GANGLIA. Includes the caudate nucleus, putamen and globus pallidus. See also LIMBIC SYSTEM.

cortex (Bot.) In some brown and red algae, tissue internal to epidermis but not central in position; in lichens, compact surface layer(s) of the thallus; in vascular plants, parenchymatous tissue located between vascular tissue and epidermis. (Zool.) (1) Outer layers of some animal organs, notably vertebrate ADRENALS, KIDNEYS and CEREBRAL HEMISPHERES. (2) Outer cytoplasm of cells (*ectoplasm*) where this is semi-solid (see CYTOSKELETON).

corticosteroids (corticoids) Steroids synthesized in the ADRENAL cortex from CHOLESTEROL. Some are potent hormones.

Divisible into *glucocorticoids* (e.g. CORTISOL, cortisone, corticosterone), and *mineralocorticoids* (e.g. ALDOSTERONE). Some synthetic drugs related to cortisone (e.g. prednisone) reduce inflammation (e.g. in chronic bronchitis, relieving airway obstruction).

corticotropin See ACTH.

cortisol (hydrocortisone) Principal glucocorticoid hormone of many mammals, humans included. (Corticosterone is more abundant in some small mammals.) Promotes GLUCONEOGENESIS, especially during starvation, and raises blood pressure. Low plasma cortisol level promotes release of *corticotropin releasing factor* (CRF) from the HYPOTHALAMUS, causing release in turn of ACTH from the anterior PITUITARY. Prevents excessive immune response.

corymb INFLORESCENCE, more or less flat-topped and indeterminate.

cosmid (cosmid vector) Hybrid VECTORS bearing the complementary single-stranded *cos* (cohesive) sites ('sticky ends') by which LAMBDA PHAGE circularizes, plus a standard plasmid replication origin and a gene for drug-resistance (e.g. tetracycline-resistance). Can carry up to 45 kb of DNA to be cloned – about three times that of phage vectors, but less than YEAST ARTIFICIAL CHROMOSOMES. The sequences separately cloned are usually fragments which have some overlap with at least one other fragment in the set, enabling CHROMOSOME WALKING.

cosmoid scale In primitive lobefin fish (see CROSSOPTERYGII), a non-placoid scale comprising layers of (from least to most superficial): lamellar bone, vascular bone, a thick layer of dentin (cosmine), and enamel. See GANOID SCALE, PLACOID SCALE.

costa (1) In some members of the BACILLARIOPHYTA (diatoms), a ridge in the silica cell wall formed by well-defined siliceous ribs. (2) The midrib, or multilayered area, of a bryophyte leaf.

costal Relating to ribs.

cost of reproduction See REPRODUCTION.

cost of meiosis The disadvantage which most (amphimictic) sexual individuals seem to incur in contributing copies of only half their genomes to any of their offspring (through meiosis) whereas greater genetic fitness would seem to come from producing parthenogenetic offspring. See SEX.

costimulatory molecules Proteins which must be present along with antigen peptide fragments on the surfaces of antigen-presenting cells in order for T CELLS to be activated. See Fig. 158, B7, IMMUNITY.

cosuppression A GENE SILENCING effect in plants in which some transgenes suppress themselves and homologous chromosome loci simultaneously.

cotyledon (seed leaf) Leaf, forming part of seed EMBRYO; attached to embryo axis by hypocotyl. Structurally simpler than later formed leaves and usually lacking chlorophyll. Monocotyledons have one, dicotyledons two, per seed; the number varies in gymnophytes. Play important role in early stages of seedling development. In non-endospermic seeds, e.g. peas, beans, they are storage organs from which the seedling draws nutrients; in other seeds, e.g. grasses, compounds stored in another part of the seed, the ENDOSPERM, are absorbed by transfer cells on the outer epidermis of the cotyledons and passed to the embryo. Cotyledons of many plants (epigeal) appear above the soil, develop chlorophyll, and photosynthesize. See SCUTELLUM.

Cotylosauria (Mesosauria) The 'stem reptiles' of the late Palaeozoic and Triassic. Limbs splayed sideways from the body; superficially rather amphibian. Probably a heterogeneous (polyphyletic) order.

countercurrent system System where two fluids flow in opposite directions, one or both along vessels so apposed to one another that exchange of contents, heat, etc., occurs resulting in the level dropping progressively in one fluid while it rises progressively in the other. It may involve active secretion, as in the *countercurrent multiplier* system in the loop of Henle in the vertebrate KIDNEY, or be passive, as in the *countercurrent*

exchange of respiratory gases in the teleost gill, the mammalian PLACENTA and the VASA RECTA. See RETE MIRABILIS and Fig. 68.

counterstaining See STAINING.

coupled oscillations Oscillations linking the abundance of two species, e.g. a predator and its prey, in which low prey abundance leads to low predator abundance. This leads to high prey abundance, which in turn, leads to high predator abundance. This again leads to low prey abundance and so on. See DENSITY-DEPENDENCE.

COV See CROSS-OVER VALUE.

coxa Basal segment of insect leg, linking trochanter and thorax.

coxal bones See PELVIC GIRDLE.

coxal glands Paired arthropod COELOMO-DUCTS. In Arachnida, opening on one or two pairs of legs; in Crustacea, a pair of coelomoducts (antennal glands) on the third (antennal) somite, or on the somite of the maxillae; sometimes both. In Onychophora, a pair in most segments. Excretory.

cpDNA Chloroplast DNA. Larger than its mitochondrial counterpart mtDNA, but like it circular. Encodes the chloroplast's ribosomal RNAs and transfer RNAs, and part of ribulose bisphosphate (RuBP) carboxylase. Nuclear DNA encodes much of the rest of chloroplast structure, and transfer of plastid DNA into the nucleus is still occurring at a rate (6×10^{-5} per generation) comparable to that of spontaneous mutation of nuclear DNA. Of the 3000–4000 genes thought to have been present in the prokaryote ancestor of chloroplasts, only about 120 remain as cpDNA (see GMO). Many cpDNA mutations affecting the chloroplast are transmitted maternally, e.g. some forms of VARIEGATION. See MATERNAL INHERITANCE.

CpG islands Regions in the genome where the CpG dinucleotide is unmethylated (see DNA METHYLATION). Of particular interest because they are associated with the 5' ends of genes and because unmethylated CpG dinucleotides are more common in microbial than mammalian genomes (see TOLL-LIKE RECEPTORS).

C₃, C₄ plants See PHOTOSYNTHESIS.

cranial nerves Peripheral nerves emerging from brains of vertebrates (i.e. within the skull); distinct from SPINAL NERVES, which emerge from the spinal cord. Dorsal and ventral roots of nerves from several segments are involved, but (unlike those of spinal nerves) these remain separate. Each root is numbered and named as a separate nerve; numbering bears little relation to segmentation, but does to the relative posteriority of emergence. There are 10 pairs of cranial nerves in anamniotes; 11 or 12 pairs in amniotes. Nerves I and II (*olfactory* and *optic* nerves) are largely sensory; III (*oculomotor*) innervates four of the six eye muscles; IV (*trochlear*) innervates the superior oblique eye muscle; V (*trigeminal*) is sensory from the head, but motor to the jaw muscles; VI (*abducens*) innervates the posterior rectus eye muscle; VII (*facial*) is partly sensory, but mainly motor to facial muscles in mammals; VIII (*vestibulocochlear*) is sensory from the inner ear; IX (*glossopharyngeal*) is mainly sensory from tongue and pharynx; X (*vagus*) is large, including sensory and motor fibres to and from viscera; XI (*accessory*) is a motor nerve accessory to the vagus; XII (*hypoglossal*) is motor, serving the tongue.

Craniata See VERTEBRATA.

cranium The vertebrate skull. See NEUROCRANIUM.

crassulacean acid metabolism (CAM) A variant of the C₄ pathway of PHOTOSYNTHESIS that evolved independently in many succulent plants (e.g. Cactaceae, Crassulaceae). Photosynthetic cells can fix carbon dioxide in the dark via phosphoenolpyruvate carboxylase forming malic acid which is stored in the vacuole. During the next light period, the malic acid is decarboxylated, and, with CO_2, is transferred to ribulose 1,5-bisphosphate (RuBP) of the CALVIN CYCLE within the same cell. CAM plants, both C₄ and C₃ pathways (like C₄ plants); differ from C₄ plants in having a temporal separation of the two pathways in the CAM plants, rather than a spatial one as in C₄ plants.

CAM plants, then, are largely dependent upon nighttime accumulation of carbon dioxide for photosynthesis because stomata are closed during the day, which retards water loss. This is an adaptation to conditions of high light intensity and water stress; efficiency of water use can be many more times greater than C₃ or C₄ plants. CAM is widespread among vascular plants, more so than C₄ plants, being reported in at least 23 families of flowering plants.

CRE (cyclic AMP response element) A short DNA sequence in the regulatory regions of genes activated by CYCLIC AMP. It binds the regulatory protein CREB as a preliminary to gene transcription.

creatine Nitrogenous compound $(NH_2.C[NH].N[CH_3].CH_2.COOH)$, derivative of arginine, glycine and methionine; reversibly phosphorylated to *phosphocreatine* (see Fig. 80), which transfers its phosphate to ADP via the enzyme *creatine kinase*. Found in muscle. Its anhydride breakdown product, *creatinine*, is excreted in mammalian urine. See PHOSPHAGEN, MUSCLE CONTRACTION.

creatinine See CREATINE.

CREB Cyclic AMP-response element binding protein. Transcription factor activated by raised calcium ion and cAMP levels, binding the cAMP-response element (see CRE) of those genes having this DNA sequence in their regulatory region and involved in many signal transduction pathways and in long-term memory mechanisms in a wide range of animals, including humans. When activated by PKA, it recruits the transcriptional coactivator CBP (CREB-binding protein) to initiate transcription.

Crenarchaeota See ARCHAEA.

creodonts Order of extinct mammals. Very varied, lasting into the Miocene, and ancestral to Fissipedia (dogs, cats, etc.). See CARNIVORA.

Cretaceous GEOLOGICAL PERIOD lasting from about 135–65 Myr BP. Much chalk deposited; anthophytes (flowering plants) and many groups of insects appeared, and became dominant. Large dinosaurs radiated, but along with ammonites and aquatic reptiles became extinct by the close. Climate tropical to subtropical throughout; Africa and South America separate. See EXTINCTION.

Crinoidea Feather stars; sea lilies. Primitive class of ECHINODERMATA. Have long, branched, feathery arms; well-developed skeleton; tube feet without suckers; usually sedentary and stalked, with mouth upwards; microphagous; most modern forms free as adults. Long and important fossil history from Ordovician onwards (providing crinoid marble).

crista (pl. cristae) See (1) VESTIBULAR APPARATUS, (2) MITOCHONDRION.

critical group Group of evidently closely related organisms, not easily categorized taxonomically. Used in the context of those apomicts which also reproduce by normal amphimictic means.

Crocodilia Sole surviving order of ARCHOSAURS. Alligators and crocodiles, appearing in the Triassic. Ancestors probably bipedal. Internal nares (CHOANAE) open far back in the mouth owing to presence of long bony FALSE PALATE. Some exhibit considerable parental care. Close affinities with birds.

Cro-Magnon man Earliest 'anatomically modern' human. See *HOMO*.

crop In vertebrates, distensible expanded part of oesophagus in which food is stored (esp. birds). In invertebrates, an expanded part of the gut near the head, in which food may be stored or digested.

Crop Genome Project Proposed project, initiated in the USA, to map the genomes of several cereals and centred on MAIZE, the quintessential American food crop. The ultimate aim is improved productivity. See RICE.

crop milk Secretion comprising sloughed crop epithelium of both sexes in pigeons, on which nestlings are fed. Production influenced by PROLACTIN, like mammalian milk.

crop rotation The practice of growing different crops in regular succession to assist control of insect pests and diseases, increase soil fertility (especially when one season's growth includes nitrogen-fixing leguminous plants), and decrease erosion.

cross The process or product of cross-fertilization. Contrast SELFING.

crossing-over Mutual exchange of homologous chromosome regions (and largely homologous DNA sequences) between chromatids during the first prophase of meiosis. It effectively involves a 'deliberate' double-strand DNA break-and-repair process. This form of general (homologous) RECOMBINATION involves enzymatic extension of the invading 3' end of broken end for thousands of nucleotides by using the appropriate strand on the recipient DNA helix as a template, while the other broken strand is similarly extended in the opposite direction. The process differs from similar events in mitotic cells because, rather than the two original DNA molecules being restored, a HETERODUPLEX joint is formed, holding two different DNA helices together (see Fig. 38). In both cases, some DNA sequences from one chromosome end up copied onto the other, which may lead to a GENE CONVERSION event if different alleles are involved (see also MATING TYPE). Responsible for the chiasmata observed in the first meiotic division, two Holliday junctions are required in order for the cross-over to be completed, and this involves RecA-type proteins. See DNA REPAIR MECHANISMS, PARASEXUALITY, SYNAPTONEMAL COMPLEX, TWIN-SPOTS, TRANSPOSABLE ELEMENTS.

Crossopterygii Order (sometimes superorder, or subclass) of OSTEICHTHYES, in the heterogeneous subclass CHOANICHTHYES. One known living form (*Latimeria*), the COELACANTH; fossil forms included ancestors of land vertebrates. Bony skeleton; paired fins, with central skeletal axes; COSMOID SCALES, or derivatives. First appeared in Devonian around 400 Myr BP, the fossil record dis-

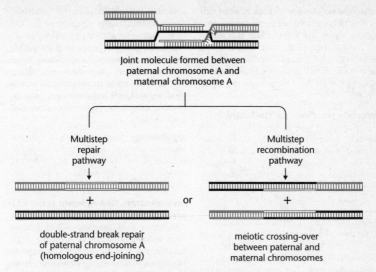

Joint molecule formed between
paternal chromosome A and
maternal chromosome A

Multistep
repair
pathway

Multistep
recombination
pathway

+   or   +

double-strand break repair
of paternal chromosome A
(homologous end-joining)

meiotic crossing-over
between paternal and
maternal chromosomes

FIG. 38 *The different resolutions of a general recombination intermediate in mitotic and meiotic cells. As shown in Fig. 144b, general* RECOMBINATION *begins when a double-strand break is generated in one double helix, followed by DNA degradation and strand invasion into a homologous DNA duplex. New DNA synthesis follows, to generate the joint molecule illustrated. Depending upon subsequent events, resolution of the joint molecule can lead either to a precise repair of the initial double-strand break (left) or to chromosome crossing-over (right). If the maternal and paternal chromosomes produced by general recombination (both in mitotic cells, where crossing-over is rare, and in meiotic cells) differ in DNA sequence in the region of new DNA synthesis illustrated here, then the sequence of the 'upper' DNA duplex in that region is converted to that of the 'lower' duplex – the terms in quotes referring to this particular diagram.*

appearing around 70 Myr BP. Differ from DIPNOI in having normal conical teeth. See RHIPIDISTIA.

cross-over value (COV, recombination frequency) The percentage of meiotic products that are recombinant in an organism heterozygous at each of two linked loci. In diploid organisms, most easily measured by crossing the double heterozygote to the double recessive. This value gives the MAP DISTANCE between the two loci, used in CHROMOSOME MAPPING. The percentage can never exceed 50%, which value would indicate absence of LINKAGE between the loci. The CHI-SQUARED TEST can give the probability that this is so.

cross-pollination Transfer of pollen from stamens to stigma of a flower of a different plant of the same species. Compare SELF-POLLINATION.

crown clade The lineage founded by the last common ancestor of a group of living organisms.

crown gall See AGROBACTERIUM.

crozier formation (hook formation) (1) Similar to formation of CLAMP CONNECTION in Basidiomycotina, but occurring in dikaryotic cells of certain Ascomycotina; a hook develops in the ascogenous hypha where conjugate nuclear division takes place and is followed by cytokinesis; crozier formation may or may not immediately precede formation of an ascus. (2) In ferns and allies, the coiled juvenile leaf or stem.

cruciate 1 o'clock–7 o'clock type of cell
A type of flagellated cell in the CHLOROPHYTA
(green algae) like the cruciate 11 o'clock–
5 o'clock type but with anticlockwise
rotation). Again, quadriflagellate cells can
also occur, in which the basal bodies of two
flagella occupy the 1 o'clock–7 o'clock positions.

cruciate 11 o'clock–5 o'clock zoid In the
CHLOROPHYTA a type of flagellated cell with
apically inserted flagella, in which when
viewed from the anterior, the basal bodies
of the two flagella are slightly displaced
anticlockwise. Thus, by analogy with a clock
face, exactly opposite flagella would have a
12 o'clock–6 o'clock configuration, while an
imaginary anticlockwise rotation would
bring about an 11 o'clock–5 o'clock configuration. The two basal bodies are linked
to four cruciate microtubular roots. This
type of configuration can also occur in quadriflagellate cells, but only for two of the
four flagella.

cruciate microtubular roots In the CHLO-
ROPHYTA, where four microtubular roots
anchor the flagella in the cell and form a
cross when seen from the apex of the cell.
The four roots generally display an x-2-x-2
arrangement, which means that two roots,
lying opposite one another, each contain
two microtubules, while the other two
roots, also lying opposite one another, comprise a varying number of microtubules.

Crustacea Class of ARTHROPODA, including
shrimps, crabs, water fleas, etc., whose
origins lie in at least the pre-Lower Cambrian, as evidenced by fossil phosphato-
copid ostracod from Shropshire, England.
Mostly aquatic, with gills for gaseous
exchange (and often nitrogenous excretion). Many segments with characteristic
BIRAMOUS APPENDAGES. Head bears single pairs
of ANTENNULES, ANTENNAE (both primarily sensory), mandibles and maxillae (both for
feeding). A pair of compound eyes common.
Trunk composed of thorax and abdomen,
often poorly distinct. Chitinous cuticle
(COPEPODS produce several million tons of
CHITIN per year) often impregnated with calcium carbonate and excretory wastes

(calcium reabsorbed prior to moulting).
Small coelom, partly represented by 'kidneys' (antennal glands, maxillary glands)
located in the head. Sexes usually separate;
development usually via a NAUPLIUS larva.
Includes hugely important microphagous filter-feeders in freshwater and marine
food webs. Major subclasses are BRANCHI-
OPODA, OSTRACODA, COPEPODA, CIRRIPEDIA,
MALACOSTRACA.

cryobiology Study of effects of very low
temperatures on living systems. Some
organisms or their parts (e.g. corneas,
sperm) can be preserved under these conditions.

**cryo-electron tomography (cryo-ET),
cryo-electron microscopy** Technique
combining flash-freezing of whole cells
with automated electron scanning (now
quickly enough to prevent damage to
structures) to convert two-dimensional data
to three-dimensional images of intact
organelles within cells, reducing the risks of
producing artifacts. It has been instrumental in revealing the structure of the apoptosome (SEE APOPTOSIS).

cryophytes Organisms growing upon ice
and snow; mostly algae, but including some
mosses, fungi and bacteria. Algae may be so
abundant as to colour the substratum, as in
'red snow', due to the green alga *Chlamydomonas nivalis*. Algae can also be frozen into
marine ice packs, possibly adapted to brine
cell environments, as well as being found
attached to the under surface of ice packs,
where, for example, in the Arctic they can
form layers 2–3 cm thick, with diatoms
being common along with small species of
dinoflagellates and cryptophytes.

crypsis A relational term, indicating that
an individual organism in a particular
environmental setting tends to be overlooked by one or more potential predators,
through its having some combination of
size, shape, colour, pattern and behaviour.
Such characters may be polymorphic, as in
the banding and shell colour polymorphism of the small genus *Cepaea*. To the
extent that such a combination can be
shown to improve the bearer's fitness by

reducing attention from predators, it is regarded as cryptic. There are similarities between crypsis and MIMICRY. See INDUSTRIAL MELANISM, SEARCH IMAGE.

cryptic species See SPECIES.

cryptochromes Blue, flavoprotein ultraviolet-A PHOTORECEPTORS, apparently ubiquitous in plants, but found also in algae and animals, mediating a variety of light responses including entrainment of CIRCADIAN RHYTHMS. The first one isolated (CRY1) came from *ARABIDOPSIS THALIANA*, following cloning of the HY4 gene whose encoded protein has partially overlapping sequences resembling both tropomyosin A and the apoproteins of type 1 microbial PHOTOLYASES – flavoenzymes catalysing light-dependent repair of certain pyrimidine dimers formed by damage through ULTRAVIOLET IRRADIATION. Not all cryptochromes have photolyase activity. The chromophore of CRY1 is fully oxidized FAD. Blue-light responses in plants include phototropisms, inhibition of axis (e.g. hypocotyl) elongation, opening of stomata, chloroplast rearrangement and expression of several genes.

cryptomonads See CRYPTOPHYTA.

Cryptophyta Cryptophytes (cryptomonads). A small algal division containing about 12 genera and 100 freshwater species and 100 marine species. Freshwater species occur in lakes, and sporadically in small ponds and puddles, particularly if the water is slightly enriched with nutrients. Marine species can be found in tidal pools, in brackish water as well as the plankton of the open ocean. In both fresh and marine waters these algae are at times a significant component of the nanoplankton. They are found in the plankton of montane and north-temperate lakes throughout the winter months. Certain species inhabit the interstitial water of sandy beaches, providing these are not exposed to heavy surf. One highly reduced cryptophyte lives as an endosymbiont in the euryhaline ciliate *Mesodinium rubrum*. This ciliate lacks a cytostome (cell mouth) and depends on the endosymbiont's photosynthesis for its nutrition.

Almost all cryptophytes are unicellular

flagellates. Some can form sessile, encapsulated (= palmelloid) stages. One genus, *Bjornbergiella*, possesses a simple, filamentous thallus. Each flagellate cell bears two flagella of unequal length apically or laterally inserted. The longer flagellum bears two rows of stiff lateral hairs, while the shorter flagellum has a single row of shorter hairs. Flagella are also covered with tiny organic scales, which bear a seven-sided rosette pattern. Cells are dorsoventrally compressed in one plane, naked and surrounded by a periplast composed of the plasmalemma, and a plate or series of plates directly under the plasmalemma. Cells also possess a deep gullet whose wall is lined by trichocysts. The flagella emerge from the cell just above and to the right of the gullet.

The chloroplasts, one to two per cell, can be blue, blue-green, reddish, red-brown, olive green, brown or yellow-brown. These colours arise because the chlorophyll is masked by accessory pigments. Chlorophylls a and c_2 are present, while the accessory pigments include (cryptophyte-) phycocyanin, (cryptophyte-) phycoerythrin, alpha-carotene and the xanthophylls, alloxanthin, crocoxanthin, zeaxanthin and monadoxanthin. The presence of alloxanthin preserved in lake sediments can be used by palaeolimnologists to detect past fluctuations in their populations. The phycocyanin and phycoerythrin fill the thylakoid lumen unlike in the RHODOPHYTA, GLAUCOPHYTA and CYANOBACTERIA, where they lie in phycobilisomes on the outside of the thylakoid. Two membranes of chloroplast endoplasmic reticulum (CER) surround each chloroplast. A nucleomorph is present in the space between the chloroplast and the CER, often lying in a depression in the surface of the pyrenoid. A double, porate membrane encloses the nucleomorph. It contains DNA and a nucleus-like structure and is interpreted as the vestigial nucleus of a photosynthetic eukaryotic endosymbiont, which was originally incorporated into the heterotrophic ancestor of the cryptophytes. Within each chloroplast thylakoids are often in pairs, forming lamellae. The pyrenoid projects out from the inner side of the chloroplast. It may or may not contain lamellae, each with thylakoids. Starch reserves are stored in the periplastidal compartment between the CER and the chloroplast envelope. The chloroplast DNA is concentrated into numerous small bodies (nucleoids) scattered throughout the chloroplast. There is often no eyespot, but where present it comprises a series of spherical globules and lies in the centre of the cell, just within the chloroplast.

Immediately below the plasmalemma, trichocysts occur and at the anterior end of the cell there is a contractile vacuole. Adjacent to the gullet are located the Maupas bodies, whose function is unknown. It is speculated that they may be involved in removal and digestion of superfluous membrane.

The nucleus is large and is located in the posterior half of the cell. Chromosome number is high, the nucleomorph within *Guillardia theta* apparently having the most gene-dense eukaryotic genome known, and including 44 overlapping genes. Mitosis is open (i.e. the nuclear envelope disintegrates), and at metaphase the chromosomes congregate forming a massive plate. Within this there are chromatin-free channels containing bundles of pole to pole microtubules. Spindle poles are flat and delimited by cisternae of rough ER; centrioles are absent. A cleavage furrow brings about cell division. Aspects of cryptophyte life cycles remain largely unknown. However, isogamous sexual fusion has been observed in one species.

cryptorchid With testes not descended from abdominal cavity into scrotum.

CSF Acronym for any of CEREBROSPINAL FLUID, COLONY STIMULATING FACTOR or CYTOSTATIC FACTOR.

c-*src* Cellular ONCOGENE whose product (c-Src) is an intracellular signalling protein kinase. Combines multiple inputs to produce a single output, analogously to integration by a neuron of multiple synaptic inputs. See DOMAIN.

ctenidia Gills of MOLLUSCA, situated in the mantle cavity. Involved in gaseous exchange, excretion, and/or filter-feeding.

Ctenophora Comb-jellies, sea-gooseberries. Subphylum of COELENTERATA; with LASSO CELLS, but no cnidoblasts. Body neither polyp nor medusa; movement by cilia fused in rows (*combs* or *ctenes*); no asexual or sedentary phase. See CNIDARIA.

cultivar Variety of plant found only under cultivation.

culture-independent PCR See POLYMERASE CHAIN REACTION.

Cupressaceae Redwoods, cypresses. A family comprising 25 to 30 genera and between 110 to 130 species of evergreen trees and shrubs (usually deciduous in *Taxodium*) that are widespread in temperate regions. Plants are resinous, aromatic and monoecious (usually dioecious in *Juniperus*).

cusp Pointed projection on biting surface of mammalian molar tooth. Each cusp is termed a *cone*. Cusping pattern may have taxonomic value.

cuticle Superficial non-cellular layer, covering and secreted by the epidermis of many plants and invertebrates (esp. terrestrial species). In plants, an external waxy layer covering outer walls of epidermal cells; in bryophytes and vascular plants comprises waxy compound, cutin, almost impermeable to water. In algae may contain other compounds. In higher plants cuticle is only interrupted by stomata and lenticels. Its function is to protect against excessive water loss as well as protecting against mechanical injury.

The arthropod cuticle contains alpha-CHITIN, proteins, lipids, and polyphenol oxidases involved in its tanning (see SCLEROTIZATION). Normally subdivisible into an outer non-chitinous *epicuticle* and an inner chitinous *procuticle* (itself comprising an outer *exocuticle* and an inner *endocuticle*). The epicuticle (1 μm thick in insects) is composed of cemented and polymerized lipoproteins, and affords good waterproofing and resistance to desiccation (less so in Onychophora), having been of utmost importance in terrestrialization by insects. The endocuticles of decapod crustaceans are highly calcified, and often thick. In many arthropod larvae and some adults the cuticle remains soft and flexible (due largely to the protein *arthropodin*) but at some hinges (e.g. insect wing bases) another protein, *resilin*, enables greater flexibility still. Tanning hardens much of the arthropod cuticle, producing SCLERITES. Hardening is also brought about by water loss as the water-soluble arthropodin is converted to the insoluble protein sclerotin.

cuticularization Process of CUTICLE formation.

cutin Heterogeneous polymer formed by ester bonds between hydroxyl groups and carboxyl groups of fatty acids, which contain either sixteen or eighteen carbon atoms. Component, with waxes and pectin polysaccharides, of plant cuticles.

cutinization Impregnation of plant cell wall with CUTIN.

cutting Artificially detached plant part used in vegetative propagation. See MICROPROPAGATION.

Cuvier, Georges (1769–1832) Professor of vertebrate zoology at the *Musée d'Histoire Naturelle* in Paris (see LAMARCK). Polymath, specializing in geology and palaeontology. Formed the premise that any animal is so adapted to its environment (or conditions or existence) that it can *function* successfully in that environment. All parts of the animal therefore had to interrelate to form a viable whole; but certain parts are relatively invariant between organisms, and may therefore have value in CLASSIFICATION. His was not, however, an evolutionary system of classification, unlike Lamarck's; it is fairly clear that the two did not enjoy a cordial relationship. Cuvier had a genius for 'reconstructing' a whole vertebrate skeleton from a single bone.

Cuvierian duct Paired major (common cardinal) vein of fish and tetrapod embryos returning blood to heart from CARDINAL VEINS (under gut) in a fold of coelomic lining forming posterior wall of pericardial cavity. Becomes part of the superior vena cava of adult tetrapods.

c-value (constant-value) The amount of DNA per haploid GENOME. Among eukaryotes, varies from 0.009 pg in the yeast *Saccharomyces cerevisiae* to 700 pg in *Amoeba dubia*, being increased by transposable elements and non-coding DNA (1 pg = 10^{-12} g). The amount appears to determine nuclear volume.

Several unicellular protists contain much more DNA than mammals, which gives rise to the C-value paradox, compounded by the fact that many sibling species have very different C-values – with no comparable morphological differences. It is non-genic DNA which accounts for this, varying in eukaryotes from perhaps 3×10^6 base pairs to 1×10^{11} base pairs (a 100,000-fold range). See REPETITIVE DNA, SELFISH DNA.

cyanelle Endosymbiotic blue-green alga, usually in association with protozoa. See GLAUCOPHYTA.

Cyanobacteria (Cyanophyta, Myxophyta) Blue-green algae. Monerans sharing general prokaryotic characteristics. Belong to the Kingdom Eubacteria. The division contains about 150 genera and 2,000 species found in many freshwater and marine habitats, on damp soil, in aerial situations, glaciers, deserts and hot springs. Most live in freshwater. In temperate lakes they generally form dense populations after the water column has become stratified. Species of *Anabaena, Aphanizomenon*, and *Microcystis* often dominate these populations. Part of their success can be attributed to an ability to use low light intensities effectively, so that they can grow below the surface, deep in the epilimnion. They have an ability to control their buoyancy by means of their gas vacuoles and so can maintain position in a water column. It is only under suboptimal conditions that they congregate at the surface, where exposure to direct sunlight may bring about photooxidation of their photosynthetic pigments. Some planktonic blue-green algae possess heterocysts; differentiated cells with a colourless interior and thick walls, which can fix atmospheric nitrogen. *Anabaena flos-aquae, Aphanizomenon flos-aquae* and *Microcystis aueruginosa* are notorious for producing toxic blooms in freshwaters. Only certain strains of these three species produce toxins. If birds, small or large animals drink water containing a toxic bloom, they develop breathing problems, severe diarrhoea and may die. *Anabaena flos-aquae* and *Aphanizomenon flos-aquae* both produce potent ALKALOID neuromuscular POISONS anatoxin and saxitoxin, respectively. *Microcystis aeruginosa* produces microcystin, a cyclic polypeptide, causing necrosis and haemorrhage of the liver.

The blue-green algae are important components of the marine picoplankton (cells 0.2–2 µm in diameter). These algae are unicellular, coccoid forms and appear to be ubiquitous in temperate and tropical oceans. In shallow water filamentous blue-green algae are important components of the epipelon where they can form thick layers on the submerged sediments (e.g. species of *Oscillatoria*). These algae are frequently dislodged from the sediment by bubbles of oxygen formed via photosynthesis or by gas escaping from the sediment, and float upwards, forming mats on the surface. This also commonly occurs in the spring in lakes where the ice is melting, the mats floating up to the under surface of the ice. Many epipelic blue-green algae are capable of movement (gliding). According to recent theories, immediately outside the MUREIN layer (see below) there are numerous microfibrils in the cell wall. These wind spirally around the filament or cell; waves propagate in rapid succession along these fibrils, imparting a rotating forward movement to the whole organism through friction between microfibrils and the substratum. Epilithic blue-green algae are also common in upper intertidal habitats (e.g. species of *Gloeocapsa* and *Pleurocapsa*). Other species live endolithically, while terrestrial blue-green algae live on damp soil, rocks, roofs and tree trunks. They play an important role as primary colonizers in the establishment of the soil flora and in accumulation of humus.

Some blue-green algae live as symbionts with or within other plants. They occur in many lichens, in the roots of *Cycas* (*Nostoc* species), in leaf cavities of the aquatic fern *Azolla* (*Anabaena*). *Mastocladus laminosus* and *Phormidium laminosum* are two species that live in hot springs, at temperatures of ~50°C, and can even tolerate temperatures up to 70°C.

In tropical countries *Anabaena* species are important for contributing nitrogen to rice plants growing in flooded paddy fields, as these algae fix atmospheric nitrogen. Dried cakes of *Spirulina platensis* have traditionally been used as food by inhabitants living around Lake Chad in North Africa. Large scale production in man-made ponds in sunny regions produced upwards of 720 tonnes of this alga in 1984. This protein-rich fodder is fed to cattle and used in health foods.

The photosynthetic pigments are located in thylakoids, which lie free in the cytoplasm. They are not stacked. Thylakoids contain chlorophyll *a* as the only chlorophyll. Cell colour is generally blue-green to violet; sometimes red or green. The accessory pigments phycocyanin, allophycocyanin and phycoerythrin mask the colour of the chlorophyll. Accessory pigments lie in hemidiscoidal or hemispherical phycobilisomes. These lie on the outer surfaces of the thylakoids. The reserve polysaccharide is cyanophycean starch. Formed in tiny granules lying between the thylakoids. Cells often contain cyanophycin granules. These are polymers of amino acids, arginine and asparagine; polyphosphate bodies are also present along with carboxysomes, which contain the primary enzyme for photosynthetic CO_2 fixation, ribulose 1,5-bisphosphate carboxylase-oxygenase (RuBisCO). The DNA lies in the centre of the cell. The central part is called the nucleoplasm. Many blue-green algae also contain plasmids (small circular DNA molecules). The structural component of the cell wall comprises murein; outside there is a lipopolysaccharide layer. Cells are often embedded in sheaths of mucilage.

Blue-green algae can reproduce vegetatively via cell division. Some filamentous species form specialized vegetative fragments called hormogonia, which serve in vegetative reproduction. Sexual reproduction is lacking, although the bacterial

PARASEXUAL processes of transformation and conjugation may bring about genetic RECOMBINATION.

Blue-green algae evolved about 3,000 Myr ago when the atmosphere contained no oxygen; photosynthesis by these organisms eventually increased atmospheric oxygen concentrations. Many extant species have the ability to photosynthesize under aerobic and anaerobic conditions (oxygenic and anoxygenic photosynthesis). Under aerobic conditions, electrons for photosystem I are derived from photosystem II, but under anaerobic conditions, in the presence of sulphur, electrons are derived from the reduction of sulphur (facultative phototrophic anaerobes).

cyanocobalamin (vitamin B$_{12}$) Cobalt- and nucleotide-containing vitamin, synthesized only by some microorganisms. In vertebrates, carried across the gut wall by a glycoprotein (*intrinsic factor*) of gastric juice. Essential for erythrocyte maturation and nucleotide synthesis (through derivative, *coenzyme B$_{12}$*), so required for nuclear division; herbivores obtain theirs from their gut flora, carnivores from their prey. Absence in diet or lack of intrinsic factor cause pernicious anaemia. See FREE RADICALS, VITAMIN B COMPLEX.

cyanome See GLAUCOPHYTA, RHODOPHYTA.

cyanophycin granules Large bodies in the cells of blue-green algae (CYANOBACTERIA) composed of protein in the form of polypeptides, usually containing aspartic acid and arginine in the form of L-arginyl-poly (L-aspartic acid). Vary in appearance, but normally full of convoluted membranes. Amount of cyanophycin varies with the growth cycle being low in exponentially growing cells, but high in those in stationary growth phase. Polypeptide is produced by a non-ribosomal enzymatic process.

Cycadofilicales   (Pteridospermae) Order of extinct, Palaeozoic gymnosperms that flourished mainly during the Carboniferous. Of great phylogenetic interest. Reproduced by seeds, but with fern-like leaves; internal anatomy combined fern-like vascular system with development of secondary wood. Micro- and mega-sporophylls little different from ordinary vegetative fronds; not arranged in cones. Compare CYCADOPHYTA.

Cycadophyta Cycads. Extant and extinct seed plants; palm-like, found mainly in the tropics and subtropics; mostly fairly large plants, some with a distinct trunk densely covered with leaf bases of shed leaves. Exhibit secondary growth from a vascular cambium; central portion of the trunk comprises a large pith. Reproductive units are more or less reduced leaves with attached sporangia that are loosely or tightly clustered into a cone-like structure near the apex of the plant; pollen and seed are produced on different plants. Sperm is flagellated and motile, but is carried in a pollen tube to the vicinity of the ovule. Comprises two classes: the Cycadopsida, which includes the extant species that first appeared during the late CARBONIFEROUS; and the Cycadeoidopsida, which includes the fossil genera. These formed one of the dominant elements of mid-MESOZOIC floras. Together with the cycads, they were so abundant that the Mesozoic era has been called the 'Age of Cycads'. Evolved in late CARBONIFEROUS or PERMIAN, reaching a zenith of development in the JURASSIC, declining dramatically to extinction in late CRETACEOUS. They appear to parallel, in a general manner, the rise and decline of the dinosaurs, which has led to the suggestion that cycadeoids and dinosaurs were to some degree interdependent; but no direct evidence exists. They resembled cycads with entire or pinnate leaves. Stems possessed a large pith and cortex, and a narrow cylinder of xylem, resembling that of extant Cycadophyta. Primary xylem was ENDARCH; secondary xylem rays arranged in radial rows. Tracheids possessed bordered pits of a circular or scalariform design. Reproductive cones BISPORANGIATE, the closed nature of which suggests a high degree of self-pollination, although insects and other animals may have chewed the cones, allowing parts to fall to the ground, where the pollen became windblown, hence disseminated.

cyclic AMP (cAMP, cyclic adenosine monophosphate) Cyclic nucleotide SECOND MESSENGER produced from ATP by the soluble (or as in some receptor cells, membrane-bound) enzyme adenylyl cyclase. cAMP-gated Ca^{2+} channels occur in vertebrate olfactory receptor ciliary membranes (see CYCLIC GMP), and it can also activate other ion channels directly in certain specialized cells; it normally exerts its effects by activating PROTEIN KINASE A (PKA), which transfers the terminal phosphate from ATP to specific serine or threonine residues of target proteins. A cAMP molecule can reach any part of a mammalian cell in 10^{-1} s. Activates lipolysis in adipose tissue, both ADRENALINE and GLUCAGON acting on cells via adenylyl cyclase membrane receptors. Extracellular cAMP causes cells of *Dictyostelium* to aggregate into a plasmodium. Contrary to previous belief, cAMP does play a role in plant cells: it can mediate the action of auxins (esp. growth regulation, cytokinins attenuating this growth). Many genes (e.g. that for SOMATOSTATIN) activated by cAMP contain a short regulatory DNA sequence called the *cyclic AMP response element* (CRE) which is bound by the regulatory protein CREB. When CREB is phosphorylated on a serine residue by PKA, it binds the transcriptional coactivator *CREB-binding protein* (CBP) which promotes transcription of these genes. Can bind to regulatory subunits of inactive enzymes (e.g. protein kinase A) to liberate catalytic subunits. For more detail, see ADENYLYL CYCLASE, CALCIUM SIGNALLING, CASCADE. See Fig. 39.

cyclic GMP (cGMP, cyclic guanosine monophosphate) Cyclic nucleotide SECOND MESSENGER produced from GTP by the soluble (or in some receptor cells, membrane-bound) enzyme guanylyl cyclase. Usually present in animal cells at about a tenth of cAMP concentration; but level can be raised by INOSITOL 1,4,5 TRIPHOSPHATE pathway. Activates G-kinase, which in turn phosphorylates target proteins. Some membrane ion channels, e.g. in vertebrate photoreceptors, are specifically gated by cGMP (see RHODOPSIN); others (e.g. in vertebrate olfactory receptor ciliary membranes) are gated equally by cAMP and cGMP. Receptor guanylyl cyclases use cGMP as an intracellular mediator just as some G protein-linked receptors use cAMP, although the binding in the former is a direct one. Increasing numbers of receptor guanylyl cyclases are being discovered, although their ligands are usually unknown: they are thus referred to as orphan receptor guanylyl cyclases. See G-PROTEIN, CYCLIC AMP, NITRIC OXIDE.

cyclic nucleotides See CYCLIC AMP, CYCLIC GMP.

cyclins Proteins of dividing eukaryotic cells, most of them accumulating and disappearing in waves during the CELL CYCLE (see Fig. 22). Cyclin D (= Cln3) is an exception in not oscillating and may be an 'initiator cyclin' helping coordinate entry into the cell cycle with cell growth. Cyclins are sometimes grouped into 'mitotic cyclins' (A and B) and 'G1 cyclins' (C, D, E and F). Oscillations are engineered in part by transcriptional control and in part by their degradation by specific UBIQUITIN ligases, enzymes which selectively bind them to ubiquitin at precise points in the cell cycle. Cyclins and CDKs appear to have multiple roles in co-ordination of the chromosome replication cycle, linking S-phase with other cell cycle events by activating DNA REPLICATION, preventing its re-replication in G2, and by allowing establishment of metaphase in mitosis – after which chromatids can separate properly in anaphase as mitotic cyclins are degraded by an ANAPHASE-PROMOTING COMPLEX (APC) and ubiquitin-dependent proteolysis. Extracellular influences such as reduced nutrient supply and removal of GROWTH FACTORS can prevent G1 cyclin breakdown and halt the cycle at G1 (see CKIs). Without degradation of mitotic cyclins, a cell is held in metaphase, breakdown returning the cell to interphase. The speed of cleavage cell cycles in frogs is affected by how fast cyclin accumulates; but this role cannot yet be universalized. Encoded by a diverse family of related genes, cyclins are often stored by way of maternal mRNA in the egg cytoplasm. They often exert their effects by binding to cyclin-

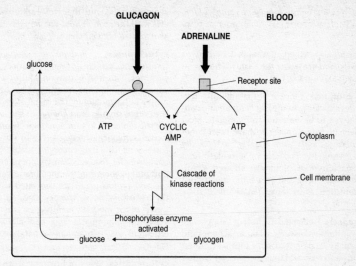

FIG. 39 *Signal transduction of glucagon and adrenaline are both mediated via* CYCLIC AMP *(cAMP). Separate membrane receptors for the hormones are involved, but both promote glycogenolysis via converging components of the kinase cascade.*

dependent kinases (CDKs). See MATURATION PROMOTING FACTOR, C-*MYC*.

cycloheximide Antibiotic, inhibiting TRANSLATION of nuclear mRNA on cytoplasmic ribosomes. Prokaryotic translation unaffected, so used to distinguish mitochondrial/chloroplast from nuclear encoding of cell protein.

cyclomorphosis A term that describes the seasonal POLYMORPHISM frequently observed in plankton organisms (e.g. dinoflagellates such as *Ceratium hirundinella*; rotifers; *Daphnia*). In its extreme form it appears to be exhibited primarily by organisms that reproduce throughout most of the year by asexual or parthenogenetic means, giving rise to genetically homogeneous clones.

cyclosis Circulation of protoplasm (cytoplasmic streaming) in many eukaryotic cells, especially large plant cells. Sometimes restricted, as in small plant cells and animal cells, to jerky movement of organelles and granules in the cytoplasm. Some eukaryotic cells (e.g. squid giant axons) use both ACTIN-based (MYOSIN-dependent) and TUBULIN-based (DYNEIN- or KINESIN-dependent) cytoskeletal systems for membrane-bound organelle transport. One suggestion is that microtubules deliver organelles rapidly anywhere from the CENTROSOME to the cell periphery, after which actin microfilaments route them to their specific terminals, such as the plasma membrane. Both mechanisms are ATP-dependent.

cyclosome See ANAPHASE-PROMOTING COMPLEX.

cyclosporin A See IMMUNOSUPPRESSANT.

cyclostomes (I) AGNATHA whose sole living representatives are lampreys and hagfishes. Eel-like but jawless, with a sucking mouth and one nostril (lampreys), or two (hagfishes); without bone, scales or paired fins. Sometimes placed in a single order (CYCLOSTOMATA), but more usually regarded as two long-separated orders (*Petromyzontiformes*, lampreys, *Myxiniformes*, hagfishes) in subclasses Monorhina and Diplorhina respectively. Lampreys are ectoparasitic on vertebrates and ANADROMOUS; hagfishes are colonial burrowers, feeding largely on polychaete worms or corpses. Lamprey larva is the AMMOCOETE. (2) A suborder of ECTO-PROCTA.

cyme Branched, flat-topped or convex INFLORESCENCE where the terminal flower on each axis blooms first.

cynodonts Lineage of mammal-like reptiles appearing in the late Permian and considered to be ancestral to mammals. Most Triassic forms appear to have been endothermic and were largely extinct by ~225 Myr BP. Presence of secondary hard palate in roof of mouth allowed simultaneous eating and breathing. Heterodont dentition, with lower jaw dominated by dentary.

cypsela Characteristic fruit of Compositae (sunflower, daisy, etc.). Like an achene (and usually so described), but formed from an inferior ovary and thus sheathed with other floral tissues outside ovary wall. Strictly, a pseudo-nut, being formed from two carpels.

cysticercus Bladderworm larva of some tapeworms. Some (e.g. *Echinococcus*) are asexual in an encapsulated cysticercus known as a *hydatid cyst*. Here, brood capsules form on the inner wall, each budding off scolices producing new hydatid cysts. Human liver may become infected with these cysts. See POLYEMBRYONY.

cystocarp Structure developed after fertilization in red algae; consisting of filaments bearing terminal CARPOSPORES produced from the fertilized carpogonium, the whole enveloped in some genera by filaments arising from neighbouring cells.

cystolith Stalked body, consisting of ingrowth of cell wall, bearing deposit of calcium carbonate; found in epidermal cells of certain plants, e.g. stinging-nettle. See STATOLITH.

cytochalasins Anti-cytoskeletal drugs binding reversibly to ACTIN monomers. Effective against APICOMPLEXA.

cytochrome c A protein usually anchored to the inner mitochondrial membrane (see Fig. 113) and involved in electron transport. But if a cell receives a signal to undergo APOPTOSIS, it leaks into the cytosol and associates with the protein Apaf-1 to generate the APOPTOSOME. See CYTOCHROMES, Fig. 45.

cytochromes System of electron-transferring proteins, often regarded as enzymes, with iron-porphyrin or (in *cyt c*) copper-porphyrin as prosthetic groups; unlike in haemoglobin, the metal atom in the porphyrin ring must change its valency for the molecule to function. Cytochrome c oxidase reduces dioxygen (O_2) to water, conserving the free energy available by coupling the redox chemistry to proton pumping, generating the membrane voltage and proton gradients associated with the CHEMIOSMOTIC HYPOTHESIS. Located in inner mitochondrial membranes, thylakoids of chloroplasts and endoplasmic reticulum. See ELECTRON TRANSPORT SYSTEM, Fig. 45.

cytogenetics Study of linking cell structure, particularly number, structure and behaviour of chromosomes (e.g. their rate of replication) to data from breeding work. Often provides evidence for phylogeny. See TAXONOMY.

cytokines A broad range of molecules which, on binding an appropriate receptor, act as signals in cell survival, growth, differentiation and apoptosis. They are used by lymphocytes in signalling to each other and to non-lymphoid cells (e.g. MACROPHAGES, epithelial cells and stromal elements), and involved in the regulation of HAEMOPOIESIS. They can exert an effect on the cell which produces them (autocrine action) or on other cells (paracrine action), often pleiotropically. Type I cytokines (having four similar α-helices) bind a family of receptors sharing certain molecular signal pathways (e.g. JAKS, STATS). Type II cytokines have different structures and include interferons (IFN-α/β and IFN-γ) as well as interleukin 10 (IL-10). All exert their effects, at least in part, via the JAK-STAT pathway (see Fig. 40). See GROWTH FACTORS, INTERFERONS, TUMOUR NECROSIS FACTOR.

cytokinesis Division of a cell's cytoplasm, as opposed to its nucleus. Distinguished, therefore, from MITOSIS and MEIOSIS. In

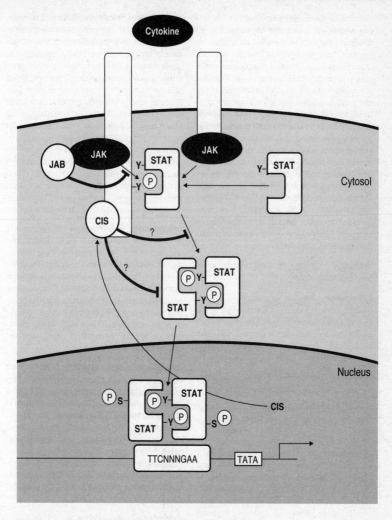

FIG. 40 *Scheme of* CYTOKINE *signalling mechanism, with CIS (cytokine-induced Src-homology 2 protein) associating with the receptor (vertical bars through the membrane) and regulated by Janus kinases (JAKs) and STATs (signal transducers and activators of transcription). The protein JAB interacts with JAKs, inhibiting their action. Inhibition is indicated by a bar at the end of an arm. See* IMMUNITY.

eukaryotic cells, onset is usually in late anaphase, when the plasma membrane in middle of cell is drawn in to form a *cleavage furrow*, formed by a contractile ring of ACTIN microfilaments; this enlarges and finally breaks through the remains of the spindle fibres to leave two complete cells.

cytokinins (phytokinins) Group of plant GROWTH SUBSTANCES recognized and named because of their stimulatory effect (requiring AUXIN) on plant cell division; but of diverse origin. Chemically identified as purines; first discovered (kinetin) was isolated from yeast, animal tissues and sweet corn kernels. Substances with similar physiological action occur in fruitlets, coconut milk and other liquid endosperms; also in microorganisms causing plant tumours, witches' brooms and infected tissues. Other growth-promoting influences include cell enlargement, seed germination, stimulation of bud formation, delay of senescence, and overcoming apical dominance. Effects are thought to involve increased nucleic acid metabolism and protein synthesis; and they appear to affect cells via both a G PROTEIN-coupled receptor and a histidine kinase receptor. Their mode of transport in plants is unresolved. See GIBBERELLINS.

cytology   Study of cells, particularly through microscopy.

cytolysins   Soluble products of macrophages and cytotoxic T CELLS, which kill ANTIGEN-PRESENTING CELLS, sometimes (as with perforins) by perforating the cell's plasma membrane, sometimes as TUMOUR NECROSIS FACTORS.

cytolysis   Dissolution of cells, particularly by destruction of plasma membranes.

cytonemes   Threads of cytoplasm forming cytoplasmic contacts between distant cells.

cytoplasm   All cell contents, including the plasma membrane, but excluding any nuclei. Comprises cytoplasmic matrix, or CYTOSOL, in which ORGANELLES are suspended, some membrane-bound and some not, plus crystalline or otherwise insoluble granules of various kinds. In amoeboid cells, there is a distinction between a semi-solid outer *plasmagel* (ectoplasm) and a less viscous inner *plasmasol*. A highly organized aqueous fluid, where enzyme localization is of paramount importance. See CELL MEMBRANES, CYTOSKELETON.

cytoplasmic ground substance   Equivalent to the CYTOSOL.

cytoplasmic incompatibility   Phenomenon in which egg cytoplasm containing the parasitic rickettsia-like microbe *WOLBACHIA* is transmitted vertically through lineages causing failure to produce progeny either unidirectionally or bidirectionally in reciprocal crosses by preventing male pronucleus disintegration. Occurs in a large variety of insects. More than one *Wolbachia* strain may be present in a host, making for complex interactions. The parasite is transmitted through infected females; but in *Drosophila simulans* it can even prevent infected males from producing offspring when mated to uninfected females, so that infected females produce more offspring. See ABERRANT CHROMOSOME BEHAVIOUR (3).

cytoplasmic inheritance   (1) Eukaryotic genetics involving DNA lying outside the nucleus, often in organelles (see cpDNA, mtDNA, PLASMID) or endosymbionts (see KAPPA PARTICLES). Patterns of inheritance from such a source characteristically fail to observe Mendelian ratios. (2) Inheritance of a cytoplasmic pattern, apparently independently of both nuclear and organelle DNA. Compare MATERNAL INHERITANCE; see ABERRANT CHROMOSOME BEHAVIOUR (3).

cytoplasmic male sterility (CMS)   Trait of higher plants (e.g. maize) determined by either a mitochondrial PLASMID gene or a mtDNA gene, and an example of MATERNAL INHERITANCE. Pollen production aborts in development and plants are therefore self-sterile. Important agronomically, since hybrid plant lines can be produced combining desirable characters from different inbred parents. See CYTOPLASMIC INCOMPATIBILITY, HALDANE'S RULE, SPECIATION, WOLBACHIA.

cytoplasmic streaming   See CYCLOSIS, MICROTUBULE.

cytosine   A PYRIMIDINE base found in the nucleic acids DNA and RNA, as well as in appropriate nucleotides and their derivatives.

cytoskeleton   Network of ACTIN microfilaments, tubulin MICROTUBULES and INTERMEDIATE FILAMENTS, much of it just beneath

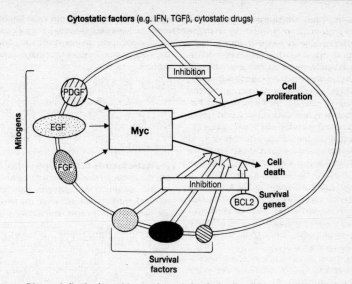

FIG. 41a *Diagram indicating how c-Myc protein can induce both cell proliferation and programmed cell death given appropriate conditions. See* CYTOSTATIC FACTORS *and* APOPTOSIS.

IFN = interferon; TGFβ = transforming growth factor β; PGDF = platelet-derived growth factor; EGF = epidermal growth factor; FGF = fibroblast derived growth factor.

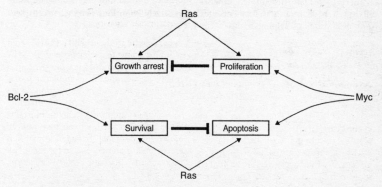

FIG. 41b *Scheme indicating how the oncoproteins Ras, Myc and Bcl-2 can result in cell proliferation only when acting in combination. Inhibition of a pathway is indicated by a bar at the end of an arm.*

the plasma membrane, conferring upon a EUKARYOTE cell (especially an animal cell) its shape and generating the spatial organization within it, providing its attachment capabilities and enabling it to move material both within it and out from it. It provides several structures involved in locomotion (e.g. in muscle contraction, amoeboid and ciliary locomotion). Concentration changes in such diffusible substrates as cyclic nucleotides and divalent ions (generated by interactions between a cell and its EXTRACELLULAR MATRIX) can alter polymerization of cytoskeletal proteins. In the myxomycotan *Dictyostelium*, phosphorylation of actin monomers by tyrosine kinase, causing depolymerization of subcortical cytoskeletal actin, occurs when starved cells round up and lose adhesion with the substratum on return to rich growth medium; dephosphorylation of actin, linked to assembly of filaments and formation of spikes, occurs when cells spread and extend pseudopodia when starved. In mammalian fibroblasts, assembly of focal adhesions and stress fibres is governed by the GTP-ase Rho (see ADHESION, Fig. I, G-PROTEIN). Rho family of GTPases Cdc42, Rac1 and RhoA regulate reorganization of the actin cytoskeleton when cells are stimulated by such signals as growth factors, and are implicated in cell shape changes which result in formation of filopodia, lamellipodia, membrane ruffles and stress fibres. Actin is the major constituent in cell surface MICROVILLI, MICROSPIKES and stereocilia, while tubulin is the main constituent in cilia and flagella. Beneath the cell surface, belt DESMOSOMES, STRESS FIBRES and CONTRACTILE RINGS all involve actin, sometimes in association with myosin or other protein filaments. Animal epithelial cells often have keratin-attachment sites (spot desmosomes) helping to link cells together. When such cells die, cross-linked keratinized cytoskeletons may form a protective surface layer, as in HAIR and NAILS. See CELL LOCOMOTION, CYCLOSIS, INTERCELLULAR JUNCTION.

cytosol The fluid and semi-fluid matrix of the cytoplasm, including the CYTOSKELETON, in which are suspended the organelles. Also defined as the high-speed supernatant obtained from cell and tissue homogenates.

cytostatic factors (CSFs) Any substance with the ability to stop cell growth, including interferons (IFNs) and transforming growth factor β (TGFβ). However, the term 'cytostatic factor' is often reserved for the complex protein (which contains as a subunit the protein product of the vertebrate c-*mos* gene, c-Mos) which holds oocyte development as metaphase of the second meiotic division, possibly by preventing CYCLIN degradation and maintaining activity of MATURATION PROMOTING FACTOR. See Figs. 21, 22 and 41, CELL CYCLE diagrams, C-*MYC*.

cytotoxic Poisonous, or lethal, to cells. See T CELL.

D

darwin   In evolutionary studies, one darwin is the change in the mean of the natural log of a morphological character divided by the elapsed time in millions of years over which the change has occurred.

Darwin, Darwinism   Charles Robert Darwin (1809–1882) began studying medicine at Edinburgh in 1825, but left after two years to study for the clergy at Cambridge. In 1831 the botanist J. S. Henslow suggested to him that he might join HMS *Beagle* on its survey of the South American coast, and in the same year he set sail on a momentous voyage lasting almost five years. In 1842 the Darwin family moved to Down House in Kent, where his most famous work *The Origin of Species by Means of Natural Selection* (published in 1859), and other influential books, were written. In 1857 Darwin received a letter from Alfred Russel WALLACE indicating that the two men were thinking along similar lines as to the mechanism of evolution – natural selection. After many years of ill health, Darwin died at Down House on 19 April 1882 (see website www.lib.cam.ac.uk/Departments/Darwin for his correspondence).

In essence, *Darwinism* is the thesis that species are not fixed, either in form or number; that new species continue to arise while others become extinct; that observed harmonies between an organism's structure and way of life are neither coincidental nor necessarily proof of the existence of a benevolent deity, but that any apparent design is inevitable given: (a) the tendency of all organisms to over-produce, despite the limited nature of resources (e.g. food) avail-able to them; (b) that few individuals of a species are precisely alike in any measurable variable; (c) that some at least of this variance is heritable; and (d) that some of the differences between individuals must result in a largely unobserved selective mortality, or 'struggle for existence', to which (a) must lead. Moreover, he argued, the 'struggle' is likely to be most intense between individuals of the same species, their needs being most similar. Darwin's metaphor 'struggle' caused much confusion, but his more abstract phrase for the mechanism of evolution, NATURAL SELECTION, has survived the test of time.

Much of the theoretical input for the theory of evolution by natural selection came from Lyell's *Principles of Geology* (1830), which Darwin took on HMS *Beagle*, and Malthus's *Essay on the Principle of Population* (1798), which he read 'for amusement' in 1838. The chief empirical influences included observations on the fauna and flora of the Galapagos archipelago, so similar to – yet distinct from – those on the mainland, the fossil armadillos and ground sloths of Patagonia, and the effects of ARTIFICIAL SELECTION on domesticated plants and animals. Darwinism argues for common genealogical links between living and fossil organisms (see also COMPARATIVE METHODS). Darwin gave an unconvincing account of the origin and nature of phenotypic variation, so essential to his theory, and had rather little to say on the strict title of his major work; but Darwinism remains the single most powerful and unifying theory in biology. See BIOGEOGRAPHY, EVOLUTION, NEO-DARWINISM.

Dasycladophyceae A Class of the CHLORO-PHYTA with about eleven extant genera and 50 species. Almost all are marine and native to tropical and subtropical coastal water; one species is found inland in brackish water. Cells of almost all members are heavily encrusted with calcium carbonate (aragonite crystals). Because of this, members of this group have often become fossilized with more than 150 fossil genera having been described. The oldest dates from the Cambrian (c. 550 Myr old) or even the Precambrian (c. 1,200 Myr old).

All thalli are siphonaceous; essentially, each mature plant represents a single, multinucleate cell. However, the plant remains uninucleate for much of its life, as meiosis of the zygote nucleus is delayed until a late stage in the development of the thallus. The principal polysaccharide in the fibrillar structure of the cell wall is a β-1,4-linked mannan; however cellulose dominates in the walls of gametangial cysts. The biflagellate gametes have two flagella and a cruciate 11 o'clock–5 o'clock configuration of the flagellar apparatus. Basal bodies overlap. Mitosis is closed, acentric, and the telophase spindle is prominent and persistent, so the telophase nucleus has a dumb-bell shape. The life cycle is interpreted as haplontic with a macroscopic haploid phase that alternates with a zygote containing one giant nucleus. This diploid stage develops into the gametophyte with meiosis of the diploid nucleus taking place followed by many mitotic divisions of the haploid nucleus. Sexual fusion is isogamous. Chloroplasts are numerous, ellipsoidal or fusiform, devoid of pyrenoids but do contain fructan and starch grains. These grains are also found in the cytoplasm.

daughter cells and nuclei Cells and nuclei resulting from the division of a single cell.

day-neutral plants Plants that flower without regard to day-length. See PHOTO-PERIODISM.

deamidation A form of POST-TRANSLATIONAL MODIFICATION of proteins at asparaginyl and glutaminyl residues, resulting in negative charges at the site of modification, which may lead to changes in the tertiary structure (and hence function) of the protein. See p53.

deamination Removal of an amino ($-NH_2$) group, frequently from an AMINO ACID, by *transaminase* ENZYMES. In mammals, occurs chiefly in the liver, where the amino group is used in production of UREA. The process is important in GLUCONEOGENESIS, where resulting carbon skeleton yields free glucose.

Decapoda (1) Order of MALACOSTRACA, including prawns and lobsters (long abdomens), and crabs (reduced abdomens). Three anterior pairs of thoracic appendages are used for feeding; remaining five are for walking (hence name) or swimming, although first and second of these may bear pincers (chelae). Fused CEPHALOTHORAX covered by a CARAPACE, which may be heavily calcified. (2) Suborder of MOLLUSCA, of the class Cephalopoda, having ten arms; squids, cuttlefish. Compare OCTOPODA.

decerebrate rigidity See RETICULAR FORMATION.

decidua Thickened and highly vascularized mucus membrane (endometrium) lining the uterus in many mammals (not in *ungulates*) during pregnancy. Some or all of the decidua comes away with the PLACENTA at birth.

deciduous (Of plants) shedding leaves at a certain season, e.g. autumn. Compare EVERGREEN.

dedicuous teeth (milk teeth, primary teeth) First of the two sets of teeth which most mammals have; similar to second (permanent) set which replaces it, but having grinding teeth corresponding only to the PREMOLARS, not to the MOLARS, of the permanent set.

decomposer Any HETEROTROPH breaking down dead organic matter to simpler organic or inorganic material. In some ECO-SYSTEMS the *decomposer food chain* is energetically more important than the *grazing food chain*. All release a proportion of their organic carbon intake as CO_2 and the heat release, as evidenced by compost heaps, can be considerable; a useful aspect is that most pathogenic bacteria and cysts, eggs, and immature forms of plant and animal parasites that may have been present, as in sewage, are killed.

An important factor in composting is the carbon to nitrogen ratio. A ratio of about 30:1 (by weight) is optimal: any higher and microbial growth slows. Because composting greatly reduces the bulk of plant wastes, it can be very useful in waste disposal.

decussation Crossing of nerve tracts (e.g. some COMMISSURES) from one side of the brain to the other. Some fibres of the optic nerves and corticospinal tracts of vertebrates cross over in this way, so that parts of the visual field from each retina are transmitted to contralateral optic tecta; each side of the body is served by its contralateral cerebral hemisphere. See OPTIC CHIASMA.

β-defensins See ANTIMICROBIAL PEPTIDES.

deficiency disease Disease due to lack of some essential nutrient, in particular a VITAMIN, trace element or essential AMINO ACID.

deforestation Permanent clear-felling of an area. Loss of some 170,000 km² of rain forest occurs every year. Loss of transpiring leaf surfaces causes much more rain to pass through the ground water, with consequent leaching of chemicals and weathering of soil and rock as roots die; potential nutrients released by decomposers (e.g. in spring) also leached. Loss of soil nitrogen (mobile nitrate ions) is very serious; other biologically important ions are leached as nutrient cycling mechanisms are uncoupled (see TROPICAL RAIN FOREST for laterites). If the land is then used for agriculture, costly fertilizers are often required to replace these ions. Apart from the expense, EUTROPHICATION of nearby waters often results. Apart from direct human impact, our indirect influence through ozone and nitrogen oxide production also leads to forest stress and eventual thinning. The loss of habitats, BIODIVERSITY, and of potentially economic resources, are major concerns. But native people are equally interested in economic development of the land as are non-indigenous people. International finance, providing national compensation, and the development of economic models recognizing intellectual property rights, may be the best long-term hope for forest conservation.

degeneration Reduction or loss of whole or part of an organ during the course of evolution, with the result that it becomes VESTIGIAL. In the context of cells (e.g. nerve fibres) it usually implies their disorganization and death. Compare ATROPHY.

degradative succession A temporal SUCCESSION of species that occurs on a degradable resource. Any dead organic material is exploited by microorganisms and detritivorous animals. Usually, there is a succession of species, as degradation of the organic matter uses up some resources and makes others available. Since heterotrophic organisms are involved in such a succession it has been referred to as heterotrophic succession. Ceases because resources are completely metabolized and mineralized, as in fungal succession on herbivore dung.

dehiscent (Of fruits) opening to liberate the seeds; e.g. pea, violet, poppy.

dehydration Elimination of water. (a) Prior to STAINING, usually achieved by soaking for up to 12 hours in successively stronger ethanol (ethyl alcohol), with at least two changes of 100%, in the preparation of tissues for microscopical examination. Failure to dehydrate properly leads to shrinkage and brittleness on embedding in paraffin, and deterioration of histological structure. See CLEARING. (b) Dry mass of soil and biological material is found after dehydration by heating to constant weight in an oven at 90–95°C.

dehydrogenase See OXIDOREDUCTASE, ENZYME.

deletion Type of chromosomal MUTATION in which a section of chromosome is lost, usually during MITOSIS or MEIOSIS. Unlike some mutations, deletions are not usually reversible or correctable by a SUPPRESSOR MUTATION. See CHROMOSOME MAPPING.

deme, -deme Denotes a group of individuals of a specified taxon – the specificity given by the prefix used. The term deme on its own is not generally advocated, but where found usually refers to a group of individuals below, at or about the SPECIES level. Thus, groups of individuals of a specified taxon in a particular area (*topodeme*); in a particular habitat (*ecodeme*); with a particular chromosome condition (*cytodeme*); within which free gene exchange in a local area is possible (*gamodene*), and of those in a gamodene which are believed to interbreed more or less freely under specified conditions (*hologamodeme*). In zoology, the term deme without prefix tends to be used in the sense of gamodeme. See INFRAS-PECIFIC VARIATION.

demography Numerical and mathematical analysis of populations and their distributions. In the human population, trends in fertility, mortality and migration (given a 90% uncertainty range of those trends in different parts of the world) seem to suggest that doubling of the global population in the 21st century (from its current ~6 billion) has only a 33% probability. Stagnation or shrinkage of population, and accompanying population ageing, are likely to be particularly dramatic in Europe. Other regional populations may triple or quadruple in the next 50 years. Because of the world 'population explosion', roughly 10% of all humans who have entered the world are now alive. *Demographic transition* occurs as populations experience reduced neonatal and infant mortality, and longer individual life expectancy; and people in an increasing number of societies worldwide voluntarily reproduce at lower levels than would seem possible, especially given the availability of resources (including financial). Wealthy families reduce their fertility earlier, and often more dramatically, than the rest of the population. In pre-transition societies, the wealthy tend to produce more offspring than average. Australia was colonized 60–53 kyr BP (earlier than the Americas and

northern Eurasia), South America probably in the last 10–20 kyr, and New Zealand in the last 2 kyr. See AGE PYRAMIDS for the human population in Europe and Asia. See EXTINCTION.

denaturation Changes occurring to molecules of globular proteins and nucleic acid in solution in response to extremes of pH or temperature, or to urea, alcohols or detergents. Most visible effect with globular proteins is decrease in solubility (precipitation from solution), non-covalent bonds giving the molecule its physiological secondary and tertiary structures being broken but the covalent bonds providing primary structure remaining intact. Solutions of double-stranded DNA become less viscous as denaturation by any of the above factors results in strand separation by rupture of hydrogen bonds. This *melting* of DNA occurs with just very small increases in temperature, and may be reversed by the same temperature drop, the *re-annealing* being used in DNA HYBRIDIZATION techniques for assessing the degree of genetic similarity between individuals from different taxa.

dendrite One of many cytoplasmic processes branching from the CELL BODY of a nerve cell and synapsing with other neurons. Several hundred BOUTONS may form synaptic connections with a single cell body and its dendrites. Most dendrites have few voltage-gated sodium channels in their membranes, so they do not conduct action potentials. However, they do conduct electrotonic current to the soma of the neuron by means of ion conduction within their fluid, and this influences the axon hillock (see IMPULSE).

dendritic cells See ANTIGEN-PRESENTING CELL.

dendrochronology Use of isotopes and annual rings of trees to assess age of tree; in fossils, used to date the stratum and/or make inferences concerning palaeoclimate. See GROWTH RING.

dendrogram Branching tree-like diagram indicating degrees of phenetic resemblance between organisms (a *phenogram*), or their phylogenetic relationships. In the latter, the vertical axis represents time, or relative level of advancement. A CLADOGRAM is a dendrogram representing phylogenetic relationships as interpreted by CLADISTICS. See TAXONOMY.

denitrification Process carried out by various facultative and anaerobic soil bacteria, in which nitrate ions act as alternative electron acceptors to oxygen during respiration, resulting in release of gaseous nitrogen. This nitrogen loss accounts in part for the lack of fertility of constantly wet soils that support nitrate-reducing anaerobes and for lowered soil fertility generally, products of denitrification not being assimilable by higher plants or most microorganisms. Bacteria such as *Pseudomonas*, *Achromobacter* and *Bacillus* are particularly important. Nitrous oxide, N_2O (which depletes ozone), and nitric oxide, NO, are sometimes by-products of denitrification. See NITROGEN CYCLE.

density-dependence Widely observed and important way in which populations of cells and organisms are naturally regulated. One or more factors act as (a) increasing brakes on population increase with increased population density, and/or (b) decreasing brakes on population increase with decreased population density. There must be a *proportional* increase or decrease in the effect of the factor on population density as density rises or falls, respectively (see COMPETITION). For example, the proportion of caterpillars parasitized by a fly must increase with increase in caterpillar density if the fly is to act as a density-dependent control factor. Since the caterpillar and fly may be regarded as a kind of NEGATIVE FEEDBACK system, some have suggested that ECOSYSTEMS might self-regulate by this sort of process (see BALANCE OF NATURE, CHAOS). CONTACT INHIBITION by cells is a form of density-dependent inhibition of tissue growth.

density gradient centrifugation Procedure whereby cell components (nuclei, organelles) and macromolecules can be separated by ultracentrifugation in caesium chloride or sucrose solutions whose densities increase progressively along (down) the centrifuge tube. Can be used to separate and hence distinguish different DNA molecules on the basis of whether or not they have incorporated heavy nitrogen (^{15}N) atoms. Cell components or macromolecules cease sedimenting when they reach solution densities that match their own buoyant densities.

dental formula Formula indicating for a mammal species the number of each kind of tooth it has. The number in upper jaw of one side only is written above that in lower jaw of the same side. Categories of teeth are given in the order: incisors, canines, premolars, molars. The formulae for the following mammals are:

	i	c		pm	m
hedgehog	3 .	1 .		3 .	3
	2 .	1 .		2 .	3
grey squirrel	1 .	0 .		2 .	3
	1 .	0 .		1 .	3
ruminants	0 .	0–1 .		3 .	3
	3 .	1 .		3 .	3
cats	3 .	1 .		2 .	0
	3 .	1 .		2 .	1
man	2 .	1 .		2 .	3
	2 .	1 .		2 .	3

dentary One of the tooth-bearing MEMBRANE BONES of the vertebrate lower jaw, and the only such bone in lower jaws of mammals, one on each side.

denticle See PLACOID SCALE.

dentine Main constituent of teeth, lying between enamel and pulp cavity. Secreted by ODONTOBLASTS (hence mesodermal in origin), and similar in composition to BONE, but containing up to 70% inorganic material. Ivory is dentine. See DENTITION.

dentition Number, type and arrangement of an animal's TEETH. Where teeth are all very similar in structure and size (most non-mammalian and primitive insectivore mammals) the arrangement is termed *homodont*; where there is variety of type and size of teeth, the arrangement is termed *heterodont*. Teeth which are replaced once only, as in mammals, are termed diphyodont; those replaced continuously, as in reptiles, are termed polyphyodont. See DECIDUOUS TEETH, DENTAL FORMULA, PERMANENT TEETH.

deoxyribonuclease See DNase.

deoxyribonucleic acid See DNA.

dependence receptor See RECEPTOR.

dephosphorylation See PHOSPHORYLATION.

deregulation Escape from normal control processes; e.g. by a gene.

dermal bone (membrane bone) Vertebrate bone developing directly from mesenchyme rather than from pre-existing cartilage (cartilage bone). Largely restricted in tetrapods to bones of CRANIUM, JAWS and PECTORAL GIRDLE. See OSSIFICATION.

Dermaptera Small order of orthopterous, exopterygote insects, including earwigs. Fan-like hind wings folding under short, stiff, forewings (resembling elytra); but wingless forms common; biting mouthparts; forceps-like cerci at end of abdomen.

dermatogen See APICAL MERISTEM.

dermatome A layer of vertebrate cells giving rise to mesenchymal connective tissue of the dorsal skin during development of a somite and originating as part of the dermamyotome which lies above the sclerotome. See MESODERM.

dermatophyte Fungus causing disease of the skin or hair of humans and other animals. Two of the most common diseases are athlete's foot and ringworm.

dermis (corium) Innermost of the two layers of vertebrate skin, much thicker than the EPIDERMIS, and comprising CONNECTIVE TISSUE with abundant collagen fibres (mainly parallel to the surface); scattered cells including CHROMATOPHORES; blood and lymph vessels and sensory nerves. Sweat glands and hair follicles project down from the epidermis into the dermis, but are not strictly of dermal origin. Responsible for tensile strength of skin. May contain SCALES or BONE. See DERMAL BONE.

Dermoptera Small order of placental mammals (one genus, *Cynocephalus*); so-called flying lemurs, although the Tertiary paromomyids provide the only fossils currently considered to lie within the order; a Chiroptera-Dermoptera clade is accepted by many. See ARCHONTA. Also called colugos. The lower incisors are each divided and comb-like. Dermopterans glide by means of the *patagium*, a hairy membranous skin fold stretching from neck to webbed finger tips and thence to webbed toes and tip of longish tail.

desert A major BIOME, characterized by little rainfall and consequently little or no plant cover. Included are the *cold deserts* of polar regions, such as tundra and areas covered by permanent snow and ice. *Hot deserts* have very high temperatures, often exceeding 36°C in the summer months. Rainfall may be less than 100 mm per year. Deserts can be extensive: the African Sahara is the world's largest; Australian deserts cover some 44% of the continent. Annual plants (desert ephemerals) are most important in these conditions, both numerically and in kind. They have a rapid growth cycle which can be completed quickly when water is available, seeds surviving in desert soils during periods of drought. Perennials that do occur are mostly bulbous and dormant for most of the time. Taller perennials are either succulents (e.g. cacti) or possess tiny leaves that are leathery or shed during periods of drought. Many succulents exhibit CRASSULACEAN ACID METABOLISM, absorbing carbon dioxide at night. See OSMOREGULATION.

desmid Informal taxon referring to two families of freshwater green algae (CHLOROPHYTA) of the Class Zygnematophyceae, Order Desmidiales. The saccoderm desmids are members of the order Zygnematales. They are basically non-filamentous algae possessing non-porate walls, and do not form a new semicell during cell division.

The nucleus is central in the cell and three chloroplast types can be found (axial and plate-like, stellate or spiral); one to several pyrenoids may be found. Sexual reproduction occurs via a conjugation tube. The *placoderm desmids* (Desmidiaceae) or 'true desmids' are distinguished from the saccoderm desmids in that each cell comprises two parts (semicells), with a much more complex, two-layered wall, perforated by a system of pores. The two semicells are usually separated by a median constriction (sinus), joined by a connecting zone (isthmus). Cells may be solitary, joined end-to-end in filamentous colonies, or united in amorphous colonies. Their taxonomy is complicated by polymorphism. Cells possess a single nucleus and two chloroplasts, one in each semicell, with one to several pyrenoids. Mucilage is secreted through pores in the cell wall and is responsible for movement and, when copiously secreted, for adherence to a substratum (commonly as epiphytes or epiliths). Barium sulphate crystals occur within a vacuole in each semicell. Sexual reproduction occurs via conjugation, but no tube is formed and the gametes are amoeboid. Ornamentation develops on the zygote walls, and meiosis precedes germination. Cell division is unique since it results in each semicell of the parent cell forming a new semicell, and becoming a new individual. As a group, desmids are INDICATORS of relatively clean, unpolluted water, with low calcium and magnesium concentrations, and an acidic pH.

desmin An INTERMEDIATE FILAMENT protein characteristic of smooth, striated and cardiac muscle cells and fibres. In sarcomeric muscle it may help to link the Z-discs together; in smooth muscle it probably serves to anchor cells together. Also a prominent component of fibres on the cytoplasmic side of DESMOSOMES.

desmosomes One kind of INTERCELLULAR JUNCTION (see Fig. 92) occurring typically where animal cells require ADHESION against stresses which would shear them. They hold cells together by rivet-like complexes in which the extracellular domains of cadher-

ins form 'Velcro-like' attachments, their intracellular domains linking to a plaque of attachment proteins which in turn anchor intermediate filaments of the cytoskeleton (especially keratins and desmins). A *belt desmosome* (adhesion belt, zonula adherens) comprises a band of contractile actin filaments surrounding E-cadherin molecules, typically near the apical surface of each epithelial cell. A sheet of cells so united may roll up to form a tube by contraction of these filaments – as in neural tube formation. *Spot desmosomes* are sites of keratin filament attachment on the inner cell membrane surface, the filaments forming a network within the cell and connected to those in other cells, spot desmosomes being paired in adjacent cells. Hemidesmosomes are plasma membrane anchorage sites for INTERMEDIATE FILAMENTS, establishing and maintaining strong adhesion between epithelial cells and the underlying basement membrane and connective tissue proteins. They contain one type of INTEGRIN heterodimer, and like spot desmosomes are rivet-like, probably transferring stress from the epithelium to underlying connective tissue via the BASAL LAMINA. See CYTOSKELETON.

desmotubule The tubule that traverses a plasmodesmatal canal, uniting the endoplasmic reticulum of two adjacent plant cells.

determinate growth (Bot.) Growth of limited duration; characteristically seen in floral meristems and leaves of plants.

determined Term applied to an embryonic cell after its fate has been irreversibly fixed. See COMPETENCE, EPIGENESIS, ORGANIZER, POSITIONAL INFORMATION, PRESUMPTIVE.

detoxification See ENDOPLASMIC RETICULUM, POISONS, TOXINS.

detritus Organic debris from decomposing organisms and their products. The source of nutrient and energy input for the *detritus food chain*. See DECOMPOSER.

Deuteromycota See MITOSPORIC FUNGI.

Deuterostomia That assemblage of coelomate animals (some call it an infragrade) in which the embryonic BLASTOPORE becomes the anus of the adult, a separate opening emerging for the mouth (see STOMODAEUM). Some fossils discovered in the CHENGJIANG FAUNA of southern China have been interpreted as basal deuterostomes and placed in a new phylum, the VETULICOLIA. See also DORSOVENTRAL AXIS-INVERSION THEORY. It thus includes the POGONOPHORA, ECHINODERMATA, HEMICHORDATA, UROCHORDATA and CHORDATA. Compare PROTOSTOMIA.

development Growth and differentiation of a multicellular organism from its earliest progenitor cell. Environmental and genetic influences are often difficult to distinguish, or are arbitrary, on account of EPIGENESIS. In animals, *mosaic development* occurs where CLEAVAGE of the egg produces blastomeres lacking the capacity to develop into entire embryos when isolated, even under favourable conditions, specific components of the embryo being absent (e.g. tunicates and echinoderms). This is often contrasted with REGULATIVE DEVELOPMENT. The cleavage divisions setting up the mosaic condition are described as determinate cleavage divisions and often depend on non-random protein distributions in the egg cytoplasm. Development is sometimes regarded as a source of variation on a par with natural selection (see EPIGENESIS, GENETIC ASSIMILATION) and is increasingly recognized as being highly modular in its genetic and molecular organization. See HOMEOTIC, LONG-GERM, MATERNAL EFFECT, MOSAIC DEVELOPMENT.

development threshold The body temperature of an organism below which no development occurs.

Devonian GEOLOGICAL PERIOD lasting from about 400–350 Myr BP. Noted for Old Red Sandstone deposits, for the variety of fossil fish, including ACTINOPTERYGII, CHOANICHTHYES and primitive Amphibia (i.e. ichthyostegids). Primitive land plants (RHYNIOPHYTA, ZOSTEROPHYTA, TRIMEROPHYTA) became extinct. Lycopods (LYCOPHYTA) first appeared; however, their diversity at this time suggests they evolved earlier, perhaps in the SILURIAN period. Progymnosperms first appeared in

the Devonian, while the earliest seed to date is from the late Devonian, some 35 Myr after the first vascular plants appeared. Ecologically, sea covered most of the land, with mountains locally.

dextrans Storage polysaccharides of yeasts and bacteria in which D-glucose monomers are linked by a variety of bond types, producing branched molecules.

dextrin Polysaccharide formed as intermediate product in the hydrolysis of STARCH (e.g. to maltose) by AMYLASES.

dextrose Alternative name for GLUCOSE.

dia- In plant tropisms, orientation of growth response at right angles to source of stimulus; e.g. diagravitropism, diaphototropism.

diabetes (1) *Diabetes insipidus*. An uncommon disorder in which a copious urine arises usually owing to a person's inability to secrete ANTIDIURETIC HORMONE. (2) *Diabetes mellitus*. The INSULIN-dependent form (IDDM, or Type I, responding to insulin treatment) generally presents early in life and results from autoimmune destruction of insulin-producing β-cells in the pancreatic islets of Langerhans; a classic symptom is presence of glucose in the urine. Twin studies indicate multifactorial inheritance. Twelve separate chromosome regions are implicated, the major disease locus *IDDM1* in the MHC on chromosome 6p21 accounts for ~40% of observed familial clustering and hence genetic risk; *IDDM2* is encoded by a minisatellite embedded in the insulin gene on chromosome 11p15. It is not inherited in a simple Mendelian way. The non-insulin dependent form of the disorder (NIDDM, Type II or maturity-onset diabetes) presents later in life and is increasing as adults become overweight and live longer. It seems to involve both impaired pancreatic insulin secretion and tissue resistance to insulin (linked to raised levels of certain free fatty acids), leading to high urination frequency and thirst. It has a strong familial element (heritability). Elevation of blood glucose (e.g. after a meal or 'sugar snack') lasts longer in the middle-aged than the young, causing β cells to secrete insulin for longer. According to the 'lipotoxicity hypothesis', insulin resistance develops when excess lipids are deposited in insulin-sensitive cells other than fat cells, likely suspects being fatty acyl CoA and/or diacyl glycerol molecules, acting through one form of PROTEIN KINASE C(PKC) to reduce activity of IRS-1, a key component of the insulin signalling pathway. See AGEING (for glycation), LEPTINS, MILK (for trigger of IDDM), OBESITY.

diacylglycerol A SECOND MESSENGER; cleavage product of phospholipase C activity upon phosphatidyl inositol bisphosphate (PIP2) of many mammalian (and maybe other) plasma membranes. Activates protein kinase C (C-kinase) when bound to Ca^{2+} and phosphatidyl-serine. Rapidly phosphorylated or cleaved to arachidonic acid. See INOSITOL 1,4,5-TRIPHOSPHATE.

diadelphous (Of stamens) united by their filaments to form two groups, or having one solitary and the others united; e.g. pea. Compare MONADELPHOUS, POLYADELPHOUS.

diad-symmetry/asymmetry See DYAD-SYMMETRY/ASYMMETRY.

diageotropism See DIAGRAVITROPISM.

diagravitropism Orientation of plant part by growth curvature in response to stimulus of gravity, so that its axis is at right angles to direction of gravitational force, i.e. horizontal; exhibited by rhizomes of many plants. See GEOTROPISM, PLAGIOTROPISM.

diaheliotropism See PHOTOTROPISM.

diakinesis Final stage in the first prophase of MEIOSIS.

dialysis Method of separating small molecules (e.g. salts, urea) from large (e.g. proteins, polysaccharides) when in mixed solution, by placing the mixture in or repeatedly passing it through a semipermeable bag or *dialysis tube*, e.g. made of cellophane, surrounded by distilled water (which itself may be removed and replaced). Small molecules will diffuse out of the mixture into the surrounding water, whereas large molecules are prevented by size from doing so. The principle underlies the design of artificial kidneys, which work by *renal dialysis*.

diapause Term used to indicate period of suspended development in insects (and occasionally in other invertebrates). In insects, usually a true DORMANCY, implying a *condition* rather than a *stage* in morphogenesis. Insect diapause can occur at any stage in development, perhaps most commonly in eggs or pupae, but usually only once in any life cycle. Whereas diapause is genetically controlled, the environmental trigger being a PROXIMATE FACTOR, 'prewarning' of alternative or associated stress, *quiescence* can occur in normal stages of a life cycle as a more immediate and short-term response to environmental fluctuations, the organism entering a state of suspense (arrest of life functions) in direct response to an environmental factor, e.g. increasing salinity, low dissolved oxygen. Such occurrences generally leave the organism more vulnerable to these and other environmental factors than would have been the case after diapause.

diapedesis Process by which neutrophils and monocytes perform EXTRAVASATION, i.e. squeeze between endothelial cells and open the basement membrane, gaining access to the tissue fluid and tissues.

diaphragm Sheet of tissue, part muscle and part tendon, covered by a serous membrane and separating thoracic and abdominal cavities in mammals only. It is arched up at rest, its flattening during inspiration reducing the pressure within the thorax, helping to draw air into the lungs. See VENTILATION.

diaphysis Shaft of a long limb bone, or central portion of a vertebra, in mammals. Contains an extended OSSIFICATION centre. See EPIPHYSIS.

diapsid Vertebrate skull type in which two openings, one in the roof and one in the cheek, appear on each side. The feature is found in many reptiles (e.g. ARCHOSAURS) and all birds.

diastase See AMYLASES.

diastema Gap in the tooth rows of some (typically herbivorous) mammals separating cropping teeth (incisors/canines) from cheek teeth. Apes typically have diastemata ('monkey-gaps') in their upper jaws into which the lower canines fit. Almost half the specimens of the early hominid *Australopithecus afarensis* have small diastemata in the upper jaw, an increasingly uncommon trait in later hominids.

diastole Brief period in the vertebrate HEART CYCLE when both atria and ventricles are relaxed, and the heart refills with blood from the veins. The term may also be used of relaxation of atria and ventricles separately; in which case the terms *atrial diastole* and *ventricular diastole* are used, and are not to be confused with true diastole. Compare SYSTOLE.

diatom See BACILLARIOPHYCEAE.

Dicer An ATP-dependent ribonuclease, a member of the Rnase III family, which cuts long double-stranded RNAs (dsRNAs) into smaller dsRNAs (small interfering RNAs, siRNAs) during RNA SILENCING. See miRNAS.

dichasium A CYME possessing two axes running in opposite directions; the type of cyme produced in those plants having opposite branching in the inflorescence.

dichlamydeous (diplochlamydeous) (Of flowers) having perianth segments in two whorls.

dichogamy Condition in which male and female parts of a flower mature at different times, ensuring that self-pollination does not occur. See OUTBREEDING, PROTANDRY, PROTOGYNY. Compare HOMOGAMY.

dichotomous venation Branching of leaf veins into two more or less equal parts, without any fusion after they have branched.

dichotomy Branching, or bifurcation, into two equal portions. *Dichotomous keys* are employed in those identification manuals where one passes along a path dictated by consecutive decisions, each choice being binary (there being just two alternatives), one route involving the strict negation of the other. See IDENTIFICATION KEYS.

Dicotyledonae See MAGNOLIOPSIDA (MAGNOLIIDAE).

Dictyoptera Insect order containing the cockroaches and mantids, a group sometimes included in the ORTHOPTERA. Largely terrestrial; wings often reduced or absent, and in general poor fliers; fore wings modified to form rather thick leathery tegmina (similar to elytra). Specialized stridulatory and auditory apparatus absent.

dictyostele An amphiphloic SIPHONOSTELE, comprising independent vascular bundles occurring as one or more rings. Present in stems of certain ferns. Individual bundles here are termed *meristeles*. See STELE, PROTOSTELE.

Dictyostelium A genus of MYCETOZOAN protists, famous for the role of the chemoattractant extracellular CYCLIC AMP on plasmodium (or slug) aggregation and as an intracellular signal for differentiation. Slug formation occurs when food runs out and starvation is imminent, as many as 100,000 previously independent cells coming together.

diencephalon Posterior part of vertebrate forebrain. See BRAIN.

differential resource utilization The use of different resources by two different species; or the use of the same resource at a different time, in a different place, or generally in a different manner. Used in the context of interspecific COMPETITION.

differentiation The process whereby cells or cell clones assume specialized functional biochemistries and morphologies previously absent. Such *determined* cells usually lose the ability to divide. Usually associated with the selective expression of parts of the genome previously unexpressed, brought on e.g. by cell contact, cell density, the extracellular matrix and molecules diffusing in it (e.g. see GROWTH FACTORS). Division of labour thus achieved is one evolutionary by-product of MULTICELLULARITY.

diffuse porous wood Secondary XYLEM in which the vessels (pores), when viewed in cross-section, are distributed fairly uniformly throughout a growth layer; or any change in size is gradual from early to late wood. Compare RING POROUS WOOD.

diffusion Tendency for particles (esp. atoms, molecules) of gases, liquids and solutes to disperse randomly and occupy available space. The rate of diffusion between two points varies as the square of the distance between them. Process is accelerated by rise of temperature, the source of movement being thermal agitation. Cells and organisms are dependent on the process at many of their surfaces and interfaces; on its own it is often inadequate for their needs. See ACTIVE TRANSPORT, FACILITATED DIFFUSION, FICK'S LAW OF DIFFUSION, OSMOSIS, WATER POTENTIAL.

Digenea Order of the TREMATODA. Includes those flukes which are usually vertebrate endoparasites as adults and mollusc endoparasites as sporocysts and rediae. Suckers simple. E.g. *Fasciola*, SCHISTOSOMA.

digestion Breakdown by organisms, ultimately to small organic compounds, of complex nutrients that are either acted upon outside of the organism (e.g. by saprotrophs), or have entered some organelle (e.g. food vacuole), or organ (enteron) or gut specialized for the purpose. Often includes the physical events of chewing and emulsification besides chemical breakage of covalent bonds by mineral acids and enzymes. Food molecules are often too large to simply diffuse across CELL MEMBRANES, and their digestion is first required. A gut forms a tube to confine ingested material, while extracellular enzymes hydrolyse it, given appropriate conditions (e.g. pH, temperature). Later stages of digestion may occur through enzymes located in the brush borders of intestinal epithelia (as with nucleotidases and disaccharides (see PANETH CELLS, MICROVILLUS). After digestion, molecules are incorporated (assimilated) into cells of the body.

Although plants lack guts, their cells can digest contained material (e.g., polysaccharides, lipids) (see LYSOSOME). Carnivorous plants (pitcher plant, sundew, Venus fly trap) capture and digest insects with enzymes secreted by the plant, which absorbs the available nutrients (e.g. nitrogenous compounds and other organic compounds as well as minerals). Digestive enzymes are synthesized in glands. Until stimulation, these enzymes are stored in enlarged vacuoles and in the wall. Although both digestion and RESPIRATION are catabolic, and digestion like some respiration is anaerobic, digestion does not release significant amounts of energy.

digit Finger or toe of vertebrate PENTADACTYL LIMB. Contains phalanges. May bear nails, claws or hooves.

digitigrade Walking on toes, rather than on whole foot (PLANTIGRADE). Only the ventral surfaces of digits used. E.g. cat, dog.

dihybrid cross A cross between two organisms or stocks heterozygous for the same alleles at the same two loci under study. A classic example was MENDEL's crossing of F1 pea plants obtained as progeny from plants *homozygous* for different alleles at two such loci. This gave his famous 9:3:3:1 ratio of phenotypes (often called the *dihybrid* ratio); but by no means all dihybrid crosses give this ratio. The phrase *dihybrid segregation* is often used to describe the production of these ratios. *Dihybrid selfing* may be possible in some hermaphrodites (e.g. peas) and monoecious plants. Contrast MONOHYBRID CROSS. See LINKAGE.

dikaryon (dicaryon) Fungal hypha or mycelium in which cells contain two haploid nuclei which undergo simultaneous division during formation of a new cell. This forms a third, dikaryotic, phase (*dikaryophase*) interposed between the haploid and diploid phases in a life cycle. Occurs in the ASCOMYCOTA, where it is usually brief, and in the BASIDIOMYCOTA, where it is of relatively long duration. The paired nuclei may or may not be genetically similar. See HETEROKARYOSIS, MONOKARYON.

Dilleniidae A subclass of the MAGNOLIOPSIDA that comprises about 25,000 species belonging to thirteen orders. This group cannot be fully characterized morphologically; however, members are more advanced than the MAGNOLIIDAE, but less advanced than ASTERIDAE. Except for about 400 species of the Dilleniales, the vast majority of this subclass are syncarpous (carpels united to form a compound pistil). Those with numerous stamens have stamens initiated in centrifugal sequence. More than one third have parietal placentation. Another third of the species are sympetalous (petals grown together or attached, at least toward the base), but only a very few have isomerous, epipetalous stamens alternate with the corolla lobes and also ovules with a single integument and a nucellus comprising a single layer of cells. The ovules of the Dilleniidae as a whole have two integuments, although transitional types can be found.

dimerization Either, formation of a dimer from two monomers (e.g. of a disaccharide from two monosaccharides); or, the aggregation of two (usually inactive) polypeptides to form a (usually active) protein – a common occurrence when extracellular ligands bind transmembrane receptor subunits. Receptors frequently form large assemblies with associated cytoplasmic SIGNAL TRANSDUCTION components. Homodimers or heterodimers are formed.

dimorphism Of members of a species, structures, etc., existing in two clearly separable forms. E.g. *sexual dimorphism*, often very pronounced, between the two sexes; *heterophylly* in some plants (e.g. water crowfoot), in which leaves in two different environments have different morphologies. See DOSAGE COMPENSATION, POLYMORPHISM.

dinoflagellate See DINOPHYTA.

Dinophyta Dinoflagellates. Important freshwater and marine planktonic algae with about 130 genera and more than 2,000 extant and 2,000 fossil species. Fossil dinoflagellates date back to the Silurian (410 Myr BP) or even the late Precambrian (600 Myr BP). They were probably the dominant primary producers in the Palaeozoic seas. About 90% of the extant species are marine or found in brackish water. This is a diverse

group of mostly unicellular algae, some biflagellate (the dinoflagellates proper), some non-motile (coccoid, filamentous, palmelloid, amoeboid). Many (50%) are devoid of chloroplasts, and are obligate heterotrophs. Many photosynthetic species are facultative heterotrophs being able to feed phagotrophically, ingesting bacteria and small planktonic algae. Although the majority of dinoflagellates are planktonic, some species are benthic living in the upper layers of marine sand, while the most important ecologically are species living as endosymbionts in tissues of invertebrates (e.g. all reef-building corals depend on cells of their endosymbiont dinoflagellates – 'zooxanthellas' and cannot grow without them). In favourable conditions some species form dense blooms, colouring the surface waters of seas and lakes a red colour (red tides) due to accumulation of carotenoid pigments (e.g. *Gymnodinium, Gonyaulax, Glenodinium, Dinophysis* and *Prorocentrum*). Although red tides primarily occur in tropical or subtropical waters they also can occur in cooler water during the spring and summer after the spring diatom populations wane. They frequently appear near coasts where nutrient concentrations are elevated. Some of these species produce TOXINS (e.g. *Ptichodiscus brevis* produces brevetoxin that kills fish in the Gulf of Mexico). Others, particularly those that become lodged in shellfish, also produce toxic red tides. The shellfish themselves may not be affected but they can be very dangerous to anyone eating them. *Protogonyaulax catenella* and *P. tamarensis* are particularly important, causing an illness called 'paralytic shellfish poisoning'. They produce the toxins saxitoxin and neosaxitoxin, and various gonyautoxins, respectively. The latter causes paralysis, which leads to death by suffocation. See POISONS.

A typical motile dinoflagellate cell comprises two parts; the upper epicone and lower hypocone, separated by a cingulum or transverse girdle. Both epicone and hypocone are normally divided into several thecal plates. A longitudinal sulcus runs perpendicular to the girdle down one side of the hypocone. Both the transverse and longitudinal flagella emerge through the thecal plates in the area where the girdle and sulcus meet on the ventral side of the cell. The transition zone of each flagellum contains two parallel discs at the base of the two central axonemal microtubules, together with one or further rings which lie below, and closer to the basal body. The longitudinal flagellum projects out behind the cell; the transverse flagellum is wave-like and closely appressed to the girdle. Both flagella bear lateral hairs. The transverse flagellum is about 2–3 times as long as the longitudinal one, but has a helical shape and comprises (a) a helical axoneme; (b) a striated strand running parallel to the longitudinal axis of the axoneme but outside the loops of the coil; and (c) a flagellar sheath enclosing both axoneme and striated strands. The transverse flagellum (left hand screw) causes forward motion, while simultaneously rotating the cell; but the longitudinal one acts as a rudder.

Chloroplasts, where present, are surrounded by three membranes but none is connected to the endoplasmic reticulum. Thylakoids are mostly united in stacks (lamellae) of three. Girdle lamellae are generally absent, and are peripheral stacks of three thylakoids. Where girdle lamellae are found within dinoflagellate chloroplasts, the latter belong to more or less reduced eukaryotic algae living endosymbiotically within the dinoflagellate cell. Contain chlorophylls *a* and c_2, (-carotene and unique xanthophylls (e.g. peridinin). Cells with endosymbiontic algae possess other accessory pigments. The chloroplast DNA is in small nodules scattered throughout the chloroplast but other kinds of nucleoids can be present within algae with endosymbionts. Various types of pyrenoids occur in chloroplasts and they may be partly penetrated by thylakoids. Eyespots may or may not be present. They are also variable; some are complex and lens-like. The principle reserve polysaccharide is starch. This is synthesized outside the chloroplast. The interphase nucleus has chromosomes almost always highly contracted and condensed, with a characteristic helicoidal, garland-like structure. This type of nucleus

has been termed mesokaryotic or dinokaryotic. The nuclear envelope remains intact during mitosis, bundles of spindle microtubules passing through the mitotic nucleus in tunnels lined by the nuclear membrane (see MESOKARYOTE). Cells usually possess a complex system of tubes called the pusule that opens to the exterior near the flagella bases. Explosively discharging trichocysts are found at the cell's surface. Around the periphery of a dinoflagellate cell is a superficial layer of flat, polygonal vesicles (thecal vesicles), which can be empty but often contain cellulose plates of varying thickness.

The dinoflagellate life cycle is haplontic with only the zygote nucleus diploid. Thickwalled resting zygotes (hypnozygotes) are usually formed. Some species form thickwalled asexual spores (hypnospores). Both are termed cysts, because often it is not known whether they have been formed asexually or sexually. The function of both is survival during less than favourable conditions. Fossilized cysts called hystrichospheres occur in rocks and sediments from the Precambrian, certainly the Silurian to the Holocene. They are important in stratigraphic correlation and are especially useful in oil exploration.

dinosaur See ORNITHISCIA, SAURISCHIA.

dioecious Unisexual, male and female reproductive organs being borne on different individuals. Compare MONOECIOUS, HERMAPHRODITE. See OUTBREEDING.

diphyodont See DENTITION.

diplanetism In some PROTISTA (e.g. some OOMYCOTA), the occurrence of a succession of two morphologically different zoospore stages separated by a resting stage. Contrast MONOPLANETISM.

diploblastic Level of animal organization in which the body is composed of two cell layers (germ layers), the outer ECTODERM and inner ENDODERM. Found only in the COELENTERATA, in which a jelly-like mesogloea separates the layers.

diploid Nuclei (and their cells) in which the chromosomes occur as homologous pairs (though rarely *paired up*), so that twice the HAPLOID number is present. Also applicable to appropriate tissues, organs, organisms and phases in a life cycle (see SPOROPHYTE). Most SOMATIC CELLS of animals are diploid (see MALE HAPLOIDY), but some of the cells of the GERMINAL EPITHELIUM engage in MEIOSIS, giving haploid products. See ALTERNATION OF GENERATIONS, DIKARYON.

diploid apogamy See APOMIXIS, PARTHENO-GENESIS.

diplomonads Group of amitochondriate protists, usually considered primitive. Nucleus attached to flagella by fibres, *rhizoplasts*. The genus *Giardia* is an obligate gut parasite (see ARCHEZOA), causing diarrhoea.

diplont, diplophase The DIPLOID stage of a LIFE CYCLE. See ALTERNATION OF GENERATIONS, HAPLONT.

Diplopoda Class (or subclass) of ARTHRO-PODA, containing millipedes. Abdominal trunk segments fused in pairs to form *diplo-segments*, each with two pairs of legs; exoskeleton calcareous; ocelli and one pair of club-like antennae present; young usually hatch with three pairs of legs, suggesting possible relations with the Insecta (see NEO-TENY). Development gradual. See CHILOPODA, MYRIAPODA.

diplospory (Bot.) Form of *apomixis* in which a (diploid) megaspore mother cell gives rise directly to the embryo. See PAR-THENOGENESIS.

diplotene Stage in first prophase of MEIOSIS.

Dipnoi (Sarcopterygii) The order of CHOANICHTHYES including lungfishes.

Diptera Two-winged (or true) flies. Large order of the INSECTA, with enormous specialization and diversity among its members. Endopterygote. The hind pair of wings is reduced to form balancing HALTERES. Head very mobile; compound eyes and ocelli present; mouth-parts suctorial, usually forming a *proboscis* and sometimes adapted for piercing; larvae legless.

disaccharide A carbohydrate comprising two monosaccharide groups joined coval-

ently by a *glycosidic bond*. The group includes LACTOSE, MALTOSE and SUCROSE.

disaster Recurrent events such as tornadoes and hurricanes may be defined as disasters. The community, or a population, is affected. These events occur sufficiently often as to leave their record in the 'genetic memory' (= adaptation) of the population. See CATASTROPHE.

disc flower Actinomorphic tubular flowers (florets) composing the central portion of the flower head (capitulum) of most Asteraceae (Compositae); contrasted with the flattened, zygomorphic ray-shaped florets on the margins of the head.

discontinuous variation See VARIATION.

discrete generations A series of generations in which each one finishes before the next begins. Commonly, however, the early life cycle stages of the succeeding generation overlap with the end of the final stages of the preceding generation (e.g., annual life cycles taking about 12 months or less to complete, although species displaying discrete generations need not be annual, since generation lengths other than one year occur).

disease resistance Any genetic defence against infection or its consequences; for example, the human sickle cell trait (see SICKLE-CELL ANAEMIA). See PLANT DISEASE AND DEFENCES, RESISTANCE GENES.

disinfectant Substance used particularly on inanimate surfaces to kill microorganisms, thus sterilizing them. Hypochlorites, phenolics, iodophores (complexes of iodine less staining, toxic and irritant than iodine solutions) and detergents all have disinfectant properties. The phenolic *hexachloro-phene* is widely used in the food industry and hospital wards to reduce pathogenic staphylococci. See ANTISEPTIC.

dispersal The spreading of individuals away from others (e.g. their parents or siblings). Dispersal may be active (walking, swimming or flying) or passive (e.g. wind-borne pollen or spores). Dispersal refers to the movement of, e.g., plant seeds from the parent, of voles from one grassland to another (usually leaving some behind) and of birds amongst an archipelago of islands in the search for suitable habitat.

displacement activity Act expressive of internal *ambivalence* and seemingly irrelevant or inappropriate to the context in which it occurs. Tends to occur when an animal is subject to opposing motivations or when some activity is thwarted.

disruptive colouration Colouration in animals tending to break up their outlines, thus avoiding visual predation.

disruptive selection See NATURAL SELECTION, POLYMORPHISM.

distal Situated away from; e.g. from place of attachment; from the head, along an antero-posterior axis; from the source of a gradient. Thus, the insect abdomen is *distal* to the head. Contrast PROXIMAL.

distribution The spatial range of a species, usually on a geographic scale but sometimes on a smaller one, or the arrangement or spatial pattern of a species over its habitat.

disturbance An event in community ecology that opens up space that can be colonized by individuals of the same or different species.

diuresis Increased output of urine by kidney, as occurs after drinking much water or taking *diuretic* drugs. See ANTIDIURETIC HORMONE.

diurnal rhythm See CIRCADIAN RHYTHM.

divergent resolution See SPECIATION.

diversity (species diversity) See BIODIVERSITY.

diverticulum Blind-ending tubular or sac-like outpushing from a cavity, often from the gut.

division Major group in the Linnean hierarchy used in classifying plants. Includes closely related classes and is the taxonomic category between kingdom and class; equivalent to *phylum* in animal classification.

dizygotic twins See FRATERNAL TWINS.

DNA (deoxyribonucleic acid) The nucleic acid forming the genetic material of all cells, some organelles, and many viruses; a major component of CHROMOSOMES and the sole component of PLASMIDS. A polymer (polynucleotide), which is formed in cells by enzymatic dephosphorylation and CONDENSATION of many (deoxyribo) NUCLEOSIDE TRIPHOSPHATES (esp. dATP, dGTP, dCTP, dTTP). The product is a long chain of (deoxyribo-)NUCLEOTIDES, bonded covalently by *phosphodiester bonds*. Duplex DNA comprises two such antiparallel strands (running in opposite directions; see Fig. 42b) but complementary in base composition, and held together in a double helix by hydrogen bonds between the complementary bases, by electronic interactions between bases, and by hydrophobic interactions. Each strand comprises a *sugar-phosphate backbone* from which the bases project inwardly. The stacking of the complementary bases in the double helix produces two grooves in the sugar-phosphate backbone termed the major and minor grooves (see Fig. 42a, DNA-BINDING PROTEINS, B-FORM HELIX). Each end of the double helix has one 5'-ending strand paired to a 3'-ending strand, where the 5' and 3' indicate which carbon atoms of the two terminal deoxyriboses are bonded to their terminal phosphate group. It is as duplex DNA that nuclear DNA is normally found, and this form is the ultimate store of *molecular information* for all cells (but single-stranded DNA BACTERIOPHAGES occur). Single-strand DNA regions are normally transient consequences of DNA replication and repair mechanisms (see DNA REPAIR MECHANISMS) and tend to form short hairpin loops (helices) by complementary base-pairing, although single-strand DNA-binding proteins (SSBs) bind cooperatively to prevent this happening (see *recA*). Unlike RNA, DNA is not hydrolysed by dilute alkali. Two of the bases abundant in the nucleotides of DNA are purines (adenine and guanine), that form hydrogen bonds with two common pyrimidines (thymine and cytosine respectively, see BASE PAIRING). This

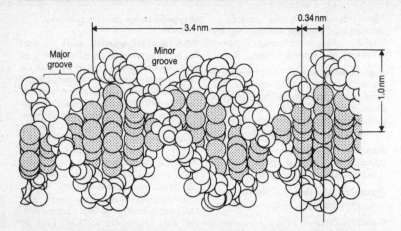

FIG. 42a *Diagrammatic section of duplex DNA, showing the minor and major grooves and approximate dimensions. The two outer helical sugar-phosphate backbones are lightly shaded and the base-pairs are darkly shaded.* (From *Introduction to Genetic Analysis (5th edn)* by Griffiths, Miller, Suzuki, Lewontin and Gelbert. Copyright © W. H. Freeman and Company, 1993. Reprinted with permission.)

ability of one strand of duplex DNA to act as template for the other enables DNA to be replicated by DNA POLYMERASE in a *semiconservative* way (see DNA REPLICATION). A combination of X-ray crystallographic and chemical data (see BASE RATIO) led J. Watson and F. Crick in 1953 to propose the three-dimensional model of duplex DNA held today. The three-dimensional shape of DNA (its 'architecture') is affected by the type of helix adopted, by DNA sequences themselves and by accessory proteins, including some TRANSCRIPTION FACTORS, which bend or kink the molecule (e.g. see CAP). Some DNA-binding proteins intercalate hydrophobic side chains within the minor groove, unstack bases, unwind the DNA and alter the direction of the helix axis without affecting the normal B-form helix of adjacent DNA. Such events are critical in regulation of TRANSCRIPTION (see DNA REPLICATION, HIGH MOBILITY GROUP PROTEINS).

AT-rich DNA contains few genes compared with GC-rich DNA, so deletions and TRANSPOSON integration within AT-rich regions are less harmful and so more widely tolerated. See SINES.

It is thought that most DNA degrades within 10^5 yr. It has been suggested that so-called *ancient* (or *fossil*) *DNA*, reportedly recovered from, e.g., insects trapped in amber (which occludes water) and dinosaurs trapped in coal, may instead be of more recent (e.g. bacterial) origin. Racemization, a chemical change converting L-forms of amino acid to their alternative D-forms, takes place at about the same rate as DNA degradation; so if amino acids in the DNA-providing material show this to even a modest degree it is likely the original DNA is long gone. More recent material, e.g. coprolitic dung from 20-Kyr ground sloths and 45-Kyr Neanderthals (mitochondrial DNA), is much more promising as a source, but indicates hydrolysis and oxidation, and modification or loss of some bases and sugars (see FOSSIL). DNA from an extinct species was first cloned in 1984, using mitochondrial DNA from the once-common zebra, the quagga (*Equus quagga*). It was shown to resemble that from the extant Burchell's zebra (*E. burchelli*). See all DNA references below, DENATURATION, GENE, GENE MANIPULATION, GENETIC CODE, PROTEIN SYNTHESIS, REVERSE TRANSCRIPTASE, RNA.

DNA arrays See NUCLEIC ACID ARRAYS.

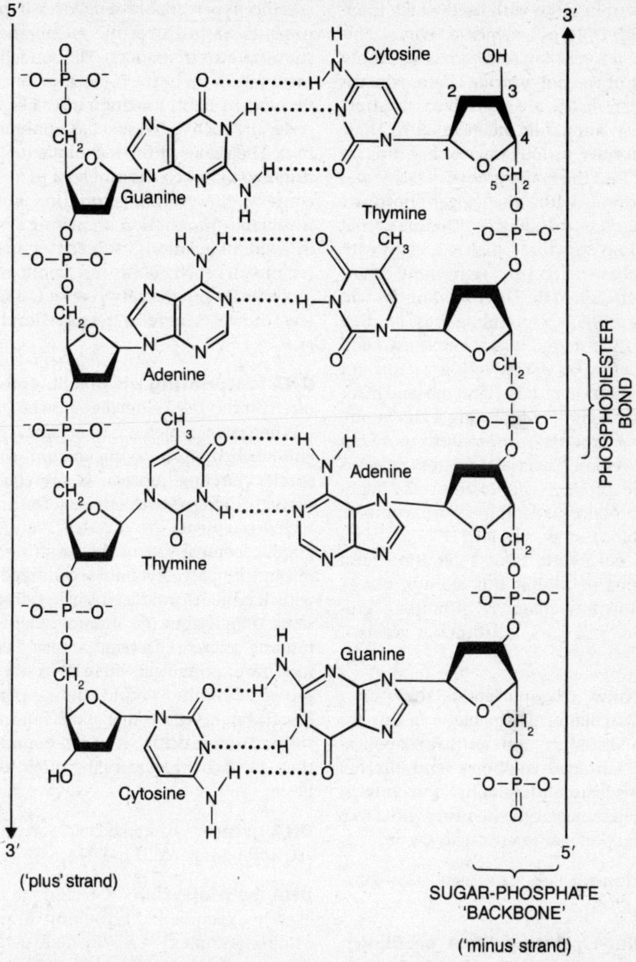

FIG. 42b *The two antiparallel strands of a short section of a DNA duplex. The strands actually twist round each other in a double helix. The dots linking the central base-pairs represent hydrogen bonds. Primed numbers (5', 3') indicate the carbon atom numbers on the deoxyribose moiety involved in the phosphodiester bonding of each chain. These carbon atoms are numbered in the deoxyribose moiety attached to the top cytosine. The left-hand chain runs from 5' to 3' top to bottom, the right-hand chain runs from 5' to 3' bottom to top.*

DNA-binding proteins Proteins (prokaryotic and eukaryotic) with DNA-binding ability. The majority function as TRANSCRIPTION FACTORS, including steroid-receptor proteins, TATA FACTOR, heat-shock proteins (see CHAPERONES), and the TUMOR NECROSIS FACTOR, p53. During DNA replication, single-strand DNA-binding proteins (SSBs) bind tightly to exposed single-stranded DNA and prevent it forming the short hairpin loops with itself which interferes with DNA polymerase activity. DNA-binding DOMAINS can fold correctly without the rest of the polypeptide chain, whereas particular *motifs* are conserved substructures that cannot fold independently. These proteins have various ways of inserting an α-helix into the major groove of DNA, and interactions with the sugar-phosphate backbone are critical for positioning so that protruding structures align correctly with specific base-pairs. A *zinc finger* motif occurs as a protrusion in the DNA-binding domain of some proteins, esp. among those binding steroids (i.e. steroid NUCLEAR RECEPTORS) and those acting as transcription factors for mRNA promoters. It is a cysteine- and histidine-rich region, complexing a zinc atom. The DNA-binding *helix-turn-helix* motif is a feature of some bacterial regulatory proteins and the products of HOMEOGENES. Replication protein A is a multi-subunit complex which binds single-stranded DNA and may initiate eukaryotic DNA replication (and p53 seems to inhibit this ability). DNA-PK phosphorylates many DNA-binding proteins. See *recA*, RNPS, HOMOLOGOUS RECOMBINATION.

DNA chips Oligonucleotide arrays on a microchip platform, capable of testing for genetic disorders and ANTISENSE applications. Technique combines light-directed synthesis (photolithography) and conventional oligonucleotide chemistry with DNA hybridization. See NUCLEIC ACID ARRAYS.

DNA cloning See POLYMERASE CHAIN REACTION, VECTOR.

DNA fingerprinting (DNA profiling)
Technique, developed in 1984, which makes use of the variable number of tandem nucleotide repeats in hypervariable microsatellite DNA sequences (see SATELLITE DNA), there being strong similarities in these between relatives compared with a random outgroup. Using very small samples of DNA, these are first digested by a RESTRICTION ENZYME then subjected to gel ELECTROPHORESIS. After removal of the fragments to a nylon membrane a small radioactive or fluorescent double-stranded DNA PROBE (with

specific hypervariable sequences it recognizes) is washed over the membrane (see SOUTHERN BLOT TECHNIQUES). The bonded fragments show up in the X-ray film, or under fluorescent light, looking rather like a bar code, and each individual has a unique pattern. The rigour of the technique has been disputed in law courts, not least in forensic contexts; however, fingerprinting provides invaluable information on mating systems in wild populations, sometimes making redundant such terms as 'monogamy', 'polygyny', 'promiscuity', etc. at the sexual level (however useful at the social level). See RFLP.

DNA footprinting DNA-binding proteins often protect DNA from digestion by DNase I. When DNA, which has been radioactively end-labelled prior to incubation with a specific binding protein, is subsequently partially digested with DNase I, unprotected fragments are digested. Autoradiographic comparison of the resulting high resolution polyacrylamide gel fragments with fragments produced by digesting the same DNA, but in the absence of all DNA-binding proteins, reveals a space ('DNA footprint') on the gel where DNA has been protected by the specific binding protein. The fragments either side of this hole reveal the position of the DNA region bound, and their subsequent sequencing often identifies it.

DNA gyrase A bacterial DNA TOPOISOMERASE, adding supercoils to DNA.

DNA hybridization A technique often used in experimental taxonomy, in which a source of duplex DNA is 'melted' (see DENATURATION) by slight temperature rise in solution, and allowed to re-anneal with either a similarly treated sample of DNA from a different source, or else commonly a single-stranded RNA sample (e.g. messenger RNA). The time taken to re-anneal, or the thermal stability of the hybrid duplex, indicates the degree of complementarity of the original strands. This can be used to indicate whether a piece of duplex DNA codes for the polypeptide translated from the messenger RNA, or perhaps to estimate the degree of relatedness of the organisms providing the original duplex DNA sources. See also DNA LIGASE, GENE MANIPULATION, *IN SITU* HYBRIDIZATION.

DNA ligase Enzyme which repairs 'nicks' in the DNA *backbone*; i.e. where the *phosphodiester bond* linking adjacent nucleotides has yielded 3'-hydroxyl and 5'-phosphate groups. Its role therefore overlaps that of some DNA POLYMERASES. Valuable for hybridization (insertion) of DNA fragments with appropriate overlapping or 'sticky' ends. See GENE MANIPULATION.

DNA loop See ENHANCER.

DNA methylation Vertebrate DNA contains a covalently modified variety of cytosine, 5-methylcytosine. Restriction enzyme Hpa II cleaves only sequence CCGG when the Cs are unmethylated indicating that, in 98% of the mammalian genome, the dinucleotide CpG occurs at about 25% of its expected frequency – and then in this methylated form. In the remaining 2% of the genome, CpG occurs at its expected frequency but is unmethylated and stable (except in the inactivated X chromosome in female placentals). These so-called 'CpG islands' are about 1 kb long and commonly occur at the 5'-ends of genes and serve to indicate the proximity of a gene. CpNpG sequences of exogenous genes occur in plants, equivalent to CpG islands in animals. Methylation can interfere with transcription, e.g. by inhibiting binding of *TRANS*-ACTING CONTROL ELEMENTS to a promoter (although not all transcription factors are so affected), or interfering with NUCLEOSOME formation. It is possible that methylation binds a gene into a non-transcribable state once trans-acting factors have initiated that state. A process known as 'ripping', first described in fungi, occurs when multiple copies of foreign DNA or duplicated copies of endogenous genes are extensively methylated and transmitted as such to daughter cells. These may subsequently be rearranged (as in *Neurospora crassa*), although this normally inactivates the DNA. A similar process has been observed in mammals and plants. This form of chromatin marking, when inherited, constitutes an EPIGENETIC system. Unmethylated CpG repeats occur far more commonly in bacterial DNA than in the DNA of higher organisms and strongly promote INFLAMMATION. This is mediated by the toll-like receptor TLR9 on APCs (see ANTIGEN-PRESENTING CELLS). See DNA REPAIR MECHANISMS, GENOME (for invertebrate/vertebrate differences), GENOMIC IMPRINTING, HETEROCHROMATIN.

DNA microarrays See NUCLEIC ACID ARRAYS.

DNA-PK (DNA-dependent protein kinase) A nuclear protein-serine/threonine kinase which phosphorylates DNA-bound substrates. An autoantigen Ku binds with free ends of double-stranded (ds) DNA and serves as the DNA binding component of DNA-PK, which phosphorylates numerous DNA-BINDING PROTEINS. DNA-PK binds to and is activated by DNA ds ends, linking V(D)J recombination (see ANTIBODY DIVERSITY) to DNA REPAIR MECHANISMS.

DNA polymerase MULTIENZYME COMPLEX which incorporates appropriate nucleoside triphosphates into a DNA chain (see DNA REPLICATION). Bacterial polymerases (DNA pols) and eukaryotic polymerases (DNA polymerase α-δ) are of several forms and require short complementary oligonucleotide RNA sequences, and several associated proteins, for *initiation*, after which they carry out chain *elongation*. See DNA REPAIR MECHANISMS, POLYMERASE CHAIN REACTION.

DNA probe A defined and fairly short DNA sequence, isotopically or otherwise labelled, which can be propagated by GENE MANIPULATION and introduced to DNA from the same or another individual (often from a different taxon) in order to detect complementary DNA sequences through DNA HYBRIDIZATION (see CHROMOSOME PAINTING, NUCLEIC ACID ARRAYS, QUANTUM DOTS). DNA probes can be used to identify embryos carrying genes for human genetic disorders such as SICKLE-CELL ANAEMIA and A-THALASSAEMIA using chorionic villus sampling (see CHORION). See also SOUTHERN BLOT TECHNIQUE.

DNA profiling See DNA FINGERPRINTING.

DNA proofreading See DNA REPAIR MECHANISMS.

DNA repair mechanisms If cells are to survive, they must receive a complete copy of their genome each time they divide. DNA damage in mammalian cells triggers one of three responses: DNA repair, cell-cycle arrest (see p53), or APOPTOSIS. General cellular DNA repair mechanisms involve enzymes in the removal of loose ends, filling-up of single-stranded regions and ligation of single-strand nicks. Some bacterial and eukaryotic DNA POLYMERASES can replace a nucleotide they insert incorrectly (mismatch proofreading and repair). This system may also be responsible for preventing recombination events between complementary DNA strands when the strands do not match very well, as with many eukaryotic multigene families, so preventing genome scrambling. Some bacterial proteins reduce DNA replication fidelity and seem to be induced when DNA damage to a cell outstrips its repair capacity. Whereas in unicells it is evidently worthwhile to repair damaged DNA, in animals it is often 'safer' to kill the affected cell than attempt repair; thus just a single DNA break is sufficient to activate, through p53, either permanent growth arrest or APOPTOSIS (see Fig. 21). DNA LIGASE then seals the phosphodiester bond. To avoid removing the nucleotide from the wrong (i.e. error-free) strand, cells methylate DNA which has been formed some while; repair enzymes thus distinguish old from new DNA, and repair only the new strand error. Mutants lacking such repair mechanisms are likely to be more susceptible to irradiating sources and to express (somatic) mutations so induced. The worst DNA damage involves double-stranded breaks (DSBs), typically radiation-or superoxide-induced which, if unrepaired, can lead to chromosomal defects, inherited disorders and cancer. In non-homologous end-joining, the broken ends are juxtaposed and simply rejoined by ligation. Ku86 knockout mice are completely defective in this kind of repair and exhibit accelerated AGEING symptoms. But in homologous end-joining, reliance is made of the fact that in diploid nuclei general RECOMBINATION (as in meiosis) can transfer nucleotide sequence information from the intact chromosome to the site of the DSB through HETERODUPLEX formation, special proteins being involved. Very few DSBs per nucleus are required to activate the huge dimeric protein kinase ATM, which seems to recognize (directly or indirectly?) the kind of chromatin restructuring characteristic of a DSB event. This then separates into its two components and initiates events leading either to DNA repair or to apoptosis. One of its target proteins is p53. In *photoreactivation*, cells of many organisms (but apparently not of placental mammals) repair radiation-damaged DNA (as from UV light) using an enzyme that functions when exposed to strong visible light. The main damage products are *pyrimidine dimers* formed by linking adjacent pyrimidines in the same DNA strand. The photoreactivating enzyme monomerizes these dimers again (see Fig. 166). See CHECKPOINTS, CROSSING-OVER, PARP, RNA SURVEILLANCE SYSTEM.

DNA repeats See REPETITIVE DNA.

DNA replication Almost universal biological processes, in which DNA duplexes are catalytically and *semiconservatively* replicated by a DNA POLYMERASE (see MULTIENZYME COMPLEX) at rates of between 50–500 nucleotides per second (but the polymerase requires a short complementary RNA primer). The duplex is first 'unzipped' by helicases (see Fig. 43a) by breaking the hydrogen bonds holding base-pairs together. The resulting Y-shaped molecule is a *replication fork* and occurs first at *replication origins* (of which there may be a hundred or more) on the chromosome. Special DNA-BINDING PROTEINS stabilize the two strands to prevent re-annealing. Appropriate DNA polymerases then move down the two single-stranded arms in a 5'-to-3' direction (see DNA), incorporating nucleotides in accordance with BASE PAIRING rules. Energy is supplied by hydrolysis of substrate nucleoside triphosphates, also catalysed by the polymerase. There are usually several simultaneous replication forks on one replicating chromosome, and newly-synthesized sections are joined up by the DNA *ligase* component of the polymerase (see DNA REPAIR MECHANISMS). Completion of DNA replication is a key event in the control of the CELL CYCLE. Restriction of initiation of eukaryotic DNA replication to S-phase of cell cycle depends on G1-phase preparation of numerous chromatin-specific replication origins. Protein clusters called origin recognition complexes (ORCs), comprising six subunits, bind chromatin throughout the cell cycle and produce a platform for the binding of other proteins during late mitosis, forming a pre-replication complex (pre-RC). Once bound to DNA ORCs usually facilitate the unwinding or distortion of adjacent DNA to provide for the entry of the DNA helicase, usually recruiting in turn additional proteins involved in the assembly of the DNA replication fork. Complexes of cyclin-dependent kinases with CYCLINS bind during mitosis and block mitotic DNA replication; S-phase cyclins are required for initiation of DNA replication and UBIQUITIN-dependent hydrolysis of inhibitors (CKIs) of S-phase cyclin/CDK complexes (in response to growth factors) must occur around START (see Fig. 43b) for DNA replication to occur. After DNA replication, during which DNA TOPOISOMER-ASES prevent tangling of the molecule, S-phase cyclin/CDK complexes block formation of the pre-RC complex in G2 needed for a further round of DNA replication. The ATR (ATM- and Rad3-related)-dependent pathway recognizes and signals many forms of DNA damage and replication problems. Damage is detected by topological changes in DNA, base mismatches and double-strand breaks leaving the DNA in different shapes and causing different protein complexes to bind and repair the mistake (see DNA REPAIR MECHANISMS).

DNase (DNAase, deoxyribonuclease) An enzyme (of which there are many forms) breaking down DNA by hydrolysis of the phosphodiester bonds of its sugar-phosphate backbone. Depending on the enzyme, it does this at either the 3'- or the 5'-end of the bond. As with peptidases, there

are *endonucleases* and *exonucleases*, cleaving respectively terminal and non-terminal nucleotides from either a single strand or from both strands of the duplex, depending on the type of DNase. Pancreatic juice contains DNases. Valuable in GENE MANIPULATION. See RESTRICTION ENDONUCLEASE.

DNA sequencing Determination of the sequence of nucleotides making up a length of DNA. RESTRICTION ENDONUCLEASES digest the strand; the fragments are isolated by gel ELECTROPHORESIS, and then the sequence can be determined by rendering the DNA single-stranded and using it as a template for DNA POLYMERASE to resynthesize the complementary strand with labelled nucleoside triphosphates, or by chemical analysis of the fragments (see Fig. 44). Nowadays, the four nucleotides can be stained with different dyes and a laser can read off the sequence as the acrylamide gel passes it.

Two methods have dominated the field since 1975, both using single-stranded DNA. In the Maxam and Gilbert (base-destruction) method, duplex 3' ends are labelled with ^{32}P, digesting the molecule. Strands of one length are then isolated, the two strands separated, from which one is isolated to form a homogeneous population. This is split into four, each part being subjected to selective destruction of a different one or two of the four bases present (G, A & G, T & C or C) in such a way that sequences of different lengths result. After separation on a gel, bands of labelled fragments are revealed by autoradiography, different for each of the four treatments. Comparison of the four band displays successively shorter bands from top to bottom of the gel, and the presence or copresence of bands indicates which base was present at that place in the strand.

In the Sanger (dideoxy) method, four tubes of single-stranded DNA are given all necessary DNA-synthesizing requirements, but a small proportion of a different dideoxynucleoside triphosphate (which stops chain elongation) is given to each tube – in addition to a short labelled primer complementary to the end of the strand and all four normal triphosphates. Each tube will

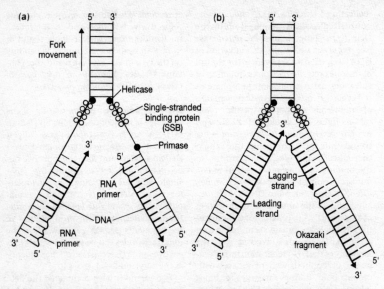

FIG. 43 (a) *Initiation of DNA synthesis by an RNA primer.* (b) *DNA synthesis proceeds by continuous synthesis on the leading strand and discontinuous synthesis on the lagging strand.* (From *Introduction to Genetic Analysis (5th edn)* by Griffiths, Miller, Suzuki, Lewontin and Gelbert. Copyright © W. H. Freeman and Company, 1993. Reprinted with permission.)

then synthesize a variety of labelled complementary strands, most shorter than normal. These are then isolated and run on an acrylamide gel, when sorting by size occurs. Comparison of the autoradiographs reveals the DNA sequence of the original (unlabelled) strand. Comparative DNA sequencing has the potential to supply phylogenetic information about the organisms concerned, only if the splitting pattern for the gene sequences corresponds to the splitting pattern for the organismal lineages (see MULTI-GENE FAMILIES). DNA from an extinct species was first cloned in 1984, using mitochondrial DNA from the once-common zebra, the quagga (*Equus quagga*). It was shown to resemble that from the extant Burchell's zebra (*E. burchelli*). For role of sequencing in population research, see SIBLING SPECIES.

A recent sequencing technique, *pyrosequencing*, is a sequence-by-synthesis approach in which pyrophosphate (PPi), released during the incorporation by DNA polymerase of natural nucleotides using a template strand, is converted to ATP and causes quantifiable light emission by luciferase. See CHROMOSOME MAPPING, PHYLOGENETIC TREES, POLYMERASE CHAIN REACTION, POSITIONAL CLONING.

DNA vaccines See VACCINE.

dolipore septum A septum characteristic of members of the dikaryotic hyphae of the BASIDIOMYCOTA. The septum flares out in the middle portion to form a barrel-shaped structure with open ends.

Dollo's Law The generalization that evolution does not proceed back along its own path, or repeat routes.

domain In general, a regionally differentiated feature. (1) Sequence of amino acids forming a functional group within a protein molecule (see Structural Classification Of Proteins website: http://scop.mrc-lmb.cam.ac.uk/scop). They are closed, globular structures supported by a closed system of hydrogen bonds and built around a well-

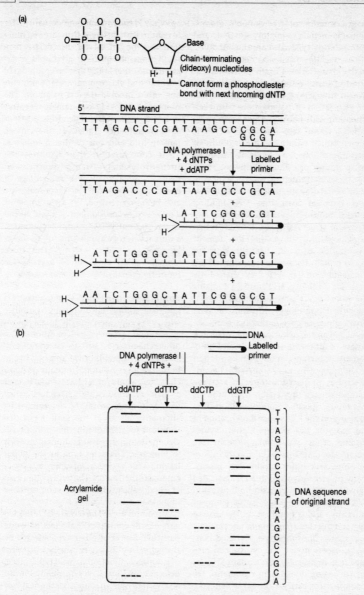

FIG. 44 *The Sanger ('dideoxy') method of* DNA SEQUENCING. *dNTP stands for any of four deoxynucleoside triphosphates. ddATP stands for dideoxyATP.* (a) *The molecular process;* (b) *the resulting gel appearance.* (From *Recombinant DNA (2nd edn)* by Watson, Gilman, Witkowski, and Zoller, 1992. Reprinted with permission of W. H. Freeman and Company.)

packed interior of hydrophobic groups. Domains include regulatory, catalytic and DNA-binding types and are the 'modules' whose exact linear assembly determines the precise characteristics of the protein. Regulation of many cellular processes involves protein interaction domains which direct the association of polypeptides with each other and with phospholipids, with small molecules and with nucleic acids. Their often modular nature and the flexibility of their interactions have probably enhanced the evolution of cellular pathways. There may be several structural ways of achieving a functional domain; thus DNA-BINDING PROTEINS have at least one DNA-binding domain, but this may take the form of any of several structural types, termed MOTIFS. Crystallography shows that the domain organization of actin, heat-shock (chaperone) protein Hsc70 and hexokinase B are remarkably similar, yet they share little sequence similarity. SH2 and SH3 domains (Src homology 2 and 3 domains, named from their homology with Rous sarcoma virus cellular oncogene product c-Src) are about 100 amino acids long and occur in a variety of otherwise structurally and functionally unrelated polypeptides (often brought about by EXON SHUFFLING). Present in various protein substrates of RECEPTOR TYROSINE KINASES, they enable these proteins to bind specific autophosphorylated domains on the RTKs once dimerization of the receptor has occurred after ligand-binding of its extracellular domain (see RECEPTOR for diagrams). SH2 binds phosphotyrosine; SH3 binds proline-rich motifs. Many proteins involved in RTK-associated signal transduction contain both SH2 and SH3 domains; but some contain one or other, in various numbers. In the case of SH3, some of these proteins may be catalytic (e.g. c-Src, phospholipase Cγl and Ras-GAP), others not. Some cytoskeletal proteins (e.g. myosin-lB) also contain SH3. (2) For packaging domains in chromatin, see POSITION EFFECT variegation. (3) In syncytia, specialized domains may occur in which some nuclei may become transcriptionally distinct from the remainder, as with myofibre nuclei in the region of a neuronal synapse. (4) Plasma membrane domains refer to regions with different structure and function from the general cell surface. They may result from phospho- or glycolipid separation from other lipids; or from groupings of intrinsic and extrinsic membrane proteins and often involve the CYTOSKELETON. The number of the AMPA glutamate receptors in one domain of brain dendritic postsynaptic membranes varies as they move rapidly into and out of this membrane. Their concentration there controls the strength of excitatory transmission between neurons and perhaps also the storage of memories in the brain. Nuclear domains are also being recognized. (5) An example of a gene expression domain is found in the *BICOID* GENE entry. (6) In some taxonomies, a formal category approximating to superkingdom in older systems. The three 'domains of life' are reputedly BACTERIA (see for further details), ARCHAEA and EUKARYA.

dominance (1) In genetics, one character is said to be completely dominant to another when it is expressed equally in the *homozygous* and *heterozygous* conditions; the other character is said to be completely RECESSIVE to it, and is only expressed in the homozygous condition. Normally, the two characters would form what Mendel termed a 'pair of contrasting characters'; they would, in other words, be determined by alternative alleles at the same locus (see MUTATION for gain-of-function mutations). Genetic dominance is not synonymous with 'commonest character type in the population': that will depend upon SELECTION, amongst other factors. The term is often used of *genes* (or alleles); but since a 'gene' can have more than one effect (see PLEIOTROPY) accuracy requires use of the term to be restricted either to the context of characters, or to one particular aspect of phenotype affected by the gene. Dominance is usually a property of a normally functional ('wild type') allele; defective (mutant) alleles are usually, but not always, recessive. In some cases the degree of dominance is altered by selection when it is an evolving property of characters (see DOMINANCE MODIFICATIONS, MODIFIER). Two characters are said to be *codominant* when the respective homozygotes are distinguishable both from each other and from the heterozygote, and where the effects of both alleles can be detected in the phenotype; two characters are said to be *incompletely dominant* to one another when the heterozygote is distinguishable from both homozygotes, but distinct effects of the two alleles in the phenotype are not recognizable.

There can be no dominance in the HAPLOID state, or in the HEMIZYGOUS condition generally.

(2) In animal behaviour, a relational property indicating one individual's priority over another in contexts where some resource (e.g. food, mate, shelter) either is, or has in the past been, contested.

(3) In ecology, out of hundreds of organisms present in a community only a relatively few species or species groups generally exert the major controlling influence by virtue of their numbers (abundance), size, production, etc.; species or species groups which largely control the energy flow as well as affecting the environment of all other species are known as the dominant species, e.g. beech trees in a beech wood. When more than one dominant species or species group occurs in a particular plant community, they are called *codominants*.

dominance modification Phenomenon whereby different populations of a species evolve different genetic backgrounds (see MODIFIER) by which phenotypic effects of the same genetic mutation are expressed as either DOMINANT or RECESSIVE. Crossing between individuals from such populations may result in breakdown of dominance, producing an unclassifiable range of phenotypes.

Donnan equilibrium The unequal distribution of ions across a membrane which occurs when the membrane is permeable to some, but not to all, of the ions. For instance, addition of potassium chloride (KCl) to one of two compartments containing distilled water and separated from another compartment by a selectively permeable membrane, will result in equal distributions of K^+ and Cl^- in the two compartments. But addition of a non-diffusible anion, A^-, with multiple negative charges, to compartment I will bring a redistribution of K^+ into compartment II and of Cl^- ions into compartment I. The resulting distribution is achieved when there is electrical neutrality within both compartments, any osmotic inequality between them being relieved by movement of water down its water potential gradient. The Donnan equilibrium is summarized by the equation:

$$[K^+]_I = [Cl^-]_{II}$$
$$[K^+]_{II} = [Cl^-]_I$$

Because cell membranes are never in steady state, the distribution of substances across them cannot be entirely predicted by Donnan equilibrium principles. Since there are always more protein molecules (large anions) within cells than in their surroundings, these principles will be important in establishing voltages across cell membranes. See NERNST EQUATION, RESTING POTENTIAL.

donor Source of material being grafted on to, or somehow inserted into, some other individual.

dopa See L-DOPA.

dopamine 3,4-**D**ihydr**o**xy**p**henylethyl-**a**mine, a monoamine intermediate in the biosynthesis of NORADRENALINE and ADRENALINE from phenylalanine and tyrosine.

Dopamine

Converted into noradrenaline in appropriate synaptic vesicles. A catecholamine NEUROTRANSMITTER of the vertebrate brain, esp. the basal ganglia, frontal cortex and limbic system – cell bodies of which are found largely in the *substantia nigra*. It is one of several neurotransmitters involved in slow synaptic transmission (see SYNAPSE). Still thought to play a critical role in motivation and reward, but now believed to play a part in drawing attention to significant or surprising events rather than directly producing feelings of pleasure or euphoria.

Abnormally low levels in the human *caudate nucleus* produce symptoms of Parkinson's disease, treatable by administration of the precursor *L*-DOPA. In schizophrenics there is increased sensitivity of dopamine receptors in the frontal cortex. The anti-Parkinsonian drug selegiline inhibits monoamine oxidase B, which breaks down dopamine in astrocytes and microglia.

dormancy (Bot.) See RODMANT. (Zool.) Term sometimes used of insect and other animal DIAPAUSE.

dormant In a resting condition. Alive, but with relatively inactive metabolism and cessation of growth. Dormancy may involve the whole organism (higher plants and animals) or be confined to reproductive bodies (e.g. resting spores such as statoblasts, fungal sclerotia, bacterial spores). May be due to unfavourable conditions, and end as these ameliorate. Many seeds (e.g. pea, wheat), though capable of germinating after harvesting, do not do so unless kept moist. On the other hand, a dormant period is part of an annual rhythm for most plants. Often has survival value (e.g. winter dormancy in deciduous trees). After vegetative growth and flowering in spring, many bulbs (e.g. snowdrop, daffodil) have a dormant period coinciding with conditions favourable to growth of other plants. This is common in plants of moist, tropical climates. Dormancy of seeds in conditions otherwise favourable to germination is common (e.g. hawthorn, the weed wild oats) and is associated with incomplete development of the embryo, impermeable seed coats, limiting entry of water and/or oxygen, inhibitors and absence of growth stimulators. Dormancy in some seeds and deciduous trees is regulated by photoperiod (see PHYTOCHROME). See AESTIVATION, DIAPAUSE, HIBERNATION.

dorsal (Zool.) Designating the surface of an animal normally directed away from the substrate; in chordates, the surface (posterior) in which the NEURAL TUBE forms, lying closest to the eventual nerve cord. In flatfish, the apparent adult dorsal surface is in fact lateral. (Bot.) Also used of leaves; synonymous with ABAXIAL.

dorsal aorta See AORTA.

dorsal lip See BLASTOPORE, ORGANIZER.

dorsal placentation Attachment of ovules to midrib of carpels in apocarpous gynoecia.

dorsoventral (dorsiventral) Term generally used to indicate some gradient or morphological feature associated with the AXIS linking the upper and lower parts of an organism or its parts. As with leaves, it often indicates some difference in structure along the axis. Compare ISOLATERAL.

dorsoventral axis-inversion theory See BODY PLANS, DEUTEROSTOMIA, PROTOSTOMIA.

dosage compensation Some animals whose SEX DETERMINATION mechanism employs differences in the sex chromosome ratio compensate for the resulting dosage imbalances of some X-linked genes (and their products) in males and females. MODIFIER loci on the X chromosome (*dosage compensators*) act either to enhance gene expression in the heterogametic sex or repress such expression in the homogametic sex. In male *Drosophila*, an early step in dosage compensation is the assembly of the non-coding rox RNAs with the male-specific lethal complex that associates with the X chromosome along its length. Uniquely in mammals, dosage compensation is achieved by inactivation of one of the two X chromosomes of females in every cell early in development. Control of transcription at the whole chromosome level seems to depend on nuclear RNA molecules, such as the large NON-CODING RNA encoded by the gene *Xist*, which form a 'cage' around the inactivated chromosome after processing in the nucleus. Dosage compensation in mammals is achieved by (but is not the same as) CHROMOSOME INACTIVATION and seems to evolve in response to loss of function of loci on the Y. Very early in eutherian development it is the paternally derived X chromosome which is inactivated while X-inactivation is random later. In marsupials, it is always the paternally derived X chromosome which is inactivated (see BARR BODY). DNA METHYLATION does not appear to be the primary signal for inactivation. See ANDROGENESIS, POSITION EFFECT VARIEGATION.

double circulation See HEART.

double fertilization The unique and probably universal condition in flowering plants (ANTHOPHYTA) whereby, from a single pollen grain, the two generative nuclei within the pollen tube fuse with different nuclei within the EMBRYO SAC of the ovule, one with the egg cell nucleus (the product of AUTOMIXIS) to form the zygote, the other with the diploid secondary endosperm nucleus to form the triploid primary endosperm nucleus. This appears to ensure that no nourishment (as endosperm) is laid down in the prospective seed until a zygote has been formed to take advantage of it.

double helix See DNA.

double recessive Individual or stock in which each of two loci involved in breeding work is homozygous for alleles bringing about expression of RECESSIVE characters. See BACKCROSS.

double-stranded RNA (dsRNA) See RNA SILENCING.

doubling time Time required for a population of a given size to double in number.

down feather See FEATHER.

Down's syndrome (Down syndrome) CONGENITAL disorder of people caused by TRISOMY of chromosome 21 (often by nondisjunction). Characterized by mental retardation, *mongoloid* facial features, simian palm and reduced life expectancy. Has a frequency of about one per 700 live births. High maternal plasma HUMAN CHORIONIC GONADOTROPHIN titre is suggestive of Down pregnancy, especially when coupled with low α-foetoprotein (AFP), which can then be confirmed by amniocentesis (see AMNION); but better diagnostic tests are being sought, one being that women carrying Down babies seem to have twice the plasma levels of inhibin A than do women with normal foetuses. See AMNION.

downstream processing See BIOREACTOR.

drive Specific causal explanations are now sought for most animal activities, so *general drive theories* of motivation have been surpassed by investigation of the control of behaviour rather than its powering. Those specific causal influences promoting an action may be regarded as a part of that activity's *specific drive mechanism*.

driving genes See ABERRANT CHROMOSOME BEHAVIOUR (4), SELFISH DNA/GENES.

Drosophila Genus of fruit flies (Diptera). Probably the best described animal genetically, and of enormous significance to studies of LINKAGE, CYTOGENETICS, SPECIATION and, most recently, developmental biology (e.g. see COMPARTMENT, HOMEOTIC GENE).

drug Any substance, e.g. neurotransmitters and hormones, capable on administration of altering the behaviour and/or physiology of an organism. See ALKALOID, PSYCHOACTIVE DRUGS, THERAPEUTIC DRUGS and (for detoxification of lipid-soluble drugs) ENDOPLASMIC RETICULUM.

drug addiction See PSYCHOACTIVE DRUGS.

drugs, therapeutic See THERAPEUTIC DRUGS.

drupe Succulent FRUIT in which the wall (pericarp) comprises an outer skin (epicarp), a thick fleshy mesocarp, and a hard stony endocarp enclosing a single seed. Commonly called a stone-fruit; e.g. plum, cherry. Compare BERRY. In some drupes the mesocarp is fibrous; e.g. in the coconut the pericarp has tough, leathery epicarp, thick fibrous mesocarp and hard endocarp enclosing the seed and forming with it the nut we buy. Compare NUT.

dry mass See DEHYDRATION.

dryopithecine Term given to a heterogeneous group of hominoids of uncertain relationship to one another and to other forms, apparently representing a hominoid radiation occurring between 17 and 12 Myr BP. Three tribes have been proposed: Afropithecini (17–15 Myr BP; thickened molar enamel, enlargement of premolars and Proconsul-like postcrania), Kenyapithecini

(15–14 Myr BP; even thicker molar enamel) and Dryopithecini (12–8 Myr BP; primitively thin molar enamel, limb bones with rounded shafts, distal humerus with rounded capitulum and deep olecranon fossa – adaptations for brachiation, greater brow ridge development). *Dryopithecus* itself remains the only European Miocene ape and may represent a sister group to the African apes/hominid clade. Retaining many plesiomorphic characters, *Dryopithecus* of Europe is generally regarded as a primitive hominoid; but dryopithecines reaching East Asia, e.g. *Sivapithecus*, are morphologically and phylogenetically more closely related to the extant genus *Pongo* (orangutan). See PROCONSUL, PONGINAE.

dsDNA Double-stranded (duplex) DNA.

dsRNA Double-stranded RNA. Not a requisite product of normal cellular gene expression but produced, at least transiently, by many viruses. Exogenous dsRNAs can act as potent triggers of POST-TRANSCRIPTIONAL GENE SILENCING. They can also be used in a wide variety of eukaryotes to suppress the expression of virtually any gene, allowing rapid analysis of the gene's function (see RNA INTERFERENCE, RNA SILENCING).

dsRNA-triggered interference A GENE SILENCING effect in which a target gene's activity is blocked by artificially provided sense and antisense RNA corresponding to that gene.

ductless gland See ENDOCRINE GLAND.

ductus arteriosus (duct of Botallo) Remnant of (homologous with) ancient vascular connection between pulmonary artery (AORTIC ARCH VI) and the systemic AORTA (aortic arch IV), persisting in amniote embryos and serving as a bypass for most blood from the right ventricle past the lungs while they are deflated and functionless. When the pulmonary circuit opens at birth the ductus closes and atrophies.

ductus Cuvieri See CUVIERIAN DUCT.

duodenum Most anterior region of small intestine of mammals, its origin guarded by the pyloric sphincter. Receives the bile duct and pancreatic duct. Characterized by alkaline-mucus-secreting *Brunner's glands* in the submucosa. So-called because it is about 12 finger-breadths long, about 25 cm, in humans; site of active digestion and absorption; like the rest of the small intestine, its luminal surface has numerous villi.

duplex Of a molecule composed of two chains or strands, usually held together by hydrogen bonds; e.g. double-stranded (duplex) DNA.

duplication Chromosomal MUTATION in which a piece of chromosome is copied next to an identical section, increasing chromosome length. Can result from non-homologous CROSSING-OVER in which two homologous chromosomes pair up imprecisely and a cross-over transfers an abnormally large piece of one chromosome to its homologue, resulting in a DELETION on one chromosome and a duplication on the other. See GENE DUPLICATION.

dura mater See MENINGES.

dyad-symmetry/asymmetry (Of a DNA region) which has symmetry/asymmetry on either side about a central pair of base-pairs. Dyad-symmetric sequences tend to attract DNA-BINDING PROTEINS with a leucine zipper MOTIF composed of homodimeric (identical) subunits whereas dyad-asymmetric sequences tend to attract those with heterodimeric (non-identical) subunits. See Fig. 117.

dye Alternative name for stain. See STAINING.

dynactin (dynein-activator complex) See DYNEIN.

dynamically fragile This describes a community that is stable only within a narrow range of environmental conditions.

dynamically robust This describes a community that is stable within a wide range of environmental conditions.

dynamic equilibrium The state of a system when it remains unchanged because two opposing forces are proceeding at the same rate.

dynamic self-organization Term recognizing that many intracellular structures are constantly forming and disappearing in a flexible way which does not depend upon pre-existing templates (see SELF-ASSEMBLY).

dynamin A GTP-binding protein catalysing several steps in CLATHRIN-dependent vesicle formation. Its microtubule-binding ability is of uncertain role.

dynein A major eukaryotic motor protein. Its microtubule-activated ATPase is responsible for axoneme motility in CILIUM and FLAGELLUM, (and left–right dynein, LRD, is required for their polarity), while cytoplasmic dynein moves membrane-bound organelles along microtubules in the opposite direction from those attached by KINESIN, towards the minus (proximal) end. May also be involved in early stages of chromosome movements of nuclear division (see KINETOCHORE). Addition of ATP or GTP prevents binding of cytoplasmic dynein to polymerizing microtubules. Dynein interacts with a complex of proteins collectively termed 'dynactin' in the positioning of nuclei and the SPINDLE, and in orientation of the latter in some animals at least. The dynein-dynactin complex forms a microtubule motor complex in mitotic polarized epithelial cells, localized on astral microtubules from prometaphase onwards.

dysgenesis See HYBRID DYSGENESIS.

dystrophic Term applied to lakes and rivers having heavily stained brown water, through receipt of large amounts of exogenous organic matter with a high humic organic content. The colour originates from bog soils or from peat at the lake margin or in the catchment through which the streams or rivers flow. See EUTROPHIC, MESOTROPHIC, OLIGOTROPHIC.

E

E1A The principal growth-promoting ONCO-PROTEIN encoded by adenovirus. See APOPTOSIS.

E2F A family of transcription factors regulating several genes controlling early events in the CELL CYCLE, notably the transition between G_0 and G_1 and S phase. Some of these genes are also regulated by the transcription factor Myc, and by other Myc family members (see *MYC* GENE). E2F-1 through to E2F-5 bind members of the 'pocket protein family', including retinoblastoma protein; but E2F-6 promotes gene silencing during G_0 by recruiting a multimeric chromatin-modifying protein complex to quiescent cells which is replaced by E2F-1 or E2F-4 in G_1, allowing regulated transcription of target genes.

ear, inner Membranous labyrinth. Vertebrate organ which detects position with respect to gravity, acceleration, and sound. Lies in skull wall (auditory capsule); impulses transmitted to brain via AUDITORY NERVE. Comprises the VESTIBULAR APPARATUS and the COCHLEA. See LATERAL LINE SYSTEM.

ear, middle Tympanic cavity. Cavity between eardrum and auditory capsule of tetrapod vertebrates (but not urodeles, anurans or snakes). Derived from a gill pouch (spiracle). Communicates with pharynx via eustachian tube, and is filled with air, ensuring atmospheric pressure is maintained on both sides of the eardrum. In it lie the EAR OSSICLES.

ear ossicles Bones in middle ear connecting eardrum to INNER EAR in tetrapod vertebrates. Instead of just the single auditory bone (stapes, see COLUMELLA) of amphibia and primitive reptiles, mammals have in addition the incus and malleus. The first retains its original attachment to the oval window of the inner ear, but here articulates via the incus with the malleus which attaches to the eardrum (tympanum). These last two bones have evolved respectively from the quadrate and articular bones of mammal-like reptiles, in which they were involved in jaw suspension. By this articulation the pressure of the stapes on the oval window is amplified 22 times compared with that of the pressure waves on the tympanum: vibrations are damped, but produce larger forces.

ear, outer (or external) That part of the tetrapod ear, absent from amphibians and some reptiles, external to the eardrum. Comprises a bony tube (*external auditory meatus*). In addition in mammals there is a flap of skin and cartilage (the pinna) at the outer opening which amplifies and focuses pressure waves upon the eardrum. Well developed in nocturnal mammals (e.g. bats).

eardrum (tympanum, tympanic membrane) Thin membrane stretching across the aperture between skull bones at the surface of the head (most anurans and turtles) or within an external meatus (most reptiles, birds and mammals). Vibrates, often aperiodically, transmitting external air pressure changes to EAR OSSICLES of middle ear cavity.

ecad (Bot.) A habitat form, showing characteristics imposed by the habitat conditions and non-generic; also called ecophenes or phenoecotypes. Compare ECOTYPE.

ecdysis Moulting in arthropods. Periodic shedding of the CUTICLE in the course of growth. In insects this includes much of the lining of the tracheal system. The number of larval moults varies (up to fourteen in APTERYGOTA); in endopterygotes there is one pupal moult (producing adult), but among insects only apterygotans moult as adults. In most crustaceans it proceeds throughout adult life. Insect ecdysis is under the control of the hormone ECDYSONE, triggered by a neuropeptide, *eclosion hormone*.

ecdysone (moulting hormone, growth- and-differentiation hormone) Hormone produced by insect *thoracic* (*prothoracic*) *glands*, and possibly also by the crustacean *Y-organ*. In insects its release is under the control of *thoracotropic hormone* produced by neurosecretory cells in the brain and released from the CORPORA CARDIACA. In crustaceans the brain neurosecretion is produced in the X-organs and transported to the *sinus glands* of the eyestalk. Its release inhibits release of moulting hormone by the Y-organs. In insects at least ecdysone induces 'puffing' of selected chromosome regions, the sequence being tissue-specific. This is associated with selective gene transcription, notably by the epidermis; but one of its major effects is to make appropriate cells sensitive to *juvenile hormone* from the CORPORA ALLATA, with which ecdysone works to bring about moulting to the appropriate developmental stage. See DIAPAUSE.

ecesis Germination and successful establishment of colonizing plants; the first stage in succession.

echidna Spiny anteater. See MONOTREMATA.

Echinodermata Phylum of marine and invertebrate deuterostomes; typically with pentaradiate symmetry as adults; an internal skeleton of calcareous plates in the dermis; TUBE FEET; nervous system typically one circular and five longitudinal nerve cords, lacking brain and ganglia; surface epithelium often ciliated, and sensory; coelom well developed, including peculiar WATER VASCULAR SYSTEM; no excretory organs; larvae typically pelagic, roughly bilaterally symmetrical, with tripartite coelom (*oligomerous*) and an often dramatic metamorphosis. Affiliations with HEMICHORDATA. Includes classes Stelleroidea (including subclasses Asteroidea, the starfish, and Ophiuroidea, the brittlestars); Echinoidea (sea urchins, etc.); Holothuroidea (sea cucumbers); Crinoidea (crinoids); and the new class Concentricycloidea (sea daisies).

Echinoidea Sea urchins, heart urchins, etc. Class of ECHINODERMATA; lacking separate arms; more or less globular in shape; mouth downwards; with rigid calcareous *test* of plates in dermis bearing spines and defensive PEDICELLARIAE; browsers and scavengers, often in enormous numbers, on sea bed.

echolocation Method used by several nocturnal, cave-dwelling or aquatic animals for determining positions of objects by reflection of high-pitched sounds. Many bats and dolphins use it, as do oil birds and the platypus.

ecodeme See DEME.

ecological memory The capacity of past states or 'experiences' to influence present or future responses of a community. Existence of memory in communities has always been explicit in ecology; it is an historical feature, and comprises all potential recruit-species that are not completely excluded because of spatial and temporary heterogeneity.

ecological isolation REPRODUCTIVE ISOLATION resulting from differences in ecology. Prezygotic isolation results if (a) individuals may be sympatric but confine mating to different habitats, or (b) they may live in different subniches of the same area and rarely, if ever, come into contact. Postzygotic isolation, which is ultimately caused ecologically, occurs when (c) individuals live in different subniches of the same area and come into contact and mate, but form poorly adapted hybrids. Data on the genetics of ecological isolation are, however, very limited. Compare SEXUAL ISOLATION.

ecological niche See NICHE.

ecology A term first used in 1869 by Ernest Haeckel deriving from the Greek οἶχος ('house' or 'place to live'). Ecology is the study of relationships of organisms, or groups of organisms, to their environments, both biotic and abiotic. It deals with three levels of concern. (i) Individual organisms and how these organisms are affected by (and how they can affect) their biotic and abiotic environment. (ii) the POPULATION (comprising conspecifics). Here one deals with the presence or absence of particular species, with their abundance or RARITY, and with trends and fluctuations in numbers. There are two approaches that can be taken: (a) one can deal with the attributes of individual organisms and consider the manner in which these combine to determine the characteristics of the population and then try to relate these to aspects of the environment; (b) one can deal directly with characteristics of populations, and try to relate these aspects of the environment. (iii) The COMMUNITY. Here one is concerned with composition and structure of communities and with the pathways taken by energy, elements, nutrients, etc., as they pass through them; i.e., with 'community function' (see ECOSYSTEM). Ecologists are versed in many interrelated disciplines; e.g. genetics, molecular biology, evolutionary and behavioural biology, physiology and increasingly study human-made or human-influenced environments (e.g. orchards, wheat fields) and the environmental consequences of human activity (e.g. all aspects of POLLUTION, GLOBAL WARMING). The discipline has become increasingly quantitative, employing modelling, computer simulation and predictive equations, even inferring past environmental changes and impacts. See BALANCE OF NATURE.

ecomorphs Unrelated organisms with similar appearance in some respect, on account (*ex hypothesi*) of the similar selection pressures (evolutionary constraints) in similar environments (e.g. drip tips on leaves of tropical rain forest trees).

ecophysiology The study of physiology and tolerance limits of species that en-

hances understanding of their distribution in relation to abiotic conditions.

ecospecies Group within a species comprising one or more ECOTYPES, whose members can reproduce amongst themselves without loss of fertility among offspring. Approximates to a *hologamodeme* (see -DEME), or to an ideal 'biological' SPECIES, and as a term is used more in botanical than in zoological contexts. See INFRASPECIFIC VARIATION.

ecosystem COMMUNITY of organisms, interacting with one another, plus the environment in which they live and with which they also interact; e.g. a lake, a forest, a grassland, tundra. Such a system includes all abiotic components such as mineral ions, organic compounds, and the climatic regime (temperature, rainfall and other physical factors). The biotic components generally include representatives from several TROPHIC LEVELS; primary producers (autotrophs, mainly green plants), macroconsumers (heterotrophs, mainly animals) which ingest other organisms or particulate organic matter, microconsumers (saprotrophs, again heterotrophic, mainly bacteria and fungi) which break down complex organic compounds upon death of the above organisms, releasing nutrients to the environment for use again by the primary producers. See BALANCE OF NATURE, FOOD CHAIN, PYRAMID OF BIOMASS.

ecotone The transition between two or more diverse communities, as between forest and grassland. Zone which may have considerable length, yet be far narrower than adjoining communities.

ecotoxicology The study of the fate and adverse effects of chemicals on ecosystems.

ecotype Term, generally applied in botanical contexts, referring to a species population exhibiting genetic adaptation to the local environment, whose phenotypic expression withstands transplantation of the plant, or its offspring, to a new environment. Ecotypes can be distinguished by morphology, physiology and phenology, and are potentially infertile with other ecotypes of the same species. Most species comprise an assemblage of ecotypes, each ranging in size from a single population to a regional group of many populations; the wider the species' range, the more ecotypes it has. The terms race, genecotype and ecological race are sometimes used synonymously for eco-type. Random genetic variants (individuals, or groups of individuals) within ecotypes are called biotypes. See INFRASPECIFIC VARIATION, ECOSPECIES. Compare ECAD.

ectexine See EXINE.

ectoderm Outermost GERM LAYER of meta-zoan embryos, developing mainly into epi-dermal and nervous tissue and, when present, NEPHRIDIA.

ectoparasite Parasite living on the outside surface of its host. See PARASITE.

ectophloic siphonostele A siphonostele with phloem external to the xylem.

ectoplasm A peripheral layer of cytoplasm in cells of some eukaryotes that is relatively rigid and non-granular (e.g. amoebae, CHAR-OPHYCEAE). See ENDOPLASM, cell CORTEX.

Ectoprocta (Polyzoa, formerly Bryo-zoa) Phylum of colonial and often poly-morphic coelomates, retaining continuity by coelomic tubes (cyclostomes) or merely by a tissue strand (ctenostomes, cheilos-tomes). Feeding (polyp) individuals do so by microphagy using a LOPHOPHORE of ten-tacles, and secreting a calcareous zooecium (together termed a ZOOID). See STATOBLAST.

ectotherm An animal whose body tem-perature is maintained by heat received from the environment. Such animals are *ectothermic*, and employ *ectothermy*. See ENDOTHERM; compare POIKILOTHERMY.

ectotrophic (Of mycorrhizas) with the mycelium of the fungus forming an external covering to the root and penetrating only between the outer cortical cells; e.g. in pine trees. See MYCORRHIZA; compare ENDOTROPHIC.

edaphic factors Environmental con-ditions determined by physical, chemical and biological characteristics of the soil.

edentata (Xenarthra) Aberrant order of eutherian Mammalia, mainly of South American history and distribution. Includes three sloths, anteaters, armadillos and extinct glyptodonts. Only anteaters are truly toothless (hence ordinal name), the others having molars at least.

Ediacara fauna See CAMBRIAN.

EDTA Ethylenediaminetetraacetic acid. A strong chelating agent of divalent metal ions (e.g. Ca^{2+}, Mg^{2+}, Zn^{2+}). Causes non-specific increase in permeability of Gram-negative bacteria and can inhibit or lyse many such species.

EEG See ELECTROENCEPHALOGRAM.

effector (1) Any structure, be it organ, cell or organelle, by which an animal responds to internal or external stimuli, often via the nervous system. Includes muscles, glands, chromatophores, cilia. Lymphocytes other than memory cells are often described as 'effector cells'. They include, e.g., T cells with cytotoxic or other functions, and anti-body-secreting plasma cells. Cnidoblasts are often regarded as *independent effectors* in that they do not seem to require stimulation from other cells (e.g. of the nervous system) for their activity. (2) The term is also applied to molecules, particularly those at the ter-mini of signalling pathways (e.g. p53).

effector cells See EFFECTOR (1).

efferent Leading away from; e.g. from the central nervous system (motor nerves), from the gills (blood vessels) or from a glom-erulus (arteriole).

egestion Removal of undigested material and associated microorganisms of the gut flora (up to 50% dry weight in man) from the anus. This material has never been inside body cells. A quite different process from EXCRETION, with which it may be con-fused. The voided material is termed *egesta*.

egg cell See OVUM.

egg membranes Few animal eggs, if any, have just a plasma membrane separating the cytoplasm from the external environment. Additional membranes are: (1) *Primary* *membranes*: the vitelline membrane, or thicker chorion. (2) *Secondary membranes*: consisting of or formed by the follicle cells around the egg. (3) *Tertiary membranes*: secreted by accessory glands, oviduct, etc., including albumen, shell membranes, egg 'jelly', etc. Protective against mechanical damage, desiccation.

elcosanoids Collective term for derivatives of 20-carbon precursor fatty acids (eicosanoic acids, e.g. arachidonic acid), including PROSTAGLANDINS, prostacyclins, thromboxanes and leukotrienes. Prostacyclins and thromboxanes tend to have opposite physiological effects. The former (formed in arterial walls) are powerful inhibitors of PLATELET aggregation, relaxing arterial walls and inducing drop in blood pressure; the latter are platelet-derived and induce platelet aggregation (see BLOOD CLOTTING), contraction of artery walls and raising of blood pressure. Leukotrienes promote INFLAMMATION, the strength depending on the fatty acid origin.

elaioplast Colourless plastid (leucoplast) in which oil is stored; common in liverworts and monocotyledons.

Elasmobranchii Subclass of CHONDRICHTHYES, appearing in the middle Devonian. Includes sharks (SELACHII), skates and rays (Rajiformes) and angel sharks (Squatiniformes). Cartilaginous skeleton; dermal denticles probably the remnants of ancestral bony placoderm armour; upper jaws independent of braincase (hyostylic jaw suspension) or in some sharks with anterior attachment to braincase (amphistylic jaw suspension). Gills border gillslits (usually five); spiracle present. Internal fertilization, male having *claspers*, modified pelvic fins, acting as intromittant organs. Tail heterocercal; teeth in rows, replacing in turn those lost. The HOLOCEPHALI (Chimaeras) form a second chondrichthyan subclass.

elastin Principal fibrous protein of the yellow fibres of animal CONNECTIVE TISSUE. Numerous in lungs, walls of large arteries and in ligaments. Highly extensible and elastic. Compare COLLAGEN.

elater (1) Elongated cell with wall reinforced internally by one or more spiral bands of thickening, occurring in numbers among spores in capsules of liverworts. Assist in discharge of spores by movements in response to humidity changes. (2) Appendage of spores of horsetails; formed from outermost wall layer, coiling and uncoiling as the air's moisture level varies; possibly assisting in spore dispersal.

electric organs Organs of certain fishes which produce electric currents by means of modified muscle cells (*electrocytes*) which no longer contract but generate ion current flow on nervous stimulation. Two basic kinds: those producing strong stunning current (e.g. electric eel, electric ray, electric catfish), and those producing currents of low voltage (e.g. in Mormyridae, Gymnotidae excepting electric eel) as a continuous series of pulses for locating prey and obstacles in muddy water, and for mate location.

electrocardiogram (ECG) Record of electrical changes associated with the HEART CYCLE, usually by means of electrodes placed on the patient's skin. Can also monitor foetal heart in the uterus.

electroencephalogram (EEG) Record of changes in electrical potential ('brain waves') provided by the cerebral cortex; detected through the skull and picked up by electrodes placed on the scalp. The waves are then amplified. Four main types: *alpha* (produced when awake, but disappearing when asleep); *beta* (appear when nervous system is active – as in mental activity); *theta* (produced in children, and in adults in emotional stress situations); *delta* (occur in sleeping adults; in awake adults they indicate brain damage).

electron microscope See MICROSCOPE.

electron spin resonance See RADIOMETRIC DATING.

electron transport system (ETS, electron transport chain) Functional chain of independent and mobile enzymes (mostly conjugated proteins, e.g. CYTOCHROMES) and associated COENZYMES (e.g.

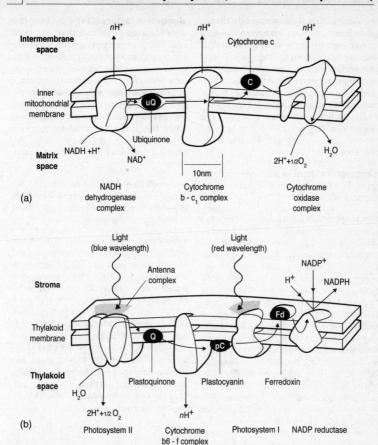

FIG. 45 *Hypothesized arrangements of electron-transporting molecules within (a) mitochondrial and (b) chloroplast membranes. Protons are extruded from mitochondria during activity but taken into thylakoids during the light reactions of photosynthesis.*

QUINONES) in an ion-impermeable membrane. Essential to the oxidoreduction chemistry of aerobic respiration and photosynthesis (see Fig. 45). It was once thought that these components were structurally ordered; but it is now thought they achieve electron transfer from one member of the chain to another, down a REDOX gradient, through random collisions. In respiration, the cytochrome and quinol oxidases achieve a sustained flow of electrons (from the initial respiratory substrates) by catalys-

ing efficient O_2-reduction. Through the exergonic electron transfer involved, energy release pumps protons across the membrane, providing a store of potential energy in the form of a proton gradient. As the protons return down their electrochemical gradient through specific channels in the membrane, they provide the *protonmotive force* needed for ATP synthesis by ATPase, itself associated with the proton channels. This process is common to the inner membranes of MITOCHONDRIA and the thylakoid system of CHLOROPLASTS. In chloroplasts the electrons are first boosted to a high energy level by photons; in mitochondria they are derived from hydrogen atoms (also the proton source) covalently bonded in electron-rich respiratory substrates. Associated coenzymes, not all intrinsic to the membrane, may include NAD, NADP, FAD, flavoproteins, plastoquinone and ubiquinone. Similar ETSs occur in bacterial membranes (e.g. see BACTERIORHODOPSIN). See SUPEROXIDES.

electrophile Electron-deficient functional group in molecules, which seeks electrons. Contrast NUCLEOPHILE.

electrophoresis Technique for separating charged molecules in buffer solution, particularly proteins, nucleic acids and their degradation products, based on different mobilities (caused by their different net charges at a given pH) in an electric field generated by direct current through the buffer. Substances for separation are usually allowed to move through a porous medium such as a gel (e.g. starch, agar, polyacrylamide) or paper (e.g. filter, cellulose acetate). Separated substances occur in bands on the medium and may be stained or identified by some labelling device, by fluorescence, by comparisons with knowns, or by removal and subsequent analysis. In *immunoelectrophoresis* antigens are placed in wells cut in agar gel. After separation of antigens by electrophoresis a trough is cut between the wells, filled with antibody, and diffusion allowed to take place. Where antigen meets appropriate antibody, arcs of precipitin form, allowing complex antigen mixtures to be compared. See IMMUNOASSAY.

electroporation (electroporosity) Process by which an electrical potential applied across (typically) plant cells in culture disrupts their membranes sufficiently to create tiny pores through which DNA in the medium can be taken up – a small fraction of which may get incorporated into the cell's genome.

ELISA Enzyme-linked immunoabsorbent assay. See IMMUNOASSAYS.

elongation factors (EFs) Conserved proteins involved in protein synthesis in all cells. Two major groups exist: EF-1 binds the amino acid component of an amino-acyl-tRNA enabling it to bind an empty A-site on a ribosome; EF-2 is involved in transfer of amino acids from activated tRNAs to the growing polypeptide. The two groups are PARALOGOUS, produced by a gene duplication event; they may serve as outcrops for each other in rooting phylogenetic trees. See EOCYTES, RIBOSOMES; compare INITIATION FACTORS.

emasculation Removal of stamens from hermaphrodite flowers before they have liberated their pollen, usually as a preliminary to artificial hybridization.

embedding Method employed in the preparation of permanent microscope slides of thin tissue sections. After DEHYDRATION and CLEARING, the material is put into molten paraffin wax (usually for 1–3 hours, with one or two changes of wax) which impregnates the tissue. After setting, the wax block is sectioned using a MICROTOME. The wax is removed by xylene, itself removed by absolute alcohol, and gradual rehydration of the section is achieved by passing for a few minutes through progressively more dilute alcohols. Staining can then proceed. In electron microscopy, Araldite® is frequently used for the embedding.

embryo (Bot.) Young plant developed from an ovum after sexual (including parthenogenetic) reproduction. In seed plants, it is contained within the seed and comprises an axis bearing at its apex either the apical meristem of the future shoot or, in some species, a young bud (the plumule), while at the other end is the root (the radicle). From the centre of the axis grow one or more seed leaves (cotyledons). (Zool.) The structure produced from an egg (usually fertilized), by generations of mitotic divisions while still within the EGG MEMBRANES, or otherwise inside the maternal body. Embryonic life is usually considered to be over when hatching from membranes occurs (or birth); in humans an embryo becomes a FOETUS when the first bone cells appear in cartilage (at about seven weeks of gestation). Of seventeen types of protoctist that evolved multicellularity (or coloniality), only three groups – those that gave rise to fungi, plants and animals – evolved the ability to form multicellular aggregates that can differentiate into different cell types.

embryogenesis Formation and development of an EMBRYO.

embryology Study of embryo development.

embryonic membrane See EXTRA-EMBRYONIC MEMBRANES.

embryonic stem cells (ES cells) Pluripotent STEM CELLS isolated directly from the INNER CELL MASS of the mammalian blastocyst. They can be maintained *in vitro* as stem cell lines or made to differentiate in a variety of ways. When microinjected into a host blastocyst, and even earlier stages (e.g. the 8-cell embryo), they contribute to all the tissue of the resulting CHIMAERA, including the GERM LINE (see Fig. 46). ES cells are useful in studies of developmental regulation and can be used to generate specific defined mutations by GENE TARGETING. Human ES cells are derived from 5-day-old human embryos (blastocysts), a process which destroys them. Many believe the main effect of ES technology on biomedical research will be less in cell-replacement therapies than in uncovering the mechanisms of genetic diseases and in generating sources of normal and impaired tissues for use in drug discovery. It is feasible to generate ES cell lines from mouse models of human disease, although the underlying causes may often be different. It has not yet (2003) proved possible to produce ES cell lines from adult humans – existing lines originate from embryos left over from *IN VITRO* FERTILIZATION, and these do not harbour the mutations causing the human diseases we should like to treat by future ES cell research. New ES cell lines would, at present, have to be generated by taking a nucleus from an adult cell and introducing it into an enucleated human oocyte (somatic cell nuclear transfer, SCNT), which would then be allowed to develop to the blastocyst stage (see CLONE).

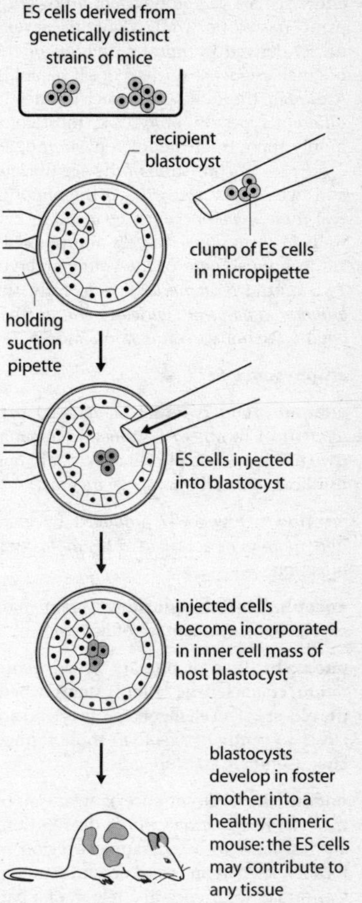

ES cells derived from genetically distinct strains of mice

recipient blastocyst

clump of ES cells in micropipette

holding suction pipette

ES cells injected into blastocyst

injected cells become incorporated in inner cell mass of host blastocyst

blastocysts develop in foster mother into a healthy chimeric mouse: the ES cells may contribute to any tissue

FIG. 46 *Making a chimeric mouse with ES cells. The cultured ES cells can combine with the cells of a normal blastocyst to form a healthy chimeric mouse, and can contribute to any of its tissues, including the germ line. Thus the ES cells are totipotent.*

There is widespread opposition to these procedures because they produce human embryos for research purposes.

embryo sac Large oval cell in the nucellus of the ovule, in which fertilization of the egg cell and development of the embryo take place. At maturity, it represents the entire female gametophyte of a flowering plant (MAGNOLIOPHYTA). Contains several nuclei derived by mitotic division of the original MEGASPORE nucleus (itself haploid). Although the number of nuclei varies in different types of embryo sac, most commonly there is, at the micropylar end, an *egg-apparatus* consisting of the egg nucleus and two others, *synergids*. At the opposite end three nuclei become separated by cell walls to form *antipodal cells* and probably aid in nourishment of the young embryo. Two central *polar nuclei* fuse to form the *primary endosperm nucleus*. For further details, see DOUBLE FERTILIZATION and Fig. 6.

emphysema See LUNG.

enamel Hard covering of exposed part (crown) of tooth; 97% inorganic material (two thirds calcium phosphate crystals, one third calcium carbonate), 3% organic.

enation Outgrowth produced by local hyperplasia on a leaf as a result of viral infection.

encephalization quotient Brain mass expressed as a percentage of body mass.

endarch Type of primary xylem maturation, characteristic of most stems, where the oldest xylem elements (protoxylem) are closer to centre of axis than those formed later. Compare EXARCH.

endemism (1) Occurrence of organisms or taxa (termed *endemic*) whose distributions are restricted to a geographical region or locality, such as an island or continent. (2) Continual occurrence in a region of a particular (endemic) disease, as opposed to sporadic outbreaks of it (epidemics).

endergonic (Of a chemical reaction) requiring energy; as in synthesis by green plants of organic compounds from water and carbon dioxide by means of solar energy. Compare EXERGONIC. See THERMODYNAMICS for more detail.

endexine Inner layer of EXINE of bryophyte spores and vascular plant pollen grains.

endocarp Innermost layer of the carpel wall, or pericarp of fruit, in flowering plants.

Frequently used to denote the 'stone' of drupes.

endocrine-disrupting chemicals (EDCs) Substances produced by human technology and which, some believe, adversely affect the endocrine system by blocking hormones but in doses too small to trigger a conventional toxic response. Suspect molecules include PCBS and cocontaminants, plasticizers and insecticides (see PEST MANAGEMENT).

endocrine gland (ductless gland) Gland whose product, one or more HORMONES, is secreted directly into the blood and not via ducts (compare EXOCRINE GLAND). The gland may be a discrete organ, or comprise more scattered and diffuse tissue. Examples include: ADRENAL, OVARY, PANCREAS, PITUITARY, PLACENTA, TESTIS, THYROID. See ENDOCRINE SYSTEM.

endocrine system Physiologically interconnected system of ENDOCRINE GLANDS occurring within an animal body. Compared to neurotransmitters, the more diffuse hormonal outputs can take more time to reach effective concentrations, and therefore require a longer physiological half life (i.e. persistence in the body). HORMONES generally exert effects over longer timescales, appropriate in growth, timing of breeding and control of blood and tissue fluid composition. Effects of peptide hormones depend as much on the distribution of membrane RECEPTORS on target cells as on the molecules secreted (see also NUCLEAR RECEPTORS). See NERVOUS SYSTEM for a further comparison of roles, and NEUROENDOCRINE COORDINATION, NEUROHAEMAL ORGAN, NEUROSECRETION.

endocrinology Study of the structure and function of the ENDOCRINE SYSTEM.

endocytosis Collective term for PHAGOCYTOSIS and PINOCYTOSIS. An essential process in much eukaryotic CELL LOCOMOTION. For receptor-mediated endocytosis, see RECEPTOR (2) and refs. there. See also EXOCYTOSIS, CELL MEMBRANES, PHAGOCYTE, COATED VESICLE, and Figs. 34, 47 and 105.

endoderm (entoderm) Innermost GERM LAYER of an animal embryo. Composed like mesoderm (when present) of cells which have moved from the embryo surface to its interior during GASTRULATION. Develops into greater part of gut lining and associated glands, e.g. where applicable, liver and pancreas, thyroid, thymus and much of the branchial system. Not to be confused with ENDODERMIS.

endodermis Single layer of cells forming sheath around the vascular region (stele), most clearly seen in roots; in some stems identifiable by its content of starch grains (the *starch sheath*). Usually regarded as innermost layer of cortex. In roots, most characteristic feature of very young endodermis is band of impervious wall material, the Casparian strip, in radial and transverse walls of cells. With age, especially in monocotyledons, endodermis cells (except PASSAGE CELLS) may become further modified by deposition of layers of suberin over entire wall surface followed, particularly on the inner tangential wall, by a layer of cellulose, sometimes lignified. Endodermis is important physiologically in control of transfer of water and solutes between cortex and vascular cylinder, since these must pass through protoplasts of endodermis cells.

endogenous clock See BIOLOGICAL CLOCK.

endolith Microorganism living in rock. See EXTREMOPHILE.

endolymph Viscous fluid occurring within the vertebrate COCHLEA and VESTIBULAR APPARATUS. These are separated from the skull wall by PERILYMPH.

endometrium Glandular MUCOUS MEMBRANE lining the uterus of mammals. Undergoes cyclical growth and regression or destruction during the period of sexual maturity. Receives embryo at IMPLANTATION. See OESTROUS CYCLE, MENSTRUAL CYCLE, PLACENTA.

endomitosis (endoreduplication) Process whereby all the chromosomes of an INTERPHASE nucleus replicate and separate within an intact nuclear membrane (which does not divide). No spindle or other mitotic apparatus found. Resulting nuclei are ENDOPOLYPLOID, the degree of ploidy some-

times exceeding 2,000. Compare POLYTENY, in which chromosomes do not separate after duplication.

endonuclease See DNAase.

endoparasite Parasite developing within its host. Includes PARASITOIDS. Contrast ECTOPARASITE.

endopelon Community of algae living and moving within muddy sediments. See BENTHOS.

endopeptidase Acid protease, hydrolyzing specific internal peptide bonds producing peptides and amino acids; e.g. PEPSIN. Compare EXOPEPTIDASE.

endophyte An organism that lives within a plant (compare EPIPHYTE). A term that is used in several ways, which gives rise to some confusion and ambiguity (e.g. bacteria and parasitic angiosperms, such as mistletoe), that live either completely or partially within plants. With respect to fungi the term is used to refer to endomycorrhizal species in plant roots and fungi occurring within gametophytes of non-flowering vascular plants. Modern usage with fungi defines fungal endophytes as those living within aerial parts of a plant but cause no symptoms of disease. With algae, endophyte refers to an alga living within spaces of other plants. See ENDOPHYTON.

endophyton Community of algae growing between cells of other plants, or in cavities within plants. Well known associations occur in some liverworts. See BENTHOS.

endoplasm (plasmasol) Cytoplasm that is granular and streams actively and lies in the interior of the cell, surrounded by a thin peripheral layer of rigid, non-granular ECTOPLASM. present in some eukaryotes (e.g. amoebae, CHAROPHYCEAE).

endoplasmic reticulum (ER) Eukaryotic cytoplasmic organelle comprising a complex system of membranous stacks (cisternae) and not unlike chloroplast thylakoids in appearance, but often being continuous with the outer of the two nuclear membranes and, like this membrane, bearing attached ribosomes on the cytosol side (when termed *rough ER*). A ribosome-free system of tubules (*smooth ER*), continuous with the cisternae, projects into the cytosol and pinches off transport vesicles. ER is not physically continuous with the GOLGI APPARATUS, but is functionally integrated with it. A large rough ER is indicative of a metabolically active (e.g. secretory) cell.

ER seems to be the sole site of membrane production in a eukaryotic cell, membrane proteins and phospholipids being incorporated from precursors in the cytosol (see SIGNAL RECOGNITION). Enzymes in the lipid bilayer pick up fatty acids, glycerol phosphate and choline and create LECITHIN, while protein components are fed into the ER lumen as they are produced at ribosomes bound to attachment sites on the cisternae. GLYCOSYLATION of newly synthesized proteins occurs within the cisternae through activity of *glycosyl transferase* located in the ER membrane. Only rough ER is involved in PROTEIN synthesis, subsequent folding of proteins and disulphide bond formation, specific cleavages of PREPROTEINS, oligosaccharide addition and assembly of quaternary structures, all occurring within the ER lumen under the direction of specific enzymes. Smooth (*transcisternal*) ER is generally a small component but from it TRANSPORT VESICLES (some of them COATED VESICLES) are budded off to carry protein and lipid to other parts of the cell. Some proteins are processed in the Golgi apparatus after the vesicles have fused there (see SNARES). Smooth ER contains enzymes involved in lipid metabolism and so is abundant in cells synthesizing cholesterol and those converting it to steroid hormones. Liver cells contain large amounts of smooth ER, in which enzymes of the cytochrome P450 family detoxify lipid soluble drugs and their metabolites (which otherwise remain in membranes) by direct reduction of carbonyl groups ($>C = O$) to hydroxyl groups ($>HC–OH$) and by conjugation of these with sulphate or glucuronic acid, rendering them water-soluble and excretable. Extra smooth ER made during a time of such drug administration seems to be removed afterwards by autophagosomes. A specialized ER region has been implicated in storage and subsequent localization of

RNAs in the *Xenopus* oocyte cytoplasm. See Figs. 20 and 74.

endopodite See BIRAMOUS APPENDAGE.

endopolyploidy The result of ENDOMITOSIS.

Endopterygota Insects with complete metamorphosis (pupal stage in life cycle) and with wings developing within the larva (see IMAGINAL DISC), although first visible externally in the pupa. Sometimes regarded as a subclass. Includes orders Neuroptera (lacewings); Coleoptera (beetles); Strepsiptera (stylopids); Mecoptera (scorpion flies); Siphonaptera (fleas); Diptera (true flies); Lepidoptera (butterflies, moths); Trichoptera (caddis flies); Hymenoptera (bees, ants, wasps). Often used synonymously with Holometbola, but see THYSANOPTERA. Compare EXOPTERYGOTA.

end organ Structure at peripheral end of a nerve fibre; usually either a RECEPTOR or a motor end-plate (see NEUROMUSCULAR JUNCTION).

endorphins Peptide NEUROTRANSMITTERS, isolated from the PITUITARY GLAND, having morphine-like pain-suppressing effects. Also implicated in memory, learning, sexual activity, depression and schizophrenia. See ENKEPHALINS, PSYCHOACTIVE DRUGS.

endoskeleton Skeleton lying within the body. Vertebrate cartilage and bone provide support, protection and a system of levers enabling manipulation of the external environment; arthropods have internal projections of their cuticle (*apodemes*) for muscle attachment; echinoderms and annelids, among other invertebrates, use a *hydrostatic skeleton* to greater or lesser extent, and these too are endoskeletons. In vertebrates comprises both *axial skeleton* (bones/cartilage around body axis, esp. vertebral column, cranium, ribs, sternum, hyoid) and *appendicular skeleton* comprising PECTORAL GIRDLE and PELVIC GIRDLE and skeletons of associated limbs/fins.

endosome Membranous organelle (see Fig. 47) to which molecules taken up by ENDOCYTOSIS are transferred prior to the

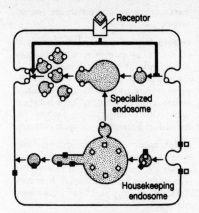

FIG. 47 *The cell-surface modification pathway and the involvement of distinct, but connected,* ENDOSOME *compartments.*

Dark squares = LDL-receptors, transferrin receptors, etc.; light squares = LDL, transferrin, etc.; circles = other (specialized) proteins, as in epithelial apical and synaptic vesicles; diamond = hormone or transmitter (e.g.).

endocytic vesicle's becoming a LYSOSOME by fusion of transport vesicles from the GOLGI APPARATUS. Vesicles bud from it to return RECEPTOR molecules to the cell surface, one function of endosomes being to sort proteins destined for different locations in the cell. There is evidence for at least two kinds of endosomal compartment: a 'housekeeping endosome' common to most if not all eukaryotic cells (e.g. fairly close to the basolateral membranes of epithelial cells and within the cell bodies and dendrites of neurons) which recycles, among others, low-density lipoprotein receptors (SEE LDL) and TRANSFERRIN receptors, and a specialized endosome (near the apical membrane of epithelial cells and within axons and nerve termini of neurons) from which vesicles leave and target apical membrane and synaptic vesicle proteins (SEE PROTEIN TARGETING, TOXIN).

endosperm Nutritive tissue surrounding and nourishing the embryo in seed plants. (1) In flowering plants (ANTHOPHYTA), formed in embryo sac by division of usually triploid endosperm nucleus after fertilization. In some seed plants (non-endospermic, exalbuminous), it is entirely absorbed by the embryo by the time seed is fully developed (e.g. pea, bean seeds); in other seeds (endospermic, albuminous), part of the endosperm remains and is not absorbed until seed germinates (e.g. wheat, castor oil). (2) Also applied to tissue of female gametophyte in conifers and related plants which is formed by cell division within the embryo sac before fertilization, outer layers persisting in the seed. Compare PERISPERM.

endospore (1) Spore formed within a parent cell; in bacteria (mostly Gram-positive species, e.g. the fermenting *Clostridium* and aerobic *Bacillus*, but also a few Gram-negative sulphate-reducers, e.g. *Thiobacillus*); a thick-walled spore – usually one per cell – resistant to heat and harsh chemicals; in Cyanobacteria, a thin-walled spore. Term also used for inner layer of spore wall. See GRAM'S STAIN. (2) Structure within the elaters of spores of the sphenophyte *Equisetum*.

endosporic Development of GAMETOPHYTE within spore wall.

endostyle Ciliated and mucus-secreting groove or pocket in ventral wall of pharynxes of urochordates, hemichordates, cephalochordates and ammocoete larvae of lampreys. The vertebrate thyroid is probably homologous with it.

endosymbiosis Symbiotic association between cells of two or more different species, one inhabiting the other, the larger being host for the smaller. In *serial endosymbiosis*, one after another such symbiotic associations may occur telescoped within the largest cell. It is believed to account for the occurrence of eukaryotic chloroplasts (ancestor a cyanobacterium), mitochondria (ancestor a purple non-sulphur bacterium, on biochemical grounds similar to present-day hydrogen-oxidizing *Paracoccus*) and, some believe, cilia. See ALGAE and EUKARYOTE for further information; also CRYPTOPHYTA, DINOPHYTA, GLAUCOPHYTA, HYDROGENOSOME, KAPPA PARTICLES, SYNTROPHY, *WOLBACHIA*. See also Fig. 48.

Bacteria   Eukarya   Archaea

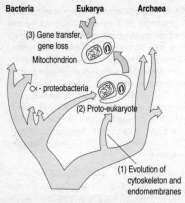

(3) Gene transfer,
gene loss
Mitochondrion

α - proteobacteria

(2) Proto-eukaryote

(1) Evolution of
cytoskeleton and
endomembranes

FIG. 48 *A widely accepted model for the endosymbiotic origin of mitochondria, and its implications for the three-domain system of biological classification. An undigested α-proteobacterium (purple bacterium) became an intracellular respiratory symbiont, and this in turn became a mitochondrion with loss of some genes and transfer of others from the α-proteobacterium to the nucleus. See* ENDOSYMBIOSIS.

endothelin A 21-amino acid peptide made by endothelial cells and ten times more potent in VASOCONSTRICTION than even angiotensin II.

endothelium Single layer of flattened, polygonal cells lining vertebrate heart, blood and lymph vessels. Mesodermal in origin, endothelial cells originate from the same stem cell as do the smooth muscle cells surrounding all blood vessels except capillaries, requiring transforming growth factor and platelet-derived growth factor. See ANGIOGENESIS, PLASMINOGEN ACTIVATORS.

endotherm (adj. endothermic) An animal whose body heat is generated through its own metabolic activities. Many are homeothermic or regionally heterothermic. Endotherms, such as birds and mammals, have far greater independence of the environmental temperature than do ectotherms, in particular allowing them to be alert and active at night when solar energy is absent. However, endotherms must consume food more quickly, and with a higher total calorific value, than ecto-therms. Contrast ECTOTHERM; see HETEROTHERMS.

endotoxins Complexes of LIPOPOLYSACCHARIDE and lipoprotein, forming the major components of the outer membrane of Gram-negative bacteria (e.g. *Salmonella typhi*, causing typhoid fever) and released on lysis of the bacterial cell. Extremely heat stable. Often complex with protein. Released during autolysis. Compare EXOTOXIN.

endotrophic (Of mycorrhizas) with mycelium of the fungus within cells of root cortex; e.g. orchids (where it may be the sole means of nutrient support, host cells digesting the hyphae). See MYCORRHIZA; compare ECTOTROPHIC.

endozoon Community of algae living within animals.

end-product inhibition (retroinhibition, feedback inhibition) The inhibition of an ENZYME, often the first in a metabolic pathway, by the last product of the pathway. Ensures against overproduction of the final product. See ALLOSTERIC, REGULATORY ENZYME. Compare CATABOLITE REPRESSION.

energy flow The passage of energy through an ECOSYSTEM from source (generally the sun), through the various TROPHIC LEVELS (within organic compounds), and ultimately out to the atmosphere as respiratory heat loss. There is about 90% loss of energy between one trophic level and the next in the grazing food chain. See PYRAMID OF BIOMASS.

***engrailed* gene** A SEGMENT POLARITY GENE, whose expression is determined by PAIR RULE GENES. Has roles in both invertebrate and vertebrate development.

enhancer DNA sequence that can activate *cis*-linked promoter utilization (i.e. promoter on same chromosome), doing so at great distances from the promoter (up to 50 kb) and at either side of it (the 3' or 5' side). Like promoters, enhancers may bind transcription factors. This may involve DNA looping caused by DNA-BINDING PROTEINS: sites normally separated on a chromo-

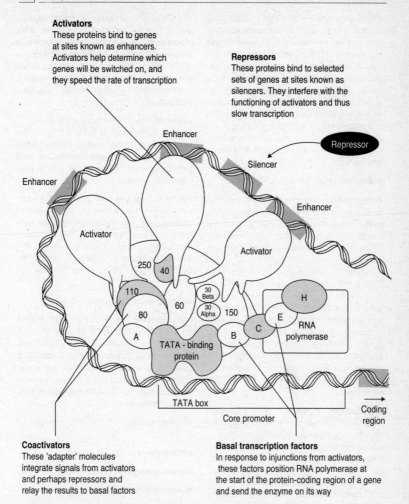

Activators
These proteins bind to genes at sites known as enhancers. Activators help determine which genes will be switched on, and they speed the rate of transcription

Repressors
These proteins bind to selected sets of genes at sites known as silencers. They interfere with the functioning of activators and thus slow transcription

Coactivators
These 'adapter' molecules integrate signals from activators and perhaps repressors and relay the results to basal factors

Basal transcription factors
In response to injunctions from activators, these factors position RNA polymerase at the start of the protein-coding region of a gene and send the enzyme on its way

FIG. 49 *The four main components in the human control of transcription and their relationship to genetic elements (ENHANCERS, TATA box and promoter).*

some are brought into close contact. Some enhancers (e.g. glucocorticoid enhancer) bind steroid-nuclear protein complexes, enabling binding in turn by RNA polymerase II and the initiation of transcription. Enhancers exhibit tissue or species specificity; e.g. some viral enhancers only function in the species in whose cells the virus

grows best. In animals (metazoans), a typical gene is likely to contain several enhancers, located in 5′ and 3′ regulatory regions as well as in introns. Each is responsible for bringing about a subset of the total gene expression pattern, often in a tissue- or cell-specific manner. Each will contain ten or more binding sites for at least three different transcription factors, usually two activators and one repressor, and some contain sequences ('tethering elements') that recruit distant enhancers (10 kb away in *Drosophila* and 100 kb in mammals) to the core promoter. This kind of long-distance regulation is not observed in yeast and may play a critical role in morphogenesis. Enhancers and other *cis*-regulatory elements of higher metazoans act in a highly structured and modular fashion, and may have been a crucial factor in the evolution of COMPLEXITY. See Fig. 49.

enhancer trap See GENE CAPTURE.

enkephalins Peptide NEUROTRANSMITTERS isolated from the thalamus and parts of the spinal cord and concerned with pain-related pathways. Morphine-like pain reducers. The substantia gelatinosa (the tip of the dorsal horn of grey matter where dorsal roots of spinal nerves enter the spinal cord) contains high levels of the endogenous opioid peptide enkephalin, thought to act as a transmitter of some dorsal root interneurons which establish presynaptic contacts with terminals of afferent neurons possessing opiate receptors. Binding of enkephalin, and probably of morphine, to these receptors reduces the production of substance P, with consequent analgesic effects. See ENDORPHINS.

enrichment culture Microbiological technique allowing selection and isolation from a natural, mixed population of microorganisms of those having growth characteristics desired by the investigator. Involves culturing on a medium whose composition is adjusted for selective growth of desired organisms, by altering nutrients, pH, temperature, aeration, light intensity, etc. Employed in bacteriology, mycology and phycology.

enterocoely Method of COELOM formation within pouches of mesoderm budded off from embryonic gut wall. Develops this way in echinoderms, hemichordates, brachiopods and some other animals.

enteroendocrine cells (EECs) Endocrine cells of the gut mucosa, e.g. in humans,

G cells of the stomach, secreting GASTRIN, and cells in the crypts of Lieberkuhn in the ileum secreting SECRETIN, CHOLECYSTOKININ or glucose-dependent insulinotropic peptide (GIP). Their differentiation in the ileum requires the transcription factor Math 1, a component of the Notch signalling pathway.

enterokinase (enteropeptidase) Enzyme (peptidase) secreted by vertebrate small intestine, converting inactive trypsinogen to active trypsin. Removes a small peptide group. Component of succus entericus. See KINASE.

enteron (coelenteron) The gut (gastrovascular) cavity within the body wall of coelenterates, having a single opening serving as both mouth and anus. May be subdivided by mesenteries (as in sea anemones); sometimes receives the gametes (as in jellyfish). May serve as hydrostatic skeleton. See ARCHENTERON.

enteropneusts See HEMICHORDATA.

enterotoxins These are EXOTOXINS that act on the small intestine, causing massive losses of fluid and resulting in diarrhoea. Produced by a variety of bacteria, including *Staphylococus aureus*, *Clostridium perfringens* and *Vibrio cholerae*. That of *Escherichia coli* is plasmid-encoded. See ADENYLYL CYCLASE, ORAL REHYDRATION THERAPY.

entomogenous (Of fungi) parasitic of insects.

entomology Study of insects.

entomophagous Insect-eating.

entomophily Pollination by insects.

Entoprocta Phylum of pseudocoelomate and mostly marine invertebrates, of uncertain relationships. Trochophore larva attaches by its oral surface; stolon grows out from the new aboral surface and produces a colony of adult individuals. These feed by ciliated tentacles which are simply folded away inside their protective cover, not withdrawn into a body cavity as in ECTOPROCTA. Excretion by protonephridia. Anus opens within tentacular ring.

entrainment Synchronization of an endogenous rhythm with an external cycle such as that of light and dark. See CIRCADIAN RHYTHM.

envelope Term applied to the two membranes surrounding the nucleus, the membranous covering of those viruses which have them (i.e. enveloped viruses), and the complex of one or more membranes and peptidoglycan forming the bacterial cell surface (see GRAM'S STAIN).

environment Collective term for the conditions in which an organism lives, both biotic and abiotic. Compare INTERNAL ENVIRONMENT.

enzyme A protein catalyst produced by a cell and responsible for the high rate and specificity of one or more intracellular or extracellular biochemical reactions. Enzyme reactions are always reversible. Almost all enzymes are globular proteins consisting either of a single polypeptide or of two or more polypeptides held together (in quaternary structure) by non-covalent bonds. By virtue of their three-dimensional conformations in solution, enzymes act upon other molecules (substrates), and thus catalyse one type of (but not necessarily just one) chemical reaction. Many enzymes are secreted as an inactive form, or ZYMOGEN, and many are inactive unless phosphorylated by a KINASE.

Their shapes provide them with one or more *active sites* (domains) which bind temporarily and usually non-covalently with compatible substrate molecules to form one or more *enzyme-substrate (ES) complexes*, catalysis occurring only during the brief existence of the complex. One or more *products* are then released as the active site is freed again to bind fresh substrate. Active sites have conformations and charge distributions which are substrate-specific and their component amino acids commonly alter their relative three-dimensional positions (termed an *induced fit*) as the substrate binds, enabling several sub-reactions involved in catalysis to proceed.

Enzymes do nothing but speed up the rates at which the *equilibrium positions* of reversible reactions are attained. In some poorly understood way, ultimately explicable in terms of THERMODYNAMICS, enzymes reduce the *activation energies* of reactions, enabling them to occur much more readily at low temperatures – essential for biological systems. Enzymes thus often by-pass the expected energetically unfavourable intermediate in a reaction. When placed in low water concentrations, many enzymes catalyse the reverse of the reaction normally promoted in the biological system.

It is now known that RNA molecules can act as catalysts of reactions, sometimes involving themselves as substrates (see SPLICING). When they involve non-self RNA molecules as substrates, as some do, they can be regarded as enzymes in the full sense (see RIBOZYMES, TELOMERE).

In general, cells can do only what their enzymes enable them to do. During both evolution and multicellular development, cells come to look and function differently from each other because they come to have different biochemical capabilities. An enzyme's presence in a cell is dictated by the expression of one or more cistrons encoding it; thus DIFFERENTIATION is understood through molecular biology (see GENE EXPRESSION).

Because enzyme molecules are generally globular proteins, their shapes and functions may be affected by pH changes in their aqueous environments (see DENATURATION). Denaturation by extremes of pH is usually reversible; not so denaturation by heat. Temperature increase will raise the rate of collision of enzyme and substrate molecules, thus increasing the rate of ES complex formation and raising the reaction rate. This is opposed by increased enzyme denaturation as the *optimum temperature* for the reaction is exceeded. Eventually the reaction ceases, sometimes only at temperatures well in excess of 100°C (see ARCHAEBACTERIA).

At any one instant, the proportion of enzyme molecules bound to substrate will depend upon the substrate concentration. As this is increased, the initial velocity of the reaction (V_o) on addition of enzyme increases up to a maximum value, V_{max} (see Fig. 50), at which substrate level the enzyme

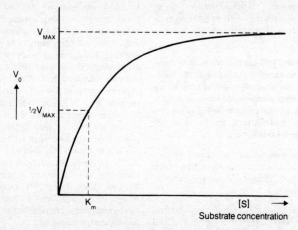

FIG. 50 *Effect of increasing substrate concentration on velocity of enzyme–substrate reaction.*

is said to be *saturated* (all active sites maximally occupied), and no further addition of substrate will increase V_o. The value of substrate concentration at which $V_o = \frac{1}{2}V_{max}$ is known as the MICHAELIS CONSTANT (K_m) for the enzyme-substrate reaction. Low K_m indicates high affinity of the enzyme for the substrate.

Some enzymes (e.g. aspartase) bind just one very specific substrate molecule; others bind a variety of the same kind (e.g. all terminal peptide bonds in the case of exopeptidases). The difference arises from the degree of *stereospecificity* of the enzyme. Many need an attached PROSTHETIC GROUP or a diffusible COENZYME for activity. In such enzymes the protein component is termed the *apoenzyme* and the whole functional enzyme-cofactor complex is termed the *holoenzyme*. Enzymes requiring metal ions are sometimes termed *metalloenzymes*, the commonest ions involved being Zn^{2+}, Mg^{2+}, Mn^{2+}, Fe^{2+} or Fe^{3+}, Cu^{2+}, K^+ and Na^+. These ions commonly provide a needed charge within an active site (but see POISONS).

Some enzymes occur as part of a MULTIENZYME COMPLEX. In nearly all cases, the shape of the enzyme alters as the ES complex forms, and this brings appropriate groups into such proximity that they are obliged to react. In so doing their electrostatic and hydro-

phobic bondings to the enzyme break, they fall away, and the enzyme returns to its original shape again. This *induced fit theory* is supported by X-ray crystallographic evidence. The suffix *-ase* often replaces the last few letters of a substrate's name to give the common name of the enzyme using it as substrate: thus sucrase digests sucrose. But an international code for enzymes recognizes six major categories of enzyme function, numbered as follows: 1. *oxidoreductases* (e.g. dehydrogenases), catalysing REDOX REACTIONS; 2. *transferases*, transferring a group of atoms from one substrate to another; 3. *hydrolases*, catalysing hydrolysis reactions; 4. *lyases*, catalysing additions to double bonds (saturating them); 5. *isomerases*, performing isomerizations; 6. *ligases*, performing condensation reactions involving ATP cleavage.

Allosteric enzymes have, in addition to an active site, another stereo-specific site to which an *effector*, or *modulator*, molecule can bind. When it does, the shape of the active site is altered so that it can or cannot bind substrate (allosteric stimulation or inhibition respectively). In this way the enzyme can be part of a fine control circuit, requiring the presence or absence of a substance – in addition to substrate presence – before enzyme activity proceeds. Some allosteric enzymes respond to two or more such modulators, permitting still finer control over timing of enzyme activity (see REGULATORY ENZYME).

Feedback (or *retro-*) *inhibition* of a biochemical pathway is often achieved by *allosteric inhibition* of the first enzyme in the sequence by the final product. The product binds non-covalently to the modulator site on the enzyme, closing the active site allosterically.

Enzyme inhibition of a simpler kind is achieved in competitive inhibition, where an inhibitor substance competes with the substrate for the enzyme's active site. The binding is reversible so that the percentage inhibition for fixed inhibitor level decreases on addition of substrate. An extremely important example of this involves probably the most abundant enzyme, *ribulose biphosphate carboxylase*, the CO_2-fixing enzyme in C_3 PHOTOSYNTHESIS, in which O_2 molecules compete with CO_2 molecules for the active site (see PHOTORESPIRATION). In *uncompetitive inhibition* the inhibitor combines with the ES complex (one piece of evidence for the latter's existence), which cannot therefore yield normal product. In *non-competitive inhibition* (a form of allosteric inhibition) the inhibitor binds at a non-active site on the enzyme and ES complex so as to deform the active site and prevent ES breakdown, a process unaffected by increasing substrate concentration, being either reversible or irreversible.

Some enzymes are *constitutive*, being synthesized independently of substrate availability; others are *inducible* (e.g. many liver enzymes), being synthesized only when substrate becomes available. The molecular biology of this is to some extent explained in GENE EXPRESSION. Some substances, e.g. steroids, PCBs, barbiturates, ethanol, and organochlorine insecticides, are known to induce enzymes non-specifically. See PSYCHOACTIVE DRUGS.

Some enzymes are located randomly in the cytosols of cells; others have very restricted distributions and may be attached to particular membranes or within the matrices of particular organelles. One effect of the latter restriction is that initial velocities of reactions (V_o) can be quite high for a substrate level that would be too low if the molecules were randomized over the whole cell. Another advantage is that incompatible reactions can be kept physically separated. Restriction of enzyme movement to two dimensions as part of a fluid membrane increases the probabilities of collision with cofactors and/or substrates (see INDUSTRIAL ENZYMES).

Methods are now available for attaching some enzymes, and even cells containing them, to insoluble support materials which *immobilize* them, holding them in place (e.g. for an industrially important reaction). The immobilizing medium may be a silica gel lattice, a collagen matrix or cellulose fibres; or enzymes can be encapsulated in beads of alginate or polymer microspheres. Advantages include recovery of the enzyme, lack of contamination of product by enzyme and sometimes greater enzyme stability at extremes of pH and temperature. Immobilization is of great service in continuous fermentations (see BIOREACTOR).

enzyme inhibition See ENZYME.

enzyme kinetics Study of the effects of substrate, inhibitor and modulator concentrations on the rate of an enzyme reaction, particularly on initial velocities (V_o). The interrelationships are normally expressed graphically, giving enzyme–substrate–inhibitor curve characteristics, one example being included in the entry for ENZYME, Fig. 50.

Eocene GEOLOGICAL EPOCH of the TERTIARY period lasting from about 56.5–35.4 Myr BP. Mammals and birds radiated extensively; initial formation of grass lands occurred. Australia separated from Antarctica, and India collided with Asia. In general, climate was mild to tropical. See Appendix.

eocytes Archaebacterial sulphobacteria, thought by some to be the sister group of eukaryotes. Evidence rests on the inclusion of a 10-amino acid sequence in their ELONGATION FACTOR-1 (EF-1) and shared with those of eukaryotes but absent from EF-1s of other archaebacteria, all true bacteria and from all EF-2 sequences.

eosinophil Vertebrate granulocytic white blood cell of myeloid origin; 10–12 μm in diameter, nucleus usually bilobed and cytoplasm filled with red-orange granules when Wright's stain used. They combat effects of histamine in ALLERGIC REACTIONS, phagocytose antigen–antibody complexes and release toxic proteins against large targets (e.g. parasitic worms). Recruited by the CYTOKINE RANTES (produced by T cells) to sites where MAST CELLS are active. Also release antihistamine, damping inflammatory responses. Migrate towards regions containing T CELL products. Capable of limited phagocytosis.

ependymal cells Cells forming the ependymal layer, lining the ventricles of the adult vertebrate brain. Now known to be neural STEM CELLS, providing neuronal and glial cell precursors.

ephemeral Plant with a short life cycle (seed germination to seed production), having several generations in one year; e.g. groundsel. *Desert ephemerals* pass the dry season as dormant seeds. Compare ANNUAL, BIENNIAL, PERENNIAL.

Ephemeroptera Mayflies. Order of exopterygote insects; with long-lived aquatic nymphs which may moult up to twenty-three times, adults living from a few minutes to a day since they have rudimentary mouthparts and neither eat nor drink. Final nymphal moult produces a unique *subimago*, which moults to produce the adult. Two pairs of membranous wings, held vertically at rest. One pair of CERCI, with or without additional third caudal prolongation.

ephyra Pelagic larval stage in life cycle of Scyphozoa (jellyfish); develops into adult medusa. Budded asexually from sessile scyphistoma.

epiblast Cell layer of avian blastodisc and mammalian blastocyst; presumptive ectoderm. Compare HYPOBLAST.

epiboly Process, observed in amphibian and other vertebrate embryos, during which the region occupied by cells of the animal half of the blastula expands over the vegetal half. In amphibians the cells migrate and roll under through the BLASTOPORE, the vegetal cells remaining as just a plug filling the blastopore.

epicotyl Upper portion of the axis of an embryo or seedling, above the cotyledons and below the next leaf or leaves. Compare HYPOCOTYL.

epidemic Large-scale temporary increase in prevalence of a disease due to a parasite or some health-related event. Once an epidemic has passed through a population, the majority of those remaining are immune to pathogens of the same strain. Compare PANDEMIC; contrast ENDEMIC.

epidermis Outermost layer of cells of a multicellular organism. (Bot.) Primary tissue, one cell thick, forming protective cell layer on surface of plant body, covered in aerial parts by a non-cellular protective CUTICLE. (Zool.) In invertebrates, often one cell thick, secreting a protective non-cellular CUTICLE. In vertebrates there is no non-cellular cuticle, and the epidermis is composed of several layers of cells; the outermost ones often undergo CORNIFICATION and die.

epididymis Long (6 m in man) convoluted tube, one attached to each testis in amniotes. Receives sperm from seminiferous tubules and houses them during their maturation, reabsorbing them if they are not ejaculated (in four weeks in man). Peristaltic contractions of the epididymides propel sperm into the sperm duct during ejaculation. Derived embryologically from the mesonephric (Wolffian) duct.

epigamic (Of animal characters) attractive to the opposite sex and therefore subject to SEXUAL SELECTION. Often concerned with courtship and mating.

epigeal (1) Seed germination in which the seed leaves (cotyledons) appear above the ground; e.g. lettuce, tomato. Compare HYPOGEAL. (2) Of animals, inhabiting exposed surface of land, as distinct from underground.

epigenesis Theory of reproduction and development deriving from Aristotle and

Genetic	Epigenetic

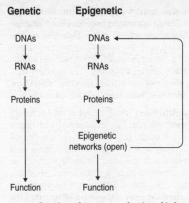

FIG. 51 *Genetic and* EPIGENETIC *theories of information processing.*

espoused by William Harvey (1651) that the parts of an embryo are not all present and *preformed* at the start of development but arise anew one after the other during it. More recently, the term 'EPIGENETICS' has come to indicate the study of developmental non-linearities and to represent an alternative to genetic determinism. Epigenesis can be viewed as those mechanisms by which DNA is contextualized, controlled and regulated to produce changing patterns of gene expression in the face of changing environmental signals. A central tenet is that DNA sequence information, by itself, contains insufficient information for determining how gene products (proteins) interact to produce any kind of mechanism. See CLONING, EPISTASIS, NATURE–NURTURE DEBATE, PREFORMATION.

epigenetics Term introduced in 1947 by the geneticist C. H. Waddington for the branch of biology which studies those causal interactions between genes and their products which bring the phenotype into being. It has two main aspects: (i) changes in cellular composition (cell differentiation, or histiogenesis) and (ii) changes in geometrical form (morphogenesis).

Essentially, epigenetics reminds us that living systems are more than the mere sum of their genes and is anti-reductionist in flavour. R. Holliday has defined it as 'the study of the mechanisms that impart temporal and spatial control on the activity of all those genes required for the development of a complex organism from the zygote to the adult'. The hierarchical nature of biological systems, so often remarked upon, may be the outcome of control processes which, although they include the genome, are not themselves genetically programmed but emerge out of the interactions of the system itself. Understanding of complex FUNCTIONS may require recourse to influences outside the genome, i.e. to epigenetic (non-linear) influences. *Epigenetic inheritance systems* are those mechanisms that store and transmit cellular information additional to those based on DNA sequences; they include CHROMATIN REMODELLING systems (see DNA METHYLATION); steady-state inheritance systems, in which a gene product acts as a positive regulator of its continued activity and is passed into daughter cells following division; and structural inheritance systems where a three-dimensional structure acts as a template for the production of an identical structure in daughter cells, as with some cortical modifications in ciliate protozoans. See Fig. 51, CELL MEMORY, COMPLEXITY.

epiglottis Cartilaginous flap on ventral wall of mammalian pharynx. The glottis pushes against it during swallowing, preventing food, etc., from entering the trachea.

epigynous See RECEPTACLE.

epinasty (Bot.) More rapid growth of upper side of an organ. In a leaf, would result in a downward curling leaf blade. The growth substance ETHENE has been implicated. Compare HYPONASTY.

epinephrine American term for ADRENALINE.

epipelon Extremely widespread community of algae occurring in all waters where sediments accumulate on to which light penetrates. The species are almost all microscopic, living on and in the surface millimetres of the sediment, being unable to withstand long periods of darkness and anaerobic conditions. Motile species exhibit endogenous vertical migration rhythm. An important algal community, particularly in shallow ponds and lakes, as well as in highly transparent oligotrophic and montane lakes. See BENTHOS.

epipetalous (Of stamens) borne on the petals, with stalks (filaments) more or less fused with the petals and appearing to originate from them.

epiphysis (1) Separately ossified end of growing bone, forming part of joint; peculiar to mammalian limb bones and vertebrae. Separated from rest of bone (DIAPHYSIS) by cartilaginous plate (epiphysial cartilage). Epiphysis and diaphysis fuse when growth is complete. See BONE, GROWTH HORMONE. (2) Synonym for PINEAL GLAND.

epiphyte Plant attached to another plant, not growing parasitically upon it but merely using it for support; e.g. various lichens, mosses, algae, ivy, and orchids, all commonly epiphytes of trees.

epiphyton Community of organisms living attached to other plants, sometimes in very large populations; well developed in aquatic habitats where algae attach to other plants.

epipsammon Community of algae found living attached to sand grains in both freshwater and marine environments. It includes very small species of diatoms and blue-green algae, and includes both motile and nonmotile species. Motile species, like those of the EPIPELON, exhibit an endogenous vertical migration rhythm; however the speed of movement is slower with epipsammic species.

episome A genetic element (DNA) that may become established in a cell either autonomously of the host genome, replicating and being transferred independently, or else as an integrated part of the host genome, participating with it in recombination and being transferred with it. Term first applied to temperate BACTERIOPHAGE, but includes PLASMIDS. See F FACTOR, TRANSPOSON.

epistasis Interaction between non-allelic genetic elements or their products, some-times restricted to cases in which one element suppresses expression of another (*epistatic dominance*). Analogous to genetic DOMINANCE. Segregation of epistatic genes in a cross can modify expected phenotypic ratios among offspring for characters they affect. See HYPOSTASIS, MODIFIER, SUPPRESSOR, MUTATION, POLYGENES, GENETIC VARIATION.

epithelium (Zool.) Sheet or tube of firmly coherent cells (see DESMOSOME) with minimal material between them, of ectodermal or endodermal origin, lining cavities and tubes and covering exposed surfaces of body; one surface of epithelium is therefore free, the other usually resting on a BASEMENT MEMBRANE over connective tissue. Its cells are frequently secretory, secretory parts of most glands being epithelial. Classified according to: height relative to breadth (e.g. *columnar, cuboidal,* or *squamous,* in order of diminishing relative height); whether the sheet is one cell thick (*simple*) or many (*stratified* or *pseudostratified*); and presence of cilia (*ciliated*). When morphologically identical tissue is derived from mesoderm, it is either ENDOTHELIUM or MESOTHELIUM. See INTERCELLULAR JUNCTION (Fig. 92) for polarity in epithelial cells. (Bot.) Layer of cells lining schizogenously formed secretory canals and cavities, e.g. in resin canals of pine.

epitope Antigen determinant. See ANTIGEN.

epitreptic behaviour Behaviour by one individual tending to cause the approach of a member of the same species (a conspecific).

epizoite Non-parasitic sedentary animal living attached to another animal. Compare EPIPHYTE.

epizoon Community of algae living attached to the outer surfaces of animals, which may range from tiny aquatic invertebrates to fishes and whales.

Epstein–Barr virus A VIRUS of the Herpes group and agent of the lymphatic disease, infectious mononucleosis, in which infected B cells become enlarged and resemble monocytes. Transmitted orally, occurring mainly in young children and more commonly in females. Cause of the cancers Burkitt's lymphoma and naso-pharyngeal carcinoma. Implicated in chromosome translocations.

equatorial plate Plane in which the chromosomes of a cell lie during metaphase of mitosis and meiosis; the equator of the spindle.

equilibrium potential Potential (voltage gradient) at which a particular ion type passes equally easily in either direction across a cell membrane. Different ions have different equilibrium potentials. See MEMBRANE POTENTIAL.

equilibrium theory A theory of community organization that focuses attention on the properties of the system at an equilibrium point (time and variation are not the central concern), to which the community tends to return after a DISTURBANCE. See NON-EQUILIBRIUM THEORY.

Equisetophyta (Equisetopsida) (formerly SPHENOPHYTA) Horsetails. This is an ancient group of plants extending back to the Devonian period, having their greatest abundance and diversity in the Palaeozoic era (c. 300 Myr BP). During the late Devonian and carboniferous periods, they were represented by the calamites: a group of trees that reached 15 m in height, with a trunk that could be more than 20 cm thick. Today there is a single genus (*Equisetum*) which comprises numerous species, widespread in moist or damp habitats (e.g. alongside streams or along the edges of woodland). Plants have jointed stems, with distinct nodes and elevated siliceous ribs. Leaves are small, whorled and fused into sheaths with their tips remaining free and tooth-like. Sporangia are borne on peltate sporophylls, which are aggregated together in a cone (strobilus) at the apex of a stem. Aerial stems arise from a branching, underground rhizome. Gametophytes are green, free-living, terrestrial and unisexual (male gametophytes are smaller than the female). Gametophytes become established mostly upon mud that has recently been flooded and is rich in nutrients. Sperm is multiflagellate, requiring water to swim to the egg. Horsetails have been used to scour pots and pans and have acquired the name 'scouring rushes'.

ergosterol In most fungi, a sterol replacing cholesterol in the cell membranes. Ergosterol inhibitors are antifungal compounds.

ergot Several ascomycotan fungi are parasitic on higher plants. The disease of plants called ergot is caused by *Claviceps purpurea*, a parasite of rye (*Secale cereale*) and other grasses. Dark spur-shaped SCLEROTIA develop in place of healthy grain in a diseased inflorescence. After the grain shatters, they overwinter in the soil. Activated by frost, they form several multicellular spore-bearing structures which contain abundant perithecia. Ascospores are shed when rye and other grasses are flowering and germinate among the flowers. The resulting mycelium gives rise to abundant conidia, which are embedded in a sticky liquid, and are spread further by insects. Fungus converts individual immature fruits into sclerotia. Ergot is a serious disease of rye because when eaten it can cause severe illness among domestic animals and humans. Ergotism, the toxic condition caused by eating grain infected with ergot, is often accompanied by gangrene, nervous spasms, psychotic delusions and convulsions. Occurred frequently during the Middle Ages, when it was known as St Anthony's Fire. Ergot contains the ALKALOID lysergic acid amide (LSA), a precursor of lysergic acid diethylamide (LSD). Some compounds in ergot (e.g. ergotamine, ergonovine) are used medicinally.

erythroblast Nucleated bone marrow cell which undergoes successive mitoses, develops increasing amounts of haemoglobin, and gives rise to a *reticulocyte*, and finally the fully differentiated RED BLOOD CELL.

erythrocyte See RED BLOOD CELL.

erythropoiesis Red blood cell formation. See HAEMOPOIESIS.

escape Cultivated plant found growing as though wild, some detrimentally to the natural ecosystems (e.g. purple loosestrife, *Lythrum salicaria*), which in certain areas of the United States and Canada is rapidly taking over wetlands, causing them to become dry.

***Escherichia coli* (*E. coli*)** Motile, Gram-negative, rod-shaped bacterium (Enterobacteriaceae) used most extensively in bacterial genetics and molecular biology. Normal inhabitant of the human colon; is usually harmless although some strains can cause disease. See BACTERIA, CHROMOSOME, JACOB-MONOD THEORY, GRAM'S STAIN.

essential amino acid See AMINO ACID.

essential fatty acid FATTY ACIDS required in the diet for normal growth. In mammals, include linoleic and gamma-linolenic acids, obtained from plant sources, without which poor growth, scaly skin, hair loss and eventually death occur. Precursors of arachidonic acid and PROSTAGLANDINS.

essentialism The view, associated in particular with Aristotle, that for any individual there is a definitive set of properties, individually necessary and collectively sufficient, rendering it the kind of individual that it is. This approach has at times been adopted in the context of the taxa used in classification, sometimes rhetorically and in opposition to evolutionary theories. See NATURAL KIND, NOMINALISM.

etaerio (Of fruits) an aggregation; e.g. of achenes, in buttercup; of drupes, in blackberry.

ethene See ETHYLENE.

ethidium bromide Substance used to 'nick' double-stranded DNA when the combination is irradiated with ultraviolet light (i.e. photochemically, not enzymatically). Ultracentrifugation can then be used to separate covalently-closed circular DNA from linear or nicked DNA. Also used to 'stain' irradiated DNA on gels.

Ethiopian Designating a zoogeographical region comprising Africa south of the Sahara. Sometimes Madagascar is treated as a separate region (the Malagasy Region).

ethology Study of animal behaviour in which the overriding aim is to interpret behavioural acts and their causes in terms of evolutionary theory. The animal's responses are interpreted within the context of its actual environmental situation. Much empirical study is designed to test models of sexual selection and of cooperativity.

ethylene (ethene) Simple gaseous hydrocarbon (C_2H_4) produced in small amounts by many plants (found in flowers, leaves, leafy stems, roots, and in some species of fungi), and acts as a plant hormone, or GROWTH SUBSTANCE. The biosynthesis of ethylene begins with the amino acid methionine, which reacts with ATP to form a compound known as S-adenosylmethionine (SAM). SAM is then split into two different molecules, one of which contains a ring consisting of three carbon atoms (1-aminocyclopropane-1-carboxylic acid, ACC). This compound is then converted into ethylene, carbon dioxide and ammonia by enzymes on the TONOPLAST. The reaction forming ACC is the step of the pathway that is affected by several treatments (e.g. high AUXIN concentration, air pollution damage, wounding). These stimulate ethylene production by plant tissues. Release of ethylene commonly inhibits auxin synthesis (negative feedback) and transport. It normally inhibits longitudinal growth, but promotes radial enlargement of tissues. The final shape and size of cells, as influenced by ethylene, are the result of its interaction not only with auxin, but also with gibberellic acid and cytokinin. Its effect upon fruit ripening has agricultural importance, e.g. to promote ripening of tomatoes picked green and stored in ethylene until marketed. Also used to ripen grapes. Ethylene promotes abscission of leaves, flowers and fruits in a variety of plant species, and is commonly used commercially to promote fruit loosening in cherries, grapes and blueberries. It is also used as a thinning agent in commercial prune and peach orchards. In MONOECIOUS flowers ethylene appears to play a major role in determining the sex of the flowers (e.g. in Cucurbitaceae, ethylene is important in sex expression and is associated with the promotion of femaleness). Plants, often the same species, have been found to employ several forms of prokaryote-like sensors for detecting gaseous ethylene. See CLIMACTERIC.

etiolation Phenomenon exhibited by green plants when grown in darkness. Such plants are pale yellow because of absence of chlorophyll, their stems are exceptionally long owing to abnormal lengthening of internodes, and their leaves are reduced in size.

eubacterials Eubacteria; a large and diverse order of BACTERIA, lacking photosynthetic pigments. Simple, undifferentiated cells with rigid cell walls, either spherical or straight rods. If motile, move by peritrichous flagella. Thirteen recognized families. Includes the important genera *Azotobacter* and *Rhizobium* (both nitrogen-fixers), *Escherichia*, etc.

eucarpic (Of fungi) with a mature thallus differentiated into distinct vegetative and reproductive portions. Compare HOLO-CARPIC.

Eucarya See DOMAIN.

eucaryote See EUKARYOTE.

euchromatin Eukaryotic chromosomal material (chromatin) staining maximally during metaphase and less so in the interphase nucleus, when it is less condensed. See CHROMOSOME, HETEROCHROMATIN.

eugenics Study of the possibility of improving the human GENE POOL. Historically associated with some extreme political tendencies and with encouragement of breeding by those presumed to have favourable genes and discouragement of breeding by those presumed to have unfavourable genes; nowadays the more humanitarian GENETIC COUNSELLING has largely replaced talk of eugenics.

Euglenophyta Euglenoids. A group of 40 genera and 800 species of algal flagellates found in most freshwater habitats, particularly in water polluted by organic waste or decaying organic matter, which can give rise to very dense populations of bloom proportion. They are also found in marine and brackish water, the open sea, in tidal areas among seaweeds, and as sand inhabitants on beaches. Brackish epipelic species can colour estuarine sediment bright green during the daytime, as countless cells migrate to the sediment surface. Shortly before submergence by the incoming tide, the cells migrate back into the top few millimetres of the sediment. Such movements are under the control of an endogenous, clock-like rhythm cued by variations in phototactic behaviour.

Most euglenoid flagellates possess chloroplasts, but many are heterotrophic. Photosynthetic species are able to supplement photosynthesis by taking up organic compounds, but there are many colourless forms (e.g. *Astasia*), which are totally dependent on heterotrophic nutrition. Most of these forms are saprophytes, while some are phagotrophic. Of these, *Peranema* and *Entosiphon* species possess a special apparatus (cytosome) for capturing and ingesting prey, which may be other algae or yeasts. Freshwater copepods can have a variety of parasitic euglenoids.

Although the vast majority of euglenoids are flagellates, some develop palmelloid stages. Two flagella covered by fibrillar hairs, together with a felt-like covering of shorter hairs, arise at the bottom of a flask-shaped invagination (ampulla), comprising a reservoir and a canal. The large contractile vacuole at the anterior end of the cell discharges its contents into the reservoir. Bright green chloroplasts are enclosed by one membrane of chloroplast endoplasmic reticulum, and within thylakoids are usually grouped in threes, forming lamellae. Chlorophylls *a* and *b* are present. Of the accessory pigments present (beta-carotene, neoxanthin and diadinoxanthin are most important. Echinenone, diatoxanthin, and zeaxanthin are also present. Chloroplast DNA occurs as tiny granules throughout the entire chloroplast. The reserve polysaccharide is PARAMYLON (a $\beta1,3$-linked glucan), which lies as granules in the cytoplasm. Within the cytoplasm is found an orange-red eyespot. It contains several droplets containing carotenoids. Surrounding the cell is a spirally arranged pellicle, comprising abutting strips of protein, and can be flexible or rigid. When flexible, euglenoids can undergo a flowing movement. These wind helically around the cell. Cells can also be surrounded by mucilage secreted by muciferous bodies lying beneath the pellicle. Some are surrounded by a rigid lorica (e.g. *Trachelomonas*). The nucleus is mesokaryotic (see MESOKARYOTE), while mitosis is closed and the nucleolus (endosome) persists throughout nuclear division.

Euglenoids probably arose from the ingestion of green algal chloroplasts by a protozoan in the Kinetoplastida (bodonids, trypanosomatids and the closely related protozoan *Isonea*). Such ENDOSYMBIOSIS would have occurred when the food vacuole membrane of the protozoan became the single membrane of the CHLOROPLAST ENDO-PLASMIC RETICULUM surrounding the two membranes of the chloroplast envelope. See NUCLEOLUS.

eukaryote (eucaryote) A term traditionally applicable to any organism in whose cell, or cells, chromosomal material is (or was) contained within one or more nuclei and separated from the cytoplasm by two nuclear membranes (the nuclear envelope). These organisms comprise the domain of life Eukarya, forming a distinct but very variable group in which highly differentiated nuclei and multicellularity have evolved several times. All extant eukaryotes seem to postdate the origin of mitochondria, although these organelles have been lost repeatedly or have degenerated into small residual organelles of unknown function. Key 'marker' genes of eukaryotes include those encoding the components of the nuclear pore complex. Some eukaryotic cells (e.g. mammalian erythrocytes, phloem sieve tubes) lose their nuclei during development; but all are distinguished from prokaryotic cells by having somewhat denser (80S) RIBOSOMES, a greater variety of membrane-bound organelles but not necessarily mitochondria (see AMITOCHONDRIATE, ARCHEZOA), their generally much larger size and the presence of the proteins ACTIN, MYOSIN, TUBU-LIN and HISTONES; though an actin-like protein has been found in *Escherichia coli*; and a possible prokaryotic homologue for tubulin (FtsZ) has also been found. Indeed, evidence

that several protein domains have been conserved since before the bifurcation of prokaryotes and eukaryotes requires us to reconsider the distinction between these grades of organization at the molecular level. But no prokaryote engages in mitosis or meiosis, and CELL LOCOMOTION in eukaryotes involves different MOTOR PROTEINS. Unlike prokaryotes, eukaryotes have sterols in their membranes. They also evolved a CYTOSKELETON, meiotic sexual life cycles, a greater diversity of intracellular PROTEIN KINASES and signalling pathways, and more extensive use of ENHANCERS for transcriptional regulation, INTRON excision and EXON shuffling. RNA SILENCING may be a eukaryotic innovation. As to the origin of eukaryotes, the 'hydrogen hypothesis' posits that, 2×10^9 years ago or more, a free-living H_2- and CO_2-producing facultatively anaerobic eubacterium (the symbiont, with the mechanisms for Krebs cycle and oxidative phosphorylation), and a strictly autotrophic, hydrogen-dependent, methanogenic ARCH-AEBACTERIUM (the host) were associated syntrophically in an anaerobic environment where CO_2 and a geological source of H_2 were present. If that H_2 supply dried up, the archaebacterium would have been dependent for its survival on the eubacterial heterotroph. Such a scenario would then favour the transfer of genes for glucose-to-pyruvate glycolysis from the eubacterium to the 'cytosol' (and thence to the chromosome) of the archaebacterium, and for eubacterial import receptor proteins for organic substrates to become incorporated in the archaebacterial plasma membrane, with subsequent transfer of their genes to the archaebacterial genome. Such a hypothetical early eukaryotic cell closely resembles the living amitochondrial hydrogenosome-containing *Trichomonas vaginalis* (see Fig. 53) and may have been one route by which proto-eukaryotes made use of the increasingly abundant O_2 produced by cyanobacteria around 2 billion years ago. By hypothesis, the archaebacterial host would not have needed a nucleus, cytoskeleton (or endocytosis) or mitosis, the requirement for hydrogen being what 'forged' the eukaryotic line out of PROKARYOTES. We now realize

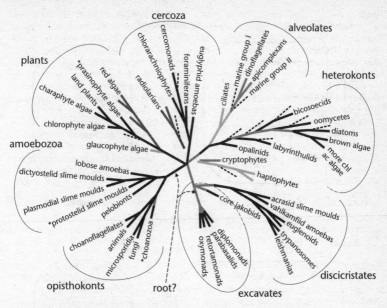

FIG. 52 *One possible phylogeny of* EUKARYOTES. * *Indicates probable paraphyletic group.*

that there exists a huge potential diversity of extremely small eukaryotes (nano-eukaryotes, 2–20 μm; pico-eukaryotes, <2 μm), distributed across the domain, overlapping bacteria (~0.5 to 2 μm) in size, and requiring the creation of several new taxa. The smallest pico-eukaryote to date is *Ostreococcus tauri*, <1 μm in diameter, but still housing a nucleus, 14 linear chromosomes, one chloroplast and several mitochondria.

Most eukaryotes can be assigned to one of eight major subgroups (see Fig. 52): (a) *opisthokonts* (having a single basal flagellum on reproductive cells and flat mitochondrial cristae); (b) *plants* (having plastids with just two outer membranes); (c) *heterokonts* (having a unique flagellum bearing hollow tripartite hairs and, normally, a second plain one); (d) *cercozoans* (amoebae with filose pseudopodia), often inhabiting shell-like tests – very elaborate in foraminiferans; (e) *amoebozoans* (mostly test-less amoebae, often with lobose pseudopodia for part of their life cycle); (f) *alveolates* (having

systems of cortical alveolae directly under their plasma membranes); (g) *discicristates* (having discoid mitochondrial cristae and, sometimes, a deep excavated ventral groove); and (h) *amitochondrial excavates* (a group lacking much biochemical support for its identity, but all AMITOCHONDRIATE and usually having a ventral excavated feeding groove).

Locating the root of the eukaryote tree has proved very challenging, but progress seems to have been made by using the derived and 'difficult-to-reverse' gene fusion between dihydrofolate reductase (DHFR) and thymidylate synthase (TS). In eubacteria, both genes are separately translated, as they are in animals and fungi – presumably, therefore, the original eukaryote condition. But plants, alveolates and Euglenozoa have instead a bifunctional fusion gene in which both enzyme activities are found in a single protein. If this is the derived condition (and assuming the fusion is monophyletic), the eukaryote root must

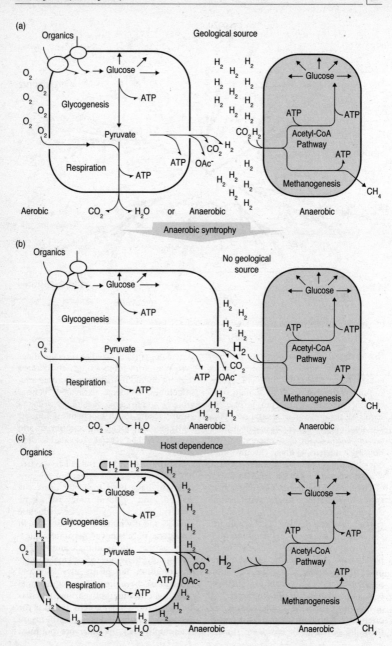

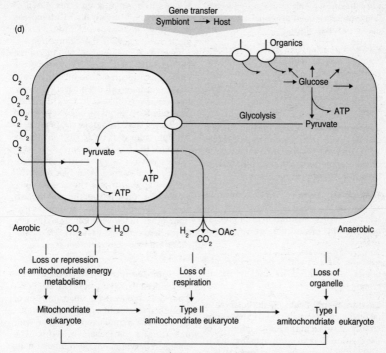

FIG. 53 *Hypothetical model deriving eukaryotic energy metabolism from the original situation (a) in which a free-living H_2- and CO_2-producing eubacterium (the symbiont) and a methanogenic archaebacterium become (b) syntrophically dependent once the geological hydrogen source is removed. Host dependence (c) is sustainable if the symbiont can obtain sufficient organic substrates. Importers would allow the host to supply its symbiont with organic matter; but selection would then dictate that the host autotrophic pathways are lost (d), with eubacterial genes moving into archaebacterial chromosomes. The situation found in the amitochondrate* EUKARYOTE *Trichomonas vaginalis may have arisen in this manner.*

lie earlier than the common ancestor of these last three. See CELL, EOCYTES, NUCLEUS, MESOKARYOTE, URKARYOTE.

euphotic zone (photic zone) Uppermost zone of lakes, seas and rivers, with sufficient light for active photosynthesis. In clear water, may extend to 120 metres.

euploid Term describing cells whose nuclei have an exact multiple of the HAPLOID set of chromosomes, there being no extra or fewer than that multiple. Thus, DIPLOID, TRIPLOID, TETRAPLOID, etc., cells are all euploid. Compare ANEUPLOID.

Euryarchaeota See ARCHAEA.

euryhaline Able to tolerate a wide variation of osmotic pressure of environment. Compare STENOHALINE, OSMOREGULATION.

Eurypterida Fossil subclass of the MEROSTOMATA, appearing in the Ordovician. Free-swimming, marine, brackish and freshwater forms; prosoma with six pairs of ventral appendages, the first being CHELICERAE, the others modified for grasping, walking and swimming. Larva resembled trilobite larva of king crab. Active predators, about two metres in length. See ARACHNIDA.

eurythermous Able to tolerate wide variations of environmental temperature. Compare STENOTHERMOUS.

eusocial Term applied generally to certain colonial insects which exhibit cooperative brood care, overlap between generations, and reproductive and sterile worker CASTES. Includes termites, bees, ants and wasps.

eusporangiate (Of sporangia in vascular plants) arising from a group of parent cells and possessing a wall of two or more layers of cells. Spore production greater than in the LEPTOSPORANGIATE type.

Eustachian tube Tube connecting middle ear to pharynx in tetrapod vertebrates. Allows equalization of air pressure on either side of eardrum. See EAR, MIDDLE; SPIRACLE.

eustele Stele in which primary vascular tissues are arranged in discrete strands around a pit.

Eustigmatophyceae Unicellular algae (about seven genera and twelve species previously classified in the Xanthophyceae) of the division Heterokontophyta that live in freshwater, or marine environments (e.g. *Nannochloropsis*) or on damp soil. Small numbers of zoospores are produced that mostly have a single emergent flagellum; but a second basal body is present, indicating a biflagellate ancestry. Where two flagella are present both are inserted near the apex of the cell. The emergent flagellum bears tripartite tubular hairs. The transition zone of the flagellum contains a transitional helix. The eyespot is prominent at the anterior end of the cell. It lies outside the chloroplast and comprises several carotenoid-containing globules not enclosed by membranes, either individually or as a group. Appressed to the cell above the eyespot is a wing-like basal expansion emergent flagellum. One or more yellow-green chloroplasts occur in each cell with chlorophyll *a* and β-carotene present with two major xanthophylls (violaxanthin and vaucheriaxanthin). Violaxanthin is the main pigment involved in light-harvesting, a role it plays in apparently no other group of photosynthetic organisms. Chloroplasts have three thylakoids per band with no girdle band beneath the chlorophyll envelope. Two membranes of chloroplast endoplasmic reticulum surround the chloroplast. In coccoid, non-flagellate cells it is continuous with the nuclear envelope. In coccoid vegetative cells, the chloroplast generally bears a stalked, angular pyrenoid on its inner side. Chloroplast DNA is organized into numerous nucleoids, which may be united to form a reticulum.

The particular combination of photosynthetic pigments, the unique type of photoreceptor apparatus and the absence of a girdle band (lamella) in the chloroplasts serve to differentiate this group of algae from other heterokontophytes. This evidence has been used to justify the classification of the Eustigmatophyceae as a division.

Eustigmatophyta Protistan algae; basically unicellular, living in freshwater or on damp soil, producing a small number of zoospores, most with a single emergent flagellum; but a second basal body is present, indicating a biflagellate ancestry. The emergent flagellum has microtubular hairs and is inserted subapically. Named after the large, orange-red EYESPOT at the anterior end of the zoospore, independent of the chloroplast (main difference from the XANTHOPHYTA). Chlorophyll *a* and β-carotene are present, with two major xanthophylls (violaxanthin and vaucheriaxanthin). Violaxanthin is the major light-harvesting pigment in the Eustigmatophyta. Chloroplasts have three thylakoids per band with no girdle band beneath the chloroplast envelope. Two membranes of CHLOROPLAST ENDOPLASMIC RETICULUM surround the chloroplast; but there is no connection between this and the outer nuclear membrane. Vegetative cells normally possess a characteristic polygonal pyrenoid, although this is absent in the zoospores.

Eutheria (Placentalia) Placental mammals. Infraclass of the MAMMALIA, and the dominant mammals today. Most of the 3,800 species occur within about six orders:

Insectivora (e.g. shrews, hedgehogs), Chiroptera (bats), Rodentia (e.g. mice, rats), Artiodactyla (e.g. deer, pigs), Carnivora (e.g. cats, dogs, weasels) and Primates (e.g. lemurs, monkeys, apes, humans). Appear in Upper Cretaceous, at time of dinosaur extinction. Connection between embryo and uterus intimate and complex; amnion and chorion present; umbilicus links embryo to chorio-allantoic PLACENTS; scrotum posterior to penis. Gestation period of varying length; newborn young more advanced developmentally than in other mammals. Great ADAPTIVE RADIATION in early Cenozoic. See PROTOTHERIA, METATHERIA.

eutrophic (Of lakes and rivers) originally introduced to describe the phytoplankton assemblages characteristic of 'lowland lakes', which received a rich source of nutrients (e.g. phosphorus, nitrogen). Thus, the original use of the term implied variation in nutrient content. Nowadays, eutrophic is used in a more general sense to describe a lake which has a high concentration of nutrients; highly productive in terms of the organic matter produced. Compare DYSTROPHIC, MESOTROPHIC, OLIGOTROPHIC.

eutrophication Usually rapid increase in the nutrient status of a body of water, both natural and occurring as a by-product of human activity. May be caused by run-off of artificial fertilizers from agricultural land, or by input of sewage or animal waste. May occur when large flocks of migrating birds collect around watering holes. Leads to reduction in species diversity as well as change in species composition, often accompanied by massive growth of dominant species. Excessive production stimulates respiration, increasing dissolved oxygen demand and leading to anaerobic conditions, commonly with accumulation of obnoxious decay and animal death. Artificial eutrophication can be slowed or even reversed by removal of nutrients at source, but may require costly sewage treatment plants. See CYANOBACTERIA.

evapotranspiration Water loss to the atmosphere from soil and vegetation. The potential evaporation may be calculated from physical features of the environment such as incident radiation, wind speed and temperature. The actual evapotranspiration will commonly fall below the potential depending on the availability of water from precipitation and soil storage. See TRANSPIRATION.

evergreen (Of plants) bearing leaves all year round (e.g. pine, spruce). Contrasted with DECIDUOUS.

evocation Ability of an inducer to bring forth a particular mode of differentiation in a tissue which is *competent*. It has been suggested that the inducer brings about release of a substance (the *evocator*) which initiates the differentiation. See INDUCTION, ORGANIZER.

'evo-devo' Evolutionary developmental biology. A term often used to gloss the principle that developmental processes can provide insight into the evolution of anatomy – most notably of BODY PLANS. See ULTRABITHORAX GENE.

evolution (1) *Microevolution*: changes in appearance of populations and species over generations. (2) *Macroevolution* or *phyletic evolution*: origins and EXTINCTIONS of species and grades (see SPECIATION).

Microevolution includes changes in mean and modal phenotype, morph ratios, etc. such as occur within populations from one generation to the next. When statistically significant changes in such variables (or the genes responsible for them) occur with time, a population may be said to evolve. Evidence suggests that African elephant populations subjected to intense poaching are evolving a tusk-less condition, and that intensively fished cod populations are evolving life cycles with earlier breeding. The case of INDUSTRIAL MELANISM in the peppered moth is well documented. Macroevolution includes large-scale phyletic change over geological time (e.g. successive origins of crossopterygian fish, amphibians, reptiles, birds and mammals), as well as extinctions of taxa within such groups. It is usually accepted that causes of evolutionary change include NATURAL SELECTION and GENETIC DRIFT, and that macroevolutionary change can be explained by the same factors that bring about microevolution. Evolution by natural selection is often regarded as 'short-sighted': intermediates cannot be explained by arguing that the end result is advantageous, SO PREADAPTATION or CO-OPTION are often invoked.

Debate has recently centred upon the rate of evolutionary change. Some biologists accept that evolution largely occurs by gradual ANAGENESIS; others stress the role of CLADOGENESIS and take the view that species persist unchanged for considerable periods of time, and that relatively rapid speciation events punctuate the fossil record (*punctuated equilibrium*) (see FOUNDER EFFECT). Darwin considered both to be possibilities. At the molecular level, controversy centres on the respective influences in evolution of random alterations in genetic material (the *neutralist* view) and of selective changes (the *selectionist* view). See MOLECULAR CLOCK. Opposed to evolutionary explanations of the composition of the Earth's fauna and flora is the group of views termed 'SPECIAL CREATIONISM', which holds that there are no bonds of genetic relationship between species, past or present. See ORIGIN OF LIFE.

Although Anaximander (6th. cen. BC), Empedocles (5th. cen. BC) and Aristotle (4th. cen. BC) all held evolutionary views of some kind, they depended more on *a priorism* than on observation and testable theory. LAMARCK is often considered the most influential evolutionary thinker prior to Charles DARWIN and Alfred WALLACE but his theory was very different from theirs. They themselves drew apart on the question of human origins and the role of sexual selection.

Evidence for common descent and the fact of evolution comes principally from molecular biology (see DNA HYBRIDIZATION, ELECTROPHORESIS, GENETIC CODE, MOLECULAR CLOCK), comparative biochemistry, comparative morphology (e.g. anatomy and embryology), geographical distributions of organisms and FOSSIL records. The modern theory of evolution (NEO-DARWINISM) derives largely from the kind of genetical knowledge which Darwin lacked, principally the occurrence of Mendelian segregation,

which helps explain how variations can be maintained in populations. Evidence for microevolution and Darwinian natural selection (amounting to his 'special theory of evolution') stems largely from population genetics (e.g. see INDUSTRIAL MELANISM), although Darwin himself drew heavily on the analogy of ARTIFICIAL SELECTION. See NATURAL SELECTION.

evolutionarily stable strategy (ESS) A heritable strategy (commonly but by no means always behavioural) which, if adopted by (expressed in) most members of a population, cannot be supplanted in evolution by an alternative (mutant) strategy. The strategy may be complex and involve a variety of different sub-responses in accordance with environmental changes, not least other organisms' behaviours. See GAME THEORY.

evolutionary taxonomy A school of biological CLASSIFICATION which makes use of both phenetic and phylogenetic data in classifying organisms. Because there is no theoretical guide as to when one approach should be used and when the other, this very influential school has been criticized by adherents of CLADISTICS. See PARALLEL EVOLUTION.

evolutionary transformation series A pair of HOMOLOGOUS characters, one derived directly from the other. See PLESIOMORPHOUS, APOMORPHOUS, CLADISTICS.

evolution of sex See SEX (2).

exact compensation DENSITY-DEPENDENCE in which increases in initial density are exactly counterbalanced by increases in death rate and/or decreases in birth rate and/or growth rate, such that the outcome is the same irrespective of initial density.

exarch Type of maturation of primary xylem in roots, in which the oldest xylem elements are located closest to the outside of the axis. Compare ENDARCH.

excretion (1) Any process by which an organism gets rid of waste metabolic products. Differs from EGESTION in that wastes removed are products of the organism's cells rather than simply undigested wastes; and from SECRETION since substances produced would generally be harmful if allowed to accumulate, and as a rule have no intrinsic value to the organism. The simplest excretory method is passive diffusion, either through the normal body surface or across organs with enlarged surface areas (gills, lungs). These may be supplemented or replaced by internal excretory organs, particularly where the body surface cannot be used. Excretory organs typically remove metabolic products from interstitial fluids (e.g. lymph, blood plasma). The gut occasionally serves as a route for excretory products, but is not an excretory organ. Nitrogenous excretion is usually in the form of ammonia (aquatic environments), urea (terrestrial environments) or uric acid (environments where water is at a premium). Common invertebrate excretory organs include FLAME CELLS, NEPHRIDIA, and MALPIGHIAN TUBULES, but in some cases (e.g. large crustaceans) excretion may be deposited in the exoskeleton, commonly to be lost during moulting. Vertebrate KIDNEYS work by filtration and selective reabsorption, and, like some invertebrate excretory organs, also have roles in OSMOREGULATION.

Excretion in plants includes GUTTATION and removal by diffusion of excess oxygen produced by photosynthesis, since oxygen may inhibit that process. Leaf fall also removes a number of metabolic wastes. (2) A substance, or mixture of substances, excreted: *excreta*.

exercise See CORONARY HEART DISEASE, MUSCLE CONTRACTION.

exergonic (Of a chemical reaction) yielding energy. See THERMODYNAMICS.

exine Outer layer of spores and pollen grains; usually divided into two main layers: an outer ectexine and an inner endexine. Often composed of SPOROPOLLENIN.

exobiology The study of extra-terrestrial life. See ORIGIN OF LIFE.

exocrine gland Any animal gland of epithelial origin which secretes, either directly or most commonly via a duct, on to an epithelial surface. See GLAND, ENDOCRINE GLAND.

exocytosis Process whereby a vesicle (e.g. secretory vesicle), often budded from the ENDOPLASMIC RETICULUM or GOLGI APPARATUS, fuses with the plasma membrane of the cell, with release of vesicle contents to exterior. Common process in SECRETION. When restricted to anterior region of cell it is an important stage in much eukaryotic CELL LOCOMOTION. Compare ENDOCYTOSIS. See SYNAPTIC VESICLES.

exodermis Layer of closely fitting cortical cells with suberized walls, replacing the withered piliferous layer in older parts of roots.

exogamy See OUTBREEDING.

exon Used in two senses: (i) any sequence of DNA represented by its RNA equivalent in mRNA (i.e. after RNA PROCESSING); (ii) any DNA sequence encoding and giving rise to a translated polypeptide sequence (often a protein DOMAIN). Exons alternate with INTRONS in most eukaryotic, and some prokaryotic, genes. See EXON SHUFFLING.

exon shuffling Recombination between exons in different genes, probably usually brought about by TRANSPOSABLE ELEMENTS. If two transposable elements of the same type insert close together, the next transposition may move the two closest ends of the separate transposons along with any exon(s) lying between them to a new site. This insertion would bring together, after intron excision, exons which had not been associated before. This could explain how modular proteins, containing DOMAINS with considerable sequence homology in other proteins, evolve.

exonuclease Enzyme which removes nucleotides one by one from the end of a polynucleotide chain. See DNase.

exopeptidase Proteolytic enzyme which removes amino acids one by one from the end of a protein molecule. Compare ENDOPEPTIDASE.

expodite See BIRAMOUS APPENDAGE.

Exopterygota   (Heterometabola) Winged insects with incomplete metamorphosis; sometimes regarded as a subclass of the INSECTA. No pupal stage. Wings develop outside the body; successive larvae (nymphs) become progressively adult-like. Includes palaeopteran orders Ephemeroptera and Odonata; orthopteroid orders Plecoptera, Grylloblattoidea, Orthoptera, Phasmida, Dermaptera, Embioptera, Dictyoptera, Isoptera and Zoraptera; and hemipteroid orders Psocoptera, Mallophaga, Siphunculata, Hemiptera and Thysanoptera. See ENDOPTERYGOTA.

exoskeleton Skeleton covering the outside of the body, or located in the skin. In arthropods (see CUTICLE), secreted by the epidermis; in many vertebrates, e.g. tortoises, armadillos, the exoskeleton consists of bony plates beneath the epidermis. Many primitive jawless vertebrates (ostracoderms) and primitive jawed vertebrates (placoderms) had body armour comprising bony skin plates and scales. The scales and denticles of modern fish are remnants of this.

exosome Versatile multienzyme complex of $3'$-to-$5'$ RNA exonucleases involved in processing eukaryotic RNAs, occurring in both nuclear and cytoplasmic matrices.

exotoxins Toxins released by a microorganism into surrounding growth medium or tissue during *growth phase* of infection. Generally inactivated by heat and easily neutralized by specific antibody. Produced mainly by Gram-positive bacteria, such as the agents of botulism, diphtheria, *Shigella* dysentery and tetanus. *Enterotoxins* are exotoxins acting on the small intestine, usually causing massive secretion of fluid and leading to diarrhoea (see ADENYLYL CYCLASE). They are produced by a variety of food-poisoning organisms, such as *Staphylococcus aureus*, *Clostridium perfringens*, *Bacillus cereus*, *Vibrio cholerae*, *Salmonella enteriditis* and *Escherichia coli*. The alga *Prymnesium parvum* forms a potent exotoxin that causes extensive fish mortalities in brackish water conditions in many countries in Europe and in Israel. See AFLATOXINS, ENDOTOXINS, POISONS.

expansins A class of protein which seems to have evolved in the land plant lineage, with functions in CELL GROWTH and other situations where the movement, adhesion and enzymatic access to wall polysaccharides are involved.

experiment The intentional manipulation of material conditions so as to elicit an answer to a question, often posed in the form: what is the effect of x on y? The aim of the experimenter is to isolate x as the only free variable, keeping constant all other variables which might affect the value of y. Values of x can then be paired off with the values of y, when changes in x are said to be the cause of any changes in y. A similar approach compares the results of two experimental situations differing in just one initial condition, which often has zero value in one of the experimental situations (called the *control*) but is allowed free range over its values in the other experimental situation (called the *experiment*). The effects of this free-ranging variable are then compared with the effect of its absence (zero value), and since it is the only independent variable, any differences in effect can be said to have been caused by changes in its value. Controlled experiments must have this comparative element. The rationale is to eliminate all possible alternative causes of effects save the one under investigation. Without such controlled experiments, the material causes of phenomena could never be ascertained. It is often assumed, not always with justification, that methods used to study biological material do not themselves affect the properties being studied.

explanation A phenomenon may be said to have been fully explained when all its component parts can be formally deduced as consequences of sets of actual initial conditions (the minor premises) satisfying the terms of whichever general law (the major premise) represents our most inclusive summary of the relevant experimental data to date. Attempts to explain biological phenomena solely in terms of the language employed in physics and chemistry exemplify what is termed *reductionism*. Most people believe this can only be achieved if terms peculiar to biology can be 'paired off' by identity or equivalence relations to terms in the physical sciences. It is highly contentious whether this can be achieved, even in principle.

explantation See TISSUE CULTURE.

exploitation competition Competition in which any adverse effects on an organism are brought about by reductions in resource levels caused by other competing organisms.

exploiter-mediated coexistence Where predation promotes the coexistence of species amongst which there would otherwise be competitive exclusion.

exponential growth Growth of cells, populations, etc., in which the numbers double during each unit time period, rate of increase depending only upon the number of individuals and their potential net reproductive rate. In other words, no competition occurs between individuals for resources nor is there any other detrimental effect of individuals upon one another. Such a situation is characteristic of the initial growth phase of microorganisms in cultures, or of organisms introduced into regions where food is not limiting and where natural controls (e.g. predators, parasites) are absent. The exponential growth curve can be defined by the equation:

$$N_t = N_o e^{(b-d)t}$$

where t is a very short time interval
     N_t is the number of individuals after time t
     N_o is the number of individuals at the beginning of the time interval
     b is the 'birth' rate during time t
     d is the 'death' rate during time t
     e is a constant, taken for convenience to be the base of Napierian logarithms, 2.718 (the exponential constant).

When cells grow exponentially in culture, the time taken for cell numbers to double is the mean generation time. Once this time is established (t minutes) the *exponential growth rate constant*, k, is defined as

$$k = 60/t$$

expressed sequence tags (ESTs) cDNA sequences which act as markers for genes expressed in particular tissues. Instead of sequencing the entire cDNA molecule, only a few hundred bases are sequenced, saving time.

expression signals The molecular signals which may need to be spliced upstream of a gene (or its cDNA equivalent) for it to be properly transcribed (initiated and terminated) and translated (i.e. expressed) in a different organism. These may take the form of an 'expression casette' in a plasmid VECTOR containing all appropriate signals, such as a strong PROMOTER (often patented) and sequences to achieve high copy number, proper ribosome binding and correct targeting of product (see PROTEIN TARGETING). Such signal sequences are needed when, e.g., a product protein has to be secreted from the host cell so that it can be easily extracted from the culture (e.g. yeast) in cell-free solution; but these signals may need subsequent removal (if not done by the host), e.g. to avoid immunogenicity. Eukaryotic genes for expression in prokaryotic hosts need a prokaryotic promoter and must be in the form of cDNA to avoid problems of RNA PROCESSING. See YEAST ARTIFICIAL CHROMOSOME.

expression system Any host cell into which has been placed a gene, or cDNA, so that it will be expressed (i.e. transcribed and translated). EXPRESSION VECTORS are often a part of such a system.

expression vector A cloning VECTOR with the required regulatory sequences to enable transcription and translation of a cloned gene, genes, or cDNA. See GENE MANIPULATION.

expressivity The level to which a given genotype is expressed in an individual's phenotype. Compare PENETRANCE. See GENOMIC IMPRINTING, MODIFIER.

extensor Muscle or tendon straightening a joint, antagonistic to FLEXOR.

exteroceptor A RECEPTOR detecting stimuli emanating from outside an animal. Compare INTEROCEPTOR.

extinction Termination of a genealogical lineage. Used most frequently in the context of species, but applicable also to populations and to taxa higher than species. Agents of 'background rate' extinction include competition, predation (e.g. see SPECIES FLOCK) and disease, alteration of habitat and random fluctuations in population size. The extinction of so many large terrestrial animals during the QUATERNARY has been interpreted as the result of climatic change, human hunting, or a combination of both. The Australian megafauna extinction (~60 vertebrate species lost) started with human arrival there between 60–53 Kyr BP. The absence of significant climate change in Australia at that time, or of similar extinctions in the New Zealand megafauna (then uncolonized by humans), seems to favour human hunting as the cause.

Although there are doubts about two of the major *mass extinctions* (some say those of the Late Ordovician and Devonian are better termed 'mass depletions'), it is generally held that there have been five such periods, when the Earth's fauna suffered extinction rates far higher than the normal background rate. These occurred in the Ordovician, the late Devonian, the late Permian (225 Myr BP), the late Triassic (190 Myr BP) and the late Cretaceous/Tertiary (K/T, 57 Myr BP). The end of the Permian saw the greatest of these mass extinctions, ~90% of all oceanic species, and $>\frac{2}{3}$ of terrestrial reptiles and amphibians, disappearing in its last several million years. Possible causes of greater than normal extinction rates include evolutionary competition, geological (e.g. vulcanism, plate tectonics, deep-sea warming) and climatic change and cometary or other impact. Victims of the K/T extinction included the dinosaurs and 60–75% of all marine species, and evidence (high iridium levels and soot in K/T boundary clays; glass fragments in the correct trajectory path; crater of correct size off Chicxulub, in the Yucatán Peninsula) indicates that meteoritic impact was responsible. Landing where it did, it would have released huge volumes of nitrogen oxides and sulphur oxides into the atmosphere,

causing severe acid rain and reducing surface temperatures. This, combined with the massive volcanic activity in the Deccan region of western India, caused all land animals larger than about 25 kg to become extinct in less than a million years; in the seas, mososaurs became extinct; and surviving animals tended to be those forming parts of detritus food chains rather than grazing food chains. Evidence for impact as the cause of other mass extinctions, notably at the end of the Permian, is increasing. One likely genetic factor in extinction as population size decreases is *inbreeding depression* (see PUNCTUATED EQUILIBRIUM). There is no evidence that any plant species has ever been driven to extinction by competition with another plant species, but plenty of evidence that animals can extinguish plant populations and entire plant species. However, the evidence for mass extinctions in plants is more problematic than for animals.

The first DNA extracted from an extinct species was achieved in 1982, for mitochondrial DNA from the hide of a once-common zebra (the quagga, *Equus guagga*) and shown to resemble that of the extant Burchell's zebra (*E. burchelli*) (see DNA).

extracellular In general, occurring outside the plasma membrane; but where a CELL WALL is present, often refers to the region surrounding this. See GLYCOCALYX.

extracellular matrix (ECM) Complex network of macromolecules lying between cells where these form tissues and colonies, and in cell walls and cuticles. Comprises mainly locally secreted proteins including, in animals, structural collagens and elastin; adhesive FIBRONECTIN and LAMININS (see ADHESION, INTEGRINS) and polysaccharides (e.g. PROTEOGLYCANS, GLYCOSAMINOGLYCANS). The ECM may contain bound growth factors and matrix metalloproteinases (MMPs) – secreted and membrane-anchored proteinases – which, by loosening up the dense meshwork of matrix molecules, may promote both normal physiological and tumorigenic invasion of the connective tissue stroma (see CANCER). Forms BASAL LAMINA in animals between epithelium and under-

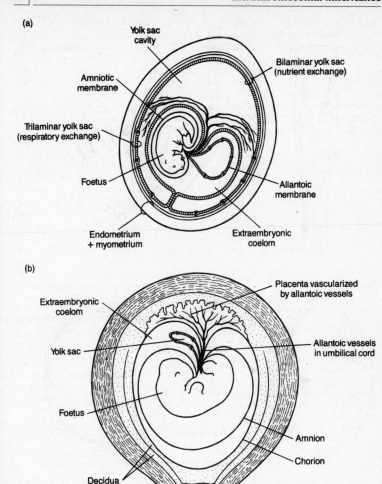

(a)

Yolk sac cavity

Bilaminar yolk sac (nutrient exchange)

Amniotic membrane

Trilaminar yolk sac (respiratory exchange)

Foetus

Allantoic membrane

Endometrium + myometrium

Extraembryonic coelom

(b)

Placenta vascularized by allantoic vessels

Extraembryonic coelom

Allantoic vessels in umbilical cord

Yolk sac

Foetus

Amnion

Chorion

Decidua

FIG. 54 *Extraembryonic membranes during development of (a) wallaby (marsupial) and (b) human. The uterus wall is outermost in both.*

lying CONNECTIVE TISSUE. Houses many local signalling molecules profoundly influencing cell division, differentiation, growth (see GROWTH FACTORS) and APOPTOSIS – the latter because detachment of cells from the stroma or adjacent cells deprives them of cadherin- and integrin-mediated SURVIVAL FACTORS which prevent it.

extrachromosomal inheritance Inheritance of genetic factors not forming part of a chromosome. Examples include PLASMID, mitochondrial and chloroplast inheritance. Inheritance of a variety of intracellular symbionts may also be regarded as extrachromosomal. See CYTOPLASMIC INHERITANCE, EPISOME.

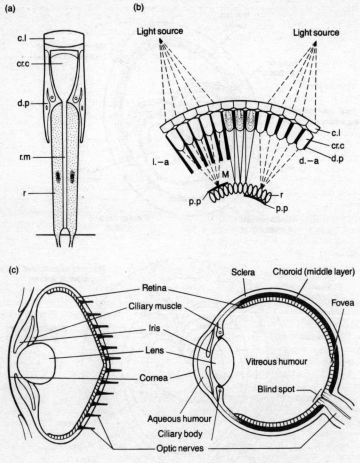

FIG. 55 *An ommatidium from an insect compound apposition eye (b) A superposition compound eye, light-adapted (l.-a) and dark-adapted (d.-a). The central three ommatidia show the proposed laminated structure of the cones; c.l. = corneal lens, cr.c. = crystalline cone, d.p. = distal pigment cell, p.p. = proximal pigment cell, r = retinula cell (photoreceptor), r.m. = rhabdome. (c) Comparison of cephalopod eye (left) with human eye (right).*

extraembryonic coelom In amniote development, the space lying between the mesoderm layers lining inner surface of the chorion and outer surface of the amnion.

extraembryonic membranes The YOLK SAC, CHORION, AMNION and ALLANTOIS of amniote vertebrates; membranes derived from the zygote but lying outside the epidermis of the embryo proper. Have played a major part in evolution of vertebrate terrestrialization. See Fig. 54.

extravasation Creeping, by diapedesis, of monocytes, some other leucocytes, and

CANCER CELLS, through the endothelium of
small blood vessels into the tissue fluid.
Involves ADHESION through INTEGRINS on cell
surfaces to selectins in the endothelium; see
INFLAMMATION.

extremophile Any microbe (mainly certain bacteria, fungi and protists) inhabiting
an extreme environment, such as a hydrothermal vent, volcanic spring, the interior
of an exposed rock, or even subsurface rocks.
Deep-living microbes inhabit oceanic and
continental crust, especially sedimentary
formations; but 'SLiMEs' (subsurface lithotrophic microbial ecosystems) exist between the mineral grains of many igneous
rocks.

One extremophile enzyme (extremozyme), Taq polymerase, is at the heart of
the POLYMERASE CHAIN REACTION procedure
and other industrial enzymes are being
actively sought from these organisms. See
PSYCHROPHILIC, THERMOPHILIC.

eye Sense organ responding to light. In
invertebrates, either a simple scattering of
light-sensitive pigment spots in the general
epithelium but more often comprising an
optic cup of receptor cells with screening
pigment cells (each functional unit an *ocellus*), lacking a refractive surface so that no
image can be formed. Although a lens may
be present, most ocelli can only differentiate
between light and dark. Nonetheless, this
enables orientation with respect to light
direction and intensity.

The basic unit of the arthropod compound eye is the *ommatidium*, comprising
a cornea lens, crystalline cone, a group of
usually 7–8 sense (retinula) cells radially
arranged around a central rhabdome
formed from their innermost fibrillar surfaces (rhabdomeres composed of microvilli), in which the light-sensitive pigment is
located, each rhabdome extending into a
nerve fibre distally.

Higher molluscan (i.e. cephalopod) eyes

(e.g. of *Octopus*) resemble those of vertebrates in complexity (see CONVERGENCE);
however, there is an ommatidium-like
organization in the retina. For details of the
vertebrate eye, see Fig. 55c and entries for
structures labelled. See also TAPETUM.

The *Pax 6* genes appear to be master control genes for eye morphogenesis in various
animal phyla, and are capable of inducing
ectopic eye development when substituted
(e.g. mouse *Pax 6* induces ectopic compound eyes in *Drosophila*), but it remains to
be seen how much else of the morphogenetic pathway is monophyletic.

eye muscles (a) *Extrinsic* (outside eyeball).
In vertebrates six such muscles rotate the
eyeball: a pair of anterior oblique and four,
more posterior, rectus muscles. Supplied by
cranial nerves III, IV and VI. (b) *Intrinsic*
(inside eyeball); see IRIS, CILIARY BODY.

eyespot (stigma) (1) Light-sensitive pigment spots of some invertebrates. See EYE.
(2) Rather a misnomer for orange to red-coloured lipid droplets or globules close to
or within the chloroplasts of some eukaryotic flagellates. The colour is imparted by
carotenoid pigments. In the green algae
(CHLOROPHYTA), for example, the eyespot is
always within the chloroplast situated
anteriorly near the flagellar bases, and the
lipid droplets (one to several layers) are
contained in the stroma between the
chloroplast envelope and the outermost
thylakoids. In other algae (e.g. DINOFLAGELLATES) the lipid droplets may be in the cytoplasm, and not surrounded by a membrane
or in a plastid-like structure, while the most
complex eyespot comprises a lens mounted
in front of a pigment cup (e.g. the dinoflagellate *Nematodinium armatum*). In flagellates, it is highly probable that the 'eyespot'
casts a shadow on the presumably light-sensitive swelling at the flagellar bases, the
flicker frequency indicating the angle of the
cell rotation with respect to the light source.
See SIGNAL TRANSDUCTION.

F

F₁ (first filial generation) Offspring obtained in breeding work after crossing the parental generation (P_1) or by selfing one or more of its members.

F₂ (second filial generation) Offspring obtained after crossing members of the F_1 generation or by selfing one or more of its members.

facial nerve See CRANIAL NERVES.

facilitated diffusion Carrier-mediated transport across CELL MEMBRANES, the transported molecule never moving against a concentration gradient. Only speeds up rate of equilibrium attainment across membrane. Examples include transport of glucose across plasma membranes of fat cells, skeletal muscle fibres, the microvilli of ileum mucosa and across proximal convoluted tubule cells of vertebrate kidneys. ATP hydrolysis is not involved. Compare ACTIVE TRANSPORT. See TRANSPORT PROTEINS.

facilitation (1) Increase in responsiveness of a postsynaptic membrane to successive stimuli, each one leaving the membrane more responsive to the next. Compare temporal SUMMATION. (2) Social facilitation. The increased probability that other members of a species will behave similarly once one member has acted in a certain way. See NERVOUS INTEGRATION.

facultative Indicating the ability to live under altered environmental conditions or to behave adaptively under markedly changed circumstances. Thus, a *facultative parasite* may survive in the free-living or parasitic mode (see MIXOTROPH); a *facultative anaerobe* may survive aerobically or anaer-

obically; a *facultative apomict* may reproduce either by apomixis or by more conventional sexual means. A *facultative annual* is a plant, which in some circumstances completes its life cycle within twelve months.

facultative mutualism The condition in which one or both species in a mutualistic association may survive and maintain populations in the absence of the other partner.

FAD (flavin adenine dinucleotide) PROSTHETIC GROUP of several enzymes (generally flavoproteins). Derived from the vitamin riboflavin and involved in several REDOX REACTIONS, e.g. as catalysed by various dehydrogenases (e.g. NADH dehydrogenase, succinate dehydrogenase) and oxidases (e.g. xanthine oxidase, amino acid oxidase). Reduced flavin dehydrogenase (FD, see ELECTRON TRANSPORT SYSTEM) can reduce methylene blue:

$$FD-FADH_2 + \text{methylene blue} =$$
$$\text{(blue)}$$

$$FD-FAD + \text{methylene blue}_{red} \, .$$
$$\text{(colourless)}$$

faeces See EGESTION.

fairy rings Fairy rings are produced by some 60 species – generally members of the BASIDIOMYCOTA. They frequently occur in lawns and grassland. Three main types exist: (i) those in which the development of sporocarps has no effect on the vegetation (e.g., *Lepista sordida*, myxomycete rings), (ii) those in which there is increased growth of vegetation (e.g., *Calvatia Agaricus praerimosus*, *Leucopaxillus giganteus*, *Calocybe* *gambosa*). Rings are started from a mycelium that grows at the outer edge of the ring. This underground dikaryotic mycelium develops radially from its original starting point marking the form and extent of the underground mycelium. Some are estimated to be up to 500 years old. The grass immediately inside the ring becomes stunted and a lighter green than outside it.

Fallopian tube In female mammals, the bilaterally paired tube with funnel-shaped opening just behind ovary, leading from perivisceral cavity (coelom) to uterus. By muscular and ciliary action it conducts eggs from ovary to uterus. Is frequently the site of fertilization. Represents part of MÜLLERIAN DUCT of other vertebrates.

false annulus Discrete grouping of thick-walled cells on the jacket of some fern sporangia, not directly influencing dehiscence.

family Category employed in biological CLASSIFICATION, below order and above genus. Typically comprises more than one genus. Familial suffixes normally end -aceae in botany and -idae in zoology. Term also employed in the wider taxonomy of proteins, e.g. the protein family encoded by *CDC* GENES, and the families of PROTEIN KINASES.

Fas A cell-surface receptor, one of the main triggers of APOPTOSIS in cells of the immune system and possibly in cancers and other noninfectious diseases. Apoptosis occurs when the ligand FasL binds Fas. Not to be confused with RAS PROTEINS or Fos (see *C-FOS*).

fascia Sheet of connective tissue, as enclosing muscles.

fasciation Coalescing of stems, branches, etc., to form abnormally thick growths.

fascicle (1) Bundle of pine leaves or other needle-like leaves of gymnophytes. (2) Now obsolete term, formerly applied to vascular bundle.

fascicular cambium Cambium that develops within a vascular bundle.

fast fibres See STRIATED MUSCLE.

fat (neutral fat) Major form of LIPID store in higher animals and some plants. Commonly used synonymously with TRIGLYCERIDE, which not only stores more energy per gram than any other cell constituent ($2\frac{1}{2}$ times the ATP yield of glycogen) but, being hydrophobic, requires less water of hydration than polysaccharide and is therefore far less bulky per gram to store. ADIPOSE TISSUE is composed of cells with little besides fat in them. Hydrolysed by lipases to yield fatty acids and glycerol. See CHYLOMICRON.

fat body (1) Organ in abdomen of many amphibia and lizards containing ADIPOSE TISSUE, used during hibernation. (2) In insects, diffuse tissue between organs, storing fat, protein, occasionally glycogen and uric acid.

fate map Diagram showing future development of each region of the egg or embryo. A series of such maps indicates the trajectories of each part from egg to adult. Construction may involve vital staining, cytological and genetic markers. Most easily constructed in cases of highly MOSAIC DEVELOPMENT.

fatty acid Organic aliphatic and usually unbranched carboxylic acid, often of considerable length. Condensation with glycerol results in ester formation to form mono-, di-, and triglycerides (fat). Commonly a component of other LIPIDS. Free fatty acids are transported in blood plasma largely by albumin (but see LIPOPROTEINS). Saturated fatty acids have no double bonds; monounsaturated have one and polyunsaturated more than one. Classification of polyunsaturated fatty acids into families is based on how many carbon atoms the last double bond is from the methyl ($-CH_3$) end of the molecule: thus *n*-3, *n*-6 and *n*-9 describe families where the last double bond is 3, 6 and 9 carbon atoms from the methyl end respectively. (See Fig. 56.) FATTY ACID OXIDATION makes an important contribution to a cell's energy release. Some types of unsaturated fatty acids stored in a cell's membranes can be released and transformed into local cell-signalling EICOSANOIDS. A diet lacking in ESSENTIAL FATTY ACIDS results in changes in the fatty acid compositions of cell membranes leading to their malfunction. Pregnant

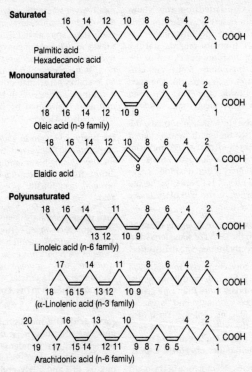

Saturated

Palmitic acid
Hexadecanoic acid

Monounsaturated

Oleic acid (n-9 family)

Elaidic acid

Polyunsaturated

Linoleic acid (n-6 family)

(α-Linolenic acid (n-3 family)

Arachidonic acid (n-6 family)

FIG. 56 FATTY ACID *structures (methyl end to left). Double bonds indicated by parallel pairs of lines; carbon atoms numbered.* (From *Human Nutrition and Dietetics* (9th edn) by J. S. Garrow and W. P. T. James. Copyright © Churchill Livingstone, 1993.)

women should ensure their intake of linoleic and α-linolenic acids is sufficient to allow proper foetal brain cell production (50% of brain mass is due to lipids), and subsequently for proper milk production. These two fatty acids are required for production of the small amounts of the long-chain polyunsaturated fatty acids arachidonic and docosahexanoic acids present in human milk and required for active neonatal brain development. Saturated fatty acids include *palmitic acid*, $CH_3(CH_2)_{14}COOH$ and *stearic acid*, $CH_3(CH_2)_{16}COOH$; unsaturated fatty acids include *oleic acid*, $CH_3(CH_2)_7CH:CH(CH_2)_7COOH$. See FATTY ACID OXIDATION, LIPASE.

fatty acid oxidation (beta-oxidation)
Prior to oxidation, fatty acids undergo a complex activation in the cytosol followed by transport across the mitochondrial membranes, whereupon the actyl group binds to COENZYME A to form a fatty acyl-CoA thioester. The terminal two carbon atoms are removed enzymatically (forming acetyl-CoA, for entry into the KREBS CYCLE) while another CoA molecule is bound to the remaining fatty acid chain. Each sequential 2-carbon removal is accompanied by dehydrogenation and production of reduced NAD for entry into the ELECTRON TRANSPORT SYSTEM of mitochondria and ATP production. See VITAMIN E.

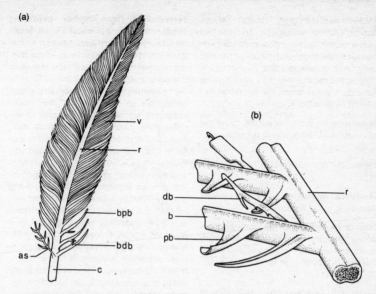

FIG. 57 *(a) Generalized contour feather; as = aftershaft, bdb = barb with distal barbule, bpb = barb with proximal barbule, c = calamous or quill, r = rachis, v = one side of vane. (b) Enlarged view of part of rachis of contour feather seen from dorsal side. Three proximal barbules (pb) have been cut short. These and the distal barbules (db) are less widely spaced than shown here. b = barb.*

feathers Elaborate and specialized epidermal productions characteristic now of birds (AVES) as a Class, but known to have evolved in dinosaurs before birds appeared. The original selective value of feathers was clearly not the provision of lift and thrust for wings during flight. It may have been, as it remains in birds today, the provision of thermal insulation, colouration, courtship, camouflage or defence.

They develop from *feather germs*, minute projections from the skin, within which longitudinal ridges of epidermal cells (*barb ridges*) form early on. On each ridge further cells of appropriate shape and position form the *barbules* after keratinization, some cell processes becoming *barbicels*, or hooks. A deep pit in the epidermis, the *feather follicle*, surrounds the bases of feathers. The first feathers are *down feathers* in which the quill is very short, a ring of barbs with minute and non-interlocking barbules sticking up from its top edge giving a soft and fluffy texture. Some follicles produce down feathers throughout life, but most are pushed out by new feathers during moulting. The adult (*contour*) feathers grow in definite tracts on the skin, with (except in penguins and RATITES) bare patches between.

Other feather types include: *intermediate feathers*, showing a combination of features of contour and down feathers; *filoplumes* (*plumulae*), which are hair-like, usually lacking vanes; *vibrissae*, stiff and bristle-like, often around the nares; and *powder down*, which is soft downy material giving off dusty particles used in feather cleaning.

The amount of keratin required to make a new set of contour feathers may cause the timing of moulting to be under strong selection pressure. Feathers may be pigmented or have a barbule arrangement which produces interference colours. See Fig. 57.

Evidence from developmental biology is very damaging to the classical view that feathers evolved from elongate SCALES. Rather, feathers are tubular, and the two planar sides of the vane (front and back) are created by the inside and outside of the tube only after unfolding from its cylindrical sheath. But the two sides of a scale develop from the top and bottom of the initial epidermal outgrowth.

fecundity Reproductive output, usually of an individual. Number of offspring produced. See FITNESS.

feedback See HOMEOSTASIS, POSITIVE FEEDBACK.

feedback inhibition See END-PRODUCT INHIBITION.

female haploidy In the phytophagous false spider mite, *Brevipalpus phoenicis*, both males and females are haploid and males infected by *WOLBACHIA* are feminized. See MALE HAPLOIDY.

femur (pl. femora) (1) Thigh-bone of tetrapod vertebrates. (2) The third segment from the base of an insect leg.

feral Of domesticated animals, living in a wild state. See ESCAPE.

fermentation Enzymatic and anaerobic breakdown of organic substances (typically sugars, fats) by microorganisms to yield simpler organic products. Pasteur showed in about 1860 that microorganisms were responsible (contrary to view of Liebig). Kuhne called the 'active principle' an ENZYME in 1878, and Buchner first isolated a fermentative cell-free yeast extract in 1897.

The term is often used synonymously with anaerobic respiration, but this is incorrect (see RESPIRATION). Classic examples include alcohol production by yeasts, and the conversion of alcohol to vinegar (acetic acid) by the bacterium *Acetobacter aceti*, a process commonly called *acetification*. Lactic acid production by animal cells is another example. In all cases the final hydrogen acceptor in the pathway is an organic compound. See BIOTECHNOLOGY.

fermenter See BIOREACTOR.

ferns See PTEROPHYTA.

ferredoxins (iron-sulphur proteins) Proteins containing iron and acid-labile sulphur in roughly equal amounts; extractable from wide range of organisms, where they are components of the ELECTRON TRANSPORT SYSTEMS of mitochondria (involved in aerobic respiration) and of chloroplasts, where they undergo reversible Fe^{2+}/Fe^{3+} transitions. Ferredoxin has several roles in the chloroplast, including donation of electrons to enzymes involved in amino and fatty acid biosynthesis. Amitochondriate eukaryotes produce pyruvate by glycolysis and those lacking HYDROGENOSOMES convert this through pyruvate:ferredoxin oxidoreductase (PFO) to reduce ferredoxin and acetyl-CoA, this last being converted to ethanol and acetate yielding 0–2 additional mol ATP per mol of glucose; but those with hydrogenosomes import the pyruvate into these organelles, whereupon PFO converts it to CO_2, acetyl-CoA and reduced ferredoxin (the ferredoxin being reoxidized by hydrogenase to yield H_2). See EUKARYOTE.

ferritin Iron-storing protein (esp. in spleen, liver and bone marrow). The iron (Fe^{3+}) is made available when required for haemoglobin synthesis, being transferred by TRANSFERRIN.

fertilization (syngamy) Fusion of two GAMETES (which may be nucleated cells or simply nuclei) to form a single cell (*zygote*) or fusion nucleus. Commonly involves cytoplasmic coalescence (*plasmogamy*) and pooling of nuclear material (*karyogamy*). With MEIOSIS it forms a fundamental feature of most eukaryotic sexual cycles, and in general the gametes that fuse are HAPLOID. When both are motile, as primitively in plants, fertilization is *isogamous*; when they differ in size but are similar in form it is *anisogamous*; when one is non-motile (and usually larger) it is termed *oogamous*. This is the typical mode in most plants, animals and many fungi. In many gymnosperms and all anthophytes neither gamete is flagellated, and a POLLEN TUBE is involved in the fertilization process. In animals, *external fertilization* occurs (typically in aquatic forms) where gametes are shed outside the body prior to fertilization; *internal fertilization* occurs (typically as an adaptation to terrestrial life) where sperm are introduced into the female's reproductive tract, where fertilization then occurs. After fertilization the egg forms a *fertilization membrane* to preclude further sperm entry. Sometimes the sperm is required merely to activate the egg (see ACTIVATION, PARTHENOGENESIS, PSEUDOGAMY). See ACROSOME, DOUBLE FERTILIZATION, *IN VITRO* FERTILIZATION.

Feulgen method Staining method applied to histological sections, giving purple colour where DNA occurs.

F factor (F plasmid, F particle, F element, sex element, sex factor) One kind of PLASMID found in cells of the bacterium *E. coli*, and playing a key role in its sexuality (i.e. inter-cell gene transfer). It encodes an efficient mechanism for getting itself transferred from cell to cell, like many drug-resistance plasmids. Rarely, the F plasmid integrates into the host chromosome (forming an *Hfr* cell), when the same transmission mechanism results in a segment of chromosome adjacent to the integrated F plasmid being transferred from donor cell to recipient. In this condition it behaves very like a λ PROPHAGE, replicating only when the host chromosome does. When this cell conjugates with a cell lacking an F particle, a copy of the *Hfr* chromosome passes along the conjugation canal, the F particle entering last (see CHROMOSOME MAPPING, TRANSDUCTION), if at all. The resulting diploid or partial diploid cells do not remain so for long since recombination (hence *high frequency recombinant*, *Hfr*, strain) between the DNA duplexes occurs and the emerging clones are haploid.

fibre (Bot.) See SCLERENCHYMA, PHLOEM. (Zool.) (1) Term applied to thin, elongated cell (e.g. nerve fibre, muscle fibre) or the characteristic structure adopted by molecules of collagen, elastin and reticulin (see also FILAMENT). (2) Component of (human) diet. The term 'fibre' has proved difficult to define, but a common interpretation is 'non-starch polysaccharides' (NSP), approximating to the sum of plant cell wall polysaccharides. Others define plant fibre as those food components resisting digestion in the small intestine – which would include among others NSP, lactose, LIGNIN (a minor component of human foods, but present in wheat bran), uronic acid and free sugars. Since resistance to digestion depends upon the amount of chewing, food processing, the time food spends in the intestine and other factors not directly related to the food itself, to define fibre as NSP has more practical value. However, some non-starch polysaccharides such as MUCILAGES and GUMS – especially in fruits and vegetables – are soluble and on the NSP definition of fibre would have to be classed as 'soluble fibre'. Rural communities tend to consume more NSP than do urban communities, but very little is known about the NSP content of tropical dietary intakes. In general, insoluble dietary fibre tends to speed the passage of faeces through the large intestine.

fibril (1) Submicroscopic thread comprising cellulose molecules, in which form cellulose occurs in the plant cell wall; (2) thread-like thickening on the inner faces of large hyaline cells in the leaf or stem cortex of the moss *Sphagnum*.

fibrin Insoluble protein meshwork formed on conversion of fibrinogen by thrombin. See BLOOD CLOTTING.

fibrinogen Plasma protein produced by vertebrate liver. See BLOOD CLOTTING.

fibrinolysis One of the homeostatic processes involved in HAEMOSTASIS. As in BLOOD CLOTTING the major inactive participant is a plasma protein, here PLASMINOGEN, which is converted to the SERINE PROTEASE enzyme *plasmin* by a variety of activators. Plasmin dissolves blood clots and removes fibrin which may otherwise build up on endothelial walls.

fibroblast Characteristic cell type of vertebrate connective tissue, responsible for synthesis and secretion of extracellular matrix materials such as tropocollagen, which polymerizes externally to form COLLAGEN. Migrate during development to give rise to mesenchymal derivatives. See FILOPODIUM.

fibroblast growth factors (FGFs) A family of secreted proteins seemingly acting as competence factors having permissive rather than concentration-related effects on cell fate; hence acting alongside MORPHOGENS.

fibroin A major protein component of silk; rich in β-pleated sheets.

fibronectin An important cell ADHESION molecule which also seems to guide cell migration in embryos. Cultured human T cells require adhesion to fibronectin in order to divide. When released by damaged tissue, fibronectin can activate the TOLL-LIKE RECEPTOR TLR4.

fibrous root Root system comprising a tuft of adventitious roots of more or less equal diameter arising from the stem base or hypocotyl and bearing small lateral roots; e.g. wheat, strawberry. Compare TAP ROOT.

fibula The posterior of the two bones (other is TIBIA) in lower part of hindlimb of tetrapods. Lateral bone in lower leg of human.

Fick's law of diffusion Law stating that the rate of DIFFUSION of gases and of solvents and solutes within a system is directly proportional to their respective concentration gradients (net movement being from higher to lower concentration). If there is a barrier to diffusion (e.g. a membrane), the rate of diffusion will be inversely proportional to the thickness of the barrier and directly proportional to the permeability of the barrier to diffusing substance. See GASEOUS EXCHANGE.

filament (1) Stalk of the STAMEN, supporting the anther in flowering plants. (2) Term used to describe thread-like thalli of certain algae and fungi. (3) Term used of many long thread-like structures or molecules. Viruses whose coat proteins produce a rod-shaped virion (e.g. TMV) are described as *filamentous*. See ACTIN, GILL, MYOSIN.

filarial worms Small parasitic nematode worms of humans and their domestic animals, typically in tropical and semitropical regions. *Filariasis* is caused by blockage of lymph channels by *Wuchereria bancrofti* (up to 10 cm long), the young (microfilariae, 200 μm long) accumulating in blood vessels near the skin. Transmitted by various mosquitoes. Cause gross swellings of legs: *elephantiasis*. Another filarian, *Onchocerus volvulus*, transmitted by blackflies (*Simulium* spp.), causes *onchoceriasis* (river blindness). The flies need water to breed, and inject the worms when they bite humans. These cause fibrous nodules under the skin and inflammation of the eye, leading to blindness. See SUPERSPECIES.

Filicales Order of PTEROPHYTA (ferns), including the great majority of existing ferns and a few extinct forms. Perennial plants with a creeping or erect rhizome, or with an erect aerial stem several metres in height (e.g. tropical tree ferns). Leaves are characteristically large and conspicuous. Sporophylls either resemble ordinary vegetative leaves, bearing sporangia on the under surface, often in groups (sori), or else are much modified and superficially unlike leaves (e.g. royal fern). Generally homosporous, prothalli bearing both antheridia and archegonia; but includes a small group of aquatic heterosporous ferns. See LIFE CYCLE.

filoplume (plumule) See FEATHER.

filopodium Dynamic extension of the cell membrane up to 50 μm long and about 0.1 μm wide, protruding from the surfaces of migrating cells, e.g. FIBROBLASTS, or growing nerve axons. Grow and retract rapidly, probably as a result of rapid polymerization and depolymerization of internal actin filaments, which have a paracrystalline arrangement. Possibly sensory, testing adhesiveness of surrounding cells. Smaller filopodia, up to 10 μm long, are termed *microspikes*. See CYTOSKELETON, CELL LOCOMOTION.

filter feeding Feeding on minute particles suspended in water (MICROPHAGY) which are often strained through mucus or a meshwork of plates or lamellae. Water may be drawn towards the animal by cilia, or enter as a result of the animal's locomotion. Very common among invertebrates, and found among the largest fish (basking and whale sharks) and mammals (baleen whales).

fimbriae Bacterial structures, resembling short flagella but not involved in locomotion. Possibly involved in adhesion (e.g. to animal tissues in pathogenic forms). Compare PILLI.

finger domain An amino acid sequence within a protein which binds a metal atom, producing a characteristic 'finger-like' conformation within the protein. Such domains tend to bind nucleic acid and may be found repeated tandemly as in *multifinger loops* (see TRANSCRIPTION FACTORS). Compare HOMOEOBOX. See NUCLEAR RECEPTORS, UBIQUITIN.

fingerprinting See DNA FINGERPRINTING.

fin ray See FINS.

fins (1) Locomotory and stabilizing projections from the body surface of fish and their allies. Include unpaired medial AGNATHAN *fin-folds* with little or no skeletal support; but the term generally refers to the medial and paired ray fins of the CHONDRICHTHYES and OSTEICHTHYES, in which increasingly extensive skeletal elements (*fin rays*) articulate with the vertebrae, and PECTORAL and PELVIC GIRDLES. The pectoral and pelvic fins are paired and are used for steering and braking. The dorsal, anal and caudal fins are unpaired and medial, opposing yaw and roll. The caudal fin (tail) is generally also propulsive (see HETEROCERCAL, HOMOCERCAL). Fins are segmented structures, seen clearly in the muscle attachments of the ray fins of ACTINOPTERYGII. (2) Paired membranous and non-muscular stabilizers along the sides of arrow worms (Chaetognatha). (3) Horizontal and muscular fringe around the mantle of cephalopods such as the cuttlefish (*Loligo*) by means of which its gentler swimming is achieved.

fish (1) General term, covering AGNATHA (jawless fish), CHONDRICHTHYES (cartilaginous fish) and OSTEICHTHYES (bony fish). Discovery of *Psarolepis* from the latest Silurian and earliest Devonian confuses our taxonomy of fishes. It has a jointed brain case like sarcopterygians and bears sarcopterygian

teeth, yet its tooth-bearing bones of the snout are actinopterygian. A large spine in front of the pectoral fin is placoderm-like, yet spines in front of its medial fins are like those of acanthodians. All this indicates that *Psarolepis* is primitive, and that placoderm-like and acanthodian-like features may be components of the primitive osteichthyan phenotype. See website http://www.fishbase.org.

(2) See CHROMOSOME PAINTING.

fission Form of vegetative reproduction, involving the splitting of a cell into two (*binary fission*) or more than two (*multiple fission*) separate daughter cells. See CELL DIVISION.

Fissipedia Suborder of CARNIVORA, including all land carnivorous mammals. Canines large and pointed; jaw joint a transverse hinge (preventing grinding); carnassial teeth often present. Includes cats (Felidae), foxes, wolves and dogs (Canidae), weasels, badgers, otters (Mustelidae), civets, genets and mongooses (Viverridae), hyaenas (Hyaenidae), racoons and pandas (Procyonidae), and bears (Ursidae).

fitness (selective value) Factor describing the difference in reproductive success of an individual or genotype relative to another. Usually symbolized by *w*. Often regarded as compound of survival, or longevity (see SURVIVORSHIP CURVES) and annual fecundity.

(1) Of individuals. Lifetime reproductive success; either 'lifetime reproductive output' (the lifetime fecundity), or the number of offspring reaching reproductive age. Both omit information on the reproductive output of these offspring, and hence on the number of grandchildren reaching reproductive age, or the number of great grandchildren doing so, and so on. Fecundity alone is therefore only one component of fitness: an individual may leave more descendants in the long term by producing fewer total offspring but by ensuring a greater probability of their survival to reproductive age (e.g. by provisioning fewer seeds with more food reserves). Likewise, natural selection will favour any heritable factor

that improves the chances of a gene's representation in subsequent generations. Thus an individual may promote future representation of its own genes, even if it leaves no offspring itself, by contributing to the fitness of close relatives. Any actions which do so contribute improve an agent's *inclusive fitness*, calculated from that individual's reproductive success plus its effects upon the reproductive success of its relatives, each effect weighted by the relative's coefficient of RELATEDNESS to the agent (see HAMILTON'S RULE). Likewise, an individual's fitness may be improved by the effects of its relatives. See HELPER, UNIT OF SELECTION.

(2) Of genotypes. Usually applied to a single locus, where the fitness value, *w*, of a genotype such as Aa is defined as $1 - s$, where s is the *selection coefficient* against the genotype. The existence of fitness differences between genotypes creates selection for the evolution of the genetic system itself. See COEFFICIENT OF SELECTION.

fixation (1) In microscopy, the first step in making permanent preparations of organisms, tissues, etc., for study. Aims at killing the material with the least distortion. Solutions of formaldehyde and osmium tetroxide often used. Some artifacts of structure usually produced.

(2) Of genes. The spread of an allele of a gene through a population until it comes to occupy 100% of available sites (i.e. until it is the only allele found at that locus). It is then *fixed* in the population.

(3) No clear definition, but usually involves either the conversion of the major atmospheric source of an element (e.g. C, N) to a less mobile compound, or the conversion of an inorganic source of an element into an organic form. Instances include conversion of molecular nitrogen to ammonia or nitrogen oxides and conversion of CO_2 to organic carbon by photosynthesis. See NITROGEN FIXATION. C-fixation occurs in photosynthesis; N-fixation occurs in soils, ponds, etc., through the action of prokaryotes (e.g. bacteria, blue-green algae).

flagellar apparatus The whole complex of flagellar basal bodies, microtubular roots and their associated structures and rhizoplasts (if any) present in a flagellate cell.

flagellar canal A narrow invagination of the plasmalemma from the floor of which a flagellum arises (e.g. in DINOPHYTA) or a canal in the cell envelope through which the flagellum runs and emerges to the exterior (e.g. in *Chlamydomonas* and other CHLOROPHYCEAE).

flagellar pit A depression in the cell surface from the bottom of which the flagella arise (e.g. PRASINOPHYCEAE).

flagellar roots Root-like structures comprising microtubules and striated fibres and connected to the flagella inside the cell. Their function is largely unknown but includes controlling the conformation of the flagellar apparatus and also anchorage within the cell.

flagellata See MASTIGOPHORA.

flagellum (1) Extension of the cell membrane of certain eukaryotic cells, comprising an internal axoneme of nine doublet microtubules surrounding two central doublets. Upon entering the cell body, the two central microtubules end at a double plate, whereas the nine peripheral doublets continue into the cell, usually picking up extra structures transforming them into triplets. Between the microtubules in the flagellum's basal region is the basal body, attached to which are the fibrillar roots. Thus, the structure is identical to that of a CILIUM (see Fig. 58), but more variable in length and generally longer. The flagellar membrane may be smooth, possessing no hairs on its surface (smooth, whiplash or acronematic flagellum); or it may have hairs on its surface (hairy tinsel or pantonematic flagellum). Two types of hairs occur: (a) *fibrous hairs* or *flimmer filaments* (5 nm thick), composed of glycoprotein, which are solid and wrap around the flagellum, increasing its surface area and efficiency of propulsion; (b) *tubular hairs* or *mastigonemes* (20 nm thick, 1 μm long), composed of proteins and glycoprotein, which are tripartite having a tapering base, a microtubular shaft and terminal filaments. Able to reverse the

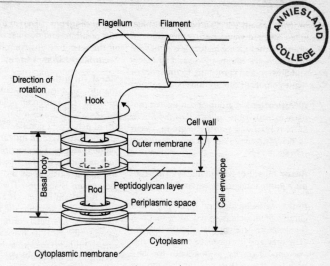

FIG. 58 *The wheel-like rotating basal body complex of the* FLAGELLUM *of* E. coli, *encoded by about thirty-five genes.* (Adapted from *The Sensing of Chemicals by Bacteria,* by Julius Adler. Copyright © Scientific American, Inc., 1976. All rights reserved.)

thrust of the flagellum. In addition to hairs, several types of scales, both inorganic and organic, can be found on the surfaces of flagella.

Flagella beat in wave-like undulations, unlike cilia, whose down-beat power-stroke is followed by an up-stroke offering less resistance. Absent from red algae and fungi; but in other algae flagella have a locomotory role, propelling organisms through the water, allowing them to undergo diurnal vertical phototactic rhythms. In plants, such as mosses, liverworts, ferns, cycads and ginkgophytes, and some algae (e.g. *Chara* and *Fucus*), flagella are only found on gametes.

(2) In some prokaryotes, a hollow membrane-less filament, 3–12 mm long and 10–20 mm in diameter, composed of helically arranged subunits of the protein *flagellin*. The attachment of the flagellum is by *hook*, *bearing* and *rotor*. The flagellum is thus in the form of a fixed helix, several often rotating in unison, powered by *proton motive force* (see BACTERIORHODOPSIN). Involved in chemotactic responses by the cell. See LOCOMOTION, PILI.

flame cell (solenocyte) Cell bearing a bunch of flickering flagella (hence name) and interdigitating with a *tubule cell* (which forms a hollow tube by wrapping itself around the extracellular space). Combined, they form the excretory units (*protonephridia*) of the PLATYHELMINTHES, nemertine worms and the ENTOPROCTA.

flatworms See PLATYHELMINTHES.

flavin Term denoting either of the nucleotide COENZYMES (FAD, FMN) derived from RIBOFLAVIN (vitamin B_2) by the enzymes *riboflavin kinase* and FMN *adenylyltransferase*. ATP hydrolysis accompanies the reactions. See FLAVOPROTEINS.

flavonoids Compounds in which two 6-carbon rings are linked by a 3-carbon unit. They are the most important pigments in floral colouration, and probably occur in all flowering plants (MAGNOLIOPHYTA), but are more sporadically distributed among the members of other groups of vascular plants; rare in algae and animals. They function in blocking far-ultraviolet radiation, which is highly destructive to nucleic acids and proteins, usually selectively admitting blue-green and red wavelengths, important in photosynthesis. One major class of flavonols which are commonly found in leaves and flowers, and contribute to the ivory or white colours of certain flowers.

flavoproteins A group of conjugated proteins in which one of the *flavins* FAD or FMN is bound as prosthetic group. Occur as dehydrogenases in ELECTRON TRANSPORT SYSTEMS.

flexor Muscle or tendon involved in bending a joint; antagonizes extensors.

flight Aerial locomotion involving either gliding or muscle-powered movement of the skeleton. Animals that glide or fly, generate drag and lift forces to produce thrust and oppose gravity, as in aquatic locomotion; but the relatively lower density and viscosity of the medium favours the maximum use of lift by all but the smallest fliers. Gliding in a still medium entails loss of altitude although there is no direct metabolic cost. Flying animals generate lift by muscle contraction, usually to move a wing in an up-and-down flapping motion (maximal at the wing tip and minimal at its base). Large birds can generate forward thrust on both upstroke and downstroke; insects do so only on the downstroke because of their lower ratio of forward velocity to vertical wing movement. Aerodynamic force is the vectorial resultant of lift and drag forces. Insect wings may have evolved from gills which formed part of an ancestral polyramous limb.

flimmer flagellum Pleuronematic or pantonematic flagellum.

flimmers (mastigonemes) See FLAGELLUM.

flora (1) Plant population of a particular area or epoch. (2) List of plant species (with descriptions) of a particular area, arranged in families and genera, together with an IDENTIFICATION KEY.

floral apex Apical meristem that will develop into a flower or inflorescence.

floral diagram Diagram illustrating relative positions and number of parts in each of the sets of organs comprising a flower. See FLORAL FORMULA, and Fig. 59.

floral formula Summary of the information in a FLORAL DIAGRAM. The floral formula of buttercup (Ranunculaceae), $k_5C_5A\infty G\infty$, indicates a flower with a calyx (K) of five sepals, corolla (C) of five petals, androecium (A) of an indefinite number of stamens and a gynoecium (G) of an indefinite number of free carpels. The line below the number of carpels indicates that the gynoecium is superior. The floral formula of the campanula (Campanulaceae), $K_5C_{(5)}A_5\overline{G_{(5)}}$, shows that the flower has five free sepals, five petals united () to form a gamopetalous corolla, five stamens, and five carpels united () to form a syncarpous gynoecium. The line above the carpel number indicates that the gynoecium is inferior (see RECEPTACLE).

floral tube Cup or tube formed by fusion of basal parts of sepals, petals and stamens, often in flowers possessing inferior ovaries. See RECEPTACLE.

floret One of the small flowers making up the composite inflorescence (Compositae), or the spike of grasses. In the former, *ray florets* are often female while *disc* florets are often hermaphrodite. See GYNOMONOECIOUS.

floridean starch Polysaccharide storage product occurring in red algae (RHODOPHYTA); somewhat similar to amylopectin.

florigen Hypothetical plant 'hormone' (see GROWTH SUBSTANCE), invoked to explain transmission of flowering stimulus from leaf, where it is perceived, to growing point.

floristic equilibrium The species composition of a flora when there is no further change generated internally or externally.

floristics Study of composition of vegetation in terms of species (FLORA) present in a particular region. Floristics aims to account for all plants of the region, with keys, descriptions, ranges, habitats and phe-

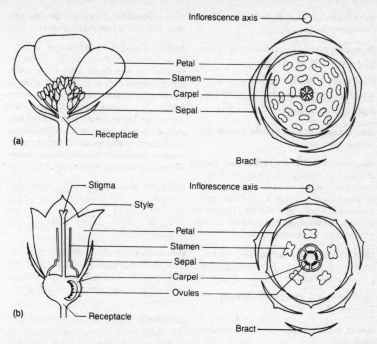

FIG. 59 *Diagrams illustrating flower structure; half-flower (median vertical section; left) and floral diagram (right). (a) Buttercup, (b) Campanula.*

nology, and to offer analytical explanations of the flora's origin and geohistorical development.

flower Specialized, determinate, reproductive shoot of flowering plants (ANTHO-PHYTA), consisting of an axis (RECEPTACLE) on which are inserted four different sorts of organs, all evolutionarily modified leaves. Outermost are SEPALS (the *calyx*, collectively), usually green, leaf-like, and enclosing and protecting the other flower parts while in the bud stage. Within the sepals are petals (the *corolla*, collectively), usually conspicuous and brightly coloured. Calyx and corolla together constitute the *perianth*. They are not directly concerned in reproduction and are often referred to as *accessory flower parts*. Within the petals are STAMENS (microsporophylls), each consisting of a filament (stalk) bearing an ANTHER, in which

pollen grains (microspores) are produced. In the flower centre is the GYNOECIUM, comprising one or more CARPELS (megasporophylls), each composed of an OVARY, a terminal prolongation of the STYLE and the STIGMA, a receptive surface for pollen grains. The ovary contains a varying number of OVULES which, after fertilization, develop into seeds. Stamens and carpels are collectively known as *essential flower parts*, since they alone are concerned in the process of reproduction. See FLORAL DIAGRAM.

flowering The production of flowers by a flowering plant. Timing of flowering tends to be more variable in annual and ephemeral species; in Britain, the least variable tend to be those flowering later in the year (May to August) and mean monthly temperature appears to be a major factor in determining flowering time for most species. See FLORIGEN, LONG-DAY PLANTS, SHORT-DAY PLANTS.

flowering plants See MAGNOLIOPHYTA (MAGNOLIIDAE). See INTRODUCTION.

fluid mosaic model Current generalized model for structure of all CELL MEMBRANES.

fluke See TREMATODA.

fluorescence in situ hybridization See CHROMOSOME PAINTING.

fluorescent antibody technique Cells or tissues are treated with an antibody (specific to an antigen) which has been labelled by combining it with a substance that fluoresces in UV light and can therefore indicate the presence and location of the antigen with which it combines.

FMN (flavin mononucleotide) A FLAVIN; derivative of riboflavin. Prosthetic group of some FLAVOPROTEINS.

focal contacts (focal adhesions) See ADHESION.

foetal membranes See EXTRAEMBRYONIC MEMBRANES.

foetus (fetus) In mammals, the stage in intrauterine development subsequent to the appearance of bone cells (osteoblasts) in the cartilage, indicating the onset of OSSIFICATION. In humans, this occurs after seven weeks of gestation. See EMBRYO.

folic acid (pteroylglutamic acid) Vitamin of the B-complex (water-soluble) whose coenzyme form (tetrahydrofolic acid, FH_4) is a carrier of single carbon groups (e.g. $-CH_2OH$, $-CH_3$, $-CHO$) in many enzyme reactions. Involved in biosynthesis of purines and the pyrimidine thymine. Very little in polished rice, but widely distributed in animal and vegetable foods. Often given to pregnant women because deficiency of folic acid is a major causal factor of megaloblastic anaemia and neural tube defects (e.g. spina bifida).

follicle (Bot.) Dry fruit derived from a single carpel which splits along a single line of dehiscence to liberate its seeds; e.g. of larkspur, columbine. (Zool.) See GRAAFIAN FOLLICLE, HAIR FOLLICLE, OVARIAN FOLLICLE.

follicle-stimulating hormone (FSH) Gonadotrophic glycoprotein hormone secreted by vertebrate anterior PITUITARY gland. Mammalian FSH stimulates maturation of both granulocytes and ovum of follicle; it increases granulocytes' production of LH receptors and of P450 aromatase (which converts androgens to oestrogens); it also promotes formation of spermatozoa in the testis. See INHIBINS, MATURATION OF GERM CELLS, MENSTRUAL CYCLE.

follicular phase Phase in mammalian OESTROUS and MENSTRUAL CYCLES, in which Graafian follicles grow and the uterine lining proliferates due to increasing oestrogen secretion.

follistatins Family of glycosylated proteins which modulate pituitary FSH release and act locally within the gonads. Usually inhibit FSH release; but also bind ACTIVINS. See INHIBINS.

fontanelle Gap in the skeletal covering of the brain, either in the chondrocranium or between the dermal bones, covered only by skin and fascia. Present in newborn babies between frontal and parietal bones of the skull; closes at about eighteen months.

food chain A metaphorical chain of organisms, existing in any natural community, through which energy and matter are transferred. Each link in the chain feeds on, and hence obtains energy from, the one preceding it and is in turn eaten by and provides energy for the one succeeding it. Number of links in the chain is commonly three or four, and seldom exceeds six. At the beginning of the chain are green plants (autotrophs). Those organisms whose food is obtained from green plants through the same number of links are described as belonging to the same TROPHIC LEVEL. Thus green plants occupy one level (T^1), the PRODUCER level. All other levels are CONSUMER levels: T^2 (herbivores, or primary consumers); T^3 and T^4 (secondary consumers; the smaller and larger carnivores respectively). At each trophic level, much of the energy (and carbon atoms) is lost by respiration and so less biomass can be supported at the next level. Bacteria, fungi and some protozoa are consumers that function in decomposition of all levels (see DECOMPOSER). All the food chains in a community of organisms make up the FOOD WEB. See BIOMAGNIFICATION, PYRAMID OF BIOMASS.

food chain magnification See BIOMAGNIFICATION.

food poisoning See EXOTOXINS.

food vacuole Endocytic vesicle produced during PHAGOCYTOSIS. See ENDOCYTOSIS.

food web The totality of interacting FOOD CHAINS within a community of organisms. For stability of food webs, see BALANCE OF NATURE. See ECOSYSTEM.

footprinting See DNA FOOTPRINTING.

foramen Natural opening (e.g. FORAMEN MAGNUM of skull, FORAMEN OVALE of foetal heart). Foramina in bones permit nerves and blood vessels to enter and leave.

foramen magnum Opening at back of vertebrate skull, at articulation with vertebral column, through which spinal cord passes.

foramen ovale Opening between left and right atria of hearts of foetal mammals, normally closing at birth (failure to do so resulting in 'hole' in the heart). While open, it permits much of the oxygenated blood returning to the foetal heart from the placenta to pass across to the left atrium, thus bypassing the pulmonary circuit (which in the absence of functional lungs is largely occluded). From there blood passes via the left ventricle to the foetal body.

foraminifera Order of mainly marine protozoans whose shells form an important component of chalk and of many deep sea oozes (e.g. *Globigerina ooze*). Shells may be calcareous, siliceous, or composed of foreign particles. Thread-like pseudopodia protrude through pores in the shell and may

or may not exhibit cytoplasmic streaming. See HELIOZOA, RADIOLARIA.

forebrain (prosencephalon) Most anterior of the three expansions of the embryonic vertebrate brain. Gives rise to *diencephalon* (thalamus and hypothalamus) and *telencephalon* (cerebral hemispheres). Also the origin of the eyestalks. Associated originally with olfaction. See Fig. 18b.

form (Bot.) Smallest of the groups used in classifying plants. Category within species, generally applied to members showing trivial variations from type, e.g. in colour of the corolla. See INFRASPECIFIC VARIATION. (Zool.) Used more or less synonymously with *morph*, to indicate one of the forms within a dimorphic or polymorphic species population.

formation See BIOME.

forming face (Of a Golgi body) the side on which cisternae of the Golgi body are assembled through fusion of vesicles derived from the endoplasmic reticulum or nuclear envelope.

formose reaction See ORIGIN OF LIFE.

fossil Remains of an organism, or direct evidence of its presence, preserved in rock, ice, amber, tar, peat or volcanic ash. Animal hard parts (hard skeletons) commonly undergo *mineralization*, a process which also turns sediment into hard rock (both regarded as diagenesis). The aragonite (a form of $CaCO_3$) of molluscs and gastropods may recrystallize as the common alternative form, calcite; or it may dissolve to leave a void. This mould may then be filled later by *replacement*, involving precipitation of another mineral (possibly calcite or silica). Partial replacement and impregnation of the original hard parts in both plants and animals by mineral salts (*permineralization*) may occur, especially if the material is porous – as are wood and bone. Fossils may occur *in situ*, or else (derived fossils) be released by erosion of the rock and subsequent reposition in new sediments. Sometimes, the fossil imprints of different locomotory styles (gaits) of the same individual animal are given different taxonomic names. *Index fossils* are those occurring only at restricted time horizons (and hence rock layers) which can thus be used to date rocks containing them in new exposures. Fossils may be dated by various methods, and provide direct evidence for EVOLUTION, as well as telling us about past conditions on Earth. See GEOLOGICAL PERIODS. For fossils of mankind's ancestors, see HOMINID.

The absence of simpler predecessors of the complex Cambrian fossils bothered Charles Darwin; and although incontestable fossils of multicellular cyanobacteria and algae have been dated to 2 billion (2×10^9) years, earlier material is more difficult to interpret. Because the fossil record is incomplete, durations of fossil lineages as estimated from the rock layers in which they occur must always underestimate the real durations, so we never really know when a lineage actually appeared or disappeared; and the uncertainties can be very large, especially when the lineage concerned is poorly or patchily represented in the fossil record. Confidence levels of tens of millions of years may have to be applied in such cases.

It is unlikely that enough sequence data will be recovered from fossil DNA (e.g. from specimens trapped in amber, where it is protected from dehydration, oxygen and bacteria) to understand the developmental biology of extinct species; and although we might conceivably recover dinosaur DNA from the Cenozoic and Mesozoic, its base sequences degenerate progressively so that CAMBRIAN DNA is presumed to be permanently lost to us (see DNA). Contamination by recent DNA is a problem, given the extreme sensitivity of the POLYMERASE CHAIN REACTION method of DNA amplification. DNA from osteocytes deep within bones may last relatively unchanged in tar pits, as from the Pleistocene. Sequences from several Siberian mammoths (10–50 Kyr) resemble those of living elephants; those from the 40 Kyr cave bear place this species as a sister group of the European brown bear. See EVOLUTION, BURGESS SHALE, INTRODUCTION.

fossil DNA See DNA.

fossorial Of animals adapted to digging, burrowing.

founder effect Effects on a population's subsequent evolution attributable to the fact that founder individuals of the colonizing population have only a small and probably non-representative sample of the parent population's GENE POOL. Subsequent evolution may take a different course from that in the parent population as a result of this limited genetic variation. Likely to occur where colonization is a rare event, as on oceanic islands, and where the colonizer is not noted for mobility. May be combined with effects of GENETIC DRIFT. It was Ernst Mayr's founder effect model of speciation that underpinned the theory of punctuated equilibrium (see EVOLUTION).

fovea (Bot.) Pit in the wall of palynomorphs, such as spores, pollen, or dinoflagellate cysts. (Zool.) Depression in retina of some vertebrates, containing no rod cells but very numerous cone cells. May lie in a circular CONE-rich region termed the MACULA LUTEA. Blood vessels absent, and no thick layer of nerve fibres between cones and incoming light as in rest of inverted retina. It is a region specialized for acute diurnal vision. Found in diurnal birds, lizards and primates, including man.

fractionation See CELL FRACTIONATION.

frameshift mutation See MUTATION.

fraternal twins Dizygotic twins who develop as a result of simultaneous fertilization of two separate ova. Such twins are no more alike genetically than other siblings. See MONOZYGOTIC TWINS.

freemartin Female member of unlike-sexed twins in cattle and occasionally other ungulates. Sterile, and partially converted towards hermaphrodite condition by hormonal (or possibly H–Y ANTIGEN) influence of its twin brother reaching it through anastomosis of their placentae.

free nuclear division Stage in development in which unwalled nuclei result from repeated division of primary nucleus.

free nuclear endosperm Endosperm in which there are many nuclear divisions without cell division (cytokinesis) before cell walls start to form.

free radical (radical) A chemical species containing at least one unpaired electron. Cobalamin, coenzyme B_{12}, provides radicals which catalyse intramolecular rearrangements involving hydrogen. The radical $NO_2^\bullet$ is produced by CIGARETTE SMOKING. For 'free radical attack theory', see AGEING; see also ORGANIC RADICALS, SUPEROXIDES.

freeze drying Method of preserving unstable substances by drying when deeply frozen.

freeze-etching Technique used in electron MICROSCOPY for examining the outer surfaces of membranes.

freeze-fracture Technique used in electron MICROSCOPY for examining the inner surfaces of membranes.

frequency-dependent selection Form of SELECTION occurring when the advantage accruing to a character trait in a species population is inversely proportional to the trait's frequency in the population. When rare, it will be favoured by selection; when common, it will be at a disadvantage compared with alternative traits. Two or more traits determined by the same genetic mechanism (locus, or loci) may thus coexist in the population in a condition of POLYMORPHISM.

front Term applied to leaf of a fern as well as divided leaves of other plants (e.g. palm).

frontal bone A MEMBRANE BONE, a pair of which covers the front part of the vertebrate brain (forehead region in man). Air spaces (*frontal sinuses*) extend from nasal cavity into frontal bones of mammals.

frontal lobe Major part of the CEREBRAL CORTEX of the primate brain, including human's. Behind frontal bone. Has numerous connections with many parts of the brain.

fructification Reproductive organ or fruiting structure, often used in the context of fungi, myxomycetes and bacteria.

fructose A ketohexose reducing sugar, $C_6H_{12}O_6$. In combination with glucose, forms sucrose (non-reducing). The sweetest of sugars.

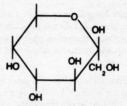

β-form in pyranose ring form

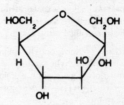

α-form in furanose ring form

Two isomers of D-fructose

fruit Ripened ovary of the flower, enclosing seeds.

fruit set The production of a ripe fruit from an immature carpel. See AUXINS, ETHYLENE.

frustule Silica elements of the diatom cell wall.

fruticose Lichen growth form where the thallus is shrub-like and branched.

FSH See FOLLICLE-STIMULATING HORMONE.

FtsA Slowly evolving (diverging) prokaryotic protein, which some believe (with bacterial heat shock protein 70) to be homologous with, and ancestral to, ACTIN.

FtsZ An essential cell division protein forming a cytokinetic ring in bacteria and archaeans; its homology with TUBULIN was first supported by sequence and functional studies but recently also by three-dimensional studies (only their core regions, which bind GTP are very similar). Used in division by CHLOROPLASTS, but not

MITOCHONDRIA. Both FtsZ and tubulin are GTPases and form protofilaments; but their functions are different and convergence cannot be ruled out.

fucoidin Polymer of α-1,2, α-1,3 and α-1,4 linked residues of L-fucose sulphated at C_4. Commercially marketed phycocolloid in cell walls and intercellular spaces of brown algae (PHAEOPHYTA).

fucosan vesicle Refractive vesicles, usually around the nucleus, containing a tannin-like compound in the brown algae (PHAEO-PHYCEAE). Also called *physodes*.

fucoserraten A sexual attractant (gamone) produced by macrogamete (egg cell) in the brown algal genus *Fucus*.

fucoxanthin Xanthophyll pigment present with chlorophylls (and other pigments) in various algal groups; e.g. PHAEO-PHYCEAE, CHRYSOPHYCEAE, PRYMNESIOPHYTA, BACILLARIOPHYCEAE.

function In one sense, the function of a component in an organism is the contribution it makes to that organism's FITNESS. Therefore it may also be the ultimate reason for that component's existence in the organism, having been selected for in previous generations. This does not exclude the possibility that a component may arise by mutation in an individual and have immediate selective value (and hence function) in that individual; but its function would not then be the reason for its existence in that individual. We need to distinguish the question 'for what end does X exist?' from the question 'what caused X to exist?' The latter requires an aetiological (historical and causal) explanation; but it is debatable whether functional explanations justify etiological claims in any clear-cut way. Functions may also emerge epigenetically (see CO-OPTION) and molecular genetics alone may be insufficient to explain their emergence. See TELEOLOGY.

functional genomics See GENOMICS.

Fungi Kingdom of eukaryotic, primarily ACELLULAR organisms that lack chlorophyll, and are hence saprotrophs, parasites or symbionts. Typical organizational unit is the cylindrical HYPHA. Although flagella are absent, modern classifications usually include fungi within the eukaryote group 'opisthokonts' since they share with other members of that group both separation of DHFR and TS genes (see EUKARYOTE and Fig. 52) and gene insertion elongation factor-1α (EF-1α). Their nutrition is absorptive (osmotrophic) and never phagotrophic. Fungi do not display an amoeboid pseudopodial phase. Hyphal walls are characterized by the presence of chitin and β-glucans. Internally mitochondria have flattened cristae and peroxisomes are nearly always present. Morphologically, fungi can be unicellular or filamentous. Filaments comprise the coenocytic (siphonaceous) haploid phase, which can be either homokaryotic or heterokaryotic. Fungi reproduce asexually or sexually, generally with a short-lived diploid phase. Included in the Fungi are the ASCOMYCOTA, BASIDIOMYCOTA, ZYGOMYCOTA and the mitosporic fungi (formerly Deuteromycota). The OOMYCOTA and HYPHOCHYTRIDIOMYCOTA are more closely allied to the algal division HETEROKONTOPHYTA than Fungi, and are sometimes placed in the kingdom Chromista.

fungicide A chemical destructive to fungi. Fungicides can be systemic (taken up systemically by plants), protective (protects an organism against infection by a fungal pathogen) or act as an erradicant (chemical applied to a substratum in which a fungus is present or used in disease control after infection has occurred). Many chemicals can act as fungicides or inhibit fungal growth (e.g., copper-, sulphur-, mercury-containing compounds, and many other inorganic compounds as well as organic compounds, antibiotics). The effective fungicide methyl bromide is being banned because of environmental concerns. See PEST MANAGEMENT.

Fungi Imperfecti See MITOSPORIC FUNGI.

funiculus (Bot.) Stalk attaching ovule to the placenta in an ovary.

furcula See WISHBONE.

fusiform initial (Bot.) Meristematic cells of the vascular cambium. Vertically elongated cells, much longer than their width, appearing flattened or brick-shaped in transverse section. Produce xylem and phloem cells having their long axes orientated vertically. Comprise the axial system of the secondary vascular tissues. See RAY INITIALS.

fusion gene The fusion of two adjacent protein-coding genes into a single one, by the deletion of the stop-signals of the first gene and the start-signals of the second gene, enabling translation to begin at the first one and carry on uninterruptedly until the end of the second one. See EUKARYOTE.

G

G1, G2, G0 Gap phases in the cell cycle.

GABA (γ-aminobutyric acid) Inhibitory NEUROTRANSMITTER in the (human) brain, potentiated by benzodiazepine tranquillizers. Derivative of GLUTAMIC ACID. It is a ligand for two different target receptors, $GABA_A$ (a ligand-gated chloride channel) and a $GABA_B$ which, linked to a G PROTEIN, may open adjacent potassium channels, close voltage-gated calcium channels, or inhibit ADENYLYL CYCLASE. See SYNAPSE.

G-actin See ACTIN.

Gaia hypothesis The view that Earth is (or resembles) a self-regulating 'superorganism': a dynamic system with a particular behaviour and clearly defined boundaries (unlike, perhaps, the concept 'BIOSPHERE'). Although it could not have evolved in a Darwinian manner, there are parallels between Gaia and a physiological system. There is a case for regarding the proliferation of life on Earth as responsible for the retention of water on the planet; even for continued plate tectonics. See BALANCE OF NATURE, HYDROLOGICAL CYCLE.

gain-of-function mutation See MUTATION.

galactose An aldohexane sugar; constituent of LACTOSE, and commonly of plant polysaccharides (many gums, mucilages and pectins) and animal GLYCOLIPIDS and GLYCOPROTEINS.

gall bladder Muscular bladder arising from BILE DUCT in many vertebrates, storing bile between meals. Bile is expelled under influence of the intestinal hormone CHOLECYSTOKININ.

gametangial contact Form of CONJUGATION in which, following growth and contact of the gametangia, nuclei are transferred from the antheridium through a fertilization or copulation tube; e.g. in oomycetes.

gametangium (Bot.) Gamete-producing cell; most commonly in the contexts of algae and fungi. However, more complex antheridia, oogonia and archegonia are sometimes cited as examples too. Compare SPORANGIUM.

gamete (germ cell) Haploid cell (sometimes nucleus) specialized for FERTILIZATION. Gametes which so fuse may be identical in form and size (*isogamous*) or may differ in one or both properties (*anisogamous*). The terms 'male' and 'female' are often applied to gametes, but serve only to indicate the sex of origin, for gametes do not have sexes. Where they differ in size it is customary to refer to the larger gamete as the *macrogamete*, and to the smaller as the *microgamete*. Sometimes plasmogamy is absent in fertilization, in which case the nuclei which fuse may be regarded as gametes. See AUTOGAMY, AUTOMIXIS, MATURATION OF GERM CELLS, OVUM, PARTHENOGENESIS, SPERM.

game theory In biology, denotes all approaches to the study of 'decision making' by living systems (usually lacking conscious overtones) in which an organism's responses to its conditions are viewed as *strategies* whose (evolutionary) goal is maximization of the organism's FITNESS. Often convenient to regard each organism as having at any time a decision procedure for responding to future circumstances in such a way as to maximize any possible payoff to itself while minimizing the pay-offs to others (but see INCLUSIVE FITNESS). Decision procedures which cannot be super-seded by rival procedures will be EVOL-UTIONARILY STABLE STRATEGIES. See ARMS RACE, OPTIMALITY THEORY.

gametocyte Cell (e.g. oocyte, spermato-cyte) undergoing meiosis in the production of gametes. Primary gametocytes undergo the first meiotic division; secondary gametocytes undergo the second meiotic division. See MATURATION OF GERM CELLS.

gametogenesis Gamete production. Fre-quently, but by no means always, involves MEIOSIS. In eukaryotes there are often hap-loid organisms and stages in the life cycle where gametes can only be produced by MITOSIS. See SPERMATOGENESIS, OOGENESIS.

gametophore In bryophytes, a fertile stalk bearing gametangia.

gametophyte In plants showing ALTERNA-TION OF GENERATIONS, the haploid (n) phase; during it, gametes are produced by mitosis. Arises from a haploid spore, produced by meiosis from a diploid SPOROPHYTE. See LIFE CYCLE.

gamma globulins (immune serum globulins) Class of globular serum pro-teins. Includes those with ANTIBODY activity, and some without.

gamone Compound involved in bringing about fusion of gametes in some brown algae (Phaeophyta); e.g. ectocarpan (in *Ectocarpus*), multifidin and aucanten (in *Cutleria*), fucoserraten (in *Fucus*).

gamopetalous (sympetalous) (Of a flower) with united petals; e.g. primrose. Compare POLYPETALOUS.

gamosepalous (Of a flower) with united sepals; e.g. primrose. Compare POLYSEPALOUS.

ganglion Small mass of nervous tissue containing numerous CELL BODIES with syn-apses for integration. CENTRAL NERVOUS SYSTEMS of many invertebrates contain many such ganglia, connected by nerve cords. In vertebrates the CNS has a different overall structure, but ganglia occur in the periph-eral and AUTONOMIC NERVOUS SYSTEMS, where they may be encapsulated in connective tissue. Some of the so-called nuclei of the vertebrate brain are ganglia.

ganglioside Type of glycolipid common in nerve cell membranes.

ganoid scale Type of SCALE found in palaeo-niscoids and primitive ACTINOPTERYGII; with a thick coat of enamel (ganoin) lying above layers of vascular bone and deeper lamellar bone (palaeoniscoids), or upon thick lamel-lar bone alone (gars and polypteriforms). Ganoin was formerly not recognized as enamel, nor cosmine as dentin, leading to a classification of these non-teleost scales into 'ganoid' and 'cosmoid' which is less justified today. See COSMOID SCALE, PLACOID SCALE.

gap genes A class of *Drosophila* segmen-tation genes (e.g. *Krüppel* (*Kr*), *hunchback* (*hb*) and *knirps* (*kni*)), mutants of which delete several adjacent segments and create gaps in the anterior-posterior pattern. As a result of signals laid down by maternal genes in oogenesis (see *BICOID* GENE, *OSKAR* GENE) they are the first zygotic genes to be expressed in *Drosophila* development. Most encode TRANSCRIPTION FACTORS whose transi-ent gradients constitute POSITIONAL INFOR-MATION in the embryo and determine patterns of expression of HOMEOTIC GENES and other patterning genes (e.g. PAIR-RULE GENES and SEGMENT-POLARITY GENES), although mutual interactions complicate the picture. The *Drosophila* gap gene *knirps* encodes a hormone receptor-like protein of the steroid-thyroid superfamily essential for abdominal segmentation, similar to ligand-dependent DNA-binding proteins of verte-brates (see RECEPTOR PROTEINS). The *Krüppel* product (K_r) and *hb* product (Hb) bind to different DNA sequences upstream of the two *hb* promotors.

gap junction See INTERCELLULAR JUNCTION.

GAPs (GTPase-activating proteins) Proteins stimulating the GTP-hydrolysing activity of specific GTPases (e.g. RAS PROTEIN-related GTPases). The GTP p190 appears to be involved in coupling signalling pathways involving Ras and Rho. See RHO PROTEINS, GEFS G-PROTEIN, RECEPTOR.

GAP protein See GAPs.

gas bladder (swim bladder, air bladder) Elongated sac growing dorsally from anterior part of gut in most of the ACTINO-PTERYGII. In fullest development (in ACANTHO-PTERYGII) acts as hydrostatic organ; but may also act as an accessory organ of gaseous exchange, as a sound producer, or as a resonator in sound reception. Opinions differ as to whether the gas bladder or the vertebrate lung is the ancestral structure; they are certainly homologous.

gaseous exchange Exchange of a respiratory gas, either between (a) an organism's external and internal environments, or (b) between a part of the organism's internal environment (e.g. its vascular system) and the tissues of its body. The surfaces of spongy mesophyll and palisade cells in a leaf, animal gills and lungs are examples of the former; the mammalian PLACENTA, involving exchange of dissolved gases between foetal capillaries within the villus, the blood sinus of the mother (see Fig. 135), tissue fluids of both organisms, and ultimately the tissues themselves, is an example of the latter. Insect tracheal systems (see TRACHEA) allow gaseous exchange to occur directly between the external air and the body cells. Animal body surfaces in contact with the external environment and specialized for gaseous exchange are termed *respiratory surfaces*. They tend to be thin and permeable to oxygen and carbon dioxide molecules, have relatively large surface areas, are usually kept moistened and regularly ventilated, and are usually richly vascularized (by capillaries), or otherwise in contact with the HAEMOCOEL (many invertebrates) or blood sinuses (e.g. the mammalian placenta). Factors affecting the relative solubilities of gases in different media (e.g. temperature), and factors affecting the steepness of concentration gradients across gaseous exchange surfaces (e.g. ventilation, countercurrent systems, and the ease of removal of respiratory gases by

specialized blood mechanisms such as CARBONIC ANHYDRASE within red blood cells) all influence the efficiency of gaseous exchange. See FICK'S LAW OF DIFFUSION.

gasohol A mixture of 80% petrol and 20% dehydrated ethanol. Has a long history of use as a fuel for cars, although in many countries petrol is cheaper than ethanol. A cheap source of fermentable carbohydrate is crucial if gasohol's economic production is to be viable. See ALCOHOLS, BIOGAS.

gastric Of the stomach. *Gastric juice* is a product of vertebrate *gastric glands*, and contains hydrochloric acid (see CARBONIC ANHYDRASE), proteolytic enzymes (see CHIEF CELLS) and mucus.

gastrin Hormone secreted by mammalian stomach and duodenal mucosae in response to proteins and alcohol. Stimulates gastric glands of stomach to secrete large amounts of gastric juice. Relaxes pyloric sphincter and closes cardiac sphincter. Oversecretion may result in gastric ulcers. See SECRETIN, CHOLECYSTOKININ.

Gastropoda Large class of the MOLLUSCA. Marine, freshwater and terrestrial. Head distinct, with eyes and tentacles; well-developed, rasping tongue (radula). Foot large and muscular, used in locomotion. Visceral hump coiled, and rotated on the rest of body (torsion) so that the anus in the mantle cavity points forward; some forms undergo a secondary detorsion. Visceral hump commonly covered by a single (uni-valve) shell. Often a trochosphere larva. Includes subclasses Prosobranchia (e.g. limpets), Opisthobranchia (e.g. sea hares) and Pulmonata (e.g. snails, slugs).

Gastrotricha Class of the ASCHELMINTHES (or a phylum in its own right), probably closely related to nematode worms. Composed of a small number of cells, these minute aquatic invertebrates have an elastic cuticle but unlike nematodes have a ciliated by acellular hypodermis. Hermaphrodite or parthenogenetic. No larval stage. See ROTIFERA.

gastrula Stage of embryonic development in animals, succeeding BLASTULA, when the primary GERM LAYERS are laid down as a result of the morphogenetic process of GASTRULATION.

gastrulation Phase of embryonic development in animals during which the primary GERM LAYERS are laid down; its onset is characterized by the morphogenetic movements of cells, typically through the BLASTOPORE, forming the ARCHENTERON. Movements may result in EPIBOLY, but frequently also *emboly* in which cells invaginate, involute and ingress.

gas vacuole Structure aiding dispersal/buoyancy comprising gas vesicles (hollow, cylindrical tubes with conical ends) found in the cytoplasm of all orders of CYANOBACTERIA (blue-green algae), except the Chamaesiphonales. Gas vacuoles do not possess a true protein-lipid membrane but rather protein-ribs or spirals arranged like the hoops of a barrel, a form of membrane which is quite rigid (gas inside is at 1 atmosphere pressure) and permeable to gases. Inner membrane surface is hydrophobic, preventing condensation on it of water droplets, restraining them by surface tension, water seeping through the pores. The outer surface is hydrophilic to maximize interfacial tension, which would otherwise collapse the vesicle. It functions in light shielding and/or buoyancy.

gated channels TRANSPORT PROTEINS of membranes, not constitutively (permanently) open to the passage of molecules, but capable of closure. *Ligand-gated channels*, such as those responding to NEUROTRANSMITTERS, open only in response to an extracellular ligand; *voltage-gated channels* (e.g. the SODIUM PUMP of nerve and muscle fibres) are dependent for opening and closure upon an appropriate membrane potential. Others may only open when concentrations of certain ions in the cell are appropriate. See IMPULSE, MUSCLE CONTRACTION.

gatekeepers Genes that control cell proliferation and death. Compare CARETAKERS; see CANCER.

Gause's Principle The idea that if two competing species coexist in a stable environment, then they do so as a result of differentiation of their realized niches. However, if there is no such differentiation or if it is precluded by the habitat, then one competing species will eliminate or exclude the other.

GEFs (guanine-nucleotide-exchange factors) Proteins activating RAS PROTEINS through exchange of bound GDP for free GTP.

gel Mixture of compounds, some commonly polymeric, having a semisolid or solid constitution. See CHROMATOGRAPHY, ELECTROPHORESIS.

gel electrophoresis See ELECTROPHORESIS.

gel filtration See CHROMATOGRAPHY.

gemma Organ of vegetative reproduction in mosses, liverworts and some fungi. Consists of a small group of cells of varying size and shape that becomes detached from the parent plant and develops into a new plant; often formed in groups, in receptacles known as gemmae-cups.

gemmule (1) Of sponges, a bud formed internally as a group of cells, which may become free by decay of the parent and subsequently form a new individual. Freshwater sponges overwinter in this way. (2) See PANGENESIS.

gene Usually regarded as the smallest physical unit of heredity encoding a molecular cell product; commonly considered also to be a UNIT OF SELECTION. The term 'gene' (coined by W. Johannsen in 1909) may be used in more than one sense. These include: a) ALLELE, b) LOCUS, and c) CISTRON. What MENDEL treated as algebraic units ('factors') or 'atoms of heredity' obeyed his laws of inheritance and were considered to be the physical determinants of discrete *phenotypic* characters. This may be called the classical gene concept (see GENETICS). In 1903, W. S. Sutton pointed out that the segregation and recombination of Mendelian factors studied in heredity found a parallel in the behaviour of chromosomes revealed by

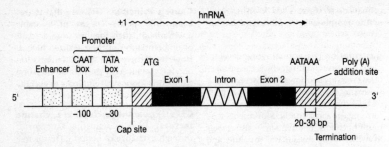

FIG. 60 *Generalized diagram of a eukaryotic Class II gene with a single intron. Enhancers may occur within or downstream of the gene. Transcription factor binding sites may also occur at the 5'-end, and occasionally the CAAT and TATA consensus sequences are absent. The region encoding hnRNA is indicated above. The cap site indicates where post-transcriptional modification occurs to the hnRNA by addition of a 'cap' (see RNA CAPPING).*

the microscope. Through the work of T. H. Morgan in the period 1910–20, chromosomes came to be regarded as groups of linked genes (or, more abstractly, of their loci), and the positions of loci and their representative alleles were first mapped on the chromosomes of *Drosophila* in this period. Morgan found that alternative genes (alleles) at a locus could mutate from one to another.

The importance of genes in enzyme production first emerged through work on the chemistry and inheritance of eye colour in *Drosophila*, and through work on AUX-OTROPHIC mutants of the mould *Neurospora crassa* by G. W. Beadle and E. L. Tatum (1941). It heralded the modern phase of genetics and molecular biology.

Genes soon became accepted as the heritable determinants of enzymes (one gene : one enzyme). However, the correspondence between the nucleotide composition of a gene and the amino acid composition of its encoded product was first revealed in variants of haemoglobin (a non-enzymic protein) and its genes, and their precise sequential correspondence (*colinearity*) was first established in detailed studies of the bacterial enzyme *tryptophan synthetase* and its gene. Great precision was by now being achieved in fine genetic mapping of BACTERIOPHAGE chromosomes using the *CIS*-TRANS TEST, and genes were soon regarded as nucleic acid sequences, mappable geogra-

phically on a chromosome, each encoding a specific enzyme, or (as in the polypeptide subunits of haemoglobin) non-enzyme protein. Subsequent work on *tryptophan synthetase* of the bacterium *E. coli* showed that two genes were required to encode this enzyme, and that their different polypeptide products associated to give the *quaternary structure* of the functional enzyme (see PROTEIN). The *functional gene concept* thus denoted a nucleic acid sequence encoding a single polypeptide chain. Nowadays, the term 'gene' is used to indicate the length of nucleic acid encoding any molecular cell product, be it a polypeptide, transfer RNA or ribosomal RNA molecule, and can usually be equated with CISTRON (but see ALLELIC COMPLEMENTATION). Class I genes (in the nucleolar organizer) encode the large rRNAs and 5.8S RNA; Class II genes encode tRNA, 5S RNA and some snRNAs. The initial RNA transcripts of all three classes usually undergo RNA PROCESSING, Class II transcripts in particular having their INTRONS spliced out. The manner in which these get excised from pre-RNA often dictates the precise cell product encoded by the gene, particularly when an intron contains a stop codon. This need not cause serious problems for gene definition, since just one promoter is usually employed – even though the final gene products may differ. But occasionally, true gene overlap does occur (see GENETIC CODE) through a shift in reading frame. One stretch of DNA (or RNA) may then be part of two genes; but even this leaves the functional gene concept unscathed. When specified by name, genes are given italics (e.g. *cdc2*, *osk*, etc.); but their protein products may be given first-letter or full capitals and are unitalicized. See Fig. 60; CHROMOSOME, PROTEIN SYNTHESIS and genetic references below.

gene amplification Process in which a small region of the GENOME of a cell is selectively copied many times while the rest remains unreplicated. Occurs in some specialized cell lines where large quantities of a particular cell product are needed rapidly. In rRNA cistrons, up to 1000 extra nucleoli may arise in amphibian oocytes in this way, with consequent large-scale ribosome production. Cistrons for rRNA are amplified in all cells with nucleoli. Gene amplification is associated with some kinds of dry resistance in cell cultures; amplification of cellular ONCOGENES is a fairly common feature of tumour cells. The whole phenomenon of selective DNA amplification is rich in theoretical interest. A powerful method of gene amplification *in vivo* (and of selective DNA in general) involves the *rolling circle*, whereby an extrachromosomal circular copy of the DNA sequence is produced which in turn produces many copies containing tandem repeats of the sequence. If these linearize and re-integrate into the chromosome, the enlarged genome will contain these identical repeat sequences, with consequent increase in copy number. Gene amplification is an excellent marker of genetic instability and commonly occurs in NEOPLASMS (see TUMOUR NECROSIS FACTORS). See GENE DUPLICATION, NUCLEOLUS, POLYMERASE CHAIN REACTION, POLYTENY.

gene bank Alternative term for GENE LIBRARY.

gene capture Method of discovering a gene's whereabouts by techniques which uncover its expression pattern. *Enhancer traps* are recombinant DNA constructs lacking cis-regulatory DNA and combining a weak promoter with a transgene (not the gene being 'captured') whose expression pattern is to be analysed *in vivo*. This gene will only be expressed if it inserts near one or more enhancer elements, and by histochemical staining for its product the expression pattern conferred by these enhancers can be observed, indicating the normal expression pattern of a nearby gene – which can then be identified by recombinant DNA techniques and analysed. The contribution to phenotype of genes so captured may then be studied by GENE KNOCKOUT techniques. See Fig. 61; GENE TRAPPING.

genecology Study of population genetics with particular reference to ecologies of populations concerned.

gene conversion Phenomenon, in eukaryotes occurring mainly at synapsis during meiosis, and thought to be a consequence of the mechanisms of homologous RECOMBINATION and DNA repair, whereby a donor DNA sequence, just a few hundred bases or perhaps a kilobase in length, is transferred from one gene to another having substantial sequence homology (usually between homologous loci, but sometimes between related sequences at non-homologous loci, notably those of dispersed MULTIGENE FAMILIES). The donor sequence is repaired back to its original form. It may be responsible for much of the diversity in some mammalian immunoglobulin production (see ANTIBODY DIVERSITY). In one model this involves 'nicking' (cutting) of a single-strand invading DNA sequence and melting (unzipping) of the invaded duplex DNA so that *heteroduplex* base pairing between the two can occur. The ousted sequence is enzymatically degraded while the invading sequence is cut and then annealed into its new position. Its original complementary strand is then used as template for its resynthesis to form the original duplex again. In another model, increasingly favoured, the recipient DNA duplex is nicked and gapped in both strands, the gap being filled by copy-synthesis using both strands of the donor duplex as templates. This would generate heteroduplexes only in regions flanking the gap. In a heterozygote of the yeast *Saccharomyces*, where conversion occurs at a rate of several per

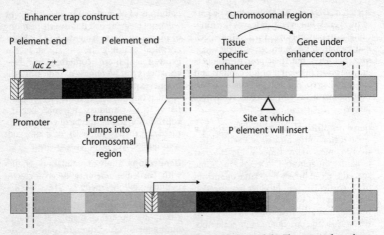

FIG. 61 *Use of an enhancer trap to capture tissue-specific genes in* Drosophila. *The essence of an enhancer trap is that it contains a weak promoter joined to a marker gene, in this case lacZ. In contrast to the reporter-gene construct there are no enhancers within the transgene construct, and so the lacZ gene will be expressed only if the transgene has inserted near enhancers. Thus, insertions of the same transgene into different regions of the genome will be associated with different tissue patterns of lacZ expression. The triangle indicates the site at which the P transgene inserted in the example.*

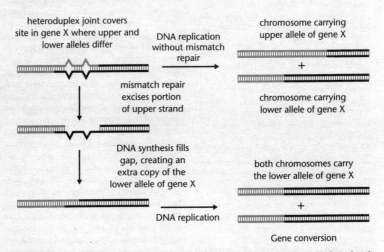

FIG. 62 *Gene conversion by mismatch correction. In this process, heteroduplex joints are formed at the sites of the crossing-over between homologous maternal and paternal chromosomes. If the maternal and paternal DNA sequences are slightly different, the heteroduplex joint will include some mismatched base pairs. The resulting mismatch in the double helix may then be corrected by the DNA mismatch repair machinery, which can erase nucleotide on either the paternal or the maternal strand. The consequence of this mismatch repair is gene conversion, detected as a deviation from the segregation of equal copies of maternal and paternal alleles that normally occurs in meiosis.*

cent per gene per meiosis, each allele can usually convert the other with about equal frequency, but examples of strongly biased conversion are known which could, in principle, lead to fixation of the favoured allele (see MATING TYPE).

Initiation of the cutting, and hence of the recombination, seems in one form of the process to occur within a gene promoter region. There is growing support for the view that the sites of heteroduplex formation (*Holliday junctions*) are responsible for much of eukaryotic crossing-over: there is about 30–50% association of gene conversion with crossing-over. Gene conversion in prokaryotes involves similar processes, although the initial alignment of homologous duplexes is less highly organized (see *recA*, *recB* and *recC*). See Fig. 62.

gene dosage Effective number of copies of a gene in a cell or organism. See DOSAGE COMPENSATION, GENE AMPLIFICATION.

gene duplication Mechanisms resulting in tandem duplication of loci along a chromosome. One of the possible evolutionary consequences of diploidy as opposed to haploidy is that with two functional representatives of a locus per cell it may not matter if one mutates and loses its original function. This is very likely also the evolutionary significance of gene duplication: one copy is free to mutate and take on a new function, the other functioning as normal (see CO-OPTION). The enzymes of the glycolytic pathway may have arisen this way from a common ancestral gene sequence, as most certainly do the various types of globin in haemoglobin. Controversy surrounds whether random fixation or positive selection of mutations occurs in the duplicate gene(s). At least one piece of evidence indicates selection operates. Gene duplication, especially in genes whose spatially restricted expressions now underpin animal development, appears to have been followed by acquisition by one or both genes of one or more additional roles. Following duplication, the alternative fate will be either that one of the genes mutates to a functionless state and is lost by genetic drift, or is stabilized by acquisition of a new role

(CO-OPTION). Acquisition of novel roles can occur by evolution of new expression sites because control regions of genes can evolve rapidly, e.g. by moving between gene neighbourhoods. See ORTHOLOGOUS GENES, PARALOGOUS GENES. Non-homologous CROSSING-OVER is one mechanism for producing gene duplication. Genome analysis indicates that many organisms contain significant percentages of duplicate genes: for *Haemophilus influenzae*, 30% of its 1,760 genes; for *Escherichia coli*, 46% of its 4,100 genes; for yeast, 14% of its 5,800 genes; in worms, flies, mice and humans there is as yet insufficient data although the majority of genes in mice and humans exist as multigene families (e.g. 2,000 or so protein kinases and perhaps 1,000 phosphatases compared with perhaps 350 protein kinases and 80 phosphatases in the nematode *Caenorhabditis elegans*). See MULTIGENE FAMILIES, GENE AMPLIFICATION, DUPLICATION.

gene expression/gene regulation The effect of those mechanisms which dictate whether or not a particular genetic element is transcribed (acts as a template for mRNA synthesis; see RNA and PROTEIN SYNTHESIS for more details). In bacteria it may best be explained by some variant of the JACOB–MONOD THEORY; in archaeons and eukaryotes this theory finds application too, although histones complicate the issue more in eukaryotes. Furthermore, unlike classical bacterial REPRESSORS, negative-acting proteins in eukaryotes do not in general bind DNA in a gene-specific manner to sterically inhibit polymerase binding; rather, they tend to exert their effects by interaction with TRANSCRIPTION FACTORS. It is common in eukaryotes to recognize two classes of regulatory phenomena involving gene expression: short-term (reversible) regulation, and long-term (often irreversible) regulation. Short-term regulation often relates to a cell's production of inducible and repressible enzymes. Steroid hormones ('effectors', see ECDYSONE) frequently bind to receptor proteins (see NUCLEAR RECEPTORS) in the cell prior to entry into the nucleus and activate transcription of selected genes. Patterns of gene expression sometimes exhibit

HOMOLOGY, where the use of REPORTER GENES and NUCLEIC ACID ARRAYS contribute greatly in PROTEOMICS generally. Long-term eukaryote regulation includes those processes involved in: a) rendering a cell DETERMINED, prior to differentiation, b) MATERNAL EFFECTS, c) the origins of facultative and constitutive HETEROCHROMATIN. Core histones of NUCLEO-SOMES have a key role in eukaryotic transcription regulation. Acetylation of the lysine residues in amino-terminal tails of particular core histones by acetyltransferases destabilizes a nucleosome, enabling DNA-binding components of the basal transcriptional machinery (which line up RNA polymerase correctly) to gain better access to promoter elements, facilitating transcription. Activator and coactivator proteins are further components in the control of transcription (see ENHANCER and Fig. 49). There are also regulators that deacetylate histones and form part of the transcription repression pathway, conserved from yeasts to vertebrates. Increasing interest centres around the roles of NON-CODING RNAS in the regulation of gene expression and of gene product activity. See CHROMOSOMAL IMPRINTING, DNA METHYLATION, EPIGENETICS, GENE CAPTURE, HISTONES and Fig. 140.

gene family See MULTIGENE FAMILY.

gene flow The spread of genes through populations as affected by movements of individuals and their propagules (e.g. spores, seeds, etc.), by NATURAL SELECTION and GENETIC DRIFT. See VAGILITY, PANMIXIS.

gene frequency Frequency of a gene in a population. Affected by MUTATION, SELECTION, emigration, immigration and GENETIC DRIFT. See HARDY–WEINBERG EQUILIBRIUM.

gene fusion See FUSION GENE.

gene gun (mechanical particle delivery) Methods involving mechanical delivery of DNA on microscopic particles into target tissues and cells. Termed 'biolistics', such methods are increasingly used to produce transgenic organisms by 'shooting' them with DNA-coated particles (see RESIST-ANCE GENES). The approach is also employed in injecting DNA VACCINES.

gene imprinting See CHROMOSOMAL IMPRINTING.

gene knock-in Refinement of GENE KNOCK-OUT technique allowing insertion of gene into strain.

gene knockout Replacement of a normal gene by a specifically mutated copy. Valuable tool for study of phenotypic effects in mice of mutants similar to those found in humans. Gene function is now commonly studied by an alternative to the rather laborious gene knockout method: RNA INTER-FERENCE. See GENE CAPTURE, HOMEOTIC GENES.

gene library (gene bank) Term given to the collection of DNA fragments resulting from digestion of a genome by a RESTRICTION ENDONUCLEASE. Each fragment is clonable by inserting it into an appropriate phage VECTOR and introducing it into an appropriate host cell (e.g. *E. coli*) for copying. cDNA libraries are libraries of those genes being transcribed in a system. See BACS, CHROMOSOME WALKING.

gene manipulation (gene modification) Set of procedures by which selected pieces (genes) of one genome (e.g. human) can be enzymatically cut out from it, spliced into a vector (e.g. a PLASMID) and inserted into an appropriate host microorganism (e.g. *E. coli*, yeast, etc.) in which it is replicated and passed on to all daughter cells of the microorganism forming a clone. A cloned gene may then be modified by site-directed mutagenesis (see MUTAGEN). Various enzymes used include RESTRICTION ENDONU-CLEASES, DNA LIGASES, etc. If the aim is mass production of the substance encoded by the transposed gene, then insertion is so arranged that transcription of the gene occurs within the clone of cells, producing large amounts of the desired substance (e.g. for insulin see Fig. 63; see also Fig. 168). Clones can be screened for their activities and selected appropriately. The recombinant DNA thus artificially produced might be hazardous unless properly contained, and strict precautions are applied in such work. Thus only non-pathogenic strains are used as hosts, or else strains that can only grow in laboratory conditions. See BIOTECH-NOLOGY, CLONE, ELECTROPORATION, GENE LIB-

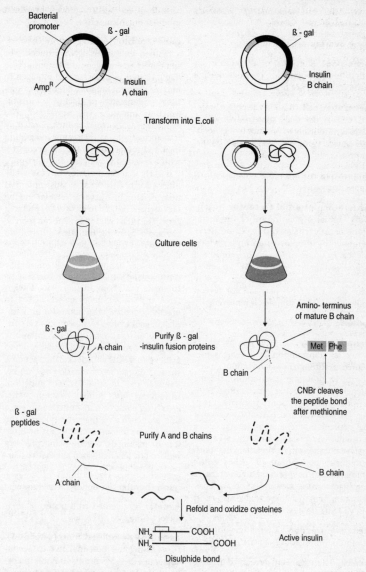

FIG. 63 *The expression of genes for human insulin A and B polypeptide chains in* E. coli. *The two genes are expressed in separate bacterial systems and emerge fused to the carboxy terminus of β-galactosidase. These prefixes are removed by cyanogen bromide (CNBr), which cleaves at internal methionines (which insulin chains lack). The A and B chains are then refolded and joined covalently to form the mature protein, insulin. See* GENE MANIPULATION.

RARY, GMO, PEST MANAGEMENT, RESISTANCE GENES, TRANSGENIC, VECTOR.

gene overlap See GENETIC CODE.

gene pool Sum total of all genes in an interbreeding population (gamodeme) at a particular time. See DEME.

generative cell (Bot.) In the pollen grain, the cell of the male GAMETOPHYTE which divides mitotically to produce two generative nuclei (gametes). See DOUBLE FERTILIZATION.

generative nucleus See DOUBLE FERTILIZATION, GENERATIVE CELL.

generator potential (receptor potential) Initial depolarization of the membrane of an excitable cell (receptor, nerve or muscle) by stimulus or transmitter, which triggers an ACTION POTENTIAL when threshold depolarization is reached. Unlike an action potential, its magnitude decreases rapidly with distance and it may last for several milliseconds after stimulus is removed.

generic (Adj.) Of GENUS.

gene silencing (1) Processes by which either (a) transcription of a gene is prevented, as through DNA METHYLATION, especially of its promoter (transcriptional gene silencing, TGS); or (b) a gene's mRNA transcript is destroyed prior to translation (e.g. RNA INTERFERENCE, and POST-TRANSCRIPTIONAL RNA SILENCING); or (c) a micro-RNA acts as a translational repressor of mRNA. (2) Larger-scale phenomena involving silencing (transcriptional prevention) of either entire chromosomes, or specific sections of them. See CHROMOSOMAL IMPRINTING, DOSAGE COMPENSATION, HETEROCHROMATIN. See also PLANT DISEASE AND DEFENCES.

gene suppression Alternative, and less common, phrase for GENE SILENCING.

genet The organism developed from a zygote. The term is used especially for modular organisms, and members of a clone, to define the genetic individual and (to contrast with RAMET) the potentially physiologically independent part that may arise from the iterative process by which modular organisms grow.

gene tagging A cloning method in which a defined exogenous DNA sequence, or sometimes an endogenous transposon (the 'tag'), is used as an insertional mutagen to mutate a gene at random, altering its function – generally producing a loss-of-function mutation (see MUTATION). The exogenous DNA could be incorporated by transformation or injection, integrating into a chromosome (if the tag segregates with a specified target gene during crosses, it is likely that the mutation was caused by integration). The DNA of this mutant is then used to construct a GENE LIBRARY and the tag can then be used to recover the mutated gene. The intact wild-type gene can then be recovered by using the DNA adjacent to the inserted tag to probe a wild-type gene library. Compare GENE CAPTURE.

gene targeting Techniques enabling us to create mutations in (i.e. place foreign DNA in) specific genes at will. Used to inactivate (create NULL MUTATION) in mammalian genes, often by HOMOLOGOUS RECOMBINATION using a targeting vector containing a disrupted gene construct; but far easier to achieve in yeast, with its smaller genome size and simpler recombination machinery. Valuable tool in study of gene function. Introduced into animal EMBRYONIC STEM CELLS, they could be transmitted to the germ line and hence to the gene pool. In the future, *gene transplacement* might be possible in mammals, a gene being precisely replaced by an engineered alternative (currently possible with yeasts). See Fig. 64.

gene therapy See HUMAN GENE THERAPY.

genetic Concerned with genes, or their effects. Compare HEREDITARY.

genetically modified food (GM food) In assessing the potential risk to a consumer of a diet containing GMOs it is important to take into account the bedrock of scientific evidence. Transgenes ought, in principle, to be digested to their constituent nucleotides and nucleosides in the same way as any other nucleic acid in the diet,

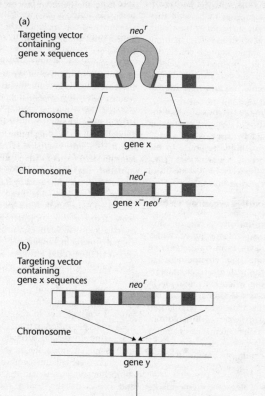

FIG. 64 (a) *Homologous recombination (gene targeting). The targeting vector, a neomycin resistance gene (neo^r) flanked by gene x sequences is shown pairing with a chromosomal copy of gene x. Homologous recombination between the targeting vector and genomic x DNA results in the specific disruption of the chromosomal copy of gene x (that is, x^- but now neo^r). (b) Non-homologous recombination (random integration). In non-homologous recombination the targeting vector is randomly inserted into the chromosomal DNA (here arbitrarily shown to be inserted into gene y). The cells are still x^+ because the targeting vector was inserted somewhere else in the genome. Most non-homologous or random insertions of the targeting vector occur through the ends of the targeting vector, presumably at an existing double-strand break in the chromosomal DNA. The darkly shaded boxes represent exons, the open boxes introns.*

although some work with *Caenorhabditis* suggests that transgenes in bacteria they eat can cause GENE SILENCING in the worm. However, it is incorrect to equate 'natural' with 'safe'. A single cup of coffee, for instance, contains natural carcinogens equal to at least one year's worth of synthetic carcinogenic residues in a normal western diet.

Recent work in the UK indicates that the ecological merits and demerits of growing GM food should be assessed on a case-by-case basis, no simple response to the matter being possible. See RESISTANCE GENES, and websites www.food.gov.uk and www.newscientist.com/hottopics/gm.

genetically modified organism See GMO.

genetic assimilation Phenomenon involving conversion of an acquired character (resulting maybe from transference of individuals from one environment to another) into one with greater HERITABILITY than it had before, where the causal mechanism involved is selection acting on the genotypes of the transferred population. The significance of the process in evolutionary terms is debatable: there is no assurance that the initial acquired character will be adaptive in the conditions bringing it about. See NATURAL SELECTION, MUTATION, PHENOCOPY.

genetic code Table of correspondence between (a) all possible triplet sequences (codons) of messenger RNA and (b) the amino acid which each triplet causes to be incorporated into protein during PROTEIN SYNTHESIS. In addition, certain triplets cause termination of polypeptide chain synthesis and occur regularly at the 3'-ends of polypeptide-encoding sequences (open reading frames). They may also arise as motivating mutations within an encoding sequence (see CODON). DNA is sometimes spoken of as a 'code', but this is shorthand. Indeed, the genetic code itself is really a *cipher*, since the amino acids in a polypeptide correspond to the letters of an alphabet rather than to words. Since more than one triplet may encode some amino acids, the code is said to be degenerate. The translation of mRNA into protein is only possible because of the

specificity of transfer RNA molecules for particular amino acids – itself the result of specificity of the enzymes which activate amino acids and bind them to appropriate tRNAs. See Fig. 65.

The code is remarkably uniform from prokaryotes to eukaryotes; but in mitochondria there are fewer codons and slightly different reading rules; in *Mycoplasma capricolum* the codon UGA is read as tryptophan rather than as a stop codon. Until the 1970s it was thought that the reading frame for translating mRNA into polypeptide never overlapped: that there were unequivocal initiation sites for any mRNA molecule. However, overlapping reading frames (overlapping genes) are widespread in viruses (e.g. ϕX174), organelles and bacteria; both same-strand and complementary strand overlaps exist. In Archaea, eubacteria and animals, including mammals, the triplet codon UGA encodes a twenty-first amino acid, called selenocysteine, which is not created post-translationally – unlike some rare amino acids found in proteins – but formed enzymatically from serine while the latter is still attached to a special transfer RNA molecule, different from that for normal serine. In certain specialized niches, such as mitochondria, the meanings of certain codons are redefined and there is now compelling evidence that in certain Archaea and eubacteria the triplet codon UAG encodes a twenty-second amino acid, called pyrrolysine.

The lack of pairing between codons and amino acids suggests the code arose by accident; but some work suggests the code could have been shaped by chemical affinities between specific base sequences and certain amino acids: sequences tending to bind a particular amino acid tend to contain codons for it; the present code also seems to be among the best of possible codes at minimizing the effect of mutations, in that single base changes are likely to substitute a similar amino acid and thus make minimal changes to the protein product. Also, mutations swapping bases of similar size are commoner than those substituting purine for pyrimidine and vice versa. During protein synthesis the first and third members of

1st position (5' end) ↓	U	C	A	G	3rd position (3' end) ↓
U	Phe	Ser	Tyr	Cys	U
	Phe	Ser	Tyr	Cys	C
	Leu	Ser	STOP	STOP	A
	Leu	Ser	STOP	Trp	G
C	Leu	Pro	His	Arg	U
	Leu	Pro	His	Arg	C
	Leu	Pro	Gln	Arg	A
	Leu	Pro	Gln	Arg	G
A	Ile	Thr	Asn	Ser	U
	Ile	Thr	Asn	Ser	C
	Ile	Thr	Lys	Arg	A
	Met	Thr	Lys	Arg	G
G	Val	Ala	Asp	Gly	U
	Val	Ala	Asp	Gly	C
	Val	Ala	Glu	Gly	A
	Val	Ala	Glu	Gly	G

FIG. 65 *Diagram of the amino acids encoded by* RNA *triplets (codons). The triplets run from the 5'-end to the 3'-end of the* RNA*. The three nucleotides for any triplet are found by taking one from the left column, one from the horizontal row and one from the right column. The amino acid indicated at their intersection is that encoded by that triplet. Thus CCC encodes proline. Three stop codons are also indicated.*

a codon are much more likely to be misread than the second, apparently making the existing code more error-proof than one in a million randomly generated codes. See WOBBLE HYPOTHESIS and Fig. 65.

genetic counselling Service, generally provided by specialists in human genetic disorders. Seeks to explain to parents with children already affected by genetic disorders the nature of those disorders, and the probability of their (and their children) having further affected offspring, and helps families to reach decisions and take appropriate action in the light of this information.

Many carriers of genetic disorders (i.e. heterozygous for the condition, but not expressing it) can be diagnosed through screening procedures, preferably of more than one tissue type (SEE RFLP, POLYMERASE CHAIN REACTION); e.g., several inherited mutations predisposing patients to cancer can now be detected early in life. The probability of someone being a carrier can often be ascertained from information about the occurrence of the disorder in close relatives and, to an increasing extent, by DNA analysis. See AMNION (for amniocentesis), CHORION, DOWN'S SYNDROME, HUMAN GENE THERAPY, *IN VITRO* FERTILIZATION.

genetic diversity See BIODIVERSITY, GENETIC VARIATION.

genetic drift (Sewall Wright effect) Statistically significant change in population gene frequencies resulting not from selection, emigration or immigration, but from causes operating randomly with respect to the fitnesses of the alleles concerned. Such random sampling error might for example occur, if, in a population of beetles, a disproportionately large number of those killed by a wandering elephant happened to be heterozygous for a recessive eye colour. The frequency of this, and maybe other, alleles could now alter significantly even though no selection had taken place. Genetic drift is expected to be of significance only in small populations, where alleles may easily go to extinction or fixation by chance alone. In large populations, effects of sampling error are usually considered to be small in comparison with those of NATURAL SELECTION. See FOUNDER EFFECT.

genetic engineering See GENE MANIPULATION.

genetic fingerprinting See DNA FINGERPRINTING.

genetic immunization See VACCINES.

genetic load The proportionate decrease in average fitness in a population relative to that of the fittest or optimal genotype, as caused principally by mutation (mutational load) and segregation (segregational load). Mutational load is the degree to which mutation impairs the average fitness of a population. J. B. S. Haldane (1937) argued that when all loci are considered, this equals the gametic mutation rate multiplied by a factor of 1 or 2 (depending on dominance) and that since partial dominance is usual the mutation load is twice the total rate per gamete. So, for mutations with large heterozygous effect, if the total mutation rate per gamete for these loci is 0.01, population fitness would be lowered by 0.02. However, adjustments have to be made to allow for mutations of minor effect. Segregational load is the loss of fitness due to segregation in polymorphic populations, where genes are maintained by HETEROZYGOUS ADVANTAGE and, as in the case of SICKLE-CELL ANAEMIA, may be high. An alternative view is that many or most polymorphisms involve selectively neutral allelic differences (see NEUTRALISM).

genetic mapping See CHROMOSOME MAPPING.

genetic marker Genes, DNA sequences, heterochromatic regions or any other distinguishing feature of genotypes, chromosomes or karyotypes which may be used to keep track of specific chromosomes, cells or individuals. Almost by definition they must exist in at least two alternative and (preferably) readily identifiable forms; when the rarest has a frequency of 1–2%, the marker is described as polymorphic (see POLYMORPHISM). Genetic markers enabled MENDEL to score discrete phenotypes and develop his theories of segregation and random assortment and provided subsequent workers with the tool to develop the principles of LINKAGE and the chromosomal theory of inheritance. Markers are now routinely used to identify transgenic cells or organisms carrying a particular VECTOR. Expression of the marker gene, carried on the same vector as another introduced gene of unrelated but useful effect, indicates which cells or organisms also harbour this other gene. If the marker gene carries drug resistance, positive selection of the cells can be achieved by administration of the drug. Marker genes closely linked to as yet unisolated genes for human genetic disorders are of enormous help in identifying cells, embryos and individuals carrying these deleterious alleles. Once isolated, these genes will be locatable by DNA PROBES. Other markers include RFLPS and some SATELLITE DNAS. A recent survey of 109 human loci (30 MICROSATELLITE and 79 RFLPs) confirms the view that most human molecular variation is homogeneous, with fluctuations consistent with random drift following a recent expansion from the (African?) parent population. The result is a mixture of clines, balanced polymorphisms and recent selective 'sweeps', but nothing resembling discrete 'races'. Local adaptations do, nonetheless, result from selection.

In developmental biology, cell markers are frequently used. These are cells whose phenotype indicates the occurrence of some genetic event, e.g. mitotic recombination or onset of gene expression, as a guide to some important developmental feature (e.g. a compartment boundary). Good cell markers do not damage their cells and are autonomous: their expression should be restricted to the cells carrying them. They should not, e.g., cause other cells to express the marker phenotype simply through leakage of product.

genetic modification See GENE MANIPULATION, GMO.

genetic redundancy Two or more genes performing the same function, inactivation of one having little or no effect on phenotype. Modelling suggests that such redundant genes may persist in populations, not least if the overlap in function is pleiotropic.

genetic resources Defined in the Convention on Biological Diversity 2001 as any material of plant, animal, microbial or other origin containing functional units of heredity of actual or potential value (humans excluded).

genetics Study of heredity and variation in biological systems. The origin of the modern, particulate, theory of inheritance is marked by the work of MENDEL, with many other contributors to this classical phase of genetics in which phenotypic ratios in breeding tests were ultimately explained in terms of chromosome behaviour. Postclassical work, largely with microorganisms and phages, directed attention towards a biochemical understanding of genetics, isolation of DNA and determination of its structure, eventually enabling GENE MANIPULATION. Population (ecological) genetics attempts to quantify the roles of selection and genetic drift in shaping the GENETIC VARIATION within populations. See GENE.

genetic screens See SCREENING.

genetic variation Occurrence of genetic differences between individuals, most commonly studied in species populations. Upon such differences, when expressed, can natural selection act. MUTATION is the ultimate source of genetic variation; but in most sexual populations MEIOSIS then results in recombination both between and within parentally-derived chromosomes, generating enormous genetic diversity among gametes and, at least potentially, among offspring phenotypes. The BREEDING SYSTEM of the population is an important consideration, affecting the relative levels of free and potential variability. *Free variability* is genetic variation which is expressed phenotypically and approximates to the proportion of homozygotes in the population; *potential variability* is genetic variation which does not express itself phenotypically and approximates to the proportion of heterozygotes in the population. See Fig. 66. Using the same MICROSATELLITE loci and RFLPS, it is estimated that the proportion of human genetic variation within local populations is 84.4%, that between local populations is ~15%, typical of other animal species, and that only a small further fraction is held between continents.

However, two or more loci often affect the same character (polygenic inheritance), and in the simplest case (where both loci are additive with respect to effect on phenotype) may generate the same phenotypes in the double heterozygote, GgHh, as in both the double homozygotes, GGhh and ggHH, thus 'protecting' the latter from selection, even though their *homozygous potential variation* contributes to the total potential variation in the population. Alleles always exert their effects against a 'genetic background' which can modify their expression.

Only through crossing, with its subsequent segregation, can a major part of the potential variability in a population be freed and become available for selection. Inbred populations, comprising all possible types of homozygotes, will harbour almost as much potential variability as outbred ones with the same genes in the same frequencies; but this variability is never freed in inbred populations and they tend to be evolutionarily static. See VARIATION, GENETIC MARKER, POLYMORPHISM, HERITABILITY.

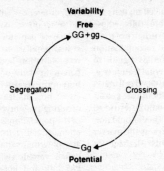

FIG. 66a *Free and potential variability. Free variability (open to direct selection) is represented by differences in phenotype between GG and gg genotypes and is converted to potential variability of Gg by crossing and is released again by segregation.*

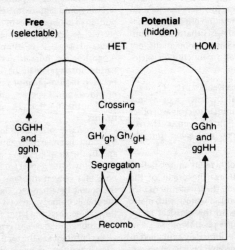

FIG. 66b *The states of variability in a system where the G and H loci are additive in effect on phenotype, so that GgHh and GGhh are phenotypically identical. The extremes of phenotype (free variability) are expressed by the double dominant and double recessive.*

gene trapping A procedure in which a REPORTER GENE integrates within an intron of a gene, in the correct orientation and reading frame, so that the upstream introns of the host gene are spliced to it. This results in production of a fusion protein including part of the host gene product fused to the product of the reporter gene. Expression of this transgene now depends upon expression of the host gene containing it,

the advantage being that since the gene trap acts as an insertional mutation, organisms may reveal altered phenotypes; and cloning of the interrupted gene is simple because the construct integrates within the gene. See GENE CAPTURE.

genitalia (Zool.) External reproductive organs. In many arthropods, especially insects, structures of male and female genitalia are often species-specific and serve as a pre-zygotic mechanism preventing hybridization.

genome The total genetic material within a cell or individual, depending upon context. A bacterial genome comprises a circular chromosome containing an upper limit of about 0.01 pg (1 picogram = 10^{-12} g) of DNA; a haploid mammalian genome contains from about 3 to 6 pg of DNA, while some amphibia and psilopsid plants may contain well over 100 pg of DNA per haploid genome. The marine chordate *Oikopleura dioica* has a minute genome (~51 Mb, ~15,000 genes). The first organism to have its genome's complete nucleotide sequence analysed (1995) was the non-infectious Rd type of the bacterium *Haemophilus influenzae* (1.8 Mb; 38% of bases G+C; see ARCHAEA). Apart from humans (see HUMAN GENOME PROJECT), animals already sequenced are the nematode *Caenorhabditis elegans*, the fruit fly *Drosophila*, mouse, rat and zebrafish. The next four candidates are: chicken, chimpanzee, honey bee and sea urchin. During 2002, a highly organized draft of 278 million base pairs of the mosquito *Anopheles gambiae*, and the genome sequence of the malarial parasite, *Plasmodium falciparum*, were published. See *ARABIDOPSIS THALIANA*, CROP GENOME PROJECT.

The considerable content of archaeal genes within some bacterial genomes suggests repeated HORIZONTAL TRANSFER of genes from that source during evolution (see HYPERTHERMOPHILE). The DNA content may or may not correlate with chromosome number, the fit being generally better in plants than in animals. Much eukaryote DNA is not part of structural genes coding for detectable cell products. The evolution of genome size is much debated. Increasing genome size would predictably bring with it increased per-genome mutation rate, if per-locus mutation rate remains constant. Yet despite the increase in genome size by several orders of magnitude from phage to fungi there appears to be no increase in per-genome mutation rate; in which case the per-locus mutation rate also varies by several orders of magnitude. The evolution-ary transitions from prokaryote to eukaryote, and from invertebrate to vertebrate, were accompanied by dramatic increases in average genome size. Perhaps a rise in inappropriate gene expression with increasing genome size was a constraint keeping genome sizes small until a mechanism for silencing such genes evolved. The uncoupling of gene transcription and translation accompanying evolution of the nuclear envelope may have enabled better filtering of such 'noisy genes', and better targeting of signals to gene expression (increasing the signal:noise ratio). Histones and nucleosomes may well have had a part to play in this. As for the invertebrate–vertebrate transition, it is known that genomes in the latter are more heavily methylated than those in the former. Invertebrates appear to methylate small portions of their genomes, apparently inactivating SELFISH DNA. Possibly, vertebrates extended this control to other parts of their genomes, reducing inappropriate gene expression. See C-VALUE, GENOME MAPPING.

genome mapping See HUMAN GENOME PROJECT for some of the methods being employed here. It should be remembered that two sorts of map are available: genetic maps based largely on linkage data (see CHROMOSOME MAPPING) and physical maps based largely on cloned DNA segments (see VECTOR, YEAST ARTIFICIAL CHROMOSOMES). Single DNA base alterations (single nucleotide polymorphisms, or SNPs) between human individuals are being analysed in an effort to discover which genes are involved in the development of disorders such as diabetes and schizophrenia, and assess predisposition to them.

genomic imprinting (parental imprinting) The heritable modification of expressivity and/or PENETRANCE of genes in offspring determined by sexual provenance – which parental sex they are inherited from. Maternally derived genes experience oogenesis whereas paternally derived genes experience spermatogenesis; consequently, maternal and paternal diploid embryos are not equivalent (diploid embryos with two egg or two sperm genomes generally fail to develop). Can be both developmentally stage-specific and tissue-specific. Implicated in an increasing number of human genetic disorders, but the molecular mechanisms are only just becoming clear. Sometimes (at least) involves DNA METHYLATION (more methylation, less expression). See CELL MEMORY, CHROMOSOMAL IMPRINTING, INSULIN/INSULIN-LIKE GROWTH FACTOR PATHWAY.

genomic library See GENE LIBRARY.

genomic remodelling See CHROMOSOME REMODELLING.

genomics The application of DNA sequencing and genome mapping to the study, design and manufacture of biologically important molecules. Applications abound in agribusiness, pharmaceutical and chemical industries; but some techniques are controversial and agreed international standards are required. See GMO.

Functional genomics comprises the techniques employed to engineer whole chromosomes (e.g. human or mouse YEAST ARTIFICIAL CHROMOSOMES) or eliminate a single gene, either in the whole organism or in just a single tissue, by homologous recombination in embryonic stem cells (e.g. in the mouse); or to insert large pieces of self or foreign DNA in transgenic organisms. Compare PROTEOMICS.

genotype Genetic constitution of a cell or individual, as distinct from its PHENOTYPE.

genus Taxonomic category, between FAMILY and SPECIES, which may include one or more examples of species. See BINOMIAL NOMENCLATURE, CLASSIFICATION.

geographical isolation The processes underlying allopatric SPECIATION models.

geological periods, epochs See Appendix. The fossil record commences in pre-Cambrian times (SEE ARCHAEAN, PROTEROZOIC) with organisms resembling bacteria and blue-green algae (see STROMATOLITES) in deposits 3 billion years old. A few green algae have been identified from an upper Precambrian formation about 1 billion years old. In Cambrian rocks, in addition to algae

and bacteria, a diverse range of aquatic invertebrate animals is found (e.g. brachiopods, trilobites and the onychophoran *Aysheaia*), including many which apparently left no descendants. Fungi and spores with wall markings like those of various land plants were also present. In the Ordovician period there was sufficient land vegetation to support such animals as burrowing worms, while the first fish-like vertebrates (ostracoderms) appeared. In this and the Silurian period, land vegetation seems to have included plants with bryophyte-like attributes, while primitive land vascular plants appeared and expanded in the late Silurian and early Devonian (420–390 Myr BP). Terrestrial arthropods have been found in early Silurian (Pennsylvania) and Upper Silurian of England (411 Myr BP): the arthropod invasion of land occurred prior to the Devonian radiation of vascular plants. But by the late Devonian both insects and vertebrates (amphibians) had appeared on land. Club mosses, horsetails and ferns (many of them tree-like) were now abundant there, providing an enormously rich flora in the Carboniferous period. Tremendous accumulations of remains of Carboniferous plants, partially decayed and subjected to intense pressure, formed the coal seams of today. The Mesozoic era was the age of the great dinosaurian reptiles, while gymnophytes were the dominant land plants. Flowering plants appeared in the Jurassic period and radiated during the late Cretaceous. However, despite an increase in species numbers, it seems that at this time they played a subordinate role in vegetation that was largely fern-dominated. The ascendancy of mammals also began with the end of the Mesozoic era. See CONTINENTAL DRIFT, and individual periods, epochs and groups of organisms.

geophytes Class of RAUNKIAER'S LIFE FORMS; plants possessing perennating buds below the soil surface on a corm, bulb, tuber or rhizome.

geotaxis TAXIS in which the stimulus is gravitational force.

geotropism See GRAVITROPISM.

germ cell Any of the cells forming a germinal epithelium, plus its cell products, the GAMETES. In vertebrates, *primordial germ cells* migrate from the early gut or yolk sac to sites of the eventual genital ridges in ovaries or testes where, after rounds of mitosis, some of their daughter cells will undergo meiosis and differentiate into gametes. See GERM PLASM, OVARY, TESTIS.

germinal epithelium See OVARY, TESTIS.

germinal vesicle Enlarged oocyte nucleus formed during diplotene of first meiotic prophase. See MEIOTIC ARREST.

germ layer One of the main layers or groups of cells distinguishable in an animal embryo during and immediately after gastrulation. Diploblastic animals have two such layers: ectoderm (outermost) and endoderm (innermost). Triploblastic animals have a third layer, mesoderm, situated between these two. Roughly speaking, ectoderm gives rise to epidermis and nerve tissue; endoderm gives gut and associated glands; mesoderm gives blood cells, connective tissue, kidney and muscle. Cartilage derives from more than one lineage. The derivations of various organs and tissues in vertebrates are indicated in Fig. 67.

germ line That cell line which, early in development of many animals, becomes differentiated (determined) from the remaining *somatic cell line*, and alone has the potential to undergo MEIOSIS and form gametes. See GERM PLASM, POLAR PLASM.

germ plasm, continuity of The germ plasm theory (WEISMANN, 1892) held that the nuclei of an individual's germ cells, unlike those of its body cells, are qualitatively identical to the nucleus of the zygote from which the individual developed. Weismann held that cellular differentiation was preceded by some loss of material from the nucleus of a cell, but that heredity from one generation to another was brought about by transference of a complex nuclear substance (germ plasm), itself part of the germ plasm of the original zygote which was not used up in the construction of the animal's body, but was reserved unchanged for the formation of its germ cells. See ABERRANT CHROMOSOME BEHAVIOUR (3).

gestation period Period between conception and birth in viviparous animals. Where there is no conception (e.g. in some cases of PARTHENOGENESIS), it would be the interval between egg maturation and birth.

ghost of competition past A term introduced by J. H. Connell in 1980 to stress that interspecific competition, acting as an evolutionary force in the past, has often left its mark on the behaviour, distribution or morphology of species, even when there is no present-day competition between them.

giant chromosome See POLYTENY.

giant fibres (giant axons) Nerve axons of very large diameter (e.g. 1 mm in squids). Occur in many invertebrates (e.g. annelids, crustacea, and nudibranch and cephalopod molluscs) and some vertebrates where rapid conduction is required (e.g. for escape) and achieved through reduced electrical resistance of the axoplasm with larger axon diameter. May be either a single enormous cell or the result of fusion of many cells. There are fewer synaptic barriers in these nerves than is usual, increasing speed of conduction.

gibberellins Class of plant GROWTH SUBSTANCES (see ISOPRENOIDS), originally isolated from the fungus *Gibberella fujkoroi* when it caused abnormal elongation of infected RICE plants in the 1930s. The best-studied is *gibberellic acid*, GA_3, which has the following structure:

Gibberellins have spectacular effects upon stem elongation in certain plants; they cause dwarf beans to grow to the same height as tall varieties, are involved in seed germination, and will substitute in many species (e.g. lettuce, tobacco, wild oats)

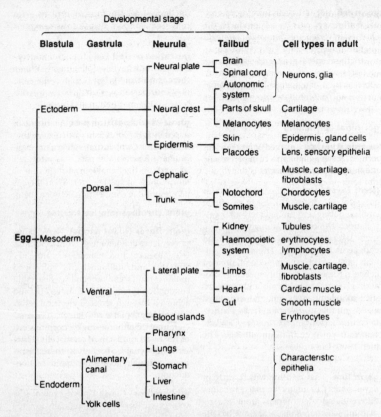

FIG. 67 *Times of production and* GERM LAYER *origins of the basic tissues and organs of a vertebrate. The diagram represents a decision-making plan for differentiation although many more decision points are actually involved. Some cell types, e.g. cartilage, have more than one derivation.*

for the dormancy-breaking cold or light requirement, promoting early growth of the embryo. Specifically, they enhance cell elongation, making it possible for the root to penetrate growth-restricting seed coat or fruit wall, which has practical application in ensuring uniformity of germinating barley in production of barley malt used in the brewing industry. Can be used to promote early seed production of biennial plants; stimulate pollen germination and growth of pollen tubes in some species; and can promote fruit development (e.g. almonds, peaches, grapes). In barley and

other grass seeds, the embryo releases gibberellins causing the aleurone layer of the endosperm to produce enzymes (e.g. α-amylase), digesting the starch store to mobilize sugars for germination. The gibberellin causes expression of the gene encoding the amylase (i.e. de-repression), the mechanism resembling that by which ECDYSONE exerts its effects. Some commercial plant growth retardants achieve their results by blocking gibberellin synthesis.

gill (Bot.) See LAMELLA, GILL FUNGI. (Zool.) Any of several organs of gaseous exchange in aquatic animals, such as the vascularized projections of the external surface (*external gills*) of many annelids and arthropods. Parapodia of polychaetes and thin-walled trunk limbs of branchiopod crustaceans increase the surface area for diffusion of dissolved gases and probably function as gills; but in some *tracheal gills* of insect larvae and pupae, these leaf-like projections, despite their rich supply of tracheae (though lacking open spiracles), absorb less oxygen than the remaining body surface. *Spiracular gills* occur in some aquatic insect pupae where one or more pairs of spiracles is drawn out into long processes and generally supplied with a PLASTRON. Vertebrate gills are either *external* (ectodermal) or *internal* (endodermal). The former occur in larvae of a few bony fish (*Polypterus* and some lungfish) and of amphibians, and are almost always soon lost or replaced by internal gills. Fish gills may serve as organs of OSMOREGULATION. See VENTILATION and Fig. 68.

gill arches Visceral arches between successive gill slits in jawed fish (typically five pairs) and larval amphibians (never more than four pairs) comprising gill bars and attendant tissues.

gill bar Skeletal support of gill slits in chordates, containing in addition blood vessels and nerves. In jawed fish typically comprises a dorsal element (*epibranchial*) and a ventral element (*ceratobranchial*), the two bent forward on each other. Five pairs are present in most jawed fish, in addition to the jaws and HYOID ARCH.

gill book Stacks of segmentally arranged vascularized leaf-like lamellae attached to the posterior faces of oscillating plates. This movement and locomotion of the limbs on which they occur (e.g. the swimming paddles of the merostomatid *Limulus*) provide ventilation of the gill book under water. See LUNG BOOK.

gill fungi Fungi belonging to the family Agaricaceae (BASIDIOMYCOTA); possessing characteristic fruiting body (BASIDIOCARP) comprising a stalk (stipe) supporting a cap (pileus), on the under-surface of which are radially arranged gills (lamellae) bearing the hymenium; e.g. mushrooms, toadstools.

gill pouch Outpushing of side-wall of pharynx towards epidermis in all chordate embryos. Precursor of gill slit in fish and some amphibia; in terrestrial vertebrates breaks through to exterior only temporarily or (as in humans) not at all. See SPIRACLE.

gill rakers Skeletal projections of inner margins of gill bars of fish, particularly elongated in those which strain incurrent water for food particles.

gill slit One of a series of bilateral pharyngeal openings of aquatic chordates. Usually vertically elongated and, in urochordates, cephalochordates and cyclostome larvae, primarily concerned with FILTER-FEEDING, but probably with additional role in gaseous exchange. In fish and most larval amphibia they have the latter role; but presence of GILL RAKERS in filter-feeding fish may again give them a nutritional role. See GILL POUCH, from which they develop.

Ginkgophyta Ginkgos, Maidenhair tree. A group of gymnosperms comprising one family (Ginkgophyceae), one extant genus (over six known from fossils), and one species *Ginkgo biloba* (Maidenhair tree, native to China) possessing active cambial growth and fan-shaped, simple, alternate to fascicled leaves with open, dichotomous venation. Petioles are equal to or exceeding the blades and resin canals are absent. Plants are deciduous, dioecious (possibly rarely monoecious) trees with a furrowed, grey coloured bark with flattened ridges. Catkin-like pollen cones are borne on spurs, while the sporophylls are distributed along the axis. Seed cones are absent; ovules and seeds are exposed and seed coats fleshy. After pollination, sperm are transported to the vicinity of an ovule in a pollen tube but they are flagellated and motile. Pollen is spherical and not winged. *Ginkgo biloba* may owe its survival in China to the persistence of trees that resprout from specialized basal swellings on the trunk, has changed very little during the last 80 Myr and is the sole survivor of a once widespread and moderately abundant group that flourished in

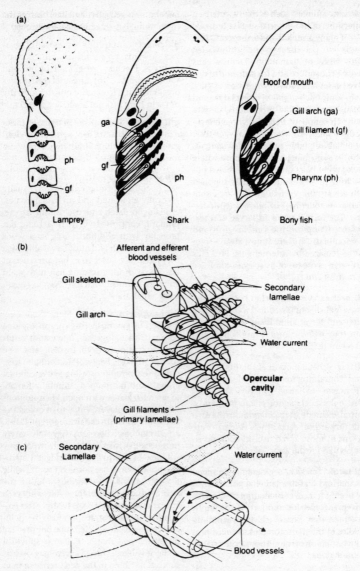

FIG. 68 *(a) The arrangement of gills in three types of fish. (b) The arrangement of secondary gill lamellae in a bony fish. (c) The countercurrent flow of blood through the lamellae.*

mid-Mesozoic times but diminished during the later Mesozoic and Tertiary. Today, there are probably no wild stands of *Ginkgo* anywhere in the world, although it is reported to occur naturally in remote mountain valleys in China's Zhejiang province. The tree was preserved in temple grounds in China and Japan from where it has been introduced into cultivation. It has been an important feature of gardens of temperate regions for more than 150 years. The unusual shape of the crown, natural resistance to disease, tolerance of pollution and yellow leaf colour in the autumn have made this a favourite street and park tree. Ovulate trees produce large quantities of seeds, which have an obnoxious odour. Canned seeds with fleshy outer coats removed are sold in ethnic markets as 'silver almonds' or 'white nuts', the gametophyte and embryo being edible. Oils from the outer coat are known to cause dermatitis in some humans. The tree is grown commercially as its leaves are used in herbal medicine.

gizzard Region of the gut in many animals, where food is ground prior to the main digestion. Walls very muscular, often with hard 'teeth' (e.g. crustaceans) or containing grit, stones, etc. (e.g. birds, reptiles).

gland An organ (sometimes a single cell) specialized for secretion of a specific substance or substances. (Bot.) Superficial, discharging secretion externally, e.g. glandular hair (lavender), NECTARY, HYDATHODE, or embedded in tissue, occurring as isolated cells containing the secretion, or as layer of cells surrounding intercellular space (secretory cavity or canal) into which secretion is discharged; e.g. resin canal of pine. (Zool.) In animals, glands are either *exocrine* (secreting onto an epithelial surface, usually via a duct), or *endocrine* (secreting directly into the blood, not via ducts). Fig. 69 illustrates one classification of exocrine glands. See APOCRINE GLAND, PARACRINE CELLS.

Glaucophyta A group of ALGAE that possesses eukaryotic cells lacking chloroplasts but harbouring 'chloroplasts' that are remarkably similar to blue-green algae and also to the chloroplasts of the red algae (RHODOPHYTA). Discovery of a thin peptidoglycan wall around their chloroplasts provides further evidence of a link with the CYANOBACTERIA. The genome size of the glaucophyte chloroplast is about ten times smaller than genomes of free-living blue-green algae. The actual size of 125,000 base pairs is similar to that of other chloroplast genomes. In its organization the glaucophycean chloroplast seems more like those of other plants than the genomes of BLUE-GREEN ALGAE (e.g. circular genomes of glaucophytes contain two 'inverted repeats', which are identical sections, containing the genes for the chloroplast ribosomal RNAs, but which are transcribed in opposite directions. The inverted repeats are separated by two sections of DNA present only in single copy, like those of plants). Such evidence indicates that the glaucophycean chloroplast or cyanelle is functional and not an endosymbiotic blue-green alga, although it did evolve from one. Moreover, in the centre of the chloroplast are polyhedral bodies (carboxysomes) that are typically found in the centre of blue-green algal cells. Thylakoids are single and equidistant as found in the blue-green algae and red algae. Pigments include chlorophyll *a* as the only chlorophyll, while accessory pigments include phycocyanin and allophycocyanin contained in phycobilisomes, which are attached to the thylakoids, and β-carotene, zeaxanthin and β-cryptoxanthin. The reserve polysaccharide is starch, formed outside the chloroplast.

The three genera in the Glaucophyta are all unicellular, dorsoventral (rounded dorsal side and flat ventral side), biflagellate algae. Flagella are unequal in length, possess two rows of delicate hairs and are inserted just below the apex of the cell. A superficial layer of vesicles, subtended by microtubules surrounds each cell. These vesicles may be empty or may contain scales or fibrillar material.

glenoid cavity Cup-like hollow on each side of pectoral girdle (on the scapula, and the coracoid when present) into which head of the humerus fits, forming tetrapod shoulder-joint.

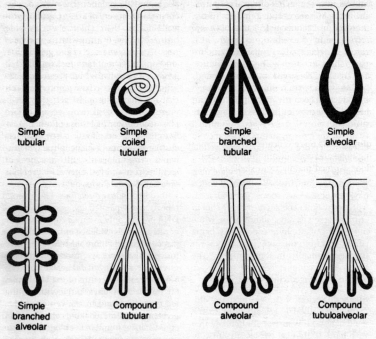

Simple tubular

Simple coiled tubular

Simple branched tubular

Simple alveolar

Simple branched alveolar

Compound tubular

Compound alveolar

Compound tubuloalveolar

FIG. 69 *Diagram illustrating the major types of animal exocrine* GLAND; *secretory portions black.*

glial cells (glia, neuroglia) Non-conducting nerve cells, performing supportive and protective roles for neurons. Include *astrocytes* (attaching neurons to blood vessels), *oligodendrocytes* (forming myelin sheaths of axons of central nervous system), SCHWANN CELLS, *microglia* (phagocytic) and *ependyma* cells (lining ventricles of the brain and cerebrospinal canal). Two-way communication between neurons and glial cells is essential for axonal conduction of impulses, synaptic transmission (see ASTROCYTES) and information processing. Signals include ion fluxes (especially intracellular calcium fluxes), neurotransmitters, ADHESION molecules and other signal molecules.

glires Superorder combining rodents and lagomorphs. This clustering may be artificial (see LAGOMORPHA).

global stability The tendency of a community to return to its original state even when subjected to a large perturbation. See GAIA HYPOTHESIS.

global warming One aspect of climatic change. Records indicate that since 1900, the Earth has warmed by 0.5°C. In years, such as 1997 and 1998, in which there is an El Niño (unusually warm surface waters in the tropical Pacific) there tends to be a higher than average global (land and sea) temperature. Computer climate models suggest that rising greenhouse gas levels should be having an even greater effect than they are on this increase (see GREENHOUSE EFFECT). In the early years of the 21st century, the sun should warm slightly as it reaches a sunspot maximum.

Since 1960 volcanic activity has resumed to its pre-1920s levels, releasing sulphate aerosols into the stratosphere. These and anthropogenic aerosols (e.g. mineral dust, industrial black carbon and those arising from burning of biomass and fossil fuels) scatter incoming solar radiation back to space. There is considerable support for the view that anthropogenic activities, moderated by variations in solar and volcanic activity, have been the main drivers of climate change during the past century. See POLLUTION.

globigerina ooze Calcareous mud, covering huge areas (about one third) of ocean floor. Formed mainly from shells of FORAMINIFERA, *Globigerina* being an important genus.

globulins Group of globular proteins, soluble in aqueous salt solutions and of wide occurrence in plants (e.g. in seeds) and animals (e.g. in vertebrate blood plasma). In humans, ELECTROPHORESIS can separate alpha$_1$-, alpha$_2$-, BETA- and GAMMA-GLOBULINS. Normal individuals belong to one of three genetic types, separable by electrophoretograms of their plasma proteins. Alpha and beta globulins transport iron, fats and fat-soluble vitamins in the blood. See ALBUMIN, TRANSFERRIN.

glomerulus (1) Small knot of capillaries covered by basement membrane and surrounded by BOWMAN'S CAPSULE, forming part of a NEPHRON of vertebrate KIDNEY. Through it small solute molecules (i.e. not cells or plasma proteins) are filtered under pressure from the blood to form the *glomerular filtrate*. (2) *Caudal glomeruli*; small tissue masses containing RETIA MIRABILIA in mammalian tails, some involved in heat conservation. See Fig. 17. (3) Nodes of the olfactory bulb of the brain, receiving inputs from olfactory cells.

glossopharyngeal nerve See CRANIAL NERVES.

glottis Slit-like opening of trachea into pharynx of vertebrates. Can usually be closed by muscles. In mammals, opens between vocal cords. See LARYNX, EPIGLOTTIS, TRACHEA.

glucagon Polypeptide hormone of vertebrates (twenty-nine amino acids), produced by alpha-cells of pancreas in response to drop in blood glucose level. Activates ADENYLYL CYCLASE in target cells (eg. liver, adipose tissue) with resultant rise in cyclic AMP in those cells and appropriate enzyme activation to ensure glycogenolysis (with release of glucose from glycogen), GLUCONEOGENESIS and lipolysis (with release of free fatty acids), restoring levels of these metabolites in the blood plasma. Also stimulates INSULIN release from beta-cells, the two hormones acting antagonistically in control of blood glucose and free fatty acid levels. Its effects are *hyperglycaemic*.

glucans Homoglycans consisting entirely of condensed glucose residues. See POLYSACCHARIDES.

glucocorticoids Steroid adrenal cortex hormones (principally CORTISOL and cortisone) concerned with normal metabolism and resistance to stress conditions (e.g. long-term cold, starvation) by promoting deposition of glycogen in liver, GLUCONEOGENESIS and release of fatty acids from fat reserves. They render blood vessels more sensitive to vasoconstrictors, thereby raising blood pressure. Stabilize lysosomal membranes, thus inhibiting release of inflammatory substances (and are hence anti-inflammatory). Undersecretion results in *Addison's disease*, oversecretion in *Cushing's syndrome*. See ACTH, CORTICOSTEROIDS.

glucokinase See HEXOKINASE.

gluconeogenesis Conversion of fat, protein and lactate molecules into glucose, notably by the vertebrate liver. It is not a reversal of GLYCOLYSIS. By appropriate enzyme activity glucogenic AMINO ACIDS (e.g. alanine, cysteine, threonine, glycine, serine) may be converted to *pyruvate*; glycerol may be converted to *glyceraldehyde-3-phosphate*; fatty acids to *acetate*. Four different enzymes, not active in glycolysis, convert pyruvate back to glucose, the final step occurring in the endoplasmic reticulum. In addition to citrate, other intermediates of the KREBS CYCLE are precursors for gluconeogenesis. See Fig. 70.

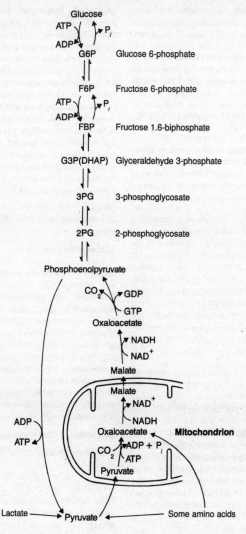

FIG. 70 *Diagram indicating role of mitochondria in reversal of glycolytic pathway during* GLUCONEOGENESIS. *The enzyme pyruvate carboxylase (pyruvate® oxaloacetate) is mitochondrial while glucose-6-phosphatase (glucose-6-phosphate® glucose) is on the inner surface of the endoplasmic reticulum membrane. Other enzymes are cytosolic. Entry points for glycerol, lactate and amino acids are indicated.*

Gluconeogenesis is stimulated by CORTISOL, THYROXINE, ADRENALINE, GLUCAGON and GROWTH HORMONE.

glucosamine Nitrogenous hexose derivative of glucose forming monomer of various polysaccharides, notably CHITIN and HYALURONIC ACID.

glucose (dextrose) The most widely distributed hexose sugar (*dextrose* in its dextrorotatory form). Component of many disaccharides (e.g. sucrose) and polysaccharides (e.g. starch, cellulose, glycogen). The isomers α glucose (starch, glycogen) and β glucose (cellulose) have important biological consequences; and phosphorylated glucose, unlike glucose, cannot readily diffuse out of cells, so that those (e.g. liver cells) which can release glucose into the blood therefore contain a glucose phosphorylase. Glucose is retained by muscle and brain, for which it is a major fuel, but not by liver, for which it is not (see GLUCONEOGENESIS). Glucose is an *aldohexose* reducing sugar, and (as glucose-1-phosphate) is the initial substrate of GLYCOLYSIS for most cells. When completely oxidized (in combined glycolysis and aerobic respiration) sufficient energy is released per glucose molecule under intracellular conditions for the generation of thirty-six molecules of ATP from ADP and P_i, in the overall equation:

$$C_6H_{12}O_6 + 6O_2 = 6CO_2 + 6H_2O$$

glucose oxidase Found as a salivary enzyme in a wide variety of herbivorous caterpillars, converting D-glucose and molecular oxygen to D-gluconic acid and hydrogen peroxide, it is used commercially as a diagnostic assay for blood glucose concentration. In this process, the hydrogen peroxide produced by glucose oxidase becomes a substrate for the enzyme peroxidase, which uses it to convert a colourless to a coloured compound whose concentration is then measured spectrophotometrically. See BIOFUEL CELLS.

glume (serile glume) Chaffy bract, a pair of which occurs at the base of grass spikelet, enclosing it.

glutamate (glutamic acid, Glu) An acidic AMINO ACID (typical pK value = 4.4), which is itself the source of most amino groups for other amino acids by TRANSAMINATION and the major excitatory NEUROTRANSMITTER in the mammalian brain (where it is released by most neurons) and spinal cord. Nearly always negatively charged at physiological pH. Maintained at high levels in animals (which lack the glutamate synthase of bacteria and plants) by transamination of α-ketoglutarate during amino acid catabolism, but produced from GLUTAMINE in brain cells by phosphate-activated glutaminase. Glutamate receptors linked directly with a membrane ion channel (ionotropic receptors) include AMPA- and NMDA-receptors, the former mediating moment-to-moment signalling, the latter mediating synaptic plasticity. Proline and arginine are derivatives of glutamate.

glutamic acid (Glu) See GLUTAMATE.

glutamine An AMINO ACID (with a polar uncharged R-group) synthesized from GLUTAMATE by glutamine synthetase (found in all organisms) by assimilation of ammonium ions (NH_4^+), a terminal amine replacing the carboxylic acid moiety in the R group of glutamic acid. Released from astrocytes into the vicinity of brain neurons, and taken up by neutral-amino-acid transporters, when it is converted to glutamate and packaged into synaptic vesicles.

glutathione (GSH) Tripeptide found in bacteria, plants and animals, formed from glutamate by ATP-dependent condensation of cysteine and glycine. Acts as a redox buffer, probably helping maintain sulphydryl groups of proteins in the reduced state.

gluten Protein occurring in wheat giving firmness to risen dough in bread-making. Those allergic to gluten suffer from *coeliac disease*, resulting in poor absorption of dietary components and in the consequent *malabsorption syndrome*. Gluten-free products are available for such people.

glycan Synonym of POLYSACCHARIDE.

glycation Abnormal protein glycosylation resulting from chronically raised blood glucose levels, as during late-onset DIABETES mellitus. Through glycation, receptor proteins may lose their responsiveness to signal molecules. See AGEING.

glyceride Fatty acid ester of glycerol. When all three –OH groups of glycerol are so esterified the result is a *triglyceride*. See FAT.

glycerol A trihydric alcohol and component of many lipids, notably of glycerides. See DIACYLGLYCEROL.

(a) GLYCEROL

(b) TRIGLYCERIDE (neutral fat)

glycocalyx (cell coat) Carbohydrate-rich region at surfaces of most eukaryotic cells, deriving principally from oligosaccharide components of membrane-bound GLYCOPROTEINS and GLYCOLIPIDS, although it may also contain these substances secreted by the cell. Role of the cell coat is not properly understood yet. See CELL MEMBRANES, PHOSPHOLIPIDS, EXTRACELLULAR MATRIX.

glycogen The chief polysaccharide store of animal cells and of many fungi and bacteria; often called 'animal starch'. It resembles AMYLOPECTIN structurally in being an α-[1,4]-linked homopolymer of glucose units, although it is more highly branched, the numerous ends of the molecule each simultaneously digestible by amylases. It can be isolated from tissues by digesting them with hot KOH solutions. As with amylopectin, it gives a red-violet colour with iodine/KI solutions. Its hydrolysis is termed *glycogenolysis*. Like starch it is osmotically inactive and therefore a suitable energy storage compound. Glycogen synthase, the enzyme integrating the synthesis of glycogen in liver and muscle, has at least ten distinct sites at which phosphates can be added by six or more different kinases, and one or more phosphatases removing them. Well-trained endurance athletes can load their muscles with ~2 kg of glycogen, eating a very high carbohydrate diet for a few days before a race while simultaneously reducing their training regime. See GLUCONEOGENESIS, GLYCOLYSIS, GLUCAGON, INSULIN. See Figs 71a and b.

glycolipid Lipid with covalently attached mono- or oligosaccharides; found particularly in the outer half of phospholipid bilayers of plasma membranes. Considerable variation in composition both between species and between tissues. All have a carbohydrate polar head end. Range in complexity from relatively simple galactocerebrosides of the Schwann cell MYELIN SHEATH to complex *gangliosides*. May be involved in cell-surface recognition. See GLYCOSYLATION, GLYCOPROTEIN, CELL MEMBRANES.

glycolysis Anaerobic degradation of glucose (usually in the form of glucosephosphate) in the cytosol to yield pyruvate, forming initial process by which glucose is fed into aerobic phase of RESPIRATION, which usually occurs in MITOCHONDRIA. Cells without mitochondria (and those prokaryotes without a MESOSOME) rely on glycolysis for most of their ATP synthesis, as do facultatively anaerobic cells (e.g. striated muscle fibres) when there is a shortage of oxygen. The pathway is illustrated in Fig. 72. It generates a net gain of two molecules of ATP per molecule of glucose used, plus reducing power in the form of two $NADH_2$ molecules. The $NADH_2$ is available for reduction of pyruvate to lactate, or of acetaldehyde to alcohol, or for fatty acid and steroid syn-

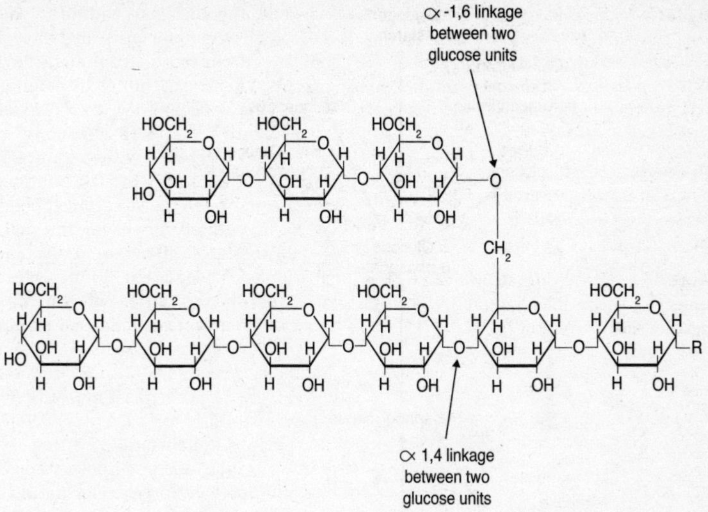

FIG. 71a *The structure of two outer branches of a* GLYCOGEN *molecule. R represents the rest of the polymer.*

thesis from acetyl coenzyme A, as occurs in liver cells.

Most significant in glycolysis is the hydrolysis of each fructose 1,6-bisphosphate molecule into two triose phosphate molecules, the remaining steps in the pathway thereby effectively occurring twice for every initial glucose-phosphate molecule used. Conversion of fructose-6-phosphate to fructose 1,6-bisphosphate is the main rate-limiting step in glycolysis, and phosphofructokinase, the enzyme involved, is a REGULATORY ENZYME, modulated by the ratio in the levels of (AMP + ADP):ATP in the cytosol so that high ATP levels inhibit glycolysis. Enzymes involved in the glycolytic pathway appear to have evolved from one ancestral enzyme by a process involving GENE AMPLIFICATION. The switch from glycolysis to GLUCONEOGENESIS is under the control of kinases and phosphatases acting on fructose-6-phosphate. See FERREDOXINS, PASTEUR EFFECT, PENTOSE PHOSPHATE PATHWAY. For *aerobic glycolysis*, see CANCER CELL.

glycoprotein Protein associated covalently at its n-terminal end with a simple or complex sugar residue. In PROTEOGLYCANS the carbohydrate forms the bulk of the molecule, with numerous long and usually unbranched GLYCOSAMINOGLYCANS bound to a single core protein. These important extracellular components contrast with cell surface glycoproteins, which generally comprise short but often complex nonrepeating oligosaccharide sequences bound to an integral membrane protein. Proteins become glycosylated (i.e. have their sugar residues added) in the ENDOPLASMIC RETICULUM and GOLGI APPARATUS. See CELL MEMBRANES, GLYCOSYLATION, GLYCOLIPIDS.

glycosaminoglycans (GAGs) Long, unbranched polysaccharides (formerly called mucopolysaccharides) of repeated disaccharide units, one member always an amino sugar (e.g. N-acetylglucosamine, N-acetylgalactosamine). They comprise varying proportions of the extracellular matrices of tissues, where they are often numerously bound to a core protein to become *proteoglycans*, e.g. HYALURONIC ACID, CHONDROITIN, HEPARIN. See GLYCOCALYX, CELL MEMBRANES.

glycoside Substance formed by reaction of an *aldopyranose* sugar, such as glucose,

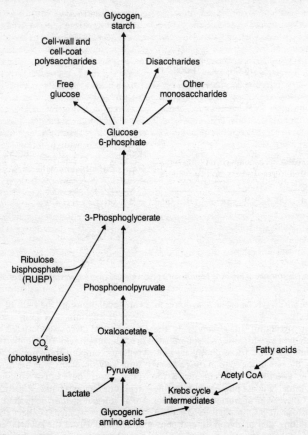

FIG. 7b *Diagram of biochemical pathways linking some non-carbohydrates to carbohydrates. Glucose-6-phosphate acts as a key branch-point.*

with another substance such that the alde- hyde moiety in the sugar is replaced by another group. Glycosidic bonds form the links between monosaccharides. Some plant glycosides, termed *cardiac glycosides*, alter the excitability of heart muscle and may be defensive; examples include *ouabain* and *digitalin*. See ANTHOCYANINS, TANNINS, ALKALOIDS.

glycosylation Bonding of sugar residue to another organic compound. GLYCOPROTEINS are formed in the lumen of rough endoplas-

mic reticulum, but may be subsequently modified in the lumen of the GOLGI APPAR- ATUS, where other amino acids of the protein may become glycosylated. Nucleotides may be glycosylated. UDP-glucose is an impor- tant coenzyme in transport of glucose, most probably in cell wall formation. GLYCOLIPIDS are also formed by glycosylation in the endoplasmic reticulum.

glyoxylate cycle Modified form of KREBS CYCLE, occurring in most plants and micro- organisms but not in higher animals, by which acetate and fatty acids can be used as sole carbon source, especially if carbohydrate is to be made from fatty acids. The cycle by-passes the CO_2-evolving steps in the Krebs cycle. The innovative enzymes are *isocitrate lyase* and *malate synthase*. In higher plants, these enzymes are found in GLYOXYSOMES, organelles lacking most of the Krebs cycle enzymes; so isocitrate must reach them from mitochondria. Plant seeds converting fat to carbohydrate are rich sources of glyoxysomes. See Fig. 73.

glyoxysome Organelles containing catalase, related to PEROXISOMES, and the sites of the GLYOXYLATE CYCLE.

GMO Genetically modified organism: one whose genome has been modified by GENE MANIPULATION. Plant breeders often work for many years without success to achieve pathogen-resistant crops using traditional methods of selective breeding. Thus, chickpea is an important crop in the Middle East, the Mediterranean and Australia but, despite years of effort by breeders, remains susceptible to the grey mould *Botrytis cenerea*. The gene for polygalacturonase-inhibiting protein (PGIP) from the raspberry plant can control the mould and attempts are now being made to splice it to a regulatory element from cauliflower mosaic virus, which keeps the gene continuously active, before inserting it into the legume. Such TRANSGENIC methods increase our chances of countering pathogens at a time when their rate of global spread is greatly increasing (see Tables 8 and 9). In 1997 about 12% of US soya bean fields were planted with genetically modified (GM) soya beans, and about 6% of corn acreage with GM corn. In 1998 the percentage for GM soya beans was 30%. The soya beans contain a transgene engineered to resist the herbicide Roundup. The corn contains a transgene from the bacterium *Bacillus thuringiensis* (*Bt*), first used in crops in 1995, encoding a protein toxic to a common pest, the European corn borer (see PEST MANAGEMENT). In the USA in 1997, just under 1 million ha (15% of the cotton crop were planted with GM cotton containing a *Bt* transgene providing resistance to the cotton boll-worm. Concern that insects feeding on these crops will become resistant to the toxin may be tempered by current attempts by the producers to transform the crops with genes from several different strains of the bacillus, each producing a slightly different toxin. RESISTANCE to such multiple toxins is less likely to develop rapidly. So-called '*Bt* maize' is a transgenic strain of maize developed to express the *Bt* gene (see BIODIVERSITY). Farmers growing *Bt* maize in the USA are obliged to plant 20% non-*Bt* varieties so that these 'refuge' areas should prevent pests from developing resistance to the insecticide. But many farmers are apparently ignoring this directive. The rationale is that resistant insects that do evolve will breed with susceptible forms living in the refuges, so diluting the trait (see GENETICALLY MODIFIED FOOD (GM FOOD)). There is increasing consumer concern about genetically altered foods, and too little public debate. In the US, critics argue that large-scale planting of GM crops will put intense selection pressure on insects, so that increased resistance (already detected in trials) will rapidly result unless adequate gene dilution occurs through matings with insects from non-GM plants grown adjacently. The potential spread of herbicide RESISTANCE GENES in pollen from GM crops to agriculture WEED species is a major concern. One solution to this is to engineer the resistance gene into the chloroplast DNA (as has been done with glyphosate-resistance in tobacco), since chloroplasts are inherited maternally in many species (so avoiding transgene escape through pollen). Adding a plastid-specific promoter to the transgene is another precautionary measure, although recent work on cpDNA (see entry) indicates that broken plastid gene sequences incorporated next to a nuclear promoter might be expressed at a very low frequency. Although farmers are advised to plant a certain proportion of non-GM plants alongside the GM varieties, the percentages involved are disputed by the manufacturers and the critics. Objections to engineering into a food plant a transgene from a totally different organism may be mollified by research aimed at modifying the plant's own disease-RESISTANCE

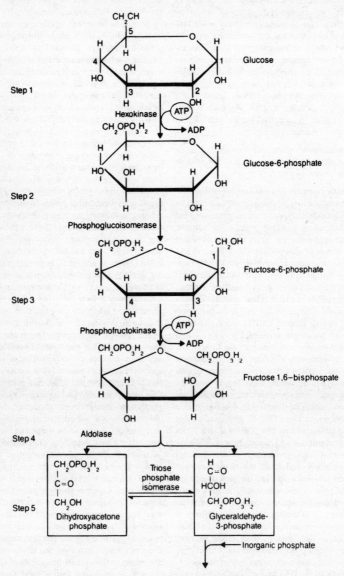

FIG. 72 *(continued on facing page) The stages of vertebrate* GLYCOLYSIS *and the enzymes involved. Because of the hydrolysis at Step 5 all later stages are represented by two molecules for every original glucose molecule.*

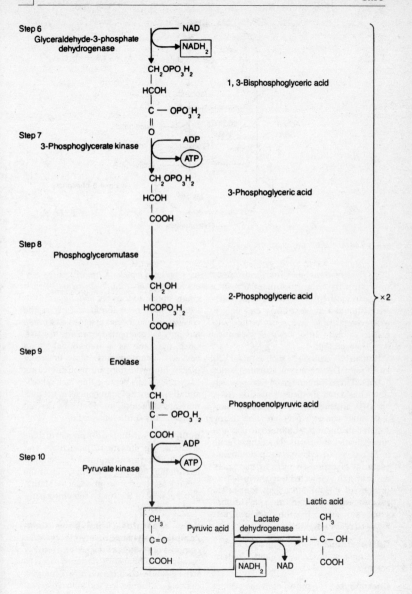

FIG. 72 (cont.)

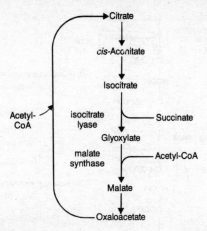

FIG. 73 *Diagram of the* GLYOXYLATE CYCLE.

GENES. Use of transgenes from indigenous species (e.g. the resurrection plant *Xerophyta viscosa* in southern Africa for drought resistance) should also be strongly encouraged wherever possible (see PLANT DISEASE AND DEFENCES and http://www.newscientist.com/hottopics/gm/).

Human blood products such as Factor VIII and Factor IX (see BLOOD CLOTTING), once extracted from human serum (an approach now considered inefficient and too risky, with HIV and hepatitis C virus transmission a possible consequence) are now being secreted in the milk of transgenic pigs and domesticated RUMINANTS. The human gene is linked to an appropriate promoter to ensure it is expressed only in mammary tissue and then injected into the male pronucleus of fertilized pig eggs (one-celled zygotes), where it inserts into a pig chromosome (probably using the cell's own DNA REPAIR MECHANISMS). See CLONE (1).

Gnathostomata Subphylum or superclass containing all jawed vertebrates. Contrast AGNATHA.

Gnetophyta Gymnosperms possessing many flowering plant-like characteristics (e.g. possess vessels in the secondary xylem, motile sperm is absent, stroboli are similar to some flowering plant inflorescences, have no resin canals). Comprise three genera (*Gnetum*, *Ephedra*, *Welwitschia*). Pollen grains similar to those of *Ephedra* and *Welwitschia* have been found as early as the PERMIAN, although that assigned to *Gnetum* has only been reported from the TERTIARY. Cited by some as a transitional group between gymnosperms and flowering plants; however, they are probably closer to gymnosperms because they lack carpels, and the presence of many nuclei in the free-nucleate gametophyte. Probably derived from an early coniferophyte line.

goblet cell Pear-shaped cell present in some epithelia (e.g. intestinal, bronchial), and specialized for production of MUCUS. As in differentiation of other epithelial cells of the gut mucosa, the transcription factor Math 1, a component of the Notch signalling pathway, is required.

Golgi apparatus (Golgi body, Golgi complex) A dynamic eukaryotic organelle, comprising a system of stacked and roughly parallel interconnecting flattened sacs (cisternae) sandwiched between two complex networks of tubules (*cis* and *trans* Golgi networks), the whole situated close to, but physically separate from, the *endoplasmic reticulum* (ER). It provides a series of membranous subcompartments through which

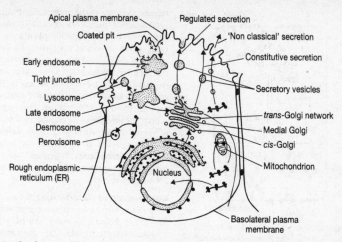

FIG. 74a *Involvement of the* GOLGI APPARATUS *in some of the protein targeting pathways in a mammalian epithelial cell. The stippled compartments are topologically equivalent to one another and to the outside of the cell.*

move components destined (some via ENDO-SOMES) for the plasma membrane, secretory vesicles and lysosomes (this requires PROTEIN TARGETING). Stacks of cisternae can move towards the 'minus' (organizing) ends of MICROTUBULES, maintaining the organelle's central position in the cell needed for its role in membrane traffic. Numbers vary from one to hundreds per cell; tend to be interconnected in animal but not plant cells, with up to 30 cisternae (normally about 6) per Golgi body. Each cisterna has a *cis* surface (towards the nucleus) and a *trans* surface (away from the nucleus). Transport vesicles (including non-clathrin-coated) from the ER arrive and fuse with the *cis* Golgi network, adding their membrane material to the cisternae and depositing GLY-COPROTEINS for processing within the cisternal lumina. Some of the oligosaccharide of the glycoprotein may be removed, while other sugar units are added to yield mature glycoproteins of different kinds – possibly vesicle-specific (SEE MAJOR HISTOCOMPATIBILITY COMPLEX). Many of these will be retained within the membrane during modification. Vesicular carriers engage in membrane traffic between components of the Golgi complex, apparently budding from one

compartment and fusing with the next, *en route* to the *trans* Golgi network, where two types of vesicles are budded off: COATED VES-ICLES (about 50 nm in diameter), and larger secretory vesicles (about 1,000 nm in diameter). Much remains to be learned about these movements, which seem to involve PHOSPHOLIPID TRANSFER PROTEINS. Possibly, different 'membrane scaffolds' are associated with budding and non-budding membrane domains, restricting the lateral movement of 'resident' Golgi proteins from non-budding regions. Non-clathrin coated vesicles are characterized by several types of COP (coatomer protein), uncoating of which is a prerequisite for fusion of vesicle with an appropriate acceptor membrane – in turn governed by soluble 'fusion proteins' which link the two membranes and small Ras-like GTPases. The fungal product brefeldin A causes uncoating of COPs and leads to disintegration of the whole organelle.

At the onset of mitosis, exocytic and endocytic membrane traffic ceases and the Golgi body fragments into tubular clusters and small vesicles (compare NUCLEUS). See LYSOSOME, CELL MEMBRANES and Fig. 74.

In plant cells, the Golgi apparatus is

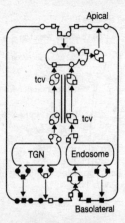

FIG. 74b *Protein targeting examples in an epithelial cell. Newly synthesized proteins of the basolateral membrane (dark squares and circles) go straight there from the trans Golgi network (TGN); proteins (dark squares) that also have an endosomal targeting sequence go to basolateral endosomes. Newly synthesized proteins destined for the apical membrane either go directly there (open circles) or go first to the basolateral membrane (open squares) prior to entry into the basolateral endosome, whence it is sorted into a transcytotic vesicle (tcv) and transported along microtubules (parallel lines) to the apical membrane. See ENDOSOME.*

involved in secretion; e.g. it synthesizes cell wall polysaccharides, which collect in vesicles pinched off from the cisternae. These secretory vesicles migrate and fuse with the plasmalemma; vesicles discharge their polysaccharide contents to the exterior, where they become part of the cell wall. In some algae, scales (both organic and inorganic) are formed in vesicles cut off from the Golgi apparatus before being transported to the cell periphery. In diatoms (BACILLARIOPHYTA), the Golgi apparatus again gives rise to translucent vesicles which collect beneath the plasmalemma, where they fuse to form the SILICA DEPOSITION VESICLE within which the siliceous cell wall is synthesized.

gonad Animal organ producing either sperm (TESTIS) or ova (OVARY). See OVOTESTIS.

gonadotrophins (gonadotropins, gonadotropic hormones) Group of ver-tebrate glycoprotein hormones, controlling production of specific hormones by gonadal endocrine tissues. Anterior pituitaries of both sexes produce FOLLICLE-STIMULATING HORMONE (FSH) and LUTEINIZING HORMONE (LH, or interstitial cell stimulating hormone (ICSH) in males); but their effects in the two sexes are different. HUMAN CHORIONIC GONADOTROPHIN (HCG) is an embryonic product whose presence in maternal urine is usually diagnostic of pregnancy. Release of FSH and LH is controlled by hypothalamic *gonadotrophic-releasing factors* (GnRFs). PROLACTIN is also gonadotrophic. See MENSTRUAL CYCLE.

Gondwanaland (Gondwana) Southernmost of the two Mesozoic supercontinents (the other being LAURUSSIA) named after a characteristic geological formation, the *Gondwana*. Comprised future South America, Africa, India, Australia, Antarctica and New Zealand (the last breaking earliest from the supercontinent, with present-day examples of a relict Gondwana-like flora and fauna). Rifts between Gondwanaland and Laurussia were not effective barriers to movements of land animals until well into the Cretaceous. By the dawn of mammalian radiation Gondwanaland had largely split into its five major continental regions, each being the nucleus of radiation for its inhabitant fauna and flora. Flora was characterized by *Glossopteris*; podocarps and tree ferns still persist in New Zealand, as do the reptile *Sphenodon* (see RHYNCHOCEPHALIA), giant crickets and flightless birds (e.g. kiwi and, up to 5,000 years ago, moas). See CONTINENTAL DRIFT, ZOOGEOGRAPHY.

gonochorism Having a chromosomally determined sex, that is either male or female. Contrast HERMAPHRODITE.

gonocytes Primordial germ cells which, in male mammals, migrate to the genital ridge and after birth into the seminiferous tubules, retaining their stem cell potential and differentiating into spermatogonial stem cells.

GPCR G PROTEIN-coupled RECEPTOR, such as pituitary thyrotropin and luteotropic receptors, among many others.

G proteins Trimeric (3-subunit) eukaryotic proteins, coupling light or hormonal activation of membrane receptor (GPCRS) to activation of a target protein (e.g. ADENYLYL CYCLASE, thereby amplifying the signal) or membrane ion channel. Some are heterotrimeric. Activity mediated by the dissociation of a G protein subunit bound to GTP. A typical vertebrate cell has (at a guess) dozens of different G proteins. Different G proteins enable channelling of signals from membrane receptors to effector molecules in the cell. Although present in plant cells, their roles are not clear. See CYTOSKELETON, GTP, RAS PROTEIN, RECEPTOR, SECOND MESSENGER.

Graafian follicle Fluid-filled spherical vesicle in mammalian ovary containing OOCYTE attached to its wall. Growth is under the control of FOLLICLE-STIMULATING HORMONE of anterior pituitary, its rupture (*ovulation*) also being a gonadotrophic effect (see LUTEINIZING HORMONE). After ovulation the follicle collapses, but theca and granulosa cells grow and proliferate forming the CORPUS LUTEUM. Androgen precursors are made by the *theca cells* of the Graafian follicle, and aromatized to oestrogens by the *granulosa cells*; in primates *theca lutein cells* of the corpus luteum make oestrogen precursors. See INHIBINS, MENSTRUAL CYCLE, OVARY.

grade A given level of morphological organization sometimes achieved independently by different evolutionary lineages, e.g. the mammalian grade. See CLADE.

gradient analysis Analysis of species composition along a gradient of environmental conditions.

graft (transplant) Artificial transfer of part of one organism to a new position, either in the same organism (*autograft*) or a different conspecific organism (*allograft*). 'Graft' and 'transplant' are almost synonyms, but transplants do not necessarily have a close union with the adjacent tissues in their new position, whereas grafts do. *Isografts* involve transfer of tissue between genetically identical (isogeneic) individuals. Allografts are the common clinical transplants. *Xenografts* (xenotransplants) involve transfer of tissue between different species

(with possible complications resulting from HORIZONTAL TRANSFER of viruses, prions, etc.). Pigs are currently the most suitable xenotransplant donors for humans; but destruction of the donor vascular endothelium often occurs within minutes of exposure to human blood. Commercial interest in overcoming this *hyperacute rejection* is great. Allograft rejection (host-versus-graft rejection) usually arises when the graft carries antigens not present in the host, T cells recognizing donor peptides in association with the MHC antigens expressed on the graft (see T CELLS). The major LYMPHOKINES involved in rejection are IL-2 and gamma-interferon, IFN_γ. See IMMUNE TOLERANCE.

graft hybrid See CHIMAERA.

graft-versus-host response Reaction of immunocompetent donor cells to recipient tissues (e.g. skin, gut epithelia, liver), often destroying them. Particularly problematic in bone transplants.

gramicidin See IONOPHORE.

Gramineae See POACEAE (grasses).

Gram-negative bacteria See GRAM'S STAIN.

Gram's stain Stain devised by C. Gram in 1884 which differentiates between bacteria which may be otherwise similar morphologically. To a heat-fixed smear containing the bacteria is added crystal violet solution for 30 s which is then rinsed off with Gram's iodine solution; 95% ethanol is applied and renewed until most of the dye has been removed (20 s–1 min). Those bacteria with the stain retained are *Gram-positive* (e.g. *Bacillus subtilis*, *Staphylococcus aureus*); those without are *Gram-negative* (e.g. *Escherichia coli*, *Agrobacterium tumefaciens*). A counterstain (e.g. eosin red, saffranin, brilliant green) is then applied, colouring the Gram-negative bacteria but not the Gram-positive ones. Differentiation reflects differences in amount and ease of access of PEPTIDOGLYCAN in the bacterial envelope. Gram-negative bacteria possess an outer membrane composed of lipopolysaccharide (LPS) held together by magnesium and calcium ions bridging negatively charged phosphosugars. Cationic (acidic) ANTIMICRO-

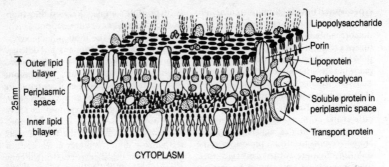

Outer lipid bilayer

Periplasmic space

25 nm

Inner lipid bilayer

Lipopolysaccharide
Porin
Lipoprotein
Peptidoglycan
Soluble protein in periplasmic space
Transport protein

CYTOPLASM

FIG. 75 *The arrangement of the double membrane envelope of Gram-negative bacteria such as* E. coli.

BIAL PEPTIDES cause displacement of the metal ions, damaging the outer membrane and allowing access by other molecules (see TOLL-LIKE RECEPTORS). Gram-positive bacteria lack this outer membrane, but have a thicker peptidoglycan coat. Among other differences, Gram-positive bacteria are more susceptible to penicillin, acids, iodine and basic dyes, and some species form resistant endospores; Gram-negative bacteria are more susceptible to alkalis, ANTIBODIES and COMPLEMENT, and do not form endospores; ampicillin is effective against some Gram-negative bacteria. See Fig. 75.

grana (sing. granum) See CHLORO-PLASTS.

grandchildless (gs) Adjective describing a group of at least eight MATERNAL EFFECT genes of *Drosophila* which when homozygous in females causes failure of regionalization of the POLAR PLASM (and polar granule production) producing sterile offspring.

granulocyte (polymorph) Granular LEU-COCYTE. Develops from MYELOID TISSUE and has granular cytoplasm. Include NEUTRO-PHILS, EOSINOPHILS and BASOPHILS.

granulosa cells See GRAAFIAN FOLLICLE, INHIBINS.

granum See GRANA.

graptolites Extinct invertebrates of doubtful affinities (possibly with either COELENTERATA or HEMICHORDATA), whose name (literally, written on stone) indicates importance as fossils, notably of shales. Upper Cambrian–Lower Carboniferous; used to subdivide the Ordovician and Silurian.

grasses See POACEAE, CEREALS.

grassland (prairie, steppe) Includes a wide variety of plant communities. Some integrade with savannas, others with deciduous woodland or desert. Unlike savannas, grasslands lack trees – except along rivers and streams. They are generally characterized by cold winters and occur over large areas in the interior portions of continents. Annual precipitation determines distribution of the grass species; e.g. typical short-grass species occur where precipitation is low, while taller species dominate in moister areas. Grassland regions have been used extensively for agriculture, particularly cultivation of cereal crops and for pasture. They were once inhabited by herds of grazing mammals associated with large predators. Such herds were widespread during Pleistocene glaciations.

gravitropism (Bot.) Orientation of plant parts under stimulus of gravity. Main stems, *negatively geotropic*, grow vertically upwards and, when laid horizontally, exhibit increased elongation of cells in growth region on lower side at tip of stem, which turns upwards and resumes its vertical position. Main roots, *positively geotropic*, grow vertically downwards and if laid horizontally exhibit increased elongation of cells on upper side of growth region, the root turning down again as a result. In shoots placed in a horizontal position, differences in GIBBERELLIN and AUXIN concentrations develop between upper and lower sides. Together, these cause lower side of the shoot to elongate more than upper side, giving the observed upward growth. When it eventually resumes vertical growth, lateral asymmetry in growth substance concentrations disappears, and growth continues vertically. In roots, growth substance asymmetries are less well understood. There is evidence that gradients of these substances within the root cap are brought about by relatively tiny movements of starch-containing plastids (amyloplasts), the 'statoliths', for when a plant is placed in a horizontal plane, these plastids move from the transverse walls of vertically growing roots and come to rest near what were previously vertically orientated walls. Then, after several hours, the root curves downwards and these plastids return to their original positions. It is not yet clear how these plastid movements translate into growth substance gradients. However, starch-less mutants of *Arabisopsis* still exhibit gravitropism, although with less accuracy and speed. Other work supports the hypothesis that although the mass of statoliths is important for the perception of gravity, there must be another mechanism capable of doing so, and protoplast pressure on the cytoskeleton and plasma membrane has received some experimental support as a factor. There is little evidence for the role of auxin (IAA) in roots; some have been unable to find it in the root cap at all. Instead, ABSCISIC ACID is found there, and has been shown to be redistributed from the cap to the root itself, and to act as an inhibitor on cells in the region of elongation on lower side of a horizontally positioned root. Recent work implicates Ca^{2+} ions, instead of auxin, as the active ingredient, Ca^{2+} ions being redistributed to the lower side of a root cap within 30 min. of gravistimulation, while substances that bind to calcium make radicles insensitive to gravity. See DIAGRAVITROPISM, PLAGIOTROPISM.

great chain of being (scala natura) View proposed by Aristotle and incorporated by Leibniz in his metaphysics and by BUFFON (for whom it was a scale of degradation, from man at the top): that there is a linear and hierarchical progression of forms of existence, from simplest to most complex, lacking both gaps and marked transitions.

green fluorescent protein (GFP) Isolated originally from the jellyfish *Aequoria victoria*, the most useful fluorescent protein in cytology, undergoing a self-catalysed reaction which produces a bright fluorescent molecule centre, which has been improved upon by site-directed mutagenesis of the original gene. Used as a reporter molecule in monitoring gene expression when its gene is linked to a promoter of a transgene or other gene whose pattern of expression is being investigated (see REPORTER GENE). Many other uses for GFP are emerging. See HORSERADISH PEROXIDASE.

greenhouse effect Effect in which short wavelength solar radiation entering the Earth's atmosphere is re-radiated from the Earth's surface in longer infrared wavelengths, and then reabsorbed by components of the atmosphere to become an important factor in heating the total atmosphere. Effect resembles heat reflection by greenhouse glass. Oxygen, OZONE, carbon dioxide, methane and water vapour all absorb in the infrared wavelengths, and increasing amounts of carbon dioxide from the combustion of fossil fuels, the destruction of tropical forest cover and the ploughing of soil are growing factors in raising the average atmospheric temperature. Geological records indicate that CO_2 levels have been considerably higher on occasions in the past 4.7 billion years than they are now (e.g. 10–20 times higher in the Devonian), and correlate with stomatal densities on plant leaves. It has been estimated that the Earth's atmosphere would now have a mean temperature of $-20°C$, and not $15°C$, if CO_2 were absent (see GLOBAL WARMING).

Methane is a much more effective greenhouse gas than is carbon dioxide and its increasing level is causing concern. Predictions include that the increase in temperature will cause an increase in the great deserts of the world, and that increased photosynthetic activity will result from increased carbon dioxide concentrations in the atmosphere. The phenomenon is alarming because the consequences are not known, existing global climate models being too unsophisticated to predict future climate patterns with any certainty; yet mankind carries on seemingly obliviously. See POLLUTION.

grey matter Tissue of vertebrate spinal cord and brain containing numerous cell bodies and dendrites of neurons, along with unmyelinated neurons synapsing with them, glial cells and blood vessels. Occurs as inner region of nerve cord, around central canal; in brain too it generally occupies the inner regions, but in some parts (e.g. CEREBRAL CORTEX of higher primates), some cell bodies of grey matter have migrated outwards to form a third layer on top of the white, axon zone.

ground meristem Primary MERISTEM in which procambium is embedded and which is surrounded by protoderm; matures to form the GROUND TISSUES.

ground tissue (Bot.) All tissues except the epidermis (or periderm) and the vascular tissues; e.g. those of the cortex and pith.

group selection Postulated evolutionary mechanism whereby characters disadvantageous to individuals bearing them, but beneficial to the group of organisms they belong to, can spread through the population countering the effects of selection at the individual level. Invoked to explain apparent reproductive restraint by individuals when environmental resources are scarce, and other cases of apparent ALTRUISM which may have a simpler explanation in terms of NATURAL SELECTION. Some hold that group selection may have played a part in the evolution of SEX, and others hold that, especially among hunter-gatherers, human social groups have been guided by an egalitarian ethic for millennia (groups of altruists having higher fitness than groups of non-altruists). Darwin was aware of the difficulty

of explaining the evolution of altruism by natural selection and proposed that natural selection could operate at more than one level of biological hierarchy. See UNIT OF SELECTION.

growth Term with a variety of senses. At the individual level, usually involves increase in dry mass of an organism (or part of one), whether or not accompanied by size increase, involving differentiation and morphogenesis. Commonly involves cell division, but cell division without increase in CELL SIZE does not produce growth (see CELL GROWTH, CELL SHAPE). Nor is uptake of water alone sufficient for growth. Usually regarded as irreversible, but ATROPHY of tissues and dedifferentiation of cells can occur. Algal and plant growth forms may be *diffuse*, where most cells are capable of division; *apical*, where a single terminal cell gives rise to cells beneath; *trichothallic*, where a cell divides forming a hair above the thallus below; *promeristematic*, where a non-dividing apical cell controls a large number of small dividing cells beneath it; *intercalary* where a zone of meristematic cells occurs forming tissue above and below the meristem; *meristodermic*, when a layer of usually peripheral cells divides parallel to the thallus surface to form tissue below the meristoderm (usually cortex). In most higher plants, growth is restricted to MERISTEMS. In animals, growth is more diffuse. Cell survival and proliferation in animals are generally under different genetic control, and disturbances in both are likely to contribute to transformation of cells into CANCER CELLS. The terms 'lytic growth' and 'lysogenic growth' are often used in the context of BACTERIOPHAGE 'life cycles'. See GROWTH CURVE, EXPONENTIAL GROWTH, ALLOMETRY, RAS PROTEINS, AUXIN, GROWTH HORMONES.

growth cone The expanded tip of a nerve axon, whose dynamic structure resembles that of a motile cell, which leads the extending axon along its appropriate path during growth. Extracellular guidance cues may attract or repel growth cones, operating either locally or at a distance. See INTEGRINS, NEUROTROPHINS.

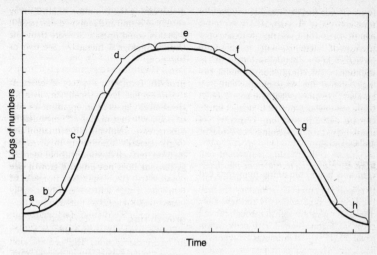

FIG. 76 *The* GROWTH CURVE *of unicellular organisms under optimal growth conditions. Phases a to h are explained in the entry.*

growth curve Many general features of the growth of a population may be indicated by the growth curve of unicellular organisms under optimal conditions for growth. (See Fig. 76.) Stages (a) to (e) represent a logistic growth curve (see EXPONENTIAL GROWTH).

(a) *Lag Phase*: latent phase, in which cells recover from new conditions, imbibe water, produce ribosomal RNA and subsequent proteins. Cells grow in size, but not in number. (b) *Phase of Accelerated Growth*: fission initially slow, cell size large. During this phase, rate of division increases and cell size diminishes. (c) *Exponential or Logarithmic Phase*: cells reach maximum rate of division. Characteristic of this phase that numbers of organisms, when plotted on a logarithmic scale, generate a straight-line slope. (d) *Phase of Negative Growth Acceleration*: food, e.g. begins to run out, waste poisons accumulate, pH changes and cells generally interfere with one another. Increase in number of live cells slows as rate of fission declines. (e) *Maximum Stationary Phase*: number of cells dying balances rate of increase, resulting in a constant total viable population. (f) *Accelerated Death Phase*: cells

reproduce more slowly and death rate increases. (g) *Logarithmic Death Phase*: numbers decrease at unchanging rate. (h) *Phase of Readjustment and Final Dormant Phase*: death rate and rate of increase balance each other, and finally there is complete sterility of the culture. See J-SHAPE GROWTH FORM.

growth factors (GFs) Diffusible molecules variously signalling the growth, division, locomotion, survival and/or differentiation of cells; e.g. COLONY STIMULATING FACTORS and some INTERLEUKINS. All bind RECEPTORS of the cell surface or EXTRACELLULAR MATRIX (ECM), which are often linked to SIGNAL TRANSDUCTION pathways via such intracellular enzymes as PROTEIN KINASES, notably the enzyme PI_3-kinase, which phosphorylates inositol phospholipids in the plasma membrane, and S6 kinase (see CELL GROWTH, MITOGEN, Fig. 151). Also initiate changes in CELL SHAPE (see CYTOSKELETON). Growth factor stimulation causes increased expression of the *myc* GENE whose product, Myc protein, increases transcription of several genes involved in a cell's metabolism and macromolecular synthesis. *T cell growth* *factor* (TCGF, interleukin 2) is required for proliferation of T cells, TCGF receptors appearing on the T cell surface in response to mitogen (antigen); *platelet-derived growth factor* (PDGF) is a chemoattractant as well as a mitogen and cell growth promoter and stimulates connective tissue and neuroglial cells (see PLATELETS); *epidermal growth factor* (EGF) stimulates many cell types to divide (see Fig. 143b); INSULIN-LIKE GROWTH FACTORS (ILGF-I and ILGF-II) collaborate with PDGF and EGF to stimulate fat and other connective tissue cells, causing permanent exit from the CELL CYCLE; *transforming growth factor* β (TGF-β, or activin) prepares many cells to respond to other GFs, stimulates the synthesis of ECM components and reversibly decreases the rate of transit through the cell cycle – a feature shared with *nerve growth factor* (NGF), which promotes axon growth and longevity of sympathetic and some sensory and CNS neurons. Growth factor antagonists include some in the extracellular medium. The *cerberus* protein (Cer) is expressed in the vertebrate anterior endoderm where it acts as a multivalent antagonist of Nodal, BMP and Wnt proteins, all of which are involved in more rostral regions during trunk development. See AGEING, ANGIOGENESIS, APOPTOSIS, CYTOKINES, INTEGRINS, NEUROTROPHINS, RETROGRADE SIGNALLING, TUMOUR NECROSIS FACTOR, Fig. 77).

growth hormone (GH, somatotrophin) Polypeptide (and most abundant) hormone produced by somatotroph cells of anterior PITUITARY. These cells proliferate and secrete on binding growth hormone-releasing factor (GRF) at their G-PROTEIN-linked receptors, a signal which is transduced via cyclic AMP. Regulates deposition of collagen and chondroitin sulphate in bone and cartilage; promotes mitosis in osteoblasts and increase in girth and length of bone prior to closure of EPIPHYSES. Transported in plasma by a GLOBULIN protein. Its release is prevented by hypothalamic SOMATOSTATIN. Induces release from liver of INSULIN-LIKE GROWTH FACTORS which mediate its effects at the cell level. Low levels in humans result in pituitary dwarfism (epiphyseal growth plates close before normal height attained); high levels prior to closure of the growth plates result in gigantism (increase in long bone length) and after closure of growth plates to acromegaly (thick bones of hands, feet, cheeks and jaws). Dwarfism and gigantism syndromes in humans have been explained by mutations in one or other genes encoding the variety of molecules regulating the hypothalamic-pituitary-target tissue signalling system. Human growth hormone (hGH) is hyperglycaemic and its excess secretion can stimulate the pancreas into continuous INSULIN secretion, leading to failure of pancreatic beta cells to produce the hormone (β-cell burn out) which would bring on type I diabetes mellitus.

growth ring (Bot.) Growth layer in secondary xylem or secondary phloem, as seen in transverse section. The periodic and seasonally related activity of vascular cambium produces growth increments, or growth rings. In early and middle summer new xylem vessels are large and produce a pale wood, contrasting with the narrower and denser vessels of late summer and autumn, which produce a dark wood. Width of these rings varies from year to year depending upon environmental conditions, such as availabilities of light, water and nutrients and temperature. See DENDROCHRONOLOGY.

growth substance (Bot.) Term used in preference to plant hormone, to include all natural (endogenous) and artificial substances with powerful and diverse physiological and/or morphogenetic effects in plants. Sites of natural production are often diffuse and rarely comprise specialized glandular organs – often just a patch of cells, commonly without physical contact. The minute quantities produced, and the notorious synergisms in their modes of action, pose profound research problems. See AUXINS, ABSCISIC ACID, CYTOKININS, ETHENE, GIBBERELLINS.

gs See GRANDCHILDLESS.

GTP (guanosine 5′-triphosphate) Purine nucleoside triphosphate. Required for coupling some activated membrane recep-

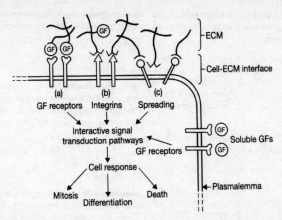

FIG. 77 *Possible interactions between* GROWTH FACTORS, *the extracellular matrix (ECM) and receptors (shown variously as structures crossing the plasmalemma).*

tors to ADENYLATE CYCLASE activity. A GTP-binding protein (G-PROTEIN) hydrolyses the GTP and keeps the enzyme's activity brief. Involved in tissue responses to various hormones (e.g. ADRENALINE), a precursor for NUCLEIC ACID synthesis and essential for chain elongation during PROTEIN SYNTHESIS. See CYCLIC GMP.

guanine Purine base of NUCLEIC ACIDS. See GTP, INOSINIC ACID.

guanosine Purine nucleoside. Comprises D-ribose linked to guanine by a beta-glycosidic bond. See GTP.

guard cells (Bot.) Specialized, crescent-shaped, unevenly thickened epidermal cells in pairs surrounding a stomatal pore (see STOMA). Changes in shape of guard cells, due to changes in their turgidities, control opening and closing of the stomata, and hence affect rate of loss of water vapour in TRANSPIRATION and amount of gaseous exchange. It is now generally accepted that turgor changes in the guard cells are caused principally by salt uptake and extrusion, although the presence of starch is held to play a role, too. Guard-cell membranes can extrude protons using an ATP-driven proton pump, potassium ions (K⁺) replacing them down an electrical gradient, thereby reducing water potential and generating turgor, while chloride ions (Cl⁻) enter to maintain pH. The rise in pH appears to favour the activity of PEP carboxylase within the chloroplasts (see PHOTOSYNTHESIS), catalysing fixation of carbon dioxide to produce oxaloacetate using pyruvate released from stored starch. The oxaloacetate is converted to malic acid, which dissociates to release malate ions, balancing the charge due to potassium ion uptake. The growth substance ABSCISIC ACID is believed to act on the guard cell membrane and bring about opening of calcium (Ca²⁺) channels there so that Ca²⁺ enters the cell. Once inside, Ca²⁺ binds a CALMODULIN and through its action as a second messenger, changes the cell's turgidity in some as yet unspecified way. Stomata can be closed by elevating external Ca²⁺ and release of Ca²⁺ from cells of a stressed root or leaf may be a relevant message. See CELL WALL.

guard hypothesis See RESISTANCE GENES.

guild A group of species that exploit the same class of environmental resources in a similar manner.

gums Polysaccharide components, some pentose-rich, of algal and plant cell walls; sometimes with oligosaccharide side chains, some residues of which may be acetylated. Include AGAR, ALGINIC ACID and CARRAGEENAN.

gustation The sense of taste. See RECEPTOR, SIGNAL TRANSDUCTION.

gut See ALIMENTARY CANAL.

guttation Excretion of water drops by plants through HYDATHODES, especially in high humidity, due to pressure built up within the xylem by osmotic absorption of water by roots. See ROOT PRESSURE.

gymnosperm A term no longer used in the formal schemes of classification, but still used informally. The term literally means 'naked-seeded plants', ovules and seeds lying exposed on their sporophyll (or analagous) surfaces, this being one of the principal characteristics of non-anthophyte seed plants. The four divisions with living species today are CONIFEROPHYTA, CYCADO-PHYTA, GINKGOPHYTA and GNETOPHYTA. See also PROGYMNOSPERMOPHYTA.

gynandromorph Animal, usually an insect, which is a genetic MOSAIC in that some of its cells are genetically female while others are genetically male. Loss of an X-chromosome by a stem cell of an insect which developed from an XX zygote thus produces a clone of 'male' tissue. Sometimes expressed bilaterally, one half of the animal being phenotypically male, the other female. Also occurs in birds and mammals. See INTERSEX.

gynandrous (Bot.) (Of flowers) having stamens inserted on the gynoecium.

gynobasic (Bot.) (Of a style) arising from base of ovary (due to infolding of ovary wall in development); e.g. white dead nettle.

gynodioecious Having female and hermaphroditic flowers on separate plants; e.g. thymes (*Thymus*). Compare ANDRODIOE-CIOUS, GYNOMONOECIOUS.

gynoecium (pistil) Collective term for the carpels of a flower; i.e. the female components of the flower. Compare ANDROECIUM.

gynogenesis Condition whereby a female animal must mate before she can produce parthenogenetic eggs. In some triploid thelytokous animals (the salamanders *Ambystoma*, the fish *Poeciliopsis*) the sperm penetrates the egg to initiate cleavage but contributes nothing genetically. See ANDRO-GENESIS, GYNOGENONE.

gynogenone Diploid embryo with two maternal genomes produced by pronuclear transplantation. See ANDROGENESIS, GYNO-GENESIS.

gynomonoecious (Bot.) With female and hermaphroditic flowers on the same plant; e.g. many Compositae. Compare ANDRO-MONOECIOUS, GYNODIOECIOUS.

gyrogonite Lime encrusted, fossilized oogonia and encircling sheath cells (nucule) of the CHAROPHYCEAE. Earliest fossils occur in the Upper SILURIAN. Gyrogonites of extant species date from the Upper CARBONIFEROUS.

gyrus A fold (convolution) on the surface grey matter of the CEREBRAL CORTEX, between which occur deep grooves, or sulci. Both gyri and sulci are functionally differentiated and identified by names.

H

habitat Place or environment in which specified organisms live; e.g. sea-shore. Compare NICHE.

habituation LEARNING, in which an animal's response to a stimulus declines with repetition of the stimulus at the same intensity. Needs to be distinguished experimentally from sensory ADAPTATION and muscular fatigue.

haem (heme) Iron-containing PORPHYRIN, acting as PROSTHETIC GROUP of several pigments, including HAEMOGLOBIN, MYOGLOBIN and several CYTOCHROMES.

haemagglutination Clumping of red blood cells due to cross-linking of cells by antibody to surface antigens. Not to be confused with BLOOD CLOTTING.

haemagglutinin Glycoprotein product of INFLUENZA virus, associated, like neuraminidase, with the encapsulating host cell membrane and involved in attachment to host cells. Antigenic variation (shift) in haemagglutinin is responsible for new epidemics of the virus, against which earlier antibodies are ineffective.

haemerythrin Reddish-violet iron-containing respiratory pigment of sipunculids, one polychaete, priapulids and the brachiopod *Lingula*. Prosthetic group is not a porphyrin, the iron attaching directly to the protein. Always intracellular.

haemocoele Body cavity of arthropods and molluscs, containing blood. Continuous developmentally with the BLASTOCOELE. Unlike the COELOM, it never communicates with the exterior or contains gametes. Often

functions as a hydrostatic skeleton. Contains haemolymph. See Fig. 35.

haemocyanin Copper-containing protein (non-porphyrin) respiratory pigment occurring in solution in haemolymph of malacostracan and chelicerate arthropods and in many molluscs. Blue when oxygenated, colourless when deoxygenated.

haemoglobin Protein respiratory pigment with iron(Fe^{2+})-containing porphyrin as prosthetic group. Tetrameric molecule, comprising two pairs of non-identical polypeptides associated in a quaternary structure, binding oxygen reversibly, forming *oxyhaemoglobin*. Occurs intracellularly in vertebrate ERYTHROCYTES, but when found in invertebrates (e.g. earthworms) is usually in simple solution in the blood, increasing its viscosity. Also found in root nodules of leguminous plants (as *leghaemoglobin*), but only if *Rhizobium* is present. Scarlet when oxygenated, bluish-red when deoxygenated. The ability of the haemoglobin molecule to pick up and unload oxygen depends on its shape in solution, which varies allosterically with local pH (see BUFFER). This in turn is a function of the partial pressure of CO_2 (see BOHR EFFECT). Haemoglobins are adapted for maximal loading and unloading of oxygen within the oxygen tension ranges occurring in their respective organisms. See Fig. 78.

Normal adult human haemoglobin (HbA) contains two α- and two β-globin chains; foetal haemoglobin (HbF) contains two α- and two γ-globin chains. The transition from foetal to adult haemoglobin production begins late in foetal life and is completed in early infancy. Some carbon

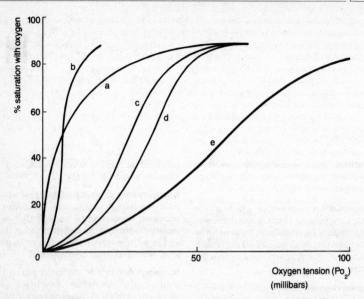

FIG. 78 *Oxygen equilibration curves for: (a) myoglobin, (b) Arenicola Hb, (c) human foetal Hb, (d) adult human Hb, (e) pigeon Hb.*

dioxide is carried by haemoglobin as *carba-mino compounds*, and one consequence of inhaling cigarette smoke is that the carbon monoxide produced binds irreversibly to haemoglobin, forming *carboxyhaemoglobin*. See MYOGLOBIN, NITRIC OXIDE, SICKLE-CELL ANAE-MIA, THALASSAEMIAS.

haemolysis Rupture of red blood cells (e.g. through osmotic shock) with release of haemoglobin.

haemophilia Hereditary disease, X-linked and recessive in humans, in which blood fails to clot owing to absence of Factor VIII (see BLOOD CLOTTING). Insertions of *LINE* transposable elements into the Factor VIII gene are responsible for this classic haemophilia (Haemophilia A). Haemophilia B (Christmas disease) is due to Factor IX (plasma thromboplastin component) deficiency.

haemopoiesis Blood formation; in vertebrates includes both plasma and cells. In anoxia, the kidney produces a COLONY STIMULATING FACTOR, *erythropoietin*, stimulat-ing red cell production in red bone marrow (see MYELOID TISSUE). In vertebrate embryos, erythropoiesis occurs commonly in yolk sac, liver, spleen, lymph nodes and bone marrow; in adults is restricted to red bone marrow (and yellow bone marrow in long bones under oxygen stress). All blood cells, myeloid and lymphoid, are ultimately gen-erated from a common multipotent bone marrow stem cell, which also gives rise to osteoclasts; but each cell type is under spe-cific control. The stem cells depend much more on contact with stromal cells of the marrow for long-term maintenance than do their progeny. Much of the plasma protein is formed in the liver, notably fibrinogen, albumen and α- and β-globulins.

haemostasis Several homeostatic mech-anisms maintaining blood in a fluid state, and within blood vessels. Includes BLOOD CLOTTING and FIBRINOLYSIS.

hair (Bot.) Trichome. (1) Single- or many-celled outgrowth from an epidermal cell; usually a slender projection composed of cells arranged end-to-end whose functions are various. Root hairs facilitate absorption of water and minerals from the soil; increased hairiness of leaves results in increased reflectance of sunlight, a lower leaf temperature, and a lower TRANSPIRATION rate – particularly important in plants of arid environments. Foliar trichomes are also used for absorption of water and minerals (e.g. bromeliads); those of the saltbush (*Atriplex*) secrete salt from leaf tissue, preventing toxic accumulation in the plant; others may act as a defence against insects (hooked hairs of some species, e.g. *Phaeolus vulgaris*, may impale insects and larvae; glandular hairs may provide a chemical defence). (2) A simple filament of the CYANO-BACTERIA, devoid of a sheath. (3) A flagellar appendage, either a fibrous solid hair or a tubular hair (see FLAGELLUM). (Zool.) Epidermal thread protruding from mammalian skin surface, composed of numerous cornified cells. A hair acts as a sensory lever, movement of the tip being magnified at the base. Each develops from the base of a HAIR FOLLICLE (its 'root') in the undersurface of which (the hair bulb), cells are produced mitotically. Hair colour depends upon the amount of MELANIN present, and with age an increasing presence of air bubbles results in total reflection of light, making hair appear white. In most non-human mammals body hair is thick enough for hair erection (by *erector pili muscles*) to have a homeostatic effect on heat retention. Nerve endings provide hair with a sensory role (see SKIN). So-called 'hairs' of arthropods are bristles. See HORN.

hair cell Ectodermal cells with modified membranes found in vertebrate VESTIBULAR APPARATUS, acting as mechanoreceptors by responding to tension generated either by a gelatinous covering layer (in maculae, cristae), or by the tectorial membrane (in cochlea). They normally bear one long true cilium (the *kinocilium*) and a tuft of several large and specialized microvilli (*stereocilia*) of decreasing length, the tip of one attached to the next longer one by protein links which open or close membrane transduction channels, these in turn being linked to

actin microfilaments within the stereocilium. Rather than using G-proteins and second messengers to achieve their sensitivity (see RECEPTOR), an ion-channel gating 'lid' is located near the tip of each pivoting stereocilium so that when the channel opens to calcium (distortion of 0.1 nm is sufficient) there is a marked dip in the stiffness-displacement curve for these organelles and a receptor potential is generated.

hair follicle Epidermal sheath enclosing the length of a hair in the skin. Surrounded by connective tissue serving for attachment of erector pili muscles. May house a SEBACEOUS GLAND. See SKIN.

hairpin (foldback) loops Palindromic duplex DNA sequences (INVERTED REPEAT SEQUENCES) which can fold back on themselves and form hairpin double-stranded structures on renaturing after being denatured. Single-stranded RNAs may also form such loops. Sometimes the degree of base-pairing required for hairpins to form is not great.

Haldane's rule J. B. S. Haldane (1922) proposed the rule that 'when in the F_1 offspring of two different animal races one sex is absent, rare or sterile, that sex is the heterogametic sex'. Evolutionary explanations for this centre on the notions that the X chromosome and autosomes are in greater disharmony in the XY (or ZW) sex; that evolution of hybrid male sterility is faster in species with XY males (driven by SEXUAL SELECTION?); and that maternal-zygotic incompatibility preferentially affects the viability of the XX sex. Study of this phenomenon will probably illuminate the genetics of SPECIATION.

halophyte Plant growing in and tolerating very salty soil typical of shores of tidal river estuaries, saltmarshes, or alkali desert flats.

hallux 'Big toe'; innermost digit of tetrapod hind foot. Often shorter than other digits. Turned to the rear in most birds, for perching. Compare POLLEX.

haltere Modified hind wing of DIPTERA (two-winged flies) concerned with maintenance of stability in flight. Comprises basal lobe closest to thorax, a stalk and an end knob. Halteres are like gyroscopes; their combined nervous input to the thoracic ganglion enables adjustment of wings to destabilizing forces.

Hamamelidae (Hamameliflorae) A subclass of the MAGNOLIOPSIDA that comprises about 3,400 species that form a coherent group of dicotyledons with more or less strongly reduced, often unisexual flowers (including nettles and beeches). These are often borne in catkins and never have numerous seeds or parietal placentas. Pollen grains are often porate and with a granular rather than columellate, infratectal structure. See INTRODUCTION.

Hamilton's rule Prediction that genetically determined behaviour which benefits another organism, but at some cost to the agent with the allele(s) responsible, will spread by SELECTION when the relation $(rb - c) > o$ is satisfied; where r = degree of RELATEDNESS between agent and recipient, b = improvement of individual FITNESS of recipient caused by the behaviour and c = cost to agent's individual fitness as a result of the behaviour. This concept of *inclusive fitness*, although associated with W.D.Hamilton, is to be found in the works of R.A. Fisher (1930) and J.B.S. Haldane (1932), to which Hamilton referred in his papers. See ALTRUISM, KIN SELECTION, PRISONER'S DILEMMA.

handling time The length of time a predator spends in pursing, subduing and consuming its prey item and then preparing itself for further prey searching. See OPTIMALITY THEORY.

haplochlamydeous See MONOCHLAMYDEOUS.

haplodiplontic (Of a LIFE CYCLE) in which both haploid and diploid mitoses occur.

haploid (Of a nucleus, cell, etc.) in which chromosomes are represented singly and unpaired. The haploid chromosome number, n, is thus half the DIPLOID chromosome number, $2n$. The amount of DNA per haploid genome is its C-VALUE. Haploid cells are commonly the direct product of MEIOSIS, but haploid mitosis is relatively common too. No haploid cell can undergo meiosis. Diploid organisms generally produce haploid gametes. In humans, $n = 23$. The false spider mite *Brevipalpus phoenicis* is the first animal to be discovered which is entirely haploid. See ALTERNATION OF GENERATIONS, MALE HAPLOIDY, POLYPLOIDY.

haploid parthenogenesis (Bot.) Development of an embryo into a haploid sporophyte from an unfertilized egg on a haploid gametophyte. See PARTHENOGENESIS.

haplo-insufficiency See MUTATION.

haplont Organism representing the HAPLOID stage of a LIFE CYCLE. Compare DIPLONT.

haplontic (Of a LIFE CYCLE) in which there is no diploid mitosis, but in which haploid mitosis does occur.

haplorhine The Suborders Haplorhini and Strepsirhini (see STREPSIRHINE) are the two sister groups of living primates. Haplorhines, lacking a RHINARIUM, include the tarsiers and the anthropoids (platyrrhines and catarrhines). See Fig. 138 and PRIMATES.

haplostele Solid cylindrical STELE in which a central strand of primary xylem is sheathed by a cylinder of phloem.

haplotype A haploid genotype. The gametes produced by a normal outbred diploid individual will be of a variety of haplotypes.

hapten Molecule binding specifically to an antigen-binding site (epitope) on an antibody molecule or lymphocyte receptor, but without inducing any immune response.

hapteron (holdfast) Bottom part of some algae, attaching the plant to the substratum; may be discoid or root-like in structure.

haptonema A thin, filamentous appendage arising near the flagella occurring in many species belonging to the algal division HAPTOPHYTA. Function not fully understood but can serve as temporary attachment to a surface and has been implicated in acquisition of food.

Haptophyta (Prymnesiophyta) A class of the algal division HETEROKONTOPHYTA where the great majority are marine with only a few occurring in freshwater. There are about 75 genera of living algae, with about 500 species. The great majority are unicellular flagellates, although some species also have amoeboid, coccoid, palmelloid or filamentous stages. Planktonic haptophytes play an important role in the world's oceans as primary producers. Almost all the planktonic species are small algae, belonging to the nannoplankton.

Haptophytes are also commonly found in nearshore or inshore phytoplankton. The importance of the coccolithophorids, a group of planktonic haptophytes that form calcareous scales has long been realized and fossil coccolithophorids are important marker fossils in stratigraphical correlation and dating of marine sediments. Fossil coccolithophorids were very diverse during the Jurassic, while there are a few occurrences in rocks dating back to the Carboniferous. They seem to have reached their peak in the Upper Cretaceous and chalk deposits of this period, which occur worldwide, comprise largely coccolith.

Recurring blooms of *Phaeocystis pouchetii* have been known for a long time in the southern North Sea, while in May 1988 a huge bloom of *Chrysochromulina polylepis* occurred around the western coasts of Sweden and southern coasts of Norway. This bloom was highly toxic to fish, invertebrates and even to epilithic seaweeds.

Flagellate cells possess two apically or laterally inserted flagella that may or may not be equal in length. Mastigonemes are not found on the flagella, although the anterior flagellum of *Pavlova* bears delicate nontubular hairs. A thin filamentous appendage, which may be short or long, called a haptonema, is located between the two flagella. Each chloroplast is enclosed with a fold of endoplasmic reticulum and where it lies against the nucleus, its own endoplasmic retuculum is continuous with the endoplasmic reticulum around the nucleus. A periplast reticulum occurs in the narrow space between the chloroplast endoplasmic reticulum and the chloroplast envelope. Thylakoids are stacked in threes to form lamellae; however, there is no girdle lamella. A pyrenoid, usually penetrated by lamellae containing two thylakoids, is found in each chloroplast. Chloroplasts contain chlorophylls a, c_1 and/or c_2. Chloroplast colour is golden brown due to accessory pigments of which the most important is fucoxanthin; other important carotenoid pigments are β-carotene, diadinoxanthin and diatoxanthin. Chloroplast DNA occurs as numerous nucleoids scattered throughout the chloroplast. When present an eyespot lies at the anterior of the cell and comprising a row or layer of small spherical globules just beneath the chloroplast envelope. Paramylon has been found in one species.

Reserve polysaccharides in vacuoles outside the chloroplast. Tiny scales or granules of organic material (cellulose) cover the cell surface. Calcified scales (coccoliths) may also be present. The scales are formed within the golgi apparatus and then secreted from the cell. They usually have a characteristic structure, each scale having radially arranged, spoke-like fibrils on the side that faces the cell, and concentrically arranged ones on the outer face. The cytoplasm of each cell is surrounded by a narrow, peripheral cisterna of endoplasmic reticulum. The haptophytes have their own characteristic type of mitosis. A heteromorphic diplohaplontic life cycle occurs in some species in which a diplod, planktonic flagellate stage alternates with a haploid, benthic filamentous stage. In other species alternation between flagellate and non-flagellate stages occurs.

haptotropism (thigmotropism) (Bot.) A TROPISM in which the stimulus is a localized contact, e.g. tendril in contact on one side with solid object such as a twig; response is curvature in that direction producing coiling around the object.

Hardy–Weinberg theorem (H–W law, principle or equilibrium) Theorem predicting for a normal amphimictic population the ratios of the three genotypes (e.g. AA : Aa : aa) at a locus with two segregating alleles, A and a, given the frequencies of these genotypes in the parent population. The theorem assumes: random (i.e. non-assortative) mating, no NATURAL SELECTION, GENETIC DRIFT or MUTATION, or immigration or emigration. Its utility is that once the parental population's genotypic ratios have been determined, their predicted ratio in the next generation can be checked against the observed values, and any departures from expectation tested for significance (e.g. by CHI-SQUARED TEST). If significant, and if all assumptions other than selection can be discounted, then this is *prima facie* evidence that selection caused the departures from expected values. If in the parental generation the frequencies of alleles A and a are *p* and *q* respectively, then the theorem states that the genotypic ratios in the next and all succeeding generations will be:

$$AA : Aa : aa$$
$$p^2 : 2pq : q^2$$

Hartig net The intercellular hyphal network formed by an ectomycorrhizal fungus upon the surface of a root.

Hatch–Slack pathway See PHOTOSYNTHESIS.

haustorium (pl. haustoria) Specialized penetrative food-absorbing structure. Occurs (1) in certain fungal parasites of plants (e.g. rusts, powdery mildews and downy mildews) at the end of a hyphal branch within a living host cell; (2) in LICHENS, commonly penetrating the algal cells; (3) in some parasitic plants (e.g. dodder), withdrawing material from the host tissues.

haversian system (osteon) Anatomical unit of compact BONE. Comprises a central *Haversian canal*, which branches and anastomoses with those of other Haversian systems and contains blood vessels and nerves, surrounded by layers of bone deposited concentrically by osteocytes and forming cylinders. Blood is carried from vessels at the bone surface to the Haversian system by *Volkmann's canals*. They produce the zoological equivalent of the 'grain' of wood, preventing spread of stress fractures and resisting bending forces parallel to the Volkmann's canals, although resulting in weakness to forces at right angles to these.

H-2 complex Mouse major histocompatibility complex. See MHC.

HCG, hCG, hcg See HUMAN CHORIONIC GONADOTROPHIN.

HDL High-density LIPOPROTEIN. The smallest of the lipoproteins (mean diameter 8 nm). In plasma, HDLs donate C and E apoproteins to CHYLOMICRONS, and exchange cholesterol ester with these particles for triglyceride. Excessive dietary calorific intake reduces HDL concentration in the plasma, while regular exercises increases it. See LDL for interactions with other lipoproteins.

heart Muscular, rhythmically contracting pump forming part of the cardiovascular system and responsible for blood circulation (see BLOOD PRESSURE). All hearts have valves to prevent back-flow of blood during contraction. Initiation of heart beat may be by extrinsic nerves (*neurogenic*), as in many adult arthropods, or by an internal pacemaker (*myogenic*), as in vertebrates and some embryonic arthropods.

Vertebrate heart muscle (CARDIAC MUSCLE) does not fatigue, and is under the regulation of nerves (see CARDIOACCELERATORY/INHIBITORY CENTRES) and hormones (e.g. ADRENALINE). The basic S-shaped heart of most fish comprises four chambers pumping blood unidirectionally forward to the gills. This *single circulation* has the route:

$$body \rightarrow heart \rightarrow gills \rightarrow body.$$

Replacement of gills by lungs in tetrapods was associated with the need for a heart providing a *double circulation* in order to keep oxygenated blood (returning to the heart from the lungs) separate from deoxygenated blood (returning to the heart from the body) (See Fig. 79). The blood route becomes:

$$body \rightarrow heart \rightarrow lungs \rightarrow heart \rightarrow body.$$

In amphibians, two atria return blood from these two sources to a single ventricle,

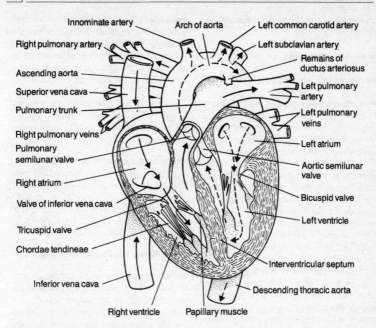

FIG. 79 *Mammalian* HEART *showing direction of oxygenated (--->) and deoxygenated (→) blood.*

and separation is limited; reptiles have a very complex ventricle with a septum assisting separation; birds and mammals have two atria and two ventricles, one side of the heart dealing with oxygenated and the other with deoxygenated blood (see HEART CYCLE). In annelids, the whole dorsal aorta may be contractile with, in addition, several vertical contractile vessels, or 'hearts'. The insect heart is a long peristaltic tube lying in the roof of the abdomen and perforated by paired segmental holes (*ostia*) through which blood enters from the haemocoele. There may be accessory hearts in the thorax. Blood is driven forwards into the aorta, which opens into the haemocoele. A similar arrangement occurs in other arthropods. The basic molluscan heart comprises a median ventricle and two atria. See CORON-ARY HEART DISEASE, PERICARDIUM.

heart cycle (cardiac cycle) One complete sequence of contraction and relaxation of heart chambers, and opening and closing of valves, during which time the same volume of blood enters and leaves the heart. Chamber contraction (*systole*) is followed by its relaxation (*diastole*) when it fills again with blood. In mammals and birds ventricular diastole draws in most of the blood from the atria; atrial systole adds only 30% to ventricular blood volume. The PACEMAKER and its associated fibres ensure that the two atria contract simultaneously just prior to the two ventricles. Atrioventricular valves open when atrial pressure exceeds ventricular and close when ventricular pressure exceeds atrial. See CARDIO-ACCELERATORY CENTRE.

heartwood Central mass of xylem tissue in tree trunks; contains no living cells and no longer functions in water conduction but serves only for mechanical support; its elements frequently blocked by TYLOSES, and frequently dark-coloured (e.g. ebony), impregnated with various substances (tannins, resins, etc.) that render it more resistant to decay than surrounding SAPWOOD.

heat-shock proteins See CHAPERONES.

heavy metals See POISONS.

hedgehog **gene** One of a family (Hedgehog family) of SEGMENT POLARITY GENES encoding the ligand Hedgehog protein (Hh), a component of cell signalling pathways. Its spatial expression during development in *Drosophila* is under the control of the *engrailed* gene and leads to maintenance of transcription of *wingless* gene in neighbouring cells and regulated segmental and IMAGINAL DISC patterning. In vertebrates, it is involved in control of left–right asymmetry, polarity of the central nervous system, somites and limbs, in spermatogenesis and other aspects of morphogenesis. At least three vertebrate genes encode Hedgehog proteins, one being *sonic hedgehog* (product Shh; see APOPTOSIS). All active *hedgehog* products bind cholesterol and are ligands of the transmembrane receptor protein, Patched. In the absence of Hh protein, Patched inhibits the activity of another receptor, Smoothened, which only mediates the Hh signal to the cell when it arrives on Patched. In *Drosophila*, absence of Hh leads to repression of Hh-sensitive nuclear genes; but in vertebrates, binding of a Hedgehog protein to Patched leads to expression of such genes. The signalling pathway in both animals is thought to involve (certainly in *Drosophila*) proteolytic digestion of an extranuclear regulatory protein by proteasomes prior to its activity as a transcriptional repressor or activator. In humans, mutant Patched receptors and resultant excessive Hedgehog signalling occur frequently in the most common form of skin cancer (basal cell carcinoma), indicating that Patched may have a role in checking skin cell proliferation. See β-CATENIN.

HeLa cell Cell from human cell line widely used in study of cancer. Original source was Helen Lane, a carcinoma patient, in 1952.

helicase An ATP-dependent enzyme which, at a replication fork, breaks the hydrogen bonds holding the two strands of duplex DNA together. Special proteins then bind each of the DNA strands and prevent re-annealing. The core transcription factor IIH (or TFIIH) has two helicase protein subunits, which facilitate the partial unwinding of duplex DNA prior to binding of the RNA polymerase II transcription machinery.

Helicobacter pylori Gram-negative spiral bacterium infecting the stomach of two-thirds of the human population and causing 7 million deaths each year from stomach ulcers and malignant cancer. Infection route is via contaminated food, especially unpasteurized milk, and infected pets.

heliophytes Class of RAUNKIAER'S LIFE FORMS.

heliotropism The diurnal movement of the leaves and flowers of many plants which are orientated either perpendicular (*diaheliotropism*) or parallel (*paraheliotropism*) to the sun's direct rays (also known as solar tracking). Unlike stem PHOTOTROPISM, leaf movement of heliotropic plants does not result from asymmetric growth; in the majority of cases, movement involves PULVINI at the bases of the leaves; sometimes the whole petiole possesses pulvinal-like properties.

Heliozoa SARCODINA of generally freshwater environments, without shell or capsule, but sometimes with siliceous skeleton, and usually very vacuolated outer protoplasm. Locomotion by 'rolling', successive pseudopodia pulling the animal over in turn. Food is caught by cytoplasm flowing over axial supports of pseudopodia. Flagellated stage common. Some are autogamous; binary fission usual.

helix-loop-helix (helix-turn-helix) The MOTIF of a class of DNA-BINDING PROTEINS, which dimerize at an interface comprising two helices joined by a loop. The proto-oncogene C-*MYC* and several genes involved in differentiation encode helix-turn-helix proteins. See HOMEOBOX.

helminth Term usually applied to parasitic flatworms, but occasionally to nematodes also.

helper An animal which helps rear the young of a conspecific to which it is not paired or mated. Commonly there are genetic bonds between the helper and its beneficiary 'family'. Many of the studies are on communal nesters in birds. Of considerable theoretical interest. See COOPERATIVE BEHAVIOUR, HAMILTON'S RULE, INCLUSIVE FITNESS, ALTRUISM.

helper cell See T CELL and Fig. 60.

hemicelluloses Heterogeneous group of long-chain polysaccharides (mostly $\beta_1 \rightarrow 4$-linked) and composed entirely of a single monomer, be it arabinose, xylose, mannose or galactose. Hydrogen-bonded to cellulose in plant CELL WALLS, especially in lignified tissue. More soluble and less ordered than cellulose; may function as food reserve in seeds (e.g. in endosperm of date seeds).

Hemichordata Now generally considered to constitute a distinct, but chordate-like, animal phylum, comprising three classes: Enteropneusta (acorn worms, e.g. *Balanoglossus*), Pterobranchia (pterobranchs) and Graptolithina (graptolites). Notochord absent, as is endostyle; dorsal neurocord confined to the mesosome and usually solid; proboscis (= collar) pore and ducts present. Development indirect, the enteropneust larva being the *tornaria*, almost identical with the young echinoderm auricularia larva and supporting a close link between the Chordata and Echinodermata. With the Tunicata and Cephalochordata, forming the informal group 'protochordates'. Enteropneusts, tunicates, cephalochordates and vertebrates form a clade. Some molecular data support a close alliance of pterobranchs and enteropneusts, and monophyly of hemichordates; other such data do not.

hemicryptophytes Class of RAUNKIAER'S LIFE FORMS.

hemidesmosome See DESMOSOME.

Hemimetabola See EXOPTERYGOTA.

Hemiptera (Rhyncota) Large order of exopterygotan insects. Includes aphids, cicadas, bed bugs, leaf hoppers, scale insects. Of enormous economic importance. Usually two pairs of wings, the anterior pair either uniformly harder (*Homoptera*) or with tips more membranous than the rest of the wing (*Hemiptera*). Mouthparts for piercing and sucking. Many are vectors of pathogens.

hemizygous Term applied to cell or individual where at least one chromosomal locus is represented singly (i.e. its homologue is absent), in which case the locus is hemizygous. Sometimes a chromosome pair bears a non-homologous region (as in the HETEROGAMETIC SEX), or all chromosomes are present singly (as in HAPLOIDY).

Hensen's node The anterior, funnel-like tip of the PRIMITIVE STREAK which induces neurectoderm to become neural plate. Consists largely of prospective notochord and considered to be the ORGANIZER of bird and mammal embryos. If transplanted ectopically, can organize a complete second embryo, a property deriving from its induction of the TGF-β superfamily factor sonic hedgehog, and in turn of nodal, on the left side of the embryo (see AXIS SPECIFICATION).

heparin GLYCOSAMINOGLYCAN product of MAST CELLS; an anticoagulant, blocking conversion of prothrombin to thrombin. Reduces EOSINOPHIL degranulation. Stored with HISTAMINE in mast cell granules, and hence found in most connective tissues.

hepatic (Adj.) Relating to the LIVER.

hepaticae See HEPATICOPSIDA.

Hepaticopsida (Hepaticae) Liverworts. Class of BRYOPHYTA, whose sporophytes develop capsule maturation and undergo meiosis before the seta elongates. Consist of a thin prostrate, or creeping to erect body (thallus), a central stem with three rows of leaves, attached to the substratum by rhizoids. Sex organs antheridia and archegonia, variously grouped; microgametes flagellated and motile. Fertilization is followed by development of a capsule, which contains elaters and spores. The latter are dispersed when, on maturity, the capsule splits to its base through four longitudinal slits. The spores germinate on being shed, most usually forming a short thalloid protonema from which new liverwort plants arise. Includes leafy and thallose species. Generally occur in moist soils, on rock, or epiphytically. Rarely aquatic. See LIFE CYCLE.

hepatic portal system System of veins and capillaries conveying most products of digestion (not CHYLOMICRONS) in cephalochordates and vertebrates from the gut to the liver. Being a portal system, it begins and ends in capillaries.

herb Plant with no persistent parts above ground, as distinct from shrubs and trees.

herbaceous Having the characters of a herb.

herbarium Collection of preserved and diverse plant specimens, usually arranged according to a classificatory scheme. Used as a reference collection for checking identities of newly collected specimens, as an aid to teaching, as a historical collection, and as data for research.

herbivore Animal feeding largely or entirely on plant products. See FOOD WEB, CARNIVORE, OMNIVORE.

hereditary (Adj.) Of materials and/or information passed from individuals of one generation to those of a future generation, commonly their direct genetic descendants. Hereditary and genetic material are not identical: an egg cell, e.g., contains a great deal of cytoplasm that is non-genetic; material passed from mother to embryo across a placenta might also be termed hereditary but not genetic. Certain EPIGENETIC systems, e.g. CELLULAR MEMORY, may be heritable independently of changes in DNA sequence.

heritability Roughly speaking, the degree to which a character is inherited rather than attributable to non-heritable factors; or, that component of the variance (in the value) of a character in a population which is attributable to genetic differences between individuals. Estimation of heritability is complex. May be regarded as the ratio of additive genetic variance to total phenotypic variance for the character in the population, where *additive genetic variance* is the variance of breeding values of individuals for that character, and where *breeding value* (which is measurable) is twice the mean deviation from the population mean, with respect to the character, of the progeny of an individual when that individual is mated to a number of individuals chosen randomly from the population.

hermaphrodite (bisexual) (Bot.) (Of a flowering plant or flower) having both stamens and carpels in the same flower. Compare UNISEXUAL, MONOECIOUS. (Zool.) (Of an individual animal) producing both sperm and ova, either simultaneously or sequentially. Does not imply self-fertilizing ability, but if self-compatible such individuals would probably avoid the COST OF MEIOSIS. Commonly, but not exclusively, found in animals where habit makes contact with other individuals unlikely (e.g. many parasites) or hazardous. Rare in vertebrates. See OVOTESTIS, PARTHENOGENESIS.

heterocercal Denoting type of fish tail (caudal fin) characteristic of CHONDRICHTHYES, in which vertebral column extends into dorsal lobe of fin, which is larger than the ventral lobe. Compare HOMOCERCAL.

heterochlamydeous (Of flowers) having two kinds of perianth segments (sepals and petals) in distinct whorls. Compare HOMOCHLAMYDEOUS.

heterochromatin Parts of, or entire, chromosomes which stain strongly basophilic in interphase. Such regions are transcriptionally inactive and highly condensed (see CHROMATIN REMODELLING). *Facultative heterochromatin* (as in inactivated X-chromosomes of female mammals) occurs in only some somatic cells of an organism and appears not to comprise repeat DNA sequences (see DOSAGE COMPENSATION). The resulting animal may thus be a MOSAIC of cloned groups of cells, each with different heterochromatic chromosome regions. *Constitutive heterochromatin* (e.g. TELOMERES, around human CENTROMERES, and CHROMOCENTRES of e.g. *Drosophila*) and consists of very high numbers of transposable elements (most of them retroelements), which may be a means of 'globally silencing' them (see NON-CODING RNAS). Heterochromatin contains a high density of chromosomal nicks, which are preferred targets of retroelement insertion, and this genome-wide repressor system has been subsequently co-opted to serve host regulatory functions (e.g. see BARR BODY). Has been implicated in initiation and maintenance of chromosome pairing (at least in *Drosophila*), and can cause POSITION EFFECT variegation. See HISTONES.

heterochrony Changes during ONTOGENY in the relative times of appearance and rates of development of characters which were already present in ancestors. Sometimes regarded as inclusive of two distinct processes: *progenesis*, in which development is cut short by precocious sexual maturity; and *neoteny*, in which somatic development is retarded for selected organs and parts. Heterochronic gene pathways comprise a cascade of regulatory genes, mutations in which cause temporal transformations in cell fates in which stage-specific events are omitted or reiterated (see miRNAs). See ALLOMETRY.

heterocyst Specialized cell of some filamentous genera of CYANOBACTERIA. Larger than vegetative cells and dependent on them nutritionally (via microplasmodesmata), with a higher respiration rate but only half the phycobilisomes; they lack photosystem II of PHOTOSYNTHESIS, have no short wavelength form of chlorophyll *a* (670 nm) but a higher concentration of P_{700} in photosystem I. Unlike vegetative cells, they appear empty and not granular under light microscopy, but develop from them by dissolution of thylakoids and storage products and production of new internal membranes and a MULTILAYERED STRUCTURE outside the cell wall. Heterocysts are involved in NITROGEN FIXATION, producing the enzyme nitrogenase. They divide only exceptionally, have a limited physiological

life (vacuolizing when senescent) and usually break off from the filament, causing a fragmentational form of vegetative reproduction.

heterodimer A protein (e.g. an INTEGRIN) composed of two different polypeptide chains (e.g. α and β) held together in quaternary structure. In homodimers, the two polypeptides are identical. See Fig. 159.

heterodont See DENTITION.

heteroduplex The double helix (duplex) formed by annealing of two single-stranded DNA molecules from different original duplexes so that mispaired bases occur within it. Heteroduplex regions are likely to occur as a result of most kinds of RECOMBINATION involving breakage and annealing of DNA, and may be short-lived, for when heteroduplex DNA is replicated any mispaired bases should base-pair normally in the newly synthesized strands (but see Fig. 62 and GENE CONVERSION). Furthermore, most organisms have DNA REPAIR MECHANISMS of greater or lesser efficiency for correcting base-pair mismatches by excision/replacement. See DNA HYBRIDIZATION, *recA*.

heteroecious (Of RUST FUNGI) having certain spore forms of the life cycle on one host species, and other forms on an unrelated host species; e.g. *Puccinia graminis* (wheat rust). Compare AUTOECIOUS.

heterogametic sex The sex producing gametes of two distinct classes (in approx. 1 : 1 ratio) as a result of its having SEX CHROMOSOMES that are either partially HEMIZYGOUS (as in XY individuals) or fully hemizygous (as in XO individuals). This sex is usually male, but is female in birds, reptiles, some amphibia and fish, Lepidoptera, and a few plants. Sometimes the XY notation is restricted to organisms having male heterogamety, female heterogamety being symbolized by ZW (males here being ZZ). See HOMOGAMETIC SEX, SEX DETERMINATION, SEX LINKAGE.

heterograft See GRAFT, XENOTRANSPLANTS.

heterokaryosis Simultaneous existence within a cell (or hypha or mycelium of

FUNGI) of two or more nuclei of at least two different genotypes to produce a '*heterokaryon*'. These nuclei are usually from different sources, their association being the result of plasmogamy between different strains. They retain their separate identities prior to karyogamy. Heterokaryosis is extremely common in coenocytic filamentous fungi. ASCOMYCOTA and BASIDIOMYCOTA have a dikaryotic phase in their life cycles (see DIKARYON). See PARASEXUALITY.

heterokonts In some taxonomies, a major EUKARYOTE group (see Fig. 52) whose members possess a unique kind of FLAGELLUM decorated with hollow tripartite hairs (stramenopiles) and have, normally, a second plain one. See HETEROKONTOPHYTA

Heterokontophyta In some taxonomies, an algal division constituting a natural group characterized primarily by similarities in the fine structure of its members and secondarily by biochemical characteristics. This division comprises at least nine algal classes (e.g. BACILLARIOPHYCEAE, CHRYSOPHYCEAE, DICTYOCHOPHYCEAE, EUSTIGMATOPHYCEAE, PARMOPHYCEAE, PHAEOPHYCEAE, RAPHIDOPHYCEAE, SARCINOCHRYSIDOPHYCEAE, XANTHOPHYCEAE). It is also recognized that this division contains not only algae but also several classes of colourless, heterotrophic protozoans, and also classes of multicellular or siphonaceous (coenocytic) fungus-like organisms. The flagella types and structure in motile cells provide evidence for this. The main heterotrophic groups are the unicellular protozoans, *Bicocoecida*, and the classes OOMYCETES, Hyphochytridiomycetes and Labyrinthulomycetes. The close link between heterotrophic and photoautotrophic heterokontophytes was confirmed by molecular sequencing, which showed similarities between the 18S rRNAs of the autotrophic chrysophycean alga *Ochromonas* and the heterotrophic oomycete *Achlya*. A phylogenetic analysis of nucleotide sequences for 18S rRNA in three different oomycetes and five algae, representing five classes of heterokontophytes supports a common phylogenetic origin for heterotrophic and photoautrophic heterokontophytes. The division contains both freshwater and marine groups (e.g. Chrysophyceae, Xanthophyceae and Eustigmatophyceae are primarily found in freshwater, while Parmophyceae, Sarcinochrysidophyceae, Dictyochophyceae, Phaeophyceae are marine). Morphologically, heterokontophytes display a great diversity of form from motile unicells, very simple multicellular types to complex multicellular thalli, with a parenchymatous or pseudoparenchymatous construction and highly specialized structures as seen in the Phaeophyceae (brown algae).

Flagellate stages of this division bear dissimilar (heterokont) flagella. There is a long pleuronematic flagellum directed forwards, and a shorter smooth (acronematic) flagellum that points backwards along the cell. The pleuronematic flagellum bears two rows of special stiff hairs composed of glycoprotein, called mastigonemes, each synthesized in cisternae of the endoplasmic reticulum. Each consists of three subunits: (a) a basal unit; (b) a tubular shaft; and (c) one or more terminal hairs. The transitional region of the flagellum between the flagellar shaft and basal body normally contains a transitional helix, although this is absent in the Bacillariophyceae, Raphidophyceae and Phaeophyceae. The chloroplast is surrounded by a fold of CHLOROPLAST ENDOPLASMIC RETICULUM (CER), and where a chloroplast lies against a nucleus the CER is often continuous with the nuclear membrane. Situated in the narrow space between the CER and the chloroplast membrane is a complex of anastomosing tubules, the periplastidial network. Thylakoids are in groups of three, forming lamellae. A girdle lamella is present in most members of the division. Chlorophylls include a, c_1 and c_2, while the principal accessory pigment is fucoxanthin (Chrysophyceae, Bacillariophyceae, Phaeophyceae and some Raphidophyceae) or vaucheriaxanthin (Xanthophyceae, Eustigmatophyceae). The main polysaccharide reserve product is chrysolaminarin, a β-1,3 linked glucan formed outside the chloroplast in special vacuoles.

The acronematic flagellum has a swelling

near its base that fits against a concave eye-spot, which is enclosed within the chloroplast. The eyespot comprises a single layer of globules containing reddish-orange pigment. The eyespot and flagellar swelling together form the photoreceptor, which is the light-perceiving organelle. One to several or many (Raphidophyceae) lie appressed to the nuclear envelope. See OOMYCOTA.

heterologous Term often used to indicate non-homology of DNA sequences.

heteromerous (Of lichens) where the thallus has algal cells restricted to a specific layer, creating a stratified appearance.

Heterometabola See EXOPTERYGOTA.

heteromorphism Occurrence of two or more distinct (heteromorphic) morphological types within a population, due to environmental and/or genetic causes. Examples include genetic POLYMORPHISM, HETEROPHYLLY, and the phenotypically distinct phases of those life cycles, particularly in the algae (e.g. the green alga *Derbesia*), with ALTERNATION OF GENERATIONS.

heterophylly Production of morphologically dissimilar leaves on the same plant. Many aquatic plants produce submerged leaves that are much dissected, while floating leaves of the same plant are simple and entire. Juvenile stages of some plants have leaves that are morphologically different from 'adult' forms. See HETEROMORPHISM, PHENOTYPIC PLASTICITY.

Heteroptera Suborder of HEMIPTERA.

heteropycnosis The occurrence of HETEROCHROMATIN. Heterochromatic regions were formerly called heteropycnotic.

heterosis See HYBRID VIGOUR.

heterosporous (Bot.) Of individuals or species producing two kinds of spore, microspores and megaspores, that give rise respectively to distinct male and female gametophyte generations. Examples are found in some club mosses and ferns and all seed plants. See ALTERNATION OF GENERATIONS, LIFE CYCLE.

heterostyly Condition in which the length of style differs in flowers of different plants of the same species, e.g. pin-eyed (long style) and thrum-eyed (short style) primroses (*Primula*). Anthers in one kind of flower are at same level as stigmas of the other kind. A device for ensuring cross-pollination by visiting insects. Compare HOMOSTYLY. See POLYMORPHISM, SUPERGENE.

heterothallism (Of algae, fungi) the condition whereby sexual reproduction occurs only through participation of thalli of two different MATING TYPES, each self-sterile. In fungi, includes *morphological heterothallism*, where mating types are separable by appearance, and *physiological heterothallism*, where interacting thalli (often termed plus and minus strains) offer no easily recognizable differences by which to distinguish them. Compare HOMOTHALLISM.

heterotherms Animals intermediate between 'pure' ectotherms and endotherms. Some insects are ectotherms when inactive, but generate and retain sufficient heat in their muscles for their temperature to be maintained considerably higher than the rest of the body. Some fish have countercurrent heat-exchange systems keeping their aerobic (red) swimming muscles warm while surface tissues are close to environmental temperature.

heterotrichous (Of algae) having a type of thallus comprising a prostrate creeping system from which project erect branched filaments. Common in filamentous forms.

heterotrophic (organotrophic) Designating those organisms dependent upon (i.e. consumers of) some external source of organic compounds as a means of obtaining energy and/or materials. All animals, fungi, and a few flowering plants are *chemoheterotrophic* (chemotrophic), depending upon an organic carbon source for energy. Some autotrophic bacteria (purple non-sulphur bacteria) are *photoheterotrophic*, using solar energy as their energy source but relying on certain organic compounds as nutrient materials. Both these groups contrast with those organisms (*photoautotrophs*) able to manufacture all their organic requirements from inorganic sources, and upon which all heterotrophs ultimately depend. Thus all herbivores, carnivores, omnivores, saprotrophs and parasites are heterotrophs. See DECOMPOSER.

heterozygous Designating a locus, or organisms, at which the two representatives (alleles) in any diploid cell are different. Organisms are sometimes described as *heterozygous for* a character determined by those alleles, or *heterozygous at* the locus concerned. Thus, where two alleles *A* and *a* occupy the *A-locus*, of the three genotypes possible (*AA*, *Aa*, *aa*), *Aa* is heterozygous while the other two are HOMOZYGOUS. See DOMINANCE.

heterozygous advantage (overdominance) Selective advantage accruing to heterozygotes in populations and which may be responsible for some balanced POLYMORPHISMS. Mutations in diploids arise in the heterozygous condition and must therefore confer an advantage in that state if they are to spread. Theory has it that if a mutation is advantageous, selection will make its advantageous heterozygous effects dominant and its deleterious effects recessive, thus giving heterozygotes a higher FITNESS than homozygotes. There are remarkably few well-documented examples in which the evidence for heterozygous advantage is conclusive. See SICKLE-CELL ANAEMIA. Compare BALANCED LETHAL SYSTEM.

hexokinase An allosteric regulatory enzyme catalysing the entry of free glucose into the glycolytic pathway (see Fig. 72) by converting it to glucose-6-phosphate (G-6-P) – in which form it cannot leave the cell via the plasma membrane. High G-6-P levels temporarily and reversibly inhibit its activity. Several isozymes of the enzyme occur in mammals, the predominant form in the liver being *glucokinase* (hexokinase D), where its activity is modulated by the pro-apoptotic protein BAD. Mutations in glucokinase are associated with some forms of Type II DIABETES. See INSULIN.

hexose Carbohydrate sugar (monosaccharide) with six carbon atoms in its molecules. Includes glucose, fructose and galactose. Combinations of hexoses make up most of the biologically important disaccharides and polysaccharides.

hibernation A mild form of TORPOR, in which body temperature does not drop below 30°C. Such 'carnivore lethargy' of bears, badgers, etc., is a response to cold or food deprivation in which heart rate drops from 60 min^{-1} to 8 min^{-1} and oxygen consumption falls to a third of normal levels. See AESTIVATION, THERMOREGULATION.

high-energy phosphates Group of phosphorylated compounds transferring chemical energy required for cell work. Depends upon their tendency to donate their phosphate group to water (to be hydrolysed) as is indicated by their STANDARD FREE ENERGIES of hydrolysis (the more negative they are, the greater the tendency to be hydrolysed). *Phosphate-bond energy* (not *bond energy*, the energy required to *break* a bond) indicates the difference between free energies of reactants and products respectively before and after hydrolysis of a phosphorylated compound.

Fig. 80 indicates tendencies for phosphate groups (shown by arrows) to be transferred between commonly occurring phosphorylated compounds of cells. High-energy phosphate bonds arise because of *resonance hybridity* between single and double bonds of the phosphorus/oxygen atoms, which renders them more stable (i.e. they have less free energy) than expected from structure alone. All common phosphorylated compounds of cells, including ADP and AMP, are resonance hybrids at the phosphate bond; but the terminal phosphate of AMP has a standard free energy less than half as negative as those of ADP and ATP.

high-mobility group (HMG) proteins Acid-soluble, non-histone nuclear DNA-BINDING PROTEINS with an HMG domain capable of bending or looping DNA in such a way as to form DNA supercoils. Some may have a role in immunoglobulin gene maturation.

high-throughput A term, used especially in the contexts of GENOMICS and PROTE-

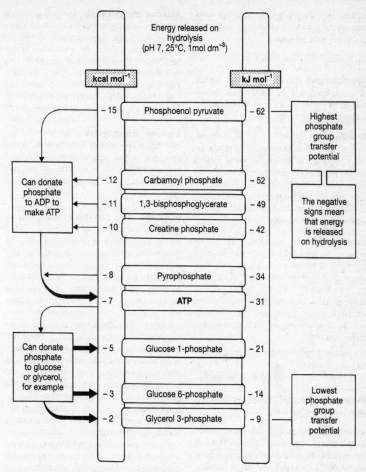

FIG. 80 *The flow of phosphate groups from substances with high transfer potential to those with low transfer potential, assuming molar concentrations of reactants and products. See* HIGH ENERGY PHOSPHATES.

OMICS, to indicate that the assays used involve the simultaneous study of very large numbers of genes, nucleic acid sequences or proteins, often requiring robotic sampling and handling techniques. See NUCLEIC ACID ASSAYS.

Hill reaction Light-induced transport of electrons from water to acceptors such as potassium ferricyanide (*Hill reagents*) which

do not occur naturally, accompanied by the release of oxygen. During it, electron acceptors become reduced. Named after R. Hill, who studied the process in 1937 in isolated chloroplasts. See PHOTOSYNTHESIS.

hilum Scar on seed coat, marking the point of former attachment of seed to funicle.

hindbrain Hindmost of the three expanded regions of the vertebrate brain as marked out during early embryogenesis, developing into the cerebellum and medulla oblongata. See BRAIN.

hip girdle See PELVIC GIRDLE.

hippocampus Part of the cerebral cortex, with a special role in learning (in which long-term POTENTIATION is involved). Affected by GLUCOCORTICOIDS; see also MEMORY.

Hirudinea Leeches. Class of ANNELIDA. Marine, freshwater and terrestrial predators and temporary ectoparasites with suckers formed from modified segments at anterior and posterior ends. Most remain attached to host only during feeding. Hermaphrodite; embryo develops directly within cocoon secreted by chitellum.

histamine Potent vasodilator formed by decarboxylation of the amino acid *histidine* and released by MAST CELL degranulation in response to appropriate antigen. Degraded by *histaminase* released by EOSINOPHIL degranulation. Increases local blood vessel permeability in early and mild inflammation. Responsible for itching/sneezing during ALLERGY.

Histamine

histiocytes Macrophages of the skin and subcutaneous tissues.

histochemistry Study of the distribution of molecules occurring within tissues, within both cells and intercellular matrices. Besides direct chemical analysis, it involves sectioning, STAINING and AUTORADIOGRAPHY.

histocompatibility antigen ANTIGEN (often cell-surface GLYCOPROTEIN) initiating an immune response resulting in rejection of an ALLOGRAFT. See HLA SYSTEM.

histogenesis Interactive processes whereby undifferentiated cells from major GERM LAYERS differentiate into tissues.

histology Study of tissue structure, largely by various methods of STAINING and MICROSCOPY.

histone acetyltransferase Any of several enzymes activating transcription by modifying particular amino-terminal tails of specific core histones by acetylating lysine residues in the histone tail domains. This destabilizes the nucleosome, allowing DNA-binding components of the basal transcriptional machinery better access to promoter elements. See GENE EXPRESSION.

histone deacetylase An enzyme removing acetyl groups from specific lysine residues in the tails of histones H3 and H4, causing gene repression by altering nucleosomal packaging (see GENE EXPRESSION).

histones Basic proteins of major importance in packaging of eukaryotic DNA. DNA and histones together comprise CHROMATIN, forming the bulk of the eukaryotic CHROMOSOME. Histones are of five major types: H1, H2A and H2B are lysine-rich; H3 and H4 are arginine-rich. H1 units link neighbouring NUCLEOSOMES while the others are elements of nucleosome structure. Among prokaryotes, bacterial cells may contain histone-like proteins; but genuine histones are found in the ARCHAEA. Histone acetylation and deacetylation by enzymes play a major role in chromatin assembly and in regulating GENE EXPRESSION (see NUCLEAR RECEPTORS and THYROID HORMONES). Of special importance are the histone methyltransferases that methylate specific lysine residues, such as lysine 9 of histone H3. This is a signal for HETEROCHROMATIN components to bind and initiate heterochromatin assembly (see CHROMATIN REMODELLING). Trichostatin A (TSA) is a reagent that inhibits histone deacetylation. See CELL CYCLE, MOLECULAR CLOCK, TRANSCRIPTION FACTORS.

HIV Human immunodeficiency virus; one of a large group of immunodeficiency viruses (IVs) widely spread among primates and other mammals. Simian immunodeficiency virus (SIV) is the closest known relative of HIV, the causative agent of AIDS (acquired immune deficiency syndrome) in humans, estimated in 2003 to be killing 3 million people per year. HIV is not an

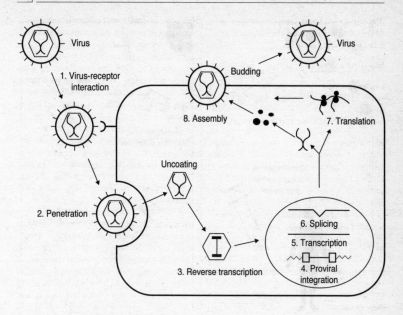

Step	Inhibition strategy
1. Virus-receptor interactions	Soluble CD4
	Intracellular anti-gp 120
2. Penetration/uncoating	Intracellular anti-gp 120
3. Reverse transcription	
4. Proviral integration	
5. Transcription	TAR decoys
	Ribozymes
	Transdominant Tat
	IFN-inducible suppressors
	Antisense RNA
6. Splicing	RRE decoys
	Transdominant Recv
7. Translation	Antisense RNA
	Ribozymes
8. Assembly	Transdominant Gag
	Gag-nuclease fusions
	CD4 trap
1 - 8. All	SELEX
	GSE

FIG. 81a *HIV 'life cycle' and possible inhibition strategies.*

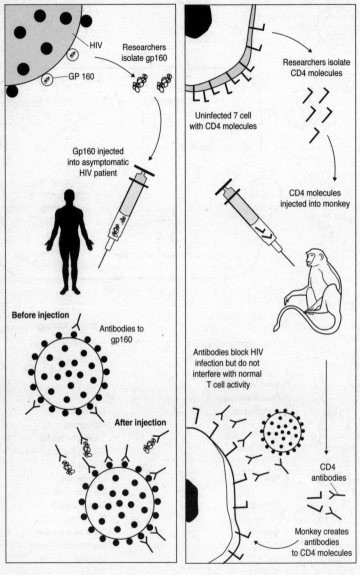

FIG. 81b *Two forms of AIDS vaccines currently being tested. On the left, isolation of the gp160 glycoprotein is followed by its injection to boost antibody titre able to bind the whole virus. On the right, antibodies to CD4 molecules on the T cells are raised in monkeys and on injection should bind to T cells and prevent HIV from entering them.*

oncogenic virus like HTLV-1 (human T cell leukaemia virus type I), but rather the first human lentivirus to be discovered.

IVs form a subgroup of the retroviruses – whose life cycle is shared with the genomic elements called *retrotransposons* (see TRANS-POSON). Each virion has a protein core surrounding the genome and an enclosed enzyme (reverse transcriptase), the whole encapsulated by a segment of host cell membrane in which viral glycoproteins are located. This glycoprotein (gp120) recognizes and binds to the ACCESSORY MOLECULE CD4, so HIV infects any CD4$^+$ cell, including T-helper (T_H) cells (see T CELL). Once bound, the membranes of HIV and the host cell fuse, releasing the infective virus core within the host cell. Immunodeficiency results, but see end of VIRUS entry.

HIV's genome comprises an RNA molecule (two copies per virion), 5–10 kb in length, housing at least three genes (*gag*, encoding core proteins; *pol* encoding viral enzymes; *env*, encoding envelope glycoproteins). On entry into the host cell, viral reverse transcriptase creates a cDNA copy of the RNA genome which is then converted into double-stranded DNA capable of inserting into the human genome and existing as a provirus for long periods; however, there is no evidence that HIVs integrate into the germ line. Eventually, productive virus synthesis occurs and new HIV particles leave the host cell encapsulated in its modified membrane. T cells producing HIV no longer divide, and eventually die. Also, because of the gp120-CD4 binding, HIV-infected cells bind to uninfected CD4$^+$ cells to produce synctia, at which point the uninfected T cells lose their immune capacity and die. Immunodeficiency results. See Fig. 81a.

One HIV-I and SIV accessory protein is Nef, a key viral protein in disease induction *in vivo*. When expressed, Nef leads to increased production of sICAM-I and sCD23 molecules from infected macrophages. sCD23 alters B cells in such a way that they in turn convert non-cell-cycling (i.e. 'resting') T cells into fertile hosts for HIV-I infection and replication – but only if B cells contacting these T cells have also been modified by sICAM-I to express cell-surface CD80 molecules. In this way, 'resting' T cells become additional reservoirs for HIV-I.

The virus is transmitted in sexual fluids and blood plasma (hence on contaminated hypodermic needles) as well as across the placenta and probably in breast milk. Perinatal infection (at the time of birth) through haemorrhage is also possible. Those receiving blood transfusions, haemophiliacs receiving blood products, and those using non-sterile needles for intravenous drug injection may all at times risk contamination, although blood products are increasingly screened/tested for HIV. Travellers to the 'Third World' should pre-empt such risks by obtaining medical insurance enabling them to fly home in emergency.

Monoclonal antibodies to HIV gp120 do not seem to hold the key to an HIV VACCINE. See Fig. 81b. HIV binds and enters CD4-bearing T cells only if they also bear a CHEMO-KINE receptor to act as a co-receptor. If chemokines bind all available receptors, or if the cell carries mutant chemokine receptors, HIV cannot infect the cell. Two chemokine receptors are occupied by different HIV strains at different stages of infection: CCR5 by the main family of HIV-1 during early stages of infection, and CXCR4 by viruses emerging as the disease progresses. Individuals who have been repeatedly exposed to HIV, but remain uninfected, have a deletion in both alleles of their CCR5 gene (i.e. they are homozygous for the deletion) and are significantly protected from HIV-1 infection without apparent side effects from the loss of normal CCR5 receptors. However, HIV-1 can often use CCR3 receptors instead, and any synthetic modified chemokine binding to such receptors would prejudice normal chemokine signalling functions. Progress is needed in designing modified chemokines that specifically bind the CCR5 receptor, and prevent HIV infection, without eliciting the normal T-cell migration that induces severe inflammation if used in high doses. Clinical trials using the drug GEM-91 are taking place (see OLIGONUCLEOTIDE).

HLA system (human leucocyte antigen system) The most important human MAJOR HISTOCOMPATIBILITY COMPLEX, located as a gene cluster on chromosome 6 and probably involving several hundred gene loci.

hnRNA (pre-mRNA) Heterogeneous nuclear RNA. See RNA PROCESSING.

hnRNP See RNA PROCESSING.

Holarctic ZOOGEOGRAPHICAL REGION amalgamating the Palaearctic and Nearctic regions.

Holliday structure (Holliday junction) Alternative term for the cross-strand exchange structure involved in some forms of HOMOLOGOUS RECOMBINATION events.

holoblastic Form of CLEAVAGE.

holocarpic (Of fungi) having the mature thallus converted in its entirety to a reproductive structure. Compare EUCARPIC.

Holocene (recent) The present, post-Pleistocene, epoch (system) of the QUATERNARY period. See Appendix.

holocentric Of chromosomes with diffuse CENTROMERE activity, or a large number of centromeres. Common in some insect orders (Heteroptera, Lepidoptera) and a few plants (*Spirogyra, Luzula*).

Holocephali Subclass of the CHONDRICHTHYES, including the ratfish *Chimaera*. First found in Jurassic deposits. Palatoquadrate fused to cranium (*autostylic* jaw suspension). Grouped with elasmobranchs because of their common loss of bone.

holocrine gland Gland in which entire cells are destroyed with discharge of contents (e.g. sebaceous gland). Compare APOCRINE GLAND, MEROCRINE GLAND.

holoenzyme Enzyme/cofactor complex. See ENZYME, APOENZYME.

hologamodeme See DEME.

Holometabola Those insects with a pupal stage in their life history. See ENDOPTERYGOTA. Compare THYSANOPTERA.

holophytic Having plant-like nutrition; i.e. synthesizing organic compounds from inorganic precursors, using solar energy trapped by means of chlorophyll. Effectively a synonym of photoautotrophic. Compare HOLOZOIC; see AUTOTROPHIC.

Holostei Grade of ACTINOPTERYGII which succeeded the chondrosteans as dominant Mesozoic fishes. Oceanic forms became extinct in the Cretaceous, but living freshwater forms include the gar pikes, *Lepisosteus*, and bowfins, *Amia*. Superseded in late Triassic and Jurassic by TELEOSTEI. Tendency to lose GANOINE covering to scales.

Holothuroidea Sea cucumbers. Class of ECHINODERMATA. Body cylindrical, with mouth at one end and anus at the other; soft, muscular body wall with skeleton of scattered, minute plates; no spines or pedicillariae; suckered TUBE FEET; bottom-dwellers, often burrowing; tentacles (modified tube feet) around mouth for feeding. Lie on their sides.

holotype (type specimen) Individual organism upon which naming and description of new species depends. See NEOTYPE, BINOMIAL NOMENCLATURE.

holozoic Feeding in an animal-like manner. Generally involves ingestion of solid organic matter, its subsequent digestion within and assimilation from a food vacuole or gut, and egestion of undigested material via an anus or other pore. Compare HOLOPHYTIC.

homeobox Conserved DNA sequence MOTIF of ~180 base pairs, encoding DNA-binding regions of many proteins, specifically HMG boxes, α1 domains and homeodomains. Most proteins containing these regions are transcriptional regulators, and a single amino acid alteration can change the promoters to which they bind, potentially altering the downstream activation pattern of genes. First identified in 1984 within several HOMEOTIC GENES of *Drosophila*, the homeobox product confers the helix-turn-helix MOTIF upon a protein, giving it its DNA-binding properties (see DNA-BINDING PROTEINS, REGULATORY GENE). Mating-type genes in ascomycete fungi control sexual development and contain homeobox motifs, and the rapidly evolving homeobox gene, *Odysseus*, is responsible for reproductive isolation (see SPECIATION) between sibling species of *Drosophila*.

homeodomain See HOMEOBOX.

homeogenes HOMEOBOX-containing genes whose protein products are all TRANSCRIPTION FACTORS; of wide (possibly universal) occurrence in animals, from cnidarians to vertebrates, but also present in other eukaryotes (plants, fungi and slime moulds, at least). Homeogenes include all homeotic and some SEGMENTATION GENES. One subset of HOMEOTIC GENES, the *Hox* (Antennapedia-like) homeogene family, is involved in encoding the relative positions of structures along the antero-posterior body axis of (possibly all) animals and has recently been suggested as the defining character (synapomorphy) of the kingdom (see ZOOTYPE). The *HOX* GENES are themselves regulated by upstream genes whose products determine the rough axial specification of eggs and early embryos, in *Drosophila* by means of a cascade of gene expression (see *BICOID* GENE). The same colinear order of expression of *Hox* gene clusters and their homologues is found in embryos as distant phylogenetically as those of fruit flies and mice: gene expression at the 3' end of each cluster on the chromosome starts in anterior embryonic regions, each next gene in the cluster in the 5' direction along the chromosome being expressed in progressively more posterior embryonic parts. Genes in a *Hox* cluster are numbered according to their positions within it. The ANTENNAPEDIA COMPLEX and BITHORAX COMPLEX in *Drosophila* form a single (split) *Hox* gene cluster. Some homeogenes contain DNA-binding domains other than the homeobox and encode transcription factors that regulate differentiation of specific tissues (e.g. the *Pit-1* gene is specific to the pituitary, controlling synthesis of prolactin and growth hormones). *Hox* gene complexes in most animals other than *Drosophila* are united into a single continuous cluster, which appears to be the primitive condition. While amphioxus has one cluster, vertebrates have four, suggesting that duplication has occurred. This may have been associated with the evolution of the NEURAL CREST. Homeotic mutations probably rarely provide raw material for evolution; rather, they indicate the genetic controls co-opted by insects in the evolution of the tagmois patterns characterizing their various orders and by vertebrates in controlling the identities of neural crest derivatives (rhombomeres, vertebrae and somites). So, although homeogenes affect axial information in both groups, the manner by which they do it is very different. See CHROMOSOMAL IMPRINTING, GAP GENES.

homeosis Term coined by William Bateson in 1894 to indicate a form of variation such that 'something has been changed into the likeness of something else'. He thought that homeotic changes provided the sort of discontinuous variation required to bolster Darwinian theory. See HOMEOTIC GENES, HOMOLOGY.

homeostasis Term given to those processes, commonly involving *negative feedback*, by which both positive and negative control are exerted over the values of a variable or set of variables, and without which control the system would fail to function.

(1) *Physiological.* Various processes which help regulate and maintain constancy of the internal environment of a cell or organism at appropriate levels. Each process generally involves: (a) one or more sensory devices (misalignment detectors) monitoring the value of the variable whose constancy is required; (b) an input from this detector to some effector when the value changes, which (c) restores the value of the variable to normality, consequently shutting-off the original input (negative feedback) to the misalignment detector. In unicellular organisms homeostatic processes include osmoregulation by contractile vacuoles and movement away from unfavourable conditions of pH; in mammals (homeostatically sophisticated) the controls of blood glucose (see INSULIN, GLUCAGON), CO_2 and pH levels, and its overall concentration and volume (see OSMOREGULATION), of ventilation, heart rate and body temperature provide a few examples. (2) *Developmental.* Mechanisms which prevent

the FITNESS of an organism from being reduced by disturbances in developmental conditions. The phrase *developmental canalization* has been used in this context. (3) *Genetic*. Tendency of populations of outbreeding species to resist the effects of artificial SELECTION, attributable to the lower ability of homozygotes than heterozygotes in achieving *developmental homeostasis*. (4) *Ecological*. Several ecological factors serving to regulate population density, species diversity, relative biomasses of trophic levels, etc., may be thought of as homeostatic. See ARMS RACE, BALANCE OF NATURE, CHAOS, DENSITY DEPENDENCE.

homeothermy Maintenance of a constant body temperature higher than that of the environment. Involves physiological HOMEOSTASIS. Characteristic of mammals and birds. Some fish are able to keep some muscles considerably warmer than the surrounding water, and there is considerable evidence that many large extinct reptiles were homeothermic. See POIKILOTHERMY.

homeotic genes (1) Term describing a *control gene* which, by either being transcribed or remaining silent during development (according to decisions between alternative pathways of DEVELOPMENT), can profoundly affect the developmental fate of a region of a plant or animal's body. They occur, at the very least, in nematodes, insects, vertebrates and plants. A hierarchical sequence of binary decisions provides clones of cells with 'genetic addresses' for differentiation. In the insect wing IMAGINAL DISC, such a decision sequence appears to be: anterior/posterior, dorsal/ventral, proximal/distal. A *homeotic mutation* is a DNA sequence change comprehensively transforming its own and its descendant cells' morphologies into those of a different organ, in insects, normally one produced by a different imaginal disc. In *Drosophila*, examples include *engrailed*, *antennapedia* and *bithorax*. Implications of homeosis for HOMOLOGY are controversial. *Antennapedia* mutants have all or parts of their antennae converted into leg structures. Since antennae are regarded as the paired appendages of the second embryonic somite, their homology with and evolutionary development from paired ambulatory appendages receives some support from this source. The same applies to the segmentally-arranged mouth-parts. Homeotic mutations have to be experimentally generated in mice by GENE KNOCKOUT techniques and generally transform axial skeletal elements, converting one vertebra into that anterior or posterior to it; others cause defects rather than transformations. See ZOOTYPE.

(2) Of organs whose positions are altered as a result of homeotic mutation. See COMPARTMENT, HOMEOBOX, IMAGINAL DISC.

home range That part of a habitat that an animal habitually patrols, commonly learning about it in detail; occasionally, identical to the animal's total range. Differs from a TERRITORY in that it is not defended, but may be geographically identical to it if it becomes a territory at some part of the year.

homing Returning accurately to the place of origin (e.g. return of salmon to the same river where they developed, after migration to the sea).

hominid A term with different meanings to different taxonomists. In the 1960s, all extent African and Asian 'great apes' were grouped together as pongids in contrast to the human (*Australopithecus*-to-*Homo*) lineage, or hominids. But bones, teeth, soft tissues and molecular evidence indicate that the 'old' Pongidae/Hominidae distinction is unnatural, modern African apes and humans sharing a much more recent common ancestor than any of them does with the orang-utan, *Pongo*. Although the term 'hominid' has lost none of its currency, 'pongid' has. But modern usage of 'hominid' tends to retain its old classificatory overtones (see 'Old Scheme', page 347), whereas a cladistic analysis gives the term far more inclusive scope, making Family Hominidae in the 'Old Scheme' equivalent to Tribe Hominini in the 'New Scheme' (see HOMININI).

The strongest current claimant for 'earliest hominid' in the old sense (still popular with physical anthropologists) seems to be *Sahelanthropus tchadensis* from the Chad

Hominid classifications	
Old Scheme	*New Scheme*
Superfamily Hominoidea	Suborder Hominoidea
Family Hylobatidae	Family Proconsulidae
Family Pongidae	Genus *Proconsul*
Subfamily Dryopithecinae	Family Hylobatidae
Genus *Proconsul*	Family Hominidae
Genus *Dryopithecus*	Subfamily Dryopithecinae
Subfamily Ponginae	Tribe Sivapithecini
Genus *Pongo*	Genus *Sivapithecus*
Subfamily Gorillinae	Tribe Pongini
Genus *Gorilla*	Genus *Pongo*
Genus *Pan*	Subfamily Homininae
Family Hominidae	Tribe Gorillini
Genus *Australopithecus*	Genus *Gorilla*
Genus *Homo*	Genus *Pan*
	Tribe Hominini
	Genus *Australopithecus*
	Genus *Homo*

basin in central West Africa, a fossil cranium dated by accompanying vertebrate fossils to 6–7 Myr BP. This is not to ignore *Ardipithecus ramidus* (~5.5 Myr; Middle Awash, Ethiopia) or *Orrorin tugenensis* (6 Myr; Lukeino, Kenya). The *S. tchadensis* fossils exhibit a unique combination of primitive and derived features copresent early in the human lineage and in animals some distance (2,500 km) from the Rift Valley. Early hominids ought to be distinguished by at least some skeletal and other adaptations to bipedalism, along with a chewing apparatus combining proportionately larger cheek than front teeth, male canines wearing only at the crown tip, and some evidence of increased brain size. If so, then the claim of *S. tchadensis* to be the earliest hominid discovered so far rests on the face, jaw and canines.

Homininae Hominid subfamily (members are hominines). Includes the African ape and human clades. See HOMINID, HOMININI.

Hominini Tribe within the hominid subfamily Homininae (members are hominins). Most distinctive attribute is bipedalism – upright walking. Short face (lacking specialized brain-cooling muzzles of baboons), small incisors and canines and tendency towards parabolic shape of den-

tary also characteristic, as is the elaboration of material culture. Probably originating *c.* 6–5 Myr BP, earliest fossils date from *c.* 3.5 Myr BP. Includes *HOMO*, and the AUSTRALOPI-THECINES – whose detailed relationships with *Homo* require further fossil material. Pre-molecular data were once thought to suggest *Ramapithecus* (see PONGINAE) as earliest fossil ape to date with human affinities; postmolecular data suggest a much more recent separation of the human lineage from apes, at perhaps as little as 7 Myr BP. Some even place African anthropoid apes (*Pan, Gorilla*) and hominins in the same clade, with sivapithecines in the sister clade.

hominoid Member of the primate super-family Hominoidea (see Fig. 82). Includes gibbons (Hylobatidae), the great apes (Pongidae) and HOMINIDS. Distinguished from other CATARRHINE primates by widening of trunk relative to body length, elongated clavicles, broad iliac blades and broad, flat back. Normally no free tail after birth, when the spinal column undergoes curvature as an adaptation to partial or complete bipedalism. A large-bodied hominoid from the Miocene of Uganda (20.6 Myr BP) is the oldest-known hominoid sharing derived features of the shoulder and vertebral

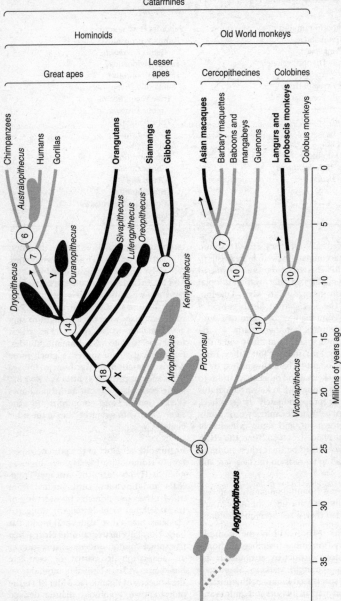

FIG. 82 A synthetic hypothesis of catarrhine and primate evolution, the branching order being well supported by molecular phylogenetic studies. Fossil genera are italicized and placed by parsimony analyses of morphological datasets. Dates for attachment points are best guesses. Species found in Africa and those in Eurasia are distinguished by line quality. Arrows indicate minimal intercontinental dispersals required to explain distributions of living and fossil species. See HOMINOID.

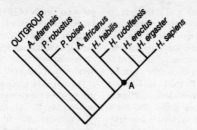

FIG. 83a *Cladogram of the hominid clade generated from a set of ninety cranial, mandibular and dental characters. Node A represents the origin of those autapomorphies distinguishing the* Homo *clade from the* Australopithecus *clade, as described in the* HOMO *entry.*

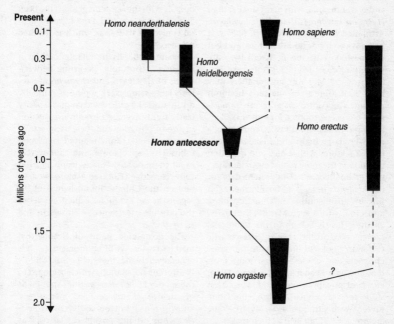

FIG. 83b *A controversial tree for later hominins. See* HOMO.

column significantly similar to those of modern apes and humans. See ANTHROPOIDEA.

Homo Primate genus within the subfamily Homininae (see HOMINID), including mankind and its immediate relatives. The three species regularly (but not universally) recognized are *H. habilis*, *H. erectus* and *H. sapiens*, only the last of which, modern man, exists today.

The genus is defined cladistically by the autapomorphies distinguishing it from *Australopithecus*, including increased cranial vault thickness and height, reduced postorbital constriction, more anteriorly situated foramen magnum, reduced lower facial prognathism, narrower tooth crowns (esp.

mandibular premolars) and reduction in length of the molar tooth row. Additional to these autapomorphies, cranial capacity (brain volume) is greater than 700 cm³, rising to in excess of 1,600 cm³; dental arcade is evenly rounded with no diastema in most individuals; pelvic girdle and limb skeletons are fully adapted for bipedalism; and hands are capable of precision grasp. The stock immediately ancestral to *Homo* is generally thought to have been AUSTRALOPITHECINE, possibly a descendant of *A. afarensis*.

In 1964 an early (1.9–1.6 Myr BP) fossil hominin (Tribe HOMININI) was reported from Olduvai, Tanzania, with brain smaller than *H. erectus* but larger than *Australopithecus*. Attributed to a new species, *H. habilis*, it had advanced jaw and dental features but was australopithecine postcranially. Subsequent more varied fossils from Lake Turkana (1.9–1.8 Myr BP) suggest there may have been more than one early hominin present. The controversial taxon *H. rudolfensis* has been proposed for some Lake Turkana and Lake Malawi fossils, combining a relatively large brain and *Homo*-like postcranial features with a face and dentition resembling the more robust australopithecines ('*Paranthropus*'). The present date for *H. rudolfensis* is 2.4–1.8 Myr BP, the taxon possibly arising during the climatic cooling event in E. Africa at 2.5 Myr BP. Culturally, all these fossils belong in the traditional Lower PALAEOLITHIC, with characteristic Oldowan tool culture comprising stone choppers and crude scrapers; some tools may also have been made from bone, teeth, etc., and possibly from wood. The oldest stone tools described are from the Gona River, Awash, Ethiopia, at 2.5 Myr BP, some arguing for *Paranthropus* as their maker.

Various fossils are thought to form intermediates between *H. habilis* and *H. erectus*. The earliest *H. erectus* fossils come from Lake Turkana, date from at least 1.7 Myr BP and indicate close temporal proximity to *H. habilis*. *H. erectus* is distinguished by larger cranial capacity (750–1250 cm³), flattened skull vault, large supraorbital torus and marked postorbital constriction, and early African fossils (1.7 Myr BP, sometimes given specific status, *H. ergaster*) indicate close temporal proximity of *H. erectus* and *H. habilis*. Evidence of Eurasian *H. erectus* now extends back to 1.75 Myr BP (Dmanisi, Georgia). The earliest evidence of the Acheulian culture associated with *H. erectus* is at Dmanisi (1.75 Myr), Africa (1.6 Myr BP) and Ubeidiya, Israel (1.5 Myr BP), the species moving into Eurasia prior to the extensions of the polar ice sheets (0.9–0.7 Myr BP). The find in 2002 of a small *H. habilis*-like skull at Dmanisi in Georgia, dated to ~1.75 Myr and placed provisionally within *H. erectus*, suggests that the ancestors of this population migrated out of Africa before the emergence of humans identified broadly with the *H. erectus* grade. It has been argued that *H. erectus* did not have the neural control of its thorax which is required for speech.

Debate surrounds the autapomorphies of *H. sapiens*, continuity occurring between fossils of *H. sapiens* and *H. erectus* at sites with large samples. *H. sapiens* is normally distinguished by increased cranial capacity (early forms average 1,166 cm³), reduction in size of posterior and increase in size of anterior teeth, and reduced muscular robusticity, esp. of lower limbs; but overlap in all these occurs. Studies using electron spin resonance (ESR, see RADIOMETRIC DATING) on the Chinese Jinniushan skull (*H. sapiens*) at 200 Kyr BP raise questions about the possible coexistence of *H. sapiens* and *H. erectus*.

The taxonomic status of Neanderthal man (*c.* 230–36 Kyr BP) is problematic. It is generally held that they derive from Middle Pleistocene populations which belonged to the species *H. heidelbergensis* (as specimens of archaic *H. sapiens* are increasingly known), possibly before 400 Kyr BP. *H. heidelbergensis* extends possibly as far back as 500 Kyr BP; but recent finds in northern Spain (Atapuerca hills) of a boy with a remarkably modern face, dated to 780 Kyr BP, could result in displacement of *H. heidelbergensis* as a direct ancestor of anatomically modern humans. *Homo antecessor*, as this specimen has been named, illustrates the complexity in interpreting an increasingly diverse fossil record. But insertion of this new species, as a putative direct ancestor of *H. sapiens*, is too radical a move for many.

Fossils come principally from Europe and the Middle East, mass spectrometry (^{230}Th/^{234}U) and ESR studies on Israeli specimens indicating Neanderthals (at Tabun) and modern *H. sapiens* (at Skhul & Qafzeh) were approximately coeval at 100 Kyr BP ($\pm$ 5 Kyr). Some French fossils suggest Neanderthals and modern humans coexisted until *c.* 36 Kyr BP, although it is not clear whether they were reproductively isolated (see MITOCHONDRIAL EVE). But sequencing of 30–100 Kyr Neanderthal mtDNA suggests that they went extinct without contributing to modern humans (but with a common ancestor perhaps 550–690 Kyr BP) and that they were a distinct species: *H. neanderthalensis*, tending to support the 'Out-of-Africa' school (see later). European Neanderthals varied anatomically, but compared with modern humans had the following: cranial capacity 1,300–1,640 cm³; enlarged occiput (esp. during Wurm glaciation); cranium with lower, flatter, crown; chin receding or absent; cheek bones large; large supraorbital torus; large nasal cavity and nasal prognathism; larger (esp. broader) incisors and canines. Postcranially, Neanderthals were short, powerfully built people; but the popular preconceptions of stupidity combined with brute animal strength require overhauling. For long, the Zhoukoudian site (500 Kyr ago) in China was thought to provide evidence of the first human-controlled use of fire, by *H. erectus*; but this is no longer certain, leaving the earliest such site at 300 Kyr, the perpetrators being *H. sapiens*. After that time, well-built hearths in open sites do appear, indicating use of temporary or semi-permanent huts and cabins. The discovery of quite complete Neanderthal bodies in burial-like position might indicate only the absence of large carnivores from the relevant caves.

The precise origins of anatomically modern humans prior to 40 Kyr BP await clarification, but an African origin at *c.* 100 Kyr BP has received recent fossil support. Controversy surrounds the origins of such symbolic behaviour as art and ornament and whether their origins were gradual or sudden. It is customary to regard the Upper Palaeolithic Aurignacian, at >43 Kyr BP in Eastern Europe, as the first such 'true' industry, coinciding as it did with the spread of anatomically modern humans into Europe. In France, however, flake-based Mousterian culture and Neanderthals themselves persisted after this date. From this time on, open-air (as opposed to cave) burials and impressive campsites become common. Anatomically modern humans probably originated in Africa (the 'Out-of-Africa' school of thought) at perhaps 130 Kyr BP, arrived in the Middle East by *c.* 100 Kyr BP (Qafzeh & Skhul) and reached Australia by *c.* 55 Kyr BP. Newly discovered Ethiopian fossils from Herto, dated to 160 Kyr BP and combining some Qafzeh-like features (wide upper face and moderately domed forehead) along with the wide interorbital breadth of more ancient African crania, suggest that they may link the earlier to the later fossils. These fossils may be our earliest record to date of what is currently thought of as modern *H. sapiens*. Y chromosome DNA polymorphism evidence suggests that all human Y chromosomes have a recent common ancestry, complementing mitochondrial studies and this, with comparative studies of nuclear DNA Alu deletions (see GENETIC MARKERS), provisionally supports the 'Out-of-Africa' school. The antecedents of Aurignacian culture may have been in Southeast Asia. This model, lacking continuity between *H. erectus* and *H. sapiens* anywhere outside Africa, stands in contrast to the largely abandoned view that different modern populations arose at different times from different *H. erectus* stocks (the 'multiregional continuity' model). In this context, the Jinniushan skull's date might be controversial. There is some archaeological evidence supporting repeated expansions of *H. sapiens* out of Africa. See Figs. 83 and 84.

homocercal Designating those outwardly symmetrical fish tails (caudal fins) in which the upper lobe is approximately of the same length as the lower. Typical of modern ACTINOPTERYGII, evolved from the HETEROCERCAL fin.

homochlamydeous (Of flowers) having perianth segments of one kind (sepals) in two whorls. Compare HETEROCHLAMYDEOUS.

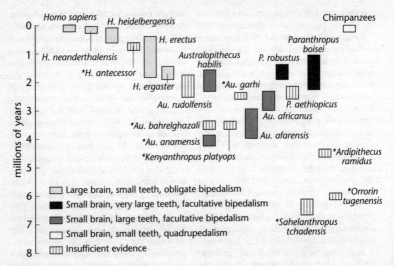

FIG. 84 *The known fossil record of hominids, including S. tchadensis, also showing ourselves (top left) and the chimpanzee (top right). Extinct species are indicated with the dates of the earliest and latest fossil evidence, but these are likely to increase and decrease, respectively, especially for the less well-known examples. Species are assigned to one of four categories, based on brain and cheek-tooth size, and inferred posture and locomotion (we are obligately bipedal; facultative bipedalism is the ability to walk or run on two legs, or as a quadruped, according to circumstances). A fifth category is for 'insufficient evidence'. The species marked with an asterisk were all unknown a decade or so ago, an indication of the paucity of evidence, until recently, of hominid evolution between 1 and 4 million years ago. This comparatively rich record contrasts with the earlier part of the hominid fossil record. There are likely to be many 'undiscovered' species in the fossil record between 7 and 4 million years ago, and in reconstructing the early stages of human evolution in particular the incompleteness of data should always be acknowledged.*

homodont See DENTITION.

homoeobox See HOMEOBOX.

homogametic sex The sex producing gametes which are uniform with respect to their SEX-CHROMOSOME complement. In mammals, this is the female sex (XX); in birds, reptiles and lepidopterans it is the male sex (ZZ). See HETEROGAMETIC SEX, SEX DETERMINATION.

homogamy Condition in which male and female parts of a flower mature simultaneously. Compare DICHOGAMY.

homograft See GRAFT.

homokaryon (homocaryon) Cell, fungal hypha or mycelium with more than one genetically identical nucleus in its cytoplasm; in fungi such nuclei commonly are haploid. See COENOCYTIC.

homologous (1) For homologous characters, see HOMOLOGY. (2) Homologous chromosomes: those capable, at least potentially, of pairing up to form BIVALENTS during first prophase of MEIOSIS, having approximately or exactly the same order of loci (but not normally of alleles). See ANEUPLOID, POLYPLOIDY, HEMIZYGOUS.

homologous recombination Also known as general RECOMBINATION. Process whereby, after a nuclease has cut two strands with the same polarity, these cut DNA segments search for and recognize homologous sequences (through mediation of proteins like RecA and single-stranded DNA-BINDING PROTEINS) and form

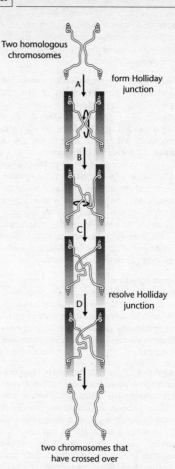

Two homologous chromosomes

A → form Holliday junction

B ↓

C ↓

D → resolve Holliday junction

E ↓

two chromosomes that have crossed over

FIG. 85 *The resolution of a Holliday junction to produce crossed-over chromosomes. Here, homologous regions of the two chromosomes have formed a Holliday junction by exchanging two strands, cutting of which would terminate the exchange without crossing-over. With isomerization of the Holliday junction (steps B and C), the original non-crossing strands become the two crossing strands, and cutting these now creates two DNA molecules that have crossed over (bottom). This type of isomerization may be involved in the breaking and rejoining of two homologous DNA double helices in meiotic general RECOMBINATION. These isomerization events can, as indicated here, occur without disturbing the rest of the chromosomes.*

cross-strand exchange structures (HOLLIDAY STRUCTURES), after which both crossing strands are cut and finally ligated. Easily employed in yeast GENE TARGETING. See Figs. 64 and 85.

homology A controversial term. In evolutionary biology denotes common descent. Two or more structures, developmental processes, DNA sequences, behaviours, etc., usually are occurring in different taxa, are said to be *homologous* if there is good evidence that they are derivations from (or identical to) some common ancestral structure, developmental process, etc. Very often a structure, etc., serving one function in one taxon has come, often with modification, to serve a different function in another, a process now termed *co-option*. Two of the EAR OSSICLES of mammals are homologous with the articular and quadrate bones of ancestral reptiles; the vertebrate THYROID gland is almost certainly homologous with the ENDOSTYLE of urochordates, etc. The term may be applicable even when, as with the vertebrate PENTADACTYL LIMB, comparable structures do not always arise from the same embryonic segments. Where structures are repeated along the organism with little or no modification, as occurs in METAMERISM, they are termed *serially homologous* structures. The term may be applied to any of a hierarchy of levels during development: i) the genes involved (which may be orthologous or paralogous); ii) patterns of gene expression; embryological origin of structures; iii) final anatomical arrangement of parts. The term should not be applied across these different levels unless the evidence supports it (e.g. HOMEOTIC GENES and their expression patterns may be homologous in two animals even though the structures they have a role in producing may not be). See ANALOGOUS.

In PHYLOGENETICS, or CLADISTICS, two characters, etc., are homologous if one (the *apomorphic character*) is derived directly from the other (the *plesiomorphic character*). The relationship is often termed special homology. Some workers in cladistics equate homology with SYNAPOMORPHY. In molecular biology, the term often indicates a significant degree of sequence similarity between DNA or protein sequences.

homology-dependent trans silencing
A GENE SILENCING effect in which an RNA or DNA trigger silences a corresponding chromosomal locus in *trans* (there being no genetic linkage to the target locus).

homonomy Characters of two or more taxa which have the same development, are found on different parts of the organism, and whose developmental pathways have a common evolutionary origin. E.g. each mammalian hair is *homonomous* with all other mammalian hair.

homoplasy In CLADISTICS, the term used to denote parallel or convergent evolution of characters.

Homoptera See HEMIPTERA.

homosporous Having one kind of spore giving rise to gametophytes bearing both male and female reproductive organs, e.g. many ferns. Compare HETEROSPOROUS. See LIFE CYCLE.

homostyly Usual condition in which flowers of a species have styles of one length, as opposed to HETEROSTYLY.

homothallism (Of algae, fungi) the condition whereby thalli are morphologically and physiologically identical, so that fusion can occur between gametes produced on the same thallus. Compare HETEROTHALLISM.

homozygous Any LOCUS in a diploid cell, organism, etc., is said to be homozygous when the two ALLELES at that locus are identical. Organisms are said to be homozygous for a character when the locus determining that character is homozygous. Homozygous mutations are normally expressed phenotypically, unless the genetic background dictates otherwise (SEE PENETRANCE). Characters which are RECESSIVE are only expressed in the homozygous (or HEMIZYGOUS) condition. See HETEROZYGOUS.

hopanoids Sterol-like molecules present in the CELL MEMBRANES of several bacteria and probably performing the strengthening role there that eukaryotic sterols do. In contrast

to sterol synthesis, molecular oxygen is not required for hopanoid synthesis, reflected by their presence in several anaerobic, as well as aerobic, bacteria.

horizontal transfer/transmission Transfers between members of different species (e.g. in xenotransplantation; see GRAFT). (1) Of pathogens, as when the parasite WOLBACHIA transfers from one dipteran species to another in host blood. (2) Of genes, as during TRANSFORMATION by DNA (see also AGROBACTERIUM, ANTIBIOTIC RESISTANCE ELEMENT). The archaeon microorganism *Thermoplasma acidophilum*, which significantly lacks all restriction endonuclease activity, seems to have acquired a high proportion of its genes from other organisms, 17% of them from the bacterium *Sulfolobus solfataricus* while a further similar percentage seems to be of bacterial origin.

hormogonium Short piece of filament, characteristic of some filamentous blue-green algae, that becomes detached from the parent filament and moves away by gliding, eventually developing into a separate filament. Several may develop from one filament.

hormone Term once applied in both botanical and zoological contexts, but now restricted to the latter (see GROWTH SUBSTANCE). Denotes any molecule, usually of small molecular mass, secreted directly into the blood by ductless glands and carried to specific target cells/organs by whose response they bring about a specific and adaptive physiological response. The term *chemical messenger* is still sometimes used in this context, SECOND MESSENGERS being molecular signals produced within, but not exported by, a cell. Neurotransmitters and neurosecretions customarily fall outside the compass of the term hormone, a distinction blurred by NEUROHAEMAL ORGANS. Hormones tend to be either water-soluble peptides and proteins, or lipid-soluble steroids, retinoids, thyroid hormones and vitamin D_3. The latter have the longer physiological half-lives and are hydrophobic, being rendered soluble by binding to specific transport proteins (see NUCLEAR RECEPTORS, TRANSCRIPTION

FACTORS). In this form they may enter nuclei to bring about selective GENE EXPRESSION and typically mediate long-term responses. Water-soluble hormones commonly bind to RECEPTOR sites on cell membranes (see ADENYLYL CYCLASE) and tend to mediate short-term responses. Examples of hormones include ADRENALINE, ECDYSONE, GASTRIN, THYROXINE, INSULIN, TESTOSTERONE and OESTROGEN. See ENDOCRINE SYSTEM, PITUITARY GLAND, PROSTAGLANDINS, CASCADE.

hormone response element (HRE) See RESPONSE ELEMENT.

horn Matted hair or otherwise keratinized epidermis of mammal, surrounding a knob-like core arising from a dermal bone of the skull. Neither the core nor the keratinized sheath is ever shed, nor do they ever branch, unlike ANTLERS.

horseradish peroxidase A haemin-containing peroxidase used in immuno-peroxidase methods. See GREEN FLUORESCENT PROTEIN.

horsetails See SPHENOPHYTA.

host (1) Organism supporting a PARASITE in or on its body and to its own detriment. A *primary* (*definitive*) *host* is that in which an animal parasite reproduces sexually or becomes sexually mature; a *secondary* (*intermediate*) *host* is that in which an animal parasite neither reproduces sexually nor attains sexual maturity, but which generally houses one or more larval stages of the parasite. (2) Organism supporting (e.g. housing) a non-parasitic organism such as a commensal. See SYMBIOSIS.

host race See INFRASPECIFIC VARIATION, SPECIATION.

host restriction See PHAGE RESTRICTION.

housekeeping genes Genes encoding enzymes for centrally important metabolic pathways, such as those for the Krebs cycle, usually expressed constitutively.

***Hox* genes** A subset of HOMEOGENES, evolving in clusters by tandem gene duplication and divergence. They encode transcription factors, and their pattern of expression (regionally, along the anterior–posterior body axis) specifies the positional identities of cells. These cells then differentiate according to the logic of their own downstream target ENHANCERS and PROMOTERS, changes in these regulatory targets being a major source of evolutionary change (see CO-OPTION, HOMOLOGY). Inappropriate expression of *Hox* genes, as by mutations in their *cis*-regulatory sequences, can cause structures to develop ectopically. See *ULTRABITHORAX* GENE, NUCLEAR RECEPTORS.

hpRNA Hairpin RNA. Self-complementary RNA which folds back on itself and forms a hairpin loop.

HTLV (human T-leukaemia virus) See HIV.

human Term indicating any HOMINID.

human artificial chromosome (HAC) Although YEAST ARTIFICIAL CHROMOSOMES have been available for some time, artificial chromosomes were first constructed and introduced into human cells in 1997. It is hoped that this approach will improve delivery of therapeutic DNA to target cell types compared with other methods of HUMAN GENE THERAPY and improve long-term maintenance of the DNA in the cell. In addition to genes, artificial chromosomes require telomeres (to maintain their ends) and a centromere (to ensure mitotic segregation to daughter cells). Making a working centromere, enabling a kinetochore to assemble there, has proved difficult: its repetitive DNA is difficult to clone. But a novel 'neo-centromere' (actually a fragment of a human centromere) lacking these repeats seems to work well. Although yeast chromosomes have defined replication origins (see DNA REPLICATION), it is not yet clear whether human chromosomes do.

human chorionic gonadotrophin (hCG, hcg) Peptide hormone produced by developing human blastocyst and PLACENTA, prolonging until four months of pregnancy the period of active secretion of oestrogens and progesterone by the CORPUS LUTEUM, which otherwise atrophies, inducing menstruation. Its presence in urine is usually diagnostic of early pregnancy (see IMMUNOASSAYS). See DOWN'S SYNDROME, MENSTRUAL CYCLE.

human demography See DEMOGRAPHY.

human gene therapy Treatment of human disease by gene transfer. The first such trial, begun in 1990, involved transfer of the adenosine deaminase gene into lymphocytes of a patient whose defect in this enzyme (leading to immune deficiency) would otherwise have proved fatal. The death in September 1999 of a young volunteer during a gene therapy trial involving adenoviruses made the climate for gene therapy technology turn very cold. The difficulty centres on getting the genes safely into the correct cells and avoiding host immune responses to vectors and to new proteins produced. Most approved gene therapy trials have involved use of retroviral vectors for gene transfer into cultured human cells that would be administered to patients (*ex vivo* gene therapy); but *in vivo* gene therapy trials employing liposomes in the transfer of plasmid DNA directly into patients' cells while still in their bodies have occurred in some cases of cancer and cystic fibrosis (CF). Liposomes are simpler than retroviruses and contain no proteins, thus minimizing host response to the vector; however they are currently inefficient at targeting cells and frequent high doses are involved. Retroviral vectors are retroviruses lacking all functional viral genes, so no viral protein is produced in infected cells. Viral replication is achieved using 'packaging cells' that produce all the viral proteins but no infective virus, so that when the DNA form of the retrovirus is introduced into these cells virions carrying vector RNA are produced that can infect target cells in the patient but cannot spread thereafter, so rather than 'infection', the term 'transduction' is used to describe this process. This results in efficient and stable targeting of cells, the main drawback being the inability of the vectors to infect non-dividing cells. There is also the risk of insertional mutagenesis (e.g. activation of ONCOGENES). Adenoviruses are also being employed as vectors, notably for transferring the normal gene (*CTFR*) into CF patients, with the lung as initial target. Proposals for human gene therapy have to pass several levels of review to ensure safety, not least ensuring there is no transmission of transgenes to the GERM LINE. Other candidate monogenic disorders for treatment by gene replacement therapies include Duchenne muscular dystrophy and the severe immunodeficiency disorder resulting from lack of adenosine deaminase (ADA). In THALASSAEMIAS, however, the problem is not to introduce a normal gene but to regulate the relative expressions of the α and β globin genes. Gene therapy against viral diseases (e.g. AIDS; see HIV) uses ANTISENSE nucleic acids in order to prevent completion of the viral life cycle. See GENETIC COUNSELLING, HUMAN ARTIFICIAL CHROMOSOMES.

Human Genome Project (HGP) By 1987, a genetic map with an average resolution of about 10 cM (see CENTIMORGAN) between markers had been achieved. Two enterprises, using slightly different protocols, have recently been involved in sequencing the entire human genome: the publicly funded Human Genome Sequencing Consortium involving about 20 laboratories around the world, and the private company Celera Genomics. The entire human genome is estimated to comprise $\sim 3.2 \times 10^9$ base pairs and the 'first draft' sequence, consisting of 92% of the euchromatic (gene-rich) sequence, was published in 2001 – some years earlier than originally expected. Only about 1.1–1.4% of the total sequence actually encodes protein – representing just 5% of the 28% of the DNA sequence that is transcribed into RNA. This leaves about three-quarters of the entire genome unsequenced – much of it consisting of REPETITIVE DNA, but including as well gaps in the euchromatic regions. The public consortium predicts there will be ~31,000 human protein-coding genes, while Celera's estimate is ~39,000 – both far lower than the earlier estimate of ~100,000. Techniques involved include the use of GENETIC MARKERS (e.g. RFLPS and SATELLITE DNA), CHROMOSOME JUMPING, POSITIONAL CLONING, DNA SEQUENCING, and production of GENE LIBRARIES. Single DNA base alterations (single nucleotide polymorphisms, or SNPs) between human individuals are being analysed in an effort to discover which genes are involved in the development of disorders such as diabetes and schizophrenia, and assess predisposition to them. The social consequences of the HGP have yet to be assessed. Leaders of the project acknowledge that there are potential ethical, legal and social spin-offs. Health insurance, e.g., might be difficult to obtain if an individual were to seek extensive surgery if details of their disorder had been required, or otherwise obtained, by the insurance company.

human placental lactogen (hPL, uCS) Hormone produced by human placenta after about five weeks of pregnancy. Major effect is to switch maternal metabolism from carbohydrate to fat utilization. INSULIN antagonist.

humerus Bone of tetrapod fore-limb adjoining PECTORAL GIRDLE proximally, and both radius and ulna distally. See PENTADAC-TYL LIMB.

humoral Transported in soluble form, particularly in blood, tissue fluid, lymph, etc. Often refers to hormones, antibodies, etc. See HUMORAL IMMUNITY.

humoral immunity Immunity due to soluble factors (in plasma, lymph or tissue fluid). Production of ANTIBODY constitutes *humoral response* to an antigen. Contrasted with *cellular response* (see IMMUNITY).

humus Complex organic matter resulting from decomposition of dead organisms (plants, animals, decomposers) in the soil giving characteristic dark colour to its surface layer. Colloidal (negatively charged), improving cation absorption and exchange and preventing leaching of important ions, thus acting as a reservoir of minerals for plant uptake; water-retention of soil also improved. See SOIL PROFILE.

hyaluronic acid Non-sulphated GLYCOSAM-INOGLYCAN of D-glucuronic acid and N-ace-tylglucosamine; found in extracellular matrices of various connective tissues.

hyaluronidase Enzyme hydrolysing hyaluronic acid, decreasing its viscosity. Of clinical importance in hastening absorption and diffusion of injected drugs through tissues. Some bacteria and leucocytes produce it. Reptile venoms and many sperm ACROSOMES contain it.

H-Y antigen (H-W antigen) Minor HISTO-COMPATIBILITY ANTIGEN encoded by locus on Y sex chromosome of most vertebrates (W sex chromosome in birds), and responsible for rejection of tissue grafted on to animal of opposite sex. Not now thought to be a product of the gene for testis-determining factor (TDF). See SEX DETERMINATION, SEX REVERSAL GENE.

hybrid In its widest sense, describes progeny resulting from a cross between two genetically non-identical individuals. Commonly used where the parents are from different taxa; but the term also has wider applicability as with *inversion hybrids* (inversion heterozygotes) where offspring are simply heterozygous for a chromosome INVERSION. Where parents of a hybrid have little chromosome homology, particularly where they have different chromosome numbers, hybrid offspring will be sterile (e.g. *mules*, resulting from horse × donkey crosses) through failure of chromosomes to pair during MEIOSIS, although one offspring sex may be partially or completely fertile. Hybrid sterility is one factor maintaining species boundaries, and selection against hybrids is a major factor in theories of SPECI-ATION. Mitosis in hybrid zygotes is unlikely to be affected by lack of parental chromosome homology but development may be thwarted by imbalance of gene products. Sometimes hybridization (especially between inbred lines of a species) may produce HYBRID VIGOUR. See ALLOPOLYPLOIDY.

hybrid dysgenesis Phenomenon in which crosses between laboratory stocks of female *Drosophila* (said to have the M cytotype) and males from wild populations (said to have the C cytotype) produce a range of surprising offspring phenotypes, e.g. sterility, high rates of point mutation and chromosomal aberration, and nondisjunction. The reciprocal crosses produce no such disorder, while the aberrant mutations themselves revert to wild-type at very high frequencies. The explanation appears to be that mobile genetic elements, and in particular *P* ELEMENTS, disrupt genomes by inserting into specific genes and rendering them inactive. The hypothesis is that *P* elements encode both a transposase (giving them mobility) and a *P*-repressor, which prevents transposase production.

hybridization (1) Production of one or more HYBRID individuals. (2) Molecular hybridization (see DNA HYBRIDIZATION). (3) See CELL FUSION.

hybridoma Clone resulting from division of hybrid cell resulting from artificial fusion of a normal antibody-producing B CELL with a B cell tumour cell. Technique involved in production of *monoclonal antibodies*. See ANTIBODY.

hybrid speciation See SPECIATION.

hybrid sterility See CYTOPLASMIC MALE STERILITY, HALDANE'S RULE, SPECIATION, *WOLBACHIA*.

hybrid swarm Continuum of forms resulting from hybridization of two species followed by crossing and backcrossing of subsequent generations. May occur when habitat is disturbed or newly colonized, as with the oaks *Quercus ruber* and *Q. petraea* in Britain. See INTROGRESSION.

hybrid vigour (heterosis) Increased size, growth rate, productivity, etc., of the offspring resulting usually from a cross involving parents from different inbred lines of a species, or occasionally from two different (usually congeneric) species. Possibly results from HETEROZYGOUS ADVANTAGE or, probably more generally, from fixation of different deleterious recessives in the inbreds.

hybrid zone Area (zone) between two populations normally recognized as belonging to different species or subspecies and occupied by both parental populations and their phenotypically recognizable hybrids. Existence of a narrow hybrid zone may indicate that the parent populations are distinct evolutionary species; a wide zone may indi-

cate that they are geographical variants of the same evolutionary species. Not to be confused with INTROGRESSION. See SEMISPECIES.

hydathode Water-excreting gland occurring on the edges or tips of leaves of many plants. See GUTTATION.

hydatid cyst Asexual multiplicative phase of some tapeworms (e.g. *Echinococcus*) within the secondary host (e.g. man, sheep, pig) in which a fluid-filled sac produces thousands of secondary cysts (brood capsules), each of which buds off a dozen or so retracted scolices. In humans the cysts may become malignant and send *metastases* around the body with sometimes fatal results. An example of POLYEMBRYONY.

hydranth See POLYP.

hydrocortisone See CORTISOL.

hydrogenase An enzyme, common in anaerobic microorganisms, which takes up or evolves gaseous hydrogen; commonly contains nickel.

hydrogen bond Electrostatic attraction forming relatively weak non-covalent bond between an electronegative atom (e.g. O, N, F) and a hydrogen atom attached to some other electronegative atom. Responsible in large measure for secondary, tertiary and quaternary PROTEIN structures, for BASE PAIRING between complementary strands of nucleic acid, for the cohesiveness and high boiling point of water.

hydrogen hypothesis See EUKARYOTE.

hydrogenosome Double-membraned (i.e. enveloped) organelle, about 1 μm in diameter, of certain anaerobic AMITOCHONDRIATE eukaryotes (e.g. the protist *Trichomonas* and some chytrid fungi) in which pyruvate is metabolized to yield 2 additional mol ATP and 2 mol each of H_2, CO_2 and acetate per mol glucose as waste products. Inner membrane folded into cristae-like projections. Sometimes considered degenerate or derived mitochondria, and although most lack a genome there is evidence for at least one hydrogenosome genome of mitochondrial descent (in

Nyctotherus ovalis, a ciliate from a cockroach hindgut). Alternatively, hydrogenosomes and mitochondria may have evolved from the same eubacterial endosymbiont. See EUKARYOTE, FERREDOXINS, Fig. 53.

hydroid Member of the HYDROZOA, in its polyp form.

hydrolase Enzyme catalysing addition or removal of a water molecule. See HYDROLYSIS.

hydrological cycle The movement of water from the oceans, by evaporation to the atmosphere. Winds distribute it over the planet's surface. Precipitation brings it down to Earth, where it may be temporarily stored in soils, lakes and ice fields. Loss occurs through evapotranspiration or as liquid flow through stream channels and ground water aquifers. Eventually, the water returns to the ocean. Movement of water, under the force of gravity, links nutrient budgets of terrestrial and aquatic communities with terrestrial systems losing dissolved and particulate nutrients into streams and ground waters. Aquatic systems lose/gain nutrients from stream flow and ground water discharge.

hydrolysis Reaction in which a molecule is cleaved with addition of a water molecule. Some of the most characteristic biochemical processes (e.g. digestion ATP breakdown and other dephosphorylations such as those in respiratory pathways) involve hydrolysis reactions. Chemically it is the opposite of a CONDENSATION REACTION.

hydrophily Pollination by means of water.

hydrophyte (1) Plant whose habitat is water, or very wet places; characteristically possessing AERENCHYMA. Compare MESOPHYTE, XEROPHYTE. (2) Class of RAUNKIAER'S LIFE FORMS.

hydroponics System of large-scale plant cultivation developed from water-culture methods of growing plants in the laboratory. Roots are allowed to dip into a solution of nutrient salts, or else plants are allowed to root in some relatively inert material (e.g. quartz sand, vermiculite) irrigated with nutrient solution. The external environment is commonly kept artificially constant.

hydrosere SERE commencing in water or otherwise moist sites.

hydrotropism TROPISM in which the stimulus is water.

hydroxyapatite Crystalline calcium phosphate. Mineral component of BONE. Used in column chromatography for eluting proteins with phosphate buffers.

5-hydroxytryptamine See SEROTONIN.

Hydrozoa Class of CNIDARIA containing hydroids, corals, siphonophores, etc. Usually there is an ALTERNATION OF GENERATIONS in the life cycle between a sessile polyp (hydranth) phase and a pelagic medusoid phase; but one or other may be suppressed. Most polyp forms (but not *Hydra*) are COLONIAL, showing division of labour between feeding and reproductive individuals. Gonads ectodermal, unlike those of SCYPHOZOA.

hymenium Layer of regularly arranged spore-producing structures in the fruit bodies (ASOCARPS and BASIDIOCARPS) of many fungi (e.g., ASCOMYCOTA, BASIDIOMYCOTA). May or may not be exposed.

Hymenoptera Large and diverse order of endopterygote insects, including sawflies (Symphyta), bees, ants, wasps and ichneumon flies (Apocrita). Fore wings coupled with hind wings by hooks. Mouthparts typically for biting, but sometimes for lapping or sucking (as in bumble bees). Ovipositor used, besides egg-laying, for sawing (sawflies), piercing or stinging. Abdomen often constricted to form a thin waist, its first segment fused with metathorax. Larvae generally legless. Bees, ants and wasps often EUSOCIAL.

hyoid arch Vertebrate VISCERAL ARCH next behind jaws. Dorsal part forms HYOMANDIBULAR; ventral part in adults forms *hyoid bone*, usually supporting tongue. Contains facial nerve; gives rise to many face muscles.

hyomandibular Dorsal element (bone or cartilage) of hyoid arch, taking part in jaw attachment in most fish (see HYOSTYLIC). Becomes columella auris (stapes) in tetrapods (see EAR OSSICLES).

hyostylic Method of jaw suspension of most modern fishes. Upper jaw has no direct connection with the braincase and the jaw is supported entirely by the hyomandibular. This widens the gape. See AUTOSTYLIC, AMPHISTYLIC.

hypermutation The introduction of point mutations at high rate into variable (*V*) region genes of ANTIBODY-producing B cells in germinal centres of lymph nodes, leading to AFFINITY MATURATION of antibodies during an immune response. The process has been termed 'somatic hypermutation' because mutation is about a million-fold higher than the rate of spontaneous mutation in other cells, and because of its occurrence in non-germ line cells. The molecular details were clarified in 2002, when it was realized that the protein activation-induced cytosine deaminase (AID), produced only in B cells which have bound antigen, converts cytosines to uracils in DNA (not RNA). DNA polymerase may then mistake uracil for thymine and introduce errors accordingly; or the uracils may be removed by uracil glycosylase, after which the gap may be filled more or less randomly; or a thymine may be inserted preferentially; or the gap may be filled and extended, generating mutations that are mostly opposite adjacent adenines and thymines. See MUTATION for *E. coli* hypermutability.

hyperparasite Organism living parasitically upon another parasite. Provoked Jonathan Swift's doggerel expressing supposed infinite regress.

hyperplasia Increase in amount of tissue resulting from cell division (each remaining the same size) as opposed to increase in cell volumes *per se* (hypertrophy).

hypersensitive response See PLANT DISEASE AND DEFENCES.

hyperthermophile An organism with a temperature optimum above 80°C, many capable of growth above the normal boiling point of water. Most belong to the Crenar-

chaeota (see ARCHAEA). Protein stability at these temperatures is often enhanced by salt bridges between amino acids, and the membranes of these organisms lack fatty acids but have hydrocarbons of various lengths (repeating units of 5-carbon phytane) bonded to glycerolphosphate. There seems to have been HORIZONTAL TRANSMISSION of archaeal genes to the genome of the hyperthermophilic bacterium *Aquifex aeolicus* (perhaps 10% of its genome). See EXTREMOPHILE, PSYCHROPHILE.

hypertonic Relational term expressing the greater relative solute concentration of one solution compared with another. The latter is *hypotonic* to the former. A hypertonic solution has a lower WATER POTENTIAL than one hypotonic to it, and has a correspondingly greater OSMOTIC PRESSURE. See ISOTONIC, EURYHALINE.

hypertrophy Often interchangeable with HYPERPLASIA. Sometimes used of enlargement of individual components of tissues, organs, etc., without increase in cell division. See REGENERATION.

hypervariable regions See ANTIBODY DIVERSITY, DNA FINGERPRINTING.

hypha (Of fungi) a tubular filament or thread of a thallus, often vacuolated. Increases in length by growth at its tip; but proteins are synthesized throughout the MYCELIUM and transported to hyphal tips by cytoplasmic streaming. May be separate (with cross-walls) or non-septate. New hyphae arise by lateral branching. See COENOCYTE.

Hyphochytridiomycota (Hyphochytridiomycetes) A group previously classified with the fungi but having closer links with the algal division HETEROKONTOPHYTA; sometimes placed in the kingdom Chromista. This group comprises about seven genera and 24 species. Thalli are eucarpic or holocarpic. Holocarpic species often have branched rhizoids. Zoospores are uniflagellate with an anterior flagellum that has mastigonemes. Species are found on algae and fungi in freshwater and soil as parasites or saprophytes, as well as on plant and insect remains as saprotrophs.

hypoblast Cell layer of avian blastodisc and mammalian blastocyst; presumptive endoderm. Compare EPIBLAST.

hypocotyl Part of a seedling stem, below the cotyledon(s).

hypodermis Layer of cells immediately below the epidermis of leaves of certain plants, often mechanically strengthened (e.g. in pine), forming an extra protective layer, or forming water-storage tissue.

hypogeal (Of cotyledons) remaining underground when the seed germinates; e.g. broad bean, pea. Contrast EPIGEAL.

hypoglossal nerve See CRANIAL NERVES.

hypogynous See RECEPTACLE.

hyponasty (Bot.) More rapid growth of lower side of an organ than upper side; e.g. in a leaf, resulting in upward curling of leaf-blade. Compare EPINASTY.

hypophysis See PITUITARY GLAND.

hypostasis Suppression of expression of a (*hypostatic*) gene by another non-allelic gene. Compare RECESSIVE. See SUPPRESSOR MUTATION, EPISTASIS.

hypothalamus Thickened floor and sides of the third ventricle of the vertebrate fore-brain (diencephalon). Its nuclei control many activities, largely homeostatic: e.g. the suprachiasmatic nucleus (SCN) is the centre of the human BIOLOGICAL CLOCK mechanism. It integrates the AUTONOMIC NERVOUS SYSTEM, with centres for sympathetic and parasympathetic control; receives impulses from the viscera. Ideally situated to act as an integration centre for the endocrine and nervous systems, secreting various RELEASING FACTORS into the PITUITARY portal system and neurosecretions into the posterior pituitary. The neurons releasing GnRHs into the pituitary originate in the olfactory epithelium and migrate into the hypothalamus during foetal development. These stimulate specific G PROTEIN-coupled membrane receptors in the pituitary, and mediate secretion of the hormone by SECOND MESSENGER amplifi-cation within the cell. Releases substances inhibiting release of releasing factors (e.g. see SOMATOSTATIN). Contains control centres for feeding and satiety – the latter inhibiting the former after feeding (see LEPTINS, OBESITY). In higher vertebrates, is a centre for aggressive emotions and feelings and for psycho-somatic effects. Contains a thirst centre responding to extracellular fluid volume; helps regulate sleeping and waking patterns; monitors blood pH and concentration and, in homeotherms, body temperature. See NEUROENDOCRINE CO-ORDINATION and Fig. 18.

hypothallus Thin, shiny, membranous adherent film at base of a slime mould (MYCE-TOZOA) fruitification.

hypothesis A temporary working explanation or conjecture, commonly based upon accumulated data, which suggests some general principle or relationship of cause and effect; a postulated solution to a problem that may then be tested experimentally. See NULL HYPOTHESIS.

hypotonic Of a solution with a lower relative solute concentration (higher WATER POTENTIAL) than another. See HYPERTONIC, EURYHALINE.

hypoxanthine Base present in small amounts in certain transfer RNA molecules,

Hypoxanthine

at the 5'-end of the anticodon and capable of base-pairing with U, C and A in the corresponding mRNA codon. A hydrolytic product of adenine, it is also formed by deamination of adenine by nitrous acid. See WOBBLE HYPOTHESIS.

hypoxia Reduced oxygen levels, typically in aquatic environments or within organisms. Cells typically respond to low oxygen tensions by increasing carbohydrate consumption (the PASTEUR EFFECT), with induction of genes encoding glycolytic enzymes. Organisms may respond by ANGI-OGENESIS, VASODILATION, ERYTHROPOIESIS and increased ventilation, shut-down of non-essential cell functions and (most severe) induction of cellular suicide by p53-mediated apoptosis. Especially when it fluctuates, hypoxia leads to bursts of production of reactive oxygen species by mitochondria (SEE SUPEROXIDES), which can lead to DNA strand breaks. Absence of oxygen is *anoxia*. See CANCER.

IAA Indole-3-acetic acid. Most common GROWTH SUBSTANCE in plants, produced in apical meristems of shoots and tips of coleoptiles. One of several AUXINS. Also synthesized by some bacteria from tryptophan.

IAN Indole-3-acetonitrile. Natural plant AUXIN.

i-band See STRIATED MUSCLE.

ichthyosaurs Extinct diapsid reptiles (Ichthyopterygia), contemporaries of the dinosaurs, whose adaptations for oceanic life made them a successful group from the Early Triassic (~245 Myr BP) to the Cretaceous (~90 Myr BP). In 1982, two nearly complete specimens of the earliest known ichthyosaur, *Utatsusaurus*, were discovered in Japan. That same year, the most complete specimen of another early ichthyosaur, *Chaohusaurus*, was found. Both resembled lizard-like animals with flippers, their skulls and limbs indicating that ichthyosaurs probably branched off from other diapsids near the separation of the two main groups of living reptiles, lepidosaurs and archosaurs. Vertebral column curved downwards to form reverse heterocercal tail; legs modified into paddles, with addition of extra digits. Fleshy dorsal fin and upper tail lobe lacked skeletal support. Jaws with homodont dentition. Became increasingly streamlined for aquatic locomotion; convergence with porpoises, etc.

ICSH Interstitial-cell stimulating hormone. See LUTEINIZING HORMONE.

identical twins See MONOZYGOTIC TWINS.

identification keys Keys used in discovering the name of a specimen are commonly constructed so as to lead the investigator through a sequence of choices between mutually exclusive character descriptions, so chosen as to eliminate all but the specimen under observation. The format is commonly DICHOTOMOUS. A disadvantage arises when not all characters of the dichotomous key are observable, as through damage or incompleteness. A *polyclave* overcomes this, placing reliance only upon characters observable in the specimen to hand. It commonly comprises a set of punchcards, each representing a different form or state that a character can take. Each species within the set dealt with by the cards is located on a master sheet and is given a unique number representing its set of character-states. When sufficient cards with character descriptions appropriate to the specimen are held up together only one punch-hole remains open, and the number corresponding to that hole is the number in the master sheet which identifies the species (or other taxonomic unit being identified).

idioblast (Bot.) Cell clearly different in form, structure or contents from others in the same tissue; e.g. cystolith-containing parenchyma cells.

idiotype Antigenic constitution of the variable (V) region of an immunoglobulin molecule. See ANTIBODY, ANTIBODY DIVERSITY.

IFNγ Gamma-interferon. See LYMPHOKINES.

IgA Monomeric or polymeric immunoglobulin, often dimeric (composed of two polypeptides). Most abundant in seromucous secretions such as saliva, milk, and such as occur in urogenital regions. See ANTIBODY.

IgD Low titre immunoglobulin found attached to B CELL membranes. Of uncertain function.

IgE Low titre immunoglobulin located on basophil and mast cell surfaces. Possibly involved in immunity to helminths; also involved in asthma and hay fever hypersensitivity. See ALLERGIC REACTION.

IGF INSULIN-LIKE GROWTH FACTOR. See GROWTH HORMONE.

IgG Class of monomeric immunoglobulin proteins with four subclasses (IgG1-4), accounting for at least 70% of human immunoglobulin titre. Each molecule contains two heavy and two light chains. The sole antitoxin class, and major antibodies of secondary immune responses (see B CELL). The only antibody class to cross the mammalian PLACENTA. See ANTIBODY.

IgM Class of large pentameric immunoglobulin molecules (five linked subunits), largely confined to plasma. Produced early in response to infecting organisms, whose surface antigens are often complex.

ileum Region of mammalian small intestine closest to colon and developing from region occupied by embryonic yolk sac; not anatomically distinct from JEJUNUM. Its numerous villi, intestinal glands (crypts of Lieberkühn) with their characteristic PANETH CELLS and GOBLET CELLS, enable it to serve as the major digesting and absorbing region of the mammalian alimentary canal.

ilium Paired bone forming dorsal part of tetrapod PELVIC GIRDLE (present in rudimentary form even in fishes) and articulating with one or more sacral vertebrae.

imaginal disc Organ-specific PRIMORDIA of holometabolous insects, derived from blastoderm and distributed mainly in larval thorax. Composed of imaginal cells. There are nineteen such discs in *Drosophila* larvae, which evaginate at metamorphosis and differentiate largely into adult epidermal structures: eyes and antennae from one pair; front legs from another pair; wings and halteres from two more pairs; genitals from a midline unpaired disc, and so on. Most discs are DETERMINED in the larval insect and remain so through artificial subculture; but *transdetermination* of cells may occur. They tend to differentiate according to their original position in embryo regardless of transplantation to new locations. Much remains to be learnt of the evolution of this mode of development. Cues for differentiation of discs into adult tissues are apparently hormonal; but the origin of the determined state appears to depend upon POSITIONAL INFORMATION, expressed as genetic addresses. See CHROMOSOMAL IMPRINTING, COMPARTMENT, HOMEOTIC GENE.

imago Sexually mature adult insect.

imbricate (Of leaves, petals, etc.) closely overlapping.

immediate-early gene Gene which is rapidly and transiently induced by a variety of extracellular stimuli. An example is *c-fos* (see ONCOGENE).

immobilization See ENZYME.

immortalized cells Most vertebrate cells stop dividing in culture after a finite number of divisions – typically 25–40. However, some cell lines (e.g. human fibroblasts) can be 'immortalized' by ensuring that the gene for the catalytic subunit of telomerase is present and switched on (see TELOMERE), so that the line is perpetuated indefinitely. But other cell types engage the checkpoint arrest mechanisms even though their telomeres remain long; but rodent cell lines can undergo mutations which inactivate these checks to division. Tumour-inducing viruses or mutagens can transform some cell lines (see CANCER, TRANSFORMATION), although the resulting cells are then abnormal. See ANTIBODY for production of monoclonal antibodies by cell fusion methods.

immune tolerance (immunological tolerance) Acquired inability to react to particular self- or non-self-antigens. Both B CELLS and T CELLS display tolerance, generally to their specific antigen classes. The concentration of antigen required to induce tolerance in neonatal B cells is 100-fold less than for adult B cells. Responsible for suppression of transplant rejection. First noticed in non-identical twin cattle which shared foetal circulations (i.e. were synchorial). For role of HLA proteins in maternal tolerance to foetus, see MAJOR HISTOCOMPATIBILITY COMPLEX; see also IMMUNOSUPPRESSION, THYMUS.

immunity Ability of animal or plant to resist infection by parasites and effects of other harmful agents. Essential requirement for survival, since most of these organisms are perpetually menaced by viruses, bacteria and fungi, or parasitic animals.

In animals there are two functional divisions of the immune system: *innate (nonspecific) immunity* and *adaptive (specifically acquired) immunity*. The former includes several barriers to pathogen entry (e.g. lysozyme, mucus, intact skin/cuticle, sebum, stomach acid, ciliary respiratory lining and commensal gut competitors) as well as nonspecific cellular responses (e.g. release of ANTIMICROBIAL PEPTIDES), inflammation resulting from activation of MACROPHAGES and NEUTROPHILS (both of which bear a variety of conserved surface receptors to the common constituents of microorganism antigens), the latter releasing several bacteriostatic proteins and peptides from preformed granules. Macrophages, especially, also release chemokines and cytokines (chemoattractant and vasoactive factors involved in INFLAMMATION).

Adaptive immune responses, absent from invertebrates, are usually slower acting than the innate immune mechanisms and involve two new classes of immune cells, T CELLS and B CELLS. However, these lymphocytes usually respond to foreign antigens only if the innate immune system has already been activated.

Adaptive immune responses, unlike innate immune responses, differ in quality and/or quantity of response on repeated exposure to antigen: the primary response to antigen takes longer to achieve significant antibody titre than does the secondary response. They include *active natural immunity*, in which the animal's MEMORY CELLS respond to a secondary natural contact with antigen by multiplication and specific antibody release (see CLONAL SELECTION THEORY); and *active induced immunity*, in which a VACCINE (see also INOCULATION) initially sensitizes memory cells. *Passive immunity* may be either natural, as by acquisition of antibodies via the placenta or colostrum in mammals, or induced, usually via specific antibodies injected intravenously. There is sometimes a distinction made between *cell-mediated* (lymphocytic and phagocytic) responses and *humoral* (antibody) responses in immunity, but it is never clear-cut: cells are involved in initiation of antibody responses and cell-mediated responses are unlikely in the complete absence of antibody. Cell-mediated immunity tends nowadays to refer to any immune response in which antibodies play a relatively minor role. For distinctions between *primary* and *secondary immune responses*, see B CELL. See ANTIGEN-PRESENTING CELL, T CELL, INFLAMMATION, and Figs. 86 and 99.

Plants lack an immune system producing specialized cells; but see PLANT DISEASE AND DEFENCES for their equally complex defence mechanisms.

immunization Process rendering an animal less susceptible to infection by pathogens, to toxins, etc. May confer active IMMUNITY through use of VACCINE; or passive immunity through injection of appropriate antibodies (in antiserum). See INOCULATION.

immunoassay A semiquantitative method of assaying the degree of genetic similarity between species involves raising antiserum to the proteins (e.g. albumin) of one of them, placing this in the central well of a plate of agar gel and placing samples of albumin of other species in adjacent wells so that diffusion may occur through the agar. From the shapes of the lines of precipitated proteins between the wells, one can infer the relative level of genetic similarity between the species (see Appendix and Fig. 87).

If an enzyme is attached to a secondary antibody (see IMMUNOFLUORESCENCE) and, by its reaction, produces a colour change, then it acts as a very specific amplifier of the antigen signal. Alkaline phosphatase is often used as such an enzyme, the technique being termed 'enzyme-linked immunoabsorbent assay', or ELISA. It is a rapid and

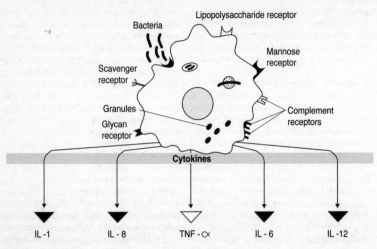

FIG. 86a *Phagocytes (e.g. macrophages) of the innate immune system recognize cell surface components of bacteria and components of complement which bind to receptors and stimulate cytokine release. IL indicates interleukin; TNF-a is tumour necrosis factor-a. IL-6 causes fever and induces acute-phase protein production. See* IMMUNITY.

very sensitive test for a target molecule, as for instance in a common test for pregnancy (SEE HUMAN CHORIONIC GONADOTROPHIN), or for various types of infection. See PROTEOMICS.

immunocompetence handicap A theory proposing that, in vertebrates at least, males 'honestly' signal their parasite resistance. It is debatable, because of testosterone's oft-assumed dual role: on the one hand its effects on secondary sexual characteristics and on the other its compromising effects on the immune system. According to the theory, only the fittest males are able to express traits attractive to females and still withstand parasitic infections. See SEXUAL REPRODUCTION.

immunoelectrophoresis See ELECTROPHORESIS.

immunofluorescence Use of antibodies, with a fluorescent marker dye attached, in order to detect whereabouts of specific antigens (e.g. enzymes, glycoproteins) by formation of antibody-antigen complexes which show up on appropriate illumina-

tion. Antibody with such a fluorescent dye attached is serving as a *primary antibody*; but if the antibody is itself unlabelled but serves as a target for a labelled *secondary antibody* then a stronger signal is achieved (see IMMUNOASSAY).

immunogen Any substance that can induce an immune response. Compare ANTIGEN.

immunoglobulin (Ig) Those members of the IMMUNOGLOBULIN SUPERFAMILY with antibody activity. See IgA, IgD, IgE, IgG, IgM.

immunoglobulin (Ig) superfamily A large glycoprotein superfamily including ANTIBODIES and their membrane-bound isotypes; T CELL RECEPTORS, lymphocyte Fc receptors, the CD2, CD4 and CD8 ACCESSORY MOLECULES and MHC molecules. All contain one or more homologous immunoglobulin-like DOMAINS about 100 amino acids long, containing two antiparallel β-pleated sheets.

immunological distance A unit indicating the closeness of relationship between two organisms, based on such evidence as

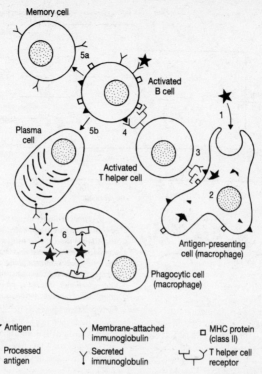

FIG. 86b *The humoral (B-cell-mediated) immune response to an infection. (1) The antigen (the infecting agent) is taken up by a cell such as a macrophage. (2) This cell processes the antigen, breaking it down into components that are then displayed on the surface of the cell. (3) The antigen is recognized by a T helper cell, which in turn (4) activates B cells that are also carrying pieces of the antigen. The activated B cells either (5a) differentiate into memory cells, which respond in future infections caused by the same agent, or (5b) become plasma cells that secrete antibody. (6) The antibodies bind to the antigen, thereby forming a complex that is destroyed by macrophages. (Adapted from The Molecules of the Immune System, by S. Tonegawa. Copyright © Scientific American, Inc., 1985. All rights reserved.)*

Species tested	Per cent reaction	Species tested	Per cent reaction
Humans (*Homo*)	100	Baboon (*Papio*)	73
Chimpanzee (*Pan*)	95	Spider monkey (*Ateles*)	60
Gorilla (*Gorilla*)	95	Ruffed lemur (*Varecia*)	35
Orang-utan (*Pongo*)	85	Dog (*Canis*)	25
Gibbon (*Hylobates*)	82	Kangaroo (*Macropus*)	8

TABLE 6 *The amount of antigen/antibody reaction produced by the indicated albumin and anti-human albumin relative to that given by human albumin and the same antiserum.*

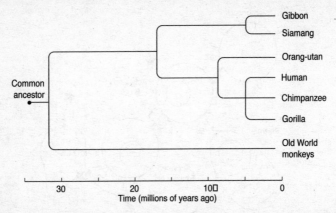

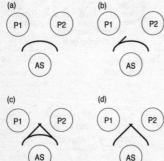

FIG. 87 *Times of divergence of Old World monkeys, apes and humans as estimated by Vincent Sarich and Allan Wilson in 1967 using an immunological approach. They assumed that Old World monkeys and hominoids separated about 30 Mya and, presuming approximate constancy of the rate of evolutionary change, this led to an estimated time of divergence of humans, gorillas and chimpanzees of about 5 Myr. For (a) to (d) see* IMMUNOASSAY.

IMMUNOASSAYS and immunoprecipitation. The smaller the value, the greater the cross-reaction and the closer the relationship.

immunological memory Process whereby B CELLS and T CELLS exhibit accelerated responses when re-exposed to antigen owing to the persistence for many years of MEMORY CELLS of both kinds. Thucidides, describing the plague of Athens in 430 BC, noted that 'the same man was never attacked twice', while the 19th-century Danish physician Panum noted that protective immunity was sustained in the absence of chronic exposure to measles virus.

immunological synapse Term coined to describe the interface between a T cell and either an ANTIGEN-PRESENTING CELL or B cell, which includes a minute gap into which cytokines, among other molecules, are secreted.

immunological tolerance See IMMUNE TOLERANCE.

immunophilins Proteins serving as receptors for the principal immunosuppressant drugs (e.g. cyclosporin A, FK506 and rapamycin). They interface with signal transduction systems in the cell, notably those involved in CALCIUM SIGNALLING and phosphorylation.

immunoprecipitation See IMMUNO-ASSAYS.

immunosuppression Complete or partial suppression of the normal immune responses, by disease (e.g. HIV infection) or human intervention. The first 'artificial' immunosuppressants were highly toxic; but introduction of less toxic *cyclosporin A* (CsA) enabled increased success in organ transplantation. Lower doses were required when using *FK506*. Both of these interfere with the calcium-dependent steps occurring when a T CELL RECEPTOR is bound by a ligand, which otherwise leads to transcription of genes for interleukin-2 and its receptor. *Rapamycin* interferes with the subsequent interleukin-2 dependent proliferation of T cell clones. One of the reasons the mammalian foetus is not rejected by the mother seems to involve downregulation of maternal NK and T cell receptors specific for paternal antigens. See IMMUNE TOLERANCE, MAJOR HISTOCOMPATIBILITY COMPLEX.

implantation (nidation) Attachment of mammalian BLASTOCYST to wall of uterus (endometrium) prior to further development, placenta formation, etc. In humans the blastocyst is small and penetrates the endometrium, passing into the subepithelial connective tissue (*interstitial implantation*). This involves breaking of JUNCTIONAL COMPLEXES between endometrial cells, and proteolytic enzymes may be secreted by the TROPHOBLAST to achieve this.

imprinting (1) Form of LEARNING, often restricted to a specific *sensitive period* of an animal's development, when a complex stimulus may appear to elicit no marked response at the time of reception but nonetheless comes to form a model whose later presentation (or something appropriately similar) elicits a higher significant response. Particularly prevalent in birds. *Filial imprinting* involves narrowing of preferences in social companion (e.g. to mother, or to artificial object, in ducklings); *sexual imprinting* involves the preferential directing of sexual behaviour towards individuals similar to those encountered early in life. (2) See CHROMOSOMAL IMPRINTING, GENOMIC IMPRINTING.

impulse An ALL-OR-NONE RESPONSE comprising an ACTION POTENTIAL propagated along the plasmalemma of an excitable tissue cell, such as a nerve axon (between NODES OF RANVIER in myelinated axons) or muscle fibre. Impulses are initiated at synapses by depolarizations of the postsynaptic membrane's resting potential (usually internally negative by some 70 mV), generally brought about by release of an excitatory neurotransmitter molecule (for general details see ACETYLCHOLINE). This opens up Na^+- and K^+-*ligand-gated channels* in the postsynaptic membrane and allows influx of Na^+ and efflux of K^+ along their electrochemical gradients; Cl^- channels remain closed if the transmitter is excitatory. Enzymic degradation of the transmitter restores these channels to their closed state, but current resulting from ion flow opens *voltage-gated* Na^+- and K^+-channels in the adjacent membrane, and the flow of ions which results causes depolarization, further current flow along the membrane and further depolarization.

Voltage-gated channels close again when depolarization in their region reaches its peak. These combined causes and effects result in propagation of an action potential away from the site of original depolarization. Because the K^+-channels open later than the Na^+-channels and stay open longer, the action potential has the characteristics shown in Fig. 88.

There is a period of less than 1 ms, when the Na^+-channels are closing and the K^+-channels are open, when the membrane is unresponsive to a depolarizing current (the *absolute refractory period*) which because it is so short enables nerves to carry the rapid succession of impulses (up to 2,500 s^{-1} for large diameter fibres, 250 s^{-1} for small diameter fibres) involved in information transfer. In addition there is a recovery period (the *relative refractory period*) after passage of an impulse during which stimuli must be of greater strength than normal to cause a propagated impulse. It lasts about 2 ms from the end of the absolute refractory period. On stimulation, individual nerve or muscle fibres respectively either conduct an impulse or contract, or they do not. There are no partial conductions or contractions because of the statistical way in which their membrane ion channels open: only when sufficient ligand-gated channels are open

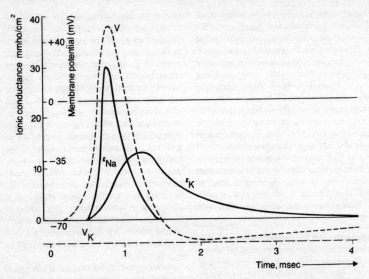

FIG. 88 *The action potential curve (V) resulting from changes in sodium and potassium conductances (^{g}Na, ^{g}K) across an axon membrane at a point on its surface during propagation of an* IMPULSE.

(the threshold level) does sufficient depolarization occur to open adjacent voltage-gated channels (the *all-or-none rule*). Their depolarizing effect in turn causes adjacent voltage-gated channels to open in a reiterated fashion (*accelerating positive feedback*). Unmyelinated nerves conduct at from 0.5–100 m. s^{-1}, increasing with diameter; myelinated axons conduct at around 120 m. s^{-1}. Rise in temperature up to about 40°C increases conduction rate.

Depolarization of muscle *sarcolemma* occurs through release of acetylcholine at NEUROMUSCULAR JUNCTIONS. Impulses are then propagated along the sarcolemma in just the same way as along nerves, but are carried inwardly to myofibrils by transverse tubules. See RESTING POTENTIAL, MUSCLE CONTRACTION, GATED CHANNELS.

inbreeding Sexual reproduction involving fertilization between gametes from closely related individuals, or in its most extreme form between gametes from the same (usually haploid or diploid) individual or genotype. Such *selfing* is not uncommon,

even obligatory, in some plants, such as first colonizers and those lacking pollinators. One end of a continuum, with OUTBREEDING at the other (see BREEDING SYSTEM). The process tends to produce homozygosity at loci (at all loci instantaneously in haploid selfing), with expected disadvantages from the expression of deleterious alleles and reduction in the level of genetic variance among offspring (see GENETIC VARIATION). R. A. Fisher explained the evolution of self-fertilizing plant populations from cross-fertilizing ones on the basis that in a stable population they will contribute three gametes rather than two to the succeeding generation: one megagamete, one self-fertilizing microgamete and one outcrossing microgamete. If the reduction in fitness due to segregation of recessive lethals is more than offset by the 50% excess in gene transmission, the 'gene for self-fertilization' will tend to increase. He reasoned that most plant species are either mainly selfing or mainly outcrossing, with few examples having a mixture of both types, and this appears to be the case. Self-fertilizers do evolve despite their being less fit than out-crossers; but they lack HETEROZYGOUS ADVAN-TAGE and are likely to become extinct in competition with them. However, many plant populations which do outbreed and inbreed (e.g. *Viola*, violets, and gynodioeci-ous species) may, by regular exposure to selection of rare alleles with recessive del-eterious effects, be 'purged' of two such alleles for each death resulting from their expression. In the normally xenophobic naked mole rat *Heterocephalus glaber* there is a rare dispersive morph (inviting com-parisons between these mammals and social insects). This morph does not mate with siblings (even the queen in oestrus), but leaves home and mates with females from another colony. Although rare, it is prob-ably common enough to enable the gene flow required to maintain the heterogeneity required for reproductive compatibil-ity between isolated populations. See ASSORTATIVE MATING, GENETIC LOAD, INBREEDING DEPRESSION, KIN RECOGNITION.

inbreeding depression Increase in pro-portion of debilitated or inviable offspring consequent upon INBREEDING. Extensively documented in captive plants and animals, examples have been slow to accumulate from wild populations.

incisor Chisel-edged tooth of most mam-mals, occurring at the front of the dentary. Primitively three on each side, of both upper and lower jaws. Gnawing teeth of rodents (which grow continuously) and tusks of elephants are modified examples. Used for nipping, gnawing, cutting and pulling. See DENTAL FORMULA.

inclusion granule Microscopically visible bodies produced in the cytoplasms of many plant and animal cells, sometimes in the nucleus, as a result of viral infection. Often consist largely of virions, which may form crystals.

inclusive fitness See FITNESS.

incompatibility (Bot.) (1) In flowering plants, the failure to set seed (i.e. failure of fertilization and subsequent embryo de-velopment) after either self- or cross-pollination has occurred. It is due to the inability of the pollen tubes to grow down the style. In tobacco (*Nicotiana alata*) and others of Family Solonaceae, RNAses both detect incompatible pollen tubes and inhibit their growth. RNAses are also implicated in the gametophytic self-incompatibility of Rosaceae, but not in Poaceae (grasses). In the Brassicaceae (including cabbages), signal transduction mechanisms similar to those in animals are involved in self-incompatibility (SI). FREQUENCY-DEPENDENT SELECTION operates on alleles at the self-incompatibility locus (the *S*-locus) here, the major component of which encodes S receptor kinase (SRK) – a serine-threonine kinase with an extracellu-lar ligand-binding domain expressed in the stigmas. The *S* allele expressed in the *Bras-sica* pollen encodes a cysteine-rich protein (SCR), *SRK* and *SCR* alleles being linked in the *S* locus. See UBIQUITIN. (2) In physiologi-cally heterothallic organisms, it is the failure to reproduce sexually in single or mixed cultures of the same mating type. Geneti-cally determined in both cases, it prevents the fusion of nuclei alike with respect to alleles at one or more loci, thus preventing inbreeding. It is analogous to negative ASSORTATIVE MATING. (3) In horticulture, inability of the scion to make a successful union with the stock. (Zool.) The cause of rejection of a graft by the host organism through an immune response. See IMMUNITY.

incomplete dominance See DOMINANCE.

incomplete flower Flower which lacks one or more of the kinds of floral parts, i.e. lacking sepals, petals, stamens or carpels.

incus One of the mammalian EAR OSSICLES, homologous with the quadrate bone of other vertebrates.

indehiscent (Of fruits) not opening spon-taneously to liberate their seeds; e.g. hazel nuts.

independent assortment (1) See MENDEL'S LAWS. (2) Events occurring in normal diploid MEIOSIS which cause one representative from each non-homologous chromosome pair to pass together into any gamete randomly, irrespective of the eventual genetic composition of the gamete. Results in random RECOMBINATION and is an important source of GENETIC VARIATION in eukaryotic populations. See ABERRANT CHROMOSOME BEHAVIOUR (*meiotic drive*).

indeterminate growth Unrestricted or unlimited growth; continues indefinitely.

indeterminate head Flat-topped INFLORESCENCE possessing sterile flowers with the youngest flowers in the centre.

index fossil See FOSSIL.

indicator, indicator species Species whose ecological requirements are well understood and which, when encountered in an area, can provide valuable information about it. In palaeolimnology, for example, certain diatoms, chrysophytes and ostracods are invaluable indicator species enabling inferences to be made about past lake environments. Absence of an indicator species (e.g. a lichen) from an area where it might be expected to occur could be symptomatic of pollution or some other environmental impoverishment.

indigenous Indicating an organism native to a particular locality or habitat.

individualistic concept The concept of the community as an association of species that occur together simply because of similarities in requirements and not as a result of a long co-evolutionary history.

indoleacetic acid See IAA.

inducible enzyme Enzyme synthesized only when its substrate is present. See ENZYME, GENE REGULATION, JACOB–MONOD THEORY.

induction In embryology, the process resulting from combined effects of EVOCATION and competence (see COMPETENT); results in production by one tissue (the *inducing tissue*) of a new cellular property in a dependently differentiating second tissue where the inducing tissue neither exhibits the resulting property nor alters its developmental properties as a result of the interaction. *Primary induction* events take place early in development; *secondary inductions* take place later in development. See ORGANIZER.

indusium Membranous outgrowth from undersurface of leaves of some ferns, covering and protecting a group of developing sporangia (a sorus).

industrial enzymes Enzymes employed in industrial and commercial applications. Most enzymes for commercial use are produced in BIOREACTORS by batch culture. Intracellular enzymes are often released by macerating the cells, filtering and concentration by gentle evaporation (without denaturing the enzymes). 'Biological' washing powders often contain a cocktail of carbohydrate- and protein-digesting enzymes, usually operating in the alkaline conditions provided by the detergent and at temperatures between 10–90°C; but generally at 40°C, so reducing energy costs. Such powders often contain a cellulase, which removes microfibrils which form on cotton during wear, thereby making it feel smooth. Sugar-releasing enzymes include the enzyme glucose isomerase, produced by continuous culture of *Bacillus coagulans* and used in an immobilized form (see ENZYME) to convert glucose to the sweeter fructose for addition to many foods and drinks. The glucose itself may be derived from corn starch, gelatinized by heat treatment and simultaneously digested by heat-tolerant microbial α-amylase and amino-glucosidase. Enzymes are often employed industrially in BIOSENSORS. See IMMUNO-ASSAYS, POLYMERASE CHAIN REACTION.

industrial melanism Occurrence, common in insects, of high frequencies of dark (*melanic*) forms of species in regions with high industrial pollution, where surfaces on which to rest are darkened by soot and where atmospheric SO_2 levels are high enough to prevent crustose lichen growth. A mutation darkening an individual will tend to be selected for (and hence come to predominate) in polluted regions since it will decrease the bearer's risk of falling prey to a visual predator; but in non-polluted parts of the species range the non-melanic form will be advantageous and occur with higher frequency. In the peppered moth, *Biston betularia*, heterozygous mutants collected in the mid 19th century were paler than they are today, providing evidence in support of the theory that DOMINANCE is an evolving property of characters in populations of species. Industrial melanism provides one of the best examples of evolution within species and of selection resulting in POLYMORPHISM; but not all melanism is necessarily adaptive against visual predation. *Thermal melanism* has been suggested in one ladybird (*Adalia bipunctata*), in which dark forms absorb more energy in regions where atmospheric soot lowers levels of incident solar radiation. They warm up earlier in the season, and gain a reproductive benefit by being mobile sooner than non-melanics. The precise roles of migration and predation on gene frequencies in industrial melanism have yet to be elucidated.

infarction See under SCLEROSIS.

inflammation Local response to injury in vertebrates; also involved in ALLERGY. Involves vasodilation and increased permeability of capillaries in damaged area due largely to release of HISTAMINE and serotonin from MAST CELLS. White blood cells, nutrients and fibrinogen enter and neutrophils are followed into the area by monocytes which become transformed into wandering macrophages for engulfing dead tissue, dead neutrophils, bacteria, etc. Fibrin forms from fibrinogen leaked into the tissues from blood, creating an insoluble network localizing and trapping invading pathogens, forming a fibrin clot preventing haemorrhage while isolating any infected region. Most inflammatory responses are usually down-regulated by circulating steroids, including CORTISOL, but are activated by interleukin-1 (see INTERLEUKINS) and are dependent upon the sequential migration of the above-mentioned effector cells from the peripheral blood to the site of injury in response to the release of various CHEMOKINES. Of the several pro-inflammatory CYTOKINES released by macrophages in response to molecules signalling infection and tissue damage (see LIPOPOLYSACCHARIDES), TUMOUR NECROSIS FACTOR (TNF) has been implicated in chronic inflammatory disorders such as rheumatoid arthritis and Crohn's disease. It is now appreciated that the vagus nerve (cranial nerve X) can reduce TNF release from macrophages by binding their nicotinic ACETYLCHOLINE receptors, which may have therapeutic implications (see NICOTINE). Some forms of Pavlovian-type conditioning, hypnosis, acupuncture and meditation, are forms of 'alternative' medicine known to reduce inflammation. See TOLL-LIKE RECEPTORS. *Pus* usually results following inflammation and comprises dead and living white blood cells and cell remains from damaged tissues. See EICOSANOIDS, NF-κB.

inflorescence Collective term for specific arrangement of flowers on an axis, grouped according to the method of branching, into: (a) indefinite, or racemose; (b) definite, or cymose. In (a), branching is monopodial, inflorescences consisting of a main axis which increases in length by growth at its tip, giving rise to lateral flower-bearing branches. These open in succession from below upwards or, if the inflorescence axis is short and flattened, from the outside inwards. The following are recognized (see Fig. 89): RACEME, whose main axis bears stalked flowers; PANICLE, compound raceme, such as oat; CORYMB, raceme with flowers borne at the same level due to elongation of the stalk (pedicel) of lower flowers, e.g. candytuft; SPIKE, raceme with sessile flowers, e.g. plantain; SPADIX, spike with fleshy axis, e.g. cuckoo pint; CATKIN, spike of unisexual, reduced and often pendulous flowers, e.g. hazel, birch; UMBEL, raceme in which the axis has not lengthened, the flowers arising at the same point to form a head with the oldest outside and youngest at the centre, e.g. carrot, cow parsley; CAPITULUM, where the axis of the inflorescence is flattened and laterally expanded, with growing point in centre, and bearing closely crowded sessile flowers (florets), the oldest at the margin and youngest at the centre, e.g. dandelion.

In (b), branching is sympodial and the main axis ends in a flower, further develop-

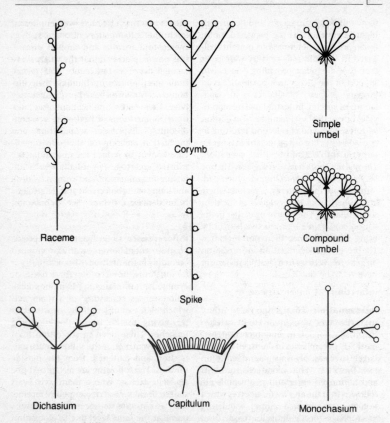

Raceme

Corymb

Simple umbel

Spike

Compound umbel

Dichasium

Capitulum

Monochasium

FIG. 89 *Diagram of different types of* INFLORESCENCE.

ment taking place by growth of lateral branches, each behaving in the same way. The CYME is described as a MONOCHASIUM when each branch of the inflorescence bears one other branch (e.g. iris), and as a DICHASIUM when each branch produces two other branches (e.g. stitchwort). Inflorescences are often *mixed* (part indefinite, part definite): a raceme of cymes.

influenza Respiratory disease caused by an enveloped single-stranded RNA virus (an orthomyxovirus; see Fig. 90), ~100 nm in diameter; the human form exists only in humans. Major coat proteins include HAEMAGGLUTININ (which binds sialic acid resi-

dues on target cells) and the enzyme neuraminidase (see ANTIGENIC VARIATION). The virus evolves by 'antigenic drift' through point mutations, and by 'antigenic shift', through recombination with another 'flu strain, often from a different vertebrate host. Type A viruses are more varied in the make-up of coat proteins than are type B viruses, both of which are transmitted through the air, typically in aerosols caused by coughing and sneezing. The virus infects mucous membranes of upper respiratory tract, and occasionally invades lungs. In children and the elderly, may be followed by bacterial pneumonia; however, recovery from 'flu is usually spontaneous and rapid.

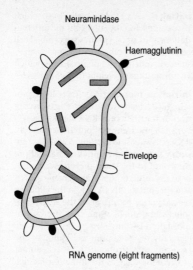

Neuraminidase

Haemagglutinin

Envelope

RNA genome (eight fragments)

FIG. 90 *Some of the* INFLUENZA *virion components.*

Often occurs in PANDEMICS (1918, 1957, 1968, 1977). The viral genes are 'segmented' (occurring in separate fragments of the RNA genome), recombination occurring between these and the fragments from any other strain of the virus simultaneously infecting the cell. Avian 'flu outbreaks are now closely monitored in case they spark a pandemic through recombination with a human 'flu strain. Vaccines against new strains take about six months to mass-produce, in which time a major pandemic could kill millions worldwide. New drugs which attack the virus directly show promise in reducing symptoms and as prophylactics. Unlike vaccines, zanamivir (Relenza), GS 4104 and GS 4071 target the active site of neuraminidase and reduce viral proliferation. It was estimated that influenza would kill half a million people in 2003. See SARS.

infradian rhythm Biological rhythm with periodicity significantly less than a day.

infraspecific variation Variation within SPECIES. It takes several different forms. Clarification depends on population structure, BREEDING SYSTEM and effectiveness of gene flow of the particular case. *Clines* refer to variable phenotypic characters whose distributions display gradients mappable geographically on to gradients in environmental conditions. Morph ratio clines occur when the ratios of different morphs change in a similarly graded way. Phenotypic plasticity may be the cause but where genetic fixation is involved the variation may be closer to ecotypic. ECOTYPES (botanical equivalent of races in ZOOLOGY; see later) involve adaptation of populations to local edaphic, climatic or biotic influences. A *form* in botany is the category within the species generally applied to members showing trivial variations from normal (e.g. in petal colour). In zoology the term is often synonymous with morph (see below), or else a seasonal variant, or used as a neutral term when it is unclear whether a species, subspecies or lesser category is appropriate. A *morph* is one form of a polymorphic species population; e.g. one BLOOD GROUP type (see POLYMORPHISM).

A *race* is a non-formal category used chiefly in zoological contexts. *Geographical races* approximate to subspecies (see later). Host races are those species populations with the same favoured hosts (if parasites) or food plants (when egg-laying, feeding, etc.); such preferences may involve various genetic and non-genetic influences. Belief in human racial subdivision is based largely on subjective impressions of physical appearance (e.g. skin colour) and is not supported by genetic or molecular data (see GENETIC MARKER). A *subspecies* is a formal taxonomic category used to denote various forms (types), commonly geographically restricted, of a polytypic species. It should ideally be used of evolutionary lineages rather than mere phenetic subdivisions of a species. Most easily applied when a population is geographically isolated from other populations of the species (e.g. on an island, mountain top). Subspecific status is often conferred on populations which are really part of a clinal series for the characters used but where intermediate populations have not been studied. In some groups (e.g. Diptera) taxonomists have dispensed with the category. A *variety* is a formal category in botany below the level of subspecies and is used of groups which differ, for various reasons, from other varieties within the same subspecies.

infundibulum (1) Outpushing, or stalk, from floor of vertebrate forebrain attaching the PITUITARY to the hypothalamus. Its terminal swelling produces the posterior pituitary (neurohypophysis); its tissues combine with those growing up from the embryonic mouth to form the combined pituitary organ. (2) Anterior end of ciliated funnel of vertebrate oviduct.

infusoria Term formerly applied to rotifers, protozoa, bacteria, etc., found in cooled suspensions of boiled hay, etc.

inguinal Relating to the groin.

inhibins Family of GROWTH FACTORS comprising dimeric polypeptides, inhibiting FSH release by the pituitary and, produced by Sertoli (see TESTIS) and granulosa cells (see GRAAFIAN FOLLICLE), acting locally within the gonads. See TGFs.

inhibition (nervous) Prevention of activation of an effector through action of nerve impulses. Some inhibitory NEUROTRANSMITTER molecules hyperpolarize rather than depolarize postsynaptic membranes at synapses thus reducing the probability of a propagated action potential at a synapse (see INHIBITORY POSTSYNAPTIC POTENTIAL). Alternatively, an inhibitory neuron may, by its activity, reduce the amount of excitatory neurotransmitter released by another neuron stimulating an effector. See SUMMATION.

inhibitory postsynaptic potential (IPSP) Hyperpolarization of a postsynaptic membrane at a synapse; brought about usually by release of a NEUROTRANSMITTER from the presynaptic membrane which fails to open ligand-gated Na^+- or K^+-channels, but instead opens Cl^--channels, making the inside of the cell more negative (polarized) than it was during the resting potential. Tends to inhibit formation of an ACTION POTENTIAL at a synapse. See SUMMATION.

initial(s) (Bot.) Cell, or cells, from which tissues develop by division or differentiation, as in APICAL MERISTEMS; a cambial cell layer with the capacity to remain meristematic indefinitely; or a cell from which an antheridium develops in bryophytes.

initiation factors (IFs) (1) Soluble proteins (often quaternary) initiating transcription or translation of RNA during PROTEIN SYNTHESIS. In transcription, IFs enable an RNA polymerase to locate its consensus PROMOTER sequence. In initiating translation, they bind to the small ribosomal subunit, or to the initiator tRNA, and enable mRNA and the initiator tRNA to join the *initiation complex* prior to arrival of the large ribosomal subunit. In some cells the rate of translation is controlled by IFs, which are generally removed prior to binding of the large ribosomal subunit. GTP is required at the same stage as the initiation factors. Compare ELONGATION FACTORS. (2) Some TRANSCRIPTION FACTORS are also described as initiation factors. (3) For initiation of DNA replication, see DNA REPLICATION.

innate immune response See IMMUNITY.

inner cell mass Group of cells formed after sinking inwardly from outer layer of the mammalian morula (blastocyst); determined by the 64-cell stage to become the future embryo rather than trophoblast.

inner ear See EAR, INNER.

innervation Nerve supply to an organ.

innexins See GAP JUNCTIONS.

innominate artery Short artery arising from aorta of many birds and mammals and giving rise to right subclavian artery (to fore-limb) and right carotid artery (to head).

innominate bone Each lateral half of the PELVIC GIRDLE when pubis, ilium and ischium are fused into a single bone as in adult reptiles, birds and mammals.

inoculation Injection of living or otherwise mildly infective pathogen into a person or domestic animal followed usually by a mild but non-fatal infection which results in the patient's immunity to the virulent pathogen. Nowadays rarely used, immunization by non-infective agents being preferred. See IMMUNITY.

inoperculate Opening of a SPORANGIUM or ASCUS by an irregular tear or plug to liberate spores.

inosinic acid (IMP) Purine nucleotide precursor of AMP and GMP. Also a rare monomer in nucleic acids where, being similar to guanine, it normally pairs with cytosine. Where it occurs at the 5'-end of an anticodon it may pair with adenine, uracil or cytosine in the 3'-end of the codon. See WOBBLE HYPOTHESIS.

inositol Water-soluble carbohydrate (a sugar alcohol) required in larger amounts than vitamins for growth by some organisms.

inositol 1,4,5-triphosphate (IP3, InsP3)
A SECOND MESSENGER produced by phospholipase C activity as a breakdown product of the minor cell membrane phospholipid phosphatidylinositol. Hydrophilic and small, IP3 diffuses into the cytosol and initiates calcium ion (Ca^{2+}) release from the endoplasmic reticulum and from intact plant cell vacuoles (where it initiates closure of STOMATA when released from its bound form within guard cells; see CALMODULIN). Another second messenger, DIACYLGLYCEROL, is a product of the same phospholipase C activity. IP3 is the common product of antennal pheromone stimulation in various insect species, and IP3-gated ion channels have been located in olfactory receptors of vertebrates and arthropods. Invertebrate PHOTORECEPTORS tend to be driven by an IP3-based signalling pathway. See CYCLIC GMP, PHOSPHOLIPID TRANSFER PROTEINS, PLECKSTRIN HOMOLOGY DOMAINS, PI3-KINASES, Fig. 91.

inquilinism See SYMBIOSIS.

Insecta (Hexapoda) Class of Phylum ARTHROPODA (see TABLE 1), whose members have a body with distinct head (6 embryonic segments), thorax (3 segments) and abdomen (11 segments; see TAGMA). Head bears one pair of antennae and paired mouthparts (mandibles, maxillae and a single fused labium); thorax bears three pairs of uniramous

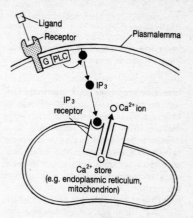

FIG. 91 *The signal pathway thought to cause release of Ca^{2+} ions into the cytosol from intracellular stores, involving* PHOSPHOLIPASE C *(PLC). See* INOSITOL 1,4,5-TRIPHOSPHATE.

legs and frequently either one or two pairs of wings on second and/or third segments; abdomen bears no legs but other appendages may be present (e.g. see CERCI; and see *ULTRABITHORAX* GENE for EVO-DEVO findings). Found as fossils from Devonian onwards. Most have a tracheal system with spiracles for gaseous exchange, and excretion by means of MALPIGHIAN TUBULES. METAMORPHOSIS either effectively lacking (APTERYGOTA), partial (EXOPTERYGOTA), or complete (ENDOPTERYGOTA). The phenomenal adaptive radiation of the group is often attributed to its diversity of mouth parts and feeding methods. However, the evolution of IMAGINAL DISCS must also be considered, enabling as it does a change of mouth parts and feeding mode within a single life cycle. More numerous in terms of species and individuals than any other animal class.

insecticide See PEST MANAGEMENT.

Insectivora Order of placental mammals (e.g. moles, shrews, hedgehogs); a primitive insect-eating or omnivorous group resembling and probably phylogenetically close to Cretaceous ancestors of all placentals. Have small, relatively unspecialized teeth (but incisors tweezer-like). Tree shrews and elephant shrews tend nowadays to be placed in separate orders, Macroscelidia and Scandentia respectively.

insertion A MUTATION involving addition of one or more nucleotides into a genome.

insertion sequence (IS, i. element) One sort of TRANSPOSABLE ELEMENT capable of inserting into bacterial chromosomes using enzymes they encode. Their ends form INVERTED REPEAT SEQUENCES. ISs may also occur in PLASMIDS (e.g. F FACTOR). Can mediate integration of plasmids into main bacterial chromosome by recombination, but cannot self-replicate; are therefore only inherited when integrated into other genomes which do have DNA replication origins. Can mediate a variety of DELETIONS, INVERSION and self-excision. See AUXOTROPH.

in situ **hybridization** A method for detecting a messenger RNA (mRNA) molecule. A label is incorporated into a DNA or RNA molecule (cDNA or cRNA) complementary to the target mRNA molecule so that it binds to it and can be detected. In 'whole mount *in situ* hybridization', this complementary molecule is then incubated with (e.g.) a whole mouse embryo, or with a tissue section. Developing the label gives a coloured product showing where the mRNA was present – at tissue or organ level resolution in whole embryos; but at tissue or cell level in tissue sections, with lower sensitivity.

InsP₃ See INOSITOL 1,4,5-TRIPHOSPHATE.

instar Stage between two ecdyses in insect development, or the final adult stage.

instinct Behaviour which comprises a stereotyped pattern or sequence of patterns; typically remains unaltered by experience, appears in response to a restricted range of stimuli and without prior opportunity for practice. Distinction between this and learnt behaviour has been blurred by research in the last two decades: attention had focused upon developmental pathways of different behavioural responses. Even learnt behaviour presumably has some heritable component; the HERITABILITY of some behavioural patterns is high, and such behaviour tends still to be termed instinctive.

insulin Protein hormone comprising 51 amino acids in two chains held together by disulphide bridges. Secreted by β-cells of vertebrate pancreas in response to high blood glucose levels, e.g. after a meal, as monitored by the β-cells themselves, and active only after removal of two amino acid sequences (pre-pro-insulin to pro-insulin, and pro-insulin to insulin conversions, occur in the GOLGI APPARATUS). Promotes uptake by body cells (esp. muscle, liver, adipose cells) of free glucose and of amino acids by muscle; essentially anabolic in its action. Is thus a *hypoglycaemic* hormone – the only one in most vertebrates – reducing blood glucose. Increases rate of fusion of glucose transporter-bearing vesicles with plasma membrane of sensitive cells; also activates their transporter-mediated glucose uptake and causes synthesis of hexokinase, which phosphorylates glucose on entry to cell, preventing its diffusion out. Also stimulates lipid biosynthesis in fat and liver cells. Low insulin levels cause the glucose transporters to accumulate once again in intracellular vesicles. The cell-surface receptor for insulin is a TYROSINE RECEPTOR KINASE. Opposed in its action by GLUCAGON, the two hormones together regulate and maintain blood glucose at appropriate levels (about 100 mg glucose/100 cm³ blood in humans) through negative feedback via the pancreas. Human insulin was the first protein to be commercially produced by GENE MANIPULATION (see Fig. 63). See DIABETES, LEPTIN.

insulin/insulin-like growth factor pathway INSULIN-LIKE GROWTH FACTORS (IGFs) are a large family of peptide hormones which bind to transmembrane receptor tyrosine kinases. A major protein which is activated as a result is PI₃-kinase, initiating a signal transduction pathway that leads to the activation or repression of particular transcription factors.

IGF-1 and IGF-2 and their subtypes are vertebrate growth-promoting hormones; and the vertebrate insulin and IGF-1 receptors are homologues. The insulin/insulin-like growth factor signal transduction pathway is a central regulator of dauer (a non-feeding stress-resistant larval state in the nematode *Caenorhabditis elegans*) and of ageing, as well as the control of CELL SIZE. Reduced activation of this pathway, or of its homologues, can greatly increase adult lifespan in animals as diverse as *C. elegans*, fruit flies and mice (see AGEING). The pathway involves PI_3-KINASES which dock on to the phosphorylated (activated) insulin receptor substrate. The gene for IGF-2 (*Igf2*), whose expression is required for human prenatal growth, is an example of an imprinted gene (see GENOMIC IMPRINTING): only the paternally derived gene is transcribed. See LIFE HISTORIES, RELAXIN.

insulin-like growth factors (IGFs, somatomedins) Liver-produced peptides with even more potent growth-promoting (anabolic) effects than INSULIN. Secreted in response to human growth hormone (hGH) by liver, muscle, cartilage and bone. Autocrine or paracrine in action. Besides stimulating protein synthesis, IGFs decrease protein catabolism and the channelling of amino acids into ATP synthesis (see KREBS CYCLE). They also stimulate release of glucose by liver (*contra* insulin) and decrease its use by most cells in respiration, enabling neurons to use glucose in times of shortage. See INSULIN/INSULIN-LIKE GROWTH FACTOR PATHWAY.

integrated pest management A term introduced in 1976 to emphasize the need for a variety of biological, chemical and cultural approaches in bringing pests below their economic thresholds in a manner that is reliable in the long term and causes as few objections as possible. Nowadays most PEST MANAGEMENT approaches incorporate this philosophy.

integrins Transmembrane cell surface ADHESION receptors by which cells bind the extracellular matrix (ECM), forming stable interaction zones within the plasma membrane by means of which the ECM can be linked to the cytoskeleton. These act as foci for protein complexes which recruit adaptor proteins and kinases and affect both cytoplasmic enzymes and signalling pathways. Many GROWTH FACTOR signalling pathways are modulated by integrins (see Figs. 77 and 92).

integument (Bot.) (Of seed plants) outer cell layer or layers of ovule covering nucellus (megasporangium) and ultimately forming the seed coat. Most flowering plants have two integuments, an inner and an outer. (Zool.) Outer protective covering of an animal, such as skin, cuticle. Integuments, or *teguments*, of tapeworms (cestodes) and flukes (trematodes) were once thought to comprise inert cuticle, but are now known to be metabolically active structures.

inteins Proteins which splice together disparate protein fragments to form an unbroken amino acid chain. *Split inteins* are 'proteins-within-proteins' which splice together proteins encoded on very different parts of chromosomes and then cut themselves out. Unlike introns, intein sequences are encoded by both DNA and its mRNA transcript and it is thought they operate like some intron splicing mechanisms in forming a loop, bringing the protein fragments together, prior to catalysing peptide bond formation between them.

intercalary (Of a meristem) situated between regions of permanent tissue, such as at bases of nodes and leaves in many monocotyledons, or at junctures of stipes and blade in some brown algae.

intercalated disc See CARDIAC MUSCLE.

intercellular Occurring between cells. Often applied to the matrix or ground substance secreted by cells of a tissue, as in CONNECTIVE TISSUES. For *intercellular fluid*, see TISSUE FLUID. See EXTRACELLULAR, INTERSTITIAL, INTRACELLULAR.

intercellular junction Any of a variety of cell–cell ADHESION mechanisms, particularly abundant between animal epithelial cells, the three commonest of which are (1) DESMO-SOMES, which are principally adhesive, (2) *gap junctions*, involved in intercellular communication, and (3) vertebrate *tight junctions*, and invertebrate *septate junctions*, occluding the intercellular space, thereby restricting movements of solutes. Vertebrate gap junctions consist of a variable hexamer of one or more different cylindrical channel proteins (connexins) with a channel diameter of 1.5 nm, coupling cells electrically (as in electrical synapses and cardiac muscle) and flipping between open and closed states, their permeability changing with calcium ion concentration. Invertebrate gap junctions consist of the protein innexins, with no primary sequence homology with connexins but having similar secondary and tertiary structures. Like PLAS-MODESMATA, gap junctions directly connect adjacent cytoplasms and allow small molecules to pass between them (see Fig. 92). Tight junctions and septate junctions (whose proteins joining adjacent cell membranes are more regularly spaced) form a continuous band around each epithelial cell and perform a selective barrier function in cell sheets, preventing diffusion of ions, etc., from one side of the sheet to the other through intercellular spaces. This is essential to proper functioning of epithelia such as intestinal mucosae and proximal convoluted tubules of vertebrate kidneys. Tight junctions also limit diffusion of membrane proteins and lipids in the outer (not the inner) lipid layers of adjacent membranes. Some of the proteins in the vertebrate tight junctions and invertebrate septate junctions show structural similarities, and mutant fruit flies (*Drosophila*) deficient in one of these, Dlg, develop epithelial tumours. This indicates that some of these junctional proteins are involved in the regulation of cell proliferation and are encoded by TUMOUR SUPPRESSOR GENES. Proteins destined for the apical and basolateral portions of polarized epithelial cells linked by tight junctions and septate junctions are sorted differently, the *trans* Golgi network sometimes, perhaps often, playing a part in this (see PROTEIN TARGETING, SNARES).

intercellular space (Bot.) In plants, air-filled cavities between walls of neighbouring cells (e.g. in cortex and pith) forming internal aerating system. Spaces may be large, making tissue light and spongy as in AERENCHYMA, occasionally harbouring algae, particularly blue-green algae.

intercostal muscles Muscles between RIBS of tetrapods which work in conjunction with the DIAPHRAGM during VENTILATION. *External intercostals* elevate ribs in quiet breathing; *internal intercostals* depress ribs aiding forced expiration.

interfascicular cambium Vascular cambium arising between vascular bundles. Compare FASCICULAR CAMBIUM.

interference competition Competition between two organisms in which one physically excludes the other from a portion of habitat and, therefore, from the resources that could be exploited there.

interferons (INFs) A group of three vertebrate glycoproteins, production of two of which (α and β) is induced in small amounts within virally infected cells (before death) in response to low doses of intracellular double-stranded RNA of a certain minimum length (some of which may be produced by chance when complementary single strands anneal). Infected leucocytes tend to produce α-interferon; infected fibroblasts tend to produce β-interferon; however, γ-interferon ('immune interferon', see INTERLEUKINS (3)) has little sequence homology with α or β and is produced by lymphocytes in response to mitogens (e.g. some growth factors) rather than to viral infection. Interferon induces within adjacent cells an *antiviral state*, apparently binding to specific cell-surface receptors and, by SIGNAL TRANSDUCTION, protecting the cell from viral infection by inducing synthesis of two enzymes which, activated by double-stranded RNA, inhibit the viral production cycle. One of these digests any single-stranded RNA in the cell, thereby killing the cell. Some viruses (e.g. adenoviruses), by producing high levels of short double-stranded RNA, circumvent the antiviral state. See RNA SILENCING.

intergradation zone See HYBRID ZONE.

interleukins (lymphokines) Soluble factors involved in communication between lymphocytes; some also produced by a

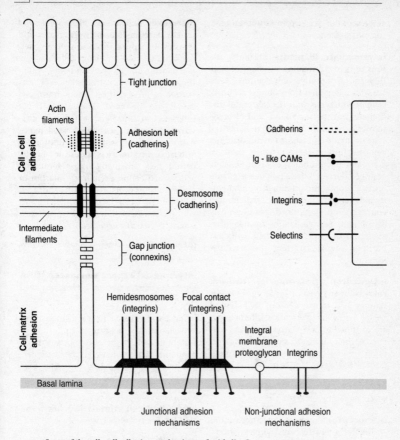

FIG. 92 *Some of the cell–cell adhesion mechanisms of epithelia. See* INTERCELLULAR JUNCTION.

variety of non-leucocytic cells (hence term a misnomer). May be involved in (1) recognition of foreign antigens by T CELLS while in contact with ANTIGEN-PRESENTING CELLS (see T CELLS for diagram); (2) amplifying proliferation of activated T cells; (3) attracting macrophages and rendering them more effective at phagocytosing microorganisms (e.g. γ-INTERFERON); and (4) in promoting HAEMO-POIESIS. B CELLS can produce lymphokines, but this seems not to be important in cell-mediated immune response. Interleukin-1 (IL-1) is a CYTOKINE activating many aspects of INFLAMMATION, boosting other cytokines

and prostaglandins and stimulating a more defensive CORTISOL release via the hypothalamic–pituitary–adrenal axis. Interleukin-2 (IL-2 or TCGF) is used clinically in the treatment of melanoma and some kidney cancers. IL-3 (or multi-CSF) promotes growth of some T cells, of pre-B cell lines and mast cells. IL-1 and IL-2 act early in immune responses and tend to attract a variety of leucocytes; IL-8, released by monocytes, attracts neutrophils during inflammation. The IL-1 receptor is a close structural relative of Toll and intracellular pathways involved in IL-1 signalling converge on NFⰔB. See COLONY-STIMULATING FACTOR, INTERFERONS.

intermediate filaments Insoluble fibrous proteins, providing mechanical stability in (apparently, only) animal cells. Variably assembled from fibrous subunits, each of which has an α-helical central 'rod domain' but specific amino- and carboxy-terminal termini. The fully polymerized state normally consists of a 10 nm-diameter cylinder (intermediate between those of actin and microtubules) of eight helically arranged tetramers, i.e. a polymer of 32 subunits. Cytoplasmic intermediate filaments include keratins (20 distinct kinds in human epithelia) and vimentin (the most widely distributed). Nuclear lamins are also intermediate filaments, but contain a nuclear transport signal targeting them to the NUCLEUS.

interneuron (internuncial neuron, relay neuron) Neuron synapsing between sensory and motor neurons in a typical spinal REFLEX arc. Vertebrate interneurons are confined to GREY MATTER of central nervous system. Afford cross-connections with other neural pathways, enabling INTEGRATION of reflexes, and learning.

internode Part of plant stem between two successive NODES.

interoceptor Receptor detecting stimuli within the body, in contrast with exteroceptor. Includes BARORECEPTORS, PROPRIOCEPTORS, pH-receptors and receptors sensitive to concentrations of dissolved O_2 or CO_2 (see CAROTID BODY, CAROTID SINUS).

interphase Interval between successive nuclear divisions, usually mitotic in proliferative cells; also preceding, and occasionally following, meiosis. Somewhat misleading term suggesting a quiescent or resting interval in the CELL CYCLE (indeed the nucleus is often termed a resting nucleus during it). On the contrary, it is the period during which most components of the cell are continuously made. Cell mass generally doubles between successive mitoses. See Fig. 22b (CELL CYCLE) for some of the molecular details.

intersexes Individuals, often sterile and usually intermediate between males and females in appearance; sometimes hermaphrodite; resulting from failure of the mechanism of SEX DETERMINATION, often through chromosomal imbalance. A FREEMARTIN is an example where hormonal causes are involved. The discovery of intersexes in *Drosophila* led to an understanding of its balance mode of sex determination. A more stringent definition of 'intersexuality' restricts its use to specific pathologies involving sex-determining genes and therefore excludes freemartinism and mutations in genes encoding androgen receptors or 5-α-reductase, which lead to androgen insensitivity. See GYNANDROMORPH, TESTICULAR FEMINIZATION.

interspecific Between species; as in *interspecific* COMPETITION.

interspersed repeat sequences DNA repeat sequences that are typically derived from TRANSPOSONS.

interstitial Lying in the spaces (interstices) between other structures, *interstitial cells* of vertebrate gonads lie either between the ovarian follicles or between the seminiferous tubules of the testis; the latter cells secrete the hormone TESTOSTERONE. See LUTEINIZING HORMONE.

interstitial cell stimulating hormone (ICSH) See LUTEINIZING HORMONE.

interstitial fluid See TISSUE FLUID.

interzonal spindle microtubules Microtubules that extend from one pole of the mitotic spindle to the other; during anaphase they appear to increase in length, thus separating the chromatids, but separation seems to be achieved more by the microtubules sliding over one another.

intestine That part of the ALIMENTARY CANAL between stomach and anus or cloaca. Responsible for most of the digestion and absorption of food and (usually) formation of dry faeces. In vertebrates, former role is often performed by the anterior *small intestine* (see DUODENUM, JEJUNUM, ILEUM), which commonly has a huge surface area brought about by a combination of (i) folds of its inner wall, (ii) VILLI, (iii) BRUSH BORDERS to the epithelial mucosal cells, (iv) SPIRAL VALVES, if present (e.g. elasmobranchs) and (v), in herbivores especially, its considerable length. The more posterior *large intestine*, or *colon*, is usually shorter and produces dry faeces by water reabsorption. The junction between the two intestines is marked in amniotes by a valve, and often a CAECUM. Products of digestion are absorbed either into capillaries of the submucosa or, in CHYLOMICRONS, into lacteals of the lymphatic system. Such digestion as occurs in the caecum and large intestine is largely bacterial.

intracellular Occurring within a cell, which generally means within and including the volume limited by the plasma membrane. Contents of food vacuoles and endocytotic vesicles, although geographically within the cell, are not strictly intracellular until they have passed through the vacuole or vesicle membrane and into the cytosol. See ENDOSYMBIOSIS, GLYCOCALYX.

intracellular immunization The efficient and stable transfer of genetic elements inhibiting viral replication into cells of the patient which are potential targets for viral replication. See RESISTANCE GENES.

intraspecific Within a species. See DEME, INFRASPECIFIC VARIATION.

introgression (introgressive hybridization) Infiltration (or diffusion) of genes of one species population into the gene pool of another; may occur when such populations come into contact and hybridize under conditions favouring one or the other, the hybrids and their offspring backcrossing with the favoured species population. See HYBRID, HYBRID SWARM.

intromittant organ Organ used to transfer semen and sperm into a female's reproductive tract. Include CLASPERS, PENIS.

intron DNA sequence lying within a coding sequence, but not usually encoding cell product, and resulting in so-called 'split-genes' (see Fig. 60); almost universal in eukaryotic genes (but not INTERFERON genes), where on average there are ten per gene;

but few examples of prokaryotic introns. Usually spliced out from RNA (pre-mRNA) during RNA PROCESSING to avoid translation into missense protein (see SPLICEOSOME). Many of the vertebrate genes cloned to date contain several or many introns. DNA from T4 phage and from fungal and plant organelles contains *self-splicing* introns, but eukaryote introns are not self-splicing. When transcribed, these fold into highly ordered tertiary structures which can catalyse self-removal and relegation of host exons without help of a spliceosome. Many also double as mRNAs from which proteins are expressed that assist self-splicing and/or mediate intron translocation to new DNA sites. At least one intron has been shown to cause mitochondrial recombination (see MITOCHONDRION). Until recently, it was widely held (the 'introns early' view) that the progenote ancestor of all living groups of organisms had introns in its DNA, for they had been found in ARCHAEBACTERIA but not in eubacteria (which were thought to have lost them). Now, however, they have been found in the Cyanobacteria (putative ancestors of chloroplasts) and those purple bacteria believed to have given rise to mitochondria (see ENDOSYMBIOSIS). This favours the alternative 'introns late' view that introns are essentially selfish elements (see SELFISH DNA/GENES) which have invaded eukaryotic genes subsequent to the evolution of the nucleus. It now looks as though eukaryotic nuclear (spliceosomal) introns evolved from group II (or self-splicing) introns, found in genomes of chloroplasts, mitochondria, cyanobacteria and proteobacteria, which self-splice without ATP hydrolysis and probably moved into the nuclear genome by retroposition from the mitochondrion, whence they had come by way of purple bacteria (see SPLICEOSOME). Some small nucleolar RNAs (snoRNAs, required for processing rRNAs) are derived from intron RNA by nucleotide excision. These splice all group II introns in the nuclear genome. Introns have short conserved sequences (splice sequences) at their ends, which are crucial for their recognition and for splicing accuracy. See RNA EDITING.

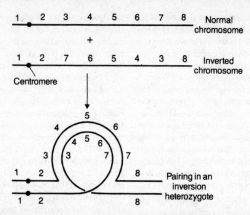

FIG. 93 *Diagram showing the effect of a large* INVERSION *upon pairing of a bivalent during meiotic prophase. Cross-overs within such an inversion hybrid (heterozygous for the inversion) result in acentric fragments and reduced fertility. Numbers indicate positions of loci.*

introrse (Of anther dehiscence) towards the centre of a flower, promoting self-pollination. Compare EXTRORSE.

intussusception (Bot.) Insertion of new cellulose fibres and other material into an existing and expanding CELL WALL, increasing its surface area. Cellulose microfibrils are interwoven among those already existing, as opposed to being deposited on top of them. Compare APPOSITION.

inulin Soluble polysaccharide, composed of polymerized fructose molecules, occurring as stored food material in many plants, such as members of the Compositae and in dahlia tubers. Absent from animals.

invagination Intucking of a layer of cells to form a pocket opening on to the original surface. Common in animal development, as during GASTRULATION.

inversion (1) A type of chromosome MUTATION in which a section of chromosome is cut out, turned through 180° and rejoined, or spliced back, to the chromosome upside down. If long enough, it results in an *inversion loop* in POLYTENE or meiotic cells heterozygous for the inversion. The inverted region is stretched in order to pair up homologously with its partner, as shown in Fig. 93.

If CROSSING-OVER occurs within the inver-

sion loop then the chromosomes which result are nearly always abnormal, having either deletions, duplications, too many centromeres or none at all (acentric). This normally results in reduced fertility. See SUPERGENE. (2) Hydrolysis of sucrose by INVERTASE to equimolar concentrations of glucose and fructose.

invert sugar Equimolar mixture of glucose and fructose, usually resulting from digestion of sucrose by INVERTASE.

invertase (sucrase) See SUCROSE.

invertebrate Term designating any organism that is not a member of the VERTEBRATA. There are many invertebrate chordates.

inverted repeat sequence DNA sequences, often lying on either side of TRANSPOSABLE ELEMENTS and, when single-stranded, run in opposite directions along the chromosome (i.e. are palindromic) and may form a HAIRPIN LOOP by folding back and base-pairing. Double-stranded inverted repeats are also found.

in vitro 'In glass' (Latin). Biological process occurring, usually under experimental conditions, outside the cell or organism; e.g. in a Petri dish.

***in vitro* fertilization (IVF)** Technique developed primarily as a treatment for human infertility. Typically, a woman is superovulated by exogenous gonadotrophins; oocytes collected are matured in a culture medium; fresh or frozen sperm are washed in a suitable medium to induce CAPACITATION and then placed in culture with oocytes; continued incubation (temperature is critical) leads to development of pre-implantation embryos, up to blastocyst stage; these may be frozen or transferred back to the uterus of the oocyte supplier, or to another woman (recipient). Success of the method varies from centre to centre and declines with the age of the individual. Several inherited disorders can now be diagnosed by genetic analysis of single cells removed from human preimplantation embryos following *in vitro* fertilization, enabling 'at risk' couples to be assured that only unaffected embryos will be replaced in the uterus, avoiding possible later pregnancy termination. Considerable debate surrounds the ethics of such procedures. See GENETIC COUNSELLING.

***in vitro* mutagenesis** Techniques which alter specific base pairs in genetic material, enabling many types of point MUTATION, deletion and insertion to be engineered at will within segments of cloned DNA. Such 'designer genes' can then be reintroduced into the organism so that their phenotypic effects may be observed. See MUTAGEN.

in vivo Biological process occurring within a living situation, e.g. in a cell or organism.

involucre A protective investment. (1) In thalloid liverworts, scale-like upgrowth of the thallus overarching the archegonia; (2) in leafy liverworts and mosses, groups of leaves surrounding the sex organs; (3) in many flowering plants (e.g. Compositae), group of bracts enveloping the young inflorescence.

involution (1) Decrease in size of an organ, e.g. thymus and other lymphoid tissue after puberty, contrasting with hyperplasia and hypertrophy. See ATROPHY. (2) Rolling over of cells during GASTRULATION, from the surface towards the interior of the developing gas-trula. (3) Production of abnormal bacteria, yeasts, etc. (e.g. in old cultures).

ion channels Membrane-spanning water-filled pores composed of protein, allowing ions across (highly selectively in eukaryotic cells, less so in prokaryotic). Not continuously open, but have 'gates' which open briefly on appropriate stimulation. Voltage-gated, ligand-gated, mechanically gated and G-PROTEIN-gated forms exist. Sodium and potassium channels (SEE IMPULSE, RESTING POTENTIAL) are examples. See RECEPTOR (2), and references there.

ionizing radiation See CHECKPOINTS, DNA REPAIR MECHANISMS, MUTAGEN, ULTRAVIOLET IRRADIATION, X-RAY IRRADIATION.

ionophore One of a range of small organic molecules facilitating ion movement across a cell membrane (usually the plasma membrane). They either enclose the ion and diffuse through the membrane (e.g. valinomycin-K^+) or form pore-channels in the lipid bilayer (e.g. gramicidin), in which case water molecules are allowed through too. Some are products of microorganisms and may have adverse effects upon cells of competing species.

ionotropic receptor Membrane RECEPTOR forming a common structure with its ion channel. Contrast METABOTROPIC RECEPTOR.

IP3 See INOSITOL 1,4,5-TRIPHOSPHATE.

iris Pigmented, muscular diaphragm whose reflex opening and closing causes varied amounts of light to fall upon the retinas of vertebrate and higher cephalopod EYES (the *iris reflex*). Contributes to depth of focus during ACCOMMODATION. Derived from fused CHOROID and retinal layers in vertebrates.

iron-sulphur proteins Non-haem iron proteins; components of the ELECTRON TRANSPORT CHAIN. The iron of these proteins is not attached to a porphyrin ring; instead, the iron molecules are attached to sulphides and to sulphurs of cysteines of the protein chain. Like CYTOCHROMES, iron-sulphur proteins carry an electron but not a proton. See QUINONE.

ischaemia (adj. ischaemic) Reduced blood supply. See NICOTINE.

ischium Ventral, back-projecting, paired bones of vertebrate PELVIC GIRDLE. They bear the weight of a sitting primate.

isidium Rigid protuberance of upper part of a lichen thallus which may break off and serve for vegetative reproduction.

islets of Langerhans Groups of endocrine cells scattered throughout the vertebrate PANCREAS; some of these cells secrete INSULIN, some GLUCAGON.

isoantigen See ALLOANTIGEN.

isobilateral (Of leaves) having the same structure on both sides. Characteristic of leaves of monocotyledons (e.g. irises), where leaf-blade is more or less vertical and the two sides are equally exposed. Compare DORSIVENTRAL.

isodiametric Having equal diameters; used to describe cell shape when length and width are essentially equal.

isoelectric point The pH of solution at which a given protein is least soluble and therefore tends to precipitate most readily. At this pH the net charge on each of the protein molecules has the highest probability of being zero, and as a result they repel each other least in solution. They also tend not to move in an electric field, e.g. during ELECTROPHORESIS.

isoenzymes (isozymes) Variants of a given enzyme within an organism, each with the same substrate specificity but often different substrate affinities (SEE MICHAELIS CONSTANT); separable by methods such as ELECTROPHORESIS. E.g. lactate dehydrogenase occurs in five different forms in vertebrate tissues, the relative amounts varying from tissue to tissue. Eastern Australian peripatids (Onychophora), formerly considered one species, have turned out on isozyme analysis to comprise several genera and about 50 species.

Isoëtaceae Quillwort family. A family of the LYCOPODIOPSIDA, comprising one genus and about 150 species of tufted, grass-like, heterosporous, perennial, from evergreen aquatics to ephemeral terrestrials found nearly worldwide. The rootstock is brown, corm-like and lobed, while the simple or dichotomous branched roots arise along a central groove separating each rootstock lobe. Leaves are simple, linear, spirally or distichously arranged, dilated toward their base and tapering to an apex. Megasporophylls and microsporophylls are usually borne in alternating cycles. Sporangia are solitary, adaxial and embedded in a basal cavity of a leaf with a velum partly to completely covering the adaxial surface of the sporangium. The megasporangium and microsporangium produce several hundred megaspores and thousands of microspores, respectively. Megagametophytes are white, endosporic and are exposed when the megaspore opens along proximal ridges. There are one to several archegonia. The microgametophyte is nine-celled and endosporic. The antheridium releases four multiflagellate spermatozoids.

isoforms Different forms of a protein, produced by POST-TRANSLATIONAL MODIFICATION.

isogamy Fusion of gametes which do not differ morphologically, i.e. are not differentiated into macro- and micro-gametes. Compare ANISOGAMY; see SEX.

isogenic (syngeneic) Having the same genotype.

isograft (syngraft) Graft between isogeneic individuals, such as identical twins, or mice of the same pure inbred line. Unlikely to be rejected. See GRAFT.

isokont (Bot.) Motile cell or spore possessing two flagella of equal length. Compare HETEROKONT.

isolating mechanisms Mechanisms restricting gene flow between species populations, sometimes of the same but usually of different species. Sometimes classified into *prezygotic mechanisms*, including any process (including behaviour) tending to prevent fertilization between gametes from members of the two populations, and *postzygotic mechanisms*, which prevent development of the zygote to maturity or render it partially or completely sterile. These mechanisms are likely to arise during geographical isolation (a non-biological isolating mechanism) of populations of the same species but they may be reinforced by selection during subsequent sympatry. Their role in sympatric speciation is under investigation. See ECOLOGICAL ISOLATION, REINFORCEMENT, SEXUAL ISOLATION, SPECIATION.

isomerase Any enzyme converting a molecule to one of its isomers, commonly a structural isomer.

isometric contraction Increase in tension during MUSCLE CONTRACTION, but without change in length. Produced artificially if contracting muscle is pinned down at each end. Compare ISOTONIC CONTRACTION.

isomorphic Used of ALTERNATION OF GENERATIONS, particularly in algae, where the generations are vegetatively identical. Compare HETEROMORPHIC.

Isopoda Order of the crustacean subclass MALACOSTRACA, containing such forms as aquatic waterlice (e.g. *Asellus*) and terrestrial woodlice (e.g. *Oniscus*). No carapace; body usually dorso-ventrally flattened. Females have a brood pouch in which young develop directly. Little division of labour between appendages. Many parasitic forms. About 4,000 species.

isoprene A lipid ($CH_2.C.CH_3CHCH_2$) which is an intermediate in the conversion of acetate to cholesterol and a monomer in the formation of polyisoprenoid polymers, such as dolichol phosphate. See ISOPRENOIDS.

isoprenoids (terpenoids, terpenes) Diverse group of compounds, many of them plant products, with some of the general properties of lipids. All contain, or are formed from, a common 5-carbon unit, *isoprene* ($CH_2.C.CH_3CHCH_2$) which is an intermediate in the conversion of acetate to cholesterol and a monomer in the formation of polyisoprenoid polymers, such as dolichol phosphate. Isoprenoids include GIBBERELLINS and ABSCISIC ACID, farnesol (stomatal regulator), squalene, STEROLS, CAROTENOIDS, turpentine, rubber and the phytol side-chain of CHLOROPHYLL. Some restrict 'terpene' to isoprenoids lacking oxygen atoms, these being pure hydrocarbons.

$$CH_3$$
$$|$$
$$\text{(head)} — CH_4 — C = CH—CH_2 — \text{(tail)}$$

Isoprene unit

Isoptera Termites (white ants). Order of EUSOCIAL exopterygote insects, with an elaborate system of CASTES, each colony founded by a winged male and female; wings very similar, elongated, membranous and capable of being shed by basal fractures. Numerous apterous forms.

isotonic (1) Of solutions having equal solute concentration (indicated by their *osmotic pressure*). Even if a cell is surrounded by an isotonic solution, it may still swell or suffer from osmotic shock (see OSMOSIS) if the solutes outside the cell are more concentrated than those within, and the membrane is permeable to them. Under these circumstances the solutes will enter the cell, lower its water potential, and cause osmotic water uptake. See HYPERTONIC, HYPOTONIC, WATER POTENTIAL. (2) See ISOTONIC CONTRACTION.

isotonic contraction MUSCLE CONTRACTION decreasing muscle length without change in tension. Compare ISOMETRIC CONTRACTION.

isotype (1) See ANTIBODY DIVERSITY. (2) Duplicate of type specimen, or HOLOTYPE.

isozyme See ISOENZYME.

iteroparous Breeding more than once in its life. In botany, polycarpic. Compare SEMELPAROUS.

J

Jacob–Monod theory An influential theory (1960) of prokaryotic GENE EXPRESSION, of considerable value in the explanation of eukaryotic gene expression. Its basic concept is that of the *operon*, a unit of TRANSLATION, comprising a group of adjacent structural genes (CISTRONS) on the chromosome, headed by a non-coding DNA sequence (the *operator*) whose conformation binds a DNA-BINDING PROTEIN (the *repressor*, or *regulator*) encoded by a regulator gene elsewhere on the chromosome (see Fig. 94). In one model, the repressor binds the operator preventing the enzyme RNA polymerase from gaining access to an adjacent DNA sequence (the *promoter*), which it must do if any of the structural genes of the operon are to be transcribed. The repressor–operator complex is stable and only broken if another molecule, the *inducer*, binds to it, in which case the inducer–repressor complex loses its affinity for the operator and transcription of the operon's cistrons results. This serves to explain prokaryotic enzyme induction, where the presence of substrate, acting directly or indirectly as inducer, is required for an enzyme to be produced by a cell. In one model of enzyme repression, the repressor does not bind to the operator until it has itself bound to some other molecule, the *corepressor* (e.g. enzyme product, or other gene product). Only then will transcription of the operon be inhibited. Mutants in the repressor gene, affecting repressor shape so that it cannot bind the operator or the corepressor, will result in CONSTITUTIVE production of the operon's cistron products. One observation which the theory helped explain was that in the bacterium *Escherichia coli* the enzymes encoded by what is now referred to as the *lac* operon were either all produced together or not produced at all. Eukaryotic promoters alone provide insufficient information for RNA polymerase to initiate TRANSCRIPTION and further REGULATORY GENE products are first required (e.g., see TRANSCRIPTION FACTORS). Eukaryotic chromosome structure has the added complication of histone-based nucleosomes (see GENE EXPRESSION), but while the operon model was originally dismissed for eukaryotes because polycistronic mRNA is rare and because of their somewhat different mode of gene expression, SYNEXPRESSION GROUPS in eukaryotes provide a remarkable parallel. See Fig. 164, NUCLEAR RECEPTORS.

JAKs Janus kinases. See CYTOKINES for involvement.

jaws Paired (upper and lower) skeletal structures of GNATHOSTOMATA certainly deriving from the third pair of VISCERAL ARCHES (comparable to gill bars) of an ancestral jawless (agnathan) vertebrate. Establishing serial homologies between jaws and gill arches has proved difficult. But it now seems likely that the innervations of hagfish and lamprey heads, and direct comparisons of the contributions of different embryonic rhombomeres to head structures in these and jawed vertebrates, will further our understanding of jaw evolution (see Fig. 95). Upper jaw (see MAXILLA) varies in its articulation with the braincase (see AUTOSTYLIC, AMPHISTYLIC and HYOSTYLIC JAW SUSPENSION). Progressive reduction in number of skeletal elements in lower jaw (see MANDIBLE) during vertebrate evolution, only the dentaries remaining in mammals. Tooth-bearing. See DENTITION.

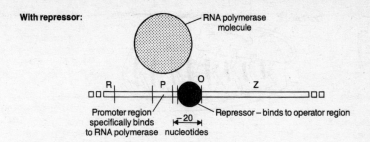

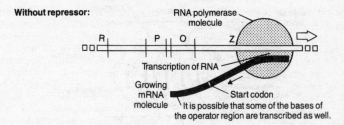

FIG. 94 *Simplified diagram indicating the effect of a repressor molecule (above) in inhibiting transcription of gene Z by RNA polymerase compared with loss of this inhibition (below) when it is removed. Gene R encodes the repressor.* © J. D. Watson: *Molecular Biology of the Gene* 3rd ed. (1976), Fig. 14-12(a). Pub. Benjamin/Cummings. See JACOB–MONOD THEORY.

jejunum Part of the mammalian small intestine succeeding the duodenum and preceding the ileum. Has larger diameter and longer villi than the rest of the small intestine, from which it is not anatomically distinct.

jellyfish See SCYPHOZOA.

JNK signalling pathway Pathway involved in regulation of many cell events, including CELL GROWTH control, TRANSFORMATION and APOPTOSIS. Some SURVIVAL FACTORS operate through this pathway. The transcription factor c-Jun is specifically phosphorylated by c-Jun N-terminal protein kinase (JNK, aka stress-activated protein kinase, SAPK), which also reportedly activates proteins p53 (see entry) and c-*myc* (see entry).

joints Articulations of animal endoskeletons or exoskeletons, the former definitive of arthropods, the latter characteristic of vertebrates where they may be either immovable (e.g. between skull bones), partly movable (e.g. between vertebrae) or freely movable (e.g. hinge joints, ball-and-socket joints). Commonly a feature of lever systems employing antagonistic muscles.

j-shape growth form Type of population growth form, in which density increases rapidly in an exponential manner and then stops abruptly as environmental resistance or other limit takes effect more or less suddenly. See GROWTH CURVES, EXPONENTIAL GROWTH.

jugulars Major veins in mammals and related vertebrates returning blood from the head (particularly the brain) to the superior vena(e) cava(e). Usually in the form of paired interior and exterior jugulars fusing on each side to form common jugulars before joining the subclavian veins and ultimately draining into the venae cavae.

junctional complex See INTERCELLULAR JUNCTION.

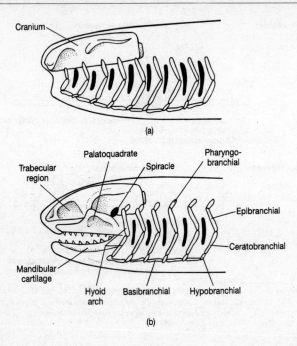

(a)

(b)

FIG. 95 *Proposed evolution of vertebrate jaw from anterior gill arches of agnathan ancestor.* (a) *agnathan,* (b) *gnathostome. The first gill arch may have had no descendant structure; the second probably gave rise to the trabecular cartilage, while a more posterior arch gave rise to the pair of jaws, for which the hyoid arch acts as a support in* (b). *The gill slit between jaws and hyoid arch was reduced to the spiracle seen in modern sharks.*

Jurassic GEOLOGICAL PERIOD of the MESOZOIC era, extending from about 205–135 Myr BP. A major part of the age of reptiles, during which the ARCHOSAUR radiation, begun in the TRIASSIC, continued. Plesiosaurs and ichthyosaurs also flourished. Mammal-like reptiles (e.g. therapsids) dwindled; true mammals were scarce. Earliest fossil birds discovered (e.g. *ARCHAEOPTERYX*) were deposited in the Upper Jurassic. Climate was generally mild, continents low with large areas covered by seas. Gymnosperms, especially cycads, prominent.

juvenile hormone (neotenin) Hormone produced by insect CORPORA ALLATA. See also ECDYSONE.

K

K See KD.

kappa particles Gram-negative bacterial species present endosymbiotically (as commensals) within cytoplasm of ciliate *Paramecium aurelia*; their maintenance requires activity of some nuclear genes of host cell. May be transferred from one host cell to another during CONJUGATION, and may produce toxins (e.g. *paramecin*) which kill sensitive *Paramecium* strains but not the producer cells (termed *killer cells*). Inheritance of the killer trait is an instance of extrachromosomal or CYTOPLASMIC INHERITANCE.

karyogamy Fusion of nuclei or their components. Feature of eukaryotic sexual reproduction. See FERTILIZATION, HETEROKARYOSIS. Compare PLASMOGAMY. For possible evolutionary origin, see SEX.

karyotype Characteristics of the set of chromosomes of a cell or organism (their number, sizes and shapes). A photograph or diagram of chromosomes, generally arranged in pairs and in order of size, is termed a *karyogram*.

Kb Kilobase; 10^3 bases.

Kd (K) Kilodalton. Having a relative molecular mass of 1,000.

keel (carina) (Zool.) Thin medial platelike projection from sternum (breast-bone) of modern flying birds (*carinates*) and bats providing attachment for wing muscles. Absent from RATITES and many flightless birds. (Bot.) (1) Ridge alongside a fold applied to coalescent lower petals of a papilionaceous corolla of the pea family. (2) In some pennate diatoms, summit of a ridge bearing the raphe, where the valve is sharply angled at the raphe.

kelp Common name given to brown algae (PHAEOPHYTA) of Order Laminariales, the main group of brown algae having important economic uses (e.g. alginic acid, green manure, fucoidin, food).

keratin Tough fibrous sulphur-rich protein of vertebrate epidermis forming resistant outermost layer of skin. α-keratin was the protein in which the α-helix was first described; but each molecule adopts a helical tertiary structure when wound around an identical α-helix to form a quaternary dimer of two identical subunits, linked by covalent sulphur bridges (a coiled coil). See CORNIFICATION, CYTOSKELETON.

ketone bodies Substances such as acetoacetate and hydroxybutyrate produced mainly in the liver from acetyl coenzyme A, itself an oxidation product of fatty acids, released for use by peripheral tissues as fuel. The metabolic pathway is termed *ketogenesis*.

kidney Major organ of nitrogenous EXCRETION and OSMOREGULATION in many animal groups of little or no homology. Its elements usually open directly to the exterior in invertebrates, but usually via a common excretory duct in vertebrates. Functional units in vertebrates, kidney tubules or *nephrons*, were probably originally paired in each trunk segment and drained through a pair of ducts, one on each side of the body (see WOLFFIAN DUCT). In higher vertebrates, anterior tubules (forming what remains of the *pronephros*) are embryonic and transitory, kidney function being normally dominated by the opisthonephros (*mesonephros* and *metanephros*), whose segmental organization is all but lost in the adult, a new excretory duct (the *ureter*) draining from the mass of nephrons into the bladder. The mesonephros is the functional kidney in adult fish and amphibians. In embryonic amniotes the two kidneys are initially mesonephric, lying in the trunk, their nephrons having glomeruli and coiled tubules; but the whole structure loses its urinary role during development and in males becomes invaded by the vasa efferentia. Each mesonephric duct gives off *ureteric buds* which grow into the intermediate mesoderm and develop into the collecting ducts, pelvis and ureter – in due course draining the *metanephros*. Each bud gives rise to a cluster of capillaries, a *glomerulus*, and a long tubule differentiating into *Bowman's capsule, proximal convoluted tubule, loop of Henle* and *distal convoluted tubule*, joining the collecting duct. Units developing from the cap tissue are termed *nephrons*, most of whose components lie in the kidney *cortex*, only the loops of Henle lying in the *medulla*, where they join the collecting ducts. Over a million nephrons may occur in each mammalian metanephros.

Hydrostatic pressure in the blood forces water and low molecular mass solutes (not proteins) out of the glomeruli into the Bowman's capsules. In mammals 80% of this glomerular filtrate is then reabsorbed across the cells of the proximal convoluted tubule by FACILITATED DIFFUSION and ACTIVE TRANSPORT into capillaries of the VASA RECTA draining the kidney. The descending and ascending limbs of the loop of Henle form a COUNTERCURRENT SYSTEM whose active secretion and selective permeabilities result in a high salt concentration in the interstitial fluid of the medulla, enabling water to be drawn back into the medulla osmotically from the collecting ducts if these are rendered permeable by ANTIDIURETIC HORMONE. The long juxtamedullary nephrons, and the overlying vasa recta, together form a countercurrent multiplier, whereby each successive change in molarity of tubule contents, although small, accentuates the original dif-

ference. Blood flow in the vasa recta is slow, enabling salt, water and urea exchanges with the interstitial fluid. Urea is also reabsorbed, but never against a concentration gradient, half being excreted on each journey through the kidneys. Kidneys play a major role in osmoregulation and help regulate blood pH by controlling loss of HCO_3^- and H^+. The remnants of glomerular filtrate from all collecting ducts comprise the URINE. See RENIN. Human patients who have undergone kidney transplants are commonly treated with the monoclonal antibody 'Orthoclone OKT-3', which specifically binds the OKT-3 antigen on the surface of T cells which otherwise mount an attack on the foreign organ. This is said to have saved the lives of 90% of the patients whose immune system would normally have rejected them.

killer T cells (cytotoxic T cells) See T CELLS. Compare NATURAL KILLER CELLS.

kinase Enzyme transferring a phosphate group from a HIGH-ENERGY PHOSPHATE compound to a recipient molecule, often an enzyme, which is thereby activated and able to perform some function. Opposed by phosphatase activity, which removes the transferred phosphate group. See PHOSPHORYLASE KINASE, PROTEIN TYROSINE KINASE, ENTEROKINASE, CASCADE.

kinesin One of any eukaryotic cell's three major motor proteins (MYOSIN and DYNEIN being the other two). Able to bind separately to microtubules and organelles and, through its ATPase activity, generate the force required to move the latter along the former (towards the *plus*, or distal, end). In contrast to myosin and dynein – also transducing proteins – binding of kinesin to its partner protein (e.g. tubulin of a microtubule) is promoted by ATP's presence, ATP being hydrolysed only while kinesin is bound. Once released, kinesin can take another step in the walk along the microtubule. Kinesin-dependent organelle movement, unlike dynein-dependent, is usually centrifugal. Plays a major role in transport within axons of neurons. See CYCLOSIS, NEURON.

kinesis (pl. kineses) Any form of orientation in which an animal's response is proportional to the intensity of stimulation and is independent of the spatial (e.g. directional) properties of the stimulus. Thus common woodlice (*Porcellio scaber*) are very active at low levels of humidity and less active at high levels, resulting in their spending more time in damp parts of the environment. Their rapid locomotion in dry places increases the probability that they will discover damper conditions. Woodlice tend to form aggregations in damp places beneath rocks and fallen logs, selection of this habitat being based entirely upon kinesis. Compare TAXIS.

kinetin A purine, probably not occurring naturally, but acting as a CYTOKININ in plants.

kinetochore Structure assembling on a CENTROMERE during MITOSIS and MEIOSIS, essential for chromosome movements during those processes. Between 10–50 microtubules attach early in prometaphase to each centromere's kinetochore. When correctly captured by the two kinetochores from opposite poles, microtubules experience tension as they start to depolymerize. Without such tension, the microtubule–kinetochore attachments collapse stochastically. Persisting attachments result in an ordered chromosome distribution to daughter nuclei. This is partly because of changes in microtubule length. When they depolymerize at the kinetochore attachment, microtubules move chromosomes towards the spindle poles, and when they polymerize, they move chromosomes away from the poles. But *in vitro* experiments indicate that microtubule-based MOTOR PROTEINS located at the kinetochores can move along microtubules in a plus-end direction (towards the growing end), remaining attached when they reach the end, and that kinetochores also maintain attachment to the shrinking (depolymerizing) ends of microtubules. These motor proteins (e.g. the KINESIN-related protein CENP-E) may also modulate the stabilities of microtubules themselves. Most of the proteins involved in mitotic and meiotic chromosome movements have not been identified. See Fig. 96.

kinetoplast Organelle present in some flagellate protozoans (the Kinetoplastida, e.g. *Trypanosoma, Leishmania*) and containing sufficient DNA for this to be visible under light microscopy when suitably stained. Apparently a modified mitochondrion; commonly situated near the origin of a flagellum.

kinetosome See CENTRIOLE.

Kingdom Taxonomic category with the greatest generality commonly employed, inclusive of divisions (phyla). Controversy has occurred over the number of kingdoms to employ. Today, most recognize five kingdoms: MONERA (prokaryotic organisms), PROTISTA, FUNGI, ANIMAL and PLANTAE.

kinin See CYTOKININ, BRADYKININ.

Kinorhyncha Class of minute marine ASCHELMINTHES (or a phylum in its own right) with superficial metameric appearance but without true segmentation. Share a syncytial hypodermal structure with GASTROTRICHA but unlike them lack external cilia. Cuticle covered with spines. Muscular pharynx similar to that of nematode worms. Nervous system a ring around the pharynx with four longitudinal nerve cords. Usually dioecious.

kin recognition Several animal studies indicate that close relatives are unattractive as mates; often, in primates, as distantly related as first and second cousins. Domestic Japanese quail choose mates of intermediate relatedness. In inbred house mice, individuals with different MHC gene products are preferred as mates; but in other mouse strains, familiar oestrous first or second cousins are preferred. Some bumble bees will only mate with non-nestmates. See INBREEDING, OUTBREEDING.

kin selection Selection favouring genetic components of any behaviour (in its broadest sense) by one organism which has beneficial consequences for another, and whose strength is proportional to the RELATEDNESS of the two organisms. Contrast GROUP SELECTION. See COOPERATIVE BEHAVIOUR, HAMILTON'S RULE.

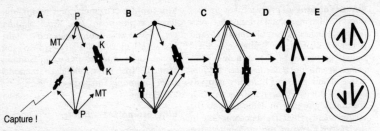

Capture !

FIG. 96 *The normal errorless events of chromosome attachment and transmission during a first meiotic division. For simplicity, chromosomes are not shown in the form of two chromatids. (a) A spindle with two poles (P), paired chromosomes with two spindle attachment sites each, and* KINETOCHORES *(K). By chance,* MICROTUBULES *(MT) growing from a pole may encounter, and be captured by, a kinetochore. The chromosome is thereby attached to that pole. (b–d) The two kinetochores of each chromosome become attached to opposite poles of the spindle, each chromosome pair separating into two chromosomes. (e) The two daughter cells each have a copy of each chromosome.*

Klebsormidiophyceae A class of algae in the division CHLOROPHYTA, restricted to freshwater and moist subaerial habitats, comprising seven genera and about 45 species that display three different levels of organization (coccoid; sarcinoid – composed of three-dimensional packets of cells; and branched or unbranched filaments). The main polysaccharide in the fibrillar structural component of the cell wall is crystalline cellulose.

Kleiber's Law See ALLOMETRY.

Klinefelter's syndrome Syndrome occurring in men with extra X-chromosome giving genetic constitution XXY. Usually results from NON-DISJUNCTION and expresses itself by small penis, sparse pubic hair, absence of body hair, some breast development (gynaecomastia), small testes lacking spermatogenesis. Long bones often longer than normal. See TURNER'S SYNDROME, TESTICULAR FEMINIZATION.

klinotaxis See TAXIS.

K_M-value See MICHAELIS CONSTANT.

knockouts See GENE KNOCKOUTS.

Korarchaeota See ARCHAEA.

Kranz anatomy (K. morphology) Wreath-like arrangement (*Kranz* being German for wreath) of palisade mesophyll cells around a layer of bundle-sheath cells, form-ing two concentric CHLOROPLAST-containing layers around the vascular bundles; typically found in leaves of C_4 plants, such as maize and other important cereals. See PHOTOSYNTHESIS.

Krebs cycle (citric acid cycle, tricarboxylic acid cycle, TCA cycle) Cyclical biochemical pathway (see Fig. 97) of central importance in all aerobic organisms, prokaryotic and eukaryotic. The pathway is cyclical since the 'end-product' (oxaloacetic acid) is also an initial substrate for the next round; and being cyclical, the pathway does not suffer from END-PRODUCT INHIBITION. The reactions themselves contribute little or no energy to the cell, but dehydrogenations involved are a source of electrons for ELECTRON TRANSPORT SYSTEMS via which ATP is produced from ADP and inorganic phosphate. In addition, some GTP is produced directly during the cycle. Cells with mitochondria perform the cycle by means of enzymes within the matrix bounded by the inner mitochondrial membrane; in prokaryotes these enzymes are free in the cytoplasm. Bulk of substrate for the cycle is acetate bound to coenzyme A as acetyl CoA (see PANTOTHENIC ACID), but intermediates of the cycle can act as substrates. Acetate is derived from pyruvate (produced by glycolysis) after this enters the mitochondrion. The MULTIENZYME COMPLEX pyruvate dehydrogenase (PDH) is the engine for this. But

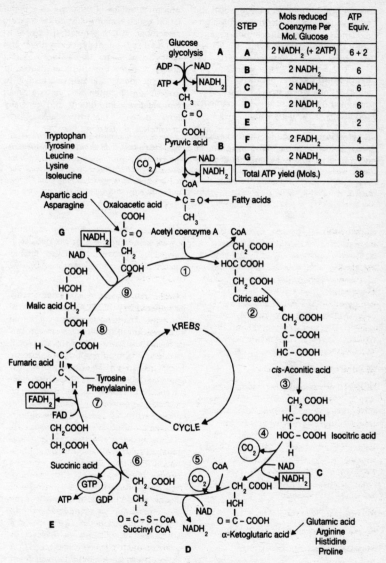

STEP	Mols reduced Coenzyme Per Mol. Glucose	ATP Equiv.
A	2 NADH$_2$ (+ 2ATP)	6 + 2
B	2 NADH$_2$	6
C	2 NADH$_2$	6
D	2 NADH$_2$	6
E		2
F	2 FADH$_2$	4
G	2 NADH$_2$	6
Total ATP yield (Mols.)		38

FIG. 97 *Diagram showing the* KREBS CYCLE *and its relationship to glycolysis and to amino acid and fatty acid input. The table indicates how one glucose molecule can provide the energy to release thirty-eight ATP molecules once the reduced coenzymes have passed their hydrogen atoms on to the respiratory chain and electron transport system.*

acetate is also a product of amino acid and fatty acid oxidation, decarboxylation and oxidation within the mitochondrion generating CO_2, acetate and reduced NAD (see also FERREDOXINS). See ATP, CALVIN CYCLE.

krummholz Region between the alpine and tree lines, where trees are dwarfed and deformed owing to severe environmental conditions, particularly wind.

K-selection Selection for those characteristics which enable an organism to maximize its FITNESS by contributing significant numbers of offspring to a population which remains close to its CARRYING CAPACITY, K. Such populations are characterized by intensely competitive and density-dependent interactions among adults, with few opportunities for recruitment by young. Adults invest heavily in growth and maintenance, have a small reproductive commitment and generally have delayed maturity. See DENSITY-DEPENDENCE, R-SELECTION.

Kupffer cell See RETICULO-ENDOTHELIAL SYSTEM.

labelling Variety of indispensable techniques for detecting presence and/or movement of certain isotopic atoms both *in vitro* and *in vivo*. Isotopes used are either radioactive (RADIOISOTOPES) or differ in their atomic masses without being radioactive. Commonly used radioactive isotopes include ^{14}C, ^{35}S, ^{32}P and ^{3}H; non-radioactive 'heavy' isotopes include ^{15}N and ^{18}O. Material containing the unusual isotope is often administered briefly (pulse-labelling), and 'chased' by unlabelled material. The time taken for label to pass through the system and the route it takes contribute to our understanding of the dynamics of biological systems and processes (e.g. cell membranes, photosynthesis, aerobic respiration, DNA and protein synthesis) as well as of molecular structure. See AFFINITY LABELLING, AUTORADIOGRAPHY, SOUTHERN BLOTTING, DNA HYBRIDIZATION, *IN SITU* HYBRIDIZATION.

labial Referring to LABIUM, or to lip-like structure.

labiate process Tube or opening through the valve wall of a diatom (BACILLARIOPHYTA) with an internal flattened tube or longitudinal slit often surrounded by two lips.

labium (Bot.) (1) Lower lip of flowers of the family Labiatae. (2) The lip subtending the ligule in the lycopod *Isoetes*. (Zool.) (1) An insect mouthpart forming the lower lip and comprising a single structure formed from a pair of fused second maxillae and bearing a pair of palps distally. Compare crustacean MAXILLIPED. (2) One member of two pairs of skin folds (*labia*) in female mammals protecting the exterior opening of the vagina.

labrum Plate of exoskeleton hinged to front of the head in some arthropod groups and serving to enclose a space (cibarium) in front of the actual mouth. Found in insects and crustaceans. Trilobites had a similar structure, but placed ventrally on the head.

Labyrinthodontia Subclass of extinct amphibians found as fossils from Upper Devonian to Upper Triassic (see GEOLOGICAL PERIODS). Generally believed to have had rhipidistian ancestors.

Labyrinthulomycota (Labyrinthulomycetes) Net slime moulds, previously classified with the fungi; however, they are more closely related to the algal division HETEROKONTOPHYTA; sometimes placed in the kingdom Chromista. The group comprises 13 genera and 42 species of aquatic, mostly marine organisms, often associated with plants and some algae; some act as pathogens. The trophic stage is an ectoplasmic network containing spindle-shaped cells that move by gliding within this network. Cells uniquely contain an invaginated organelle at the cell surface, which connects the plasma membrane to the network membranes (i.e. bothrosome). Biflagellate zoospores are produced which have mastigonemes on one of the flagella. Sexual reproduction is known for some species.

lac operon DNA sequence in the genome of the bacterium *Escherichia coli* comprising an *operator*, *promoter*, and three structural genes, transcribed into a single mRNA, encoding enzymes involved in lactate uptake and metabolism. See JACOB–MONOD THEORY.

lactation Process of milk production by mammary glands. Involves hormone activity, notably of PROLACTIN, itself released by hypothalamic *prolactin releasing factor* reflexly secreted during sucking at nipples. Oestrogens and progesterone promote prolactin secretion but inhibit milk secretion; however after delivery the female sex hormone levels in the blood drop abruptly and this inhibition is removed. Sucking at the nipples reflexly releases OXYTOCIN from the posterior pituitary causing muscle contraction in the breast alveoli and milk letdown. See HUMAN PLACENTAL LACTOGEN.

lacteal Lymph vessel draining a VILLUS of vertebrate small intestine. After digestion, reconstituted fats are released into the lacteal as CHYLOMICRONS.

lactic acid A carboxylic acid, CH_3 $CHOHCOOH$, produced by reduction of pyruvate during anaerobic respiration (see GLYCOLYSIS). Many vertebrate cells can produce lactate, notably muscle and red blood cells. After transport in the blood it may be converted back by the liver to glucose during GLUCONEOGENESIS. Probably responsible for sensation of muscle fatigue. See OXYGEN DEBT. Also formed in metabolism of many bacteria (e.g. from lactose in souring of milk), as well as in fungi (e.g. Deuteromycotina), commonly in association with alcohol or acetaldehyde.

lactose Disaccharide found in mammalian milk. Comprises galactose linked to glucose by a $\beta.1-4$ glycosidic bond. A reducing sugar. See LACTOSE INTOLERANCE.

lactose intolerance Underproduction by intestinal mucosal cells of lactase, required for lactose digestion after intake of milk; most frequent in the elderly and in Afro-Caribbean peoples. Undigested lactose in the lumen retains water and encourages bacterial fermentation in the colon, with resulting cramps, gas production and diarrhoea. Lactase is required in early life, but its production usually declines in adolescence, except in populations where milk has remained an important component in the adult diet.

lacuna (Zool.) Any small cavity, such as those containing bone or cartilage cells.

lacustrine Inhabiting lakes.

Lagomorpha Order of placental mammals including pikas, rabbits and hares. Gnawing herbivores differing from rodents chiefly in having two pairs of incisors as opposed to one pair in the upper jaw, the second pair being small and functionless. Classified as 'ruminants' in the Old Testament, and placed within Rodentia by Linnaeus. Study of orthologous protein sequences suggests that the group is more closely related to Primates and Scandentia than to rodents and that the clustering of rodents and lagomorphs into Superorder Glires may be based upon symplesiomorphies and not synapomorphies.

Lamarck, lamarckism Jean Baptiste de Lamarck (1744–1829) was a French natural philosopher who united a wide range of scientific interests under general principles. Coined the term *biology* in 1802, and worked in the newly-created National Museum of Natural History as professor of the zoology of the lower animals, although his previous work had been largely botanical. First to classify animals into invertebrates and vertebrates and to use such taxa as Crustacea and Arachnida. Lamarck's evolutionary views owed something to BUFFON'S interest in modification of organisms by changes in their environments – changes which for Lamarck altered an organism's needs and habits. He held (as did Geoffroy St Hilaire) that continuous spontaneous generation was required to restock the lowest life forms which had evolved into more complex ones.

For Lamarck the mechanism of evolutionary change was environmental: organs which assist an organism in its altered conditions are strengthened (e.g. the giraffe has acquired its high shoulders and long neck by straining to reach higher and higher into trees for leaves); others progressively atrophy through disuse. Such acquired characters, he thought, were then inherited. But there were and are no clear examples of such inheritance; nor does this theory account satisfactorily for evolutionary stability (stasis). It would require a theory of inheritance completely at variance with that receiving experimental support today. See CUVIER, DARWIN, WEISMANN, NEO-LAMARCKISM, MUTATION.

lambda (λ) phage and repressor A temperate phage (see BACTERIOPHAGE) and mobile genetic element which can undergo either lytic growth or enter lysogenic growth. The decision to move to lysogeny after infection is enhanced by λ repressor concentration, while transfer from lysogenic growth to the lytic cycle depends on repressor destruction. The λ repressor is a dimeric protein encoded by λ phage which prevents entry of further λ and whose transcription is activated by phage cII protein. It prevents binding of RNA polymerase and shuts off early phage promoters, repressing phage genes involved in the lytic phase, switching development into the lysogenic mode. The λ repressor promoter region is itself blocked by Cro protein, when the phage lytic cycle commences. Joining of the λ phage DNA with the bacterial chromosome is achieved by a lambda integrase, encoded by the virus. See TRANSPOSONS, VECTOR.

lamella Any thin layer or plate-like structure. (Bot.) (1) One of the spore-bearing gills in the fruiting body of a mushroom or related fungus, attached to the underside of the cap (pileus) and radiating from centre to margin. (2) One of a series of double membranes (thylakoids) within a chloroplast which bear photosynthetic pigments. (3) In bryophytes, a thin sheet of flap-like plates of tissue on dorsal surface of the thallus or leaves. (Zool.) One of the concentric layers of hard calcified material of compact bone forming part of a HAVERSIAN SYSTEM.

Lamellibranchia See BIVALVIA.

lamellipodium Sheet-like extension, or flowing pseudopodium, of the leading edges of many vertebrate cells during locomotion, some forming attachments with the substratum, others carried back in waves (ruffling, producing a *ruffled border*). Many give rise to microspikes. See CELL LOCOMOTION, LECTIN.

lamina (1) Sheet or plate; flat expanded portion of a leaf or petal. (2) In brown algae (PHAEOPHYTA), expanded leaf-like portion of the thallus.

lamina propria Loose connective tissue of a MUCOUS MEMBRANE (e.g. of gut mucosa), binding the epithelium to underlying structures and holding blood vessels serving the epithelium.

laminarin Storage polysaccharide product of the brown algae (PHAEOPHYTA), occurring as oil-like liquid contained in a membrane-bound sac surrounding the pyrenoid but outside the CHLOROPLAST. Comprises β-1,3 linked glucans containing 16 to 31 residues.

laminar placentation Attachment of ovules over surface of carpel.

laminins Glycoproteins of the eukaryotic EXTRACELLULAR MATRIX containing many epidermal growth factor (EGF) repeats and implicated, like other such proteins, in developmental processes and stimulation of cell division. Laminins provide chemical signals, leading axons along growth routes in the embryo, notably on the surfaces of glial and neuroepithelial cells. They also initiate the construction of a BASAL LAMINA by some cultured epithelial cells.

lamins See INTERMEDIATE FILAMENTS, NUCLEUS.

lampbrush chromosome BIVALENTS during diplotene in some vertebrate (notably amphibian) oocytes in which long chromatin loops, which are transcriptionally very active, form at right angles to the chromosome axis and become covered with newly transcribed RNA. Not certain that much of this RNA acts as mRNA in protein synthesis; but if it is functional, the high transcriptional activity is presumably an adaptation to serving a relatively large cell from a single nucleus. See CHROMOSOME, GENE AMPLIFICATION.

lampreys See CYCLOSTOMES.

lamp shells See BRACHIOPODA.

land use ecology The study of the role and impact of ecological (or environmental) factors on land use patterns, productivity and management, in the past as well as the present.

large intestine See COLON.

larva Pre-adult form in which many animals hatch from the egg and spend some time during development; capable of independent existence but normally sexually immature (see PAEDOGENESIS, PROGENESIS). Often markedly different in form from adult, into which it may develop gradually or by a more or less rapid METAMORPHOSIS. Often dispersive, especially in aquatic forms. In insects especially, the phase of greatest growth in the life cycle. Examples include AMMOCOETE, CATERPILLAR, LEPTOCEPHALUS, NAUPLIUS, TADPOLE, TROCHOPHORE, VELIGER.

Larvaceae See UROCHORDATA.

larynx Dilated region at upper end of tetrapod trachea at its junction with the pharynx. 'Adam's Apple' of humans. Plates of cartilage in its walls are moved by muscles and open and close glottis. In some tetrapods, and most mammals, a dorso-ventral and membranous fold (vocal cord) within the pharynx projects from each side wall, vibrations of these producing sounds. Movement of the cartilage plates alters the stretch of the cords and alters pitch of sound. See SYRINX.

lasso cell Cell type characteristic of CTENOPHORA, whose tentacles are armoured with these sticky thread-cells for capturing prey. Do not penetrate prey, unlike nematocysts of cnidarians.

latent period (reaction time) Time between application of a stimulus and first detectable response in an irritable tissue. The latent period for a striated muscle fibre may be as little as 2 milliseconds, from initiation of an action potential, its propagation along the transverse tubules, release of calcium ions from the sarcoplasmic reticulum, their diffusion to the troponin molecules, activation of myosin cross-bridges which then bind to actin, and the generation of force.

lateral inhibition (1) See APICAL DOMINANCE. (2) Processes by which the CELL POLARITIES of adjacent cells are altered so that they come to follow different FATES during development. Small stochastic initial asymmetries are sometimes involved.

lateral line system (acoustico-lateralis system) Sensory system of fish and aquatic and larval amphibians whose receptors are clusters of sensory cells (*neuromast organs*) derived from ectoderm, found locally in the skin or within a series of canals, or grooves on head and body. Neuromasts resemble cristae of the VESTIBULAR APPARATUS of higher vertebrates, having a gelatinous cupula but lacking otoliths; they are probably homologous structures. Pressure waves in the surrounding water appear to distort the neuromasts, sending impulses via the vagus nerves on either side, where they associate with the *lateral lines* themselves – an especially pronounced pair of these sensory canals, one running the length of each flank. The head canals are served by the facial nerves.

Structures termed the 'lateral line system' exist in larval and pupal Trichoptera (caddis flies), comprising a linear and bilateral series of filaments, lamellae and setae emerging from the abdominal surface. Their function is unclear.

lateral meristems Meristems giving rise to secondary tissue; the vascular cambium and cork cambium.

lateral plate See MESODERM.

lateral transfer of genes See HORIZONTAL TRANSFER/TRANSMISSION.

laterite See TROPICAL RAIN FOREST, DEFORESTATION.

latex Fluid product of several flowering plants, characteristically exuding from cut surfaces as a milky juice (e.g. in dandelions, lettuce). Contains several substances, including sugars, proteins, mineral salts, alkaloids, oils, caoutchouc, etc.; rapidly coagulates on exposure to air. Function not clearly understood, but may be concerned in nutrition and protection, as well as in healing wounds. Latex of several species is

(a)   (b)

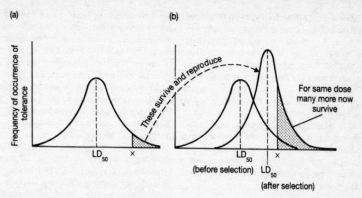

FIG. 98 *Artificial selection and the development of insecticide resistance.* (a) *Tolerance distribution before selection;* (b) *tolerance distribution after selection (unwittingly) by agricultural practices, showing differences between parental and progeny populations.*

collected and used in manufacture of several commercial products, the most important being rubber.

Latimeria See COELACANTHINI.

Laurussia One of two great Upper Carboniferous land masses, the other being GONDWANALAND, formed by the breakup of Pangaea. Comprised what are now North America, Greenland, Europe and Asia. Originally straddling the Equator, it gradually moved northwards by plate tectonics. For much of the Jurassic and Cretaceous most of Europe was covered by the Tethys and Turgai Seas, the latter only drying up to link Europe and Asia about 45 Myr BP. Separation of Gondwanaland and Laurussia was completed by early Cretaceous (130 Myr BP) with the result that much of the later radiation of dinosaurs took place in Laurussian continents but not in Gondwanaland.

laver General name given to the edible dried preparation made from the red alga *Porphyra* (Rhodophyta).

LD$_{50}$ (Of pesticide, insecticide, etc.) the dose required to kill 50% of the pest population. It is generally found that after a pesticide's application, owing to selection for greater pesticide resistance, the population's LD$_{50}$ increases and the dose required to achieve

the same effect as on initial application has to be raised. (See Fig. 98.)

LDL (low-density lipoprotein) Spherical complexes (density 1.019–1.063 g/cm^3) of phospholipid and specialized protein (apolipoproteins, or apoproteins) found in blood plasma. Because the apoproteins are amphipathic (one surface contains hydrophilic amino acid residues and the other hydrophobic ones) they can bind phospholipid by hydrophobic interactions. The core of the particle contains non-polar CHOLESTEROL esters (55%) and TRIGLYCERIDE (20%). LDL binds cell surface LDL receptors, where they cluster in coated pits (see Fig. 34, COATED VESICLES). As a result, cholesterol is provided to cells throughout the body and excess is delivered to the liver for recycling or excretion as bile acids. LDL receptors are synthesized in response to a fall in the free cholesterol within a cell. High density lipoproteins (HDLs) remove excess cholesterol from body cells and transport it to the liver for elimination. A high HDL:LDL blood ratio is indicative of reduced risk of CORONARY HEART DISEASE caused by atheroma (see SCLEROSIS). Endothelial cells of the artery oxidize LDLs as they are transported into the artery wall and this oxLDL may transform monocytes into macrophages (see SCLEROSIS (2)). See CHYLOMICRON, LIPOPROTEIN, RECEPTORS.

_l_-dopa   (_l_-dihydroxyphenylalanine)
Intermediate in the pathway from phenylalanine to noradrenaline and immediate
precursor of the brain neurotransmitter
DOPAMINE. Dopamine is deficient from the
caudate nuclei of the brain in patients with
Parkinson's disease. While neither oral nor
intravenous dopamine reaches the brain,
_l_-dopa does so and is of widespread clinical
use in treating parkinsonism, at least in the
short term. Transplantation of embryonic
adrenal medulla (production site for _l_-dopa)
into the caudate nuclei of sufferers from
parkinsonism is under clinical test but raises
ethical issues.

leaf Major photosynthesizing and transpiring organ of bryophytes and vascular plants;
those of the former are simpler, non-
vascular and not homologous with those of
the latter, which consist usually of a leaf
stalk (_petiole_), attached to stem by LEAF BASE,
the leaf blade typically lying flattened on
either side of the main vascular strand, or
midrib. The lamina is often lobed or toothed,
possibly reducing the mean distance water
travels from main veins to sites of evaporation, helping to cool the leaf. Hairy leaves
probably reduce insect damage, as may latex
channels, resin ducts, essential oils, tannins
and calcium oxalate crystals. Some leaves
produce hydrogen cyanide when damaged.
During evolution, leaves have become
modified to serve in reproduction, either
sexually (as SPOROPHYLLS), or asexually (as
producers of propagating buds). Leaves of
higher plants usually have a bud in their
axils and are important producers of GROWTH
SUBSTANCES. See COTYLEDON, CUTICLE, LEAF
BLADE, PHYLLOTAXIS.

leaf base (phyllopodium) Usually the
expanded portion of the leaf, attached to
the stem.

leaf blade (lamina) Thin, flattened and
flexible portion of a LEAF; major site of PHOTO-
SYNTHESIS and TRANSPIRATION, for which it is
admirably adapted. May be simple (comprising one piece) or compound (divided
into separate parts, leaflets, each attached
by a stalk to the petiole). A typical dicotyledon leaf presents a large surface area,
photosynthesizing cells being arranged
immediately below the upper epidermis
allowing maximum solar energy absorption. The blade is provided with a system
of supporting veins bringing water and
mineral nutrients and removing photosynthetic products. Has a system of intercellular spaces opening to the atmosphere
through STOMATA, permitting regulated
gaseous exchange and loss of water vapour;
its thinness reduces diffusion distances
for gases, keeping their concentration
gradients steep, so increasing diffusion
rates. Its large surface area and transpiration rate help cooling. Evergreen leaves generally have blades twice as thick as
deciduous ones, but neither can offer too
great a wind resistance. See KRANZ ANATOMY,
MESOPHYLL.

leaf gap Localized region in vascular cylinder of the stem immediately above point of
departure of LEAF TRACE (leaf trace bundle)
where the parenchyma rather than vascular
tissue is differentiated. In some plants where
there are several leaf traces (leaf trace
bundles) to a leaf, these are associated with
a single leaf gap. Leaf gaps are characteristic
of ferns, gymnosperms and flowering plants
(anthophytes).

leaflet   Leaf-like part of a compound
leaf.

leaf scar   Scar marking where a leaf was
formerly attached to the stem.

leaf sheath   Base of a modified leaf, forming a sheath around the stem (e.g. in grasses,
sedges).

leaf trace   (1) Vascular bundle extending
between vascular system of stem and leaf
base; where more than one occurs, each
constitutes a leaf trace. (2) Vascular supply
extending between vascular system of stem
and leaf base, consisting of one or more
vascular bundles, each known as a _leaf trace
bundle_.

learning   Acquisition by individual animal
of behaviour patterns, not just as an
expression of a maturation process but as a
direct response to changes experienced in
its environment. Various forms of learning

include CONDITIONING, HABITUATION and IMPRINTING. Insight learning, in which an animal uses a familiar object in a new way to solve a problem creatively, may be a form of instrumental conditioning: actions may come to be selected because their consequences form part of a route to obtaining a goal. Once clearly contrasted with INSTINCT, but it is now realized that these are not mutually exclusive categories of process. Learning ability is a clear example of adaptability, and of the adaptiveness of behaviour. Methods of monitoring brain activity, including those of individual neurons, in behaving animals, including humans, are being developed with increasing pace. See MELANOCYTE-STIMULATING HORMONE.

lecithin See PHOSPHOLIPID.

lectins Globulin proteins and glycoproteins cross-linking cell-surface carbohydrates and other antigens, often causing cell clumping (SEE AGGLUTINATION). May act as antigens themselves, as do the mitogenic lymphocyte-stimulators *phytohaemagglutinin* (*PHA*) and *concanavalin A* (*ConA*). In plants, may provide toxic properties of seeds; some are secreted extracellularly and recognized by receptors with LEUCINE-RICH REPEATS after binding to microbial signals. Are also involved in artificial CAPPING of specific cell-surface components. See MACROPHAGE.

lectotype Specimen or other component of original material, selected to serve as a nomenclatural type when no HOLOTYPE was designated at the time of publication, or as long as it is missing. See ISOTYPE, NEOTYPE, SYNTYPE.

leghaemoglobin See HAEMOGLOBIN.

legume (1) A pod; fruit of members of the Family Leguminoseae (peas, beans, clovers, vetches, gorse, etc.). Dry fruit formed from single carpel that liberates its seeds by splitting open along sutures into two parts. (2) Used by agriculturists for a particular group of fodder plants (clovers, alfalfa, etc.) belonging to the Leguminoseae. Important in crop rotation, having symbiotic nitrogen-fixing root nodule bacteria. See *RHIZOBIUM*, COMPLEMENTARY RESOURCES.

lemma Lower member of pair of BRACTS surrounding grass flower, enclosing not only the flower but also the other bract (PALEA).

lemurs See DERMOPTERA (flying lemurs); and PRIMATES.

lens Transparent, usually crystalline, biconvex structure in many types of eye, serving to focus light on to light-sensitive cells. In vertebrate eyes, constructed of numerous layers of fibres of the protein *crystallin*, arranged like layers of an onion, normally enclosed in connective tissue capsule and held in position by suspensory ligaments, absorbing potentially damaging light of wavelength < 400 nm. See ACCOMMODATION, EYE.

lentic (Of freshwaters) where there is no continuous flow of water, as in ponds, lakes. Compare LOTIC.

lenticel Small raised pore, usually elliptical, developing on woody stems in portions of the peridium in which CORK CAMBIUM is more active than elsewhere. Lenticel tissue possesses numerous intercellular spaces which begin forming during the development of the first peridium, and in the stem, typically appear below stomata. Their function is to allow for gaseous exchange between the interior of the stem and the atmosphere. They are also formed upon the surfaces of some fruits (e.g. apples, pears) and roots.

lentivirus Any of a group of 'slow' retroviruses inducing neurological impairment and chronic pneumonias. Include visna virus of sheep and HIV.

Lepidoptera Butterflies and moths. Endopterygote insect order. Two pairs of large membranous wings, with few crossveins, and covered with scales; larva a CATERPILLAR, usually herbivorous and sometimes a defoliator of economic importance. Adults feed on nectar using highly specialized and often coiled proboscis formed from grooved and interlocked maxillae. There is no simple way to distinguish all moths from all butter-flies, but in Europe any lepidopteran with club-tipped antennae, flying in the day, and capable of folding its wings vertically over its back, is a butterfly.

Lepidosauria Dominant subclass of living reptiles; possess DIAPSID skulls and overlapping horny scales covering the body. Includes the orders Rhyncocephalia (*Sphenodon*, the tuatara) and Squamata (lizards, snakes, amphisbaenids).

leptin Hormone produced by ADIPOSE TISSUE. Expression and function of leptin and its receptor induce satiety (feeding reduction) and are essential for avoidance of OBESITY, insulin-resistant DIABETES and infertility. It is thought that circulating leptin levels rise with increasing obesity, receptors in the HYPOTHALAMUS being its target. This signalling activates pathways which normally lead to a reduction in food intake, promote additional heat release from fat and alter insulin secretion and action. Falling leptin levels, however, seem to initiate excessive feeding, as well as reductions in both reproductive output and thyroid hormone levels (both energy-conserving). Leptin administration increases the frequency of action potentials in the anorexigenic POMC (proopiomelanocortin)-type neurons in the arcuate nucleus of the HYPOTHALAMUS, where its effects may be mediated by neuropeptide Y, injection of which causes typical symptoms of leptin deficiency, repetitive injection leading to obesity. But other leptin targets in the brain undoubtedly exist. In most obese individuals (human and rodent), OBESITY is leptin-resistant, although leptin has been shown to increase fatty acid consumption by its oxidation in skeletal muscle, through activation of AMP-activated protein kinase (AMPK). AMPK phosphorylates acetyl CoA carboxylase, inactivating it and preventing it from producing the lipid malonyl CoA, which inhibits transport of fatty acids into mitochondria for beta-oxidation (see FATTY ACID OXIDATION). So, leptin might operate by increasing mitochondrial uptake of fatty acids in insulin-dependent tissues, and although many obese individuals are leptin-resistant, others do lose weight in response to its administration. Clinically, it may also reduce fat levels in liver, heart and pancreatic beta cells (see DIABETES mellitus), which may have been its original selective advantage.

leptocephalus Oceanic larva of European eel. Migrates over 2,000 miles across Atlantic from breeding site near West Indies (Sargasso Sea) to European fresh waters, where it becomes adult. Transparent.

leptoma Thin area in wall of gymnophyte pollen, through which the pollen tube emerges.

leptome (leptoid) Photosynthate-conducting cell, approaching phloem in structure and function, found in bryophytes (esp. Bryales).

leptosporangiate (Of vascular plant sporangia) arising from a single parent cell and possessing a wall of one layer of cells. Spore production is low in comparison with EUSPORANGIATE type.

leptotene Stage in first prophase of MEIOSIS during which chromosomes are thin and attached at both ends to the nuclear membrane. DNA has already replicated but each chromosome appears as one thread, sister chromatids being closely apposed.

lethal loci Gene loci which, when homozygous for a certain allele, bring about death. Cuenot, in 1904, first detected such a locus in mice: the dominant yellow coat colour is determined by the allele A^Y, and homozygous embryos, A^YA^Y, do not survive.

leucine-rich-repeat (LRR) A repeat MOTIF in an intracellular DOMAIN of certain plant and animal RECEPTORS. In animals, LRR is found in the TOLL-LIKE RECEPTORS of the innate immune system. In plants, it is found in a large subfamily of proteins with intracellular serine/threonine kinase domains triggering signalling cascades, many of which are involved in defence against pathogenic microbes and some plant–

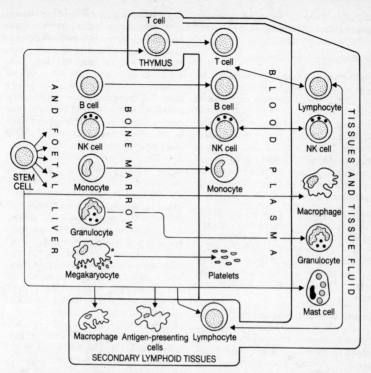

FIG. 99 *Origins of cells of the immune system, all derived from the haemopoietic stem cell. Granulocytes include neutrophils, basophils and eosinophils. NK = natural killer cell. For details, see separate entries and LYMPHOID TISSUE, MYELOID TISSUE. Monocytes which leave the blood at a site of inflammation transform into wandering macrophages.*

microbe symbioses (see PLANT DISEASE AND DEFENCES, RESISTANCE GENES).

leucine zipper One kind of DNA-BINDING PROTEIN which must dimerize before binding, doing so along the interface formed by several leucine residues spaced seven amino acids apart towards the C-terminal ends of each of the two polypeptides (the structural MOTIF). Some proto-oncogenes (e.g. C-*FOS*) encode leucine zipper proteins. Heterodimers of leucine zipper proteins increase the range of DNA sites which can be bound, and the strength of binding. See Fig. 117.

leucocyte (white blood cell) Nucleated blood corpuscle lacking haemoglobin.

Includes *granulocytes* (neutrophils, eosinophils, basophils), with granules in their cytoplasm, developing typically from MYELOID TISSUE, and *agranulocytes* (lymphocytes and monocytes), lacking cytoplasmic granules and developing typically from LYMPHOID TISSUE. Lymphocytes are of two kinds: T cells and B cells (see also IMMUNITY). The monocytes in tissue fluid are called *wandering macrophages*, and these and neutrophils are the major phagocytic leucocytes. Natural killer cells may also be of lymphoid origin. See specific cell types, and Fig. 99.

leucoplast Colourless plastid found in cells of plant tissues not normally exposed to light. Includes AMYLOPLASTS, storing starch, ELAIOPLASTS, storing oil, and ALEURO-PLASTS, storing protein.

leucosin See CHRYSOLAMINARIN.

leukaemia Malignant overproduction by MYELOID TISSUE of white blood cells, crowding out normal red cell- and platelet-producing lines (leading to poor blood clotting) and resulting in a lack of mature and normal white cells. Death can result not so much directly from the cancer as from these indirect effects. Treatment by X-rays and chemotherapy may result in partial or complete remission; myeloid transplants sometimes required.

leukotrienes (LTs) Lipid product of arachidonic acid removal from membrane phospholipids and its subsequent modification. They constrict bronchial airways, stimulate chemotaxis of white blood cells and mediate INFLAMMATION by increasing permeabilities of small blood vessels. See MAST CELL.

Leydig cells (interstitial cells) Groups of relatively scanty cells in the interstices between seminiferous tubules of the vertebrate testis responsible for steroid production (especially TESTOSTERONE). Have unusually large smooth endoplasmic reticulum. Testosterone output is synergistically enhanced by PROLACTIN and LUTEINIZING HORMONE. See MATURATION OF GERM CELLS.

L-form bacteria Bacteria lacking cell walls.

LH See LUTEINIZING HORMONE.

libriform fibre Xylem fibre having thick walls and greatly reduced pits.

lice See SIPHUNCULATA (sucking lice), MALLOPHAGA (biting lice), and PSOCOPTERA (book lice).

licensing A process which ensures that chromatin is available for a further round of DNA replication only after its passage through mitosis. See CELL CYCLE.

lichen A symbiotic association between a fungus (mycobiont) and a green alga (phycobiont), a cyanobacterium, or both, which develops into a unique morphological form quite distinct from either partner. Lichens are a biological, and not systematic, group. They are unique in that in many cases, although not all, the life form and behaviour differ from that of the isolated partners. The mycobionts do not usually occur free-living and appear to be responsible for the overall form of the thallus and fruiting structures, but there appears to be an interplay between both the mycobiont and phycobiont to produce the final form. The lichens are polyphyletic in origin. About 19% of all fungi and 42% of all members of the Ascomycota are lichenized. Lichenization also occurs in a few members of the Basidiomycota. The number of algal genera in lichenized associations is small with only 40 genera represented, 15 of these being members of the CYANOBACTERIA. About 90% of lichen thalli have as their phycobiont species of CHLOROPHYTA (green algae) (e.g. *Trebouxia*, *Pseudotrebouxia*, *Cephaleuros*, *Coccomyxa*, *Myremecia* and *Trentepohlia*). Cyanobacterial genera in lichen associations include *Calothrix*, *Nostoc*, *Gloeocapsa*, *Scytonema* and *Stigonema*. Most lichens have only one species of alga, although there are a few with two.

Morphologically, lichen thalli may be: (i) foliose, having several well-defined stratified layers including an upper cortex, an algal layer, a medulla and a lower cortex. They have two growth forms, either lobed and leaf-like and attached to a substratum by numerous rhizines or (less common) circular in outline and attached by a single central chord; (ii) crustose, having an upper cortex and a medulla in direct contact with a substratum via rhizoidal hyphae. They form a thin, flat crust on a substratum; (iii) squamulose, having a structure intermediate between a foliose and crustose thallus, comprising numerous small lobes or squamules. Internally, there are no lower cortex or rhizines; (iv) fruticose, having a radially symmetrical, cylindrical-to-flattened thallus in cross-section, hollow or solid, and either erect or pendulous. The outer cortex is quite thick, the algal layer thin and scattered, and neither rhizines nor a lower cortex are present (*Cladonia* has a squamulose thallus but a more conspicuous secondary fruticose thallus called a podetium). The fruticose thallus is the most complex and largest, and (v) gelatinous, having no stratified layers other than a thin upper and lower cortex and a medullary layer of loosely interwoven hyphae and scattered algae. When the algae are scattered at random the thallus is said to be homeomerous. In contrast, if the algae are distributed in a distinct layer below the upper cortex the thallus is said to be heteromerous. There are a few filamentous lichens where the algal filaments predominate (e.g. *Coenogonium*, *Cystocoleus*, *Racodium*).

Non-sexual reproduction occurs through: (i) fragmentation of the lichen thallus; as well as (ii) through formation of small or widespread areas of cortical breakdown (soralia), which contain minute powdery propagules called soredia. These are clusters of algal cells together with some fungal hyphae; (iii) through development of numerous small, simple or branched-coralloid, cortical papillae called isidia. The fungus may also reproduce asexually by production of conidia. The phycobiont reproduces mainly by cell division; zoospores or aplanospores have been reported very rarely. Sexual reproduction is confined to the mycobiont; structures formed are commonly ascocarps (apothecia or perithecia). Ascospores are formed in asci and discharged. On germination new lichen individuals are formed if the algal partner happens to be present, but in its absence the fungus dies. Where several dual propagules of the same or different lichens commence growing close together intermixing and formation of interspecific or intergeneric hybrids can occur.

In the lichen relationship, the fungus obtains organic carbon from the alga. Those with nitrogen-fixing blue-green algae (e.g. *Nostoc*) also obtain nitrogenous products. Lichen fungi form a close network of hyphae around the algal cells, and commonly form haustoria, which penetrate the algal cells. Other fungi form specialized organs called appresoria that lie along the surface of the algal cells penetrating them by means of specialized pegs. Growth of many lichens is slow (0.1–10 mm yr^{-1} increase in colony diameter). Some mature lichens may approach 4,500 years or more. Lichens dominate the flora in large areas of mountain and Arctic regions, where few other plants can exist. They occur on bare soil, rocks, tree-trunks and fences. One species, *Verrucaria serpuloides*, is a permanently submerged marine species found on rocky shores. Lichens play an important role in primary colonization of bare areas. In the Arctic, certain lichens are a valuable food source (e.g. Iceland moss, Reindeer moss). Others produce dyes (e.g. *Rocella* provides litmus), and are sources of medicines, poisons, cosmetics and perfumes. Many are sensitive to gases such as sulphur dioxide, probably due to the sensitivity of the phycobiont, and so can be used as indicators of air POLLUTION. They have been used to monitor fallout from nuclear tests.

life Complex physico-chemical systems whose two main peculiarities are (1) storage and replication of molecular information in the form of nucleic acid, and (2) the presence of (or in viruses perhaps merely the potential for) enzyme catalysis. Without enzyme catalysis a system is inert, not alive; however, such systems may still count as biological (e.g. all viruses away from their hosts). Other familiar properties of living systems such as nutrition, respiration, reproduction, excretion, sensitivity, locomotion, etc., are all dependent in some way upon their exhibiting the two above-mentioned properties.

Living systems also have an evolutionary history. Whatever the ORIGIN OF LIFE may have been, all *existing* life forms derive from living antecedents. The earliest living system would have been very different from any modern life form, particularly so in their genetic systems (modes of storage and implementation of molecular information).

life cycle Progressive series of changes undergone by an organism or a lineal succession of organisms from fertilization to death of the stage producing the gametes beginning an identical series of changes. As do phenotypes, they display adaptation to

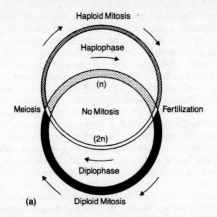

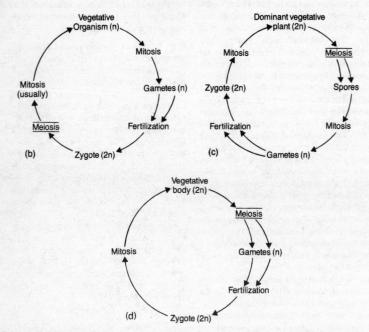

FIG. 100 (a) *Scheme indicating the various ways in which mitosis and meiosis can be included within a eukaryotic sexual* LIFE CYCLE. *Both fertilization and meiosis may or may not be followed by mitosis. The situation in which no mitosis at all occurs (the central route) is rarely if ever found: cell number would be increased solely by meiosis.* (b) *The typical haplontic life cycle (e.g. Mucor, Spirogyra, Chlamydomonas).* (c) *The typical haplodiplontic life cycle (e.g. mosses, liverworts, ferns, conifers, flowering plants). Spores may be produced homosporously or heterosporously.* (d) *The typical diplontic life cycle (e.g. Fucus, most animals).*

the external environment; and within a species there is often more than one way in which a generation may be completed, not least through variation in the modes of reproduction. In eukaryotes, this sometimes involves an ALTERNATION OF GENERA-TIONS (*metagenesis*) between sexual and asexual stages. The simplest form of life cycle (found in both prokaryotes and eukaryotes) is purely vegetative, involving repeated binary fission, budding, vegetative cell division or cell division coupled with fragmentation of the filament or colony. Next in complexity are ASEXUAL cycles (again found in both prokaryotes and eukaryotes), involving production of spores (e.g. aplanospores, conidia, zoospores) which do not engage in fertilization and were formed other than by meiosis.

With the evolution of eukaryotic sexual reproduction, the stage was set for the evolution of diploidy – probably when two haploid cells combined to form a diploid zygote. Presumably, this zygote divided by meiosis to restore the haploid condition; however, some zygotes divided mitotically instead, with meiosis occurring later. In animals, this delayed meiosis results in production of gametes. In eukaryotes sexual life cycles may be classified according to whether or not haploid mitosis and/or diploid mitosis occur (see Fig. 100). Such life cycles may either be (a) *haplontic*, in which only haploid cells divide mitotically, the diplophase being represented by a single nucleus (e.g. some green algae such as *Spirogyra*, *Chlamydomonas*; fungi such as *Mucor*); (b) *diplontic*, in which all haploid cells are the immediated products of meiosis (e.g. most animals; the brown alga *Fucus*), or (c) *haplodiplontic* (or *diplohaplontic*), in which both haploid and diploid mitoses occur. Haplophase and diplophase may be more or less equally prominent (e.g. the green alga *Ulva*), or one may be dominant (haplophase in mosses and liverworts; diplophase in ferns and seed plants). In most haplodiplontic cases (yeasts are somewhat variable) gametes are produced by mitosis, and it is a mitotic product of the zygote that undergoes meiosis.

It is usual with plants and algae to refer to the haplophase and diplophase as the *gametophyte* and *sporophyte* respectively. Multicellular haploid organisms that appear alternately with the diploid forms are found in plants, and among the PROTISTA in some brown, red and green algae, in two closely related genera of the CHYTRIDIOMYCOTA and in one or more other groups. Such organisms display alternation of generations. Even in mosses and ferns these phases are practically distinct individuals. The gametophyte has become increasingly restricted during evolution, represented in flowering plants by just the microspore, megaspore and their greatly reduced mitotic products. Although the development of gametophytic heterospory, begun in ferns, has been taken far further in flowering plants, it is the sporophyte that has become the dominant phase from ferns to seed plants, indicating a probably greater robustness and adaptability of the diploid plant in terrestrialization.

Most animal life cycles are diplontic; but in some forms (e.g. polychaetes) the products of meiosis may undergo mitosis to produce gametes. In most cases of MALE HAPLOIDY, sperm are formed mitotically, although in bees they are formed by unipolar meiosis. Life cycles of animal parasites often involve POLYEMBRYONY as well as two or more hosts; similarly some fungal parasites may require more than one host and produce several spore types (e.g. rust fungi), while life cycles of periodical cicadas (Hemiptera) may take as long as 17 years to complete, nymphs taking this long to reach maturity. See PARTHENOGENESIS.

life histories An organism's life history is more than just its LIFE CYCLE. As early as 1930, the statistician and biologist R.A. Fisher argued that demographic traits could be treated as aspects of an organism's phenotype, and that they could be explained as strategies for survival, coexistence and for competition in the environment as much or as little as could morphology, physiology and behaviour. Thus, longevity, senescence and resource allocation are but three of the components of an organism's life cycle which impinge upon its life history. See OPTIMALITY THEORY.

lifespan See AGEING, LIFE HISTORIES.

ligament Form of vertebrate CONNECTIVE TISSUE joining bone to bone. *Yellow elastic ligaments* consist primarily of elastic fibres and form relatively extensible ligaments joining vertebrae, and true vocal cords; *collagenous ligaments* by contrast consist largely of parallel bundles of collagen and resist extension.

ligase ENZYME catalysing condensation of two molecules and involving hydrolysis of ATP or another such triphosphate. DNA LIGASE is much used in GENE MANIPULATION, as well as forming part of DNA REPAIR MECHANISMS.

lignin Complex polymeric molecule composed of phenylpropanoid units associated with CELLULOSE (as *lignocellulose*) in CELL WALLS of sclerenchyma, xylem vessels and tracheids, making them strong, rigid and impermeable to water (see SUBERIN). After cellulose, lignin is the most abundant plant polymer, forming 20–30% of the wood of trees.

lignocellulose Major chemical component of wood. Valuable resource, not least of energy as when converted to methane or alcohol in techniques of BIOTECHNOLOGY. See CELLULOSE, LIGNIN.

ligule (1) Membranous outgrowth arising (a) from junction of leaf blade with leaf sheath in many grasses, (b) from base of leaves of certain lycopods. (2) Flattened corolla of ray flower in Compositae.

likelihood A quantity proportional to the probability of the data (or probability density), given specific values for all parameters in the model. The likelihood function provides a means of estimating the parameters of the model. Parameters associated with the global maximum of the likelihood function are 'maximum likelihood estimates'. 'Maximum likelihood' is an optimality criterion in phylogenetics (see OPTIMALITY THEORY).

Liliidae (Liliforae) A subclass of the LILIOPSIDA comprising about 50,000 species with more than 80% belonging to two families, the Orchidaceae (about 25,000 species to which can be added about 70,000 artificial hybrids made by growers throughout the world) and Liliaceae. Characteristically, flowers are showy (there are exceptions) with sepals that are petaloid. They have intensively exploited insect-pollination. A few members are arborescent; some have broad, net-veined leaves, and some have vessels throughout the shoot as well as in the root. Plants are often strongly mycotrophic. See INTRODUCTION.

Liliopsida (Liliidae) (monocotyledons, monocots) The smaller and less diverse group of flowering plants (MAGNOLIOPHYTA) comprising about 50,000 species occurring in a wide range of habitats. The Liliopsida are considered to comprise five subclasses: the Alismatidae, Arecidae, Commelinidae, Zingiberidae and LILIIDEAE. The monocots are predominately herbaceous plants. Less than 10% of all monocots are woody. Those that are woody characteristically have an unbranched (or sparingly branched) stem with a terminal crown of large leaves and mostly belong to a single family, the Arecaceae (Palmae). The difference in habit is partly a reflection of the complete absence of typical cambium in monocots. The majority of monocots are smaller plants but include many important food plants (e.g. cereals, fodder grasses, bananas) and ornamentals (e.g. orchids, grasses). Monocot seed germinates to form one cotyledon, or the embryo is sometimes undifferentiated. Leaves have mostly parallel venation and plants lack an intrafascicular cambium (i.e. vascular bundles 'closed'); no cambium of any sort is present. The vascular bundles of the stem are generally scattered or in two or more rings. Floral parts, when of definite number, are typically borne in sets of three or six, seldom four and almost never five (carpels often fewer). Pollen is typically uniaperturate or of uniaperturate-derived type. Monocot root systems are wholly adventitious. It is widely agreed that monocots are derived from primitive dicots and that the monocots must therefore follow rather than precede the dicots in any proper lineage sequence. See INTRODUCTION.

limb bud Circular bulge on the neurula surface during vertebrate embryonic development, the somatic mesenchyme cells of which will proliferate and differentiate to form the limb elements. A ridge along its anterior margin is a major signalling centre for the developing limb, keeping the mesenchyme dividing and generating the anterior–posterior limb axis (digit polarity).

limbic system The brain's 'emotional system'. Structures in the medial limb of the brain, often with looped and complex connections, all ultimately projecting into the HYPOTHALAMUS. Essential to memory. Analyses information and relays to the nucleus accumbens, the limbic part of the corpus striatum (see BASAL GANGLIA). The amygdala seems to provide affective (value and emotional) content to experience and is involved in conditioning the animal to its environment (or context). The septum links the amygdala to the hypothalamus. The hippocampus and the temporal lobe are major sites for long-term memory storage, the former receiving inputs from the inferior temporal region of the cortex. Its efferent pathway is the fornix, which in turn provides input to the hypothalamus. The parahippocampal gyrus, cingulate gyrus and underlying cortex of the hippocampal formation form what is called the *limbic lobe* of primitive cortex around the brainstem, linking the limbic system with the autonomic nervous system. Thus, the amygdala receives CATECHOLAMINE and SEROTONIN projections from the brainstem. See Fig. 101.

limiting factor Any independent variable, increase in whose value leads to increase in the value of a dependent variable. Ideally, values of other independent variables should be held constant while this relationship is examined. In plots of dependent against independent variables, the latter are *limiting* only while there is a linear or near-linear relationship to the plot. Thus in the plot of initial velocity of an enzyme reaction (V_o) against substrate concentration, *S*, the latter is limiting only until the enzyme begins to be saturated with substrate (see ENZYME). In ecology, the term applies to any variable factor of the environment whose particular level is at a given time limiting some activity of an organism or population of organisms; e.g. temperature may limit photosynthesis when other conditions would favour a higher rate; growth of planktonic diatoms may become limited by depletion of dissolved silica. The *principle of limiting factors* says: 'When a process is dependent on more than one factor, the rate of the overall process is limited by the factor in the shortest supply relative to the demand for it.'

limiting similarity See COMPETITIVE EXCLUSION PRINCIPLE.

limnology Study of fresh waters and their biota.

Limulus See MEROSTOMATA.

Lincoln index Equation used to estimate populations of mobile animals after a mark–release–recapture programme. A large number of animals are first captured, marked in a non-injurious way and released back to mix into the population. The ratio of marked to unmarked individuals obtained in a second capture can then be used to estimate population size, as follows:

$$N = (n_1 \times n_2)/n_3$$

where N = population estimate, n_1 = number originally marked and released, n_2 = number captured on second occasion, n_3 = number recaptured (i.e. marked individuals recaught). Assumptions include no immigration/emigration or selection (as in HARDY–WEINBERG THEOREM).

LINEs (*L1* elements) Long interspersed nuclear elements (compare SINES), targeting AT-rich regions of the host DNA, LINEs are TRANSPOSONS of the non-retroviral retrotransposon family. Rarely mobile, their repeat sequences contribute a significant proportion to the human genome and their reverse transcriptase activity is believed to be responsible for most reverse transcription in the human genome, including the retrotransposition of the non-autonomous SINES. Rare insertions occur into the gene

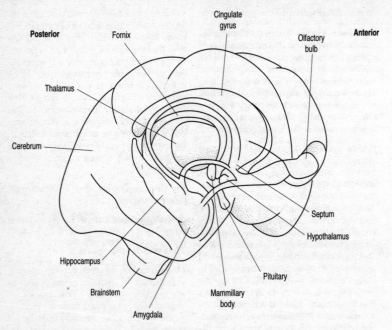

FIG. 101 *Anatomical components of the* LIMBIC SYSTEM.

encoding Factor VIII (see BLOOD CLOTTING), leading to one form of haemophilia. Similar elements occur in other mammals, insects and yeast mitochondria. Compare *Alu*.

lineage Evolutionary lineages are defined and delineated by the identification, first, of large-scale synapomorphies (shared derived features) shared by sister groups and representing states derived from the primitive features of all members of the CLADE; and then of successively more restrictive synapomorphies separating off smaller clades until, finally, the uniquely distinctive features of the group terminating the sequence (i.e. the autapomorphies) are identified. The resulting nested set of branchings reveals the phylogenies (i.e. lineages) involved.

linkage Two or more loci (see LOCUS), and their representative genes, are said to be linked if they occur on the same chromosome. Such loci normally occur in the same linear sequence on all homologous chromosomes, so that chromosomes form *linkage groups*.

Contrary to the second of MENDEL'S LAWS of inheritance, eukaryotic genes which are linked (forming part of a chromosome) will tend to pass together into nuclei produced by MEIOSIS and are not randomly assorted. This is very important, because whereas there is an equal statistical probability of all combinations of unlinked genes ending up in a given nucleus after meiosis (simply through the behaviour of non-homologous chromosome pairs), genes which are linked can only be prevented from passing together into nuclei by CROSSING-OVER, or by some chromosome MUTATION such as translocation, which separates them (see LINKAGE DISEQUILIBRIUM). The events of meiosis create new linkage groups and thereby new sets of phenotypic characters in organisms upon which SELECTION can act. Without linkage (i.e. without chromosomes) selection could never alter the probabilities with which characters appear together in different individuals of a population, and adaptation and evolution in so far as we understand them could not have occurred. Linkage makes possible the selection of *coadapted combinations* of alleles of different gene loci, which will tend to be transmitted together. Because rates of crossing-over between loci can be altered by selection (see MAP DISTANCE), adaptive combinations of genes, and of their resulting character combinations, can be preserved rather than disrupted by meiosis. See CHROMOSOME MAPPING, RECOMBINATION, SEX LINKAGE, SUPERGENE.

linkage disequilibrium Occurrence in a population of two or more loci so tightly linked that few of the theoretically possible gene combinations are found. E.g., if loci A and B are represented by just two alleles each in the population (*A* and *a*; *B* and *b*) yet only individuals with genotypes *AABB* and *aabb* are found in any frequency, then there is linkage disequilibrium between the two loci. Without crossing-over, and without selection for and against any of the genotypes, one expects linkage equilibrium ($AB + ab = Ab + aB$) to be achieved eventually. See LINKAGE, SUPERGENE.

linkage map See CHROMOSOME MAP.

linkage mapping Genetic mapping. See CHROMOSOME MAPPING.

Linnaeus, Carolus (adj. Linnean) Swedish naturalist (1707–88), physician and originator of modern system of BINOMIAL NOMENCLATURE. In his *Systema Naturae Fundamenta Botanica* he assigned to every known product of nature (including minerals) its place in one great system of classification; indeed, discovery of the natural method of such classification was for him the naturalist's task. He thought he had achieved this for species and genera but only partially for classes and orders. Nature was for him imbued with divine *economy*: nothing happened unnecessarily, nor did any created form ever become extinct, although fossil evidence clearly forced him to consider this possibility. He was a little perplexed by the discovery of hybridization between species, and in later works considered that new

species might arise in this way; but fundamentally his thinking was by today's standards conservative and non-evolutionary. His chief critic was BUFFON. Today, his rules of nomenclature and his taxonomic categories result in a hierarchy without biological meaning. The current approach is to find ways of constructing branching, tree-like, evolutionary LINEAGES in which origins, rather than phenotypes, are the deciding factor in taxonomy. See GREAT CHAIN OF BEING.

linnean hierarchy Arrangements of organisms into TAXA; forming nested sets, with KINGDOM the highest taxonomic category and SPECIES the lowest.

linoleic acid ESSENTIAL FATTY ACID of mammals, making up to 20% of the total fatty acid content in their triglycerides.

lipase Any of a class of enzymes catalysing hydrolysis of esters of fatty acids and other lipids, often regulated by CYCLIC AMP; they convert triglycerides (fats) into fatty acids, monoglycerides and glycerol. Present in pancreatic juice and succus entericus of vertebrates. *Lipoprotein lipase* on surfaces of adipose tissue cells is largely responsible for hydrolysis of triglycerides within CHYLOMICRONS in blood plasma.

lipid A See LIPOPOLYSACCHARIDE.

lipids Wide and heterogeneous assemblage of organic compounds having in common their solubility in organic solvents such as alcohol, benzene, diethyl ether, etc. Include fats, waxes and oils, steroids and sterols (e.g. cholesterol), glycolipids, phospholipids, terpenes, fat-soluble vitamins, prostaglandins, carotenes and chlorophylls. Tests for lipids include dissolving in alcohol and adding a few drops of distilled water, when a creamy emulsion of lipid in water forms on the surface; in histochemistry, Sudan IV and oil red O stain lipids red. See Fig. 102 and MICELLE.

lipogenesis FATTY ACID synthesis.

lipolysis Hydrolysis of lipids, particularly of triglycerides. See ADIPOSE TISSUE, LIPASE.

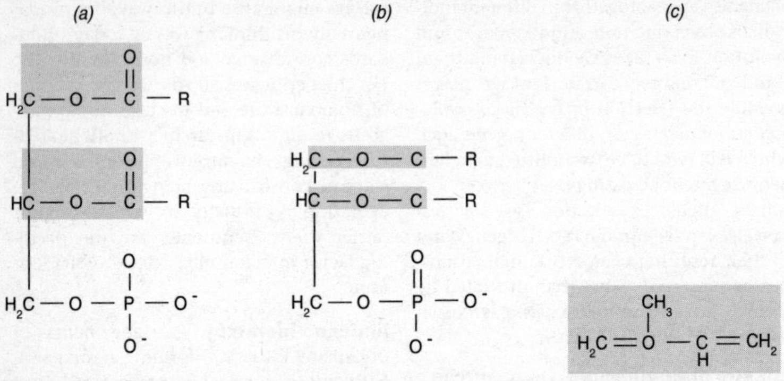

FIG. 102 *Chemical bonds in* LIPIDS: (a) *the ester linkage as found in lipids of bacteria and eukaryotes;* (b) *the ether linkage of lipids from archaeons; and* (c), *isoprene, from which the hydrophobic side chains (R) of archaeal lipids derive. By contrast, bacterial and eukaryotic lipids have fatty acid R-groups.*

lipopolysaccharide The dominant feature of the surfaces of Gram-negative bacteria (see GRAM'S STAIN), these molecules have in common a lipid region (Lipid A), a core polysaccharide, and an O-specific chain which usually distinguishes the bacterial serotype. *Salmonella typhimurium* and *Escherichia coli* have so many of these molecules in their surface membrane that the cell surface is effectively covered by specific O-chains. The lipid A complex is responsible for the toxicity of ENDOTOXINS while the polysaccharide complex makes the molecule water-soluble. See Fig. 75, ANTIMICROBIAL PEPTIDES, MACROPHAGE, TOLL-LIKE RECEPTOR.

lipoprotein Micellar complex of protein and lipids. *Apolipoproteins*, protein components of these complexes, transport otherwise insoluble lecithin, triglyceride and cholesterol in vertebrate blood plasma. The transport lipoprotein particle comprises an outer lipid bilayer with specific conjugated protein components, within which the transported molecules are either free or esterified to bilayer fatty acids. There is a complex turnover and interaction of these particles, CHYLOMICRONS, low-density lipoprotein (LDL), high-density (HDL) and very high-density (VHDL) forms increasing relative to the amount of protein in the particle. Regular exercise (physical conditioning) raises blood HDL, which seems to counteract impact of CHOLESTEROL in heart disease, decrease triglyceride levels and improve lung function. See Fig. 103, LIPASE, ADIPOSE TISSUE, LIVER, COATED PIT.

liposome Artificially produced spherical lipid bilayers, 25 nm or more in diameter, which can be induced by segregate out of aqueous media. Their selective permeabilities to organic solutes and the subsequent reactions which can occur within them have led some to propose a similar structure as a cell prototype. See COACERVATE, CHYLOMICRON, HUMAN GENE THERAPY.

list Cellulose extension of cell wall in some armoured dinoflagellates, usually extending out from the cingulum and/or sulcus.

Lister, Joseph (1827–1912) English surgeon and Professor of Surgery at Glasgow. Motivated by the work of PASTEUR, he appreciated that the putrefaction of wounds was a bacterial fermentation. His use of carbolic (phenol) as a DISINFECTANT and ANTISEPTIC in surgery greatly reduced postoperative infection rates.

lithophyte Plant found growing on rocks.

lithosere SERE originating on exposed rock surface.

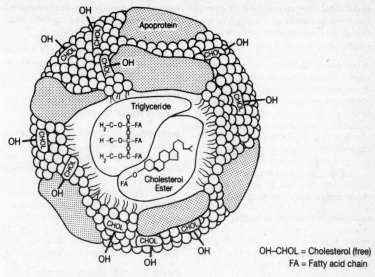

FIG. 103 *A human plasma* LIPOPROTEIN. *Diameter depends on the class of lipoprotein, but excepting larger chylomicrons would be ~40 nm.*

lithotroph Autotrophic bacterium (e.g. nitrifying or sulphur bacterium) obtaining its energy from oxidation of inorganic substances (e.g. sulphur, iron) by inorganic oxidants: the terminal hydrogen acceptor in respiration is always inorganic.

littoral Inhabiting the bottom of sea or lake, near the shore. From shore to 200 m in the sea; to 6–10 m in lakes, depending upon the extent of rooted vegetation.

liver Gland, usually endodermal in origin and arising as a diverticulum of gut. Livers in different phyla are not homologous. In vertebrates its main glandular function is production of BILE, which leaves via the hepatic duct for storage in the gall bladder. Has a wide capability enzymatically, much of which is inducible. Structural unit is the *lobule*, a roughly hexagonal block of cuboidal cells (*hepatocytes*) supplied at its corners with products of digestion via factors of the hepatic portal vein, and with oxygen by branches of the hepatic artery. Blood leaves the lobule at its centre through a vessel leading to the hepatic vein. Bile leaves through *bile canaliculi*, also at corners of lobules. As a homeostatic organ it is second to none, involved in production of GLYCOGEN from monosaccharides and fat and its subsequent storage; absorption and release of blood glucose, absorption of CHYLOMICRONS, deamination of amino acids, urea production, GLUCONEOGENESIS, raising of blood temperature, production of prothrombin, fibrinogen, albumin and other plasma proteins, storage of vitamins A, D, E and K, red blood cell breakdown, detoxification of some poisons and steroid hormone conversion to cholesterol.

liverworts See HEPATICOPSIDA.

LMPs (low molecular mass proteins) See PROTEASOME.

locule Compartment, cavity or chamber. In ASCOMYCOTA, chambers containing ASCI; in flowering plants (ANTHOPHYTA), cavity of the ovary where ovules occur; in diatoms (BACILLARIOPHYTA), a chamber within the frustule having a constricted opening on one side and a VELUM on the opposite side.

loculicidal (Bot.) Describing dehiscence of multilocular capsule by longitudinal splitting along a dorsal suture (midrib) of each carpel; e.g. iris. Compare SEPTICIDAL.

locus (gene locus) Position on homologous chromosomes occupied, normally throughout a species population, by those genes which determine the state of a particular phenotypic character (see GENE), e.g. a locus for eye colour in *Drosophila*. This position may be determined by CHROMOSOME MAPPING, various occupants of a locus in a given population being referred to as ALLELES. Relative positions of loci within an individual's cells may be altered by some chromosome MUTATIONS. See LINKAGE.

lodicules Reduced perianth of grass flowers; two small scale-like structures below ovary which at time of flowering swell up, forcing open enclosing bracts (pales), exposing stamens and pistil.

log phase (logarithmic phase) See GROWTH CURVES.

lomasome Membranous evagination of the plasmalemma of a fungal cell or hypha, occurring singly or in groups and situated between the rest of the plasmalemma and the wall material. Consists of membranous tubules, vesicles or parallel sheets lying in a matrix. May play a role in normal development of either the plasmalemma or wall.

lomentum Type of leguminous fruit, constricted between seeds and breaking into one-seeded portions when ripe; e.g. in bird's foot trefoil.

long-day plants Plants that flower when exposed to dark periods less than a critical length and therefore flowering primarily during the summer. See PHOTOPERIODISM, SHORT-DAY PLANTS, DAY-NEUTRAL PLANTS, PHYTOCHROME.

long-germ Mode of DEVELOPMENT in higher insects (e.g. Diptera) in which all body segment primordia are established at the blastoderm stage, segments becoming visible along the entire embryo soon after gastrulation. Compare SHORT-GERM.

long shoot Main branch in some gymnophytes (gymnosperms) bearing short dwarf shoots; e.g. in *Pinus*, *Ginkgo*.

long-term potentiation One of the best-studied models of memory at the cellular level. See SYNAPTIC PLASTICITY.

looped domain See CHROMOSOME.

loop of Henle See KIDNEY.

lophophore Hollow ring of ciliary feeding tentacles surrounding the mouth and (strictly) containing an extension of the coelom, as in Ectoprocta, Brachiopoda, Phoronidea and enteropneusts. All except the last of these *lophophorate* animals appear on current evidence to be protostomes, the lophophores of enteropneusts apparently being convergent. Term also applied more loosely to any ring of oral tentacles, as in some polychaete worms and entoproctans.

lorica Envelope surrounding some flagellated (e.g. *Trachelomonas*, *Dinobryon*) and ciliate (e.g. *Cothurnia*) protists, fitting loosely enough to allow the cell to move fairly freely – unlike shells and tests. Generally proteinaceous, of glycosaminoglycan (generally described as 'chitin', 'pseudochitin' or 'tectin') or cellulose.

loss-of-function mutation See MUTATION.

lotic Of freshwaters where there is continuous flow of water, e.g. streams, rivers. Compare LENTIC.

low molecular mass proteins (LMPs) See PROTEASOME.

LUCA The last universal common ancestor of existing taxa.

luciferase, luciferin See BIOLUMINESCENCE.

Ludwig effect See entry on NATURAL SELECTION.

lumbar vertebrae Bones of the lower back region, lacking rib attachments and situated between thoracic and sacral vertebrae.

lumen (1) Cavity within tube (e.g. within blood vessel, gut) or within sac. (2) Cavity within cell wall of plant cell, from which the protoplast has been lost.

lung Sac-like organ of gaseous exchange, invariably with moist inner surface. (1) In vertebrates, they arise as a diverticulum of the pharynx (see GAS BLADDER) and were present in fish prior to the rise of amphibians, serving probably as an adaptation to drought and/or poorly oxygenated water (see DIPNOI). Generally paired, and in most higher tetrapods internally subdivided into bronchi, bronchioles and alveoli, they lie surrounded by coelomic membranes that in mammals will later form fluid-lubricated pleural cavities separating lungs from thorax (see MEDIASTINUM). Lungs are relatively small in birds, where AIR SACS are the major respiratory surfaces. Lungs lack muscles of their own and are ventilated by rib muscles of the trunk and by the diaphragm (in mammals). Only in birds does air actually circulate through lungs; elsewhere air is tidal and terminates in richly vascularized alveoli, providing in all ~140 m² for gaseous exchange. The lung's (pulmonary) blood circulation has evolved in tetrapods towards the DOUBLE CIRCULATION of birds and mammals. Stretch receptors in a lung's connective tissue walls feed back to RESPIRATORY CENTRES controlling ventilation. *Asthma* is a chronic inflammatory problem, causing overproduction of mucus and reduction in diameters of smaller bronchi and bronchioles (through contraction of surrounding muscle) in response to an allergen. *Bronchitis* is an inflammatory response involving hypertrophy of glands and goblet cells lining the bronchi. Atmospheric pollution may be a contributory factor, excessive mucus being produced as sputum. Cigarette smoking, which produces a cocktail of inflammatory substances, is the major cause of chronic bronchitis, lasting from months to years (see NICOTINE). In *emphysema*, the alveolar walls disintegrate, resulting in large air spaces which do not close during exhalation. Reduced surface area for gaseous exchange and loss of lung elasticity, making exhalation very hard work, are effects. Major causes are long-term irritation of the air passages, as from smoke and other air pollution. Cigarette smoke deactivates the enzyme α-1-antitrypsin, which would otherwise inhibit the proteases and elastases which attack the alveoli. (2) In molluscs, the lung is most advanced in pulmonate gastropods where it is a specialization of the mantle cavity and opens to the air via a valved pneumostome. Muscles ventilate the chamber.

lung book Organ gaseous exchange in some air-breathing arachnids (e.g. in scorpions and, in conjunction with tracheae, some spiders). Consists of leaf-like projections sunk into pits opening through a narrow pneumostome to the air. Four pairs occur in scorpions, two pairs or one pair in some spiders, none in others. Gaseous exchange is by diffusion.

lungfish See DIPNOI.

lutein A xanthophyll pigment.

luteinizing hormone (LH, interstitial cell-stimulating hormone, ICSH) A glycoprotein GONADOTROPHIN, produced by the anterior PITUITARY under influence from hypothalamic releasing factor (GnRF). Its main function in males is to stimulate interstitial cells of the testis to produce TESTOSTERONE, which in turn shuts off GnRF release. In females, brings about ovulation (see MENSTRUAL CYCLE), and transforms the ruptured Graafian follicle into a corpus luteum with subsequent progesterone production.

Lycopodiaceae Club moss family. An ancient family containing about 400 or more species of terrestrial, epilithic or epiphytic plants with horizontal stems present or absent. When present they are mainly protostelic (actino- or plectostelic in some species) and can be on the surface of a substratum or subterranean or form stolons. Erect shoots are simple or branched, usually leafy. Lateral shoots may be present or absent. Sporangia are solitary, adaxial near the leaf base or axillary; sporophylls are unmodified, photosynthetic to much modified non-photosynthetic, reduced and aggregated in strobili. The plants are mostly homosporous and spores vary from being trilete, thick walled and pitted, to small-grooved, rugulate or reticulate. Gametophytes are subterranean and non-photosynthetic or surficial and photosynthetic. See SELAGINELLACEAE, ISOËTACEAE, LYCOPODIOPHYTA.

Lycopodiophyta (Lycophyta (Lycopodiopsida)) Club mosses. Plants that have an evolutionary lineage extending back to the Devonian period. The progenitors of the Lycopodiophyta were almost certainly Zosterophyllophyta. Lycopodiophytes split very early on into two major groups. The first remained herbaceous, the second the lepidodendroids (or tree ferns) became woody and tree-like and were the dominant plants of coal-forming forests of the Carboniferous period (e.g. Lepidodendron). Tree ferns became extinct in the Permian period (c. 280 Myr BP). Living lycopodiophytes (lycopods) are divided into three families: the Lycopodiaceae, Selaginellaceae and Isoëtaceae. In general plants are small and evergreen, with upright or trailing stems that bear numerous small leaves. Sporangia are borne singly in axils of the leaves. Sporophylls occur in groups at intervals along the stem forming terminal cones. They are either homosporous (e.g. *Lycopodium*) with small mycotrophic prothalli wholly or partly subterranean or heterosporous (e.g. *Selaginella*), with reduced prothalli remaining largely enclosed by the spore wall. See LIFE CYCLE, LYCOPODIACEAE, SELAGINELLACEAE, ISOËTACEAE.

lymph Clear fluid within vessels of LYMPHATIC SYSTEM and derived from TISSUE FLUID and resembling it in composition, except in those lymph vessels between the gut and venous system after a meal, when it is generally 'milky' due to presence of CHYLOMICRONS. Since capillaries are impermeable to them, proteins secreted by tissues into the tissue fluid tend to gain access to the blood indirectly via the lymph. Lymphocytes enter as it passes through LYMPH NODES, but macrophages can enter from tissue fluid.

lymphatic (lymphoid) system System of LYMPH-containing vessels and the organs producing and accumulating lymphocytes

linked by these vessels in vertebrates (see Fig. 104). The 'second' circulatory system in these animals. Comprises a blindly-ending meshwork of highly permeable endothelial *lymph capillaries* (resembling blood capillaries, but having non-return valves) permeating most body tissues (not the nervous system) and joining to form larger vessels (usually not larger than 2–3 mm diameter), resembling veins, with non-return valves but thinner connective tissue walls. Finally joins the venous system, usually near the heart. TISSUE FLUID drains into lymph capillaries and is slightly concentrated by loss of some water and electrolytes as it passes along the system, under the same forces as achieve venous return. Most of the large *lymph trunks* enter the left *thoracic duct* which opens into the venous system close to or at the left subclavian vein, but some also enter from a right thoracic duct near the right subclavian vein. See LYMPH NODE, LYMPHOID TISSUE.

lymph heart Enlarged part of lymphatic vessel with muscular pulsating wall. Present in many vertebrates, but not in birds or mammals.

lymph nodes Ovoid structures on lymph vessels of mammals and to a lesser extent of birds (but not other vertebrates); up to 25 mm long in humans, comprising connective tissue framework of capsule and inner extensions (trabeculae) supporting successive internal parenchyma tissues: (a) the cortex of B CELLS, (b) the paracortex of T CELLS and (c) the medulla of T and B cells. Afferent lymphatic vessels join the capsule, where valves ensure that lymph passes progressively towards the medulla, where an efferent lymph vessel rejoins the lymphatic circulation. As it passes through the node, lymph is processed by fixed macrophages of the RETICULO-ENDOTHELIAL SYSTEM lining the sinuses, while lymphocytes respond to antigens from adjacent ANTIGEN-PRESENTING CELLS. Each node has its own blood supply, and both B cells and T cells respond by clonal expansion in a way dependent upon the type of antigenic stimulation. Compare SPLEEN. See LYMPHOMA.

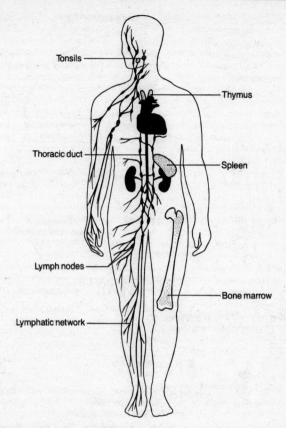

FIG. 104 *Diagram of the human* LYMPHATIC SYSTEM *and its associated lymphoid organs and tissues (labelled).*
Adapted from N. Staines, J. Brostoff and K. James, *Introducing Immunology*, Gower Medical Publishing, 1985, courtesy of the authors and publishers.

lymphocyte One of two kinds of vertebrate white blood cell (LEUCOCYTE), confined to the blood system. Most are *small lymphocytes* (diameter 6–8 μm), agranular with high nuclear : cytoplasmic ratio; but *large lymphocytes* are larger (diameter 8–10 μm) and granular, with low nuclear : cytoplasmic ratio and with azurophilic granules (staining with azure dyes) in the cytoplasm; some may serve as NATURAL KILLER CELLS. Small lymphocytes (B CELLS and T CELLS) are involved in both humoral and cell-mediated IMMUNITY.

Lymphocytes can move by amoeboid locomotion but most are non-phagocytic. They develop in LYMPHOID TISSUE. See Fig. 99.

lymphoid tissue Vertebrate tissue (e.g. bursa of Fabricius in birds, THYMUS in mammals) in which LYMPHOCYTES, DENDRITIC CELLS and NATURAL KILLER CELLS develop. Lymphocytes are produced in *primary lymphoid tissue* (thymus, embryonic liver, adult bone marrow) and migrate to *secondary lymphoid tissue* (spleen, lymph nodes, unencapsulated

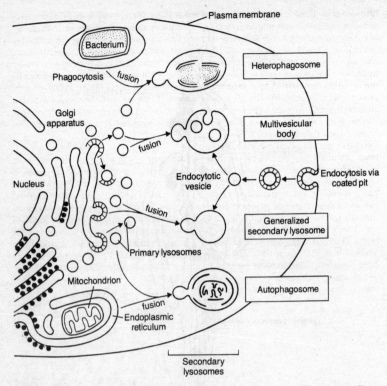

FIG. 105 *Diagram showing the source of primary LYSOSOMES and their fusions with other organelles to form secondary lysosomes. Ladder-like rings indicate coated vesicles.*

lymphoid regions of gut submucosa, respiratory and urogenital regions), where ANTIGEN-PRESENTING CELLS and mature T CELLS and B CELLS occur. The kidney is a major secondary lymphoid organ in lower vertebrates. Tissues in animals without close chordate affinities may nonetheless form regional aggregates of phagocytes which are loosely termed lymphoid. See MYELOID TISSUE.

lymphokines Any soluble factors produced by lymphocytes; those produced by T CELLS often termed INTERLEUKINS. Release of lymphokines (notably gamma-INTERFERON, IFN$_\gamma$) by T cells activates macrophages to destroy intracellular parasites and some cancer cells. NATURAL KILLER CELLS also release lymphokines.

lymphoma Any tumour composed of lymph tissue. *Hodgkin's disease* is a malignant lymphoma of reticulo-endothelial cells in lymph nodes and other lymphoid tissues.

lysigenous (Of secretory cavities in plants) originating by dissolution of secreting cells; e.g. oil-containing cavities in orange peel. Compare SCHIZOGENOUS.

lysis Destruction of cells through damage to or rupture of plasma membrane, e.g. by osmotic shock. In bacteria, may be brought about by infection by BACTERIOPHAGE. See HYDROLYSIS.

lysogenic bacteria, lysogeny See BACTERIOPHAGE.

lysosomes Diverse membrane-bound vacuolar organelles, forming integral part of eukaryotic intracellular digestive system (see Fig. 105). Contain large variety of hydrolytic enzymes (e.g. acid phosphatases and hydrolases) which operate at pH ~5 within the organelle maintained by a membrane proton pump. Substances for digestion entering the cell's geography by endocytosis (heterophagy) become part of the lysosomal system when they fuse with *primary lysosomes* (0.5 μm diameter), which have not been involved in hydrolytic activity, budded from the GOLGI APPARATUS. The vacuoles (*heterophagosomes*) formed by this fusion are sites of digestion analogous to the gut, and similar digestion products diffuse through specific channels or carriers into the cytosol through the lysosomal membrane. Undigested remains may persist in 'residual bodies' for some time. Autophagy often involves organelles being wrapped up in a membranous vacuole, probably originating from smooth endoplasmic reticulum, and its subsequent fusion with primary lysosomes to form *autophagosomes*. These are characteristic of cells involved in hormone-related developmental reorganization. Autophagosomes and heterophagosomes are types of *secondary lysosome*, and may be considerably larger than primary lysosomes. Some extracellular enzyme secretion results from fusion of primary lysosomes directly with the plasma membrane. See Fig. 74.

lysozyme Class of ENZYME catalysing hydrolysis of (glycosaminoglycan) walls of Gram-positive bacteria, leading to rupture and death of remaining protoplast. Secreted by skin and mucous membranes and found in tears, saliva and other body fluids of mammals; also in egg white. Provides one innate immune response to bacteria. Lysozyme was among the first proteins whose three-dimensional molecular structure was elucidated.

M

macrocyst See MYCETOZOA.

macroevolution The origins and evolution of higher taxa (i.e. above species). See EVOLUTION.

macrogamete (megagamete) See GAMETE, ANISOGAMY.

macromolecule Molecule of very high molecular weight, characteristic of biological systems, e.g. proteins, nucleic acids, polysaccharides, and complexes of these.

macronucleus (meganucleus) See NUCLEUS.

macronutrients Substances required in large amounts for plant growth; e.g. nitrates, phosphates, sulphates, calcium and magnesium.

macrophage Phagocytic cell of vertebrate connective tissue but not typically of the blood itself. They are crucial effector cells of the innate immune system, releasing a broad array of cytokines which integrate both this and the adaptive immune response (see IMMUNITY). Included here are *wandering macrophages* derived from monocytes (see LEUCOCYTE), and more static macrophages (*histiocytes*) dispersed throughout connective tissue but capable of migrating towards a site of infection. Phagocytes of the RETICULO-ENDOTHELIAL SYSTEM are rather more specialized macrophages, but all are important as ANTIGEN-PRESENTING CELLS, capable of collecting immune intelligence about pathogens and activating a sensitized helper T CELL. Macrophages endocytose particles or soluble glucose-conjugated compounds that are bound by the broad-specificity mannose receptor (a LECTIN); but they also have a receptor for lipopolysaccharide, a component of Gram-negative bacterial outer membranes, which stimulates synthesis of such CYTOKINES as interleukins-1,6 and 12, and TUMOUR NECROSIS FACTOR – all of which induce production of ACUTE PHASE PROTEINS. Cell debris at the site of a wound or infection includes self-antigens, heat-shock protein 60 and mitochondrial and nuclear material which can bind to the TOLL-LIKE RECEPTORS TLR4 and TLR2 and activate the macrophage to ingest dying cells. Macrophages have one form of nicotinic acetylcholine receptor on their cell surface (see NICOTINE), with implications for INFLAMMATION. See Fig. 106.

macrosporangium See MEGASPORANGIUM.

macrospore See MEGASPORE.

macrosporophyll See MEGASPOROPHYLL.

macula (1) See FOVEA. (2) Small receptor in walls of the utricle and saccule of tetrapod VESTIBULAR APPARATUS providing information about position of the head in relation to gravity when the animal is moving or at rest. Lying in planes perpendicular to one another, they comprise an epithelium containing support cells and HAIR CELLS whose stereocilia are embedded in a gelatinous layer on which lie *otoliths* of calcium carbonate which move the gelatinous layer, so pulling on the stereocilia. Impulses generated travel along CRANIAL NERVE VIII to the MEDULLA. Otoliths have an inertia which makes them stay at relative rest or movement as the body respectively moves or comes to rest. See STATOCYST.

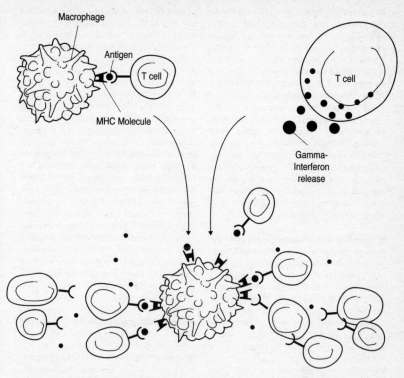

FIG. 106a *Chemical messengers (cytokines) produced by the immune system, such as gamma-interferon, improve the immune response to antigens. Here, gamma-interferon increases the level of antibody being presented by* MACROPHAGES *to T cells, enhancing their activity.*

magnetic resonance imaging (MRI) Spectroscopic technique in clinical diagnosis (formerly called *nuclear magnetic resonance imaging*), in which protons in tissues respond to a pulse of radio waves (~300 MHz) while they are being magnetized (field ~10 tesla). Because protons have spin, each can adopt an orientation either parallel to, or antiparallel to, the field. Since these states differ in energy, when photons are beamed through the material changes in the immediate environment of each nucleus can be detected as changes in resonance frequency. These can be colour-imaged to build up a two- or three-dimensional picture of cellular chemistry on the screen. The technique is used to detect tumours and artery-clogging plaques (see SCLEROSIS), changes in brain chemistry, etc. It is not recommended for pregnant women and cannot be used on people with artificial pacemakers or metal joints.

Magnoliophyta (angiosperms, or flowering plants) Division of the plant kingdom that forms a natural group. Seed plants whose ovules are enclosed in a carpel, and with seeds borne within fruits. Vegetatively diverse; characterized by flowers; pollination usually by insects, but other modes (e.g. anemophily) have evolved in several lines. Gametophytes are much reduced. The male gametophyte is initiated by a pollen grain (microspore) comprising two non-motile gamete nuclei and a tube cell nucleus; each associated with a little cytoplasm in the pollen tube. The female gametophyte develops entirely within the wall of the megaspore, which at maturity is a large cell containing eight nuclei, the embryo sac. Characteristic double fertilization. The Magnoliophyta comprise two groups that have been referred to as the dicotyledons (dicots) and monocotyledons (monocots), or equivalent names with latinized endings; however, they are not considered as formal taxa. They are now considered as the subclasses MAGNOLIOPSIDA and LILIOPSIDA, these names being recommended in the International Code of Botanical Nomenclature. See Fig. 107 and INTRODUCTION to dictionary.

Magnoliopsida (Magnoliidae) (dicotyledons, dicots)

The larger and more diverse of the two groups (classes) of flowering plants (MAGNOLIOPHYTA), comprising about 165,000 species found growing in a wide range of habitats. The Magnoliopsida are considered to comprise six subclasses, the MAGNOLIIDAE, HAMAMELIDAE, CARYOPHYLLIDAE, DILLENIIDAE, ROSIDAE and ASTERIDAE. About half the species are more or less woody-stemmed and many are trees, usually with a deliquescently branched trunk (with a central axis melting away irregularly into a series of smaller branches). Other examples include potatoes, cabbages and such ornamentals as roses, primulas, clematis and snapdragons. A primary root system developed from a radicle is common; others have an adventitious root system. Upon germination the seed produces two cotyledons (seldom one, three or four or the embryo seldom undifferentiated). True leaves are mostly net-veined. An (intra)fascicular cambium is usually present and the vascular bundles of the stem are usually borne in a ring that encloses a pith. The floral parts, when a definite number, typically are borne in sets of five, less often four, seldom three (carpels often fewer). Pollen is typically triaperturate (with three apertures) or of triaperturate-derived type, except in a few of the more archaic families. See INTRODUCTION.

maize (corn) *Zea mays*, a domesticated CEREAL and the second largest crop worldwide, showing extreme morphological divergence from its apparent wild ancestor, teosinte (*Zea mays* ssp. *parviglumis*, Family POACEAE), apparently brought about by changes in just five gene regulatory regions. Economically, the most important plant to carry out C4 PHOTOSYNTHESIS. Has a huge genome ($\sim 3 \times 10^9$ base pairs) and seems to be a segmental allotetraploid: DNA sequence data suggests that one of its two ancestral diploids was more closely related to modern SORGHUM than to the other ancestor, the polyploid event occurring at least 11.4 Myr BP. Although domestication might have been expected to reduce genetic variation in maize, it is actually more genetically varied than a lot of wild plants – many teosinte alleles apparently contributing to its gene pool. Its genome exhibits much vigorous TRANSPOSABLE ELEMENT activity. May form the focus of proposed Crop Genome Project, in which genomes of several cereals would be mapped. See STAPLES.

major histocompatibility complex (MHC) Mammalian gene complex of several highly polymorphic linked loci encoding glycoproteins involved in many aspects of immunological recognition, both between lymphoid cells and between lymphocytes and antigen-presenting cells. Initially detected as the region encoding antigens involved in graft rejection, the most important region in humans (the HLA system) is located on the sixth chromosome pair; in mice the comparable region (the H-2 complex) is located on the seventeenth chromosome pair. *Class I* MHC molecules occur in low levels on most body cells, but abundantly on T CELLS; *Class II* MHC molecules are constitutive of lymphoid dendritic and B CELL membranes and inducible on macrophages. Both Class I and Class II molecules initially insert into the endoplasmic reticulum. There, the former bind intracellular antigenic peptide fragments prior to export to the cell surface via the Golgi apparatus; but the latter enter a type of Golgi vesicle destined to fuse with endocytic vesicles and bind antigenic peptides of extracellular

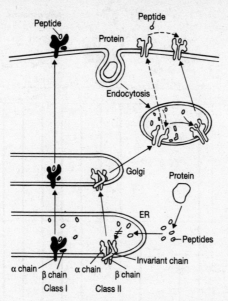

FIG. 106b *Two pathways for antigen presentation. The Class Iα and β₂m chains (shaded) are inserted into the endoplasmic reticulum (ER) where they associate non-covalently and bind peptide. Class I-peptide complexes are exported via the Golgi apparatus to the cell surface – most of the peptides being cleaved from proteins in the cytosol. Class IIα and β chains are inserted into the ER but associate with each other and the invariant chain which blocks peptide-binding until it is removed after processing in the Golgi apparatus and reach the cell surface, where they bind extracellular peptides for recognition by appropriate T cells.*

origin prior to export to the cell surface for T cell surveillance. Several genes in the MHC appear to encode proteins involved in supply of peptide fragments to Class I and Class II molecules. See ACCESSORY MOLECULES, IMMUNITY.

As for the evolutionary origin of the MHC, large-scale MHC sequencing efforts are expected to provide the most substantial database yet for testing molecular evolutionary hypotheses. Sequences encoding the peptide-binding regions are currently the focus of much population genetic research, e.g. testing NEUTRALISM. The traditional explanation for MHC diversity is HETEROZYGOUS ADVANTAGE, heterozygotes presenting a wider range of foreign peptides to T cells. But the class I *HLA-G* product expressed on foetal trophoblast cells seems to protect the foetus from maternal NK cell attack in the decidua (maternal tolerance to foetus). Surveys also reveal that a significant excess of human couples with a history of spontaneous abortions share antigens for one or more *HLA* loci: when mother and foetus differ at *HLA* loci, there is an apparent requirement for an immune response to occur before implantation can take place. Alternative explanations invoke SEXUAL SELECTION (mate choice) or parasite resistance (SEE SEXUAL REPRODUCTION). Some of the best examples of strong associations between MHC loci and these factors come from birds, e.g. when selection reduces diversity through advantage conferred by a specific haplotype. However, MHC sequence *HLA-B*5301*, a product of micro-recombination, is thought to confer human resistance to malaria. Resistance to hepatitis B virus in West African humans has been

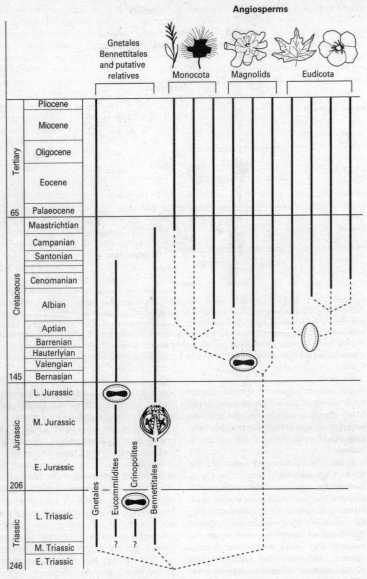

Angiosperms

FIG. 107 *A simplified phylogenetic tree showing the relative stratigraphic ranges of selected magnoliophyte groups and likely close relatives. The bennettitalian flower-like reproductive structure illustrated is* WILLIAMSONIELLA. *Monocolpate pollen is the basic magnoliophyte condition retained by liliopsids and magnoliopsids.*

shown to involve heterozygous advantage at *HLA* loci, enhancing diversity. The major genetic locus (*IDDM1*) implicated in development of type I DIABETES mellitus lies in the MHC on chromosome 6p21, accounting for ~35% of observed familial clustering. See PROTEASOME, HUMAN GENOME PROJECT for SNPS.

Malacostraca Largest subclass of CRUSTACEA, including prawns and shrimps (Amphipoda), crabs and lobsters (Decapoda) and woodlice (Isopoda). Thorax with eight segments, covered by carapace (not in isopods or amphipods); abdomen usually with six. Abdominal appendages on all segments, usually functioning in swimming. Thoracic appendages serve for locomotion and sometimes for feeding, and often bear gills. Compound eyes stalked.

malaria A widespread and debilitating human disease, caused by a protozoan parasite (*Plasmodium* spp., see APICOMPLEXA) injected by mosquitoes of the genus *Anopheles*. Of the many forms of *Plasmodium*, *P. falciparum* alone infects 5–10% of the human population and is the most commonly fatal, killing two million people annually. Parasite life cycle includes asexual multiplicative stages in human liver and erythrocytes, fertilization in the mosquito gut lumen, with subsequent asexual multiplication in its wall. Work on a vaccine has been held up by ANTIGENIC VARIATION of the parasite. Rather than attempt to eradicate *Plasmodium*, which is widespread in wild game, we probably ought to develop vaccines to boost human immunity. Mosquitoes are now resistant to many insecticides, as is the parasite to prophylactic drugs (e.g. chloroquine, mefloquine); but drug combinations interfering with plasmodial folic acid metabolism have been effective, although global drug-resistance is now a problem. However, use of OLIGONUCLEOTIDES in treatment may side-step this problem. The antimalarial agent triclosan is potent against *in vivo* blood stages of *P. falciparum*, targeting Fab I, an enzyme component of the parasite's fatty acid biosynthesis pathway. Short-lasting oil on surfaces of mosquito breeding pools, use of mosquito nets impregnated with permethrin at night and mosquito-repellant creams are all of some help. In Africa, ~250 children die of malaria every hour. Visitors to malarial areas need to seek advice and take appropriate precautions. Natural resistance to malaria is conferred by one haemoglobin variant (see SICKLE-CELL ANAEMIA) and probably also by MHC sequence *HLA-B*5301*, a product of microrecombination. During 2002, a highly organized draft of 278 million base pairs of the mosquito *Anopheles gambiae* and the genome sequence of *P. falciparum* were published. See DNA PROBE, SICKLE-CELL ANAEMIA, THALASSAEMIA.

male haploidy Arrhenotoky. Condition in which males arise from unfertilized eggs, females from fertilized ones. Males have no father; females have two parents, but only three grandparents. Males are cytologically haploid, producing gametes by mitosis (see LIFE CYCLE). Universal in hymenopteran insects; found also in some scale insects (Coccoidea) and mites, among other animal groups. Form of PARTHENOGENESIS. See FEMALE HAPLOIDY, HAMILTON'S RULE, THELYTOKY.

male sterility See CYTOPLASMIC MALE STERILITY.

malleus See EAR OSSICLES.

Mallophaga Biting lice; bird lice, feather lice. Exopterygotan insects; ectoparasitic on birds and occasionally on mammals. Flattened, with tarsal claws and reduced eyes. Cannot pierce skin (unlike SIPHUNCULATA), but bite small particles of FEATHERS or hair. Include hen louse, *Menopon*.

malpighian corpuscle In vertebrate KIDNEY, a glomerulus and its associated Bowman's capsule. See NEPHRON.

malpighian layer Innermost layer of epidermis of vertebrate skin, next to dermis. An active region of MITOSIS, non-stem cells being pushed to the surface, ageing and becoming cornified. Its cells often contain the pigment melanin, darkening the skin as a protection against ultraviolet light.

malpighian tubules Long, blind-ending, slender tubes lying in the haemocoeles of most myriapods and terrestrial insects and arachnids (where independently evolved), but not Onychophora. Involved in excretion and osmoregulation. In insects, open into intestine near junction of hindgut and midgut. Water, nitrogenous waste and salts enter tubule lumen, probably by active secretion. The hindgut then selectively reabsorbs water and useful metabolites, leaving a precipitated concentrate rich in uric acid salts which then leaves with the faeces. See COXAL GLANDS.

maltose Disaccharide sugar composed of two glucose molecules bonded in an α[1,4]glycosidic linkage. Formed by the enzymatic degradation of starch by AMYLASES. Occurs e.g. in germinating seeds such as barley, and during digestion in animals.

Mammalia Vertebrate class originating in Early or Middle Jurassic; present forms distinguished by presence of body hair and MILK secretion, generally via mammary glands. Homeothermic; with a diaphragm used in lung ventilation. Lower JAW composed of a single pair of bones (dentaries), unlike therapsid reptilian ancestors, presumed also to have been homeotherms. DENTITION heterodont. Only the left systemic arch remains (see AORTA, DORSAL). Three bones form the EAR OSSICLES in each middle ear. Mostly of small size until the Tertiary. Three extant subclasses: oviparous PROTOTHERIA (monotremes, e.g. spiny anteaters, duck-billed platypus); viviparous METATHERIA (marsupials, e.g. opossums, wombats, koalas, kangaroos) and EUTHERIA (placentals, in 16 extant orders). Mammals from the Mesozoic are known principally from their teeth, and these often show convergence. A recently discovered 120 Myr-old fossil (*Jeholodens jenkinsi*) from Liaoning, China, is remarkable in being complete. Classified as a triconodont, it exhibits a chimaera-like combination of reptile-like pelvic girdle and back leg articulation alongside more mobile and typically mammalian front leg articulation. Several extinct orders. One possible phylogeny is illustrated in Fig. 108a. Debate surrounds the origin(s) of the TRIBOSPHENIC DENTITION thought to distinguish the marsupial-placental clade from that containing the monotremes. It is not yet clear from fossils whether it evolved before or after the break-up of Pangaea 160 Myr ago, or whether it evolved more than once, by convergence. Application of BAYESIAN STATISTICS to DNA sequences suggests that placentals can be arranged into four major groups, one of which groups together rodents, lagomorphs and primates. It also suggests that placentals arose little more than 100 Myr ago during the break-up of GONDWANA. See MAMMAL-LIKE REPTILES.

mammal-like reptiles (Therapsida) Reptilian subclass radiating from the mid-Permian and becoming extinct at the end of the Triassic. Sluggish herbivores and active carnivores, the latter with elbow and knee swung towards (tucked into) the body allowing rapid four-footed gait. The herbivores may have formed herds. Small Mesozoic therapsids gave rise to the earliest mammals, contemporaries of the ruling archosaurs for several million years. See CYNODONTS, REPTILIA.

mammary glands Milk-producing glands, peculiar to the ventral surfaces of female viviparous mammals (similar structures are present in monotremes). Rudimentary in males, unless abnormal hormonally. Develop from epidermis and resemble sweat glands, consisting of a branching series of ducts terminating in secretory alveoli during pregnancy. See MILK, PROLACTIN.

mandible (1) Of vertebrates, the lower JAW. (2) Of insects, crustacea and myriapods, one of the first pair of MOUTHPARTS ('jaws'), usually involved in biting and crushing food and heavily sclerotized. See MAXILLA.

mandibular arch See VISCERAL ARCHES.

mannan Polysaccharide, composed of mannose units and occurring in cell walls of some algae and yeasts.

mannitol A 6-carbon sugar alcohol universally present in the brown algae (PHAEOPHYTA) in which D-mannitol is the accumulation product of photosynthesis

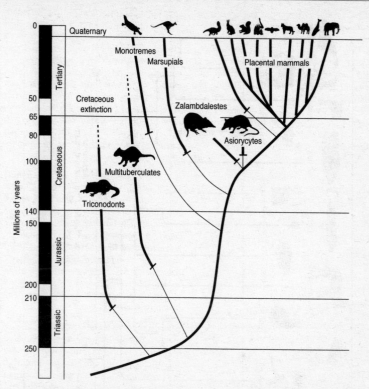

FIG. 108a *Postulated evolutionary branching events for some mammalian groups. Thick lines above bars indicate age ranges of groups, as inferred from fossil record; other thick and thin lines indicate hypothesized branching times. Multituberculate position is controversial: the group might be connected to the node leading to monotremes, marsupials and placentals.*

(up to about 25% of the dry weight of some *Laminaria* species in the autumn).

mannoglycerate Saccharide storage product in some red algae (RHODOPHYTA).

mantle (Bot.) (1) A dense mass of fungal hyphae surrounding a root. (2) That part of the diatom valve that bends away at 90°. (Zool.) Surface layer of visceral hump of MOL-LUSCA, secreting shell. Flaps of the mantle enclose the mantle cavity. Similar structure is found in brachiopods.

map distance A measure of the frequency of CROSSING-OVER between two linked chromosome marker loci, equal to the percentage recombination in meiotic products, provided there are no double or multiple cross-overs between the markers. The further apart geographically two loci are on a chromosome the more will their apparent distance (as derived from cross-over values) be foreshortened by double and multiple cross-overs between them. Map distances between loci are reduced (underestimated) to the extent that such cross-overs occur between them, so that long map distances are best derived from summation of short distances, using intervening markers. Suppression of all crossing-over between two

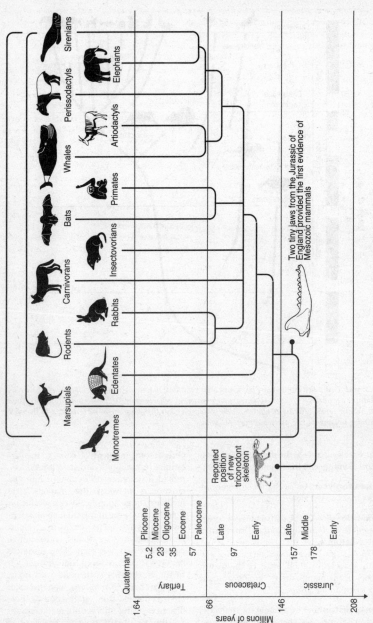

FIG. 108b *Putative timescale of mammalian evolution. The new triconodont fossil described in the text has been placed near the base of the crown lineage,* MAMMALIA.

loci (see SUPPRESSOR MUTATION) would effectively render them indistinguishable from a single complex locus. This might have selective advantages. See SUPERGENE.

MAPKs (MAP-kinases) Mitogen-activated protein kinases. MAPKs must be phosphorylated on both a serine and a threonine residue (separated by a single amino acid) to become fully activated, events achieved by a MAP-kinase-kinase (MAPKK) – which is MEK in the mammalian Ras signalling pathway (see Fig. 151). MAPKK is itself activated through phosphorylation by a MAP-kinase-kinase-kinase (MAPKKK), which is Raf in that same pathway – itself activated by activated Ras. See NEUROTROPHINS.

marginal meristem MERISTEM located along margin of leaf primordium and forming the blade.

marginal placentation Attachment of ovules to carpel margin.

marginal value theorem A proposed theoretical model relating to the decision of a predator to leave (move from) patches of prey which it has depleted. It states that the predator should leave all patches at the same rate of prey extinction, namely the maximum average overall rate for that environment as a whole.

marker See GENETIC MARKER.

marsupial See METATHERIA.

marsupium Pouch of many marsupials and spiny anteater (*Echidna*, PROTOTHERIA). Fold of skin supported by epipubic bone of pelvic girdle, forming pouch containing mammary glands or similar structures, into which newborn (or eggs in echidnas) are placed. Young marsupials attach there to teats, or lick milk in the case of teatless *Echidna*.

masseter muscle Muscle running from inside the zygomatic arch and inserting on the outside and rear of the mammalian lower jaw, pulling jaws forward and up (protraction) and assisting in side-to-side motions. Compare TEMPORALIS MUSCLE.

mass extinction See EXTINCTION.

mass flow Hypothesis for explaining TRANSLOCATION of material in phloem of vascular plants.

mast cell Granular LEUCOCYTE derived from MYELOID TISSUE, often associated with mucosal epithelia. When not in connective tissue, dependent upon T CELLS for reproduction. Granules in cytoplasm contain SEROTONIN, HEPARIN, HISTAMINE and TUMOUR NECROSIS FACTOR-α; released when allergen crosslinks the IgE molecules bound to plasma membrane. Involved in many allergies, but also in immunity to parasites. See BASOPHIL, EOSINOPHIL.

mastigoneme See FLAGELLUM.

Mastigophora Often employed as a class of PROTOZOA, but including both holophytic, holozoic and facultative forms. Usually refers to flagellated protozoans, both *zoomastigophorans* such as trypanosomes and *phytomastigophorans* such as euglenoids. Close affinities of flagellated and pseudopodial protozoans have resulted in some authorities uniting them in the Class Sarcomastigophora. Likely that all these are grades rather than clades. See FLAGELLUM.

masting Production, in some years, of especially large crops of seeds by most conspecific trees (e.g. European beech, *Fagus sylvatica*) and shrubs in the same geographical area, with a dearth of seeds produced in the years in between. Most often observed in tree species that suffer from generally high intensities of seed predation. Especially significant is that the chances of escaping seed predation are typically higher in mast years than in other years. The individual seed predators are satiated in mast years, and the populations of predators cannot increase in abundance rapidly enough to exploit the glut. By the time they have increased (usually the following year), glut has given way to famine. The years of seed famine are essential for the recovery of the trees as production of a mast crop makes great demands on the trees' internal resources. European beech do this.

mastoid process Part of mammalian

auditory capsule (periotic bone) containing air spaces communicating with middle ear.

maternal effect (m. influence) Instances in which some aspect of phenotype of normal sexually produced offspring is controlled more by the genotype of maternal parent than by its own. Distribution of cytoplasmic molecules in the egg, under the control of maternal genes, is one notable example in that it may have a marked effect upon CLEAVAGE and later development (see CYCLINS, *STRING* gene, *OSKAR* gene). For instance, during oogenesis of the *Drosophila* egg, maternal mRNA products of the *BICOID* GENE are retained at the anterior pole of the egg, and the eventual protein translation product disperses some way along the egg, forming the kind of gradient required of a positional signal in the POSITIONAL INFORMATION theory of development. See AGEING, GERM PLASM, GRANDCHILDLESS, MATERNAL INHERITANCE.

maternal gene Term sometimes used specifically for a gene responsible for a MATERNAL EFFECT.

maternal inheritance Characters inherited through the female line only. Not all genes are transmitted via nuclear chromosomes (see cpDNA, mtDNA, PLASMIDS). Evolution of anisogamy means that the female parent may contribute more genetically to its offspring than does the male parent. See MATERNAL EFFECT, CYTOPLASMIC INHERITANCE.

mating type In e.g. algae and fungi, used to designate a particular genotype with respect to compatibility in sexual reproduction. Gametes are identical in appearance and referred to as plus (+) and minus (–), rather than as male and female. In the budding yeast *Saccharomyces cereviseae*, a single locus (*MAT*) determines whether a haploid cell is of a mating type **a** or α. Each new generation switches its mating type, and it turns out that this involves a form of GENE CONVERSION in which unexpressed copies (so-called *cassettes*) of the two alternative mating type alleles (i.e. **a** and α) lie one on either side of the expressed form and that the expressed form is degraded each generation and replaced by the alternative allele, which persists as a flanker (see CONJUGATION and MITOCHONDRION). A short *cis*-acting enhancer achieves this switch by stimulating genetic recombination across the whole left arm of chromosome III, catalysed by the enzyme HO endonuclease. In the fission yeast *Schizosaccharomyces pombe*, cells of the two mating types (P and M) differ only in their mating potential and respond to pheromone of the opposite type by extending a conjugation tube towards the source. RECEPTORS for the pheromones **a**-factor and α-factor in *S. cereviseae* and P-factor and M-factor in *S. pombe* all belong to the RHODOPSIN family of receptors. See CELL DIVISION, HETEROTHALLISM, HOMEOBOX, SEX.

maturation of germ cells Processes normally taken to include the cell division and differentiation involved in the production of functional gametes from their germ mother cells.

During *oogenesis* in humans, oocyte development is arrested at first meiotic prophase in the newborn female (see CYTOSTATIC FACTOR), most oocytes being surrounded by follicle cells, some already organized in developing follicles; but these degenerate before puberty. At puberty, FOLLICLE STIMULATING HORMONE restarts follicle development, the surrounding thecal cells secreting oestrogens which reduce FSH production but increase the number of FSH receptors on the thecal cells, so that as FSH level drops only that follicle with most receptors can bind the remaining FSH; the others die. Follicle cells contribute small molecules to the oocyte via gap junctions, enabling it to synthesize macromolecules; but CYCLIC AMP is also transferred and inhibits resumption of meiosis by phosphorylating key oocyte proteins (see MATURATION PROMOTING FACTOR). The mid-cycle surge of pituitary LUTEINIZING HORMONE disrupts the gap junctions causing a drop in cAMP and onset of the second meiotic division arrested at second meiotic metaphase, in which state the oocyte may be fertilized. The follicle enlarges and ruptures in response to LH and internal prostaglandin, releasing the secondary oocyte (see Fig. 109a).

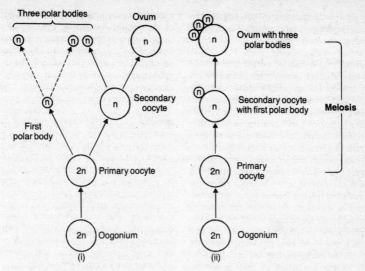

FIG. 109a *Diagram illustrating vertebrate oogenesis (i) with polar bodies shown distinct from oocyte; (ii) with polar bodies remaining attached to oocyte, as normally occurs.*

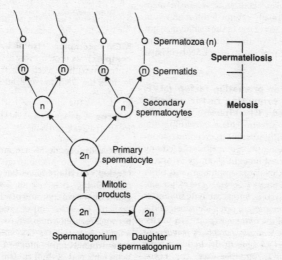

FIG. 109b *Diagram illustrating vertebrate spermatogenesis.*

Spermatogenesis in the TESTIS (Fig. 109b) is also under pituitary control, LH (= ICSH) being the more important hormone, but acting synergistically with FSH. ICSH causes LEYDIG CELLS to release testosterone, some of which probably stimulates the germinal epithelium of the seminifrous tubules to divide. ICSH also affects the SERTOLI CELLS (see ACTIVIN, INHIBINS), influencing the rate at which spermatozoa are released into their tubule lumen. Spermatogonia remain attached to each other after mitosis, as do their primary and secondary spermatocytes and spermatids (forming a *syncytium*). Meiosis takes about 24 days to complete in man and most sperm differentiation (*spermiogenesis*, or *spermateliosis*) occurs after meiosis is completed, spermatids losing excess cytoplasm and developing an ACROSOME and, from one of the two centrioles, a flagellum – all the while surrounded and protected from immune attack by their Sertoli cell (gametes are antigenically unlike body cells). The testis is an immunoprivileged site, evading the immune system. Spermiogenesis takes about nine weeks in man. Sperm normally cannot fertilize an ovum until they have been in the epididymis, nor until they have undergone CAPACITATION in the female reproductive tract. See MENSTRUAL CYCLE.

maturation promoting factor (MPF, M-phase promoting factor, growth-associated HI kinase) A mitosis-inducing protein kinase governing the G2-M transition in the CELL CYCLE and responsible for resumption of meiosis (i.e. ending meiotic arrest brought about by CYTOSTATIC FACTOR) in ovulated amphibian and other vertebrate oocytes. Several of its major subunits are CYCLINS, but its catalytic subunit is the phosphoprotein pp34, identified as the product of the *cdc2* gene (p34^{cdc2}) in the fission yeast *Schizosaccharomyces pombe* and of the *CDC28* gene in the budding yeast *Saccharomyces cerevisiae* (see CDC GENES, YEASTS). Probably ubiquitous in eukaryotes, p34^{cdc2} has an ATP-binding domain and is inhibited by phosphorylation of a tyrosine residue in that domain by *wee1* kinase, among other tyrosine kinases. MPF is acti-

vated by the tyrosine phosphatase product of the (fission yeast) *cdc25* gene, when it triggers chromosome condensation, breakdown of the nuclear lamina and formation of the mitotic spindle (see Fig. 22 in the CELL CYCLE entry). Some of these events may result directly from MPF-activation, others from subsequent regulatory CASCADES.

maxilla (1) One of the DERMAL BONES of the vertebrate upper jaw (and of the face in man) carrying all the upper teeth, except incisors. (2) Sometimes used for the whole vertebrate upper jaw. See PREMAXILLA. (3) One of a pair of MOUTHPARTS in insects, crustaceans and myriapods, lying beneath the mandibles and articulating with the head capsule by a ball-and-socket joint. Sclerotized. In insects, bears a jointed sensory palp. In crustaceans, tends to be a phyllopodium, and may serve in feeding and gaseous exchange (see BIRAMOUS APPENDAGES).

maxillule Paired crustacean mouthpart, lying behind the mandibles and in front of the maxillae. Usually a phyllopodium (see BIRAMOUS APPENDAGES), passing food to the mouth.

MCM complex (minichromosome complex) Six proteins whose interactions with chromatin as a cell exits from mitosis licenses the DNA for replication. See DNA REPLICATION.

meatus A passage; e.g. external auditory meatus (see EAR, EXTERNAL).

mechanoreceptor See RECEPTOR.

Meckel's cartilage Paired bar of cartilage, one forming each side of lower jaw of gnathostome embryo, and of adult elasmobranch. In most other vertebrates it becomes reduced, and partly ossified, as the *articular bone* (see EAR OSSICLES), which in non-mammals forms hinge of lower jaw. Represents part of third VISCERAL ARCH.

meconium Contents of mammalian foetal intestine; derived from glands discharging into gut, and from swallowing of amniotic fluid during process of excretion.

median Situated in or towards the plane dividing a bilaterally symmetrical organism or organ into right and left halves.

mediastinum Space between the two pleural cavities containing the heart in its pericardium, aorta, trachea, oesophagus, thymus, etc.

medulla Central part of an organ. (Bot.) Central core of usually parenchymatous tissue in stems where vascular tissue is in cylindrical form; functions in food storage. May also occur in some roots where the central tissue develops into parenchyma instead of xylem. Also refers to innermost region of the thallus in lichens and in some brown and red algae. (Zool.) (1) Central part of an animal organ, typically where outer part is termed the *cortex*. See ADRENAL GLAND, mammalian KIDNEY, LYMPH NODE. (2) Abbreviation of MEDULLA OBLONGATA.

medulla oblongata (medulla) Most posterior part of vertebrate brain. See BRAINSTEM and Fig. 18.

medullary ray Thin vertical plate of parenchyma tissue, one to several cells wide, running radially through the STELE. May be either *primary*, passing from pith (medulla) to cortex between primary vascular bundles, or *secondary*, formed from cambium during secondary thickening, ending blindly in secondary xylem and phloem. Since these latter have no connection with pith (medulla), they are sometimes termed *vascular* (*phloem* or *xylem*) rays. The function of medullary rays is in storage and radial translocation of synthesized (organic) materials.

medullated nerve fibre A myelinated NERVE FIBRE.

medusa Free-swimming form of the CNIDARIA. Shaped like bell or umbrella, swimming by rhythmic contractions of circular muscle in the rim and/or sub-umbrella, producing a jet-propulsion effect through the water and involving concentrations of nerve cells (ganglia) and formation of nerve tracts or rings (two in jellyfish). Relatively thick mesogloea. Produced by budding, they themselves are sexual. Absent from life cycles of corals, sea anemones and some hydrozoans (e.g. *Hydra*).

megagamete See GAMETE, ANISOGAMY.

megagametophyte In heterosporous plants, the female gametophyte; located within the ovule of seed plants.

megakaryocyte Large, highly polyploid bone marrow cell giving rise to PLATELETS by a kind of pinching-off, or cellular autotomy.

meganucleus (macronucleus) See NUCLEUS.

megaphyll Type of leaf possessing a branched system of veins and with a LEAF TRACE generally associated with one or more LEAF GAPS in STELE of stem. Usually associated with siphonostelic vascular system in stem. Characteristic of ferns and seed plants. Compare MICROPHYLL.

megasporangium (macrosporangium) SPORANGIUM in which MEGASPORES are formed; i.e. a meiosporangium of heterosporous plants, producing usually one to four megaspores. In seed plants, an OVULE.

megaspore (macrospore) Larger of two kinds of spore (meiospore) produced by heterosporous ferns; the first cell of the female gametophyte generation of these and of seed plants. It becomes the EMBRYO SAC in flowering plants (ANTHOPHYTA).

megaspore mother cell Diploid cell (meiosporocyte) in which MEIOSIS will occur, contained within the megasporangium (within nucellus of ovule); produces one or more megaspores. See DOUBLE FERTILIZATION.

megasporophyll (macrosporophyll) Leaf or modified leaf, bearing MEGASPORANGIA. In flowering plants, the carpel. Compare MICROSPOROPHYLL.

meiosis (reduction division) Process whereby a nucleus divides by two divisions into four nuclei, each containing half the original number of chromosomes, in most cases forming a genetically non-uniform haploid set (see Fig. 110). A necessary aspect of eukaryotic sexual reproduction, for without it fertilization would usually double the chromosome number every generation.

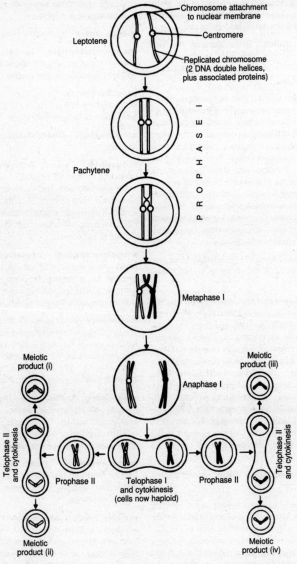

FIG. 110 *The diagram shows the behaviour of one pair of homologous chromosomes during* MEIOSIS. *One cross-over results in a chiasma, visible at metaphase I. From this stage onwards the two chromosomes are shown black or white to indicate origins of segments.*

Meiosis ensures that all gamete nuclei from diploid parents contain a haploid set of chromosomes. It also ensures, in sexually outbred populations at least, wide GENETIC VARIATION between offspring. This results from genetic RECOMBINATION, both random and non-random.

The first meiotic division is initiated by DNA replication (S phase), so that each chromosome comes to comprise two sister chromatids, as in mitosis; but at the start of prophase the sister chromatids of each chromosome remain tightly together, their early appearance (*leptotene*) giving the impression of unreplicated chromosomes. They remain attached to the nuclear membrane. Chromosomes then pair up homologously (*zygotene*), as a synaptonemal complex develops between them holding them together (*synapsis*). Each such pair is termed a *bivalent*; but where there is little homology (e.g. between sex chromosomes as in the heterogametic sex of most vertebrates) synapsis is only partial. Completion of synapsis is followed by shortening and thickening of the bivalents (*pachytene*), during which interval, often lasting for days, CROSSING-OVER occurs between non-sister chromatids, producing recombination between homologous chromosomes. The synaptonemal complex then dissolves (*diplotene*) and the two homologous chromosomes fall apart, except where crossing-over holds them together at one or more visible *chiasmata*. Chromosomes begin to unwind (decondense) and may commence RNA transcription again. At *diakinesis*, any RNA synthesis ceases (but see LAMPBRUSH CHROMOSOMES); bivalents shorten, thicken and detach from the nuclear membrane. For the first time they appear as four distinct chromatids, linked at their centromeres and chiasmata.

After first meiotic prophase, the nuclear membrane disintegrates and at *first metaphase* bivalents lie on the midline of the cell, between the two poles. To commence *first anaphase* the spindle fibres, attached to kinetochores of the centromeres, then pull the two members of each bivalent to opposite poles, the sister chromatids appearing somewhat more 'splayed out' than at mitotic anaphase (see Fig. 96). Chance governs which pole each chromosome of a bivalent moves to, ensuring random recombination between non-homologous chromosomes. Each of the two sets of chromosomes produced is HAPLOID, although each dyad has chromatids of mixed parental origin. A short *first telophase* and *interphase* may follow; or the second meiotic division proceeds at once. Nuclear membranes normally reform, chromosomes decondensing, then recondensing, at a brief *second prophase*. The nuclear envelope breaks down, a spindle forming either parallel to (e.g. plant megaspores), or at right angles to, the first. *Second metaphase, anaphase* and *telophase* pass quickly, resembling mitotic phases except that non-identical sister *chromatids* separate, lying on the metaphase plate until the centromeres holding them together separate at anaphase. Again it is a matter of chance which of two poles a chromatid moves to. Nuclear membranes form around the four haploid chromosome sets to complete meiosis. See LIFE CYCLE, SPERMATOGENESIS, OOGENESIS and MATURATION OF GERM CELLS for vertebrate gametogenesis. See COST OF MEIOSIS, POLYPLOID.

meiospore Spore produced by meiosis.

meiotic arrest See MATURATION PROMOTING FACTOR.

meiotic drive See ABERRANT CHROMOSOME BEHAVIOUR.

Meissner corpuscles Moderately rapid-acting touch (velocity) receptors in dermis of vertebrate hairless skin.

MEK See MAPKS.

melanins Pigments of the amniote skin, product of tyrosine metabolism, including oxidation to dopa and dopaquinone (red) and then via an indole-quinone to yield melanin. Produced by melanocytes. The most noticeable pigments in hair and skin are: *eumelanin* (brown-black), *phaeomelanin* yellow/red (brownish) and *erythromelanin* (the pigment of red hair). See ALBINISM.

melanism See INDUSTRIAL MELANISM.

melanocyte The pigment cells of the amniote skin, in humans located around hair bulbs and in the basal layer of the epidermis, producing MELANINS. Especially in fish, amphibians and many reptiles, the pigment granules (melanosomes) in the cells can alter their positions by moving along microtubules in response to neural or hormonal stimuli, affecting the overall colour pattern of the body surface.

melanocyte-stimulating hormone Peptide hormone of the *pars intermedia* of the vertebrate anterior PITUITARY GLAND causing darkening of the skin in lower vertebrates by contraction of melanin in melanophores (see CHROMATOPHORE). In humans it appears to raise the general excitability of neurons in the central nervous system, probably affecting learning. Release inhibited by hypothalamic factor.

melanomas Cancerous outgrowths of melanocytes.

melanophore See CHROMATOPHORE.

melatonin See PINEAL GLAND.

membrane See CELL MEMBRANES, MUCOUS MEMBRANE, SEROUS MEMBRANE.

membrane attack complexes Complexes formed in a cell surface membrane by a cascade reaction from late COMPLEMENT components, near the site of C3 activation. They disturb the lipid bilayer, make the cell membrane leaky and cause osmotic rupture of the cell. Such complexes can even destroy enveloped viruses. See IMMUNITY.

membrane bone See DERMAL BONE.

membrane potential Electrical potential across a cell membrane. All plasma membranes have such voltage gradients, the inside negative with respect to the outside. Most pronounced in animal excitable tissues. See RESTING POTENTIAL.

membranous labyrinth Vertebrate inner ear, comprising the VESTIBULAR APPARATUS and, in higher vertebrates, the COCHLEA.

memory cell Both B cells and T cells may be involved in immunological memory, and differ from their respective effector cells in not requiring persistence of antigen for their continued existence, although it seems that memory T cells require more regular restimulation with antigen than do memory B cells. Both exhibit clonal expansion (see CLONAL EXPANSION THEORY) when appropriately triggered by their specific antigen after initial antigen-presentation (see ANTIGEN-PRESENTING CELL). Two functional classes of memory T cell seem to exist: 'central' memory cells circulating through the lymphoid system but in greater numbers and with a more rapid response rate than naïve antigen-specific T cells; and 'effector', or 'tissue' memory T cells that leave the blood and penetrate the tissues before a pathogen has a chance to disseminate. See B CELLS, T CELLS and IMMUNITY for overview.

menarche The initiation of MENSTRUAL CYCLES.

Mendel, Johann (Gregor) Austrian experimental biologist (1822–84) of peasant stock. After study at Olomouc University, he entered Brno monastery in 1843 as it provided opportunities for continued academic work. His experimental work involved production of pea stocks (*Pisum* is self-fertile), pure-breeding for one or more characters; stocks which could be selfed and crossed in large numbers, reciprocally when necessary, whose seeds and offspring could be scored for several pairs of contrasting characters used. He carried out both monohybrid and dihybrid crosses. Since, as is now realized, the pairs of characters studied were determined by unlinked loci, he was able to obtain offspring in ratios enabling the subsequent formulation of laws of inheritance. His greatest conceptual innovation was to regard heritable factors determining characters as atomistic and material particles which neither fused nor blended with one another – a conclusion inescapable in the light of the experimental results he obtained. Mendel was elected Abbot of Brno monastery in 1868. His experimental results were confirmed in the first decade of the 20th century, after the 're-discovery' of his laws.

Mendelian heredity (M. inheritance, Mendelism) The view, expressed here in modern terminology, that in eukaryotic genomes alleles segregate (separate into different nuclei) during MEIOSIS, after which any member of a pair of alleles has equal probability of finding itself in a nucleus with either of the members of any other pair (if the loci are unlinked). As a result of chromosome behaviour during meiosis and fertilization, and of dominance and recessiveness among characters, ratios of characters among offspring phenotypes are predictable, given knowledge of the parental genotypes.

MENDEL WAS unaware of the genetic role of chromosomes (see GENE, WEISMANN), and studied inheritance of variation determined by unlinked allelic differences of major effect. He was also unaware of POLYGENIC INHERITANCE and LINKAGE, both of which are liable to cause departures from Mendelian ratios in breeding work. Linkage provides a clear exception to the law of independent assortment (SEE MENDEL'S LAWS). Mendelian ratios may also be distorted by MUTATION, SEX-LINKAGE, MEIOTIC DRIVE, CYTOPLASMIC INHERITANCE, MATERNAL EFFECT, EPISTASIS and by SELECTION among embryos or gamete types. MALE HAPLOIDY will also result in distortions; but even here, as with sex-linkage, alleles behave in a basically Mendelian way. It is simply that the genetic system produces non-Mendelian ratios in breeding work.

Mendel's Laws This account of the laws uses terminology which Mendel did not employ himself.

First Law, of Segregation: during meiosis, the two members of any pair of alleles possessed by an individual separate (segregate) into different gametes and subsequently into different offspring, neither having blended with nor altered the other in any way while together in the same cell (but see GENE CONVERSION). The law asserts that alleles retain their integrities (barring mutation) during replication from generation to generation. None of the cells normally produced by meiosis contains two alleles from any locus.

Second Law, of Independent Assortment: asserts that during meiosis all combinations of alleles are distributed to daughter nuclei with equal probability, distribution of members of one pair having no influence on the distribution of members of any other pair. It holds, in effect, that during meiosis random reassortment occurs between alleles at different loci. Thus if one locus is represented by the two alleles A and a, while another locus is represented by alleles B and b, then all four haploid nuclei AB, aB, Ab and ab will be formed in equal frequency by meiosis. Mendel's second law is refuted by LINKAGE. Genes at linked loci tend to retain their linear sequences on chromosomes during meiosis and therefore tend to be inherited as blocks (but see CROSS-OVER VALUE).

Mendel's first law is a consequence of the behaviour of all chromosomes during meiosis. His second law is a consequence of the independent behaviour of non-homologous chromosomes during meiosis. See MENDELIAN HEREDITY.

meninges Three membranous coverings of the vertebrate brain, spinal cord, and spinal nerves as far as their exits from between the vertebrae. Innermost is the vascular *pia mater*, separated from the *arachnoid* by the fluid-filled arachnoid spaces in which a fine web of fibres and villi reabsorb the CEREBROSPINAL FLUID back into the venous system. Outermost membrane, the *dura mater*, is a thick, fibrous membrane lining the skull and separated from the arachnoid below it by the dural sinus draining blood from the brain. The pia mater supplies capillaries to the ventricles of the brain.

menopause Dramatic and fairly rapid endocrine changes which signal the end of a WOMAN'S MENSTRUAL CYCLES, often anticipated by raised FSH levels during the follicular phase of the cycles and involving a decrease in the pool of follicles. Suspected link with alteration in rhythmicity of hypothalamic suprachiasmatic nucleus pacemaker (SCN, see BIOLOGICAL CLOCK) and its effect on gonadotrophin, and hence on LH, release.

menstrual cycle See Fig. III. Modified oestrous cycle of catarrhine primates; characterized by sudden breakdown of uterine endometrium, producing bleeding (*menstruation*), and by absence of a period of heat (oestrus) in which the animal is particularly sexually receptive. Controlled and coordinated nervously and hormonally, in particular by the hypothalamus and pituitary gland and the gonads. A mature human cycle has the following main sequence of events: a hypothalamic releasing factor (FSH-RF) causes release of FOLLICLE-STIMULATING HORMONE from the pituitary, initiating growth of one GRAAFIAN FOLLICLE in the ovaries and its consequent increased production of oestrogens, especially oestradiol. High oestrogen levels eventually inhibit release of both hypothalamic FSH-RF and LH-RF, decreasing FSH and LH output; but about two days prior to *ovulation* a marked rise in output of pituitary LUTEINIZING HORMONE occurs. Through a positive feedback mechanism, rising levels of oestrogens induce LH release (via LHRF) from the pituitary, which has increased sensitivity at this time. A mid-cycle surge of LH induces rupture of the Graafian follicle (ovulation) with release of the egg and development of the Graafian follicle into a CORPUS LUTEUM, which secretes oestradiol and increasing amounts of progesterone, which maintains the vascularization of the uterine endometrium begun by oestrogens. Progesterone is secreted only so long as small amounts of pituitary LH maintain the corpus luteum, about 10–12 days in women unless pregnancy occurs. With atrophy of the corpus luteum, progesterone and oestradiol levels drop sharply, causing *menstruation* – the shedding of the uterine lining each month in the absence of pregnancy. If there is no pregnancy, the cycle repeats immediately as hypothalamic FSH-RF is released again, having been inhibited by the negative feedback effect of steroid sex hormones. If pregnancy occurs, maintenance of the corpus luteum is sustained for a short period by LH, then by hCG (HUMAN CHORIONIC GONADOTROPHIN) secreted by the implanted blastocyst and later by the placenta. See CONTRACEPTIVE PILL, MATURATION OF GERM CELLS.

mericarp Single-seeded portion of SCHIZOCARP.

meristele Individual vascular unit of DICTYOSTELE.

meristem Localized region of active mitotic cell division in plants, from which permanent tissue is derived. New cells formed by activity of a meristem become variously modified to form characteristic tissues of the adult (e.g. epidermis, cortex, vascular tissue, etc.). A meristem may have its origin in a single cell (e.g. in ferns), or in a group of cells (e.g. in flowering plants). The principal meristems in latter group occur at tips of stems and roots (APICAL MERISTEMS, or growing points), between xylem and phloem of vascular bundles (CAMBIUM) in cortex (CORK CAMBIUM), in young leaves and (e.g. in many grasses) at bases of internodes (intercalary meristems). Meristems may also arise in response to wounding.

meristematic activity State of active mitosis in a MERISTEM.

meristoderm Outer meristematic cell layer in certain brown algae.

meroblastic See CLEAVAGE.

merocrine gland Gland whose cells secrete their product while remaining intact: no portion is pinched off (see APOCRINE GLAND), nor do cells have to disintegrate in order to release their product (see HOLOCRINE GLAND). Examples include vertebrate salivary glands and exocrine cells of the pancreas.

meroplanktonic Term describing organisms that spend part of their life cycle in the plankton and part in the benthos as a resting stage.

Merostomata Class of aquatic arthropods (formerly the order Xiphosura of the class Arachnida). Includes the extinct EURYPTERIDA. The only living forms (king crabs, e.g. *Limulus*) have a broad cephalothorax (pro-

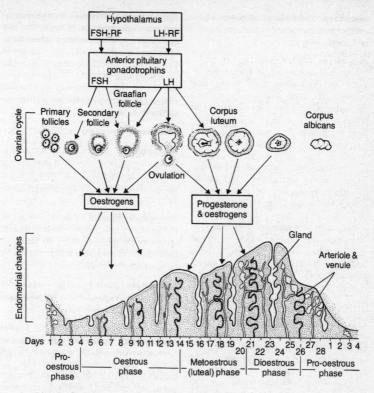

FIG. 111 *Diagram illustrating the hormonal and endometrial changes in the human* MENSTRUAL CYCLE *when no pregnancy occurs.*

soma) covered dorsally by carapace which covers limbs and in which is located a pair of eyes. Chelicerae chelate. Pedipalps resemble walking limbs. Gnathobases (spinal basal segments of legs) function as mandibles, which are absent. Opisthosoma represented by fused tergites forming a single dorsal plate, with a long caudal spine hinged to its posterior border. Gill books serve for gaseous exchange. The trilobite larva of *Limulus* has chelicerae and lacks antennae; however, the chelicerate arthropods could have been derived from a pro-trilobite. See ARTHROPODA.

mesarch Type of maturation of primary xylem from a central point outwards;

i.e. the oldest xylem elements (protoxylem) are surrounded by later-forming metaxylem.

mesencephalon See MIDBRAIN.

mesenchyme Embryonic mesoderm comprising widely scattered tissue giving rise to connective tissue, blood, cartilage, bone, etc.

mesentery (1) Double-layered extension of the peritoneum attaching stomach and intestines to the dorsal body wall. Contains blood, lymph and nerve supply to these organs. (2) Vertical partitions of body wall of anthozoans (sea anemones) forming compartments within the enteron.

mesocarp Middle layer of mature ovary wall, or pericarp; between the exocarp and endocarp. See FRUIT.

mesoderm Middle germ layer of triploblastic animals, coming to lie between ectoderm and endoderm from gastrulation onwards (see GERM LAYER for mesodermal derivatives). In animals with a large coelom it is separable into an inner *splanchnic mesoderm* forming outer covering of digestive tract and its diverticula, and an outer *somatic mesoderm* from which, when present, develop the somatic skeleton and musculature, kidneys and gonads. Part of the mesoderm commonly becomes divided during development into a series of blocks, or *somites*, which form the basic developmental units of segmented coelomates. In vertebrates the somites, which are dorsal and on either side of the neural tube, undergo differentiation into lateral *dermatome* and medial *myotome* with *sclerotome* around the neural tube and notochord. These will give rise respectively to dermis, striated muscle and vertebral column and adjacent rib components. Vertebrate mesoderm undivided into somites forms unsegmented *lateral plate mesoderm*, giving rise to splanchnic and somatic mesoderm. The *extraembryonic mesoderm* of amniotes is mesoderm that has spread out to cover the trophoblast and forms the CHORION.

mesogene development Development of stomata where GUARD CELLS and subsidiary cells share a common parental cell.

mesogloea Layer of jelly-like material between ectoderm and endoderm of coelenterates. Merely a non-cellular collagenous membrane in hydrozoans such as *Hydra*, but much enlarged, thickened and fibrous in jellyfish (scyphozoans) and in these and some other groups comes to contain cells derived from the two tissue layers. It is not itself a germ layer.

mesokaryote (dinokaryote) Term used to describe the nucleus of the DINOPHYTA and EUGLENOPHYTA, where the chromosomes remain condensed throughout the mitotic cycle. Nucleolus persistent, dividing by pinching in two; nuclei large, the nuclear envelope remaining intact during the entire mitotic cycle (mitosis closed); chromosomes present attach to the nuclear membrane and not the spindle microtubules. In Dinophyta, few basic proteins (histones) are associated with the DNA. The term is used to denote the evolutionary position of dinoflagellates and euglenoids between prokaryotes and eukaryotes.

mesonephros See KIDNEY.

mesophilic (Of microorganisms, e.g. *Escherichia coli*) with optimum growth temperature (measured by generations per hour) in the 30–40°C range. Compare PSYCHROPHILIC, THERMOPHILIC.

mesophyll Internal tissue of LEAF BLADE; differentiated into upper *palisade* and lower *spongy* mesophyll, the latter cells with generally fewer chloroplasts and separated by large air spaces. See LEAF.

mesophyte Plant found growing under average conditions of water supply. Compare HYDROPHYTE, XEROPHYTE.

mesosome Infolded region of plasma membrane of bacterial cell, containing electron transport system, attaching to the circular chromosome and involved in initiation (and termination?) of chromosomal replication. Generally located near newly-forming cell wall in binary fission. Very probably homologous with mitochondrial cristae. See LOMASOME.

mesothelium Epithelium-like layer covering vertebrate SEROUS MEMBRANES; flattened squamous cells derived from mesoderm, and therefore not strictly epithelial.

mesotrophic (Of lakes and rivers) which are neither nutrient-rich nor nutrient-deficient (i.e. between OLIGOTROPHIC and EUTROPHIC). Category based upon so-called phytoplankton indices which involves comparing ratios of eutrophic to oligotrophic species. Compare DYSTROPHIC.

Mesozoic Geological era extending from 225–70 Myr BP; includes Triassic, Jurassic and Cretaceous periods. See GEOLOGICAL PERIODS.

messenger RNA (mRNA) Single-stranded RNA molecule, translated on ribosomes into a polypeptide. Produced (transcribed) by RNA polymerase, only one of the two DNA strands being read by the polymerase and acting as template (see PROTEIN SYNTHESIS). The transcription product (hnRNA in eukaryotes) generally contains some base sequences not coding for any part of the eventual polypeptide, these being cut out enzymatically (see RNA PROCESSING, INTRON) before the processed mRNA is attached to a ribosome. The amino acid sequence of the polypeptide reflects the mRNA nucleotide sequence, as dictated by the GENETIC CODE. Most mRNA molecules are broken down enzymatically by nucleases, usually within minutes (several hours in mammalian cells), but some eukaryotic cells lacking nuclei have relatively long-lived mRNA, as do some egg cells. Eukaryotic mRNA is commonly preserved from exonuclease degradation by a poly(A) tail added to the 3′ end, and by a 7-methylguanosine 'cap' added to the 5′ end, during transcription. The poly(A) tail enables researchers to distinguish mRNA from other forms of RNA in the cell. Slowed rates of deadenylation and decapping by enzymes contribute to mRNA stability. See RT-PCR.

metabasidium Cell in which meiosis occurs in members of the BASIDIOMYCOTA. Compare PROBASIDIUM.

Metabola See PTERYGOTA.

metabolic pathway Sequence of reactions, each catalysed by a different enzyme, leading to the formation of one or more functional products. Pathways may be linear (e.g. GLYCOLYSIS) or cyclical (e.g. KREBS CYCLE), and a by-product of one may serve as a substrate in another. See REGULATORY ENZYMES.

metabolic rate The rate at which energy and materials are transformed within an organism and exchanged between the organism and its environment. Whole organism metabolic rate scales with the $\frac{3}{4}$-power of body mass and increases exponentially with temperature (see ALLOMETRY, POWER LAWS, Q_{10} EFFECT), which governs metabolism through its effects on enzyme activity. A general model has recently been shown to explain the scaling of whole organism metabolic rate, B, with body mass, M, where

$$B \propto M^{3/4}$$

so that mass-specific metabolic rate

$$B/M \propto M^{-1/4}$$

a quarter-power scaling based on the fractal-like constitution of exchange surfaces and distribution of networks in plants and animals. See BASAL METABOLIC RATE.

metabolism Sum of the physical and chemical processes occurring within a living organism; often intended to refer only to its enzymic reactions. May be regarded as comprising *anabolism* (build-up of molecules) and *catabolism* (breakdown of molecules). Frequently used in context of a particular class of compounds within an organism, e.g. fat metabolism. The term METABOLIC RATE is often used rather loosely, more or less synonymously with respiratory rate, the level of oxygen consumption in aerobes being an indicator of the general metabolic activity of the organism. See METABOLITE.

metabolite Substance participating in METABOLISM. Some are intermediary compounds of biochemical pathways; others are taken in from the environment. Primary metabolites are those required for or produced by metabolic reactions essential to the life of the organism; SECONDARY METABOLITES are those required for, or produced by, reactions not essential to continuance of the organism's life.

metabotropic receptor Membrane RECEPTOR acting indirectly on an ion channel by activating a second messenger and other molecular mechanisms within the cell. Contrast IONOTROPIC RECEPTOR.

metacarpal bones Rod-like bones of forefoot of tetrapods articulating with carpals (wrist-bones) proximally and finger bones (phalanges) distally. Usually one corresponding to each digit. Compare METATARSAL BONES. See PENTADACTYL LIMB.

metacarpus Region of tetrapod forefoot containing metacarpal bones. Palm region of man.

metacentric See CENTROMERE.

metachromatic granule (polyphosphate body, volutin granule) Spherical granule containing stored phosphate found in cells of blue-green algae (CYANOBACTERIA). Absent from young cells or cells grown in phosphate-deficient medium; prominent in older cells.

metachromatic stain See STAINING.

metachronal rhythm Pattern of beating adopted by groups of cilia, segmented parapodia of polychaetes and arthropod limbs, in which each unit (cilium, parapodium, limb) is at a slightly different stage in the beat cycle from those on either side of it. Result is a smooth progression of waves of beating along the units. Often occurs in forms of locomotion, microphagous feeding and exchange of water over a respiratory surface.

metagenesis See ALTERNATION OF GENERATIONS.

metamere One of the repeating modules of antero–posterior bodily organization in animals with METAMERIC SEGMENTATION. See PARASEGMENT.

metameric segmentation (metamerism) See SEGMENTATION.

metamorphosis Process during, and as a result of, which an animal undergoes a comparatively rapid change from larval to adult form. Under hormonal control, it is most notable in the life histories of many marine invertebrates, the majority of insects (especially ENDOPTERYGOTA) and of AMPHIBIA. Often requires destruction of much larval tissue (see LYSOSOMES) and changes in gene expression. Of enormous evolutionary and ecological significance. See CORPORA ALLATA, THYROXINE.

metanephros See KIDNEY.

metaphase Stage in MITOSIS and MEIOSIS.

metaphloem Primary PHLOEM formed after the protophloem.

metaphyton Algae present between epiphytic algae, occurring as a loose collection of non-motile or weakly motile forms, lacking ready means of attachment to a substratum.

metaplasia Transformation of one kind of adult cell into another not usually found in that part of the body. May occur during tumour formation.

metarterioles Minute branches of arterioles (8–18 μm in diameter). Generally give rise to capillary beds which may be opened and closed by precapillary sphincters. See CAPILLARY.

metastasis Movement of infectious microorganisms or malignant CANCER CELLS, usually via bloodstream or lymph, from one focus of growth to another part of the body where they set up further foci, in the case of cancer cells often by attachment to the endothelium of a capillary. See EXTRAVASATION.

metatarsal bones Rod-like bones of tetrapod hind-foot (forming metatarsus), usually one corresponding to each digit, articulating with ankle bone (tarsal) proximally and toe bones (phalanges) distally. See PENTADACTYL LIMB.

Metatheria Marsupials. Subclass or infraclass of MAMMALIA containing just one order (Marsupialia). Originated in early Cretaceous; short gestation period and very undeveloped offspring at birth; a protective pouch (marsupium) in most forms investing the mammary glands of the female and housing the young while they complete development. Restricted to Australian region (including New Guinea) and the New World; fossil marsupials appear to be absent from Europe and Mongolia. Decline in diversity during late Cretaceous. Many resemble placental mammals of various orders, exhibiting CONVERGENCE. See PLACENTA.

metaxylem Primary xylem formed after the protoxylem. Cell elongation is generally complete, or nearly so, by the time of its production.

Metazoa In older classifications, an informal group denoting 'multicellular animals'. The term is made somewhat redundant if, as generally nowadays, animals are regarded as multicellular by definition. See ANIMALIA.

methanogen An organism (e.g. see ARCHAEBACTERIA) producing methane. Methanogenic bacteria are obligate anaerobes, most using CO_2 as their terminal electron acceptor in fermentation, although methanol, formate, methyl mercaptans, acetate and methylamines are also used. Major environments for methane production include marshes, swamps, rice paddies and the rumen.

methanotrophs Bacteria which use methane as an electron donor and sole source of carbon. Many are aerobic and inhabit freshwater and soils, and are of industrial importance in manufacture of SINGLE CELL PROTEIN. But methanotrophic archaea are now known to be crucial in utilizing and preventing methane which moves upwards through the ocean floor from entering the atmosphere and promoting global warming. These archaea live in clusters around bacteria which use sulphate ions (SO_4^{2-}), possibly live mutualistically with the archaea, consuming such products of their metabolism as molecular hydrogen and carbon compounds. It is estimated that there are more than 10^{13} tons of methane buried in the ocean floor: twice that of all known coal, oil and other fossil fuels.

5-methylcytosine See DNA METHYLATION.

7-methylguanosine See MESSENGER RNA.

MHC See MAJOR HISTOCOMPATIBILITY COMPLEX.

MHC restriction See T CELL.

micelle Particle of colloidal size, normally spherical, which in aqueous medium has a hydrophilic exterior and hydrophobic interior. Commonly consists of a phospholipid monolayer, sometimes with associated protein, within which non-polar materials may be housed and transported. Several storage polysaccharides (e.g. amylose, amylopectin) form hydrated micelles rather than true solutions within cells. See BILE, CHYLOMICRON, LIPOPROTEIN, LIPOSOME.

Michaelis constant (Michaelis–Menten constant) For a given enzyme–substrate reaction, the value (K_M) of the substrate concentration at which the initial velocity of the reaction (V_O) is half maximal. A low value indicates high affinity of enzyme for substrate. Allosteric enzymes do not strictly have K_M values. See ENZYME.

microaerophilic (Of organisms, e.g. microorganism) requiring gaseous oxygen, but at a level lower than atmospheric (e.g. *Rhizobium*; see NITROGEN FIXATION).

microarrays See NUCLEIC ACID ARRAYS.

microbe Microscopic organism of any type, whether prokaryotic or eukaryotic. Many are pathogenic.

microbial loop Term referring to the cycling of elements in the surface waters of the ocean in which fixed carbon is dissolved when microorganisms die, then used by other microbes which may in turn be consumed by others, eventually finding its way to the zooplankton. When these in turn die, some of their biomass sinks to the ocean floor, locking away the fixed carbon.

microbiology Branch of biology dealing with MICROORGANISMS.

microbodies Term often used to indicate PEROXISOMES and glyoxysomes (see GLYOXYLATE CYCLE).

microdissection Technique used in operating upon small organisms whereby the specimen is dissected while viewed through a microscope. Instruments are normally manipulated by mechanical means. Micromanipulation is often used where living cells or their nuclei are removed or inserted.

microevolution Evolution within a species. Some would include SPECIATION. See EVOLUTION.

microfibrils Sheaths of glycoprotein complexes, including fibrillin, which bind to the elastin core in elastic FIBRES and are essential to their integrity.

microfilament See ACTIN, CYTOSKELETON.

microfossil Microscopic FOSSIL; includes spores, pollen grains, algae, tracheids, pieces of plant cuticle, animals, etc.

microgamete See GAMETE, ANISOGAMY.

microgametophyte One of two types of gametophyte produced by HETEROSPOROUS plants; develops from microspore. In ferns, comprises a *prothallus* (confined to the microspore in club-mosses like *Selaginella*); in seed plants, comprises microspore within wall of pollen grain, plus its mitotic products, including pollen tube and its nuclei.

micrograph Photograph of an image obtained during either light or electron microscopy.

micrometre (micron) Unit of length often used in microscopy. Symbol μm; 10^{-3} mm; 10^{-6} m. Many bacterial cells are approximately 1 μm in length. 1 μm = 1,000 nm = 10,000 Å.

micron See MICROMETRE.

micronucleus See NUCLEUS.

micronutrient Substance required by an organism from its environment for healthy growth, but only in minute amounts. Includes TRACE ELEMENTS and VITAMINS.

microorganisms Informal term of little taxonomic value. Include the viruses, prokaryotes, microscopic fungi and protists, all of which are of great intellectual, economic and cultural significance. They tend to concentrate at interfaces, such as those between the lithosphere, hydrosphere and atmosphere, where nutrients and energy can fluctuate.

microphagy Methods employed by animals in feeding upon particles which are small in relation to their own size. Such feeding tends to occur continually, frequently by sieving of particles from water by such devices as baleen plates (baleen whales), by ciliary-mucus devices in

pharyngeal gill slits (urochordates, cephalochordates) and other gills (bivalve molluscs); by lophophores; by trapping particles in bristle-fringed trunk limbs (many crustacea), or even cytoplasmic nets (heliozoans, foraminiferans).

microphyll Type of leaf usually but not always small, possessing very simple vascular system comprising single branched vein. LEAF TRACE is not associated with a LEAF GAP in stele of stem. Associated with PROTOSTELIC vascular system. Characteristic of club mosses (LYCOPHYTA), horsetails (SPHENOPHYTA) and related forms. Compare MEGAPHYLL.

micropropagation One form of BIOTECHNOLOGY, in which whole plants can be raised from single cells. Naked PROTOPLASTS can be cultured to produce callus which can be treated to give whole plants. Apical meristems may be cultured to produce virus-free plants. Technique is used with tomatoes, potatoes, raspberries, strawberries, etc. Increasingly important as a tool in the conservation of threatened species, notably orchids and other plants for which there is horticultural demand.

micropyle Canal formed by extension of integument(s) of ovule beyond apex of nucellus; recognizable in a mature seed as a minute pore in seed coat through which water enters at start of germination.

microsatellite DNA See SATELLITE DNA.

microscope, microscopy Microscopes are instruments employing lenses to produce magnified images, and hence fine detail, of objects too small to observe clearly with the naked eye. Earliest *light microscopes* (visible light as the transmitted medium) appeared in about 1590 in Holland and had two glass lenses. In these *compound microscopes* one short focal length lens, the eyepiece lens, produces a magnified apparent image of a real image produced by a second short focal length lens, the objective lens, placed closer to the object. Total magnification is the product of the magnifications produced by these two lenses. Hooke (1665) first recorded cells in cork, and Leeuwenhoek first noted infusoria (1676) and bacteria

(1683), which were first stained by von Gleichen (1778). In the 1870s E. Abbe and C. Zeiss produced the first oil-immersion objective lens enabling a good image of up to ×1500 magnification. Achromatic and aplanatic objective lenses (the finest available to this day) were developed by J. J. Lister, those of 1886 from Zeiss being of a very high quality.

The thinner the material being observed, the greater the clarity of image; hence the improvement of EMBEDDING, cutting and sectioning methods to match the evolution of the microscope. Light reflected from a point object cannot be recombined again to form another true point, but only a disc of light. When discs representing adjacent object points overlap detail is lost. The *resolving power* of a microscope, its ability to distinguish fine detail, is proportional to the wavelength of the transmitted medium. Resolving power depends on the quality of the objective lens, since the eyepiece lens cannot add to what the objective lens has failed to pick up. Visible light has a wavelength of about 0.5 µm and the best resolving power (even using visible light of the shortest wavelength) is about 0.45 µm. Objects closer together will not be resolved as more than one object. Where an object (e.g. a typical cell) is transparent, with features differing only in refractive index, light rays will emerge with different phase relations depending on the paths they have taken. The resulting image has uniform brightness, but the technique of *phase contrast microscopy* makes use of its phase differences so as to produce the image that would have been seen had these been amplitude differences.

Electron microscopes use wave properties of electrons fired through the object held in a vacuum. In transmission electron microscopes, electrons are accelerated through the microscope by a large voltage (up to 1,000 kV), those passing through the object hitting a screen, which fluoresces giving an image. The electrons are focused by electron magnets. The wavelength of electrons is inversely related to the voltage used, but even in low voltage apparatus is about 0.005 nm – four orders of magnitude (10^4)

less than that of a light wave. Resolving power of the order of 1 nm for biological material (less for crystals) is achievable. This is about 200 times better than in light microscopy. Materials for observation are commonly first fixed (see FIXATION) and then embedded in Araldite® prior to sectioning by an ultramicrotome, giving sections 20–100 nm thick. The dangers of so generating artefacts are well known and are avoided as far as possible. Biological material contains few heavy atoms and consequently its electron-scattering ability is poor, but can be improved by soaking the object in, or spraying it with, a salt of a heavy metal ('staining' it). In *scanning electron microscopy* the electron beam causes the object to emit its own electrons. These can be used to produce an image which is built up as the electron beam scans the specimen (rather in the way a television picture is produced). Scanning tunnelling electron microscopy (STM) enables visualization of atomic-scale images and even chemical modification at the nanometre level. Micrographs produced have a three-dimensional quality, but lower resolution than in transmission microscopy. One technique often employed for looking at hydrophobic interiors of membranes is *freeze-fracture electron microscopy*, in which cells are frozen in ice to the temperature of liquid nitrogen (–196°C) and then fractured, the plane of fracture tending to pass through the middles of lipid bilayers. Exposed fracture surfaces are then shadowed with platinum and carbon followed by digestion of the organic content (*freeze-etching*), leaving a platinum 'replica' which can be examined with the electron microscope. This is particularly useful for detecting where proteins are located in membranes. In electron microscopy, the specimen is inevitably in a vacuum and exposed to high temperature electron bombardment; hence it is impossible to view living material, specimens requiring cooling during viewing.

In *confocal microscopy*, which has had an enormous impact in biology, laser light sources are scanned across the specimen and electrons emitted from fluorescent chromophores are captured by sensitive digital detectors. A variable pinhole aperture allows light emanating from only a single focal plane to reach the detector. The apparatus is sensitive to the square of the light intensity, not to the intensity itself, and this discriminates against light that is out of focus. The technique enables high spatial (three-dimensional) resolution and permits optical sectioning so that it is possible to record fluorescent images showing the chromophore distribution in just one focal plane notably inside optically dense objects (e.g. embryos). See also CRYO-ELECTRON TOMOGRAPHY.

microsomes Products of homogenization of endoplasmic reticulum. Rough microsomes are closed vesicles with ribosomes attached to their outer surfaces; smooth microsomes lack ribosomes and may derive from plasma membrane as well as from smooth endoplasmic reticulum.

microspike Small FILOPODIUM.

microsporangium Sporangium in which microspores are formed; borne on a microsporophyll. In seed plants, a *pollen sac*.

microspore Smaller of two kinds of spore produced by heterosporous plants (e.g. ferns, seed plants); first cell of microgametophyte generation in these plants, developing after meiosis from the *microspore mother cell* (microsporocyte). In seed plants, becomes a pollen grain. Compare MEGASPORE.

Microsporidia Obligate intracellular parasites of several eukaryotes, structurally simple apart from having a unique and complex infection apparatus. Although they are thought to lack mitochondria and peroxisomes (see AMITOCHONDRIATE), the presence of nuclear genes encoding orthologues of typical mitochondrial heatshock Hsp70 proteins is evidence that they have lost these organelles secondarily; and the discovery of this protein in the human microsporidial *Trachipleistophora hominis*, and within minute organelles (~50 × 90 nm) with double membranes, suggests that eukaryotes are loath to lose their mitochondria, even when its primary role in aerobic respir-

ation has been lost, and that the taxon is related to fungi rather than representative of a basal eukaryote.Genealogies making use of α- and β-tubulin sequences also relocate them as highly derived fungi rather than as early eukaryotes. Cause gastrointestinal illness. See EUKARYOTE.

microsporophyll Leaf, or modified leaf, that bears microsporangia. In flowering plants, a STAMEN. Compare MEGASPOROPHYLL.

microtome Machine for cutting extremely thin sections of tissue, etc. For light microscopy the material is first fixed and either frozen or embedded in paraffin wax, sections cut with a steel knife usually being 3–20 μm thick; for electron microscopy, fixing is followed by embedding in a resin such as Araldite®, sections cut with the glass or diamond knife of an *ultramicrotome* being 20–100 nm thick.

microtubular root Bundles of one to many parallel microtubules that extend down into the cell from the flagellar basal bodies.

microtubule One of the essential protein filaments of the CYTOSKELETONS of probably all eukaryotic cells, and of their cilia, flagella, basal bodies, centrioles and mitotic and meiotic spindles. Microtubules originate at MTOCs. Each microtubule consists of a hollow cylinder, about 25 nm in diameter, made up of 13 protofilaments of the protein TUBULIN. Each protofilament consists of globular tubulin molecules polymerized together. Like ACTIN filaments, microtubules grow and depolymerize at different rates at their two ends. Magnesium ions are required for formation from tubulin dimers; GTP hydrolysis is required for assembly of dimers into filaments, the energy apparently being stored until required for disassembly (but GTP-like subunits at the end of a microtubule protect it from depolymerization). Microtubule roles include guiding organelle and chromosome movement in the cell (see KINETOCHORE), causing cell elongation by their own elongation and involvement in ciliary and flagellar beating. Spindle microtubules are of four classes: astral microtubules contact the cell cortex,

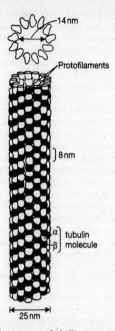

14 nm

Protofilaments

8 nm

α β tubulin molecule

25 nm

FIG. 112 *Arrangement of tubulin monomers (α and β globular protein molecules) to form a* MICROTUBULE, *cross section (top) and side view (below). Compare diameter with width of a bacterial envelope (see* GRAM'S STAIN*).*

orienting the spindle; pole-to-pole microtubules overlap in the middle of the spindle with those growing from the other pole (attached motor proteins controlling spindle length); a third class interacts with chromosome arms via chromokinesin proteins and pushes chromosomes to the middle of the spindle; and a fourth class captures kinetochores, linking chromosomes to the spindle. Various associated proteins play modifying roles in the behaviour of these microtubules. See Figs. 96 and 112, SPINDLE.

microvillus Minute finger-like projection from the surfaces of many eukaryotic cells, particularly animal epithelia involved in active uptake (e.g. small intestine, kidney proximal convoluted tubule) where, several thousand strong, they constitute the *brush*

borders observed in electron micrographs. Each microvillus is about 1 μm long and 0.1 μm in diameter. About 40 ACTIN microfilaments run along its length, supported by accessory proteins (e.g. alpha-actinin, fimbrin). They may be capable of retraction and extension, possibly through sliding of actin over myosin (see MUSCLE CONTRACTION). Brush borders may increase area of plasma membrane available for absorption 25-fold and house ENZYMES, TRANSPORT PROTEINS, etc. *Stereocilia* are long, thick microvilli, about 4 μm in length (see HAIR CELL), and the rhabdomeres or arthropod compound eyes are also composed of microvilli (see EYE). See Fig. 24d, Fig. 92.

midbrain See BRAIN.

middle ear See EAR, MIDDLE.

middle lamella Layer of intercellular material cementing together the primary walls of adjacent cells. Contains calcium pectate. See CELL WALL.

midgut (1) In vertebrate development, a somewhat arbitrary division of the endodermal layer in contact with remnants of the yolk sac and giving rise to part of duodenum, rest of small intestine and upper part of large intestine. (2) Cylinder of endodermal epithelial cells of the arthropod gut (mesenteron). Not lined by cuticle; secretes a *peritrophic membrane*, enclosing food and separating and protecting the epithelium from abrasion. Most digestion occurs here, with glycogen storage and secretion of urate (see MALPIGHIAN TUBULES). Commonly bears one or more pairs of diverticula (midgut or hepatic caeca), increasing its surface area.

migration The mass directional movements of large numbers of a species from one location to another. This term applies to intercontinental journeys of birds, the transatlantic movements of eels, the to-and-fro movements of shore animals following the tidal cycle and the vertical movements of plankton in water columns.

mildew (1) Plant disease caused by a fungus, producing superficial, powdery or downy growth on host surface. (2) Fungus causing such a disease. (3) Often used synonymously with MOULD.

milk (1) Complex aqueous secretion of MAMMARY GLANDS, with which mammals suckle their young. Composition varies, but usually rich in suspended fat, in protein (mostly *casein*) and sugar (mainly *lactose*). In addition, minerals, vitamins and antibodies (including antibacterial IgA in human milk, but not in cow's milk) and (again in humans but not in cows) the important iron-binding protein *lactoferrin* inhibiting bacterial growth in the baby's intestine. Because the newly born human baby's gut is fairly porous, introduction of foreign proteins (e.g. in cow's milk) can be absorbed unchanged into its circulation where they may provoke an immune response. Bottle-fed babies are much more prone to allergies in later life. It may also be that antibodies produced against one such protein in cow's milk may cross-react with cells in the islets of Langerhans in the infant pancreas and trigger insulin-dependent DIABETES. Milk production is under PROLACTIN control, and its ejection involves OXYTOCIN release during the *sucking reflex*. See PASTEURIZATION, PARTURITION. (2) See CROP MILK.

milk teeth See DECIDUOUS TEETH.

millipedes See DIPLOPODA.

mimicry Relational term, indicating that the signal component of some mutually beneficial evolved signal–response pairing between two organisms (one signalling, the other responding) has been simulated or employed by a third party to its own advantage, often to the detriment of one or both of the original parties.

In *Müllerian mimicry* (after F. Müller) two or more organisms independently derive protection from predation, for example by tasting repellent; but they also benefit mutually through convergent evolution of similar warning (aposematic) colouration and pattern, predators learning by association to avoid both after tasting one. In *Batesian mimicry* (after H. W. Bates) an edible or otherwise relatively defenceless organism (the *mimic*) gains protection from predation by resemblance to a distasteful, poisonous

or harmful organism, termed the *model*. In this case, the benefit gained will depend in part upon the ratio of availabilities of models and mimics, predators tending to require reinforcement of the learnt association of colour/pattern and disagreeableness. Some species (e.g. certain butterflies) may be Batesian mimics of more than one model (see POLYMORPHISM). In *aggressive mimicry*, the mimic feeds on one of the original parties in the original signal–response pairing. Some instances of CRYPSIS (e.g. stick or leaf insects) may appear to overlap the criteria of mimicry given above; but it can be argued that crypsis does not involve manipulation of a true signal (as opposed to background noise) in the service of a new function. Many flowering plants, notably several orchids, achieve pollination through mimicry of signals (e.g. visual, olfactory) attracting specific insect pollinators. Mimicry has provided much support for Darwinian evolution. See SEARCH IMAGE.

minimal medium Medium for growing microorganisms, containing ample quantities of all minimal nutritional requirements of the wild-type organism. See PROTOTROPH, AUXOTROPH.

minisatellite DNA See SATELLITE DNA.

Miocene Geological epoch; subdivision of TERTIARY period. Forests receded, grasslands spread along with radiation of grazing animals and apes. Climate was moderate, while extensive glaciation occurred in the Southern Hemisphere. See Appendix.

miracidium Ciliated larval stage of endoparasitic trematode platyhelminths (flukes). Commonly bores into snail, where it develops into a sporocyst. See TREMATODA.

miRNAs (microRNAs) A large family of NON-CODING RNAS.

mismatch proof-reading and repair See DNA REPAIR MECHANISMS.

missense mutation See MUTATION.

mites See ACARI.

mitochondrial DNA See mtDNA.

mitochondrial Eve Human mitochondrial DNA (see MtDNA) is mostly or entirely inherited via the female line (matrolinearly) and mutates at about ten times the rate of nuclear DNA. Attempts have been made to estimate the times of divergence of different human populations by assuming a direct proportionality between the sum of the differences in their mtDNA and the time since they diverged. But human mtDNA from a variety of geographical populations (so-called 'races') is remarkably uniform, although more diverse in African populations, indicating recency of common origin from an original ancestress, probably African, dubbed 'mitochondrial Eve'. Controversy surrounds the statistical methods employed to analyse mtDNA data, and mitochondrial Eve's date currently ranges from 400–60 Kyr BP. It is even disputed that she was of African origin. Mitochondrial DNA amounts to only one part in 4×10^5 of the total per human cell; the rest was inherited from other contemporaries of the mitochondrial Eve. It would be an error to assume that just one female founded all the non-African members of *Homo sapiens*. Work on human DRB1 (immunity) polymorphisms suggests that perhaps 10^5 'contemporaries of mitochondrial Eve' are implicated in maintaining these polymorphisms at their present level. However, it is anticipated that greater knowledge of human mtDNA and more refined analytic techniques will lead to greater precision in these matters. There is no evidence from mtDNA for genetic input from Neanderthals to modern humans, but this does not preclude their hybridization (see *HOMO*).

mitochondrion (chondriosome) Cytoplasmic organelle of all eukaryotic cells engaging in aerobic RESPIRATION, and the source of most ATP in those cells. Vary in number from just one to several thousand per cell. Most electron micrographs of mitochondria show them as either cylindrical (up to 10 μm long and 0.2 μm in diameter) or roughly spherical (0.5–5 μm in diameter). Have two membranes, the outer smooth and generally featureless, the inner invaginating to produce *cristae*, generally at

right angles to long axis of the mitochondrion. Mitochondria from very active tissues have large numbers of tightly-packed cristae, those from relatively anoxic cells have few cristae. Mitochondria increase 5- to 10-fold in resting skeletal muscle which has been stimulated to contract for long periods. They are a site of acetylcholine production and their outer membranes house proteins (e.g. BCl-2) involved in the control of APOPTOSIS. The matrix within the inner membrane is often granular and contains ribosomes (of prokaryotic type), several copies of a circular DNA molecule (see MtDNA) and other protein-synthesizing components, several enzymes (e.g. those for the KREBS CYCLE, part of the UREA CYCLE, FATTY ACID OXIDATION and monoamine oxidation) and variable amounts of calcium and phosphate ions. Enzymes for haem synthesis (for cytochromes and haemoglobin) occur in the mitochondrial matrix.

Inner membrane is the site of the ELECTRON TRANSPORT SYSTEM, the ATP-synthesizing complex and enzymes involved in fatty acid synthesis. Enzymes in outer membrane include those involved in oxidation of adrenaline and serotonin, and others engaged in phospholipid metabolism. The two membranes differ in their permeabilities: outer freely permeable to small and medium-sized molecules and ions; inner permeable only to small uncharged molecules such as O_2 and undissociated H_2O and impermeable to glucose and $NADH_2$, but with various TRANSPORT PROTEINS permitting exchange of ATP and ADP, Krebs cycle intermediates and the accumulation of pyruvate, calcium and phosphate ions against concentration gradients, all coupled to electron transport (see ACTIVE TRANSPORT). The ATP *synthetase* complex ($F_0F_1ATPase$) is also embedded in inner membrane, comprising at least nine different polypeptides in two separable units: five make a spherical head ($F_1ATPase$) which can catalyse ATP synthesis; remainder comprise a proton channel (F_0) through which protons re-enter the mitochondrion after ejection by the electron transport system (see BACTERIO-RHODOPSIN and Fig. 113).

New mitochondria arise by growth and

division of existing mitochondria (resembling binary fission in bacteria). Mitochondrial fusion is involved in the production of 'giant mitochondria' during spermatid maturation, e.g. in *Drosophila*. Many of their proteins are encoded by genes in the nuclear genome, in particular the enzymes of the Krebs cycle and fatty acid oxidation, much of the electron transport system, the outer membrane proteins and DNA and RNA polymerases. In the slime mould *Physarum*, mitochondrial fusion and genome recombination can occur between *mif*$^+$ (plasmid-containing) and *mif*$^-$ strains. Most derived mitochondria contain the plasmid responsible. Yeast mitochondrial genomes may engage in recombination, intron $\omega+$ being commonly responsible, and nearly all mitochondria end up $\omega+$. Both the *mif* plasmid and $\omega+$ intron appear to be examples of SELFISH DNA/GENES. Compare F FACTOR. Mitochondrial GENETIC CODE differs from both 'nuclear' and bacterial codes in that the triplet UGA codes for tryptophan and is not a stop codon. Other codons may also have different meanings, even between mitochondria from different organisms.

Mitochondria of most C_3 plants of temperate latitudes do not engage in aerobic respiration during daylight, this being restricted to the hours of darkness. Instead these plants engage in the seemingly wasteful process of PHOTORESPIRATION, which is greatly reduced or absent in C_4 plants of tropical origin.

Mitochondria are generally regarded as having originated by ENDOSYMBIOSIS from a purple bacterium. Mitochondria in some protists harbour strange anaerobic ATP-producing pathways involving fumarate or nitrate as electron acceptors. See AMITOCHONDRIATE, EUKARYOTE, HYDROGENOSOME.

mitogen Substance triggering MITOSIS, most commonly by overcoming the braking mechanisms normally blocking progress through the CELL CYCLE, especially by acting on the G_1 phase to suppress Cdk activity. One of the first mitogens discovered was platelet-derived GROWTH FACTOR, which is both a mitogen and a growth factor (NB the term 'growth factor' does not imply that a substance has mitogenic activity). Mitogens often act through signalling cascades involving MAPKs (see Fig. 151). The important PROMOTER serum response element (SRE) lies upstream of many mitogen-inducible genes and is targeted by many extracellular signal molecules (see SERUM).

mitosis (karyokinesis) Method of nuclear division (M-phase of the CELL CYCLE) which produces two daughter nuclei, genetically identical to each other and to the original parent nucleus. Commonly accompanied by division of parental cytoplasm (cytokinesis) around the daughter nuclei to produce two daughter cells (cell division). It is the usual method by which nuclei are replicated during the growth, development and repair of multicellular organisms and coenocytic fungi. Prokaryotes lack nuclei and have no mitosis, replicating and segregating their DNA by a different process. Body cells (and nuclei) not undergoing mitosis are typically in INTERPHASE. The first stage of mitosis (*prophase*) commences with the first appearance of condensed chromosomes, each consisting of two identical sister chromatids held together by a CENTROMERE. In contrast to their behaviour in first meiotic prophase, homologous mitotic chromosomes do not normally pair up (see Fig. 114).

During *prophase* a spindle of MICROTUBULES assembles outside the nucleus. If centrioles are present, they have replicated in S-phase (see CELL CYCLE) and form foci for origins of spindle fibres. The two ASTERS get pushed apart by growth of microtubules. In *metaphase* the nuclear membrane disintegrates and spindle microtubules attached to the KINETOCHORES of the chromosome centromeres bring chromosomes into one plane (the 'metaphase plate') half way between the two spindle poles. Each chromosome lies with its long axis at right angles to the axis of the spindle. Metaphase may be quite lengthy. A surveillance mechanism, the 'spindle assembly checkpoint', monitors the spindle and blocks sister chromatid separation until these are attached to opposite poles of the spindle. Special proteins (Mad2 and Bub) seem to detect kinetochores with

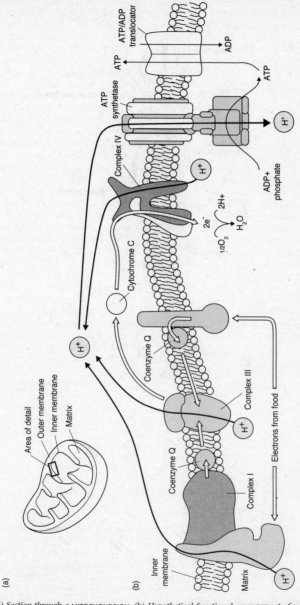

FIG. 113 (a) *Section through a* MITOCHONDRION. (b) *Hypothetical functional arrangement of proteins in the mitochondrial inner membrane.*

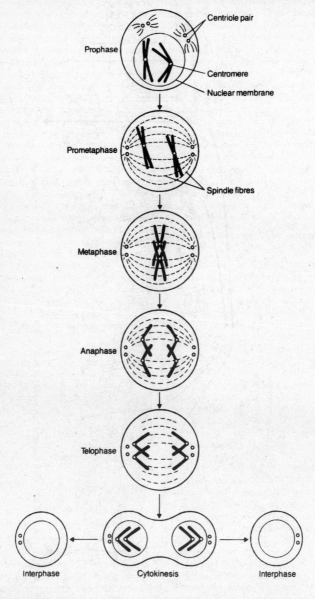

FIG. 114 *Diagrams showing the behaviour of one pair of homologous chromosomes in animal* MITOSIS.

unattached microtubules and feed this information to the cell cycle machinery (see ANAPHASE-PROMOTING COMPLEX). *Anaphase* begins when kinetochores separate, the two sister chromatids of a chromosome being carried towards the opposite spindle poles, which move further apart. Each chromatid is now a chromosome in its own right. A few minutes later, once chromosomes have reached the poles, a new nuclear membrane appears around each group of chromosomes, which decondense. Nucleoli reappear, constituting *telophase*. Mitosis is now complete and two daughter cells are normally produced by cytokinesis. Mitosis normally takes between a half and three hours. See FISSION, CLEAVAGE, MESOKARYOTE, MICROTUBULES (for involvement of four microtubule classes). Compare MEIOSIS.

mitospore Spore produced as a result of mitosis. Contrast MEIOSPORE.

mitosporic fungi Formerly Deuteromycota (Fungi Imperfecti); these are asexual fungi comprising about 2,600 genera and 15,000 species. It is an artificial assemblage and comprises known mitosporic fungi, which have not yet been correlated with any meiotic states (i.e. sexual stage is unknown). As these fungi lack a sexual stage they lie outside the remainder of the fungal classification, which is based heavily upon sexual reproduction. The mitosporic group is an artificial grouping. Some species may be secondarily sexual, others possibly never possessed a sexual stage or during the course of evolution it became lost and its functions have been replaced by PARASEXUALITY. These fungi are generally believed to be members of the Ascomycota or Basidiomycota, with the largest number in the former division. Mitosporic fungi that have been correlated with teleomorphs in the Ascomycota and Basidiomycota can be termed the anamorphs or anamorphic stages of those groups. Where correlation has been made between the anamorphic and teleomorphic stages the name of the holomorph (the whole fungus in all its correlated states) is that of the teleomorph. Classification within the mitosporic fungi is based on characteristics such as the manner in which the spores (conidia) are formed, conidial characteristics, absence of a sexual stage and absence of any meiotic or mitotic reproductive structures (agonomycetes, mycelia sterilia). Three classes are recognized: (i) Hyphomycetes – mycelial forms that bear conidia on separate hyphae or aggregations of hyphae but not inside discrete sporodochial conidiomata; (ii) Agonomycetes – mycelial forms that are sterile but may form chlamydospores, sclerotia and other vegetative structures, and (iii) Coelomycetes – forms bearing conidia in pycnidial, pycnothyrial, acervular, cupulate or stromatic conidiomata.

Mitosporic fungi are cosmopolitan in their distribution. Many are saprophytic in soil; others are plant and animal parasites (e.g. causing such diseases as ringworm and athlete's foot); some occur in flowing water. Many moulds are included, some commercially important, e.g. *Penicillium roquefortii* and *P. camembertii* in cheese production; *Aspergillus oryzae* in production of soy paste and in the brewing of saki; *A. oryzae* and *A. soyae* used together to produce soy sauce ('shoyu'). Citric acid is produced commercially from colonies of *Aspergillus* grown under very acid conditions. Mitosporic fungi grow prolifically on artificial media, and are widely used in genetic, biochemical and nutritional research. Many produce ANTIBIOTICS.

mitotic arrest The interruption of mitosis by the spindle assembly checkpoint. See MITOSIS, ANAPHASE-PROMOTING COMPLEX, p53.

mitotic recombination See RECOMBINATION, TWIN-SPOTS.

mitral valve Valve comprising two membranous flaps and their supporting tendons between the atrium and ventricle on left side of the hearts of birds and mammals. See HEART.

mitrates Small fossils (usually a few millimetres long) from marine sediments laid down 500–360 Myr ago and difficult to interpret because they resemble nothing alive today. Their calcite skeletons indicate links with echinoderms, but a water vascular system is only dubiously present and penta-radiate symmetry is lacking. The 'head' has two distinct sides, one made of large plates and the other convex with a patchwork of smaller plates. The 'tail' is segmented. They may have links with chordates or hemichordates.

mixed tissue Tissue containing more than one cell type, all originating from same group of stem cells in development. XYLEM and PHLOEM are examples.

mixotroph PHOTOAUTOTROPH capable of utilizing organic compounds in the environment; *facultative heterotroph* or *facultative parasite* may also refer to a mixotroph.

mobile element See TRANSPOSABLE ELEMENT.

modification Methylase activity preventing or reversing host restriction (see PHAGE RESTRICTION). Specific bases of the phage DNA within target site of the restriction endonuclease are methylated. Nonheritable change.

modifier Gene capable of altering expression of another gene at a different locus in the genome. Selection for such modifiers can establish a genetic background within which a new mutation at a locus will consequently be expressed to a greater or lesser extent. Modifiers are thus strong candidates for control of DOMINANCE and recessiveness of characters. See DOSAGE COMPENSATION, EPISTASIS, GENOMIC IMPRINTING, PENETRANCE.

modular organism An organism that grows by repeated iteration of its parts (e.g. leaves, shoots, branches of a plant, polyps of a coral or bryozoan). Such organisms are almost always branched, though the connections between branches may separate or decay and separated parts may in most cases then become physiologically independent (e.g. *Lemna minor* – duckweed). See RAMET, UNITARY ORGANISM.

modular proteins See DOMAIN (1), MOTIF, PROTEIN COMPLEXES.

modulator (Of enzyme) see allosteric ENZYME, ATP.

molar Crushing back tooth of mammal, without milk tooth predecessor (unlike premolar). Usually has several roots and complicated pattern of ridges and projections on grinding surface. See DENTITION.

mole See INSECTIVORA.

molecular activity (turnover number) Number of substrate molecules catalysed per minute under optimal conditions by a single enzyme molecule (or active site). Carbonic anhydrase has one of the highest molecular activities: 36×10^6 min^{-1} $molecule^{-1}$.

molecular clock The more important it is for the amino acid sequence of a protein to be conserved, the more slowly does that sequence evolve (the more conservative it is) through evolutionary time. For instance, the discarded peptide fragments formed when fibrinogen is converted to fibrin have no biological import and their sequence evolves rapidly, whereas haemoglobin and especially HISTONES are very conservative. Can be used to estimate the time since any two evolutionary lineages diverged, using proteins from living species as representatives of those lineages; however, it has to be remembered that the 'clock' runs faster for neutral sequences than for conserved ones. Nucleotide substitutions which result in a change in amino acid sequence (non-synonymous substitutions) will tend to be under-represented compared to those (synonymous, or silent) substitutions which cause no amino acid alteration (through degeneracy of the GENETIC CODE). This is because the effects of the former are prone to weeding-out by selection (but see NEUTRALISM), although nonsense mutations (synonymous, but producing start or stop codons) would also be subject to selection. Interestingly, the rate of synonymous nucleotide substitution varies from gene to gene.

The 'relative-rate test' is devised to test whether molecular sequences do evolve at a constant rate in different lineages. The rate of substitution would be the number of substitutions per site per year, given by the number of substitutions between two

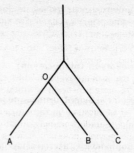

FIG. 115 *Figure illustrating the tree used in the relative-rate test as described in the* MOLECULAR CLOCK *entry.*

homologous sequences (K) divided by $2T$, T being the time taken for the two sequences to diverge from the ancestral sequence, usually inferred from palaeontological data (see RADIOMETRIC DATING). It involves comparing sequences from species A and B with those from a third (outgroup) species, C, whose lineage is known to have diverged earlier than the common ancestor (O in Fig. 115) of A and B. A and B have had the same time to accumulate differences since their divergence at O (call the number of differences K_{OA} and K_{OB} respectively). Assuming that sequence differences are cumulative, i.e. that substitutions gained in the divergence of C from O can be added to those gained in the transition from O to B to arrive at the number of substitutions between C and B, and that the rates of accumulation of differences are the same in the two lineages, it should be possible to determine the rate of substitution between A and B directly from the value of the sequence difference $K_{AC} - K_{BC}$.

However, there is considerable controversy at present over the interpretation of sequence data, although more confidence is usually placed in DNA data than protein data when it comes to inferring times of lineage divergences since DNA sequences contain more information than protein sequences. This is because replacement of an amino acid by another may require more than one non-synonymous substitution at the DNA level. Another sense of the phrase

'molecular clock' occurs in molecular studies of AGEING. See mtDNA, MITOCHONDRIAL EVE, MUTATION RATE.

molecular machines See PROTEIN COMPLEXES.

mollicutes General name for group of primitive gliding bacteria distinguished by possession of the smallest known cells, as little as 0.1 μm in diameter. Cells lack cell walls, but possess ribosomes and all other protein-synthesizing machinery, encoded by a very small genome of fewer than 650 genes (about 20% of a bacterial genome). Some are saprotrophic; others are pathogens of plants and animals, including humans. Known by a variety of names, including *pleuropneumonia-like organisms* (PPLOs), after the disease caused in cattle by the first member to be described. Six genera to date, about 50 of the 60 species belonging to *Mycoplasma*. Their ribosomal RNA sequences suggest they form a natural group, probably of bacterial origin. *Mycoplasma-like organisms* (MLOs) have been isolated from over 200 plant species and have been implicated in more than 50 plant diseases, often with symptoms of yellowing or stunting; they appear to be confined to sieve-tubes and to be passed passively from one sieve-tube member to another through sieve-plate pores.

Mollusca Phylum of bilaterally symmetrical, unsegmented invertebrates. Largely aquatic; coelom restricted to spaces around heart and within kidneys and gonads. Characteristically soft-bodied, with anterior head (rudimentary in Bilvalvia); commonly with rasping tongue (radula), large muscular ventral foot (modified to 'arms' in Cephalopoda); with viscera usually in a hump dorsal to the foot. Outer layer of visceral hump called the *mantle*, usually covered by a shell. Mantle typically enclosing a *mantle cavity* in which usually lie the gills (ctenidia), anus and opening of kidney and reproductive ducts. Heart typically present; circulation often includes both open (haemocoele) and closed components. Nervous system of cords and ganglia (see CENTRAL NERVOUS SYSTEM). Development mostly by spiral CLEAVAGE, often with trochophore larva. Includes Monoplacophora, Amphineura (chitons), Gastropoda (snails, slugs), Scaphopoda, Bivalvia (clams, oysters, etc.) and Cephalopoda (octopus, cuttlefish, squid). Enormous radiation among the phylum. About 80,000 species living; numerous fossils back to the Cambrian.

monadelphous (Of stamens) united by their filaments to form tube surrounding the style; e.g. in lupin, hollyhock. Compare DIADELPHOUS, POLYADELPHOUS.

Monera The KINGDOM including all PROKARYOTES.

mongolism See DOWN'S SYNDROME.

monoamine oxidase Enzymes (e.g. on the outer membrane of mitochondria) oxidatively removing the amine group of monoamines converting, e.g., noradrenaline to a glycoaldehyde. See MITOCHONDRION.

monocarpic (semelparous) (1) Having one carpel per flower, as in Leguminoseae. (2) (Of plants) reproducing only once and then dying. Contrast POLYCARPIC.

monochasium See INFLORESCENCE.

monochlamydeous (haplochlamydeous) (Of flowers) having only one whorl of perianth segments.

monoclonal antibody (mAb) See ANTIBODY.

monocolpate Referring to a pollen grain with one furrow or groove through which pollen tube will emerge.

Monocotyledonae See LILIOPSIDA and INTRODUCTION.

monocyte Largest of the kinds of vertebrate LEUCOCYTE. Can differentiate into MACROPHAGES (see LDL) and are the source of numerous GROWTH FACTORS, notably PDGF and interleukin-1. See also MYELOID TISSUE and Fig. 99.

monoecious (1) (Of plants) having both male and female reproductive organs on the same individual; in flowering plants, having unisexual, male and female, flowers on the same plant, e.g. hazel. See DIOECIOUS, HERMAPHRODITE. (2) (Of animals) see HERMAPHRODITE.

monokaryon (monocaryon) A fungal cell, hypha or mycelium in which there are one or more genetically identical haploid nuclei. Characteristic of the early phase in life cycles of many of the BASIDIOMYCOTA. Compare DIKARYON, HETEROKARYON.

monophyletic Of a taxon or taxa originating from and including a single stem species (either known or hypothesized) and either including the whole CLADE so derived (a holophyletic taxon), or else excluding one or more smaller clades nested within it (a paraphyletic taxon).

monoplanetism In some PROTISTA (e.g. some oomycota), where a zoospore of only one type is produced. Contrast DIPLANETISM.

monoploid Term used, in contrast to 'HAPLOID', to indicate the number of chromosomes (x) derived from each separate parental species' gametes produced by hybrids (typically allopolyploids). Thus, in modern wheat (*Triticum aestivum*), x = 7, although the haploid number n = 21 (i.e. three sets of 7 chromosomes).

monopodium An axis produced and increasing in length by apical growth; e.g. trunk of a pine tree. Compare SYMPODIUM.

monosaccate Refers to a type of pollen grain with a single air bladder.

monosaccharides Simple sugars, with molecules often containing either five carbon atoms (pentoses such as ribose, $C_5H_{10}O_5$) or six (hexoses such as glucose and fructose, $C_6H_{12}O_6$). Some, notably glucose and its amino derivatives, are monomers of biologically important POLYSACCHARIDES and GLYCOSAMINOGLYCANS. As carbohydrates their empirical formula approximates to $C_x(H_2O)_x$. Trioses (3-carbon sugars) such as phosphoglyceraldehyde and phosphoglyceric acid are important intermediates of many biochemical pathways (e.g. see GLYCOLYSIS, PHOTOSYNTHESIS).

monosomy Abnormal chromosome complement, one chromosome pair in an otherwise diploid nucleus being represented by just a single chromosome. An instance of ANEUPLOIDY. May arise by NON-DISJUNCTION. The norm in the heterogametic sex of those animals with the XO/XX mode of SEX DETERMINATION.

monosulcate Refers to a pollen grain with one furrow or groove on distal surface.

Monotremata Monotremes. The only order of the mammalian subclass PROTOTHERIA. Molecular systematics links monotremes with marsupials (see Fig. 108b).

monozygotic twins Twins developing from two genetically identical cells, cleavage products of a single fertilized egg which become completely separate (failure to do so may give rise to Siamese twins). They will be of the same sex. The nine-banded armadillo regularly produces a genetically identical litter and this amounts to a form of asexual reproduction. See FRATERNAL TWINS, FREEMARTIN, POLYEMBRYONY.

morph One form of INFRASPECIFIC VARIATION; one of the *forms* of a species population exhibiting POLYMORPHISM, generally of a markedly obvious kind.

morphactins Group of synthetic compounds derived from fluorenecarboxylic acid, that reduce and modify plant growth. Internally, the orientation of the SPINDLE axes of dividing cells is altered; most striking external effect is development of dwarf, bushy habit due to shortening of the internodes and loss of apical dominance (see AUXINS). Other effects include inhibition of phototropism and geotropism, of seed germination, and of lateral root development.

morphogen Any molecule controlling cell fate in a concentration-dependent manner. Controversy surrounding their putative existence has resolved in their favour although a prime candidate, RETINOIC ACID, has turned out not to be a morphogen. The strongest candidate morphogens are members of the transforming growth factor-β (TGF-β), Hedgehog (Hh) and Wnt families of secreted proteins. The factors creating and shaping morphogen gradients (e.g. endocytosis, antagonistic proteins which associate with them and prevent binding to receptors) vary from case to case and even differ for the same morphogen at different stages of embryo development. See *BICOID* GENE, MATERNAL EFFECT.

morphogenesis The generation of form and structure during development of an individual organism. *Morphogenetic movements* of large numbers of cells during ontogeny, being particularly pronounced during GASTRULATION. One aspect of such movements, if they are to result in appropriate cell location, is ADHESION between cells.

Considerable interest surrounds the theory that an animal's form develops through a hierarchical series of decisions determining the fates of cells and clones derived from them as cell number increases (see PHYLOTYPIC STAGE, ZOOTYPE). The cues enabling these decisions to be made are believed to result in PATTERN FORMATION, and take the form of molecules (*morphogens*) activating or suppressing gene expression. See COMPARTMENT, POSITIONAL INFORMATION.

Convergent extension is the term given to the movement of cells during gastrulation and neurulation of animal embryos (e.g. during germ band extension in *Drosophila*, gut elongation in echinoderms, as well as in ascidians, teleosts, birds and mammals). It involves a sheet of cells moving from the surface to the interior and then intercalating to form a longer, narrower line of cells, often extending in a plane at right angles to the initial movements. The polarization of intercalating cells during convergent extension involves WNT SIGNALLING. See Fig. 116.

morphology (1) Study of the form, or appearance, of organisms (both internal and external). Anatomy is one aspect of morphology. *Comparative morphology* is important in evolutionary study. (2) The actual 'appearance' of an organism, including such diverse features as behaviour, enzyme constitution and chromosome structure. Any and all detectable feature(s) of an organism.

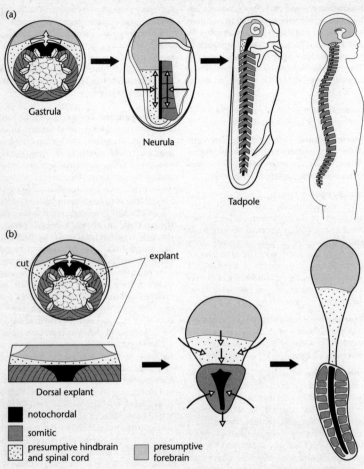

(a)

Gastrula

Neurula

Tadpole

(b)

cut

explant

Dorsal explant

■ notochordal

■ somitic

▦ presumptive hindbrain and spinal cord

▨ presumptive forebrain

FIG. 116 *Convergent extension movements elongate the anterior–posterior axis of the vertebrate body plan. The notochordal (black) and somitic (dark grey) tissues turn inside and converge (narrow) and extend (lengthen) in the gastrula and neurula stages of the frog embryo (a). The overlying presumptive hindbrain and spinal cord (stippled) tissues converge and extend coordinately but on the outside of the embryo. These movements push the head away from the tail and elongate the body axis of the tadpole. Similar movements elongate the body axis of mammals. (b) Cultured explants of the same tissues also converge and extend, showing that convergent extension movements are driven by internal forces. See* MORPHOGENESIS.

morphospecies Species recognized by a museum taxonomist using conventional criteria. These may harbour morphologically indistinguishable sibling species. See SPECIES.

morula Animal embryo during CLEAVAGE, at the stage when it is a solid mass of cells (blastomeres). Stage prior to the BLASTULA. Human embryos implant at this stage.

mosaic (1) Organism comprising clones of cells with different genotypes derived, however, from the same zygote (unlike CHIMAERAS). See HETEROCHROMATIN, OVOTESTIS. (2) (Bot.) Symptom of many virus diseases of plants; patchy variation of normal green colour, e.g. tobacco mosaic.

mosaic development See DEVELOPMENT.

mosses See BRYOPSIDA.

motif Term used to indicate conserved regions of developmental genes, or of the protein amino acid sequences they encode (more conserved than base sequences), which correspond to functional DOMAINS. Examples include homeodomain (encoding the HOMEOBOX), zinc finger, LEUCINE ZIPPER, basic HELIX-LOOP-HELIX, and paired domain (a conserved domain encoding the paired box). Some developmental genes have only one motif, while others have more. Thus, vertebrate *Pax* genes contain a homeobox and the paired domain. Many proteins' biochemical activities or cellular functions, e.g. those of transcription factors and enzymes, are recognizable from their motifs alone. See Fig. 117.

motor end-plate See NEUROMUSCULAR JUNCTION.

motor neurone (motor nerve) Neurone carrying impulses away from central nervous system to an effector (usually a muscle or gland). Efferent neurone. In vertebrates, motor neurones leave spinal cord via the ventral horn of a spinal nerve.

motor protein A protein which, through its ability to hydrolyse an appropriate energy-rich molecule (e.g. ATP, GTP), is able to generate the force required to move objects within cells, or the cells themselves – acting as a molecular transducer. Examples in eukaryotes are MYOSIN, DYNEIN and KINESIN. RNA polymerases act as motor proteins, pulling DNA through during transcription.

motor unit One motor neurone, its nerve terminals, and the skeletal muscle fibres innervated by them. There may be 200 or more muscle fibres in each vertebrate motor unit in large muscles of the leg or trunk.

mould (1) Any superficial growth of fungal mycelium. Frequently found on decaying fruit or bread. (2) Popular name for many fungi.

moulting (1) For moulting in arthropods, see ECDYSIS. (2) The periodic shedding of FEATHERS in birds, and of hair in mammals.

mouthparts Paired appendages of arthropod head segments, surrounding the mouth and concerned with feeding and sensory information. Evolved from walking limbs. Similarities in mouthparts between different arthropod groups are often due to convergence rather than homology. Arachnids have a pair of CHELICERAE and PEDIPALPS but no mandibles; crustaceans a pair each of MANDIBLES, MAXILLULES, MAXILLAE and MAXILLIPEDS; insects a pair of mandibles and maxillae, a LABIUM and a LABRUM; centipedes a pair of mandibles and two pairs of maxillae; millipedes a pair of mandibles and maxillae. Mouthparts show most adaptive radiation in insects, commonly being modified for piercing and/or sucking (see PROBOSCIS). See Fig. 118.

M-phase See CELL CYCLE.

MRCA Most recent common ancestor. See CLADISTICS.

MRSA Methicillin-resistant *Staphylococcus aureus*. Strains of this bacterium arose from acquisition of a novel β-lactam-resistant penicillin-binding protein gene on a large DNA fragment of unknown source. It has since been transferred to *S. epidermidis* (MRSE) and other coagulase-negative staphylococci. MRSA is a major global problem, often harbouring multiple resistance genes. See ANTIBIOTIC RESISTANCE ELEMENT.

mtDNA Mitochondrial DNA. The mitochondrial genome almost always consists of a circular duplex, always less than 100 kilobases in length (16.5 kb in humans). In animals, generally contains less than 20 structural genes, including some for mitochondrial ribosomal RNAs, transfer RNAs and perhaps for some subunits of a few of the respiratory enzymes. Mitochondrial DNA evolves rapidly, with a mutation rate

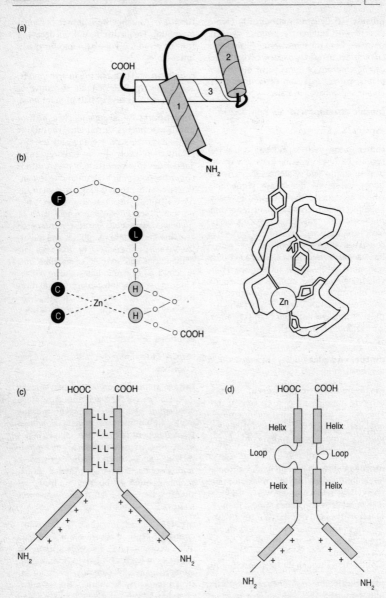

FIG. 117 *Structural models of:* (a) *the helix-turn-helix DNA-binding domain;* (b) *the zinc finger DNA-binding domain;* (c) *the leucine zipper DNA-binding domain;* (d) *the helix-loop-helix DNA-binding domain. See* MOTIF.

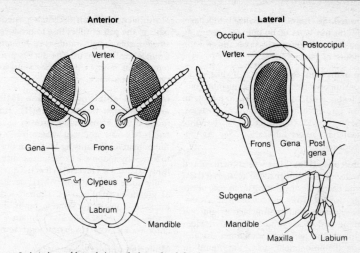

FIG. 118 *Anterior and lateral views of a locust head showing positions of* MOUTHPARTS *and head areas.*

up to ten times that of nuclear DNA. Once thought to exhibit strict MATERNAL INHERIT- ANCE, but some paternal inheritance (esp. in mussels, *Mytilus*) has been shown. Because it exists in many more copies ($500-10^3$) per cell than does nuclear DNA (2 copies), mtDNA is much more likely to survive intact in long-dead specimens, e.g. *HOMO*. mtDNA resembles a eubacterial genome in miniature, that of the flagellate *Reclino- monas* containing 62 protein-coding genes, 25 more than its nearest competitor (the mtDNA of the bryophyte *Marchantia*). Mutations in mtDNA have been linked to heart failure, diabetes, AGEING and certain degenerative diseases such as ALZHEIMER'S and motor disturbances. See CYTOPLASMIC MALE STERILITY, MITOCHONDRIAL EVE, MITOCHON- DRION, PLASMID, SPECIES FLOCKS.

MTOC **(microtubule-organizing centre)** The cell's nucleation centre for MICROTUBULES, duplicated along with nuclear DNA replication during S-phase of the CELL CYCLE. In most animals the CENTROSOME is the MTOC; however, the microtubules are not directly nucleated by the centriole itself. It is not currently known how they do originate, although CYCLINS A and B are involved. Immunofluorescence work will be of value here. The fungal MTOC is the spindle pole body.

Mu One mobile DNA element. See TRANSPOSON.

muciferous body (= mucilage body) Vesicles found in some algal cells (e.g. CHRY- SOPHYCEAE, HAPTOPHYTA, RAPHIDOPHYCEAE, DINOPHYTA) that contain granular mucilage bound by a single membrane and lie beneath the plasmalemma and often dis- charge upon stimulation. In the EUGLENO- PHYTA cells are permanently coated with a thin slime layer produced by the muciferous body, and those species with large mucifer- ous bodies eject their contents upon irri- tation and produce a copious slime layer around the cell, which in *Euglena gracilis* comprises glycoprotein and polysac- charides.

mucilage Slimy fluid containing complex carbohydrates; secreted by many plants and animals. Swells in water, and is often involved in water retention.

mucilage canal Elongate cells in cortex of some brown algae (Order Laminariales) and cycads that may be involved in conduc- tion of mucilage.

mucilage hairs Specialized mucilage-producing hairs of bryophytes, occurring most commonly near leaf axils and growing points of gametophores.

mucins (mucoproteins) Jelly-like, sticky or slippery GLYCOPROTEINS. Formed by complexing a GLYCOSAMINOGLYCAN to a protein, the former contributing most of the mass. Some provide an intercellular bonding material, others lubrication. See PEPTIDO-GLYCAN.

mucopeptides Major constituents of cell walls of blue-green algae (CYANOBACTERIA) and bacteria, constituting up to 50% of the dry weight. They are GLYCOSAMINOGLYCANS, with chains of alternating N-acetylglucosamine and N-acetylmuramic acid residues linked by peptides sometimes including diaminopimelic acid; not found in eukaryotes.

mucopolysaccharide See GLYCOSAMINO-GLYCAN.

mucoproteins See MUCINS.

mucosa Se MUCOUS MEMBRANE.

mucous membrane General name for a moist epithelium and, in vertebrates, its immediately underlying connective tissue (*lamina propria*). Applied particularly in context of linings of gut and urogenital ducts. Epithelium is usually simple or stratified, often ciliated; usually contains goblet cells secreting mucus.

mucus Slimy secretion containing MUCINS, secreted by goblet cells of vertebrate mucous membranes. Some invertebrates produce similar sticky or slimy fluids.

Müllerian duct Oviduct of jawed vertebrates (i.e. excluding Agnatha). Generally paired, except in birds. A mesodermal tube, arising embryologically from mesothelium of coelom in close association with WOLFFIAN DUCT, and opening at one end by a ciliated funnel into coelom, joining the cloaca (or its remnant in placental mammals) at the other. Muscular and ciliary movements pass eggs down the tube; where fertilization is internal, sperms pass up it. In marsupials and placentals, gives rise to *Fallopian tube*, *uterus* and *vagina*. In all placentals, posterior parts of the pair of tubes fuse to produce a single median vagina and, as in humans, a single median uterus. The duct is just a remnant at most in males. See MÜLLERIAN INHIBITING SUBSTANCE.

Müllerian-inhibiting hormone (MIH, anti-Müllerian hormone, AMH) Hormone of the transforming growth factor family (see TGFS) produced by Sertoli cells of the human foetus causing Müllerian ducts to regress by apoptosis. Exogenous addition leads to partial sex reversal in foetal X X rats, while male rats lacking AMH develop as pseudo-hermaphrodites. Human locus involved is on chromosome 19, but is only expressed if foetal cells also contain a Y chromosome. See TESTIS-DETERMINING FACTOR.

Müllerian mimicry See MIMICRY.

Müller's ratchet Process, first recognized by H. J. Müller, whereby in asexual populations no line can ever evolve to contain less mutational load than is present in the line currently with the least mutational load. The ratchet is supposed to move on one notch as each least mutationally loaded line in turn becomes extinct through genetic drift, leaving a more mutationally loaded line as the new least loaded one as a new mutant allele drifts to fixation. It has been suggested that one advantage of sex (recombination) may have been to slow the ratchet and provide time for back and compensatory mutations to appear.

multiaxial Having an axis with several apical cells that give rise to nearly parallel filaments; found in many red algae (RHO-DOPHYTA).

multicellularity Occurrence of organisms composed not merely of more than one cell, but of cells between which there is (a) considerable division of labour brought about by cell DIFFERENTIATION, and (b) a fairly high level of intercellular communication. Increase in cell number must be accompanied by intercellular ADHESION, and by itself is of relatively little evolutionary significance. However it provides opportunities for division of labour and specialization between cells, which may enable further increase in size, providing scope for improved competitiveness in a variety of respects, and exploitation of fresh niches. Evolved multicellular organization depends upon epigenetic inheritance mechanisms (e.g. DNA METHYLATION and CELLULAR MEMORY) for maintenance of the DETERMINED and differentiated state. Sponges (PORIFERA) display considerable communication between cells (see AMOEBOCYTES), although this does not take the form of a nervous system. Occasionally used rather loosely of groups of cells between which there is little differentiation and for which the terms colonial (e.g. *Volvox*) or filamentous (e.g. *Spirogyra*) are more appropriate. Social organization of prokaryotic MYXOBACTERIA and eukaryotic MYCETOZOA may well be termed multicellular. The earliest fossils of multicellular organisms are currently those of certain algae from Canadian rocks dated to 1.2 billion (1.2×10^9) years BP, although the earliest animals (in old terms, 'metazoans') so far discovered date from the Ediacaran beds of Newfoundland, Canada (see CAMBRIAN). Disputed structures, thought by some to be worm casts, are found in sandstones of south-western Australia. See ACELLULAR, COLONY, COENOBIUM.

multienzyme complex Complex and compound molecule usually containing several enzyme subunits performing different stages in a biochemical pathway along with intrinsic cofactors (prosthetic groups). The DNA replicating enzymes (DNA *polymerase complex*) of cells, including those coded by viral genomes, form such complexes, as do *pyruvate dehydrogenases* of mitochondria. Numerous examples occur in eukaryotic biosynthetic pathways for amino acids, folate, fatty acids, purines and pyrimidines. See PROTEASOMES, PROTEIN COMPLEXES, RIBOSOMES.

multifactorial inheritance Any pattern of inheritance where variation in a particular aspect of phenotype is dependent upon more than one gene locus (in this sense synonymous with *polygenic*) but often also, especially in some human clinical disorders, upon particular environmental conditions which tip the balance in favour of expression: enough genes may predispose one to having the disease, but exposure to some environmental factor may be necessary before actual symptoms of the disease are presented. Allelic differences involved are usually of varying but small individual effect. Sex differences may also be involved. Examples include schizophrenia, diabetes and hypertension. At least 12 separate chromosome regions have been implicated in the development of type 1 DIABETES mellitus. See HUMAN GENOME PROJECT.

multifinger loop See TRANSCRIPTION FACTORS.

multigene families Genes with considerable base sequences in common and thought to have descended from a single ancestral gene through GENE DUPLICATION and modification. The duplication events (as when gene *A* gives rise in one lineage to gene A) that produce gene families occur in individuals and, unlike speciation events, do not match the cladistic events in the lineage of organisms carrying those genes; so subsequent evolutionary divergence of the gene copies does not correspond to the splitting patterns of the organismal lineages. Descendants of the species in which the gene duplication occurred will possess both *A* and A. Such genes are PARALOGOUS and, by themselves, misrepresent phylogeny. Multigene families include HOMEOBOX- and PAIRED BOX-containing genes, genes for globins and NUCLEAR RECEPTORS, the IMMUNOGLOBULIN SUPERFAMILY, and genes whose products contain multifinger loops.

multilayered structure (Of flagella) comprising a more or less rectangular body attached to the anterior end of the single broad band of microtubules in the CHAROPHYCEAE and in the spermatozoids of lower land plants. It lies directly beneath the basal bodies of the flagella and comprises four layers, the layer closest to the plasmalemma containing microtubules of the root. Beneath are two electron-dense layers, the bottom-most layer comprising small microtubules. May be present in the Micromonadophyceae (CHLOROPHYTA) but absent in the Chlorophyceae and Ulvo-phyceae.

multinucleate Cell or hypha containing many nuclei. See ACELLULAR, COENOCYTE, SYN-CYTIUM.

multiple allelism Simultaneous occurrence within a species population of more than two alleles (*multiple alleles*) at a given gene locus. A common phenomenon, contributing not only to continuous variation of a character in the population but sometimes also, as in inheritance of human BLOOD GROUPS, to discontinuous variation. Common at loci responsible for INCOMPATI-BILITY in plants. See SUPERGENE.

multipolar neuron See NEURON.

multiregional model For this model of human evolution, see *HOMO*.

Multituberculata Extinct group of proto-therian mammals, possibly the sister group of the therians. Mainly of late Jurassic–Cretaceous times, but by extending into the late Eocene (~35 Myr BP) its span far exceeds that of any other mammalian order. They had chisel-like incisors and multicuspate cheek teeth (as opposed to the three or fewer of therian mammals). Probably the first her-bivorous mammals; somewhat rodent-like in form and dentition. Some were medium-sized. Coprolitic fossils from northern China (~55 Myr BP) reveal multituberculate hair. See PANTOTHERIA.

multivalents Associations of more than two chromosomes, joined by chiasmata, at meiosis in POLYPLOIDS (seen most clearly at first metaphase). Where homology extends to more than just pairs of chromosomes (as it does particularly in AUTOPOLYPLOID cells) synapsis will probably result in crossing-over between any homologous regions present. When such multiply chiasmate chromosomes move apart at anaphase the result may be a chain or ring of chromo-somes, and since multivalents do not always disjoin regularly, autopolyploids often have a proportion of meiotic products with unbalanced (aneuploid) chromosome numbers giving loss of fertility. See ALLOPOLY-PLOID.

mureins Group of MUCINS found in bac-terial cell walls.

muscarine An ALKALOID which, like nic-otine, mimics acetylcholine action at cer-tain cholinergic junctions (*muscarinic junctions*). Affects target organs of parasym-pathetic nervous system (e.g. vagus nerve); its effects being blocked by atropine, those of nicotine by curare.

Musci See BRYOPSIDA.

muscle Any of a spectrum of animal tissues of mesenchymal origin, from fibroblast-like cells through to skeletal muscle fibres. Smooth and striated muscle are found in all animal phyla from the Coelenterata upwards. Contractile role of muscle is attributable to a proliferation of protein microfilaments from cell cortex to cell interior. See STRIATED MUSCLE, SMOOTH MUSCLE, CARDIAC MUSCLE.

muscle contraction The force-generating response of muscle to stimulation. May be *isotonic* (when muscle shortens during con-traction) or *isometric* (when there is no change in muscle length during contrac-tion). Molecular mechanism of contraction is probably similar in smooth, cardiac and striated muscle but is best understood in the last. Myofibrils in a resting striated muscle fibre consist of very precise arrangements of filaments of proteins ACTIN and MYOSIN (see STRIATED MUSCLE). In order for contrac-tion to occur, several conditions must be met. Currently accepted account (*sliding filament hypothesis*) holds that force genera-tion can only occur when actin and myosin filaments make contact and form the com-plex ACTOMYOSIN. This they can only do in the presence of intracellular calcium ions (Ca^{2+}). But the tubular sarcoplasmic reticu-lum has CALCIUM PUMPS which accumulate Ca^{2+}, keeping it scarce. These pumps are shut off when ACTION POTENTIALS reach the terminal cisternae of the sarcoplasmic retic-ulum where they make contact with inwardly folded transverse tubules (T-tubules) of the fibre's outer membrane (see Fig. 119). Only then does Ca^{2+} flow from the cisternae, surrounding the myofilaments. These ions then bind to *troponin* molecules

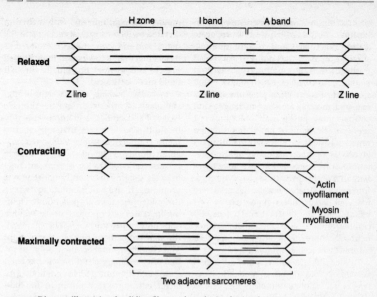

FIG. 119 *Diagram illustrating the sliding filament hypothesis of striated* MUSCLE CONTRACTION.

attached to actin filaments and in turn cause a shift in positions of *tropomyosin* molecules also attached to actin. These unmask sites on actin to which myosin can bind, forming actomyosin. But the 'heads' on myosin filaments will only bind to actin after they have been primed to do so by hydrolysis of ATP, which each head catalyses through its own ATPase activity. Only when ATP has been hydrolysed (i.e. before contraction occurs) is the myosin head ready to attach to an actin filament, changing its own shape (allosterically) so as to pull the actin filament past it. It can then release its bound ADP and inorganic phosphate, hydrolyse another ATP molecule (which requires magnesium ions, Mg^{2+}) and swing back ready for another power stroke. Repeated cycles of this sequence result in the actin filament sliding along relative to myosin; and since in muscle cells actin filaments are bound to Z-discs, the H-zone in the middle of the contracting sarcomere will narrow and darken and the I-bands disappear: the sarcomere has then fully contracted.

Striated muscle contracts with an ALL-OR-NONE RESPONSE. It relaxes only if ATP is available to free myosin from actin and if Ca^{2+} ions are actively removed from sarcoplasm by terminal cisternae, which they are in the absence of action potentials along the sarcolemma (and transverse tubules). In the absence of ATP the actomyosin complex persists and *rigor mortis* occurs.

There is limited availability of ATP in sarcoplasm, and the immediate source of energy for its resynthesis is another HIGH-ENERGY PHOSPHATE compound, the PHOSPHAGEN *creatine phosphate (phosphocreatine)*. If muscle is forced to contract repeatedly the amplitude of contractions will decrease and ultimately fail altogether when energy resources are expended. Muscle, like nerve, has brief absolute and relative *refractory periods* (see IMPULSE) after an action potential has passed a point on its membrane. During these the membrane becomes repolarized. The relative refractory period of striated muscle is much shorter than that of cardiac muscle, but longer than that of nerve. See TETANUS. Vertebrate striated muscle is of two types: slow-contracting *tonic fibres*, often involved in postural control, which do not propagate action potentials, and require multiple stimulation to contract doing so in a graded (rather than an all-or-nothing) manner; and *phasic*, or *twitch, fibres*, which may be either slow- or fast-contracting. Slow phasic fibres (e.g. red muscle) are slow to fatigue and contract more quickly than tonic fibres, but more slowly than fast phasic fibres (white muscle). Fast phasic fibres rely mainly upon glycolysis or upon oxidative phosphory-lation for ATP production, oxidative fibres fatiguing more slowly than glycolytic ones. Invertebrate striated muscle has relatively few, but larger, fibres than that of ver-tebrates and the innervation is sparse by comparison. Unlike vertebrate motor neurons, those of invertebrates may be inhibitory as well as stimulatory.

Striated muscle will respond to a suc-cession of rapid stimuli. In moderate activ-ity oxygen supply to muscle is adequate to prevent build-up of lactic acid, but if work rate exceeds oxygen supply an OXYGEN DEBT arises. Prior to this, the oxygen stored in MYOGLOBIN will have been used up. *Muscle tone* (*tonus*) is the sustained low-level reflexly controlled background muscle con-traction without which muscle flabbiness and poor responsiveness result. MUSCLE SPINDLES maintain this tone, which is variable.

Smooth muscle contractions are either *tonic* or *rhythmic*. Some smooth muscle is self-exciting (see MYOGENIC). In tonic con-tractions slow waves of electrical activity pass across adjacent cells, which are often fused to form a syncytium. Bursts of action potentials then cause the muscle to contract in functional blocks when the waves reach a sufficient intensity. Parasympathetic stimulation is responsible for tonus in much of the vertebrate gut. Rhythmic contrac-tions of smooth muscle in walls of tubular organs (*peristaltic* contractions) are also due to rapid spikes of action potentials; but the initiating signal here is usually stretching of the tube wall, which elicits contraction either directly or reflexly via local nerve plexuses (SEE AUERBACH'S PLEXUS).

Cardiac muscle contracts and relaxes rapidly and continuously, with a rhythm dictated by its intrinsic pacemaker and the neural and endocrine influences upon it. It has a longer refractory period than striated muscle and so does not fatigue through build up of lactic acid.

Anaerobic training (e.g. weight-lifting) stimulates synthesis of muscle proteins (skeletal and cardiac) and improves ability in short-term activities that rely on iso-metric contraction. *Aerobic training* (e.g. jog-ging) improves blood flow to muscle because raised CO_2 levels relax arteriole muscles, dilating them, and angiogenesis is promoted. It also builds endurance, where isotonic contractions are paramount. Train-ing increases resting heart stroke volume and pulse rate (up to 7-fold increase), through sympathetic nervous stimulation. Mitochondrial numbers increase 5- to 10-fold after resting skeletal muscle has been stimulated to contract for long periods. Any factor reducing oxygen tension in the blood (e.g. smoking) would tend to negate these positive training effects. See CORONARY HEART DISEASE, GLYCOGEN (for loading), LIPOPROTEIN, NICOTINE.

muscle spindle Stretch receptor (proprio-ceptor) of vertebrate muscle. Each com-prises 3–10 intrafusal muscle fibres partially enclosed in a spindle-shaped connective tis-sue capsule, surrounding which are normal extrafusal muscle fibres. Different motor neurons, arising in the grey matter of the spinal cord, innervate the two types. When muscle is stretched, sensory nerve endings in its spindles fire and send impulses via the spinal cord to the cerebellum and cerebral cortex. When muscle contracts these impulses are inhibited. Responsiveness of the spindle can be altered by central control from the reticular formation of the mid-brain: each spindle can be made to contract even when its muscle is not contracting, thus enabling regulation of the rate of muscle contraction in response to different loads. Increased spindle firing due to muscle stretch will result in reflex muscle contrac-tion and reduced spindle firing. The nerve circuitry is quite complex and involves inhibitory and excitatory loops.

mushroom Common name for the edible fruiting bodies of fungi belonging to the Agaricaceae (BASIDIOMYCOTA).

mutagen Any influence capable of increasing MUTATION rate. Usual effect is chemical alteration, addition, substitution or dimerization of one or more bases or nucleotides of the genetic material (DNA or RNA). Chemical mutagens include: nitrous acid, which deaminates adenine to hypoxanthine and cytosine to uracil; base analogues such as 5-bromouracil and 2-amino purine, which are incorporated into DNA but have slightly different base pairing properties; acridine dyes, which become lodged in the DNA helix and interfere with its replication, causing an extra base to be inserted (frameshift mutation); ALKYLATING AGENTS, formaldehyde, 1-nitropyrene, dinitropyrenes (see POLLUTION), mustard gas and aflatoxin. ETHIDIUM BROMIDE causes nonenzymatic (photochemical) nicking of double-stranded DNA. Physical mutagens include: ULTRAVIOLET IRRADIATION, X-RAY IRRADIATION and beta and gamma irradiation. *In vitro* mutagenesis is now routinely used to modify existing gene products. In site-directed mutagenesis, a synthetic oligonucleotide is used to achieve GENE CONVERSION in a plasmid prior to introduction into its *Escherichia coli* host. Alternatively, wild-type sequences can be removed from a plasmid and the desired mutant sequence (cassette) ligated in instead. See AMES TEST, CARCINOGEN, *IN VITRO* MUTAGENESIS.

mutant An individual, stock or population expressing a MUTATION.

mutation Alteration in the arrangement, or amount, of genetic material of a cell or virus. They may be classified as either *point mutations*, involving minor changes in the genetic material (often single base-pair substitutions), or *macromutations* (e.g. deletions), involving larger sections of chromosome. Mutations often occur during DNA replication, may involve TRANSPOSABLE ELEMENTS and TRANSPOSONS and commonly involve enzyme activity. Precise definition of the term is difficult if one wishes to exclude such phenomena as CROSSING-OVER and other forms of recombination (see, e.g., EXONS, ANTIBODY DIVERSITY).

The effects of macromutations on chromosome structure are often visible during mitosis and meiosis and include INVERSION, TRANSLOCATION, DELETION, DUPLICATION (see TRANSPOSABLE ELEMENT) and POLYPLOIDY. NON-DISJUNCTION produces chromosome imbalance (e.g. aneuploidy). Whereas point mutations commonly cause amino acid substitutions in the polypeptide encoded by the gene and often have minor effects on gene function, macromutations (especially if they involve deletions) commonly lead to syndromes of abnormalities which seriously reduce FITNESS, and are more often lethal. Many point mutations fail to result in amino acid substitutions because the GENETIC CODE is degenerate. Others, though altering amino acid sequence, either have little effect or only partially inactivate gene function (as in *leaky* auxotrophs of fungi and bacteria). But a single amino acid replacement in a critical position can abolish an enzyme's activity. One type of radical effect that can follow from a single base-pair substitution is the creation of STOP CODON within the open reading frame, usually rendering the translation product useless. Mutations creating stop codons (*nonsense* mutations) can often be phenotypically suppressed by compensating mutations in the anticodon sequence of tRNA molecules (*suppressor mutations*); but see NONSENSE-MEDIATED DECAY. *Frameshift mutations* are nucleotide additions and deletions not involving a multiple of three base-pairs, which move the 'reading frame' of tRNA to the left or right during the translation phase of PROTEIN SYNTHESIS. These usually have drastic effects on gene function. *Mis-sense mutations* arise when a DNA sequence transcribed into a codon specifying one amino acid is altered so that it specifies another. The effect depends upon the importance to the final product of the original amino acid. One such is a cause of ALZHEIMER'S DISEASE. Most mutations lead to partial or complete loss of functional gene-product and are recessive to wild-type alleles, normal gene product normally being produced in amounts sufficient for function even when the allele is present in single dose. However, *loss-of-function mutations* are those causing the normal allele (now present in single dose) to produce inadequate product for normal phenotype (*haplo-insufficiency*, or *haplo-lethality*) and therefore tend to have a dominant effect on phenotype. *Null mutations* are extreme cases, where all function is lost. *Gain-of-function mutations*, such as constitutively active alleles, tend either to cause overproduction of normal gene-product or production of novel or toxic gene products and tend also to have dominant effects on phenotype – esp. if key regulatory genes. Gain-of-function mutations are often expressed in tissues where that gene is not usually transcribed. *Mutation 'hot spots'*, consisting largely of trinucleotide repeat sequences, have been identified in numerous human disease-causing genes. Unstable, they arise during mitosis or meiosis and can increase stochastically in copy number, producing somatic or germ-line mosaicism. See DOSAGE COMPENSATION.

A distinction may be made in some multicellular organisms between mutations occurring in body cells (somatic mutations), and those occurring in GERM LINE cells. Only the latter can normally be inherited and play a part in evolution (but see POLYPLOIDY).

By themselves, mutations do not normally direct the path of evolution: there is no 'mutation pressure'; rather, they provide the heritable variation upon which selection may act. It is usually accepted that mutations are random with respect to requirements of the cells or organisms in which they arise; but debate is developing as to whether or not some bacterial populations can respond to selection pressure by increasing the frequency of mutation rates for favourable genes. It now looks as though, when *E. coli* is subjected to unfavourable conditions (e.g. lactose starvation), it enters a 'hypermutable' state in which mutations arise randomly with respect to the adaptive benefit of the bacterium, although some such mutations may actually be adaptive. Some mutations are caused by the insertion of retrotransposons (see TRANSPOSABLE ELEMENTS). See MUTAGEN, GENE CONVERSION, GENETIC VARIATION, MOLECULAR CLOCK, NEO-LAMARCKISM, REPLICA PLATING.

mutation load The extent to which mutation impairs average population fitness. 'Other factors being equal', if a new mutation exhibits partial dominance, with large heterozygous effects, calculation shows that the mutation load is twice the allele's mutation rate per gamete (i.e. 2μ). If recessive, it is equal to μ. Mutation load is proportional to mutation rate, not to the selective disadvantage of individual mutant alleles. See MULLER'S RATCHET.

mutation rate The frequency with which MUTATIONS arise in populations of organisms or in tissue culture. Experiments suggest that an average of about one base-pair changes 'spontaneously' per 10^9 base-pair replications. If proteins are, on average, encoded by about 10^3 base-pairs then it would take about 10^6 cell generations before the protein contained a mutation. See MOLECULAR CLOCK, MUTAGEN.

mutual interference Interference amongst predators leading to a reduction in the consumption rate of individual predators, which increase with predator density. See PSEUDOINTERFERENCE.

mutualism See SYMBIOSIS.

mycelium Collective term for mass of fungal hyphae constituting vegetative phase of the fungus; e.g. mushroom spawn. Some bacteria (e.g. Actinobacteria) have mycelial vegetative forms.

Mycetozoa A subgroup within the amoebozoan group of eukaryotes, comprising protostelid, plasmodial and dictyostelid slime moulds. See MYXOMYCOTA for more traditional taxonomy.

***myc* gene, Myc protein** The *myc* gene is a dominant oncogene encoding the oncoprotein Myc, whose deregulated expression drives cell proliferation since one of its normal roles is to promote CELL CYCLE progression. Myc increases transcription of several CELL GROWTH-promoting genes. See C-MYC.

mycobiont Fungal partner of a lichen; usually member of ASCOMYCOTA. Compare PHYCOBIONT.

mycology Study of fungi.

mycophage Fungal PHAGE.

mycoplasmas See MOLLICUTES.

mycoprotein A fibrous meat substitute produced by growing the fungus *Fusarium graminearum* on a glucose-and-mineral solution in continuous fermenters (see BIO-REACTOR). Sold under brand names, such as QUORN. See SINGLE-CELL PROTEIN.

mycorrhiza A mutualistic, non-pathogenic, or weakly pathogenic, association of a fungus and the roots of a higher plant. Mycorrhizae are found in most groups of vascular plants. In fact they occur in the majority, perhaps up to 85% of plant species. Two main types of mycorrhizae occur: (i) Endomycorrhizae, which are most common, occurring in about 80% of all vascular plants. The mycorrhiza most widely distributed through the plant kingdom, and geographically, is a type of endo-infection formed by members of the Zygomycota of the order Glomales and is termed vesicular-arbuscular. In this association the penetrating hyphae form finely branched haustorial branches (arbuscules) or coils (pelotons) and vesicles. Other associations include the orchid-basidiomycete and ericoid-ascomycete associations in which the fungal hyphae enter the cortical cells of the root, enveloped by the plasmalemma of the host; (ii) Ectomycorrhiza, in which relationships are species-specific involving certain trees and shrubs, including beech, willow, and pine, as well as a few tropical trees. In fact ectomycorrhizal relationships are characteristic of temperate and boreal forest trees with different species of basidiomycetes (e.g. species of *Amanita*, *Boletus*, *Russula* and some ascomycetes, e.g., *Tuber*). In this relationship the fungus develops a sheath around the root surface from which hyphae extend outward into the soil and inwards between the cortical cells with which they interface with a HARTIG NET. The mycelium in the soil plays a highly important role transferring phosphorus and nitrogen to the plant, replacing the function of root hairs, which are frequently absent. See Fig. 120.

mycosis Disease of animals caused by fungal infection; e.g. ringworm.

mycotrophic (Of plants) having MYCO-RRHIZAS.

MyD88 An adaptor protein activated in TOLL-LIKE RECEPTOR pathways and containing both a TIR domain and a death domain. On association with a TLR, it recruits members of the interleukin-1 receptor-associated kinase family through death domain–death domain homophilic interactions.

myelin sheath Many layers of membrane of SCHWANN CELL (in peripheral nerves) or of oligodendrocyte (central nervous system) wrapped in a tight spiral round a nerve axon forming a sheath preventing leakage of current across the surrounded axon membrane except at *nodes of Ranvier*. Cell membranes comprising the myelin sheath contain large amounts of the glycolipid galactocerebroside. Such *myelinated neurons* conduct faster than equivalent non-myelinated ones as current 'jumps' from one node to the next. See IMPULSE.

myeloid tissue Major site of vertebrate HAEMOPOIESIS, restricted except in embryo to red bone marrow (e.g. in ribs, sternum, cranium, vertebrae, pelvic girdle). Mesodermal in origin. Myeloid stem cells give rise to cell lineages distinct from those of LYMPHOID TISSUE. Products include MONOCYTES and descendant MACROPHAGES, MAST CELLS, BASO-PHILS, NEUTROPHILS, EOSINOPHILS, MEGAKARYO-CYTES and their PLATELETS. In embryos, liver and spleen are principal myeloid sites.

myeloma Malignant cancers of MYELOID TISSUE; causing anaemia; especially in the middle-aged and elderly. See CANCER CELL.

myoblast Precursor cells of vertebrate SKEL-ETAL MUSCLE fibres. Can divide but eventually fuse to form typical multinucleate syncytia of differentiated skeletal muscle tissue.

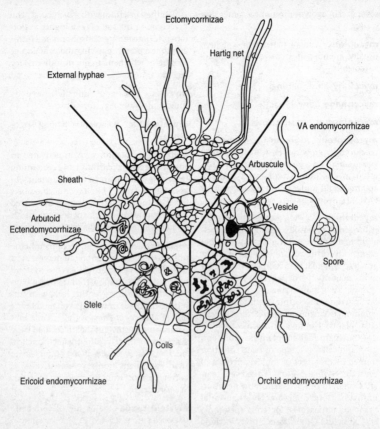

Ectomycorrhizae

Hartig net

External hyphae

VA endomycorrhizae

Arbuscule

Sheath

Vesicle

Arbutoid
Ectendomycorrhizae

Spore

Stele

Coils

Ericoid endomycorrhizae

Orchid endomycorrhizae

FIG. 120 MYCORRHIZAE *are composed of septate or non-septate fungi within a root, but also exploring the surrounding soil. In ectomycorrhizae, a septate fungus forms a sheath around the root ('sheath-forming mycorrhizae') and grows between cortical cells forming the so-called hartig net. Endomycorrhizae involve non-septate Glomales (VA mycorrhizae) or septate fungi (ericoid and orchid mycorrhizae). In both cases, the fungus penetrates the cortical cells where it forms respectively arbuscules or coils. Ectendomycorrhizae formed by basidiomycetes in some Ericales (e.g.* ARBUTUS) *share features of both ecto- and endomycorrhizae, i.e. a sheath and intracellular coils.*

myofibril Structural unit of STRIATED MUSCLE fibres, several to each fibre.

myogenic (Of muscle tissue) capable of rhythmic contraction independently of external nervous stimulation as a result of presence of PACEMAKER. CARDIAC MUSCLE always has such a pacemaker and SMOOTH MUSCLE may have.

myoglobin Conjugated protein of vertebrate striated and cardiac muscle fibres. Single polypeptide, whose iron-haem prosthetic group binds molecular oxygen. Its oxygen equilibrium curve lies well to the left of HAEMOGLOBIN'S, enabling it to load and unload its oxygen at lower oxygen partial pressures. It gives up its oxygen only when the muscle is under anoxic conditions. See

Fig. 78, and see SATELLITE DNA for minisatellite DNA.

myosin A protein found in the majority of eukaryotic cells. At least two classes of myosin exist, a single-headed tail-less variety (*myosin I*) involved in CELL LOCOMOTION, and a two-headed, tailed variety (*myosin II*) involved in MUSCLE CONTRACTION. Each filament of myosin II is thicker than those of ACTIN, comprising a tail composed of two heavy α-helices of about 134 nm in length by which it interacts with other myosin filaments, and two pairs of light chains forming a pair of knob-like heads at each end containing actin-binding ATPase sites essential for ACTOMYOSIN formation, as in muscle contraction. Its presence in STRIATED MUSCLE is responsible for A-bands. Myosin I molecules, on the other hand, are single-headed and instead of the tail have a non-α-helical domain housing a second actin-binding site which is ATP-independent (i.e. lacks ATPase activity). Molecules of cytoplasmic myosin (myosin I) can assemble into dynamic thick filaments, in contrast to the more static thick filaments formed on assembly of myosin II. Cytoplasmic myosins establish the cleavage furrow before onset of cytokinesis and generate the force required for constriction of the contractile ring – cells lacking thick cytoplasmic myosin filaments become large and multinucleate. Phosphorylation of cytoplasmic myosin by cdc2 kinase appears to inhibit its activity until the appropriate stage of the CELL CYCLE (after nuclear division) and cdc2 kinase is itself inhibited. Cytoplasmic myosins can apparently interact with phospholipid bilayers of membranes and may mediate membrane-dependent cell movements which occur in the absence of myosin II. They have been located in microvilli and *Drosophila* eye rhabdomeres. There are several unconventional myosins. See the Myosin Home Page: www.mrc-lmb.cam.ac.uk/myosin/myosin.html. See Fig. 121.

myotome See MESODERM.

Myriapoda Class of ARTHROPODA including two important subclasses: carnivorous Chilopoda (centipedes), and herbivorous DIPLOPODA (millipedes). Have long bodies of many segments, and distinct heads bearing one pair of antennae, mandibles and at least one pair of maxillae (see MOUTHPARTS). Terrestrial, with tracheal system. Centipedes have one pair of legs per segment and flattened bodies; millipedes have two pairs of legs per segment (the result of segment fusion) and cylindrical bodies.

myxamoeba Naked cell characteristic of vegetative phase of slime fungi (MYCETOZOA) and some simple fungi; capable of amoeboid locomotion.

myxo- Prefix indicating mucus or 'slime', notably on cell surfaces (e.g. see MYXOBACTERIA, MYCETOZOA). Thus *myxoviruses*, such as the influenza virus, invade the mucous membrane of the respiratory tract.

Myxobacteria Small group of rod-shaped bacteria, many of which exhibit gliding movement in contact with a solid surface and by delicate, flexible, cell wall. In many, vegetative cells mass together (particularly when starved) to form minute fruiting bodies in which cells differentiate into durable spores. Some use lactones as signalling molecules. Fruiting myxobacteria have the most complex behaviour and life cycles of any prokaryotes. The gliding forms are related to the purple bacteria group. Compare eukaryotic MYCETOZOA.

myxoedema Human condition in which metabolic rate is abnormally low, fuels being stored rather than consumed. Lethargy results. Symptom of hypothyroidism (see THYROID HORMONES).

Myxomycetes See MYCETOZOA.

Myxomycota Slime moulds; acellular slime moulds (Kingdom PROTISTA) comprising about 74 genera and 719 species of free-living organisms that can be unicellular or plasmodial. They can be non-flagellate in single or multi-celled phagotrophic stages. Characteristically, the mitochondrial cristae are tubular. Sporocarps have single to many spores that germinate to form uni- or biflagellate cells. Spore walls are cellulosic

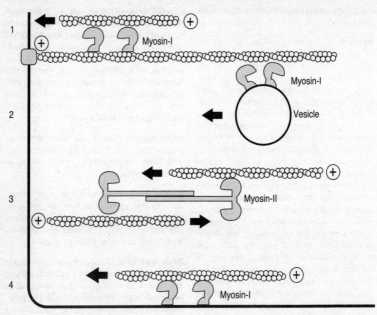

FIG. 121 *Possible roles for* MYOSIN *I and myosin II in a typical eukaryotic cell. Short heads of myosin I bind other filaments or membranes, moving one actin filament relative to another (1), a vesicle relative to an actin filament (2), or an actin filament and membrane relative to each other (4). Small antiparallel-assemblies on myosin II molecules can slide filaments over each other, mediating local contractions in an actin filament bundle (3). In all cases, the head end 'walks' towards the plus end of the actin filament it contracts.*

or chitinous. Two classes are recognized: (i) Myxomycetes and (ii) Protosteliomycetes.

The Myxomycetes (true slime moulds, e.g. *Dictyostelium*) comprise 60 genera and 690 species found on old wood or other plant material undergoing decomposition that form a plasmodium. This is multinucleate, coenocytic and saprotrophic. The plasmodium is motile, typically fan-shaped with flowing protoplasmic tubules thicker at their base, and then spreading out, branching and becoming thinner toward the periphery. As it moves, bacteria, yeast cells, fungal spores and small particles of decaying plant and animal debris are engulfed, and subsequently digested. Plasmodial growth continues until the food source is exhausted. Then internal differen-

tiation occurs and a fruiting structure (sporangium) is produced or under poor conditions the plasmodium can become a resting body or sclerotium, and the swarm of myxamoebae (see above) become microcysts. Sporangia are stalked or sessile and often brightly coloured. Spores develop in the sporangium after meiosis, therefore, the plasmodium is diploid. They are produced in masses with a persistent or evanescent peridium. Spores germinate to produce biflagellate cells or myxamoebae (swarm cells). Amoeboid and flagellated cells are readily interconvertible.

Protosteliomycetes (protostelid slime moulds) comprise 14 genera and 29 species found on dead or decaying plant material, especially bark and on dung. These slime moulds comprise amoeboid cells with filose pseudopodia. They are plasmodial but not with the protoplasmic movement shown by the Myxomycetes. Cells may or may not bear flagella. Sporocarps have simple, often delicate stalks. Spores may be formed singly or several together in some species germinating to form eight haploid flagellate spores. See SLIME FUNGI for debate.

myxophycean starch Storage polysaccharide of the CYANOBACTERIA, having a similar structure to glycogen; occurs as granules (α-granules) whose shape varies among species from rod-shaped granules, to 25 nm particles, to elongate 31–67 nm bodies.

Myxophyta See CYANOBACTERIA.

N

NAD (nicotinamide adenine dinucleotide, coenzyme I) See Fig. 122. Dinucleotide COENZYME, derivative of NICOTINIC ACID, required in small amounts in many REDOX REACTIONS where oxidoreductase enzymes transfer hydrogen (i.e. carry electrons), as in KREBS CYCLE and ELECTRON TRANSPORT SYSTEM (ETS) in respiration.

NAD^+ usually functions in oxidations. In the Krebs cycle the oxidized form of the coenzyme (NAD^+) receives one hydrogen atom from isocitrate (under the influence of isocitrate dehydrogenase) to become NADH, while another hydrogen atom from the isocitrate becomes a proton:

$$\text{isocitrate} + NAD^+ = \text{ketoglutarate} + CO_2 + NADH + H^+$$

NAD dehydrogenase, which contains the flavoprotein FMN and is an important part of mitochondrial ETS, then transfers both the proton and the hydrogen from reduced NAD to its FMN component, releasing NAD^+ for re-use:

$$NADH + H^+ + FMN = NAD^+ + FMNH_2$$

By this means, electrons can be collected as NADH from a variety of sources and funnelled into the ETS. So NADH is an energy source for ATP synthesis, through its link to the ETS, and as the mitochondrion is impermeable to NADH, its electrons are first transferred to glycerol 3-phosphate, which crosses the outer mitochondrial membrane and is reoxidized to dihydroxyacetone phosphate on the inner membrane, electrons passing there to FAD (so entering the ETS) while dihydroxyacetone phosphate diffuses back to the cytosol.

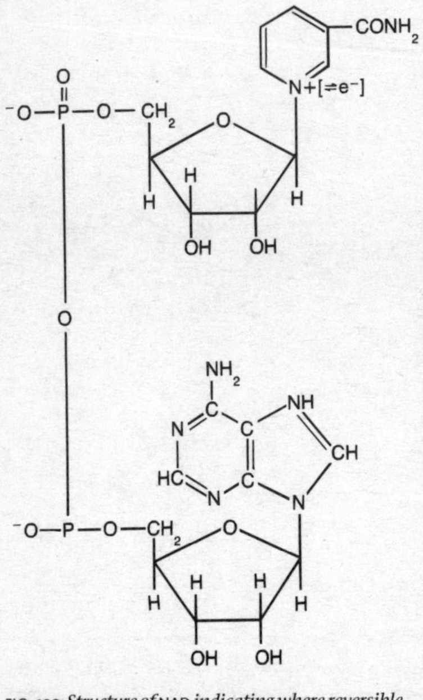

FIG. 122 *Structure of* NAD *indicating where reversible reduction occurs, giving NADH.*

NADP (nicotinamide adenine dinucleotide phosphate, coenzyme II) A dinucleotide coenzyme generally used (as NADPH) in reductions; a phosphorylated NAD, and like it involved in several dehydrogenase-linked redox reactions, but usually acting as an electron donor. Not as abundant in animal cells as NAD. Involved more in synthetic (anabolic) reactions than in breakdown. Major role in electron transport during PHOTOSYNTHESIS.

nail Hard, keratinized epidermal cells of tetrapods covering upper surfaces of tips of digits, forming flattened structures, often for arboreal grasping locomotion. *Claws* are generally tougher still, narrowing and curving downwards at their tips. The mitotic epithelium of the *nailbed* is protected by a fold of skin. See CYTOSKELETON.

nanobacteria Extremely small organisms, taxonomically problematic, as little as 50 nm in diameter. See ARCHAEA.

nanobiotechnology A term whose scope includes all commercial efforts to employ nanotechnology in the life sciences (e.g. see http://www.nano.gov/). Somewhat arbitrarily, 'nano' implies that the sizes of structures involved are in the range ~1–100 nm ('micro' implies a size range of ~100 nm to 1 μm). The physical sciences study quantum phenomena and some extraordinary properties of structures in this size range, including electrical conductivity. A vast impetus has been given to the possibility of a productive nanotechnology by the discovery of fullerenes and construction of nanotubules; but cells have their own 'nanomachines' in the form of the diverse PROTEIN COMPLEXES that populate them and which often arise through SELF-ASSEMBLY. There is also a rapidly expanding range of novel tools being developed in the physical sciences, including nanoparticles for use as probes (see QUANTUM DOTS) and new optical systems for imaging, with possible applications in molecular and cell biology. Scanning probe devices, for instance, have been very successful in permitting the application of forces to single molecules. 'Nanoparticles' already employed in the life sciences include fluorescent particles labelled with antibodies (see IMMUNOASSAYS) and magnetic particles for use in MAGNETIC RESONANCE IMAGING.

nanometre Unit of length (nm). 10^{-9} metres; 10 Ångstroms; one thousandth of a MICROMETRE. Formerly called a millimicron.

nanoplankton See PLANKTON, PHYTO-PLANKTON.

narcotics See OPIOIDS.

nares (sing. naris) Nostrils of vertebrates. *External nares* open on to surface of head; *internal nares* are the CHOANAE. Usually paired.

nasal cavity Cavity in tetrapod head containing olfactory organs, communicating with mouth and head surface by internal and external NARES respectively. Lined by mucous membrane. See PALATE.

nastic movement (In plants) response to a stimulus that is independent of stimulus direction. May be a growth curvature (e.g. flower opening and closing in response to light intensity), or a sudden change in turgidity of particular cells causing a rapid change in position of an organ (e.g. leaf). These movements are classified according to nature of the stimulus; e.g. *photonasty* is a response to alteration in light intensity, *thermonasty* to change in heat intensity, *seismonasty*, to shock. Compare TROPISM.

natural genetic competence The ability of cells to bind to and take up exogenous DNA. May be important in horizontal transfer of genes (between different taxa). Reported in a wide variety of Gram-negative and Gram-positive bacteria. DNA taken up can replace homologous regions of the cell's chromosome. See TRANSFORMATION.

natural killer cell (NK cell) The equivalent, during the innate immune response, of the cytotoxic T cell in the adaptive immune response (see IMMUNITY). Unlike those T cells, NK cells are always armed and ready to induce APOPTOSIS in foreign, virus-infected or otherwise compromised cells. Activated by interleukin-12 (IL-12), they release cytotoxic granules when a receptor on their surface binds surface carbohydrates of virus-infected cells, infection having rendered them deficient in Class I MHC molecules which would normally have prevented the binding (recognition of Class I usually turns off NK cells). NK cells have two kinds of surface receptors: those that activate their cytotoxic response and those that inhibit it. They produce the pro-phagocytic IFN-γ (see INTERFERONS, INTERLEUKINS) when viral proteins interact with one of the activating receptors, but can also recognize stress-induced and constitutive self ligands. See Fig. 99.

natural kind A term filling a need experienced by some who seek to justify a classificatory system on objective, non-arbitrary grounds. Such people hold that objects would 'sort themselves' into categories, or kinds, if all facts about them were known. At such a time one could predict the kind to which an organism belonged, given knowledge only of those facts about it that *materially caused* it to be a member of that kind. This predictability presupposes a degree of theoretical sophistication which might be unattainable given the epigenetic nature of biological systems. So, even if natural kinds do exist in biology, we might never be in a position to know which they are. See CLASSIFICATION.

natural selection (selection) Most widely accepted theory concerning the principal causal mechanism of evolutionary change ('descent with modification'); propounded by Charles DARWIN and Alfred Russel WALLACE. The theory asserts that, given diversity (both genetic and phenotypic) among individuals making up a species population, not all individuals in the population at time t_0 will contribute equally to the make-up of the population at a subsequent time t_1. To the extent that this is due to the effects of heritable differences upon individuals, natural selection has occurred.

Confusion arises over the use of Herbert Spencer's phrase 'the survival of the fittest'. Individual organisms do not survive through geological time (unlike some evolutionary lineages), but what they inherit and pass on does: that is, GENES (see UNIT OF SELECTION). The theory of natural selection asserts that the genetic composition of an evolutionary lineage will change through time by non-random transmission of genes from one parental generation to the next, a non-randomness ('selection') due solely to the fact that not all gene combinations are equally suited to a given environment, and that consequently individuals differ in their biological (Darwinian) FITNESS. Constraints

upon phenotype from the environment, which produce this differential gene transmission, are termed '*selection pressure*'. It is commonly assumed that all regular components of a species' phenotype have been favoured by natural selection, but evolution may sometimes result from causes other than natural selection (see GENETIC DRIFT).

When a character, especially a POLYGENIC one, is under *directional* selection (*orthoselection*) in a population, it undergoes an incremental or decremental shift in its mean value with time. When alternative phenotypes (e.g. red-eye and white-eye) at one character mode (here eye-colour) are each favoured in the same population, selection is *disruptive* and the population may become polymorphic for that character mode. When a particular (mean) character state is favoured, selection is *stabilizing*. *Balancing selection*, resulting from opposing selection forces, probably maintains many balanced polymorphisms, e.g. by frequency-dependent selection (see POLYMORPHISM) and may also occur when an allele is (i) favoured at one developmental stage but selected against at another; (ii) favoured in one sex and selected against in another; (iii) favoured at one time but selected against at another. Or it may occur when different genotypes in a population exploit different environmental resources within a habitat, reducing competition and elimination by selection (the *Ludwig effect*). See Fig. 123, SEXUAL SELECTION, ARMS RACE.

natural taxon Taxon comprising organisms more closely related to one another than to organisms in any other taxon at the same taxonomic level.

nature–nurture debate Debate surrounding the relative extents to which phenotypic characters are genetically or environmentally determined. See, e.g., HERITABILITY, OBESITY, PHENOTYPIC PLASTICITY.

nauplius Larval form of many crustaceans. Oval, unsegmented, bearing three pairs of appendages. It approximates to the 'head' of the eventual adult, successive segments

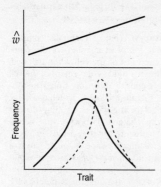

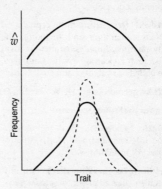

Directional selection: linear selection for higher or lower phenotypic values, detected by an association between the mean of a trait and fitness. Directional selection increases (positive) or decreases (negative) the trait mean.

Stabilizing selection: concave non-linear selection against extreme phenotypes, detected by a negative relationship between the second moment of a distribution and fitness. Stabilizing selection decreases the variance of a trait.

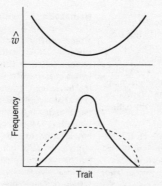

Disruptive selection: the opposite of stabilizing selection - convex non-linear selection against intermediate phenotypes detected by a positive relationship between the second moment of a distribution and fitness. Disruptive selection increases the variance of a trait.

FIG. 123 *Modes of* NATURAL SELECTION. *Selection is a covariance between phenotypes and their expected relative fitnesses (ŵ), here categorized by its impact on phenotypic distributions both as a function (above) and as within-generation change in phenotypic distributions (below).*

being added in an anterior direction from the rear as it develops.

nautiloid Designating those cephalopod molluscs resembling and including the pearly nautilus, *Nautilus*. Numerous fossil forms, first appearing in the Cambrian. Have coiled or straight chambered shell.

ncRNAs See NON-CODING RNAs.

Neanderthal man See *HOMO*.

Nearctic Zoogeographical region consisting of Greenland, and North America southwards to mid-Mexico.

necridium (separation disc) Cell in a blue-green alga filament (CYANOBACTERIA), whose death results in formation of a HORMOGONIUM.

necromass The weight of dead organisms usually expressed per unit area or volume of water. This term is sometimes used to include the dead parts of living organisms (e.g. bark and heartwood of trees, hair and claws of animals).

necrosis Relatively uncontrolled process of cell/tissue death usually following gross perturbation to the cellular environment and loss of plasma membrane integrity causing water and ions to pass down their respective osmotic and chemical gradients with cell and organelle swelling followed by cell rupture. The causes are typically pathological. Contrast APOPTOSIS.

necrotrophic (Of parasites) which kill and can then continue to live on their host; e.g. the fungus *Botrytus fabi*, a parasite of beans. Contrast BIOTROPHIC; see PLANT DISEASE AND DEFENCES.

nectary Fluid-secreting gland of flowers. *Nectar* contains sugars, amino acids and other nutrients attractive to insects, especially in insect-pollinated flowers.

negative feedback See HOMEOSTASIS.

negative staining Those methods employed in both light and electron MICROSCOPY by which only the background is stained, the unstained specimen showing up against it.

nekton Swimming animals of pelagic zone of the sea or lake. Includes fishes and whales. See BENTHOS, PLANKTON, PELAGIC.

nematocyst (Bot.) Structurally complex organelle found in members of the DINOPHYTA, fired out of the cell when irritated, resulting in a sudden movement of the cell in the opposite direction from the discharge. Include *trichocysts*, which discharge straight, tapering rods many times longer than the charged trichocyst (the actual benefit is obscure, but they could be a mechanism for quick escape or might spear a naked intruder); *cnidocysts*, which are explosive projectiles somewhat similar to nematocysts of coelenterates (anemones, *Hydra*). More complex is a *nematocyst–taeniocyst* complex occurring in the dinoflagellate *Polykrikos*; on discharge, this complex has a long thread attached to a conical structure at one end; probably used to capture prey which is then digested by the dinoflagellate. (Zool.) Inert stinging capsule produced by a cnidarian CNIDOBLAST.

Nematoda Roundworms (eelworms, threadworms). Abundant and ubiquitous animal phylum (or class of the ASCHELMINTHES). Unsegmented, triploblastic and pseudocoelomate. Circular in cross-section, with a characteristic undulating, or thrashing movement. Elastic CUTICLE of collagen acts as antagonist of the unique longitudinal muscles (no circular muscle) during swimming. Muscle cells have contractile bases adjacent to cuticle and non-contractile 'tails' lying in the vacuolated parenchymatous pseudocoelom. Gut with suctorial pharynx; anus terminal. Nervous system of simple nerve cords, ganglia and anterior nerve ring. Excretory system intracellular, consisting of two longitudinal canals. Cilia absent; sperm amoeboid. Sexes usually separate. Includes free-living and parasitic forms. FILARIAL WORMS of man and domestic animals, and root eelworms (e.g. of potato) are of great economic importance. Each species has its own number of cells, and there are usually four moults.

Nematomorpha Small phylum (or aschelminth class) of thin, elongated worms resembling nematodes but lacking excretory canals and with a brain linked to a single ventral nerve cord. Their larvae bore into insects. Like nematodes, have only longitudinal muscles in body wall. See ASCHELMINTHES for more general properties.

Nemertina (Nemertea, Rhynchocoela) Ribbon, or proboscis, worms. Small phylum of mostly marine worms with platyhelminth-like characteristics. Differ in having tube-like gut with mouth and anus (gut entire), peculiar proboscis, simpler reproductive system and a circulatory system.

neo-darwinism Brand of DARWINISM, current since early decades of 20th century, which combines Darwin's theory of evolution by natural selection with MENDELIAN HEREDITY and post-Mendelian genetic theory. Accounts, more successfully than Darwin was able, for the origin and maintenance of variation within populations. The combination has to some extent resolved problems surrounding the nature and origins of species (see SPECIATION).

Neogea See NEOTROPICAL REGION.

neogene Collective term for the Miocene, Pliocene and Pleistocene epochs. See GEOLOGICAL PERIODS.

Neognathae Largest superorder of NEORNITHES, including all non-fossil and non-ratite species. Many orders. See PALAEOGNATHAE.

neo-lamarckism View, generally discredited, that acquired characters may be inherited. Notoriously espoused by the Stalinist biologist T. D. Lysenko (mainly for crops and domesticated animals). In another episode this century, Austrian biologist P. Kammerer tried to demonstrate the phenomenon in midwife toads and sea squirts. More recently still, genetic transfer of acquired immunity in rats has been alleged. Experimental support for the view has generally been inconclusive, and sometimes even fabricated. See LAMARCK, GENETIC ASSIMILATION, MUTATION.

Neolithic (New Stone Age) Phase of human history, commencing approximately 10,000 years BP, and succeeding the PALAEOLITHIC, during which domestication of animals and cultivation of plants first occurred. Production of sophisticated stone tools (sometimes by mass-production), arrowheads, fine bone ornaments, etc., took place. Farming began to supersede hunter-gathering ecologically. First detected in Mesopotamia, from where it spread.

neopallium See CEREBRAL CORTEX.

neoplasm Tumour, or cancerous growth. Malignant if invasion or METASTASIS occurs, or is likely to occur; otherwise BENIGN. See CANCER CELL.

Neornithes See AVES.

neoteny Retardation of somatic development. Form of HETEROCHRONY, often confused with PROGENESIS. Involves a slowing in rate of growth and development of specific parts of the body relative to (especially) the reproductive organs, although frequently accompanied by delayed onset of sexual maturity. Usually results in retention of juvenile features by otherwise adult animal. Formerly used in broad sense of PAEDOMORPHOSIS. Examples include the amphibian *Ambystoma* (axolotl); development of combat and display structures in some social mammals (e.g. mountain sheep, giraffe, African buffalo); increased brain size in slowly-growing mammals with small litter sizes and intense parental care, e.g. some cat species and, arguably, humans. A correlation exists between presence of neotenic attributes and conditions favouring K-SELECTION.

Neotropical Region ZOOGEOGRAPHICAL REGION consisting of South and most of Central America. Compare NEARCTIC.

neotype Specimen chosen as a replacement of the HOLOTYPE when that is lost or destroyed.

nephridium Tubular organ present in several invertebrate groups (platyhelminths, nemerteans, rotifers, annelids, some molluscan larvae and *Amphioxus*), developing independently of the COELOM as an intucking of the ectoderm. Lumen often formed by intracellular hollowing-out of nephridial cells, closing internally (*protonephridium*) or acquiring an opening, the nephrostome, into the coelom (*metanephridium*). Protonephridia end internally either in FLAME CELLS or SOLENOCYTES. Adult annelids usually have metanephridia, often replacing larval protonephridia, and sometimes these have ciliated funnels and resemble COELOMODUCTS (which they are not). Nephridia open either to outside of the body (via *nephridiopores*), or into the gut. Carry excretory products, wafted by cilia or flagella; may have an osmoregulatory role; occasionally carry gametes.

nephron Functional unit of vertebrate KIDNEY.

neritic Waters over the continental shelf. Depth varies from just zero at low tide to as much as 200 m at the outer edge of the shelf. Much of the region is less than 80 m deep, and is well within the PHOTIC ZONE (i.e. euphotic). These waters are far more productive than the open ocean; it is over the continental shelves that the great fish and mammal migrations occur. Neritic waters differ from oceans in being less constant in chemical and physical features; the further inshore, the more variable these become.

Nernst equation Equation asserting that, for any diffusible ion, X, the equilibrium potential across a membrane depends on the absolute temperature, the valency of the ion, and the ratio of the ion's concentrations on the two sides of the membrane:

$$E_X = \frac{RT}{zF} \ln \frac{[X]_O}{[X]_I}$$

where R is the gas constant, T the absolute temperature, F is the Faraday constant, ln is the logarithm to base e, z is the valency of the ion X, and $[X]_O$ and $[X]_I$ are the chemical activities of ion X outside and inside the membrane, respectively. E_X is expressed in volts.

It is thus possible to calculate the theoretical membrane RESTING POTENTIAL if the ratios of the internal and external concentrations of all relevant ions are known. See DONNAN EQUILIBRIUM.

nerve (Bot.) Narrow thickened strip of tissue found running the length of the middle of a moss leaf (costa). (Zool.) Bundle of motor and/or sensory NEURONS and GLIAL CELLS, with accompanying connective tissue, blood vessels, etc., in a common connective tissue sheath, or *perineurium*. Each neuron conducts independently of its neighbours. *Mixed nerves* contain both sensory and motor neurons. Nerves may be nearly as long as the whole animal and contain thousands of (usually myelinated) neurons. See CRANIAL NERVE, SPINAL NERVE, NERVOUS SYSTEM.

nerve cell See NERVOUS SYSTEM, NEURON, GLIAL CELLS.

nerve cord Rod-like axis of nervous tissues forming, usually, a longitudinal through-conduction pathway and integration centre for both sensory and motor information and forming a major element of the CENTRAL NERVOUS SYSTEM. Usually linked anteriorly to the BRAIN. In segmented animals especially, a major route for REFLEX ARCS. In invertebrates the nerve cord or cords are often *nerve chains*, composed of linked ganglia. In vertebrates it forms the SPINAL CORD. Nerve cords in annelids and arthropods are usually ventral and paired; in chordates they are single, dorsal and usually hollow (see NEURULATION).

nerve ending Structure forming either the sensory or motor end of a peripheral neuron. If the former, may comprise free nerve endings or a RECEPTOR end organ; if the latter, usually consists of a motor end-plate (see NEUROMUSCULAR JUNCTION).

nerve fibre Axon of a NEURON, and its MYELIN SHEATH if present. Diameters vary from 1–20 μm in vertebrates up to 1 mm in GIANT FIBRES of some invertebrates. Nerve fibres commonly branch towards their termini into small-diameter 'twigs'.

nerve impulse See IMPULSE.

nerve net Network of neurons, often diffusely distributed through tissues, making up all or most of the nervous system of coelenterates and echinoderms and a large component of peripheral nervous systems of hemichordates. Cells may fuse to form a syncytium or form specialized synapses capable of transmitting in both directions. In some coelenterates, especially motile ones, there is often division of labour between two nerve nets, one (*through-conduction net*) conducting faster and more unidirectionally than the other. Nerve nets characteristically conduct away slowly in all directions from a point of stimulation, temporal SUMMATION and FACILITATION involving an increasing area of excitation with increased stimulus strength. Found in gut walls of some arthropods, molluscs and vertebrates.

nerve plexus Diffuse network of neurons and/or ganglia. In vertebrates, *brachial plexi* and *sacral plexi* are associated with limb movements and consist of anastomosing spinal nerves. *Solar plexus* is the collective term for a number of ganglia in the coeliac/anterior mesenteric region connected to the sympathetic nerve chain by the splanchnic nerves.

nervous integration Process whereby sensory inputs, often from more than one source and modality, either give rise to unified motor responses or are stored under some principle of association. SYNAPSES are the basic physical units of integration, providing for SUMMATION of excitatory and inhibitory potentials. Non-synaptic membranes of the postsynaptic cell may integrate through ADAPTATION, ACCOMMODATION or by their refractory periods (SEE IMPULSE). Ganglia, nerve 'nuclei' and brains all depend on synaptic connections for integration. Reflex arcs are major sites of nervous integration, *relay neurons* often communicating sensory information to the brain where even quite complex behaviour is often reflexly coordinated. Quite simple muscular activity often requires fairly complex nervous integration, in vertebrates employing feedback from MUSCLE SPINDLES, after which the CEREBELLUM coordinates this information and transmits it to cerebral motor areas. In segmented invertebrates this role is often performed by segmental ganglia, largely independently of the brain. All forms of LEARNING involve integration of neural information, as do long- and short-

term memory. See HYPOTHALAMUS, NEURO-ENDOCRINE COORDINATION.

nervous system Complement of nervous tissue (NEURONS, NERVES, RECEPTORS and GLIAL CELLS) serving to detect, relay and coordinate information about an animal's internal and external environments and to initiate and integrate its effector responses and activities. Present in all animals except sponges, developing from the ectodermal GERM LAYER. Characteristic mode of information carriage lies in patterns of nerve IMPULSES transmitted along neurons, NEUROTRANSMITTERS relaying the impulse pattern at SYNAPSES. Presence of synapses enables some nervous integration to occur; this is more marked the more advanced the nervous system. Nervous systems may be relatively simple, as in NERVE NETS, but even here there is a tendency towards *through-conduction pathways* enabling rapid transfer of impulses from one region to specific, and often distant, parts. Thus invertebrate and vertebrate NERVE CORDS mediate local reflexes, via *segmental* and *spinal nerves* respectively. Further, ganglia often serve as integration centres. The nerve cord(s) and BRAIN make up the CENTRAL NERVOUS SYSTEM (CNS), the *peripheral nervous system* comprising most of the nerves conducting impulses towards and away from it (in vertebrates, these include the spinal and CRANIAL NERVES). The AUTONOMIC NERVOUS SYSTEM forms an additional visceral motor circuit in vertebrates, integrated anatomically and functionally with the CNS.

Compared with ENDOCRINE SYSTEMS, nervous systems provide for *reception* of more specific environmental information and, through integration centres simultaneously receptive to inputs from a wide variety of sources (see SYNAPSE), can elicit responses (generally more complex, preprogrammed and integrated) far more quickly. There are no structures equivalent to sense organs or integration centres in endocrine systems, which rely upon unidirectional transport of solutes via the blood system (see HORMONES). It is the distribution patterns of receptor sites on the membranes of target cells which are instrumental in responses to many hormonal messages, diffusely broadcast as they are when compared to the pin-point accuracy of nerve impulses. Nervous systems function via rapid, multi-directional and highly integrated through-conduction pathways. Different endocrine cells use different hormones to signal to target cells; but many nerve cells respond to the same transmitter yet still convey specific information. Hormones are much more diluted than transmitters, so must be effective at lower titres. Not surprisingly, rapid responses to environmental changes are integrated via the nervous system, whereas seasonal and circadian responses often involve the endocrine system. See NERVOUS INTEGRATION, NEUROENDOCRINE COORDINATION.

neural arch Arch of bone resting on centrum on each VERTEBRA, forming tunnel (neural canal) through which spinal cord runs.

neural coding The information content of action potentials propagated along neurons, and the information content of their absence. It has long been held that mean firing rates are the necessary and sufficient mechanism underlying neural information processing ('rate coding'), and that neuron firing rates not only code peripheral sensory and motor events but also mediate central cognitive processes. This belief is now under scrutiny as evidence accumulates that, since large neuron populations are involved in any behaviour, the temporal structure of spike activity in neural assemblies may provide extra avenues for coding information.

neural connectivity The ability of neurons to extend their axons and connect to others with a high degree of specificity. It seems to derive from the general developmental programme of the organism, in which neuronal identity is translated into axonal behaviour. See NEUROGENESIS, NEUROTROPHINS.

neural crest Band of embryonic vertebrate ectoderm on both sides of the developing neural tube, giving rise to dorsal root ganglia, CHROMAFFIN CELLS, SCHWANN CELLS, and other cell types indicated in Fig. 67 for GERM LAYER. Sometimes dubbed the 'fourth germ layer' on account of their importance in vertebrates (distinguishing vertebrates from protochordates and invertebrates), neural crest cells often attain their final positions after lengthy migrations (see CELL LOCOMOTION).

neural network A connectionist model of how the brain (for instance) computes non-digitally to perform learning and cognitive tasks. Competitive and co-operative interactions between the neuron-like processing units employed in the model are such that each unit has an activity level, or state, normally ranging between 0 and 1, and each connection between units is weighted. The total input to each unit is then the sum of the states of those units with which it connects, each multiplied by a corresponding connection weight (equivalent to synaptic strength) which varies with the use made of the connection. Each unit's state is thus a smooth, but non-linear, function of this total input and the weights encode all long-term knowledge in the system. ARTIFICIAL INTELLIGENCE is an attempt to examine cognitive processes by use of computer programs and other information-processing devices.

neural plate Flat expanse of chordate ectodermal tissue, the first-formed embryonic rudiment of the nervous system. Will sink and round up to form neural tube.

neural spine Spine of bone which may be produced from top of NEURAL ARCH, running up between dorsal muscles of each side and serving for their attachment. Successive spines are usually bound by ligaments.

neural tube Hollow dorsal tube of chordate embryonic nerve tissue formed by rolling up of neural plate, *neural folds* so produced fusing in the mid-dorsal line, a process termed *neurulation* (see DESMOSOMES). The epidermis then fuses above the neural tube. Failure to roll up and fuse gives rise to *spina bifida* and, in its most extreme form, *anencephaly*. The tube expands in front to form the brain and its ventricles, the narrower more posterior part forming the spinal cord; thereby produces the central nervous system and peripheral motor neurons.

neurite Term for any of the dendrites and axons of a NEURON.

neuroblast Embryonic and presumptive nervous tissue cell.

neurocranium The part of the skull surrounding brain and inner ear, as distinct from the part composing JAWS and their attachments, the splanchnocranium.

neuroendocrine cells See NEUROSECRETORY CELLS.

neuroendocrine coordination Combined and integrated activities of the nervous and endocrine systems; involved in many physiological and behavioural responses to internal and external signals in multicellular animals.

Moulting (ECDYSIS) in some insects is initiated nervously but is mainly hormonally controlled; copulation and ovulation in many female mammals are integrated so that hormonal influences bring about OESTROUS behaviour (involving nervous control); the ensuing cervical stimulation initiates nervous release of gonadotrophic releasing factors (GnRFs), bringing about ovulation and increasing the probability of fertilization. The vertebrate hypothalamus and pituitary illustrate the close anatomical and physiological links between the nervous and endocrine systems. For instance, a sudden fall in air temperature around a mammal induces shivering. This involves sensory input from skin to the hypothalamus, whose output (along vagus nerve) causes adrenaline release from adrenals followed by rapid rhythmic contractions in appropriate body muscles and rise in body temperature. See NEUROHAEMAL ORGAN, NEUROSECRETORY CELLS.

neurogenesis Nerve cell formation. Different kinds of nerve cells are made in different parts of the nervous system, each type depending on the competence of the progenitor cell – determined by molecules, such as transcription factors, intrinsic to the cell. In *DROSOPHILA*, two of these (Hunchback and Krüppel) are sufficient and necessary for their continued expression in a neural cell lineage (see CELL MEMORY) and, interestingly, are expressed in the same temporal order in entirely different neural cell lineages at different times in development – the same sequence being used to generate different cell types in each lineage.

neurogenic (Of muscle) requiring neural stimulation before it will contract. Contrast MYOGENIC.

neuroglia See GLIAL CELL.

neurohaemal organ Organ lying outside the nervous system and storing secretion from numbers of NEUROSECRETORY CELLS, releasing it into the blood. Particularly widespread in arthropods. Insect CORPORA CARDIACA, and *sinus glands* in the eyestalks of some crustaceans are examples.

neurohumour See NEUROTRANSMITTER.

neurohypophysis See PITUITARY GLAND.

neuromast organ See LATERAL LINE SYSTEM.

neuromodulators Substances, e.g. catecholamines and cholecystokinin which, unlike classical neurotransmitters, bind non-channel-linked receptors on a postsynaptic membrane of a SYNAPSE and set in motion complex intracellular responses with onset latency of several seconds and duration of a few minutes or more (neuromodulation).

neuromuscular junction Area of membrane between a motor neuron and the muscle cell membrane forming a SYNAPSE between them. The area of muscle membrane under the nerve is the *motor endplate*. In the fruit fly *Drosophila*, spontaneous release of glutamate transmitter from the presynaptic membrane is required to change the random distribution of its receptors on the postsynaptic membrane into an ordered clustering at the junction; but in mice this is not the case for acetylcholine receptors. Each SKELETAL MUSCLE fibre commonly receives just one terminal branch of a neuron. Each nerve impulse releases a 'jet' of acetylcholine, and resulting small depolarizations of the endplate summate in a graded way to a critical threshold level (about –50 mV internal negativity) at which an action potential is generated and travels along the muscle (see IMPULSE, MUSCLE CON-TRACTION).

neuron (neurone, nerve fibre) Major cell type of nervous tissue, specialized for transmission of information in the form of patterns of IMPULSES. Nucleus and surrounding cytoplasm comprise the *cell body* (*perikaryon*, or *soma*), which may be the site of multiple synaptic connections with other nerve fibres. In non-receptor cells numerous projections from the cell body, *dendrites*, provide a large surface area for synaptic connections with other neurons, while one or more other regions of the cell body (*axon hillocks*) extend into long thin *axons* carrying impulses away from the cell body to other neurons or to effectors, making contacts via SYNAPSES and usually through secretion of neurotransmitters (see Fig. 124). Axons and dendrites are collectively known as *neurites*, and their outgrowths are mediated in part by GLIAL CELL contacts and secretions. Mice lacking the protein kinesin KIF2A develop abnormally long and re-branching collateral branches (spines) to their neurons in the hippocampus (see ALZHEIMER'S DISEASE). Kinesin KFI2A appears to depolymerize microtubules at the edge of the cell, and in its absence these seem to cause the edge to push out and so initiate outgrowths. See AMPHOTERIN.

The membrane of the axon hillock region commonly has the lowest threshold for production of ACTION POTENTIALS and is often the site of their origin. Most neurons have one axon (*monopolar*); others have two (*bipolar*), or several (*multipolar*). Axons are also termed *nerve fibres*, and may or may not be covered in a MYELIN SHEATH. Myelinated fibres conduct faster than unmyelinated ones. Unmyelinated fibres, lacking the electrical insulation of myelinated fibres, are normally located in association centres of the nervous system (see GREY MATTER).

neuronal spines Tiny protrusions from dendrites of nerve cells, constituting the receiving parts of synapses. Considered to be the most basic functional units in the brain, their large number (~10^4 per neuron) helping to explain the huge memory-storage capacity of the cerebral cortex.

neurophysin See NEUROSECRETORY CELLS.

neuropil(e) Mass of interwoven axons and dendrites within which neuron cell bodies of the central nervous system are embedded.

Neuroptera Order of ENDOPTERYGOTE insects including alderflies and lacewings. Two similar pairs of membranous wings which, when at rest, are held up over body. Biting mouthparts. Larvae carnivorous, often taking insect pests (lacewing larvae eat aphids). Mostly terrestrial; some aquatic (e.g. alderfly larvae). Alderflies (Metaloptera) include some of the most primitive endopterygotes.

neurosecretory (neuroendocrine) cells Cells present in most nervous systems, combining ability to conduct nerve impulses with terminal secretion of hormones which travel via the blood and act on target cells. Do not transmit impulses to other cells. Distinction between these and ordinary neurons is blurred as the latter secrete NEUROTRANSMITTERS, although not into the blood. Examples include cells in the brains, CORPORA CARDIACA, CORPORA ALLATA and THORACIC GLANDS of insects, and cells in the vertebrate HYPOTHALAMUS. Hormones produced are transported inside the axon, and then usually in the blood, by proteins termed *neurophysins*. See NEUROHAEMAL ORGAN, CHROMAFFIN CELL.

neurotransmitters Low molecular mass substances released in minute amounts from SYNAPTIC VESICLES at interneural, neuromuscular and neuroglandular SYNAPSES. May be excitatory (depolarizing postsynaptic membrane) or inhibitory (hyperpolarizing postsynaptic membrane). Neuropeptides (e.g. ENDORPHINS, ENKEPHALINS, corticotropin releasing factor and CHOLECYSTOKININ) are often derivatives of POLYPROTEINS and comprise the largest family. One widespread transmitter is ACETYLCHOLINE, but L-glutamate (GLUTAMIC ACID) serves at the majority of mammalian excitatory synapses and at insect and crustacean neuromuscular junctions. Other transmitters include ADRENALINE, NORADRENALINE and DOPAMINE (monoamine derivatives of the amino acid

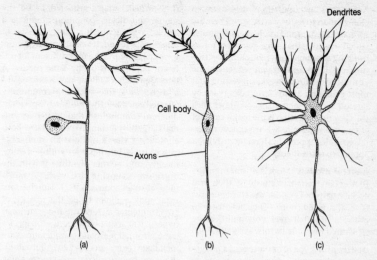

FIG. 124 *Schematic diagram of three types of* NEURON. *(a) Monopolar (many motor neurons and interneurons of higher vertebrates and many sensory neurons); (b) bipolar neuron (many sensory neurons); (c) multipolar neuron (most vertebrate interneurons and motor neurons). Impulses are normally considered as travelling from dendrites to axons, although this has not always been shown.*

tyrosine), SEROTONIN (monoamine derivative of the amino acid tryptophan), aspartate, glycine, GABA and NITRIC OXIDE. Inactivation of transmitter may be by extracellular enzyme or through removal by an uptake carrier and subsequent transport into neurons and glial cells. Some transmitters (e.g. small brain peptides) act in a PARACRINE mode, influencing several cells in the local region. Some are hydrolysed in the synaptic cleft; others are pumped back into the releasing cell. Some bind non-channel-linked RECEPTORS to initiate intracellular enzyme CASCADES. See IMPULSE, PRESYNAPTIC INHIBITION, PSYCHOACTIVE DRUGS.

neurotrophins A family of signal proteins which generally activate and ligate the cell surface Trk receptors of neurons, which have specific tyrosine kinase activities and operate via the PI3-KINASE and MAPK signalling pathways. The first discovered was nerve growth factor (NGF), which promotes survival of specific sets of neural crest-derived sensory neurons and is produced by the tissues these neurons innervate (see SURVIVAL FACTORS). It increases the activity of GROWTH CONES and has been shown to be transported in a retrograde manner, even complexed to its receptor, from nerve terminals to the neuron cell body, although it is debated whether this alone constitutes RETROGRADE SIGNALLING. Neurotrophins regulate the outgrowth of axons and dendrites (see NEURAL CONNECTIVITY) and can prune individual dendrites selectively. Some even modulate synaptic events (see NEUROMODULATORS), among them brain-derived neurotrophic factor (BDNF) which has transmitter-like properties and is implicated in long-term potentiation (see SYNAPTIC PLASTICITY).

neurula Stage of vertebrate embryogenesis after most gastrulation movements have ceased, manifested externally by presence of NEURAL PLATE. Stage ends when NEURAL TUBE is complete. See PRIMITIVE STREAK.

neuter Organism lacking sex organs, but otherwise normal.

neutralism (1) The thesis that gene

frequencies in populations owe more to chance than to NATURAL SELECTION. There are several molecular and mathematical studies which suggest that any selective forces bringing about changes in molecular structure (e.g. DNA and protein sequences) are at most exceedingly weak, suggesting that most evolution at this level takes place by random GENETIC DRIFT. This contrasts with the view that selection is the major cause of allelic substitutions. See MOLECULAR CLOCK. (2) Neutralist approaches have recently been applied to NICHE theory.

neutral models Models of communities that retain certain features of their real counterparts but exclude the consequences of biotic interactions. They are used to evaluate whether real communities are structured by biotic factors. See NICHE.

neutrophil Type of GRANULOCYTE, a phagocytic polymorph, developing in bone marrow from a common granulocyte/macrophage progenitor cell. They release CYTOKINES and like other granulocytes circulate in the blood for only a few hours before migrating out of capillaries into the tissues where they survive but a few days. See IMMUNITY.

newt See URODELA.

nexin links Protein links present between the nine peripheral doublets in the AXONEME of a flagellum.

NF-𝑥B Nuclear factor 𝑥B. Actually, a family of vertebrate transcription factors which form a variety of homo- and heterodimers and, under stimulation of the cell by appropriate microbial lipopolysaccharide, interleukins or tumour necrosis factor, are translocated to the nucleus and activate antimicrobial genes. DNA-binding motifs for NF𝑥B are found in promoters and enhancers of more than 60 genes known to be activated during INFLAMMATION. The dimers are bound by I𝑥B proteins, holding them inactive within large cytoplasmic PROTEIN COMPLEXES from which they are released by I𝑥B ubiquitinization on activation of TNF-α or IL-1 signalling pathways.

niche (ecological niche) Originally

(J. Grinnell, 1914) considered to be the spatial and dietary conditions, biotic and abiotic, within which a particular type of organism is found. It was appreciated these were complex and possibly different for different populations of the same species. A later approach (C. Elton, 1927) was to regard a niche as a role within a community enacted equivalently by different species in different communities and defined for animals largely by feeding habits and size. Both approaches view a niche as an immutable 'place' (in a broad sense) within a community, and neither identifies it with the organism occupying that place. A much more abstract approach (G. E. Hutchinson, 1944) took a niche to be the totality of environmental factors (the n-dimensional hyperspace) acting on a species (or species population). If temporal considerations are included (e.g. nocturnality/diurnality), then the COMPETITIVE EXCLUSION PRINCIPLE may be expressed thus: realized niches do not intersect. This approach defines a niche strictly with respect to (in terms of) its occupant, and with respect to a set of continuous (as opposed to discrete) axes, i.e. axes upon which all other niches are also defined. Modern niche theory is concerned particularly with resource competition between species. See SPECIES.

However, neutrality models which attempt to explain the distribution, abundance and co-existence of species, do so by taking into account only the births and deaths of individuals, overall population sizes, species numbers and origins, while assuming dispersal to be randomized. Ignoring individual ADAPTATION, and hence Darwinism, the models predict ecosystems and patterns remarkably similar to those we find in nature. See NEUTRALISM.

nicin Peptide produced by bacterium *Lactobacillus lactis*, killing its competitors. Binds Lipid II of bacterial cell membranes, the same target as vancomycin (see ANTIBIOTIC RESISTANCE ELEMENT). Might replace vancomycin as a broad-spectrum antibiotic; effective in low concentrations.

nicking (Of DNA) cutting of a strand of DNA duplex by an endonuclease.

nicotinamide adenine dinucleotide See NAD.

nicotinamide adenine dinucleotide phosphate See NADP.

nicotine Psychoactive ALKALOID, and mimic of acetylcholine, present as an inducible defence compound in the leaves of *Nicotiana tabacum* (tobacco), and hence in cigarettes, snuff, chewing tobacco and some gums. Induces both physiological and psychological dependence (addiction) and, as with all stimulant drugs, depression can follow use. Stimulates nicotinic ACETYLCHOLINE receptors, particularly those in ganglia of the AUTONOMIC NERVOUS SYSTEM, raising blood pressure, heart rate, release of ADRENALINE and tone and activity of the gastrointestinal tract. Reduces sensory output from muscle spindles, reducing muscle tone (partial cause of feeling relaxed). Nicotine has powerful immunosuppressive and inflammation-suppressing effects, one consequence being that it may protect against such inflammatory diseases as ulcerative colitis, Parkinson's disease and even Alzheimer's disease. It can also reduce fever and protect against otherwise lethal infection by influenza virus – these effects attributable to its ability to bind to the α7 subunit of the acetylcholine receptor of MACROPHAGES. A general stimulant of the central nervous system, raising behavioural activity; may cause vomiting and nausea. In large doses, induces tremors and convulsions. Induces release of ANTIDIURETIC HORMONE, causing fluid retention. Reduces weight gain. With the carbon monoxide present in cigarette smoke, increases atherosclerosis (see SCLEROSIS) and thrombosis (clotting) in coronary arteries; but if these are already atherosclerotic, causes cardiac *ischemia* (reduced blood supply) when oxygen supply fails to meet demand brought on by its stimulatory effect on heart muscle, inducing angina or myocardial infarction (heart attack). There is a 5- to 19-fold increased risk of death from coronary heart disease in smokers compared with non-smokers – worse still in diabetic smokers or those with already high blood pressure. Apart from the effects of nicotine in cigarette smoke, smoking is also the major cause of lung cancer in men and women and is a major cause of bladder cancer. Intake of alcohol while smoking exacerbates the risk of mouth cancers. At least 23 of the more than 2,000 compounds in cigarette tar are carcinogenic. Nicotine crosses the placenta, and the carbon monoxide a mother inhales in cigarette smoke reduces oxygen supply to the foetus. Smoking increases spontaneous abortion and stillbirth and probably results in reduced average body mass of neonates – with potentially irreversible effects on the intellectual and physical abilities of the child. See CAFFEINE, CORONARY HEART DISEASE.

nicotinic acid (niacin) VITAMIN of the B-complex, lack of which is part of the cause of *pellagra* in man (symptoms being dermatitis and diarrhoea). Yeast, fortified white bread and liver are good sources, but it can also be synthesized in the body from the amino acid tryptophan, of which milk, cheese and eggs are good dietary sources. Contributes structural components to coenzymes NAD and NADP. Synthesized by many microorganisms.

nictitating membrane Transparent membranous skinfold (third eyelid) of many amphibians, reptiles, birds and mammals, lying deeper than the other two; often very mobile. Moves rapidly over the cornea independently of the other two, if these move at all, cleaning it and keeping it moist. Often used under water.

nidation See IMPLANTATION.

nidicolous Of those birds which hatch in a relatively undeveloped state (naked, blind) and stay in the nest, being tended, for some time after hatching. Compare NIDIFUGOUS.

nidifugous Of those birds which hatch in a relatively advanced and mobile state and are capable of leaving the nest immediately and of searching for food, often assisted by one (usually) or both parents. Compare NIDICOLOUS.

nit Egg of human louse; cemented to hair.

nitric oxide (NO) Neurotransmitter and muscle relaxant produced by the enzyme NO synthetase (NOS) on deamination of arginine. Binds avidly to haem groups, e.g. HAEMOGLOBIN in the lungs, whereupon it is released in oxygen-poor blood vessels along with oxygen – a shuttle service helping stabilize blood pressure. The thiol groups of haemoglobin's two cysteines bind NO and prevent the haems from inactivating it. This so-called 'S-nitrosylation' is believed to play an important role in post-translational modification of other proteins. Raises a cell's CYCLIC GMP level. When produced by damaged endothelial cells, promotes vasodilation and oxidizes low-density lipoprotein (see LDL, SCLEROSIS (2)). Overactive microglia are known to produce nitric oxide, which can enter nearby neurons and cause production of free radicals, which disrupt cellular structures (see SUPEROXIDES). NO can inhibit synthesis of complex I enzyme in mitochondria, resulting in energy deficit, FREE RADICAL production and a decrease in antioxidants (see PARKINSON'S DISEASE). See OZONE, PLANT DISEASE AND DEFENCES.

nitrification Conversion of ammonium ions (NH_4^+) to nitrite and nitrate ions (NO_2^-, NO_3^-) by chemotrophic soil bacteria. The oxidation of ammonia is an energy-yielding reaction, and such energy released in this process is used by the bacteria (*Nitrosomonas*) to reduce carbon dioxide.

$$2NH_3 + 3O_2 \rightarrow 2NO_2^- + 2H^+ + 2H_2O$$

Nitrite is toxic to plants, but rarely accumulates in soil. *Nitrobacter* oxidizes nitrite to form nitrate ions, again with release of energy.

$$2NO_2^- + O_2 \rightarrow 2NO_3^-$$

Because of nitrification, nitrate is the form in which almost all nitrogen is absorbed by plants. Low soil temperatures and pH greatly reduce the rate of nitrification. See NITROGEN CYCLE.

nitrogen cycle Circulation of nitrogen atoms, brought about mainly by living organisms. Inorganic nitrogenous compounds (chiefly nitrates) are absorbed by autotrophic plants from soil or water and synthesized into organic compounds. These autotrophs die and decay or are eaten by animals, and the nitrogen, still in the form of organic compounds (e.g. proteins, nucleic acids), returns to the soil or water via excretion or by death and decay. Ammonifying and nitrifying bacteria then convert them to inorganic compounds (see AMMONIFICATION, NITRIFICATION). Some nitrogen is lost to the atmosphere as nitrogen gas by DENITRIFICATION. A great deal (about 140–700 $mg.m^{-2}.yr^{-1}$, more in fertile areas) is exacted from the atmosphere by N-fixing bacteria and blue-green algae (see NITROGEN FIXATION). Lightning causes oxygen and nitrogen to react, producing oxides of nitrogen which react with water to form nitrate ions, adding on average approx. 35 $mg.m^{-2}.yr^{-1}$ of nitrogen to the soil.

The nitrogen cycle is intimately linked to the CARBON CYCLE. Human activity now fixes more atmospheric nitrogen gas into biologically available forms each year than all natural processes combined, causing a possible sink for excess atmospheric CO_2 through stimulation of plant growth in N-limited ecosystems. The Haber–Bosch method of transforming atmospheric N_2 into ammonia, thus creating nitrogen fertilizer, was in large measure responsible for the improvement in crop yields of the 20th century, but it has dramatically altered the global nitrogen cycle, doubling its nitrogen input over the past hundred years.

nitrogen fixation Incorporation of atmospheric nitrogen (N_2) to form nitrogenous organic compounds. The majority (about 10^8 tonnes globally per year) are produced by nitrogen-fixing prokaryotes. Of the various groups of nitrogen-fixers, the symbiotic bacteria are by far the most important in terms of the total nitrogen fixed; however, species of free-living bacteria and blue-green algae (CYANOBACTERIA) can also fix atmospheric nitrogen. The most common nitrogen-fixing symbiotic bacterium is *RHIZOBIUM*, which invades roots of leguminous plants (e.g. alfalfa, *Medicago sativa*; clovers, *Trifolium*; peas, *Pisum sati-*

vum; and beans, *Phaeolus*). The beneficial effects upon soil fertility from growing leguminous plants has been recognized for centuries. The bacteria use carbohydrates supplied via the host phloem, providing in turn nitrogenous products to the host.

Blue-green algae are the only photosynthetic prokaryotes producing oxygen, which can readily inactivate the nitrogen-fixing enzyme NITROGENASE; similar to nitrogenase of the anaerobic bacterium *Clostridium*; dissimilar to the nitrogenase of the aerobic bacterium *Azotobacter*, where the nitrogenase is stable in the presence of oxygen. With respect to their nitrogen fixing abilities, cyanobacteria can be divided into two groups: (a) filamentous forms producing HETEROCYSTS, which are the site of nitrogen fixation; lacking photosystem II, they cannot evolve oxygen, yet by cyclic phosphorylation they form the necessary ATP for nitrogen fixation; and (b) non-filamentous blue-green algae that fix atmospheric nitrogen in the dark, when there is no production of nitrogenase-inhibiting oxygen by photosynthesis. Nitrogen-fixing blue-green algae are important in symbiotic relationships between liverworts, ferns and gymnosperms. The relationship between the blue-green alga *Anabaena azollae* and the small aquatic, free-floating fern *Azolla* is important because heavy growths are allowed to develop in rice paddies. As the fern dies after being shaded out by the rice, nitrogen is released for use by the rice plants themselves. *Trichodesmium*, a N-fixing colonial aerobic cyanobacterium, occurs throughout oligotrophic tropical and subtropical oceans and fixes nitrogen gas while evolving photosynthetic oxygen. Its activities probably represent a major input to the marine, and global, nitrogen cycle.

Mutualistic SPIROCHAETES (mainly *Treponema* spp.) in the hindguts of termites contribute up to 60% of total termite nitrogen biomass in wood-feeding forms. It is suspected that free-living spirochaetes also contribute significantly to N-fixation in zooplankton, the cordgrass rhizosphere and Antarctic ice pools. Other N-fixing mutualisms occur between the fungus *Actinomyces* and alders (*Alnus* spp.). See NITROGEN CYCLE.

NK cell See NATURAL KILLER CELL.

NMR (nuclear magnetic resonance) See MAGNETIC RESONANCE IMAGING.

nociception Experience of pain. See ENKAPHALINS, PSYCHOACTIVE DRUGS.

nodal bract Modified leaf-like appendage emanating from a NODE on a stem, cone or fruiting apex.

node (1) (Bot.) Part of plant stem where one or more leaves arise. (2) (Zool.) See HENSEN'S NODE.

node of Ranvier Exposed region of axon of NERVE FIBRE, where the MYELIN SHEATH is absent between Schwann cells. Only here does current flow through the membrane during passage of an IMPULSE along the myelinated nerve, so that these axons have much faster impulse transmission than non-myelinated axons. The conduction is sometimes referred to as saltatory, since it leaps from one node to the next.

nominalism Approach to CLASSIFICATION which denies the reality of categories employed, holding instead that they are artificial constructs employed for convenience or for naturalistic reasons. Compare ESSENTIALISM.

non-coding RNAs (ncRNAs) First identified in 1960 because of their high expression, and soon identified by labelling and polyacrylamide separation. They do not function as mRNAs, tRNAs or rRNAs, but variously have increasingly appreciated and crucial roles in transcription and chromosome modelling (e.g. the *Xist* product in DOSAGE COMPENSATION; and some CHROMOSOMAL IMPRINTING), RNA processing and modification (see RNA SILENCING), mRNA stability, protein stability and transport. In bacterial contexts, the term 'small RNAs' (sRNAs) has been preferred, although there is no uniformity in the use of this term.

ncRNAs range in size from 18 to 25 nucleotides (nt) in the case of microRNAs (miRNAs), ~100–200 nt for bacterial sRNAs (often involved as translational regulators), and up to >10,000 nt for those higher eukaryote ncRNAs involved in GENE SILENCING.

MicroRNAs (miRNAs) form part of an evolutionarily conserved eukaryotic system of RNA-based gene regulation, about which much remains unknown. Their nucleotide sequences indicate they are breakdown products of larger precursors transcribed from non-protein-coding genes, some being processed from partially duplexed precursors (dsRNA) by DICER. They are negative regulators of protein synthesis, either by forming complexes which block the translation process (animals) by binding to the 3'-untranslated regions of their mRNAs, or by promoting specific mRNA degradation (plants). The targets of most of the currently known miRNAs (over 200 in mammals) await discovery, and although some miRNA genes are uniformly expressed, suggesting general roles in gene regulation, others such as *lin-4* and *let-7* are temporally regulated and tissue-specific in *Drosophila* and *Caenorhabditis elegans* (as are other miRNA loci in plants). This and the fact that miRNA expression is often cell type-specific points to a function in developmental control. Indeed, in the fruit fly *Drosophila*, the *bantam* locus encodes a miRNA functioning to suppress APOPTOSIS and promote cell division in development. The term 'small heterochromatic' RNAs (shRNAs) has been applied to certain dsRNAs which resemble Dicer cleavage products that correspond to centromeric repeats and recruit heterochromatin.

Small interfering RNAs (siRNAs) downregulate gene expression by binding complementary mRNAs and either initiating their elimination by sequence-specific cleavage (see RNAi, RISC) or curtailing their translation. They are employed by cells as an antiviral defence mechanism. In flies and humans, they are 21–23 nt long with two-nucleotide, 3'-overhanging ends.

non-disjunction Chromosomal MUTATION resulting in failure of either (a) the two members of a bivalent to separate during the first meiotic anaphase, or (b) the two sister chromatids of a chromosome to separate at second meiotic anaphase. Results in aneuploidy: resulting cells have either one too many or one too few chromosomes. Trisomies and monosomies may arise this way. In man, may cause DOWN'S SYNDROME. See TELOMERE.

non-equilibrium theory In community ecology, concerned with the transient behaviour of a system away from any equilibrium point, it specifically focuses attention on time and variation. See EQUILIBRIUM THEORY.

nonsense-mediated decay (NMD) A highly conserved eukaryotic pathway targeting and destroying RNAs containing premature 'STOP' codons. See RNA SURVEILLANCE SYSTEM.

nonsense mutation See MUTATION.

non-synonymous sites (non-synonymous sequence changes) DNA sequence changes in coding regions that cause amino acid sequence changes (see GENETIC CODE).

noradrenaline (norepinephrine) Generally excitatory catecholamine NEUROTRANSMITTER of sympathetic nervous system, produced at postganglionic termini; in the brain, from cell bodies located in the brainstem terminating in the cortex, limbic system, hypothalamus and cerebellum. Implicated in maintaining AROUSAL in human brain, while dreaming, hunger, thirst, emotion and sexual behaviour may all involve its release. A link between noradrenaline depletion and clinical depression has gained much experimental support, relevant circuits originating in the brainstem and projecting to the limbic system. Like ADRENALINE, a tyrosine derivative; also secreted by the adrenal gland – but in smaller quantities.

norepinephrine See NORADRENALINE.

Northern blotting A technique following the same principles as SOUTHERN BLOTTING except that RNA is separated according to

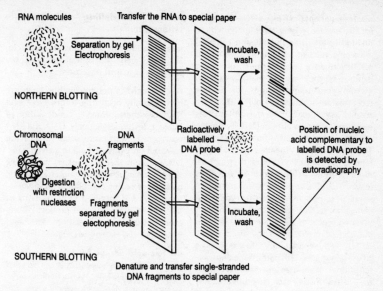

RNA molecules   Transfer the RNA to special paper

Separation by gel
Electrophoresis

Incubate,
wash

NORTHERN BLOTTING

Chromosomal
DNA

DNA
fragments

Radioactively
labelled
DNA probe

Position of nucleic
acid complementary to
labelled DNA probe
is detected by
autoradiography

Digestion
with restriction
nucleases   Fragments
separated by gel
electophoresis

Incubate,
wash

SOUTHERN BLOTTING

Denature and transfer single-stranded
DNA fragments to special paper

FIG. 125 *A comparison of the techniques of Northern and Southern blotting.*

size in a denaturing gel prior to being blotted onto a solid support. The mRNA transcripts can then be detected by hybridization with a DNA PROBE, the intensity of the radioactive signal indicating mRNA abundance. See Fig. 125.

nostrils See NARES.

notochord Rod of vacuolated tissue enclosed by firm sheath and lying along long axis of chordate body, between central nervous system and gut. Present at some stage in all chordates. In most vertebrates, occurs complete only in embryo (or larva) but remnants may persist between the vertebrae, which obliterate it. Found in larval and adult cephalochordates, larval urochordates, and perhaps adult hemichordates. In these latter forms it acts skeletally, antagonizing the myotomes. Usually regarded as mesodermal, and with it forming the *chorda-mesoderm*.

Notogea Australian ZOOGEOGRAPHICAL REGION.

nucellus Tissue between megaspore and integument(s) of gymnosperm or flowering plant ovule; part of sporophyte. Generally provides nutrition for embryo sac.

nuclear magnetic resonance imaging (NMRI) See MAGNETIC RESONANCE IMAGING.

nuclear pore See NUCLEUS.

nuclear proteins (1) Proteins forming part of the nuclear envelope, pore complexes and nuclear lamina. (2) Proteins imported into the nucleus. They bear a short targeting sequence of amino acids (often basic: the nuclear localization signal) required to open the pore complex, ATP-dependently. Such signals, apparently permanently attached, may also be recognized by receptor proteins which carry imported proteins through the pore. These proteins include enzymes, nuclear RNPs, transcription factors (e.g. E2F, p53) and their modulators (e.g. retinoblastoma protein, Rb). See TUMOUR SUPPRESSOR GENE.

nuclear receptors DNA-BINDING PROTEINS (of the *nuclear receptor superfamily*) which, on binding an intracellular ligand, can bind a specific nuclear chromatin region and inhibit/enhance transcription of target genes (i.e. they are latent TRANSCRIPTION FACTORS); but many ligands for nuclear receptors are unknown – hence these are termed *orphan nuclear receptors*. All nuclear hormone receptors bind to DNA as dimers. *HOX* GENES are regulated by nuclear receptors. There are at least 35 members in the superfamily, in at least two classes: thyroid hormone/retinoic acid/vitamin D receptors (class 1) and steroid hormone receptors (class 2). Class 1 receptors can recognize a palindromic DNA sequence comprising an inverted repeat of the sequence AGGTCA, but each also recognizes the direct repeat of AGGTCA separated by either a 3-, 4- or 5-base pair spacer. This permits specific transcriptional responses to each receptor. Indeed, isoforms of each encoding receptor gene exist which, given the ability of some receptors to form heterodimers (e.g. with any of three 9-cis-retinoic acid-binding *retinoid X receptors*, *RXR*) each with its own activity once bound to a given PROMOTER. This further enlarges the variety of responses. Retinoic acid-responsive elements, to which heterodimers bind, are known as RAREs. When THYROID HORMONE (see Fig. 162) binds its TR/RXR receptor (already bound to chromatin), histone acetyltransferases are recruited and transcription is induced. In the absence of thyroid hormone and retinoic acid, nuclear hormone receptors for them repress the transcription of a series of genes needed for cellular differentiation. Class 2 receptors first bind their steroid ligand, then on entering the nucleus must dimerize with a similar steroid-receptor complex in order to activate transcription. Auxiliary proteins further enhance the DNA-binding properties of class 2 receptors, forming a possible third class of receptor. These proteins, and the cell surface receptors activating them, are absent from fungi and plants; and the majority of pathways involving them in mammals appear to occur in invertebrates. See ENHANCER, TRANSACTIVATION.

nuclear remodelling See CHROMATIN REMODELLING.

nuclease Enzyme degrading a nucleic acid. See DNase, RNase, RESTRICTION ENDONUCLEASE. Some occur within LYSOSOMES.

nucleic acid A polynucleotide. There are two naturally occurring forms, DNA and RNA, polymers formed by condensation of nucleotides by, respectively, DNA and RNA polymerase. Both forms occur in all living cells, but only one of them in a given virus. DNA (except in RNA viruses) carries the store of molecular information (the genetic material, a major component of CHROMOSOMES); RNA is involved in deciphering of this information into cell product during transcription and translation. See GENE, NUCLEOPROTEIN, PROTEIN SYNTHESIS.

nucleic acid arrays Microarray slides, or chips, which are systematically dotted with DNA from thousands of genes that can serve as probes for detecting which genes are active in different types of cells. The DNA is tagged with a fluorescent marker and DNA or RNA in the sample bonds to complementary sequences on the chip, the chip putting out its results on a computer file as a record of glowing dots of different intensities. This is transforming studies of GENE EXPRESSION and studies of biological COMPLEXITY. See DNA PROBE, YEAST TWO-HYBRID SYSTEM.

nucleic acid hybridization See DNA HYBRIDIZATION.

nucleohistone See CHROMATIN, CHROMOSOME.

nucleoid An aggregation of DNA, found in prokaryotes and in the mitochondria and chloroplasts of eukaryotes, that somewhat resembles a nucleus but lacks a nuclear envelope and histones.

nucleolus Spherical body, occupying up to 25% of nuclear volume, one or more of which stain with basic dyes in interphase nuclei. Its size reflects its level of activity. Consists of decondensed chromatin containing the chromosome-specific and highly amplified tandem gene sequences for ribosomal RNA (*nuclear organizers*), along with their transcription products (ribosomal RNA sequences; see RNA POLYMERASE), certain RNA-binding proteins and associated ribosomal proteins, all involved in the early phase of ribosome formation. Nucleoli self-assemble in association with the ribosomal DNA in response to transcription of pre-rRNA, sequestering all the components needed for ribosome synthesis (see RIBOSOME, SELF-ASSEMBLY) and their formation and continued existence require ongoing transcription of the rDNA loci, so that they usually dwindle and disappear early in nuclear division, but reappear as rRNA synthesis recommences in telophase. Much of the nucleolar RNA and protein seem to be carried by chromosomes while the nucleolus itself is disassembled. In the EUGLENOPHYTA, nucleoli (= endosomes) persist through the mitotic cycle. See GENE AMPLIFICATION.

nucleomorph An organelle closely associated with a chloroplast, which contains DNA and is enclosed by a double membrane envelope. Found in the CRYPTOPHYTA, CHLORARACHNIOPHYTA and some dinoflagellates. The nucleomorph is interpreted as the vestigial nucleus of a photosynthetic, eukaryotic endosymbiont.

nucleophiles Functional groups in molecules which are rich in electrons and capable of donating them. Contrast ELECTROPHILE.

nucleoprotein Complex of nucleic acid and associated proteins forming a CHROMOSOME. In eukaryotes the proteins include histones. See CHROMATIN.

nucleosides The base-ribose moieties of NUCLEOTIDES. Common naturally-occurring forms in RNA are adenosine, guanosine, cytidine and uridine, while *deoxy* forms (deoxyribose instead of ribose) of all but uridine occur in DNA, deoxythymidine replacing uridine.

nucleoside triphosphate (nucleotide diphosphate) Molecule comprising a NUCLEOSIDE to which three phosphate groups are bound, usually at the 5′ hydroxyl of the pentose. ATP, GTP, CTP and UTP are substrates in RNA synthesis, while the *deoxy* forms dATP, dGTP, dCTP and dTTP are substrates in DNA synthesis (see DNA REPLICATION).

nucleosome Fundamental packing unit of a eukaryotic CHROMOSOME, comprising an octomeric HISTONE core and 146 base pairs of DNA wound around it. Transcription of eukaryotic RNA occurs on DNA organized as nucleosomes, which may reduce spurious transcription in response to insertion of transposable elements.

nucleotide A molecule comprising either a purine or pyrimidine base bonded to either a phosphorylated ribose or (in the case of a deoxynucleotide) deoxyribose moiety. A phosphorylated nucleoside. Nucleotides and deoxynucleotides comprise the monomers of RNA and DNA respectively, in which they are linked by phosphodiester bonds. In *cyclic nucleotides* such as cyclic AMP, the phosphate moiety forms a diester bond with the 3′ and 5′ hydroxyls of the ribose. Nucleotides are incorporated into nucleic acids by polymerases which use nucleoside or deoxynucleoside triphosphates as substrates. See DNA REPLICATION.

nucleus (1) The organelle in eukaryotic cells making about 10% of the cell volume and containing the cell's CHROMOSOMES (see Fig. 20). In the interphase nucleus, individual chromosomes occupy discrete compartments, termed *chromosome territories*. A eukaryotic cell may have no nucleus, or one or more; but each such cell ultimately derives from one which was nucleated.

Nuclei vary in size, even growing and diminishing within the same cell, tending to be largest in actively synthetic (e.g. secretory) cells. Each nucleus has an *envelope* of two membranes punctuated by *pore complexes* of 30–100 nm internal diameter (nuclear pores) where the inner and outer membranes are continuous.

Nuclear pore complexes (NPCs, see Fig. 126) are involved in active transport of molecules into and out of the *nucleoplasm* within the envelope, and have an iris-like

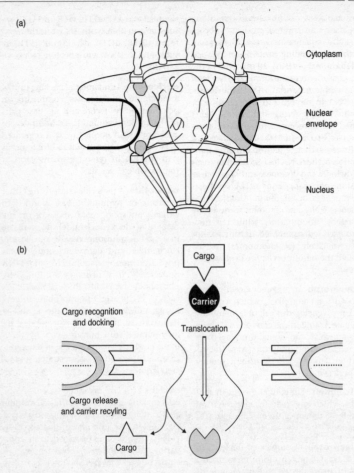

FIG. 126 *Models of the nuclear pore complex: (a) schematic and (b) diagrammatic. See* NUCLEUS. *(b) A unifying model for nucleocytoplasmic transport. This model emphasizes the role of shuttling carriers. It is well supported for protein import but largely speculative for RNP export. The NPC (in black) is shown schematically, embedded in the nuclear envelope (grey). Cytoplasmic fibrils and baskets are omitted and, as a consequence, cytoplasm and nucleosides sides are not defined.*

variable aperture; they are able simultaneously to transport protein into the nucleus (shuttle carriers and signal transduction involved) and RNA out of it (see BCl-2). The inner nuclear membrane has attachment sites for INTERMEDIATE FILAMENTS (chiefly *lamins*) constituting the thin *nuclear lamina*, which support the CHROMATIN fibres during interphase. Lamins are believed to be the target for kinases whose activity in the presence of cyclin initiates the CELL CYCLE (see esp. MATURATION PROMOTING FACTOR). Other intermediate filaments surround the cytoplasmic surface of the outer membrane. Phosphorylation of the lamins at mitotic and meiotic prophase in higher EUKARYOTES (e.g. not diatoms, dinoflagellates or fungi) is associated with disintegration of the nuclear membrane, while their dephosphorylation at anaphase causes vesicles of nuclear membrane to assemble around chromosomes and gradually fuse to reform the nuclear envelope. The outer membrane of the envelope is usually continuous in places with the ENDOPLASMIC RETICULUM. Nucleoplasm is regionally differentiated into domains (e.g. NUCLEOLUS, COILED BODY) and contains enzymes, PROTEASOMES, RNPS, HIGH-MOBILITY GROUP PROTEINS and other proteins. Some of these are involved in transcription of ribosomal RNA from multiple copies of rRNA genes located, often tandemly as *nucleolar organizers*, on particular chromosomes. The nucleus cannot synthesize proteins, but can import them from the cytosol through pore complexes if they have appropriate peptide SORTING SIGNALS.

If properly stained, chromosomes are visible in interphase nuclei as thin grains of chromatin. Interphase nuclei contain relatively decondensed chromatin. This condenses (shortens and thickens by coiling) during nuclear division to give chromosomes visible in light microscopy. Ciliates and some other protozoa contain two nuclear classes: a *macronucleus* concerned with the vegetative 'day-to-day' running of the cell, and one or more *micronuclei* involved in CONJUGATION. The macronucleus, which contains much of the genome in multiple and often fragmented copies, divides amitotically (see ABERRANT CHROMOSOME BEHAVIOUR), the micronucleus by mitosis. Mature eukaryotic cells in which the nucleus has disintegrated (e.g. mammalian red blood cells) may rely on 'long-lived' messenger RNA for protein synthesis. Nuclear transplantation has revealed much about cell differentiation, not least that the nucleus generally releases information to the cytoplasm (in the form of mRNA) only when 'instructed' by it to do so, as for example by GROWTH FACTORS or by steroid hormones complexed to NUCLEAR RECEPTORS. The nuclear envelope may have originated in response to mitochondrial endosymbiosis, ensuring the separation of these two genomes. The nuclear envelope provides an additional level of control over transcription (and RNA PROCESSING) compared with prokaryotes. This is because many transcription factors only cross the envelope when released from amino acid sequences holding them in the cytoplasm, or from association with other proteins which have first to be targeted and digested by the proteasome system.

(2) One of several anatomically distinct aggregations of nerve cell bodies (ganglia) in the vertebrate BRAIN. The *lateral geniculate nuclei* within the THALAMUS, for example, contain cell bodies of neurons of the two optic nerves. (3) The central part of an atom of a chemical element, containing at least one proton and usually at least one neutron.

nucleus accumbens Most basal portion of the striatum (see BASAL GANGLIA).

nude mice Developmentally abnormal mice, lacking thymus (*athymic*). Used in immunological work.

null hypothesis Hypothesis constructed so as to predict from it a hypothetical set of experimental results (*expected results*) which would obtain under specified experimental conditions were the hypothesis true, and which may be compared with *observed results* derived from an actual experiment operating under those conditions. If there is a significant discrepancy between the two sets of results (see CHI-SQUARED TEST), then it is probable that the null hypothesis is incorrect. Testing explicit null hypotheses is reckoned to be less methodologically problematic than corroborating positive hypotheses, since they avoid the problems inherent in all inductive generalizations and are often more precisely stated. Because a null hypothesis is usually expressed largely in ignorance of the factors affecting the results of the experiment it is often stated in a way that suggests that varying the one under investigation makes no difference to those results. See EXPERIMENT.

nullisomy Abnormal chromosome complement in which both members of a chromosome pair are absent from an otherwise diploid nucleus. Type of ANEUPLOIDY;

viable only in a polyploid (e.g. wheat, which is hexaploid).

null mutation A MUTATION eliminating the activity of a wild-type GENE. See GENE TARGETING.

numerical phyletics Erection of an inferred phylogeny of living organisms through numerical analysis of characteristics. See PHYLETICS.

numerical taxonomy Erection of classifications by numerical analysis of characteristics. See CLASSIFICATION.

nurse cells (Bot.) In some liverworts, sterile cells among the spores lacking any special wall thickening and frequently disintegrating before spores mature. (Zool.) For Sertoli cells, see TESTIS.

nut Dry indehiscent, single-seeded fruit, somewhat similar to an achene but the product of more than one carpel and usually larger, with a hard woody wall; e.g. hazel nut. A small nut is called a nutlet.

nutation (circumnutation) Spiral course pursued by apex of plant organ during growth due to continuous change in position of its most rapidly growing region; most pronounced in stems, but also occurs in tendrils, roots, flower stalks, and sporangiophores of some fungi.

nutrition Any process whereby an organism obtains from its environment the energy and atoms required for maintenance, growth, reproduction, etc. Either AUTOTROPHIC or HETEROTROPHIC.

nyctinasty Response of plants to periodic alternation of day and night; e.g. opening and closing of many flowers, 'sleep movements' of leaves. Related to changes in temperature and light intensity. See PHOTONASTY, THERMONASTY.

nymph Larval stage of exopterygote insect. Resembles adult (e.g. in mouthparts, compound eyes), but is sexually immature. Wings absent or undeveloped.

obesity Overweight. Obesity is most readily the result of long-term over-consumption of energy, with resultant excessive increase in body mass. However, there may be contributory malfunction in the individual's energy homeostasis (see ADIPOSE TISSUE, DIABETES, LEPTINS). The dis-covery of two peptides, orexins A and B, produced in the lateral hypothalamus in increased amounts during starvation brings the known number of such selectively expressed molecules to five. Orexin admin-istration to the brain brings about excessive feeding. Conversely, leptin administration increases the frequency of action potentials in the anorexigenic POMC (proopiomelan-ocortin) type of neurons in the arcuate nuc-leus of the hypothalamus. Although it is possible that massive obesity may result from mal-signalling by a single (e.g. leptin) receptor, it is unlikely that obesity will com-monly be monogenic. It will probably far more commonly be multifactorial (i.e. not only polygenic, but also diet- and stress-related). Published tables of 'desirable weights' converge on the range between 20–25 on 'Quetelet's Index' (QI), also known as *body mass index* (BMI, see Fig. 127). Each person's position on the index is calculated by dividing their weight (kg) by the square of their height (m). Positions on the table below a QI of 20 are regarded as 'under-weight'. Values above 25 lie within one of three bands of 'overweight' and a QI value of 30 or above is commonly regarded as 'obese'.

obligate Term indicating some type of restriction in an organism's way of life, from which it cannot depart and survive. Used particularly in the contexts of parasites, aerobes and anaerobes. Contrast FACUL-TATIVE.

occipital condyle Bony knob at back of skull, articulating with first vertebra (the atlas). Absent in most fish, whose skulls do not articulate with vertebral column. Single in reptiles and birds, double in amphibians and mammals.

occiput (occipital region) (1) In ver-tebrates, an arbitrarily delimited region of the skull and head in the neighbourhood of the occipital condyle. (2) In insects, a plate of exoskeleton forming back of the head.

oceanic Inhabiting the sea where it is deeper than 200 metres. Compare NERITIC.

ocellus General term for several types of *simple eye*, as found in some coelenterates, flatworms, annelids, insects (the only eyes of larval endopterygotans), arachnids and other arthropods. Usually incapable of image-formation. See EYE.

ochre mutation A 'stop' mutation; UAA triplet of messenger RNA. The ochre codon forms the normal terminus of the coding regions of many codon. See GENETIC CODON, AMBER MUTATION.

octamer-binding proteins Proteins which bind a highly conserved 8-base pair sequence (octamer) found in every Ig pro-moter so far studied and within the IgH enhancer. Some at least of these proteins contain a HOMEOBOX motif similar to the homeodomain common to *Drosophila* regu-latory proteins encoded by SEGMENTATION

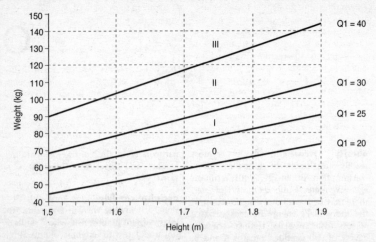

FIG. 127 *Height–weight variation. The region O indicates the desirable range of weight for height in adults. Below this area is underweight, and grades of overweight are indicated by the bands I, II and III, which begin at Quetelet's Index values of 25, 30 and 40 respectively. See OBESITY.*

GENES. Also present in adenovirus origin of replication.

octopamine One of the monoamine family of NEUROTRANSMITTERS.

oculomotor nerve Third CRANIAL NERVE of vertebrates. Supplies four of the extrinsic eye muscles and, by neurons of parasympathetic system, via the ciliary ganglion, intrinsic eye muscles of ACCOMMODATION and pupil constriction. A ventral root.

Odonata Dragonflies and damselflies. Order of exopterygote insects with aquatic nymphs. Carnivorous as nymphs and adults, with biting mouthparts, hinged on an extensible 'mask' in the nymph. Large, compound eyes; two pairs of similar wings, folded over the back in damselflies (Zygoptera) and horizontally in dragonflies (Anisoptera). Some fossil (Carboniferous) forms had wingspans in excess of half a metre.

odontoblasts Cells lying in pulp cavities of vertebrate teeth, sending processes into adjacent dentine, which they help form.

odontoid process See ATLAS.

oedema Swelling of tissue through increase of its TISSUE FLUID volume. Can occur during INFLAMMATION, through blockage of vessels of LYMPHATIC SYSTEM returning tissue fluid to venous system (see FILARIAL WORM) and through low levels of blood proteins causing poor return of tissue fluid through CAPILLARIES, as in protein malnutrition (kwashiorkor).

oesophagus Anterior part of gut, between pharynx and stomach. Usually concerned simply in peristaltic passage of food to stomach. May contain a CROP.

oestradiol, β-oestradiol Most potent OESTROGEN in female mammals.

oestrogens Group of STEROIDS, both naturally-produced ovarian hormones and artificial homologues of them. Synergists of GROWTH HORMONE. High blood levels inhibit release of gonadotrophic releasing factors (GnRFs) from the HYPOTHALAMUS, but while increasing cause the LH surge that initiates mammalian ovulation. The most potent natural oestrogen in humans is β-oestradiol; two others are oestrone and oestriol. Produced by theca cells of developing GRAAFIAN FOLLICLES of the ovary and, after ovulation, by granulosa cells; adrenal cortices also produce small amounts.

Main roles of oestrogens in female mammals are: development of MAMMARY GLANDS (synergistically with PROGESTERONE), regulation of fat deposition, softening and smoothing of skin texture (in humans), regulation of uterine endometrium, again with progesterone (see MENSTRUAL CYCLE), promotion of flattening and widening of PELVIC GIRDLE and GROWTH spurt at puberty (all *secondary sexual characteristics*). Effects on sexual behaviour depend on the species involved, but may cause 'maternal behaviour' if injected into males.

Oestrogens enter target cell nuclei in combination with specific cytoplasmic receptors, annealing to specific chromosome regions and initiating mRNA transcription (see GENE EXPRESSION). Breast cancer incidence is transiently increased by pregnancy, and when oestrogens are used as oral contraceptives or in hormone replacement therapy (HRT), but permanently lowered by late menarche, early menopause, early first childbirth and by increased offspring production. Endometrial cancer rate increases with HRT and both endometrial and ovarian cancers are less common in oral contraceptive users. They are major components of the CONTRACEPTIVE PILL. See NUCLEAR RECEPTORS.

oestrous cycle Reproductive cycle of short duration (usually 5–60 days) occurring in sexually mature females of many non-human mammals, in the absence of pregnancy. Regulated by endocrine (often neuroendocrine) interactions involving the HYPOTHALAMUS, pituitary and ovaries. The phases in the cycle include: *oestrus*, *metoestrus*, *dioestrus* and *pro-oestrus*. Compare human MENSTRUAL CYCLE.

oestrus Follicular phase of the OESTROUS CYCLE during which sexual desire and attractions of the female may be heightened, leading to copulation (see NEUROENDOCRINE CO-ORDINATION); less so in some primates. In cats and dogs, periods of oestrus (when females are 'on heat') occur two or three times per year, separated by long periods of *anoestrus*.

oidium (pl. oidia) Haploid spore of Ascomycota and Basidiomycota, also known as

an arthrospore, produced asexually by a primary mycelium.

oils Triglycerides (see LIPIDS) which are liquids at ambient temperatures.

olecranon process Bony process on mammalian ulna, extending below elbow joint and employed in attachment of muscles straightening limb.

olfaction Detection of odours; sense of smell. Involves, at least in mammals, specific G PROTEIN-linked olfactory RECEPTORS, present on the surfaces of modified cilia extending from these cells, odorants activating their intracellular adenyl cyclase components on binding and initiating the sodium ion entry which depolarizes the receptor cell. See PHEROMONES.

olfactory nerve First CRANIAL NERVE of vertebrates, running from olfactory area of cerebral cortex to olfactory bulb (see OLFACTORY ORGANS).

olfactory organs Organs of smell. In vertebrates, consist of sensory epithelia in nasal cavities, whose cells respond to molecules dissolved in their moist mucous membranes. *Olfactory cells* are bipolar neurons whose dendrites synapse with branches of the olfactory nerve in a mass of grey matter termed an *olfactory bulb* within the braincase. The olfactory nerve travels to olfactory regions of cerebral cortex. See TASTE BUD, VOMERONASAL ORGAN.

oligo- Prefix indicating 'few'. Thus oligosaccharides, oligopeptides, oligonucleotides, referring to relatively short-chain polymers (i.e. each contains relatively few, perhaps up to 20 or so, condensed monomers).

Oligocene GEOLOGICAL EPOCH, subdivision of the TERTIARY period lasting from 35.4–23.3 Myr BP, during which the Alps and Himalayas rose, South America separated from Antarctica, and volcanoes erupted in the Rocky Mountains. Earth's atmosphere began to cool during the Oligocene, its mean temperature dropping by ~15°C, causing succulent rain forest vegetation to be replaced by low-nutrient siliceous grasses over much of the planet. Non-ruminant herbivorous mammals (e.g. brontotheres, hyracodons) were increasingly replaced by ruminants (e.g. camels, deer and cattle). Many genera of modern plants evolved in the Oligocene, and deposits from Fayum in Egypt indicate the presence of simian primates with resemblances to New World monkeys (see Fig. 138).

Oligochaeta Class of annelid worms which includes earthworms. See ANNELIDA.

oligodendrocytes Myelin-forming GLIAL CELLS of the central nervous system.

oligonucleotides Laboratory-prepared oligonucleosides ('oligos', or aptamers) are proving useful in clinical trials to treat viral infections, cancer, Alzheimer's disease and MALARIA. Oligos bind complementary nucleic acid sequences and form double-stranded complexes when designed to target mRNA, or triple-stranded complexes when targeting duplex DNA. Either way, oligos can very selectively close down expression of a specific gene. One ANTISENSE oligo under trial is designed to prevent expression of *gag* protein, crucial for HIV replication (see HIV). Another targets the mRNA encoding the enzyme dihydrofolate reductase-thymidylate synthase (DHFR-TS) specific to the malarial parasite *Plasmodium falciparum* (i.e. slightly different from the human form of the enzyme) required for its pyrimidine synthesis. Such treatment might sidestep the parasite's global resistance to the drug pyrimethamine, which also blocks this enzyme. Besides their probable clinical value, oligos will likely have great application as probes (e.g. see DNA PROBE) of important cellular processes.

oligopeptide See PEPTIDE.

oligosaccharide Carbohydrate formed by joining together monosaccharide units (4–20) in a chain by links termed *glycosidic bonds*, formed enzymatically. Are intermediate digestion products of POLYSACCHARIDES, and some form side chains of GLYCOPROTEINS and GLYCOLIPIDS. See GLYCOSYLATION.

oligotrophic (Of lakes and rivers) originally introduced to describe the phytoplankton assemblages characteristic of montane lakes, which were believed to be deficient in combined nitrogen and phosphorus. Nowadays used in a more general sense to describe lakes and rivers that are unproductive in terms of organic matter formed; nutrient status of the water and nutrient supply are very low. Compare EUTROPHIC, MESOTROPHIC, DYSTROPHIC.

ommatidium See EYE.

omnivore Animal eating both plant and animal material in its diet. See CARNIVORE, HERBIVORE.

onchoceriasis 'River blindness'. See FILARIAL WORMS.

onchosphere (hexacanth) Six-hooked embryo of tapeworms (CESTODA). Develops from egg, usually while still in proglottis. Will bore through gut wall of secondary host and is then carried in blood to host tissues, in which it lodges and develops into the CYSTICERCUS.

oncogene A gene (or gene locus) in which mutation, loss of function or other reduction of expression induces neoplasia (transformation of a cell into a CANCER CELL). Oncogenic viral genes (v-*onc*) generally occur in retroviruses; a cellular oncogene (c-*onc*) is sometimes termed a proto-oncogene (proto-*onc*). They are generally referred to by three italicized letters indicating the retrovirus in which they were first identified, most v-*onc*s starting in a host cell genome and getting into the viral genome by a transduction-like event after which they can exert their effect in a host cell which becomes infected by the retrovirus. A v-*onc* sequence is normally very like its c-*onc* homologue but often lacks the latter's introns, suggesting its origin by reverse transcription from cellular mRNA. Several oncogenic viruses interfere with particular SIGNAL TRANSDUCTION pathways, mimicking different signals (e.g. producing G1 cyclins) which induce CELL CYCLE activation in quiescent G0 cells and transforming them. Most oncogenes (e.g. c-*abl*, c-*fos*) encode PROTEIN KINASES, involved in the regulation of the cell cycle, and may exert their effects in mitotic cells, meiotic cells or both; others direct the transition out of the cell cycle and into the differentiation pathway.

Most retroviral oncogenes originate as normal cellular (host) genes (proto-oncogenes or c-*oncs*) that are picked up by the retrovirus by a transduction-like process and subsequently undergo mutation (e.g. for c-*myc*, see APOPTOSIS). Some oncogenes may be expressed when TRANSPOSABLE ELEMENTS insert adjacent to them in the genome. See CYTOSTATIC FACTORS, MITOGEN, SUPPRESSOR MUTATION.

oncogenic virus See ONCOGENE, VIRUS.

oncoproteins Proteins, such as TYROSINE KINASES and RAS PROTEINS, encoded by ONCO-GENES. See E1A, MYC.

one gene–one enzyme hypothesis See GENE.

ontogeny The whole course of an individual's development, and life history. Compare PHYLOGENY. See RECAPITULATION.

Onychophora Velvet worms. Small group of animals (e.g. *Peripatus*) with annelid and arthropod affinities, sometimes regarded as a distinct phylum, but here placed as a class within the ARTHROPODA. Represent an early stage of *arthropodization*. Body segmented with soft, unjoined cuticle 1 μm thick and permeable to water; body wall with longitudinal and circular smooth muscles; head not demarcated, bearing a pair of long, mobile *pre-antennae*, simple jaws and oral papillae. Limbs fleshy, unsegmented, with terminal claws; probably evolved from parapodial appendages. The heart has ostia. Coelom replaced in adult by haemocoele. Gaseous exchange by tracheae and spiracles. Excretory system by segmentally repeated *coxal glands*, similar to those of arachnids and crustaceans but with ciliated excretory ducts. The pair of ventral nerve cords lack segmental ganglia but have numerous connectives. Eyes simple. Forest-dwelling, generally nocturnal predators. About 70 species. Some aquatic Cambrian fossils, such as *Aysheaia* and *Opabinia*, show affinities. See TARDIGRADA.

oocyte Cell undergoing MEIOSIS during OOGENESIS. *Primary oocytes* undergo the first meiotic division, *secondary oocytes* undergo the second meiotic division. See MATURATION OF GERM CELLS.

oogamy Form of sexual reproduction involving production of large non-motile gametes, which are fertilized by smaller motile gametes. Extreme form of ANISOGAMY, occurring in all metazoans and some plants.

oogenesis Production of ova, involving usually both meiosis and maturation. See MATURATION OF GERM CELLS and Fig. 109a.

oogonium (1) (Bot.) Female sex organ of certain algae and fungi, containing one or several eggs (OOSPHERES). (2) (Zool.) Cell in an animal ovary which undergoes repeated mitosis, giving rise to oocytes. Compare SPERMATOGONIUM.

Oomycota (Oomycetes) A group previously classified with the fungi but having closer links to the algal division HETEROKON-TOPHYTA; sometimes placed in the kingdom Chromista. The Oomycota comprise about 95 genera and 694 species that are cosmopolitan and widespread in terrestrial or aquatic, freshwater or marine. Both saprophytic and parasitic species occur. Some of the latter are economically important on higher plants. Probably the best known plant pathogenic genus is *Phytophthora* ('plant-destroyer') with species causing widespread destruction of many crops including pineapples, tomatoes, strawberries, onions, apples, tobacco and citrus fruits. The best known species is *P. infestans*, which caused the great potato famine in Ireland in 1846–7, and caused devastating losses when transported on fruits and vegetables from Mexico to Canada and the northwestern USA in the early 1990s. Another species of *Phytophthora* is causing mayhem among wild oaks in California. Of the water moulds, the genus *Saprolegnia* includes both saprophytic and parasitic species, causing disease of fish and fish eggs.

The thallus can be unicellular or mycelial. The individual hyphae are coenocytic, and mainly aseptate. The assimilative stage is diploid. Most reproduce asexually through production of zoospores. These are biflagellate and bear dissimilar flagella; one is smooth or has fine hairs, the other bears mastigonemes. The cell walls comprise a glucan-cellulose, rarely with minor amounts of chitin, thus they are markedly different from the cell walls of fungi. Sexual reproduction is oogamous. The oogonium contains many eggs (oospheres), and the antheridium contains many male nuclei. The zygote is thick-walled (the oospores).

oosphere (Bot.) Large, naked, spherical non-motile macrogamete (egg); formed within an OOGONIUM.

oospore Thick-walled resting spore formed from a fertilized OOSPHERE.

opal mutation A 'stop' mutation. A UGA mRNA triplet. See AMBER MUTATION.

open bundle Vascular bundle in which a vascular cambium develops.

open reading frame (ORF) A DNA sequence has six possible reading frames: three in each direction (see GENETIC CODE). In any such sequence, however, an ORF starts with an initiation codon (usually ATG) and ends with a stop codon (TAA, TAG or TGA). That is, it is uninterrupted by stop codons. Computers can scan DNA sequences 'looking for' ORFs (potential genes), normally more than 100 codons long. RNA will contain complementary ORFs. See YEAST TWO-HYBRID SYSTEM.

operator See JACOB–MONOD THEORY.

operculum (Bot.) Lid or cover; in certain fungi, part of cell wall; in mosses, a multicellular apical lid that opens the sporangium. (Zool.) (1) Cover of gill slits of holocephalan and osteichthyan fishes, and of larval amphibia. Contains skeletal support. (2) Horn or calcareous plate borne on back of foot in many gastropod molluscs, brought in over the body when animal withdraws into its shell. (3) A second EAR OSSICLE of many urodele and anuran amphibians; flat plate fitting into oval window of inner ear. May replace the stapes.

operon See JACOB–MONOD THEORY.

Ophidia Snakes. Suborder of the Order SQUAMATA (sometimes a separate order). Limbless reptiles, with exceptionally wide jaw gape due to mobility of bones. Eyelids immovable, nictitating membrane fused over cornea; no ear drums.

Ophiuroidea Brittlestars. Class of ECHINO-DERMATA; star-shaped, with long, sinuous ambulatory arms radiating from clearly delineated central disc; mouth downwards; well-developed skeleton of articulating plates; tube feet without suckers; no pedicillariae; easily break up by autotomy; scavengers.

ophthalmic Alternative adjective to optic.

opiates See ENDORPHINS, ENKEPHALINS.

opines Tumour-specific modified amino acids produced in a host cell (e.g. AGROBAC-TERIUM) under viral gene direction.

opioids Any natural or synthetic drug exerting effects on the body similar to those induced by morphine (the major analgesic derived from the opium poppy). They include the narcotics (poppy-derived pain-relievers morphine and codeine), heroin, and neuroactive analgesic peptides (ENDOR-PHINS, ENKEPHALINS). See PSYCHOACTIVE DRUGS.

Opisthobranchia Subclass of gastropod molluscs, in which shell and mantle cavity are reduced or lost. Body bilateral and slug-like. Sea hares (e.g. *Aplysia*), other nudibranchs, and the actively swimming pteropods.

opisthosoma One of the TAGMATA of chelicerate arthropods (Merostomata and Arachnids). Comprises the trunk segments, devoid of walking limbs; often (e.g. in scorpions) separable into an anterior *mesosoma* and posterior *metasoma*.

opsins Proteins belonging to the super-family of G-PROTEIN-coupled RECEPTORS which, bound to 11-*cis*-retinal, act as vertebrate visual pigments (see RHODOPSIN, ROD CELL, CONE).

opsonins Proteins coating foreign particles, such as surfaces of pathogenic microorganisms, rendering them more susceptible to ingestion by phagocytic leucocytes (*opsonization*). Include ACUTE PHASE PROTEINS, components of COMPLEMENT and some ANTIBODIES.

optic chiasma Structure formed beneath vertebrate forebrain by those nerve fibres of the right optic nerve crossing to left side of brain and those of left side crossing to the right side. In most vertebrates all the fibres cross; in mammals about 50% on each side do so. See BINOCULAR VISION, DECUSSATION.

optic nerve Second CRANIAL NERVE of vertebrates; really part of brain wall. See RETINA.

optimal diet model Mathematical formulation of prey selection that would theoretically provide the best energy returns to the consumer. Such a model is used to assess actual performance.

optimality theory Very generally the theory that, through natural selection, the behaviours of organisms are such that they tend to the most cost-effective use of time and available resources, given environmental circumstances. Despite charges of being vacuous and truistical, and although similar statements may be deducible from some formulations of Darwinian theory, their heuristic value has been great, prompting detailed and testing studies which have greatly enhanced understanding of adaptations involved in such diverse fields as foraging behaviour, diet choice, habitat selection and competition in animals, and reproductive behaviour more generally. See GAME THEORY.

The procedure for optimality testing is to specify an optimality criterion, define alternative strategy sets and estimate the pay-offs for each strategy – so determining which is optimal under the specified conditions. The justification for using optimality theory in evolution is that, in general, natural selection is expected to maximize inclusive fitness – the appropriate optimality criterion. In deciding which of two evolutionary trees is best (see BAYESIAN STATISTICS), four optimality criteria are widely used: (a) maximum parsimony (the tree requiring fewest character state changes is the better alternative); (b) maximum LIKELIHOOD (the tree maximizing the likelihood under the assumed evolutionary model is the better); (c) minimum evolution (the tree having the smaller sum of branch lengths is better); (d) least squares (the tree showing best fit between estimated pairwise distances, and the corresponding pairwise distances obtained by summing paths through the tree, is the better). See GAME THEORY.

optimum similarity The level of similarity between competing species, which, if either exceeded or reduced, would lead to a lowering of fitness for individuals in at least one species. See COMPETITIVE EXCLUSION PRINCIPLE.

oral rehydration therapy (ORT) A treatment for replacing salt and water lost in cases of severe diarrhoea (e.g. in acute gastroenteritis) which does not require intravenous drip. A combination of sugar and salt is taken by mouth, enabling glucose-linked sodium absorption across the villi even though heavy intestinal epithelial sodium secretion is occurring through the cells' binding of enterotoxin. The Na^+/glucose co-transporters in the epithelial brush border membranes appear to co-transport water molecules into the cells; but any uptake of solutes by epithelial cells would improve their osmotic uptake of water molecules, rehydrating the patient.

orbit Cavity or depression in vertebrate skull, housing eyeball.

ORC (origin recognition complex) A collection of proteins which bind sequentially to chromatin, probably at origins of DNA replication, remaining associated throughout the CELL CYCLE. Its presence is required for the binding of other replication proteins but does not, apparently, regulate the timing of S-phase onset. See CELL CYCLE, DNA REPLICATION.

Orchidaceae An important family of flowering plants belonging to the LILIOPSIDA (LILIIDAE) (monocots) that comprises about 25,000 species and more than 70,000 artificial hybrids made by growers throughout the world (more are being added each year). Orchids are found worldwide in the mountains or the tropics, in temperate regions and even in the Arctic where seven species have been found. Ecologically diverse; in the UK many are confined to chalk downlands, others to fens or acid moors. Many are increasingly threatened by habitat disturbance and by collectors. The majority of orchids are herbaceous but several are vines. More than half grow as epiphytes, clinging to the surfaces of other plants, a tree or shrub and derive their nutrients from materials dissolved in rainwater that seeps over them. Orchids have developed special vegetative features (e.g. swollen stems called pseudobulbs, succulent leaves and aerial roots) to survive dry periods. Orchids are grown and appreciated by countless people throughout the world for their fascinating flowers. Every orchid flower is alike in having six parts: three coloured sepals, two petals and a third specialized petal called the lip. This surrounds the reproductive parts, which are embedded in a central column. Their great attraction lies in the tremendous variation in flower size and shape, colour and markings, scent, texture and the length of time flowers last. Orchid seed is dust-like and dependent upon ectomycorrhizal fungi, sometimes even for food, but can grow only where the fungal symbiont finds suitable conditions. This dust-like orchid seed is well adapted for distribution by wind and lodging in the bark of trees. The requirement for the presence of a symbiotic fungus hampered early attempts at raising orchids from seed; however, the advent of MICROPROPAGATION has lead to an enormous increase in the number of plants in cultivation and a consequent reduction in their price.

Order Taxonomic category; inclusive of one or more families, but itself included by the class of which it is a member. See CLASSIFICATION.

ordination Mathematical system for categorizing communities on a graph so that those that are most similar in species composition appear closest together.

Ordovician GEOLOGICAL PERIOD, second of the PALAEOZOIC era, lasting approximately from 505–440 Myr BP. Period starts with the first major extinction event. Molluscs diversified while plants possibly invaded the land. The first fungi appeared. Climate was mild; seas were shallow and continents flat. See Appendix.

organ Functional and anatomical unit of most multicellular organisms, consisting of at least two tissue types (often several) integrated in such a way as to perform one or more recognizable functions in the organism. Examples in plants are roots, stems and leaves; in animals, liver, kidney and skin. Organs may be integrated into functional SYSTEMS. Sometimes it is debatable where the limits of organs are; thus the stomach is often regarded as an organ, but on occasions so is the whole alimentary canal.

organ culture By partial immersion in nutrient fluid, the growth and maintenance *in vitro* of an organ after removal from the body. The organ must be small (usually embryonic) since nutrients must diffuse into it from outside. See TISSUE CULTURE.

organ of Corti See COCHLEA.

organelle Structural and functional part of a cell, distinguished from the cytosol. Often membrane-bound (e.g. nucleus, mitochondria, chloroplasts, endoplasmic reticulum); sometimes not (e.g. ribosomes). The plasma membrane is itself an organelle. Those in cytoplasm are termed *cytoplasmic organelles*, in contrast to the nucleus. *De novo* synthesis of membrane-bound organelles is never required: they grow by division of pre-existing organelles, transmitted maternally. Techniques for their study include CELL FRACTIONATION, CRYO-ELECTRON TOMOGRAPHY, MICROSCOPY. See Fig. 20 and Table 5.

organic farming Principally, those methods of agriculture which avoid the use of herbicides, insecticides, artificial fertilizers, and intensive animal husbandry, and are therefore more reliant on natural or ecologically sustainable procedures. Use of genetically modified organisms has recently brought public opprobrium, and is probably excluded from organic farming. Many past civilizations over-farmed the land, notably in north Africa at the end of the Roman Empire. However, a sediment record indicates that one ancient people, the Incas of the Peruvian highlands, practised ecologically sensitive conservation techniques such as canal building, terracing and possibly tree planting (nitrogen-fixing alders, *Alnus acuminata*) to restore degraded farmland – the same soil-sparing practices as are used there today. Illegal woodcutting and burning were punishable by death in preconquest Inca territories. After the conquest by Europeans, both terraces and trees went into decline.

organic radicals Organic molecules with unpaired electrons which are often transient intermediates in enzyme reactions, including those involving electron transfer, isomerization, oxidation, oxygenation, reduction, DNA repair and synthesis of antibiotics and DNA. Their participation in these reactions requires binding of the enzyme by coenzymes, many of which (e.g. THIAMINE pyrophosphate, TPP) have previously been thought of as involved solely in polar (non-radical) reactions. These are thought to donate an electron to some non-toxic acceptor while picking up the unpaired electron from the organic radical, so rendering it safe. See FREE RADICAL.

organizer Any part of an embryo which can induce another part to differentiate. Classic examples are the dorsal lip of the BLASTOPORE, and HENSEN'S NODE. See EVOCATION, INDUCTION, POSITIONAL INFORMATION.

organotrophic See HETEROTROPHIC.

Oriental ZOOGEOGRAPHICAL REGION comprising India and Indo-China south to WALLACE'S LINE.

origin of life Major hurdles to the origin of life-forms resembling even the simplest known cells are the production of (a) a genetic system which is (b) sufficiently discrete from other such systems to enable its collection of properties to be a unit in reproductive competition with them. These two requirements would seem to put a premium on simple genetic (information) systems engaging in, at least initially, autocatalysis, with subsequent expansion of the stored information to code for molecules which do not themselves store information but instead form part of the structure of the system.

Even the simplest living systems today are far too complicated to resemble at all closely the earliest forms of life. Three major features of present living systems (construction from complex, often polymeric, organic compounds; catalysis by proteins; storage of molecular information in the form of nucleic acid) are likely to be the end results of evolution by natural selection between systems which were originally far simpler and lacked all these features. The following account is highly conjectural, but raises some of the issues being debated.

For long it was thought that the early Earth's atmosphere consisted, locally if not universally, of reducing molecules (ammonia, methane, nitrogen and water vapour), and attempts to create likely precursor molecules of life (amino acids, bases) by passing electrical sparks through such gaseous mixtures were rewarding. However it is probable that four thousand million years ago the sun was weaker in terms of total thermal output than it is now, and more active in terms of ultraviolet output. Implications for the Earth's early atmosphere and the origin of life are that: (a) without an atmospheric GREENHOUSE EFFECT, the surface temperature of the Earth would have been too low to avoid ocean freezing, and that (b) incident ultraviolet radiation on the Earth's atmosphere would have resulted in photolysis of methane and ammonia, releasing free hydrogen out of the atmosphere, with probable retention of carbon dioxide and atomic and molecular nitrogen. The carbon dioxide could have provided the greenhouse effect: the early oceans were not frozen.

Adenine, one of the bases in RNA and DNA, can be produced by oligomerization of hydrogen cyanide (HCN). RNA has become a lively contender in the 'origin' debate on account of its catalytic ability (see RIBOZYME, RNA WORLD). It is now clear that RNA is capable of self-catalysis (see AUTOCATALYST). HCN may have been formed from methane present as a trace gas in the early mantle. Hydrothermal vents (undersea hot springs) provide reducing compounds which, by permeating rock, could form complex molecules, cooled sufficiently by the water to reduce the likelihood of their dissociation. These, in association with the self-assembling crystalline organization provided by such clays as *kaolinites* and *illites*, could come to form polymers supported by the lattice of the crystal. An alternative site for the earliest stages of life may have been the surface of *pyrite* (FeS$_2$) close to hydrothermal vents, making the content of the atmosphere at the time less of an issue. However, if these early processes occurred on the surfaces of water droplets in clouds, atmospheric content would have been paramount. The significance of having a surface such as clay crystals or pyrite, rather than a solution, for life's origin is that the former achieve less entropy loss. The evolution of non-enzymatic biochemical cycles could have been important in life's origin; e.g. the *formose reaction* in which sugars are produced autocatalytically from formaldehyde. In addition, activated nucleotides can condense into chains 20–40 nucleotides long in the absence of templates. Although some of the most primitive organisms alive today (HYPERTHERMOPHILES) inhabit hot water, some researchers favour cool pools or even cold oceans as the location for cell origins. Indirect evidence includes the fact that the adenine- and uracilcontaining nucleotides of ribosomal RNA form weak bonds and that the G–C bond holds up better in hot conditions. However, one computer model, extrapolating from the rRNA make-up of 40 very widely differing living organisms to a hypothetical common ancestor, proposes an ancestral rRNA with only moderate G–C bonds.

The first membranes might have arisen by self-assembly, forming spherical vesicles in much the same way that some phospholipids do when mixed with water. Such lipid vesicles might then have accumulated any organic molecules soluble in their membranes and if these polymerized inside they might not have been able to escape. Continued polymerization within might have put the vesicles under strain, promoting their enlargement by incorporation of further membrane components, as can happen with artificial LIPOSOMES. Rupture of the spheres followed by renewed growth of the rupture products would have achieved a simple form of reproduction.

In the past, RNA may have had wider catalytic ability than it has now (see TELOMERE). If RNA were among early polymers, formed perhaps on clay crystals trapped within membrane species, and were it to catalyse amino acid polymerization, then protein synthesis might gain independence from clay crystals, RNA also supporting the protein. If some of these proteins were to associate with the membrane in such a way as to stabilize it and even promote entry of building materials, selection could have favoured those systems which in replicating retained catalytic production of the membrane proteins by RNA; that is, which achieved the reproducible link between genotype and phenotype. Something like a simple cell ('RNA life') would have been formed, and interestingly a motif that binds ATP in solution has been located on some cellular RNAs. DNA could eventually have been synthesized off the RNA (retroviruses do this today), the resulting duplex providing a more stable store of information. See EUKARYOTE.

ornithine cycle (urea cycle) See UREA.

Ornithiscia Extinct order of ARCHOSAURS ('ruling reptiles'), lasting throughout Jurassic to late Cretaceous. 'Bird-hipped' and herbivorous dinosaurs. Some early forms were bipedal. Both the pubes and ischia of the PELVIC GIRDLE pointed downwards and backwards. Included stegosaurs (e.g. *Stegosaurus*), ankylosaurs and ceratopsids (e.g. *Triceratops*). Compare SAURISCHIA.

ornithology Study of birds.

ornithophily Pollination by birds.

orphan receptors Receptors for which no ligand has yet been found. For orphan nuclear receptor superfamily, see NUCLEAR RECEPTORS; for orphan guanylyl cyclase receptors, see CYCLIC GMP.

orthogenesis Theory, prevalent in early decades of 20th century, that the evolutionary path of a lineage can acquire an 'impetus', or 'inexorable trend' carrying it in a direction independently of selective constraints imposed by the environment. It was once suggested that the genetic material itself might somehow be responsible. Now discredited.

orthognathous 'Short-faced' mammals, lacking muzzles. Hominid evolution has been accompanied by the evolution of increasingly orthognathous faces from PROGNATHOUS ones.

orthologous If an ancestral species carries gene *A* and, after speciation, so does each descendant species, then as the species diverge so might the gene *A* sequences in the different lineages. Such gene sequences in the two species are said to be orthologous. As these alterations reflect the splitting (cladistic) event, the information has phylogenetic significance (e.g. the large and small subunit rRNA genes). Gene sequences brought about by GENE DUPLICATION and subsequent divergence within a lineage would not be of phylogenetic significance and would be PARALOGOUS.

Orthoptera Large order of exopterygote insects, including locusts, crickets and grasshoppers. Medium or large insects, usually with two pairs of wings (sometimes absent), front pair narrow and hardened, hind pair membranous; hind legs usually large and modified for jumping. Stridulatory organs; usually involving wings being rubbed together (*alary*), or inner face of hind femur being rubbed against a hardened vein on forewing (*femoro-alary*). Paired auditory organs (*tympana*) either on anterior abdomen or tibias of front legs.

orthoselection Directional selection. See NATURAL SELECTION.

osculum Large opening through which water leaves the body of a sponge (PORIFERA), having entered through *ostia*.

oskar genes (osk.) A group of MATERNAL EFFECT genes in *Drosophila*, whose mRNA transcription products become localized in the posterior pole of the developing oocyte. Embryos from females mutant for any of these genes develop normal heads and thoraxes, but lack abdomens. The *osk* protein product activates transcription of the *nanos* gene, whose product in turn activates *knirps* transcription and represses *hunchback* translation (an example of PLEIOTROPY). See *BICOID* GENE, EICOSANOIDS.

osmium tetraoxide Substance (OsO_4) which, in aqueous solution, blackens fat and is often used to demonstrate myelin sheaths of neurons. Used as a fixative in light and electron microscopy. Sometimes called osmic acid.

osmophilic body Lipid-containing granules, appearing dark when treated with osmium fixation.

osmoregulation Any mechanism in animals regulating (a) the concentration of solutes within its cells or body fluids, and/or (b) the total volume of water within its body.

Each major inhabited environment, freshwater, marine and terrestrial, poses its own osmotic problems for organisms. Cells approximate to sea water in WATER POTENTIAL, but have lower water potentials than freshwater and far higher water potentials than air at normal humidities. Freshwater protozoans and sponges use CONTRACTILE VACUOLES to expel osmotic water from their cells; the SODIUM PUMPS of most animal cell membranes are important in limiting the cell's internal ion concentration. Protection from osmotic uptake of water and from desiccation is achieved by a body surface covered with an impermeable CUTICLE or SKIN, but there are usually soft parts (e.g. gut, gills) through which water enters. Adaptive behaviour may reduce osmotic dangers; thus limpets adhere closely to rocks at low tide; marine mussels and periwinkles retract into shells. *Euryhaline* animals tolerate wide fluctuations in water potentials of their surroundings, possibly through a limited ability to swell in hypotonic conditions and by a reduction in intracellular levels of some organic compounds, or inorganic ions such as sodium and chloride. Some engage in active ion uptake (e.g. of sodium) if the external medium becomes dilute enough to cause sodium loss by diffusion. Unless urine can be made hypotonic to body fluids, inorganic ions will be lost by this route, and a kidney which reabsorbs ions from the excretory fluid will reduce this loss. ANADROMOUS animals, such as many cyclostomes and some fish, automatically take up water in freshwater but produce a copious hypotonic urine, active ion uptake by the gills replacing urinary loss. On return to marine conditions they lose water to the sea and correct this by drinking sea water: sodium, potassium and chloride are absorbed by the intestine and excreted across the gills (see SODIUM PUMP).

Restriction of water loss on land is associated with evolutionary change from excretion of ammonia (ammonotely) to urea (ureotely) and uric acid (uricotely). Amphibian metamorphosis includes a change from ammonotely to ureotely; land reptiles produce a urine hypotonic to body fluids, but because of uricotely (uric acid being insoluble requires only incidental water in its removal) the total volume produced is not great. Evolution of the AMNIOTIC EGG would have been less likely without the uricotely of reptilian and bird embryos. Marine birds and reptiles do not regularly drink sea water, but they absorb much salt from their food and excrete it through *salt glands* above the orbits of the eyes.

Vertebrate KIDNEYS, especially mammalian, can produce a urine which is variably copious and hypotonic to body fluids or sparse and hypertonic, depending upon the water balance of the body. The hypothalamus monitors blood concentration, and any increase above the homeostatic norm results in release of ANTIDIURETIC HORMONE by the posterior PITUITARY. This causes uptake of water by the collecting ducts and reduced volume of hypertonic urine. Moreover, the hypothalamic *thirst centre* now

initiates drinking, restoring blood concentration. Sudden loss of blood volume (or pressure) causes *aldosterone* release from the ADRENAL cortex (see ANGIOTENSINS). This increases potassium excretion and sodium retention by the kidney, tending to make extracellular (tissue) fluid more concentrated than body cell fluid, lowering its water potential and drawing water out from the cells to help restore blood volume. After a blood meal, the bug *Rhodnius* disposes of excess fluid by release of a *diuretic hormone* from its thoracic ganglia. This accelerates active secretion of sodium ions from the gut, chloride and water following them into the haemocoele, where they are dealt with by MALPIGHIAN TUBULES. Osmoregulation may therefore involve structural, physiological and behavioural adaptations.

osmosis The net diffusion of water across a selectively permeable membrane (permeable in both directions to water, but varyingly permeable to solutes) from one solution into another of lower WATER POTENTIAL. The *osmotic pressure* of a solution is the pressure which must be exerted upon it to prevent passage of distilled water into it across a semipermeable membrane (one impermeable to all solutes, but freely permeable to solvent), and is usually measured in pascals, Pa (1 Pa = 1 Newton/m^2). A solution's *osmotic potential* is always negative, the more so the greater the concentration of solute particles. The plasmalemma is selectively (i.e. not semi-) permeable and permits selective passage of solutes (see CELL MEMBRANES), as does the *tonoplast* in plant cells. See AQUAPORINS.

The osmotic pressure of a solution depends upon the ratio between the number of solute and solvent particles present in a given volume: 1 mole of an undissociated (non-electrolytic) substance dissolved in 1 dm^3 of water at 0°C has an osmotic pressure of 2.26 MPa (megapascals). But a solution of 0.01 mole of sodium chloride per dm^3 of water has almost twice the osmotic pressure of a solution of 0.01 mole of glucose per dm^3 of water. This is because sodium chloride in water dissociates almost completely into Na$^+$ and Cl$^-$ ions (giving twice the particle number) whereas glucose does not dissociate in water. Solutions of equal osmotic pressure are *isotonic*; one with a lower osmotic pressure than another is *hypotonic* to it, the latter being *hypertonic* to the former.

When cells produce polymers from component monomers they may dramatically reduce the osmotic pressure of their cytoplasm; hence polymers (e.g. starch, glycogen) make good storage compounds because they are *osmotically inactive*. Glucose is far more *osmotically active* than is starch.

Osmosis is a physical process of great importance to all organisms, affecting relations with their environments as well as between their component cells. If an animal cell takes up water osmotically its plasma membrane will eventually rupture (*osmotic shock*; see ISOTONIC); but the SODIUM PUMP reduces this problem under normal physiological conditions. Cells with cell walls are normally prevented from osmotic rupture. Those which lose water osmotically shrink and become *plasmolysed* (see PLASMOLYSIS). Water relations of cells and organisms are generally discussed in terms of water potential. See TURGOR, DIFFUSION, OSMOREGULATION.

osmotic activity, osmotic pressure and osmotic shock See OSMOSIS.

ossicle A very small bone, or calcified nodule.

ossification Formation of bone. May occur by replacement of CARTILAGE (*endochondral ossification*) or by differentiation of non-skeletal mesenchyme (*intramembranous ossification*), forming DERMAL BONES. In the former, cartilage first calcifies and hardens to form *ossification centres*. The cartilage cells die and those in the periochondrium develop into *osteoblasts* and lay down a ring of bone to form the *periosteum*, which enlarges as blood vessels invade and osteoblasts enter vacant lacunae to develop into *osteocytes*. The bone lengthens and thickens, finally becoming mineralized. Cartilage remains as *articular cartilage* forming JOINTS at the bone ends, and in the *growth plate*

in the bone EPIPHYSIS. See Fig. 128, DIAPHYSIS, GROWTH.

Osteichthyes Bony fishes (usually regarded as sister group to CHONDRICHTHYES). Largest vertebrate class. Some fossils come from the Upper Silurian but a considerable radiation had occurred by the Lower to Middle DEVONIAN. Includes subclasses the lobe-fins, CHOANICHTHYES (= SARCOPTERYGII) and ACTINOPTERYGII (which itself includes the TELEOSTEI). Other fish groups (e.g. agnathans, placoderms, acanthodians) contain bone; but the osteichthyans are the only jawed fishes with bony vertebrate and gill arches and with paired fins, but lacking bony plates on head and body. ACANTHODII are sometimes regarded as a subclass of the Osteichthyes, and do not really fall outside the above criteria. For PSAROLEPSIS, see FISH.

osteoblast Cell type responsible for formation of calcified intercellular matrix of BONE.

osteoclast Multinucleate cell which breaks down the calcified intercellular matrix of BONE. Remodelling of bone shape by such activity accompanies bone GROWTH.

Osteostraci Extinct order of AGNATHA (Monorhina: single external nostril). Head covered by strong bony shield. Heterocercal tail. See OSTRACODERMI.

ostiole Pore in fruit bodies of certain fungi, and in conceptacles of brown algae through which, respectively, spores or gametes are discharged.

ostium (1) One of the many inhalant pores of sponges. See PORIFERA. (2) Opening in the arthropod heart, into which blood flows from the haemocoele.

Ostracoda Subclass of CRUSTACEA. Most are a few mm long. Aquatic. CARAPACE bivalved, closable by adductor muscles, and completely covering body. FILTER FEEDING. Trunk limbs few and small; food gathered by head appendages. About 2,000 species. Shells of ostracods are calcified, and become preserved in lake sediments, particularly under calcareous or saline conditions (where the

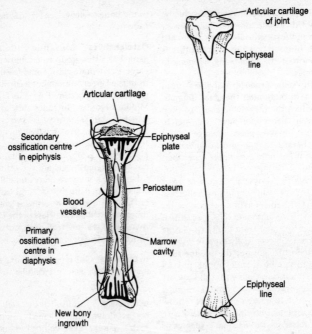

FIG. 128 *Final stages in* OSSIFICATION *of a human tibia. The epiphyseal line represents the bony remains of the once actively dividing cartilaginous epiphyseal plate (growth plate).*

action by carbonic acid is minimal). Therefore, they are most abundant in marl-sediments; rare in the typical organic lake sediment. Useful INDICATOR organisms and used in palaeolimnological studies because they are particularly sensitive to the concentration of dissolved solids in lake water, each species having a distinct range of tolerance, from very dilute to extremely concentrated (saline). They are also sensitive, to a lesser degree, to water depth and permanence of the lake or stream, as well as to climate in terms of temperature and precipitation.

Ostracodermi Group of fossil AGNATHA (Diplorhina: external nostrils paired) found from Upper Silurian to mid-Devonian, but probably present from Cambrian. Covered in bony armour; up to 50 cm in length. Lacked paired appendages. Similarities with cyclostomes. Some may have had mouthparts, but not jaws. Vertebrae so far

absent from fossils; notochord apparently the skeletal support. See OSTEOSTRACI.

otic Concerning the ear. See reference under AUDITORY.

otocyst See STATOCYST.

otolith Granule of calcium carbonate in the vertebrate MACULA. See also STATOCYST.

outbreeding Effect of any mechanism which tends to ensure that gametes which fuse at fertilization produce zygotes with a higher degree of heterozygosity than they otherwise would. Darwin realized that many plants (e.g. orchids, primroses) are structurally adapted to cross-pollination, and subsequent work on the genetics of incompatibility mechanisms, preventing self-fertilization, has shown that selection must favour such plants under certain conditions. Mechanisms promoting outbreeding including having separate sexes (see DIOECIOUS), DICHOGAMY, HETEROSTYLY, INCOMPATIBILITY mechanisms and, commonly in animals, the social structure of the breeding group (e.g. incest taboos in humans). Advantage of outbreeding in sexual populations is that it serves to free genetic variability in the population to selection and thereby reduces evolutionary stagnation of the population (see GENETIC VARIATION). Compare INBREEDING.

outcrossing (Bot.) Pollination between (normally genetically) different plants of same species.

outgroup A taxon used to root a PHYLOGENETIC TREE. If, e.g., we aim to infer such a tree for three groups of organisms (G1, G2 and G3) thought to be clades, any outgroup taxon would be one which diverged from G1, G2 and G3 before they diverged from each other.

ovarian follicle See GRAAFIAN FOLLICLE.

ovary (1) (Bot.) Hollow basal region of a carpel, containing one or more ovules. In a flower which possesses two or more united carpels the ovaries are united to form a single compound ovary. (2) (Zool.) Main reproductive organ in female animals, producing *ova*. In invertebrate development eventual GERM CELLS often become distinct at early cleavage. In vertebrates a pair of ovaries develops from mesoderm in the roof of the abdominal coelom; they become invaded by cells from the endoderm, but it is debatable whether these or mesodermal cells developing *in situ* become the *primordial germ cells*; however, the ovary epithelium (*germinal epithelium*) invaginates and some of its cells begin to undergo *oogenesis* (SEE MATURATION OF GERM CELLS). Each *oogonial* cell becomes surrounded by other cells to form a *follicle*. The ovaries produce various sex hormones, for details of which in mammals see MENSTRUAL CYCLE. See TESTIS.

overcompensating density-dependence Density-dependence which is so intense that increases in initial density lead to reduction in final density.

overdominance See HETEROZYGOUS ADVANTAGE.

overexploitation The exploitation of (i.e. removal of individuals or biomass from) a natural population at a rate greater than the population is able to match with its own recruitment. This tends to drive the population towards extinction.

overgrowth competition Competition amongst sessile organisms where the mechanism is the growth of one organism over another, preventing the efficient capture of light, suspended food or some other resource.

overturn Complete mixing of a body of water, from surface to bottom; the breakdown of STRATIFICATION. Results from various external factors.

oviduct (Zool.) Tube carrying ova away from ovary (or from the coelom into which ova are shed) to the exterior. See MÜLLERIAN DUCT.

oviparity Laying of eggs in which the embryos have developed little, if at all. Many invertebrates are *oviparous*, as are the majority of vertebrates, including the prototherian mammals. See OVOVIVIPARITY, VIVIPARITY.

ovipositor Organ formed from modified paired appendages at hind end of abdomens of female insects, through which eggs are laid (*oviposition*). Consists of several interlocking parts. Frequently long (e.g. in ichneumon wasps), and capable of piercing animals or plants, permitting eggs to be laid in otherwise inaccessible places. Stings of bees and wasps are modified ovipositors.

ovotestis Organ of some hermaphrodite animals (e.g. the garden snail, *Helix*; sea bass, Serranidae) serving as both ovary and testis. In the citrus tree pest *Icerya purchasi* (cottony cushion scale) diploid ('female') individuals transform into functional self-fertilizing hermaphrodites, possessing an ovotestis in which the centre is haploid and testicular, the cortex diploid and ovarian. They are thus chromosomal MOSAICS. In order to achieve the haploid interior, one set of chromosomes is eliminated in some of the early ovotestis cells (see ABERRANT CHROMOSOME BEHAVIOUR (3)).

ovoviviparity Development of embryos within the mother, from which they may derive nutrition, but from which they are separated for most, or all, of development by persistent EGG MEMBRANES. Examples include many insects, snails, fish, lizards and snakes. See OVIPARITY, VIVIPARITY.

ovulation Release of ovum or oocyte from mature follicle of vertebrate ovary. See MENSTRUAL CYCLE for hormonal details.

ovule Structure found in seed plants which develops into a SEED after fertilization of an egg cell within it (see DOUBLE FERTILIZATION). In gymnophytes (gymnosperms), ovules are unprotected; in flowering plants (anthophytes), they are protected by the MEGASPOROPHYLL, which forms a closed structure (CARPEL) within which they are formed singly or in numbers. Each ovule is attached to carpel wall by stalk (funicle) which arises from its base (chalaza). A mature flowering plant ovule comprises a central mass of tissue (the NUCELLUS) surrounded by one or two protective layers (integuments) from which the seed coat is ultimately formed. Integuments enclose the nucellus except at the apex, where a small passage (the micropyle) permits entry of water and oxygen during germination. Within the nucellus is a large oval cell, the EMBRYO SAC, developed from the megaspore and containing the egg cell.

ovum Unfertilized, non-motile, egg cell. In many animals it is an OOCYTE. Product of the OVARY.

oxidative phosphorylation Process by which energy released during electron transfer in aerobic RESPIRATION is coupled to production of ATP. See BACTERIORHODOPSIN, NAD, MITOCHONDRION.

oxidoreductases Major group of ENZYMES; catalyse REDOX REACTIONS.

oxygen debt Amount of oxygen needed to oxidize the LACTIC ACID produced in the anaerobic work done by muscle during vigorous exercise, and to resynthesize *creatine phosphate* used (see PHOSPHAGEN). Until then, oxygen intake remains above normal while lactic acid remains in the muscle and blood. May be oxidized to pyruvic acid, converted to carbohydrate by the liver in GLUCONEOGENESIS, or neutralized by blood bicarbonate BUFFER and excreted as *sodium lactate* by the kidneys.

oxygen dissociation curves (oxygen equilibrium curves) See HAEMOGLOBIN.

oxygen quotient See Q_{O_2}.

oxyhaemoglobin See HAEMOGLOBIN.

oxyntic cells (parietal cells) Hydrochloric acid-secreting cells of the gastric pits of vertebrate stomachs. They achieve this through their CARBONIC ANHYDRASE activity.

oxytocin Oligopeptide hormone secreted by *pars nervosa* of PITUITARY glands of birds and mammals. Involved in contraction of alveoli of mammary glands during expression of milk, and in promoting smooth muscle contraction of uterus during coitus, and during PARTURITION.

ozone A gas existing as a layer in the stratosphere, 20–40 km above Earth, formed when short wavelength ultraviolet (uv) light splits a molecule of oxygen gas (O_2) into its two atoms high in the stratosphere. These reactive atoms combine with oxygen molecules to form ozone molecules, O_3. Ozone gas absorbs uv of a longer wavelength and protects life on Earth from damaging effects of ULTRAVIOLET RADIATION. Ozone in the lower atmosphere (troposphere) appears to be generally increasing, largely through the rising levels of nitrogen oxides (NOx), hydrocarbons and hydrogen peroxide from which O_3 is produced through photochemical reactions, high light and temperature promoting production. It is an efficient greenhouse gas (see GREENHOUSE EFFECT), absorbing infrared radiation. Ozone damage to European herbaceous plants occurs when levels exceed 2.5 times normal, shifting sucrose distribution from roots to shoots and increasing drought-dependent stress, possibly through reacting with ethylene to produce damaging free radicals. Ozone in both the Arctic and the Antarctic is destroyed because the low temperatures in the upper atmosphere permit cloud formation containing crystals of NITRIC OXIDE, which absorb chlorine compounds during the winter; but the spring warmth and sunshine release these chlorine compounds which are then further broken down by sunlight into reactive chlorine molecules. These go on to destroy ozone, and although chlorine is normally mopped up by nitrogen dioxide (NO_2), in polar clouds the latter tends to form nitric acid instead. See ACID RAIN, CFCS.

P

P₁ Parental generation in breeding work. Their offspring constitute the F_1 generation.

p53 An oncosuppressor protein encoded by the *p53* gene (see TUMOUR NECROSIS FACTOR), involved in cell death caused by HYPOXIA, and in APOPTOSIS. It directs cells either towards delaying their division, or towards apoptosis. Functionally inactivated in many human cancers, it plays a major part in preventing expansion of mutated somatic cells. Damage to DNA is detected by p53-based checkpoint controls, reversibly arresting the cell cycle at the G1/S transition and allowing DNA REPAIR MECHANISMS to offset mild DNA damage before genomic DNA is replicated. However, p53 induces apoptosis in the face of extensive damage to DNA, and also has a role in the spindle checkpoint mechanism such that CELL CYCLE arrest (mediated by induction of the Cdk-inhibitor p21) occurs in response to aberrant mitotic spindles. In normal cells, DNA-damaging agents activate p53, which then upregulates the expression of the cyclin-dependent kinase (CDK) inhibitor p21 and proapoptotic proteins containing the BH3 domain. As a result of this suppression of cyclin/CDK activity the *Rb* gene product, Rb, is activated (see TUMOUR SUPPRESSOR GENE) and suppresses DEAMINATION of Bcl-x$_L$. This inhibits the ability of BH3-containing proteins to induce apoptosis. By contrast, tumour cells that lack Rb suffer deamidation of Bcl-x$_L$ by DNA-damaging agents with the result that the proapoptotic activities of BH3-containing proteins are not suppressed and tumour cell apoptosis results. Embryonic fibroblasts from *p53*-null mice overproduce centrosomes, while mice with increased p53 activity exhibit premature senescence (see AGEING). Bcl-2 can bind the p53-binding protein p53-BP2, and this may explain how Bcl-2 can interfere with translocation of p53 from the cytosol to the nucleus. See Fig. 129, CANCER CELLS, ULTRA-VIOLET IRRADIATION.

pacemaker (1) Groups of modified cardiac muscle cells in *sinus venosus* of vertebrate HEART. In mammals and birds, forms *sinuauricular* (*sinoatrial*) *node* and lies in right atrial wall near superior vena cava. Cells have wandering membrane potentials, with a built-in tendency to depolarize (the so-called pacemaker potential), and the cell with the fastest intrinsic rate of depolarization leads the rest to fire with it simultaneously. The electrical current generated spreads to adjacent *Purkinje fibres*, and from there to atrial muscle cells, which contract together. The intrinsic pacemaker discharge is increased by sympathetic nerves from the thorax and decreased by branches of the vagus (parasympathetic system). See CARDIOACCELERATORY and CARDIO-INHIBITORY CENTRES. (2) A *myogenic pacemaker* occurs in the smooth muscles of vertebrate gut; longitudinal and circular layers each have one.

pachytene See MEIOSIS.

Pacinian corpuscles Very rapid-adapting pressure receptors in subcutaneous adipose tissue of both hairy and hairless vertebrate skin. Also found in joint capsules, tendons, etc.

paedogenesis Form of HETEROCHRONY, in which reproductive organs undergo accelerated development relative to rest of body, giving larval maturity. The alternative term

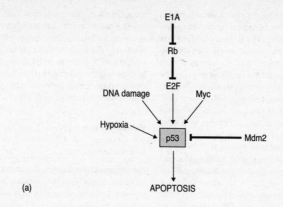

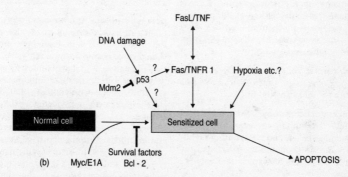

FIG. 129 *Two models for the relations between oncogenes and various death signals. (a) A conventional model in which p53 acts as the transducer of a variety of diverse signals. (b) In this model, it is proposed that growth-deregulating mutations sensitize cells to apoptotic triggers. The models are simplified and not intended to encompass the total complexity of signalling pathways.*

PROGENESIS has been advocated. See NEOTENY, PAEDOMORPHOSIS, PARTHENOGENESIS.

paedomorphosis Evolutionary displacement of ancestral features to later stages of development in descendant organisms, either through PROGENESIS or NEOTENY. See HETEROCHRONY.

pain A subjective experience due to nociception. For analgesics, see PSYCHOACTIVE DRUGS.

pair-rule genes A class of at least eight segmentation genes, particularly studied in *Drosophila*, mutants which bring about repetitive deletion of specific parts of alternating segments. The fact that they are expressed in seven or eight stripes during cellularization of the *Drosophila* BLASTODERM is a key event in the PATTERN FORMATION process. Their expression patterns determine PARASEGMENT boundaries.

The earliest to be expressed in development (primary pair-rule genes) are those directly controlled by the prior expression and distribution of GAP GENES, and in *Drosophila* are *hairy* (*h*), *even-skipped* (*eve*) and *runt* (*runt*). Their PROMOTERS are recognized by gap gene products, and variations in the promoter 'strengths' for different gap gene products seem to enable pair-rule genes to respond to their varying (often very low) concentrations. Primary pair-rule gene products then act as regulators (whether positive or negative is not certain) of secondary pair-rule genes such as *fushi tazaru* (*ftz*) and *paired* (*prd*), producing in the end fourteen stripes (bands) of product. The result of this mutually interactive sequence of gene activation and repression is that blastoderm cells emerge with different combinations of gene product that can now serve as reference points for further pattern formation, particularly in defining the domains of the SEGMENT-POLARITY GENES *engrailed* (*en*) and *wingless* (*wg*). The pair-rule segmentation mechanism may not exist in some SHORT-GERM insects. See Fig. II for some of these relationships.

Palaearctic ZOOGEOGRAPHICAL REGION, consisting of Europe, north Africa, and Asia south to Himalayas and Red Sea.

palaeoanthropology Combination of studies attempting to reconstruct social structures, beliefs and behaviour of prehistoric human (hominid) populations from their artefactual and fossil remains. See *HOMO*.

palaeobotany (paleobotany) Study of fossil plants.

Palaeocene GEOLOGICAL EPOCH; earliest division of the TERTIARY period, during which the early insectivorous mammals and primates appeared. Climate was mild to cool. Wide, shallow continental seas largely disappeared.

palaeoecology The study of relationships between past organisms and the environment in which they lived. Largely concerned with reconstruction of past ecosystems. Ecology and palaeoecology have similar aims, and invoke many of the same biological principles; however, there

are differences since past ecosystems cannot be examined directly. Inferences have to be made from fossils. Thus, palaeoecology is limited to the study of those past organisms whose fossils preserve or of those that can be detected chemically. Further, the fossil or sedimentary record can be altered through diagenesis, transportation and redeposition.

Palaeogene Collective term for Palaeocene, Eocene and Oligocene epochs. See GEOLOGICAL PERIODS.

Palaeognathae Ratites. Superorder of the subclass NEORNITHES. Includes birds with a reduced breastbone, and which are therefore secondarily flightless. Examples are: ostrich, rheas, cassowaries, tinamous, kiwis, and the extinct moas and elephant birds. Their feathers lack barbs. Probably represent a GRADE rather than a CLADE. See NEOGNATHAE.

palaeolimnology (paleolimnology) A branch of LIMNOLOGY concerned with the interpretation of past limnology from changes that occurred in the ecosystem of the lake, and their probable causes. Palaeolimnologists are interested in understanding how the lake functioned; how this aquatic ecosystem responded to external changes in climate, catchment area changes, fire, volcanism, tectonic events and anthropogenic influences (e.g. clearing of forests and establishment of settlements; agriculture and industry). It is through the palaeolimnological approach that long-term trends can be ascertained; thus palaeolimnology provides an historic perspective and is today an important approach in studies of ACID RAIN, EUTROPHICATION, and climate change.

Palaeolithic An archaeological term denoting the Old Stone Age of human prehistory. For over a century, Eurasian material has been classified into Lower, Middle and Upper Palaeolithic, and African material into Early, Middle and Late Stone Age. The Lower Palaeolithic was characterized by the hand-axe and complementary pebble tools/chopping tools; the Middle Palaeolithic usually lacked these large artefacts but contained many tools classified on their probable functions as scrapers; the Upper Palaeolithic technology included long, thin parallel-sided blade-like objects, while artistic and decorative material also appeared. RADIOMETRIC DATING methods have taken the Upper Palaeolithic back to >40,000 yr BP; uranium-series dates for carbonates, electron-spin resonance dates for mammal teeth and thermoluminescence dates for sediments and burnt flint have taken the Middle Palaeolithic (in Europe, Mousterian) industries back to >200,000 yr BP, while the Lower/Middle Palaeolithic division probably took place between 250,000–200,000 yr BP, the former traditionally extending back to European colonization at least 700,000 yr BP. It was long held that Middle Palaeolithic culture was associated with Neanderthals (see *HOMO*) while Upper Palaeolithic artefacts were the province of Cro-Magnon (modern humans). But Neanderthal remains from Kebara and modern crania and skeletons from Qafzeh (both in Israel) are all associated with Middle Palaeolithic tools. Again, outside Europe, there is no clear-cut artefactual boundary-line between Middle and Upper Palaeolithic, since blades and tools traditionally associated with the Upper Palaeolithic sometimes lie beneath strata with Mousterian culture. This lends support to those classifying tools less on appearance (typology) than on function. See PLEISTOCENE.

palaeontology Study of FOSSILS and evolutionary relationships and ecologies of organisms which formed them.

Palaeozoic Earliest major geological era. See GEOLOGICAL PERIODS.

palate Roof of the vertebrate mouth. In mammals and crocodiles the roof is not homologous with that of other vertebrates; a new (false) palate has developed beneath original palate, by bony shelves projecting inwards from bones of upper jaw. In mammals, bony part of false palate (*hard palate*) is continued backwards by a fold of mucous membrane and connective tissue, the *soft palate*.

palatoquadrate (pterygoquadrate) Paired cartilage or cartilage bone forming primitive upper jaw (as in CHONDRICHTHYES and embryo tetrapods). See AUTOSTYLIC JAW SUSPENSION, HYOSTYLIC.

palea (superior palea, pale) Glume-like bract of grass spikelet on axis of individual flower which, with the LEMMA, it encloses.

palindromic Reading the same forwards and backwards. Some DNA sequences are palindromic, a notable discovery being that the long arm of the human Y chromosome (Yq) has eight huge, but imperfect, palindromic sequences comprising a total of 5.7 Mb, or one quarter of the male-specific region. See INVERTED REPEAT SEQUENCE, INSERTION SEQUENCE.

palisade See MESOPHYLL.

pallium See CEREBRAL CORTEX.

palmelloid (Of algae) describing an algal colony comprising indefinite number of single, non-motile cells, embedded in mucilaginous matrix.

palps Paired appendages of many invertebrates, on the head or around the mouth. In polychaete annelids, tactile (on head); in bivalve molluscs, ciliated flaps around mouth generating feeding currents; in crustaceans, distal parts of appendages carrying mandibles (locomotory or feeding); of insects, parts of first and second MAXILLAE, sometimes olfactory.

palynology See POLLEN ANALYSIS, PALAEOECOLOGY.

pancreas Compound gland of vertebrates, in mesentery adjacent to duodenum; endocrine and exocrine functions. Secretes *pancreatic amylase*, *lipase*, *trypsinogen* and *nucleases* from its ACINAR CELLS in an alkaline medium of sodium hydrogen carbonate, promoted by CHOLECYSTOKININ and SECRETIN, and by the vagus (cranial nerve X), when the gastric phase of digestion is complete. The two major pancreatic hormones are INSULIN and GLUCAGON, secreted respectively from β and α cells of the islets of Langerhans. See DIGESTION.

pancreatin Extract of pancreas containing digestive enzymes.

pancreozymin See CHOLECYSTOKININ.

pandemic Worldwide EPIDEMIC; e.g., of INFLUENZA in 1918.

Paneth cells Actively mitotic cells at the bases of the crypts of Lieberkuhn of the ILEUM, secreting lysozyme and capable of limited phagocytosis. As they move up the sides of the crypts (taking up to five days), they differentiate into enteroendocrine cells, goblet cells and absorptive cells. These last have brush border enzymes in their microvilli and carry out the final stages of digestion; but some are sloughed off into the lumen and break open there to release enzymes. They also secrete angiogenin into the gut lumen, where it has bactericidal activity against intestinal microbes.

Pangaea The most recent supercontinent on Earth, starting to form ~330 Myr BP and reaching its maximum extent in the Late Permian (250 Myr BP). Its ultimate break-up, ~175 Myr BP, was preceded by and included widespread volcanic activity. Formed by the collision of GONDWANALAND and LAURUSSIA. See EXTINCTION.

pangenesis Theory adopted by Charles DARWIN to provide for the genetic variation his theory of natural selection required. Basically Lamarckian, it supposed that every part of an organism produced 'gemmules' ('pangenes') which passed to the sex organs and, incorporated in the reproductive cells, were passed to the next generation. Modification of the body, as through use and disuse, would result in appropriately modified gemmules being passed to offspring. Severely criticized by Darwin's cousin, Francis Galton. Empedocles held a similar view, but Aristotle rejected it.

panicle Type of INFLORESCENCE.

panmixis Result of interbreeding between members of a species population, no important barriers to gene flow occurring within it. The whole population represents one GENE POOL. Panmictic animal populations tend to have high VAGILITY.

pantothenic acid VITAMIN of the B-complex; precursor of COENZYME A (CoA), a large molecule comprising a nucleotide bound to the vitamin. See KREBS CYCLE.

Pantotheria Extinct order of primitive therian Jurassic mammals. Small and insectivorous. Have therian features: large alisphenoid on wall of braincase, and triangular molar teeth. Contemporaries of MULTITUBERCULATA.

papilla Projection from various animal tissues and organs. *Dermal papillae* project from dermis into epidermis of vertebrates, providing contact (and finger-print patterns); in feather and hair follicles, papillae provide blood vessels. *Tongue papillae* increase surface area for taste buds in mammals; they are cornified for rasping in cats, etc.

pappus Ring of fine, sometimes feathery, hairs developing from calyx and crowning fruits of the Family Compositae (e.g. dandelion). Act as parachute in wind dispersal of fruit.

parabiosis Surgical joining together of two animals so that their blood circulations are continuous. Each member of the pair is termed a *parabiont*. Often employed to monitor humoral influences in behaviour and development (e.g. in insect moulting).

paracrine cell (1) Cell secreting a local chemical signal which is rapidly taken up at receptor site on another cell, but destroyed or inactivated within about 1 mm from release site. Includes several cells secreting PROSTAGLANDINS, MAST CELLS and nerves of the sympathetic system (see VARICOSITY). (2) Any cell secreting a substance having an effect upon another cell. The substances secreted would be described as having a *paracrine action*.

paradox of the plankton Phrase coined by G. E. Hutchinson as a result of observing that 10–50 phytoplankton species commonly co-exist in the same body of water at any one time. Ecological principle of competitive exclusion predicts that relatively homogeneous environments, such as the photic zone of lakes and oceans, should contain very few species which possess similar ecological requirements. Since all phytoplankton are essentially photoautotrophs, with similar needs, competition for resources, especially nutrients, should result in the elimination of all but one species (best able to use the limiting resources).

This 'paradox' can be explained in terms of each species having its own nutrient-utilization characteristics resulting in each different species being limited simultaneously by a different nutrient. Direct competition is avoided, and potentially as many species can co-exist as there are limiting nutrients. Grazing by zooplankton may also improve the chances for co-existence among planktonic algae. Other explanations stress that the photic zone is not homogeneous, since gradients in light intensity, nutrients and temperature occur providing spatial heterogeneity.

parallel evolution Possession in common by two or more taxa of one or more characteristics, attributable to their having similar ecological requirements and a shared genotype inherited from a common ancestor: the common characteristics would be HOMOLOGOUS. In CLADISTICS, no distinction is made between parallel evolution and CONVERGENCE.

parallel venation Pattern of leaf venation, where principal veins are parallel or nearly so; characteristic of MONOCOTYLEDONAE.

paralogous Homologous gene sequences originating by GENE DUPLICATION and (often) subsequent divergence within a lineage. They therefore lack phylogenetic significance. See ELONGATION FACTORS; compare ORTHOLOGOUS.

paramutation Heritable change in the expression of one (paramutable) allele when heterozygous with another (paramutagenic) allele at the same locus. May explain TRANSGENE silencing (co-suppression) which sometimes occurs when a transgene insertion reduces its own level of expression, as well as those of homologous host alleles, which is sometimes accompanied by DNA METHYLATION.

paramylon (Of algae) storage polysaccharide composed solely of β-1,3 linked glucans in the EUGLENOPHYTA, XANTHOPHYTA and PRYMNESIOPHYTA (e.g. *Pavolva mesolychnon*). Occurs outside the chloroplast as water-soluble, single-membrane-bound inclusions of various shapes/sizes.

parapatry Where the ranges of two populations of the same or different species overlap they are SYMPATRIC; but if the ranges are contiguous, i.e. if they abut for a considerable part of their length but do not overlap, the distributions show parapetry. Distributions of several organisms formerly regarded as single species have been shown to consist of several SIBLING SPECIES or SEMISPECIES with parapatric distributions, as in frogs of *Rana pipiens* species group.

paraphyletic Term describing taxon or taxa originating from and including a single stem species (known or hypothetical) but excluding one or more smaller CLADES nested within it. E.g. if, as is commonly accepted, flowering plants arose from a gymnosperm ancestor, then gymnosperms are a *paraphyletic group* since the group does not include all descendants of a common ancestor. Such taxa are not permitted in CLADISTICS but are much used by adherents of EVOLUTIONARY TAXONOMY. See CLADE, MONOPHYLETIC.

paraphysis (Bot.) Sterile filament, numbers of which occur in mosses and certain algae, interspersed among the sex organs, and in the hymenia of certain fungi (ASCOMYCOTA, BASIDIOMYCOTA).

parapodium Paired metameric fleshy appendage projecting laterally from the body of many polychaete annelid worms (especially errant polychaetes). Usually comprises a more dorsal *notopodium* and a ventral *neuropodium*, each with bundles of chaetae and endowed with a supporting chitinous internal *aciculum* to which muscles moving the parapodium are attached.

parasegment Unit in insect development which corresponds not to a morphological segment but to the posterior COMPARTMENT of one segment and the anterior compartment of the next most posterior segment. See HOMEOBOX, PAIR-RULE GENES (whose expressions determine parasegment boundaries), SEGMENT POLARITY GENES.

parasexual cycle See PARASEXUALITY.

parasexuality Fungal life cycle (*parasexual cycle*) discovered by G. Pontecorvo and J. A. Roper in 1952 which includes the following: occasional fusion of two haploid heterokaryotic nuclei in the mycelium to form a diploid heterozygous nucleus; mitotic division of this nucleus during which crossing-over (*mitotic crossing-over*) occurs, then restoration of haploidy to the nucleus by either mitotic NON-DISJUNCTION, or some form of chromosome extrusion, removing a haploid set of chromosomes (see ABERRANT CHROMOSOME BEHAVIOUR). The non-sexual spores produced differ genetically from the parent mycelium. It accounts for the variation of pathogenicity in certain fungal pathogens, and has enabled genetical studies to be made using members of the DEUTEROMYCOTA.

parasite One kind of symbiont. Organism living in (*endoparasite*) or on (*ectoparasite*) another organism, its *host*, obtaining nourishment at the latter's expense. Metabolically dependent upon their hosts, as are carnivores, herbivores, etc. Distinction between herbivorous caterpillar and ectoparasitic fluke is not clear-cut. Endoparasites generally display more, and more specialized, adaptations to parasitism than do ectoparasites, and often include both primary and secondary HOSTS in the life cycle. *Obligate parasites* cannot survive independently of their hosts; *facultative parasites* may do so. *Partial parasites*, or *hemiparasites* (e.g. mistletoe, *Viscum*), are plants which photosynthesize and also parasitize a host. Sometimes relationships between members of the same species are parasitic (e.g. males of some angler fishes live attached to the female and suck her blood). Placental reproduction shares features with parasitism, as

do forms of viviparity in which young emerge causing death of the parent (e.g. in the midge *Miastor* and many aphids and water fleas). See SYMBIOSIS, MALARIA, PARASITOID.

parasitoid Insects, and some other animals, which introduce their eggs into another animal, in which they grow and develop in a slow and controlled manner using the host's resources without killing it. The parasitoid wasp *Cotesia congregata* lays hundreds of eggs in each large caterpillar of its host, the tomato and tobacco pest *Manduca sexta*, and by consuming the haemolymph the larvae avoid killing their host until well after emergence. During development, the wasp's larvae are protected from immune attack by causing apoptosis (APOPTOSIS) of the host's granulocytes. This is due to viral infection of the host and to a covering of viral protein on the early larvae, the virus (polydnavirus) being injected by the female wasp along with her eggs. Unlike the situation in long-established parasitic associations, parasitoids usually cause the deaths of their hosts and are often used in commercial biological control (e.g. *Encarsia* on whitefly).

parastichy See PHYLLOTAXY.

parasympathetic nervous system See AUTONOMIC NERVOUS SYSTEM.

parathyroid glands Endocrine glands of tetrapod vertebrates, usually paired, lying near or within THYROID depending on species. Arise from embryonic gill pouches, and produce PARATHYROID HORMONE. Removal produces abnormal muscular convulsions within a few hours.

parathyroid hormone (PTH, parathormone) Polypeptide hormone of parathyroid glands operating with vitamin D and CALCITONIN in control of blood calcium levels. Injection releases calcium from bone and raises blood Ca^{2+} level, inhibiting further parathyroid hormone release, apparently via direct negative feedback on parathyroid glands. Also reduces Ca^{2+} excretion by the kidneys. Deficiency produces muscle spasms.

paratype Any specimen, other than the type specimen (HOLOTYPE) or duplicates of this, cited with the original taxonomic description and naming of an organism.

Parazoa Subkingdom of the ANIMALIA, containing the phylum PORIFERA (sponges).

parenchyma (Bot.) Simple tissue, comprising parenchyma cells occurring in the primary plant body commonly as continuous masses in the cortex of stems, in roots, in the stem pith, in the leaf mesophyll, and in the flesh of fruits. Parenchyma cells also occur as vertical strands in the primary and secondary vascular tissues, as well as horizontal strands called rays in the secondary vascular systems. Cells are thin-walled, often almost as broad as they are long, and remain alive and capable of dividing at maturity. Some have a secondary wall in addition to the primary wall. Those with just primary walls, because of their ability to divide, play an important role in regeneration and wound healing; it is these cells from which adventitious structures (e.g. adventitious roots) arise. Other functions may include photosynthesis, secretion, storage, water movement and transport of food materials. See TRANSFER CELL. (Zool.) Loose tissue of irregularly shaped vacuolated cells with gelatinous matrix, and forming a large part of the bodies of some invertebrate groups, notably platyhelminths and nematodes.

***par* genes** See CELL POLARITY.

parietal (Bot.) Referring to peripheral position, as in chloroplasts of some algal cells located near the cell's periphery. (Zool.) *Parietal bones* lie one on each side of the vertebrate skull, behind and between the eye orbits. See also PINEAL GLAND.

parietal eye See PINEAL EYE.

parietal placentation (Bot.) Attachment of ovules in longitudinal rows on carpel wall.

Parkinson's disease Neurodegenerative disease causing loss of coordinated motion, tremors, difficulty initiating movements, loss of balance, unpredictable 'freezing'. Non-motor symptoms such as excessive sweating and other disturbances of the autonomic nervous system, depression and, eventually, dementia. Damage to the substantia nigra, and of their links with the BASAL GANGLIA, accounts for most symptoms. In the 1960s, the drug L-dopa (laevodopa) was developed to compensate for the lack of DOPAMINE in the brain, but is not a cure-all and loss of sensitivity to the drug commonly follows after a few years. Discovery of fatty acid oxidation products in the substantia nigra suggested that free radicals might be oxidizing cell membranes: Parkinson's patients tend to lack the brain antioxidants which would protect membranes from free radical attack, and tend also to have raised brain iron levels, associated with free radical formation (see SUPEROXIDES, NITRIC OXIDE). See CHAPERONES, ALZHEIMER'S DISEASE.

Parmophyceae A class of algal division HETEROKONTOPHYTA or sometimes considered as an order of the Chrysophyta (Parmales); members of the marine nannoplankton (e.g. *Pentalamina*). Characterized by coccoid, $2-5.5$ µm diameter cells, having siliceous walls comprising round and triradiate plates that abut, edge to edge in Antarctic and subarctic Pacific oceans.

PARP Poly(ADP-ribose) polymerase. An enzyme present in all animals (metazoans) where it has been looked for, and involved in several chromatin transactions (e.g. nucleosome decondensation and telomere elongation), DNA-REPAIR MECHANISMS, APOPTOSIS, DNA replication and perhaps genomic remodelling during development. It is activated by binding specifically to nicked DNA and uses NAD to add long chains of ADP-ribose residues to target proteins, such as histones and transcription factors.

parsimony In scientific reasoning, the 'principle of parsimony' is a rule to the effect that one should not introduce more causes, entities or changes to an existing explanatory hypothesis than are necessary to accommodate the data. In this way, one promotes, or maximizes, the LIKELIHOOD of the hypothesis. As a rule of thumb, the most parsimonious explanation of events is the most persuasive. See BAYESIAN STATISTICS, OPTIMALITY THEORY, PHYLOGENETIC TREES.

parthenocarpy (Bot.) Development of fruits without prior fertilization. Occurs regularly in banana and pineapple (which are therefore seedless). Can be induced by certain auxins in unfertilized flowers, e.g. those of tomato.

parthenogenesis Development of an unfertilized gamete (commonly an egg cell) into a new individual. One of a spectrum of forms of uniparental SEXUAL REPRODUCTION.

A parthenogenetic egg cell (or nucleus) may become diploid either through nuclear fusion, or through a restitution division (see RESTITUTION NUCLEUS). Sometimes cleavage products of a haploid egg may undergo fusion, producing a diploid embryo. Phenomenon includes non-gametic forms of AUTOMIXIS and in animals is a common cause of MALE HAPLOIDY. In animals, THELYTOKY (absence of males) enables rapid production of offspring without food competition from males. *Cyclical parthenogenesis* (as in some aphids and flukes) involves a combination of thelytoky and bisexual fertilization. The crustacean *Daphnia* switches from production of parthenogenetic eggs to bisexually produced resting eggs on the basis of multiple environmental stimuli, including a maternal food effect and photoperiod – information on both being transmitted to, and influencing the seasonal production of resting eggs in, their offspring. In some aphids thelytoky prevails in summer, males only appearing in autumn or winter when fertilization occurs. In the midges *Miastor* and *Heteropeza*, larvae possess functional ovaries enabling progenetic reproduction by automixis, adults not appearing for generations; some larval flukes are progenetic. Some instances of thelytoky (automictic, or meiotic, thelytoky) involve meiotic egg-production, and two of the four meiotic products sometimes fuse to restore diploidy; in others (apomictic, or ameiotic, thelytoky), mitosis produces the egg cells. In some cases diploidy may be restored by ENDOMITOSIS after meiosis.

In the protozoan *Paramecium*, fusion may occur of two of the micronuclei produced meiotically from the cell's parent micronucleus (*automixis*) but no new individual is produced.

Since development of unfertilized eggs can occur, on rare occasions, in many species (e.g. *Drosophila* and grasshoppers), and be induced artificially in many others (see CENTRIOLE), it is still surprisingly rare, especially since it avoids the COST OF MEIOSIS. Thelytokous forms seem to be liable to early extinction compared with their bisexual relatives, probably through progressive homozygosity (see GENETIC VARIATION).

Haploid egg cells have been recorded developing parthenogenetically in plants, but this seems to have had little evolutionary impact. Unreduced (diploid) gametophytes may arise either from an unreduced megaspore (*diplospory*) or from an ordinary unreduced somatic cell of the sporophyte (*apospory*). Both are genetically equivalent to apomictic (ameiotic) parthenogenesis in animals. In dandelions (*Taraxacum*) the megaspore mother cell undergoes meiosis, the first division producing a restitution nucleus, the second producing two cells, each unreduced, from one of which the 8-nucleate embryo sac is produced. See GYNOGENESIS, PARTHENOCARPY.

partial refuges Areas of prey habitat in which their consumption rate by predators is less than the average for the habitat as a whole, as a result of the predator's behavioural responses to the prey's spatial distribution.

parturition Expulsion of foetus from uterus at end of pregnancy (term) in therian mammals. In man and other primates it appears that the foetus determines pregnancy length by initiating the release of prostaglandins, principally PGF2a, from the endometrium (and later from the myometrium) of the uterus which initiates labour through its effects on smooth muscle contraction in the uterine wall. Oxytocin has also been implicated in the onset of labour. Successful transition from intrauterine to extrauterine life requires the previous maturation of lungs, the laying down of fat and carbohydrate reserves and the onset of maternal lactation (see PROLACTIN). Foetal adrenal glands are involved in much of this regulation, becoming more active as term approaches.

Parvovirus Single-stranded DNA VIRUS which attacks rapidly dividing cells.

passage cells Cells of the ENDODERMIS, typically of older monocot roots, opposite protoxylem groups of stele, remaining unthickened and with casparian strips only after thickening of all other endodermis cells. Allow transfer of material between cortex and vascular cylinder.

passerines Members of largest avian order, the Passeriformes. Perching birds, characterized by having large first toe directed back, the other three forward. Includes most of the common inland birds. See NEOGNATHAE.

Pasteur, Louis (1822–95). French chemist and microbiologist; professor of chemistry at the Sorbonne, but worked mostly at the Ecole Normale in Paris. Became director of the Institut Pasteur, Paris, in 1888. Championed the view that fermentation was a *vital* rather than a *simple* chemical process, as against the chemical theory of Liebig and Berzelius. Already aware, with others, that yeasts were associated with alcoholic fermentations, he demonstrated presence of microorganisms in other fermentations. In 1858 he demonstrated fermentation in the absence of organic nitrogen, destroying the chemical theory. Pasteur's many experiments supported the germ theory of fermentation, as against the theory of spontaneous generation, and its implications were appreciated by Joseph Lister in his work on antisepsis in the 1860s. In 1879, Pasteur's assistant Emile Roux carried out important work on the attenuation of cholera bacilli in chickens and later, on the use of rabbit spinal cords in attenuating rabies virus. These efforts were not always given due credit by Pasteur. Pasteur came to accept the role of microorganisms in disease, showing that attenuated forms of bacteria produced by serial culture could be used in inoculation to immunize the host. His vaccines against anthrax and rabies, like Jenner's earlier ones against smallpox, were instrumental in establishing the germ theory of disease. See PASTEURIZATION, VIRCHOV.

Pasteur effect Phenomenon whereby onset of aerobic respiration inhibits glucose consumption and lactate accumulation in all facultatively aerobic cells, conserving substrates. Depends upon the allosteric inhibition of glycolytic enzymes by high intracellular ATP to ADP ratio. See ATP, GLYCOLYSIS.

pasteurization Method of partial sterilization, after Louis PASTEUR, who discovered that heating wine at a temperature well below its boiling point destroyed the bacteria causing spoilage without affecting its flavour. Widely used to kill all pathogenic bacteria in food, without achieving complete sterility. Potential pathogens include tubercle bacteria (*Mycobacterium tuberculosis*) and the intracellular rickettsia-like *Coxiella burnetti*, the most heat-resistant pathogen of milk and cause of Q fever. In the 'holder' method, milk is heated to 63–65°C for 30 min. In the HTST or 'flash pasteurization' method, milk is heated to 72°C for at least 15 sec followed by rapid cooling, delaying fermentation by the coliform bacteria always present in raw milk (most being killed). UHT treatment (ultra-high-temperature sterilization) involves heating milk to 132.2°C for at least one second, giving it a keeping time of several months without much flavour diminution. Coliforms present in milk after pasteurization are derived from endospores, or have contaminated it subsequently. Compare TYNDALLIZATION.

patch dynamics The concept of communities as comprising a mosaic of patches within which abiotic disturbances and biotic interactions proceed.

patchy habitat A habitat within which there are significant spatial variations in suitability for the species under consideration.

patella Kneecap. Bone (sesamoid bone) over front of knee joint in tendon of extensor muscles straightening hind limb.

Present in most mammals, some birds and reptiles.

patenting In mid 1991, The National Institutes of Health, USA, applied to patent some DNA sequences they had found to be useful in human gene mapping. The application was opposed on the grounds that it would impede the flow of scientific information and obstruct international cooperation, and eventually failed because of cost. A second application for DNA patenting by a private company was turned down on the grounds that the research had been done with public money. But in 1997 the US Patent Office agreed to allow patenting of gene strands with no known biological function if they were likely to have industrial use. The result has been that big academic labs and companies have been sequencing and patenting sequences in order to secure private investment. In order for DNA sequences to be distinguished from their natural counterparts (which cannot be patented), a patent application must state that the invention has been purified or isolated or is part of a recombinant molecule or vector. Some pharmaceutical companies, and the NIH, are compiling public databases of sequences to ensure public access. Donor consent is a concern in human DNA patenting: formal consent has not been obtained for all human DNA sequences in data banks and cell lines. In 1999, the European Patent Office lifted a four-year moratorium on applications for patents on plants and animals.

paternal effect Sperm provide important developmental information in some species which cannot be compensated for by the egg. In *Drosophila*, males homozygous for the allele $ms(3)K81^1$ produce motile sperm that enter the egg but cause embryonic lethality, independently of the mother's genotype. Some of these mutations affect the structure of the zygote mitotic spindle (see MITOTIC ARREST).

pathogen Disease-causing parasite, usually a microorganism. See VIRULENCE.

pathology Study of diseases or diseased tissue.

patristic (Of similarity) due to common ancestry. See CLADISTICS.

pattern, pattern formation Describing phenomena whereby cells in different parts of embryo become locked into different spatio-temporal developmental pathways, coordinated in such a way as to produce a viable multicellular system. Early animal embryos engage in a hierarchy of decisions involving progressive steps in *regional specification*, in which POSITIONAL INFORMATION is imparted to cells, which then respond to it (see AXIS). Cells of a particular histological type may have arrived at their condition via alternative, *non-equivalent* routes. Parts of an embryo acquire different DETERMINED states through regionalization processes. See MATERNAL EFFECT, COMPARTMENT, PAIR-RULE GENES.

Pax5 A gene regulatory protein involved in production of mature B cells from progenitors. In its absence, these progenitors can differentiate into other blood cell types.

Pax-6 A gene regulatory protein crucial to vertebrate EYE development.

PCBs (polychlorinated biphenyls) Persistent and toxic pollutants, toxicity varying with the position and number of chlorine atoms bound to the biphenyl core. PCBs appear to interfere with plasma transport of vitamin A and thyroxine (whose structure is not dissimilar).

PCR See POLYMERASE CHAIN REACTION.

PDE Acronym for PHOSPHODIESTERASE.

peat Accumulated dead plant material which has remained incompletely decomposed owing, principally, to lack of oxygen. Occurs in moorland, bogs and fens, where land is more or less completely waterlogged; often forms a layer several metres deep. Of local value as fuel for burning.

peck order See DOMINANCE (2).

pectic compounds Acid polysaccharide carbohydrates, present in CELL WALLS of unlignified plant tissues; comprise pectic acid and pectates, pectose (propectin) and pectin. Form gels under certain conditions.

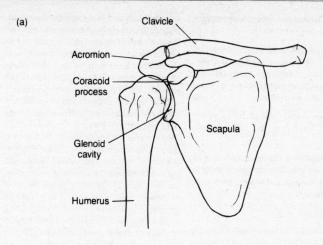

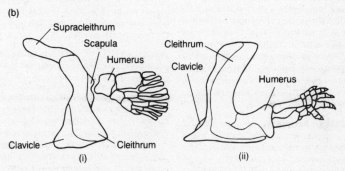

FIG. 130 (a) *Human shoulder region, viewed from the front.* (b) *Pectoral girdle and fin of* (i) *crossopterygian fish and* (ii) *early fossil amphibian. One side only shown in each case.*

Principal components are galacturonic acid, galactose, arabinose and methanol. Form the basis of fruit jellies.

pectoral fin See FINS.

pectoral girdle (shoulder girdle) Skeletal support of vertebrate trunk for attachment of fins or forelimbs; in fish, attaches to skull. Primitively, a curved bar of cartilage or bone on each side of body, fusing ventrally to form a hoop transverse to long axis, incomplete dorsally. Each bar bears a joint with fin or limb (see GLENOID CAVITY). Components are: *scapulae* dorsal to the

joint, *coracoids* ventrally. Additional dermal bones are the *cleithra* in fish and primitive tetrapods, and *clavicles*, usually on the ventral side. In mammals each clavicle joins a scapula at a process of the coracoid, the *acromion*. Scapulae do not articulate with the vertebral column or ribs (compare PELVIC GIRDLE). In tetrapods, coracoids and clavicles join mid-ventrally to the sternum. See Fig. 130.

pedicel (Bot.) Stalk of individual flower of an inflorescence. (Zool.) Narrow tube-like 'waist' of many hymenopteran insects.

pedipalps Second pair of head appendages of ARACHNIDA. See MOUTHPARTS.

peduncle (1) Stalk of an inflorescence. (2) An extensible pseudopod used to catch and either suck out or engulf a prey organism (another alga or a protist) found in some members of the DINOPHYTA.

Peking man Early form of *Homo erectus* from China; formerly termed *Sinanthropus*. See *HOMO*.

pelagic Inhabiting the mass of water of lake or sea bottom (see BENTHOS). Pelagic animals and plants are divided into PLANKTON and NEKTON.

Pelecypodia See BIVALVIA.

p **element** A kind of TRANSPOSABLE ELEMENT found in the fruit fly *Drosophila* and responsible for HYBRID DYSGENESIS in crosses between P-strain male and M-strain female flies. 0.5–1.4 kb in length, they are flanked by inverted repeats 31 base pairs long. They originate through deletions within larger *P factors*, a few copies of which occur in *P strains* of the fly. Appropriately injected into M-strain embryos they can be used as gene vectors, the resulting fly's germ line tending to acquire the gene. See COPIA.

pellicle (= periplast) An outer skin-like layer comprising mostly protein found around cells of some unicellular algae (RAPHIDOPHYCEAE, EUGLENOPHYTA, CRYPTOPHYTA). In the DINOPHYTA it is an additional wall layer or envelope made of cellulose and sporopollenin-like material, lying below the theca.

peltate (Bot.) Circular, with the stalk inserted in the middle; as of some leaves, hairs and horsetail and cycad sporangiophores.

pelvic fin See FINS, PELVIC GIRDLE.

pelvic girdle (hip girdle) Skeletal support for attachment of vertebrate hindlimbs or pelvic fins. In fish (see Fig. 131), a pair of curved bars of bone or cartilage embedded in the abdominal muscles and connective tissue, fused to form a mid-ventral plate, articulating with the fins but not with the vertebral column. In tetrapods, the ventral plate ossifies from two centres on each side: the *pubis* anteriorly, and *ischium* posteriorly. A large rounded socket (*acetabulum*) receives the head of the femur on each side where these two bones join a third and dorsal element, the *ilium*, which unites with one or more sacral vertebrae to form a complete girdle around this region of the trunk, giving rigid support to hindlimbs for locomotion. Pelvic girdle structure varies in tetrapod classes. In mammals, the ilium extends anteriorly towards the SACRUM, while the pubis and ischium have moved posteriorly, hardly reaching the acetabulum. In most mammals, many reptiles and *Archaeopteryx* the pubes articulate or fuse mid-ventrally to form the *pubic symphysis*, consisting in humans of fibrocartilage between the two *coxal bones* (fused ilium, pubis and ischium on each side). In monotremes and marsupials a pair of *prepubes* reaches forward from the pubes to form a body wall support. Compare PECTORAL GIRDLE.

pelvis (1) The PELVIC GIRDLE. (2) Lower part of the vertebrate abdomen, bounded by the pelvic girdle. (3) The *renal pelvis*. See KIDNEY.

penetrance A dominant character determined by an ALLELE is either always expressed in any individual where it occurs (*completely penetrant*), or is expressed in some individuals but not in others (*incompletely penetrant*). Once a character finds expression, it may be expressed to varying degrees in different individuals (*variable expressivity*). Possible factors affecting penetrance and expressivity include the genetic background (see MODIFIER) and environmental influences during development. See EPIGENESIS.

penicillins See ANTIBIOTIC, MITOSPORIC FUNGI.

penis Unpaired intromittant organ of male mammals, some reptiles and a few birds (especially of those mating on water). In mammals, contains the terminal part of the urethra.

pentadactyl limb The type of limb found in tetrapod vertebrates. Evolved as an adaptation to terrestrial life from the paired fins

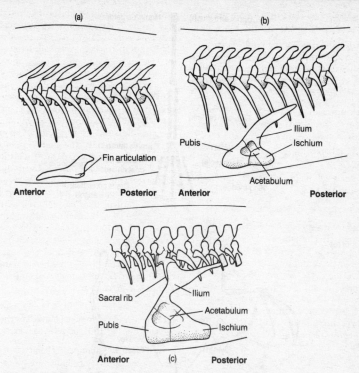

(a)

Fin articulation

Anterior   **Posterior**

(b)

Ilium
Pubis
Ischium
Acetabulum

Anterior   **Posterior**

(c)

Sacral rib
Ilium
Acetabulum
Pubis
Ischium

Anterior   **Posterior**

FIG. 131 *Pelvic girdles of* (a) *fish,* (b) *early tetrapod and* (c) *later tetrapod with attachment of girdle to enlarged sacral rib.*

of crossopterygian fishes. The basic plan is illustrated in Fig. 132. Many modifications occur through loss or fusion of elements, especially in the terminal parts. The earliest known tetrapods, from the Upper Devonian, had more than five digits, raising the possibility that the pentadactyl condition is not primitive for the clade. See Fig. 132.

pentastomids (tongue worms) Phylum Pentastoma. Bilaterally symmetrical, flattened, pseudocoelomate and parasitic worms. 90% of known species parasitize reptiles and all parasitize vertebrates. Separate sexes; fertilization internal. Development via three 'larval' stages, the first with three pairs of lobed, leg-like and unjointed appendages.

pentosans Polysaccharide (e.g. xylan, araban) whose monomers are PENTOSE units (e.g. xylose, L-arabinose). Include many HEMICELLULOSES, although these more often contain non-pentose units as well.

pentose Monosaccharide with five carbon atoms in molecule, e.g. ribose and deoxyribose (important constituents of NUCLEIC ACIDS), and ribulose. Found in various plant polysaccharide chains, e.g. pectin, gum arabic.

pentose phosphate pathway (phosphogluconate pathway, hexose monophosphate shunt) An alternative route to GLYCOLYSIS for glucose catabolism, involving initially conversion of glucose-phosphate to phosphogluconate. Some intermediates

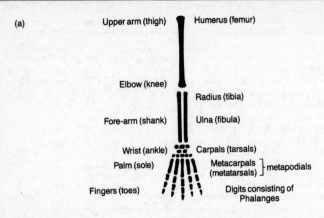

(a)

Upper arm (thigh)	Humerus (femur)
Elbow (knee)	
Fore-arm (shank)	Radius (tibia)
	Ulna (fibula)
Wrist (ankle)	Carpals (tarsals)
Palm (sole)	Metacarpals (metatarsals) } metapodials
Fingers (toes)	Digits consisting of Phalanges

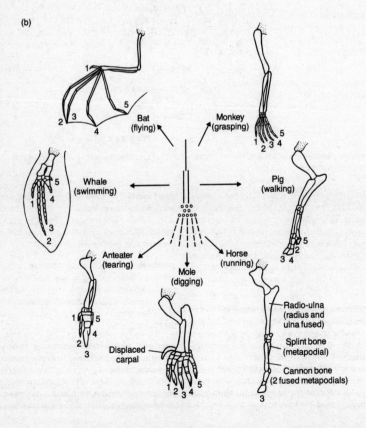

(b)

Bat (flying)

Monkey (grasping)

Whale (swimming)

Pig (walking)

Anteater (tearing)

Mole (digging)

Horse (running)

Displaced carpal

Radio-ulna (radius and ulna fused)

Splint bone (metapodial)

Cannon bone (2 fused metapodials)

are the same as those of glycolysis. Generates extramitochondrial reducing power in the form of NADPH, important in several tissues (e.g. adipose tissue, mammary gland, adrenal cortex) where fatty acid and steroid synthesis occur from acetyl coenzyme A. Ribose-phosphate is one intermediary in the pathway, and may be used for nucleic acid synthesis or 'for glucose production from CO_2 in some plants. Other monosaccharides are intermediates, and can be fed into the pathway and linked with glycolysis.

pepsin Vertebrate ENDOPEPTIDASE, secreted as the zymogen pepsinogen by chief cells of gastric pits and spontaneously formed from it below pH 5 (intramolecular process), when it is active.

peptidase One of a group of enzymes hydrolysing peptides into component amino acids. Also sometimes used of proteolytic enzymes in general. See ENDOPEPTIDASE, EXOPEPTIDASE.

peptide (1) A compound of two or more amino acids (strictly, amino acid residues), condensation between them producing PEPTIDE BONDS. Shorter in length than a polypeptide. Besides being intermediates in protein digestion and synthesis, many peptides are biologically active. Some are hormones (e.g. OXYTOCIN, ADH, MELANOCYTE-STIMULATING HORMONE); some are vasodilatory. (2) Sometimes employed in the context of any chain of amino acids, however long.

peptide bond Bond between two amino acids resulting from combination of an amino group ($-NH_2$) attached to α-carbon of one amino acid and the carboxyl group ($-COOH$) attached to the α-carbon of another amino acid. The bond formed ($-NH-CO-$) involves the elimination of one molecule of water, and is a *condensation reaction*. See PROTEIN, AMINO ACID.

peptidoglycan (mucoprotein) Rigid component of cell walls of prokaryotes

only, e.g. of virtually all eubacteria. Comprises chains of glycosaminoglycan (N-acetylglucosamine linked to N-acetylmuramic acid) crosslinked by a tetrapeptide consisting of L-alanine, D-alanine, D-glutamic acid and either lysine or diaminopilemic acid (DAP). Also called murein, mucopeptide, mucocomplex. See GRAM'S STAIN, MUCINS.

peptone Large fragment produced by initial process of protein hydrolysis.

perennation (Of plants) surviving from year to year by vegetative means.

perennial Plant that continues its growth from year to year. In herbaceous perennials, the aerial parts die back in autumn and are replaced by new growth the following year from the underground structure. In woody perennials, permanent woody stems above ground form starting point for each year's growth, a characteristic enabling some species (trees, shrubs) to attain large size. Compare ANNUAL, BIENNIAL, EPHEMERAL.

perforation plate (Bot.) End wall of a vessel element containing one or more holes or perforations. See XYLEM.

perforins Proteins released from granules by cytotoxic T CELLS, disrupting the integrity of the target cell membrane by forming channels there, in the presence of calcium ions. See COMPLEMENT.

perianth (Bot.) (1) Outer part of flower, enclosing stamens and carpels; usually comprising two whorls. Differentiated in Dicotyledonae as an outer green CALYX and an inner COROLLA, the latter usually conspicuous and often brightly coloured. In Monocotyledonae, usually no differentiation of calyx and corolla, both whorls looking alike. (2) In leafy liverworts, a tubular sheath surrounding the archegonia and, later, the developing sporophyte.

FIG. 132 (a) *Diagram of skeleton of pentadactyl limb of vertebrate, giving names of parts of fore-limb, and (in brackets) of hind-limb. On left, common name of whole part; on right, names of bones.* (b) *Diagram indicating the adaptive radiation of vertebrate limb from a basic archetype.*

periblem See APICAL MERISTEM.

pericardial cavity Space enclosing the heart. In vertebrates, is coelomic and bounded by a double-walled sac, the *pericardium*. In arthropods and molluscs, is haemocoelic, supplying blood to heart.

pericarp (Bot.) Wall of an ovary after it has matured into a FRUIT. May be dry, membranous or hard (e.g. achene, nut), or fleshy (e.g. berry).

perichaetium Distinct whorl of leaves surrounding sex organs in mosses.

periclinal (Bot.) (Of planes of division of cells) running parallel to the surface of the plant. See Fig. 23; compare ANTICLINAL.

pericycle (Bot.) Tissue of the vascular cylinder lying immediately within the ENDODERMIS, comprising parenchyma cells and sometimes fibres.

periderm (Bot.) Cork cambium (phellogen) and its products; i.e. cork and secondary cortex (phelloderm).

perigene development (Bot.) Development of GUARD CELLS in which the guard cell mother does not give rise to subsidiary cells.

perigonium (Bot.) The antheridium, together with associated perigonial leaves or bracts (surrounding the antheridium), borne on a specialized branch (perigonial branch), in mosses and liverworts.

perigynous See RECEPTACLE.

perikaryon Alternative term for CELL BODY of neurone.

perilymph See VESTIBULAR APPARATUS.

perineum Region between anus and urogenital openings of placental mammals.

periosteum Sheath of connective tissue investing vertebrate bones, and to which tendons attach. Contains osteoblasts, and white and elastic fibres.

Peripatus See ONYCHOPHORA.

peripheral nervous system See NERVOUS SYSTEM.

periplasmic space Space between the two cell membranes of Gram-negative bacteria (see Fig. 75) or an organelle envelope (e.g. of chloroplast or mitochondrion). Proteins with the correct signal sequence can be targeted to this region.

periplastidial compartment The narrow space between the chloroplast and the chloroplast endoplasmic reticulum in the HETEROKONTOPHYTA, CRYPTOPHYTA and CHLORARACHNIOPHYTA.

periplastidial reticulum (= periplastidial network) A network of interconnected tubules that occur in the periplastidial compartment between the chloroplast and the chloroplast endoplasmic reticulum in the HETEROKONTOPHYTA and HAPTOPHYTA. The reticulum may serve to transport polypeptides from the nucleus to the chloroplast.

perisperm Nutritive tissue surrounding the embryo in some seeds; derived from the NUCELLUS. Compare ENDOSPERM.

Perissodactyla Order of eutherian mammals containing the odd-toed ungulates (e.g. horses, tapirs, rhinoceroses). Walk on hoofed toes, the weight-bearing axis of foot lying along the third toe, which is usually larger than the others (some of which may have disappeared): horses have very large third, but minute second and fourth, toes. Tapirs have four toes on front feet, three on hind feet. Rhinoceroses have three toes on each foot. See ARTIODACTYLA.

peristalsis Waves of contraction of smooth muscle, passing along tubular organs such as the intestines. Serve to move material from one end of the tube to the other.

peristome (Bot.) Fringe of pointed appendages (teeth) around the opening of dehiscent moss capsule, concerned with spore liberation. (Zool.) Spirally twisted groove leading to cytostome in some ciliate protozoans.

perithecium Rounded, or flask-shaped, fruiting structure (ASCOCARP) of some members of the ASCOMYCOTA and LICHENS;

with an internal hymenium of asci and paraphyses, and opening via a pore (ostiole) through which ascospores are discharged.

peritoneum   Epithelium (serous membrane) lining posterior coelomic cavity (*abdominal, perivisceral* and *peritoneal cavities*) of vertebates, around the gut. That covering the gut and other viscera is termed *visceral*; that lining the wall of the cavity is *parietal*. The MESENTERY carrying blood vessels to and from the viscera is also peritoneal.

perivisceral cavity   The vertebrate coelomic cavity lined by PERITONEUM.

permanent teeth   Second of the two successive sets of teeth of most mammals, replacing DECIDUOUS TEETH. See DENTITION.

permanent wilting point   The condition of a soil in which water is sufficiently unavailable to cause plants growing in it to wilt irrecoverably.

permease   A TRANSPORT PROTEIN, or carrier, assisting in transport of a solute down its concentration gradient across a cell membrane.

Permian   Last GEOLOGICAL PERIOD of the PALAEOZOIC era, 286–251 Myr BP. In Britain, Permian and TRIASSIC marine faunas are often united as New Red Sandstone (*Permo-Trias*). Among vertebrates, palaeoniscoid fish dominated and holosteans appeared. Reptiles diversified and included pelycosaurs and their replacements the THERAPSIDA, but there were few representatives of the dominant MESOZOIC reptiles. Conifers, cycads and ginkgos evolved, and the earliest forest types declined. Global climate was arid; extensive glaciation occurred in the Southern Hemisphere. At its close there was a considerable mass EXTINCTION of marine fauna; a rapid overturn of deep anoxic oceans which introduced toxic levels of CO_2, and possibly H_2S, into surface waters has been held responsible. See Appendix.

peroxiredoxins   See NITRIC OXIDE.

peroxisome   MICROBODY containing CATALASE, especially in vertebrate liver and kidney cells, and those plant cells involved in PHOTORESPIRATION or BETA-OXIDATION. Catalase uses hydrogen peroxide (itself produced by oxidative enzymes using molecular oxygen inside the organelle) to remove hydrogen atoms from substrates, so oxidizing (detoxifying) phenols, formic acid, formaldehyde and ethanol (oxidized to acetaldehyde). Biogenesis not well understood. All their component enzymes, including those involved in β-oxidation of fatty acids, are encoded by nuclear DNA and imported post-translationally, apparently by PROTEIN TARGETING. See GLYOXYLATE CYCLE, SUPEROXIDES.

perturbation approach   In community ecology, an experimental approach in which artificial disturbances are used to unravel species interactions.

pesticide   See PEST MANAGEMENT.

pest management   Five major strategies of pest control are employed, each dependent for its effectiveness upon the ecological 'strategy' of the pest organism (see PESTS). They are: pesticide control; BIOLOGICAL CONTROL; cultural control (where agricultural or other practices are used to change the pest's habitat); breeding for pest resistance in cultivated organisms, and sterile mating control, where pest populations are variously sterilized to reduce their reproductive rates. These main approaches, and the types of pest against which they are employed, are indicated in Table 7. See INTEGRATED PEST MANAGEMENT.

WEED seeds often contaminate a crop after harvesting, reducing its economic value and founding the next weed problem when the crop seeds are sown. Wild oats (*Avena fatua, A. barbata*) and cleavers weed (*Galium aparine*) often make up 2–8% of cereal seed sown. CROP ROTATION is successful in controlling wild oats, which is not affected by broad-leaf herbicides (since it is a grass, like the wheat crop). The aim is not to eradicate the pest, but to reduce it to a level where it will do no economic damage. Crop rotation commonly involves alternating cereals with a nitrogen-fixing legume (such as clover, alfalfa, or some other member of the pea family). Soil pests, such as the nematode

	r-pests	*Intermediate pests*	*k-pests*
Pesticides	Early widescale applications based on forecasting	Selective pesticides	Precisely targeted applications based on monitoring
Biological control		Introduction or enhancement of natural enemies	
Cultural control	Timing, cultivation, sanitation and rotation	→ ←	Change in agronomic practice, destruction of alternative hosts
Resistance	General, polygenic resistance	→ ←	Specific, monogenic resistance
Genetic control			Sterile mating technique

TABLE 7 *Principal control methods appropriate for different pest strategies.*

worms which attack potatoes, can often be controlled this way. Crop rotation will only be effective when a pest is thereby unable to colonize successive crops. So-called 'volunteer populations' of weeds often arise after harvest and can act as 'green bridges' – hosts for pests and pathogens while crop plants are absent.

By sowing early or late, it may be possible to avoid the egg-laying period of a pest, or the vulnerable stages in plant growth may have passed by the time the pest reaches 'pest proportions'. *Autumn sowing* of cereals in Europe has reduced the risk of aphid damage to seedlings to a minimum, while the dense growth of an autumn-sown crop also reduces weed problems the following spring. Autumn sowing does increase the risk from wheat bulb fly, but late sowing of spring wheat reduces this risk. *Early harvesting* may prevent crops from being infested with a pest, and early maturing varieties are one of the major goals of plant breeders. As with early/late sowing, the principle behind this is to desynchronize the growth of the crop from the life cycle of the pest. Thus, infestation by maize weevil is reduced by prompt harvesting of the maize (= corn) crop before the adult beetles emerge from the stored crop.

Trap crops can be used to lure pests off crop plants. This involves growing plants which the pests prefer to the crop. They are often planted every fifth row, or as a peripheral band around the crop, and saves insecticide use. *Intercropping* is similar, two crops being planted together. Scented plants like onions can protect carrots from carrot root fly which is normally attracted by the scent of carrots, but is overwhelmed by the scent of onion. In Thailand, ground nuts are intercropped with maize. On hot days, locusts seek the shade of the ground nuts, where they may be eaten by ducks! Such ingenuity is often successful, and avoids use of pesticides. Intercropping of cauliflower with narrow rows of barley (*Hordeum*) considerably reduces infection by mosaic virus. Reflective plastic sheets interfere with aphid, thrips and other insect vision behaviour based upon ultraviolet, disrupting the spread of viruses; and clear plastic covering raises the temperature of soil to about 50°C, which can kill some fungal spores, weed seeds and nematodes before crops are planted.

Stubble burning, now strictly controlled in Britain (dangerous close to roads and habitation), is still widely used elsewhere to destroy both weeds and animal pests. Pesticides include herbicides (e.g. the growth-regulating phenoxy acid 2,4-D; the photosynthesis inhibitor paraquat); FUNGI-CIDES (e.g. the seed dressings benomyl and organomercurials); and insecticides. Non-specific (broad spectrum) forms include chlorinated hydrocarbons and some organophosphates (see ACETYLCHOLINE) and carbamates. Narrow spectrum (bio-rational) insecticides include some other organophosphates and organochlorides, synthetic hormone-like growth regulators, and 'semiochemicals' such as pheromones which alter the insect's behaviour. Organochlorines (e.g. DDT) are fat-soluble, and an animal on a restricted diet (when fat is used) will experience raised plasma levels, to toxic values, of these insecticides. Two common problems associated with such pesticide use are BIOMAGNIFICATION and pest RESISTANCE. Use of microorganisms in BIOLOGICAL CONTROL includes commercial applications of the Gram-positive soil bacterium *Bacillus thuringiensis* (*Bt*) (a biopesticide now forming over 90% of the global market). *Bt* spores contain an endotoxin which is rendered soluble by the high pH of the insect mid-gut, epithelial microvilli being rendered more permeable to cations so that the cells rupture. Paralysis and death follow. Some see potential in the agents of CYTOPLASMIC INCOMPATIBILITY as biopesticides.

pests Species whose existence conflicts with human profit, convenience, or welfare. Some cause serious nuisance; their injuriousness is well established and their control is either a social or economic necessity. Pest status commonly arises from (a) entry of species into previously uncolonized regions; (b) some change in the properties of previously unproblematic species; (c) changes in human activities, bringing contact with species to which there was previous indifference; (d) increase in abundance of a species, with resulting nuisance value.

Pest status may be interpreted in terms of ecological strategies wrought by different selection pressures to which pest species are exposed; thus: *r-pests*, where R-SELECTION influences are uppermost; *k-pests*, where K-SELECTION influences dominate; and *intermediate pests*, lying somewhere between these two. See PEST MANAGEMENT, WEEDS.

PET (positron emission tomography) Technique in which a positron-emitting radioactive substance is introduced into the body; gamma rays then released by collision of positrons with electrons are used to construct a computerized image (a PET scan) which may be helpful in detecting metabolic changes in healthy and diseased organs (esp. heart, brain).

petal One of the parts forming corolla of flower; often brightly coloured and conspicuous. See FLOWER.

petiole Stalk of a LEAF.

Peyer's patches Patches of secondary LYMPHOID TISSUE in submucosa of amniote intestines.

pH A quantitative expression denoting the relative proton (hydrogen ion, H^+) concentration in a solution. The pH scale ranges from 0–14: the higher the pH value, the lower the acidity. A pH of 7 indicates a neutral solution. Defined as $-\log [H^+]$, where $[H^+]$ = concentration of protons, expressed as $g.dm^{-3}$. In consequence, a solution of pH 6 has ten times the H^+ concentration of a solution of pH 7. It is important that the pH of cells and body fluids is kept within acceptable values, one reason being the effect of pH change on the shapes of globular protein molecules. See PROTEIN, BUFFER.

Phaeophyceae Brown algae. A class of the algal division HETEROKONTOPHYTA (Kingdom PROTISTA) containing about 265 genera and between 1,500 and 2,000 species. This group is almost exclusively marine; only four freshwater genera are known (*Heribaudiella, Pleurocladia, Bodanello, Sphacelaria*); several marine forms are found in brackish water and salt marshes. Most brown algae live attached to rocks (epilithic algae) on seacoasts; others are epiphytic and occur in the intertidal or subtidal zones. Brown algae dominate in the Northern Hemisphere in colder waters. A species of the genus *Sargassum* is exceptional, as it is pelagic, accumulating in large amounts in the Sargasso Sea. Fossils indisputably linked to the brown algae have been recovered from Miocene deposits in California (*c.* 7–15 Myr old). Since these fossils belong to some of the most advanced orders of the division, the class as a whole must be much older. Putative phylogenetic relationships derived from sequences of nucleotides in 5S ribosomal RNA molecules indicate that the brown algae and diatoms diverged from each other rather recently around 200 Myr BP. They appear to have arisen from Sarcinochrysidalean algae.

All brown algae are multicellular with the morphological complexity of the thallus varying enormously, from microscopic branched filaments to foliose plants many metres in length and possessing a complex anatomy. Brown algae range from minute (<1 mm long) filaments to very large (60–70 m long) thalli with a root-like holdfast and a stem-like stipe bearing branched or unbranched leaf-like blades or laminae, often with one or more air bladders and relatively complex internal structure. Complex thalli show several growth forms (e.g. diffuse, apical, trichothallic, promeristem, intercalary, meristoderm). Chloroplasts are generally discoid and are golden brown – deriving this colour from large concentrations of the accessory xanthophyll fucoxanthin. Chlorophylls a, c_1 and c_2, as well as β-carotene and violaxanthin, are also present. The pyrenoid is stalked or at least protrudes from the chloroplast. Surrounding the pyrenoid outside the chloroplast endoplasmic reticulum is a membrane-bound sac containing the main storage product chrysolaminarin. Mannitol is, however, the accumulation product (up to 25% dry weight) of some species of Laminaria in the autumn. There are two membranes of chloroplast endoplasmic reticulum, usually continuous with the outer membrane of the nuclear membrane in the Ectocarpales but apparently discontinuous in the Dictyotales, Laminariales and Fucales. A periplastidial reticulum (network of interconnected tubules) is present in the narrow space between the nuclear envelope and the two membranes of the chloroplast envelope. Chloroplasts have three thylakoids per band and a girdle lamella is also present beneath the chloroplast envelope. Chloroplast DNA is organized in a ring-shaped nucleoid. Cell walls are composed of a network of cellulose microfibrils stiffened by calcium alginate.

This forms the structural component of the cell wall. An amorphous matrix component comprises fucoidin and mucilaginous alginates. Alginates are non-toxic and are used extensively in industry (e.g. ice cream and beer manufacture) because of their colloidal properties, while fucoidin is a complex sulphated polysaccharide containing in addition to the monosaccharide fucose, varying proportions of the monosaccharides galactose, mannose, xylose and glucuronic acid.

Mitosis is semi-closed, the nuclear envelope breaking down in late anaphase. The spindle collapses in early telophase and as it collapses, the daughter nuclei move back towards each other and lie close together. The spindle is formed inside the nucleus between two polar pairs of centrioles; microtubular asters assemble around the centrioles, and radiate out into the cytoplasm. Two types of reproductive structures occur in the brown algae: (a) unilocular, single-celled sporangia releasing haploid zoospores after meiosis (meiosporangium), which give rise to the gametophyte generation, and produce gametes; (b) plurilocular sporangia, each cell of which produces a single motile cell, functioning either as a gametangium (producing haploid gametes) if borne upon the gametophyte, or as a sporangium (producing diploid zoospores) if borne upon a sporophyte. Motile cells, always gametes or zoospores, have two dissimilar flagella inserted laterally. The long flagellum is pleuronematic, bearing tripartite hairs (mastigonemes). The shorter flagellum is posteriorly directed and is acronematic (except the Fucales where the posterior flagellum of the sperm is longer than the anterior one). Flagellate cells usually possess a typical heterokont photoreceptor apparatus, consisting of a flagellar swelling on the posterior flagellum and an eyespot contained within the chloroplast; however, no eyespot apparatus is present in most genera in the Laminariales.

Sexual fusion can be isogamous (Sphacelariales), anisogamous (Cutleriales) or oogamous (Desmarestiales, Fucales, Laminariales, Dictyotales, Durvillaeales). The life cycle is generally diplohaplontic, and can be isomorphic or heteromorphic. The orders Fucales and Durvillaeales have a diplontic life cycle with gametic meiosis.

phage Viruses infecting bacteria are termed BACTERIOPHAGES; those infecting fungi are termed *mycophages*.

phage conversion Phenomenon whereby new properties may be conferred upon host cells when infected by temperate phage. Each cell receiving the prophage also acquires the new property. See BACTERIO-PHAGE.

phage restriction When phage are grown upon bacteria of one strain and then upon another, their titre may drop in the second host strain. They are then said to be *restricted* by the second host strain. Due to degradation of the phage DNA by host RESTRICTION ENDONUCLEASES. See MODIFICATION.

phagocyte Cell which can ingest particles from its surroundings (phagocytosis), forming vacuole composed of the plasma membrane in which the material lies. The vacuoles may then fuse with LYSOSOMES to form *heterophagosomes*. Receptor sites on phagocytes of the immune system recognize common components of bacterial cell surfaces, allowing ingestion of complement-coated pathogens, and such receptor binding probably initiates vacuole formation in phagocytes generally. Many protozoans are phagocytic, as are dendritic cells, NEUTROPHILS, MONOCYTES and MACROPHAGES which engulf bacteria and clumped antigens. See Fig. 86, LEUCOCYTE, IMMUNITY.

phagocytosis One form of ENDOCYTOSIS in which large particles/cell debris are taken up into large endocytic vesicles (*phagosomes*) which seems to occur by a 'membrane-zippering' process in which adhesion between engulfing cell and particle occurs progressively until the latter is completely surrounded by the cell's pseudopodia, in which actin and actin-binding proteins accumulate – sometimes with CLATHRIN. The resulting endocytic vesicle (food vacuole) is normally converted to a heterophagosome as lysosomes fuse with it, degrading the contents to particles smaller than 1 Kd in size.

The resulting concentration gradients of amino acid, sugars, etc., drive diffusion through specific carriers or channels into the cytoplasm. See also POTOCYTOSIS.

phagosome For autophagosomes and heterophagosomes, see LYSOSOME.

phalanges Bones of vertebrate digits (fingers and toes). Each finger has 1–5 phalanges (more in whales) articulating end-to-end in a row, the proximal of each row forming a joint with a *metacarpal bone*. See PENTADACTYL LIMB.

Phanerophytes Class of RAUNKIAER'S LIFE FORMS.

Phanerozoic Geological division lasting from approx. 600 Myr BP to the present. Initiated by appearance of metazoan fossils. During it atmospheric oxygen level rose from about 0.1 of present levels to that of the present.

pharynx (1) The vertebrate gut between mouth (buccal cavity) and oesophagus, into which opens the glottis in tetrapods and gill slits in fish. In man and other mammals is represented by throat and back of nose; partly divided by soft palate into upper (nasal) section and lower (oral, or throat) section. Contains sensory receptors setting off swallowing reflex. The GAS BLADDER and EUSTACHIAN TUBE, where found, also open into it. See GILL POUCH. (2) Part of the gut into which gill slits open internally in urochordates and cephalochordates.

phase contrast See MICROSCOPE.

phasic receptor Rapidly-adapting RECEPTOR, e.g. touch receptor in skin, in which a steady stimulus may evoke only a few impulses. Does not provide information about the length of stimulus, but responds to rate of change of stimulus intensity, e.g. at the stimulus onset or offset. Compare TONIC RECEPTOR.

Phasmida Order of exopterygote insects containing stick insects and leaf insects.

phellem See CORK.

phelloderm Tissue formed by the CORK CAMBIUM; a cork skin. The cells of the phelloderm are living at maturity, lack SUBERIN, and resemble PARENCHYMA cells of the cortex, and can be distinguished from cortical cells by their inner position in the radial rows of other peridium cells.

phellogen (cork cambium) Meristematic cells producing CORK.

phenetics Grouping of organisms into taxa on the basis of estimates of overall similarity, without any initial weighting of characters. Diagrams which result (*phenograms*) are devoid of necessary phylogenetic implications although may be interpreted phylogenetically. Phenetics is a branch of numerical taxonomy. See CLASSIFICATION, PHYLOGENETICS.

phenocopy Environmentally induced alteration in the phenotype of an organism (or cell), commonly resulting from abnormal developmental conditions. Mimics effect of a known mutation but is non-heritable. See GENETIC ASSIMILATION. Contrast TRANSDETERMINATION.

phenogram See PHENETICS.

phenology Study of periodicity phenomena in plants, such as timing of flowering in relation to climate.

phenotype Total appearance of an organism, determined by interaction during development between its genetic constitution (GENOTYPE) and the environment. Different phenotypes may result from identical genotypes, but it is unlikely that two organisms could share all their phenotypic characters without having identical genotypes. Shared presence of a character in two organisms does not necessitate identical genotypes with respect to that character. See DOMINANCE, PENETRANCE, ECOTYPE, PHENOTYPIC PLASTICITY.

phenotypic plasticity Extent to which phenotype may be modified by expression of a particular genotype in different environments. Its often adaptive nature is illustrated by HETEROPHYLLY in some members of *Ranunculus* (subgenus *Batrachium*) and by *Polygonum amphibium*. Spectral quality of light allows many plants to 'gauge' canopy density and potential competition from neighbours. PHYTOCHROMES seem to interact with at least some plant growth substances (esp. gibberellin) to form complex light-transduction pathways that include several regulatory genes and account for the considerable plasticity at the phenotypic level, including rates of branching and vertical stem elongation, and timing of flowering. In animals, the phenomenon includes some cases of CASTE determination in insects and of SEX DETERMINATION in diverse groups. Any genotype probably has a characteristic degree of developmental plasticity (see HOMEOSTASIS). In plants, dwarf, prostrate, thorny and succulent forms are often produced when RAMETS of a cloned genotype are grown in different conditions; genetic fixation (by selection) of an altered phenotype, preadapted to the conditions inducing it, may not be uncommon in wild populations although evidence is rather sparse. In a broad sense, animal LEARNING could be included here; cell differentiation takes the range of the concept below the level of the individual. Most living organisms exhibit a special kind of plasticity to temperature changes, including a *heat-shock response* (see CHAPERONES). See CLINE, DEME, ECOTYPE.

phenylketonuria Recessive human genetic disorder. The enzyme converting dietary phenylalanine to tyrosine is deficient, causing excretion of phenylpyruvate (or phenylalanine) in urine. Intellectual impairment common and epileptic attacks occur in about 25% of cases. Tendency to lighter hair and skin pigmentation than average. Can be detected soon after birth; a diet low in phenylalanine reduces symptoms.

pheromones Chemical substances which, when released into an animal's surroundings, influence the behaviour or development of other individuals of same species. Include sexual attractants in many insect species. Worker and queen bees produce several different pheromones, each with its own effect; one deer produces pheromones from at least seven sites on its body, each with a different social function.

phloem Principal food-conducting tissue in vascular plants. Mixed tissue, containing parenchyma and occasionally FIBRES, besides sieve elements, the main conducting cells (SIEVE CELLS in nonanthophytes, SIEVE-TUBE MEMBERS in anthophytes), and their COMPANION CELLS. Substances transported include sugars, amino acids, some mineral ions and growth substances. Phloem may be *primary* or *secondary* in origin, the former frequently being stretched and destroyed during elongation of the plant. Protoplasts of adjacent sieve elements connect via groups of narrow pores, the *sieve areas*, concentrated on the overlapping ends of these long, slender cells. Where the pores are large, the sieve area is termed a *sieve plate*. See MASS FLOW, VASCULAR BUNDLE, XYLEM.

phlorotannins Polymers of phloroglucinol (1,3,5-trihydroxybenzene), known only from the Phaeophyceae (brown algae). Phlorotannins display properties of tannins, precipitating proteins from solution and binding metal ions. Like tannins, they also have an astringent taste.

Phoronidea Small phylum of marine worm-like animals, unsegmented and coelomate, living in tubes (tubicolous) of chitin which they secrete. Resemble ECTO-PROCTA in habit and appearance, but are only superficially colonial. Feed by ciliated LOPHOPHORES. Planktonic larva resembles a trochophore. Vascular system contains haemoglobin.

phosphagen Any of several HIGH-ENERGY PHOSPHATE compounds which act as reservoirs of phosphate-bond energy in the cell, a KINASE transferring their phosphate to ADP, forming ATP. Include *phosphocreatine* and *phosphoarginine*. Nerve and muscle, especially, contain phosphagens. Creatine and arginine become phosphorylated when ATP concentration in the cell is high, the reverse occurring when the cell's ATP to ADP ratio is low. Phosphoarginine is characteristic of invertebrates, phosphocreatine of vertebrates. Both occur in echinoids and hemichordates. See CREATINE.

phosphatases Enzymes splitting phosphate from organic compounds. Compare KINASE.

phosphate-bond energy See HIGH-ENERGY PHOSPHATE.

phosphatides See PHOSPHOLIPIDS.

phosphatidic acid (phosphatidate, PA) Diacylglycerol 3-phosphate. The simplest phosphoglyceride and produced, along with a free headgroup, when phospholipase D hydrolyses membrane phospholipids. Believed to serve as a SECOND MESSENGER, as does DIACYLGLYCEROL.

phosphatidyl inositol (PI) A membrane lipid (see PHOSPHOLIPIDS) unique in undergoing multiple reversible phosphorylations to produce a variety of inositol phospholipids. Two distinct signalling pathways are based on this: (a) the IP_3 pathway which depends on PLC-β or PLC-γ activity (see INOSITOL 1,4,5-TRIPHOSPHATE), and (b) that depending upon PI_3-KINASE (phosphatidylinositol 3-kinase) activity.

phosphocreatine See PHOSPHAGEN, CREATINE.

phosphodiesterase (PDE) Any enzyme capable of hydrolysing cyclic AMP to (non-cyclic) 5'-AMP. Some also hydrolyse the phosphodiester backbone (sugarphosphate backbone) of nucleic acids.

phospholipase C (PLC) A G-PROTEIN-linked enzyme in the inner phospholipid layer of perhaps most eukaryotic plasmalemmas, releasing INOSITOL 1,4,5-TRIPHOSPHATE and DIACYLGLYCEROL (both SECOND MESSENGERS) from phosphatidylinositol 4,5-bisphosphate located in the same layer. Exists in two isozymic forms: PLC-β and PLC-γ. See PHOSPHATIDYLINOSITOL, RECEPTOR, TASTE BUD.

phospholipids (phospholipins, phospholipoids, phosphatides) Those LIPIDS bearing a polar phosphate end, commonly esterified to a positively charged alcohol group. Major components of cell membranes and responsible for many of their properties. There are two main groups: (a) the *phosphoglycerides*, derivatives of

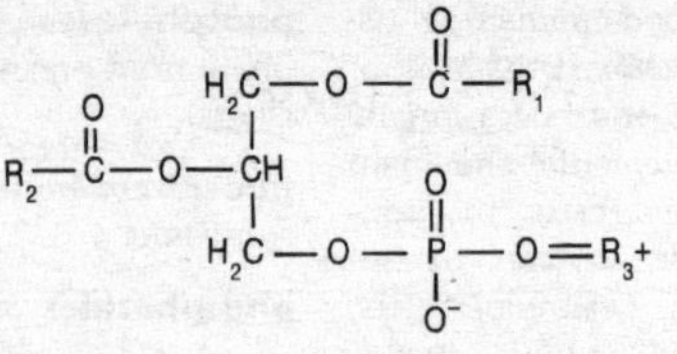

FIG. 133a *A generalized* PHOSPHOLIPID. *R_1 and R_2 represent fatty acid chains. R_3 represents one of several groupings.*

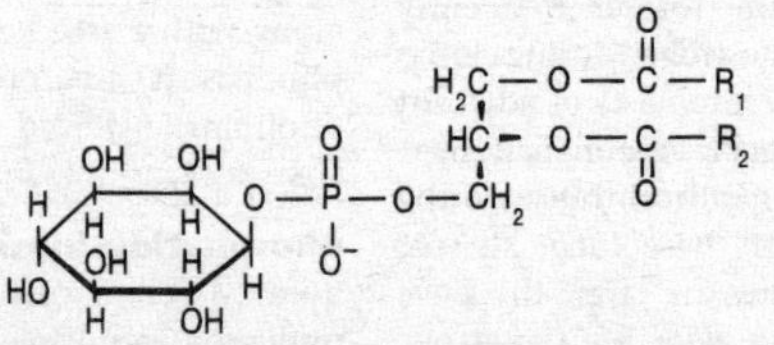

FIG. 133b *Phosphatidyl inositol.*

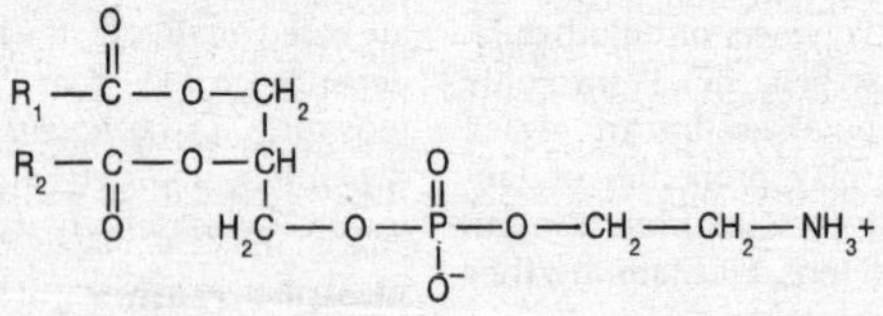

FIG. 133c *Phosphatidyl ethanolamine.*

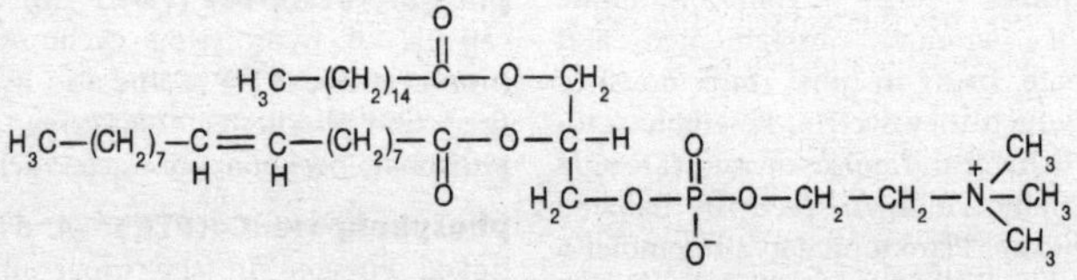

FIG. 133d *A phosphatidyl choline (= lecithin), with palmitic and oleic acid residues as the fatty acid chains.*

phosphatidic acid, include *lecithin* (phosphatidylcholine), *cephalin* (phosphatidylethanolamine) and phosphatidylinositol. Several important proteins are anchored in membranes by links between their C-terminal portions and phosphatidylinositol via a mannose-rich glycan and ethanolamine. Lecithin, an important cell membrane component (and surfactant in vertebrate lungs), is a component of BILE, helping to render cholesterol soluble; (b) the *sphingolipids*, containing the basic *sphingosine* instead of glycerol, forming components of plant and animal cell membranes, e.g. *sphingomyelins* (not to be confused with *myelin*) of liver and red blood cell membranes. Some inositol phospholipids act as SECOND MESSENGERS (see INOSITOL 1,4,5-TRIPHOSPHATE). See also CELL MEMBRANES, CEREBROSIDES, Fig. 133.

phospholipid transfer proteins (PLTPs) Diffusible cytosolic eukaryotic proteins catalysing *in vitro* energy-dependent transfer of lipids between membrane bilayers. Some are very specific in

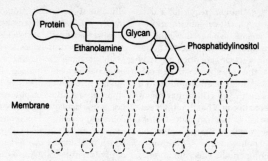

FIG. 133e *Figure indicating how a phospholipid molecule (here phosphatidylinositol) can form an attachment site for a membrane protein, in this case via a glycan and ethanolamine linker. Such attachments are common, notably in the glycocalyx.*

their catalysis (e.g. PCTP, for phosphatidyl-choline), some less so (e.g. PITP, for phosphatidylinositol), others much less so, carrying most PLs, glycolipids and cholesterol. Work *in vivo* suggests that in some yeasts the ratio of PITP/PCTP may affect GOLGI APPARATUS secretory competence. See CHOLESTEROL.

phosphoproteins Proteins with one or more attached phosphate groups. The milk protein *casein* has them attached to serine residues. See PHOSPHORYLASE KINASE.

phosphorylase Enzyme which transfers a phosphate group, often from inorganic phosphate ions, on to an organic compound which thereby becomes *phosphorylated*. Glycogen and starch phosphorylases are enzymes involved in the mobilization of carbohydrate reserves, forming glucose-phosphate.

phosphorylase kinase A Ca^{2+}/calmodulin-dependent PROTEIN KINASE (see CALMODULINS) activating a PHOSPHORYLASE by transfer of a phosphate group from ATP. Often AMP-dependent or Ca^{2+}-dependent (see CALMODULIN). May exert its effect by allosteric change in the protein adjacent to the phosphorylase in a multiprotein complex. One of many regulatory enzymes in cells, activating glycogen breakdown (see PROTEIN KINASE A).

phosphorylation Transfer of a phosphate group by a PHOSPHORYLASE to an organic compound. Usually ATP-dependent, this produces compounds which are highly reactive in water with other organic molecules in the presence of appropriate enzymes, when their phosphate is transferred in turn and energy made available for work. The most important energy-transfer system in metabolism. See KINASE, HIGH-ENERGY PHOSPHATE, OXIDATIVE PHOSPHORYLATION, PHOTOPHOSPHOR-YLATION and ATP.

photic zone Upper portion of a lake, river or sea, sufficiently illuminated for photosynthesis to occur.

photoautotroph See AUTOTROPHIC.

photoauxotroph Photosynthetic organism requiring an external vitamin source.

photoheterotroph See HETEROTROPHIC.

photoperiodism A biological response to changes in the ratio of light and dark in a 24-hour cycle. In plants, although flowering is the best known example, many other responses are photoperiodic and regulated by the photoreversible pigment PHYTO-CHROME. CIRCADIAN RHYTHMS are thought to be fundamental to photoperiodism.

With respect to flowering, plants may be grouped into three categories: (a) *short-day plants*, flowering in early spring or autumn (fall), requiring a dark period exceeding a critical length; (b) *long-day plants*, flowering mainly during summer, requiring a dark period less than a critical length; and (c) *day-neutral plants*, where flowering is unaffected by photoperiod. Different populations within a species are often precisely adjusted to the photoperiodic regimes where they live. Some plants require only a single exposure to the critical day–night cycle in order to flower. The stimulus is perceived by the leaves and transmitted (probably by some growth substance, or *florigen*) to growing points where flowering is initiated. Photoperiodic effects in animals include initiation of mating in aphids, fish, birds and mammals. See PROXIMATE FACTOR.

photophile Literally, light-loving; light-receptive phase of a CIRCADIAN RHYTHM, lasting about 12 hours. Compare SKOTOPHILE.

photophosphorylation Coupling of phosphate with ADP to produce ATP, using light energy absorbed in PHOTOSYNTHESIS. See BACTERIORHODOPSIN.

photopigment Proteins which undergo structural changes on absorbing light and initiate the biochemistry of vision; e.g. those in outer segment membranes of vertebrate ROD CELLS and CONES.

photoreactivation See DNA REPAIR MECHANISMS.

photoreceptor (1) (Bot.) The photoreceptor contains a light-sensitive pigment and is situated either in a flagellar swelling (Heterokontophyta, Euglenophyta) or in a specialized area of the plasmalemma overlying the eyespot (Chlorophyta). (2) In general, any light-sensitive RECEPTOR organ (e.g. EYE), cell (e.g. ROD CELL) or molecules (e.g. CRYPTOCHROME, PHOTOTROPIN, PHYTOCHROMES, RHODOPSIN). See BIOLOGICAL CLOCK, CYCLIC GMP, INOSITOL 1,4,5-TRIPHOSPHATE.

photoreceptor apparatus The whole complex of the EYESPOT (stigma) and photoreceptor in flagellated algal cells.

photorespiration Type of very active non-mitochondrial respiration occurring in conditions of high light intensity, reduced CO_2 levels and raised O_2 levels in temperate plants carrying out C_3 PHOTOSYNTHESIS; usu-

ally absent (or low) in tropical C_4 plants. Involves oxidation of carbohydrates, takes place in PEROXISOMES, and yields neither ATP nor $NADH_2$; thus appears very wasteful. Main substrate, *glycolic acid*, is derived from oxygenation of ribulose bisphosphate (RuBP) through competitive inhibition of chloroplast ribulose bisphosphate carboxylase by molecular oxygen (SEE ENZYME). Glycolic acid is oxidized by molecular oxygen in peroxisomes to yield hydrogen peroxide, which is destroyed by catalase. Up to 50% of photosynthetically fixed carbon may be reoxidized to CO_2 during photorespiration, lessening the efficiency of C_3 photosynthesis.

photosynthesis The light-dependent synthesis of organic carbon from inorganic molecules occurring in CHLOROPLASTS, cells of blue-green algae, prochlorophytes, green sulphur bacteria, purple sulphur bacteria, and purple non-sulphur bacteria, in the presence of one or more types of light-trapping pigments (notably chlorophylls). Chloroplasts, blue-green algal and prochlorophycean cells contain chlorophyll *a* (occurs in all photosynthetic eukaryotes) as the major light-trapping pigment. Depending upon the group, other chlorophylls may or may not be present, e.g. chlorophyll *b* in the bryophytes, vascular plants, green and euglenoid algae, and prochlorophytes. When a molecule of chlorophyll *b* absorbs light, the excited molecule transfers its energy to a molecule of chlorophyll *a*, which then transforms it into chemical energy. Because chlorophyll *b* absorbs light of different wavelengths from chlorophyll *a*, it extends the range of light that can be used for photosynthesis. In other groups such as the CHRYSOPHYTA, BACILLARIOPHYTA and PHAEOPHYTA, chlorophyll *c* serves the same function as chlorophyll *b*. Chlorophylls present in bacteria differ in several ways from chlorophyll *a* (e.g. chlorobium chlorophyll *a* in the green sulphur bacteria; bacteriochlorophyll in purple sulphur and purple non-sulphur bacteria), but they have the same basic structure. Other accessory pigments include carotenoids and phycobilins. Photosynthesis is the route by which virtually all energy enters ecosystems. More than 150 billion metric tons of sugar are produced on a global scale per year.

It was Joseph Priestley (1771) who first reported that 'vegetation could restore air'; then F. F. Blackman (1905) demonstrated that photosynthesis was a two-step process, while B. Van Niel showed that purple sulphur bacteria reduced carbon to carbohydrates but released no oxygen. These bacteria require H_2S for their photosynthetic activity.

$$CO_2 + H_2S \xrightarrow{\text{light}} (CH_2O) + H_2O + 2S$$

Sulphur accumulated inside the bacterial cells. In the 1930s, Van Niel made the brilliant extrapolation to plants, proposing that water, not carbon dioxide, was split in photosynthesis. This was proven years later using ^{18}O isotopes:

$$CO_2 + 2H_2{}^{18}O \xrightarrow{\text{light}} (CH_2O) + H_2O + {}^{18}O_2$$

Thus, water acts as an electron donor in plants, algae, CYANOBACTERIA and prochlorophytes.

Photosynthesis occurs in two stages: (a) a light-trapping stage (*light phase*), in which light energy (photons) initiates photochemical reactions on pigment molecules, producing energy-rich compounds (ATP and NADPH) with the release of molecular oxygen (plants, algae, blue-green algae and prochlorophytes only); (b) a light-independent stage, in which enzymes located off the pigment molecules use products of the light phase and incorporate (fix) CO_2, using the atoms of its (inorganic) molecules to synthesize more organic molecules. (Since the carbon-fixing enzyme is not itself dependent upon light, this is often called the *dark phase*. It occurs in the light unless experimentally contrived so as not to.) Energy for this carbon-fixation comes from ATP and NADPH produced momentarily earlier in the light phase.

Chloroplasts are thought to have evolved through ENDOSYMBIOSIS from prokaryotic cells resembling blue-green algae. Bacterial photosynthesis differs from others in not releasing molecular oxygen, for the pigment system (photosystem) needed to utilize

hydrogen atoms in water is lacking, and these usually anaerobic bacteria use alternative hydrogen donors (see AUTOTROPHIC).

In blue-green algae, prochlorophytes and chloroplasts, photons are absorbed on thylakoid membranes by chlorophylls and accessory pigments arranged in the form of ANTENNA COMPLEXES. Energy is passed by resonance transfer between pigments until it reaches the reaction centre of one of two types of photosystem, each containing a specific form of chlorophyll a. Light energy enters photosystem II where it is trapped by the reaction centre, P_{680}, either directly or indirectly via one or more of the pigment molecules. When P_{680} is excited a pair of excited electrons (excitons) are donated to an organic receptor molecule, designated as 'Q' because of its ability to quench the loss of energy by fluorescence of excited P_{680} (PLASTO-QUINONE). The electrons lost are replaced by a pair from a water molecule as protons and molecular oxygen are released into the thylakoid space (see Fig. 26). This light-dependent oxidative splitting of water molecules is termed *photolysis*. Manganese is an essential cofactor for the oxygen-evolving mechanism. In the other photosystem, photosystem I, chlorophyll a_{700} donates a pair of electrons to another organic receptor which, assisted by a FERREDOXIN molecule, reduces NADP to release NADPH into the chloroplast stroma. Electrons lost are replaced by those from P_{680} after they have passed along an ELECTRON TRANSPORT SYSTEM which includes cytochromes, iron-sulphur proteins, quinones, chlorophyll and the copper-containing protein plastocyanin. Energy released drives protons from the stroma across the thylakoid membrane into the thylakoid spaces, diffusing out through an ATPase in the membrane and releasing sufficient energy for synthesis of ATP from ADP and inorganic phosphate. The whole electron and proton flow sequence is called *non-cyclic photophosphorylation*. After the electrons have left P_{700} of photosystem I they may short-circuit back again via some of the electron transport molecules. This produces no NADPH, but proton-pumping and ATP formation do occur; this is termed *cyclic photophosphorylation*. See Fig. 45. The dark phase of photosynthesis differs between so-called C_3 and C_4 plants. In the former (including most temperate plants; about 85% of all plant species), CO_2 is fixed by the enzyme *ribulose bisphosphate carboxylase/oxidase* (RuBisCO), acting as a carboxylase. The two substrates are CO_2 and ribulose 1,5-bisphosphate (RuBP), and the product is the 3-carbon (hence C_3) compound 3-phosphoglyceric acid (PGA). PGA is not energetic enough for further metabolism, but is converted using ATP and NADPH from the light phase to glyceraldehyde-3-phosphate. This can be used to recycle RuBP (the Calvin cycle – after its discoverer, the Nobel laureate Melvin Calvin; see AUTOCATALYST) to synthesize starch, or sucrose. If RuBisCO functions as an oxidase then RuBP serves as a substrate for PHOTORESPIRATION, and photosynthesis is less efficient. RuBisCO in higher plants such as rice is 100 times more likely to pick up CO_2 than O_2, but the effect is concealed by the far higher relative concentration of O_2. So 20–50% of carbon fixed by photosynthesis is lost by photorespiration. Although not a handicap when O_2 levels were low at the dawn of oxygenic photosynthesis, RuBisCO's weakness (competitive inhibition by O_2) was exposed as oxygen levels rose. Some diatoms and red algae have more specific RuBisCOs than other plants, often with threefold greater efficiency at fixing CO_2.

In C_4 plants (e.g. many cereals and the rice grass *Spartina*), the first product of CO_2-fixation is 4-carbon oxaloacetate, produced by phosphoenolpyruvate carboxylase (PEP carboxylase), which uses phosphoenolpyruvate as its other substrate. In this Hatch–Slack pathway (named after the two Australians who played key roles in its elucidation), oxaloacetate is further reduced to malate or changed by addition of an amino group to aspartate. These reactions occur in the mesophyll cells of the leaf, and the malate (or aspartate) moves from them to bundle-sheath cells surrounding the vascular bundles (see KRANTZ ANATOMY). Here malate is decarboxylated to yield CO_2 and pyruvate, the CO_2 entering the Calvin cycle as substrate for RuBP-carboxylase while pyruvate reacts with ATP to form more PEP

molecules. Such plants can generally photosynthesize at temperatures far higher than C_3 plants, and generally have a far lower CO_2 COMPENSATION POINT. The C_4 cycle imposes a high energy cost on the plant and is only cost effective at the higher temperatures which C_4 plants tolerate (hence no winter maize crop). The C_4 pathway has evolved over 30 times and two C_3 plants (tobacco, *Nicotiana tabacum*, and celery, *Apium graveolens*) have recently been shown to be able to assimilate carbon from malate and CO_2 using RuBisCo, and to have C_4-type decarboxylases in their vascular bundle cells. This enables the plants to conserve respiratory carbon products, known to be transported in xylem sap. Research is under way to modify non-functioning equivalents of C_4-type genes in rice and replace them with their active counterparts in maize. See AEROBIC ANOXYGENIC PHOTOSYNTHESIS, BACTERIORHODOPSIN, CRASSULACEAN ACID METABOLISM, CARBON CYCLE.

photosystem See PHOTOSYNTHESIS and Fig. 45.

phototaxis TAXIS in which the stimulus is light. In the alga *Chlamydomonas*, the genetics of the process has been clarified.

phototrophic See AUTOTROPHIC.

phototropism TROPISM in which light is the stimulus, e.g. the bending of a stem of an indoor plant towards a window, brought about by increased cell elongation in the growth region of the shaded side. Has been shown to be caused by unequal distribution of the growth substance AUXIN, which migrates from the light side to the dark side of the shoot, particularly in light of wavelengths of 400–500 nm. Thus a pigment absorbing blue light mediates the effect; evidence suggests that it is a flavin pigment.

phragmoplast (Bot.) A structure composed of the interzonal (continuous) spindle microtubules, together with various other microtubules around the periphery of the spindle. The phragmoplast is parallel to it, and is involved in the assembly of a cell plate, and hence also in the formation of a new cell wall, in the plane of the spindle equator. See PHYCOPLAST.

phycobiliprotein Water-soluble pigments located on (CYANOBACTERIA, RHODOPHYTA) or inside (CRYPTOPHYTA) thylakoids of the chloroplasts of the three aforementioned algal divisions. Coloured proteins (chromoproteins) in which the prosthetic group (non-protein part of the molecule or chromophore) is a tetrapyrrole (bile pigment) known as phycobilin. The prosthetic group is tightly bound by covalent linkages to its protein portion of the molecule (apoprotein). There are two apoproteins, α and β. The major chromophore of phycocyanin and allophycocyanin (blue pigments) is phycocyanobilin, and that of phycoerythrin is phycoerythrobilin. In addition in B- and R-phycoerythrin there is the chromophore phycourobilin. Each phycobiliprotein usually comprises a basic aggregate of three molecules of α-apoprotein and three molecules of β-apoprotein with the chromophore attached to the apoproteins. Phycobiliproteins have to be assembled into proper sequences and attached to the thylakoid membranes by means of linker polypeptides. These function as ACCESSORY PIGMENTS.

phycobiont Term referring to the algal partner in a lichen.

phycocolloid A polysaccharide colloid formed by an alga; e.g. CARRAGEENAN formed by various red algae.

phycocyanin See PHYCOBILIPROTEINS.

phycoerythrin See PHYCOBILIPROTEINS.

phycology Study of ALGAE.

Phycomycetes In older classifications, all lower fungi; included the distantly-related Mastigomycotina and ZYGOMYCOTINA.

phycoplast (Bot.) An assembly of microtubules at the equator of the spindle during telophase. These microtubules are orientated with their long axes in the equatorial plane, at right angles to the spindle axis. Cell division takes place in the plane of the phycoplast (e.g. through formation of a wall via a cell plate). The phycoplast functions in ensuring that the cleavage furrow passes between the daughter nuclei. See PHRAGMOPLAST.

phyletic evolution Term encompassing both the origins of higher taxa and the rate of taxonomic evolution (the rate of change of species number in a clade). Often linked to the notion of 'key innovations' in evolution, difficult though these are to test. Species selection (not to be confused with group selection) is the result of differential survival rates or speciation rates of different species: long-term ('unforeseen') consequences of the evolution of short-term selectively advantageous traits (i.e. of microevolution). Phyletic gradualism and 'punctuated equilibrium' are two extremes in the interpretation of a continuum of the process. HOMEOTIC GENES have been invoked in the origins of some key innovations.

phyllid Flattened leaf-like appendage in bryophytes.

phylloclade See CLADODE.

phyllode Flat, expanded petiole replacing blade of leaf in photosynthesis.

phyllotaxy (phyllotaxis) (Bot.) Arrangement of leaves on the stem: whorled, opposite or spiral. In spiral phyllotaxy, a line connecting attachment points of successive leaves forms a spiral; individual leaves regularly positioned within this spiral. In the most simple, truly alternate, arrangement, leaves are 180° apart, and passage from one half leaf to that precisely above it involves one circuit of the stem and two leaves – a phyllotaxy of 1/2. Various forms of phyllotaxy occur, such as 1/3, 2/5, 3/8, 5/13, etc. (a Fibonacci series), each fraction representing an angle made by successive leaves with the stem (looking vertically downwards). At the end of each spiral, a leaf is directly above the one at the beginning. Looking down on a stem, these points of superimposition are identified as vertical rows of leaves known as *orthostichies*. At the growing point, though leaf primordia are spirally arranged, orthostichies do not occur; but looking down at the apex, the primordia are arranged in a series of descending curves, or *parastichies*, some clockwise. Parastichy in an apex may become orthostichy in a mature shoot by straightening during elongation.

phylogenetics Approach to biological CLASSIFICATION concerned with reconstructing PHYLOGENY and recovering the history of speciation. This is possible when speciation is coupled with, and does not proceed faster than, character modification. PHYLOGENETIC TREES so produced should represent the hypothetical historical course of speciation and be open to rigorous testing. The system in ascendancy today is CLADISTICS. Where appropriate, the techniques include comparative anatomy and embryology; DNA protein (and cytochrome) sequencing; DNA HYBRIDIZATION; immunodiffusion and immunoelectrophoresis. See MOLECULAR CLOCK; compare PHENETICS.

phylogenetic tree Diagram representing the history of a group of organisms, in terms of ancestors, descendants, and the relationships between these, implicitly including current evolutionary hypotheses and often information about the times at which taxa are thought to have existed. Phylogenetic trees therefore include much hypothetical and inferential content absent from CLADO-GRAMS. Estimates of absolute time derive from the fossil record and may be contentious; rates of evolutionary change are likewise uncertain; ancestor–descendant links are usually hypothetical. Phylogenetic trees may include ecological data relating to adaptations, which cladograms always exclude. Molecular phylogenies, perhaps more than others, require ROOTING, and the aim is to find the optimal tree (maximizing parsimony): we assume that evolution follows the line of fewest changes and build this into decisions when interpreting genealogies (see HOMOLOGY). DNA SEQUENCING rarely involves more than a small fraction of the genome, and comparisons between taxa often estimate the *bootstrapping proportion* (BP), which is the number of times a given node in a phylogenetic tree is supported over the total number of nucleotide samplings carried out, low BPs indicating low support for individual branches and providing scope for introduction of a new

taxon to change the entire topology of a tree. See, e.g., Fig. 107, OUTGROUP.

phylogeny Evolutionary history. Genealogical history of a group of organisms, in practice represented by its hypothesized ancestor–descendant relationships. Compare ONTOGENY. See RECAPITULATION.

phylotypic stage In an animal's development, the stage at which the precursor to the general body plan characteristic of the phylum to which it belongs is expressed; or, the stage at which all major body parts are represented in their final relative anatomical positions as undifferentiated cell groups (immediately after the principal morphogenetic tissue movements). Such stages have been proposed as: the tailbud stage (for vertebrates); the fully segmented term band stage (for insects); the fully segmented ventrally closed stage (for leeches) and the completion of most embryonic cell divisions (for nematodes). See ZOOTYPE.

phylum Taxonomic category often restricted to the animal kingdom; includes one or more CLASSES and is included within a KINGDOM in the taxonomic hierarchy. Corresponds to the category DIVISION in botany. See PHYLOTYPIC STAGE.

physiological saline See RINGER'S SOLUTION.

physiological time A measure combining both time and temperatures and applied to ECTOTHERMIC and POIKILOTHERMIC organisms, reflecting the fact that growth and development in particular are dependent upon environmental temperature and therefore require a period of time-temperature rather than simply time for their completion.

physiology Study of processes, many either directly or indirectly homeostatic, that occur within living organisms; in multicellular organisms, includes interactions between cells, tissues and organs and all forms of intercellular communication, both energetic and metabolic.

phyto- Prefix indicating a botanical context.

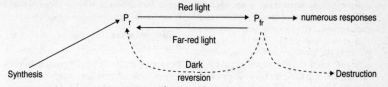

FIG. 134 *A summary of some transformations of* PHYTOCHROME. *Dashed lines indicating dark reversion and destruction do not seem to occur with type 2 P_{fr} molecules.*

phytoalexins Complex, low-molecular weight, antimicrobial organic compounds (often phenolic) produced by plants in response to infection and that inhibit further growth by the pathogen. Infected cells die, and they and the immediately surrounding cells produce phytoalexins. They are strictly localized, rather non-specific in their action and are produced as a reaction to a variety of stimuli. Compare PHYTOANTICI-PINS. See PLANT DISEASE AND DEFENCES.

phytoanticipins Preformed, rather than inducible, infection inhibitors of plants. A phytoanticipin of one plant may be a PHYTO-ALEXIN in another. See PLANT DISEASE AND DEFENCES.

phytochrome Plant chromophore-containing protein pigment existing in two alternative, interconvertible, forms. The chromophore (similar to phycobilin) absorbs the light that causes phytochrome responses, and two major types of phytochrome (types 1 and 2) exist, each with P_r and P_{fr} forms (see Fig. 134). The kinase activity of plant phytochromes has been confirmed, although they have serine/threonine kinase activity instead of the expected histidine kinase activity, and this could underlie their role in the long-term regulation of gene activity. There is now strong evidence for photoactivated nuclear translocation of phytochromes and this research field is rapidly moving. P_r absorbs (is *receptive* to) red light (660 nm) and P_{fr} absorbs far-red light (730 nm), conversion from one to the other apparently modifying auto-phosphorylation sites. When a P_r molecule absorbs a photon it is converted in milliseconds to a molecule of P_{fr}, the reverse happening when P_{fr} absorbs a photon. P_{fr} acts primarily by changing membrane permeability (in seconds to minutes), similar to blue-activated CRYPTOCHROME. Red-light treatment appears to activate auxin's effects on proton extrusion from the cytoplasm into the wall, a brief far-red treatment reversing this. The $P_r : P_{fr}$ ratio is instrumental in breaking DORMANCY in some seeds, in anthocyanin development and plastid differentiation. It is also crucial in several aspects of photomorphogenesis, e.g. branching of stems and stem elongation (see PHENOTYPIC PLASTICITY).

Phytochrome-like molecules have been discovered in the non-photosynthetic eubacteria *Deinococcus radiodurans* and *Pseudomonas aeruginosa*. In the first of these it functions as a light-regulated histidine kinase, helping to protect it from bright visible light. See PHOTORECEPTOR.

phytogeography See PLANT GEOGRAPHY.

Phytophthora See OOMYCOTA.

phytoplankton Algal plankton, comprising a diverse range of cell size and cell volume, including non-motile and motile unicells and colonies. Organisms complete all or the majority of their life cycle suspended in the open water. Phytoplankton can be arbitrarily classified according to size (e.g. microplankton, 60–500 μm diameter; nanoplankton, 5–60 μm diameter pico-plankton (= ultraplankton); ultraplankton, 0.5–5 μm diameter), and can be subdivided into several categories (*Euplankton* = Holo-plankton, the permanent planktonic assemblage of organisms which completes its life cycle suspended in the water; *meroplankton*, includes those species which spend part of each season resting on the bottom sediments; *pseudoplankton* = tychoplankton, the assemblage of casual species derived from other habitats – quite common in lakes and rivers, especially after the storms). Further, the prefixes *limno-*, *potamo-* and *heleo-* are used to refer to the plankton of lakes, rivers and ponds. There is also a flora, both in marine and freshwaters associated with flocs of organic detritus up to 1 mm or more in diameter. Algae, not planktonic taxa but benthic taxa associated with surfaces, can be found living in this unusual site and have been termed the *detritiplankton*.

phytosociology See PLANT SOCIOLOGY.

PI3-kinases (phosphatidylinositol 3-kinases) Several forms of this enzyme occur. The one activated by receptor tyrosine kinases (RTKs) binds to their phosphotyrosines at its SH_2 DOMAINS. Another binds activated G-protein-linked receptors and then binds activated Ras PROTEINS, by which it is then activated. Activated PI3-kinases mainly phosphorylate phosphatidylinositol at the 3 position of the inositol ring, generating the lipids $PI(3,4)P_2$ or $PI(3,4,5)P_3$ which serve as docking sites for intracellular proteins. PI3-kinase may be activated by antigen receptors on B CELLS, when these docking sites recruit PLC-β and the kinase BTK which signal the B cells to proliferate or survive. The PI3-K signal transduction pathway is one of the main signalling routes by which CELL GROWTH is controlled. See NEUROTROPHINS.

pia mater The innermost of the MENINGES.

pileus Cap-like part of fruiting body of fungi of Basidiomycotina (e.g. mushrooms), bearing the hymenium on its undersurface.

pili (sing. pilus) Proteinaceous filaments protruding the cell wall of mainly Gram-negative bacteria; shorter and straighter than flagella. The exact function is not known; however, they may serve as a bridge for transfer of DNA, or may pull conjugating cells together or may have an adhesive role, attaching to membrane surfaces. Pili (fimbriae) also occur protruding from the cell walls of some blue-green algae (CYANOBAC-

TERIA) but differ in several characteristics from those of bacteria. See F-FACTOR.

piliferous layer That part of the root epidermis bearing root hairs.

Pinaceae Pines. A family comprising ten genera and about 200 species of trees (occasionally shrubs), almost entirely confined to the Northern Hemisphere. Plants are evergreen (annually deciduous in *Larix*), resinous, aromatic and monoecious. Lateral branches are well-developed and similar to the leading (long) shoots or reduced to well-defined short (spur) shoots (*Larix*, *Pinus*). Twigs are terete, sometimes clothed by persistent primary leaves or leaf bases. Leaves (needles) are simple, shed singly (except whole fascicles shed in Pinus), and are alternate and spirally arranged but sometimes twisted so as to appear 1- or 2-ranked or fascicled, linear to needle-like. Leaves are sessile to short petiolate. Pollen cones mature and shed annually. They can be solitary or clustered. Sporophylls overlap and bear two abaxial microsporangia (pollen sacs). Pollen is 2-winged and spherical (less commonly with wings reduced to a frill or not winged, (e.g. *Larix*, *Pseudotsuga*). Seed cones mature and are shed in one to three seasons or are long-persistent. The cones of certain species remain on the tree and closed for several to many years until stimulated to open, which is often due to fire (e.g. *Pinus banksiana*, *P. contorta*). Seed cone scales overlap with two inverted adaxial ovules, so two seeds are produced per scale. An elongate terminal wing partially decurrent on the seed body is present.

Pines are of major economic importance as producers of the majority of the world's softwood timber, and as sources of pulpwood, tar, turpentine, essential oils and other forest products. Species are also grown as ornamentals (e.g. species of *Abies*, *Pinus*, *Picea*, *Larix*, *Pseudotsuga* and *Tsuga*).

pineal eye See PINEAL GLAND.

pineal gland (epiphysis) Small mass of nerve tissue attached to roof of third ventricle of the vertebrate midbrain; loses all nervous connection with the brain but innervated by sympathetic nervous system.

In amphibia, primitive reptiles (e.g. *Spheno-don*, the tuatara) and some snakes, pineal cells form a *parietal eye* (median eye) lying within a parietal foramen of the skull, with a lens-like upper epithelium and retina-like lower part. In higher vertebrates, the pineal organ has a glandular structure, and secretes *melatonin* which inhibits gonadotrophins and their effects. Melatonin production is inhibited by exposure of the animal to light. SEROTONIN, another pineal product, and melatonin demonstrate inverse circadian rhythms in their production. See Fig. 18.

pinna (pl. pinnae) (Bot.) A primary division, or leaflet, of a compound leaf or frond. (Zool.) See EAR, OUTER.

pinnate Form of branching which occurs at uniform angles from different points along a central axis, all in one plane, as in a feather.

Pinnipedia Eutherian order (or suborder of the order CARNIVORA) containing specialized and aquatic mammals: seals (Phocidae), sea-lions (Otariidae) and walruses (Odobenidae). Limbs are broad flippers, with webbed feet; tail very short. In true seals, hind limbs are fused with tail.

pinocytosis Fluid-phase uptake by eukaryotic cells during ENDOCYTOSIS. *Macropinocytosis* accompanies membrane ruffling induced in many cell types when stimulated by GROWTH FACTORS and involves Rho-family GTPases signalling pathways in the protrusion of >1 μm actin-driven membrane extensions which then collapse back onto the cell membrane generating large endocytic vesicles (macropinosomes) that sample quite large volumes of the extracellular fluid. Other pinocytic uptakes occur through clathrin-COATED VESICLES (~120 nm) and caveolae (~90 nm), and others by clathrin- and caveolin-independent endocytosis. See Fig. 105.

pioneer species Organisms with adaptations enabling them to reach and colonize barren land (e.g. newly emerged islands or land subjected to recent CATASTROPHE). In many ecological contexts, lichens and mosses tend to be the first producers. Pioneer plants would have precocious reproductive abilities (including vegetatively), effective seed dispersal (e.g. in bird egesta) and self-compatibility. Animals (e.g. aphids, some arachnids) would generally be capable of aerial dispersal. A pioneer community is eventually established. As an ecological strategy, being a pioneer tends towards the '*R*-SELECTION' end of the r/K selection spectrum. See SUCCESSION.

pistil Either each separate carpel (in *apocarpous gynoecia*), or two or more fused carpels (in *syncarpous gynoecia*). Each typically comprises an ovary, style and stigma.

pistillate (Of flowers) naturally possessing one or more carpels but no functional stamens; also called carpellate. See STAMINATE.

pith Ground tissue located in the centre of stem or root, within the vascular cylinder. Usually comprises PARENCHYMA. See STELE.

Pithecanthropus See *HOMO*.

pits Small, sharply defined depressions in wall of plant cell where the secondary wall is completely absent, permitting easier passage of material between adjacent cells. Usually occur over *primary pit fields* (where plasmodesmata are concentrated), but also in their absence. Often coincide with pits in the walls of adjacent cells, separated from them by a *pit membrane* comprising a middle lamella and a very thin layer of primary wall on either side, together forming a *pit pair*. Such *simple pits* connect living cells together and also occur in stone cells and some fibres. In *bordered pits*, characteristic of xylem vessels and tracheids, the pit cavity is partly enclosed by over-arching of the cell wall (the *pit border*) and the pit membrane may possess a central, thickened impermeable *torus*, closing the pit aperture if the pit membrane is displaced laterally.

pituitary gland (hypophysis) Small but essential vertebrate gland, lying in a depression of sphenoid bone of skull and communicating with the hypothalamus (part of the diencephalon of the forebrain) by a stalk-like *infundibulum* (pituitary stalk). A composite gland, comprising an anterior lobe (*adenohypophysis*) deriving from pharyngeal ectoderm and secreting the bulk of pituitary hormones, and a posterior lobe (*neurohypophysis*) deriving from ectoderm of the hypothalamus and containing the termini of hypothalamic neurosecretory nerve axons.

The adenohypophysis releases its hormones into capillaries, drained by the hypophysial vein, under commands (*releasing factors*) from the hypothalamus which reach it via a pituitary portal system. There are six endocrine cell types of the adenohypophysis. Hormones secreted by the most anterior region (*pars distalis*) include GROWTH HORMONE, PROLACTIN, THYROID-STIMULATING HORMONE, GONADOTROPHINS and ACTH. Where present, an intermediate region (*pars intermedia*) secretes MELANOCYTE-STIMULATING HORMONE (MSH). ACTH and MSH are cleavage products of proopiomelanocortin (POMC). The neurohypophysis, neurosecretory in function, releases the hormones OXYTOCIN and ANTIDIURETIC HORMONE (see NEUROSECRETORY CELLS for *neurophysins*).

The pituitary forms an integral link in many homeostatic feedback circuits in the body. See HYPOTHALAMUS and Fig. 18.

PKA See PROTEIN KINASE A.

PKB See PROTEIN KINASE B.

PKC See PROTEIN KINASE C.

PKG Cyclic GMP-dependent protein kinase.

placenta (Bot.) That part of the ovary wall to which the ovules are attached and where they remain until mature. The arrangement of the placenta and ovules varies among different groups of flowering plants. May be *parietal* (ovules borne on the ovary walls or on extensions of it); *axile* (ovules borne upon a central column of tissue in a partitioned ovary with as many locules as carpels); *free central* (ovules borne upon a central column of tissue not connected by partitions with the ovary wall); *basal* (single ovule occurs at the very base of a unilocular ovary). Such differences are important in flowering plant classification. (Zool.) A temporary organ, consisting of both embryonic

and maternal tissues, within the uterus of therian mammals and several other viviparous animals, which enables embryos to derive soluble metabolites by diffusion and/or active transport from the maternal blood supply. In placental mammals the organ is connected to the embryo by the *umbilical cord* and has an essential role in the immunological protection of the embryo. See Fig. 54.

In mammals (see Fig. 135), foetal components of the placenta derive initially from the TROPHOBLAST, connected with the embryonic bloodstream either through its contact with the YOLK SAC (*vitelline placentation*) or ALLANTOIS (*allantoic placentation*). In the latter, the outer covering is termed the *allantochorion*, a term also used of placentae in which blood vessels, but little else, derive from the allantois (as in humans). In both vitelline and allantoic placentation, trophoblastic villi push out from the surface of the chorion and invade the uterine lining (*endometrium*; see DECIDUA), greatly increasing the surface area (14 m^2 in humans) for exchange of solute molecules. In marsupials, vitelline placentation serves the entire, brief, intrauterine life; in eutherians it is soon replaced by allantoic placentation, and sometimes only certain restricted zones of the allantochorion (termed *cotyledons*) form villi and participate in exchanges. Functions of the mammalian placenta include: (a) allowing passage of small molecules from uterine capillaries, or the intervillous blood sinuses (depending upon the type of placentation), into the capillaries of the umbilical vein. These include O_2, salts, glucose, amino acids and small peptides (all by active transport), simple fats, some antibodies (see IgG) and some vitamins (A, C, D, E and K), but in addition psychoactive (some dependency-inducing) drugs such as cocaine, CAFFEINE and NICOTINE; (b) removal of embryonic excretory molecules, notably CO_2 and urea, by diffusion from the umbilical artery into the maternal circulation; (c) storage of glycogen, and its conversion to glucose if foetal glucose levels drop; (d) hormone production, substituting for maternal ovaries in production of the sex hormone OESTROGENS (largely oestriol) and PROGESTER-

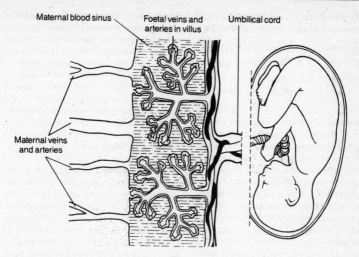

Maternal blood sinus — Foetal veins and arteries in villus — Umbilical cord

Maternal veins and arteries

FIG. 135 *Diagram of human* PLACENTA *and its relationship to the foetus. Left side of diagram is magnified compared to right side.*

ONE (primates especially), and in humans producing HUMAN CHORIONIC GONADOTROPHIN, HUMAN PLACENTAL LACTOGEN and such 'pituitary' hormones as ACTH and TSH; (e) production of ENDORPHINS, and (f) prevention of red blood cell exchange between mother and foetus, avoiding AGGLUTINATION. (The human placenta is exceptional in allowing a mutual exchange of leucocytes and blood platelets. See PARTURITION.)

Placentalia 'Placental' mammals. See EUTHERIA.

placentation (Bot.) Manner in which the ovules are attached in an ovary. See PLACENTA. (Zool.) The type of arrangement of the PLACENTA in mammals.

Placodermi Class of fossil bony fish, some large predators, mainly from the Devonian period. Bony shields on head and front part of body. AUTOSTYLIC JAW SUSPENSION; paired fins; heterocercal tail. Possibly related more to Chondrichthyes than to Osteichthyes or Acanthodii. For *Psarolepis*, see FISH.

placoid scale (denticle) Tooth-like scale, characteristic of the CHONDRICHTHYES and completely covering elasmobranch fish.

Base bone-like, but made largely of dentine, with a pulp cavity and enamel-like coating (usually *vitrodentine*). Probably represent last remnants of the bony armour covering the early vertebrate body (SEE PLACODERMI). Those fish with denticles have similar teeth. Compare GANOID SCALE, COSMOID SCALE.

plagioclimax Any plant community whose composition is more or less stable and in equilibrium under existing conditions but which, as a result of human intervention, has not achieved the natural CLIMAX; e.g. grassland under continuous pasture.

plagiosere Succession of plants deflected into a new course through human intervention. Compare PRISERE.

plagiotropism Orientation of a plant part by growth curvature in response to gravitational stimulus, such that its axis makes an angle other than a right-angle with the line of gravitational force. E.g. exhibited by branches of a main root which makes an acute angle with the vertical. Also used in a wider sense to mean the orientation of an organ so that its axis makes a constant angle with the vertical, and includes DIAGEO-TROPISM as a special type. See GEOTROPISM.

planarians Free-living members of the PLATYHELMINTHES. Class Turbellaria.

planation Dichotomies in two or more planes, flattening out into one plane, e.g. a leaf.

plankton Organisms, algae and animals kept in suspension by water turbulence and dispersed more by such water movements than by their own activities. Most are small, non-motile or motile unicells or colonies, occurring particularly within the photic or euphotic zone. Ecologically and economically of great importance, e.g. providing food for fish and whales. The open water, whether freshwater or marine, always contains particulate matter kept in suspension by water movement, which is known collectively as seston, and this is divisible into non-living (abioseston) and living (bioseston); the components of the bioseston are the PHYTOPLANKTON and ZOOPLANKTON. Compare BENTHOS.

planont (planospore) See ZOOSPORE.

Plantae Kingdom comprising eukaryotic organisms (mostly autotrophs) usually with an embryonic stage and clearly defined cellulose-containing cell walls. Plantae include the BRYOPHYTA (liverworts, mosses, hornworts), PSILOPHYTA (psilopsids), LYCO-PHYTA (lycophytes), SPHENOPHYTA (horsetails), PTEROPHYTA (ferns), CONIFEROPHYTA (conifers), CYCADOPHYTA (cycads), GINKGOPHYTA (*Ginkgo*), GNETOPHYTA (gnetophytes) and ANTHOPHYTA (flowering plants). All are adapted for life on land; their ancestors were specialized green algae (CHLOROPHYTA), and modern classifications include as plants all and only those eukaryotes whose plastids have just two outer membranes (see EUKARYOTE and Fig. 52). Plant classification is very controversial and new data continually arise and require interpretation. The 'Deep Green' group of taxonomists, for instance, using DNA sequences, would prefer to see the major split within the flowering plants placed elsewhere than between monocotyledons (Liliopsida) and dicotyledons (Magnoliopsida).

Some of them associate their findings from these genome studies not with cotyledon number but with a fundamental morphological distinction in pollen type: has it a single pore or furrows? Reproduction in plants is primarily sexual, with cycles of alternating haploid and diploid generations, the former becoming reduced during evolution in the more advanced members. Structural differentiation into e.g. leaves, roots and support organs occurred during evolution of land plants.

plant disease and defences It is estimated that up to 40% of plant productivity in Africa and Asia, and about 20% in the developed world, is lost to pests and pathogens – one third to microbes and the rest to insects and nematodes. Most of this damage occurs to full-grown plants, after most of the water needed for crop growth has been invested; so reducing crop losses to pests and pathogens effectively increases water economy. Most plants are resistant to most pests and pathogens (see ARMS RACE), and the use of GMOS – although controversial – could enhance this, its great advantage over traditional selective breeding being that resistance genes in one plant (or non-plant) species can be engineered into the genome of another without problems of breeding incompatibility. See Table 8.

Structural features providing immunity in plants include a waxy surface preventing wetting and consequent pathogen development, thick cuticles preventing entry of germ-tubes of fungal spores, or production of PHYTOALEXINS. Plants may respond systemically by accumulation of SALICYLIC ACID which is, however, inactivated by glycosylation.

Plants attacked by pathogens rapidly deposit callose (a β-1,3 glucan) at wound sites, especially in cell wall appositions (papillae) forming beneath infection sites. This constitutes their most prominent physical defence against penetration by pathogens, which can be divided into those that kill the host and digest/absorb the contents (necrotrophs) and those that require it as a living host in order to complete its life cycle (biotrophs). Plant chitinases are believed to

Disease	Hosts	Geographical distribution
Fungal		
Late blight	Potato, tomato	Spreading worldwide
Downy mildew	Corn, sorghum	Spreading out of SE Asia
Rust	Soybean	Spreading from SE Asia and Russia
Karnal bunt	Wheat	Pakistan, India, Nepal, Mexico, USA
Monillia pod rot	Cocoa	South America
Rust	Sugarcane	The Americas
Blast	Rice	Asia
Viral		
African mosaic	Cassava	Africa
Streak disease	Maize, wheat, sugarcane	Africa
Hoja blanca (white tip)	Rice	The Americas
Bunchy top	Bananas	Asia, Australia, Egypt, Pacific Islands
Tungro	Rice	SE Asia
Golden mosaic	Bean	Caribbean Basin, Florida, Central America
Plum pox virus	Stone fruits (peaches, etc.)	Europe, India, Syria, Egypt, Chile
High Plains virus	Cereals	Great Plains, USA
Bacterial		
Leaf blight	Rice	Japan, India
Wilt	Banana	The Americas

TABLE 8 *Some emerging plant pathogens.*

defend against fungal infections and appear to be evolving in response to inhibitors (small carbohydrates or proteins) produced by pathogens. Plant leaf cells often store TANNINS, while seeds also often produce ANTI-MICROBIAL PEPTIDES.

Many plant pathogens produce aggressive, Avr, proteins encoded by their avirulence (*avr*) genes. Host defences include R proteins, encoded by their RESISTANCE GENES, which engage the Avr proteins before they do serious harm. The resulting defence responses include calcium influx to the cell, alkalinization of the extracellular space, PROTEIN KINASE activation, production of reactive oxygen intermediates (e.g. H_2O_2 and SUPEROXIDES) and NITRIC OXIDE, and changes in transcriptional activation brought about by signalling pathways. These include the salicylic acid (SA), ethyl-ene and jasmonate signalling pathways, which often regulate resistance and hypersensitive cell death. The *hypersensitive response* is a common feature of many resistance reactions, and involves a microbially infected plant cell undergoing programmed cell death (or APOPTOSIS) sufficiently rapidly to prevent spread of the pathogen to adjacent cells, and may be a response that is invoked only if specific cellular stress signal thresholds are reached. It is not an inevitable response to infection, and in necrotrophic infections might even benefit the pathogen.

Most plant viruses have their genetic material as single-stranded RNA (ssRNA) and replicate their genomes by means of a virally encoded RNA-dependent RNA polymerase, which produces a double-stranded RNA (dsRNA) intermediate. Plant viruses containing sequences homologous to nuclear-expressed genes bring about silencing of the targeted genes by degradation of its mRNA transcript by a nuclease homologue of Dicer-1 (see RNA SILENCING).

Many plant transposons seem to be inactivated by *de novo* DNA methylation, a feature of POST-TRANSCRIPTIONAL GENE SILENCING (PTGS) and transcriptional GENE SILENCING (TGS). PTGS can spread systemically through a plant, both via plasmodesmata and between cells, possibly by translocation of specific mRNA. However, some viruses produce proteins that disable PTGS and prevent this.

Volatiles (aromatic compounds) released by plants, especially in response to damage by herbivores, include terpenes (see SEMIOCHEMICALS). These often attract predators or parasitoids of the feeding herbivore, or repel further egg-laying females. Volatile signalling between plants sometimes involves methyl jasmonate, notably by the sagebrush *Artemisia tridentate*; but although this compound has been associated with reduced herbivore damage in neighbouring plants, it may also restrict competitor growth. However, in the lima bean, *Phaseolus lunatus*, volatile terpenoids released by plants infested with herbivorous spider mites can cause expression of several defence-related genes in neighbouring uninfested plants. Too little experimental work has been done in the field however to assess the significance of these effects. The ELECTROPHILE and volatile 2(*E*)-hexanol, produced by many trees and shrubs, is a biocidal molecule which can induce the accumulation of sesquiterpenoid PHYTOALEXINS in wounded cotton and stress-related gene expression in *Arabidopsis*. Reactive electrophilic compounds are probably crucial in microbial disease and possibly contribute to programmed cell death in the hypersensitive response (HR). Their study could well lead to innovations in PEST MANAGEMENT. Grafting stems from one strain of plant onto the rootstock of another which is resistant to soil pathogens is one ploy to avoid infection, long known to grape growers and now being adopted by tomato growers.

plant geography Study of plant distributions, past and present, and of their causes. Includes such fields as geology, palaeontology, plant genetics, evolution and taxonomy. See ZOOGEOGRAPHY.

plantigrade Mode of walking involving the central surface of the whole foot, i.e. the metacarpus or metatarsus and digits. Humans are plantigrade. Compare DIGITIGRADE, UNGULIGRADE.

plant sociology (phytosociology) Study of plant COMMUNITIES comprising the vegetation, their origins, compositions and structures.

planula Solid free-swimming ciliated larva of most classes of Cnidaria and a few of the Ctenophora. Composed of an outer ectoderm and inner endoderm, the latter formed by migration of cells into the interior.

plaque (1) A clear area in a bacterial culture on nutrient agar caused by localized destruction of bacterial cells through BACTERIOPHAGE activity. Similarly applied to a zone of lysis caused by a virus in animal tissue culture. Counting plaques provides a simple technique for estimating virus concentration in a fluid applied to the culture. (2) Protein can deposit abnormally as plaques in organs such as the brain, as in ALZHEIMER'S DISEASE, PARKINSON'S DISEASE and Creutzfeldt–Jakob disease (see PRION); the term is also used of the atheroma found in arteries during atherosclerosis (see SCLEROSIS). See CHAPERONES.

plasma See BLOOD PLASMA.

plasma cell One kind of mature B CELL, active in antibody secretion.

plasmagel The gel-like state of the outer cytoplasm of many cells, as distinct from the more fluid inner cytoplasm (*plasmasol*) with which it is interconvertible. May be highly vacuolated in planktonic protozoans. Involved in several forms of CELL LOCOMOTION, notably *amoeboid locomotion*.

plasmalemma Outer membrane of a cell. See CELL MEMBRANES.

plasmid A piece of symbiotic DNA, mostly in bacteria but also in yeast, not forming part of the normal chromosomal DNA of the cell and capable of replicating independently of it. Plasmids carry a signal situated at their replication origin dictating how many copies (the copy number) are to be made, and this number can be artificially increased. It constitutes a form of *extra-chromosomal DNA*, sometimes housing genes encoding enzymes (e.g. conferring drug resistance) of critical value to the host cell or organism. Sometimes confer no recognizable phenotype on the host cell (*cryptic plasmids*). Those that may be considered to do so include the bacterial sex factor (F FACTOR) and some forms of PHAGE (*temperate phages*, e.g. P1, in their non-temperate phase). A segment (T-DNA) of the tumour-inducing plasmid (T$_i$) can be transferred from the bacterium *AGROBACTER-IUM* to plant cells at a wound site. Some bacteria contain 'mega'-plasmids, at 580 kb larger than some entire bacterial genomes (e.g. that of *Mycobacterium genitalium*). The bacterium *Borrelia burgdorferi*, agent of Lyme disease, contains both circular and linear plasmids. These bear genes normally found on bacterial chromosomes and have undergone recombination with the bacterial chromosome. Some plasmids, if not all, can be transferred from one cell to another by conjugation (see F FACTOR), TRANSDUCTION and TRANSFORMATION. Often act as vehicles for TRANSPOSABLE ELEMENTS such as ANTIBIOTIC RESISTANCE ELEMENTS. Plasmids which can pick up such transposable elements are termed *R plasmids*. Another type of plasmid, *Col plasmids*, are harboured by many bacterial cells and enable them to produce nutrient-harvesting proteins which effectively starve competing strains; e.g. the iron-scavenging colicin produced by one strain of *E. coli* tends to inhibit other strains of the species (see TRANSFERRIN). The plasmid also renders cells which contain it immune to the colicins. The implications for public health of R plasmids, particularly their mobility between bacteria of different species and genera, are serious. Antibiotic resistance (selected for in non-pathogenic intestinal bacteria by too frequent adminis-

tration of antibiotics) can be transferred to pathogenic bacteria (e.g. *Salmonella* and *Shigella*). Such resistance has been recorded in sewers and polluted rivers and is a strong reason for restricting antibiotic administration to essential cases. See CYTOPLASMIC INHERITANCE.

plasminogen activators Proteases secreted by endothelial cells, activating plasminogen to release the enzyme plasmin and enabling endothelial cells to digest the basal lamina of their capillary and invade new tissue (part of capillary increase, angiogenesis). Generally enable cells to dissolve their attachments and become mobile. Human urokinase plasminogen activator (of Ly-6 protein superfamily) is present on leucocytes and some non-lymphoid cells. Genetically engineered tissue plasminogen activator (t-PA) may be injected to disperse clots in THROMBOSIS.

plasmodesmata (sing. plasmodesma) Extremely fine cytoplasmic tubes (about 20–40 nm in diameter), which pass through the cell walls of living plant cells, and are lined by a plasma membrane of the two adjacent cells. Running through the centre of a plasmodesma is a fine channel, the desmotubule, linking the endoplasmic reticula of adjacent cells. Some gene regulatory proteins, and some mRNAs, are known to pass through plasmodesmata and to spread from cells with a specific mRNA into others that have not (see SYMPLAST). These somehow overcome the size exclusion mechanism normally preventing passage of molecules >800 daltons by binding to plasmodesmata components – as do some virus particles. Many plasmodesmata are formed during cell division as strands of tubular endoplasmic reticulum become trapped within the developing cell plate. They may occur throughout the cell wall, or may be aggregated in the primary pit fields or the membranes between pit-pairs.

plasmodium (1) Generic name of malarial parasites. See MALARIA. (2) Vegetative stage of slime-fungi (MYCETOZOA). Multinucleate, amoeboid mass of protoplasm bounded by a plasma membrane but lacking definite shape or size. See COENOCYTE, SYN-CYTIUM.

plasmogamy Union of two or more proto-plasts, typically of gametes, unaccompan-ied by union of any nuclei they may have. Compare KARYOGAMY.

plasmolysis Shrinkage of a cell's proto-plasm (in plants, away from cell wall) when placed in a hypertonic solution, as a result of osmosis. Compare TURGOR.

plastid endoplasmic reticulum See CHLOROPLAST ENDOPLASMIC RETICULUM.

plastids Small, variously shaped, or-ganelles in the cytoplasm of plant and algal cells, one to many per cell. Each plastid is surrounded by an envelope of two mem-branes and contains a system of internal membranes, pigments and/or reserve food material, a more or less homogeneous ground substance (stroma) in which ribo-somes of a prokaryotic type may be present. Arise either by division of existing plastids or from proplastids, and are classifiable into CHLOROPLASTS (sites of photosynthesis, and the chlorophyll and carotenoid pigments, as well as phycobiliproteins – in some algae), CHROMOPLASTS, non-pigmented LEUCOPLASTS, and PROPLASTIDS. See PLASTOGENE.

plastoquinones Mobile, lipid-soluble car-riers of two electrons (PQ_B) in the thylakoids of chloroplasts, picking these up from cyto-chrome b_6 along with protons from the stroma, becoming reduced (PQH_2) and then donating the electrons and protons to iron-sulphur proteins: a similar role to the reduced quinine, Q, in mitochondria. See Fig. 45.

plastron (1) Flat horny scutes on underside of body of a turtle or tortoise, connected to the carapace above by a bony bridge. (2) An air store forming thin film over bodies of several aquatic insects (some beetles and heteropteran bugs), communicating with the spiracles and held by hydrofuge bristles or scales. Can act as a 'physical gill' if there is adequate dissolved oxygen, enabling the insect to remain under water.

platelets (thrombocytes) Small cell-like blood-borne fragments (3–4 μm long) shed from megakaryocytes in vertebrate bone marrow and essential to normal BLOOD CLOT-TING. Discoid in shape, they contain secretory granules (containing SEROTONIN and Ca^{2+}), express Class I MHC products and receptors, Fc receptors for IgG (see ANTI-BODY) and IgE, and receptors for factor VIII (see BLOOD CLOTTING). Platelets adhere to injured endothelial cells and then release chemoattractants for leucocytes, platelet-derived GROWTH FACTOR (PDGF) and acti-vators of COMPLEMENT as well as increasing permeability of capillaries (see INFLAM-MATION). Platelets are destroyed by RETICULO-ENDOTHELIAL SYSTEM after about ten days. See EICOSANOIDS.

Platyhelminthes (flatworms) Invert-ebrate phylum, containing bilaterally symmetrical triploblastic acoelomates, dorso-ventrally flattened, with only one opening to the gut (when present), and without a blood system. Include planarians (Turbellaria), flukes (Trematoda) and tape-worms (Cestoda). They have bulky paren-chymatous tissue derived from mesoderm, FLAME CELLS and complex hermaphroditic reproductive organs.

platypus Duck-billed platypus (*Ornitho-rhynchus*). See PROTOTHERIA.

Platyrrhines Members of the anthropoid Infra-order Platyrrhini (Ceboidea, New World monkeys). Probably isolated from other primates since the Eocene. Charac-terized by broad nasal septum; 36 teeth (incl. 3 premolars in each jaw quadrant); often have prehensile tails. Include marmosets, tamarins, howler monkeys. See ANTHRO-POIDEA, CATARRHINE.

PLC See PHOSPHOLIPASE C.

Pleckstrin homology domains (PH domains) Modular protein DOMAINS enabling a molecule to bind to charged headgroups of specific phosphorylated inositol phospholipids (e.g. INOSITOL 1,4,5-TRIPHOSPHATE) in the cell membrane during signal transduction, enabling the molecule to interact with other signal mol-ecules recruited to the membrane. See PRO-TEIN NETWORKS.

Plecoptera Stoneflies. Small order of exopterygote insects with aquatic nymphs, long antennae, biting mouthparts and weak flight. Two pairs of wings, held flat over back at rest; hind pair usually larger. Adults tend to feed on lichens and unicellular algae. Good pollution indicators.

plectenchyma Tissue composed of fungal hyphae; either *prosenchyma*, in which component hyphae are loosely woven and recognizable as such, or *pseudoparenchyma*, comprising closely-packed cells which can no longer be distinguished as hyphae and resemble parenchyma of higher plants.

plectostele A protostele, split into many plate-like units.

pleiomorphism (pleomorphism) Form of POLYMORPHISM in which there are several different forms in the LIFE CYCLE; e.g., in the life cycles of rust fungi, where there is a succession of different spore forms.

pleiotropy The ability of allelic substitutions at a gene locus, or a cell product, to affect or to be involved in the development of more than one aspect of phenotype. For any given allele some effects may be dominant, others recessive (see DOMINANCE). A gene locus may have pleiotropic effects if it encodes an enzyme or regulatory protein whose product is involved in several biochemical pathways, as commonly occurs in development (compare POLYGENES). The widespread occurrence of pleiotropy makes it difficult to measure the true FITNESS of a gene, because the most obvious phenotypic differences may not be those upon which the selection coefficient of a gene rests. Several SECOND MESSENGERS, whose targets may not be restricted to a single molecule, have pleiotropic effects. PHYTOCHROME and many PROTEIN KINASES are pleiotropic.

Pleistocene Geological epoch; first of the Quaternary period (see GEOLOGICAL PERIODS). An upper value of duration would be 2.5 Myr–11,000 yr BP. Characterized by four major glacial advances (altogether more than two dozen glacial advances and retreats occurred) separated by interglacial periods lasting tens of thousands of years.

The epoch saw rapid hominid evolution (see HOMO), the extinction of many large mammals and birds, the appearance of elephants, cattle and the modern horse. The climate fluctuated from cold to mild, and the final uplift of many mountain ranges occurred.

plerome See APICAL MERISTEM.

plesiomorphic In PHYLOGENETICS, describing the original pre-existing member of a pair of homologous characters, the evolutionary novelty being the *apomorphic* member of the pair. See CLADISTICS.

Plesiosauria Extinct suborder of marine euryapsid reptiles (Order Sauropterygia), prominent in the Mesozoic. Typical length 3 metres; long-necked, short-bodied, with limbs modified into powerful paddles. Derived from similar *nothosaurs*, they were not dinosaurs.

pleura Serous membranes lining PLEURAL CAVITY and covering the lung surfaces in mammals. Normally only a thin fluid layer separates the pleura from body wall.

pleural cavity Coelomic space surrounding each lung in a mammal, separated from rest of perivisceral coelom by the diaphragm. Pleural cavities are separated from each other by the mediastinum and the pericardium.

Pleurastrophyceae A class of the algal division CHLOROPHYTA whose members can be coccoid, sarcinoid or branched filaments. At present six genera are included in the class (*Trebouxia, Myrmecia, Friedmannia, Pleurastrosarcina, Pleurastrum* and *Microthamnion*). The genera *Trebouxia* and *Myrmecia* occur widely in subaerial habitats, living free on the bark of trees. They also form the phycobionts in many lichen symbioses; the fungi in these partnerships are usually members of the Ascomycota. These algal cells are used by the fungus as photosynthetic organisms; or else the association represents controlled parasitism rather than symbiosis, since about 70–80% of the carbon fixed by the algae during photosynthesis is transferred to the fungus. There is no clear advantage to the algae which fend off the fungal attack by producing antibiotic substances (phytoalexins). *Friedmannia* was isolated from soil from the Negev desert in Israel.

Within the cell walls of *Trebouxia* and *Myrmecia* is a sporopollenin-like compound; this is considered an adaptation to a subaerial existence, hindering desiccation. Chloroplasts are either parietal and cup- or band-shaped or they may be central and stellate. Each contains a single pyrenoid or pyrenoids may be absent. Flagellated cells are strongly compressed and have four microtubular roots in a cruciate arrangement. Basal bodies show a distinct overlap. A distinctive type of mitosis and cytokinesis characterizes this class. Cytokinesis is brought about by a cleavage furrow, which grows from one side only. Mitosis is semiclosed and the telophase spindle is not persistent. Reproduction occurs through production of non-flagellate, walled cells (autospores) or as naked flagellate cells (zoospores or gametes).

pleurocarpus Growth form in some mosses, where the gametophore is multibranched and creeping, with the sporophyte borne upon a very short lateral branch.

pleuropneumonia-like organisms (PPLOs) See MYCOPLASMAS.

plexus See NERVE PLEXUS.

Pliocene The last epoch of the TERTIARY period, during which the first HOMINOIDS appeared. Deserts formed. The climate became cooler, and widespread glaciation occurred in the Northern Hemisphere. Much uplift and mountain formation occurred; Panama joined North and South Americas.

ploidy The number of haploid chromosome sets in a nucleus, or cell. A haploid nucleus has a ploidy of 1; a diploid nucleus 2, a triploid nucleus 3, etc. The ploidy of a eukaryotic cell influences its volume, approximately proportionately. See ANEUPLOID, POLYPLOID.

PLTP See PHOSPHOLIPID TRANSFER PROTEINS.

plumule (Bot.) In a plant embryo, the apical meristem of the shoot terminates the epicotyl (the stem-like axis above the cotyledons). In some embryos the epicotyl bears one or more young leaves. The epicotyl, together with these young leaves, is called a plumule. (Zool.) Down feather of nestling birds, persisting in some adult birds between contour feathers. See FEATHER.

plurilocular Possessing many chambers, e.g. describing gametangia and mitosporangia in brown algae (PHAEOPHYCEAE).

pluripotency Property of a *pluripotent* cell which can give rise to many cell phenotypes of the organism to which it belongs when suitably challenged, but lacks complete TOTIPOTENCY. Pluripotent cell progeny eventually become DETERMINED and show the same phenotype in different environments.

pluteus Larva of echinoid (sea urchin) or ophiuroid (brittle star). Ciliated and planktonic.

pneumatocyst Hollow region of stipes of some brown algae, containing gas helping to keep the alga afloat.

pneumatophore (Bot.) Negatively geotropic specialized root branch, produced in large numbers by some vascular plants growing in water of tidal swamps, e.g. mangrove; grows into the air above the water and contains well-developed intercellular system of air spaces communicating with the atmosphere through pores on the aerial portion. (Zool.) Gas sac of a siphonophoran coelenterate.

Poaceae Grasses (Gramineae a permitted, but somewhat outdated alternative family name). A family of annual or perennial herbs or sometimes woody plants of considerable size (e.g. Bamboo), comprising more than 4,000 species. Pollen that appears to represent the Poaceae enters the fossil record in the Palaeocene and major diversification of the grasses may be more recent, but was evidently well under way in the Miocene. Of enormous ecological and economic importance. See CEREALS, SAVANNA.

pocket protein family A protein family

which includes the retinoblastoma gene product pRb, a tumour suppressor protein which is crucial in the G1/S-phase transition in the CELL CYCLE.

Pogonophora Phylum of benthic and tubicolous worm-like animals; mostly very thin, but attaining large sizes near hydrothermal oceanic vents. Deuterostomes, the coelom in three parts (*oligomerous*), with a tentacular hydrostatic system comparable to the lophophore of phoronids. Lacking a gut, absorbing dissolved organic compounds across their tentacles.

poikilothermy Condition of any animal whose body temperature fluctuates considerably with that of its environment. Characteristic of invertebrates, and vertebrates other than birds and mammals along living forms. See ECTOTHERM, HOMEOTHERMY.

poisons Substances having harmful or adverse effects on living organisms, their toxicity sometimes varying with conditions, e.g. concentration. Poisons affecting enzymes directly include cadmium and mercury, which replace zinc in metalloenzymes such as alkaline phosphatase. Cadmium, zinc, lead and mercury react irreversibly with the thiol (–SH) groups of non-active sites of enzymes, esp. in the nervous system. Lead and mercury also bind to thiol and phosphate groups on membranes, reducing lipid solubility thereby affecting transport of solutes and their resulting concentration gradients across the membrane. Several crop plant yields are limited by aluminium toxicity of acid soils (40% of arable land worldwide). Some overcome this by producing oxalic acid, forming Al-oxalate complexes. Metallothioneins and phytochelatins are thought to be involved in cadmium and copper tolerance. In many pathogen infections, it is the toxins they produce rather than the presence of the pathogen *per se* that do the damage. Toxins may be released extracellularly (EXOTOXINS), some of them (ENTEROTOXINS) acting on the small intestine to cause massive fluid secretion; or be cell-bound and released in bulk only when cells lyse (ENDOTOXINS).

Some plant and bacterial toxins (A toxins) enter mammalian cells by receptor-mediated endocytosis. Some, such as diphtheria, tetanus, botulinum and anthrax toxins pass from acidified endosomes to the cytosol. Others, such as Shiga toxin, Shiga-like toxins, cholera toxin, and the plant toxins ricin and abrin, seem to be transported from endosomes via the Golgi to the endoplasmic reticulum before reaching the cytosol, where they bind enzyme substrates and turn off crucial pathways (but see ADENYLYL CYCLASE; also ENDOTOXINS, EXOTOXINS). Lead in petrol fumes also reduces production of haemoglobin and haem-containing respiratory enzymes. Poisons may be enzyme inhibitors: thus, amanitin (toxin of death cap fungus) binds irreversibly to RNA polymerase. Some organochlorine insecticides are non-specific enzyme inducers. Dinitrophenol makes the inner mitochondrial membrane leaky to protons and prevents formation of the proton gradient; and silicic acid (derivative of silica) ruptures lung macrophage lysosomes so that released enzymes kill the cell and surrounding tissues (silicosis). Aflatoxins (produced by the fungus *Aspergillus flavus*) induce tumours, esp. in grain-eating birds.

Neurotoxins include the channel toxin tetrodotoxin (TTX) from puffer fish, and saxitonin from dinoflagellates (see DINOPHYTA for others), which bind voltage-gated sodium channels and prevent sodium current. Tetraethylammonium (TEA) and 4-amino pyridine block most potassium channels. The ω-conotoxins from cone snails block calcium channels, while conatokins from the same source are non-competitive NMDA (glutamate) receptor antagonists. Toxins preventing presynaptic transmitter release include β-bungarotoxins from cobra venom, and notoxin from tiger snakes. Post-synaptic receptor toxins include acetylcholine (ACh) receptor agonists and antagonists. Muscarine and pilocarpine are agonists, while atropine (belladonna) blocks the muscarinic receptors (see ACETYLCHOLINE). NICOTINE is an agonist of nicotinic ACh receptors. Eserine (physostigmine) and DFP (diisopropylfluorophosphate – a nerve gas) inactivate acetylcholinesterase. For detoxification, see ENDOPLASMIC RETICULUM. See also CYTOTOXIC, PEST MANAGEMENT, TERATOGENS, VIRULENCE.

polar body One of up to three small haploid nuclei, usually with a covering of cytoplasm, extruded from primary and secondary OOCYTES during MEIOSIS in animals. The first polar body may undergo the second meiotic division, producing three polar bodies in all. They usually play no further genetic role, and get lost or broken down in early development.

polarity (1) Having a difference (morphological, physiological, or both) between the two ends of an AXIS, e.g. shoot/root, head/tail. Characteristic of most organisms, unicellular and multicellular, and of many cells within organisms. (2) See CELL POLARITY.

polar nucleus One of two nuclei that migrate to centre of megagametophyte of a flowering plant, fusing to form the central fusion nucleus prior to DOUBLE FERTILIZATION.

polar plasm (pole plasm) Cytoplasm at the posterior pole of the eggs of insects (e.g. *Drosophila*), the vegetable pole of eggs of nematodes (e.g. *Parascaris*), and fertilized frog eggs, which contains regionally specific molecules (proteins, mRNAs) usually as a result of localization by directional transport along MICROTUBULES; through specific attachment sites on mRNAs and/or by pole-specifying interactions of follicle cells with the oocyte. Involved in differentiation of GERM LINE cells from somatic cells. See MATERNAL EFFECT, POSITIONAL INFORMATION.

pollen analysis (palynology) Techniques for reconstructing past floras and climates using pollen grains (and other spores), especially those preserved in lake sediments and peat. Their resistance to decay and distinctive sculpturing enable quantitative and qualitative estimates of past species abundance, typically within a timescale of a few decades up to millennia.

pollen grain Microspores of seed plants, containing a mature or immature micro-gametophyte. Outer EXINE coat sculptured. See MICROSPORANGIUM.

pollen tube Tube formed on germination of a pollen grain, which transports the two 'male' nuclei to the egg. In flowering plants, each penetrates tissues of the stigma and style, entering the ovary and growing towards an ovule, where it passes through the micropyle, penetrating the nucellus and rupturing at the tip of the embryo sac to set free two nuclei. See AUXIN, CUTIN, DOUBLE FERTILIZATION, MICROGAMETOPHYTE.

pollex 'Thumb' of pentadactyl forelimb. Often shorter than other digits. Compare HALLUX.

pollination Transfer of pollen (by wind, water, insects, birds or other animals) from the anther where they were formed to a receptive stigma. Not to be confused with any later fertilization which may result. Forest-floor plants have been shown to be pollinated by different insects depending on the temperature of the microhabitat; insects with high thorax temperatures (e.g. bees) are more likely to pollinate in cool, shady parts than are those with lower ones (e.g. many flies).

pollinium Coherent mass of pollen grains, as in orchids; generally transferred as a unit in pollination.

pollution Anthropogenic and non-anthropogenic contamination of the environment by chemicals, and sometimes by heat, to the extent that existing habitats are threatened or populations of organisms are endangered. Unfortunately, criteria for pollution are often more subjective; e.g. some ecologically insensitive voices would suggest that if no one plans to swim close to a sewage outlet, we should question whether it is polluted. If we are also concerned for organisms other than ourselves, the criteria for pollution change. *Atmospheric pollution* of the troposphere includes reactive intermediates in the oxidation of volatile organic compounds (VOCs) and oxides of nitrogen (NO_x), the hydroxyl radical during the day, and the nitrate radical (NO_3) during the night. NO_x increases are associated with the increase in combustion of fossil fuels during the industrial period (see also SUPER-OXIDES). OZONE (O_3), a source of the hydroxyl radical, is a contributor day and night. OH, NO_3 and O_3 play a central role in the production and fate of airborne toxics, mutagenic aromatic hydrocarbons (e.g. malaoxon from atmospheric oxidation of the widely used organophosphate insecticide, malathion) and fine particles. These last have been shown to contain toxics which cause frameshift mutations in bacterial assays and human cell MUTAGENS, notably nitro-derivatives of polycyclic aromatic hydrocarbons (e.g. 1-nitropyrene, dinitropyrenes). The chemistry of SO_2 and acid deposition (see ACID RAIN) is closely linked with the abundance of the reactive species, OH (see also CFCS, PCBS). In addition to their effects on GLOBAL WARMING, aerosols can act as cloud condensation nuclei, and when more droplets form within a given cloud each droplet is smaller, preventing droplets from reaching the size needed for precipitation as rain (see GREENHOUSE EFFECT).

Over 1,500 substances have been implicated in *freshwater pollution*, and a classification of freshwater pollutants might include: pathogenic organisms; biodegradable organic waste (see BIOCHEMICAL OXYGEN DEMAND); water-soluble inorganic ions; plant nutrients (artificial fertilizers, organic breakdown products; see EUTROPHICATION); insoluble and soluble organics (see BIO-DEGRADABLE); suspended solids; radioactive substances (e.g. naturally, during rock erosion; or resulting/leaking from nuclear tests, nuclear power stations and uranium mines); and thermal pollution (e.g. near cooling towers of power stations), which reduces dissolved oxygen levels.

Marine pollution is typically associated with oil spillage, but terrestrial organic PESTI-CIDES also undergo BIOMAGNIFICATION there. Suspended solids (e.g. from sewage outfalls) do reduce oxygen availability for, and fertility of, crustaceans and fish. See CARBON CYCLE, CARCINOGEN, LICHENS, NITROGEN CYCLE.

polyadelphous (Of stamens) united by their filaments into several groups, as in St John's wort. Contrast MONADELPHOUS, DIADELPHOUS.

polyadenylation See MESSENGER RNA.

polycarpic (iteroparous) (Of plants) reproducing more than once. Contrast SEMELPAROUS. Compare ANNUAL, PERENNIAL.

Polychaeta Class of ANNELIDA, including ragworms, tubeworms, fanworms and lugworms. Marine; free-swimming, errant, burrowing or tube-dwelling. Clearly segmented; typically a pair of PARAPODIA per segment. Usually a well-marked head, often with jaws and eyes. Large coelom, usually divided by septa. Separate sexes; fertilization external; larva a trochophore. Compare OLIGOCHAETA, HIRUDINEA.

polycistronic Nucleic acid sequences which encode two or more different polypeptides. Common in bacteria, but rare in eukaryotes. See the somewhat different concept of a POLYPROTEIN.

polyclave See IDENTIFICATION KEYS.

polyclone Well-defined anatomical region (e.g. insect wing COMPARTMENT) consisting of several entire clones of cells.

polyembryony (Bot.) The simultaneous occurrence of more than one embryo within a seed. Can occur when more than one egg cell is fertilized within a single ovule, as in most gymnosperms, where female gametophytes produce several archegonia each (generally only one embryo survives). Alternatively, as in orange seeds, one embryo is formed sexually and others by vegetative budding of the nucellus (which acts like a multiple suspensor); or again, the pro-embryo may branch early in its growth. (Zool.) Asexual formation of more than one embryo from each zygote produced; sometimes arises by budding at some early (or larval) stage of development. Common in some parasites. See CERCARIA, HYDATID CYST, MONOZYGOTIC TWINS, PROGENESIS.

polygenes Genes, at more than one locus, variations of which in a particular population have a combined effect upon a particular phenotypic character (said to be determined *polygenically*, or to be a *polygenic* *character*). Allelic substitutions at such loci tend to have little individual effect upon phenotypic differences between individuals, the normal situation in much of developmental biology, where expression of several different genes is required for production of a character. Such *polygenic inheritance* increases the likelihood that a character will exhibit continuous VARIATION in the population. Human height is an example. Compare PLEIOTROPY, EPISTASIS.

polygenic inheritance See POLYGENES.

polymer A very large molecule comprising a chain of many similar or identical molecular subunits (monomers) joined together (polymerized). In biochemistry, this invariably involves condensation reactions. Proteins, polysaccharides and nucleic acids are characteristic biopolymers. Fatty acids are polymerized in CUTIN.

polymerase Enzyme joining monomers together to form a polymer; e.g. DNA POLYMERASE, RNA POLYMERASE, starch synthase.

polymerase chain reaction (PCR) Very adaptable process for isolating and amplifying a specifically desired DNA sequence (see Fig. 136). Artificially mutated DNA can be amplified this way, as can samples of forensic interest and amber-trapped DNA (see FOSSIL). Double-stranded DNA is first separated into its two complementary strands by heating to 72°C, while two short single-stranded DNA primers known to anneal upwards of the 3'-end of the DNA sequence to be amplified are added to either strand after the reaction mixture has cooled. A key to the process is that DNA POLYMERASE requires such a priming sequence to extend the whole complementary strand in a 5'-wards direction, and *Taq* polymerase from the extreme thermophilic bacterium *Thermus aquaticus* is added (it can withstand the high temperatures involved in the later steps and consequently allows automation of the process). *Taq* extends the complementary

strands beyond the position of the primer-binding site on the other template while the reaction is cool, and reheating separates the new DNA duplexes. On cooling, the primers are added again (four binding-sites are now available); *Taq* is added again and extension proceeds but is restricted to the desired target sequence on account of the limiting primers on the template strands. The whole process is repeated for many generations, producing over 10^6 double-stranded target molecules after 32 generations of the cycle. If the process is started with a large number of template molecules then fewer cycles are required (with less mutation risk) to produce sufficient amplified DNA for VECTOR cloning.

By sampling whole environments, notably soil, for their DNA we now realize that such 'pooled DNA samples', when analysed by PCR, can reveal the existence of unnoticed taxa. In the case of bacteria, so few bacterial species have been cultured that such culture-independent PCR (ciPCR) methods have more than doubled the estimated number of prokaryotic phyla and have revealed the existence of many bacteria-sized eukaryotes (nano- and pico-eukaryotes). These methods bypass the need to identify, much less isolate, the new organisms.

When we wish to discover the sequences of DNA flanking a known sequence (for which no primers are available), we use the same primers as used to determine the known sequence, but in such a way that they face in opposite directions. This is achieved by using a ligase to circularize dsDNA which includes the known sequence and its flanking sequences, adding the primers, and then carrying out normal PCR.

Traditional PCR methods use 'thermal cyclers', which have high thermal masses so that one cycle might take in excess of 90 minutes to complete. New microfabricated systems rely on reduction in thermal mass

FIG. 136 POLYMERASE CHAIN REACTION *procedure*. (From *Recombinant DNA (2nd edn)* by Watson, Gilman, Witkowski, and Zoller. Copyright © James D. Watson, Michael Gilman, Jan Witkowski and Mark Zoller, 1992. Reprinted with permission of W. H. Freeman and Company.)

Amplification of target sequence

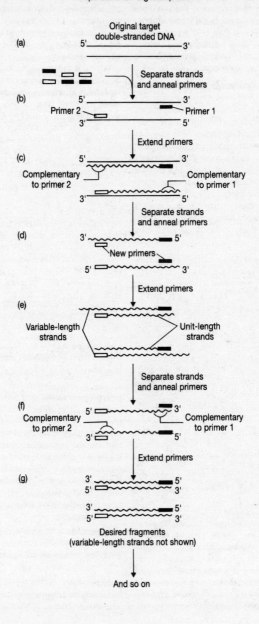

(a) Original target double-stranded DNA

5' ——————————— 3'

Separate strands and anneal primers

(b) 5' ——————————— 3'

Primer 2

3' ——————————— 5' Primer 1

Extend primers

(c) 5' ——————————— 3'

Complementary to primer 2

Complementary to primer 1

3' ——————————— 5'

Separate strands and anneal primers

(d) 3' ~~~~~~~~~~ 5'

New primers

5' ——————————— 3'

Extend primers

(e) Variable-length strands

Unit-length strands

Separate strands and anneal primers

(f) 5' ——————————— 3'

Complementary to primer 2

Complementary to primer 1

3' ——————————— 5'

Extend primers

(g) 3' ~~~~~~~~~~ 5'
5' ——————————— 3'

3' ~~~~~~~~~~ 5'
5' ——————————— 3'

Desired fragments (variable-length strands not shown)

And so on

to facilitate rapid cooling and heating, and continuous flow of very small fluid volumes through different zones held at specific temperatures. PCR can thus be integrated with reverse transcription within a single microchip, 20 cycles of DNA amplification lasting only 5 minutes. See RT-PCR.

polymorph (polymorphonuclear granulocyte) Any of several types of LEUCOCYTE with granular cytoplasms and lobed nuclei, produced in bone marrow; forming 60–70% of human blood leucocytes. Also found in tissue fluid and lymph (they can adhere to and penetrate capillaries). Predominantly phagocytic (esp. neutrophils), but include EOSINOPHILS, BASOPHILS and MAST CELLS. PLATE-LETS also have granular cytoplasms, but are not polymorphs. Neutrophils degranulate into their PHAGOSOMES; eosinophils, basophils and mast cells degranulate by exocytosis.

polymorphism Major category of discontinuous VARIATION within species. (1) *Genetic polymorphism*. Simultaneous occurrence in a species population of two or more discontinuous forms in such a ratio that the rarest of them could not be maintained solely by recurrent mutation. This excludes reproductively non-overlapping seasonal forms of a species (e.g. some PLEIOMORPHISMS), rare mutants, geographical races and all continuous variation. Some polymorphisms are maintained by NEUTRALISM, although balancing selection increases the probability that the genes involved will persist over time, which also reduces the size of the population required for their persistence. These polymorphisms generally involve clear genetic alternatives between individuals in the population, as controlled by SUPERGENES, INVERSIONS or loci where allelic substitutions tend to bring about marked differences in phenotype (so-called 'major genes'). The term is also used of single nucleotide polymorphisms (see HUMAN GENOME PROJECT). Examples of polymorphism include many forms of Batesian MIMICRY, human BLOOD GROUPS, the MHC system, some forms of INDUSTRIAL MELANISM and HETEROSTYLY, and many forms of SEX DETERMINATION. Enzyme and DNA polymorphisms, however, are often quite cryptic. The term *balanced polymorphism* describes polymorphism maintained by a balance of selective agencies whose effects tend to cancel each other out (e.g. see FREQUENCY-DEPENDENT SELECTION).

Where a mutant gene increases in frequency in a population it may eventually replace its alternative allele(s). Such a situation is termed *transient polymorphism*, but may never go to complete fixation of the new mutant. The spread of melanic forms is an example. Relative fitnesses of different *morphs* can sometimes be calculated using the HARDY–WEINBERG THEOREM. See HETERO-ZYGOUS ADVANTAGE, NEUTRALISM.

(2) *Non-genetic polymorphism*, where distinct forms of a species occur but without the definitional rigour of the cases above. Environmental influences play a large part in insect CASTE polymorphisms. Genetic factors (e.g. haplo-diploidy) may be important, but nutritional influences (e.g. in hymenopterans) and pheromones (e.g. in termites) are instrumental in developmental decisions. The term polymorphic is also applied to the different ZOOIDS in siphonophore and ectoproct colonies.

polynucleotide Long-chain molecule formed from a large number of nucleotides (e.g. NUCLEIC ACID).

polyp (hydranth) Sedentary form of the CNIDARIA. Cylindrical stalk-like body attached to substratum at one end, mouth surrounded by tentacles at the other. Many bud asexually; some are sexual. Polyp-like stages in the scyphozoan life cycle (*scyphistomas*) produce ephyra larvae (which develop into sexual medusae) by strobilation. Most hydrozoans and anthozoans have (or are) polyps. See POLYPIDE.

polypeptide Molecule consisting of a single chain of many amino acid residues linked by PEPTIDE BONDS. See PROTEIN.

polypetalous (Of flowers) having petals free from one another; e.g. buttercup. Compare GAMOPETALOUS.

polyphosphate granules Phosphate storage granules found in cells of blue-green algae (CYANOBACTERIA).

polyphyletic Term for a group of taxa (species, genera, etc.) when, despite their being classified together as one taxonomic category, it is thought that not all have descended from a common ancestor which was also a member of the group. Such a taxon forms a GRADE rather than a CLADE. If the classification is to correspond with phylogeny, the group should be split into two or more distinct taxa. See CLADISTICS, HOMOLOGY, MONOPHYLETIC.

polyphyodont See DENTITION.

polypide Polyp-like individuals, forming a COLONY of ectoprocts or entoprocts.

polyploidy Condition in which a nucleus (cell, individual, etc.) has three or more times the HAPLOID number of chromosome sets characteristic of its species (or ancestral species). The haploid number is usually represented by the letter n, so a *triploid* is commonly represented as $3n$, a *tetraploid* as $4n$, a *pentaploid* as $5n$, and so on.

Tetraploidy can arise in otherwise normal diploid somatic tissue, both plant and animal, through failure of replicated chromosomes to separate in MITOSIS. This can be induced artificially by exposure to a compound such as colchicine, which inhibits formation of the spindle apparatus in mitosis through its disruptive effect upon microtubule formation; or by heat treatment. Such AUTOPOLYPLOID cells are much more likely to become part of the germ line in plants than in animals; but there may be some reduction in fertility due to unequal segregation of chromosomes at meiosis. However, some autotetraploid plants (e.g. cocksfoot grass, *Dactylis glomerata*; purple loosestrife, *Lythrum salicaria*) are fully fertile. ALLOPOLYPLOID organisms are more likely to be fully fertile since each chromosome has only one fully homologous partner to pair with. Crossing and backcrossing between these and normal (usually diploid) individuals often produces a *polyploid complex*, resulting in gene INTROGRESSION. Most

instances of plant APOMIXIS involve polyploid species, giving them an 'escape from sterility'. Polyploidy is comparatively rare in animals, probably because most animals are bisexual and obligate cross-fertilizers. A tetraploid ($4n$) animal could only produce sterile triploid ($3n$) offspring, if any at all. Only a few amphibians and fish among vertebrates seem to be polyploid, possibly because of their method of SEX DETERMINATION. See also HYBRIDIZATION.

In plants, the range of variation is often narrower in allotetraploids than in a related diploid because each gene is duplicated. Polyploids are often self-pollinated, even when related diploids are mainly cross-pollinated, which reinforces the decrease in variability.

Polyploids are of interest because of the examples they may provide of 'instantaneous' SPECIATION, and have been extremely important in the evolution of certain groups of plants. One of the most important polyploid (hexaploid) species is wheat, e.g. bread wheat, *Triticum aestivum*, which has 42 chromosomes. It was derived some 8,000 years ago following the spontaneous hybridization of a cultivated wheat possessing 28 chromosomes with a grass of the same group with 14 chromosomes, followed by chromosome doubling (see Fig. 137).

Among plant polyploids are many of our most important crops including, besides wheat, bananas, cotton, potatoes, sugar cane and tobacco. Many garden flowers are polyploids, e.g. chrysanthemums, day lilies. See ENDOMITOSIS, POLYTENY.

polyprotein Large protein from which different functional peptides may be cleaved enzymatically. Neuropeptides are cleavage products from a polyprotein synthesized in the neuron cell body. See POST-TRANSLATIONAL MODIFICATION.

polysaccharide (glycan) Carbohydrate produced by condensation of many monosaccharide subunits to form polymers, forming long, often fibrous, molecules which are poorly soluble, osmotically inactive, and lacking a sweet taste.

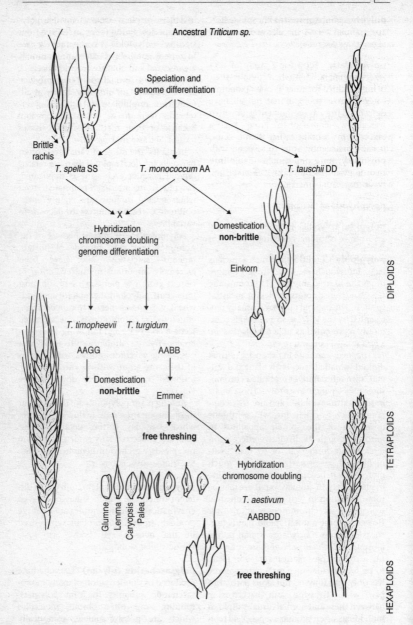

Ancestral *Triticum sp.*

Speciation and
genome differentiation

Brittle
rachis

T. spelta SS   *T. monococcum* AA   *T. tauschii* DD

X

Hybridization
chromosome doubling
genome differentiation

Domestication
non-brittle

Einkorn

DIPLOIDS

T. timopheevii   *T. turgidum*

AAGG   AABB

↓   ↓

Domestication
non-brittle

Emmer

free threshing

Glumne
Lemma
Caryopsis
Palea

X

Hybridization
chromosome doubling

T. aestivum

AABBDD

↓

free threshing

TETRAPLOIDS

HEXAPLOIDS

Extremely important structurally (e.g. CELLULOSE, CHITIN and GLYCOSAMINOGLYCANS) and as energy reserves (e.g. STARCH, GLYCOGEN and INSULIN).

polysepalous (Of a flower) having sepals free from one another; e.g. buttercup. Compare GAMOSEPALOUS.

polysome (polyribosome) See RIBOSOME.

polysomy Condition in which one chromosome (rarely more) in a nucleus is represented more frequently than the remaining chromosomes. In a *trisomy* (see DOWN'S SYNDROME) one chromosome is present three times in the nucleus as opposed to twice for other chromosomes. The nucleus would have a ploidy of $2n + 1$ (where n = haploid number). Polysomy can occur in nuclei which are otherwise of any ploidy. See NON-DISJUNCTION, POLYPLOID.

polyspermy (Zool.) Entry of more than one sperm nucleus into an ovum at fertilization. Occurs normally only in very yolky eggs (e.g. shark, bird) and commonly in insects; only one nucleus fuses with egg nucleus, the rest taking no part in development. May upset normal development in other eggs. See FERTILIZATION, PSEUDOGAMY.

polystelic (Bot.) Possessing more than one STELE, comprising more than one (sometimes many) vascular bundles.

polyteny Condition of some chromosomes, nuclei, cells, etc., in which many identical parallel copies of each chromosome are formed by repeated replication without separation, and lie side-by-side forming thick cable-like chromosomes. Best studied in salivary gland cells of the fruit-fly *Drosophila*, these and LAMPBRUSH CHROMOSOMES are often termed *giant chromosomes*, occurring only in cells which have lost the power to divide mitotically (permanent interphase). They tend to be longer than normal on account of being less coiled.

Close pairing of sister strands may produce transverse 'banding', which enables both the detection of chromosome inversions and study of chromosome 'puffing' (see ECDYSONE). May occur in secretory tissues where increase in cell size rather than number has occurred. Compare ENDOMITOSIS, POLYPLOIDY.

polythetic (Of classification) a system where membership of a taxon depends upon possession of a large number of characters in common.

polytopic Designating taxa occurring in two or more separate areas.

polytypic Designating species which occur in a variety of geographical forms, or subspecies. See INFRASPECIFIC VARIATION.

polyunsaturated fatty acid A FATTY ACID molecule with more than one double bond in its carbon chain.

Polyzoa (Bryozoa) See ECTOPROCTA, ENTOPROCTA.

pome 'False fruit', the greater part of which is developed from the receptacle and not the ovary; e.g. apple, pear. The edible fleshy part represents the receptacle, and the core represents the ovary.

Ponginae Subfamily of Hominidae (see HOMINID), including the Miocene fossil *Sivapithecus* (= *Ramapithecus*, 12.5–7 Myr BP) and the tribe Pongini (orang-utan). *Sivapithecus* is known from Indo-China to Turkey, seemingly from subtropical deposits, and was apparently an arboreal quadruped. Until fairly recently, it was thought that certain specimens (initially attributed to a distinct genus *Ramapithecus*, now placed in this subfamily) represented in early hominid line diverging from a common ape-human stock about 14–12 Myr BP. It is now thought that such a branching occurred far more recently (6–5 Myr BP?) and that the thick

FIG. 137 *The evolution of wheat, showing patterns of mutation, hybridization and chromosome doubling (adapted from Feldman, 1976). A brittle rachis makes harvesting difficult and non-free-threshing grain produces poor quality flour. Another important change has been in the number of grains in each spikelet. See* POLYPLOIDY.

dental enamel of *Ramapithecus* (shared with AUSTRALOPITHECINES) is better considered a primitive feature of hominoids and hence of no taxonomic value in discriminating putative hominids. A large assemblage of ramapithecine fossils discovered in China in the 1980s indicates that in its facial characters it resembled the modern orang-utan with, *inter alia*, narrow interorbital distance, broad high zygomatic region and large size discrepancy between upper incisors. There is no clear evidence that the group were evolving towards bipedalism; the hallux (big toe) was opposable and capable of powerful gripping. The pongine clade of hominoids seems to have branched off earlier than the separation of humans and anthropoid apes. The molecular evidence (DNA HYBRIDIZATION, protein sequencing and immunological distancing) from living forms favours this view (see HOMINID); but controversy still surrounds the affiliations of DRYOPITHECINES.

pons Floor of fourth ventricle of mammalian brain, lying in front of medulla oblongata and under cerebellum. Connects cerebrum to cerebellum and houses respiratory centres (see VENTILATION).

population Group of conspecific individuals, commonly forming a breeding unit, sharing a particular habitat at a given time. See DEME, INFRASPECIFIC VARIATION.

population ecology The study of the variations in time and space in the sizes and densities of populations and of the factors causing those variations.

population regulation A tendency in a population for some factor to cause density to increase when it is low and to decrease when it is high. See DENSITY-DEPENDENCE.

population vulnerability analysis An analysis, generally applied to populations or species in danger of extinction, of the chances of extinction.

pore complex See NUCLEUS.

Porifera Sole phylum of Subkingdom PARA-ZOA. Sponges. Have an apparently loose diploblastic organization lacking both nerve and clear muscle cells, and comprising barely discrete tissues, there being no basement membranes. Basically vase-shaped, the outer 'epithelium' of flattened contractile cells (*pinacocytes*) with amoeboid locomotion; inner layer of cells consists of CHOANOCYTES. Internal skeleton may be of either calcareous (Class Calcarea) or siliceous (Class Demospongia) spicules, and frequently contains fibres of the protein spongin. Pinacocytes may develop into *porocytes* whose intracellular lumens form the surface pores (ostia) through which water enters propelled by flagella of the food-collecting choanocytes, and after entering the central cavity (spongocoel) leaves through one or more oscula at the top. Complexity of internal chambering increases from *ascon*, via *syconoid* and *sicon*, to *leucon* levels of organization. A gelatinous MESOGLOEA between the two main layers contains amoeboid cells (see AMOEBOCYTES) whose varied functions include spicule secretion, gamete production and transport of sperm or food. Locomotion confined to flagellated larvae. See MULTICELLULARITY.

porphyrin Prosthetic group of several conjugated protein pigments; nitrogen atoms of tetrapyrrole nucleus often coordinated to metal ions (magnesium in chlorophyll, iron in haemoglobin and cytochromes).

portal vein Blood vessel connecting two capillary beds, as in *pituitary portal system* and *hepatic portal system*.

positional cloning (chromosome landing) A map-based cloning technique in which genes responsible for alternative character traits, determined either monogenically or polygenically, can be mapped to positions adjacent to DNA segments (e.g. RFLP or MICROSATELLITE markers) that have already been cloned. Such a marker is then used to screen the library and pick out the clone containing the gene, obviating the need for CHROMOSOME WALKING. Of great value in plant work, where there is often a huge genome to map, and in the HUMAN GENOME PROJECT.

positional information Embryological theory (perhaps currently out of favour) indicating those signals which enable cells in a developing metazoan to respond as though they appreciated their spatial positions in the embryo. Would normally involve specification of information as to where each cell lay on both antero-posterior and dorso-ventral axes. Such information would give cells their *positional values* and ultimately determine their fates. *Positional signals* are graded and can produce multiple outcomes, unlike *inductive signals*, which produce all-or-none responses and result in a single cell state. It is likely cells retain their positional values even when differentiated, making even cells of the same type (e.g. osteocytes) non-equivalent if they lie in different body regions. A cell might pick up and 'remember' different cues at different times prior to differentiation, such as chemical gradients (see MORPHOGEN) or those arising from key cell contacts. A cell's position on the various axes of the body might be given in sequence using a form of digital framework, evidence for which is emerging from, among others, studies of the genetics of *Drosophila* and mouse development (see HOMEOGENE, ZOOTYPE). When the vertebrate body plan has been laid out in gastrulation, and the *HOX* GENES have been regionally expressed, further migrations occur of neural crest, sclerotome, myotome and angioblast cells which in turn bring new inductions and organogenesis. Compare PREPATTERN. See AXIS, *BICOID* GENE, GAP GENES, PAIR-RULE GENES, PATTERN, SEGMENT-POLARITY GENES.

position effect Occurrence of phenotypic change resulting not from gene mutation as such but from a change in position of a piece of genetic material. Of two kinds. Stable position effects result from insertion of genes in the germ-line (e.g. inversion, translocation and crossing-over). *Position effect variegation* (PEV) produces mosaic phenotypes and can occur when a euchromatic gene finds itself close to a piece of HETEROCHROMATIN (see CHROMOSOMAL IMPRINTING) or distal euchromatin. It involves the transcriptional 'silencing' ('shutting-off') of normal eukaryotic genes in the vicinity of heterochromatic regions

of the chromosome and is mediated by the action of a complex of heterochromatin proteins, including HP1, which repackage the chromatin into heterochromatic domains. A few genes (heterochromatic genes) actually depend upon a heterochromatic environment for their proper expression. Other loci often act as modifiers of PEV, suppressing when in single dose and enhancing when in double dose. ONCOGENES and TRANSPOSONS can have PEV-like effects, in the latter case especially when they insert in telomeric regions.

position effect variegation See POSITION EFFECT.

positive feedback Pathways ensuring that the occurrence of an event makes its reoccurrence more likely. Although such feedback loops can generate instability, and are therefore short-lived, they are crucial to several biological processes, such as the self-recruitment of clotting factor enzymes during BLOOD CLOTTING, depolarization of a membrane during generation of an IMPULSE, the surge in oestrogen just prior to ovulation in the MENSTRUAL CYCLE, and several signalling pathways involved in pattern formation and some forms of CELL MEMORY.

posterior (Bot.) Of lateral flowers, part nearest main axis. See FLORAL DIAGRAM. (Zool.) Situated away from head region of an animal, or away from region foremost in locomotion (e.g. dorsal surface in bipedal animals such as humans).

postorbital bar Bony strut formed, in primates at least, by the union of the ascending process of the jugal with a descending process of the frontal bone, providing support for the outer margin of the eyeball and ensuring the eye is surrounded by a complete bony ring, which may be an adaptation to arboreal life, although it is widespread in ungulates; possibly also serves for transmission of forces during mastication of food. Presence in tree shrews (Scandentia; see ARCHONTA) does not in itself indicate a close phylogenetic relationship of these mammals with primates. Absent in fossil *Hyracotherium* but present in *Equus*, present in several mammalian orders, convergent evolution being probable in these diverse groups. Lacking in 'archaic primates', the plesiadapids. In the tarsiers and higher primates (anthropoids) a postorbital septum closes off the back of the orbit, separating it from the temporal fossa and is composed of the same bones in the two groups.

post-transcriptional gene silencing (PTGS)
A GENE SILENCING effect in which some viruses, transgenes or RNAs trigger the post-translational degradation of homologous cellular RNAs. Strongly implicated as an antiviral mechanism in plants, in which the effect spreads intercellularly. A similar PTGS process and intercellular spreading has been noted in the nematode *Caenorhabditis elegans* after extracellular delivery of dsRNA, and loss of function of genes required for RNA-interference results in activation of multiple transposable elements in the germ line.

post-transcriptional modification
Changes occurring to some tRNA and rRNA transcripts prior to translation. May involve removal of nucleotides by hydrolysis of phosphodiester bonds or chemical modification of specific nucleotides, as in methylation of thymidine or pseudouridylation of uridine in rRNA transcripts. Compare POLYPROTEIN. See RNA EDITING, RNA PROCESSING.

post-transcriptional RNA silencing
A broad group of phenomena in eukaryotes, including RNA SILENCING and POST-TRANSCRIPTIONAL GENE SILENCING in which transcription of the silenced gene is unimpaired, yet the mRNA transcript fails to reach the normal cytoplasmic concentration. A PPD (PAZ and Piwi domain) protein is required for all post-transcriptional RNA silencing in every organism where the function of this family has been tested.

post-translational modification
Changes occurring to protein structure after it has been produced on the ribosome. Post-translational modification of a protein is often responsible for their existing in the cell in multiple 'isoforms', which increases the complexity of the proteome without increasing actual genome size. See also

NITRIC OXIDE for S-nitrosylation. These may include (a) selective cleavage of the protein, sometimes removing inhibitory sequences (see POLYPROTEIN, INSULIN) or sorting signals (see PROTEIN TARGETING); (b) selective inactivation, possibly by ligand-binding; (c) covalent addition of further groups (e.g. glycosylation, addition of methyl, nucleotidyl, fatty acyl and myristoyl groups; see GOLGI APPARATUS); (d) involvement with other polypeptides to produce functional quaternary structure (e.g. HAEMOGLOBIN); (e) phosphorylation or dephosphorylation of selected tyrosine, serine or threonine residues by kinases and phosphorylases respectively, critical to progress through the CELL CYCLE (e.g. see MATURATION PROMOTING FACTOR) and to the events of SIGNAL TRANSDUCTION; (f) see DEAMIDATION.

potentiation (1) Process whereby a given cellular response is achieved at a lower stimulus threshold. At SYNAPSES, a short burst of action potentials arriving at the presynaptic membrane can cause a greatly enhanced response to action potentials in the postsynaptic cell (lasting up to weeks in long-term potentiation), while earlier these had no effect. (2) See SYNERGISM.

potocytosis Process enabling eukaryotic cells to take up and concentrate small molecules and ions prior to entry to the cytosol at special plasma membrane sites termed caveolae. These invaginated ~50 nm vesicles on the membrane's surface bear anchored proteins which form clusters and potocytosis begins when the caveolae pinch off and the anchored proteins internalize, the ions/molecules taken up being either (a) bound to a specific receptor within the caveola; (b) enzymatically converted within the caveola; (c) bound to carrier molecules ferrying them between cells and the caveola. Inside the small caveolae, they achieve a high concentration driving their diffusion out of the caveola into the cytosol.

power laws Two variables, X and Y, are associated by a power law relation whenever it is true that $Y = aX^b$, or by its logarithmic equivalent $\log Y = \log a + b \log X$, where the parameter a is often known as the scaling coefficient and *b* as the scaling exponent. See ALLOMETRY.

PPLO See MYCOPLASMAS.

P-protein (phloem-protein) Proteinaceous substance found in cells of flowering plant phloem, particularly in sieve-tube members, where it may block or hinder TRANSLOCATION through sieve pores.

Prasinophyceae A class of the algal division CHLOROPHYTA containing 16 genera and about 180 species of free-living marine, brackish water and freshwater, green flagellates. Minute forms (e.g. *Bathycoccus* species), 0.5 to a few micrometres in diameter, are an important component of the picoplankton of the oceans. These organisms, together with tiny coccoid cyanophytes, prochlorophytes, heterokontophytes and haptophytes, probably contribute greatly to oceanic primary production. This is an ancient class of green algae with fossils of cysts similar to those formed by the extant genus *Pterosperma* being found in the PRECAMBRIAN (*c.* 1200 Myr BP).

Prasinophytes are also known as photosynthetic endosymbionts of various heterotrophic organisms (e.g. *Tetraselmis convolutae* is an endosymbiont of the intertidal turbellarian worm *Convoluta roscoffensis*). Other tiny prasinophytes have been found in various radiolarians as photosynthetic endosymbionts. One species lives as an endosymbiont within the large vacuole of the phagotrophic dinoflagellate *Noctiluca*.

Cells and flagella are covered by one or more layers of more or less elaborate organic scales. These are produced in the Golgi apparatus. One to eight, apically or laterally inserted flagella can be present. Flagella usually arise from the base of a depression. Flagellar root systems vary from cruciate to unilateral. In this prasinophytes differ from the other classes of the Chlorophyta, each of which is characterized by a particular type of root architecture. Distinctive rhizoplasts are usually present and the basal bodies are unusually long. Mitosis is open or closed and cytokinesis is brought about by a cleavage furrow or by a cell plate of vesicles within a phycoplast.

Asexual reproduction is common; however, a freshwater species of the genus *Nephroselmis* has a haplontic life cycle. This alga is heterothallic and isogamous. The zygote develops into a hypnozygote (i.e. a thick-walled resting stage) which, after a period of dormancy, undergoes meiosis and produces four meiospores. Each gives rise to cells of the new haploid generation.

preadaptation Term serving to explain how genes, structures, etc. come to perform new roles, for which they were not originally selected. Thus, if it is the case that bird feathers evolved from scale-like structures originally selected for their effect in thermoregulation, their subsequent selective advantage in terms of flight would be a case of 'preadaptation to flight'. In genetics especially, the term 'CO-OPTION' is now preferred, since it lacks the semantic overtones of 'foresight'.

Precambrian See ARCHAEAN, PROTEROZOIC.

precapillary sphincter Sphincter muscle around metarteriole where it gives rise to capillary bed. See CAPILLARY.

precipitin Antibody combining with, and causing precipitation of, a soluble antigen. Such precipitations form the basis of *in vitro* procedures for separating and identifying antigens, often on gels (*precipitin reaction*).

precocial Those young of mammals and birds which hatch or are born in a partially independent condition. Compare ALTRICIAL, NIDICOLOUS, NIDIFUGOUS.

preformation Doctrine, generally accepted in Europe in 17th and 18th centuries, that all parts of the adult are already perfectly formed at the beginning of development. 'Ovists' believed these were present only in the ovum; 'spermatists' believed they were only in the sperm. Compare EPIGENESIS.

premaxilla Dermal bone forming front part of upper jaw in most vertebrates, bearing teeth (incisors in mammals). Forms most of upper beak in birds.

premolar One of the crushing cheek teeth of mammals, anterior to the molars and posterior to the canines (or incisors). Unlike molars, has a predecessor in the milk teeth; usually has more than one root and a pattern of ridges on biting surface. See DENTAL FORMULA.

pre-mRNA, pre-rRNA, pre-tRNA See RNA PROCESSING, RIBOSOMES.

prepattern A theoretical model of animal DEVELOPMENT in which a prior spatial deployment of molecular gene regulators is 'read' by the cells of an embryo (rather than decoded, as with gradient models), the patterns of deployment dictating the gene activities and thereby the paths of differentiation of the cells. In the simplest form of the model, the 'plan' of the embryo would be laid out at the time of fertilization. More reductionist than the POSITIONAL INFORMATION model, the chief difficulties of the model arise in explaining regulative development (see REGULATION) and increases in spatial complexities of embryos. It also fails to account satisfactorily for the origin of prepattern itself. See EPIGENESIS.

preprotein Proteins which, normally destined for secretion or membrane-integration, are maintained in an unfolded conformation (see CHAPERONES) and carry amino-terminal signal sequences which are subsequently removed to give the mature protein (see PROTEIN TARGETING).

pressure potential See WATER POTENTIAL.

presumptive Term indicating that a particular cell or cell group in the early animal embryo will normally differentiate in a particular way. Thus, 'presumptive epidermis' indicates that a group of cells so described will normally become epidermal cells, even prior to their becoming DETERMINED.

presynaptic inhibition Suppression of release of an excitatory NEUROTRANSMITTER occurring where the terminal of one neuron makes synaptic contact with the presynaptic ending of a second neuron which itself excites a third. The first neuron controls (modulates) the secretion of excitatory transmitter by the second, but through neurotransmitters rather than neuromodulators. It is often the control of calcium ion (Ca^{2+}) influx into the second neuron which reduces the quantity of transmitter it releases. *Presynaptic facilitation* increases the amount of transmitter released by the presynaptic cell, usually by prolonging the Ca^{2+} current into it; but the same neural circuitry is involved. See SEROTONIN, SYNAPTIC STRENGTH.

Priapulida Phylum of burrowing marine worms of dubious affinity. Possibly pseudocoelomate. A distended proboscis, everted with great force, is used to anchor the worms while the body is drawn up over it.

Pribnow box Nucleotide sequence in prokaryote PROMOTER regions, located about six bases upstream of a transcribed region. Contains the sequence TATAATG. Also called a (-10) region, because of the invariant T residue at base 10 upstream from the start of the transcribed region. However, many promoters whose expression is controlled by SIGMA FACTORS other than the normal vegetative one have different consensus sequences in different positions upstream of the actual transcription start point. See TATA BOX.

prickles Short pointed epidermal outgrowths from a plant surface, as of roses and brambles. Compare SPINES.

primary endosperm nucleus Nucleus resulting from fusion of one generative nucleus with the two polar nuclei in a flowering plant embryo sac. Usually triploid, its mitotic products produce the endosperm. See DOUBLE FERTILIZATION.

primary immune response See B CELL.

primary pit field See PITS.

primary production The total organic material synthesized in a given time by autotrophs of an ecosystem. See PRODUCTIVITY.

primary structure See PROTEIN.

Primates Members of the mammalian Order Primates. Placentals, including today lemurs, lorises and tarsiers (Prosimii), and the ANTHROPOIDEA (simians). There is an ongoing reinterpretation of primate origins based on new fossils and on calculations on the effectiveness with which fossils sample past primate species. Primate origins may be c. 85 Myr BP (Middle Cretaceous) instead of the more widely held c. 65 Myr BP (Lower Palaeocene). Unusually complete recent fossil finds from the Late Palaeocene/Early Eocene of Wyoming indicate that the archaic primates, or plesiadapiforms, were already quite diversified both in size (20 g to 4 kg) and in range of mobility. The superordinal relationships of primates remain unclear (SEE ARCHONTA). Clavicle retained (lost in many mammalian orders), with shoulder joint permitting freedom of movement in all directions, and an elbow joint allowing rotation of forearm; five digits on all limbs, with generally enhanced mobility and usually opposability of thumb (pollex) and big toe (hallux); claws modified into flattened nails; reduced snout; binocular vision, great alertness, acuity and colour vision; brain enlarged relative to body, with expansion of cerebral cortex; two mammary glands only, and young usually born singly. Most of these features are adaptations to arboreal living (most primates are treedwellers). There have been independent moves towards terrestrialization, notably in the HOMINOID line. Among possible fossil antecedents of extant primates, plesiadapids now seem closer to colugos (Dermoptera); adapids have traditionally been linked to lemurs and lorises but some authors now suggest links with simians; omomyid relationships are highly controversial but have traditionally been linked to tarsiers – but some place tarsiers as the sister group of simians. See Fig. 138. See AUSTRALOPITHECINE, HOMINID, *HOMO*, PONGID, POSTORBITAL BAR.

primitive Referring to a characteristic present in an early ancestral organism and as a relatively unmodified HOMOLOGY in living forms. A term indicating that a structure, taxon, etc. is regarded as similar in appearance to a stem line. Such a line may have given rise through adaptive radiation to diverse and often more specialized modifications of form. Has temporal rather than ecological implications. For some difficulties, see FISH.

primitive groove See PRIMITIVE STREAK.

primitive streak Longitudinal thickening at gastrulation in early (disc-like) reptilian, avian and mammalian embryos caused by cell migration from posterior epiblast towards the centre. The streak narrows, thickens and extends anteriorly, marking the antero-posterior embryo axis. As cells converge on the streak, a depression (the *primitive groove*) forms within it and serves as a BLASTOPORE through which cells migrate into the blastocoele to form endodermal and mesodermal tissue. The anterior funnel-like tip of the streak (Hensen's node) serves as the amphibian dorsal blastopore lip, cells passing through it anteriorly to form head mesoderm and notochord. During late gastrulation, the streak regresses posteriorly as notochord and mesoderm are laid down progressively from anterior to posterior beneath the presumptive ectoderm, inducing NEURAL PLATE formation and onset of NEURULATION. Posteriorly, Hensen's node forms the anal region (see DEUTEROSTOMIA).

primordial meristem See PROMERISTEM.

primordium (Bot.) An immature cell or group of cells that will eventually give rise to a specialized structure. Thus, a leaf primordium will become a leaf.

prion (proteinaceous infectious particle) Malformed version of a normal cellular protein which apparently 'replicates' by recruiting normal proteins to adopt its form, being capable of infecting other cells of the same, or a different, organism. Normal prion protein, PrP^C, contains lengthy coils of α-helices whereas the pathological protein, PrP^{Sc}, or PrP^{pres}, has the same amino acid sequence but folded into several β-pleated sheets, which aggregate to form PLAQUES (see PROTEIN, Fig. 139). Diseases such as scrapie in sheep, Creutzfeld–Jacob disease (CJD) in

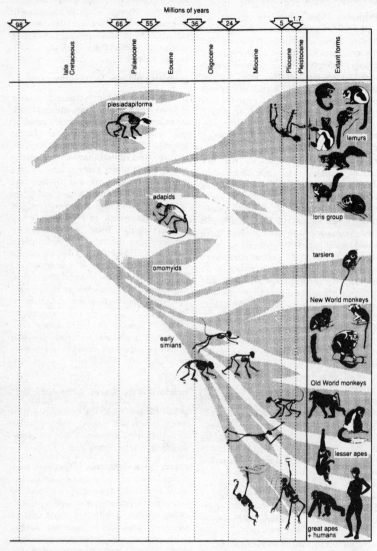

FIG. 138a *The main groups of living* PRIMATES *and their possible phylogenies. The fossil forms included are those with relatively complete skeletons, indicating the gaps which still exist in the fossil record. Illustration by Lucrezia Beerli–Bieler, reproduced with the kind permission of* Nature *and Prof. R. D. Martin.*

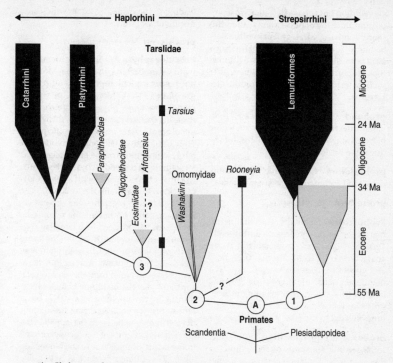

← ——————— Haplorhini ——————— → ← — Strepsirrhini — →

FIG. 138b *Cladogram of putative phylogenetic relationships of the more familiar taxa of Eocene and Oligocene* PRIMATES, *including extant haplorhines and strepsirrhines. Temporal calibration suggests that the haplorhine–strepsirrhine cladogenesis was pre-Eocene.*
A = Order Primates; 1 = Strepsirrhini; 2 = Haplorhini; 3 = Anthropoidea.

humans and bovine spongiform encephalopathy (BSE, or 'mad cow disease') in cattle are caused by plaques formed by misfolded and protease-resistant PrP. See CHAPERONES.

prisere Primary sere. Complete natural succession of plants, from bare habitat to climax. Compare PLAGIOSERE.

prisoner's dilemma A game with two players, each with the option to co-operate (C) or be selfish and defect (D). Using Robert Axelrod's payoff values, if both players cooperate they each get R = 3 points, if both defect they each get P = 1 point, while a player defecting gets T = 5 points if the other cooperates, getting S = 0. Defecting is therefore always rewarding, especially if the players (animals) meet only once; but it is

not always as rewarding as mutual cooperation, especially if the players meet regularly. Assessing payoff in terms of biological fitness – crudely, numbers of offspring – defectors should outcompete cooperators. If the prisoner's dilemma is iterative, i.e. if the probability of another round of the game is constant, one successful strategy appears to be *tit-for-tat*, in which a player cooperates for its first move and in succeeding moves simply copies what the other player did for its last move. If this is an evolutionarily stable strategy, it is difficult to see how altruistic behaviour can evolve other than when the terms of HAMILTON'S RULE are satisfied. However, it has been suggested that if an animal adopts the strategy 'always cooperate in response to cooperation but forgive defection on a proportion (one third) of occasions', then this more forgiving strategy may ultimately be more rewarding to both participants (and be evolutionarily stable) since it allows for a proportion of inevitable stochasticities such as errors of signal recognition. See ALTRUISM.

proband Propositus. Index case. Individual affected by a genetic disorder, whose presentation of it enables a pedigree for the disease to be drawn up.

probasidium Cell in which karyogamy occurs in BASIDIOMYCOTA. See METABASIDIUM.

probe DNA See DNA PROBE.

Proboscidea Elephants. Placental mammal order, probably derived during early Tertiary from the CONDYLARTHRA. Today characterized by trunk (formed from nose and upper lip), large size (largest terrestrial mammals), massive legs, greatly lengthened incisors (tusks) and huge grinding molars, only two pairs of which are in use at one time. Related to hyraxes.

proboscis (pl. proboscides) (Zool.) Suctorial feeding apparatus of some dipteran flies and most Lepidoptera. In the former most of the mouthparts contribute to its formation, and the labella contains food channels (pseudotracheae). In the latter it is formed largely from maxillae (mandibles being absent). (Bot.) A snout-like projection at the anterior end of spermatozoids of *Fucus* (PHAEOPHYCEAE) and *Vaucheria* (XANTHOPHYCEAE), whose shape is maintained by MICROTUBULES.

procambium Tissue of narrow elongated cells in vascular plants, grouped into strands, differentiating just behind growing points of stems and roots and giving rise to vascular tissue. See APICAL MERISTEM.

procaryote See PROKARYOTE.

procaspases Inactive molecules whose cleavage or conformational change, often mutual, is initiated by adaptor proteins which bring them close enough together for their interactions to result in active CASPASES,

leading to the caspase cascades which are a feature of APOPTOSIS.

prochloron See PROCHLOROPHYTA.

Prochlorophyta Prokaryotic organisms (with Cyanobacteria, in Kingdom Monera), but possessing chlorophylls *a* and *b* (thus resembling the chloroplasts of higher plants), lacking phycobiliproteins, and releasing oxygen during photosynthesis, as in eukaryotic algae and autotrophic plants. The carotenoids differ from eukaryotic algae but resemble those of the blue-green algae. Structurally, a prochlorophyte cell is similar to a blue-green algal cell, with a peptidoglycan (murein) layer in the cell wall. The photosynthetic pigments are bound to thylakoids that lie free in the cytoplasm; thylakoids are grouped into lamellae containing two to several thylakoids, as found in chloroplasts of the Chlorophyta (green algae) and higher plants. Among the accessory pigments are β-carotene and several xanthophylls, the main one being zeaxanthin as in the CYANOBACTERIA. DNA has a diffuse distribution, occurring throughout the cell in regions between the lamellae that are rich in ribosomes; it is not concentrated in the centre of the cell as in the blue-green algae. The storage product is starch-like cyanophycin, a nitrogen reserve characteristic of the blue-green algae. It comprises a polymer of the amino acids arginine and asparagine. Carboxysomes (polyhedral bodies), containing RUBISCO, are also present. Among the prochlorophytes there are three known genera. *Prochloron* and *Prochlorococcus* are unicellular, coccoid, marine forms, while the other contains but one species, the filamentous *Prochlorothrix hollandica*, which occurs in freshwater. The genus *Prochloron* contains several coccoid forms living as extracellular symbionts within tropical and subtropical colonial ascidians (sea-squirts). *Prochloron* is an obligate symbiont, and the association appears also to be obligate for the host ascidians, with products from photosynthesis being utilized by the host animals. The coccoid *Prochlorococcus* is abundant as a member of the marine picoplankton in dimly lit deeper waters about 50–100 m below the surface of the open ocean. *Prochlorothrix hollandica* gives rise to massive blooms in shallow eutrophic lakes of The Netherlands, where it was discovered. When discovered, *Prochloron* was thought to represent the missing link between prokaryotic algae and green chloroplasts. However, it seems that both *Prochloron* and *Prochlorothrix* are closely related to the Cyanobacteria. They appear to belong within lineages of the blue-green algae judging from genetic data, although such data are incomplete. Comparisons of nucleotide sequences (of 16S ribosomal RNA and of a gene for a subunit of ribosomal RNA polymerase) in *Prochloron*, *Prochlorothrix* and *Prochlorococcus* indicate that these three genera are unrelated, except that all belong to the blue-green algae. Therefore, prochlorophytes are polyphyletic and none is specifically related to chloroplasts and the Prochlorophyta is an artificial division of green algae belonging to the prokaryotic realm of the Eubacteria.

Proconsul Genus of E. African Miocene primate (Suborder Hominoidea, Family Proconsulidae; 22–17 Myr BP); skull and dentition (e.g. thin molar enamel) resemble ancestral catarrhine, but postcranial features place it in the hominoid clade. At least two species, one small (*c.* 9 kg) and one larger (26–38 kg). Arboreal and quadrupedal, with a completely opposable thumb. Possibly ancestral to the later DRYOPITHECINE apes. A maxilla from earlier East African deposits (27–24 Myr BP) has been tentatively placed within the genus. It would constitute the earliest fossil hominoid yet discovered.

proctodaeum Intucking of embryonic ectoderm meeting the endoderm of the posterior of the alimentary canal and forming anus or cloacal opening.

procumbent (Of teeth) jutting forward, as in ape incisors.

productivity (production) Rate at which solar energy and carbon dioxide are absorbed and utilized in photosynthesis. The *gross primary production* of an ecosystem is the total amount of organic matter produced by its autotrophs, and is usually

recorded in kilojoules per hectare per year. *Net primary production* is gross primary production less that used by plants in respiration and represents food potentially available to consumers in the ecosystem. See STANDING CROP.

proembryo In seed plants, a group of cells formed by initial divisions of the zygote, which by further development differentiates into the suspensor and embryo proper.

progenesis Form of HETEROCHRONY in animals, leading to PAEDOMORPHOSIS through precocious sexual maturity and a shortening of the developmental pathway, often resulting in a 'juvenilized' morphology. Believed to have been a source of important new taxa during evolution. Most successful paedomorphs seem to have been small progenetic larvae, e.g. six-legged myriapod larva (ancestral to insects) and the urochordate tadpole (ancestral to vertebrates). Several gall midges have progenetic larvae; wingless aphids and many parasites are progenetic. Progenesis tends to be favoured by *R*-SELECTION. Compare NEOTENY.

progesterone Steroid hormone secreted by CORPUS LUTEUM of mammalian ovary, and made by PLACENTA from maternal and foetal precursors. For timing of secretion, and roles in human female, see MENSTRUAL CYCLE. Broken down by LIVER. See CONTRACEPTIVE PILL.

progestogen Any substance with progesterone-like effects in the female mammal.

proglottis (pl. proglottides) Segment-like unit of the tapeworm body which, when mature, leaves the gut of the primary host in the faeces. Normally each develops male reproductive organs first and later loses these as female parts develop. Proglottides are not true segments since they are budded off from the anterior of the worm (from the scolex). This does not amount to asexual reproduction (but see HYDATID CYST). See CESTODA.

prognathous 'Long-faced', the jaws protruding considerably in front of the brain case. In primates, an ape-like character;

hominids tend towards ORTHOGNATHOUS 'short' faces.

programmed cell death (PCD) A term that is being increasingly substituted by the term APOPTOSIS. Plant biologists tend to retain it, however, on the grounds that apoptosis is one form of PCD, defined by its special molecular and cytological features.

Progymnospermophyta Progymno-sperms. A group of fossil plants of the late PALAEOZOIC era, which had characteristics intermediate between those of the trimero-phytes and gymnosperms. Morphologically some were similar to early seed ferns, while others were large trees with leafy branches. Reproduced by freely dispersed spores, and produced secondary xylem remarkably similar to that of gymnosperms. Unique in that they also produced secondary phloem. Progymnosperms and Palaeozoic ferns may have evolved from the more ancient trimerophytes, from which they differed mainly by having a more complex vascular system. Another suggestion is that the ferns evolved from progymnosperms. The most important evolutionary advance of pro-gymnosperms over trimerophytes is the presence of a bifacial vascular cambium (producing both secondary xylem and phloem), characteristic of seed plants. Mor-phological evidence suggests that gymno-sperms evolved from the progymnosperms upon evolution of the seed, but whether the seed evolved only once or several times is unknown; nor is it certain from which group of progymnosperms the gymno-sperms evolved.

prokaryote (procaryote) Traditionally, this 'superkingdom' of life included eu-bacteria, cyanobacteria, archaebacteria, prochlorophytes and mycoplasmas. Recent classifications based on 16S and 18S ribos-omal RNA sequences often distinguish two domains within the prokaryotes: BACTERIA and ARCHAEA. Eleven major subgroups of bac-teria and two archaean subgroups are often recognized, the archaeans including Cren-archaeota (hyperthermophilic sulphur-dependent prokaryotes) and Euryarchaeota (methanogenic, extremely halophilic pro-karyotes). They are typically unicellular or filamentous, and small (normally only up to 3 μm in diameter). Their DNA is not housed within a nuclear envelope (see NUCLEOID), and no prokaryotic cell is descended from such a nucleated cell. Within the plasma membrane, often folded and convoluted within cell interior, lies the rest of the cytoplasm, containing smaller ribosomes than those of EUKARYOTES, along with granular inclusions. The traditional classificatory dichotomy into prokaryotes and eukaryotes has to be revised given the discovery of tubulin-like, actin-like and histone-like molecules in archaeons (these proteins formerly being considered the sole preserve of eukaryotes; see FtsZ). Their GEN-ETIC CODE is remarkably similar to that of eukaryotes (see ARCHAEA for details of tran-scription, etc.). Cell division commonly lags behind chromosome replication, so that prokaryotic cells commonly contain at least two chromosomes, each consisting of DNA and non-histone proteins, often attached for a time to the plasma membrane. Micro-tubules, mitochondria and chloroplasts are absent, but they possess structures that function similarly (MESOSOME, CHROMATO-PHORE). See CELL WALL.

DNA reassociation methods indicate that prokaryote diversity in aquatic environ-ments is three to four orders of magnitude less than in sediments and soils, while com-munity fingerprinting techniques using fluorescence microscopy and the POLYMERASE CHAIN REACTION also indicate far lower numbers of taxa there.

prolactin (lactogenic hormone, luteo-tropic hormone, ITH) A protein go-nadotrophic hormone secreted by the vertebrate anterior pituitary gland. In mam-mals it promotes secretion of progesterone by the corpus luteum and is involved in LACTATION. Commonly prolonged in prein-dustrial societies, where it can lead to birth spacings of 3–4 years through disruption of the MENSTRUAL CYCLE (amenorrhoea). In pigeons stimulates crop milk production and is necessary for maintenance of incu-bation (which itself stimulates its secretion) by both sexes.

prolegs Stumpy unjointed appendages on ventral surface of abdomen of caterpillar.

proliferation Growth by active cell division.

promeristem Extreme tip of an APICAL MERISTEM comprising actively dividing, but as yet undifferentiated, cells.

pro-metaphase Stage in the CELL CYCLE when the nuclear envelope disintegrates; usually regarded as the termination of prophase.

promoter One type of CIS-ACTING CONTROL ELEMENT. About 100 base pairs long, it lies a short distance upstream of the coding part of the gene of which it forms a part, but downstream of the OPERATOR, and is the region to which an RNA polymerase molecule binds (often accompanied by one or more TRANSCRIPTION FACTORS) to initiate transcription. Promoters of genes which transcribe large amounts of product (strong promoters) are similar to one another (share similar CONSENSUS SEQUENCES) and have a TATA BOX about 30 base pairs upstream from where transcription begins. Most eukaryotic promoters contain core promoter sites, which are diverse; but the TATA box is their most prominent element and the transcription factor TFIID binds the core promoter during formation of the RNA polymerase II transcription initiation complex. ENHANCERS regulate the promoter's rate of transcription. The retinoblastoma gene product (see TUMOUR SUPPRESSOR GENE) is known to repress transcription of some cellular promoters by binding appropriate transcription factors. See GENE for diagram.

pronation Position of fore-limb, or rotation towards it, such that fore-foot (hand) is twisted through 90° relative to the elbow, the radius and ulna being crossed. In this way the fore-foot points forwards (the natural position for walking in many tetrapods). Humans and other primates can untwist the fore-arm (*supination*).

pronephros See KIDNEY.

pronuclei The two haploid nuclei present in a zygote prior to their fusion, which completes fertilization. They may move closer on microtubules, and in mammals remain distinct until their membranes have broken down just prior to the first CLEAVAGE division. See *WOLBACHIA*.

pro-oestrid See OESTROUS CYCLE and MENSTRUAL CYCLE.

propagule (Bot.) A dispersive structure, such as a seed, fruit, gemma or spore, released from the parent organism.

prophage Non-infectious phage DNA, integrated into a bacterial chromosome and multiplying with the dividing bacterium but not causing cell lysis except after excision from the chromosome, which can be *induced* by certain treatments of the cell. See BACTERIOPHAGE.

prophase First stage of MITOSIS and MEIOSIS.

proplastid Minute, self-reproducing, colourless or pale green, undifferentiated PLASTIDS occurring in meristematic cells of roots and shoots. They are the precursors of the more highly differentiated plastids (e.g. CHLOROPLASTS). Development of a proplastid may become arrested by a lack of light, then it may form one or more prolamellar bodies; semicrystalline bodies comprising tubular membranes. Such plastids are called etioplasts. Upon returning to the light the membranes of the prolamellar bodies develop into thylakoids.

propositus See PROBAND.

proprioceptors (1) Receptors involved in detection of position and movement including MUSCLE SPINDLES, organs of balance in the vertebrate VESTIBULAR APPARATUS, and campaniform sensillae occurring in all parts of the cuticle of the insect body subject to stress (especially in joints, at wing and haltere bases). Do not usually show sensory ADAPTATION. (2) In a wide sense, any receptor detecting changes within the body other than those caused by substances taken into the gut and respiratory tract; e.g. deep pain receptors and BARORECEPTORS; STATOCYSTS in invertebrates.

prop root Adventitious supportive root, arising from the stem above soil level.

prosencephalon See FOREBRAIN.

prosenchyma See PLECTENCHYMA.

Prosimian Of the primate suborder Prosimii. Primitive, nocturnal and mostly arboreal. ORBITS not totally enclosed in bone at rear, and less frontally positioned than in anthropoids (hence poorer stereoscopic vision). Olfactory lobes larger than in anthropoids; snout longer, with a wet muzzle (rhinarium). Incisors and canines of lower jaw form a dental comb/scraper; diet more insectivorous than anthropoids, and brain size smaller relative to body. Neither frontal bone nor two halves of lower jaw fused. Tend to have 2–3 young and leave them in nests.

prosoma One of the tagmata of CHELICERATA, a combined head and thorax. Comprises one pre-oral segment and five post-oral segments, the latter bearing walking limbs. See OPISTHOSOMA.

prostaglandins (PGs) Family of 20-carbon fatty acid derivatives continuously synthesized (in mammals at least) by most nucleated cells from precursor phospholipids of the plasma membrane. Rapidly degraded by enzymes on release; but if appropriately triggered a cell will increase its output of PG, raising local levels and influencing both itself and its neighbours. Like hormones, they may be released into the blood but, unlike them, are only effective over short distances. Promote contraction of smooth muscle, platelet aggregation, inflammation and secretion. Some bind membrane G PROTEIN-linked receptors and exert their effects by altering intracellular cyclic AMP levels. Studies on knockout mice suggest that PGs have a far wider role than previously thought, including preparation of the endometrium for pregnancy and blastocyst implantation. Of great clinical potential in alteration of blood pressure, in bronchodilation and constriction, inducing labour, reducing gastric secretion, etc. Also called local or tissue hormones. See EICOSANOIDS, PARTURITION.

prostate gland Gland of male mammalian reproductive system lying below the urinary bladder and around part of the urethra. Secretes alkaline fluid comprising up to a third of semen volume. Its effect on vaginal pH assists sperm motility. Rich source of PROSTAGLANDINS in humans.

prosthetic group Non-protein group which when firmly attached to a protein results in a functional complex (a conjugated protein). Many respiratory pigments (e.g. HAEMOGLOBIN) are conjugated proteins, while many enzymes require prosthetic groups (some of them metal ions). The carbohydrates and lipids in glycoproteins and lipoproteins are prosthetic groups of their proteins. DNA is the prosthetic group of the histones in chromatin. Compare COENZYME.

protamines Basic proteins of low molecular weight (about 5,000), lacking prosthetic groups, containing many arginine and lysine residues, and found in association with DNA. See CHROMOSOME.

protandry (1) In flowers, the situation in which anthers mature before carpels. See DICHOGAMY. (2) In animals, the condition in sequential hermaphrodites in which sperm are produced prior to eggs (e.g. in many tapeworms).

protease Enzyme that digests protein by hydrolysis of peptide bonds. Many proteases require metal ions for their activity (metalloproteases), and the zinc-binding family includes reprolysins and astacins as subfamilies. These are important in exerting control over morphogen activity in development. See PROTEOLYSIS.

proteasomes ATP-dependent MULTI-ENZYME COMPLEXES (2 megadaltons, 26S), variably composed of several subunits and located within both cytoplasm and nucleus, involved in selective protein degradation. Barrel-shaped, with a hole down its centre. Proteasomes almost always act solely on proteins that have been targeted for destruction, perhaps because they are misfolded or contain oxidized or other abnormal amino acids, by covalent attachment of several copies of the protein UBIQUITIN. The immune

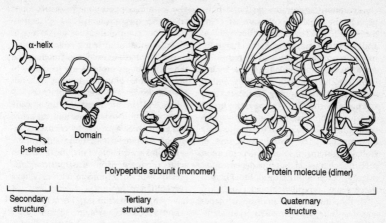

α-helix

β-sheet

Domain

Polypeptide subunit (monomer)

Protein molecule (dimer)

Secondary structure

Tertiary structure

Quaternary structure

FIG. 139 *The 'ribbon model' method of representing* PROTEIN *structure. A domain is indicated, both isolated and as a part of a polypeptide subunit of the functional protein molecule.*

system relies on specialized proteasomes (immunoproteasomes) which degrade viral or cancer cell antigens on the cell surface prior to integration with MHC Class 1 molecules and presentation on the cell surface for cytotoxic T CELL surveillance (see MAJOR HISTOCOMPATIBILITY COMPLEX).

protein Polymer of very large or enormous molecular mass, composed of one or more polypeptide chains, and whose monomers are AMINO ACIDS, joined together (in condensation reactions) by PEPTIDE BONDS. In addition, some have covalent 'sulphur bonds' formed by oxidization between two cysteine radicals in the polypeptide. Proteins include structural, often fibrous, forms (e.g. collagen) and active, often globular, forms (e.g. see ENZYMES, TRANSCRIPTION FACTORS).

The potential variety of polypeptides is infinite: there are 20 common amino acids, so joining any two together would give 400 (= 20 × 20) possible *dipeptides*. Biological polypeptides are often several hundred amino acids long, so few of the possible polypeptides actually occur in organisms (see PROTEIN SYNTHESIS). Several proteins (e.g. ACTIN, TUBULIN) form filaments of polypeptide molecules. The molecular mass which results (as in the protein coat of tobacco

mosaic virus) may exceed 40 million daltons. See BIURET REACTION, Fig. 139.

Despite considerable overlap between the following structural categories within proteins, they remain didactically useful. Each polypeptide has a *primary structure*: the number and sequence of its amino acids. There is an amino-terminal ($-NH_2$) end and a carboxy-terminal ($-COOH$) end to the molecule. During its production on a ribosome a polypeptide commonly assumes a corkscrew-like ALPHA-HELIX (α-helix) as its *secondary structure*, due to hydrogen bonding between the hydrogen atom attached to the nitrogen in one amino acid radical and the oxygen attached to a carbon atom three radicals along the chain. Other *intra*molecular hydrogen bonds may contribute to the secondary structure, as with antiparallel folding of the molecule back along itself in the same plane, hydrogen atoms of one side being linked to oxygen atoms of the side parallel to it. Such β-pleated sheets occur in many globular (spherical) protein molecules and commonly link several polypeptides of the same type (as in fibroin of silk, and some other fibrous proteins; see PLAQUE). Regions of α-helix and β-pleated sheets, along with less clearly organized (i.e. more random) stretches of the amino acid chain, may all contribute to the three-dimensional configuration (*tertiary structure*) of a polypeptide, the α-helical portions often being thrown into folds by electrostatic attraction and repulsion resulting from the charge distribution on the amino acid R-groups (see AMINO ACID). Proteins often have hydrophilic and hydrophobic segments, and in such membrane-spanning species as RECEPTOR proteins the former tend to lie in the aqueous regions on either side of the membrane while the latter lie in the membrane's lipid interior. The tertiary structures of globular proteins in solution depend upon pH (since charges on R-groups depend upon pH) and upon sulphur (disulphide) bonds. ENZYMES are, typically, globular proteins. Their functions depend upon their shapes and are affected by changes in pH and temperature (see DENATURATION, ISOELECTRIC POINT). Many important proteins (*conjugated proteins*) are formed by covalent union of a non-protein radical with the protein molecule, forming a hybrid molecule (e.g. HAEMOGLOBIN, CYTOCHROMES). The temperature stability of a protein's tertiary structure may be enhanced by *salt bridges*, in which charges on amino acids are bridged by Na$^+$ or other cations (see HYPERTHERMOPHILES). The *quaternary structure* of a protein is the shape adopted when two or more polypeptide chains associate (sometimes covalently) to produce the functional protein molecule. Both the non-enzyme haemoglobin and the enzyme lactate dehydrogenase (see ISOENZYME) are proteins formed by quaternary association between four polypeptides of two different kinds. Indeed, functional enzyme molecules often consist of two or more different polypeptide subunits. For more on protein structure see ALLOSTERIC, DNA-BINDING PROTEINS, DOMAIN, RECEPTOR, REGULATORY ENZYMES.

Fibrous proteins often have major structural roles (e.g. in CYTOSKELETONS, CONNECTIVE TISSUE, STRIATED MUSCLE, CHROMOSOMES). This is more obviously true in animals than in plants, although plant CELL WALLS contain structural GLYCOPROTEINS. All cells depend upon catalytic activities of enzymes, and have TRANSPORT PROTEINS in their membranes. Generally, RESPIRATORY PIGMENTS are conjugated proteins. Proteins also form major components of ANTIBODIES and other glycoproteins, and of LIPOPROTEINS. They commonly act as BUFFERS (e.g. in blood plasma), and being colloids reduce the WATER POTENTIALS of cells and intercellular fluids. Although proteins are insoluble in lipid solvents, globular proteins (but not fibrous) dissolve in water and dilute salt solutions. Proteins are digested hydrolytically by PROTEOLYTIC ENZYMES and mineral acids, and are usually separable by CHROMATOGRAPHY or ELECTROPHORESIS. Only autotrophic organisms are capable of making the amino acid components of proteins from inorganic precursors. Two or more proteins may have very different amino acid sequences yet have similar structures and functions: we cannot yet predict a protein's structure from its amino acid sequence unless the sequence is clearly related to that of another protein of known structure. An estimated third of eukaryotic cell proteins are phosphorylated by PROTEIN KINASES, some of them involved in events leading to the strategic tethering of regulatory proteins within cells at lipid docking sites (see PI3-KINASES). See PROTEIN ENGINEERING; PROTEIN FAMILY.

protein chips See PROTEOMICS.

protein complexes It has long been appreciated that some proteins must form stable complexes before they function (e.g. ribosomes, and proteasomes, which are MULTIENZYME COMPLEXES), but it is now appreciated that many important cellular functions involve more transient protein complexes acting as 'molecular machines' (e.g. see PROTEIN SYNTHESIS, NF-κB). Thus, Arp2/3 is a complex of seven proteins that causes the polymerization of actin, and many receptors – both intracellular and at the cell surface – are also transient protein complexes. Often, scaffolding proteins (e.g. TAB1, TAB2) are required to organize the relevant proteins into complexes. These scaffolding proteins often form complexes themselves, and some may activate MAPKS.

Many cell-signalling pathways share proteins promiscuously, components having different roles in different partnerships (e.g. the involvement of CYTOCHROME C in apoptosis and mitochondrial electron transport; see APOPTOSOME, SIGNAL TRANSDUC-TION). The overall shapes of protein complexes are more likely to have been conserved over evolutionary time than are the precise compositions of the different proteins in the complex.

The phosphotyrosine-binding (SH_2) and proline-binding (SH_3) domains of many intracellular signalling proteins enable them to act as ADAPTORS, cross-linking proteins into networks.

protein engineering Techniques used in improving naturally occurring enzymes for human use. Isolation of the gene encoding a polypeptide by specific endonucleases is followed by site-specific MUTAGENESIS (e.g. changing a particular glycine to an alanine or a lysine to an arginine may improve an enzyme's thermal stability). Likewise, increasing the number of disulphide (–SH) bridges (e.g. by one where the native molecule only has one) increases an enzyme's rigidity. Hope of progress in protein design depends upon increasing our understanding of how proteins fold to achieve their native state. Mimics of the enzymes chymotrypsin and trypsin have been designed in which short peptide sequences from their active sites are strung together between oligoglycine loops, the products hydrolysing alkyl ester and peptide bonds on a par with native enzymes. A different goal might be to replace the non-renewable resources so often used in manufacture by renewable engineered proteins providing fibrous and structural materials.

protein family A molecular taxon containing proteins with >50% amino acid sequence identity. Protein superfamilies contain proteins with 50% amino acid sequence identity. See Structural Classification Of Proteins website: http://.scop.mrc-lmb.cam.ac.uk/scop/.

protein kinases (PKs) Enzyme transferring a phosphate group from ATP to an intracellular protein (often also an enzyme), increasing or decreasing its activity. The PROTEIN TYROSINE KINASE family phosphorylate specific tyrosine residues on the target protein; the serine/threonine protein

kinases (e.g. PROTEIN KINASE A) phosphorylate specific serine or (less often) threonine residues respectively. These generally form parts of RECEPTOR protein kinases; but there are also non-receptor protein kinases (often associated with the inner surface of the plasma membrane) and these often contain SH_2 and SH_3 DOMAINS. The PI3-KINASES phosphorylate lipids rather than proteins, but can indirectly activate PROTEIN KINASE B. Calcium/calmodulin-dependent protein kinases (CaM-kinases) phosphorylate serines or threonines in proteins, as do PKA and PKB, and many of the effects of Ca^{2+} in animal cells are mediated by this group of enzymes (see CALCIUM SIGNALLING), and some also phosphorylate gene regulatory proteins such as CREB. See *cdc* GENES, CELL CYCLE, SIGNAL TRANSDUCTION and Fig. 151, INCOMPATIBILITY MECHANISMS, PKC, and specific protein kinase entries below.

protein kinase A (PKA) A CYCLIC AMP-dependent protein kinase, found in all animal cells, and important target of cAMP. It exists in a variety of isoforms and mediates many of cAMP's effects. Comprising two catalytic and two regulatory subunits, its regulatory subunits dissociate from the molecule when PKA binds cAMP. Typically bound by anchoring proteins which attach its regulatory subunits to a membrane or to the cytoskeleton, these often also bind other kinases and some phosphatases to form a signalling complex. The regulatory subunits also prevent the enzyme from operating when cAMP is scarce, and phosphatases ensure PKA's activity is brief. Activated PKA very rapidly phosphorylates enzymes involved in muscle glycogen breakdown (see Fig. 39) while inhibiting glycogen synthesis, but it is far slower to bring about transcriptional changes (see e.g. SOMATO-STATIN). As with PKB, the response within a cell to PKA depends upon which target proteins are also present.

protein kinase B (PKB, or Akt) A protein kinase activated by many GROWTH FACTORS and SURVIVAL FACTORS, with glucose-dependent but strongly anti-apoptotic effects. It contains a PH (Pleckstrin homology) domain which directs it to the plasma membrane when PI3-KINASE is activated by a signal molecule. Molecules of $PI_{(3,4,5)}P_3$ resulting from PI3-kinase activity bind PKB molecules, inducing an allosteric change allowing it to be activated by PDK1, whereupon it leaves the membrane and phosphorylates a number of proteins, one of which (BAD) is thereby inactivated and prevented from exerting its normally pro-apoptotic influence in the cell. In this way, PKB promotes cell survival – which it also does by inhibiting other cell death activators, sometimes by preventing transcription of their encoding genes. It stimulates glucose transport into the cell, recruitment of hexokinases into mitochondria and glycolysis (see CELL GROWTH). As with PKA, the response of a cell to PKB depends upon which target proteins are present. See APOPTOSIS.

protein kinase C (PKC) A calcium-dependent serine/threonine protein kinase, existing in several isozymic forms, many activated by diacylglycerol. It may be recruited from the cytosol to the plasma membrane, attaching there to diacylglycerol after phospholipase C-β has cleaved $PI_{(4,5)}P_2$ (phospatidylinositol 4,5-bisphosphate). An initial Ca^{2+} influx caused by the resulting INOSITOL 1,4,5-TRIPHOSPHATE (IP_3) brings about this translocation. See DIABETES.

protein networks Interactions between the many PROTEIN COMPLEXES (protein machines); other protein–protein and protein–nucleic acid interactions within the cell are so complex that they can be considered to form functional networks. See DOMAIN, PROTEIN SYNTHESIS.

protein sorting See PROTEIN TARGETING.

protein synthesis Proteins are manufactured by cells on RIBOSOMES, which involves joining together in the correct sequence possibly hundreds of amino acid molecules. Cells produce particular proteins either all the time (constitutively) or as and when required (SEE CELL GROWTH, GENE EXPRESSION). Sequence of amino acids forming primary structure of a protein is encoded in the sequence of nucleotides of the genetic material, usually DNA (in some viruses it is RNA). When a piece of DNA becomes involved in protein synthesis, an RNA POLYMERASE first breaks the hydrogen bonds holding the two DNA strands together, then uses one of the strands as a template on which to incorporate the nucleotides making a complementary RNA molecule, in an order dictated by BASE PAIRING rules. Once produced (the process is called transcription), this pre-RNA molecule is commonly modified by splicing (see RNA PROCESSING) to form a shortened intron-free messenger RNA molecule, to which a cap is then attached just upstream of the nucleotide triplet AUG, which is the start codon and encodes methionine (see RNA CAPPING). After transcription, the nuclear enzyme poly-A polymerase adds to most eukaryotic mRNAs a poly-A (poly-adenylyl) tail about 50–200 A-residues in length, improving the stability of the mRNAs and protecting them from ribonuclease degradation. This is bound in the cytosol by the RNA-binding protein, poly(A) tail-binding protein, which somehow determines the final length of the tail. This mRNA molecule passes into the cytoplasm (via the nuclear pore apparatus in nucleated cells) and triggers a ribosome to assemble upon it at the AUG codon nearest the 5′-end of the mRNA molecule.

Free amino acids are not assembled directly into protein but are first loosely bound to an *activating enzyme*, so-called because it (a) hydrolyses an ATP molecule, providing energy for (b) attachment of the activated amino acid to one of a small number of specific transfer RNA (tRNA) molecules, of which there is a pool in the cell. The result of this ATP-dependent catalysis is a pool of amino acyl-tRNAs from which protein synthesis proceeds.

Under the influence of INITIATION FACTORS, each ribosome draws amino-acyl-tRNAs bound to ELONGATION FACTORS from its surroundings in an order determined by the nucleotide sequence of the mRNA molecule to which it attaches. AUG is the CODON for the amino acid methionine, and the newly-assembling ribosome already has tRNA-methionine bound to it (see P-SITE). This tRNA base-pairs with the AUG codon of the mRNA by hydrogen bonds with a triplet of nucleotides exposed at one end of the molecule: its anticodon triplet. Then another tRNA molecule gets bound by the ribosome (see A-SITE), but only if its anticodon base-pairs with the codon next to AUG in the mRNA. It brings with it its own attached amino acid. This codon–anticodon hydrogen bonding provides the essential working principle of the GENETIC CODE.

The ribosome will continue to draw in appropriate tRNA molecules, joining each of their amino acids together to form a growing polypeptide chain. This chain elongation requires hydrolysis of two GTP molecules per hydrogen bond formed. Each of the tRNA molecules is released from the ribosome once it has donated its amino acid load.

As it draws in each amino acyl-tRNA in turn, the ribosome moves one codon further along the mRNA molecule, towards the 3'-end of the molecule. At an appropriate stop codon (termination codon) on the mRNA, translation ends and the ribosome releases the completed polypeptide. The poly-A tail mRNA is not translated and the N-terminal methionine is cleaved off. Downstream from the termination codon in eukaryotes, a variable nucleotide sequence is usually added containing the sequence AATAAA about 10–30 nucleotides upstream from the site of poly-A addition. Mutations here can cause transcription past the normal termination region (as in some THALASSAEMIAS).

This ribosomal phase of protein synthesis is called *translation*. Each mRNA molecule is simultaneously the site of attachment of many ribosomes (see POLYSOME), and when each ribosome reaches the stop codon it releases another identical polypeptide molecule (see PROTEIN TARGETING). Both chloroplasts and mitochondria make some of their own proteins but, as with prokaryotic systems in general, their start codon (AUG) binds N-formylmethionine rather than methionine. Protein synthesis is energy-dependent, each amino acid incorporated into the polypeptide requiring hydrolysis of three HIGH-ENERGY PHOSPHATE bonds. Several ANTIBIOTICS stop protein synthesis, during either transcription or translation. Much POST-TRANSLATIONAL MODIFICATION of proteins occurs in the GOLGI APPARATUS.

In eukaryotes, protein synthesis requires a network of multicomponent cellular machines, each complex interfacing with the others both physically and functionally (see Fig. 140). Transcription is coupled with pre-mRNA processing (splicing) through the assembly platform provided by the carboxy-terminal domain (CTD) of the RNA polymerase II large subunit for factors involved in capping, splicing and polyadenylation. Coupling also occurs between the splicing machinery (SPLICEOSOMES), transcription initiation and elongation factors, and between the machinery of nonsense-mediated decay (NMD) and mRNA export. In NMD, pre-mRNAs with premature termination codons are targeted for degradation in the cytoplasm by exon junction-marking proteins, which remain attached to the mRNA and stop the ribosome from moving along it, so triggering rapid mRNA decay. NMD also occurs in the nucleus, although the pathway is less understood.

protein targeting (protein sorting)

During synthesis (co-translationally) or after synthesis (post-translationally), proteins may move from one cellular compartment to another, crossing the hydrophobic barrier of a membrane's phospholipid bilayer. Such behaviours are made possible by amino acid sequences (sorting signals) comprising small parts of the protein's structure and effectively targeting its destination in the cell. Signal sequences (signal peptides) are hydrophobic, occur at the amino-terminal end and direct the ribosome to the endoplasmic reticulum (ER). They also target the protein through the ER before they are cleaved off by small RNA SIGNAL RECOGNITION particles, e.g. 7SL-RNA. Some proteins cross the ER membrane completely; others (e.g. membrane RECEPTORS) do so only partly, remain integrated within it and end up in the plasmalemma or the membrane of another organelle. Likewise,

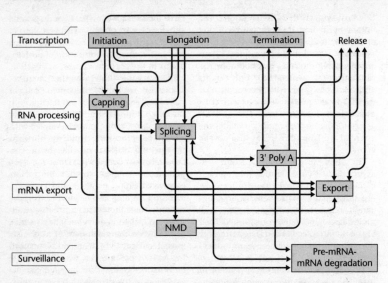

FIG. 140 *A complex network of coupled interactions in eukaryotic gene expression. The major steps of gene expression are indicated on the left and each stage in transcription is shown along the top in the thick black arrow. 'Release' indicates release of the mature mRNA from the site of transcription. The steps in RNA processing, mRNA export, nonsense-mediated decay (NMD) and RNA degradation are shown below the arrow. The downward and upward pointing black arrows indicate physical and/or functional coupling between the two steps in gene expression. Each arrow is documented by published studies showing a physical and/or functional interaction between one or more components of gene expression machines.*

the several proteins involved in INTERCELLULAR JUNCTIONS must also be very carefully targeted. Because protein translocation beyond the ER is achieved via budded-off vesicles, the ER's is the only membrane a protein ever has to cross, even if destined for export from the cell (likewise some viruses). *Transit sequences* are amino-terminal and amphiphilic, targeting the protein into chloroplasts, mitochondria, etc. *Retention signals* keep a protein within a compartment despite the bulk flow of other proteins. *Signal (targeting) patches* are formed from different amino acid sequence regions which, through the protein's tertiary and quaternary structures, come to form part of the surface conformation of the molecule after folding in the ER (compare B cell and T cell epitopes). Localization of cytoplasmic mRNA transcripts, by microtubules or by actin, is effectively another method of protein targeting if the translation process is restricted as a result. See Figs. 74 and 150, ENDOSOME, GLYCOSYLATION, SNARES.

protein tyrosine kinase (PTK) Those PROTEIN KINASES phosphorylating tyrosine residues in their target protein substrate (often an enzyme). They are enzymes providing a central switch mechanism in cellular SIGNAL TRANSDUCTION pathways, often involved in cell fate determination. The RECEPTOR TYROSINE KINASES (RTK) form intracellular components of cell-surface transmembrane receptors to growth factors and hormones, being activated when ligand binds the extracellular domain, causing autophosphorylation of the receptor's tyrosine residues, which are subsequently bound by SH2 domains of non-receptor tyrosine kinases and other proteins involved in signal transduction (see DOMAINS). Activation of a cell's PTKs can induce it to divide and migrate (e.g. fibroblasts treated with PDGF, see GROWTH FAC-TORS) or differentiate (neuronal precursors treated with NGF). Non-receptor protein tyrosine kinases (NRPTKs) include c-Src and c-Yes proteins, phosphorylated forms of which are actively involved in M-phase of the CELL CYCLE. Over-expression or gain-of-function MUTATIONS in RTK genes can cause cancers (see CANCER CELLS).

Proteobacteria (purple bacteria) A large diverse division of Gram-negative BAC-TERIA having in common certain 16S rRNA sequences; includes *H. pylori*, *H. influenzae* and *E. coli*.

proteoglycans Class of acidic GLYCO-PROTEINS found in varying amounts in extra-cellular matrices of animal tissues, notably connective tissues, as well as attached to cell membranes where they can act as co-receptors and play important roles in binding secreted signal proteins during PATTERN FORMATION. Contain more carbohydrate than protein. One forms, with collagen, the rubbery material in cartilage preventing bone ends from grating together. As with MUCINS, a carbohydrate-free area serves for cross-links between the protein chains to produce aggregation. The fibrous polysac-charide HYALURONIC ACID serves as a chain along which many of these proteoglycan molecules align themselves. See PEPTIDO-GLYCAN, CHITIN.

proteolysis Protein hydrolysis, as achieved by mineral acids and – more widely and importantly – by proteases (proteolytic enzymes). Apart from its importance in nutrient digestion, proteolysis is essential to regulation of cellular protein concentration and distribution, in which PROTEASOMES and UBIQUITINS are involved. See ANAPHASE-PROMOTING COMPLEX, CELL CYCLE.

proteolytic enzyme Any enzyme taking part in breakdown of proteins, ultimately to amino acids. Include PEPSIN, TRYPSIN, CHYMOTRYPSIN, PEPTIDASES and intracellular CATHEPSINS. See PROTEOLYSIS.

proteome, proteomics A proteome is the complete protein complement expressed by an organism or cell. Proteomics is the use

of techniques aimed at determining the current and changing protein status of cell systems and typically involves two-dimensional gel electrophoresis and mass spectrometry. These complement DNA-sequence and RNA-expression analyses, offering a more direct measure of cellular responses than do measurement of nucleic acid status. Nowadays, proteomics covers much of functional genomics (see GENOMICS) other than large-scale study of protein structure ('structural genomics'). Protein modifications due to POST-TRANSLATIONAL MODIFICATION are not apparent from DNA sequence data.

A 'protein chip' is a surface on which a variety of 'bait' proteins (e.g. antibodies) are immobilized in an array format (compare NUCLEIC ACID ARRAYS), so that when the surface is probed with the sample of interest only those proteins which bind to the relevant antibodies stay bound to the chip (see IMMU-NOASSAYS for comparable ELISA).

Proterozoic Geological division (eon) between the end of the ARCHAEAN eon (*c.* 2,500 Myr BP) and onset of Phanerozoic (*c.* 543 Myr BP), from which it is separated by a discrete event: the radiation of eucoelo-mate animals. During it, atmospheric oxygen levels rose to about one tenth of present levels. Limestones became abundant for the first time, often containing STROMATOLITES. Trace fossils permit recognition of perhaps three uppermost Proterozoic biozones, the best-known being the Ediacara fauna (soft-bodied) from Australia (590–560? Myr BP). Phytoplanktonic protists (acritarchs) and Cyanobacteria are quite well distributed from perhaps 630 Myr BP onwards. Chitin-ous sabelliditid worms and shelly fossils occur at or just below the Proterozoic/Cam-brian boundary. There may have been two or more great global glacial periods in the late Proterozoic. See CAMBRIAN for 'Cambrian explosion'. See Appendix.

prothallus Independent gametophyte stage of ferns and related plants. Small, green, parenchymatous thallus bearing antheridia and archegonia, showing little differentiation. Usually prostrate on the soil surface, attached by rhizoids. May be subterranean and mycotrophic. See CAMBRIAN.

prothoracic gland See ECDYSONE.

prothrombin See BLOOD CLOTTING.

Protista (Protoctista) Kingdom comprising eukaryotic and mesokaryotic unicellular or multicellular organisms. Nutritional modes are diverse, and include ingestion, photosynthesis and absorption. True sexuality is present in the majority of divisions (except euglenoids). Most protists have a stage bearing flagella (undulipodia). Includes all the eukaryotic and mesokaryotic algae, protozoa, slime moulds, and Oomycota and Chytridiomycota. Some protists are AMITOCHONDRIATE.

Protoctista See PROTISTA.

Protochordata Informal group, comprising HEMICHORDATA, TUNICATA, and CEPHALOCHORDATA.

protogyny (1) The condition in flowers (termed *protogynous*) whose carpels mature before their anthers, as in plantains. See DICHOGAMY. (2) The condition in sequentially hermaphrodite animals in which first eggs are produced, then sperm. See PROTANDRY.

protonema Branched, multicellular, filamentous or (less commonly) thalloid structure, produced on germination of a bryophyte spore, from which new plants develop as buds.

protonephridium See NEPHRIDIUM.

proton pump See CHEMIOSMOTIC THEORY, CYTOCHROMES.

proto-oncogene (cellular oncogene) See ONCOGENE.

protoplasm Cell contents within and including the plasma membrane but usually taken to exclude large vacuoles, masses of secretion or ingested material. In most eukaryotic cells it includes, besides the CYTOPLASM, one or more nuclei. Prokaryotic cells lack nuclei. Cell walls, if present, are nonprotoplasmic. Each protoplasmic unit constitutes a *protoplast*.

protoplast (Bot.) Actively metabolizing part of a cell (its PROTOPLASM), as distinct from cell wall. Equivalent to the whole cell in zoology. In BIOTECHNOLOGY, the use of tissue culture has involved plant regeneration and MICROPROPAGATION from protoplasts (the cell wall having been removed by enzymatic digestion). Protoplasts are also used in plant genetics; thus (a) protoplasts from different plants can be induced to fuse together to produce interspecific or, in some cases, intergeneric hybrid cells, while (b) the Ti PLASMID from the crown gall bacterium *Agrobacterium tumefaciens*, or some other DNA injection technique, is used to introduce specific genes into protoplasts. Tissue culture is then used to regenerate plants from these individual protoplasts.

protopodite See BIRAMOUS APPENDAGE.

protostele Simplest and most primitive type of STELE, comprising a central core of xylem surrounded by a cylinder of phloem. Present in stems of some ferns and club mosses and almost universal in roots. In a *haplostele*, xylem forms a central rod; in an *actinostele*, xylem is ribbed and appears star-shaped in transverse section; in a *plectostele*, xylem is in several parallel, longitudinal strips embedded in the phloem.

Protostomia Those coelomate metazoans (sometimes termed an infragrade) in which the blastopore develops into the mouth of the adult, cleavage tends to be determinate, and the coelom tends to form by SCHIZOCOELY. Includes annelids, arthropods, molluscs and, usually, those phyla with LOPHOPHORES. Compare DEUTEROSTOMIA; see DORSOVENTRAL AXIS-INVERSION THEORY.

Prototheria Mammalian subclass, of which only the monotremes (six species) survive. Includes extinct orders MULTITUBERCULATA, Triconodonta and Docodonta, and the extant Order Monotremata. The latter comprise the duckbill, or platypus (*Ornithorhynchus*), of Australia and Tasmania, and the spiny anteaters (*Tachyglossus, Zaglossus*) of Australia, Tasmania and New Guinea. Fossil forms have been found only in Australia, and have not so far pre-dated the Pleistocene. All species have hair and mammary glands and are homoiothermic; but all have the reptilian features of egg-laying (ovipary), retention of separate coracoid and interclavicle bones in the PECTORAL GIRDLE, and epipubic bones attached to the PELVIC GIRDLE. Brain size in relation to body size is lower than in placental mammals, resembling marsupials in this respect. A CLOACA is present. Echidnas incubate in a pouch (marsupium); duckbills incubate in a nest. See Fig. 108.

prototroph Any microorganism (esp. bacterium, fungus) expressing the normal (wild-type) phenotype with respect to its ability to synthesize its organic requirements when grown on nutritionally unsupplemented (i.e. *minimal*) medium. Contrast AUXOTROPH.

protoxylem The first elements to be differentiated from procambium; extensible. Described as *endarch* when internal to the later-formed metaxylem (as in roots), and as *mesarch* when surrounded by metaxylem (as in fern stems).

Protozoa Phylum, or subkingdom, of the PROTISTA, comprising unicellular and colonial animals of varied form. Generally subdivided into four classes: SARCODINA (amoebae, radiolarians, foraminifera), MASTIGOPHORA (flagellates), CILIATA (ciliates and suctorians) and APICOMPLEXA, formerly Sporozoa (e.g. coccidians, gregarines); but some of these at least represent grades rather than clades. The first two are sometimes united as the Sarcomastigophora. Reproduction commonly by binary or multiple FISSION, but sometimes by CONJUGATION. Ubiquitous, inhabiting aquatic and damp terrestrial habitats. Several of them are serious pests of humans and their domestic animals (e.g. see MALARIA).

Protura Order of the APTERYGOTA. Minute (0.5–2.5 mm long) whitish insects lacking antennae and eyes. Inhabitants of moist soils, leaf litter, etc. Stylet-like mandibles for piercing. Metamorphosis consists of addition of three abdominal segments and development of genitalia. No evidence that the adult moults.

proventriculus Anterior part of the bird stomach, where digestive enzymes are secreted, posterior part being the GIZZARD. Used synonymously with gizzard in crustaceans and insects.

provirus Viral genomes which are integrated into the host cell chromosome and most of the time remain unexpressed (*latent*). Bacteria harbouring bacteriophages which do this are termed *lysogenic*. See BACTERIOPHAGE.

proximal Situated relatively near to a point of attachment or origin. Compare DISTAL.

proximate factor Explanations in biology are mostly either proximate or ultimate. Former are characteristically mechanistic and indicate how some outcome or change is intelligible in terms of antecedent causes. The latter are teleological, rendering phenomena intelligible in terms of probabilities of future states of affairs. Thus, change in photoperiod may be cited as the proximate factor causing change of winter coat colour in stoats, etc.; but better camouflage may be cited as an ULTIMATE FACTOR responsible for the change, bringing an increase in individual fitness.

Prymnesiophyta See HAPTOPHYTA.

pseudoalleles Two mutations in the same cistron which give rise to different phenotypes when in the *cis* and *trans* conditions respectively. In the *CIS-TRANS TEST* they fail to complement one another.

pseudoautosomal inheritance Because there is a homologous region of the human X and Y chromosomes, at which there is obligatory crossing-over every meiosis, alleles in this region are effectively uncoupled from the unique regions of these chromosomes and therefore exhibit 'pseudoautosomal inheritance'. See SEX CHROMOSOMES.

pseudocoelom (pseudocoel) Fluid-filled cavity between body wall and gut with, however, an entirely different origin from true COELOM. The pseudocoelom is a persistent blastocoel, lacking a definite mesoderm lining. In some cases, as in nematode worms, it may be filled with vacuolated mesenchyme cells. Pseudocoelomate invertebrates include ASCHELMINTHES, ENDOPROCTA and possibly priapulid worms. Their interrelationships are unclear, but it is likely that all pseudocoelomate animals have had progenetic origins (see PROGENESIS), and that they are polyphyletic and represent a grade rather than a clade.

pseudogamy Phenomenon where fertilization is required for development of sexually-produced offspring which derive all their genes from their maternal parent. In the grass *Poa*, and in *Potentilla*, apomictic plants produce perfectly functional pollen and fertilization precedes seed development, but only fertilization of the endosperm nucleus occurs, the egg cell nucleus remaining unfertilized. See GYNOGENESIS, PARTHENOGENESIS.

pseudogene A DNA sequence which, despite being largely homologous to a transcribed sequence elsewhere in the genome, is not transcribed. Some probably arise through gene duplication (*processed* pseudogenes); others, more numerous, must have originated by reverse transcription of mRNA and insertion into the gene, since they have lost their introns and are dispersed in location. The gene cluster for human haemoglobin contains two pseudogenes (designated by the symbol ψ in front of the gene symbol). See REPETITIVE DNA, TRANSPOSON.

pseudointerference Pattern of declining predator consumption rate with increasing predator density, reminiscent of the effects of mutual interference, but resulting from the aggregate response of the predator. See MUTUAL INTERFERENCE.

pseudoparenchyma See PLECTENCHYMA.

pseudopodium Temporary protrusions of some cells (in some sarcodine protozoans, e.g. *Amoeba*; MACROPHAGES) involved in amoeboid forms of CELL LOCOMOTION and food capture. See CYTOSKELETON for its role in *Dictyostelium*; PHAGOCYTE.

pseudopregnancy State resembling pregnancy in female mammal, but in absence of embryos. Due to hormone secretion of CORPUS LUTEUM, and occurs in species where copulation induces ovulation (e.g. rabbit, mouse), but when such copulation is sterile, or when normal oestrous cycle includes a pronounced luteal phase (e.g. bitch).

Pseudoscorpiones Arachnid order containing the pseudoscorpions: minute scorpion-like animals lacking tail and sting and common in soil where they are predatory, using pincer-like pedipalps.

pseudouridine See POST-TRANSCRIPTIONAL MODIFICATION.

Psilophyta Whisk ferns. Two living genera (*Psilotum* and *Tmesipteris*). *Psilotum* is tropical and subtropical in distribution; *Tmesipteris* is restricted to Australia, New Caledonia, New Zealand and other south Pacific islands. Sporophytes are very simple and may represent a reduction from a fern-like ancestor. *Psilotum* is unique among living vascular plants in lacking both roots and stems; the sporophyte comprises an underground rhizome system with many rhizoids, and a dichotomously branched aerial portion with small scale-like outgrowths. The main absorptive system is the mass of fungal hyphae that penetrates the cortex and interweaves among the rhizoids. The stele is PROSTELIC, with a central, irregular mass of xylem surrounded by the phloem, which, in turn, is enclosed by several layers of parenchymatous cells that comprise the pericycle. The rhizomatous system can form minute, multicellular, ovoid gemmae, which arise at the tips of rhizoids, and are filled with starch, and readily break free. The chlorophyllous aerial portions are about 20–30 cm in length, and have a conspicuous cuticle; the epidermal cells lack chloroplasts and rhizoids, but stomata are present. Chloroplasts occur in a narrow band of outer cortical cells, and this forms the main photosynthetic tissue. Inside the cortex, and forming the main supportive tissue, is a layer of sclerenchyma cells. The endodermis forms the innermost layer of the cortex and surrounds the stele. The xylem is exarch, and comprises scalariform and pitted tracheids in the metaxylem; and annular and helical tracheids in the protoxylem. The phloem is between the endodermis and the xylem. Sporangia (or synangia) are borne upon the uppermost branches; spores germinate giving rise to bisexual gametophytes which resemble portions of the rhizome. The gametophyte usually lacks vascular tissue. Antheridia and archegonia are intermixed over the entire surface of the gametophyte. Sperm of *Psilotum* are multiflagellate, and require water to reach the egg. Sporophyte is initially attached to the gametophyte by a foot, which absorbs nutrients from the gametophyte. It eventually becomes detached from the foot. *Tmesipteris* is epiphytic, with larger, leaf-like appendages and a rhizomatous system similar to that of *Psilotum*, but it often has collenchymatous thickenings in the cortex.

P-site Binding site of ribosome for the tRNA corresponding to the START CODON of mRNA (usually AUG, but sometimes GUG) and for the tRNA linked to the growing end of the polypeptide chain (hence P for peptidyl) during PROTEIN SYNTHESIS. The P-site receives this tRNA from the A-SITE. See RIBOSOME.

Psocoptera Booklice (psocids). Order of exopterygote insects. Small, some wingless. Biting mouth parts. Feed on fragments of animal and vegetable matter and paste of book-bindings.

psychoactive drugs Drugs altering mood or behaviour. Some are useful in treating neuropsychological disorders. Their actions may be explained through their effects on the synthesis, release, activity or metabolism of a particular NEUROTRANSMITTER. Non-selective central nervous system *depressants* include barbiturates, ethyl ALCOHOL and general anaesthetics; *antianxiety agents* include benzodiazepines (e.g. valium); *psychomotor stimulants* include cocaine, CAFFEINE and NICOTINE; *antidepressants* include tricyclic antidepressants (e.g. amitriptyline) and monoamine oxidase inhibitors; natural *narcotic analgesics* (pain-killers) are opioids and include morphine and its analogues (see ENDORPHINS, ENKEPHALINS), while synthetic (chemically unrelated) analgesics include

methadone (Dolophine); *antipsychotic* drugs include reserpine; *psychedelics and hallucinogens* include atropine and scopolamine (anticholinergic, blocking ACh receptors), mescaline, MDMA, amphetamines (resembling NORADRENALINE), and the SEROTONIN mimics psilocybin and psilocin (the very potent active ingredients of 'magic mushrooms') and lysergic acid diethylamide (LSD). The active component of marijuana (derivative of the hemp plant, *Cannabis sativa*), tetrahydrocannabinol (THC), is difficult to classify. It has effects on several neurotransmitter receptors, none of which clearly dominates. It may, as with general anaesthetics, increase neuron membrane fluidity; it blocks ACh receptors (possibly inducing mild analgesia), but its analgesic effect is possibly exerted through inhibition of ADENYLYL CYCLASE. Its euphoric effects may involve dopamine and serotonin receptors and in high concentrations its hallucinogenic action may be similar to a mild LSD experience. The synthetic amphetamine derivative MDMA ('Ecstasy') has been shown to produce irreversible destruction of serotonin neurons in monkeys and rats. In high doses it may be neurotoxic. Pain relievers (analgesics) include: aspirin (stops prostaglandin release at site of injury); codeine (blocks nerve impulses inside brain and spinal cord) and paracetamol (stops prostaglandin production in brain rather than site of injury). *Narcotics* are pain-killing derivatives of the opium poppy, e.g. morphine. Heroin is a semisynthetic opiate produced by chemical modification of morphine. *Addiction* to narcotic drugs (i.e., their compulsive abuse) is basically a learning process. An addicted person has a high tendency to relapse after withdrawal of the drug, and often experiences tolerance to the drug. This is a progressive decrease in responsiveness to the drug, sometimes caused by induction of ENZYMES which break it down. Somebody physically dependent on a drug needs it in order to function normally and suffers withdrawal symptoms on its removal. The nucleus accumbens, one part of the BASAL GANGLIA, has neural connections with the mesolimbic dopaminergic system (MDS, between the midbrain and higher processing centres) and, together, these pathways form the basis for the rewarding effects of cocaine and amphetamine. The limbic system also projects into the nucleus accumbens, providing information modulated by the MDS and relayed along two pathways: one passes through the brainstem into systems controlling movement; the other passes back to the limbic cortical centres associating environmental cues with reinforcers (drugs) in learning. It is possible that there is such a common neural pathway for all addictive drug action. See Fig. 141

psychrophilic (Of microorganisms, e.g. *Flavobacterium*) with an optimum growth temperature, measured by generations per hour, in the 5–15°C range. Compare MESO-PHILIC, THERMOPHILIC.

Pteridophyta In older plant classifications, a division containing spore-bearing (as opposed to seed-bearing) tracheophytes: Psilophyta, Lycophyta, Sphenophyta and Pterophyta.

pterobranchs Sessile, tubicolous, and often colonial, hemichordates (see HEMI-CHORDATA). Lophophore present dorsally on the collar, the arms forming a cone radiating away from the ventral mouth and engaged in muco-ciliary suspension feeding. Pharyngeal perforations allow excess water brought in with food to exit. Larva of the trochophore type. Asexual budding widespread.

Pterodactyla See PTEROSAURIA.

Pterophyta The ferns. Relatively abundant as fossils since the Carboniferous period; not known from the Devonian period. About 12,000 extant species, roughly two thirds in tropical regions, the other third in temperate regions. Display great diversity of form and habit, but are almost entirely terrestrial. The fern plant, the sporophyte, is the dominant generation; the gametophyte is the prothallus. Most temperate ferns comprise an underground siphonostelic rhizome which produces new leaves each year; roots are adventitious; leaves (fronds) are the megaphylls. Unique among seedless vascular plants in possessing metaphylls. All but a few ferns are homosporous, with sporangia variously placed and commonly borne in clusters called sori. Fern heterospory is restricted to two specialized groups of aquatic ferns. Spores of most homosporous ferns germinate to give rise to free-living gametophytes which develop into prothalli; these are monoecious, antheridia and archegonia developing on the ventral surface. Sperm are motile and require water to swim to the eggs. The sporophyte resulting from fertilization is initially dependent upon the gametophyte, but growth is rapid and it soon becomes independent. See LIFE CYCLE.

FIG. 141 *Where* PSYCHOACTIVE DRUGS *may act in the brain.*

(a) *Cocaine and amphetamine exert their primary effects on the pathway running between the midbrain and the nucleus accumbens that uses dopamine as its chemical messenger, or neurotransmitter. Other drugs of abuse may also affect the release of dopamine, either directly or indirectly. This system is thought to mediate the rewarding effects of a wide variety of drugs, including cocaine, nicotine and alcohol, and also some of the unpleasant symptoms of psychological withdrawal. Other neurotransmitter systems arising in the brainstem and basal forebrain send projections widely all over the brain. They use noradrenaline, 5-hydroxytryptamine and acetylcholine, and these too may mediate the effects of drugs of abuse or withdrawal from them, as indicated in the boxes around this diagram of a rat's brain.*

(b) *The nucleus accumbens also receives important inputs from regions in the cerebral cortex that are involved with emotional behaviour, learning and memory. These include the amygdala, the hippocampus and the prefrontal cortex, a group of structures loosely known as the 'limbic system'. The nucleus accumbens, in its turn, sends its information to regions in the brainstem concerned with movement, thereby controlling behavioural responses. In this way, the nucleus accumbens can act as an interface between emotional and motivational processes and behaviour.*

Key: Bzd = benzodiazepines; ? = conflicting data on this issue; ?? = little or no data, but considerable speculation on effects described.

(a)

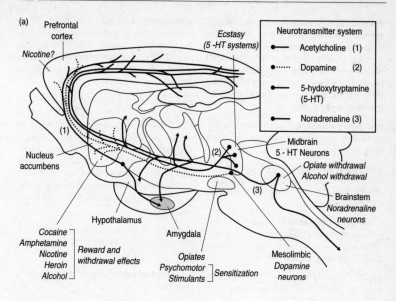

Prefrontal cortex

Nicotine?

Ecstasy
(5 -HT systems)

Neurotransmitter system

● Acetylcholine (1)

● Dopamine (2)

● 5-hydoxytryptamine
(5-HT)

● Noradrenaline (3)

(1)

Nucleus accumbens

(2)

Midbrain
5 - HT Neurons

Opiate withdrawal
Alcohol withdrawal

(3)

Brainstem
Noradrenaline
neurons

Hypothalamus

Amygdala

Mesolimbic
Dopamine
neurons

Cocaine
Amphetamine
Nicotine   *Reward and*
Heroin   *withdrawal effects*
Alcohol

Opiates
Psychomotor   *Sensitization*
Stimulants

(b)

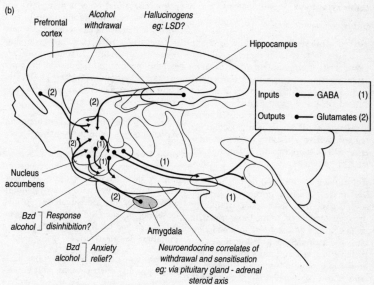

Prefrontal cortex

Alcohol
withdrawal

Hallucinogens
eg: LSD?

Hippocampus

(2)

(2)

Inputs ● GABA (1)

Outputs ● Glutamates (2)

(2)

(1)

(1)

(1)

Nucleus accumbens

(1)

(1)

(2)

(1)

Bzd ⎤ *Response*
alcohol ⎦ *disinhibition?*

Amygdala

Bzd ⎤ *Anxiety*
alcohol ⎦ *relief?*

Neuroendocrine correlates of
withdrawal and sensitisation
eg: via pituitary gland - adrenal
steroid axis

Pterosauria Pterodactyls. Extinct order of ARCHOSAURS, originating in late Triassic and disappearing at end of Cretaceous. Winged reptiles, fourth finger of the fore-limb greatly elongated and supporting a membrane. Hind-limbs feeble. Long tail tipped by a membrane and acting as a rudder. Wing spread up to 16 metres.

pterygoid muscles Muscles moving the vertebrate mandible, arising on the PTERYGOID PROCESSES.

pterygoid processes Wing-like projections of the inferior part of the SPHENOID BONE, to which some of the muscles moving the mandible attach.

Pterygota (Metabola) Insect subclass including all but the APTERYGOTA, some are secondarily wingless (e.g. fleas). Includes the ENDOPTERYGOTA and EXOPTERYGOTA.

ptyalin An AMYLASE present in saliva of some mammals, including humans.

pubic symphysis See PELVIC GIRDLE.

pubis See PELVIC GIRDLE.

puff (Balbiani ring) Swelling of giant chromosome (e.g. from dipteran salivary gland cell) normally regarded as representing regions being actively transcribed and consisting of many strands of decondensed DNA and its associated messenger RNA. Some recent work indicates some transcription of these chromosomes in regions lacking puffs, and some puff presence where transcription is lacking. See POLYTENY, ECDYSONE.

pulmonary (Adj.) Relating to the LUNG. Vertebrate *pulmonary arteries* (where present, derived from sixth AORTIC ARCH) carry deoxygenated blood to the lungs from the right ventricle of the heart in crocodiles, birds and mammals (or from the single ventricle of lungfish and amphibians) to the lung capillaries; in tetrapods, *pulmonary veins* return oxygenated blood to the left atrium.

Pulmonata Order of GASTROPODA. Lung develops from mantle cavity; e.g. snails, slugs.

pulp cavity Internal cavity of vertebrate tooth or denticle, opening by a channel to the tissues in which it is embedded. Contains *tooth pulp* of connective tissue, nerves and blood vessels, with odontoblasts lining dentine wall of cavity. See TOOTH.

pulse (Zool.) Intermittent wave of raised pressure passing rapidly (faster than rate of blood flow) from heart outwards along all arteries each time the ventricle discharges into the aorta. The increased pressure dilates the arteries, and this can be felt. Each pulse experiences resistance from the elastic walls of the arteries. See ARTERY. (Bot.) Seed of leguminous plant (Fabaceae), rich in proteins; e.g. soya bean.

pulvinus Joint-like thickening at base of leaf petiole (or petiolule of leaflet), playing important role in its movement.

punctuated equilibrium See EVOLUTION, CLADISTICS.

pupa (chrysalis) Stage between larva and adult in life cycle of endopterygote insects, during which rearrangement of body parts (METAMORPHOSIS) occurs, involving development of IMAGINAL DISCS. In some pupae (*exarate*) the appendages are free from the body; in many others (*obtect*) they are glued to the body by a larval secretion. Pupae of most culicine mosquitoes are active.

pupil Opening in IRIS of vertebrate and cephalopod EYE, permitting entry of light.

pure line Succession of generations of organisms consistently homozygous for one or more characters under consideration. Initiated by crossing two appropriately homozygous individuals or, in plants, by selfing one such individual. Such organisms *breed true* for the characters, barring mutation. See INBREEDING.

purine One of several nitrogenous bases occurring in NUCLEIC ACIDS, nucleotides and their derivatives. Synthesis involves addition of glycine, and two transaminations from glutamine, to a ribose-phosphate precursor. By far the commonest are ADENINE and GUANINE; rarer *minor purines* including

methylation products of these. See CYTOKIN-INS, PYRIMIDINE.

Purkinje fibres Modified CARDIAC MUSCLE fibres forming the bundle of His conducting impulses from PACEMAKER, and the fine network of fibres piercing the myocardium of the ventricles. See also HEART CYCLE. Not to be confused with *Purkinje cells*, which are large neurons of the cerebellar cortex.

puromycins Antibiotics interrupting the translation phase of protein synthesis in both prokaryotes and eukaryotes by their addition to growing polypeptide chain, causing its premature release from the ribosome. Compare ACTINOMYCIN D. See ANTI-BIOTIC.

purple bacteria (Proteobacteria) The largest, and most physiologically diverse, group of BACTERIA. Many are phototrophic, but many are non-phototrophic. Five main subgroups (α–ε). See PHOTOSYNTHESIS, ELEC-TRON TRANSPORT SYSTEM.

pusule A sac-like structure found in members of the DINOPHYTA, that opens via a pore into the flagellar canal. Like contractile vacuoles, has an osmoregulatory function, but is structurally more complex.

putrefaction Type of largely anaerobic bacterial decomposition of proteinaceous substrates, with formation of foul-smelling amines rather than ammonia.

pycnidium Flask-shaped structure in which conidia are formed in some members of the DEUTEROMYCOTA and ASCOMYCOTA.

pycnium Flask-like to variously shaped structure located subepidermally, wherein pycniospores or spermatia and special receptive hyphae develop. Produced by rust fungi (BASIDIOMYCOTA, Order UREDINALES). Pycniospores are exuded in a sugary liquid and are dispersed by insects. Those of one pycnium must be transferred to receptive hyphae of another to initiate the dikaryotic phase. See UREDINALES, AECIUM, UREDIUM, TELIUM.

pylorus Junction between vertebrate stomach and duodenum. Has a sphincter muscle within a fold of mucous membrane closing off the junction while food is digested in the stomach.

pyramid of biomass Diagram, pyramidal in form, representing the dry masses in each TROPHIC LEVEL of a community or food chain. The standing crops of populations are normally summated within each trophic level. Not all biomass has the same energy content, however. See ENERGY FLOW, PYRAMID OF NUMBERS, PYRAMID OF ENERGY.

pyramid of energy Diagram representing the energy contents within different TROPHIC LEVELS of a community or food chain. Generally pyramidal in form, it is difficult to represent the DECOMPOSERS, and does not easily indicate the unavailability (to the next trophic level) of storage if standing crops are used. See PYRAMID OF BIOMASS, PYRAMID OF NUMBERS.

pyramid of numbers Diagram, pyramidal (sometimes inversely) in form, representing the numbers of organisms in each TROPHIC LEVEL of a community or food chain. All organisms are equated as identical units, a massive tree being equivalent to a flagellate cell. See PYRAMID OF BIOMASS.

pyrenoid A proteinaceous region embedded in chloroplasts, or which lies beneath its surface, and contains the enzyme ribulose 1,5-bisphosphate carboxylase-oxygenase (RuBisCO). This catalyses the primary dark reaction of PHOTOSYNTHESIS, involving the fixation of CO_2 into carbohydrate. Occur in every eukaryotic algal group, and are considered a primitive characteristic. The pyrenoid is denser than the surrounding stroma and may or may not be traversed by thylakoids. In green algae (CHLOROPHYTA), they are associated with starch synthesis and surrounded by starch deposits.

pyridoxine One of the water-soluble vitamins in B6; precursor of *pyridoxal phosphate*, a PROSTHETIC GROUP associated with transaminase enzymes of mitochondria and/or the cytosol. Required by, among others, yeasts, bacteria, insects, birds and mammals.

pyrimidine One of a group of nitrogenous bases occurring in NUCLEIC ACIDS, nucleotides and their derivatives. Synthesis involves a step in which carbamoyl phosphate and aspartate form carbamoylaspartate prior to ring closure. The three commonest are cytosine, thymine and uracil. Methylated forms also occur; they are termed *minor pyrimidines*.

pyrosequencing See entry on DNA SEQUENCING.

Pyrrophyta See DINOPHYTA.

pyruvate dehydrogenase complex See KREBS CYCLE.

pyruvate : ferredoxin oxidoreductase (PFO) See FERREDOXINS.

pyxidium Type of CAPSULE.

Q

Q$_{10}$ effect The increase in the rates of enzymic reactions with an increase in temperature of 10°C (up to denaturation of the enzymes); also known as the *temperature coefficient*. Within physiological limits the Q$_{10}$ value of chemical reactions, and physiological processes, is 2–3. Purely physical processes (e.g. diffusion) usually have Q$_{10}$ values closer to 1. Metabolic rates in most animals whose body temperature varies show a Q$_{10}$ of 2–3 for every 10°C rise in ambient temperature. Animal BIOLOGICAL CLOCKS are temperature insensitive, with Q$_{10}$s of 1.

QO$_2$. Oxygen quotient Rate of O$_2$ consumption, often of whole organisms or tissues. Often expressed in $\mu l.mg^{-1}.hr^{-1}$.

quadrat A sampling unit used to assess density, frequency or biomass of organisms. Both non-destructive and destructive measures may be taken using this sampling unit. Quadrats are commonly square frames placed on the substrate so that organisms enclosed by it can be counted. Point quadrats are frames with vertical rods inserted in a horizontal row, often ten placed at 5 cm intervals, so that organisms touched by the rods can be counted. Quadrats may be arranged subjectively to include representative areas, for example areas with special features (e.g. the species being studied); or they may be placed at random over an area. Each sampling position must have an equal chance of being chosen, and it is common practice to obtain these sites by using pairs of random numbers at distances along two axes at right angles to each other. A variant of this type (random walk) may involve walking on a compass bearing for a random number of paces, sampling, changing direction and repeating this procedure. One problem with random sampling is that the cover of the areas with sampling points is not regular, some areas being under-sampled and others over-sampled. Sampling using a grid, however, results in a regular arrangement but limits the statistical tests that can be applied to the resulting data. A restricted random approach combines some advantages of both random and regular approaches. Here the study area is subdivided and each subdivision sampled at random. Then each point in the subdivision has an equal chance of being sampled and data are suitable for statistical analysis. Quadrats can also be used to investigate long-term changes in vegetation. See SIMPSON'S INDEX, SPECIES RICHNESS, BIODIVERSITY. Compare TRANSECT.

quadrate Cartilage-bone of posterior end of vertebrate upper jaw. Develops within the PALATOQUADRATE and attaches to neurocranium, in most cases articulating with lower jaw. In mammals becomes the incus (see EAR OSSICLES).

qualitative inheritance Inheritance in which the genetic VARIATION between members of a population results in recognizably distinct phenotypes; e.g. POLYMORPHISMS. Contrast QUANTITATIVE INHERITANCE.

quantasomes Granules occurring on inner surfaces of THYLAKOIDS of CHLOROPLASTS; thought to be basic structural units involved in light-dependent phase of photosynthesis rather than artefacts of electron microscope preparations.

quantitative inheritance Inheritance in which the genetic VARIATION between members of a population results in phenotypes differing in degree, rather than kind. Contrast QUALITATIVE INHERITANCE.

quantum In 1905, Albert Einstein proposed the particle theory of light in which light is considered to be composed of particles of energy (photons or quanta of light). The energy of a photon (quantum of light) is not the same for all kinds of light; instead it is inversely proportional to the wavelength.

quantum dots (QDs) Fluorescent semiconductor nanocrystals: nanometer-scale semiconductor crystallites. When conjugated to DNA and encapsulated within a layer of biocompatible material, such as a micelle, they can be used as *in vitro* probes to highlight specific complementary DNA sequences. They have been shown to be non-toxic and cell autonomous when injected into frog tadpoles. See NANOBIO-TECHNOLOGY.

Quaternary The GEOLOGICAL PERIOD comprising the last 2.6 Myr of Earth history: the Pleistocene and Holocene epochs. Data from the Turkana region of East Africa suggest that there at least the fauna was steadily adapting towards grassland vegetation from woodland, driven by a global cooling and African drying. There appears to have been a 50–60% turnover in Turkana mammal species between 3–2 Myr BP, with a 30% increase in diversity. This is the period during which our genus, *HOMO*, appeared.

quaternary structure See PROTEIN.

quelling A COSUPPRESSION phenomenon in the mould *Neurospora crassa*.

quiescence See entry on DIAPAUSE.

quiescent centre Area at the tip of the root apical meristem where rate of cell division is lower than in surrounding tissue.

quinine An important alkaloid extracted from the bark of *Cinchona* species (Family Rubiaceae) used in the treatment of malaria.

quinoa *Chenopodium quinoa* (Family Chenopodiaceae) is an important grain crop in the Andes, and elsewhere in South America.

quinone Important small hydrophobic components of the ELECTRON TRANSPORT CHAIN. Unlike CYTOCHROMES and IRON-SULPHUR PROTEINS, quinones (e.g. COENZYME Q, ubiquinone) carry the equivalent of a hydrogen atom. By alternating electron transfer between components that carry or do not carry a proton with the electron, protons can be moved across the membrane, setting up proton gradients. Most quinone molecules are not attached to proteins and diffuse rapidly in the plane of the membrane. See CHEMIOSMOTIC THEORY.

quorum sensing A prokaryotic phenomenon in which assemblages of bacteria (e.g. BIOFILMS) activate target genes when diffusible signal molecules reach threshold concentrations – often in a density-dependent manner. This form of communication can induce swarming, or a change in colony morphology. See BIOLUMINESCENCE.

R

race See INFRASPECIFIC VARIATION, ECOTYPE.

raceme Kind of INFLORESCENCE.

rachis (1) Main axis of an INFLORESCENCE. (2) The axis of a fern leaf (frond), from which the pinnae arise. (3) In compound leaves, the extension of the petiole corresponding to the midrib of an entire leaf.

RAD51, Rad 51 See *recA* entry.

radial cleavage (bilateral cleavage) See CLEAVAGE.

radial micellation (Bot.) The radial orientation of the cellulose microfibrils in the guard walls of stomata (see STOMA), which allows the guard cells to lengthen while preventing them from expanding laterally. The common wall at the ends of the guard cells remains almost constant in length during opening and closing of the stomata. Therefore, increase in TURGOR PRESSURE causes the outer (dorsal) walls to move relative to their common walls. As this occurs, the radial micellation transmits the movement to the wall bordering the stomatal opening, and the pore opens.

radial section Longitudinal section cut parallel to the radius of a cylindrical body (e.g. a stem or root).

radial symmetry Capable of bisection in two or more planes to produce halves that are approximately mirror images of each other. Characteristic of bodies of coelenterates and echinoderms; and of many flowers (see ACTINOMORPHIC) and some algae, e.g. the CENTRIC DIATOMS, when viewed in valve view. Compare BILATERAL SYMMETRY; see AXIS.

radical Root of embryo seed plants. See SUPEROXIDES for *free radical*.

radioactive labelling See LABELLING.

radioisotope (radioactive isotope) See LABELLING, RADIOMETRIC DATING.

Radiolaria Order of marine planktonic sarcodine protozoans, lacking shells but with a central protoplasm comprising chitinous capsule and siliceous spicules perforated by numerous pores through which spines project between vacuolated and jelly-like outer protoplasm. Between the spines radiate out branching pseudopodia exhibiting cytoplasmic streaming. Many house yellow symbiotic algal cells (zooxanthellae). *Radiolarian oozes* cover much of the ocean floor and are important in flint production.

radiometric dating Methods employed to measure the amount of an isotope produced by radioactive decay, or the amount of the radioisotope itself. By assuming that the proportion of the radioactive isotope to the stable isotope is the same now as when the sediment was laid down, and that no subsequent addition or dilution has occurred, the proportion of the radioactive isotope or its product remaining today is a function of the time that it has had to decay at its constant known rate (the *half-life*). Carbon-14 is the most commonly used radiometric dating technique in Quaternary palaeoecology, as it has a suitable decay rate (half-life = 5,568 years), which allows dating to be made back to *c*. 40,000 yr BP. Accelerator mass spectrometry (AMS) has provided further analytical improvements in the ^{14}C method, since samples containing 1 mg or less of elemental carbon can be analysed.

Previous methods have required nearly 1,000 times that amount. Dating of recent materials using ^{14}C is difficult because of the relatively large errors in the measurements, and the large amounts of ^{14}C-deficient carbon that have been introduced into the atmosphere through burning of fossil fuels since the industrial revolution. So other isotopes have to be used, such as caesium-137 (^{137}Cs) or lead-210 (^{210}Pb). The former was introduced into the atmosphere as a result of nuclear testing in 1954. Since then, the amount of ^{137}Cs in undisturbed lake sediments reached maximum levels in 1963, since when there has been a decline. ^{210}Pb is more frequently used and can date sediments up to 150 years old. Radium decays in soils to radon-222, which escapes into the atmosphere, where it decays to ^{210}Pb. The latter enters a lake in precipitation, and eventually becomes incorporated into sediments. This is called unsupported ^{210}Pb. Supported ^{210}Pb is produced within the sediments, and is assessed by measuring its ^{226}Ra grandparent. The excess ^{210}Pb over the expected ^{210}Pb gives the amount of unsupported ^{210}Pb. The half-life of ^{210}Pb is 22.26 years. By assuming either a constant initial concentration in the sediments which have accumulated at a constant rate, or by assuming a constant rate of ^{210}Pb supply to the sediments, the age of the sediment can be determined. The second model provides more accurate dates. The thorium-228 : thorium-232 ratio can be used to date sediments up to 10 Myr in age. In contrast, the method applicable to material from the early part of the PLEISTOCENE is the potassium/argon (K/Ar) method, by means of which the earliest HOMINID remains from East Africa have been indirectly dated. The K/Ar method depends upon the decay of ^{40}K to the non-radioactive inert gas argon, and is particularly applicable to volcanic rocks in which the K/Ar clock was set to zero when they were formed. The half-life of ^{40}K is very long (1,300 Myr) and in most cases the lower limit of the K/Ar dating is $c.$ 250,000 yr BP. Uranium-series dating methods are applicable over the time range from a few thousand to $c.$ 10^6 years, and can be applied to a wide range of materials. It is an open system using disequilibrium in the ^{238}U-^{234}U-^{230}Th (thorium) decay series.

Although not strictly radiometric, the following two techniques are now invaluable in dating. For material up to $c.$ 1 Myr old, thermoluminescence (TL) and electron spin resonance (ESR) techniques are employed. TL depends upon the fact that heating and sunlight repair the damage to crystalline materials (e.g. flint) caused when natural radiation in them displaces electrons. The process of accumulating radiation damage then starts again, and the date of a burnt flint sample can be obtained when the material is heated to 500°C or more to release the energy of the electrons displaced since the first burning. ESR employs microwave radiation to measure the accumulated amount of radiation damage in crystalline materials (carbonates, tooth enamel), since it causes displaced electrons to emit a signal proportional to the damage.

radius One of two long bones (the other being the ulna) in the tetrapod fore-limb. Articulates with the side of the fore-foot bearing the thumb. See PENTADACTYL LIMB.

radula 'Tongue' of molluscs; a horny strip, continually renewed, with teeth on its surface for rasping food. Found in Amphineura, Gastropoda and Cephalopoda; absent in Bivalvia, which are microphagous. May be modified for boring in some species. Pattern of teeth helpful in identification.

RAGE (receptor for advanced glycation end-products) A cell surface receptor protein of the immunoglobulin superfamily which binds potentially damaged glycated proteins. See AGEING.

Ramapithecus (Sivapithecus) See PONGINAE.

ramet An offshoot or module in some plants and modular invertebrates, formed by vegetative growth, and actually or potentially capable of independent physiological existence (e.g. tubers of a potato plant, the polyps on a colonial hydroid). See GENET.

raphe (1) In seeds formed from anatropous ovules, a longitudinal ridge marking position of the adherent FUNICLE. (2) Elongated slit (or slit pair) through valve wall of some pennate diatoms, involved in movement of cell over the substratum.

Raphidophyceae A class of the algal division HETEROKONTOPHYTA that contains nine genera found in both freshwater and marine habitats. In freshwater these algae are found in pools on bogs (e.g. *Goniostomum*) or above barren sandy substrata. In the marine environment, species such as *Chattonella* and *Fibrocapsa* form summer blooms along the Japanese coast causing severe disruption to fish farming. All members of this class are unicellular flagellates that lack a cell wall. Cells are large (50–100 μm), dorsoventrally constructed, with a curved dorsal side and a flatter ventral side. The latter is traversed by a shallow, longitudinal groove. The two flagella are inserted near the apex of the cell but not laterally. They arise from the bottom of a small funnel-shaped gullet, which lies on the ventral side of the cell just below the apex. The pleuronematic flagellum points forwards; mastigonemes are tripartite and formed within cisternae of the endoplasmic reticulum. The acronematic flagellum points backwards along the cell lying in the ventral groove. No eyespot or photoreceptor is present. Chloroplasts are elliptical, numerous and yellow-green or yellow-brown, and often somewhat flattened where they press against one another. Pigments include chlorophylls a, c_1 and c_2; the main accessory pigments are β-carotene, diadinoxanthin, vaucheriaxanthin and heteroxanthin, while in marine species they are β-carotene and fucoxanthin. A ring-shaped nucleoid is present in each chloroplast. Trichocysts are found beneath the cell surface; spherical mucilage vesicles may also be present. The Golgi apparatus sits like a cap on top of the nucleus. The latter lies in a capsule of relatively viscous fluid surrounded by more fluid cytoplasm. The periplastidal reticulum lies between the chloroplast envelope and the chloroplast endoplasmic reticulum.

Raphidophyta Chloromonads (Kingdom Protista). Biflagellated algae (some become palmelloid, e.g. *Vacuolaria*) possessing one anterior (commonly hairy) and one posterior (smooth) flagellum. Two membranes of chloroplast endoplasmic reticulum are present, and within the chloroplasts are chlorophylls a and c. Many cells possess mucocysts in the peripheral cytoplasm, which can burst out of the cell, releasing a fibrous material. Freshwater species are green in colour and common in the EPIPELON. Marine taxa are yellowish due to the carotenoid fucoxanthin, and are a common component of algal blooms in red tides in coastal waters. Sometimes included with the XANTHOPHYTA, due to similar flagellation and pigmentation; but mucocysts are not found in xanthophytes.

RAR RETINOIC ACID receptor; e.g. RARα, RARβ, RARγ.

RAREs **R**etinoic **A**cid-**R**esponsive **E**lements on DNA, to which NUCLEAR RECEPTOR heterodimers (RAR-RXR) bind.

rarity A growing body of evidence suggests that locally rare (and geographically restricted) species tend to share some or all of the following characteristics compared with commoner taxa: (i) lower levels of self-incompatibility, (ii) tendency towards asexual reproduction, (iii) lower overall reproductive effort, and (iv) poorer dispersal abilities.

***Ras* genes** Oncogenes initially discovered through their ability to cause RAT sarcomas (see CANCER CELL). Mutations in *ras* in mammalian cells tend to be oncogenic and are frequently associated with human cancers. Encode RAS PROTEINS.

Ras proteins Oncoproteins encoded by *RAS* GENES which bind GTP and GDP and act as GTP-activated switches in SIGNAL TRANSDUCTION pathways in virtually all metazoan cells studied to date, and as key regulators in growth of eukaryotic cells (see GAPS). In one such signalling cascade, used to transduce both mitogenic and differentiation signals, both receptor and non-receptor TYROSINE KINASES lie upstream of Ras (see Fig. 143b). Often located on the cytoplasmic surface of the plasma membrane, Ras often requires an intermediary protein complex (including the GAP, Ras-activator or Sos) which binds both the activated receptor and Ras. Involved in transduction of mitogenic signals from cell periphery to the nucleus. See DOMAIN for SH2 and SH3.

ratites See PALAEOGNATHAE.

Raunkiaer's life forms System of vegetational classification based on position of perennating buds in relation to soil level, indicating how plants survive the unfavourable season of their annual life cycle. The following are recognized: *phanerophytes*: woody plants whose buds are borne more than 25 cm above soil level (many trees and shrubs); *chamaerophytes*: woody or herbaceous plants whose buds are above soil level but less than 0.25 m above; *hemicryptophytes*: herbs with buds at soil level, protected by the soil itself or by dry dead portions of the plant; *geophytes*: herbs with buds below soil surface; *heliophytes*: herbs whose buds lie in mud; *hydrophytes*: herbs with buds in water; *therophytes*: herbs surviving the unfavourable season as seeds.

ray (Bot.) Tissue initiated by CAMBIUM and extending radially in secondary xylem and phloem. Mainly parenchymatous, but may include tracheids in the xylem. (Zool.) See FINS (for ray fins); MEDULLARY RAY.

ray flower See DISC FLOWER, ASTER.

ray initials (Bot.) One of two types of meristematic cells of the vascular cambium; horizontally elongated or squarish cells that produce horizontally orientated ray cells, which form the vascular rays or radial system. The rays, composed largely of parenchyma cells, are variable in length. See FUSIFORM INITIAL.

RdRP RNA-directed RNA polymerase. Converts single-stranded RNA into double-stranded RNA. Some viruses encode their own. Apparently not encoded by the human genome. RdRP also has a surveillance role in *Neurospora* by silencing unpaired DNA during meiosis – unpaired genes probably being foreign DNA sequences, such as transposons. See RNA SILENCING.

reaction centre (Bot.) The chlorophyll molecule of a photosystem capable of using energy in the photochemical reaction. See PHOTOSYNTHESIS.

recA Gene of the bacterium *E. coli* whose product, the RecA protein, is involved in promoting general (i.e. homologous) DNA RECOMBINATION in the genome, and has ATPase activity in the presence of single-stranded DNA as well as catalysing DNA base-pairing and strand annealing (assimilation). RecA protein binds to single-stranded DNA and anneals it to any complementary sequence in a double-stranded (duplex) DNA in such a way as to replace one of the two DNA strands of the original duplex, forming a HETERODUPLEX. Humans and mice contain at least seven RecA homologues. RecA has 50% homology to one of these, the RAD51 (or Rad51) protein, molecules of which bind cooperatively to form a helix around DNA and stretch it out of the B-form into a form 1.5 times its length so that homologous DNA sequences can pair up. Rad51 repairs broken replication forks during S phase of the CELL CYCLE, and two of its accessory proteins are BRCA1 and BRCA2, mutations in whose genes are associated with greatly increased risk of breast CANCER.

recapitulation See BIOGENETIC LAW.

recB, recC CISTRONS of the bacterium *E. coli* whose combined products make up the RecBC enzyme, which initiates DNA-unwinding at any free duplex end and has nuclease activities. Once bound to a free duplex end (not normally present in *E. coli*) RecBC proceeds to unwind the duplex, but wherever the enzyme encounters the DNA strand sequence 5'-GCTGGTGG-3' (termed *Chi*), it cuts the strand, leaving it exposed to binding by RecA protein (see *recA*). For this reason, recombination is promoted in *Chi* regions, of which there may be one thousand in the *E. coli* genome.

recent See HOLOCENE.

receptacle (thalamus, torus) Apex of flower stalk, bearing flower parts (perianth, stamens, carpels). Its relation to the gynoec-

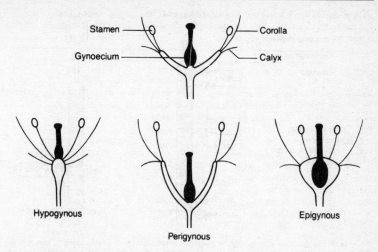

FIG. 142 *Diagrams of different types of floral* RECEPTACLE, *showing the position of the gynoecium (black) relative to other flower parts.*

ium determines whether carpels are *inferior* or *superior*. When carpels are at the apex of a conical receptacle and other flower parts are inserted in turn below, the gynoecium is superior and the flower *hypogynous* (e.g. buttercup, Fig. 142). When carpels are at the apex (centre) of a concave receptacle with other flower parts borne around its margin, the gynoecium is superior and the flower *perigynous* (e.g. rose, upper and middle diagrams in Fig. 142). When receptacle completely encloses carpels and other flower parts arise from receptacle above, the gynoecium is inferior and the flower *epigynous* (e.g. apple, dandelion, Fig. 142). In this condition, carpel walls are intimately fused with the receptacle wall. (2) Describing the shortened axis of the INFLORESCENCE (capitulum) in Compositae. (3) In some brown algae (e.g. Fucales, Phaeophyta), the swollen thallus tip containing conceptacles.

receptor (1) Sensory cell responding to some variable feature of an animal's internal or external environment by a shift in its membrane voltage. Sometimes (see Fig. 143C), as in *primary receptors* (e.g. MUSCLE SPINDLES), only part of the cell is sensory and generates action potentials with a frequency related to the stimulus intensity; the rest of the cell is axonal, transmitting signals long distances. In *secondary receptors* the altered membrane voltage initiates action potentials in a synapsing neuron by bringing about voltage changes (*receptor*, or GENERATOR POTENTIALS) in the postsynaptic membrane of the neuron. Information about the duration and rate of change of stimulus intensity is relayed by tonic receptors and PHASIC RECEPTORS, respectively. The transduction event in all receptors is the production of a receptor potential on receipt of the signal (see POTENTIATION, SIGNAL TRANSDUCTION). Receptors may be *interoceptors* or *exteroceptors* and classified by modality into *chemoreceptors*, *mechanoreceptors*, *photoreceptors*, etc. See HAIR CELL, SENSE ORGAN.

(2) Membrane receptor molecules (receptor sites). Exquisitely adapted proteins. One classification distinguishes: (a) *channel-linked* (*ionotrophic*) *receptors*, associated with ion channels. Either ligand-gated (e.g. the ACETYLCHOLINE receptor) or voltage-gated (e.g. the sodium channel). The action signalled by the stimulus is effected directly by the receptor molecule itself (unlike the *non-channel-linked*, or *metabotropic*, *receptors* below). One such class of receptor molecule

(a)

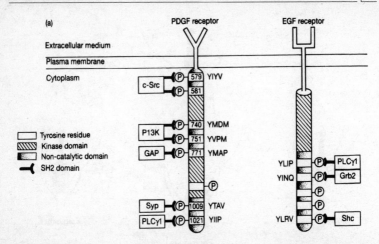

FIG. 143a *PDGF and EGF transmembrane receptor molecules with their kinase domains, non-catalytic domains and tyrosine residues (some numbered in PDGF, indicating the amino acid residue involved). Once phosphorylated, these tyrosine residues may bind specific SH2-bearing substrates, such as c-Src itself, GTPase-activating protein (GAP) of the GTPase Ras and phospholipase Cγ1 (PLCγ). See* RECEPTOR *and specific references there.*

(*seven transmembrane helix receptors*, or *heptahelicals*) spans the membrane seven times with α-helices, and includes photoreceptors (e.g. BACTERIORHODOPSIN), olfactory receptors and receptors for several neurotransmitters, all of them G-PROTEIN-linked and members of the rhodopsin superfamily of proteins; (b) *endocytic receptors*, such as the LDL and TRANSFERRIN receptors, initiate clathrin-coated pit formation (see COATED VESICLE and Fig. 34) during receptor-mediated ENDOCYTOSIS. Neutrophils, macrophages and other mononuclear phagocytic white cells have cell-surface Fc receptors so that once ANTIBODY is bound to pathogen, the Fc end can bind the phagocyte; (c) *growth factor receptors*, such as those for insulin (see DIABETES), epidermal GROWTH FACTOR and those used in intercellular signalling, e.g. CD4 and CD8 (see ACCESSORY MOLECULES) span the membrane, linked covalently or by other direct coupling to an intracellular effector molecule, e.g. an enzyme such as a tyrosine kinase (when termed *catalytic receptors*; see RECEPTOR TYROSINE KINASES). Binding of ligand here usually results in an ALLOSTERIC shift in

the receptor, initiating one or more signal cascades associated with the control of the CELL CYCLE, gene expression and cell differentiation. Receptors frequently form large assemblies with associated cytoplasmic SIGNAL TRANSDUCTION components (see Figs. 134a and b). Some of these, *dependence receptors*, initiate APOPTOSIS in the absence of their ligand, as occurs when the cell is poorly positioned as a result of developmental errors.

receptor potential See GENERATOR POTENTIAL.

receptor proteins See RECEPTOR (2) and reference there.

receptor serine/threonine kinases Transmembrane receptors found, with distinctions, in both the plant and animal kingdoms (see PROTEIN KINASE). All have an intracellular serine/threonine kinase domain and an extracellular ligand-binding domain. Unlike animals, plants seem to depend more on these receptors than they do on the RECEPTOR TYROSINE KINASES.

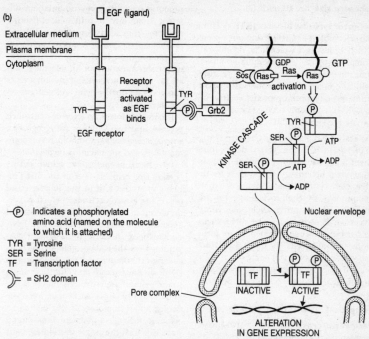

FIG. 143b *The involvement of EGF* RECEPTOR *in one of its signal transduction pathways involving the G-protein Ras. Ras-activation is required for proper functioning of many growth and differentiation factors, such as epidermal growth factor (EGF), shown here, insulin, PGDF, NGF, T cell receptor and many cytokines. Grb2 is also known as Sem5. Compare Fig. 151.*

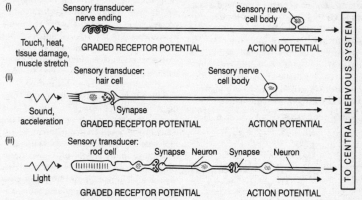

FIG. 143c *Different types of sensory cell. (i) Primary receptor; (ii) and (iii) secondary receptors. In all three cases, a graded receptor potential evoked in the sensory transducer is translated into an action potential firing frequency.*

receptor site See RECEPTOR (2).

receptor tyrosine kinases (RTKs) Those PROTEIN TYROSINE KINASES having one DOMAIN which binds an extracellular ligand, one domain crossing the membrane bilayer (transmembrane) and one or more intracellular domains for phosphorylating tyrosine radicals of specific proteins. Each family of RTKs has probably evolved by GENE DUPLICATION. See Figs. 143 and 151.

recessive (Of phenotypic characters) only expressed when the genes determining them are HOMOZYGOUS (complete recessivity). When heterozygous, the allele for the character is either 'silent', giving rise to no cell product, or its effect is masked by the presence of the other allele. Sometimes 'recessive' is used to describe alleles themselves, but since many gene loci are pleiotropic it would need to be made clear which aspect of phenotype was being described as recessive (see MUTATION for loss-of-function mutations). Compare DOMINANCE. See PENETRANCE.

reciprocal altruism See ALTRUISM.

reciprocal cross (1) Cross between two hermaphrodite individuals, in which the male and female sources of the gametes used are reversed. (2) Crossing operation between stocks of two different genotypes, where each stock is used in turn as the source of male and female gametes. Employed when testing for SEX-LINKAGE (also sex-limited and sex-controlled inheritance), where one sex has a greater influence than the other in determining offspring phenotype. When reciprocal crosses give very different results (e.g. in F1 or F2) the character studied is likely to be sex-linked or under cytoplasmic control. See CYTOPLASMIC INHERITANCE.

recombinant DNA DNA whose nucleotide sequence has undergone alteration as a result of incorporation of, or exchange with, another DNA strand. Such DNA occurs naturally as a result of CROSSING-OVER during RECOMBINATION, and also during recombinant DNA techniques employed during GENE MANIPULATION.

recombination Any process, other than point mutation, by which an organism produces cells with gene combinations different from any it inherited. Offspring resulting from such recombinant cells are *recombinant offspring*. A major source of GENETIC VARIATION, its effectiveness is dependent upon mutation for initial gene differences, from which recombination events can generate further gene and genome rearrangements.

Recombination is an enzyme-controlled process, and two varieties are recognized: *general recombination* (homologous recombination), and *site-specific recombination*.

(a) *General recombination*, which occurs both during crossing-over in the first prophase of MEIOSIS and in the double-strand break DNA REPAIR MECHANISM which is predominant in MITOSIS, begins when a special endonuclease cuts both strands of a DNA double helix. Because it has two DNA-binding sites, the RecA protein, or one of its eukaryotic homologues, can hold both a DNA double strand and another single strand simultaneously, enabling it to undertake a synapsis reaction between the duplex and a homologous single strand – achieved through a transient three-strand stage in which the single strand base-pairs with bases that flip out from the major groove of the duplex molecule (see Fig. 42a). Several RecA molecules line up (see Fig. 144a) along the three-strand intermediate as elongation of the resulting heteroduplex region takes place. This *branch migration* process, which is ATP-dependent, results in one strand of the original duplex being ousted while the originally unpaired homologue takes its place. This leads to an exonuclease pruning back the 5' end to leave the 3' end protruding and unpaired. This end then 'searches' for a homologous DNA helix with which to base-pair, which it finds on the other chromosome of the pair. If the other broken end behaves likewise, a joint molecule is formed (see Fig. 144b) and, depending upon later events, will result either in restoration of the two starting DNA helices after repair of the break, or in a crossover event in which a heteroduplex is formed which holds together two different helices. The different outcomes seem to depend upon whether or not a set of

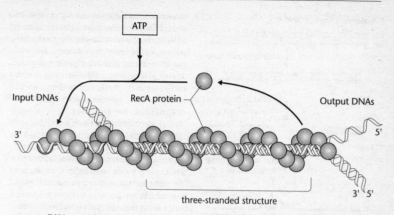

FIG. 144a *DNA synapsis catalysed by the RecA protein. In vitro experiments show that several types of complex are formed between a DNA single strand covered with a RecA protein and a DNA double helix. First a non-base-paired complex is formed, which is converted through transient base-flipping to a three-stranded structure as soon as a region of homologous sequence is found. This complex is unstable because it involves an unusual form of DNA, and it spins out a DNA heteroduplex plus a displaced single strand from the original helix. Thus the structure shown in this diagram migrates to the left, reeling in the 'input DNAs' while producing the 'output DNAs'. The net result is a DNAs strand exchange identical to that in (b).*

meiosis-specific proteins is involved, because if it is then two Holliday junctions form on the HETERODUPLEX, a different pair of strands joining up at the two junctions – producing a crossover. GENE CONVERSION is usually regarded as a straightforward consequence of general recombination (see also *meiotic drive*, in ABERRANT CHROMOSOME BEHAVIOUR (4)).

(b) *Site-specific recombination* involves mobile genetic elements a few hundred to tens of thousands of nucleotide pairs in length. Site-specific enzymes break and rejoin, often reversibly, two DNA double helices at specific sequences. The DNA may be viral (when between-cell mobility may also be involved) or involve TRANSPOSABLE ELEMENTS (when movement occurs within the genome of only one cell). Much of the REPETITIVE DNA of vertebrate chromosomes is a relic of these mobile elements, most of which have now mutated and lost their mobility. But their translocation processes, often integrating them into functional genes, may give rise to spontaneous mutations in many organisms, including

humans, and contribute to evolutionary change. The enzymes mediating site-specific recombination recognize much shorter and more specific nucleotide sequences than those involved in general recombination and there is no requirement for extensive DNA homology. Two kinds of site-specific recombination are recognized: *transpositional site-specific recombination*, in which a mobile genetic element (see TRANSPOSONS) is excised from a part of the genome and integrates at a non-homologous region elsewhere – without heteroduplex formation; and *conservative site-specific recombination*, in part so called because it is reversible, and in part because the original DNA sequences of both donor and recipient are conserved by means of a short HETERODUPLEX joint. This type therefore requires a short DNA sequence to be identical on both the donor and recipient DNA.

In bacterial TRANSFORMATION and TRANSDUCTION (examples of *homologous* recombination) and some eukaryotic gene transfers, homologous DNA duplexes first align, the donor duplex undergoes DENATURATION

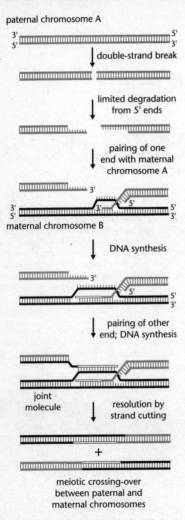

paternal chromosome A

double-strand break

limited degradation from 5' ends

pairing of one end with maternal chromosome A

maternal chromosome B

DNA synthesis

pairing of other end; DNA synthesis

joint molecule

resolution by strand cutting

+

meiotic crossing-over between paternal and maternal chromosomes

FIG. 144b *General recombination in meiosis. As indicated, the process begins when an endonuclease makes a double-strand break in a chromosome. An exonuclease then creates two protruding 3' single-stranded ends, which find the homologous region of a second chromosome to begin DNA synapsis. The joint molecule formed can eventually be resolved by selective strand cuts to produce two chromosomes that have crossed over, as shown.*

(separates into its two strands), and one strand invades the host duplex, aligning with the host strand having the greater base-pairing conformity. It is then nicked while the host strand without a partner is nicked at two places, donor DNA getting inserted by ligases in its place. The evidence for this comes from electron microscopy and CHROMOSOME MAPPING. See GENE MANIPULATION. Recombination can occur between mitochondrial genomes, as in yeast and the slime mould *Physarum* (see MITOCHONDRION).

During the generation of ANTIBODY DIVERSITY during B CELL maturation, genes from different parts of a chromosome are brought together in such a way that an RNA transcript is produced which effectively 'omits' the intervening DNA sequences. Mitotic recombination between homologous chromosomes is very rare, but can be induced by X-RAY IRRADIATION. It can lead to production of two daughter cells homozygous for different alleles and for which the parent cell was heterozygous. See TWIN SPOTS.

recombination nodule See SYNAPTONEMAL COMPLEX.

recombination value Alternative for CROSS-OVER VALUE.

rectum Terminal part of intestine, opening via anus or cloaca and commonly storing faeces. In insects, often reabsorbs water, salts and amino acids from the 'urine' (see MALPIGHIAN TUBULES); some insect larvae have tracheal gills in the rectum, while larval dragonflies also eject water forcibly from the rectum for propulsion. Ectodermal in origin (see PROCTODAEUM).

red blood cell (red blood corpuscle, erythrocyte) Most abundant vertebrate blood cell; generated in bone marrow, usually from RETICULOCYTES. Contains many molecules of HAEMOGLOBIN loading and unloading molecular oxygen (and carbon dioxide to a much lesser extent) and serving as a blood BUFFER. Mammalian erythrocytes are flattened, circular, biconcave discs (about 8 μm diameter in humans), lacking nuclei, mitochondria and most internal membranes. Tend to be larger and oval in shape in other vertebrates, retaining a nucleus. Damaged by passage through capillaries, they last about four months in humans (judged by radioactive tracers) before being destroyed by the liver's RETICULOENDOTHELIAL SYSTEM. Their surface antigens specify BLOOD GROUP. Their membrane SODIUM PUMPS regulate cell volume, but hypotonic solutions cause osmotic swelling and rupture, leaving erythrocyte membranes as *ghosts*. The important enzyme CARBONIC ANHYDRASE catalyses the reversible reaction:

$$H_2O + CO_2 \overset{\longrightarrow \text{ tissues}}{\underset{\text{lungs} \longleftarrow}{\longleftrightarrow}} H^+ + HCO_3^-$$

enabling rapid exchanges of gases in the lungs and body tissues. Role of erythrocytes in CO_2 transport is primarily to generate HCO_3^- ions for carriage in the plasma and to reconvert them back to CO_2 molecules in the lungs, where exhalation occurs. However, about 23% of CO_2 carried in human blood is in the form of *erythrocytic carbaminohaemoglobin*, which breaks down in the lungs to release CO_2 again. See HAEMOPOIESIS.

redia One of the larval types in endoparasitic Trematoda, developing asexually from the sporocyst and from other rediae (see POLYEMBRYONY). Often parasitic in snails, developing into cercariae.

redox reactions Oxidation-reduction reactions, in biology generally catalysed by enzymes. Involve transfer of electrons from an electron donor (reducing agent) to an electron acceptor (oxidizing agent). Sometimes hydrogen atoms are transferred, equivalent to electrons, so that dehydrogenation is equivalent to oxidation. Respiration involves many *redox pairs*, one member donating, the other accepting, electrons, determined by their relative *standard oxidation-reduction potentials*. See ELECTRON TRANSPORT SYSTEM.

Red Queen hypothesis The concept that 'all progress is relative', as applied in biology to evolutionary theory. It is, essentially, the view that in evolutionary conflicts there tends to be an ARMS RACE between adaptive changes which evolve in the conflicting parties: the emergence of an advantage in one party creates selection pressures favouring the evolution of counter-measures in the other party, and so on iteratively. This kind of endless feedback is a predictable consequence of Darwinism and has strong empirical support. See ANTIBIOTIC RESISTANCE ELEMENT, PLANT DISEASE AND DEFENCES.

reducing sugar A sugar capable of reducing ferric or cupric ion in solution (the basis of Fehling's reaction), as indicated by BENEDICT'S TEST. Depends upon presence of potentially free aldehyde or ketone group. Most monosaccharides are reducing sugars (and all are in weakly acid solution), as are most disaccharides except sucrose.

reduction division First division of MEIOSIS, the chromosome numbers of the daughter cells produced being half that of the parent cell.

reductionism See EXPLANATION.

reflex Innate (inherent) and often invariant neuromuscular animal response to an internal or external stimulus, usually with little delay involved. In its simplest form, mediated by a *reflex arc* involving input along a sensory neuron of a spinal nerve to the CENTRAL NERVOUS SYSTEM (CNS) and output along a motor neuron of the same or a different spinal nerve to an effector (muscle, gland). Frequently in vertebrates an association (relay) neuron intervenes in the CNS, acting to transmit impulses via white fibres to the brain. This may enable perception and learning, as in CONDITIONED REFLEXES. Vertebrate reflexes which involve only the spinal component of the CNS are termed *spinal reflexes*. Examples of innate reflexes in humans include blinking, sneezing, coughing, ventilation, regulation of heart rate and pupil diameter (the iris reflex), the thirst reflex (see HYPOTHALAMUS) and complex postural reflexes involving the CEREBELLUM during walking and running. See Fig. 145, NERVOUS INTEGRATION, NERVOUS SYSTEM.

refractory period The time interval during which, subsequent to the passage of an

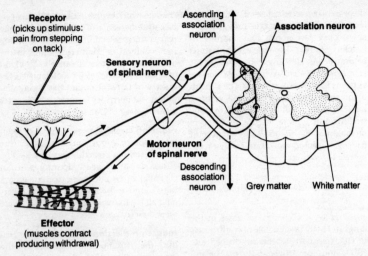

FIG. 145 *Diagram of section of spinal cord showing nervous pathways involved in a spinal* REFLEX *on one side. Ascending and descending association neurons permit involvement of higher and lower body regions.*

impulse past a point on its membrane, an excitable cell (neuron, muscle fibre) is unable to conduct another impulse. No impulse can be conducted, however strong the stimulus, until the inactivated Na+ channels have returned allosterically to their resting states, this interval being the *absolute refractory period*. The *relative refractory period* is the interval succeeding one impulse when the membrane will conduct a second impulse, but only if the excitatory depolarization is suprathreshold – i.e. larger than the normal threshold – and coincides with the time interval when voltage-gated potassium (K^+) channels are still open after inactivated Na^+ channels have returned to their resting states. Larger diameter axons have shorter absolute refractory periods (~0.4 ms) than do small diameter axons (up to 4 ms), enabling them to conduct impulses at higher frequencies. See IMPULSE, MUSCLE CONTRACTION.

refugium Locality (e.g. a tableland or mountain *nunatak*) which has escaped drastic alteration following climatic change (in this case glaciation), in contrast to the region as a whole. Refugia usually form centres for RELIC species populations or communities.

regeneration Restoration by regrowth of parts of body which have been removed, as by injury or AUTOTOMY. Commoner in lower than higher animals, but extensive in some planarians (where new individuals can be regenerated from small body fragments), in the polychaete *Chaetopterus* (which can regenerate a whole animal from one segment) and in crustaceans (which can replace limbs) and echinoderms (which replace arms). Urodeles, such as newts and salamanders, are the only vertebrates which can regenerate amputated limbs, a plasticity achieved by reprogramming mature cells at the site of amputation. Involves the re-establishment of local tissue differentiation. Some examples require long-range interactions and extensive cell movement (*morphallaxis*), while others involve short-range interactions and extensive growth (*epimorphosis*). Vertebrate embryonic limb regeneration involves production of a BLASTEMA. If a large part of an organ is removed (e.g. of liver, pancreas) the remaining organ commonly returns to normal size. This is *compensatory hypertrophy.* Removal of a mammalian kidney usually results in compensatory hypertrophy of the one remaining. Compare REGULATION. Very common in plants, occurring, e.g. in higher plants by growth of dormant buds, formation of secondary meristems and production of adventitious buds and roots. Exploited on a large scale in plant propagation. Includes wound-healing.

regulation Ability of animal embryo to compensate for disturbance (e.g. involving removal, addition or rearrangement of cells) and to produce an apparently normal individual. Eggs may be *regulative*, meaning that removal of some early CLEAVAGE products has no effect on the eventual animal, other than decreasing its size. Embryonic regulation is a broad term, including *twinning* (production of two complete animals from bisected embryo), *fusion* (fusion of more than one embryo to give a single giant embryo), *defect regulation* (removal of part of the early embryo does not disturb DEVELOPMENT), and *inductive reprogramming* (see INDUCTION, where a grafted inducer causes surrounding parts to pursue a developmental route different from that expected from the FATE MAP). Compare MOSAIC DEVELOPMENT.

regulative development See REGULATION.

regulator gene Genetic elements first identified in prokaryotes (see JACOB–MONOD THEORY) by their control of the co-expression of a set of genes. On the basis of sequence comparisons and structural considerations of the proteins they encode it is now thought that, despite limited amino acid homology, some prokaryotic regulatory proteins do have homologues in eukaryotes. Several show remarkable evolutionary conservation at the structural level, from the bacterium *Escherichia coli* to the fruit fly *Drosophila* and to humans (see HOMEOBOX, HOMEOTIC, *CDC* GENES). A *regulatory element* is any genetic element, or its product (commonly protein), involved in the control of GENE EXPRESSION and includes repressors, ENHANCERS and PROMOTERS.

regulatory element (control element) Any gene PROMOTER or ENHANCER.

regulatory enzymes Enzymes specifically involved in switching metabolic pathways on or off. Include (a) ALLOSTERIC enzymes, where activity is modulated through the non-covalent binding of a specific molecule (e.g. see ATP, ADP, GLYCOLYSIS) at a site other than the active site; (b) *covalently-modulated enzymes*, which alternate between active and inactive forms as the result of other enzyme activity (e.g. see KINASE). Regulatory enzymes are exquisitely responsive to alterations in the metabolic needs of the cell.

regulatory proteins Proteins which activate, suppress or otherwise modulate gene expression and metabolic pathways. Usually allosteric, they may in turn be modulated by phosphorylation or covalent modification (e.g. by *S*-nitrosylation). Important functional classes include REGULATORY ENZYMES, caspases (see APOPTOSIS) and TRANSCRIPTION FACTORS.

reinforcement (1) Physiological. Process whereby presentation of a stimulus (the *reinforcer*) to an animal just after it has performed some act alters the probability or intensity of future performances of that act when repeated either in isolation or in the presence of the stimulus. *Positive reinforcers* increase the probability and/or intensity of response; *negative reinforcers* decrease its probability and/or intensity. See CONDITIONING. (2) Evolutionary. The process by which two allopatric populations that have already evolved some postzygotic ISOLATING MECHANISM while apart then undergo selection for increased SEXUAL ISOLATION on becoming sympatric. It need not require the pre-occurrence of hybrid sterility or inviability, although it may result when allopatric populations evolve behavioural or ecological differences which cause hybrids to be maladapted.

relatedness Incestuous and non-diploid matings excluded, the degree of relatedness between two conspecific individuals, male A and female B, is found by first locating their nearest common ancestor(s) and then counting the number of generations passed by moving from B to the common ancestor(s) and then on to A (the *generation*

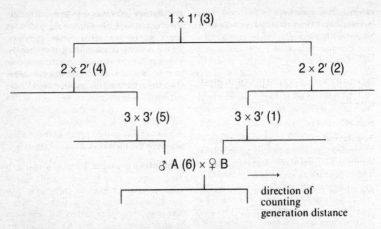

FIG. 146 *Diagram to show the calculation of* RELATEDNESS *between two first cousins.*

distance). This number is the power to which $\frac{1}{2}$ is first raised and then multiplied by the number of nearest common ancestors. For two second cousins, the calculation is as shown in Fig. 146. Generation distance is given by numbers in parentheses from B to A, i.e. 6. There are two nearest common ancestors of A and B, namely 1 and 1', so degree of relatedness, $R = 2 \times (\frac{1}{2})^6 = \frac{1}{32}$, meaning that two second cousins (such as A and B) are expected on average to share 1 in 32 of their genes. Relatedness is a meaningless concept prior to the origin of molecular replication. See HAMILTON'S RULE.

relaxin Gonadal peptide hormone (produced in corpus luteum and testes) involved in pubic symphysis relaxation and preparturition cervical softening and dilation in some female mammals, but apparently not humans. It belongs to the insulin/insulin-like growth factor group family of peptides but seems to operate via G protein-coupled receptor, raising levels of cyclic AMP. It regulates the growth and remodelling of reproductive tissues during late pregnancy and inhibits both spontaneous and oxytocin-induced uterine contractions. In humans and other primates it peaks in the first trimester and appears to be as crucial as progesterone in preparation of the uterine stroma for implantation (decidualization)

and specifically induces expression of vascular endothelial growth factor (VEGF; see ANGIOGENESIS). See INHIBINS, OVARY.

releaser A standard external stimulus evoking a stereotyped response; commonly applied in the context of INSTINCT. A *social releaser* emanates from a member of the same species as the respondent. Releasers include the *sign stimuli* (e.g. claw-waving in fiddler crabs, postures adopted by courting birds, etc.) which have been so valuable in ETHOLOGY.

releasing factor (RF, releasing hormone, RH) Substance stimulating release of hormone into blood. The *locus classicus* for their production is the vertebrate HYPOTHALAMUS, several of whose neurosecretions initiate hormone release from the anterior pituitary.

relic (1) Surviving organism, population or community characteristic of an earlier time. (2) A *relic (relect) distribution* of fauna or flora is one representing the localized remains of an originally much wider-ranging distribution such as those organisms, now confined to mountain tops, which were far more widespread in glacial times (see REFUGIUM).

renal Pertaining to KIDNEY.

renin Enzyme produced by juxtaglomerular cells of kidney when blood pressure falls. See ANGIOTENSINS.

rennin Enzyme secreted by stomachs of young mammals; converts soluble caseinogen to the insoluble protein casein, which coagulates and therefore takes longer to leave the stomach. *Rennet* is impure rennin.

repeat-induced gene silencing (RIGS) A localized *cis*-acting GENE SILENCING effect where regions of tandemly repeated sequence are silenced, often without homologous silencing elsewhere in the genome.

repetitive DNA Coding sequences may constitute a rather small percentage of a genome (<5% in humans), while repetitive DNA sequences may account for much more (at least 50% in humans, see C-VALUE). These repeats are of five main kinds: (a) TRANSPOSON-derived repeats, also known as interspersed repeats (see *Alu*, LINES, SINES); (b) inactive and partially retroposed copies of cellular genes, including some which encode proteins and small structural RNAs, usually referred to as PSEUDOGENES; (c) simple sequence repeats ('minisatellites' and 'microsatellites'), in which a short nucleotide sequence, or k-mer, is repeated (e.g. $(A)_n$, $(CA)_n$, $(CGG)_n$); (d) segmental duplications, where blocks of 10–300 kb have been copied from one part of the genome to another; and (e) blocks of tandemly repeated sequences (SATELLITE DNA), as occur at CENTROMERES, TELOMERES, short arms of acrocentric chromosomes and ribosomal gene clusters (see NUCLEOLUS). Some regions of a genome may be very dense in repeats while others, possibly because of an abundance of *cis*-regulatory elements, which cannot tolerate insertion by mobile genetic elements, are nearly devoid of them.

It is usually held that insertion of a repetitive element into a gene is deleterious, but many translated repetitive elements, especially L1 and Alu, are found in proteins. These seem to originate not by insertion into the original open reading frames but through alternative splicing (see RNA PROCESSING), which can sometimes extend or truncate the coding region. Nevertheless, many neurological and neurodegenerative disorders, including Huntington's disease and fragile X syndrome, derive from lengthening tracts of TRINUCLEOTIDE REPEATS or slightly longer sequences (see also DIABETES mellitus for role of minisatellite). See GENE DUPLICATION, INVERTED REPEAT SEQUENCE, ONCOGENES, RECOMBINATION.

replacing bone Synonym for cartilage bone. See OSSIFICATION.

replica plating Method employed to isolate mutant microorganisms sensitive to a component of, or deficiency in, a growing medium and which, by failing to grow on it, are difficult to detect. The diluted organism is plated on to a solidified medium suitable for growth and colony formation. An adsorbent material (e.g. filter paper) is then pressed lightly on to the medium surface to pick up cells from each colony, and then pressed on to fresh solidified medium with the additive or deficiency. The distribution of resulting colonies on this new medium can be compared with the original distribution, any gaps indicating sensitive strains. See AUXOTROPH.

replication origin See DNA REPLICATION.

replicon Nucleic acid molecule containing a nucleotide sequence forming a replication origin, at which replication is initiated. Usually one per bacterial or viral genome, but often several per eukaryotic chromosome.

reporter gene A gene used in studies of the regulatory control of another gene but whose phenotypic effects are easier to assay because its protein product (e.g. GREEN FLUORESCENT PROTEIN) is more easily monitored. Commonly the reporter gene is spliced immediately downstream of the regulatory elements of the other gene, when detection of its phenotype indicates the other gene is being expressed too. The actual site of insertion of a reporter gene may be unimportant, as when its role is simply to report the expression of another gene which itself had been spliced downstream and adjacent to the regulatory elements of the gene whose expression was being monitored. The pattern of expression of the reporter gene mimics that of the gene studied. Reporter genes (e.g. the *lacZ* gene of *E. coli*, which encodes β-galactosidase) can also be used to find enhancers. Commonly the reporter gene, with a weak promoter and two or more putative enhancer DNA regions, is incorporated into a plasmid bordered by P elements so that it can be inserted into the genome by P element transformation. If the reporter gene is expressed in a tissue, then the DNA did act as an enhancer (since weak promoters cannot induce transcription without enhancer help) and the exact enhancer sequence can be obtained by testing smaller and smaller pieces of the DNA in the same reporter gene assay. See GENE CAPTURE, GENE TRAPPING.

repression (1) (Of enzyme) see ENZYME feedback inhibition. (2) (Of gene) see JACOB–MONOD THEORY.

repressor molecule See JACOB–MONOD THEORY, REGULATORY GENE, LAMBDA REPRESSOR.

reproduction The production of new individuals, physically independent of their parents. See ASEXUAL, SEXUAL REPRODUCTION.

reproductive allocation The proportion of an organism's available resource input that is allocated to reproduction over a defined period of time. Often this is the proportion of an organism's biomass or volume that is reproductive tissue.

reproductive cost The decrease in survivorship and/or growth rate, and therefore the decrease in the potential for future reproduction, suffered by an individual as a result of increasing its current allocation to reproduction.

reproductive isolation Prevention of gene flow between sympatric populations, mechanisms usually being classified as *prezygotic* and *postzygotic* (see ISOLATING MECHANISMS). Although the process of speciation is complete when two divergent lineages are reproductively isolated, the evolution of reproductive isolation may continue between them. See ECOLOGICAL ISOLATION, SEXUAL ISOLATION, SPECIATION.

reproductive output The number of offspring produced by an individual or population.

reproductive rate The number of offspring produced by an organism per unit time, or over a defined time period.

reproductive value The expected relative contribution of an individual to its population, by reproduction, now and in the future.

Reptilia Reptiles. Traditionally considered a vertebrate class, including the first fully terrestrial tetrapods. Probably a MONOPHYLETIC group, whose members lay AMNIOTIC EGGS, have horny or scaly skin and metanephric kidneys. Stem reptiles (Cotylosaurs, Subclass Anapsida) evolved in the Lower Carboniferous, or even earlier, and it has been generally held that they had affinities with the fossil amphibian group, the anthracosaurs. They had solid skulls lacking fenestrations (modern turtles and tortoises are representatives). Synapsida had a single fenestration low on side of skull and included pelycosaurs. Therapsida (MAMMAL-LIKE REPTILES) appeared from the mid-Permian onwards; some were large and herbivorous, others smaller and carnivorous. They were dominant in the Permian fauna and had varied cranial morphology. By the close of the Triassic, therapsids had become very mammal-like, as indicated by the *cynodonts* (dog-sized), which gave rise to mammals. The great Mesozoic reptilian radiation produced three subclasses: Euryapsida (marine reptiles such as ichthyosaurs and plesiosaurs), Archosauria (ARCHOSAURS), including dinosaurs and crocodiles, and Lepidosauria, including extinct crocodile-like forms as well as lizards and snakes (see SQUAMATA). Modern forms are poikilothermic (ectothermic), but some extinct archosaurs were probably homoiothermic (endothermic). Cladistic biologists would regard the term 'Reptilia' as representing a grade, so it does not exist in their classifications. Instead, the AMNIOTE clade includes ARCHOSAURS and SYNAPSIDS. For *Sphenodon*, see RHYNCHOCEPHALIA.

residual reproductive value The expected relative contribution of an individual to its population, by reproduction, for all stages of its life cycle subsequent to the present.

resilience The rate at which a community returns to its former state after it has been disturbed.

resin duct Relatively large intercellular spaces lined by thin-walled parenchyma cells secreting resin into the duct. Wounding, pressure, frost and wind damage can stimulate their formation. Resin apparently protects the plant from attack by decay fungi and bark beetles. Found in conifers.

resistance (1) (Of pests, etc.) the spread within a pest population of genes (often latent in the population) through selection for their ability to destroy or otherwise reduce the effect of some artificial pesticide. See LD$_{50}$, GMO, PLASMID, RESISTANCE GENES.

(2) Term sometimes (confusingly) implicating the non-adaptive IMMUNE responses of organisms. But see PHYTOALEXINS.

(3) In ecology, the ability of a community to avoid displacement from its present state by DISTURBANCE.

resistance genes Plant–pathogen interactions, especially if the pathogen is biotrophic, are contingent upon the interactions between ligands encoded by a pathogen's avirulence (*avr*) genes and the receptors to those ligands encoded by the host's resistance genes (*R* genes). *R* gene products are proteins (R proteins) which are either cytoplasmic or cell-surface (transmembrane) receptor serine/threonine kinases. The largest subset comprises those with a nucleotide-binding site and leucine-rich repeat sequences ('LRR proteins') whose variability is the consequence of selection for polymorphic ligand-binding sites. The LRR proteins are probably intracellular, and either receptors for secreted ligands encoded by a pathogen's avirulence (*avr*) genes, or just one component of a PROTEIN COMPLEX which forms the functional receptor. The other R protein classes are transmembrane receptors binding ligands on the pathogen surface. Activation of any of these R-proteins probably triggers a chain of events in a SIGNAL TRANSDUCTION (and

innate immune) response, reducing the pathogen's virulence. One current model, the *guard hypothesis*, is that LRR proteins bind other functionally important proteins, which they 'guard' from *avr* products which target them. On binding their *avr* ligand, however, they dissociate and initiate a defence pathway. Alternatively, the R protein might bind and initiate the resistance pathway only after the *avr* product has bound the target guardee. In this way, R proteins could monitor whether a cellular protein is under attack from a pathogen *avr* protein (SEE PLANT DISEASE AND DEFENCES). Resistance involves a hypersensitive reaction such that the plant can isolate the pathogen at the site of infection and prevent its spread through the plant. It appears that avirulence genes are involved in the production of a fully virulent infection. Cloning resistance genes is expected to be a valuable tool in crop husbandry (see MAIZE). A wild species of RICE, *Oryza longistaminata*, was the source of a gene (*Xa21*) conferring resistance to a bacterial blight, called *Xoo*. The gene was tracked down by map-based cloning, and gold pellets coated with the gene and fired into the rice plant transformed some of them into blight-resistant forms, a resistance which was found to be heritable. See GMO.

Genes which prevent viral replication are also 'resistance genes'. These might include genes for the chemokine co-receptors for HIV infection.

resolution, resolving power See MICROSCOPE.

resource Whatever is consumed by one organism and, as a result, becomes unavailable to another (e.g. food, water, nutrients). When limiting, results in COMPETITION. Light may be a requirement but, although sometimes limiting, is not strictly a resource.

resource depletion zone The region around a consumer in which the availability of a RESOURCE is reduced (e.g. the zone surrounding the absorbing surface of a root from which nutrients and water are absorbed).

resource partitioning Differential use by organisms of resources such as food or substration. Thus different species of ants eat differently sized seeds, and the two sexes of many birds of prey (females usually being larger) feed on differently sized prey. See CHARACTER DISPLACEMENT.

respiration (1) (*Internal, tissue* or *cellular* respiration.) Enzymatic release of energy from organic compounds (esp. carbohydrates and fats) which either requires oxygen (*aerobic respiration*) or does not (*anaerobic respiration*). Anaerobic respiration is sometimes used as a synonym of FERMENTATION, but strictly this is incorrect since respiration always involves electron transport along cytochromes, and fermentation does not. Anaerobic respiration involves some final acceptor for electrons other than oxygen; e.g. nitrate or Fe^{2+} (see LITHOTROPH). Among eukaryotic cells only those with mitochondria, and among prokaryotes only those with mesosomes, can respire aerobically. All cells can respire anaerobically, using enzymes in their cytosol; but not all obtain sufficient energy release for their needs this way. A high proportion of the energy released during respiration is coupled to ATP synthesis (50% during anaerobic respiration in erythrocytes; 42% of aerobic respiration involving mitochondria). The major anaerobic pathway in cells is GLYCOLYSIS. This is a metabolic 'funnel' into which compounds from a variety of original sources may be fed (e.g. from proteins, fats, polysaccharides). The energy-rich products are ATP (from ADP) and NADH (from NAD); pyruvate formed is either catabolized further in mitochondria (see KREBS CYCLE) or converted to lactate or alcohol. The actual ATP-generating steps of aerobic respiration occur during passage of electrons along the ELECTRON TRANSPORT SYSTEM. (2) (*External respiration*.) See VENTILATION. (3) See PHOTORESPIRATION.

respiratory centres See VENTILATION.

respiratory chain Alternative for ELECTRON TRANSPORT SYSTEM.

respiratory movement See VENTILATION.

respiratory organ Animal organ specialized for gaseous exchanges of oxygen and carbon dioxide. Include LUNGS, LUNG BOOKS, GILLS, GILL BOOKS and the arthropod TRACHEAL SYSTEM. See VENTILATION.

respiratory pigment (1) Substance found in animal blood or other tissue, involved in uptake, transport and unloading of oxygen (and/or carbon dioxide) in solution, always by weak and reversible bonds. Some visibly colour the blood (HAEMOGLOBIN, HAEMERYTHRIN, CHLOROCRUORIN); but MYOGLOBIN is not blood-borne, and does not noticeably colour the tissues where it is found. (2) Substances (e.g. CYTOCHROMES, FLAVOPROTEINS) found in all aerobic cells (in insufficient concentrations to colour them), and involved in the ELECTRON TRANSPORT SYSTEM of aerobic respiration.

respiratory quotient (RQ) Ratio of volume of carbon dioxide produced to oxygen consumed by an organism during aerobic respiration. The theoretical RQ value for oxidation of carbohydrates is 1; for fats 0.7; for protein 0.8. Cells can oxidize more than one of these at a time, making *in vivo* results problematic, although data on levels of nitrogen excreted can be helpful. Interconversion of carbohydrates to fats can also affect RQ values. With people on a normal diet an RQ of 0.82 is expected.

respiratory surface Part of an animal's body surface specialized for GASEOUS EXCHANGE. Gills, lungs and insect tracheoles are regular examples, and although such exchange normally occurs between the animal and its external environment, the mammalian PLACENTA is one organ where the exchange occurs between two internal environments. Although the surfaces of spongy mesophyll and palisade cells of a leaf are involved in exchange of gases, they are not often regarded as respiratory surfaces because the context of the exchange is normally considered to be photosynthesis.

response Change in organism (or its parts) produced by change in its environment. Usually adaptive. See IRRITABILITY.

response element That component of a gene regulatory sequence (often within a PROMOTER) which binds a specific repressor, activator, or other regulatory molecule such as a steroid hormone-receptor complex (see NUCLEAR RECEPTORS). See CRE, NFϰB.

resting nucleus (resting cell) See INTERPHASE.

resting potential (membrane potential) Electrical potential across a cell membrane when not propagating an impulse. All plasma membranes exhibit such voltage gradients (inside negative with respect to the outside) generated by TRANSPORT PROTEINS, the more important in neuron and muscle membranes being the K$^+$ leak channel, which lets potassium ions flow back out of the cell along their electrochemical gradient until internal negativity of -75 mV (nerve axons) is achieved – the potential retarding loss of K$^+$ to the extent that its *equilibrium potential* results. Much less important is the Na$^+$/K$^+$ ATPase pump (SODIUM PUMP), which contributes slightly to the resting potential. The flow of ions needed to set up a resting potential is rapid and minute, leaving ion concentrations practically unaffected. Membrane potentials of 20–200 mV internally negative can be generated, depending on the species and cell type. See DONNAN EQUILIBRIUM, IMPULSE.

restitution nucleus (restitution meiosis, restitution division) Inclusion within the same nuclear membrane of all the chromatin after meiotic chromosome or chromatid separation has occurred, giving an unreduced egg which may develop by automictic PARTHENOGENESIS (e.g. in dandelions, *Taraxacum*). See AUTOMIXIS.

restoration ecology The science concerned with the deliberate colonization and re-vegetation of derelict land, especially after major damage (e.g. from strip mining) and after land has been released from agricultural use.

restriction endonucleases (restriction enzymes) Class of nucleases originally extracted from the bacterium *E. coli* (where it digests phage DNA but leaves the cell DNA intact). Type I restriction enzymes bind to a recognition site of duplex DNA, travel along the molecule and cleave one strand only, about 75 nucleotides long, apparently randomly. Type II are more valuable in GENE MANIPULATION and cleave the duplex at specific target sites at or near the binding site. Their naming uses the first letter of the genus and the first two letters of the specific name of the organism (host) from which they were derived, with host strain identified by a full or subscript letter. Roman numerals then identify the particular host enzyme if there are more than one. Thus *Eco*RI is the first restriction enzyme from *E. coli* strain RY13, and *Hind*II comes from *Haemophilus influenzae* Rd. Target sites for these nucleases are very specific. What has been termed the *restriction-modification system* is an epigenetic bacterial methylation marking system in which a special enzyme methylates host DNA, usually at an A or C residue, in a sequence-specific manner so that the bacterial genome is distinguished from foreign parasitic (e.g. viral) DNA, which is not methylated and is consequently degraded by the host's restriction enzymes. The term 'restriction' stems from the discovery that phage propagated in one strain of bacterium grew poorly (i.e. was restricted) in another strain, and vice versa – caused by degradation of phage DNA by the endonuclease. See DNA FINGERPRINTING, MODIFICATION, PHAGE RESTRICTION, RESTRICTION MAPPING.

restriction mapping Method employed in CHROMOSOME MAPPING, usually of organelle and viral genomes. Complete nucleotide sequencing may result.

rete mirabilis (pl. retia mirabilia) Network of arterioles running towards an organ and breaking up into capillaries adjacent to, but in the opposite direction from, a similar set of capillaries returning blood to the venous system. Exchanges between the two networks form a COUNTERCURRENT SYSTEM, passively increasing local value of some variable (e.g. metabolite concentration, temperature). They occur for example in teleost gills and GAS BLADDERS and in feet of birds which stand in cold water or on ice.

reticular fibres Fine branching collagen-like protein fibres (*reticulin*) forming an extracellular network in many vertebrate connective tissues and holding tissues and organs together. Stains selectively with silver preparations. Abundant in BASEMENT MEMBRANES under many epithelia, sarcolemma around muscle and around fat cells. Forms framework for lymph corpuscles in LYMPHOID TISSUE.

reticular formation (reticular activating system) Nerve cells scattered throughout the vertebrate BRAINSTEM, some forming nuclei, receiving impulses from the spinal cord, cerebellum and cerebral hemispheres and returning impulses to them. Stimulation of some of these cells produces AROUSAL in unanaesthetized animals, the formation playing an important role in waking, attentiveness and, in higher vertebrates, consciousness. A *decerebrate* animal, whose brain is sectioned between basal ganglia and reticular formation, exhibits the effects of overactive stimulatory and underactive inhibitory mechanisms of the reticular formation, producing *decerebrate rigidity*.

reticulate thickening Internal thickening of a wall of a xylem vessel or tracheid in the form of a network.

reticulin See RETICULAR FIBRES.

reticulocyte Immature, non-nucleated, mammalian RED BLOOD CELL. Develops from nucleated *myeloblasts* and is released into the blood during very active HAEMOPOIESIS. Cytoplasm basophilic, due to RNA.

reticulo-endothelial system (RES) Tissue macrophages either circulating in blood, dispersed in connective tissue or attached to capillary endothelium. Of MYELOID origin, these cells carry (F_c) receptors for IgG antibodies, complement and lymphokines. Attack foreign or tumour cells either by ingesting them or by lysis after adherence to them. Abundant in liver sinusoids (*Kupffer cells*), where e.g. they ingest worn erythrocytes, in kidney glomeruli (*mesangium cells*), alveoli of lung, attached to capillaries of brain (*microglial cells*), SPLEEN and LYMPH NODE sinuses.

reticulum (1) Second compartment of the RUMINANT stomach, receiving partially digested material which has been in the rumen, rechewed, and swallowed a second time. Water is here pressed out before the food passes to the abomasum. (2) See ENDOPLASMIC RETICULUM.

retina Photosensitive layer of vertebrate and cephalopod EYES, non-sensory in region of ciliary body and iris. Contains ROD CELLS and CONES, both of which synapse with intermediary neurones before impulses leave via the optic nerve. Also present are blood vessels and glial cells; in vertebrates light must traverse these tissues (the ganglion layer) before impinging upon photoreceptors, an arrangement termed an *inverted retina*. By contrast in cephalopod retinas, derived from external ectoderm, light impinges first upon photoreceptors. Vertebrate retinas arise as outpushings of the brain, photoreceptors having a pigmented layer behind them abutting the choroid. This is because the neural tube rolls up during vertebrate development, so that the receptors are on the inside. See FOVEA, BLIND SPOT.

retinoblastoma A childhood cancer in which neural precursor cells in the retina differentiate into tumour cells. See TUMOUR SUPPRESSOR GENE.

retinoic acid (RA, vitamin A acid) Naturally occurring oxidation product of vitamin A, once considered by some to be an endogenous vertebrate MORPHOGEN, in addition to playing a part in the growth and maintenance of epithelia and inhibiting growth of some malignant cells. It now appears that the true morphogen in limb development is produced by the zone of polarizing activity, which is itself induced by retinoic acid. In chick limb buds, where retinol can serve as a precursor, a gradient of RA levels provides the *positional information* specifying anterior–posterior digit pattern in the wing; Hensen's node (see PRIMITIVE STREAK) and derivative midline structures can induce a second body axis and digit pattern duplications when grafted into chick wing bud, and is a source of retinoic acid. With BMPS, RA is involved in the identity of facial regions in birds. An RA receptor (RAR) protein, once bound to its ligand, then binds to specific DNA regions within the nucleus (see NUCLEAR RECEPTORS).

retinoids Diterpenoids derived from a monocyclic parent compound and containing five C–C double bonds and a functional group at the terminus of the acyclic portion. As ligands of NUCLEAR RECEPTORS, they are required for normal development and growth. VITAMIN A (retinol), retinene (retinal), RETINOIC ACID and 3,4 didehydroretinoic acid (ddRA) are important animal examples; BACTERIORHODOPSIN is a prokaryote retinoid.

retinol (vitamin A) See RHODOPSIN, VITAMIN A.

retrograde signalling Phenomenon by which a GROWTH FACTOR binds to receptors on nerve terminals and then transmits a survival signal back to the cell body, which may be located a metre away. The NEUROTROPHIN NGF is certainly transported retrogradely from the neuron terminal to the cell body, although apparently this is not necessary for retrograde signalling. It seems that its receptor – TrkA – may be the signal, either in a vesicle or in the form of a wave of concentration.

retrograde transport Protein transport from the cell membrane via the endosome, *trans*-Golgi network, Golgi and endoplasmic reticulum, in opposite direction from normal anterograde transport. See GOLGI APPARATUS, TOXIN.

retroinhibition See END-PRODUCT INHIBITION.

retrotransposons (retroposons) See TRANSPOSONS.

retrovirus See VIRUS, REVERSE TRANSCRIPTASE, TRANSPOSON.

re-uptake For re-uptake of transmitters, see NEUROTRANSMITTER.

reverse genetics A somewhat unhelpful term, indicating those techniques by which a target gene is deliberately mutated in order to discover its function within the organism. The starting point might be a protein of interest. The encoding gene is then identified and mutated (see *IN VITRO* MUTAGENESIS) and transferred with an appropriate regulatory protein into a cell where it becomes integrated into a chromosome, becoming part of the genome. All the cell's descendants will then contain the mutant gene and its effects can be observed, and if the cell is a fertilized egg then all the organism's cells, including germ line cells, will be transformed. See also GENE KNOCKOUTS.

reverse transcriptase A DNA POLYMERASE of retroviruses (initially Rous sarcoma virus), synthesizing complementary single-strained DNA (cDNA) on a single-stranded RNA template. Its inhibition could prevent retroviral infection. cDNA formed may be cloned and screened to detect differences in mRNAs between cells, mRNA acting as template for reverse transcriptase. See GENE MANIPULATION, REPETITIVE DNA, TRANSPOSABLE ELEMENTS, VIRUS.

RFLP (restriction fragment length polymorphism) In 1979, DNA SEQUENCING by A. Jeffreys on the β-globin gene of human haemoglobin revealed an unexpected high sequence variation between individuals of approximately one base substitution every few hundred base pairs. These were effectively a new kind of GENETIC MARKER since restriction enzymes – by cleaving DNA at very specific base sequences (restriction sites) – cut DNAs from different people into fragments of different lengths. Such new cleavage sites turned out to be polymorphic and be revealed as RFLPs by SOUTHERN BLOTTING. GENETIC MAPPING of mammalian (e.g. human) chromosomes has been transformed by this technique, since markers were previously far too isolated and rare for conventional CHROMOSOME MAPPING through linkage analysis. Mutations removing or adding a restriction site may be used as markers, too. Studies of genetic RELATEDNESS in both human and wild populations have also been greatly enhanced. However, novel restriction sites are not ideal markers since they are insufficiently polymorphic, i.e. they normally exist in only two alternative forms. Southern blotting is also difficult to automate. See DNA FINGERPRINTING.

R$_f$ value In paper chromatography, the ratio between the distance moved by a particular solute compared with the distance moved by the solvent. Each solute has a characteristic R$_f$ value for a given solvent. See CHROMATOGRAPHY.

***R* genes** Plant resistance genes. See PLANT DISEASE AND DEFENCES.

Rhabdovirus A bacilliform VIRUS.

Rhesus system See BLOOD GROUP.

rhinarium Area of moist, glandular and almost hairless skin surrounding the nostrils in most mammals. Among primate mammals, lemurs and lorises have one, but tarsiers and simians do not. See HAPLORHINE, STREPSIRHINE.

Rhipidistia Traditionally, an order or suborder of crossopterygians, appearing in early Devonian but extinct by end of Palaeozoic. Teeth with folded dentine and cranium divided into anterior and posterior parts. Many (osteolepids) were well covered in thick scales and adapted to life at the water's edge. Cladistic analysis of new fossils recommends placing coelacanths and traditional rhipidistians within the Sarcopterygii (see CHOANICHTHYES), with tetrapods within the new more inclusive clade Rhipidistia.

rhizine (Of lichens) a bundle of hyphae attaching a lichen thallus to its substratum.

Rhizobium This bacterial genus and the related *Bradyrhizobium* are Gram-negative motile rods. Infection of the roots of leguminous plants leads to formation of root nodules in which gaseous nitrogen is converted to combined nitrogen (see NITROGEN FIXATION). Receptors involved in mutualistic plant–microbe interactions belong to the leucine-rich repeat (LLR) receptor subfamily, as do the TOLL-LIKE RECEPTORS of animals, despite the fact that the last common ancestors of these organisms existed ~1.8 billion years ago. Different legumes are infected by different strains of rhizobia. Initial penetration is by the root hairs, but *Nod factors* excreted by the bacterium induce curling of the root hair and formation of a cellulose tube (infection thread) by which bacteria gain access to other root cells. Plant cell division is promoted by the *Nod factors* generating the nodule. Although molecular oxygen is required to release the energy for N-fixation, the nitrogenase enzyme itself is inactivated by O_2. Interaction between the two organisms results in leghaemoglobin production, which binds O_2 and keeps its level low in the nodule. Independence from nitrogenous fertilizers in agriculture reduces POLLUTION of water tables, rivers, etc. There is considerable reciprocal symbiotic signalling between rhizobia and their legume host, both to allow entry of the bacteria and to permit subsequent nitrogen fixation. Some of the genes for these are carried on plasmids, and there is evidence of HORIZONTAL TRANSFER among rhizobia and the legume rhizosphere, as well as to other non-symbiotic bacteria in fields (e.g. the closely related *AGROBACTERIUM*). A few legume species form rhizobial N-fixing nodules on their stems.

rhizoid Single- or several-celled hair-like structure serving as a root. Present at bases of moss stems and on undersurfaces of liverworts and fern prothalli, as well as in some algae and fungi.

rhizome More or less horizontal underground stem bearing buds in axils of reduced scale-like leaves. Serves in perennation and vegetative propagation; e.g. mint, couch grass, *Scirpus*.

rhizomorph Root-like strand of hyphae produced by some fungi, transporting food materials from one part of the thallus to another, and increasing in length by apical growth. May be internally differentiated and complex.

rhizophore A form of aerial adventitious root (e.g. in *Selaginella*); or, the dichotomously branching underground axis of some Carboniferous trees (e.g. *Lepidodendron*).

rhizoplane Root-surface component of the RHIZOSPHERE.

rhizoplast See DIPLOMONADS.

Rhizopoda Protozoan class, or subclass of the SARCODINA.

rhizosphere Zone of soil immediately surrounding root, and modified by it. Characterized by enhanced microbial activity and by changes in the ratios of organisms compared with surrounding soil.

Rho A family of small GTP-ases, encoded by *rho* genes. Rho GTPase is a key regulator of the actin CYTOSKELETON and of CYTOKINESIS. See G-PROTEIN, ENDOSOME, ADHESION. Contrast RHO FACTOR.

Rhodophyta The Red Algae. A division comprising 500–600 genera and 5,000–5,500 species, the majority of which are marine, and live attached to rocks, attached to animals (epizoic), epiphytic upon *Zostera* (eelgrass) and brown algae. Only about 150 species from about 20 genera occur in freshwater. Red algae have been found living at greater depths than any other photosynthetic organisms (off a sea-mount off the Bahamas, red algal crusts grow at a depth of 268 m, where only about 0.001% of light incident upon the sea's surface remains available for photosynthesis). In the lower part of the intertidal zone, rocks are often covered with red algae. In tropical regions limestone-encrusted red algae are important in formation of reefs and sediments (e.g. members of the Corallinales). Fossil remains of calcified red algae are known from the Cambrian (600 Myr ago). Subsequently, red algae have been major contributors to marine carbonate sediments. A unicellular species belonging to the genus *Porphyridium* has been found living as a photosynthetic endosymbiont within large tropical foraminifera. A few red algae are non-photosynthetic, obligate parasites of other red algae, and are often reduced and pustular.

The red algae represent one of the most ancient lineages of eukaryotes. The absence of flagella (including the male gamete) perhaps indicates that these algae diverged from other primitive eukaryotes before the evolution of the typical eukaryotic flagellum. Cell walls generally contain cellulose, which forms the fibrillar component (some have xylan), and amorphous mucilages (polysaccharides; e.g. agar and carrageenan). Chloroplasts are usually stellate, with a central pyrenoid in the morphologically simple red algae; commonly discoid in others. Thylakoids are not stacked but lie equidistant and singly within the chloroplasts. One or two thylakoids are present around the periphery of the chloroplast running parallel to the chloroplast envelope. The only chlorophyll is *a* but the green of the chlorophyll *a* is masked by the red accessory pigment phycoerythrin. Phycocyanin also occurs in the chloroplasts. Both are located in hemispherical or hemidiscoidal bodies (phycobilisomes) on the surfaces of the thylakoids. Chloroplast DNA is present as small nucleoids scattered throughout the chloroplast. The most important storage product is a polysaccharide called floridean starch. Grains are formed in the cytoplasm, next to the chloroplast envelope. Mitosis is closed (see MESO-KARYOTE) and the telophase spindle is persistent. The mitotic nucleus is surrounded by its own envelope as well as by perinuclear endoplasmic reticulum. Commercially significant in production of CARRAGEENAN, laver bread, and sizes for paper manufacture.

rhodopsin (visual purple) Light-sensitive pigment of ROD CELLS, whose molecules contain a protein component (*opsin*) whose prosthetic group, the chromophore II-*cis* retinal (vitamin A, a component of CONE pigments), undergoes isomerization to *trans* retinal in light, becoming free of opsin in the process. Opsin seems to undergo an allosteric change, closing Na^+ channels in the rod cell membrane previously kept open in the absence of light by CYCLIC GMP, terminating release of neurotransmitter. An intermediary (perhaps Ca^{2+}) probably links these events. Photons may initiate calcium release from the discs in the outer segment of the cell, while rhodopsin's shape change possibly starts a CASCADE of reactions reducing intracellular cyclic GMP levels. Conversion of *trans* retinal to *cis* retinal occurs in absence of light and involves reduction by reduced NAD, *retinol* being an intermediary in the regeneration. See BACTERIORHODOPSIN, SIGNAL TRANSDUCTION.

rho factor (ρ) ATP-dependent hexamer of identical polypeptides, associating temporarily with some mRNAs during their termination of transcription (at a *termination signal*). Appears to help pull transcript away from the RNA polymerase. See SIGMA FACTOR. Contrast RHO.

rhombencephalon See HINDBRAIN.

rhombomeres Swellings dividing the vertebrate rhombencephalon (see BRAIN) metamerically into smaller true COMPARTMENTS (polyclones respecting boundaries), demarcated by grooves. Regulatory genes are expressed in specific sets of rhombomeres (e.g. *Hox-2.9* is expressed only in rhombomere 4).

Rhynchocephalia Order of lepidosaurian reptiles, related to eosuchians. Now represented solely by the tuatara (*Sphenodon*), restricted to New Zealand (a RELIC from GONDWANALAND), where it was in danger of extinction. Appeared in the Mesozoic, but never a dominant group. Lizard-like with many primitive features, e.g. lack of external ear opening, a pineal eye (see PINEAL GLAND) and lack of an intromittant organ. See REPTILIA.

Rhyncota See HEMIPTERA.

Rhynie chert Devonian, Aberdeenshire (Scotland), chert formed when silica-rich volcanic waters swamped a terrestrial community, 380–370 Myr BP. Famous for fossils of the early land plant *Rhynia*.

Rhyniophyta Fossil plant division containing the earliest-known vascular plants understood in detail dated back to the late SILURIAN period, at least 420 Myr ago. They became extinct in the mid-DEVONIAN period (*c.* 380 Myr ago). Earlier vascular plants go back at least another 15 Myr. Seedless, they comprised a simple dichotomously branched axis bearing terminal sporangia.

Differentiation into stems, leaves and roots is not apparent, and they were homosporous. *Rhynia* is the best known example; leafless, but possessing a rhizome with water-absorbing rhizoids. The internal structure resembled extant vascular plants, having an epidermis surrounding the photosynthetic cortex, while the centre of the axis contained the xylem. See TRIMEROPHYTA, ZOSTEROPHYTA. *Rhynia* was probably a marsh plant; its aerial parts were stems about 20–50 cm long, 3–6 mm thick, and covered by a cuticle, bearing stomata and serving as the photosynthetic organs.

rhytidome See BARK.

riboflavin (vitamin B₂) Precursor of FMN and FAD (prosthetic groups of CYTOCHROMES). An orange-red photoreceptive pigment in the zygomycote mould *Phycomyces*. Insufficient intake in humans causes inflammatory lesions in corners of mouth and blocking of sebaceous glands of nose and face. Possibly involved in phototropism. See VITAMIN B COMPLEX.

ribonuclease See RNase.

ribonuclease proteins See RNPs.

ribonucleic acid See RNA.

ribose A pentose (5-carbon) monosaccharide; $C_5H_{10}O_5$. Component of RNA and its nucleotide triphosphate precursors, etc.

ribosomal RNA (rRNA) See RIBOSOMES.

ribosomes Non-membranous, but often membrane-bound, organelles of both prokaryotic and eukaryotic cells, of chloroplasts and mitochondria. Sites of PROTEIN SYNTHESIS, each is a complex composed of roughly equal ratios of *ribosomal RNA* (rRNA) and 40 or more different types of protein. Prokaryotic ribosomes are slightly smaller than eukaryotic, with a sedimentation rate (Svedberg number, S) of 70S as opposed to 80S. Each is composed of one large (50S or 60S) and one smaller (30S or 40S) sub-particle, forming ribosomes with diameters of approx. 29 nm in prokaryotes and 32 nm in eukaryotes respectively. Each sub-particle is formed from one to three species of rRNA plus associated proteins.

A functional ribosome has one conformational groove to fit the growing polypeptide chain (see A-SITE) and another for the messenger RNA (mRNA) molecule. Although ribosomes from different groups of organisms (bacteria, yeast, wheat, rats) have different compositions, they all have the same shape of gap between the large and small subunit to permit entry of tRNA. INITIATION FACTORS are required for complexing of the small ribosomal subunit with the initiator tRNA (see P-SITE) and mRNA prior to binding of the large ribosomal subunit. Several ribosomes on an mRNA strand form a *polysome*.

Ribosomes are not attached to membranes in prokaryotes, chloroplasts and mitochondria, but are commonly membrane-bound in eukaryotic cells (esp. actively secreting ones), forming rough ENDOPLASMIC RETICULUM, as well as being attached to outer surface of outer nuclear membrane. The origin of most eukaryotic rRNA is the NUCLEOLUS (see also RNA POLYMERASES), a single large pre-rRNA transcript being cleaved into one each of 18S, 5.8S and 25/28S rRNA subunits and processed to the mature rRNAs by a series of endonuclease cleavages and exonuclease refinements. The 18S rRNA molecules form the small ribosomal subunit; the 5.8S and 25/28S rRNAs are joined by a 5S rRNA subunit made in the nucleoplasm to form the larger ribosomal subunit, the rRNA including at least one RIBOZYME, a possible remnant of the RNA WORLD. Ribosomal proteins, translated in the cytoplasm under directions from over 80 rRNA genes and imported back into the nucleus, bind the rRNAs and one another to produce the two ribosomal subunits. These ribosomal proteins assemble with the rRNAs and undergo extensive covalent nucleotide modifications, the sites being selected by snoRNAs (see SNORNAS entry). Actively growing yeast cells contain approx. 200,000 ribosomes for which approx. 16 million ribosomal protein molecules must be synthesized each cell cycle. One ribosomal subunit exits each nuclear pore every three seconds in such cells, in which 60% of all nuclear transcription is devoted to ribosome synthesis. If a critical ribosomal protein is produced in excess, it binds its own mRNA and inhibits further translation. Ribosomal proteins associate with these rRNAs in the nucleolus and seem to be exported as pre-ribosomal particles (RNPs) through the nuclear pore complexes into the cytosol. Prokaryotic rRNA types are 5S, 16S and 23S and are cut from a 30S pre-rRNA transcript. The gene for the 16S subunit evolves very slowly and yet has been shown to be remarkably diverse (see ARCHAEA). See CELL FRACTIONATION.

ribozyme Any catalytically active RNA molecule. Their discovery in 1981 has widened the extension of the term 'enzyme' beyond proteins. Several ribozymes are self-splicing introns (see SPLICING), causing speculation as to their possible roles as intermediates in the evolution of biological systems from prebiotic ones. The peptide bonds formed on RIBOSOMES are catalysed by a ribozyme – almost, but not quite, autocatalytically. See COENZYME, ORIGIN OF LIFE, RNA WORLD, TELOMERE.

ribs (costae) Vertebrate skeletal struts developing from cartilage within the connective tissue sheets (myocommata) separating successive blocks of segmental body musculature. Articulating with the vertebrae, they provide for muscle attachment. So-called *true ribs* also articulate ventrally with the sternum, those articulating ventrally with each other but not with the sternum are *false ribs*, while those articulating with nothing ventrally are *floating ribs*. See INTERCOSTAL MUSCLES.

rice The ancient origin of rice (*Oryza sativa*) is unclear, but probably it was in southeastern Asia. Some very early Chinese and Indian writings mention rice cultivation for more than 5,000 years in that part of the world. Recent work by Japanese and Chinese archaeologists suggests that rice may first have been cultivated along the middle Yangtze River in Hubei and Hunan provinces of central China 11,500 years BP, pre-dating the earliest known agriculture in west Asia. Rice is most important in the tropics and semi-tropics and has significant production in some temperate regions. The cultivated plant, customarily grown as an annual, is a semi-aquatic annual grass that grows erect. It has narrow parallel-veined leaves with spikelets borne on a loose panicle and each contains one flower enclosed by a lemma and palea. The flower has six stamens and one ovary. Because the rice grain has the inner leaves of the flower fused to the pericarp, the milling process differs from that used with WHEAT. The leaves (the husk) are first removed, producing the brown rice with the bran attached to the grain. The bran is then removed in a rice mill, said to 'pearl' the grain, leaving it polished. This leads to substantial VITAMIN B and FIBRE losses.

Rice can be classified in several ways. Based upon chemical characteristics of the starch and grain aroma, rice can be classified into three groups: (i) waxy or glutinous types characterized by a starchy endosperm containing no amylose; (ii) common types characterized by more or less translucent non-glutinous starch with the endosperm containing 25% amylose and 75% amylopectin; and (iii) aromatic or scented types that are grown in India and southeast Asia. Groups i and ii comprise more than 90% of the total rice grown.

Rice is also classified on maturity. The rate of maturity is genetically controlled and is measured by the number of days required for plants to reach 50% heading. Another classification is based on its cultured adaptation as 'japonica' or 'indica' types. The former cultivars are usually short-grain and adapted to a temperate climate, while the latter are long-grain and tropical. Hybridization has led to a breakdown of this traditional classification. Cultivars can also be classified as lowland rice (continuously or pond-flooded) or upland rice (non-irrigated or irrigated but not continuously pond-flooded). Finally, rice cultivars can be classified on the basis of kernel characteristics into short-, medium- or long-grain types. In April 2002, draft genome sequences of the *japonica* and *indica* strains were published; and later that year two chromosomes of the former were fully sequenced.

Successful rice production is dependent upon ample water availability. Therefore, important production regions lie along great rivers and deltas. Asia produces more than 90% of the world rice crop, major producers being China, India, Indonesia, Bangladesh, Thailand, Japan and Burma (Myanmar). In the Southern Hemisphere, Brazil is an important producer.

Rice grows well on moist soils. For best yields, the fields are flooded for most of the growing season; therefore, clay, clay-loam or silty clay-loam soils are frequently used to conserve water by minimizing seepage losses. Soil pH ranges between 5.0 and 7.5 for best results. In most Asian countries, the centuries of simple, non-mechanical methods of flooding, terracing, levelling and other wetland rice cultural operations have resulted in soils with common characteristics. The immediate soil surface is oxidative and the subsoil strongly reducing (anaerobic). These soil conditions create what is termed a rice paddy. Deep water rice responds to submergence by transcription of expansin genes, with subsequent rapid stem elongation (SEE CELL GROWTH). The hormone GIBBERELLIN has similar effects.

Wild rice (*Zizania palustris* var. *interior*) is an aquatic grass not closely related to common rice (*Oryza sativa*). It has been used in the Great Lakes region of Canada and the United States as a food by the indigenous Indians for more than 300 years. During each summer wild rice produces a flower stalk that elongates and emerges from the water. It bears female flowers at the top of a spike with many drooping male flowers between them. This arrangement almost guarantees cross-pollination between plants. By late summer the plants are 90–120 cm tall and very leafy. Ripening of the grains is spread over about ten days or more. The traditional way of harvesting is performed from a boat by bending the stalks over and into this boat and beating the kernels off with a flail. Such a process and the plant's seed-shattering tendencies results in considerable amounts of seed (probably more than half) being lost into the water. This portion provides seed for the following year's crop.

ricin See TOXINS.

rickettsiae Group of bacterium-like prokaryotes occurring naturally in tissues of arthropods (e.g. fleas, lice, ticks), which can transmit them to mammals where they can produce fatal diseases (e.g. typhus in humans). Some can be transmitted via food as well, notably those causing Q fever and a form of encephalitis, both in untreated goat's milk.

Ringer's solution Physiological saline; aqueous solution, containing sodium, potassium and calcium chlorides, employed in maintaining cells or organs alive *in vitro*. Always appropriately buffered, usually with bicarbonate or phosphate (SEE BUFFER).

ring chromosome Chromosome formed by the loss of the two telomere-containing tips and subsequent joining of the broken ends to form a ring-like structure, which may or may not contain the centromere. Acentric rings usually get lost in subsequent cell divisions, but centric rings may be passed on. The condition is linked with ionizing radiation. In humans, individuals heterozygous for rings often exhibit mental retardation. Crossing-over and subsequent separation at first meiotic anaphase often results in disruption of the ring, while sister chromatid exchange in somatic cells may produce double-sized dicentric chromosomes. If the problem occurs during embryogenesis, cells may be totally or partially aneuploid for the chromosome, producing a mosaic. Growth retardation, variable degrees of mental retardation and possibly infertility result.

ring-porous wood Secondary XYLEM in which the vessels (pores) of the early wood, when viewed in cross-section, are distinctly larger than those of the late wood and form a well-defined zone or ring.

ring species Species characterized by circular or looped geographical distributions (see Fig. 147), adjacent populations interbreeding on the two arms of the loop but not where arms overlap, i.e. those populations presumably furthest from the original area of spread. See SPECIATION, SPECIES, TRANSITIVITY.

riparian Inhabiting the banks of a river.

ripping See DNA METHYLATION.

RISC The RNA-induced silencing complex, which binds small interfering RNAs (siRNAs) during RNA SILENCING after their production by the endonuclease DICER. The complex contains a distinct endonuclease which uses the sequences encoded by the antisense siRNA strands to find and destroy mRNAs complementary to them.

ritualized behaviour Behaviour which has become stereotyped in evolution through its role in communication. Many display and threat sequences are ritualized and some DISPLACEMENT ACTIVITIES have become so through selection as information-bearers.

RNA (ribonucleic acid) Nucleic acid class, differing from DNA in being usually either single-stranded or looped, in containing ribose not deoxyribose, and in that uracil replaces thymine. Synthesized by RNA POLYMERASES from nucleoside triphosphates (ATP, GTP, CTP, UTP); with 5'- and 3'-ends to the molecule, as in DNA (see DNA).

The three RNA types, *messenger RNA* (mRNA), *transfer RNA* (tRNA) and *ribosomal RNA* (rRNA), are all involved in PROTEIN SYNTHESIS, in both prokaryotic and eukaryotic cells, but in eukaryotes especially, RNA PROCESSING accompanies mRNA production. In RNA viruses, RNA is sometimes double-stranded, serving as genetic material (see REVERSE TRANSCRIPTASE, CENTRAL DOGMA); in some RNA viruses the RNA is transcribed into RNA and in others (*retroviruses*) it is reverse-transcribed into DNA. tRNA molecules fold back upon themselves by complementary base-pairing to form double-stranded 'stems' and single-stranded 'loops'. A loop at one end bears a specific nucleotide triplet (the *anticodon*) while the 3'-end of the molecule carries a tRNA-specific amino acid – both essential for protein synthesis to proceed by means of a GENETIC CODE. Ribosomal RNA subunits associate with protein molecules to form

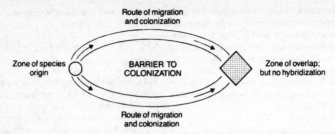

FIG. 147 *Ring species.*

RIBOSOMES. All tRNA and rRNA molecule types are encoded by DNA (see GENE), and there are many more of these molecules per cell than there are of mRNA. Eukaryotic rRNAs are of four densities: 28S, 18S, 5.8S and 5S, while the three prokaryotic rRNAs are 23S, 16S and 6S. The genes for all these tend to be highly conserved and are often employed in phylogenetic studies. Some RNA molecules have catalytic activity (see ORIGIN OF LIFE, RIBOZYMES and Table 9) The growing realization that RNA combined some of the properties of proteins and DNA led to the RNA WORLD hypothesis for the origin of life, and there is much interest in the roles of NON-CODING RNAS, e.g. in RNA interference (see RNA SILENCING).

RNA-binding proteins Proteins with at least one RNA-binding motif (RBM), encoded by members of the *RBM* gene family. Involved in RNA transport and processing. See RNPS, PROTEIN SYNTHESIS, TELOMERE.

RNA capping Eukaryotic RNAs are processed in nuclei prior to release into the cytosol. This involves attachment of a *cap* of 7-methylguanosine triphosphate to their 5'-end. Ribosomes recognize this cap and commence translation at the AUG codon nearest to the cap, finishing at the first stop codon, ensuring that translation is usually monocistronic. See CODON, RNA PROCESSING, PROTEIN SYNTHESIS.

RNA editing Process changing the information content of transcribed RNA from that in an encoding gene's DNA. Occurs in mitochondrial, chloroplast and nuclear compartments. Differs from RNA PROCESSING

Type of RNA*	Function
mRNA	Transfers information from genes to protein-synthesizing machinery
tRNA	Carries activated amino acids for protein synthesis
rRNA	Protein synthesis
U1, U2, U4/6, U5 snRNAs	mRNA splicing
ncRNA	Transcription, chromosome modelling, RNA processing, mRNA stability
siRNA	down-regulate gene expression
M1 RNA	Catalytic unit RNaseP
Telomerase RNA	Template for telomere synthesis
Primer RNA	Initiation of DNA replication
7S RNA	Part of protein secretory complex
ATP	Carrier of energy-rich bonds
Coenzyme A	A key molecule in intermediate metabolism

* mRNA = messenger RNA; nc = non-coding RNA; rRNA = ribosomal RNA; siRNA = small interfering RNA; snRNA = small nuclear RNA; tRNA = transfer RNA.

TABLE 9 *Table indicating some functions of RNA and ribonucleotides.* (Modified from *Recombinant DNA 2/E* by Watson, Gilman, Witkowski, and Zoller. Copyright © James D. Watson, Michael Gilman, Jan Witkowski and Mark Zoller, 1992. Reprinted with permission of W. H. Freeman and Company.)

in that nucleotides are added, substituted or deleted between genomic DNA and mRNA stages of protein synthesis to ensure the correct reading frame is produced for translation. Sometimes involves interruption of an open reading frame by conversion of coding triplet into a stop codon. Editing sites are identified in pre-mRNA by formation of double-stranded RNA formed when intron sequences hybridize with complementary sequences on either side of the site. Enzyme dsRAD (double-stranded RNA ADENOSINE DEAMINASE) acts on such regions of dsRNA to catalyse the conversion of adenosine (A) to inosine (I), which is read subsequently as guanosine (G), so altering the information encoded by the RNA and, possibly, the amino acid sequence in the protein product of a gene. There is speculation that INTRONS may sometimes alter the probability of RNA editing in adjacent exons. RNA editing might alter splicing during RNA processing, codon composition, translation initiation or termination. It may also affect the reading frame, mRNA stability or structure, and RNA secondary structure. RNA editing in RECEPTOR protein pre-mRNA can lead to changes in signal transduction. In the SEROTONIN receptor, this is thought to contribute to mood swings, loss of appetite and seizures. See Z-FORM HELIX.

RNAi RNA INTERFERENCE. See RNA SILENCING.

RNA-induced silencing complex See RISC.

RNA interference (RNAi) A form of RNA SILENCING occurring in the cells of most, if not all, multicellular eukaryotes and an effective method of GENE SILENCING. It ensures that a specific mRNA molecule is destroyed and has great potential as a tool in biological research and can be tailored for the suppression of any gene, possibly even viral ones. See Fig. 148.

RNA life See ORIGIN OF LIFE, ANTIBIOTICS, TELOMERE.

RNA polymerases Enzymes producing RNA from ribonucleoside triphosphates. Unlike DNA polymerases they do not require a polynucleotide primer. Three types occur in eukaryotic cells, *polymerase I* making large ribosomal RNAs, *polymerase II* transcribing structural genes (introns and exons), *polymerase III* making small RNAs such as tRNAs and rRNAs. Each RNA polymerase must interact with transcription factors, each class having a distinct set (although the TATA FACTOR is common to all three sets). See RNA PROCESSING, Figs. 157 and 164.

RNA processing mRNA transcription within the nucleus produces RNAs of various sizes (*precursor RNA*, *heterogeneous RNA*) which are modified (processed) before passage to the cytosol for translation on ribosomes. The $5'$-end of the molecule is first capped (see RNA CAPPING) and then has a long poly-AMP sequence bound to the $3'$-end, which may facilitate the rest of processing and passage to the cytosol. Major feature of nuclear processing is the excision from pre-mRNA of non-coding INTRON sequences by SPLICEOSOMES. This is achieved by cutting these sections out using a PHOSPHODIESTERASE, and then splicing the transcript. This may rejoin one encoding region (exon) to another that is not its official nearest neighbour in the pre-mRNA. Alternatively, an exon may get cut out. This *alternative splicing*, especially common in higher eukaryotes, can produce two forms of the same protein at different times in development or in different cell types (see *BITHORAX* COMPLEX). Specific protein splicing factors, related to those in the spliceosome, control these splicing events. In B cells, the heavy chains of the cell surface IgM produced early in development differ at their $3'$-ends from the secreted form produced later by the differentiated plasma cell – this is because an exon incorporated in the B cell mRNA is replaced in the plasma cell mRNA by two shorter exons encoding fewer hydrophobic amino acids, allowing the IgM to cross the plasma membrane. Alternative splicing plays a crucial part in SEX DETERMINATION in *Drosophila*. Ribosomal and transfer RNAs are derived from pre-rRNA and pre-tRNA, in both prokaryotes and eukaryotes (see RNPS) and is even more complex than mRNA processing.

Alternative splicing of RNA occurs when the primary mRNA transcript is spliced in different ways so that from a single gene different proteins may be produced at translation – often in tissue-specific ways. Involved *inter alia* in sex determination in *Drosophila* and in production of cell-surface as opposed to secreted forms of ANTIBODY. See EXON, EXON SHUFFLING (for contrast), POST-TRANSCRIPTIONAL MODIFICATION, REPETITIVE DNA, RNA EDITING, RNA SPLICING.

RNase (ribonuclease) Any of several enzymes which hydrolyse RNA by breaking their phosphodiester bonds.

RNA silencing A group of POST-TRANSCRIPTIONAL RNA SILENCING mechanisms, including RNA interference (RNAi, in animals, see Fig. 148), post-transcriptional gene silencing and cosuppression (in plants) and 'quelling' in fungi, but absent from bacteria and Archaea. Spoken of as the 'genome's immune system' and forming a conserved eukaryotic antiviral defence mechanism, it involves sequence-specific targeting and degradation of RNA and combines both specificity against foreign nucleic acids with the ability to amplify and raise a massive response against them. The RNAi pathway is also involved in higher order chromatin remodelling, not least the packaging of DNA into HETEROCHROMATIN.

Although it was initially thought that introduction of ANTISENSE or additional sense mRNA into a cell could (surprisingly in the case of sense mRNA) suppress expression of the corresponding cellular gene, it now appears that in both cases it was co-introduction of contaminating double-stranded RNA (dsRNA) that did so. It seems to be induced by the presence of double-stranded RNA (dsRNA) or by viruses. Antisense small interfering RNAs (siRNAs) are usually involved as intermediaries, while dsRNA seems normally to act either as a trigger or as an intermediate in the response; for when dsRNA is cleaved into 'primary' siRNAs, these then seem to act (in *Caenorhabditis*, *Arabidopsis*, *Neurospora* and *Dictyostelium*, but apparently not in *Drosophila* or humans) as amplifiers which form complexes with proteins and then bind to complementary target mRNAs, whereupon either or both of two events seem to occur (not certain at the time of writing). In one, RNA-dependent RNA polymerase, RdRP synthesizes complementary RNA in a 5' direction using the siRNA–protein complexes as primers and target mRNAs as transcription templates, producing dsRNA sequences which are then cut by Dicer to release 'secondary' siRNAs which in turn complex with specific proteins and target complementary RNAs for a further round of dsRNA synthesis by RdRP (hence amplifying the signal). RNA interference in flies and humans seems to involve only 'primary' siRNAs, RdRP apparently not being involved. In the other process, siRNAs bind target mRNAs and induce RISC binding so that the mRNA is degraded by this dsRNase endonuclease.

In human cytoplasm, dsRNAs longer than 25 base pairs activate the potent interferon and RNA-dependent protein kinase (PKR) antiviral pathways, leading to non-sequence-specific effects that can include apoptosis. Slightly shorter cytoplasmic dsRNAs can enter sequence-specific RNAi pathways, mediating destruction of target mRNAs.

dsRNA is only one trigger of RNA silencing. Introduction of a transgene, in some plants at least, may lead to RNA silencing of the single-stranded transgene transcript, although it is not known which of many possible factors leads to this mRNA being regarded as aberrant, however the chief candidate for an aberrant RNA sensor is RdRP. Viral infection is another trigger for RNA silencing since the life cycles of some RNA viruses include dsRNA intermediates. The trigger for RNA silencing in DNA virus life cycles has not been identified at the time of writing. Transposons and repetitive DNA sequences are also kept in check by RNAi-like mechanisms, *Caenorhabditis* defective in the *mut-7* gene showing increased transposition.

In plants, transcriptional silencing can be triggered by the introduction of transgenes generating dsRNA corresponding to the transgene's promoter, and involves methylation of the homologous transgene pro-

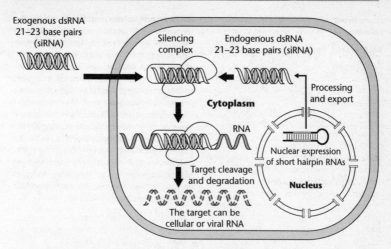

FIG. 148 *RNA interference, indicating how it is thought to occur in mammalian cells. Short double-strand RNA molecules (dsRNAs) can be introduced into cells from the outside, or are produced within the cell nucleus from longer precursors – typically RNAs that fold back on themselves by base-pairing to form 'hairpin' loops. These are then cleaved to produce shorter RNAs, containing 21–23 base pairs, that are then exported to the cytoplasm. Here the short dsRNAs (small interfering RNAs, or siRNAs) assemble with proteins into silencing complexes. Targets that perfectly match one strand of the dsRNAs are cleaved and destroyed, and include messenger RNA transcripts of certain genes, such as those of viruses that have entered the cell.*

moter in the nucleus. A new branch of horticulture involves *pathogen-derived resistance* (PDR), in which transgenes derived from a viral pathogen are used to transform plants. Once RNA silencing of the transgene is established, all RNAs with homology to the transgene are degraded. Furthermore, plant tissues far from the infected site also show specificity in RNA silencing, suggesting that a mobile signal spreads from cell to cell via plasmodesmata. In plants, at least, many viral pathogens encode proteins that suppress RNA silencing, indicating a host–pathogen evolutionary ARMS RACE, and there appears to be at least one conserved mechanism for suppression of RNA silencing for plant and animal viruses. See GENE SILENCING, PLANT DISEASE AND DEFENCES.

RNA splicing The joining together of RNA sequences on either side of intron sequences after the latter have been removed during RNA PROCESSING. Alternative RNA splicing is responsible for the production of the various immunoglobulin-like N-CAMs (neural cell adhesion molecules), the splicing steps taking place in the SPLICEOSOME, catalysed by small nuclear RNAs (snRNAs).

RNA surveillance system Mechanisms which detect and eliminate defective mRNAs and pre-mRNAs, the best understood being nonsense-mediated decay (NMD). NMD detects misplaced STOP codons, i.e. located more than 50 nucleotides upstream of an intron. Nuclear proteins bind to the exon–exon boundaries after intron excision and subsequent RNA SPLICING. Interactions with exon junction-marking proteins are further elaborated in the cytoplasm as further proteins are recruited, these complexes only persisting into the translation process on the ribosomes if the mRNA is to be degraded. Nuclear degradation of defective pre-mRNAs also occurs, although less is known about it. See DNA REPAIR MECHANISMS, PROTEIN SYNTHESIS.

RNA world There is a well-accepted view that the first life forms lived in an 'RNA world', in which not only did RNA store genetic information in primitive cells, but it also catalysed the reactions required by these early life forms – roles subsequently performed by DNA and proteins. Much of the evidence for this comes from studies of what RNA currently does in cells. Thus, some RNAs are catalytic: some ribosomal RNAs of the ciliate *Tetrahymena* are involved in splicing operations which cut out an INTRON within themselves (self-splicing), a process also occurring in fungal mitochondrial and bacteriophage pre-mRNAs; small nuclear RNAs (snRNAs) are catalytically involved in RNA SPLICING; the catalytic sites in SPLICEOSOMES are entirely composed of RNA; and ribozyme replication requires only an RNA-dependent RNA polymerase – a function (i.e. autocatalytic RNA replication) which may be performable, at least in part, in the absence of protein enzymes by a recently artificially mutated and selected polymerase ribozyme. Primer RNA is required for initiation of DNA synthesis; mRNA, tRNA and rRNA are all required for protein synthesis; 7S RNA is part of the protein secretory complex; RNA molecules can also catalyse the synthesis of the nucleotide 4-thiouridine from 4-thiouracil and activated ribose, one of the key reactions by which nucleotide constituents of nucleic acids are formed – the formation of the glycosidic bond joining sugar to base. From nucleic acid populations with randomized sequences, researchers have employed 'Darwinian selection' in a test-tube, to cull sequences that bind RIBOZYMES. These sequences, *doppelgängers* of their extinct counterparts, may give clues to the construction and testing of models of the RNA world. Ribose nucleotides are also present in ATP, an energy-rich molecule present in all cells, and in coenzyme A, universally crucial in intermediate metabolism. See Fig. 149, ORIGIN OF LIFE, RIBOSOME.

RNome The RNA equivalent of the PROTEOME.

RNPs (ribonuclear proteins) Eukaryotic complexes of protein and RNA located especially within nuclei, but also in the cytoplasm; some are restricted to the nucleus. The proteins of heterogeneous RNPs (hnRNPs) are among the most abundant in nuclei and bind pre-mRNAs while these are processed (spliced) into mRNAs (see SPLICEOSOME). Several hnRNPs shuttle between nucleus and cytoplasm. Small nuclear RNPs (snRNPs) are found with hnRNPs on those pre-mRNAs transcribed specifically by RNA polymerase II in the nucleus and occur in high concentration in the *coiled body*, a domain within the nucleus (see INTRONS). See PROTEIN SYNTHESIS, TELOMERE.

rod cell Highly light-sensitive secondary receptor of vertebrate RETINA, providing monochromatic vision even in very dim light. Its outer segment, closest to the choroid, contains parallel-stacked membranous discs with embedded molecules of *cis-retinal* (see RHODOPSIN), responding even to single photons. New discs are added to base of outer segment at 1–3 per hour; renewal is needed as old ones slough off at tip and get phagocytosed by pigment epithelial cells, which also store vitamin A. Attached to this by a cilium-containing restriction is the mitochondrion-rich *inner segment* which leads to the nuclear region (cell body) with its synaptic processes. Much integration of rod cell output is achieved by horizontal, bipolar, and AMACRINE CELLS before ganglion cells lead off via the optic nerve. Each primate retina contains about 120 million rods, few of them in the fovea. See SCOTOPIC VISION, CONE.

Rodentia Most widespread and numerous of mammalian orders. Placentals, they include rats, mice, capybaras, squirrels, beavers and porcupines. Herbivores or omnivores; one pair of large, continually growing, chisel-like incisors in both upper and lower jaws (see DENTITION, LAGOMORPHA). Large diastema; grinding molars.

rolling circle See GENE AMPLIFICATION.

root The vascular plant organ that usually grows downwards into the soil, anchoring the plant and absorbing water and nutrient

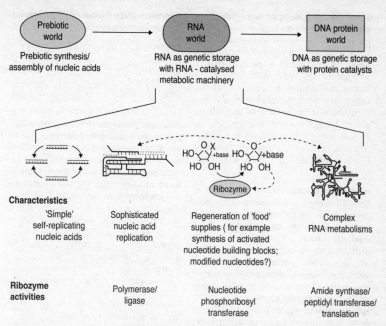

Characteristics

| 'Simple' self-replicating nucleic acids | Sophisticated nucleic acid replication | Regeneration of 'food' supplies (for example synthesis of activated nucleotide building blocks; modified nucleotides?) | Complex RNA metabolisms |

Ribozyme activities

| | Polymerase/ ligase | Nucleotide phosphoribosyl transferase | Amide synthase/ peptidyl transferase/ translation |

FIG. 149 *Biochemical epochs in the RNA WORLD. Early nucleic-acid or non-nucleic-acid replicators gave rise to faster and more faithful mononucleotide or polynucleotide polymerases. As foodstuffs for replicators were exhausted, an evolutionary advantage would have accrued to organisms that evolved the ability to generate new building blocks (for instance, using the thiouridine synthetase identified by Unrau and Bartel). At this stage, ribozymes would have possessed the chemical sophistication to modify nucleotide or oligonucleotide precursors. Modified nucleotides could have improved all extant catalysts and fostered the evolutions of more sophisticated catalysts, such as ribosomal RNA. The advent of neither cells nor energy metabolism is explicitly indicated, as either innovation would have yielded an evolutionary advantage irrespective of when it occurred.*

mineral salts. The first root originates in the embryo and is the primary root. In dicotyledons and gymnophytes, this root becomes a TAPROOT, it grows vertically downwards, giving rise to lateral roots along the way. In monocotyledons, the primary root is normally short-lived, and the root system develops from ADVENTITIOUS roots, and their lateral branches give rise to a FIBROUS root system. Roots cannot be distinguished from stems on the basis of their position with respect to the soil; some plants have roots wholly above the ground, and others have underground stems (e.g. RHIZOMES). Externally, roots principally differ from stems in not bearing leaves or buds; they possess at their tips a protective layer of cells, the ROOT CAP. Internally, roots differ from stems in having their vascular tissue in a central region and with protoxylem exarch. Many roots are important storage organs, and some of these are economically important (e.g. potatoes, carrots, parsnips, sugar beet).

root cap Cap of loosely arranged cells covering the apex of the growing point of a root; protects the apical meristem, and aids the root in penetrating the soil. As the root grows in length and the root cap is pushed forward, the cells of the periphery of the root cap are sloughed off, and these cells form a slimy covering around the root.

Thus they lubricate its passage through the soil. As quickly as cells are sloughed off they are replaced by the apical meristem. This slimy substance is a highly hydrated polysaccharide (probably a pectin), that is secreted by the outer root cap cells. It accumulates in the Golgi apparatus vesicles, which fuse with the plasmalemma, and then release the slime into the wall; it then passes through the wall emerging on the outside as droplets. In addition to a protective role, the root cap plays an important role in controlling the response of the root to gravity.

root hair (pilus) Tubular outgrowth of the root epidermal cell, possessing a thin, delicate wall in intimate contact with soil particles. Root hairs are produced in large numbers behind the region of active cell division at the root tip, forming the piliferous layer. They enormously increase the root's absorbing surface area, taking up ions and water. They are continually being replaced from new root tissue formed by the apical meristem. Water and dissolved ions move from the root hairs through the cortex, the endodermis and the pericycle into the xylem, and then upwards through the root and stem and finally into the leaves. This may occur apoplastically or symplastically. See TRANSPIRATION, TRANSPIRATION STREAM.

rooting (Of evolutionary tree) determination of the branch point in an evolutionary tree which represents the common ancestor of all the branch tips. Generally requires an OUTGROUP – a group that branched off before the common ancestor of the ingroup (the organisms being studied). See ELONGATION FACTORS.

root nodules Small swellings on roots of leguminous plants (e.g. pea, bean, clover) produced through infection by nitrogen-fixing bacteria. See NITROGEN FIXATION.

root pressure Pressure under which water passes from living root cells into the xylem; demonstrated by exudation of liquid from the cut end of a decapitated plant. May continue for long periods, arising through maintenance of water potential gradients by active transport of solutes. See CAVITATION, GUTTATION, TRANSLOCATION.

rootstock An unfortunate name for what is largely a short, stout, almost upright underground stem, with its upper end flush with the soil surface. Serves for vegetative propagation and found in herbaceous perennials.

Rosidae The largest subclass of the MAGNOLIOPSIDA in terms of the number of families but about the same size as the Asteridae in terms of the number of species with more than 60,000. Members of the Rosidae are more advanced than the MAGNOLIIDAE, but less advanced than the Asteridae. The Rosidae represent a natural group derived from the ancestral Magnoliidae, rather than by definitive distinguishing characters. Compound leaves with distinct, articulated leaflets are common, and although most members of the Rosidae are polypetalous, a considerable number are sympetalous and some are apetalous or even amentiferous. Parietal placentation is rare but other types are well represented. See INTRODUCTION.

Rotifera Abundant and widespread phylum (or class of ASCHELMINTHES); microscopic unsegmented pseudocoelomates, all aquatic and many capable of producing resistant sexual eggs. Parthenogenesis common; one group (bdelloids) thelytokous. Feed usually by ciliated trochal disc; gut entire, commonly with a muscular pharynx and chitinous 'jaws'. Excretion involves flame cells. Nervous system very simple. Elastic cuticle covers most of body. Extraordinary in being of approx. protozoan size, but at the 'organ' level of organization.

roundworms See NEMATODA.

R proteins See RESISTANCE GENES.

rRNA (ribosomal RNA) See RIBOSOMES.

r-selection Selection tending to operate in scattered and transient habitats, favouring organisms with ability to colonize rapidly, make use of short-lived resources, and complete their reproduction rapidly. Such organisms are selected for high r-values (r = intrinsic rate of increase); are clearly exploiters of their environments, have a tendency to PROGENESIS, and are at opposite end of range of strategies from organisms subject to *K*-SELECTION.

RT-PCR Reverse transcription polymerase chain reaction. A very sensitive amplification technique, enabling detection of a very few copies of an mRNA molecule from a sample. mRNA is isolated from a tissue or organ and then reversely transcribed to produce cDNA molecules, of which many copies are then made by the polymerase chain reaction.

RTKs RECEPTOR TYROSINE KINASES.

rubber See LATEX.

RuBisCO (ribulose bisphosphate carboxylase-oxidase) The major carbon-fixing enzyme in PHOTOSYNTHESIS and probably the most abundant protein on Earth.

ruderal (Of a plant) growing in disturbed soil, and so often in waste near human habitation.

Ruffini ending Slow-adapting pressure receptor in dermis and hairless vertebrate skins responding to both vertical pressure and skin stretching; often direction-specific.

ruffled membrane See LAMELLIPODIUM.

rumen Diverticulum of RUMINANT oesophagus; the chamber of the 'stomach' in which storage and initial digestion of food occur.

ruminant Artiodactyl (Suborder Ruminantia) characterized by very complex herbivorous stomach and feeding method. Grazed food is swallowed into the RUMEN and mixed with mucus, undergoing partial and anaerobic digestion by cellulase from a symbiotic bacterial flora. Some products are absorbed here; but 'pulp' can be returned to the mouth as 'cud' for chewing. After this (often done at leisure), the bolus is returned to the RETICULUM and omasum, where water is removed. The *abomasum* is the true stomach, in which further digestion occurs. In some ruminants, food can be seen regularly travelling up and down the oesophagus in the neck. *Rumination* is the act of remasticating reticular and ruminal ingesta. It is preceded by regurgitation into the mouth, followed by expression of excess water by the tongue and swallowing of this water. Renewed chewing then occurs. Ruminants are often on low-protein diets and compensate for this by excreting urea into the rumen, either by absorption directly from the blood or by its excretion in saliva. In either case, the urea reaches the rumen, where it is quickly converted by bacteria to ammonia and enters the general pool of rumen nitrogen from which microbial proteins are synthesized. These microbes are then digested as a source of free amino acids. Include deer, antelopes, giraffes, cattle, sheep and goats.

runner A STOLON that roots at its tip and forms a new plant which is eventually freed from connection with the parent through decay of the runner. Also used horticulturally of the daughter plant itself.

rust (1) A disease caused by one of the Uredinales of the class Teliomycetes (BASIDIOMYCOTA). (2) One of the UREDINALES. (3) A disease with rusty symptoms.

S

Saccharomyces Genus of YEAST fungi (ASCO-MYCOTA, Endomycetales), including organisms used in the production of bread and alcoholic drinks.

sacculus See VESTIBULAR APPARATUS.

sacral vertebrae Vertebrae of tetrapod lower back, articulating by rudimentary ribs with pelvic girdle. Just one occurs in most amphibia; two or more are fused in other vertebrae. See SACRUM.

sacrum Group of fused sacral vertebrae, ilia of the PELVIC GIRDLE being united to some or all of them. See SYNSACRUM.

sagittal A plane dividing a body, or organs, into left and right portions.

sagittal crest Medial bony ridge on top of some primate skulls (notably *Australopithecus robustus*, gorilla, chimpanzee) anchoring the TEMPORALIS MUSCLE on both sides.

St Anthony's fire See ERGOT.

salicylic acid (salicylate) Acetylsalicylate (aspirin) acetylates the terminal amino group of prostaglandin synthase, inhibiting its cyclooxygenase activity and thus inactivating it. Salicylate is formed as a result. Aspirin is anti-inflammatory because of its effect in blocking the first step in prostaglandin synthesis. Said to counter infection by stopping blood platelets from clumping in infected regions, but it may have direct antibacterial effects as well in reducing their output of fibronectin-binding protein used to bind bacteria to host cells. See PLANT DISEASE AND DEFENCES.

saliva Fluid secretion of salivary glands (labial glands in insects). In terrestrial vertebrates contains mucus and, in some of these and insects generally, contains amylases (e.g. *ptyalin* in mammals). *Invertases* occur in some insects (e.g. worker bees). Anticoagulants are present in salivas of blood-sucking leeches, insects and vampire bats, while those of the last two groups may contain pathogens.

salivary gland chromosomes See POLYTENY.

Salmonella Genus of bacterium, causing typhoid and paratyphoid; spread in food and water.

saltatory conduction See NODE OF RANVIER.

salt bridges See PROTEINS.

salt marsh The shores of protected portions of some estuaries are often occupied by a tidal salt marsh. These broad, flat areas extend from the water's edge to the uplands, at the extreme limit of tidal influence. Vegetation includes both flowering plants and algae.

SAM Abbreviation for S-adenosylmethionine, a compound produced in the biosynthesis of ETHYLENE.

samara Simple, dry, one- or two-seeded indehiscent fruit, the pericarp bearing winglike outgrowths (e.g. sycamore).

saprobiont Alternative for SAPROTROPH.

saprophyte See SAPROTROPH.

saprotroph (saprophyte, saprobe) Organism obtaining organic matter in solution from dead and decaying organisms. Usually involves secretion of extracellular enzymes on to the material, followed by absorption of the digestion products. Common in bacteria and fungi engaged in decay; of enormous importance in recycling and making available carbon dioxide, nitrates, phosphates and other nutrients via the DECOMPOSER food chain.

saprozoic (Of animals) taking up nutrients from the environment in solution; e.g. gut parasites utilizing the host's digestion products, some free-living flagellate protozoans.

sapwood Outer region of xylem of tree trunks, containing living cells (e.g. xylem parenchyma, medullary ray cells); functions in water conduction and food storage, as well as providing mechanical support. Usually distinguished from HEARTWOOD by its lighter colour.

Sarcinochrysidophyceae A class in the algal division HETEROKONTOPHYTA whose members live in estuarine and coastal marine waters. Their thalli are simple and multicellular, either filamentous or thalloid. They grow in the upper intertidal zone; only a few species are planktonic (e.g. *Ankylochrysis, Sarcinochrysis*). They possess a typical heterokontophyte cell structure (endoplasmic reticulum) surrounds the chloroplast, Golgi bodies having their forming faces appressed to the nuclear envelope, girdle lamella in the chloroplast, mastigonemes synthesized in the cisternae of the endoplasmic reticulum). Motile cells also display characteristic heterokontophyte features. They have several similarities to, and differences from the PHAEOPHYCEAE, and are interpreted as survivors of an ancestral group of algae from which the members of the Phaeophyceae arose.

Sarcodina Protozoan class characterized by PSEUDOPODIA. Some bear flagella in part of life cycle, supporting unification with flagellate protozoans in the single class Sarcomastigophora. Most are free-living, but some amoebae (e.g. *Entamoeba*) are parasitic. Many secrete shells. Includes subclasses Rhizopoda (amoebae, FORAMINIFERA) and Actinopoda (RADIOLARIA, HELIOZOA).

sarcoma See CANCER CELL.

sarcoplasmic reticulum Endoplasmic reticulum of muscle fibres; especially important in STRIATED MUSCLE in regulating calcium ion level in myofibril environment. Forms sac-like *terminal cisternae* around transverse tubule system, giving triad appearance in section. See MUSCLE CONTRACTION and Fig. 119.

Sarcopterygii (Choanichthyes) In older classifications, the lobe-finned fishes; of which there are four living genera: the three lungfishes (see DIPNOI) and the coelacanth (see COELACANTHINI). However, a cladistic classification would include as sarcopterygians all lobe-finned fishes with all their phenomenally successful descendants (including the land vertebrates), totalling over 24,300 living species.

SARS (severe acute respiratory syndrome) A pneumonia-like disease caused by a coronavirus (see Fig. 170), whose genome, published in May 2003, differs from those of the three previously known coronavirus subfamilies, indicating a long and independent evolution – probably in an animal host. It is important to discover the source from which it spread to humans (in southern China at the end of 2002) so that the likelihood of further outbreaks can be reduced. The virus appears to be remarkably invariant, unlike the influenza virus and HIV, thought to result from the method by which its genetic material is copied from several templates, reducing the chance that any mutation will become entrenched in the viral population. Worldwide, the death rate from SARS appears to be ~10%, but for those over 65 years the rate is over 50%. The ultimate cause of death is unclear, as by the time most lung damage has been done, the concentration of virus circulating in the blood has already peaked. A response overload of cytokines may be a factor in fatality. The drug glycyrrhizin, from liquorice roots, appears to be effective in ridding cultured cells of the virus. Other drugs being studied deactivate the viral proteinase that activates the viral proteins required for replication. The most promising VACCINE research seems to be on modified cat coronaviruses that have had non-essential but disease-causing genes removed.

satellite DNA Tandem repeats of a simple DNA sequence, which form minor 'peaks' (on account of its peculiar base composition and hence density) after DNA fragmentation and DENSITY GRADIENT CENTRIFUGATION. In general, any such large sequences are not transcribed and occur in HETEROCHROMATIN, as, e.g., associated with centromeres. In some mammals, a single satellite may comprise >10% of the genome, perhaps even a whole chromosome arm (see SELFISH DNA). They seem to evolve rapidly and can shift position within the karyotype. The human MYOGLOBIN locus was shown by A. Jeffreys in 1985 to contain an unusual tandem repeat DNA sequence, based on a core repeat of just 10–20 base pairs, producing a 'minisatellite'. It turned out that individuals varied in the exact number of core repeats, and that this variation was polymorphic (hence VNTR, for variable number of tandem repeats). It has provided a useful GENETIC MARKER in DNA FINGERPRINTING. Smaller repeats still, involving di-, tri- and tetranucleotides ('microsatellites') are also often highly polymorphic for repeat number, can be amplified by PCR, and seem to be more random in distribution through the genome than minisatellites, providing an almost ideal tool for GENETIC MAPPING and kinship studies. The human genome contains numerous $(CA)_n$ dinucleotide microsatellites, and considerable success has been achieved in combining maps from RFLPs and $(CA)_n$ repeats (see HUMAN GENOME PROJECT).

saturated fatty acid A FATTY ACID with no double bonds in its carbon chain.

Saurischia Extinct order of lizard-hipped reptiles forming (with ORNITHISCHIA) the dinosaur archosaurs. Distinguished by pubes pointing downwards and forwards and ischia pointing downwards and backwards (see PELVIC GIRDLE). Included *theropods* (bipedal carnivores, e.g. *Tyrannosaurus, Allosaurus*) and herbivorous *sauropodomorphs* (e.g. *Apatosaurus* (= *Brontosaurus*), *Diplodocus*). The last two were the largest land

animals yet discovered, a semi-aquatic life supporting their bulk.

savanna Grasslands with scattered broadleaved deciduous and evergreen trees, occurring singly or in small clumps. Sometimes trees dominate; others are dominated by shrubs. Savanna trees are generally deciduous, dropping their leaves during the dry season. Savannas are transitional regions between evergreen tropical rain forests and deserts, covering vast areas of East Africa, or in the regions between the prairies and the temperate deciduous forests, and between the TAIGA and the PRAIRIES in North America and Russia. They also cover much of eastern Mexico, Cuba, southernmost Florida, Australia, and parts of South America.

In savannas, both annual rainfall and temperatures vary widely seasonally. Also, light levels at ground level are high because of the sparse distribution of trees. Perennial grasses are common, along with bulbous plants. Trees of the savanna usually possess thick bark; seldom more than 15 m tall; nearly all deciduous, flowering when leafless.

The existence of savannas depends to a large degree upon periodic burning.

scaffolding proteins See PROTEIN COMPLEXES.

scalariform thickening Internal thickening of cell wall of xylem vessel or tracheid, taking the form of more or less transverse bars, suggestive of the rungs of a ladder.

scale (Bot.) In some algae, an external element covering the cell and sometimes the plasmalemma of the flagella. They may be inorganic or organic in composition, with intricate ornamentation. Inorganic scales are commonly siliceous as in the CHRYSOPHYTA, Class Synuraceae, and possess species-specific ornamentation. Such scales can provide invaluable information concerning past environmental conditions. Since such species occur widely in oligotrophic, acidic lakes they are being used as indicators of lake acidification (see ACID RAIN). Some organic scales have an inorganic covering of calcium carbonate deposited upon their surface (e.g. PRYMNESIOPHYTA, Order Coccosphaerales; see COCCOLITH). In vascular plants, a plate-like outgrowth.

(Zool.) (1) An epidermal structure characteristic of most vertebrate groups, usually plate-like or tooth-like, and serving for protection. Of limited taxonomic value in fishes (see COSMOID SCALE, GANOID SCALE, PLACOID SCALE). Teleost scales are either *cycloid* (concentric, lamella, and non-calcified) or *ctenoid* (as cycloid, but with a fringe of projections along its posterior margin). Some fish, notably eels and trout, have scales so small and/or so deeply embedded, that they appear to be scale-less. Early AMPHIBIA were scaled, as are many reptiles (e.g. SQUAMATA), in some of which bony as well as horny scales may be present. Birds have scaly legs; some mammals (e.g. rodents) have scaly tails. The production and deployment of vertebrate scales is termed *squamation*. (2) In some insects (e.g. butterflies and moths), one of numerous hollow sac-like ectodermal secretions attached to wing membranes, containing air or pigment. Their distributions may produce interference patterns, providing colouration.

scale insects (mealy bugs) Hemipterans (Superfamily Coccoidea); females wingless and scale-like, gall-like or covered in waxy secretion. Males usually with one pair of wings and vestigial mouthparts. Important economic PESTS; cottony cushion scale (*Icerya purchasi*) damages citrus fruits, others coconuts. Control by coccinellid beetles (ladybirds) is usually successful.

Scandentia Mammalian Order (tree shrews). See ARCHONTA.

scanning electron microscopy See MICROSCOPE.

scape Leafless flowering stem arising from ground level (e.g. dandelion, daffodil).

Scaphopoda Small class of bilaterally symmetrical marine molluscs (tusk shells, e.g. *Dentalium*), somewhat intermediate between bivalves and gastropods. The tubular shell opens at both ends, while the reduced foot is used in burrowing. Ctenidia absent; larva a trochosphere.

scapula Dorsal component of vertebrate PECTORAL GIRDLE; the shoulder-blade of mammals.

Schistosoma Blood fluke (see TREMATODA) of Africa (esp. Egypt), S. America and China causing *bilharzia* (+ schistosomiasis) in man. Smaller female lives attached to male within a groove of his body; eggs laid in abdominal blood vessels penetrate walls (causing most damage) and pass to bladder. Larval miracidia hatch in urine diluted on contact with freshwater and burrow into snail, in which rediae develop and from which cercariae emerge and penetrate skin of bathing/rice-planting people, debilitating over 200 million worldwide. Three common species infect humans. Education, support of irrigation engineers, molluscicides and medically prescribed drugs (e.g. antimony-containing) are all involved in prevention/cure.

schizocarp Dry fruit formed from a syncarpous ovary that splits at maturity into its constituent carpels forming several partial fruits, which are usually single-seeded, resemble ACHENES, and are called *mericarps*; e.g. hollyhock, mallow, geranium.

schizocoely Mode of formation of the COELOM in most annelids, in arthropods, molluscs and higher chordates. The coelom arises *de novo* as a cavity in the embryonic mesoderm. Compare ENTEROCOELY.

schizogenous (Bot.) (Of structure) formed by splitting due to separation of cells; e.g. oil-containing cavities in leaves of St John's wort. See LYSIGENOUS.

Schwann cell Specialized vertebrate GLIAL CELL, derived from NEURAL CREST, ensheathing axons of peripheral nervous system and responsible for formation there of the MYELIN SHEATH. See NODES OF RANVIER.

scion Twig or portion of twig of one plant grafted on to the stock of another.

sclera (sclerotic) The white of the vertebrate EYE. Coat of fibrous or cartilaginous tissue covering whole eyeball except for cornea.

sclereid See SCLERENCHYMA.

sclerenchyma Tissue, composed of sclerenchyma cells, which may form in any or all parts of the primary and secondary plant bodies that provide mechanical support to plants. Cells often lack protoplasts at maturity, and possess thick cell walls, with often lignified secondary walls. Two major types of sclerenchyma cells are found: (1) *fibres*, which are typically very elongated cells with tapering ends, occurring singly or variously grouped in strands or bundles. Economically important fibres include hemp, jute and flax, which are derived from the stems of dicotyledons. From monocotyledons comes manila hemp extracted from leaves; and (2) *sclereids*, which occur singly or in groups in the GROUND TISSUE, and are of variable shape, often branched, and relatively short cells. They form the seed coats of seeds, shells of nuts, the stone or ENDOCARP of stone fruits, and of pears. Compare COLLENCHYMA.

sclerite Region of arthropod CUTICLE in which the exocuticle is fully differentiated and *sclerotized*. Between two such sclerites there is usually a region of flexible membranous (unsclerotized) cuticle where exocuticle is undeveloped, allowing for articulation at joints. The body wall of a typical arthropod segment is divisible into four sclerotized regions: dorsal tergum, ventral sternum and a lateral pleuron on each side. Sclerites of the tergum are *tergites*, those of the sternum are *sternites* and those of each pleuron are *pleurites*.

scleroproteins Group of insoluble proteins forming major components of CONNECTIVE TISSUES. Include COLLAGEN, KERATIN and ELASTIN.

sclerosis (1) Hardening, as occurs in vertebrate tissues after injury. Involves deposition of COLLAGEN. (2) *Atherosclerosis* is the principal contributory process in the pathogenesis of gangrene and myocardial and cerebral *infarction* (death of a tissue area due to interruption of blood supply) as well as being a disease in its own right. Involves deposition of CHOLESTEROL and triglycerides in arterial walls (abnormal LDL receptors predispose individuals towards premature atherosclerosis). Normally a protective response to damage to endothelium and smooth muscle of arterial walls; involves inflammation and formation of fibro-fatty and fibrous lesions (*atheroma*) which, when excessive, become the disease – several GROWTH FACTORS (e.g. TGF β, PDGF) being involved. Endothelial cells, by producing NITRIC OXIDE (NO), tend to prevent platelet aggregation and promote vasodilation. Macrophages normally scavenge potentially toxic materials; but their uptake of too much oxidized LDL (oxLDL, produced by endothelial cells) can cause them to metamorphose into passive lipid-laden *foam cells*, which adhere to the artery's smooth muscle to form a fatty streak which may become artery-blocking plaque (atheroma) and/or produce cytokines and growth factors which attract monocytes and T cells and instruct the artery wall cells to multiply. This combination may be fatal. The intracellular bacterium *Chlamydia pneumoniae* has been found in 70% of atheromatous lesions in blood vessels, although its role in pathogenesis is unknown.

sclerotic See SCLERA.

sclerotium (Of fungi) compact tissue-like mass of fungal hyphae, often possessing a thickened rind, varying in size from that of a pin's head to that of a man's head; capable of remaining dormant for long, perhaps unfavourable, periods, and commonly giving rise to fruiting bodies; e.g. ERGOT.

sclerotization Tanning of arthropod CUTICLE. Involves cross-linking of protein by quinones (tyrosine derivatives). In some insects (e.g. *Calliphora* larvae), a quinone precursor is released onto the cuticle surface via pore canals and subsequently enzymatically oxidized by a phenol oxidase, tanning the protein cuticulin in the epicuticle and diffusing inwards to tan the outer procuticle to produce exocuticle. This is a major cause of cuticles becoming hard and brittle.

sclerotome See MESODERM.

scolex Part of tapeworm attached by suckers and hooks to gut wall of host; sometimes called the head. Proglottides are budded off behind it.

Scorpiones Order of ARACHNIDA containing scorpions. Pedipalps form large pincers, chelicerae small ones. End part of abdomen a segmented flexible tail bearing sting. Viviparous; terrestrial. See PSEUDOSCORPIONES.

scorpion flies See MECOPTERA.

scotophile See SKOTOPHILE.

scotopic vision Dark-adaptation of eye. Decrease in threshold of sensitivity of the ROD CELLS with increasing length of time in the dark. Involves enzymatic resynthesis of RHODOPSIN, increase in sensitivity of cells proceeding faster than the resynthesis. See EYE.

scramble competition The most extreme form of overcompensating DENSITY-DEPENDENCE is the effect of intraspecific COMPETITION on survivorship, where all competing individuals are so adversely affected that none of them survives.

screening, screens (1) Genetic screens. Methods resulting in the successful search for those individuals whose phenotypes indicate that they are genetic mutants in a gene, or genes, of interest. In non-human studies, these may discriminate between individuals on the basis of some aspect of behaviour. The same phenotypic trait may be caused by mutations in different genes, and COMPLEMENTATION tests can reveal whether this is so. If a set of genes is involved in bringing about the normal (wild type) phenotype, then it is useful if the order in which the genes operate can be worked out, and this requires finding mutations in each of the genes separately. REPORTER GENES are used to reveal when and where a single gene is expressed and NUCLEIC ACID MICROARRAYS can allow us to discover the expression patterns of several genes at once. (2) In microbiology, mutants are often screened by selection on media which permit only their phenotypes to grow, an important practice in GENE MANIPULATION to confirm the presence, or co-presence, of genes in recombi-

nant clones. Mutants which are non-selectable are often screened by REPLICA PLATING. (3) See GENETIC COUNSELLING.

scrotum Pouch of skin of perineal region of many male mammals, containing testes (at least during breeding season) and keeping them cooler (by 2°C in humans) than body temperature, without which sperm formation is impaired. Female spotted hyaenas have a pseudoscrotum, important in formation of dominance hierarchy.

scutellum Single cotyledon of a grass embryo, specialized for absorption of the endosperm.

scyphistoma Polyp stage in life cycles of SCYPHOZOA, during which the ephyra larvae are budded off by a kind of strobilation.

Scyphozoa Class of CNIDARIA containing jellyfish. Life cycle exhibits ALTERNATION OF GENERATIONS. Medusoid stage the rather complex adult jellyfish (gonads endodermal; enteron of four pouches, with complex connecting canals). Fertilized eggs develop into ciliated planula larvae which settle to become SCYPHISTOMAS.

sea anemones See ACTINOZOA.

sea cucumbers See HOLOTHUROIDEA.

seals See PINNIPEDIA.

search image Proposal by L. Tinbergen in 1960 that predators perform a highly selective 'sieving operation' on the visual stimuli reaching their retinas so improving their abilities to locate new (unfamiliar) cryptic prey when they are common, ignoring them when rare (see APOSTATIC SELECTION). The modern use of the phrase implies that predators individually undergo a perceptual change that enhances their detection of *familiar* prey items. Research today concerns analysing the possible role of alternative explanations of the data, such as that predators alter their search rate rather than acquire a search image in detecting such prey. The results will have consequences for traditional theories on the origin and maintenance of CRYPSIS and MIMICRY.

searching efficiency The instantaneous probability that a given predator will consume a given prey (also called the 'attack rate').

seaweeds Red, brown or green ALGAE living in or by the sea.

sebaceous gland Mammalian holocrine skin gland opening into hair follicle. Secretes oily lipid-containing *sebum* that helps to waterproof fur and epidermis. Epidermal in origin but projecting into dermis.

secondary consumer A carnivore that eats herbivores (= primary consumers). See FOOD CHAIN.

secondary growth (secondary thickening) In plants, growth derived from secondary or lateral meristems, the vascular or cork cambia. Results in increase in diameter of gymnosperm and dicotyledonous stems and roots, providing additional conducting and supporting tissues for growing plants and, in most cases, makes up greater part of mature structure. Rare in monocotyledons.

secondary meristem (Bot.) Region of active cell division that has arisen from permanent tissues, e.g. cork cambium (phellogen), wound cambium. See PRIMARY MERISTEM.

secondary metabolites A diverse array of chemically unrelated compounds such as alkaloids, quinones, essential oils (including terpinoids), glycosides (including carcinogen compounds and saponins), flavonoids and raphides (needle-like crystals of calcium oxalate). The presence of such compounds can characterize families or groups of flowering plants. Such chemicals function in decreasing or restricting the palatability of the plants in which they occur, or in causing animals to avoid them. This ability to produce and then retain such compounds in their cells and tissues is an important evolutionary step, providing such plants with a biochemical defence from most herbivores. Many fruits and 'vegetables' synthesize secondary metabolites affecting their colour, scent and flavour and such 'phytochemicals' often attract or repel

herbivory. Several brassicas (e.g. cabbage, broccoli, kohlrabi) contain the vacuolar enzyme myrosinase which is released on cell damage and converts the secondary metabolites known as glycosinoloates into isothiocyanate which in turn induces detoxifying enzymes in the human liver and may reduce the potentially harmful effects of carcinogens. Some fungi produce secondary metabolites which are highly toxic and also carcinogenic, while the siphonaceous green alga *Halimeda* contains compounds that significantly reduce feeding by phytophagous fish. In microbiology, secondary metabolites are not produced during the organism's primary growth phase (not essential for growth/reproduction) but when the culture enters its stationary phase. They usually demand a large number of complex enzymic steps. Often of industrial interest: ANTIBIOTICS are examples. See COEVOLUTION.

secondary sexual characters Characters in which the two sexes of an animal species differ, excluding gonads, their ducts and associated glands. In humans include mammary glands, subcutaneous fat deposition, shape of pelvic girdle, voice pitch, mean body temperature, mass and extent of muscle development. See OESTROGENS, TESTOSTERONE.

secondary wall Innermost layer of plant CELL WALL, formed by some cells after elongation has finished; possesses highly organized microfibrillar structure.

second messenger Organic molecules and sometimes metal ions, acting as intracellular signals, whose production or release usually amplifies a signal such as a hormone, received at the cell surface. Some hormones bind to the cell membrane and activate an enzyme there to generate the second messenger. Alternatively, the ligand may be a non-hormone which opens or closes a GATED CHANNEL affecting membrane permeability to an ion. Calcium ion (Ca^{2+}) concentration is extremely important in control of many cell functions (SEE CALCIUM PUMP, CALCIUM SIGNALLING, CALMODULINS). First organic molecule hailed as a second messenger was

CYCLIC AMP, but others include DIACYLGLYCEROL and INOSITOL 1,4,5-TRIPHOSPHATE. See CASCADE.

secretin Polypeptide hormone of intestinal mucosa, produced there in response to acid chyme from stomach. Inhibits gastric secretion, reduces gut mobility but stimulates secretion of alkaline pancreatic juice, bile production by liver and intestinal secretion. See CHOLECYSTOKININ, GASTRIN.

secretion (1) Production and release from a cell of material useful either to it or to the organism of which it is a part. Material is commonly packaged into *secretory vesicles* which bud off from the GOLGI APPARATUS and fuse with the cell's apical plasmalemma (see MEROCRINE GLAND); other methods are employed by APOCRINE, AUTOCRINE and HOLO-CRINE GLANDS. Plant cell walls are secreted. (2) *Active secretion*. Process whereby a substance (often an ion) is actively pumped out of a cell against its concentration gradient, as in the ascending loop of Henle in the vertebrate kidney, and in fish gills. See ACTIVE TRANSPORT.

secretory vesicles/secretory granules Specialized organelles mediating regulated exocytotic protein secretion, containing a set of the soluble and membrane-bound proteins which are initially synthesized in the rough endoplasmic reticulum and transported to, and through, the GOLGI APPARATUS. PROTEIN TARGETING takes place during assembly of the secretory vesicle matrix and membrane.

sedimentation coefficient See SVEDBERG UNIT.

seed Product of fertilized ovule, comprising an embryo enclosed by protective seed coat(s) derived from the integument(s). Some seeds (castor oil, pine) are provided with food material in the form of ENDOSPERM tissue surrounding the embryo, while in other (*non-endospermic*) seeds food material is stored in the cotyledons (e.g. pea). The seed habit is the culmination of an evolutionary development involving, in sequence, heterospory, reduction of a free-living female gametophyte generation dependent on water for fertilization, and its retention within the tissues of the sporophyte by which it is protected and supplied with food. It occurred early in geological history, and in more than one group of plants. Of equal importance was the reduction of the male gametophyte to the pollen grain and its pollen tube, again avoiding the need for water in fertilization. Independence from water in sexual reproduction increased immensely the range of environments open for colonization. Biological advantages of the seed habit include: continued protection and nutrition of the embryonic plant during its development, provision of dispersal mechanisms, provision of food to tide over critical periods of growth of the embryo, and its establishment as an independent plant after the seed is shed. Plants almost always free themselves from systemic diseases on passing through the phase of seed production. Passage through the sexual cycle 'cleans' the cytoplasm of viruses and cytoplasmic variants. See LIFE CYCLE.

seed bank, seed banking A seed bank is the population of viable dormant seeds that accumulates in and on soil, and in sediments under water. Seed banking involves the preservation of plants in 'suspended animation' as seeds by: collecting from those growing in the wild, cleaning them and screening by X-ray to check they contain fully developed embryos, drying to reduce moisture content to equilibrium with 15% relative humidity and finally storing in clusters at −20 to −40°C. For seeds so far investigated, this is expected to preserve 86% of seeds for an average of 200 years. Many endangered floras and commercial varieties will hopefully be banked this way. See BIODIVERSITY.

segmentation (1) Alternative term for CLEAVAGE of an egg. (2) In zoology, commonly synonym for *metameric segmentation* or *metamerism*, to indicate production of a body plan of repeating organizational units, variably distinguishable, along the antero–posterior body axis. Metamerism most marked in annelids and arthropods (see TAGMOSIS) but vertebrates exhibit it in the segmental organization of nervous system and muscles, most clearly in the embryo (see MESODERM). Has often become reduced or wholly lost in evolution, notably in molluscs and echinoderms. Debate surrounds role of the HOMEOBOX in control of segmentation. See HOMEOBOX, COMPARTMENT.

segmentation genes Genes dividing the early *Drosophila* embryo into a series of repetitive segmental primordia along the anterior–posterior axis, mutations in which cause loss of parts of (even of entire) segments. GAP GENES, PAIR-RULE GENES and SEGMENT POLARITY GENES form the three classes to date. For a diagram of one proposed cascade of development gene regulation involving these, see *BICOID* GENE. The gene *ENGRAILED* is expressed in metameres of both the fruit fly *DROSOPHILA* and the chordate AMPHIOXUS. This is evidence that segmentation was present in the common ancestor from which the insect and chordate lineages diverged. See HOMOLOGY.

segmented (1) For zoological contexts, see SEGMENTATION. (2) In the context of a viral genome, see INFLUENZA.

segment polarity genes A class of segmentation genes, studied particularly in *Drosophila*, mutants in which cause not only a loss of the same part in each segment, but also a mirror-image duplication of the remaining part to be produced in its place. In *Drosophila* at least, domains of mRNA transcript expression of segment polarity genes are dependent largely upon prior distributions of GAP GENE expression domains (e.g. see *HEDGEHOG* GENE). For instance, *engrailed* (*en*) and *wingless* (*wg*) express as stripes of transcript in the early embryo, *wg* stripes representing the posterior limits of PARASEGMENTS, while *en* stripes represent anterior limits. It is likely that some at least of these genes encode transmembrane proteins involved in translating extracellular signals into changes in cell fate. Compare PAIR-RULE GENES. See POSITIONAL INFORMATION.

segregation (1) See MENDEL'S LAWS. (2) Process whereby alleles usually present together in somatic cells of an organism separate into different cells during meiosis in the germ line. Such alleles are said to segregate.

segregation distorter (Sd) See ABERRANT CHROMOSOME BEHAVIOUR (4).

seismonasty (Bot.) Response to non-directional shock stimulus; e.g. rapid folding of leaflets and drooping of leaves in *Mimosa pudica* when lightly struck or shaken.

Selachii Elasmobranch Order containing modern sharks. Members have five to seven gill openings on each side of head and an upper jaw not fused to skull. Group appeared in the Jurassic and is characterized by the *rostrum* (snout) that hangs over the mouth. Includes largest living fishes (over 20 metres in whale sharks, *Rhinedon*). Rays and skates (Batoidea) are sometimes also included. See ELASMOBRANCHII.

Selaginellaceae Spike moss family. An ancient family dating to the Carboniferous period related distantly to the LYCOPODIACEAE and ISOËTACEAE comprising one genus, *Selaginella*, and 700 species worldwide, but primarily found in the tropics and subtropics. Plants are herbaceous, annual or perennial, sometimes remaining green over winter. Stems are leafy, branch dichotomously and can be regularly or irregularly forked or branched. Stems are protostelic (sometimes with many protosteles or meristeles), siphonostelic or actino-plectostelic. Rhizophores (modified leafless shoots bearing roots) may be present or absent. Sporophylls are monomorphic or adjacently different, slightly or highly differentiated from vegetative leaves. Solitary sporangia are borne on short stalks in the axils of the sporophylls and open via distal slits. Plants are heterosporous; megaspores (one, two or four in number) are large; microspores are numerous and minute.

selectins A small family of ADHESION molecules with a characteristic sequence of protein motifs comprising a lectin-related amino-terminal domain, an EGF-like repeat unit and two or more domains related to consensus sequences of several COMPLEMENT receptors. They seem to help regulate leucocyte binding to endothelium at sites of INFLAMMATION. See INTEGRINS.

selection See ARTIFICIAL SELECTION, NATURAL SELECTION.

selection coefficient (s) See COEFFICIENT OF SELECTION, FITNESS.

selection pressure See NATURAL SELECTION.

selector genes Genes involved in the final decision of cell fate, leading to production of a specific cell type.

self antigen Cell surface markers to which an animal does not initiate an immune response. See AUTOIMMUNE REACTION, IMMUNITY, IMMUNE TOLERANCE.

self-assembly Usually reserved for biological structures which become assembled from components without help of enzymes or 'scaffolds' which do not form part of the functional structure. Some VIRUSES (e.g. tobacco mosaic virus, TMV) assemble within the host cell from separate nucleic acid and capsomere protein molecules. Several other structures are either fully self-assembling (e.g. bacterial ribosome) or require surprisingly few enzyme steps or accessory proteins acting as 'jigs' in assembly. Some viruses, membranes, cilia, mitochondria and myofibrils are in this second category. It is thought that some steps in these more complex cases of partial self-assembly require appropriate timing, and are irreversible if disassembly is imposed upon them. See DYNAMIC SELF-ORGANIZATION, NUCLEOLUS.

self-fertilization Fusion of micro- and macrogametes from the same individual. See INBREEDING.

selfing Self-fertilizing; in higher plants, involves prior self-pollination.

selfish DNA/genes If the UNITS OF SELECTION are genes, or perhaps smaller-sized replicable hereditary material, a consequence appears to be that any mutation promoting their transmission into the next generation would automatically be selected for – at least in the short-to-medium term. Such promotion might involve simply the 'hijack-ing' (a kind of molecular parasitism) of normal genetic mechanics and machinery, which selection might be unable to prevent. It has been suggested that evidence for such 'self-promotion' of genes, even in the face of harmful phenotypic effects they may have on their bearers (uncoupling morphological and molecular evolution), is to be found in the C-VALUE paradox, mif^+ plasmid and ω^+ intron (see MITOCHONDRION), *T* HAPLO-TYPES, INTRONS, CHROMOSOME DIMINUTION, GENE CONVERSION and driving genes (see ABERRANT CHROMOSOME BEHAVIOUR (3) and (4)), TRANSPOSABLE ELEMENTS. Some of this 'selfish DNA' is now more properly regarded as mutualistic (e.g. see *Alu*, EXON SHUFFLING). There is now talk of 'intranuclear warfare' between genetic elements, even between whole chromosomes.

self limitation A process where intraspecific competition leads to a reduction in reproduction and/or survival at higher densities.

self-pollination Transfer of pollen from anther to stigma of same flower, or to stigma of another flower of same plant. See INBREEDING.

self-replication See AUTOCATALYST.

self-restriction See T CELL.

self-sterility (Of some HERMAPHRODITES) inability to form viable offspring by self-fertilization. See INCOMPATIBILITY.

self-thinning The progressive decline in density that accompanies and interacts with the increasing size of individuals in a population of growing individuals; e.g. in a maturing forest.

semelparous Reproducing only once in its lifetime. Contrast ITEROPAROUS.

semicell See DESMID.

semen Sperm-bearing fluid produced by the testes and accessory glands (e.g. PROSTATE GLAND, SEMINAL VESICLES and Cowper's gland in mammals), particularly of animals with internal fertilization.

semicircular canals Component of vertebrate VESTIBULAR APPARATUS detecting (directional) acceleration of the head.

seminal vesicle (1) Organ of lower vertebrates and of some invertebrates (e.g. earthworms, insects) that stores sperm. (2) Diverticulum of the *vas deferens* (sperm duct) of male mammals, whose alkaline secretions lower the pH of the semen and counteract vaginal/uterine acidity.

seminiferous tubules See TESTIS.

semiochemicals The plant's equivalent of PHEROMONES: volatile signal molecules released when under herbivore attack. Some attract parasitoids or predators in the host's defence – as when maize which is being eaten by caterpillars of *Spodoptera* releases volatiles which attract the parasitoid wasp *Cotesia*. Parasitized caterpillars eat fewer leaves than unparasitized ones. In the broad bean, *Vicia faba*, the composition of the mixture of volatiles released is known to depend upon the species of aphids feeding on its leaves, the parasitoids attracted being able to distinguish the different semiochemical compositions. Plant–plant communication is also known to involve volatile production (see PLANT DISEASE AND DEFENCES).

semispecies Populations in the process of acquiring mechanisms isolating one from the other reproductively. Regarded therefore as *incipient species* in the sense of the biological species concept (see SPECIES). Semispecies are generally expected to exhibit greater genetic differences than do geographical subspecies (see INTRASPECIFIC VARIATION).

senescence See AGEING.

sense organ Group of sensory RECEPTORS and associated non-sensory tissues specialized for detection of one sensory modality (e.g. light in the EYES, sound in the EARS, etc.).

sensitivity Ability to respond to environmental change. Regarded as a property common to all life forms. Sometimes referred to as 'irritability'.

sensitization Process rendering an organism or cell more reactive to a specific ANTIGEN or antigenic determinant. See IMMUNITY, ALLERGIC REACTION.

sensory transduction The general process of converting an environmental energy change into a RECEPTOR POTENTIAL.

sepal Component member of the calyx of dicotyledonous flowers; usually green and leaf-like. Cultures of pollinated flowers of several species on media containing only sucrose and mineral salts indicate that sepals have a role in nitrate assimilation and synthesize and export AUXIN, GIBBERELLINS and CYTOKININS to the developing fruit. See FLOWER.

septate junction Functionally, the invertebrate equivalent in many respects of vertebrate tight junctions. See INTERCELLULAR JUNCTIONS.

septicidal (Bot.) Describing the dehiscence of multilocular capsules by longitudinal splitting along septa between the carpels, separating the carpels from one another, e.g. St John's wort. See LOCULICIDAL.

septum Partition or wall. The structure concerned is said to be SEPTATE.

sere Particular example of plant SUCCESSION. Seres originating in water are referred to as HYDROSERES; those arising under dry conditions as XEROSERES, of which those developing upon exposed rock surfaces are known as LITHOSERES.

serial homology See HOMOLOGY.

serial sections Series of successive microtome sections, from which three-dimensional structure can be built up.

serine kinases See PROTEIN KINASES.

serine proteases (serpins) Family of homologous proteolytic enzymes (see GENE DUPLICATION) including several involved in digestion (e.g. chymotrypsin, trypsin) and BLOOD CLOTTING (e.g. thrombin). Implicated in the genesis of POSITIONAL INFORMATION during *Drosophila* development. See CASCADE.

serology The study of ANTIGEN-ANTIBODY REACTIONS *in vitro*.

serosa Outermost layer of most parts of vertebrate GUT. A SEROUS MEMBRANE.

serotonin (5-hydroxytryptamine, 5-HT) L-tryptophan-derived NEUROTRANS-MITTER of vertebrate brain; esp. of *raphe nuclei* of PONS and MEDULLA, and of PINEAL GLAND (precursor of MELATONIN). General inhibitor of activity – apparently in opposition to NORADRENALINE; tricyclic antidepressant drugs probably potentiate neurotransmission of both by blocking their neuronal uptake (see serotonin-like PSYCHOACTIVE DRUGS). Hyperpolarizes post-synaptic membranes and activates *phosphofructokinase* in liver (see GLYCOLYSIS), both mediated by cyclic AMP. Many studies have found that aggressive animals, humans included, have lower levels on average of a serotonin metabolite in their cerebrospinal fluid, believed to reflect lower serotonin levels in the brain. This suggests that aggressive behaviour is linked with reduced serotonin levels; but there are at least 14 different serotonin receptors in the human brain, which makes clarification of its roles very difficult, and at least one of these is a presynaptic receptor which decreases the level of serotonin produced (PRESYNAPTIC INHIBITION) yet tends to quell aggression when bound by ligand. See also PLATELETS, RNA EDITING, BLOOD CLOTTING.

Serotonin

serotype An antigenically distinct variety within a bacterial species.

serous membranes Mesothelial layers overlying deeper connective tissue and lining the vertebrate coelomic spaces (pericardial, perivisceral, pleural, peritoneal cavities).

Sertoli cells See TESTIS, MATURATION OF GERM CELLS.

serum See BLOOD SERUM.

sesamoid bone Bone (e.g. patella) developing within tendon of vertebrate, particularly where tendon operates over ridge of underlying bone.

sessile (1) (Of animals) living fixed to the substratum, e.g. sponges, corals, barnacles, limpets, tunicates. (2) Lacking stalks, e.g. eyes of some crustaceans. Opposite of *pedunculate* in plants.

seta (Bot.) Stalk of sporagonium in mosses and liverworts; in algae, a stiff hair, bristle or bristle-like process – an elongated hollow cell extension. (Zool.) Invertebrate epidermal bristle, consisting solely of cuticular material (CHAETA), or of a hollow projection of cuticle enclosing epidermal cell or its part (e.g. in insects).

Sewall Wright effect See GENETIC DRIFT.

sex (1) Often used as synonym of SEXUAL REPRODUCTION. (2) Used in several senses. *Germ cell* distinguishes individuals by their abilities to produce gametes of particular morphological types, namely, microgametes (sperm, generative nuclei, etc.), or macrogametes (eggs, egg cells, etc.). Males (with 'male' sex organs) produce the former; females (with 'female' sex organs) produce the latter. HERMAPHRODITES produce both, either simultaneously or sequentially. *Genetic sex* concerns an individual's genotype in so far as it bears on SEX DETERMINATION; *phenotypic sex* relates to anatomical appearances normally associated with one or other sex (e.g. SECONDARY SEXUAL CHARACTERS), sometimes distinguished from *behavioural sex*, where the two sexes behave in distinctive ways. *Hormonal sex*, identified by the particular hormonal production from an individual's sex organs, is determined by sex organ appearance and physiology rather than by genotype (see SEX-REVERSAL GENE). *Brain sex* refers to distinctive anatomical differences between the brains of the two sexes, itself sometimes causally related to behavioural sex.

The *origin and evolution of sex* are problematic, but the processes common to all forms of genetic recombination may have evolved from cellular DNA REPAIR MECHANISMS or as an antidote to driving genes (see ABERRANT CHROMOSOME BEHAVIOUR (4)). Some hold that since the only advantages of sexual reproduction over asexual seem to be long-term ones, some kind of GROUP SELECTION is necessary to account for prevalence of the former. Others argue that natural selection alone can account for eukaryotic sexual reproduction, despite the COST OF MEIOSIS. One view is that a parasitic gene (see SELFISH DNA/GENES) might force a cell to fuse with another (plasmogamy), leave a copy of itself in this second cell, and then force the two cells to split apart (see MITOCHONDRION). The evolution of nuclear fusion (karyogamy) might then have involved selection to avoid the deleterious consequences of a nucleus bearing a recessive 'mating type' allele being condensed, envacuolated and eliminated from the cytoplasm – as happens in *Physarum* (see MATING TYPE). Another view is that genes for sexual mechanics evolved during an ARMS RACE between parasites and hosts in which host genes for parasite-resistance oscillate out of phase with parasite genes for undermining host resistance. Computer simulations suggest that whereas in asexual host populations the resistance genes go to extinction as parasite anti-resistance genes increase, sexual populations can maintain these genes at low frequency. These genes may be reused in response to future parasite advances whereas asexual populations would have to 'wait' for the gene to mutate again, which may be too late. A further idea is that each time numbers get depressed by parasites in an asexual host population, it becomes more susceptible to accumulation of mutations by MÜLLER'S RATCHET, eventually failing to replace itself and going extinct. Moderate effects on hosts by parasites combined with reasonable rates of mutation could make sex evolutionarily stable against repeated invasion by asexual and parthenogenetic host clones. It is interesting that cytoplasmic pathogens appear to be cleared by the formation of SEEDS. See RECOMBINATION.

Separation of sexes (*dioecism*) has the effect of reducing the potential number of individuals between which fertilization may occur (see INCOMPATIBILITY).

sex chromatin See BARR BODY.

sex chromosome Chromosome having strong causal role in SEX DETERMINATION, usually present as a homologous pair in nuclei of one sex (the HOMOGAMETIC SEX) but occurring either singly (or with a partial homologue) in those of the other sex (the HETEROGAMETIC SEX). In cases of partial homology one chromosome (Y) is often much smaller than the other (X), and both may resemble autosomes. In birds, where females are heterogametic, males are sometimes given the genotype ZZ and females ZW. Sometimes sex-determining loci are situated on such a short region of a single chromosome pair as to make sex chromosomes indistinguishable in appearance. In species with an XX/XY system (e.g. humans) sex chromosomes usually pair up at meiosis and form a bivalent; in mammals, there may be crossing-over and chiasma formation between homologous (often sub-terminal) regions. In humans, the X and Y chromosomes normally recombine (even in male meiosis) at two Y-terminal pseudoautosomal regions which flank a 'male-specific region' (MSY) comprising 95% of the Y chromosome length. The human MSY includes at least 156 transcribed units, all in euchromatic regions, and half of these probably encode proteins – only 27 of which seem to be distinct proteins or protein families. Twelve of these are known to be expressed throughout the body while 11 are expressed mainly or exclusively in the testes. So-called 'X-degenerate sequences' in the MSY region are intron-bearing single-copy gene or pseudogene homologues of 27 different X-linked genes, with between 60% and 96% nucleotide sequence identity with their X chromosome homologues. Their presence is consistent with the prevailing view that human X and Y chromosomes, like those of other animals, evolved from an ordinary pair of autosomes, at which time they seem to have experienced incremental suppression of crossing over. Inversions in the Y chromosome may also have suppressed crossing over with the X chromosome. Decay and subsequent obliteration of the non-recombining Y chromosome regions were probably followed by upregulation of genes on homologous X chromosome regions and subsequent X chromosome inactivation (see DOSAGE COMPENSATION). Human Y chromosomes have a remarkable amount of PALINDROMIC DNA. Organisms with both types of sex organ combined in one individual (e.g. monoecious plants, hermaphrodites) lack specialized sex chromosomes. For nonparity in offspring phenotypes involving maternally- and paternally-derived X chromosomes, see CHROMOSOMAL IMPRINTING, HETEROCHROMATIN. See also AUTOSOME, HEMIZYGOUS, SEX LINKAGE.

sex-controlled character See SEX-LIMITED CHARACTER.

sex determination Control of occurrence of, and differences between, sexes (see SEX). Where male and female differentiation occurs within a single individual (e.g. a HERMAPHRODITE), it is commonly restricted to the sex organs. See GYNANDROMORPH, FREEMARTIN.

Bisexual (dioecious) species often have genetic sex-determining mechanisms. These are occasionally cytoplasmic (see CYTOPLASMIC INHERITANCE), sometimes depend upon the individual's ploidy (see MALE HAPLOIDY), but most commonly involve a pair of SEX CHROMOSOMES.

In some cases (e.g. many gonochorist species of fish, amphibians and reptiles) one or a few loci on an unspecialized chromosome pair may be responsible for sex determination through segregation of a pair or a few pairs of alleles (resembling some INCOMPATIBILITY mechanisms). In cases of sex organ differentiation within a single hermaphrodite individual (e.g. some moss and fern gametophytes) the mechanism involved resembles normal cytological DIFFERENTIATION. Sex-determining loci may not be restricted to sex chromosomes, even where these are differentiated (e.g. see SEX REVERSAL GENE). But where sex is determined by a differentiated sex chromosome pair it may be absence of a second X or presence of a Y that results in sex differentiation. In humans it is the latter. Individuals with KLINEFELTER'S SYNDROME (XXY) are male while those with TURNER'S SYNDROME (XO) are female. In mammals, the Y-chromosome encodes a testis-determining factor (see TESTIS for effect of *Sry* gene) which causes embryos with a Y-chromosome to develop testes and become males, while those lacking a Y-chromosome develop ovaries and become females. The locally acting signal, Wnt-4, is required for MÜLLERIAN DUCT development in both sexes; but it is crucial for sexual development in female mammals as it suppresses Leydig cell development, *Wnt-4* mutant females being masculinized while mutant males develop normally (see also MÜLLERIAN INHIBITING SUBSTANCE). But in the fruit fly *Drosophila* the X-chromosome is female-determining while the autosomes collectively are male-determining, an individual's sex resulting from the balance, or ratio, between the number of sets of autosomes and X-chromosomes. This was established by artificial production of INTERSEXES which are triploid for their autosomes but diploid (XX) for their sex chromosomes. 'Superfemales' were diploid for their autosomes but triploid for sex chromosomes (XXX); 'supermales' were triploid for their autosomes but hemizygous for an X-chromosome (XO). So an XY/XX sex chromosome system may in itself tell us little about the sex-determining mechanism involved. In *Drosophila*, whatever measures the ratio of sex chromosomes to autosomes also controls the details of alternative splicing (see RNA PROCESSING) of two gene transcripts (*Sxl* and *tra*); for in males the 'default' pattern) the pre-RNA of each contains an intron bearing a stop codon which makes for functionless gene product, whereas in females these introns are excised. The *tra* gene product is required for female development, which it does by regulating further RNA splicing events. The nematode *Caenorhabditis elegans* has a similar 'balance' sex-determining mechanism. The worm *Bonellia* (Echiuroidea) has environmental sex-determination, albeit with genetic involvement. If a larva develops independently it becomes female; if it is influenced by pheromones produced by the adult female's proboscis it develops into a male. In some reptiles (e.g. turtles, crocodilians) sex is strongly determined by the temperature at which the embryo develops.

sexduction (F-duction) See TRANSDUCTION, F FACTOR.

sex factor See F FACTOR.

sex hormones See OESTROGENS, TESTOSTERONE.

sex-limited character (sex-controlled character) Character determined by a genetic element expressed differently between the sexes, commonly as a result of hormonal differences. *Pattern baldness* in humans is a character expressed in men who are either homozygous or heterozygous for the allele, but only in those women who are homozygous for it. Usually caused by CHROMOSOMAL IMPRINTING. See SEXUAL SELECTION.

sex linkage A character is said to be sex-linked if it is determined by a genetic element occurring either (a) on an X-chromosome where the method of SEX DETERMINATION involves an XX/XO system, or (b) on any non-homologous region of a pair of SEX CHROMOSOMES where the method involves an XX/XY system. Most readily detected by an appropriate RECIPROCAL CROSS. In an XX/XY system, if the character in question is recessive it will appear in the heterogametic sex with the same frequency (say, n) as the chromosome bearing the element; hence it will appear in the homogametic sex with the square of that frequency (i.e. n^2). In humans, red-green colour blindness and haemophilia are such examples. *X-linked* characters cannot be transmitted from father to son since a father contributes a Y-chromosome to his male offspring. A mother may be a carrier, in which case half her male offspring would tend to be affected. Dominant sex-linked characters would be expressed in hemizygous, heterozygous and homozygous conditions.

sex pili See PILI, F FACTOR.

sex ratio Proportion of males to females in a species population, usually expressed as the number of males per 100 females. *Primary sex ratio* is assessed immediately after fertilization; *secondary sex ratio* is assessed at birth (or hatching); *tertiary sex ratio* is assessed at maturity. The sex ratio among the offspring of a particular female may be subject to fluctuations, as with queen bees and other female eusocial insects. This is the subject of much theorizing and investigation. The view that sex ratios are subject to Darwinian selection has received recent confirmation in a bird. In the Seychelles warbler (*Acrocephalus sechellensis*), female offspring can reduce their 'cost' to their parents by helping raise her later offspring, but only in favourable habitats, where food is abundant. Pairs in high quality territories produce 86% female offspring, those in low quality territories produce only 23% female offspring.

sex reversal, sex-reversal gene *SRY* in humans, *Sry* in mice. Dominant autosomally determined character of mice causing the somatic gonadal tissues of females to develop as testis. This then secretes testosterone and brings about male phenotype despite the presence of female genotype. See TESTIS, TESTIS-DETERMINING FACTOR.

sex-role reversal Occurrence, notably in some bird species (e.g. some Arctic wading birds), where females are more brightly coloured than males and display for mates, sometimes leaving them to incubate the offspring alone.

sexual dimorphism Occurrence of populations where individuals differ in respect of two distinct sets of phenotypic characters, sex itself being one of them. Some cases may be attributable to SEXUAL SELECTION.

sexual imprinting The acquisition of mate preferences by learning.

sexual isolation A form of prezygotic REPRODUCTIVE ISOLATION resulting when two populations undergoing SEXUAL SELECTION diverge in male traits and female preferences. Compare ECOLOGICAL ISOLATION.

sexual reproduction A misleading phrase, for the essential processes of sex (gene RECOMBINATION and/or transfer) and reproduction (production of new individuals) are quite distinct, despite common association in LIFE CYCLES. For example, during CONJUGATION in protozoa such as *Paramecium*, sex and reproduction are only tenuously linked and are completely uncoupled during AUTOMIXIS in the *Paramecium aurelia* species complex. Again, nuclear fusion producing the second endosperm nucleus in ANTHOPHYTA is a sexual process (automixis), yet that nucleus contributes nothing genetically to the future offspring (reproduction).

Gene transfers here include only those between existing cells, excluding cell division by mitosis, binary fission and budding – none of which is sexual; but it includes CONJUGATION, FERTILIZATION, PLASMID transfer and viral infection (and even some forms of GENE MANIPULATION). Excision of plasmids and their incorporation within the genome of an unrelated recipient species raises questions about the validity of SPECIES definitions relying on the concept of a shared GENE POOL. In prokaryotes, lacking MEIOSIS and FERTILIZATION, sexual processes commonly involve a specialized form of conjugation (e.g. see F FACTOR). In eukaryotes, the distinctive features of sexual LIFE CYCLES are meiosis and fertilization; but gametes are not invariably produced by meiosis, and fertilization is not always followed by mitosis. The genetic recombination normally associated with sexual processes tends to promote GENETIC VARIATION among offspring (but see INBREEDING). This is generally considered adaptive in unstable or patchy environments. However, meiosis has a potentially disruptive effect upon coadapted gene complexes. W.D. Hamilton and others proposed in 1990 that the main selective advantage enabling retention of sexual reproduction and outcrossing despite the COST OF MEIOSIS is that polymorphism and frequency-dependent selection at loci contributing to parasite recognition (e.g. the MHC system) restrict the loss of fitness due to disease, a model which is still debated (e.g. see IMMUNOCOMPETENCE HANDICAP). See APOMIXIS, SEX.

sexual selection Form of selection, generally contrasted with NATURAL SELECTION although also proposed by DARWIN. Results from the exercise of mating preferences (by either sex, but most commonly by females) in favour of individuals expressing certain genetically determined characters. As a result, genes for these characters tend to spread through the population, being expressed as SEX-LIMITED CHARACTERS. These could be defined as 'attractive' to the choosing sex.

This is a form of natural selection, a male's fitness being enhanced by his expression of 'attractive' characters, a female's fitness being improved by contributing genetically both to male offspring likely to have these characters and to female offspring likely to be genetic carriers of them. Sexual selection is often invoked to explain cases of extreme SEXUAL DIMORPHISM (e.g. where males are huge compared with females), and such dimorphisms do tend to evolve where males are polygynous. Sexual selection may enhance the directional component of selection, but its role in the spread of genes which have undramatic phenotypic effects is the subject of debate and experimentation. See MAJOR HISTOCOMPATIBILITY COMPLEX, SEX, SPECIATION.

SH2 and SH3 domains (*src* homology 2 and 3 domains) Modular binding DOMAINS located on several crucially important intracellular signalling proteins and enabling its possessor to bind a particular structural MOTIF on the interacting protein or lipid. The resulting interactions, which may involve several molecules linked together (see PROTEIN COMPLEXES, PROTEIN NETWORKS), are a fundamental feature of the signalling pathways involved in SIGNAL TRANSDUCTION. See RECEPTOR; Figs. 143a and 143b.

Shannon–Weiner diversity index See BIODIVERSITY.

shared homologue A character found in two or more taxa whose most recent common ancestor also had (or can be inferred to have had) it. See SYMPLESIO-MORPHY, SYNAPOMORPHY.

sharks See SELACHII.

shell No clear-cut definition; includes many hardened animal secretions having protective and/or skeletal roles, often attached only to part of body surface. Structures termed shells are found in many animal groups, e.g. the SARCODINA, MOLLUSCA, BRACHIOPODA, in CRUSTACEA (a *carapace*) and in echinoid echinoderms (a *corona*).

shiga toxin See TOXINS.

shivering See THERMOGENESIS.

short-day plants See PHOTOPERIODISM.

short-germ Mode of DEVELOPMENT in some lower insect groups (e.g. Orthoptera) in which only the most anterior head structures are defined before or during gastrulation. Compare LONG-GERM.

shotgun sequencing The commonest method of sequencing an informational polymer (e.g. protein or DNA molecule) currently employed, involving its initial fragmentation into large pieces, after which partial sequences are then determined and a computer used to piece together the whole sequence, identifying overlapping sequences and using them to guide the assembly. See CONTIG.

shoulder girdle See PECTORAL GIRDLE.

shuttle vector See VECTOR.

siblings (sibs) Brothers and/or sisters; offspring of the same parents.

sibling species Very closely related SPECIES differing only in minor respects, or appearing identical, but in fact reproductively isolated. Separation is often important where one of the species is vector to an economically important parasite (as in *Simulium* and *Anopheles* spp.), when DNA probes may be used to distinguish them. Mitochondrial DNA SEQUENCING from the two echolocating types of European pipistrelle bat (*Pipistrellus pipistrellus*) indicates genetic divergence warranting their reclassification as two sibling species. Individual sibling species can have very wide geographical distributions, and isolates from various parts of the world (e.g. in protozoa) have been shown to be capable of mating with each other. See ISO-ENZYMES, SUPERSPECIES.

sickle-cell anaemia A hereditary, genetically determined disorder affecting many newborn African and American negroes and others of tropical climates where malaria is, or has recently been, endemic. Caused by homozygosity for allele Hb^S, producing a single amino acid substitution in the β-chain of the normal haemoglobin molecule determined by allele Hb^A. Individuals heterozygous for the allele (Hb^{A}Hb^S, producing the sickle-cell trait) are resistant to the most serious form of MALARIA (subtertian malaria caused by *Plasmodium falciparum*; see HETEROZYGOUS ADVANTAGE), and allele Hb^S therefore attains a high frequency in malarial areas. However, the homozygote Hb^{S}Hb^S suffers from sickle-cell anaemia, so-called because red blood cells have a sickle shape even at high blood oxygen levels and block capillaries, becoming phagocytosed and causing anaemia and other symptoms. Sickle cells have much shorter lives than normal cells, contributing to anaemia. An example of NATURAL SELECTION in humans. See THALASSAEMIAS, GENETIC LOAD.

sieve area Area of sieve element wall containing clusters of pores through which protoplasts of adjacent sieve elements are connected. See PHLOEM, SIEVE PLATE.

sieve cell See PHLOEM.

sieve plate End-wall of SIEVE TUBE MEMBER, where pores are larger than in lateral sieve area; highly differentiated.

sieve tube Series of SIEVE TUBE MEMBERS, arranged end-to-end and interconnected by SIEVE PLATES. Functions in transport of food materials (e.g. sucrose, amino acids). Lacking nuclei at maturity. See TRANSLOCATION.

sieve tube member Component of a SIEVE TUBE. Elongated, unlignified tube-like cells found primarily in flowering plants, but also in some brown algae (Laminariales); typically associated with a COMPANION CELL. See PHLOEM.

sigma factor (σ factor) Protein component of prokaryotic RNA POLYMERASES binding loosely to the core enzyme and restricting mRNA transcription to one of the two DNA strands and appropriate promoter region (see JACOB–MONOD THEORY).

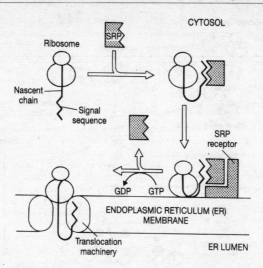

FIG. 150 *Scheme showing the dual roles of the eukaryotic* SIGNAL RECOGNITION PARTICLE *(SRP) in both chaperoning an unfolded nascent chain through the cytosol and targeting it to the SRP receptor in the ER membrane.*

Core RNA polymerases (lacking sigma factor) tend to start transcription randomly, on either DNA strand. The σ factor tends to dissociate itself from the RNA polymerase after a few RNA nucleotides have been incorporated, when its place may be taken by elongation factors involved in chain elongation and termination. For eukaryotic sigma factor equivalents, SEE TATA FACTOR, TRANSCRIPTION FACTORS. See also RHO FACTOR, PRIBNOW BOX.

signal recognition As a SIGNAL REGION of a PREPROTEIN comes away from the ribosome on which it was translated, it is bound by a specific elongated signal recognition particle (SRP) comprising a 7S RNA and six different polypeptides. Binding can slow down or stop the translation process. The SRP targets the nascent polypeptide–ribosome complex to the endoplasmic reticulum by its interaction with a receptor in the membrane. The SRP then leaves the polypeptide–ribosome complex as the signal regions contact the translocation mechanism in the ER membrane. See Fig. 150.

signal regions (signal sequences, leader sequences) See PROTEIN TARGETING, SIGNAL RECOGNITION, PLASMIDS.

signal transduction The processes by which a cell responds to an external signal (be that molecular or a form of energy), all of which involve the 'leading across' (transduction) of information from outside the cell to bring about changes in the signals within. This commonly involves a cell's binding a signal molecule (ligand) at one of its cell-surface receptor molecules (see RECEPTOR (2)), dimerization of bound receptors, and subsequent amplification of the signal at or close to the membrane, activating molecular pathways and culminating in changes in, e.g. GENE EXPRESSION, CELL CYCLE or CELL FATE. Figs. 143b and 151 indicate how the G PROTEIN Ras mediates between signal and enzyme activation. Methylation of membrane receptors alters the strength of the signal they pass to the G protein phosphorylation cycle. Receptors are said to transduce the actions of ligands. Alternatively, some signal molecules cross the cell membrane and bind intracellular NUCLEAR

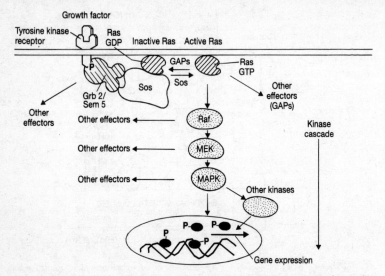

FIG. 151 *A* SIGNAL TRANSDUCTION *pathway involving a growth-factor-specific receptor tyrosine kinase whose autophosphorylation on binding its ligand enables it to recruit exchange factors, such as Sos, once an appropriate exchange control factor (here Grb2, also known as Sem5) has bound the tyrosine kinase. Such a complex can now activate Ras, turning it from inactive Ras-GDP to active Ras-GTP. Active Ras can then initiate a tyrosine kinase cascade, the ultimate signals entering the nucleus as phosphorylated transcription factors regulating gene expression. As indicated, several of the proteins involved (many of them oncoproteins) are pleiotropic in their effects. Compare Fig. 143b.*

RECEPTORS before activating gene expression. Signal molecules include traditional HORMONES, PHEROMONES and NEUROTRANSMITTERS, but have expanded widely to include GROWTH FACTORS, CYTOKINES, etc. Much research also involves bacterial (e.g. BACTERIORHODOPSIN), protistan and plant signal transduction mechanisms. The alga *Chlamydomonas* has a RHODOPSIN-regulated signal transduction chain involving its 'eyespot', flagellum base and Ca^{2+}, controlling its behaviour to light, so it is of interest that much plant-cell signal transduction seems to be calcium-linked (see CALMODULIN). Although much recent research highlights the molecular signals arriving at cell surfaces (see ACCESSORY MOLECULES, MAJOR HISTOCOMPATIBILITY COMPLEX, INCOMPATIBILITY, INTERFERONS), work on transduction mechanisms in photoreceptors and mechanoreceptors is also active (see HAIR CELL, RECEPTOR

(I), ROD CELL, SH2 AND SH3 DOMAINS, TOLL-LIKE RECEPTORS).

It was initially thought that a signal, such as a hormone binding to its receptor, initiated a response via a simple, linear chain of adaptor and effector proteins. But it is now apparent that these proteins do not work independently but rather as fleeting molecular machines (see PROTEIN COMPLEXES), the modular components of which are shared between pathways. The Alliance for Cellular Signalling (AfCS) has been founded to explore this territory (see www.signalling-gateway.org). See CANCER.

silicoflagellates Members of the CHRYSOPHYTA (DICTYCHALES) possessing cells with an external silicified skeleton. Originated in the Cretaceous. Fossil forms often found in calcareous chalks, with members of the

PRYMNESIOPHYTA. Today, these algae form an important part of the phytoplankton of colder seas.

silicula See SILIQUA.

siliqua (silique) Special type of capsule found in cabbages and related plants (Family Cruciferae). Dry, elongated fruit, formed from an ovary of two united carpels and divided by central false septum into two compartments (locules). Dehiscing by separation of carpel walls from below upwards, leaving the septum bounded by persistent placentas (replum), with seeds adhering to it. The *silicula* has a similar structure, but is short and broad; e.g. honesty, shepherd's purse.

silk Fibroin (a β-keratin protein) produced by modified salivary glands of silkworms (silkmoth larvae) and by spiders, among other arthropods. Has small amino acid R-groups, while anti-parallel polypeptide molecules hydrogen-bond to form very stable β-pleated sheets.

Silurian Period of the PALAEOZOIC era (*c.* 440–400 Myr BP). The period commences with a major EXTINCTION event. The first fossil plants occur, and the first jawed fish appear. Other animal fossils include scales of ostracoderm fishes, corals, crinids, trilobites, brachiopods and graptolites. The climate was generally mild, and existing continents generally flat. See GEOLOGICAL PERIODS.

simian Informal term; belonging to the primate suborder ANTHROPOIDEA.

Simpson's Index (of diversity) An index of species diversity (see SPECIES RICHNESS) which attempts to overcome the problem that easily overlooked species, and those represented by just one or a few individuals, may cause diversity to be underestimated. The formula used is:

$$D = \frac{N(N-1)}{\Sigma n(n-1)}$$

where D = Simpson's diversity index, N = total number of individuals in sample, n = number of individuals of each species in sample, Σ = sum of.

It is often difficult to be certain of the number of plant individuals present, since plants often grow vegetatively and it is often impracticable to count them. The number of 'hits', using point QUADRANT data, is often acceptable.

Sinanthropus (Peking man) See HOMO.

SINEs Short interspersed nuclear elements (e.g. see *Alu* entry; compare LINES). Short TRANSPOSONS (100–400 base pairs) harbouring a polymerase III promoter (usually derived from tRNA sequences), but no proteins. They probably use the LINE machinery for transposition and mostly share the 3' end with a resident LINE element (although *Alu* does not). The human genome contains three monophyletic SINE families: the active Alu, and the inactive MIR and Ther2/MIR3. Mice have both tRNA- and 7SL-derived SINEs. Many SINE species seem to be transcribed under conditions of stress, the resulting RNAs binding specific PROTEIN KINASES, so blocking their ability to inhibit protein translation and so promoting translation under stress.

single cell protein (SCP) Any microorganism (e.g. bacterium, yeast) which can be cultured on a commercial scale in a fermenter, dried and sold as a food source. The bacterium *Methylophilus methylotrophus* can utilize methanol, mineral nutrients and a mixture of air and gaseous ammonia to produce a cattle food cake ('Pruteen') rich in protein and free amino acids, vitamins and lipids. The blue-green alga *Spirulina*, eaten for centuries in the tropics, is now commercially harvested in the warm water of cooling plants, with a huge yield of dry weight and protein. Eukaryotic SCP has less nucleic acid content than does prokaryotic and is therefore more suitable as a human food source, since excess nucleic acid causes health problems. See BIOREACTOR.

single-stranded DNA (ssDNA) See DNA.

sinoatrial node See PACEMAKER.

sinus (Bot.) Space or recess between two lobes of leaf or other expanded organ. (Zool.) A *blood sinus* is an expanded vein, particularly of selachian fish; sometimes also used of the HAEMOCOEL. *Nasal sinuses* are air-filled spaces within some facial bones of mammals lined by mucous membrane and communicating with the nasal cavity.

sinusoid Type of CAPILLARY.

sinus venosus Chamber of vertebrate HEART, between veins and auricle(s). Thin-walled; absent from adult birds and mammals. See PACEMAKER.

siphonaceous (siphonous) Coenocytic; e.g. members of several orders of algae, which are filamentous, sac-like or tubular, without cross-walls.

Siphonaptera Order of endopterygote insects. Fleas. Secondarily wingless, with legless detritus-feeding larvae; exarate pupa enclosed within a cocoon. Mouthparts consist of long serrated mandibles, short palped maxillae and reduced palped labium. Laterally compressed, with legs adapted for running between body hair and for jumping. Include human flea *Pulex irritans* and rat flea (transmitter of bacterial plague) *Xenopsylla cheopis*.

siphonophora Order of CNIDARIA containing complex and polymorphic colonial animals. Colonies often form by budding from original medusoid individual.

siphonostele (solenostele) Type of STELE containing hollow cylinder of vascular tissue surrounding a pith. See ECTOPHLOIC and AMPHIPHLOIC.

Siphunculata (Anoplura) Sucking lice. Wingless exopterygotan insects, ectoparasitic on mammals. Eyes (and ocelli) reduced or absent; mouthparts modified for piercing and sucking. Thorax of fused segments. The human louse *Pediculus humanus* can carry *Rickettsia prowazeki*, distributed via the insect's faeces and causing epidemic typhoid fever.

Sirenia Order of placental mammals. Manatees, dugongs (seacows). Aquatic (coastal and river-dwelling) and herbivorous, with transversely expanded tail, front legs modified as flippers and vestigial hindlegs. One pair of mammary glands. Not closely related to cetaceans.

siRNAs Small interfering RNAs. See NONCODING RNAS.

sister group A species or higher monophyletic taxon believed to be the closest genealogical relative of a given taxon, excluding the species ancestral to both taxa. Two taxa forming sister groups are thus thought to share an ancestral species not shared by any other taxon.

site-directed mutagenesis (targeted mutagenesis) See GENE TARGETING, HOMOLOGOUS RECOMBINATION, MUTAGEN, POLYMERASE CHAIN REACTION.

site-specific recombination See RECOMBINATION.

Sivapithecus See PONGINAE.

skates See ELASMOBRANCHII.

skeletal muscle See STRIATED MUSCLE.

skeleton Any component of an animal's anatomy serving for muscle attachment and/or transmission of restoring forces extending a contracted muscle. See ENDOSKELETON, EXOSKELETON.

skin (cutis) Vertebrate organ covering most of body surface, comprising *epidermis* (ectoderm) above, produced by the MALPIGHIAN LAYER, with varying amounts of underlying connective tissue of *dermis* (mesoderm) and *subcutaneous fat* (adipose tissue). Often impervious to water, but permitting gaseous exchange in modern amphibia. Produces, variously, SCALY HAIR, FEATHERS, NAIL and sometimes bone (see DERMAL BONE). Glands located here include mucus glands (fish, amphibia), and sebaceous and sweat glands (mammals). May be variously pigmented (see CHROMATOPHORE) and contain several kinds of receptors in the dermis (e.g. pain, pressure, temperature). Attachment to underlying organs by loose connective tissue helps it move over them and return elastically. In amphibia the attachment is particularly loose. See CORNIFICATION, KERATIN.

skotophile Literally, dark-loving; dark-receptive phase of circadian rhythms, lasting about 12 hours. See PHOTOPHILE.

sliding filament hypothesis Hypothesis that the apparently diverse activities of MUSCLE CONTRACTION, CYCLOSIS, CYTOPLASMIC STREAMING, and various kinds of CELL LOCATION all depend fundamentally upon energy-dependent sliding of microfilaments of ACTIN over MYOSIN. Similar mechanism, involving MICROTUBULES of tubulin, explains the beating of eukaryotic flagella (see CILIUM), and organelle and chromosome movement.

slime fungi, slime moulds The term 'slime mould' has in the past been applied to two rather different groups of organism, often, though by no means always, regarded as a division of the kingdom Fungi: the acellular, or plasmodial, slime moulds (formerly Myxomycetes), and the cellular slime moulds, or dictyostelids. More recently, and somewhat controversially, three groups of slime moulds (dictyostelid, plasmodial and protostelid slime moulds) have been placed in the amoebozoan group of eukaryotes (MYCETOZOA), while the acrasid slime moulds are placed in the discicristates group (see EUKARYOTE). See MYXOMYCOTA for a more traditional account.

slow fibres See MUSCLE CONTRACTION.

small intestine See ILEUM, JEJUNUM.

small RNAs (sRNAs) See NON-CODING RNAS.

smooth muscle (involuntary muscle) Contractile tissue (generally called *visceral muscle* in vertebrates) comprising numbers of elongated spindle-shaped cells (sometimes syncytial) lacking transverse striations or other obvious ultrastructure. Position of the one nucleus per cell varies widely. Responsible among much else for peristaltic movement of food along gut, of blood along some contractile vessels (esp. invertebrates) and in regulation of vertebrate blood pressure (see VASOMOTOR CENTRE). Often arranged in *muscle coats* as bundles of cells averaging 100 µm in diameter, frequently within vertebrate connective tissue but occurring widely in both invertebrates and vertebrates. Among the latter principally in visceral, vascular and

other locations where hollow organs occur (e.g. uterus). Cell lengths vary from 30–450 µm; diameters from 2–6 µm. Cells in the muscle bundle are separated by a basement membrane of glycosaminoglycans, glycoproteins, collagen and elastic fibres about 60 nm wide; cell contacts include tight junctions and gap junctions. Proteins involved in its contraction are basically those of striated and cardiac muscle; but the myosin filaments are not as regularly arranged, while the actin filaments are more randomly distributed, lying along the longitudinal axis of the cell parallel to the myosin filaments; some of them terminate via accessory proteins at the cell membrane.

In vertebrates, almost always under control of AUTONOMIC NERVOUS SYSTEM, often with intrinsic PACEMAKERS. Hormones may alter the threshold for contraction. The sympathetic nervous system is generally excitatory on vascular smooth muscle but inhibitory on gut muscle; the parasympathetic system may be either excitatory or inhibitory, depending on the organ. See MUSCLE CONTRACTION.

smuts Members of the only order (Ustilaginales) within the fungal class Ustomycetes (Division BASIDIOMYCOTA). Important plant pathogens, often of their flowers, producing masses of black, sooty spores.

SNAREs Soluble *N*-ethyl-maleimide-sensitive factor-attachment protein receptors. A complex of proteins bridging two membranes at the point of their fusion, enabling the mixing of their lipids. They may be involved in defining the specificity of vesicle targeting (e.g. between the endoplasmic reticulum and Golgi apparatus). At least 35 SNARE genes are expressed in mammalian species. See CELL FUSION, INTERCELLULAR JUNCTIONS, PROTEIN TARGETING.

S-nitrosylation See NITRIC OXIDE.

snoRNAs (small nucleolar RNAs) A group of eukaryotic non-coding RNAs directing 2'-O-ribose methylation and pseudouridylation of rRNA, tRNA and ncRNAs, forming base pairs with sequences close to the site to be modified. See RIBOSOMES.

snRNAs (small nuclear RNAs) A group of eukaryotic non-coding RNAs involved in splicing pre-mRNAs during RNA PROCESSING.

SNPs Single nucleotide polymorphisms. See HUMAN GENOME PROJECT.

snRNP See RNPs.

social facilitation An increase in the rate of an aspect of an animal's activity with increasing presence of conspecifics, especially if they are performing that behaviour. Thus, consumption rate may rise with increasing consumer density, and can result from the increased time available for feeding as less time is required for vigilance against predators. Yawning in humans is socially facilitated.

society (Bot.) Minor climax community within a consociation, arising as a result of local variations in conditions and dominated by species other than the consociation dominant.

sociobiology The systematic study of the biological basis of all social behaviour. See, for instance, BATEMAN PRINCIPLE.

sodium pump ACTIVE TRANSPORT mechanism present in plasma membranes of most animal cells, consuming an estimated third of a cell's ATP production in pumping sodium ions (Na^+) out of the cell and potassium ions (K^+) into it in the ratio 3:2. It plays a minor role in establishment of a cell's RESTING POTENTIAL, but helps regulate cell volume by casting out Na^+ which would tend to enter along its electrochemical gradient, adding to the negative osmotic potential of the cell and drawing water in. The cell's internal electrical negativity prevents chloride ions (Cl^-) from entering and having the same effect. With the origin of the Na,K-ATPase, which pumps more sodium out than potassium in and so helps maintain the cell's osmotic balance, early eukaryotes could shed their cell walls with osmotic impunity. The pump is blocked by external *ouabain*, and animal cells may therefore swell or burst if this or other inhibitors of ATP synthesis or hydrolysis are added. It is indirectly responsible for glucose and amino acid uptake by cells since it creates a sodium gradient necessary for Na^+-based symports (see TRANSPORT PROTEINS).

soil profile Series of recognizably distinct layers or horizons visible in a vertical section through the soil down to the parent material. Study of soil profiles yields valuable information on the character of soils. See Fig. 152.

solar plexus See AUTONOMIC NERVOUS SYSTEM, NERVE PLEXUS.

solonostele See SIPHONOSTELE.

solute potential See WATER POTENTIAL.

soma See CELL BODY, SOMATIC CELL.

somatic cell Body cell; any cell of multicellular organism other than gametes. Mutations in somatic cells do not generally play significant role in evolution, being unlikely to be passed to further generations in gametes. However, asexual budding may produce new individuals with copies of somatic mutations, and some plants may generate polyploid microspores and megaspores from polyploid somatic precursor cells. See GERM PLASM.

somatic embryogenesis (Bot.) The production of an embryo from a single somatic (vegetative) plant cell. It represents a versatile tool for clonal propagation in such plants as corn, wheat and sorghum. See TOTIPOTENCY, TISSUE CULTURE.

somatic mutation A MUTATION in a somatic, as opposed to a GERM LINE, cell. If the mutation produces a recognizable phenotype, the patch of cells inheriting the mutation is called a *mutant sector*. They tend to have greater effects the earlier they occur in development. See AFFINITY MATURATION, HYPERMUTATION, ULTRAVIOLET IRRADIATION.

somatic nervous system That division on the peripheral nervous system which excludes the AUTONOMIC NERVOUS SYSTEM. Its sensory and motor connections with the central nervous system are principally routed through spinal nerves.

somatomedins See INSULIN-LIKE GROWTH FACTORS.

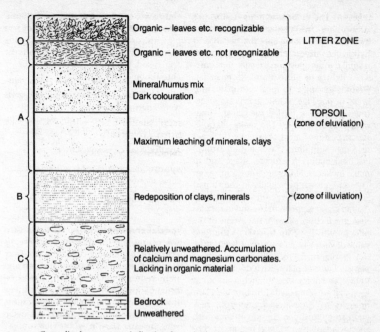

FIG. 152 *Generalized* SOIL PROFILE *indicating the major horizons. Not all soils have all these and their relative thicknesses vary greatly.*

somatostatin (growth hormone inhibiting hormone, GHIH) Hypothalamic *release-inhibiting factor* (RIF) preventing release of growth hormone releasing factor (GHRF) and hence of GROWTH HORMONE. Some pancreatic cells (delta cells) also release somatostatin, here inhibiting release of pancreatic hormones. The gene encoding this peptide has a CRE sequence in its regulatory region, and its transcription is brought about relatively slowly under the influence of raised CYCLIC AMP levels.

somatotrophin See GROWTH HORMONE.

somite See MESODERM.

***sonic hedgehog* gene** See *HEDGEHOG* GENE.

soredia Organs of vegetative reproduction in LICHENS. Minute clusters of algal cells surrounded by fungal hyphae, formed in large numbers over lichen surface and dispersed by various agencies (e.g. wind, water, insects).

sorghum This is a common catch-all name applied to all plants of the genus *Sorghum*, which includes numerous grain and foliage types. However, the plants do have markedly different characteristics and uses. Sorghum is believed to have originated in Africa. Grain sorghums are known by several names (e.g. durra, Egyptian corn, great millet, or Indian millet; in India sorghum is known as jowar, cholum or jonna; in the United States different types of grain sorghum are known as milo, kafir, hebari, feterita, shallu and kaoliang). In both Africa and Asia sorghum is one of the major CEREAL grains. Resistant to heat and drought, sorghums have effective and extensive root systems, adaptations to warm and semiarid regions.

Sorghums have been grouped into four

types: (i) The grain sorghums (Caffrorum Group) are the non-saccharine plants, including milo, kafir, feterita, hegari and hybrid derivatives, among others. The grain is similar in composition to MAIZE, but somewhat lighter in protein and lower in fat. Grain is ground into meal and made into bread or porridge. Whole grains are sometimes popped or puffed for cereal. Grain sorghums are also used to make dextrose, starch, paste and alcoholic drinks. (ii) Sweet or forage sorghums (Saccharatum Group) are used mainly for forage, silage and to make molasses. Stalks are juicy and sweet. These are grown principally in the United States and South Africa. (iii) Grass sorghum (Sudan grass) is grown in the Sudan for pasture, green chop, silage or hay. Some of the new hybrids have up to 90% of the food value of corn silage for dairy or beef cattle. (iv) Broom corn (Technicum Group) is a panicled woody plant, with dry pith, little foliage, and fibrous seed branches 30–90 cm long that are often made into brooms. Plant breeders have developed numerous new grain sorghum hybrids and cultivars with specific adaptation to local regions or conditions.

sorting signals See PROTEIN TARGETING.

Southern blotting Very sensitive method for detecting presence among restriction fragments of DNA sequences complementary to a radiolabelled DNA or RNA probe sequence. After initial separation by agarose gel electrophoresis, restriction fragments are denatured to form single-stranded chains and then trapped in a cellulose nitrate filter on to which the probe suspension is poured. Hybridized fragments are detected by autoradiography, after washing off excess probe. Used in DNA FINGERPRINTING. See NORTHERN BLOTTING, Fig. 125.

Sox **genes** A family of genes encoding transcription factors, one of which – SOX9 – is required for the conversion of mammalian undifferentiated gonads into testes (SEE TESTIS-DETERMINING FACTOR). They encode high-mobility group (HMG) domain DNA-BINDING PROTEINS, which act in a cell-specific manner and have roles in the regulation of embryonic development and the determination of cell fate. In certain cases they pair off with partner proteins prior to targeting their gene. The prototype of *Sox* genes is the *Sry* (sex-determining region of the Y chromosome) gene.

spacer DNA DNA separating one gene from another, often not transcribed itself. Common where there is GENE AMPLIFICATION (e.g. between ribosomal RNA genes).

spadix Kind of INFLORESCENCE.

spathe Bract enclosing inflorescence of some monocotyledons (e.g. *Arum*).

spatial summation See ALL-OR-NONE RESPONSE.

special creationism The view that each *kind* of organism (present and past) owes its existence to a unique creative act of God. Many influential scientists of the 17th and 18th centuries (Ray, Burnet, Whiston) believed science would vindicate a literal interpretation of the biblical creation story, as did many in the 19th (Buckland, Sedgwick, Agassiz). Darwin, whose views oppose special creationism, thought he had refuted this theory, holding that an intelligent Creator would have distributed organisms according to their physiological requirements rather than their taxonomy. Contrast EVOLUTION.

specialized Having special adaptations to a particular ecological niche which often result in wide divergence from the presumed ancestral form. Such *specializations* evolve and may result in niche restriction.

speciation The origin of SPECIES. If species are not real ('objective') entities, speciation cannot be regarded as a real process. If, however, species are lineages with temporal continuity, then speciation is whatever generates independence of lineages from one other. See REPRODUCTIVE ISOLATION.

The major models of speciation are the *allopatric* and *sympatric* models. In the former a parent species becomes physically separated into daughter populations by geography, restricting (or eliminating) gene flow between non-overlapping populations. In sympatric models, a parent species differentiates into lineages in the absence of any physical restriction on gene flow.

Some phenotypic characters bear directly on speciation. Shape of genital aperture, timing of breeding, degree of assortative mating, compatibility of pollen and stigma, degree of developmental homeostasis – all are aspects of phenotype subject to genetic variation. They are therefore responsive to disruptive or directional selection, and can alter cohesion of the population as an evolutionary unit or lineage. Debate surrounds whether speciation is adaptive, or merely a stochastic process which perhaps selection cannot prevent. Allopatric speciation, at least, seems to involve non-selective splitting of lineages. In time, the daughter populations may simply gain sufficient distinctive genetic and phenetic characteristics for taxonomists to recognize separate species. Sympatric speciation models more often invoke the adaptiveness of species formation. But some insist that in either model speciation is *de jure* incomplete until sufficient overlapping of daughter populations has occurred for the biological species concept to be applied. It is often said that after overlap, selection against hybrids either promotes lineage independence (known as 'reinforcement') or is too weak to prevent collapse of population identities; but much will hinge on the species concept, phenetic or biological, being employed; but REINFORCEMENT (see CHARACTER DISPLACEMENT) is undergoing a popularity renaissance, recent work indicating, e.g., that ASSORTATIVE MATING can best explain the increased distinctness of populations of the European flycatchers *Ficedula hypoleuca* and *F. albicollis* where their ranges overlap (see HYBRID).

Major sympatric models involve disruptive selection on a deme already polymorphic for an ecological requirement (e.g. food plant, oviposition site), and assortative mating in favour of individuals sharing that requirement. Its occurrence in the wild is slowly gaining acceptance.

Hybrid speciation, the formation of new species after hybridization between two previously existing species, is probably far commoner in plants than in animals, although few cases have been demonstrated. It is a form of sympatric speciation. Hybrids between distantly related plant taxa are far commoner than they are in animals; but in the case of the sunflower *Helianthus anomalus*, the parents are thought to have been the related *H. annuus* and *H. petiolaris*. At least three chromosome breaks, three fusions and one duplication seem to have been involved in the evolution of the *H. anomalus* karyotype. These rearrangements preclude successful introgression between the parent species through hybridization between either of them and *H. anomalus*.

Stasipatric speciation postulates that a widespread species may generate internal daughter species whose chromosomal rearrangements play a primary role in speciation (through reduced fecundity or viability of individuals heterozygous for the rearrangements). Daughter species might then extend their ranges at the expense of the parent species and might hybridize where ranges abut, although resulting offspring would be less fertile.

Polyploidy can be a form of 'instantaneous' speciation, since it may result in the complete reproductive isolation of an individual from other gene pools. Allopolyploids are more significant here than autopolyploids, permitting regular bivalent formation during meiosis as well as originating through combination of genomes from different species. They breed true for their hybrid character. Moreover, crosses of allotetraploids to their diploid progenitors give sterile triploids, preventing backcrossing and gene flow. They are widespread in plants, where vegetative propagation of sterile hybrids can enhance colonization prior to allopolyploidy, successful meiosis and improved fertility. See POLYPLOIDY.

Only recently, and then only rarely, has the genetics of speciation been dissected at the molecular level. A 'speciation gene' might be any type of gene, or even an extranuclear factor (e.g. the symbiont *WOLBACHIA*). Nor are such elements necessarily involved in reproductive isolation, since they may have accumulated subsequent to the speciation event itself. Sibling species in the process of attaining reproductive isolation are likely to be most informative and, at least in dioecious animals, the incompatibility usually follows HALDANE'S RULE in that sterile hybrids are usually males. In the sibling species *Drosophila simulans* and *D. mauritiana*, at least 120 widely scattered genes account for hybrid male sterility in the two species, some apparently through sperm protein divergence. One X-linked locus, *Odysseus*, accounts for 40% loss of male fertility when introgressed into the wrong genetic environment. Expressed in the reproductive tissues, it is a homeogene; but its homeobox shows remarkable variation in its encoded amino acid sequences between four species in the *D. melanogaster* clade, apparently driven by directional (in this case, sexual) selection. What has been dubbed '*sperm competition*' between males of *D. melanogaster* includes the evolution of male peptides in the ejaculate which increase a female's egg-laying rate but reduce her sexual appetite. Another male protein kills off competitor sperm, and in doing so poisons females. Females may in turn secrete proteins that inactivate the male proteins, or reduce sperm numbers, or counter the loss of sexual appetite. It is suggested that the speed with which such 'tricks' can evolve in the laboratory supports the view that 'barriers to fertilization' between incipient species can also evolve quickly. Intersexual competition may be a major engine of speciation. So, genetic differentiation between two populations may not be sufficient for speciation. Other processes, such as adaptation and SEXUAL SELECTION, may be crucial in the establishment of initial barriers to gene flow. It may be just a single trait which breaks down gene flow between incipient species, an occurrence uncorrelated with genetic divergence within the species. In studies (albeit from a narrow range of species) where genes have been involved in isolation, the majority have involved postzygotic isolating factors. The recent explosion of new species of cichlid fishes in the African great lakes was until recently explained by fluctuations in water

levels isolating populations in satellite lakes around Lake Victoria (thought to have been dry 12,400 yrs ago). But current work indicates that sexual selection by females for male colour, allopatric and sympatric speciation, have all been involved in the formation of the 500 cichlid species. Turbidity of Lake Victoria's waters by human activity, leading to misidentification of males by females, appears to be breaking down the species barriers.

species Term used both of a formal taxonomic category ('the species') and of taxa exemplifying it (particular species). In the system of BINOMIAL NOMENCLATURE, taxa with species status are denoted by Latin binomials, each species being a member of a genus. The naturalist John Ray, writing of plants in 1672, did not entirely avoid circularity in stating that the true criterion of species as taxa is that they are 'never born from the seed of another species and reciprocally': the cross-sterility criterion. To a special creationist like Ray such cross-sterility was to be expected (compare the nominalism of BUFFON). For Darwin, all degrees of sterility should be discoverable if sterility *barriers* (the quite different, evolutionary, concept of ISOLATING MECHANISMS) take time to evolve. See SPECIATION.

Darwin's writings sometimes reveal nominalism on the species question, to be expected on the view that species evolve gradually. At other times he drew a clear distinction between *taxa* and *categories*, and although doubtful about the possibility of defining the category 'species' (i.e. the taxonomic unit) he was in no doubt that taxa with specific status actually existed. He did not espouse ESSENTIALISM with regard to species.

Zoologists find the criterion of reproductive isolation especially valuable in demarcating species in the wild. As such, the *biological species concept* includes as species groups of populations which are phenotypically similar and reproductively isolated from other such groups, but which are actually or potentially capable of interbreeding among themselves. Problems in its application arise (a) when reproductive isolation from other populations admits of degrees, being incomplete over part of a species range, (b) with RING SPECIES, (c) with obligately asexual species (*agamospecies*), (d) with animals which, although sexual, lack males (obligate THELYTOKY), and (e) with ANAGENESIS. The biological species concept also fails to incorporate a historical dynamic into its account, thereby ignoring that species are genealogically unique. The majority of fungi, plants and marine invertebrates broadcast their gametes widely, making it impossible to establish reproductive isolation in the wild. So complexes of phenetic characters (the stock-in-trade of museum taxonomists) usually serve to identify such species. But SIBLING SPECIES pose problems, and classification rests here upon cytological techniques (e.g. use of DNA PROBES) rather than external morphology, although reproductive isolation is often the original clue to their existence. Some believe it may eventually be possible to identify an individual animal as a member of its species by reading the DNA sequence of just one of its genes (that for mitochondrial cytochrome oxidase I) from a tissue sample, using a hand-held device in the manner of a 'bar code'.

Recently favour has developed for ecological and evolutionary species concepts. The former fix a classification to independently existing environmental states or niches (difficult to isolate independently of the organisms which occupy them), equating speciation with niche change but leaving open the extent of change required; the latter stress the genealogical uniqueness of species (useful in asexual and thelytokous forms), but emphasize the cladistic (branching) nature of speciation at the expense of gradual anagenesis. Most species probably comprise two or more *subspecies* or *races* (see INFRASPECIFIC VARIATION, SEMISPECIES) and are said to be *polytypic*.

Inability to find a unified species concept is no disgrace, reflecting the variety of reproductive systems and the dynamic state of biological material. It may be argued that intraspecific sterility barriers caused in insects by the endosymbiont *WOLBACHIA* have little to do with speciation; and

Darwin's finches hybridize regularly without threat to the number of species. It is probably no accident that the species concept does not figure prominently in biological theory: species may be best regarded not as NATURAL KINDS (such as elements in chemistry) but as individuals, each historically unique and irreplaceable once extinct. If species are individuals then species names are proper names, so that properties of species would describe but not define them. Definitions of taxa would then be *necessary* in philosophical terms, even though there is no list of necessary properties for any biological taxon. See LINEAGE, PARAPATRY, TRANSITIVITY.

species diversity index See BIODIVERSITY.

species flock A species-rich, narrowly endemic and ecologically diverse group of organisms, usually considered to have radiated rapidly in geological terms from one, or more than one, very closely related stem species. They are sometimes quoted as *prima facie* candidates for sympatric speciation, but insufficient evidence is usually available to ascertain the mode of speciation which produced them. Examples include cichlid fish in African lakes, Galapagos finches, honeycreepers and fruitflies on Hawaii. Mitochondrial DNA (mtDNA) analysis on the fishes of Lakes Malawi and Victoria support monophyletic origins for these flocks against the alternative theory that they represent polyphyletic products of numerous invasions by already-differentiated taxa. Debate surrounds the involvement of SEXUAL SELECTION in speciation within such flocks. Sadly, introduction of the nile perch (*Lates niloticus*) into Lake Victoria has already led to the disappearance of about 200 endemic species.

species group Informal taxonomic category used in preference to such formal categories as subgenus or infragenus.

species richness The number of species within a habitat, or in a community of organisms. Indices of species richness often include relative abundances of species. Botanical sampling methods may involve QUADRATS. There are consistent trends in species richness across many habitats within latitudinal belts and in relation to altitude, topographic relief, island size and location, etc. See BIODIVERSITY, SIMPSON'S INDEX.

specific dynamic effect (SDE, post-prandial thermogenesis) The increase above basal metabolic rate (by up to 10–20%) brought on by the consequence of food ingestion. Heat is released by smooth muscle contractions of the gut wall, from hydrolysis of glycosidic and peptide bonds, and from active transport of digestion products. Deamination and urea production also contribute. SDE amounts to 31% of BMR for diets of meat, 13% for lipids and 6% for carbohydrate.

spectrins A family of cytoskeletal proteins (including spectrin, dystrophin and α-actinin) all of which seem capable of forming antiparallel homodimers or heterodimers. Several are involved in linking ACTIN filaments together, and appear to be linked to one another by hinges enabling extension and retraction.

sperm See SPERMATOZOID, SPERMATOZOON.

spermateliosis See MATURATION OF GERM CELLS.

spermatheca (seminal receptacle) Organ in some female or hermaphrodite animals which receives and stores sperm from the other mating individual.

spermatid Haploid animal cell resulting from the meiotic division of a secondary spermatocyte. Undergoes extensive cytoplasm loss and condensation of its nucleus during spermiogenesis (SEE MATURATION OF GERM CELLS).

spermatium Non-motile male sex cell in the RHODOPHYTA (red algae), and some fungi belonging to the ASCOMYCOTA and BASIDIOMYCOTA.

spermatocyte (Bot.) Cell which becomes converted into a spermatozoid (without intervention of cell division). See SPERMATID. (Zool.) Cell undergoing meiosis during sperm formation. *Primary spermatocytes* undergo first meiotic division; *secondary*

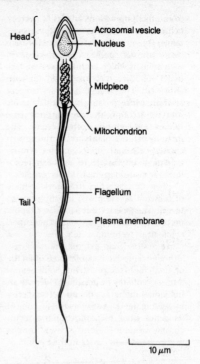

FIG. 153 *Diagrammatic illustration of a* SPERMATOZOON.

spermatocytes undergo second meiotic division. See MATURATION OF GERM CELLS.

spermatogenesis Formation of spermatozoa. See MATURATION OF GERM CELLS.

spermatogonium Cell within testis, commonly lining seminiferous tubules, which either divides mitotically to produce further spermatogonia or else gives rise to primary spermatocytes. See TESTIS, MATURATION OF GERM CELLS.

spermatophore Small packet of sperm produced by some species of animals with internal fertilization, e.g. many crustaceans, snails, mites, scorpions, *Peripatus*, newts, etc. Those of cephalopod molluscs are very complex and pump seminal fluid into the female by syringe-like structures.

Spermatophyta In some classifications, the Division containing all seed-bearing vascular plants. See ANTHOPHYTA, GNETOPHYTA, GINKGOPHYTA, CYCADOPHYTA, CONIFEROPHYTA.

spermatozoid (antherozoid) (Bot.) Small, motile, flagellated microgamete.

spermatozoon (Zool.) Microgamete (see Fig. 153), usually motile, produced by testes; commonly flagellated but amoeboid in nematodes and aschelminthes. Sperm of *Drosophila* are about 2 mm in length. Most use lipids as fuel; complex sperm store glycogen. See ACROSOME, MATURATION OF GERM CELLS.

sperm competition See SPECIATION.

spermiogenesis See MATURATION OF GERM CELLS.

spermogonium (spermagonium) Flask-shaped or flattened, hollow structure in which spermatia are formed.

S-phase The phase in the eukaryotic CELL CYCLE during which DNA is replicated. Several DNA tumour virus oncoproteins can induce S-phase entry in quiescent cells (see ONCOGENE).

Sphenodon See RHYNCHOCEPHALIA.

sphenoid bone A bone lying in the middle of the base of the skull, articulating with all the other skull bones and holding them together. See PTERYGOID PROCESSES.

Sphenophyta See EQUISETOPHYTA, INTRODUCTION.

spheroplast Cell whose wall material has been largely but not entirely removed (protoplasts have theirs entirely removed). Often employed in DNA CLONING (e.g. bacterium, yeast).

spherosome Single, membrane-bound, spherical structure in plant cell cytoplasms. Many contain lipids and are apparently centres of lipid synthesis and accumulation.

sphincter Ring of smooth muscle in wall of tubular organ, opening of hollow organ, etc., whose contractions and relaxations close and open the tube or aperture; e.g. pyloric and cardiac sphincters of stomach, anal sphincter.

sphingolipids See PHOSPHOLIPIDS.

sphingomyelin See PHOSPHOLIPIDS.

sphingosine See PHOSPHOLIPIDS.

spiders See ARANEAE.

spike Indeterminate inflorescence in which main axis is elongated and flowers are sessile.

spikelet Basic unit of grass inflorescences; small group of grass flowers.

spinal column See VERTEBRAL COLUMN.

spinal cord The part of the vertebrate CENTRAL NERVOUS SYSTEM lying within the vertebral canal, protected by the vertebral column, and consisting of a hollow cylinder of mixed nervous tissue (derived from the NEURAL TUBE) with walls of relatively uneven thickness. Paired and segmental spinal nerves leave it on each side between vertebrae. Contains both grey and white matter, former usually H-shaped in cross-section with the *cerebrospinal canal* running through the 'cross-bar', the latter surrounding the grey matter, the whole covered by the MENINGES. Carries sensory and motor information via ascending and descending tracts of white fibres; also provides for reflex arcs (intra- and intersegmental, ipsi- and contralateral). Continuous with the medulla oblongata of BRAINSTEM. See Fig. 145.

spinal nerves Peripheral nerves arising from the vertebrate spinal cord between vertebrae; typically one on each side per segment. Each has a dorsal (afferent) root and a ventral (efferent) root, which typically fuse on emergence from the vertebral column. Compare CRANIAL NERVES.

spindle (1) MICROTUBULE complex appearing during mitoses and meioses. Has two sources: regions around the CENTRIOLES, and the KINETOCHORES. Microtubules from the former extend to the equator of cell where they apparently overlap and generate sliding forces which often push the poles of the cell apart; kinetochores somehow pull the centromeres of sister chromatids towards opposite poles. Microtubules (*polar*) developed from the centriolar region have their fast-growing (+) ends away from the pole; those from the kinetochore have their (+) ends attached to the kinetochore. While a cilium-like DYNEIN has been implicated in the sliding of polar microtubules, the origin of the force pulling chromatids to the poles is not clear. For spindle positioning, see CELL POLARITY. (2) See MUSCLE SPINDLE.

spindle attachment See CENTROMERE.

spines (Bot.) Modified leaves or leaf parts, each forming a pointed structure; common in XEROPHYTES, e.g. cacti. Compare PRICKLES. (Zool.) Alternative for VERTEBRAL COLUMN. See also NEURONAL SPINES.

spinnerets Maximum of three pairs of silk-spinning organs on the posterior opisthosomas of spiders (ARANEAE); most probably modified legs. The silk, which hardens on emergence, may be used in construction of egg cocoon, feeding web, and cords wrapped around prey trapped in the web. When released as a long line it may also provide sufficient wind resistance to lift the spider into the air for its dispersal.

spiracle (1) Reduced first gill slit of many fish. Dorsally situated, its small size results from the connection formed between mandibular and hyoid arches for firm attachment of jaws, the spiracle lying between these arches. In most living bony fish the spiracle is closed up; the gill pouch of embryo tetrapods representing the spiracle develops into the cavity of the middle ear and Eustachian tube. See VISCERAL ARCHES. (2) External opening (stigma) of insect and other TRACHEAE. Often contains valves to regulate water loss and gaseous exchange.

spiral cleavage See CLEAVAGE.

spiral thickening Internal thickening of wall of xylem vessel or tracheid, in form of a spiral. Occurs in cells of protoxylem and, while providing mechanical support, permits longitudinal stretching as neighbouring cells grow.

spiral valve Spiral fold of mucous membrane projecting into the intestine of some fish, notably elasmobranchs, ganoids and Dipnoi. Probably serves to increase surface area for absorption.

spirillum Any long, coiled or spiral bacterium; e.g. a SPIROCHAETE. Also the name of a bacterial genus (*Spirillum*). See Fig. 10.

spirochaetes Elongated, spirally twisted, unicellular bacteria with thin, delicate walls; up to 500 μm long (large for a bacterium); motility by a helical wave along the cell. Some are free-living, some parasitic and pathogenic (e.g. *Treponema pallidum*, causing syphilis). Some are nitrogen-fixers (see NITROGEN FIXATION).

spleen Largest mass of LYMPHOID TISSUE, lying in mesentery of stomach or intestine of jawed vertebrates. Unlike LYMPH NODES, perfused by blood rather than lymph. Important lymphocyte and PLASMA CELL reservoir, and component of the RETICULO-ENDOTHELIAL SYSTEM, its cells phagocytosing worn red blood cells and platelets. A store of red blood cells, sympathetic contraction of smooth muscle squeezing them into the circulation (often giving a 'stitch') in emergency. See HAEMOPOIESIS.

spliceosomes Nuclear particles formed from a complex of five small nuclear RNAs (the snRNAs U1, U2, U4, U5 and U6) and many (50–100?) proteins (snRNPs); on a par with RIBOSOMES in structural complexity. The snRNAs bind in sequence along with snRNPs at an intron on the pre-mRNA (of a Class II GENE), fold the pre-mRNA and catalyse two transesterification reactions to yield mRNA and intron products. This is ATP-dependent – not on account of the splicing chemistry itself but because the assembly and rearrangements of the spliceosome require the breakage of RNA–RNA interactions and the formation of new ones, both between pre-mRNA and snRNPs and between the snRNPs themselves. Built into this process is very high quality control of events, not least through the formation of an active catalytic site only after the assembly and rearrangement of the splicing components on the pre-mRNA has occurred. This catalytic site is composed entirely of RNA molecules, and not of proteins (see RNA WORLD). There are similarities between spliceosomes and certain self-splicing introns in organelles of plants and fungi. Different splicing machines are involved in the removal of different types of intron (e.g. AT–AC introns and GU–AG introns). See INTRONS (for possible origin of snRNAs), RNPs.

splicing See INTRON, RNA PROCESSING.

split gene Gene with at least one INTRON sequence embedded within it.

sponge See PORIFERA.

spongy mesophyll See MESOPHYLL, LEAF.

spontaneous generation The view that life can arise from non-life, independently of any parent. In the sense of the ORIGIN OF LIFE, this may be regarded as a scientifically respectable view but it has been held in the past that many individual organisms arise abiogenically from e.g. fermenting broth, rotting meat, etc. (see BIOGENESIS). In the late 17th century, Francesco Redi's controlled experiments demonstrated that maggots do not appear spontaneously from rotting meat but instead grow from eggs laid in the flesh by flies (and see PASTEUR). In the 19th century, endorsement of spontaneous generation by Lamarck and Geoffroy St Hilaire resulted in its commonly being associated in France with any evolutionary theory. See ABIOGENESIS, LAMARCK, SPECIAL CREATIONISM.

sporangiole (Of fungi) small sporangium containing only one or a few spores (sporangiospores), and lacking a columella.

sporangiophore (Of fungi) hypha bearing one or more sporangia. sometimes morphologically distinct from vegetative hyphae.

sporangium Organ within which are produced asexual spores; typically in fungi and plants.

spore A single or several-celled reproductive structure (propagule) that detaches from the parent and gives rise, directly or indirectly, to a new individual; a general term. Spores are usually microscopic, of many different types, produced in various ways. Spores can be formed asexually or

sexually. They may be motile (zoospore) or non-motile (aplanospore; conidium); thin- or thick-walled, they often serve for very rapid increases in the population, as when produced in enormous numbers and distributed far and wide by wind, water, animals, etc. Others are resting spores, enabling survival through unfavourable periods. Spores occur in all plant groups, algae, fungi, bacteria, Cyanobacteria, protozoans and other protists. The term sexual spore usually indicates a spore that can engage directly in fertilization; less commonly it indicates a spore produced by meiosis or fertilization. Asexual spores, therefore, either do not engage in fertilization, or are not produced by meiosis. See MICROSPORE, MEGASPORE.

spore mother cell (Bot.) Diploid cell giving rise by meiosis to four haploid spores or nuclei.

sporic meiosis Pattern of ALTERNATION OF GENERATIONS where the spores are produced by meiosis and develop into multicellular gametophytes before the gametes are produced; occurs in many algae, all bryophytes and vascular plants.

sporocyst (1) Cyst of some SPOROZOA. (2) Stage in life cycles of many flukes (see TREMATODA); lacks mouth and gut; can produce daughter sporocysts or rediae (see POLYEMBRYONY).

sporodochium (Of fungi) a cluster of conidiophores arising from a stroma or mass of hyphae.

sporogonium Spore-producing structure of liverworts and mosses that develops after fertilization; the sporophyte generation of these plants.

sporophore (Of fungi) general term for a structure producing and bearing spores; e.g. a sporangiophore, conidiophore (simple sporophore), or mushroom (complex sporophore).

sporophyll Leaf, bearing sporangia. In some plants, indistinguishable from ordinary leaves except by presence of sporangia, e.g. in bracken fern; in others, much modified and superficially quite unlike ordinary leaves, e.g. stamens and carpels of flowering plants.

sporophyte Spore-producing diploid phase in LIFE CYCLE of a plant (see ALTERNATION OF GENERATIONS). Arises by union of sex-cells produced by haploid gametophytes.

sporopollenin Tough substance of which the exine (outer wall) of spores and pollen grains is composed. One of the most resistant organic substances known, not affected by hot hydrofluoric acid or concentrated alkali. It consists of complex polymers with an empirical formula $[C_{90}H_{142}O_{36}]$. Formed by oxidative polymerization of carotenoids and their esters.

Sporozoa Protist taxon, renamed API-COMPLEXA.

sport (rogue) Individual exhibiting the effect of a mutation, often an unusual or rare one.

spot desmosome See DESMOSOMES.

springtail See COLLEMBOLA.

Squamata Order of the REPTILIA (Subclass Lepidosauria) containing lizards (Lacertilia), snakes (Ophidia), amphisbaenids (Amphisbaenia) and the tuatara (RHYNCHO-CEPHALIA). Males generally have unique paired copulatory organs. In *lizards*, mandibles are joined at a symphysis but, as in snakes, there is a freely movable quadrate bone to which lower jaw is hinged. *Snakes* lack eardrums and movable eyelids and their eyes are covered by transparent eyelids. The jaws are exceptionally mobile and only loosely attached to the skull; mandibles are only loosely attached anteriorly by elastic ligaments and laterally by skin and muscles, with great relative mobility. Snakes feeding on prey large enough to struggle are usually constrictors (e.g. boas, pythons, anacondas) or venomous (e.g. cobras, mambas, coral and sea snakes, pit vipers and true vipers). The group is apparently undergoing rapid speciation. *Amphisbaenids* are legless burrowing forms, mostly under 1.5 m in length, with a single tooth in midline of upper jaw fitting into a space in lower jaw (effective nippers). Head shape blunt or flattened. Eardrums absent and eyes rudimentary. The Squamata have bodies covered in horny epidermal scales.

squamation See SCALE (Zool.).

squamosal A MEMBRANE BONE of skull, in mammals taking over from the quadrate the articulation of lower jaw (dentary). See EAR OSSICLES.

squamous cells Animal cells, commonly forming an epithelium, which are flattened and resemble paving stones in shape. Cells of the inside of the cheek form a squamous epithelium; those lining blood vessels are squamous endothelium. See ENDOTHELIUM, EPITHELIUM.

squamulose Lichen growth form which is similar to the FOLIOSE form, but with numerous small, loosely-attached thallus lobes or *squamules*.

Sry genes See TESTIS.

ssDNA Single-stranded DNA.

ssRNA Single-stranded RNA.

stable equilibrium In an ecological context, a level of a population or populations or of resources, which is returned to after slight displacements from that level.

stable limit cycles A regular fluctuation in abundance, the path of which is returned to after slight displacements from the path.

staining Treatment of biological material with chemicals (dyes, stains) that only colour specific organelles or parts of a structure, thus providing contrast, as between nucleus and cytoplasm, mitochondria and other organelles, or between cell wall and cytoplasm. For use in light MICROSCOPY, most stains are organic compounds (dyes) comprising a negative and positive ion. In acid stains, the colour arises from the organic anion; in basic stains it arises from the organic cation, while neutral stains are mixtures of both acid and basic stains. Stains are applied to a biological material by its immersion in a stain solution. *Vital staining* refers to staining of living tissue when no damage to the tissue occurs (e.g. neutral red stains vacuoles and granules in many cells;

Janus green stains mitochondria specifically). *Non-vital staining* refers to the staining of dead tissue. Various modified techniques are required for staining different types of tissues for light microscopy; e.g. *counterstaining* (double staining), where two stains are used in sequence so as to stain different parts of the specimen. See DEHYDRATION, FIXATION.

Staining techniques may give quantitative data, as when the amount of stain taken up is proportional to the amount of stained component, and methods exist (microspectrophotometry) to measure through the microscope the amount of stain present at a given site within a cell. When the structure stained takes on a colour different from that usually produced by the stain (or dye), the staining is said to be metachromatic.

Stains in electron microscopy are not the coloured stains of light microscopy; rather they contain heavy metal atoms (e.g. lead, uranium) in forms that combine with chemical groups characteristic of specific structures in the cell. Presence of such atoms permits fewer electrons to pass through the specimen, so that an image is produced. 'Stained' (electron-dense) structures appear darker than their surroundings. *Negative staining* for electron microscopy is used to examine three-dimensional and surface aspects of cell structure. Specimens are not sectioned; rather they are placed directly on a thin plastic film, covered with a drop of solution containing heavy metal atoms and allowed to dry, leaving the specimen with a thin layer of electron-dense material.

stamen Organ of FLOWER which forms microspores (shed after development as pollen grains); a microsporophyll; comprises stalk or filament bearing anther at apex. Anther comprises two lobes united by a prolongation of the filament connective, and in each lobe there are two pollen sacs (microsporangia) producing pollen.

staminate FLOWER which has stamens but no functional carpels, i.e. is male. See PISTILLATE.

staminode Sterile stamen; one that does not produce pollen.

standard free energy See THERMODYNAMICS.

standing crop Biomass per unit area (or per unit volume) at any one time. Not equivalent to biomass productivity, which is a *rate* measure. Standing crop values are often given in terms of energy content and commonly relate to populations or trophic levels.

stapes Stirrup-shaped mammalian EAR OSSICLE, representing columella auris of other tetrapods and the hyomandibular of fish. See COLUMELLA.

staple Any plant food, plant or animal, that is more than 10% protein by weight can be classed as a *cultural staple*. Apart from meat, they include wheat, soy beans, many nuts, legumes, some strains of MAIZE and RICE, SORGHUM, and certain others. See CEREAL, COMPLEMENTARY RESOURCES, CROP GENOME PROJECT.

starch A complex insoluble polysaccharide carbohydrate of green plants, one of their principal energy ('food') reserve materials. Formed by polymerization (condensation) of several hundred glucose subunits ($C_6H_{10}O_5$) and easily broken down enzymatically into glucose monomers (see AMYLASE). Comprises two main components: *amylose* and *amylopectin*. The former consists of straight chains of $\alpha[1,4]$-linked glycosyl residues and stains blue-black with iodine/KI solution; the latter contains in addition some $\alpha[1,6]$ branches in its molecules and stains red with iodine/KI solution. Found in colourless plastids (LEUCOPLASTS) in storage tissue and in the stroma of CHLOROPLASTS in many plants. Formed into grains, laid down in a series of concentric layers. See DEXTRIN, STATOLITH.

starch sheath Innermost layer of cells of cortex of young flowering plant stems, containing abundant and large starch grains; considered homologous with ENDODERMIS, it may sometimes lose starch and become thickened as an endodermis at later stage.

starfish See ASTEROIDEA.

start codon See PROTEIN SYNTHESIS, GENETIC CODE.

statismospore In fungi, a spore that is not forcibly discharged. See BALLISTOSPORE.

statoblasts Resistant internal buds with chitinous shells produced by some ECTO-PROCTA asexually and capable of withstanding unfavourable conditions, as during winter. They break open in spring to produce new colonies.

statocyst (otocyst) Mechanoreceptor and/or position receptor evolved independently by several invertebrate groups (Cnidaria, Platyhelminthes, Crustacea) and vertebrates (see MACULA). Typically a fluid-filled vesicle containing granules of lime, sand, etc. (*statoliths*), which impinge upon specialized setae (*statolith hairs*) and stimulate sensory cells as the animal moves. Resulting nerve impulses initiate reflexes which often serve to right the animal after it has been turned upside down. See STATOLITH.

statocyte Plant cell containing one or more STATOLITHS.

statolith Gravity sensor in plant cells. Perception of gravity probably involves the sedimentation of amyloplasts (starch-containing plastids) within specialized cells of the shoot and root. Such cells are to be found in cells contiguous to or surrounding the vascular bundles all along the shoot. In roots, in contrast, they are localized in the root cap. How the gravity sensors, and their movement, is translated into hormonal gradients is not fully understood, although calcium may play a key role in roots through its control of GROWTH SUBSTANCE transport.

statospore (stomatocyst) Resting spore characteristic of the CHRYSOPHYTA, produced asexually or sexually, and characterized by its siliceous wall, which may or may not be ornamented. Similar spores are produced by some members of the XANTHOPHYTA.

STATs (signal transducers and activators of transcription) Latent cytosolic transcription factors activated by certain (Janus family) tyrosine kinases after cellular stimulation by interferons and other CYTO-KINES (see Fig. 40 for mode of action). More than 30 cytokines and hormones activate the Jak-STAT pathway by binding to cytokine receptors, and all STATs have a SH2 DOMAIN enabling them to dock onto specific phosphotyrosines on certain activated RTKS.

stele (vascular cylinder) (Bot.) Cylinder or core of vascular tissue in centre of roots and stems, comprising xylem, phloem, pericycle, and in some steles, pith and medullary rays; surrounded by endodermis. Structure of the stele differs in different groups of plants. See DICTYOSTELE, PROTOSTELE, SIPHONOSTELE, VASCULAR BUNDLE.

stem Normally aerial part of axis of vascular plants, bearing leaves and buds at definite positions (nodes), and reproductive structures, e.g. flowers. Some are subterranean (e.g. rhizomes) but these, like all stems, are distinguished externally from roots by the occurrence of leaves (scale leaves on rhizomes) with buds in their axils, and internally by having vascular bundles arranged in a ring forming a hollow cylinder, or scattered throughout tissue of the stem, with the protoxylem most commonly endarch.

stem cells Undifferentiated cells, either embryonic (see EMBRYONIC STEM CELLS) or adult (e.g. ependyma cells lining the ventricles of the BRAIN, and haemopoietic cells of bone marrow), which divide to produce one stem cell, and another which can pass along a specific differentiation pathway in response to local signals (e.g. GROWTH FACTORS). One possibility that it is crucial to exclude before one concludes that any cell is capable of differentiating into a wide variety of cell types (i.e., is a multipotent stem cell) is that it has not fused with any other cell during its history and so acquired additional molecules and properties. It is now known that ependymal (neural stem) cells of the central nervous system are capable of giving rise to both neurons and glial cells and do persist in the adult mammal brain. Use of adult stem cells in tissue and organ remodelling would certainly have fewer ethical overtones than would use of embryonic stem cells. Interestingly, homologous genes are known to regulate stem cell maintenance in *DROSOPHILA* and *ARABIDOPSIS*.

stem group Organisms that preceded, coexisted with and may have post-dated the common ancestor of the crown group of life and with it formed the total group which began with the origin of life.

stenohaline Unable to tolerate wide variations in environmental salinity.

stenopodium See BIRAMOUS APPENDAGE.

stenothermous (-thermic) Unable to tolerate wide variations in environmental temperature. Compare EURYTHERMOUS.

stereocilium Specialized microvillus. See HAIR CELL.

sterigma (Of fungi) a minute stalk bearing a spore or chain of spores.

sterile (1) Unable to produce viable gametes and/or sexual offspring, unlike normal individuals. (2) Free from microorganisms. See ANTISEPTIC, AUTOCLAVE, DISINFECTANT.

sternum (1) Breast bone. Tetrapod bone lying ventrally in mid-chest to which ventral ends of most ribs are attached. Attached anteriorly to PECTORAL GIRDLE. (2) Cuticle on ventral side of each segment on an arthropod, often forming a thickened plate. See TERGUM.

steroids Chemically similar but biologically diverse groups of LIPIDS originating from *squalene*. Include bile acids, vitamin D, adrenal cortex and gonadal hormones, active components of toad poisons and digitalis. Saturated hydrocarbons, with seventeen carbon atoms in a system of rings, three 6-membered and one 5-membered, condensed together. *Sterols* (e.g. CHOLESTEROL and ergosterol) form a large steroid subgroup, having a hydroxyl group at C_3 and an aliphatic chain of eight or more carbon atoms at C_{17}. Insects are incapable of producing sterols. See Fig. 154, ISOPRENOIDS.

sterols See STEROIDS.

STH Somatotropic hormone. See GROWTH HORMONE.

stick insects See PHASMIDA.

sticky ends See DNA LIGASE, TELEOMERE.

stigma (Bot.) (1) Terminal portion of the style; surface of the carpel which receives the pollen. (2) See EYESPOT. (Zool.) Rare alternative name for insect SPIRACLE.

stimulus Any change in the internal or external environment of an organism intense enough to evoke a response from it without providing the energy for that response. See IRRITABILITY, ADAPTATION, HABITUATION.

stipe Stalk. (1) Of fruit bodies of certain fungi, e.g. BASIDIOMYCOTA. (2) Of the thallus of seaweeds (e.g. *Laminaria*), the organ connecting the holdfast system and the photosynthetic laminae or blades.

stipule Small, usually leaf-like, appendage found one on either side of leaf stalk in many plants, protecting axillary bud; often photosynthetic.

stock Part of plant, usually comprising the root system together with a larger or smaller part of the stem, on to which is grafted a part of another plant (the *scion*). See GRAFT.

stolon Stem growing horizontally along the ground, rooting at nodes (e.g. strawberry runner).

stoma (pl. stomata) (Bot.) An opening or pore in the plant epidermis, regulated by specialized cells called GUARD CELLS (see for details of mechanics); especially in the leaves, and through which gaseous (including water vapour) exchange occurs. Stomata are often associated with epidermal cells that differ in shape from the ordinary epidermal cells (termed subsidiary or accessory cells). Stomata may occur upon both sides of a leaf, but usually they are more numerous on the under-surface. They may occur only on the upper leaf surface of aquatic flowering plants, whose leaves float upon the water surface; immersed leaves usually lack stomata entirely. Leaves of XEROPHYTES tend to have a greater density of stomata compared to other plants, allowing for a higher rate of gaseous exchange during the

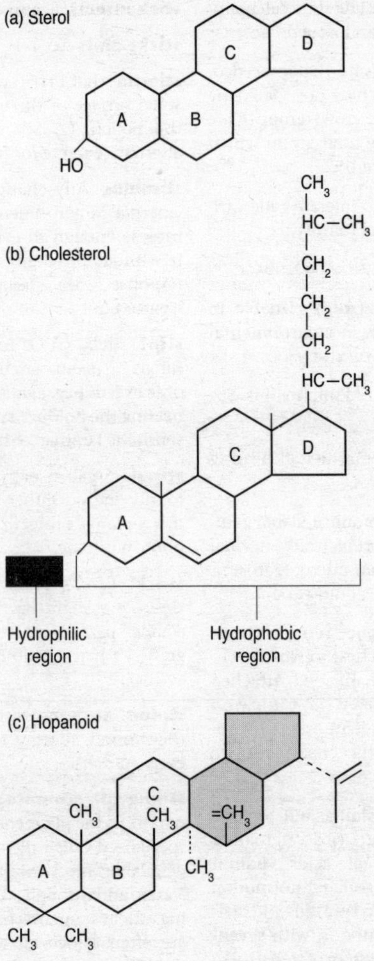

FIG. 154 (a) *The general structure of a sterol; (b) the structure of cholesterol; (c) the structure of the hopanoid diploptene. Sterols are found in the membranes of eukaryotes, while hopanoids are found in those of some prokaryotes. See* STEROIDS.

infrequent periods of favourable water supply. However, these stomata are sunken (sometimes hairy) depressions on the lower leaf surface – adaptations to reduce water loss. Dicotyledon leaves generally have a scattered arrangement of stomata, whereas the stomata in monocotyledons are arranged in rows parallel with the longitudinal axis of the leaf.

Since the industrial revolution, decreases in stomatal densities have paralleled increases in atmospheric CO_2 concentration, a correlation which also holds for fossil material over the past 400 Myr. The

Arabidopsis gene *HIC* (high carbon dioxide) encodes a negative regulator of stomatal development that responds to CO_2 concentration, mutant *hic* plants exhibiting up to a 42% stomatal density increase in response to a doubling of CO_2 concentration. See RADIAL MICELLATION, TRANSPIRATION.

stomach Enlargement of anterior region of ALIMENTARY CANAL. In vertebrates it follows oesophagus and usually has thick walls of SMOOTH MUSCLE to churn food and lining mucosal cells secreting mucus, pepsinogen and hydrochloric acid (see GASTRIC). Cardiac and pyloric sphincters can close the ends during churning. See GASTRIN, RENNIN, RUMINANT.

stomium Place in wall of fern sporangium where rupture occurs at maturity, releasing spores.

stomodaeum Intucking of ectoderm meeting endoderm of anterior part of ALIMENTARY CANAL, forming mouth.

Stone Age Term denoting PALAEOLITHIC and NEOLITHIC.

stone cell Type of sclereid. See SCLERENCHYMA.

stoneflies See PLECOPTERA.

stop codon See PROTEIN SYNTHESIS, GENETIC CODE.

stratum corneum Outer layer of epidermis of vertebrate skin. Cells undergo CORNIFICATION (keratinization) and die, becoming worn off.

Strepsiptera (stylopids) Small endopterygote insects; females endoparasitic; males free-living, short-lived, with large metathorax, anterior wings haltere-like, hind wings large and fan-shaped. Females degenerate, apodous and larviform, enclosed in persistent larval cuticle. Several forms parasitize hymenopterans. Larvae emerge from host and probably find new hosts by waiting on flowers or through contact in nest. Probable affinities with the Coleoptera.

strepsirhine Any primate mammal with a rhinarium. The justification for Suborder Strepsirhini (lemurs and lorises) is debated, although both this and Suborder Haplorhini (see HAPLORHINE) are often employed.

Streptococcus Genus of non-spore-producing, Gram-positive bacterium, forming long chains. Many are harmless colonizers of milk; some commensal in the vertebrate gut; but the *pyogenic* group are human pathogens, some producing haemolysins, destroying erythrocytes. The *viridans* streptococcal group lives usually non-pathogenically in the upper respiratory tract, but can cause serious infections (some chronic), and may cause bacterial arthritis.

streptokinase A protein product of certain bacteria (genus *Streptococcus*) which activates profibrinolysis (plasminogen) and so has indirect fibrinolytic activity (see FIBRINOLYSIS). This may assist invasiveness. Used clinically in the treatment of THROMBOSIS.

streptomycin ANTIBIOTIC inhibiting translation of mRNA on prokaryotic, but not eukaryotic, ribosomes and can be used to distinguish these translation sites. Streptomycin resistance is conferred upon prokaryotes by plasmid-borne transposon. See ANTIBIOTIC RESISTANCE ELEMENT, CHLORAMPHENICOL.

stress hormones See ADRENAL GLAND.

stretch receptor See MUSCLE SPINDLE, PROPRIOCEPTOR.

striated muscle (skeletal/voluntary/striped muscle) Contractile tissue, consisting in vertebrates of large elongated muscle fibres formed by fusion of MYOBLASTS to form syncytia. The cytoplasm of each fibre is highly organized, producing conspicuous striations at right angles to its long axis, and contains numerous longitudinal fibrils (*myofibrils*), each with alternating bands (A, anisotropic; I, isotropic), H zones and Z discs caused by distributions of ACTIN and MYOSIN myofilaments and of α-ACTININ (see Fig. 155). The cross-striations of a whole fibre result from similar bands lying side by side. Each fibre is bounded by a *sarcolemma* (plasmalemma and basement membrane),

(a)

Thin myofilament

Thick myofilament

Mitochondrion

Sarcolemma

T tubule to outside of cell

Sarcoplasmic reticulum

Sarcomere

Z line

Z line

T tubule

I band

A band

H zone

1 micron

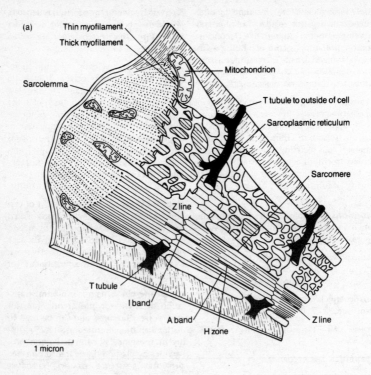

(b)

Capillary

Fibre

Fascicle

Neurone

Epimysium

Perimysium

Sarcolemma
of fibre

Arteriole or
venule

100 microns

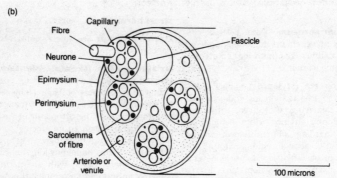

FIG. 155 (a) *Diagram of the ultrastructure of a vertebrate* STRIATED MUSCLE *fibre showing a complete sarcomere and two adjacent parts of sarcomeres. Several such fibres together, with appropriate connective tissue, form a muscle fascicle several of which in turn comprise a striated muscle as shown in (b). (b) Four striated muscle fascicles and surrounding connective tissue forming a small muscle.*

the plasmalemma of which is deeply invaginated into the fibre forming *transverse tubules* (T-system), generally between the Z discs and H zones in insects but over the Z discs in vertebrates. These bring membrane depolarizations right into the fibre, ensuring uniform contraction. Mitochondria abound between myofibrils. The endoplasmic reticulum is modified to form a confluent system of sacs (*sarcoplasmic reticulum*) controlling calcium ion concentration.

On stimulation, a striated muscle fibre contracts by shortening and thickening. Numbers of mitochondria increase 5–10 fold in resting skeletal muscle which has been stimulated to contract for long periods. Fibres are bound together by connective tissue to form muscle tissue, and bring about locomotion by moving the skeleton, to which they are attached in vertebrates by tendons.

Three kinds of vertebrate skeletal muscle fibres are found: (a) *slow twitch fibres (slow phasic fibres, slow oxidative fibres)*, the smallest diameter skeletal muscle fibres, appearing dark red because of their large quantity of myoglobin and rich blood capillary supply. Their numerous large mitochondria generate ATP aerobically, making them resistant to fatigue and capable of prolonged sustained contractions over many hours. The ATPase in the myosin heads hydrolyses ATP relatively slowly, giving the fibres a low contraction velocity (~100–200 ms). They take longer than fast twitch fibres to reach maximum tension despite being better at utilizing oxygen; (b) *fast oxidative-glycolytic fibres*, intermediate in diameter between slow twitch fibres and fast twitch fibres, containing a rich capillary supply and much myoglobin and so, like slow twitch fibres, also appearing dark red. With high intracellular glycogen levels, they can also generate ATP by glycolysis. The ATPase in the myosin heads hydrolyses ATP rapidly (hence 'fast' fibres), making their contraction velocity (100 ms) faster than in slow twitch fibres, causing tension to be reached more quickly; (c) *Fast twitch fibres (fast phasic fibres, fast glycolytic fibres)*, with the largest diameter of all skeletal muscle fibres, containing the greatest number of myofibrils,

fast twitch fibres generate the most powerful contractions. Their high glycogen and low myoglobin content, relatively scanty blood supply and small numbers of mitochondria indicate their adaptation to generate ATP by glycolysis. The ATPase in their myosin heads hydrolyses ATP rapidly, adapting them for anaerobic activities of short duration (contrast slow twitch fibres). Anaerobic exercise leads to increase in mass of these fibres by hypertrophy. See MUSCLE CONTRACTION, CARDIAC MUSCLE, NEUROMUSCULAR JUNCTION, SMOOTH MUSCLE.

striatum See BASAL GANGLIA.

string gene A gene in *Drosophila* whose product is homologous to yeast cdc2 protein (see *CDC* genes) and positively regulates MATURATION PROMOTION FACTOR. Translated from maternal mRNA stored in the oocyte for the first thirteen cell division cycles of the embryo, after which maternal mRNA is degraded and nuclear *string* mRNA is produced.

stRNAs Small temporal RNAs. A group of microRNAs, cleaved from precursor RNA by the enzyme Dicer, with effects on developmental timing, discovered in the nematode *Caenorhabditis elegans*. See NON-CODING RNAS.

strobilation Process of transverse fission which produces proglottides from behind the scolex of a tapeworm and ephyra larvae from the jellyfish scyphistoma (Scyphozoa). Regarded as a method of asexual reproduction in the latter, but less commonly in the former. The whole ribbon-like chain of tapeworm proglottides may be referred to as a strobila.

strobilus (Bot.) Cone. Reproductive structure comprising several modified leaves (sporophylls), or ovule-bearing scales, grouped terminally on a stem.

stroma (Bot.) (1) Tissue-like mass of fungal hyphae, in or from which fruit bodies are produced. (2) Colourless matrix of the CHLOROPLAST, in which grana are embedded. (Zool.) Intercellular material (matrix), or connective tissue component of an animal organ.

stromal cells (1) Fibroblasts and other connective tissue components of the bone marrow, capable of binding growth factors during HAEMOPOIESIS. Some release interleukin-7, required for B CELL development. (2) Cells migrating to central portion of the developing KIDNEY, producing growth factors permitting continued growth and differentiation of ureteric buds.

stromatolites Macroscopic structures produced by certain blue-green algae (CYANOBACTERIA) where there is deposition of carbonates along with trapping and binding of sediments. Predominantly hemispherical in shape, they possess fine concentric laminations produced by growth responses to regular (often daily) environmental change. Fossil stromatolites occur from early Precambrian (more than 3,000 Myr BP) to the Recent period.

style Slender column of tissue arising from top of ovary and through which pollen tube grows.

subarachnoid space Area between the arachnoid and the pia mater, filled with cerebrospinal fluid. See MENINGES.

subcloning Procedure involving removal and transfer of cloned DNA fragment from one vector to another.

subcutaneous Immediately below dermis of vertebrate skin (i.e. the *hypodermis*). Such tissue is usually loose connective tissue, blood vessels and nerves and generally contains fat cells (see ADIPOSE TISSUE). In many tetrapods, also includes a sheet of striated muscle (*panniculus carnosus*) to move skin or scales.

suberin Complex mixture of fatty acid oxidation and condensation products present in walls of cork and most endodermis cells, rendering them impervious to water.

suberization Deposition of SUBERIN.

subspecies Formal taxonomic category used to denote the various forms (types), usually geographically restricted, of a polytypic SPECIES. See INFRASPECIFIC VARIATION.

substitutable resource Two resources are said to be perfectly substitutable when either can wholly replace each other (e.g. seeds of wheat or barley in the diet of chickens). It does not imply that the two resources are as good as each other. Compare COMPLEMENTARY RESOURCES.

substrate (1) Substance upon which an ENZYME acts. (2) Ground or other solid object on which animals walk or to which they are attached. (3) Material on which a microorganism is growing, or solid surface to which cells in tissue culture attach.

subtidal Zone in sea or ocean extending from low-tide mark to edge of continental shelf.

succession A progressive change in species composition of a community of organisms, e.g. from initial colonization of a bare area (*primary succession*) or an already established community (*secondary succession*), towards a largely stable, climax, community. See AUTOGENIC SUCCESSION, ALLOGENIC SUCCESSION, AUTOTROPHIC SUCCESSION, DEGRADATIVE SUCCESSION, SERE.

succulent Type of xerophytic plant which stores water within its tissues and has a fleshy appearance (e.g. cacti).

succus entericus (intestinal juice) Digestive juice (pH about 7.6 in humans) containing enterokinase, peptidases, nucleases, sucrase, etc., all secreted by the glandular *crypts of Lieberkühn* between intestinal villi. Completes hydrolysis of food molecules begun higher in the gut. About 2–3 litres per day secreted in humans. See DIGESTION.

sucker Sprout produced by roots of some plants, giving rise to a new plant.

sucrase (invertase) See SUCROSE.

sucrose (cane sugar) Non-reducing disaccharide, comprising one glucose and one fructose moiety, linked between C_1 of glucose and C_2 of fructose. Abundant transport sugar in plants. Digested by the enzyme *sucrase (invertase)* and dilute mineral acids to glucose and fructose: invertase is anchored in the periplasmic space (between a plasma membrane and cell wall) of the yeast *Saccharomyces cerviseae*. See INVERTASE.

Suctoria Predatory ciliates, ciliated only in larval stage. See CILIATA.

sulcus (Bot.) (1) Longitudinal furrow, as in groove containing trailing flagellum in dinoflagellates; (2) thin furrowed area of pollen wall, notably in cycads and *Gingko* pollen. (Zool.) See GYRUS.

sulphur bonds (disulphide bonds) See PROTEINS.

summation Additive effect at synapses when arrival of one or a few presynaptic impulses is insufficient to evoke a propagated response but a train of impulses can do so. Termed *temporal summation* when impulses arrive at the same synapse, *spatial summation* when at different synapses of the same cell. In conjunction with nervous INHIBITION and FACILITATION it enables fine control over an animal's effector responses. See NERVOUS INTEGRATION, SYNERGISM, TRANSMITTER.

superantigens Antigens of bacterial and viral origin defined as having a potent stimulatory effect on T CELLS because of their ability to bind to whole classes of T cell receptor. Unlike normal antigens, they are not processed and associate with MHC Class II molecules outside the peptide binding groove and recognition is not MHC-restricted.

supergene Group of gene loci with mutually reinforcing effects upon phenotype that have come (through selection) to lie on the same chromosome, increasingly tightly linked so as to be inherited as a block (e.g. supergenes for the heterostyle or homostyle system in primroses (*Primula*) and shell colour and banding pattern in the snail (*Cepaea*)). See POLYMORPHISM.

superior ovary See RECEPTACLE.

supernumerary chromosome (accessory chromosome) Chromosome additional to normal karyotype of the species (B chromosome in botany). Either not homologous, or only partially homologous, with members of normal karyotype. In some populations of a species most of the individuals carry such chromosomes; in others, frequency is low. Most are heterochromatic. Their presence does not seem to affect markedly the individual's appearance. Geographical distribution often non-random.

superoxides During mitochondrial respiration, incomplete reduction can cause up to 4% of molecular oxygen (O_2) to be released as the superoxide anion, $O_2^{-\bullet}$, which can react on ferrous and other metal catalysts to form the more reactive $OH^\bullet$. The metaloenzyme superoxide dismutase, found in different forms in the cytosol and mitochondria, converts two superoxide radicals into hydrogen peroxide and oxygen:

$$O_2^{-\bullet} + O_2^{-\bullet} + 2H^+ \rightarrow H_2O_2 + O_2.$$

PEROXISOMES remove the H_2O_2. Superoxides are produced within the lysosomes of monocytes and macrophages, where they are antimicrobial. All such free radicals are potentially damaging oxidizing agents, and vitamin E, located within the phospholipid bilayer of many cell membranes, plays a major antioxidant role in protecting polyunsaturated fats and other membrane components from such attack. Diets deficient in vitamins E and C (also an antioxidant), and environmental factors producing free radicals (e.g. $NO_2^\bullet$ from CIGARETTE SMOKING; see CAROTENOIDS, POLLUTION) can cause oxidative stress leading to long-term degenerative tissue damage (see also PARKINSON'S DISEASE).

superspecies Informal taxonomic category usually applied to allopatric arrays of species where evidence suggests common ancestry and where the species are sufficiently similar. Such species complexes tend to be discoverable with difficulty due to cryptic distinguishing characteristics between the species (usually sibling species), such as *Anopheles gambiae* and *Stimulium damnosum* complexes. See DNA PROBE.

supertramp species Species with good colonizing ability that arrive earliest in a new habitat.

supination See PRONATION.

suppressor mutation (1) (*Intragenic.*) A mutation (nucleotide addition or deletion) at a site in a chromosome sufficiently close to a prior nucleotide deletion or addition to restore the reading frame and thus suppress the effect of the original mutation. (2) (*Intergenic.*) A mutation at one locus which prevents the expression of a mutation at another locus. (3) (Uncommonly) a mutation preventing local or complete crossing-over in meiotic cells.

Some *suppressor genes* seem to be responsible for preventing oncogenic phenotype in normal cells; their loss may result in ONCOGENE expression. See ABERRANT CHROMOSOME BEHAVIOUR (1), MUTATION, HYPOSTASIS.

suprachiasmatic nuclei (SCN) Bilateral region of anterior hypothalamus in mammals (two nuclei), comprising about 20,000 neurons. With circadian oscillatory patterns of protein abundance, phosphorylation, interactions and subcellular location – probably with functionally distinct groups of cells. The 'master' BIOLOGICAL CLOCK of mammals.

suprarenal gland See ADRENAL GLAND.

surrogate resource A resource, not itself in limited supply, which is competed for because of the access it provides to some other resource, which is or may become limited in supply.

survival factors, survival signals Cell surface ligands which promote cell survival by suppressing APOPTOSIS and in whose absence a cell becomes apoptotic, ensuring that cells only survive where they are needed (see EXTRACELLULAR MATRIX). Binding of these signal molecules to some cell surface receptors leads to activation of various protein kinases (e.g. PROTEIN KINASE B), inactivating BAD and releasing BCL-2. Alternatively, as in *Drosophila*, a survival factor may inhibit apoptosis by phosphorylating and inactivating inhibitors of inhibitors of apoptosis (IAPs). The tumour necrosis factor TNF-α is an important protein which simultaneously activates two signalling pathways: a cell-survival pathway regulated by the transcription factor NF$\varkappa$B, and a cell-death pathway associated with the JNK SIG-

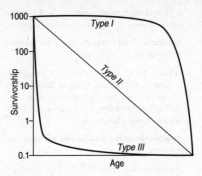

FIG. 156 *Hypothetical standard* SURVIVORSHIP CURVES (*after Pearl, 1928*).

NALLING PATHWAY cascade. Nerve growth factor (see NEUROTROPHINS) and COLONY STIMULATING FACTORS are survival factors. Compare GROWTH FACTORS.

survival value Characters and genes are said to have positive survival value if they increase an individual's FITNESS, and (less commonly) negative survival value if they decrease it. Neutral survival value is attributable to characters and genes with no effect on fitness. See NATURAL SELECTION.

survivorship The probability of a representative newly born individual surviving to various ages. See SURVIVORSHIP CURVES, Fig. 156.

survivorship curves Graphs of numbers of individuals in a cohort plotted up the y-axis (usual units log 10 I_x, where I_x is the proportion of the original cohort still alive) from the time they become independent until the time they die, plotted against time (x-axis). Pearl's (1928) classification identifies three commonly occurring curves (I, convex; II, flat; III, concave), representing the consequences of different age-dependent mortalities. See Fig. 156, AGE PYRAMIDS.

suture Area of fusion of two adjacent structures. (1) (In flowering plants) the line of fusion of edges of carpel is known as a ventral suture. Mid-rib of carpel is known as the dorsal suture, not implying any fusion of parts to form it but to distinguish it from the ventral (true) suture. (2) Junction between the irregular interlocking edges of adjacent skull bones, or between plates of hardened cuticle of exoskeleton. Suture-lines occur on shells of ammonites, marking edge of a septum with side-wall of shell. (3) (Surgical) to sew a wound together.

SV40 See VIRUS.

Svedberg unit The unit (S), related to sedimentation coefficient, used to indicate the time taken for large molecules, small organelles, etc., to reach an equilibrium level when ultracentrifuged in water at 20°C. The unit (S) is defined as a sedimentation coefficient of 1×10^{-13} seconds. Sedimentation coefficient is given by the equation:

$$s = \frac{dx/dt}{\omega^2 x}$$

where x = the distance of the sedimenting boundary from the centre of rotation (cm), t = time (s) and ω = angular velocity (radian s⁻¹). The S-value of an object is termed its *Svedberg number*. See RIBOSOMES.

swarm spore See ZOOSPORE.

sweat Aqueous secretion of mammalian SWEAT GLANDS containing solutes in lower concentrations than in blood plasma, to which it is hypotonic. In humans, contains 0.1–0.4% sodium chloride, sodium lactate and urea; about one litre is lost per day in temperate climates (up to 12 litres in hot dry conditions when salt and water supply permit). The urea and lactate may be regarded as excretory, but may also reduce colonization by skin microflora. Its production, along with vasodilation of skin capillaries, removes body heat as latent heat of vapourization of water. See HOMEOTHERMY.

sweat glands Epidermal glands, projecting into mammalian dermis, releasing SWEAT when stimulated by sympathetic nervous system. (1) *Apocrine sweat glands* are simple, branched tubular glands of (mainly) armpits, pubic region and areolae of breasts. Secretion more viscous than from eccrine sweat glands. (2) *Eccrine sweat glands* are simple, coiled tubular glands. Widely distributed; rows of them open via pores between the papillary ridges (responsible for fingerprints) on hands and feet of many mammals. Absent from cats and dogs except between toes. See MAMMARY GLANDS.

swim bladder See GAS BLADDER.

symbiont A symbiotic organism. See SYMBIOSIS, MYCOBIONT, PHYCOBIONT.

symbiosis The living together in permanent or prolonged close association of members (*symbionts*) of usually two different species, with beneficial or deleterious consequences for at least one of the parties. Included here are: *commensalism*, where one party (the commensal) gains some benefit (often surplus food) while the other (the host) suffers no serious disadvantage; *inquilinism*, where one party shares the nest or home of the other, without significant disadvantage to the 'owner'; *mutualism*, where members of two different species benefit and neither suffers (symbiosis in a restricted sense; e.g., see LICHENS, MYCORRHIZA, *RHIZOBIUM*) and *parasitism*, where one party gains considerably at the other's expense (see PARASITE). In *amensalism*, one party is harmed while the other is unaffected. Some include certain intraspecific relationships within symbiosis.

Symmetrodonta Order of extinct mammals (Infraclass Trituberculata). Small Mesozoic forms; insectivorous. Possibly ancestral to PANTOTHERIA.

sympathetic nervous system See AUTONOMIC NERVOUS SYSTEM.

sympatric Of populations of two or more species, whose geographical ranges or distributions coincide or overlap. From an ecological and genetic viewpoint, often more valuable to distinguish between sympatry as defined above and *effective* sympatry. Where adults are settled in different geographical regions but gametes, developmental stages, etc., are widely dispersed, there may be effective sympatry despite adult distribution. In lice, several species may inhabit the same host but settle in well-defined and species-specific anatomical regions. They are sympatric, yet ALLOTOPIC. See ALLOPATRIC, SYNTOPIC, SPECIATION.

sympetalous See GAMOPETALOUS.

symphysis Type of joint allowing only slight movement, in which surfaces of two articulating bones (both covered by layer of smooth cartilage) are closely tied by collagen fibres. Symphyses occur between centra of vertebral column. For *pubic symphysis*, see PELVIC GIRDLE.

symplast Interconnected protoplasts and their plasmodesmata which effectively result in the cells of different plant organs forming a continuum (e.g. from root hairs to stele). See TRANSLOCATION.

symplesiomorphy Any character which is a SHARED HOMOLOGUE of two or more taxa, but which is thought to have occurred as an evolutionary novelty in an ancestor earlier than their earliest common ancestor. Compare SYNAPOMORPHY. See PLESIOMORPHIC, CLADISTICS.

sympodium Composite axis produced, and increasing in length, by successive development of lateral buds just behind the apex. Compare MONOPODIUM.

symport See TRANSPORT PROTEINS.

synangium Compound structure formed by lateral union of sporangia; present in some ferns.

synapomorphy Term used in phylogenetics (see CLADISTICS) to denote what are otherwise termed *homologous* characters in many writings on evolution. Any character which is a SHARED HOMOLOGUE of two or more taxa and is thought to have originated in their closest common ancestor and not in an earlier one. Synapomorphy has been used in a more inclusive sense than has homology, characters involved including, e.g., geographical ranges. See HOMOLOGY, SYMPLESIO-MORPHY.

synapse Region of functional contact between one neuron and another, or between it and its effector. Commonly a gap (*synaptic cleft*) of at least 15 nm occurs between the apposed plasma membranes,

the direction of impulses defining *presynaptic* and *postsynaptic* membranes. An individual neuron commonly receives 1,000–10,000 synaptic contacts with 1,000 or more other neurons, most synapses being between axon terminals of the stimulating neuron and dendrites of the receiving neuron, but axon–axon and dendrite–dendrite synapses occur. Most synapses are chemical (humoral), involving release of neurotransmitter via *synaptic vesicles* in response to Ca^{2+} entry at the presynaptic membrane: the more Ca^{2+} entry, the more transmitter release (see ACETYLCHOLINE, IMPULSE); others are electrical, impulses passing without delay from one neuron to another via gap junctions (see INTERCELLULAR JUNCTION). Chemical synapses, although encountering greater resistance due to diffusion time of transmitter, do permit NERVOUS INTEGRATION.

There are two categories of chemical transmission between neurons: fast and slow synaptic transmission. About half of the fast brain synapses are excitatory and involve glutamate as their transmitter, the remaining half being inhibitory, involving GABA. *Fast synaptic transmission* occurs in <0.001 s, attributable to the ability of fast-acting transmitters to open ligand-gated ion channels in the post-synaptic membrane. *Slow synaptic transmission* takes place over hundreds of milliseconds to minutes and is much more complex in its operation, involving biogenic amines (e.g. DOPAMINE), peptides and amino acids serving rather like neurotransmitters (NEUROMODULATORS) but operating instead through non-ion channel-linked receptors via second messengers and protein kinases. Most PSYCHOACTIVE DRUGS exert their effects at synapses, many binding to specific receptors. $GABA_A$ receptors mediate rapid INHIBITION and are ligand-gated chloride channels acted on by GABA, by benzodiazapine tranquillizers (e.g. Valium, Librium) and by barbiturates – each binding a different site on the same receptor protein and apparently operating by allowing reduced amounts of GABA to open the channel (see POTENTIATION). Volatile anaesthetics (e.g. ethanol) appear to operate by inhibiting some ion channels while potentiating others at synapses. Initiation of synapse formation (synaptogenesis) in mice has been shown to involve proteins called neuroligins, and even kidney cells expressing the neuroligin genes can trigger early synapse formation in neurons next to them. See IMMUNOLOGICAL SYNAPSE, NEUROMUSCULAR JUNCTION, SUMMATION, SYNAPTIC PLASTICITY.

Synapsida Amniote clade including mammal-like reptiles and mammals. They have a single temporal fenestra bounded above by the upper temporal bar (squamosal-postorbital bones).

synapsis Pairing-up of chromosomes. See MEIOSIS.

synaptic plasticity The capacity of neurons to modulate the strength of their synaptic connections in response to their cellular experience and environment. Contributes to a variety of physiological and pathological processes in the adult brain, including learning, memory storage, age-related memory loss, tolerance to and dependence upon psychoactive drugs, and epilepsy. See SYNAPTIC STRENGTH.

synaptic strength The effectiveness with which a synapse functions, e.g. in releasing its transmitter and modifying the postsynaptic membrane. These are aspects of SYNAPTIC PLASTICITY. *Synaptic depression* is the decrease in synaptic strength which normally accompanies prolonged trains of action potentials and can be brought about by release of nitric oxide. In the peripheral nervous system, terminal SCHWANN CELLS (those at the neuromuscular junction) monitor synaptic activity by detecting neuron-glial signalling molecules (ATP, adenosine) released from the neuron in association with neurotransmitter, integrate the activity of the synapse and balance the strength of the connection by regulating transmitter release from the presynaptic neuron. Increased synaptic strength may be brought about by signalling pathways that increase cytoplasmic levels of calcium ions (Ca^{2+}) in the terminal Schwann cells. Similarly, in the brain, ASTROCYTES ensheath synaptic junctions and use purinergic receptors for neuron-glial signalling. A rise in astrocytic cytoplasmic Ca^{2+} levels is associated with changes in synaptic strength in adjacent synapses, both in culture and in intact retinas and associated in turn with rise in extracellular ATP concentrations. See CELL MEMORY, PRESYNAPTIC INHIBITION, SYNAPSE.

synaptic vesicle Vesicle produced by the Golgi apparatus of a nerve axon, in which NEUROTRANSMITTER is stored prior to release into synaptic cleft. The membrane components of synaptic vesicles are thought to derive from the plasma membrane after delivery there from the Golgi apparatus, and regenerated after exocytosis by local recycling from the plasma membrane endocytotically, probably refilling with transmitter in the neuron terminal, for which the membranes have special carrier proteins. See COATED VESICLES.

synaptonemal complex Ladder-like (zip-like) protein complex holding chromosome together during synapsis of MEIOSIS, chromosome loops pointing away from it. Within it lie *recombination nodules*, apparently determining where CHIASMA formation occurs. The complex dissolves at diplotene.

syncarpous (Of the gynoecium of a flowering plant) with united carpels (e.g. tulip). See FLOWER.

synchronous culture Culture (of microorganisms or tissue cells) in which, through suitable treatment, all cells are at any one time at approximately the same stage of development, or of the CELL CYCLE. In mammalian tissue culture, cells in mitosis (M phase) round up and can be separated from others by gentle agitation to start a synchronous culture. They will rapidly enter G1.

syncytium Animal tissue formed by fusion of cells, commonly during embryogenesis, to form multinucleate masses of protoplasm. May form a sheet (as in mammalian TROPHOBLAST) or cylinder (STRIATED MUSCLE), or network of more or less discrete cells linked by intercellular bridges (as in mammalian spermatogenesis; see MATURATION OF

GERM CELLS). Smooth muscle is occasionally syncytial. See ACELLULAR, SYMPLAST, DOMAIN.

synecology Ecology of communities as opposed to individual species (autecology).

synergidae (synergids) Two short-lived cells lying close to the egg in mature EMBRYO SAC of flowering plant ovule.

synergism Interaction of two or more agencies (e.g. hormones, drugs), each influencing a process in the same direction. Their combined effects are either greater than their separate effects added (*potentiation*) or roughly the sum of their separate effects (*summation*). Compare ANTAGONISM.

synexpression groups Sets of eukaryotic genes sharing a complex spatial (and maybe temporal) expression pattern in more than one tissue and functioning in the same process. It appears that in eukaryotes, relatively few genes are organized into operons (see JACOB–MONOD THEORY); instead, synexpression of genes is probably achieved by TRANS-ACTING CONTROL ELEMENTS. See MODULAR ORGANISM.

syngamy See FERTILIZATION.

syngeneic Strains of organisms produced by repeated inbreeding, each pair of autosomes being identical.

syngenesious (Of stamens) united by their anthers; e.g. Compositae.

syngraft See ISOGRAFT.

synonymous sites (synonymous sequence changes) DNA sequence changes in coding regions that do not change amino acid sequences (due to degeneracy of the GENETIC CODE).

synovial membrane Connective tissue membrane forming a bag (*synovial sac*) enclosing a freely movable joint, e.g. elbow joint, being attached to the bones on either side of the joint. Bag is filled with viscous fluid (*synovial fluid*) containing glycoprotein, lubricating the smooth cartilage surfaces making contact between the two bones.

synteny Sharing of the same linear order of genes on the chromosomes (or parts of chromosomes) between different taxa (whether species or higher taxa) subsequent to their evolutionary divergence. Genome sequencing has revealed, for instance, that of the 27,000–30,500 protein-coding genes in the mouse, *Mus musculus*, 96% lie within syntenic regions of the mouse and human chromosomes although the nature and extent of this 'conservation of synteny' differ considerably among chromosomes.

syntopic Two or more organisms are syntopic if they share the same habitats (or microhabitats) within the same geographical range. They are *microsympatric*. This may be the expression of different phenotypes within a single species. See SYMPATRIC.

syntrophy Mutual dependence of cells on one another's products for their nutritional needs. Not uncommon between different bacteria, as between methanogens and hydrogen-producing bacteria in anaerobic marine sediments and deep in the Earth's crust (see EXTREMOPHILE). The hydrogen is used by the methanogens as a metabolite, which oxidize it to methane (CH_4) using carbon dioxide as an electron acceptor. The low hydrogen tension generated by the hydrogen consumption is required thermodynamically for the complete fermentation of carbohydrates to acetate and H_2 by the hydrogen producing bacteria. In colonies of such syntrophic pairs of bacteria, partners are arranged so that cells of one species have neighbours belonging to the other. See ENDOSYMBIOSIS.

syntype Each of several specimens used to describe a new species when a single type specimen was not chosen. See HOLOTYPE.

syrinx Sound-producing organ of birds, containing typically a resonating chamber with elastic vibrating membranes of connective tissue (VOCAL CORDS); situated at point where trachea splits into bronchi. Compare very different LARYNX of mammals, which in birds lacks vocal cords.

system (organ system) Integrated group of ORGANS, performing one or more unified functions, e.g. NERVOUS SYSTEM, VASCULAR SYSTEM, ENDOCRINE SYSTEM.

systematics Term often used as synonym of TAXONOMY, but sometimes used more widely to include also identification, practice of classification, nomenclature. See BIOSYSTEMATICS, CLASSIFICATION.

systemic Generally distributed throughout an organism.

systemic arch Fourth AORTIC ARCH of tetrapod embryo, becoming in adult main blood supply for the body other than the head. In Amphibia both left and right arches persist in the adult; in birds only the right; in mammals only the left (the aorta).

systems biology See COMPLEXITY.

systole (1) Phase of HEART CYCLE when heart muscle contracts. (2) Phase of contraction of contractile vacuole.

T

tachycardia Abnormally rapid heart rate. See BETA-BLOCKER.

tactile (Adj.) Referring to the sense of touch. Sense organs include many surface pressure receptors.

tadpole Term given to aquatic larval stages of tunicates (UROCHORDATA) and of most modern amphibians. Metamorphoses into the adult.

tagma (pl. tagmata) Functional and anatomical regionalizations of the arthropod body associated with division of labour; thus appendages with similar functions tend to be grouped on adjacent segments. Tagmata include *head*, *thorax* and *abdomen*, but each class within the phylum has a characteristic pattern of *tagmosis*. Insects have distinct head, thorax and abdomen; crustaceans generally lack a clear head–thorax distinction; arachnids have PROSOMA and OPISTHOSOMA, etc. Tagmosis in post-head region is closely related to mode of locomotion. See ARTHROPODA.

tagmosis See TAGMA.

taiga The northern coniferous forest or boreal forest that extends over vast areas of Russia, Scandinavia and North America. Taiga is the Russian word for vegetation found in this BIOME. This forest is primarily evergreen, except in large areas of north-eastern Siberia, where larches (*Larix*), which are deciduous conifers, are dominant. Common trees of the taiga include *Picea* (spruce), *Abies* (fir), *Populus* (poplar) as well as *Larix*; shrubs include *Salix* (willows), *Betula* (birch) and *Ledum* (Labrador tea). The members of all these genera of trees and shrubs are ectomycorrhizal. Herbaceous vegetation is dominated by mosses and lichens. Flanked by montane forests, savannas or grassland to the south, grading unevenly into tundra to the north. The northern limits are determined by the severity of the arctic climate. Characterized by severe winters, with most precipitation falling during the summer. Evaporation rates are low, consequently lakes, bogs and marshes are common. More than 75% of the northern taiga is underlain by permafrost (permanent ice), usually within less than a metre of the surface, thus even though precipitation is low, the ground is usually moist, since the water cannot percolate through the soil. Fire occurrence is common and results in generally warmer, more productive areas for 10–20 years. Soils are generally highly acidic and low in nutrients.

tandem repeats See REPETITIVE DNA.

tangential section Longitudinal section cut at right angles to radius of cylindrical structure (e.g. root or stem).

tannins Group of complex astringent substances occurring widely in plants, dissolved in vacuoles. Particularly common in tree bark, unripe fruits, leaves and galls; also found in animals (e.g. insect cuticles). Include phenols and hydroxy acids or glycosides. Some plant tannins deter herbivores; others have roles in ALLELOPATHY and still others protect against microbial attack (see PLANT DISEASE AND DEFENCES). Employed in the tanning process, which is usually enzymatic (see CUTICLE). Commercial leather production involves cross-linking collagen molecules in skins by treatment with benzoquinone, and a similar process occurs naturally.

tapetum (tapetum lucidum) (Zool.) Reflecting layer of vertebrate choroid (retina in some teleosts), especially in nocturnal forms and deepwater fish. Generally contains guanine crystals reflecting light back through the retina, increasing its stimulatory effect but reducing visual acuity. The pigment melanin is commonly expanded over the crystals in bright light. Most prosimians, except certain diurnal or crepuscular lemurs and *Tarsius*, possess a tapetum although it is well developed in the diurnal ring-tailed lemur (*Lemur catta*). It is absent from all haplorhine primates (tarsiers, monkeys, apes and humans). Its presence in some diurnal mammals may indicate simply the retention of a feature originally developed by a nocturnal lineage, or possibly that it retains some residual role in diurnal forms, such as improving avoidance of nocturnal predators – although one might expect all lemurs to be similarly subjected to this. See CHROMATOPHORE. (Bot.) In vascular plants, a layer of cells, rich in reserve food, surrounding a group of spore mother cells; e.g. in fern sporangium, pollen sacs of the anther. Gradually disintegrate and liberate their contents, which are absorbed by developing spores.

tapeworm See CESTODA.

taphonomy The attempt to define all those circumstances, physical, chemical and biological, in which fossils are deposited and preserved. It includes the forensic study of tooth marks on bones.

tap root Primary root of a plant, formed in direct continuation with root tip or radicle and forming prominent main root, directed vertically downward, bearing smaller lateral roots; e.g. dandelion. Sometimes swollen with food reserves; e.g. carrot, parsnip. See FIBROUS ROOT.

Tardigrada Order of minute aquatic arthropods of uncertain status, but unique among these in having a terminal mouth. The four pairs of stumpy appendages are simple and lobopod (resembling ONYCHO-PHORA), ending in claws. Clear circulatory and respiratory systems lacking. Most pierce and suck plant juices. Resistant to desiccation, living among mosses, etc.

tarsal bones (tarsals) Bones of proximal part of tetrapod hind-foot (roughly the ankle). Primitively 10–12 bones in a compact group, reduced during evolution by fusion and/or loss. There are seven in humans, one (calcaneum) forming the heel. They articulate proximally with tibia and fibula, distally with metatarsals. See PENTADACTYL LIMB.

tarsus (1) Region of tetrapod hind-foot containing TARSAL BONES (approx. the ankle). Compare CARPUS. (2) A segment (fifth from the base) of an insect leg.

taste bud Vertebrate taste receptor, consisting of group of sensory cells, usually located on a papilla of the TONGUE, but in aquatic forms often widely scattered (especially in barbels around fish mouths). Buds are often (e.g. in humans) localized in regions sensitive to salt, sweet, bitter and acid (sour) tastes; other taste modalities are detected in fish. Taste stimuli are diverse, and complex mechanisms are involved in their transduction. Salts and acids interact directly with apical ion channels and depolarize taste receptor cells; but sugars, amino acids and most bitter-tasting compounds bind membrane receptors, usually bound to G PROTEINS and second messenger systems. The G protein *gustducin* (related to the transducins in the vision phototransduction process) is expressed exclusively in taste-receptor cells and activates phosphodiesterase, decreasing cAMP levels but raising calcium intake and release from intracellular stores. It is certainly involved in transducing bitter stimuli, but also seems to have a role in transducing sweet stimuli. Bitter stimuli activate PHOSPHOLIPASE C (PLC) and release diacylglycerol (DAG) and $InsP_3$. Sugar stimuli activate adenylyl cyclase and raise cAMP levels, in turn activating PKA and closing potassium channels. Synthetic sweeteners stimulate PLC to produce DAG and $InsP_3$.

TATA box Short nucleotide consensus sequence in eukaryote PROMOTER sequences bound by RNA polymerases II & III, about

25–30 base pairs upstream of transcribed sequence (hence –25 to –30). An AT-rich region, commonly including the interrupted sequence TATAT . . . AAT . . . A. Recognized and bound by TATA-binding protein, TBP (see TATA FACTOR). See ENHANCER, PRIBNOW BOX, TRANSCRIPTION FACTORS, Fig. 164.

TATA factor (TATA-binding protein, TBP, transcription factor IID) Eukaryotic protein with key role in PROMOTER recognition by RNA polymerase II, where its binding to the TATA BOX is the first step in assembly of the multiprotein transcription initiation complex (see Fig. 157). Also required for specific transcription by RNA polymerases I and III, even when promoter regions lack a TATA box. Its C-terminal 180 amino acids are thought to have some sequence homology to prokaryotic SIGMA FACTORS. The CHROMATIN-REMODELLING nucleosomal ATPase ISWI has been identified as a key molecule in erasing the TBP from association with the nuclear matrix during the nuclear remodelling involved in nuclear transplantation of adult nuclei into *Xenopus* eggs. See CLONE (I). See TRANSCRIPTION FACTORS.

taxis (pl. taxes) A form of orientation in which an animal heads directly towards or away from a source of stimulation. Thus, the maggot larva of a house fly (*Musca domestica*) shows negative phototaxis: when it has finished feeding, it leaves the food and goes to a dark place where it pupates, crawling directly away from a light source. The maggot head is equipped with light-sensitive receptors capable of registering different light intensities, and as the maggot crawls along it moves its head from side to side testing the intensity of light on each side of its body. If the intensity is greater on the right than on the left, the maggot is less likely to turn its head to the right and therefore tends to change course and crawl to the left. This successive comparison of intensity of a stimulus on one side of the body and then the other is called *klinotaxis*, and is not uncommon. See CHEMOTAXIS. Compare KINESIS.

taxol (paclitaxel) Naturally occurring anti-cancer drug (source: the Pacific yew,

Taxus brevifolia), a derivative of which is approved in treatment of metastatic breast cancer. Commercial production is being explored.

taxon The organisms comprising a particular taxonomic entity, e.g. a particular class, family or genus. Members of a *particular* species form a taxon, but the TAXONOMIC CATEGORY species does not.

taxonomic category A category, formal or informal, used in CLASSIFICATION. Formal categories include Kingdom, Phylum (Division), Class, Order, Family, Genus and Species. Instances of these (e.g. Phylum Arthropoda) are taxa, not taxonomic categories. Informal categories include superspecies and race. See INFRASPECIFIC VARIATION.

taxonomy Theory and practice of CLASSIFICATION. Classical taxonomy is concerned with morphology (including cytological, biochemical, behavioural), and may involve weighting phenetic characters in some scheme of relative taxonomic importance or value. *Numerical taxonomy* dispenses with such weighting and involves computerized analysis of data obtained from observing whether or not organisms being compared have or do not have any of the 'unit characters' involved in the comparison. Unit characters have an 'all-or-none' nature, being either present or absent. The data tend to arrange themselves into sets or phenons, which can then be organized into a DENDROGRAM. Relationships between organisms which can be thus evaluated are termed *phenetic* if determined by overall morphological similarity between organisms, or *cladistic* if they depend upon community of descent. In *experimental taxonomy*, breeding work and field experiments may be used to clarify the taxa to which organisms belong.

T cells (T lymphocytes) Lymphocytes which travel from the bone marrow via the blood and enter the thymus, after which they enter the circulation again and settle in spleen and lymph nodes. Unlike B CELLS, T cells do not recognize native antigen conformation directly, but only in association with self-antigens of the MAJOR HISTO-

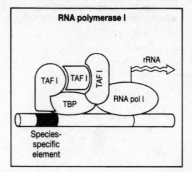

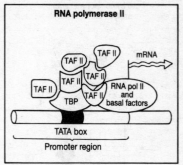

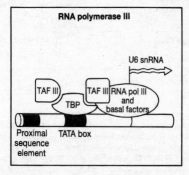

FIG. 157 *Distinct multiprotein complexes with a common subunit, the TATA-binding protein (TBP), participate in specific promoter recognition by the three eukaryotic RNA polymerases. For clarity, only a single type of minimal promoter is shown for each RNA polymerase.*

COMPATIBILITY COMPLEX (see also ANTIGEN-PRESENTING CELLS). There are two major subsets of T cell: those expressing the ACCESSORY MOLECULE CD4 and those expressing CD8. The former are mainly *T helper cells* (T_H); the latter are generally effector cells recognizing and destroying infected cells (i.e. cytotoxic (T_C) cells). T_H cells recognize a specific MHC-antigen complex on the surface of a B cell and then induce its maturation and proliferation into specific antibody-secreting (plasma) cells (see HIV). While in the thymus a T cell 'learns' during *T cell maturation* both to treat its body's Class I molecules as 'self-antigens' and to recognize as foreign a specific EPITOPE of a 'non-self antigen' when this is bound to MHC molecules. T cells do not produce antibody, but antibody production by B cells often requires T cell help. T CELL RECEPTORS (see Fig. 159) bind antigen on the surfaces of other cells only after it has been degraded or otherwise processed by that cell, and only after it has become physically associated with molecules of the MHC. Two classes of effector helper T cells can be distinguished, T_H1 and T_H2, on the basis of the CYTOKINES they secrete. T_H1 cells secrete gamma-interferon and tumour necrosis factor alpha, activating macrophages to kill microbes within their phagosomes. T_H2 cells will secrete interleukins 4, 5, 10 and 13, mainly defending against extracellular pathogens such as parasitic protozoa and worms. They also activate B cells to make most classes of antibody. *MHC restriction*

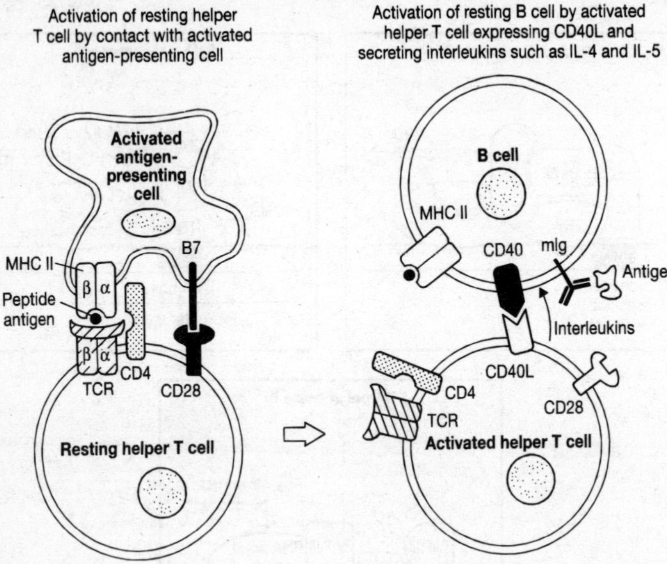

Activation of resting helper
T cell by contact with activated
antigen-presenting cell

Activation of resting B cell by activated
helper T cell expressing CD40L and
secreting interleukins such as IL-4 and IL-5

FIG. 158 *Cell contact-dependent signals in the initiation of an immune response. In the first step the resting helper* T CELL *contacts an antigen-presenting cell, such as a macrophage, presenting peptide antigens in the context of MHC Class II molecules. A second contact-dependent signal is provided by the B7 molecule, expressed by antigen-presenting cells that have themselves previously been activated and transmitted to the T cell by CD28. In the second step the activated T cell, which now expresses CD40L, provides an initial activating signal to a resting B cell via CD40. Binding of antigen to mIg on the B cell is not essential for activation but enhances the response. The interleukins, such as IL-4 and IL-5, secreted by the activated T cell enhance B cell proliferation, differentiation to a phenotype with a higher rate of antibody secretion and immunoglobulin class switching to IgG and IgE.*

refers to the process during T cell maturation in the thymus when an individual's T cells come to recognize, and be activated by, antigen only when presented in physical association with that individual's particular MHC molecules, often in conjunction with certain ACCESSORY MOLECULES. *Self-restriction* occurs when T cells preferentially recognize foreign antigens when bound to MHC molecules encountered during their own development in the thymus. *Cytotoxic T cells (T_C)* recognize tumour or virus-infected cells by their surface antigens in combination with their MHC markers, and will kill them (see CYTOLYSINS, NATURAL KILLER CELLS). Other T cells (*macrophage-activating cells*) produce LYMPHOKINES which promote macrophage activity. *Suppressor T cells (T_S)* specifically

suppress the immune response, probably through affecting antigen-presenting cells and/or through more direct interactions with T_H or B cells. See IMMUNITY, Figs. 86 and 158.

T cell receptor (TCR) T CELLS need to recognize a wide variety of antigens, doing so through the cooperation of a membrane receptor (TCR) and specific ACCESSORY MOLECULES. The TCR comprises one α and one β polypeptide (TCR-2) or one γ and one δ polypeptide (TCR-1). The genes encoding the receptor resemble those for antibodies and comprise variable and constant regions (see ANTIBODY DIVERSITY, Fig. 159) and, as in that system, production of the TCR repertoire involves both germ line diversity and

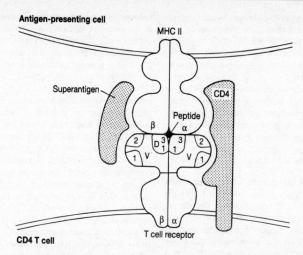

Antigen-presenting cell

MHC II

Superantigen

CD4

Peptide

β   α

2   3   3   2
  D 1   1
1   V   V   1

β   α

CD4 T cell   T cell receptor

FIG. 159 T CELL RECEPTOR *interaction with an MHC Class II molecule presenting either a peptide in the antigen-binding groove or a superantigen. 1, 2 and 3 indicate the three hypervariable regions of the T cell receptor α and β chains. The receptor interacts with a Ras protein on the inner surface of the T cell membrane to mediate further T cell response. See Fig 143.*

gene rearrangements. The β and δ chains are encoded by V, D and J segments, while α and γ chains are encoded by V and J segments only. Antigen recognition by the T cell receptors involves their binding that antigen (often a peptide fragment) when presented on another cell's surface stably bound to a protein encoded by the organism's MAJOR HISTOCOMPATIBILITY COMPLEX. Antigen-binding by the TCR activates a PROTEIN TYROSINE KINASE and generation of phosphatidyl-derived second messengers (see INOSITOL 1,4,5-TRIPHOSPHATE).

T cell restriction Alternative for MHC-restriction of T CELL.

***TDF* (testis-determining factor)** Called *TDF* in humans, *Tdy* in mice. See TESTIS.

T-DNA A DNA segment of a Ti plasmid which can be transferred by recombination into DNA of the host organism (usually a plant). See *AGROBACTERIUM*, PLASMID.

tectum (optic tectum) Dorsal region of vertebrate midbrain (see BRAIN). A correlating region integrating vestibular, tactile and visual information and initiating or modifying reflex motor responses. Except in mammals, it is the primary visual centre. See Fig. 18.

teeth See DENTITION, DECIDUOUS TEETH, PERMANENT TEETH.

tegument Animal body surface. Alternative for INTEGUMENT.

teleology Biological discourse becomes teleological when reference is made to function, purpose and design, as when structure is related to function. In recent history, vitalists, organismic biologists and others have stressed that biological systems demand a functional interpretation, but tended to explain this goal-directedness or adaptedness by a retroactive (teleological) causation peculiar to living systems 'pulling' towards perfection. As Aristotle probably intimated, and we tend to believe, the goal-directedness of the vast bulk of biological structure and behaviour is more easily interpreted as the outcome, or expression, of an organizational complexity far greater than that of normal inert objects.

It is largely attributable to DARWIN that, since the mid-19th century, an account of adaptation has been available which trivializes classical teleology and interprets contexts in which teleological terms are used as inevitable consequences of biological reproduction, accompanied as that is by selective transmission of heritable variation from one generation to the next as organisms compete for finite environmental resources. The concept of teleology is difficult to explicate without circularity. See FUNCTION, PROXIMATE FACTOR, ULTIMATE FACTOR, ADAPTATION.

Teleostei Higher bony fishes (about 20,000 living species). A fish taxon of uncertain category (e.g. suborder, series) but within Subclass ACTINOPTERYGII of Class OSTEICHTHYES (bony fishes). Probably arose from a holostean stock in the Mesozoic, but became abundant in the Cretaceous. Vertebral axis turns upwards in the tail, which is superficially symmetrical; paired fins are small; scales are usually rounded, without ganoid covering, thin and bony. Among living teleosts soft-rayed fish (e.g. salmon, herring, carp) are regarded as relatively primitive, spiny-rayed fish, which include the ACANTHOPTERYGII (e.g. perches), as more progressive. See FINS.

telium A structure in which are produced teliospores of the rust fungi (BASIDIOMYCOTA, Order Uredinales). Teliospores are two-celled, dikaryotic, overwintering spores that will germinate the following spring; however, prior to germination, the two haploid nuclei fuse to form a diploid nucleus; meiosis takes place upon germination in the two short cylindrical basidia that emerge from the two cells of the teliospore. Septa form between the resultant nuclei, which migrate into the sterigmata, and basidiospores develop. See AECIUM, UREDIUM, PYCNIDIUM.

telocentric Of chromosomes, with the CENTROMERE at one end (terminal).

telolecithal Of eggs (e.g. amphibian) with yolk more concentrated at one end and a marked POLARITY along the axis of yolk distribution.

telome One of the distal branches of a dichotomized plant axis; a morphological unit in a primitive vascular plant. Telomes are regarded as the basic units from which the diverse types of leaves and sporophylls of vascular plants have evolved.

telomere Originally, term used by cytogeneticists for an end of a eukaryotic chromosome. Such ends lacked any tendency to fuse together spontaneously (i.e. they were 'non-sticky'). They also prevent chromosomes from being degraded. Telomeres appeared never to become incorporated within the chromosome (e.g. by terminal inversion). It is now known that they are examples of simple tandem DNA repeat sequences (regular or irregular), replicated by a specific *telomerase* enzyme (in a $5' \rightarrow 3'$ end direction), which is a specialized reverse transcriptase providing the template for new telomeres. Telomerase is an enzyme composed of both RNA and protein, and as such may be an evolutionary link between systems using RIBOZYMES only and those using purely protein enzymes (SEE ORIGIN OF LIFE). The ability to replicate without loss the $5'$-ends of each DNA strand in the chromosome is a requirement, given the normal machinery of semiconservative DNA replication – which works in a $5' \rightarrow 3'$ direction – and the need of cellular DNA POLYMERASES for an RNA primer. It is eukaryotic chromosome ends that are described as telomeres, in contrast to the various termini of linear viral, non-nuclear plasmid and mtDNA genomes. Telomeric DNA is regionally organized into non-nucleosomal chromatin (the *telosome*) which can interact with the envelope of the NUCLEUS and probably facilitates chromosome pairing during early MEIOSIS. A telomere-binding protein is required for segregation of homologous chromosomes during first meiotic prophase, its absence resulting in meiotic non-disjunction. Telomeres are composed of HETEROCHROMATIN and may be involved in transcriptional regulation and nuclear architecture. Human telomeres have been cloned in YEAST ARTIFICIAL CHROMOSOMES. Erosion of telomeric DNA (perhaps 50–100 nucleotides from each of its telomeres per division) is known to be a feature of all human somatic cells, first noticed in blood cells, but much less so in germ line cells. Somatic cells usually lack telomerase activity, and when telomeres reach a critically short length its absence causes cells to undergo *replicative cell senescence*, in which they stop dividing and develop chromosomal instabilities. Such a mechanism could have evolved as a way of preventing uncontrolled proliferation of abnormal cells, as in CANCER; or it might serve to limit the total sizes of tissues and organs, being a 'measuring stick' for counting the number of cell divisions undergone (the 'telomere-clock' model; see BIOLOGICAL CLOCK). Although there is experimental support for this from tissue culture, not all cells in a tissue stop dividing when they lack functional telomeres. Human telomeres comprise up to 10,000 repeats of the sequence GGGTTA (i.e. they are G-rich) and cells which have had artificial additions to this repeat sequence miss their 'senescence cue' and continue to divide. See PRION.

telophase Terminal stage of MITOSIS or MEIOSIS, when nuclei return to interphase.

telson Hindmost segment of arthropod abdomen, forming sting of scorpions and part of tail fan of lobsters and their allies. In insects present only in the embryo.

temperate Of BACTERIOPHAGE that can insert its genome into that of its host so that it is replicated with it. See also EPISOME.

temperature coefficient See Q_{10}.

temperature-sensitive mutant Mutant form expressed only under certain temperature conditions, the wild-type phenotype being expressed at others. Normal function occurs within the permissive temperature range (commonly low), malfunction within the restrictive temperature range (commonly high). Many *cdc* mutations (see CDC GENES) are temperature-sensitive. Commonly due to heat-labile gene product. See CHAPERONES.

template In nucleic acids, the strand used by a polymerase to build a new and complementary polynucleotide strand.

temporalis muscle Muscle originating on temporal bone of mammalian skull, inserting on the dentary; elevates and retracts mandible and assists in its side-to-side movements. Compare MASSETER MUSCLE. See SAGITTAL CREST.

tendon Cord or band of relatively inextensible vertebrate connective tissue attaching muscle tissue to another structure, often bone. Consists almost entirely of closely packed collagen fibres with rows of fibroblasts between. With high tensile strength and coefficient of elasticity.

tendril Stem, or part of leaf, modified as a slender branched or unbranched thread-like structure; used by many climbing plants for attachment to a support, either by twining around it (e.g. pea, grape vine) or by sticking to it by an adhesive disc at the tip (e.g. virginia creeper). See HAPTOTROPISM.

teratogens Substances, e.g. thalidomide, which cross the placenta and damage the developing foetus. See POISONS, TOXIN.

teratoma Growth of cell mass from an unfertilized egg within mammalian ovary (or of germ cells in the male's testis) which becomes disorganized and uncontrolled. Many differentiated cell types may be represented, all mixed up. Teratomas may become malignant (*teratocarcinomas*). See PARTHENOGENESIS.

tergite A SCLERITE composing an arthropod TERGUM.

tergum Thickened plate of cuticle on dorsal side of arthropod segment.

termination See PROTEIN SYNTHESIS.

termite See ISOPTERA.

terpene Unsaturated LIPID consisting of multiples of ISOPRENE (a 5-carbon hydrocarbon), sesquiterpenes having three such units. Includes vitamins A, E and K, carotenoids and many odorous substances in plants. *Squalene* (see CHOLESTEROL) is another. Rubber and guttapercha are polyterpenes. See Fig. 102.

territoriality The establishment by an animal or animals of an area from which other individuals are partially or totally excluded.

territory Area or volume of habitat occupied and defended by an animal, or group of animals of same species. Territories commonly arise during or prior to breeding, with an important spacing role in many vertebrate and arthropod populations. Compare HOME RANGE.

Tertiary GEOLOGICAL PERIOD, lasting from abut 70–3 Myr BP. With Quaternary it comprises the Cenozoic era. Often called the Age of Mammals on account of the radiation which occurred during it. Birds dominated the air throughout, almost all major modern groups being present early in the period.

testa Seed coat. Protective covering of embryo of seed plants, formed from the integument(s); usually hard and dry.

test cross Cross between an individual of unknown genotype (or one heterozygous at a number of loci) and another homozygous for recessive characters at loci of interest. Offspring ratios are then used to deduce the genotype of the unknown, and/or the linkage relationships of its heterozygous loci. Not necessarily a BACKCROSS. See CHROMOSOME MAPPING.

testicular feminization Syndrome of some mammals, including humans. Due to presence of the *Tfm* allele at an X-linked locus, individuals chromosomally male (XY) are phenotypically female since male genital (Wolffian) ducts fail to develop, lacking cell membrane receptors capable of binding androgens (testosterone and its derivatives). Uterus absent or rudimentary; gonads usually abdominal or inguinal testes. For role of *Sry* gene, see TESTIS.

testis Sperm-producing organ of male animal; in vertebrates producing also 'male' sex hormones (androgens) and derived in part from *genital ridges* of coelomic epithelium in dorsal abdomen adjacent to the mesonephros, forming the testis cortex. There is a 35 Kb sex-determining region (*SRY* in humans, *Sry* in mice) on the Y-chromosome expressed briefly in the Sertoli cell precursors, which interact with other gonadal somatic cells to cause testis production and block ovary development (see MÜLLERIAN INHIBITING SUBSTANCE). Germ cells then develop as egg or sperm upon signals from somatic cell neighbours. Subsequent male sexual differentiation is a consequence of the hormonal products of the testis. It is very likely that the testis-specific *Sry* gene product is a TRANSCRIPTION FACTOR. Amoeboid *primordial germ cells* invade the ridges from the endoderm of the yolk sac and as the ridges hollow out, primitive sex cords develop from mesenchyme to form *seminiferous tubules* (see Fig. 160), to which primordial germ cells attach. Tubules are separated by septa and discharge sperm into the EPIDIDYMIS via ducts (*vasa efferentia*). Each testis is attached to the abdominal wall by a ligament (the *gubernaculum*) and under testosterone influence is drawn into the scrotum through a canal (the inguinal canal) as body elongation occurs. Endocrine cells (*interstitial cells*, or *Leydig cells*) between the tubules secrete androgens, most potent being TESTOSTERONE. See MATURATION OF GERM CELLS, WOLFFIAN DUCT, OVARY.

testis-determining factor (TDF) Protein produced by the *SRY* gene on the short arm of the Y chromosome in male human foetuses and by the *Sry* gene in mice. Triggers the gonadal ridges to begin developing into testes. Other autosomal and X-linked gene products are required for complete testis development. The gene sequence is very variable among mammals, apart from an included conserved HMG box which encodes a DNA-binding nuclear factor of 79 residues and shared by the SOX family of transcription factors (see *SOX* GENES). Most mutations in SRY that cause SEX REVERSAL are located within it. See MÜLLERIAN-INHIBITING HORMONE, TESTIS.

testosterone The main vertebrate androgen ('male' sex hormone). Anabolic steroid produced largely by the Leydig cells of the testes from cholesterol or acetyl coenzyme A, and in smaller quantities by the adrenal cortex. Responsible for maintaining testes and for male growth spurt at puberty by stimulation of longitudinal bone growth and deposition of calcium. Closes epiphyses

(a)

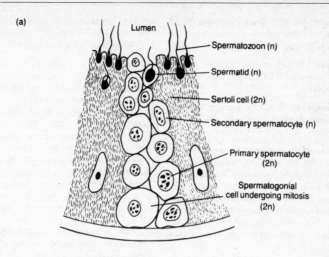

Lumen

Spermatozoon (n)

Spermatid (n)

Sertoli cell (2n)

Secondary spermatocyte (n)

Primary spermatocyte (2n)

Spermatogonial cell undergoing mitosis (2n)

(b)

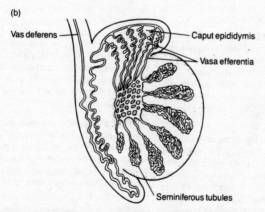

Vas deferens

Caput epididymis

Vasa efferentia

Seminiferous tubules

FIG. 160 (a) *Section through a vertebrate seminiferous tubule indicating the various cell stages involved in the production of spermatozoa.* (b) *Section through mammalian* TESTIS.

a few years after puberty, arresting growth. Promotes protein synthesis (hence muscle development), sexual behaviour and spermateliosis (SEE MATURATION OF GERM CELLS). In humans promotes hair growth in pubic, axillary, facial and chest regions and enlargement of laryngeal cartilage and voice deepening.

tetanus (1) Disease caused by toxin from anaerobic spore-forming bacterium *Clostridium tetani*. Increasing muscle spasms make opening of jaw difficult (hence lock-jaw). Progressive convulsions may be fatal, often by asphyxia or exhaustion. (2) Sustained muscle contraction resulting from nervous stimulation at a rate too great to allow muscle relaxation. Results from over abundance of calcium ions in sarcoplasm. See MUSCLE CONTRACTION.

tetrad Group of four haploid cells or nuclei produced by meiosis, while they are adjacent.

tetraploid Of nuclei, cells, individuals, having four times ($4n$) the haploid number of chromosomes. See POLYPLOID.

tetrapod Four-limbed vertebrate (see Fig. 161). Includes amphibians, reptiles, birds and mammals. Secondary loss of one or both pairs of limbs, or modification into wings, flippers, etc., has occurred in some taxa. Footprints apparently derived from true tetrapods date from at least the middle Upper Devonian. See PENTADACTYL LIMB, RHIPIDISTIA.

tetraspores In some red algae, the four spores formed through meiosis in a tetrasporangium. Borne usually on a free-living and diploid sporophyte.

tetrodotoxin (TTX) Highly specific puffer fish toxin, blocking voltage-gated Na^+ channels in striated muscle membrane and causing paralysis. Some marine dinoflagellates make a toxin (saxitonin) with the same effect. Valuable tools for studying muscle membranes and counting channel density.

TGFs (transforming growth factors) Some members of the TGF-β family of GROWTH FACTORS are signalling molecules in many developmental processes in many different organisms, and in adults have a role in bone reconstruction, wound healing and neuronal plasticity. Evidence seems to indicate that they induce secondary signals over short distances. If they act as diffusible MORPHOGENS in embryos, they do so at levels below those currently detectable. One TGF, myostatin, may be a muscle-specific growth inhibitor. See ACTIVIN, INHIBINS.

thalamus (Bot.) Receptacle of flower. (Zool.) Part of telencephalon of vertebrate forebrain, forming roof and/or lateral walls of third ventricle and composed largely of grey matter. Organized into nuclei relaying sensory information from spinal cord, brainstem and cerebellum to cerebral cortex. See HYPOTHALAMUS and Fig. 18.

thalassaemias Inherited disorder whereby the two globins (α and β) involved in haemoglobin production are synthesized in unequal amounts. Normally presents as a heterozygous disease in the Mediterranean region and Asia. In parts of Italy, over 10% of the population are carriers of the α-globin gene and the α^+-thalassaemias are the commonest known human genetic disorders, affecting up to 80% of some populations. This homozygous condition is geographically associated with high incidence of childhood MALARIA infection by the nonlethal *Plasmodium vivax*, which may then act as a 'natural vaccine', inducing cross-species protection against subsequent severe *P. falciparum* malaria. In β^+-thalassaemia, the β-globin is in low amounts, sometimes because too little of the β-globin mRNA precursor leaves the nucleus for the cytoplasm, sometimes because of a mutation of codon 39 to a termination codon (SEE PROTEIN SYNTHESIS). In β°-thalassaemia, no β-globin is made at all. When homozygous, thalassaemia causes severe haemolytic anaemia and is usually fatal in childhood. Se RFLP, GENETIC COUNSELLING, SICKLE-CELL ANAEMIA.

Thallophyta In older classifications, a Division of plant kingdom housing prokaryotes and simple plant-like eukaryotes possessing simple vegetative bodies (thalli). Included bacteria and cyanobacteria, algae, fungi, lichens and slime fungi. Term now largely abandoned.

thallus Simple, vegetative plant-like body, lacking differentiation into root, stem and leaf. Unicellular or multicellular, comprising branched or unbranched filaments; or more or less flattened and ribbon-shaped. Monostromatic when one cell thick.

T haplotype An extended genetic entity on chromosome 17 of the mouse (*Mus musculus*) which, compared to the normal homologue, contains four non-overlapping inversions within which only infrequent crossing-over occurs. Most contain recessive

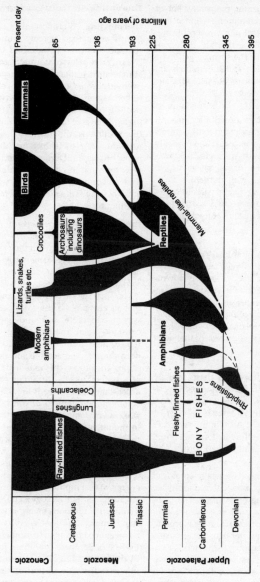

FIG. 161 *Diagram indicating approximate* TETRAPOD *phylogenies. The width of a group's entry indicates its approximate species abundance.*

lethal mutations acting prenatally. But in the heterozygous condition (+/t) in males, t is transferred to 99% or more of offspring: t-bearing spermatids always 'kill' their meiotic partners (so in the t/t male genotype they kill each other and the animals are sterile). See SELFISH DNA/GENES, ABERRANT CHROMOSOME BEHAVIOUR.

Thecodontia 'Stem' order of ARCHOSAURS, arising in early Triassic and extinct by its close, giving rise to DINOSAURS and probably birds (see AVES). South African fossil *Euparkeria* is representative, having several DIASPID characteristics, but with numerous small teeth on both pterygoid and palatine bones and other primitive features absent from later archosaurs. Probably at least partially bipedal.

thelytoky Form of animal PARTHENOGENESIS. Males are either very rare, effectively without a genetic role, or entirely absent. There are about 1,000 thelytokous animals from a large number of taxa, but only approx. twenty-five of these are vertebrates (four fish, two salamanders and about twenty lizards). See *WOLBACHIA* for thelytoky induced by infection.

theory Explanatory hypothesis, usually firmly founded in observation and experiment. They tend to have more consequences than do hypotheses, being of wider scope, and are tested by examining whether their consequences (predictions) are borne out by observation and experiment.

therapeutic drugs Clinically employed DRUG substances, used either prophylactically (preventing onset of disease) or in reducing disease symptoms. Include ANTIBIOTICS; non-narcotic ANALGESICS, e.g. aspirin, ibuprofen, acetaminophen; local ANAESTHETICS, e.g. procaine (Novocain); antiepileptic drugs, e.g. clonazepam; antiparkinsonian drugs, e.g. L-dopa; drugs for spasticity, e.g. baclofen; anti-MALARIA drugs; VACCINES, etc. The vectors used in attempts at GENE THERAPY might also be termed drugs. See PSYCHOACTIVE DRUGS.

Therapsida See MAMMAL-LIKE REPTILES.

thermocline Stratification of lakes and oceans with respect to temperature during summer months, characterized by upper layer of more or less uniformly warm, circulating, fairly turbulent water (*epilimnion*) overlying deeper, cold and relatively undisturbed region (*hypolimnion*). Between the two is a region of steep temperature drop (*metalimnion*, or *discontinuity layer*). The thermocline is the plane or surface of maximum rate of temperature drop with respect to depth. During autumn, epilimnetic temperatures decrease, water density rises and mixing (*overturn*) of water and nutrients results in a more even temperature distribution.

thermodynamics Classical thermodynamics deals with *closed systems* – those which do not interact with their surroundings in terms of energy or matter. By contrast, living systems (in so far as they are alive) are *open systems*, interacting in both ways.

The first law of thermodynamics states that in any process the total energy of a system and of its surroundings remains constant, even though energy may be transformed from one form to another (transduced). The second law states that during any process the combined *entropy* (S) of a system and its surroundings (its disorder, randomness) tends to increase until equilibrium is attained, at which point no work can be done. The tendency for entropy in the universe to be maximized could be thought of as the 'driving force' of all chemical processes. Living cells exist in states of thermodynamic non-equilibrium, and in different steady states, in which rates of energy/matter input from the surroundings equal their output. Biology is concerned primarily with reactions taking place at constant temperature (*isothermally*) but heat changes accompany even isothermal reactions. If under these conditions heat is lost to the surroundings the system is said to lose *enthalpy* (H) and the reaction is said to be exothermic. Absorption of heat characterizes endothermic reactions. The form of energy capable of doing work in a system under constant temperature and pressure is its *free energy* (G), whose value is the key to predicting the direction of a chemical reaction. These relationships may be summarized:

$$\Delta G = \Delta H - T\Delta S$$

where ΔG = the free energy change of the system, ΔH = its change in enthalpy, T = absolute temperature and ΔS = its change in entropy. Reactions are *exothermic* if ΔH is negative, *endothermic* if ΔH is positive; they are *exergonic* (and may do work) if ΔG is negative, and *endergonic* (not doing work) if ΔG is positive. Spontaneous reactions are characterized by a loss of free energy (exergonic). Endergonic reactions cannot proceed spontaneously.

The free energy change of any reaction at constant temperature and pressure involves a fixed constant for that reaction known as the *standard free energy change* of the reaction (ΔG_o), giving the loss of free energy when the reaction is allowed to go to equilibrium starting with certain standard conditions – in particular, when all reactants and products are present at 1.0 molar concentrations. It is a measure of the difference between the sum of free energies of products and of reactants and is related to the equilibrium constant K (since $\Delta G_o = -RT \ln K$). Free energy of a reaction varies with concentrations of reactants and products, affecting its probability of occurrence. A chemical reaction only occurs if its ΔG is negative in sign, and the maximum amount of work it then does equals this decrease in free energy.

thermogenesis Processes leading to heat release within the body. Apart from general heat release (e.g. by liver, brain, heart) through normal activity, they include shivering (involuntary thermogenesis) and non-shivering modes. Human babies cannot shiver, but non-shivering thermogenesis by their brown ADIPOSE TISSUE compensates for this. Shivering occurs in large insects prior to flight and in birds and mammals. It involves alternate contractions of antagonistic muscles in response to cold. See THERMOREGULATION, SPECIFIC DYNAMIC EFFECT.

thermoluminescence (TL) See RADIOMETRIC DATING.

thermonasty (Bot.) Response to a general, non-directional, temperature stimulus; e.g. opening of crocus and tulip flowers with temperature increase.

thermoneutral zone The range of environmental temperatures for an endotherm over which it has to exert minimum metabolic effort to maintain a constant body temperature. See BASAL METABOLIC RATE.

thermophilic (Of microorganisms, e.g. *Thermus*) with optimum growth temperature (measured by generations per hour) in the 55–65°C range. Compare MESOPHILIC, PSYCHROPHILIC.

thermoregulation Processes adjusting heat generation to heat loss from the body; such as behavioural control (e.g. posture, moving to and from burrow); control of blood flow to the skin, sweating, panting; or ACCLIMATIZATION. Vertebrate core thermoreceptors occur in the hypothalamus (e.g. mammals) or spinal cord (e.g. birds) and surface 'hot' and 'cold' exteroceptors occur in the skin. Fall in core temperature, detected principally by the hypothalamus (the 'thermostat'), initiates shivering and non-shivering THERMOGENESIS, vasoconstriction (by sympathetic stimulation) in the surface tissues and contraction of erector pili muscles, raising feathers and fur, so trapping a thicker layer of heat-insulating air at the surface of the body. It may also induce thyroxine release, raising BASAL METABOLIC RATE (see THYROID HORMONES), a response also brought about by sympathetic stimulation of ADRENALINE release. Rise in core temperature shuts down these hypothalamic responses and may invoke evaporative heat loss by sweating and/or panting. Humans can increase the skin's evaporative water loss from 100 mg m^{-2} min^{-1} to 23,000 mg m^{-2} m^{-1} by sweating. Panting is adopted by several mammals and birds, and many of the latter can gular flutter, when the moist throat region is fluttered. Some large long-legged birds even urinate on their legs when heat stressed.

In order to reduce the temperature gradient between the body surface and the environment, some endotherms allow the temperature of peripheral tissues to drop below the body's core temperature. This reduces heat loss, but at the risk of freezing the body surface. Penguin flippers and whale tail flukes have large surface areas which would lose heat rapidly were it not for countercurrent heat exchange between arterial and venous blood in which a central 'warm' artery is surrounded by a network of 'cooler' veins with blood flowing in the opposite direction returning from the skin, the effect of heat exchange being to conserve as much heat as possible within the core circulation. Such a system has been located in the human arm. See TORPOR.

therophytes Class of RAUNKIAER'S LIFE FORMS.

theropods Bipedal carnivorous dinosaurs (see SAURISCHIA). Many new fossils support the view that birds are their descendants. A furcula (wishbone) has been discovered in a theropod believed to be close to birds and typical theropod characters, such as an enlarged claw on digit 2 of the foot, have been found in early birds. Fossils of an Early Cretaceous theropod from China, the chicken-sized *Sinosauropteryx prima*, have branched filaments in the integument. Some suggest that these may have been proto-feathers.

thiamine (vitamin B) Vitamin precursor of coenzyme *thiamine pyrophosphate* (TPP), serving enzymes transferring aldehyde groups during decarboxylation of α-keto acids (such as pyruvate in mitochondria), and formation of α-ketols. Deficiency causes beri-beri in man, and polyneuritis in birds. See VITAMIN B COMPLEX.

thigmotropism See HAPTOTROPISM.

thinning line The line on a plot of log (mean individual weight) against log (density) along which SELF-THINNING populations of a sufficiently high biomass tend to progress and beyond which they cannot pass.

thoracic duct Main mammalian lymph vessel, receiving lymph from trunk (including lacteals) and hindlimbs and running up the thorax close to vertebral column, discharging into left subclavian vein (humans), or another major anterior vein. Often paired in fish, reptiles and birds.

thorax (1) In terrestrial vertebrates, part of body cavity containing heart and lungs (i.e. the chest); in mammals, clearly separated from abdominal cavity by diaphragm. (2) In arthropods, body region between head and abdomen; often not clearly separable, or tagmatized. In adult insects comprises three segments, each typically with a pair of walking legs and one or two of them commonly bearing a pair of wings. See TAGMA.

thread cell See CNIDOBLAST.

threadworms See NEMATODA.

threonine kinases See PROTEIN KINASES.

threshold Critical intensity of a stimulus below which there is no response by a tissue (e.g. nerve, muscle).

thrombin Proteolytic enzyme derived from prothrombin during sequence of reactions involved in BLOOD CLOTTING, converting fibrinogen to fibrin.

thrombocytes See PLATELETS.

thromboplastin (thrombokinase) Phospholipid and protein, liberated from PLATELETS and damaged tissues on wounding, initiating the CASCADE of reactions resulting in BLOOD CLOTTING.

thrombosis Formation of a blood clot too easily, or otherwise inappropriately, in an unbroken blood vessel. Such a clot is a *thrombus*. Clinical treatment may include injection of thrombolytic (clot-dissolving) substances such as STREPTOKINASE or tissue PLASMINOGEN ACTIVATOR (t-PA). See BLOOD CLOTTING, CORONARY HEART DISEASE.

thrombus See THROMBOSIS.

thylakoid Flattened membranous vesicle containing CHLOROPHYLL pigments; site of photochemical reactions in PHOTOSYNTHESIS. Vary in form and arrangement between different groups of organisms. Their membranes have a high concentration of unsaturated fatty acids, making them very fluid and protecting them against damaging cold spells, thereby increasing the thermotolerance of photosynthesis beyond 35°C.

thymidine Nucleoside comprising the base THYMINE linked to ribose by a glycosidic bond.

thymine Pyrimidine base of the NUCLEOTIDE thymidine monophosphate, a monomer of DNA but not RNA. Also a component of thymidine di- and triphosphates.

thymus gland Bilobed vertebrate organ containing primary LYMPHOID TISSUE, usually situated in the pharyngeal or tracheal region. Originates from gill pouches (in mammals, the third pouches) or gill clefts. In mammals lies in thorax (mediastinum), the gill pouch cells mixing with lymphocytes. Attains maximum size at puberty, slowly diminishing afterwards. Responsible for maturation of T CELLS, and possibly with an endocrine function. The thymus produces various hormones (e.g. thymosin) promoting T cell proliferation and maturation. As the mammalian body ages, more thymic tissue is replaced by adipose tissue and thymic hormones wane, reducing T cell production and immunological competence.

thyroglobulin (TGB) Transport protein for thyroid hormones produced by thyroid gland. Stored there and carried in blood plasma.

thyroid gland Vertebrate endocrine gland, either single (e.g. in mammals) or paired (e.g. most amphibians and birds), in neck region. Derived from endoderm of pharyngeal floor; probably homologous with ENDOSTYLE of cephalochordates, etc. Comprises sac-like *thyroid follicles* in which THYROID HORMONES (thyroxine and triiodothyronine) are stored, colloidally complexed with THYROGLOBULIN, surrounded by parafollicular cells (secreting CALCITONIN). Inadequate dietary iodine causes enlargement of the thyroid (goitre) due to raised TSH secretion (see THYROID STIMULATING HORMONE).

thyroid hormones In addition to the hormone CALCITONIN, most vertebrate thyroids produce small iodine-containing hormones as derivatives of the amino acid tyrosine. These seem to act like steroid hormones, passing through cell membranes and binding to NUCLEAR RECEPTORS, the complex then binding specific chromosomal DNA sites to bring about GENE EXPRESSION. See Fig. 162.

Thyroxine (T_4) is produced in greater quantities than *triiodothyronine* (T_3), but is about a quarter as potent. About one third of circulating T_4 is converted to T_3 in lungs and liver. Both are homeostatically controlled, and if circulating hormone levels or blood temperature are too low, the hypothalamus releases thyrotropin releasing factor (TRF) into the anterior pituitary portal system, stimulating release of TSH (THYROID STIMULATING HORMONE), which, in turn, releases bound and unbound hormones from the thyroid. These increase BASAL METABOLIC RATE (especially in the liver) and oxygen consumption, uncoupling electron transport from ATP synthesis in liver mitochondria. Respiration then releases heat into the blood (raising its temperature) rather than generating ATP.

They have general catabolic effects upon fat and carbohydrate, promoting gluconeogenesis; but they stimulate protein synthesis. Are important in tissue growth and development, and in metamorphosis of amphibian tadpoles. Deficiency in growing vertebrates causes reduced nerve and organ development; presence promotes neural activity, heart rate, blood pressure. Dietary intake of iodine (in humans 100–400 µg day^{-1}) is required for their normal production.

thyroid stimulating hormone (TSH, thyrotropin) Protein product of anterior pituitary gland, stimulating thyroid production and release of thyroxine (T_4) and triiodothyronine (T_3). Secretion influenced by TSH-releasing factor (TRF) from the hypothalamus. Negative feedback occurs between levels of (a) (T_4) and (T_3) and (b) TSH and TRF in blood. Fall in T_3 and T_4 raises outputs of TRF and TSH; rise of the last two shuts off TRF and TSH release.

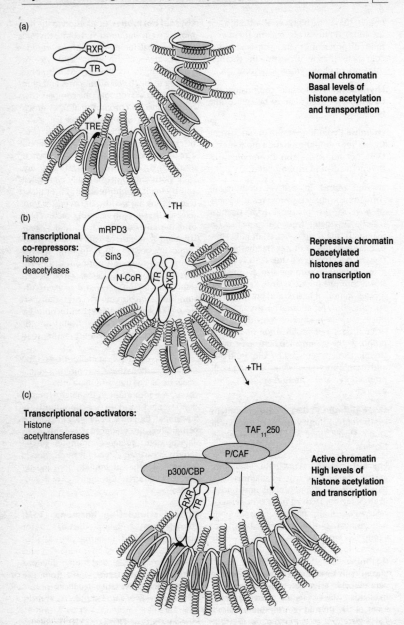

(a)

Normal chromatin
Basal levels of
histone acetylation
and transportation

-TH

(b)

**Transcriptional
co-repressors:**
histone
deacetylases

Repressive chromatin
Deacetylated
histones and
no transcription

+TH

(c)

Transcriptional co-activators:
Histone
acetyltransferases

Active chromatin
High levels of
histone acetylation
and transcription

thyrotropin-releasing hormone Hypothalamic neuropeptide stimulating thyrotropin (THYROID STIMULATING HORMONE) and prolactin release by the anterior pituitary.

thyroxine (T₄) See THYROID HORMONES.

Thysanoptera Thrips. Order of minute exopterygote insects, most with piercing mouthparts for sucking plant juices (and occasionally aphids). Have prepupal and pupal stages despite being exopterygote. Of economic importance as transmitters of disease, and ecologically in studies of population regulation.

Thysanura Silverfish. See APTERYGOTA.

tibia (1) Shinbone; the more anterior of the two long bones below knee of tetrapod hindlimb, the other being the fibula. See PENTADACTYL LIMB. (2) Fourth segment from base of an insect's leg.

ticks See ACARI.

tight junction See INTERCELLULAR JUNCTION, SEPTATE JUNCTION.

tiller (Of grasses) a side shoot arising at ground level.

TIRs (1) Terminal inverted repeat DNA sequences, characteristic of Class II TRANSPOSABLE ELEMENTS. (2) Toll/interleukin-1 receptor/resistance signalling domains, found on several R proteins (see RESISTANCE GENES) and in proteins in host defence pathways in both the plant and animal kingdoms. They govern the dimerizations of TOLL-LIKE RECEPTORS (TLRs) and associations between TLRs and TIR domain-containing adaptors. TIR domains allow molecules such as MyD88 to associate with the intracellular tail of a TLR and so connect these cell surface receptor molecules to other molecules in the cell that propagate the signal further.

tissue Association of cells of multicellular organism, with a common embryological origin or pathway and similar structure and function. Often, cells of a tissue are contiguous at cell walls (plants) or cell membranes (animals) but occasionally the tissue may be fluid (e.g. blood). Cells may be all of one type (a *simple tissue*, e.g. squamous epithelium, plant parenchyma) or of more than one type (a *mixed tissue*, e.g. connective tissue, xylem, phloem). Tissues aggregate to form organs. See ACELLULAR.

tissue culture (explantation) A technique for maintaining fragments of animal or plant tissue or individual cells from the organism. A sterile bathing medium (commonly RINGER'S SOLUTION in the case of animal tissues or cells or a culture medium for plant tissues or cells) surrounds the cells or tissues, while appropriate temperature, pH and nutrient levels are maintained and waste products removed. With respect to plants, tissue culture is one of the most important procedures in BIOTECHNOLOGY, and this is feasible only because of an understanding of how hormones function, and regulate development. The greatest impact of tissue culture is in the area of plant multiplication (clonal propagation). There are three different methods of tissue culture propagation, each taking advantage of the control of development that can be accomplished with plant hormones, and include: (a) SOMATIC EMBRYOGENESIS: (b) shoot multiplication, where growing shoot tips are exposed in culture to a high enough concentration of CYTOKININ, not only to sustain shoot growth, but also to stimulate lateral bud development. Commonly used for multiplying ferns, orchids and woody plants; and (c) callus and/or protoplasts, where cytokinin is used to promote formation of growing shoots in unorganized masses of parenchymatous tissue, which is known as

FIG. 162 *Three-step model for transcriptional regulation by the* THYROID HORMONE *receptor:* (a) *precise rotational positioning of the DNA sequence containing the thyroid-hormone response element (TRE) on the surface of the histone octamer, allowing the thyroid-hormone receptor (TR/RXR) access to chromatin;* (b) *the unliganded receptor recruits a deacetylase complex to augment repression;* (c) *subsequent addition of thyroid hormone (TH) leads to recruitment of acetyltransferases that can disrupt repressive histone–DNA interactions.*

callus; once shoots form they can be treated with AUXIN to initiate root formation. Protoplasts can also be used for plant regeneration, after cell walls have been removed by enzymatic digestion, in a technique known as protoplast fusion; protoplasts from differing plants are allowed to fuse together to produce a hybrid cell. The first plant developed using this technique was a hybrid of tobacco. Intergeneric hybrid plants have also been developed using protoplast fusion (e.g. between the potato and tomato).

tissue fluid (interstitial fluid) Fluid derived from blood plasma by filtration through capillaries in the tissues. Differs from blood chiefly in containing no suspended blood cells and in having lower protein levels. Bathes tissue cells in appropriate salinity and pH and acts as route for reciprocal diffusion of dissolved metabolites between them and blood. Most of the water filtered from capillaries is reabsorbed by them osmotically; that which remains (along with some solutes) enters the LYMPHATIC SYSTEM, when it is termed *lymph*. See OEDEMA.

titre Usually, a relative measure of the amount of antibody in a fluid, usually serum. The greatest dilution capable of producing a particular detectable antibody–antigen reaction is its titre. This assay technique can be used to compare titres in different samples. See PRECIPITIN.

T killer cells See T CELLS.

TMV Tobacco mosaic virus. See VIRUS.

TNF See TUMOUR NECROSIS FACTOR.

TNFRs Tumour necrosis factor receptors.

toadstool Common name for fruiting bodies of fungi (other than mushrooms) belonging to the Agaricaceae (BASIDIOMYCOTA). Often wrongly assumed that all are poisonous, although some are.

tocopherol Vitamin E. See VITAMINS.

tolerance (1) Tolerance to a drug involves a progressively decreasing responsiveness to it. Either there is increasing synthesis of drug-metabolizing enzymes, or receptors

(e.g. in brain) adapt to the drug's continued presence. (2) See IMMUNE TOLERANCE.

Toll Invertebrate transmembrane receptor, encoded by a maternal gene, and first identified in *Drosophila*, where it is a major player in the innate immune system. On binding its ligand (of microbial origin) it initiates a signalling pathway leading to production of antimicrobial compounds (see ANTIMICROBIAL PEPTIDES) and is probably a conserved component of an ancient animal defence mechanism (see IMMUNITY, TOLL-LIKE RECEPTORS). It seems to have been co-opted in evolution to serve in dorsoventral patterning where its ligand in early development is the protein spätzle, which once bound causes another maternal gene product, the transcription factor Dorsal, to enter nearby nuclei and organize the dorsoventral axis of the embryo.

Toll-like receptors (TLRs) Vertebrate receptors on the surfaces of macrophages and dendritic cells (APCs, see ANTIGEN-PRESENTING CELLS) and epithelia lining the gut and nasal passages and involved in both innate and adaptive immune responses. When they detect the presence of pathogens, they trigger an intracellular signalling pathway culminating in various immune responses. TLRs transduce the broadly distinct pathogen-specific molecular patterns on viruses, bacteria and fungi to elicit intracellular signals which are either dependent or independent of the Toll/IL-1 receptor (TIR) domain-containing adaptor proteins MyD88, TIRAP or TRIF and induce a core set of stereotyped responses, especially INFLAMMATION, by activating the transcription factor NFκB (which regulates the genes for the inflammatory CYTOKINES tumour necrosis factor-α and interleukin-1), mitogen-activated protein kinases (MAPKs) and c-Jun N-terminal kinase (JNK). However, TLRs can also induce specific effector responses to particular microbial infections, and differences in gene expression result when different TLRs are bound.

The first mammalian TLR to be identified was TLR4, whose activation is required for B7 and cytokine production by APCs. TLR3 detects the presence of double-stranded

RNAs (e.g. some viral genomes) and similar structures and is vital in warding off viruses. The protein TRIF (or Trif) transduces signals from activated TLR3 but is also essential for transducing signals from the important but non-mutable bacterial ligand LIPOPOLYSAC-CHARIDE, via TLR4, for beta-interferon production. TLRs are characterized in animals by LEUCINE-RICH-REPEATS. Some of the plant resistance genes (*R* genes) have homologues with Toll-like receptor genes (SEE PLANT DISEASE AND DEFENCES).

Deriving their name from the earlier discovered invertebrate receptor, TOLL, the several kinds of TLR recognize a limited, but highly conserved and common, set of pathogen-associated molecular patterns carried on pathogen surfaces, likened to recognition 'bar-coding'. Certain DNA motifs, such as unmethylated CpG islands – more common in bacterial than in mammalian DNA – are also targets. In this way, macrophages and dendritic cells effectively classify the invader by these signatures, enabling them to initiate appropriate transcriptional responses.

tone (tonus) See MUSCLE CONTRACTION.

tonic receptor Slow-adapting RECEPTOR, continuing to respond with action potentials throughout unchanging stimulus. Conveys information about the duration of a stimulus. Compare PHASIC RECEPTOR.

tonoplast (Bot.) The membrane that surrounds the VACUOLE in plant cells. It plays an important role in the active transport of certain ions into the vacuole, and their retention there.

tonsils Masses of non-encapsulated LYMPHOID TISSUE in mouth or pharynx of tetrapods, lying close underneath mucous membrane. Humans have a pair of palatine tonsils at the junction of mouth and pharynx, and a single pair of pharyngeal tonsils (adenoids) at the back of the nose. Sites of lymphocyte production. See PEYER'S PATCHES.

tonus (tone) See MUSCLE CONTRACTION.

tooth A modified part of the vertebrate dermal skeletal materials occupying the oral cavity, usually attached to the jaws.

Although cyclostomes have horny tooth-like structures, teeth are unknown from fossil and living lower chordates and agnathans.

The exposed portion of a tooth is covered in ENAMEL, beneath which occurs the DENTINE occupying the bulk of the tooth volume. The tooth is attached to the jaw by CEMENT. In its simplest form, as in many fish and reptiles, a tooth is conical in shape. More complex teeth have the upper surface flattened to form a *crown*, adapted for crushing or chewing. The tooth interior is the *pulp cavity*, containing soft tissues with associated blood vessels and nerves. One or more *roots* occur at the base, for attachment to the jaw. In mammals, the roots each contain a *root canal*, surrounded by dentine, through which blood vessels and nerves pass from the jaw. See DENTITION, Fig. 163.

topoisomerase Any enzyme affecting the topology of DNA, e.g. coiling or relaxing it, preventing tangling during replication. DNA gyrase is one such enzyme, which introduces negative supercoils. All such modifications to DNA involve transient breakage of DNA strands, *topoisomerase I* producing single-strand nicks while *topoisomerase II* forms transient double-strand nicks. See DNA REPLICATION.

tornaria Ciliated planktonic larval stage of the HEMICHORDATA. Its development by radial cleavage and ENTEROCOELY suggests links with echinoderms.

torpor A deep hypothermia (body temperature, T_b, drops below 30°C) in which metabolism, respiratory rate and heart rate are depressed, accompanied by reduced response to external stimuli. Metabolic rate, in particular, may fall to one twentieth or one hundredth of its normal value. It may be part of a daily cycle of temperature change, as it is in deer mice and some mountain hummingbirds, and is more profound than the mild torpor, often termed HIBERNATION, in which T_b does not drop below 30°C.

torus (1) Receptacle of a FLOWER. (2) Central thickened portion of pit membrane in bor-

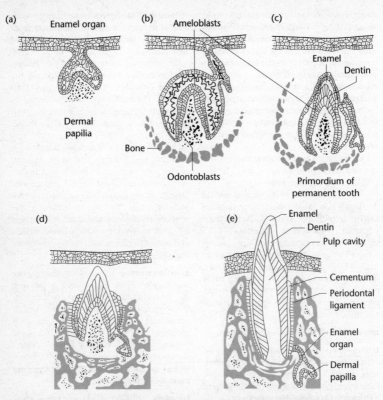

FIG. 163 *Mammalian tooth development. (a) Enamel organ (from the epidermis) and dermal papilla (from the dermis) appear. (b) Ameloblasts are the source of tooth enamel and form from the enamel organ. Odontoblasts are the source of dentine and form from the dermal papilla. Bone appears and begins to delineate the socket in which the tooth will reside. (c) The primordium of the permanent tooth appears. (d) Tooth growth continues. (e) The deciduous tooth erupts and is anchored in the socket by a well-established periodontal ligament. The enamel organ and dermal papilla of the permanent tooth primordium will not begin to form the tooth until shortly before the deciduous tooth is lost.*

dered PITS of conifers and some other GYMNO-SPERMS.

totipotency (Bot.) The inherent capability of a single cell to provide the genetic programme required to direct the development of an entire individual. See SOMATIC EMBRYOGENESIS.

toxin See ANTITOXINS, DINOPHYTA, ENDO-TOXINS, EXOTOXINS, POISONS, TOXOIDS.

toxoid Toxin modified so as to retain its antigenicity (its epitopic domain) but to lose its toxicity. Used in VACCINES against human diphtheria and tetanus.

Toxoplasma A coccidian protist parasite and cause of toxoplasmosis (a zoonosis) which is a serious and sometimes fatal illness. Domestic cats are a main source of infection, although in Europe meat consumption is the biggest single risk factor. The disease is opportunistic and associated with the immunosuppression caused by HIV.

trabeculae Slender bars of tissue lying across cavities. (Zool.) They form the mesh-like interiors of spongy BONE, commonly growing along stress lines. (Bot.) Slender rod-shaped supporting structures found in various plant organs: (1) radially elongated endodermal cells, each becoming a row of cells by division, that support the stele in stems of the lycopod *Selaginella*, linking stele to cortex; (2) membranes, comprising rows of cells, traversing interior of sporangia in the lycopod *Isoetes*; (3) rod-like outgrowths of cell wall lying across lumen of tracheids.

trace element Element required in minute amounts for an organism's normal healthy growth. A *micronutrient*. Often a component or activator of an ENZYME. *Essential* plant trace elements (without which death eventually ensues) include zinc, boron, manganese, copper and molybdenum. Fluorine is essential for hardening tooth enamel. Iodine is a component of thyroxine and triiodothyronine. For cobalt, see CYANOCOBALAMIN. Compare VITAMIN.

tracer Term sometimes used for an isotopic label. See LABELLING.

trachea (1) Vertebrate windpipe. A single tube, covered and kept open by incomplete rings of cartilage; with smooth muscle in its wall, and a ciliated lining. Leads from glottis down neck, branching into two bronchii. (2) In most insects and some other arthropods (e.g. woodlice), hollow tubes of epidermis and cuticle conducting air from spiracles directly to tissues. The finest branches (*tracheoles*, diameter about 0.1 µm) are intracellular. Both tracheae and tracheoles may be permeable but tracheoles are more important in gaseous exchanges. See VENTILATION, ECDYSIS.

tracheary element General term for the water-conducting cell in vascular plants. Comprise TRACHEIDS and VESSEL members.

tracheid Non-living XYLEM element, characteristic of vascular plants other than flowering plants. Formed from a single cell, it is elongated with tapering ends and thick, lignified and pitted walls: an empty firm-walled tube running parallel to long axis of organ in which it lies, overlapping and in communication with adjacent tracheids by means of pits. Functions in water conduction and mechanical support. Primitive compared with a VESSEL.

tracheole See TRACHEA.

Tracheophyta A division in older classifications which regard possession of vascular tissue as of greater taxonomic significance than the seed habit. Would include all vascular plants, thus comprising the majority of the world's terrestrial vegetation.

training See MUSCLE CONTRACTION.

trait A particular phenotypic character, as opposed to character mode. Thus eye colour in *Drosophila* is a character mode, whereas red eye colour is a character trait. Traits can be *autosomal* or *sex-linked*, and determined either by a single locus or polygenically. Sometimes, in heritable disorders, *trait symptoms* are contrasted with *disease symptoms*, in which case the former are usually milder and occur in heterozygotes while the latter are more severe and occur in homozygotes (see SICKLE-CELL ANAEMIA).

***trans*-acting control elements** Regulatory genetic elements (e.g. REPRESSOR genes and genes for other PROMOTER-binding proteins) whose products are not limited in their binding to the DNA molecule of which they form a part. See TRANSCRIPTION FACTORS. Contrast *CIS*-ACTING CONTROL ELEMENTS.

transactivation Interaction of steroid receptors with *TRANS*-ACTING CONTROL ELEMENTS on DNA to stimulate transcription of hormonally responsive genes. See NUCLEAR RECEPTORS.

transaminase Enzyme carrying out TRANSAMINATION.

transamination Transfer by a *transaminase* of an amino group from an amino acid (often GLUTAMATE) to a keto acid (e.g. pyruvate, ketoglutarate), producing respectively a keto acid and an amino acid. See PYRIDOXINE.

trans-configuration (Of mutations) see *CIS-TRANS TEST*.

transcriptase An RNA POLYMERASE. Compare REVERSE TRANSCRIPTASE.

transcription Production of RNA molecules by an RNA POLYMERASE enzyme, using a DNA strand as a template. Approximately 1% of all human genes are estimated to undergo transcription from both DNA strands, i.e. have ANTISENSE expression, and the figure may be considerably higher. All three classes of RNA are produced by transcription processes. The control of transcription, both positive and negative, involves *cis*-regulatory elements such as PROMOTERS and ENHANCERS as well as *trans*-acting molecules, most of them DNA-BINDING PROTEINS, such as TRANSCRIPTION FACTORS (see also NUCLEAR RECEPTORS), or their binding proteins (e.g. ACTIVATORS). Transcription is usually regarded as the key indicator of GENE EXPRESSION. It may require prior modification of the NUCLEOSOME architecture (or topology) of a chromosome. See CHROMOSOMAL IMPRINTING. Transcriptional repressors include Mad proteins which antagonize transcriptional activation by recruiting HISTONE deacetylases to target sites, e.g. promoters, on DNA.

transcription factors (TFs) Regulatory proteins determining the efficiency with which RNA polymerases bind to DNA PROMOTER regions during transcription. *Transcription activators* are involved in positive, and *transcription repressors* in negative control of gene expression. The DNA-binding ability of these proteins often comprises the common structural motif of an α-helix, a short turn and a second α-helix (the 'helix-turn-helix'). The first helix interacts with the DNA backbone while the second lies in a major groove of the DNA molecule. Many eukaryotic transcription factors contain in addition tandemly repeated amino acid sequences each producing a metal-binding (often Zn^{2+}) finger-like protrusion in the functional molecule. These finger sequences (*multifinger loops*) often include identical or near-identical (*consensus*) subsequences which are thought to be conserved and homologous. Some may have oncogenic properties if their amino acid sequence is altered by mutation in the encoding gene. Several eukaryotic transcription activators have been shown to have one nucleic acid-binding region and a second 'activating region' binding additional transcription factors or perhaps RNA polymerase itself. See Fig. 157 (TATA BOX) for the various complexes of TATA-binding protein, transcription activating factors (TAFs) and RNA polymerase required for transcription of different gene classes.

Since the 1990s it has become apparent that the research fields of SIGNAL TRANSDUCTION and the control of GENE EXPRESSION are inseparable. Transcription regulation depends upon a very large number of proteins assembled into the RNA POLYMERASE, including general transcription factors. But site-specific transcription factors form a much larger group of proteins which recruit coactivators to initiate gene-specific transcription, often several at one ENHANCER. See CREB, DNA-BINDING PROTEINS, GENE EXPRESSION, SIGMA FACTORS, references there, and Fig. 164.

transcriptome The collection of genes expressed, or transcribed, from genomic DNA by a cell. Unlike the genome, it is highly dynamic (see GENE EXPRESSION). And changes in multigene expression patterns give us clues about regulatory mechanisms and the biochemical pathways involved in cellular functions. See NUCLEIC ACID ARRAYS.

transdetermination Change brought about in groups of cells (e.g. insect imaginal disc cells) which when subsequently cultured differentiate into a structure normally produced by cells from a different lineage within the organism (e.g. a different disc). The effect resembles mutation in a stem cell, but occurs to a group of cells which shifts from one heritable state to another as a result of environmental influences. See HOMEOTIC MUTATION.

transduction (1) Process in which usually a BACTERIOPHAGE picks up DNA from one bacterial cell and carries it to another, when the DNA fragment may become incorpor-

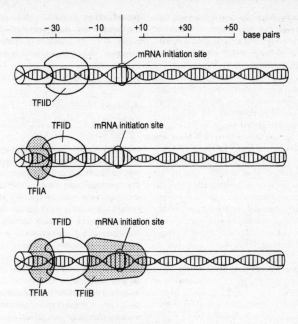

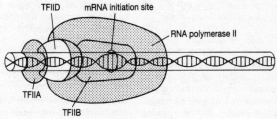

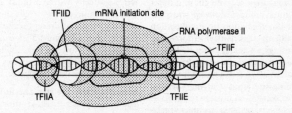

FIG. 164 *Development of active eukaryotic transcription initiation complex for Class II genes. TFIID = TATA factor or TATA-binding protein (TBP). Some notations have all TFs replaced by TAF (= transcription activating factor).*

ated into the bacterial host's genome. Two basic types: (a) *generalized transduction*, where the phage DNA-packaging mechanism picks up 'by mistake' any phage-sized fragment of chromosomal DNA, which can be integrated by homologous RECOMBINATION into the recipient genome after injection into the cell by the phage apparatus; (b) *specialized transduction*, in which, on induction, integrated phage DNA (e.g. λ) genome is imprecisely excised from the chromosome, carrying adjacent chromosomal DNA with it; since the phage is generally integrated at a specific site in the chromosome, only a few bacterial genes can be transduced this way. A similar process, not usually regarded as transduction, occurs in *F-mediated sexduction* (see F FACTOR) in which an F factor (a plasmid, not a phage) carries bacterial DNA from one cell to another. All three types are employed in CHROMOSOME MAPPING. See HUMAN GENE THERAPY. (2) (Of energy) the conversion of energy from one form to another. In this sense, chloroplasts and muscle fibres are two among many biological transducers. (3) For amplification of signals by a cell (e.g. all sensory cells), see SIGNAL TRANSDUCTION.

transect A form of systematic sampling where samples are arranged linearly and usually contiguously. They are commonly used in investigations of gradients of change. Another use is in the study of pattern. The *line transect* is a rapid but insensitive technique suitable for investigating transitions in vegetation due to some factor showing a gradient (e.g. height of a water table). A line is arbitrarily placed across the zoning of the vegetation (in the direction of the gradient). It is marked in some way and recordings are taken along its length. Line transects may be used to collect more precise information (e.g. percentage cover) by recording the length of line covering the species concerned in large numbers of parallel transects. In dense, shrubby vegetation this technique can be used when it is impracticable to use QUADRATS. The *belt transect* is an arrangement of quadrats where they are placed in rows touching one another.

transfection In prokaryotes, reserved for the uptake by transformation of naked phage DNA to produce a bacteriophage infection. See TRANSFORMATION for use in eukaryotes.

transferase Enzyme (e.g. acyl transferases, transaminases) transferring a non-hydrogen functional group from one molecule to another. Coenzymes are often involved.

transfer cells Specialized parenchyma cells possessing ingrowths of the cell wall and producing unusually high surface-to-volume protoplast ratios. Believed to play an important role in transfer of solutes between cells over short distances. Exceedingly common; probably serving a similar role throughout the plant. Occur in association with XYLEM and PHLOEM of small veins in cotyledons and leaves of many herbaceous dicotyledons; in xylem and phloem of leaf traces, at nodes in both dicots and monocots; in various tissues of reproductive structures, e.g. embryo sacs, endosperm; in various glandular structures, e.g. nectaries and glands of carnivorous plants, and in absorbing cells of haustoria of dodder (*Cuscuta*).

transfer efficiency The efficiency with which energy is passed through various steps in the trophic structure of a COMMUNITY.

transferrin Iron-binding β-globulin protein transferring iron from reticuloendothelial cells in bone marrow to immature red blood cells, where it binds and passes iron (Fe^{2+}) into the cell. Cells short of iron produce transferrin receptor molecules for import of iron-bound transferrin to endosomes, whose low pH frees the iron while transferrin bound to its receptors returns to the cell membrane. See FERRITIN.

transfer RNA See RNA, GENETIC CODE, PROTEIN SYNTHESIS.

transformation (1) Change in certain bacteria (occasionally other cells) which, when grown in presence of killed cells, culture filtrates or extracts from related strains, take up foreign DNA and acquire characters encoded by it. Transformed cells retaining this DNA may pass it to their offspring. When cultured mammalian or other animal cells take up foreign DNA the term *transfection* is often used. Discovery implicated DNA as a genetic material. See ANTIBIOTIC RESISTANCE ELEMENT, NATURAL GENETIC COMPETENCE. (2) Any heritable alteration in the properties of a eukaryotic cell. In tissue culture, usually refers to the conversion of a non-cancerous cell to a CANCER CELL ('neoplastic transformation'), as by retroviruses or carcinogens. Common properties of transformed cells include increased blebbing of the plasma membrane and increased mobility of proteins within it; reduced adhesion of the cell to surfaces ('contact inhibition' in the context of adjacent cells), tending to make them rounder in appearance; increased extracellular proteolysis on account of increased release of plasminogen activator; growth by proliferation to a high cell density, with consequent 'immortality'; reduced growth factor requirement, and tendency to cause tumours when injected into susceptible host animals. Malignant transformation leads to an immortalized cell which may then acquire tumorigenicity. It is the phenotypic result of several successive gene lesions affecting the function and control of ONCOGENES, tumoursuppressor genes and their pathways. These include: (a) acquisition of deregulated cell growth caused by disruption of the retinoblastoma (RB) pathway; (b) neutralization of a p53-dependent apoptotic response normally brought on by inappropriate CELL CYCLE events; and (c) maintenance of telomere length and function, often by telomerase activity.

transformation series See EVOLUTIONARY TRANSFORMATION SERIES.

transforming growth factors See TGFs.

transgene Any gene introduced by GENE MANIPULATION from one organism into an organism of a different species (i.e. xenogeneic).

transgenic Describing an organism whose normal genome has been altered by introduction of a gene (transgene) by a manipulative technique (microinjection of DNA into egg or whole organism; use of plasmid or virus-based DNA VECTOR), often involving introduction of DNA from a different species. The foreign DNA may then integrate into the host genome. Such organisms are potentially valuable in plant and animal husbandry (see GENE MANIPULATION, GMO). The technique is yielding results in the genetics of animal development. See TRANSDUCTION.

transition Any error in DNA replication during which one pyrimidine nucleotide is inserted in lieu of another, or one purine nucleotide in lieu of another. Contrast TRANSVERSION.

transition threshold The basic reproductive rate of a parasite that is a necessary condition for a disease to spread.

transitivity A logical relation of importance in the context of SPECIES definitions. The relational property of populations '*interbreeds with . . .*' is transitive if, when population A interbreeds with population B and population A interbreeds with population C, then population B interbreeds with population C. RING SPECIES indicate that '*interbreeds with . . .*' is not transitive at the population level and so cannot do the job which the biological species concept demands of it.

translation Ribosomal phase of PROTEIN SYNTHESIS.

translocation (1) (Bot.) Long-distance-transport of materials (e.g. water, mineral ions, photosynthates) within a plant. Absorption of mineral ions from the soil solution occurs through the root epidermis to the endodermis symplastically. The ions travel from the epidermal cells to the first layer of cortical cells through plasmodesmata in the epidermal-cortical cell walls. Radial movement of the ions continues in the cortical SYMPLAST from cell to cell (protoplast to protoplast) via plasmodesmata, through the endodermis and into parenchyma cells of the vascular cylinder by diffusion. Most inorganic (mineral) ions are taken up by active transport (support for this hypothesis comes from the fact that

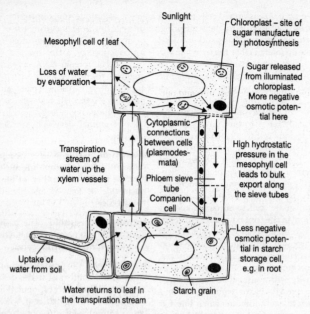

Sunlight

Mesophyll cell of leaf

Chloroplast – site of sugar manufacture by photosynthesis

Loss of water by evaporation

Sugar released from illuminated chloroplast. More negative osmotic potential here

Cytoplasmic connections between cells (plasmodesmata)

Transpiration stream of water up the xylem vessels

High hydrostatic pressure in the mesophyll cell leads to bulk export along the sieve tubes

Phloem sieve tube

Companion cell

Uptake of water from soil

Less negative osmotic potential in starch storage cell, e.g. in root

Water returns to leaf in the transpiration stream

Starch grain

FIG. 165 *Illustration of the pressure-flow (mass flow) hypothesis of phloem* TRANSLOCATION.

mineral uptake is an energy-requiring process). In fact, ion transport from the soil to the xylem requires two carrier-mediated membrane events: first at the plasmalemma of the epidermal cells; second, secretion into xylem vessels at the plasmalemma of the vascular parenchyma cells. Active transport of ions across the plasmalemma may result in electrochemical gradients being established. When this occurs, a voltage difference (the transmembrane potential) develops across the membrane. The hydrogen ion (H^+) is one of the most important cations involved; with pumping of H^+ ions to the outside, a negative potential develops on the inside of the cell, which will attract positively charged ions (e.g. Na^+, K^+). Once mineral ions are in the xylem vessels, they are rapidly distributed upward and throughout the plant in the TRANSPIRATION STREAM. Movement of organic molecules (e.g. sugars, amino acids) and some inorganic ions occurs upwards and downwards in the phloem from leaves to storage organs to

regions of active growth (or storage), the pattern changing with the season or the stage of development. Rates of movement in the phloem ($60–120$ cm h^{-1}) greatly exceed normal sucrose diffusion rates in water. Assimilate movement is said to follow a source-to-sink pattern. The *pressure-flow hypothesis* (see Fig. 165) offers the simplest and most widely accepted explanation for long-distance assimilate transport in the phloem. According to this hypothesis, assimilates move from a source to a sink along an osmotic turgor pressure gradient. In the leaves (the source), sugars produced by photosynthesis are actively secreted into sieve tubes. This active process is termed phloem loading and creates a more negative water (and osmotic) potential in the sieve tube, causing water entering the leaf in the transpiration stream to move into the sieve tube by osmosis. Sugars are transported passively (by mass flow) to a sink, where they are unloaded from the sieve tube. The removal of sugar results in a less negative water (and osmotic) potential in the sieve tube at the sink, so water moves out of the sieve tube. Thus the sieve tubes appear to play a passive role in the movement of sugars; however, active transport is involved in the loading and possibly unloading of sugars and other compounds into and out of the sieve tubes at the sources and sinks. Evidence suggests the phloem loading at the source is provided by a proton pump (energized by ATP and mediated by ATPase at the plasmalemma), involving a sucrose-proton co-transport (symport) system. Metabolic energy required for this is expended by the companion cells or parenchyma cells bordering the sieve tubes. Loading occurs across the plasmalemma of the companion cells; but some sieve tubes are capable of loading themselves, the site of active transport being their own plasmalemmas. See TRANSPIRATION. (2) Form of chromosomal MUTATION in which an excised chromosome piece either rejoins the end of the same chromosome or is transferred to another.

transmitter See NEUROTRANSMITTER.

transpiration Loss of water vapour from plant surfaces. More than 20% of the water taken up by roots is given off to the air as water vapour, most of that transpired by plants coming from the leaves. Differs from simple evaporation in taking place from living tissue and in being influenced by the plant's physiology. Most occurs through STOMATA, and to a much lesser extent through the leaf cuticle. Transpiration rates are affected by levels of CO_2, light, temperature, air currents, humidity and availability of soil water. Most of these affect stomatal behaviour, whose closing and opening are controlled by changes in the turgor pressure of guard cells (see RADIAL MICELLATION), in turn correlated with the potassium ion (K^+) levels inside them. So long as stomata are open (during exchange of gases between leaf and atmosphere), loss of water vapour to the atmosphere must occur and, although for a healthy plant this is inevitable, it is harmful when excessive and causes wilting and even death. See TRANSPIRATION STREAM, EVAPOTRANSPIRATION.

transpiration stream The TRANSLOCATION of water and dissolved inorganic ions from roots to leaves via xylem, caused by TRANSPIRATION. Water enters plant through root hairs, much of it moving along a WATER POTENTIAL gradient to the leaves via the XYLEM. This gradient is via both the SYMPLAST and APOPLAST (see ROOT HAIR). The theory of water movement is known as the cohesion–tension theory or cohesion–adhesion–tension theory. Water within xylem vessels is under considerable tension because polar water molecules cling together in continuous columns pulled by evaporation above. Because of the cohesiveness of water, this tension is transmitted all the way down the stem to the roots, so that water is withdrawn from the roots, pulled up the xylem, and distributed to those cells losing water to the atmosphere. The actual loss makes the water potential of the roots negative, thus increasing their ability to extract water from the soil. Adhesion of water molecules to the walls of the xylem, and to the walls of the leaf and root cells is just as important for the rise of water as cohesion and tension. Cell walls along which the water moves are very effective water-attracting surfaces, and take advantage of water's adhesiveness. Thus this amplifies the expression of cohesiveness. Evidence for this theory of water movement is that (1) water possesses sufficient tensile strength to withstand such tensions when contained within narrow diameter conduits, (2) water in the xylem is under tension, (3) flow begins in the uppermost branches of trees, and (4) the trunk shrinks slightly at the beginning of water movement. If the water column breaks, CAVITATION results. See TRANSLOCATION.

transplantation (1) Artificial transfer of part of an organism to a new position either in the same or a different organism. Practically synonymous with *grafting*, but no close union with tissues in new position is implied. See GRAFT for more details. (2) For nuclear transplantation, see NUCLEUS.

transport proteins (carrier proteins) Proteins intrinsic to some cell membrane (not just plasmalemma), mediating either passive diffusion, FACILITATED DIFFUSION or

ACTIVE TRANSPORT of a solute across it. All appear either to traverse the membrane or be part of a structure which does. Some exhibit enzyme-like properties, e.g. saturation and competitive inhibition. Thus, active uptake by roots of potassium is inhibited by rubidium, and of calcium by vanadium; uptake of glucose by neurons is inhibited by deoxyglucose. Reversible allosteric changes in the protein are probably responsible for carriage, and for actively transported molecules these conformational changes are in part linked to ATP hydrolysis. Transport proteins are either *uniports* (taking one solute type one way), *symports* (transport of one solute depending upon co-transport of another; e.g. pyruvate entering with H^+ ions through a symport in the inner mitochondrial membrane; or co-transport of Na^+ and glucose in the kidney proximal convoluted tubule) or *antiports* (carriage of one solute into the cell depends upon simultaneous transport of another out of the cell, e.g. SODIUM PUMP). See BACTERIORHODOPSIN, GATED CHANNELS, IMPULSE, IONOPHORE.

transport vesicles Membranous vesicles which bud from the membranes of donor cell compartments and fuse with acceptor membranes, which they are thought to recognize, at least in part, by the specific binding of proteins on the vesicles to receptors on target membranes. Such vesicles include those budded from the ENDOPLASMIC RETICULUM and GOLGI APPARATUS (e.g. see SYNAPTIC VESICLE).

transposable elements (TEs) See TRANSPOSONS.

transposition The process whereby DNA elements move about from one site in the genome (the donor site) to a second site (the target site), mediated by transposase enzymes. Similar events are catalysed by retroviral integrases such as that inserting HIV-I (see HIV). It has had a profound effect on genomic evolution (see TRANSPOSONS).

transposons DNA sequences (transposable elements) with the capacity to move about within a genome, inserting site-specifically into host DNA but without

requiring extensive DNA homology in order to do so. Because of its rather modest sequence specificity, the transposase enzyme encoded by these mobile genetic elements recognizes a DNA sequence at the two ends of the transposon and can then insert it into many host genome sites. This site-specific mobility and RECOMBINATION may lead to deletions in adjacent DNA, or its movement to another site, and often produces considerable REPETITIVE DNA in its wake, although much of this derives from ancient elements. Three groups are commonly recognized: retroviral and non-rectroviral transposons are *Class I transposable elements*; DNA-only transposons are *Class II transposable elements*.

(a) *Retroviral-like transposons* (retroposons, retrotransposons) have direct (i.e. not inverted) long terminal repeat sequences (LTRs) at each end, encode reverse transcriptase, and move via an RNA intermediate under the direction of a promoter in the LTR. They are then reversely transcribed back into DNA before insertion into the genome under the direction of an encoded integrase. Both DNA transposons (see below) and LTR retroposons appear to have become completely or almost completely inactive in the hominid lineage, but not in the mouse lineage. Very abundant in eukaryotes, retrotransposons can comprise over 70% of nuclear DNA (as in MAIZE). The *TyI* element in yeast is well studied and resembles a retrovirus without a protein coat.

(b) *Non-retroviral transposons*, which form a large component of the human genome, have a poly-A sequence at the 3′ terminal end of their RNA transcript and no identifying sequence at the 5′ end; they encode reverse transcriptase and endonuclease and move via an RNA intermediate, often directed by a neighbouring promoter. The endonuclease makes a single-strand nick in the host DNA and the reverse transcriptase uses the 3′ end of the break as a primer for synthesis of a single-strand DNA copy of the transposon RNA which thus becomes directly linked to the host DNA. In some undiscovered way, the transposon is then integrated as a DNA helix at the original nick site. They mediate a wide range of genetic effects in their host, notably inactivation or change of expression of genes, and sometimes chromosomal rearrangements (mutations) such as inversions, translocations and deletions. Their insertion into, or adjacent to, another gene often blocks its expression, but insertion into a gene promoter region may increase that gene's expression. If recombination occurs between copies of a transposable element at different sites in a chromosome, then *non-homologous crossing-over* sometimes occurs, possibly a source of reciprocal translocations, deletions, duplications or GENE AMPLIFICATION.

(c) *DNA-only transposons*. The major class in prokaryotes, having short terminal inverted repeat sequences (at each end), they encode their own transposase and move as DNA – either excising and leaving the original site, or replicating before doing so, and then inserting through the transposase activity in a 'cut-and-paste' fashion.

All three classes of transposon are present in animals, e.g. *Drosophila* and humans (see *Alu* entry, LINE, REPETITIVE DNA). At least 47 human genes have been identified as probably deriving from transposons, mostly from DNA-only transposons. Activity of the LINE element LINE1 may include EXON SHUFFLING as well as reverse transcription of genic mRNAs, which can occasionally give rise to functional processed genes in addition to functionless PSEUDOGENES. A few genes also use transcriptional terminators contributed by LTR transposons and some regulatory elements derive from repeat elements. See ANTIBIOTIC RESISTANCE ELEMENT, TRANSPOSABLE ELEMENTS.

transverse division Cell division in the plane of the long axis of a plant (or plant organ). See Fig. 23.

transverse process Lateral projection, one on each side, of neural arch of tetrapod vertebra, with which the head of the rib articulates.

transverse section Section cut perpendicular to longitudinal axis of an organism.

transversion Any error in DNA replication during which a purine nucleotide is substituted for a pyrimidine, or vice versa. Contrast TRANSITION.

Trematoda Flukes. Class of parasitic PLATYHELMINTHES. Ectoparasites (Monogenea) or endoparasites (Digenea) of vertebrates. Mouth and gut are retained, adults having a thick epidermal cuticle. All have a sucker around mouth. Adult *digeneans* live in liver, gut, lung or blood vessel of primary host and may cause serious disease (see SCHISTOSOMA). Ciliated larvae (miracidia) pass out of the host with egesta and parasitize snails (secondary host) in which they reproduce asexually (see POLYEMBRYONY). Sporocysts then give rise to REDIAE and these to CERCARIAE. *Monogeneans* normally have a large multiple central sucker for attachment, *Polystomun* alone among them being endoparasitic (frog bladder).

Trentepohliophyceae This is a small class, with one order and four genera, in the algal division CHLOROPHYTA. Species of Trentepohlia grow on rocks or tree trunks or leaves.

triacylglycerol Alternative term for TRIGLYCERIDE.

Triassic (Trias) GEOLOGICAL PERIOD, lasting from about 251–205 Myr BP. Permian and Triassic are sometimes united as the Permo-Trias. Gymnosperms and ferns dominated the forests; the first dinosaurs and mammals appeared. Continents were mountainous and were united in a single landmass, with large areas being arid.

tribe Taxonomic category between subfamily and genus, used to indicate a group of closely related genera. Tribe names normally end -eae in botany and -ini in zoology.

tribosphenic dentition A combination of triangular upper molar teeth bearing a large cusp which locks into a pestle-like basin on lower molars, and notched shearing crests on the sides of the upper molars which scissor against those of the lower molars. This action combining shearing and grinding has for long been thought to distinguish the mammalian clade containing marsupials and placentals from that containing the monotremes. See MAMMALIA.

tricarboxylic acid cycle (TCA cycle) See KREBS CYCLE.

trichogyne A receptive, sometimes elongate projection extending from the female sex organ that receives the male gamete prior to fertilization in some fungi (e.g. ASCOMYCOTA, BASIDIOMYCOTA), red algae (RHODOPHYTA), green algae (CHLOROPHYTA) and LICHENS.

trichome In blue-green algae (CYANOBACTERIA), a filament comprising a uniserate or multiserate chain of cells. See also HAIR.

trichomonads See ARCHEZOA.

Trichoptera Caddis flies. Order of moth-like, weakly-flying endopterygote insects. Two pairs of membranous wings covered in bristles (hairs) rather than scales. Mandibles reduced or absent in adult. Larvae aquatic, often encased, with biting mandibles and important as herbivores, carnivores and as prey items. Some are indicators of unpolluted water.

Triconodonta Extinct mammalian order (Lower Cretaceous–Upper Triassic), apparently lacking descendants. Up to four incisors in each half-jaw; molars typically with three sharp conical cusps in a row. Cat-sized or smaller; possibly derived from therapsids. Brain small.

tricuspid valve Valve between atrium and ventricle on right side of bird or mammalian heart. Consists of three membranous flaps preventing backflow of blood into the atrium on ventricular contraction. Compare MITRAL VALVE.

trigeminal nerve Fifth vertebrate CRANIAL NERVE.

triglyceride (triacylglycerol, neutral fat) Ester formed by condensation of three fatty acid molecules to the trihydric alcohol glycerol. Includes fats, waxes and oils. See FAT, GLYCEROL.

Trilobita Extinct class of marine arthropods, abundant from Cambrian–Silurian, surviving until Permian. The conservative body, oval and depressed, comprised the following tagmata: shield-like *cephalon* with one pair of antennae and paired sessile compound eyes, a *trunk* of freely movable segments terminating in the united segments of the *pygidium*. Along length of body, longitudinal grooves separated large pleural plates (covering the biramous limbs) from a higher middle region on each trunk segment. Average length 5 cm, but some reached 0.6 m. Probably benthic, feeding on mud and suspended matter. See ARTHROPODA, MEROSTOMATA, Fig. 15.

Trimerophyta Primitive fossil vascular plants whose main axis formed lateral branch systems that dichotomized several times; lacked leaves, and some of the smaller branches terminated in elongate sporangia, while others were vegetative. The cortex was wide, with cells having thick walls; in the centre was the vascular strand. Homosporous, the trimerophytes probably evolved directly from the RHYNIOPHYTA and seem to represent ancestors of ferns, progymnosperms and perhaps horsetails. They appeared in the early DEVONIAN period *c.* 395 Myr BP but became extinct at the end of the mid-Devonian, *c.* 20 Myr later.

trinucleotide repeats REPETITIVE DNA trinucleotides, associated with clinical disorders. Huntington's disease (HD) is caused by tandem repetition of a codon for glutamine in the human *HD* gene near the tip of the short arm of chromosome 4. Normal individuals have between six and 39 of these CAG repeats, but individuals with HD have 36–180 repetitions, and the greater the expansion, the earlier the disease onset. The protein encoded by the normal *HD* allele is a 3144-amino acid protein, *huntingtin*, apparently implicated in synaptic plasticity through association with clathrin-coated membranes along dendrites and involved in both presynaptic and postsynaptic functions. Fragile X syndrome, the most common form of familial mental retardation, is sex-linked and affects most males who inherit the affected X chromosome. The *FMR1* locus is very close to the chromosomal constriction associated with fragile X syndrome, and varies in length because of the variable number of repeats of the trinucleotide CGG. Only with its transmission through females do increases in copy number of the triplet sequence occur. Twelve other human disorders are associated with lengthening of trinucleotide repeats, while at least one is associated with a pentanucleotide repeat. Repeat instability may lead either to massive or to small-scale expansion. Misalignment of long tracts of such simple repeats may lead to deletions or expansions during DNA replication, altering the topology of the DNA – leading, for instance, to the formation of hairpin loops. Faulty DNA REPAIR MECHANISMS may then promote repeat instability.

triplet repeats See TRINUCLEOTIDE REPEATS.

triploblastic Animals with a body organization derived from three GERM LAYERS (ectoderm, endoderm, mesoderm). Includes all metazoans except coelenterates, which are diploblastic.

triploid Nuclei, cells or organisms with three times the haploid number of chromosomes. A form of POLYPLOIDY. Triploid individuals are usually sterile due to failure of chromosomes to pair up during meiosis. See ENDOSPERM.

trisomy Nucleus, cell or organism where one chromosome pair is represented by three chromosomes, giving a chromosome complement of $2n + 1$. May result from NONDISJUNCTION. See DOWN'S SYNDROME.

trochanter (1) Any knob for muscle attachment on the vertebrate femur. Three occur on mammalian femurs, largest in humans being conspicuous at hip joint. (2) Segment lying second from base of an insect leg.

trochlear nerve Fourth vertebrate CRANIAL NERVE.

trochophore (trochosphere) Ciliated larval stage of polychaetes, molluscs and rotifers. Usually planktonic, developing by spiral cleavage. The main ciliary band (*prototroch*) encircles the body in front of mouth (preorally). Commonly a second ring (*telotroch*) surrounds the anus. Coelom arises by schizocoely, not enterocoely. Two protonephridia may be present.

trophic cascade hypothesis The controversial hypothesis that food webs are more strongly influenced 'from above' (e.g. by predation) than 'from below' (e.g. by nutrient supply). Nutrient input, it is agreed, sets the potential productivity of lakes; but deviations from this potential are held to be due to food web effects. Nutrient and food web effects are complementary but act on different timescales. Species variations at the top of the food chain (top predator) bring about differences in selective predation, the effects of which cascade through the zooplankton and phytoplankton to influence the ecosystem dynamics.

trophic level Theoretical term in ecology. One of a succession of steps in the transfer of matter and energy through a COMMUNITY, as may be brought about by such events as grazing, predation, parasitism, decomposition, etc. (see FOOD CHAIN). For theoretical and heuristic purposes, organisms are often treated as occupying the same trophic level when the matter and energy they contain have passed through the same number of steps (i.e. organisms) since their fixation in photosynthesis. Primary producers, herbivores, primary, secondary and tertiary carnivores and decomposers all commonly figure as trophic levels in the analysis of ecosystems. Different developmental stages and/or sexes within a species may occur in more than one trophic level, and so some consider trophic levels to be convenient abstractions lacking even approximate material counterparts. The number of trophic levels in a community is thought to be limited by inefficiency in ENERGY FLOW from one trophic level to the next; however, food chains are no longer in tropical communities, where energy input is high, than they are in Arctic communities, where energy input is low.

trophic structure The organization of a COMMUNITY described in terms of energy flow through its various trophic levels. See TROPHIC LEVEL.

trophoblast Epithelium surrounding the mammalian BLASTOCYST, forming outer layer of chorion and becoming part of the embryonic component of the PLACENTA or of the EXTRA-EMBRYONIC MEMBRANES. In humans, produces HUMAN CHORIONIC GONADOTROPHIN and HUMAN PLACENTAL LACTOGEN.

tropical rain forest Forests found in three major regions of the world, covering only 7% of the land surface but containing more than half the total number of living species. Largest is in the Amazon basin of South America, with extensions into coastal Brazil, Central America, eastern Mexico and some islands in the West Indies; the second is in the Zaire basin of Africa, with an extension along the Liberian coast; the third extends from Sri Lanka and eastern India to Thailand, the Philippines, the large islands of Malaysia and a narrow strip along the northeast coast of Queensland in Australia. Neither water nor low temperatures are limiting factors for photosynthesis, and rainfall averages from 200–400 cm. yr^{-1}. It is the richest biome in terms of both plant and animal diversity, it being not unusual for 1 km^2 in Central or South America to contain several hundred bird species, and many thousands of butterfly, beetle and other insect species. A total of 700 tree species were identified in ten one-hectare forest plots in Borneo. The trees are evergreen, characterized by large leathery leaves. There is a poorly developed herbaceous layer on the forest floor because of low light levels; but there are many vines and epiphytes at higher levels. Most hypotheses seeking to explain the coexistence of large numbers of plant species emphasize niche partitioning or disequilibrium processes (i.e. changing environmental conditions). Equilibrium explanations for such coexistence, with a single limiting resource, probably too often ignore the relevance of concentration gradients. Many of the animals inhabit tree tops. Little accumulation of organic debris occurs because decomposers rapidly break it down, released nutrients being quickly absorbed by mycorrhizal roots or leached from the soil by rain (very little in undamaged forest).

Tropical rain forest now forms about half the forested area of the planet, but is in process of being destroyed by human activities. Many of the soils of tropical rain forests are conditioned by high and constant temperatures and abundant rainfall and are relatively infertile (*latosols* – red clays largely leached of their nutrients). When the forest is cleared, this leaching process accelerates and the soils either erode or form thick, impenetrable crusts (*laterites*) on which cultivation is impossible. It is estimated that by the end of this century most of the tropical rain forests (with huge untapped genetic resources) will have disappeared to produce fields which will be completely useless to agriculture within a very few years and then turn to desert, with considerable effect on global climate. Some areas deforested for agriculture or cattle farming, once abandoned, become recolonized by vegetation and fauna in secondary succession. After initial pioneer-dominated vegetation, longer-lived tree species gain the upper hand. Species richness may become high after just 50 years, although different taxa are involved from those in the mature forest. Mature tropical rain forests may be among the most fragile ecosystems in the world. See BIODIVERSITY, DEFORESTATION.

tropism Response to stimulus (e.g. gravity, light) in plants and sedentary animals by growth curvature; direction of curvature determined by direction of origin of the stimulus. See NASTIC MOVEMENT.

tropomyosin See MUSCLE CONTRACTION.

troponin See MUSCLE CONTRACTION.

true predator A predator that kills other organisms (their prey) more or less immediately after attacking them, killing several or many over the course of its lifetime.

truffle Fungal genus (ASCOMYCOTA, Order Tuberales) that produces a subterranean fruiting body; prized as a gastronomic delicacy.

trumpet hyphae Drawn-out sieve cells, wider at the cross-walls than in the middle, in some brown algae (Order Laminariales). Active transport of photosynthates (mostly mannitol) through sieve cells occurs in this Order.

trunk (Zool.) (1) In arthropods lacking clear demarcation between thorax and abdomen, e.g. myriapods, the entire body region behind the head. (2) In enteropneusts and pterobranchs, the entire body posterior to the proboscis and collar. (3) In vertebrates, the body regions overlying the thoracic, lumbar and sacral vertebrae; i.e. in humans, the body region between the base of the neck and the groin. (Bot.) The main stem of a tree.

trypanosomes Flagellated protozoans containing a kinetoplast. Parasites in blood of vertebrates and their bloodsucking invertebrates (e.g. insects, leeches) which act as vectors. The genus *Leishmania* parasitizes vertebrate lymphoid/macrophage cells, causing visceral leishmaniasis (kala-azar) in humans; *Trypanosoma* is a vertebrate blood parasite, different species causing sleeping sickness and Chagas' disease in humans and nagana in domesticated cattle. Notorious for ANTIGENIC VARIATION, making search for vaccines against them almost pointless.

trypsin Protease enzyme secreted in its inactive form (trypsinogen) by vertebrate pancreas and converted to active form by ENTEROKINASE. Converts proteins to peptides at optimum pH 7–8.

trypsinogen Inactive precursor of TRYPSIN.

tryptophan An aromatic AMINO ACID, essential in the human diet and, like phenylalanine and tyrosine, a product of phosphoenolpyruvate and erythrose 4-phosphate addition, with subsequent derivation, in bacteria and plants. SEROTONIN is a tryptophan derivative.

tsetse fly Genus (*Glossina*) of dipteran fly, sucking vertebrate blood and acting as vector for trypanosomes (spreading sleeping sickness).

TSH See THYROID STIMULATING HORMONE.

tuatara See RHYNCHOCEPHALIA.

tube-feet (podia) Soft, hollow, extensile and retractile echinoderm appendages,

responding to pressure changes within WATER VASCULAR SYSTEM. In some groups (starfish, sea urchins) they end in suckers and are locomotory; in other groups (brittle stars, crinoids) suckers are absent. Those around mouth may be used for feeding (sea cucumbers, brittle stars). Ciliated in crinoids, forming part of food-collecting mechanism.

tuber Swollen end of RHIZOME, bearing buds in axils of scale-like rudimentary leaves (stem tuber), e.g. potato; or swollen root (root tuber), e.g. dahlia. Tubers contain stored food materials and are organs of vegetative propagation.

Tubulidentata Aardvarks. Order of fossorial termite-eating African mammals with long snouts and tongues; once regarded as edentates but now considered related to the extinct CONDYLARTHRA. One family (Orycteropidae) and species (*Orycteropus afer*). Large digitigrade mammal, with large ears and strong, digging claws. Skin nearly naked. No incisors or canines in the permanent dentition; cheek teeth rootless, growing throughout life. Arrangement of dentine around tubular pulp cavities gives the Order its name.

tubulin A characteristic eukaryotic protein. Filamentous (F) tubulin is formed by GTP-dependent polymerization of globular (G) protein monomers, each of these being a heterodimer ($\alpha\beta$) of two tubulin subunits whose genes are in turn multiply allelic. All α subunits appear to be able to copolymerize with all β subunits. Thirteen parallel tubulin filaments make up each of the composite cylindrical MICROTUBULES which play a major role in the eukaryotic cytoskeleton. β-tubulin has 81% homology among organisms as distant phylogenetically as the unicell *Giardia* and humans; there is also 20% homology with the GTP-binding protein FtsZ of prokaryotes, supported by structural and functional studies. The longer a microtubule has been in existence, the greater the proportion of its lysine residues become acetylated and its tyrosine residues removed from the carboxyl terminus, indicating the age of the microtubule. Complete modification takes several hours and is thought to affect the binding of microtubule-associated proteins (MAPs).

tumor See TUMOUR.

tumorigenesis A complex process producing a pathological 'organ' in which host and tumour cells act together in the production of an entity which invades, spreads and ultimately kills its host. In this process, the mature tumour cell must acquire the capacity to instruct the host to grow new blood vessels, evade host immune recognition and alter the microenvironment so that it forms metastases even under less permissive growth conditions. See CANCER CELLS, TRANSFORMATION (2). See TUMORIGENESIS.

tumour (neoplasm) Swelling caused by uncontrolled growth of cells; may be malignant (see CANCER CELLS) or benign.

tumour necrosis factors (TNFs) Misleading term, deriving from the nineteenth-century observation that visible tumours sometimes regressed dramatically if the patient suffered a simultaneous acute infection. This results from an effect of TNFs on the endothelial linings of blood vessels within certain tumours. But TNFs also have powerful systemic effects, causing 'systemic' or 'endotoxic' shock in high doses. One TNF group comprises CYTOLYSINS produced by cytotoxic T CELLS and macrophages; others, from MYELOID cells, cause many tumour cells to be engulfed by macrophages and activate blood polymorphs and macrophages – critical in the elimination of intracellular pathogens (e.g. *Listeria monocytogenes*, mycobacteria and *Leishmania*) from macrophages. TNFs have direct effects on the hypothalamus and pituitary, and one or both TNF receptors (members of the nerve GROWTH FACTOR receptor superfamily) are located on most human cell types. TNF-α, produced by cells of the innate immune system, such as MACROPHAGES and MAST CELLS in response to injury or infection, is a multifunctional CYTOKINE with effects on INFLAMMATION, sepsis, lipid and protein metabolism, HAEMOPOIESIS, angiogenesis, host resistance to parasites and malignancy. As with IL-I (see INTERLEUKIN), TNF-α binding activates NF-ϰB, turning on at least 60 known mammalian genes involved in inflammatory responses. The TNF protein product p53 (see CANCER CELLS, HYPOXIA, APOPTOSIS) is a transcription factor required for transcription of cyclin-dependent kinase (cdk) inhibitors (see CKIs). Raised TNF levels, through illness, may trigger dormant integrated HIV to become activated.

tumour suppressor gene (TSG) A gene normally restraining a cell from excessive proliferation, by involvement of its product in the transduction of an anti-mitotic signal. Mutants in, or loss of, these genes seem to provoke cancerous behaviour only if the system controlling cell division is already disturbed. One such gene is *p53* (see p53, CANCER CELLS, ULTRAVIOLET IRRADIATION). The retinoblastoma gene *Rb* is an example, and its product (RB, or p105) plays a key role in cell proliferation by binding cellular transcription factor domains (notably on E2F) and modulating their behaviour positively or negatively (see PROMOTER). The unphosphorylated form of RB, which binds E2F, may suppress entry to GI of the cell cycle (see Fig. 22) unless it is first bound by transforming proteins encoded by certain viral ONCOGENES. Cells with poorly phosphorylated RB tend to be differentiated or non-dividing. Other TSG products are secreted, or are membrane-associated SIGNAL TRANSDUCTION proteins. See SUPPRESSOR MUTATION, ADHESION.

tundra Treeless region at the farthest limits of plant growth. A vast biome, occupying about one fifth of the planet's surface, and best developed in the northern hemisphere, being found mostly north of the Arctic circle. The Arctic tundra essentially comprises a huge band across Eurasia and North America, with alpine tundra, more closely related to adjacent mountain forests, extending southward in the mountains. In general, permafrost (permanent ice) lies within a metre of the surface and ground conditions in the tundra are usually moist since water is unable to drain through the soil due to the ice. The soils are acid to neutral, low in nutrients and with available nitrogen in very short supply. Evaporation is low due to the high moisture content of the air at low temperatures. In contrast, some tundra regions are so dry they constitute true polar deserts. For plant survival, the mean temperature must be above freezing for at least one month per year; the growing season may be only two months per year. Plants include low shrubs, e.g. birch (*Betula*), willow (*Salix*), blueberry (*Vaccinium*), and Labrador tea (*Ledum*), perennial herbs, mosses and lichens. Few annuals are found. Many of the plants, including several grasses and sedges, are evergreen, which enables them to initiate photosynthesis immediately conditions of light, temperature and moisture allow. Vegetative reproduction is characteristic of many perennials, and much of the biomass of tundra plants is underground, consisting not only of roots but also of rhizomes.

tunica corpus Organization of shoot of most flowering plants and a few gymnophytes, comprising one or more peripheral layers of cells (*tunica layers*) and an interior (*corpus*). Tunica layers undergo surface growth and corpus undergoes volume growth.

Tunicata See UROCHORDATA.

Tupaiidae The family of modern tree shrews (Order Scandentia). See ARCHONTA.

Turbellaria Class of PLATYHELMINTHES. Mostly aquatic and free-living. Epidermis ciliated, especially ventrolaterally, for locomotion. Gut a simple or branched enteron, with ventral mouth and protrusible pharynx.

turbidity Level of cloudiness of a suspension, commonly of cells. Often used as indicator of density of cells in suspension.

turgor (1) State of plant cell in which cell wall is rigid, stretched by an increase in volume of vacuole and protoplasm during absorption of water. The cell is described as *turgid*. An essential feature of mechanical support of the plant tissues, loss of turgor (when water loss exceeds absorption) being followed by wilting. Aids expansion of young cells in growth. Compare PLASMOLYSIS. See WATER POTENTIAL. (2) This 'swelling pressure' also occurs in gels formed from highly

hydrophilic glycosaminoglycans (e.g. in CARTILAGE matrix) which can resist compressive forces of several hundred atmospheres.

turgor pressure The hydrostatic pressure that develops within a plant cell through osmosis and/or imbibition. Plant cells tend to concentrate relatively strong solutions of salts within their vacuoles along with sugars, amino and other organic acids. Consequently, plant cells continually absorb water by osmosis and this creates their inner hydrostatic pressure which, in turn, keeps the cell rigid (turgid). This is a major role of the vacuole and its membrane (tonoplast). When turgid, the turgor pressure is equal and opposite to the pressure of the cell wall (wall pressure). See WATER POTENTIAL.

turion Detached winter bud by means of which many aquatic plants survive the winter; e.g. *Myriophyllum*.

Turner's syndrome Human genetic disorder caused by absence of a sex chromosome, leaving one X chromosome per cell (XO). NON-DISJUNCTION is often the cause. Individuals usually have webbing of the neck, narrowing of the aorta, and reduced height. Gonads undifferentiated secreting no hormones; so no menstrual cycle or pubic hair occur. Abnormal XO conditions also occur in mice, horses and rhesus monkeys. Human females with Turner's syndrome who retain the paternal X chromosome tend to achieve better cognition and social adjustment than do those inheriting the maternal X chromosome. CHROMOSOMAL IMPRINTING affects X-chromosome genes associated with Turner's syndrome. See SEX DETERMINATION.

turnover number See MOLECULAR ACTIVITY.

twin spots A form of tissue mosaicism in which two tissues of different phenotype lie adjacent to one another against a tissue background of a third (commonly wild-type) phenotype. Caused by mitotic crossing-over between homologous chromosomes in a stem cell heterozygous for recessive alleles at two linked loci (e.g. +sn/

y+) which normally complement to give wild-type tissue (background) but in this case give patches of singed (sn) and yellow (y) tissue phenotypes resulting from recessive homozygosity after the cross-over event and subsequent mitosis of mutant daughter cells.

twitch fibres See MUSCLE CONTRACTION.

two-hybrid system See YEAST TWO-HYBRID SYSTEM.

tylose (tylosis) Balloon-like enlargement of membrane of a PIT in wall between a xylem parenchyma cell and a vessel or tracheid, protruding into the cavity of the latter and blocking it; wall may remain thin or become thickened and lignified. Tyloses occur in wood of various plants, often abundantly in heartwood of trees, and may be induced to form by wounding. Often occur in the xylem below developing abscission layer before leaf-fall.

tympanic cavity See EAR, MIDDLE.

tympanic membrane (tympanum) See EAR DRUM.

tyndallization Fractional sterilization devised by the physicist John Tyndall (1820–93). Method involved lengthy immersion of the material in boiling brine (over 100°C), killing vegetative microorganisms, followed by a period in which unkilled spores were allowed to germinate, followed by further immersion in boiling brine. This could be repeated two or three times, gradually eliminating all bacterial spores. Compare PASTEURIZATION.

type specimen See HOLOTYPE.

tyrosine kinase See PROTEIN TYROSINE KINASE.

U

Ubiquinone See COENZYME Q.

ubiquitins A family of small highly conserved proteins involved in the degradation of most cellular proteins, initially by binding to them covalently, which targets them for rapid hydrolysis by a (26S) proteolytic complex which contains a 20S degradative core termed a PROTEASOME. Binding of ubiquitins to other proteins requires ubiquitin activation by the E1 enzyme and transfer to one of several ubiquitin-conjugating enzymes (E2). Accessory proteins (E3) associate with this to form an E2–E3 complex (ubiquitin ligase), the E3 component binding the degradation signals on proteins targeted for proteolysis. E2 forms a multiubiquitin chain in which each ubiquitin molecule in the complex is linked by a lysine residue to the previous one, this complex being what is recognized by a receptor on a proteasome. Cells govern the ubiquitination processes which underpin the CELL CYCLE by an enzyme complex (SCF, a group of E3s) which uses its variable binding heads to recognize proteins for degradation with exquisite control. Ubiquitins are emerging as key regulators of eukaryotic mRNA synthesis, not only by ubiquitin-dependent destruction of transcription factors by the proteasome but also by proteasome-independent processes. See CYCLINS.

ulna Bone of tetrapod fore-limb, lateral to radius. Articulates distally with carpus and proximally with humerus, above which it may protrude as the olecranon process ('funny bone'). Serves in attachment of main extensor muscle of fore-limb. May fuse with radius or, particularly in mammals, be lost altogether. See PENTADACTYL LIMB.

ultimate factor Factor considered to be a cause of some biological structure or phenomenon, whose presence it serves to explain in terms of future states of affairs rather than (mechanistic) antecedent conditions. ADAPTATION and ONTOGENY require consideration of future conditions. Compare PROXIMATE FACTOR. See TELEOLOGY.

***Ultrabithorax* gene (*Ubx* gene)** A HOMEOTIC gene of the bithorax complex (BX-C) of *Drosophila* expressed in cells forming the adult haltere (posterior thoracic segment), whose loss of function causes the haltere to be replaced with a wing (homologous structure of middle thoracic segment).

In *Drosophila*, haltere evolution may have depended upon the gradual acquisition of binding sites for Ubx protein in the *cis*-regulatory DNAs of certain 'growth genes', such as *wingless* and *decapentaplegic*. This protein also represses expression of another gene, *Distalless* (*Dll*), required for limb formation. One theory is that the Ubx protein product of the gene could function as an activator of limb development in onychophorans and that an alanine-rich repression domain was acquired when these and arthropods diverged. This domain now appears to mediate constitutive repression of limb development in insect abdominal segments but in crustaceans acts only in a conditional fashion. In crustaceans, changes in the pattern of *Ubx* expression seem to have occurred in conjunction with a change in expression of another Hox gene, *Sex combs reduced* (*Scr*) in changing swimming limbs into feeding appendages. See EVO-DEVO, HOMEOGENE.

ultracentrifuge High-speed centrifuge, devised in 1925 by Svedberg, capable of sedimenting (and separating) particles as small as protein or nucleic acid molecules. See CELL FRACTIONATION, SVEDBERG UNIT.

ultradian rhythm Biological rhythm with periodicity longer than a day.

ultramicrotome See MICROTOME.

ultrastructure Fine structure of a cell or tissue, as obtained by instruments with higher resolving power than the light MICROSCOPE.

ultraviolet irradiation Ultraviolet light (uv) has wavelengths in the band between visible light and X-rays, i.e. ~13–397 nm. UVA lies in the region 400–320 nm, while UVB lies in the range 320–300 nm. Solar radiation is high in these wavelengths (see OZONE), produced by emission from excited atoms and ions and so generated by discharge tubes. It is capable of ionizing atoms and molecules, of splitting molecules into free radicals (some biologically harmful) and of initiating polymerizations. In human skin, converts ergosterol into VITAMIN D. Because nucleic acids absorb energy in the uv range of the electromagnetic spectrum, DNA absorption peaking at ~260 nm, this can result in production of the thymidine dimers, chain breakage and cytosine hydration (see Fig. 166), with consequent impairment of DNA replication unless repaired (see DNA REPAIR MECHANISMS, MELANIN, CANCER CELLS). UVB light causes mutations in DNA at sites where two pyrimidines are adjacent (i.e. C-C, C-T and T-T). About two-thirds of such changes are C-to-T substitutions; about 10% are C-C → T-T double substitutions. Such mutations in the *p53* gene lead to under-expression of p53 protein in affected cells, and since p53 is required for programmed cell death in DNA-damaged cells, such cells will not die but undergo unrestrained growth, leading to nonmelanoma skin cancer. The protein

bcl-2 protects cells against UVB irradiation damage.

Ulvophyceae A class of the algal division CHLOROPHYTA, containing some 35 genera and 265 species. The great majority of these algae are marine or brackish water forms (e.g. *Enteromorpha*); only a few species are found in freshwater (e.g. species of *Ulothrix*). Morphologically this class contains unicellular, multicellular and siphonocladous non-flagellate green algae. Bi- or quadriflagellate cells (zoospores, gametes) are formed that have a cruciate, 11 o'clock–5 o'clock configuration of the flagellar apparatus; there is a distinct overlap of the flagellar bases. Mitosis is closed and the telophase spindle persistent. In vegetative cells cytokinesis is brought about through in-growth of a cleavage furrow, often with addition of Golgi vesicles, which coalesce with the furrow. Microtubules are often associated with the leading edge of the furrow. Plasmodesmata are not present in the cross walls. One cup-shaped or band-like parietal chloroplast is present in each cell; one to several pyrenoids are present. Known life cycles are either haplontic, in which the zygote is the only diploid cell or isomorphic diplohaplontic. Thick-walled hypnozygotes are not formed. Sexual reproduction is isogamous or anisogamous.

umbilical cord Connection between ventral surface of embryo of placental mammal and placenta. Consists mainly of allantoic mesoderm and blood vessels (umbilical artery and veins), covered by amniotic epithelium. Surrounded by amniotic fluid, it usually breaks or is bitten through at birth. See EXTRAEMBRYONIC MEMBRANES.

undercompensating density-dependence DENSITY-DEPENDENCE in which death rate increases, or birth or growth rates decrease, less than the initial density increases, so that increases in the initial density still lead to (smaller) increases in final density.

undulipodium Alternative name for eukaryotic FLAGELLUM.

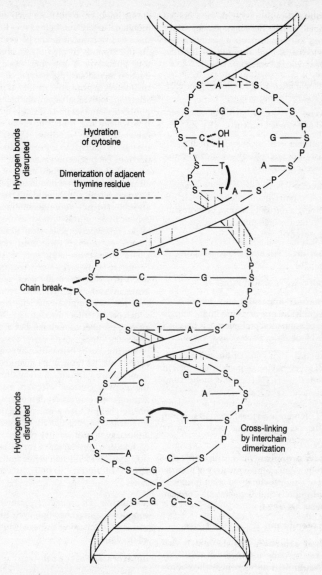

FIG. 166 *Diagram showing alterations in DNA caused by* ULTRAVIOLET IRRADIATION. *The formation of thymidine dimers is the change most likely to do damage to living cells and viruses. S-P-S-P is the sugar-phosphate backbone; A, T, G and C are the four bases; hydrogen bonds are shown on pale horizontal bars. Thymidine dimers are shown as curved bars. (Adapted from* Ultraviolet Radiation and Nucleic Acid *by R.-A. Deering. Copyright © Scientific American, Inc., 1962. All rights reserved.)*

ungulate Non-formal term for a hoofed mammal; usually herding and adapted for grazing. Runs on firm, open ground on tips of digits (*unguligrade* locomotion). Includes ARTIODACTYLA and PERISSODACTYLA.

unguligrade See UNGULATE. Compare DIGITIGRADE, PLANTIGRADE.

uniaxial (Of algae) having an axis composed of a single filament.

unicellular (Of organisms or their parts) consisting of one cell only. Compare ACELLULAR, MULTICELLULARITY.

unilocular Possessing a single chamber whose contents divide into many zoids, e.g. describing a mitosporangium or more usually a meiosporangium in brown algae (PHAEOPHYCEAE).

uniparous Producing one offspring at birth.

uniport See TRANSPORT PROTEINS.

uniramous appendage Appendage of myriapods and insects, consisting of a single series of segments (podomeres) as opposed to two series as in a BIRAMOUS APPENDAGE.

unisexual (Bot.) (1) (Of flowering plants and flowers) having stamens and carpels in separate flowers. Can be either monoecious or dioecious. (2) (Of algae) having only one type of gametangium formed on a plant. (Zool.) (Of an individual animal) not producing sperm or eggs but both. Compare HERMAPHRODITE.

unitary organism An organism that proceeds by a determinate pathway of DEVELOPMENT of a tightly canalized adult form (e.g. all arthropods and vertebrates). Compare MODULAR ORGANISM.

unit membrane See CELL MEMBRANES.

unit of selection Whatever it is that NATURAL SELECTION ultimately distinguishes between and whose frequencies thereby not only change but may as a result be causally responsible for changing frequencies of characters in the population. These causal chains would reflect our best theories at the time. Controversy surrounds this issue.

Strong initial contenders for such units are individual organisms themselves. However, since selection discriminates between these on the grounds of phenotype, and since total phenotype is not what sexual reproduction reproduces, phenotypic characters themselves might better qualify as units of selection, individual organisms being composites of character traits. Strictly, however, it is usually not traits but their material causes (physical encodings, GENES) which pass from parent to offspring in reproduction. Yet if selection acts at the level of classical Mendelian genes, many of which are pleiotropic, then as dominance relations between genetically determined characters are subject to background genetic modifiers, genetic backgrounds as much as individual alleles (alternative genes at a locus) may qualify as units of selection. However, alleles themselves may not be durable enough to be the units, since crossing-over, when it occurs, normally does so within genes. Instead, the genetic unit favoured by selection may simply be any length of DNA which persists intact through a greater than average number of meioses. But a mutant molecular 'trick' (e.g. GENE CONVERSION) producing such better-than-average persistence would tend to be favoured automatically by selection, thereby disrupting the normal control selection exerts on genotype via expression of mutants in the phenotype: for such a mutant might have no phenotypic effect at all. In conclusion, a plurality of entities may qualify as units of selection, producing a range between neo-classical Mendelian genes and 'selfish' DNA. See FITNESS, ALTRUISM, SELFISH DNA/GENES.

univoltine (Of animals, typically insects) completing one generation cycle in a year and then diapausing through winter, the adults dying.

unsaturated fatty acid A fatty acid with at least one double bond in its carbon chain.

uracil Pyrimidine base occurring in the nucleotide uridine monophosphate (UMP), whose radical is an integral component of RNA but not of DNA. The UMP radical, derived from UTP, base pairs with AMP and is incorporated into RNA by RNA POLYMERASE.

urea Main nitrogenous excretory product, $CO(NH_2)_2$, of ureotelic animals, produced by liver cells from deaminated excess amino acids via the *urea cycle*, part of which occurs in mitochondria. Degradation product of both pyrimidines and, via uric acid, purines (especially fish, amphibians). Elasmobranch fish have a high urea content in body fluid, enabling water to enter osmotically through soft body surfaces. See Fig. 167.

uredinales Rust fungi. A group of fungi belonging to the class Teliomycetes of the BASIDIOMYCOTA, comprising 164 genera and about 7,000 species that are cosmopolitan on flowering plants and ferns. Most are obligate parasites causing major diseases but several species have been maintained in culture. Many rusts are of great economic importance, causing extensive damage to crops throughout the world, e.g. *Puccinia graminis*, the cause of black stem rust of wheat and other cereals. The mycelium of these fungi has no clamp connections and is frequently intercellular (frequently with haustoria). It is limited to parts of leaves or other aerial organs of the host. No basidiocarps are produced. Rusts form up to five spore states. Traditionally spore terminology was based on morphology. Spore types include spermatia, which are monokaryotic gametes produced in pycnia. Aeciospores are unicellular, non-repeating vegetative spores (usually resulting from dikaryotization and usually associated with pycnia) which are produced in aecia. They germinate to give rise to a dikaryotic mycelium. Urediniospores are formed during the summer (summer spores, red rust spores) and are repeating vegetative spores, which give further urediniospores or teliospores. The urediniospores are borne on a dikaryotic mycelium in uredinia. Amphispores, or resting urediniospores, are formed by some rust fungi and possess thick and dark-coloured cell walls. Teliospores (winter spores, black rust spores) are basidia-producing spores produced in telia. Typically, a teliospore is a resting spore, two or more celled, sessile or pedicellate but not deciduous. Basidio-

spores are haploid, unicellular, thin-walled spores borne on two- or four-celled basidia after meiosis and liberated from sterigmata by abjection. The life cycles of rusts can be complex; species can be heteroecious (require two different hosts to complete a life cycle) or autoecious (require only one host); species can be homo- or heterothallic.

uredinium Rust-coloured, linear streaks on the leaves and stems of wheat, containing dikaryotic urediniospores in rust fungi (BASIDIOMYCOTA, Order Uredinales). These spores are capable of reinfection. See AECIUM, PYCNIDIUM, TELIUM.

ureotelic (Of animals) whose main nitrogenous waste is UREA, such as elasmobranchs, adult amphibians, turtles and mammals. Compare AMMONOTELIC, URICOTELIC.

ureter Duct conveying urea away from kidney. In vertebrates, usually restricted to duct of amniotes leading from metanephric kidney to urinary bladder. Develops as outgrowth of WOLFFIAN DUCT. See CLOACA.

urethra Duct leading from urinary BLADDER of mammals to exterior; joined by vas deferens in males, conducting both semen and urine. See PENIS.

uric acid Major urine breakdown product of adenine and guanine; also the form in which nitrogen from excess amino acids is excreted in *uricotelic* animals. Almost insoluble in water, it is an adaptive nitrogenous waste in environments where water is at a premium.

uricotelic (Of animals) whose main nitrogenous waste is URIC ACID; e.g. snakes, lizards, a few mammals (some desert rodents), embryonic and adult birds, terrestrial gastropods, and insects (see MALPIGHIAN TUBULES).

urinary bladder See BLADDER.

uriniferous tubule Convoluted tube forming bulk of each *nephron* (excluding Malpighian corpuscle) of vertebrate KIDNEYS.

urkaryote Term reserved for protoeukaryotes, organisms (presumably single-celled) from which eukaryotes evolved after

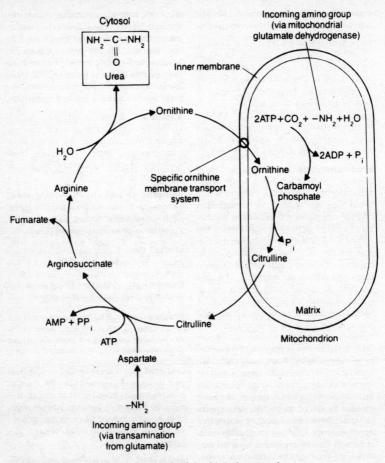

FIG. 167 *Diagram to show the involvement of mitochondria in the* UREA *cycle.*

endosymbiotic incorporation of whichever prokaryotes gave rise to mitochondria, chloroplasts and perhaps some other membranous structures now included within the category of organelles. See ENDOSYMBIOSIS.

Urochordata (Tunicata) Subphylum of the CHORDATA (see PROTOCHORDATA). Sea squirts, etc. Adults sedentary (some free-swimming forms, the Larvaceae, are probably progenetic), feeding microphagously by a ciliary-mucus method employing an ENDOSTYLE, with pharyngeal gill slits, reduced nervous system, no notochord. Larvae active and tadpole-like, with well-developed nervous system and notochord. The draft genome sequence of the tunicate *Ciona intestinalis* (160 million base pairs) was published in late 2002. As a deuterostome, but not a vertebrate, its evolutionary interest lies mainly in theories of vertebrate origins. *Ciona* has just under 16,000 genes, including single copies of genes which are absent from protostomes such as *Caenorhab-*

ditis and *Drosophila*, but which have diversified into large gene families in vertebrates, notably those involved in cell signalling and developmental regulation. *Ciona* appears to retain a primitively simple gene complement such as might have been found in the last common ancestor of sea squirts and vertebrates. The larvacean *Oikopleura* has a very small genome (72 million base pairs) with an overall organization similar to that of the nematode *Caenorhabditis*, suggesting life history similarities. It develops rapidly, living for only 4 days compared with *Ciona* living for 3 months. See PROGENESIS.

Urodela Order of AMPHIBIA. Newts and salamanders; with elongated bodies, tail and short limbs. Mostly confined to moist terrestrial habitats and/or ponds and lakes. Tadpoles resemble adults more than do those of frogs and toads. *Ambystoma*, a salamander, is an example of NEOTENY.

urostyle Rod-shaped bone (representing fused caudal vertebrae) in anuran vertebral column; important in rigidifying posterior part of the animal during jumping. Flanked by two ilia, with which it articulates.

Ustilaginales Smut fungi. An order in the fungal class USTOMYCETES (BASIDIOMYCOTA) comprising 50 genera and about 950 species. Smut fungi cause much economic loss of crop plants. Like the rust fungi, they do not form complex basidiocarps but do have dikaryotic hyphae and basidia which are septate and form spore masses. Thick-walled probasidia (ustilospores) are formed in sori on living plants; metabasidia (promycelia) is non-septate or transversely septate, producing ballisto- or statismospores. The Ustilaginales have a mycelial parasitic phase and are yeast-like when cultured *in vitro*. Smut fungi occur typically as host-specific endophytes parasitic on flowering plants, especially grasses and sedges (except for two species of *Melanotaenium* which are parasitic on *Selaginella*). Sori are usually limited to the ovary, anthers, inflorescence or leaves and stem of the host, though the root is attacked by *Entorrhiza*. The mycelium is delicate. Hyphal cells are uni- or binucleate which may ramify throughout the host plant or only be at the points of infection. The hyphae are generally intercellular, frequently with haustoria and sometimes with clamp connections but dolipore septa are lacking. Smut fungi may form conidia on the host's surface, although in most genera only the smut spores (chlamydospores, brand spores, resting spores, pseudospores, teliospores, ustospores, ustilospores) and basidiospores are formed. Ustilospores are generally exposed as a dark powder when mature. Mature spores are diploid and uninucleate. At germination meiosis takes place and a promycelium (basidium, hemi- or metabasidium, germ tube) having four or more basidiospores (sporidia, sporidola, 'conidia', promycelial spores) is produced.

Ustomycetes Smut fungi class of the Basidiomycota comprising 63 genera and about 1,060 species of fungi bearing basidiospores in sori and frequently forming a yeast-like phase. See USTILAGINALES.

uterus (womb) Muscular expansion of MÜLLERIAN DUCT of female mammals (except monotremes) in which embryo develops after implantation. Usually paired, each connecting to a Fallopian tube; single in humans, through fusion of lower part. Connected by vagina to exterior, controlled by sex hormones (see MENSTRUAL CYCLE), oxytocin and prostaglandins; enlarging at sexual maturity or in breeding season. See CERVIX. Glandular lining membrane (endometrium, decidua) nourishes early embryo. Smooth muscle in its wall greatly increases during pregnancy; its contraction ultimately expels embryos and their placentae (PARTURITION).

utriculus See VESTIBULAR APPARATUS.

UTRs Untranslated regions of RNA.

V

vaccine Material producing in an animal an immune reaction and an acquired IMMUNITY to a natural microorganism (*immunization*; compare INOCULATION). Almroth Wright (1861–1947) first showed that dead, rather than living (typhus), bacteria could produce immunity. BCG vaccine (against tuberculosis) contains live, attenuated (weakened) tubercle bacilli, while vaccination against smallpox has eliminated the disease. The ideal vaccine is one in which a single administration at birth provides lifetime protection from multiple diseases. Efforts at producing vaccines for the Third World centre upon genetically engineering live attenuated vaccine vectors (*multi-vaccine vectors*) that immunize against many antigens. Viral vaccines have the advantage of expressing naturally folded, glycosylated, assembled and secreted antigens in eukaryotic cells as well as stimulating cytotoxic T CELLS and antibody production. Immunity to poliomyelitis, yellow fever, measles and mumps has been achieved with 'live' attenuated virus; but antibodies to the numerous strains of HIV and influenza virus cross-react minimally or not at all (SEE PASTEUR). Bacterial vaccines immunize for longer periods and can provide local immunity, e.g. in the gut. In mice, engineered BCG vaccine containing *Mycobacterium tuberculosis* with a plasmid expression VECTOR for β-galactosidase, tetanus toxins and HIV proteins among others, has been shown to elicit antibody and cell-mediated responses; however, such recombinant vaccines may not provide immunity from pathogens showing ANTIGEN VARIATION. Some work suggests that the technique of *genetic immunization* may offer hope: gold

microprojectiles coated in plasmids containing a gene encoding the antigenic protein (linked to a suitable PROMOTER) are injected into the recipient animal to elicit antibody. Even 'naked' DNA injected into animals can result in expression of the transgene and responses by both cytotoxic T cells and B cells. Antibody production against INFLUENZA virus antigens following intramuscular injection of plasmid DNA encoding viral antigens was first detected in 1993. Traditional vaccines have failed to achieve immunity to tuberculosis, chlamydia, papilloma virus and malaria; nor are there any effective anti-tumour vaccines, although one approach undergoing clinical trials involves injecting into cancer patients dendritic cell precursors which have presented cancer cell antigens on their surfaces, the hope being to raise host immune responses to the tumour. There are currently human clinical trials of DNA vaccines against HIV-1 and the causes of MALARIA and hepatitis B, among others. There are also hopes expressed that there will be vaccines against cocaine and other drug addictions. Unmethylated CpG repeats (SEE DNA METHYLATION) are added as ADJUVANTS in vaccines based on naked DNA. DNA-based vaccines centre upon the use of sequences encoding and expressing an antigen stimulating division and response of appropriate T cells (SEE GENE GUN).

Some believe that vaccines may offer us the opportunity for 'VIRULENCE management' of pathogens, driving them towards mildness. This possible evolutionary effect is most often illustrated by the case of the bacterial disease diphtheria. The pathogen exists in two forms: a harmless form found in the throat, where it takes up passing nutrients, and a dangerous form which produces a cell-rupturing toxin and causes a dispersal-promoting cough. Endangering its host's life is a risky strategy for a pathogen, and diphtheria vaccine targets the toxin. Where there has been a good vaccination programme, death rates from diphtheria have fallen dramatically as the harmless forms have become more common and the virulent forms rare; but where programmes have faltered, as currently in Russia, the virulent strain is making a comeback. It is suggested that successful vaccination programmes, by targeting the toxin rather than the organism, selectively increase the cost of virulence to the organism. Others argue that unless a vaccination programme is 100% effective, virulent pathogens will ultimately find some unvaccinated individuals, in which they would be 'nastier than before'. See TOXOID.

vacuolar membrane (Bot.) See TONO-PLAST.

vacuole Membrane-bound regions within the cytoplasm. In plant cells the membrane is termed the TONOPLAST. The immature plant cell typically has numerous small vacuoles, which increase in size and fuse into a single vacuole as the cell enlarges. A vacuole may take up as much as 90% of a plant cell's volume, with the cytoplasm forming a thin peripheral layer closely pressed against the cell wall. A direct consequence is the development of turgor pressure and maintenance of tissue rigidity. Within the vacuole the cell sap comprises water, with other components depending upon the species of plant, as well as its physiological state. Salts and sugars are usually present, along with, in some species, dissolved proteins. Ions (e.g. Ca^{2+}) accumulated in the cell sap in concentrations far in excess of those in the surrounding cytoplasm, even to the extent that crystals form, e.g. calcium oxalate crystals are particularly common (SEE INOSITOL 1,4,5-TRIPHOSPHATE). The cell sap is slightly acidic, and very acidic in citrus fruits. Vacuoles are important storage sites for various metabolites (e.g. malic acid in CAM plants); they also remove toxic

secondary products, and are often the sites of pigment deposition (e.g. anthocyanins), as well as being involved with the breakdown of macromolecules, and recycling of their components within a cell. Entire organelles (ribosomes, mitochondria, plastids) may enter a vacuole where they are broken down (see LYSOSOME). Vacuoles are absent from the cells of bacteria and CYANO-BACTERIA. See CONTRACTILE VACUOLE, PUSULE, FOOD VACUOLE, WATER POTENTIAL.

vagility Distance in a straight line between either (a) an individual's birth place and the place where it dies, or (b) its site of conception and the place where it gives rise to a new zygote. Populations or demes of low mean vagility have more probability of recruiting rare homozygous mutants.

vagina Duct of female mammal connecting uteri with the exterior via short vestibule. Usually single and median due to fusion of lower part of MÜLLERIAN DUCT in the embryo. Receives the male's penis during copulation. Lined with stratified non-glandular epithelium, which may undergo cyclical changes during oestrous cycle under the influence of sex hormones.

vagus nerve Large, mixed (sensory and motor), vertebrate CRANIAL NERVE (CN X) innervating much of the gut, ventilatory system (including fish gill muscles) and heart. Emerges within skull from the medulla, running back on each side as an abdominal ramus, the main parasympathetic route. ACCESSORY NERVE is basically a motor element of the vagus, emerging further back along the spinal cord. Vagi tend to terminate in plexi, such as in the sympathetic ganglia of the solar plexus. See AUTONOMIC NERVOUS SYSTEM, INFLAMMATION.

valve (Bot.) (1) One of several parts into which a CAPSULE separates after dehiscence by longitudinal splitting. (2) One of two halves of diatom cell wall (frustule). (Zool.) (1) (Of heart) flap or pocket of vertebrate heart wall at entry and exit of each chamber, preventing back-flow of blood (e.g. MITRAL VALVE, TRICUSPID VALVE). (2) (Of veins, lymphatics) pocket-like flap in birds and mammals, preventing back-flow.

vancomycin An ANTIBIOTIC, currently administered only when most others have failed to clear the bacterium. It prevents the growing peptidoglycan chains from cross-linking to other chains. See ANTIBIOTIC RESISTANCE ELEMENT.

Vanilla Genus of orchid, the seed pods of which are the natural source of the popular flavouring of the same name. Extracted from the dried fermented seed pods. Originally used by the Aztecs in what is now Mexico. Cultivated now in Madagascar and other islands in the western Indian Ocean. Synthetically produced vanilla flavouring (*vanillin*) is now more widely used.

variation Phenotypic and/or genotypic differences between individuals of a population. *Continuous* (*quantitative*) characters are those exemplified throughout the phenotypic range, tending to be determined by POLYGENES. *Discontinuous* (*qualitative*) characters are those unrepresented in all parts of the phenotypic range (a lack of intermediates), tending to be *polymorphic*, determined by genetic 'switch-mechanisms' (see POLYMORPHISM). Controversy over the relative influence of heredity and environment in producing phenotypic differences fuels the NATURE-NURTURE DEBATE. Problems arise in obtaining acceptable control populations to test hypotheses. By starting with genetically uniform material (e.g. by cloning or repeated inbreeding) it is often possible to compare phenotypes produced under different environmental regimes and to estimate HERITABILITY. See EPISTASIS, GENETIC VARIATION, PHENOTYPIC PLASTICITY, INFRASPECIFIC VARIATION.

varicosity (Of some motor nerve axons) a swelling, releasing the neurotransmitter noradrenaline via synaptic vesicles – not at specialized active zones on the membrane (as at synapses) but seemingly randomly over its surface. An axon may have several varicosities where it passes close to smooth muscle, the transmitter acting locally on G-PROTEIN-linked RECEPTORS.

variegation Irregular variation in colour of plant organs (e.g. leaves, flowers) through suppression of normal pigment develop-ment in certain areas. Commonly due to TRANSPOSABLE ELEMENTS, or to chloroplast-(esp. cpDNA-)based mosaicism, or maybe a disease symptom (e.g. mosaic diseases due to viral infection).

variety Taxonomic category involving groups of plants or animals of less than species-rank. Some botanists view plant varieties as equivalent to subspecies, while others consider them divisions of subspecies. See INFRASPECIFIC VARIATION.

vasa recta Capillaries taking blood into the KIDNEY medulla from the efferent arterioles of juxtaglomerular nephrons (those whose loops of Henle extend into the inner zone of the medulla – about 15% of nephrons in humans). Blood flow is slow, totalling *c.* 10% of kidney blood volume at any time. Originally isotonic with a cortical fluid, blood in descending vasa recta becomes increasingly concentrated as water leaves by osmosis and small solutes enter down concentration gradients. At a hairpin bend the vasa recta turn upwards and flow towards the cortex, where water is reabsorbed into the blood from the peritubular fluid and solutes leave by diffusion. This produces a passive COUNTERCURRENT SYSTEM of exchange which preserves the medullary solute gradient (higher in the inner medulla than the outer). Water tends to leave the medulla via the vasa recta because of the colloid osmotic pressure due to their contained proteins.

vascular (Bot.) Adjective pertaining to any plant tissue or region comprising or giving rise to conducting tissue (e.g. XYLEM, PHLOEM, VASCULAR CAMBIUM).

vascular bundle Longitudinal strand of conducting (vascular) tissue comprising essentially XYLEM and PHLOEM; unit of stelar structure in stems of gymnophytes and flowering plants and occurring in appendages of leaves (e.g. veins in leaves). In gymnophyte stems, arranged in a ring surrounding the pith; in monocotyledons, scattered throughout stem tissue. May be (a) *collateral*, with the phloem on same radius as xylem and external to it (typical condition in flowering plants and gymnophytes); (b) *bicollateral*, with two phloem groups, external and internal to xylem, on same radius (uncommon, occurring in some dicotyledons, e.g. marrow); (c) *concentric*, with one tissue surrounding the other. The latter are *amphicribral* when phloem surrounds the xylem, as in some ferns, and *amphivasal* when xylem surrounds phloem, as in rhizomes of certain monocotyledons. Vascular bundles are further described as *open* when cambium is present, as in dicotyledons, and *closed* when cambium is absent, as in monocotyledons.

vascular cambium (Bot) One of two lateral meristems in vascular plants which produces secondary tissues that comprise the secondary plant body. Cells of the vascular cambium are highly vacuolated and occur in two forms: (1) *fusiform initials*, vertically elongate cells; (2) *ray initials*, squarish cells. The vascular cambium arises from procambial cells that remain undifferentiated between the primary xylem and primary phloem, as well as from PARENCHYMA of the interfascicular regions. It produces secondary xylem and secondary phloem through periclinal divisions of the cambial initials and their derivatives (i.e. the cell plate that forms between the dividing cambial cell's initials is parallel to the stem or root surface). Cambial initials divided off towards the outside and inside of the root or stem become phloem or xylem cells, respectively. Thus, a long continuous radial row of cells is formed, extending from the cambial initial outwards to the phloem and inwards to the xylem. Xylem and phloem cells produced by fusiform initials have their long axes orientated vertically and form the axial system of the secondary vascular tissues, while those cells produced by the ray initials form the vascular rays or radial system. That portion of the cambium arising within the VASCULAR BUNDLES is the fascicular cambium, and that arising in the interfascicular region is the interfascicular cambium. In woody stems, the formation of secondary xylem and secondary phloem results in the formation of a cylinder of secondary vascular tissues, with rays extending radially throughout this cylinder. More secondary xylem than secondary phloem is formed. See CORK CAMBIUM.

vascular cylinder See STELE.

vascular plant Any plant containing a VASCULAR SYSTEM. See TRACHEOPHYTA.

vascular system (Bot.) See STELE. (Zool.) System of fluid-filled vessels or spaces in animals where transport and/or hydrostatics is involved; e.g. BLOOD SYSTEM, LYMPHATIC SYSTEM, WATER VASCULAR SYSTEM.

vas deferens (pl. vasa deferentia) One of a pair of muscular tubes (one on each side) conveying sperm from testis to exterior. In male amniotes the vasa lead from the epididymis to the cloaca or urethra. See WOLFFIAN DUCT, VAS EFFERENS.

vas efferens (pl. vasa efferentia) Tube (many on each side), developing from a mesonephric tubule and conveying sperm from the testis to the epididymis in male vertebrates. See KIDNEY, TESTIS, WOLFFIAN DUCT.

vasoactive Having an effect upon rate of blood flow or tissue fluid production; e.g. affecting dilation or permeability of blood vessels. See HISTAMINE, INFLAMMATION, SEROTONIN, VASOMOTOR CENTRE.

vasoconstriction Narrowing of blood vessel diameter (common arteriole) through smooth muscle contraction. Brought about in skin and abdomen by adrenaline, but more generally by vasoconstrictor nerves of sympathetic nervous system. ANTIDIURETIC HORMONE and ANGIOTENSINS may also be involved. See VASOMOTOR CENTRE, ENDOTHELIN.

vasodilation Increase in blood vessel diameter (commonly arteriole), often through reduction in its sympathetic nervous stimulation, by cholinergic nervous stimulation, or (in heart and skeletal tissue) by adrenaline. HISTAMINE and KININS are vasodilators associated with inflammatory response. See CAPILLARY, VASOMOTOR CENTRE.

vasomotor centre Group of neurons in vertebrate medulla (see BRAINSTEM) whose reflex outputs control muscle tone of arteriole walls, producing normal VASOCONSTRIC-TION (vasomotor tone), but increasing this tone when required by raised sympathetic output. Inputs include those from BARO-RECEPTORS influencing cardiac centres.

vasopressin See ANTIDIURETIC HORMONE.

V(D)J recombination See ANTIBODY DIVERSITY.

vector (1) Organism (animal, fungus) housing parasites and transmitting them from one host to another, commonly acting as a host itself. Insects, ticks and mites often act as vectors, many transmitting parasites to man, crops and domestic animals. Female *Anopheles* mosquitoes transmit *Plasmodium*, cause of MALARIA; the mosquito *Aedes aegypti* transmits the viral cause of dengue fever, and blackflies (*Simulium* spp.) transmit FILARIAL WORMS. (2) A vehicle, e.g. a drug-resistance PLASMID or modified BACTERIO-PHAGE, which has its own replication origin and is capable of use for insertion and cloning of other pieces of DNA. *Cloning vectors* are those in which the foreign DNA is cloned prior to further study, e.g. for construction of GENE LIBRARIES. These were first constructed by cloning DNA fragments into a λ phage vector; but only fragments of about 15 kb can be packaged into this phage head. *Cosmid vectors* are cloning vectors carrying the cos sites (which leave GC-rich 'cohesive ends' after specific endonuclease treatment) from λ phage, plus a plasmid origin of replication and a drug-resistance gene. These enable cloning of fragments up to about 45 kb in length. For cloning of fragments over 100 kb in length, YEAST ARTIFICIAL CHROMOSOMES are employed. EXPRESSION VECTORS are those in which the foreign DNA is not just cloned, but also expressed (see EXPRESSION SIGNALS), the com-

bined vector-and-host system being the EXPRESSION SYSTEM. *Shuttle vectors* are those that can be replicated in both a bacterium (e.g. *E. coli*, in which they are propagated) and a eukaryote (e.g. *Saccharomyces*) in which they can be studied. They must contain selectable markers and replication origins that function in each organism, and a yeast chromosomal replication origin (an *autonomously replicating sequence*, or ARS) often forms part of the plasmid vector. See Fig. 168 and PCR.

vegetal pole (vegetative pole) Point on surface of an animal egg furthest from nucleus; usually in the yolkiest end. Compare ANIMAL POLE.

vegetative reproduction (v. propagation) (Bot.) Reproduction involving detachment of some part of the organism other than a spore (e.g. gemmae, rhizomes, bulbs, corms, tubers) and its subsequent development into a complete organism. See ASEXUAL, LIFE CYCLE, MICROPROPAGATION. (Zool.) See ASEXUAL. Compare APOMIXIS.

vein (Bot.) Vascular bundle forming part of the conducting and supporting tissue of a leaf or other expanded organ. (Zool.) (1) Blood vessel (smaller diameter than venous sinus) carrying blood back from capillaries to heart. In vertebrates, has smooth interior of ENDOTHELIUM surrounded by variable degrees of smooth muscle and connective tissue, but always much less than a corresponding artery and with a larger internal diameter. Non-return valves ensure unidirectional blood flow, low pressure being provided mainly by skeletal muscle contractions. (2) Fine tubes of toughened cuticle within insect wings, providing support. May contain tracheae, nerves or blood.

FIG. 168 *Expression* VECTOR *technique employed to ensure that a previously cloned cDNA sample is expressed within an* E. coli *cell so that the protein encoded by the original mRNA (which acted as a template for cDNA production) is produced in large amounts. An* E. coli *promoter is used for the expression of foreign protein in the cell. IPTG inactivates the repressor once large enough numbers of* E. coli *have been grown. (From Recombinant DNA 2/E by Watson, Gilman, Witkowski and Zoller. Copyright © James D. Watson, Michael Gilman, Jan Witkowski and Mark Zoller, 1992. Reprinted with permission of W. H. Freeman and Company.)*

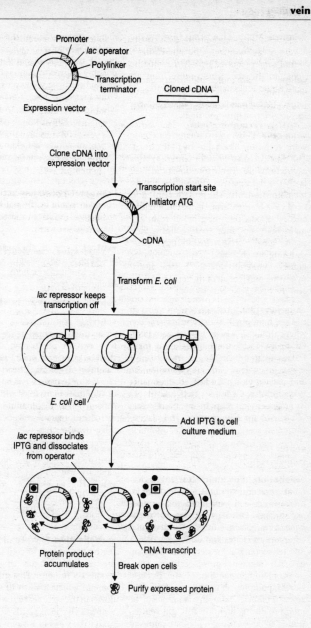

velamen Multiple epidermis covering the aerial roots of some tropical epiphytic orchids, aroids and some terrestrial roots.

veliger Planktonic molluscan larva, developing from trochophore and with larger ciliated bands; adult organs present include foot, mantle and shell.

vena cava Either of two major vertebrate veins, returning blood directly to the heart. The usually paired superior (anterior) vena cava returns blood from fore-limbs and head (only the right persists in many mammals, including humans). The posterior (inferior) vena cava is single and median in lungfish and tetrapods, returning blood from the main body posterior to fore-limbs. Supplants renal portal veins in phylogeny, shunting much blood formerly entering kidney glomeruli directly to the heart, bypassing much or all of the renal filtration and the liver. See CUVIERIAN DUCT.

venation (Bot.) Arrangement of veins in the leaf mesophyll. Termed *netted venation* where arranged in branching pattern (largest dicot vein often extending along main leaf axis as midrib); or *parallel venation* (most monocots) where veins are of similar size and parallel along the leaf, much smaller veins forming a complex network. (Zool.) Arrangement of veins in an insect wing; often used for identification and classification.

venter (Bot.) Swollen basal region of archegonium, containing egg cell.

ventilation Rhythmic breathing movements of an animal's muscles and skeleton, increasing gaseous exchange across respiratory surfaces. Often termed *external respiration*. In large active insects the abdomen commonly vibrates and undergoes reversible 'telescoping' as the sternum and tergum of each segment move, causing compression and expansion of the tracheal system, pumping air in and out of the spiracles. Thoracic pumping may also be involved during flight. In fish, gill ventilation is achieved using a pressure pump in front and a suction pump behind. Amphibians at rest on land employ throat movements to renew air in the mouth, where gaseous exchange occurs; when active, a buccal force-pump ventilates the lungs, possibly assisted by flank movements. Bird ventilation is very complex (see AIR SACS). Only mammals have a DIAPHRAGM (stimulated by phrenic nerves from the cervical spinal cord) which, with INTERCOSTALS, ventilates the lungs by changing thorax volume during inspiration and expiration under the control of the medulla (see BRAINSTEM). Here an *inspiratory centre* receives inputs from stretch receptors in the lungs (via the vagus), from CAROTID BODIES and from an inhibitory *expiratory centre* in the pons. Both centres are directly sensitive to gaseous tensions in the blood.

ventral Generally the surface resting on, or facing, the substratum. (Bot.) See ADAXIAL. (Zool.) In chordates, body surface furthest from nerve cord.

ventral aorta See AORTA, VENTRAL.

ventricle (1) Chamber of vertebrate heart; either single or paired (tendency to separate into two in phylogeny). Thicker-walled than ATRIUM. See HEART, HEART CYCLE. (2) Chamber of molluscan heart responsible for blood flow to tissues. (3) One of the cavities of vertebrate brain filled with CEREBROSPINAL FLUID. A pair in the cerebrum, another in the rest of the forebrain, and one in the medulla. See EPENDYMAL CELLS.

venule Small vertebrate blood vessel, intermediate in structure and position between capillary and vein. Most have non-return valves. Highly permeable, comprising layer of endothelium with coat of collagen fibres. Larger venules also have smooth muscle in their walls.

vernalization (Bot.) Promotion of flowering by application of cold treatment; hormones, cold and day length interact to modify the response. First studied in cereals, some of which if sown in spring will not flower in the same year but continue to grow vegetatively. Such plants (winter varieties) need to be sown in autumn of the year preceding that in which they are to flower, and contrast with spring varieties which, planted in the spring, flower in the same year. By vernalization, winter cereals can be sown and brought to flower in one season. Seed is moistened sufficiently to allow germination to begin, but not enough to encourage rapid growth, and when tips of radicles are just emerging the seed is exposed to a temperature just above 0°C for a few weeks. Seed thus vernalized acquires properties of seed of winter varieties; sown in the spring, it produces a crop in summer of same year. Even after vernalization, plant must be subjected to suitable photoperiod for flowering to occur. In some plants, gibberellin treatment can substitute for cold exposure, while vernalization effects can be reversed in some plants by exposure to high temperature in anaerobic conditions.

vernation Way in which leaves are arranged in relation to one another in the bud.

vertebra See VERTEBRAL COLUMN.

vertebral column (spinal column) Segmentally arranged chain of bones or cartilage near vertebrate dorsal surface, surrounding and protecting the spinal cord. In most vertebrates, vertebrae form a hollow rod attached anteriorly to skull, largely replacing the notochord. Each vertebra straddles the junction of two embryonic somites, a central mass (*centrum*) replacing the notochord, and an arch above (*neural arch*) enclosing the spinal cord. A similar arch below (*haemal arch*), or ventral projections, often enclose major axial blood vessels (aorta, posterior cardinal veins/vena cava). Projections from vertebrae are for muscle attachment. Vertebrae join at their centra (see SYMPHYSIS), often also by projections of neural arch; only restricted movement is possible between any two. Fish vertebrae are very similar from skull to tail; but in tetrapods there are usually these regional differences: atlas and axis; *cervical vertebrae*, with very short ribs; *thoracic vertebrae*, bearing main ribs; *lumbar vertebrae*, without ribs; *sacral vertebrae*, attached by rudimentary ribs to pelvic girdle; and tail (*caudal*) vertebrae. See BONE, CARTILAGE.

Vertebrata (Craniata) Major subphylum of CHORDATA. Cyclostomes, fish, amphibians, reptiles, birds and mammals. Differ from non-vertebrate chordates in having a skull, vertebral column and well-developed brain. All but agnathans have paired JAWS. During its development, the vertebrate head is derived from two cell populations that are absent from cephalochordates: NEURAL CREST and neurogenic placodes. The former contributes to many of the bones, muscles and nerves of the head and face while the latter contribute to most of the sensory organs and their associated nerves. Different *HOX* GENES are expressed along the anterior–posterior axis, each directing the cell fate for cells in different positions along it, their *cis*-regulatory regions being crucial for the spatial development of this singularly vertebrate structure. See ZOOTYPE. Classes include AGNATHA, PLACODERMI, ACANTHODII, CHONDRICHTHYES, OSTEICHTHYES, AMPHIBIA, REPTILIA, AVES and MAMMALIA.

vesicle Any of a variety of membranous spheres involved in transport or storage within eukaryotic cells. See GOLGI APPARATUS, PHAGOCYTOSIS, PINOCYTOSIS.

vessel (trachea) Non-living element of xylem comprising tube-like series of cells arranged end-to-end, running parallel with long axis of the organ in which it lies and in communication with adjacent elements by means of numerous pits in side-walls. Components of a vessel, *vessel members*, are cylindrical, sometimes broader than long, with large perforations in end walls, and function in TRANSLOCATION of water and mineral salts, and in mechanical support. They have evolved from tracheids, principal features of this evolution being elimination of multiperforate end walls to form a single large perforation, reduction or disappearance of taper of end walls and a change from a long, narrow element to one relatively short and wide. Generally thought to be more efficient conductors of water than tracheids because water can flow relatively unimpeded from one vessel member to another through the perforations. With few exceptions, confined to flowering plants.

vestibular apparatus (v. organ) Complex part of the *membranous labyrinth* forming part of inner ear of most vertebrates generating sensory input for reflexes of balance and general posture. Lying on either side within otic region of the skull, each comprises a closed system of cavities forming sacs and canals lined by epithelium and filled with fluid *endolymph* resembling tissue fluid. Major sac-like parts are the *utriculus* (utricle) and the *sacculus* (saccule), the latter normally lower, each containing a MACULA with HAIR CELLS and otoliths. These respond to head tilt and acceleratory movements; but turning movements are detected by the *semicircular canals*, three of which are generally present on each side. Two canals lie in vertical planes at right angles, the third horizontal. Each has a swollen *ampulla* containing a sensory *crista* with hair cells embedded in a jelly-like *cupula* which swings to and fro under influence of endolymph in the canals, pulling the hair cells whose combined outputs along the AUDITORY NERVE register turning movements in all spatial planes. The COCHLEA is the other component of the membranous labyrinth.

vestibulocochlear nerve (auditory nerve, acoustic nerve) Eighth vertebrate cranial nerve, innervating the inner ear. Essentially a dorsal root of facial nerve, relaying impulses from hair cells of the cochlea (via cochlear branch) and from vestibular apparatus (via vestibular branch).

vestigial organ An organ whose size and structure have diminished over evolutionary time through reduced selection pressure. Only traces may remain, comparative anatomy providing evidence of phylogeny. Remains of pelvic girdles in snakes and whales indicate tetrapod ancestries. See APPENDIX (not vestigial in humans), EVOLUTION.

Vetulicolia A putative fossil phylum from the CHENGJIANG FAUNA of southern China. Its members are a few centimetres long and have in common a bipartite body whose anterior half is voluminous and sac-like with a large opening (the mouth?) and five pairs of lateral openings interpreted as phar-yngeal gill slits. One structure being interpreted as a candidate endostyle, until now unique to chordates, and in a similar position to those of tunicates and amphioxus, has also been found. The posterior portion of the vetulicolian body is divided into seven limbless segments, which in other respects give a very arthropod-like appearance to the fossils. Others find in the fossils a resemblance to tunicate tadpole larvae. See Fig. 169.

vibrissiae Stiff hairs or feathers, usually projecting from face. Tactile.

vicariance biogeography See BIO-GEOGRAPHY.

villus (1) Finger-like projection from the lining of the small intestines of many vertebrates. Their great number (up to 4–5 million, each up to 1 mm long in humans) gives the mucosa a velvet-like appearance. Each is covered by epithelium whose brush borders increase the surface area for absorption and digestion 25-fold (see MICROVILLUS, DIGESTION). Each contains blood vessels and a lacteal, extending and retracting by means of smooth muscle. (2) For chorionic (trophoblastic) villi, see PLACENTA.

Virchow, Rudolph German medical microscopist (1821–1902) who did most to dispel the notion, current from the time of Hippocrates, that disease resulted from imbalance in body 'humours'. Replaced it by a cell-based theory, arguing that cells are derived only from other cells (*omnis cellula a cellula*), broadening and deepening CELL THEORY.

virion See VIRUS.

viroid Smallest known agents of infectious disease; smaller than the smallest viral genome, and lacking capsids (protein coats). Known only from plants, they comprise small, single-stranded molecules of RNA that replicate autonomously in susceptible cells. The RNA may be a closed circle or a hairpin-shaped structure open at one end, and have 300–400 nucleotides. Viroids are found almost exclusively in the nucleus of infected cells, replicating themselves by forming a complementary strand that acts

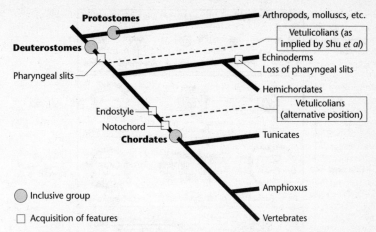

FIG. 169 *The deuterostomes, and possible places for vetulicolians in the evolutionary scheme of events. Shu and colleagues found vetulicolians – as axial deuterostomes – phylogenetically outside all extant deuterostomes. But they also suggested that the fossils have an endostyle, a structure generally seen as characteristic of a more exclusive deuterostome group, the chordates, and implying that the vetulicolians have a more 'crownward' position in the tree of life. As interpreted by Shu and colleagues, the fossils lack a notochord and so cannot be regarded as chordates. But vetulicolians could be their immediate sister taxon.*

as a template. Presumably host enzymes catalyse the process. It is suggested that viroids may cause their symptoms by interference with gene regulation in the infected host cells. They have been identified as responsible for some very important plant diseases.

virulence The level of pathogenicity engendered by a pathogen towards its host. Many plant pathogens, and bacterial pathogens of humans, produce specialized proteins which are required for, or enhance, their virulence in the host. Some deliver 'virulence factors' ('type III' proteins; e.g. haemolysins, hyaluronidase, collagenase, streptokinase) into host cells, in bacteria often by injection. Evasion of host defences by *Yersinia pestis* (cause of bubonic plague), includes a secreted protein tyrosine phosphatase which blocks phagocytic uptake, and a protein causing actin microfilament disruption. In one evolutionary model, there is a 'trade-off' between how fast a pathogen reproduces and how easily it can infect new hosts, and there is a common belief that evolution favours mildness,

allowing host and parasite to enjoy a 'peaceful co-existence'. Pathogens transmitted by host contact may be selected for reduced virulence, since mobile hosts aid their spread; but vector-transmitted pathogens may be spread as well, or better, by incapacitated hosts, and here selection may favour increased virulence. Health workers' hands may be unwitting selectors for increased virulence in human pathogens. See ARMS RACE, ATTENUATION, RESISTANCE GENES, VACCINES.

virus One of a group of minute infectious agents (20–300 nm long and/or wide), unable to multiply except inside living cell of a host, of which they are obligate parasites and outside of which they are inert. They are biological systems, since each contains molecular information in the form of nucleic acid which is transcribed and replicated within the host cell. But they are not normally regarded as 'living', since none exhibits enzyme activity away from its host cell. Unlike cells, they almost never contain both RNA and DNA: human cytomegalovirus particles were the first to be recognized as containing both DNA and mRNA. They

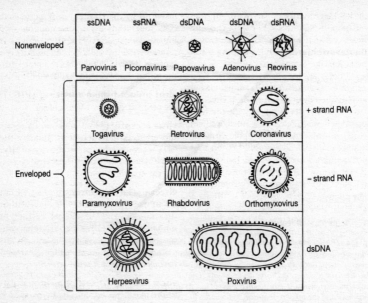

FIG. 170 *Illustration of the important families of animal viruses. ss = single-stranded; ds = double-stranded. See* VIRUS *for clarification.*

pass through filters which trap bacteria and are resolvable only by electron microscopy. When inside a cell, may pack together in a crystalline condition through symmetry and regularity of their capsids. The fully formed virus particle (*virion*) contains its nucleic acid or nucleoprotein core either within a naked coat of protein (*capsid*) consisting of protein subunits (*capsomeres*) or within a capsid enveloped by one or more host cell membranes acquired during exit from the cell and often modified, as by addition of specific glycoproteins. Capsid symmetry may be *helical* (capsomeres forming a rod-shaped helix around the nucleic acid core, e.g. tobacco mosaic virus, TMV); or *icosahedral* (*spherical*, capsomeres forming a 20-sided structure), capsomere number ranging from 12 (X174) to 252 (adenovirus). Some viruses (*phages*) infect fungi (*mycophages*) or bacteria (BACTERIOPHAGES).

Viruses may be classified in terms of the type of nucleic acid core they contain. In RNA viruses this may be either single-stranded (e.g. picornaviruses, causing polio and the common cold) or double-stranded (e.g. reovirus, causing diarrhoea); likewise DNA viruses (e.g. single-stranded in X174 phage and parvovirus; double-stranded in adenovirus, herpesvirus and poxvirus). RNA viruses are of three main types: + *strand RNA viruses*, whose genome serves as mRNA in the host cell and serves as template for a minus (–) strand RNA intermediate; – *strand RNA viruses*, which cannot serve directly as mRNA but rather as templates for mRNA synthesis via a virion transcriptase; and *retroviruses*, which are + strand and can serve as mRNAs but on infection immediately act as templates for double-stranded DNA synthesis (which immediately integrates into the host chromosome) via a contained or encoded REVERSE TRANSCRIPTASE. Human T-lymphotropic viruses (HTLVs) are single-stranded retroviruses and can cause acquired immune deficiency syndrome (AIDS). Any human *immunodeficiency virus* (see HIV) forms part of the *lentivirus* subgroup of RNA retroviruses. Parvovirus invades only rapidly dividing cells, e.g., of villi, and heart cells in young animals. Orthomyxoviruses include those causing INFLUENZA. See Figs. 90 and 170.

While bacteriophages inject their genomes into the host cell, animal viruses are engulfed by endocytosis or, if encapsulated by a membrane, fuse with the host plasmalemma to release the nucleoprotein core into the cell. HIV and poliovirus are among many with VIRUS RECEPTORS on host cell membranes. Once inside, the genome is commonly initially transcribed by host enzymes; but virally encoded enzymes usually then take over. Host cell synthesis is generally shut down, the viral genome replicated, and capsomeres synthesized prior to assembly into mature virions (see SELF-ASSEMBLY). The virus commonly encodes a late-produced enzyme, rupturing the host plasma membrane (the *lytic phase*) to release the infective progeny; but a viral genome may become integrated into the host chromosome and be replicated along with it (a *provirus*). Many eukaryotic genomes have proviral components. Sometimes this results in *neoplastic transformation* of the cell (see ONCOGENE, CANCER CELL) through synthesis of proteins normally produced only during viral multiplication. DNA tumour viruses include *adenoviruses* and *papovaviruses*; RNA tumour viruses are capsulated and include some retroviruses (e.g. Rous sarcoma virus). The DNA of viruses invading eukaryotic cells often contains ENHANCERS activated by host cell proteins. Viruses may be used as genetic tools in TRANSDUCTION. They may have evolved from PLASMIDS that came to encode capsid proteins. Host cells respond to viruses without human assistance in many ways (e.g. see PLANT DISEASE AND DEFENCES, RNA SILENCING, INTERFERONS), and attacking viral genomes is a long-evolved host defence mechanism. Recent work on a cellular protein called APO-BEC3G shows that it blocks replication of retroviruses (including HIV) by deaminating cytidine in the RNA genome and its DNA reverse transcript, leading to increase in the mutation rate of G → A in newly synthesized HIV DNA. Remarkably, APOBEC3G is

taken up by the HIV virion in permissive cells and leads to hypermutation and impaired replication of the virus, which has responded in this ARMS RACE by producing a protein, Vif, which prevents its uptake into the virion. See IMMUNITY, VACCINES.

virus-induced gene silencing (VIGS) See GENE SILENCING.

virus receptors Cell ADHESION receptors subverted as virus receptors, notably the CD4 ACCESSORY MOLECULE in the case of HIV, and ICAM-1 in the case of rhinoviruses causing the common cold.

viscera The large organs housed within an animal's coelomic cavities (i.e. within the thoracic and abdominal cavities of humans). The adjective *visceral* is often contrasted with *somatic*, where the latter in this context relates to structures lying external to the coelomic cavities.

visceral arch (1) One of a series of partitions in fish and tetrapod embryos on each side of the pharynx. In fish, lying between mouth and spiracle and between this and adjacent gill slits; in tetrapod embryos, between corresponding gill pouches; (2) skeletal bars (cartilage or bone) lying in these partitions in fish (see GILL BAR). First (mandibular) arch is modified to form the jaw skeleton (PALATOQUADRATE and MECKEL'S CARTILAGE); the second (HYOID ARCH), behind the spiracle, commonly attaches jaws to the skull, each remaining (branchial) arch lying behind a functional gill slit.

visceral pouch See GILL POUCH.

visceral purple See RHODOPSIN.

vitalism Metaphysical doctrine with early roots, popular in a variety of forms during 19th century. Opposed to alternative extreme of scientific materialism. Underlying most vitalisms was the conviction that life was more than mere complex chemistry, otherwise science would subject even human activity to deterministic explanations. Ignorance of biochemical principles made such phenomena as growth and development mystifying, employing terminology beyond physics or chemistry.

Such causal agencies as 'entelechies' and 'vital forces' were invoked for really baffling phenomena but added nothing to understanding. Vitalists, often religiously inspired, eschewed Darwin's mechanistic philosophy.

vital staining See STAINING.

vitamin Organic substance not normally synthesized by an organism, which it must obtain from its environment in minute amounts (a micronutrient). ESSENTIAL AMINO ACIDS are required in larger amounts, and are not vitamins. *Provitamins*, closely related precursors of some vitamins, may occur in the diet. Absence of a vitamin from the diet for sufficient time gives symptoms of a resulting *deficiency disease*, although it is often unclear why particular symptoms occur. A vitamin for one organism may not be so for another.

Classified into fat-soluble (lipid) and water-soluble forms. For humans the only lipid vitamins are A, D, E and K, all stored in the liver and CHOLINE; while all water-soluble vitamins are converted to coenzymes, accounting for the small amounts needed. See particular vitamin entries.

vitamin A 11-*cis*-retinal, the lipid prosthetic group of the protein opsin in visual purple. Its deficiency affects all tissues, but the eyes are most readily affected. Young animals, lacking a liver store of the vitamin, are most affected by deficiency, which causes *xerophthalmia* ('dry eyes') in human infants and young children. See RHODOPSIN.

vitamin B complex Several water-soluble vitamins (currently 12). B_1 (aneurine, or thiamine) prevents beri-beri. Group includes BIOTIN, CYANOBALAMIN (B_{12}), FOLIC ACID, NICOTINIC ACID, PANTOTHENIC ACID, PYRIDOXINE (B_6) and RIBOFLAVIN (B_2).

vitamin C See ASCORBIC ACID.

vitamin D Small group of fat-soluble (lipid) vitamins of humans and some other animals, deficiency causing rickets. Some dietary in origin, but also synthesized in the skin under ultraviolet light. One form (D_2, *ergocalciferol*) derives from provitamin ergosterol; another (D_3, *cholecalciferol*) derives from provitamin dehydrocholesterol. Both act as hormones (see NUCLEAR RECEPTORS). People receiving insufficient sunlight can supplement their vitamin D levels by eating liver, particularly fish liver oils. The main circulating form of vitamin D in animals, *25-dihydroxycholecalciferol*, is very active and derived by modification of cholecalciferol in the liver. Like other forms of the vitamin it promotes uptake of calcium ions in the ileum and is required for calcification of bones and teeth. See CALCITONIN.

vitamin E (tocopherol) Group of fat-soluble vitamins obtained principally from plant material (seed oils, wheat germ oil) but also found in dairy produce. Deficiency produces infertility in male and female rats, and probably in most vertebrates, kidney degeneration and other general wasting symptoms. They seem to prevent oxidation of highly unsaturated fatty acids (and their polymerization in cell membranes) in presence of molecular oxygen.

vitamin K Fat-soluble vitamins (K_1 and K_2), required for liver synthesis of a substance (proconvertin) required for prothrombin production, and so for conversion of fibrinogen to fibrin during BLOOD CLOTTING. Produced by many plants (K_1), and by microorganisms (K_2), including those in the gut.

vitelline membrane See EGG MEMBRANES.

vitreous humour Jelly-like semitransparent substance filling cavity of the vertebrate eye behind the lens.

viviparity (vivipary) (Zool.) Reproduction in animals whose embryos develop within the female parent and derive nourishment by close contact with her tissues. Involves PLACENTA in placental mammals; occurs also in some reptiles, amphibians, elasmobranchs, waterfleas (cladocerans), aphids and tsetse flies. See R-SELECTION. (Bot.) Having seeds germinating within the fruit, e.g. mangrove; or producing shoots (e.g. bulbils) for vegetative reproduction, instead of inflorescences; e.g. some grasses.

VNTR (variable number of tandem repeats) See SATELLITE DNA.

vocal cords See LARYNX, SYRINX.

Volkmann's canals See HAVERSIAN SYSTEM.

voltage-gated ion channels Those cation ION CHANNELS whose openings and closings depend upon the voltage across the membrane in which they lie. Propagation of action potentials along excitable cell plasma membranes (neuron and muscle) depends upon voltage-gated sodium and potassium channels, while voltage-gated calcium channels open when the action potentials reach the nerve terminal (see IMPULSE, MUSCLE CONTRACTION). The sodium channels take different allosteric conformations: closed, inactivated and open (see REFRACTORY PERIOD). The action potentials in certain muscle, egg and endocrine cells depend upon voltage-gated calcium channels rather than sodium channels. The proteins forming the channels exhibit a variety of ISOFORMS.

voltage clamp Technique whereby the potential difference across a membrane is kept steady so that movements of ions across it can be measured. Employed a great deal in nerve and muscle research.

voluntary muscle Alternative term for STRIATED MUSCLE.

volva Cup-like fragment of the universal veil at base of stipe in some members of the BASIDIOMYCOTA (Order Agaricales).

vomeronasal organ (Jacobson's organ) Specialized part of the olfactory system of numerous tetrapods, detecting stimuli from food; persisting in most mammals where they sometimes open via a secondary palate into the mouth cavity. In some reptiles (lizards and snakes) they open to the palate independently of the choanae (internal nares) and in snakes detect particles in the air brought to them by the flicking tongue.

vulva (1) Mammalian external genitalia, comprising two pairs of fleshy tissue folds, the *labia*. (2) The external opening of the ovary in nematode worms, comprising 22 cells but deriving from just four cells in *Caenorhabditis elegans*, one of them inducing the other three in development. Develops at the last larval stage.

W

Wallace, Alfred Russel British naturalist and traveller (1823–1913). According to T. H. Huxley, his short essay on the mechanism of evolution, received by Charles DARWIN from the Moluccas in June 1858, 'seems to have set Darwin going in earnest' on writing his *The Origin of Species* . . . Their joint paper of 1858, read at the Linnean Society, was the first publicized account of the theory of natural selection. At first, Wallace held that human evolution could be explained by this theory, but later departed from Darwin on this, believing a guiding spiritual force necessary to account for the human soul. Wallace considered SEXUAL SELECTION to be less important in evolution than did Darwin, holding that, unlike Darwin, it had no role in the evolution of human intellect. His contributions to zoogeography were of great importance (see WALLACE'S LINE), and his book *Darwinism* (1889) attributes a wider role for natural selection in evolution than did Darwin's works.

Wallace's Line Boundary drawn across the Malay Archipelago, separating Arctogea and Notogea (formerly between Australian and Oriental ZOOGEOGRAPHICAL REGIONS). Not a precise demarcation line, lying along the edge of the south-east Asian continental shelf, between Bali and Lombok, Borneo and Celebes, the Philippines and the Sangai and Talaud Islands. Named by T. H. Huxley after the zoogeographical work of WALLACE (see DARWIN).

wall pressure (Bot.) Pressure exerted against the protoplast by plant cell wall; opposed to turgor pressure. See WATER POTENTIAL.

warm-blooded See HOMEOTHERMY.

warning colouration See APOSEMATIC, MIMICRY.

water potential Term indicating the net tendency of any system to donate water to its surroundings. The water potential, ψ_ω, of a plant cell is the algebraic sum of its wall pressure (pressure potential), ψ_p, and of its osmotic (solute) potential, ψ_o:

$$\psi_\omega = \psi_p + \psi_o.$$

Since the water potential of pure water at atmospheric pressure is zero pressure units by definition, any addition of solute to pure water reduces its water potential and makes its value negative. Water movement in nature is therefore always from a system with higher (less negative) water potential to one with lower (more negative) water potential. The units normally employed are *megapascals* (MPa; 1 Pa $= 1$ Newton/m^2), or *millibars* (1 mbar $= 10^2$ Pa). Standard atmospheric pressure, equal to the average pressure of air at sea level, is 1.01 bar.

The term has wide applicability. The water potential of any cell is its net tendency to donate water to its surroundings, one with a higher water potential than an adjacent cell tending to donate water to it. Plant cells in equilibrium with pure water have zero water potential and will be fully *turgid* (see TURGOR PRESSURE). The volume of water entering or leaving a plant cell during equilibrium attainment is negligible and does not alter the cell's solute potential.

When applied to non-cellular systems such as solutions separated by a selectively permeable membrane, water potentials dictate the net direction in which water moves by osmosis, since net movement of water is always from higher to lower water potential values. Osmotic effects can therefore be subsumed under the wider category of water potential effects, for membranes may or may not be involved in phenomena explained in terms of water potential. It is therefore valuable in predicting water movement between cells, between cells and soil solutions and between cells and the atmosphere. It reduces to one the number of units employed in analysing the TRANSPIRATION STREAM in plants. Compare OSMOSIS.

water vascular system System of water-filled canals forming tubular component of echinoderm coelom, usually with direct connection (the *madreporite*) with the exterior. A circumoesophageal water ring internal to the skeleton usually gives off radial water canals which pass through pores in the skeleton wherever *tube-feet* protrude. Each tube-foot is expanded by a muscular ampulla internal to the skeleton and is coordinated with others in feeding and locomotion. See TUBE-FEET.

waxes FATTY ACID esters of ALCOHOLS with high molecular weights. Occur as protective coatings of arthropod CUTICLE, skin, fur, feathers, leaves and fruits, often reducing water loss. Also found in beeswax and lanolin.

W-chromosome An alternative notation for the Y-chromosome in organisms where the heterogametic sex is female. See SEX CHROMOSOME.

weed The concept of what is or what is not a weed is not precisely defined because it has both biological and sociological elements. Biologically, weeds are plants having the ability to colonize, inhabit and thrive in continually disturbed habitats, most especially in areas that are repeatedly affected by human activity. Sociologically, weeds are plants growing where someone wishes they would not. This can be regarded casually as simply a plant out of place. Weeds that thrive in areas of continual disturbance may be regarded as belonging to two intergrading components, RUDERALS and AGRESTIALS. Weeds occur in all growth forms

and in many lifestyles. The majority are flowering plants sharing the following characteristics, which tend to suggest *R*-SELECTION. Seed germination requirements can be fulfilled in many different environments. Germination is often discontinuous (e.g. chickweed) and internally controlled. Seeds have a great longevity. Once germinated the plants grow rapidly to flowering, and one can have continuous seed production for as long as conditions permit. Prolific amounts of seed are produced, and plants have the ability to produce some seed in unfavourable environments. Plants are usually self compatible but not completely autogamous or apomictic. Weedy plants have developed strategies for both short- and long-term dispersal and, if perennial, many display vigorous vegetative regeneration or can regenerate from fragments. They are also good interspecific competitors (e.g. rosette formers, choking growth or allelochemical production). Weeds often provide habitats for insect and fungal pests. Leguminous weeds can add nitrogen to the soil through N-fixing bacteria, and all increase soil fibre content on ploughing in. Most troublesome and aggressive weeds are those that are foreign or alien that have invaded or been introduced from elsewhere in the world. In North America *Lythrum salicaria* (purple loosestrife) entered via seaport cities along the east coast, as well as through garden introductions and is now a prominent and aggressive weed of wetlands in both Canada and the northern United States. See GMOs.

Weismann, August Theoretical biologist (1834–1914); one of the first to appreciate that chromosomes are the physical bearers of hereditary determinants and to forge a comprehensive theory of heredity. With Wilhelm Roux, he advocated the theory of MOSAIC DEVELOPMENT. Best remembered for his theory of the continuity of GERM PLASM. Prolific coiner of terms and entities.

Wernicke's area Region of the human temporal cortex concerned with the analysis of incoming sounds (hence with understanding?) and located in the dominant hemisphere. Projections lead from Wernicke's area to BROCA'S AREA.

Western blotting (immunoblotting) Similar procedure to NORTHERN BLOTTING (see Fig. 116), but for separating proteins rather than mRNAs.

whalebone See BALEEN.

whales See CETACEA.

wheat Although 'wheat' is commonly employed to refer to the modern bread wheat, *Triticum aestivum*, the term refers to a complex of several species, two of which are diploid, nine tetraploid, and four hexaploid. Different varieties are employed for different purposes; e.g. for pasta, biscuits and bread. Bread wheat has a complex history, involving non-conscious as well as conscious human artificial selection and both hybridization and POLYPLOIDY (see for details) have been involved. Einkorn wheat (*Triticum monococcum*, whose dated remains have been found in Swiss lake dwellings and at Jarmo, Iraq, at 9000 yr BP) produces only a single fruit in each spikelet, but although giving poor yield grows well on poor soils unsuitable for most other crops. The many varieties of bread wheat may be subdivided into 'spring wheats' (sown in early spring and harvested four months later) and 'winter wheats' (sown in autumn, germinating before winter begins, and harvested the following mid-summer). Wheat can have a slightly higher protein content (up to 16%) than RICE (up to 11%) and MAIZE (up to 12%). The major part of the wheat crop is milled to produce flours used as ingredients in other foods. Wheat's insoluble proteins produce GLUTEN when flour is kneaded with water, and the elastic properties of gluten make bread production possible. See CEREAL, COMPLEMENTARY RESOURCES.

white blood cell See LEUCOCYTE.

white matter Myelinated nerves forming through-conduction pathways of vertebrate BRAIN and SPINAL CORD, with associated glia and blood vessels. Compare GREY MATTER, to which it normally lies superficially.

whorl Three or more leaves, flowers, sporangia or other plant parts at one point on an axis or node.

wild rice See RICE.

wild type Originally denoting the type (genotype or phenotype) most commonly encountered in wild populations of a species, but commonly applied to laboratory stock from which mutants are derived. As in natural populations, the laboratory wild type is subject to change. In genetics, term is usually employed in contexts of alleles and particular phenotypic characters. Such alleles are normally represented by the symbol + rather than by a letter, a genotype homozygous for a wild type allele being given by either ++ or +/+. Wild type characters are by no means always genetically dominant (see DOMINANCE).

wilting Condition of plants in which cells lose turgidity and leaves, young stems, etc. droop. Results from excess of transpiratory water loss over absorption. Excessive wilting may be stressful, even irreversible.

wings See FLIGHT.

wishbone (furcula) Fused clavicles and interclavicles, diagnostic of birds. Present in *ARCHAEOPTERYX*.

Wnt signalling *Wnt* genes form a family of conserved vertebrate genes which encode secreted glycoproteins (Wnt proteins) modulating cell fate and behaviour in embryos by activating receptor-mediated signal pathways (see SIGNAL TRANSDUCTION). Their functions include determination of the dorsoventral embryonic AXIS. See SEX DETERMINATION.

wobble hypothesis Hypothesis, due to F. H. C. Crick, that the nucleotide at the 5'-end of a tRNA anticodon plays a less important role than the other two in determining which tRNA molecule will align by base-pairing with a codon triplet in mRNA. Helps to explain pattern of degeneracy observed in the GENETIC CODE, which is restricted to the 3'-ends of codons. See HYPOXANTHINE, INOSINIC ACID.

Wolbachia Rickettsia-like bacterial genus found as parasite of, at least, nematodes, mites, crustaceans, flies, mosquitoes and some wasps. Transmissible in egg cytoplasm, but not sperm. Alters the reproductive potential of its host: feminization of genetic males in isopod crustaceans; parthenogenesis (induced thelytoky) in wasps (see ABERRANT CHROMOSOME BEHAVIOUR (3)) and interstrain CYTOPLASMIC INCOMPATIBILITY in a large variety of insects. One consequence of cytoplasmic incompatibility caused by *Wolbachia* in the wasp *Nasonia vitripennis* is that it alters the latter's normal pattern of reciprocal centrosome inheritance. Controversy exists as to whether *Wolbachia* is on the evolutionary path to becoming a cell organelle.

Wolffian duct (archinephric duct) Vertebrate kidney duct; one on each side. Develops in all vertebrates (initially in both sexes) from region of pronephros, becoming the mesonephric duct. In male amniotes, conveys both urine and semen to the exterior (opening into the cloaca). In male amniotes it becomes the vas deferens (vasa efferentia and epididymis representing mesonephric tubules which now lead into it from the testis). In all adult amniotes an outgrowth from lower end of the Wolffian duct develops into the URETER, invading the metanephric kidney. Hormonal influences from the gonads determine whether the Wolffian duct or the MÜLLERIAN DUCT will persist and differentiate in the adult.

womb See UTERUS.

wood See XYLEM.

woodlice See ISOPODA.

worker See CASTE.

X

Xanthophyceae A class in the algal division HETEROKONTOPHYTA comprising about 100 genera and 600 species of primarily freshwater algae. Unicellular and colonial forms are found in the plankton of lakes and ponds; much rarer in the marine environment. Many species live on damp soil.

xanthophyll CAROTENOID pigment; each an oxygenated derivative of carotenes.

X chromosome See SEX CHROMOSOMES, CHROMOSOMAL IMPRINTING.

X chromosome inactivation (XCI) The silencing mechanism eutherian mammals use to equalize the expression of X-linked genes between males and females early in development. See DOSAGE COMPENSATION.

xenogeneic Originating from a (genetically) different species. Compare ALLOGENEIC, AUTOGENEIC. See XENOTRANSPLANTS.

xenograft See GRAFT.

Xenopus laevis The claw-toed frog of Africa. The most commonly used amphibian in developmental work.

xenotransplants (xenografts) Animal organs and tissues employed in TRANSPLANTATION in lieu of appropriate human material. Such is the demand for human organs that as few as 5% of the organs needed in the USA ever become available. Xenotransplants offer a ready supply of organs and some advantages over human material in that the operation can be planned in advance while the organ is fresh and undamaged, and appropriate pre-treatment of the patient can be organized. Pathogens tend to be species-specific, and xenotransplants may help avoid recurrence of disease in the graft. It may also be possible to manipulate the donor animal genetically or biochemically so as to reduce the risk of rejection or even produce specific benefit to the patient. The preferred animal source is currently the pig which, unlike primates, can be genetically manipulated to express heterologous genes, and is also less likely to be a source of lethal viruses.

The innate IMMUNITY of the host most likely to cause rejection of xenografts involves (i) pre-existing ANTIBODIES (*xenoreactive natural antibodies*) recognizing cell-surface antigens of the donor vascular endothelial cells (*hyperacute rejection*, HAR); (ii) COMPLEMENT and (iii) NATURAL KILLER CELLS. Host adaptive immune responses most likely to cause organ rejection are T CELL-mediated responses, which are also most likely to be a problem in free tissue grafts in extracellular spaces. If HAR can be prevented, the transplant may be subject to delayed xenograft rejection (*acute vascular rejection*, AVR) in which the foreign endothelium becomes activated to cause BLOOD CLOTTING and INFLAMMATION and the latter, especially, will induce T cell responses. *Chronic rejection* is a major cause of allograft organ rejection, the relative roles of immune responses and structural damage to the transplant being hotly debated. There is always a risk of zoonosis infection in xenotransplants and it is to be hoped that screening for such pathogens as viruses and prions will be adequate and supported by basic research. See ALLOGRAFT.

xeromorphy (Bot.) Possession of morphological characters associated with XERO-PHYTES.

xerophyte Plant of arid habitat, able to endure conditions of prolonged drought (e.g. in DESERTS), due either to a capacity during the brief rainy season for internal water storage, used when none can be obtained from the soil, plus low daytime transpiration rates associated with stomatal closure (e.g. in such succulents as cacti); or ability to recover from partial desiccation (e.g. such desert shrubs as the creosote bush). As long as water is freely available, it seems that non-succulent xerophytes transpire as much as, or more freely than, mesophytes but unlike the latter they can endure periods of permanent wilting during which transpiration is reduced to minimal levels. Features associated with this reduction, in different xerophytes, include dieback of leaves covering and protecting perennating buds at soil surface, shedding of leaves when water supply is exhausted, heavily cutinized or waxy leaves, closure or plugging of the stomata (which are sunken or protected), orientation or folding of leaves to reduce insolation, and a microphyllous habit reducing the possibility of drought necrosis of the mesophyll. See HYDROPHYTE, MESOPHYTE.

xerosere SERE commencing on a dry site.

Xiphosura See MEROSTOMATA.

X-ray diffraction (X-ray crystallography) Technique where a narrow beam of X-rays is fired through a crystalline source (e.g. of DNA or protein), the arrangement of atoms within the crystal being interpreted from the X-ray diffraction pattern produced (usually by computer reconstruction). Loose quasicrystalline structures such as cell walls and cell membranes are sometimes analysed this way.

X-ray irradiation A form of ionizing electromagnetic radiation with beneficial effects in microbiological sterilization and food preservation (although here some toxicity may result) but also a potentially hazardous mutagen in that it breaks chromosomes and causes developmental abnormalities (teratologies). There is no evidence for congenital abnormalities caused by diagnostic (clinical) X-ray irradiation treatment. Mammalian DNA damage caused by mild X-ray irradiation may block passage through two CELL CYCLE 'checkpoints', namely the restriction point (START) and the entry to mitosis at G_2-M. Moderate X-ray irradiation in S phase causes failure of new DNA replication origins to develop which would otherwise allow DNA repair before S phase continues. Mild uv and X-ray irradiation elevate the level of protein p53 (see CANCER CELL) by increasing its half-life, inducing G_1 arrest of the cell cycle.

xylem Wood. Mixed vascular tissue, conducting water and mineral salts taken in by roots throughout the plant, which it provides with mechanical support. Of two kinds: *primary*, formed by differentiation from procambium and comprising protoxylem and metaxylem, and *secondary*, additional xylem produced by activity of the cambium. Characterized by presence of tracheids and/or vessels, fibres and parenchyma. In mature woody plants, makes up bulk of vascular tissue itself and of entire structure of stems and roots. See TRANSPIRATION STREAM.

Y

YAC See YEAST ARTIFICIAL CHROMOSOME.

Y chromosome See *HOMO*, SEX CHROMOSOMES.

yeast artificial chromosome (YAC) A cloning VECTOR used for cloning huge fragments of DNA within yeast cells. Each contains, in addition to selectable markers and a cloning site, an *autonomously replicating sequence* (ARS, derived from yeast chromosomal DNA), a yeast centromere (for stability) and a yeast TELOMERE at each end (enabling replication as linear chromosome-like molecules). YACs behave like normal yeast chromosomes and replicate each time the cell goes through mitosis. The DNA for cloning is produced by digestion with rare cutting endonucleases, enabling production of 'YAC libraries' (see GENE LIBRARY). It is possible to identify overlapping sequences, even those making up entire human chromosomes. Human telomeres have been cloned within YACs, and YACs of >1 Mbp are now common. But, unfortunately, libraries of such large inserts often contain chimaeric clones. The technique of CHROMOSOME WALKING has been largely superseded by that of POSITIONAL CLONING. YACs may offer scope for human gene therapy, providing an unaffected allele to replace the affected one by homologous recombination: the large homologous regions would be advantageous. See BACTERIAL ARTIFICIAL CHROMOSOMES, HUMAN ARTIFICIAL CHROMOSOME, HUMAN GENE THERAPY, GENOME MAPPING.

yeasts Widely distributed unicellular members of the ASCOMYCOTA, which reproduced asexually by fission or by budding, rather than by spore formation. Sexual reproduction occurs when either two cells or two ascospore units unite and form a zygote. The zygote may also produce diploid buds or it may function as an ASCUS, undergoing meiosis and producing four haploid nuclei; a mitotic division may then take place, and then within the zygote wall, which is now the functional ascus, cell walls are formed around each of the nuclei so that either four or eight ascospores are formed. Yeasts are of great economic importance because of their ability to ferment carbohydrates, breaking down glucose to produce ethanol (ethyl alcohol) and carbon dioxide in the process. The brewing, winemaking and baking industries depend upon this capacity, with brewers and vintners utilizing yeasts to produce alcohol, and bakers utilizing yeasts to produce carbon dioxide. Most of the yeasts important in the production of wine, beer, cider and sake are strains of just one species, *Saccharomyces cerevisiae*, although other species also play a role. The complete genome of *Saccharomyces cerevisiae* was published in 1997, and was the first eukaryote genome to be sequenced. Some species of yeasts are important human pathogens, causing such diseases as thrush and cryptococcosis. Others, especially *S. cerevisiae*, are important laboratory organisms for genetic research, and some are used as hosts in DNA CLONING techniques of BIOTECHNOLOGY. Others are used commercially as sources of proteins and vitamins. See MATING TYPE.

yeast two-hybrid system A powerful functional assay and generic method for studying potential protein–protein interactions. OPEN READING FRAMES (ORFs) are fused to the DNA-binding or the DNA-activating domains of the gene GAL4 situated upstream of a REPORTER GENE such that if the proteins encoded by the ORFs interact within the yeast nucleus, the reporter gene is switched on due to the close proximity of the two domains. In large-scale situations, the strategy has been used where yeast clones containing ORFs as fusions to DNA-binding or -activation domains are arrayed on a grid and the ORFs to be tested (for reciprocal fusion ability) are screened against the entire grid to identify activating clones. Such a method has been employed to detect potential protein–protein interactions in more than one species. The inherent artificial nature of the assay has given rise to criticism, and methods to determine the biological relevance of potential interactions identified need to be developed. See NUCLEIC ACID ARRAYS, PROTEIN COMPLEXES.

yolk Store of food material, mainly protein and fat, in eggs of most animals. See POLARITY, CENTROLECITHAL.

yolk sac Vertebrate EXTRAEMBRYONIC MEMBRANE, conspicuous in elasmobranchs, teleosts, reptiles and birds. Contains yolk and hangs from ventral surface of the embryo. Has outer layer of ectoderm, inner layer (usually absorptive) of endoderm, with mesoderm containing coelom and blood vessels between. A gut diverticulum; yolk usually communicating with the intestine. In mammals it is normally devoid of yolk, forming part of the CHORION; in marsupials, it forms an integral component of the PLACENTA. As yolk is absorbed, the yolk sac is withdrawn, eventually merging into the embryo. It is the source of primordial GERM CELLS in mammals.

Y-organ Pair of epithelial endocrine glands in the heads of malacrostracan crustaceans, anterior to the brain. Appear to secrete an ECDYSONE-like hormone involved in moulting.

Z

Z-DNA DNA sequences in the form of a Z-FORM HELIX.

zeaton A CYTOKININ.

zeitgeber Any environmental entraining cue in a CIRCADIAN RHYTHM. Commonly involves light or temperature.

Z-form helix A left-handed form of double helical DNA containing the alternating same-strand sequence dG-dC. It zigs and zags (hence the name); although found in crystals of synthetically prepared DNA, its presence in naturally occurring DNA is uncertain. The Z-DNA structural motif is stabilized by transcription-dependent negative supercoiling, and such sequences seem to occur towards the 5' end of genes rather than the 3' end (see DNA METHYLATION).

zinc finger See DNA-BINDING PROTEINS, MOTIF.

Zingiberidae A subclass of the LILIOPSIDA comprising two orders (Bromeliales and Zingiberales) and about 3,800 species. Characteristically sepals are well-differentiated from the petals and are often green and herbaceous, sometimes petaloid in texture. The endosperm is typically starchy and mealy, with compound starch grains. Leaves are narrow and parallel-veined, or equally often with a broad expanded, petiolate blade and a characteristic pinnate-parallel venation. Members of the Bromeliales (bromeliads) are wholly American (mainly tropical and subtropical) except for a single species from western tropical Africa (*Pitcairnia feliciana*). Members of the Zingiberales are almost exclusively tropical with most occurring in moist forests or in wet habitats, but some (including ginger) are adapted to seasonally dry habitats.

Zinjanthropus Genus of hominid containing the species *Z. boisei*, now renamed *Australopithecus boisei*. See AUSTRALOPITHECINE.

Z lines See STRIATED MUSCLE.

zona pellucida Glycoprotein membrane around mammalian ovum, disappearing before implantation. Secreted by cells of the GRAAFIAN FOLLICLE.

zonation The characteristic distribution of species along environmental gradients; e.g. of plants and animals from the uppermost splash zone to the low water spring tide level of a rocky seashore. Zonation is determined by gradients of biotic and abiotic factors, and the levels to which organisms are adapted to particular combinations of such factors over time. Levels of competition and predation are often influential in the locations of the distribution zones.

zonula adherens One kind of DESMOSOME.

Zoochlorella Symbiotic green algae assignable to the CHLOROPHYTA (e.g. *Chlorella*, *Oocystis*); both freshwater and marine. See ZOOXANTHELLAE.

zoogeographical regions (faunal provinces) Subdivisions of world surface into regions identified by differences in their dominant animals. They are: (a) Arctogea (Europe, Asia, Africa, Indochina and North America). It includes the Palaearctic (Asia, Europe and North Africa), Oriental (Indochina) and Nearctic (Greenland and North America south to Mexico) regions; (b) Neogea (Central and South America), equivalent to the Neotropical region; (c) Notogea (Australasia), normally taken to include Australia, Tasmania, New Zealand, eastern Indonesia and Polynesia. The island of Madagascar is often considered sufficiently distinct faunally to be regarded as a minor region: the Malagasy region. Botanical equivalents are known as *floral regions*. See WALLACE'S LINE, ZOOGEOGRAPHY.

zoogeography Study of, and attempt to interpret, global distributions of animal taxa. Takes into account CONTINENTAL DRIFT and other major geological processes (e.g. orogenies, ice ages and other factors affecting sea level), as well as aspects of ecology and evolution (e.g. adaptive radiation, competition, speciation, geographical isolation, dispersal and vagility). It is largely due, for example, to isolation since the Cretaceous of the fauna of Notogea (see ZOOGEOGRAPHICAL REGIONS) on the Australian plate that its present fauna is so distinctive. Likewise, separation of Malagasy region from the rest of Africa prior to higher primate radiation results in a relict lemur population persisting there today. Island faunas, as Darwin and Wallace knew, have high levels of endemism, particularly if there is considerable separation from mainland faunas. See PLANT GEOGRAPHY, GONDWANALAND, LAURUSSIA.

zooid Member of a colony of animals (chiefly ectoprocts and entoprocts) in which individuals are physically united by living material. See COLONY.

zoology Branch of biology dealing specifically with animals. There is considerable overlap, however, with botany.

zoonosis Any infection of a human by a pathogen whose source is a reservoir of a non-human animal population.

zooplankton Animal members of plankton.

zoosporangium (Bot.) Sporangium in which zoospores are formed.

zoospore (swarm spore) Naked spore produced within a sporangium (zoosporangium); motile, with one, two or many flagella; present in certain algae and oomycotan protists.

zootype A proposed defining character (synapomorphy) of Kingdom Animalia. Animals would, on this suggestion, be those eukaryotes whose genomes contain the POSITIONAL INFORMATION system of *Hox* gene clusters, colinearly expressed along the chromosome and in their positions within the embryo during development (see HOMEOBOX). The stage at which the zootype is most clearly displayed appears to be the PHYLOTYPIC STAGE for that animal phylum.

Zooxanthellae Algae varying in colour from golden, yellowish, brownish to reddish, living symbiotically in a variety of aquatic animals; includes algae assignable to the BACILLARIOPHYTA, CHRYSOPHYTA, DINOPHYTA, CRYPTOPHYTA; especially important in coelenterates of coral reefs. See ZOOCHLORELLA.

Zoraptera Small order of very small exopterygote insects (metamorphosis slight) of Subclass Orthopteroidea. Occur under bark, in humus, etc. Many species are dimorphic, apterous or winged, although wings can be shed.

Zosterophyta Extinct, seedless, vascular plants found in strata from the early to late DEVONIAN period, from *c.* 408–370 Myr ago. Zosterophytes were leafless, dichotomously branched and possibly aquatic plants, whose aerial stems possessed a cuticle. Stomata were present but confined to the uppermost branches. Named because of their general resemblance to modern seagrasses (e.g. *Zostera* – a marine flowering plant which resembles grasses). The sporangia were globose or kidney-bean-shaped, and were borne laterally upon short stalks. Plants were HOMOSPOROUS and probably ancestors of the lycophytes.

zwitterion Ion which has both positively and negatively charged regions. All AMINO ACIDS are zwitterions, although their charge distributions are much affected by pH.

Zygnematophyceae A class of the algal division CHLOROPHYTA comprising only freshwater algae (a few members are terrestrial). About 50 genera and between 4,000 and 6,000 species are known; non-branching, filamentous and unicellular forms are found with the majority being unicells (e.g. *Cosmarium, Staurastrum*). Fossil hypnozygotes dating back to the Carboniferous (*c.* 300 Myr BP) have been found for three extant genera, *Spirogyra, Mougeotia* and *Zygnema*. The hypnozygote wall is particularly resistant to decay as it contains a sporopollenin-like compound. The cell wall's fibrillar structural component is crystalline cellulose in the form of microfibrils. Cells are uninucleate with the chloroplasts displaying several forms (e.g. spiral, axial plate or two massive chloroplasts, which are often more or less stellate). Pyrenoids are found within the chloroplasts. Mitosis is semi-closed and there is a persistent telophase spindle. Cytokinesis is brought about through in-growth of a cleavage furrow, together in many instances with a cell plate, which is formed within a phragmoplast. Plasmodesmata are not present in the cross wall that is deposited. Flagellate cells are never formed. Sexual conjugation takes place by conjugation involving fusion of two amoeboid gametes. The LIFE CYCLE is haplontic and includes formation of a thick-walled hypnozygote.

zygomatic arch Bony arches formed by the temporal processes of the zygomatic bones (malars, 'cheekbones', with part of the outer wall and floor of the orbits) and the zygomatic processes of the temporal bones; for attachment of MASSETER MUSCLES. The TEMPORALIS MUSCLES pass under it.

zygomorphic BILATERAL SYMMETRY of flowers.

Zygomycota A division in the Fungi comprising two classes, the Zygomycetes and Trichomycetes. There are about 173 genera and 1,056 species, which are saprophytes or parasites. Sexual reproduction characteristic in Zygomycetes, involving two gametangia which fuse via conjugation, with a zygosporangium (zygote) resulting. Inside the zygosporangium, the gametes, which are just nuclei, fuse to form one or more diploid nuclei. Asexual reproduction occurs through production of non-motile sporangiospores borne within a sporangium. Most species possess coenocytic (multinucleate) hyphae, within which the cytoplasm can often be seen streaming rapidly. Commercially important in production of organic acids, pigments, fermented foods, alcohols and modified steroids. One most important group of the Zygomycetes is the order Glomales, members of which occur in MYCORRHIZAL associations. The thallus of members of the Trichomycetes (many of which parasitize arthropods) is simple or branched, attached to the host cuticle or to the digestive tract by a holdfast. Asexual reproduction in this class occurs through sporangiospores or arthrospores. Sexual reproduction is known in order Harpellales and one genus in the Eccrinales.

zygospore Thick-walled resting spore; product of conjugation in the ZYGOMYCOTA and some green algae. Also used for product fertilization in the isogamous *Chlamydomonas* and its relatives.

zygote Cellular product of gametic union. Usually diploid.

zygotene Stage in the first prophase of MEIOSIS.

zygotic meiosis Meiosis occurring during maturation or germination of a zygote.

zymogen Any inactive enzyme precursor (e.g. pepsinogen, trypsinogen or prothrombin). Activation to the functional enzyme generally involves excision of a portion of polypeptide. The several zymogens in secretory vesicles of pancreatic ACINAR CELLS are termed *zymogen granules*, although some active enzyme is contained there too. Zymogen granules fuse with the cell apex under influence of acetylcholine or cholecystokinin.

Appendix

Eon	Era	Period	Epoch
Phanerozoic	Cenozoic	Quaternary	Holocene
			Pleistocene
		Tertiary	Pliocene
			Miocene
			Oligocene
			Eocene
			Palaeocene
	Mesozoic	Cretaceous	
		Jurassic	
	Palaeozoic	Triassic	
		Permian	
		Carboniferous	
		Devonian	
		Silurian	
		Ordovician	
		Cambrian	
Proterozoic	Neoproterozoic		
	Mesoproterozoic		
	Paleoproterozoic		
Archaen			
Hadean			

Unless otherwise stated, arrows in right-hand column refer to current earliest appearances in the fossil record

Myr since start	Some major evolutionary events
0.011	← Extinctions of large mammals by man
1.64	← *Homo*
5.2	← Hominids
	Global cooling, reduced rainfall, savanna spreads
	← Old world monkeys
23.3	← Apes (*Proconsul*); global warming. Spread of African mammals into Eurasia
35.4	Rapid global cooling
	← Old world simian primates
56.5	← Ungulates
	Equable global climate; high rainfall
65	← Mass extinction (including dinosaurs)
	← Primates
	Major teleost radiation
135	← Flowering plants in ascendancy
	Cycads in decline; conifers continue
	← *Archaeopteryx*
	Dinosaurs dominate land; plesiosaurs and ichthyosaurs dominate seas
205	← Mammals
	← Mass extinction
251	← Dinosaurs
	← Great Mass Extinction
	First occurrence of archosaurs and cynodonts
286	Mososaurs briefly dominate seas
	Giant club mosses, horsetails and cycads dominant
350	Pangea forming
	← Land vertebrates; seed plants
400	Placoderms flourish
440	
505	
	← Jawed vertebrates
543	▼ Cambrian Explosion
1000	Ediacara sandstones
1600	
2500	← Gunflint stromatolitic cherts
	← Earth's atmospheric oxygen level rises
4000	← Possible origin of life on earth
~4500	